THE MANULEAE

A TRIBE OF SCROPHULARIACEAE

O. M. HILLIARD

Edinburgh University Press

© Olive Hilliard, 1994
Edinburgh University Press
22 George Square, Edinburgh

Printed and bound in
Great Britain by
The Alden Press Ltd, Oxford

A CIP record for this book is
available from the British Library

ISBN 0 7486 0489 8

Contents

Acknowledgments

I am grateful to the directors of the following herbaria, not only for sending much material on loan, but also for permitting me to hold it for considerable periods of time, and often donating duplicates to the Royal Botanic Garden Edinburgh: Botanische Garten und Botanisches Museum Berlin-Dahlem (B), The Natural History Museum, London (BM), Bolus Herbarium, University of Cape Town (BOL), Botanical Institute, University of Coimbra (COI), Herbarium Universitatis Florentinae (FI), Moss Herbarium, University of the Witwatersrand (J), Royal Botanic Gardens, Kew (K), Herbarium V.L. Komarov Botanical Institute of the Russian Academy of Sciences St Petersburg (LE), Herbário, Centro do Botânica, Instituto de Investigaçao Científica Tropical (LISC), Botanische Staatssammlung München (M), Missouri Botanical Garden (MO), Compton Herbarium, Cape Town (NBG), New York Botanical Garden (NY), Laboratoire de Phanérogamie, Muséum National d'Histoire Naturelle (P), National Botanical Institute, Pretoria (PRE), Botany Department, Swedish Museum of Natural History (S), South African Museum herbarium (SAM), National Herbarium and Botanic Garden Zimbabwe (SRGH), Stellenbosch Herbarium (STE), School of Botany, Trinity College Dublin (TCD), Botanical Museum, Uppsala University (UPS), Naturhistorisches Museum Wien (W), National Herbarium of Namibia, Windhoek (WIND), Institut für Systematische Botanik, Universität Zurich (Z). I was able to visit The Natural History Museum, Botanical Museum, University of Copenhagen, Royal Botanic Gardens Kew, Linnean Society of London, Botany Department, Swedish Museum of Natural History, and Botanical Museum, Uppsala University, and thank the directors and staff there for making these visits so pleasant. Professor Nordenstam (S) later hunted for types for me, as did Dr Sven Snogerup (Lund); Dr Moberg (UPS) examined specimens in Thunberg's herbarium and sent photographs. The directors of the herbaria in Florence (FI) and Geneva (G-DC) also sent photographs of types. Dr Rourke (NBG) provided map blanks.

I am especially grateful to Mrs A. Batten, Mrs H. Bokelmann, Mrs Paine, Mr H.C. Taylor and Mr J. Vlok for the trouble they took in collecting specimens and seed from widely scattered sites in Cape Province; Mrs Batten helped further with photographs and drawings of plants. Mrs Bean (BOL) sent a number of interesting collections.
Dr C.L. Argue, University of Minnesota, examined the pollen of a great many species and generously wrote a brief synopsis and provided beautiful electron micrographs for inclusion here.

I also owe much to the technical skills and horticultural assistance of staff at the Royal Botanic Garden Edinburgh, and thank them all. However, several people need special mention. Mrs Evelyn Turnbull did most of the scans and also sectioned seeds; Miss Debbie White spent hours in the darkroom; Mrs Maureen Warwick gave technical help with the figures in her own time; she also scanned scales; Mrs Mendum (Mary Bates) drew figs 50–73 to illustrate the salient features of the genera, and with great skill painted four colour plates, working mostly from dried specimens and slides. Dr Kwiton Jong provided all the information on chromosomes and also gave me technical help with photography. Mr B.L. Burtt examined specimens for me in the Van Royen herbarium (Leiden) and the Lamarck herbarium (Paris); I thank him too for helpful criticism and interminable proof-reading.

I acknowledge with thanks funding from the Appleyard Trust (administered by the Linnean Society) for the preparation of the colour plates, from the Sibbald Trust (Royal Botanic Garden Edinburgh) for technical assistance, and both the Sibbald Trust and the Annals of Botany Co. for contributions towards the cost of publication.

My work has depended entirely on the research facilities I have been afforded at the Royal Botanic Garden Edinburgh and for these I express my warmest thanks to the Regius Keeper, Professor David Ingram, who has also contributed to publication by permitting preparation of camera-ready copy in-house. This, together with the design and lay-out of the book, has been in the capable hands of Mrs Norma Gregory, who has worked with untiring energy and great good humour and who has also been responsible for all necessary arrangements with Edinburgh University Press. I could not have been more fortunate in my collaborator.

REVISION OF MANULEAE

Bentham (1876) gave recognition to the tribe Manuleae and, acquainted with little more than a few of the Cape species, placed therein eight genera and 116 species. Nearly one hundred and twenty years later, the present study accepts 17 genera and 344 species. In the intervening period, generic concepts have fluctuated and one reason for their instability has been reliance on rather superficial characters for their definition. Here it is shown that features of indumentum, stamens, placenta, stigma and seed-coat provide important and reliable criteria, and their use has helped a far more comprehensible pattern of genera to emerge.

The one fundamental character that distinguishes Manuleae from the allied tribe Selagineae (the erstwhile family Selaginaceae, sometimes even included in Globulariaceae) is the multi-ovulate ovary, for Selagineae have but one ovule in each ovary-loculus. Because species have been described superficially, without dissection of the ovary, many mistakes have been made: species of Manuleae described in Selagineae and vice versa. These have had to be corrected, but their very existence has led to the development of a strong body of evidence insisting that these two tribes are very closely related and certainly not to be placed in separate families.

SYNOPSIS OF MANULEAE

The distinguishing characters of the tribe Manuleae are the aestivation of the corolla lobes (posticous lobes external in the bud), the synthecous anthers, and the 4- to many-seeded septicidal capsule opening further by short loculicidal splits. Most of the genera have a lingulate stigma with marginal stigmatic papillae and a dorsal nectariferous gland at the base of the ovary.

Members of Manuleae are very nearly confined to Africa south of the Sahara: one species of *Camptoloma* is endemic to Grand Canary, and one species of *Jamesbrittenia* extends from Egypt and Sudan to India.

The tribe embraces 17 genera: *Manulea* L., *Sutera* Roth, *Lyperia* Benth., *Camptoloma* Benth., *Jamesbrittenia* O. Kuntze (the last three long included under *Sutera*), *Zaluzianskya* F.W. Schmidt, *Polycarena* Benth., *Phyllopodium* Benth., *Glekia* Hilliard, *Trieenea* Hilliard, *Melanospermum* Hilliard (the last three including many species originally described under *Phyllopodium* and *Polycarena*), *Reyemia* Hilliard, *Glumicalyx* Hiern (originally monotypic but now including several species originally described under *Zaluzianskya*), *Strobilopsis* Hilliard & Burtt, *Tetraselago* Junell (all species originally described under *Selago*), *Antherothamnus* N.E. Br. and *Manuleopsis* Thellung. The principal differences between the genera are summarized in Table 1.

Antherothamnus and *Manuleopsis*, two very closely allied monotypic genera sometimes placed in Cheloneae and more recently in Freylinieae Barringer (Barringer, 1993), have all the essential characters of Manuleae (see further under *Antherothamnus*). The species are shrubs up to 2.5m tall, confined to Namibia, the arid northern Cape, Botswana, Zimbabwe and the western Transvaal. In *Antherothamnus*, the fifth (posticous) stamen is represented by a very well developed staminode (filament present, anther wanting). The geographical distributions of the genera and species of Manuleae suggest that one is looking at relicts of once much more widely distributed groups.

Antherothamnus and *Manuleopsis* possibly represent the least specialized members still extant.

Table 1. Summary of the principal characters of the genera of Manuleae.

	Habit[a]	Stem hairs[b]	Cymes[c]	Bract[d]	Calyx lobing[e]	Clavate hairs[c]	No. of stamens[f]	Posticous filaments[g]	Stigmatic surface[h]	No. of ovules[i]	Nectary[j]	Seed shape[k]	Seed length[l]	Seed colour
1. *Antherothamnus*	1	2	+	–	1	+	4½	+	+	6–12	–	2	c.1.4	brown
2. *Manuleopsis*	1	2	+	–	1	+	4	+	+	∞	–	1	0.8	brown
3. *Camptoloma*	3, 4	2	+	–	1	+	4	+	+	∞	–	3	0.4–0.5	brown
4. *Jamesbrittenia*	1, 2, 3, 4	2, 3	–	–	1	+	4 (2½ in 1 sp.)	+	+	∞	–+	4A, 4B	0.3–1	reddish brown, greyish brown
5. *Lyperia*	3, 4	2	–	–	1	+	4, 2½	+	+	∞	+	5	1–1.5	black
6. *Sutera*	3, 4	2, 3	+	–	2	+	4	–	–(+ in 2 spp.)	∞	–	6	0.5–1	amber, pallid, blue
7. *Manulea*	2, 3, 4	2, 3	+	–	2	+	4 (2 in 1 sp.)	–	–	∞	+	7	0.3–1	pallid, blue
8. *Melanospermum*	3, 4	1, 2	–	–+	2	+	4	+	–	∞	+	8	0.25–0.75	black
9. *Polycarena*	4	2	–	+	2	+	4, 2	+	–	4–∞	+	9	0.75–1 (–1.5)	pallid
10. *Glekia*	2	1	–	+	2	+	4	–	–	9–16	+	10	1–1.25	pallid
11. *Trieenea*	2, 3, 4	1, 2	–	–+	1	+	4	+	–	2–14	+	11	0.5–1	pallid, amber, greenish blue
12. *Phyllopodium*	3, 4	1, 2	–	+	1, 2	+	4 (2, 3 when selfing)	+	–	few–∞	+	12A, 12B	0.4–1	amber, pallid grey-mauve blue
13. *Zaluzianskya*	2, 3, 4	1, 2	–	+	3	+	4, 2	+	–	∞	+	13A, 13B	0.4–2	pallid, mauve
14. *Reyemia*	4	1, 2	–	+	3	+	2½	+	–	∞	+	14	0.8–1	pallid
15. *Glumicalyx*	2, 3	1	–	–	1, 2	–	4	+	–	7–∞	+	15	0.8–1	brown
16. *Strobilopsis*	3	1, 2	–	–	2	+	4, 3, 2	+	–	1–6	+	16	1.4	pallid → dark brown
17. *Tetraselago*	3	1	–	+	3	+	4	+	–	2	+	17	1	amber → black

a 1, shrub; 2, shrublet; 3, perennial; 4, annual.
b 1, eglandular; 2, stalked glandular; 3, sessile glands.
c +, present; –, absent.
d +, adnate; –, free.
e 1, nearly to base; 2, ± halfway; 3, toothed.
f ½, staminodes.
g +, decurrent; –, not decurrent.
h +, ± terminal; –, lingulate.
i In each loculus.
j –, ± annular; +, lateral gland.
k Numbers refer to fig. 1.
l In mm.

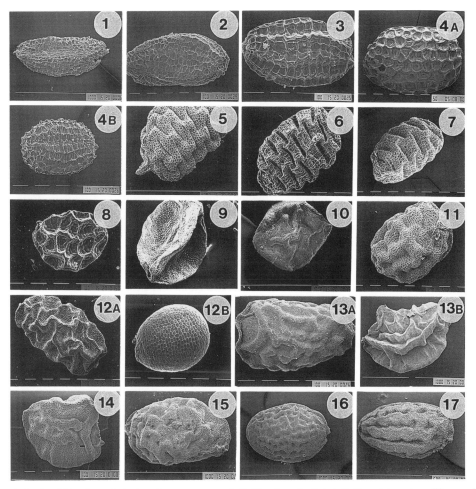

Fig. 1. Seeds of Manuleae. 1, *Antherothamnus pearsonii*; 2, *Manuleopsis dinteri*; 3, *Camptoloma lyperiiflorum*; 4A, *Jamesbrittenia microphylla*, 4B, *Jamesbrittenia pallida*; 5, *Lyperia tristis*; 6, *Sutera foetida*; 7, *Manulea cheiranthus*; 8, *Melanospermum swazicum*; 9, *Polycarena capensis*; 10, *Glekia krebsiana*; 11, *Trieenea taylorii*; 12A, *Phyllopodium multifolium*, 12B, *Phyllopodium pumilum*; 13A, *Zaluzianskya villosa*, 13B, *Zaluzianskya inflata*; 14, *Reyemia nemesioides*; 15, *Glumicalyx montanus*; 16, *Strobilopsis wrightii*; 17, *Tetraselago natalensis*. Scale bars: 1, 6, 9, 10, 13B, 15, 16 = 1mm; 17 = 0.5mm; 2, 3, 4B, 6, 7, 8, 11, 12, 13, 14 = 0.1mm; 4A = 0.05mm.

Sutera, in the broad sense of Hiern (1904) and others, is heterogeneous. In this revision, *Sutera* is taken as congeneric with only *Chaenostoma* Benth. and *Sphenandra* Benth.; it comprises 49 species, mostly perennial. *Lyperia*, lectotypified by *L. tristis* (L.f.) Benth., comprises six species, characterized, inter alia, by relatively large, black bothrospermous seeds (externally with transversely elongated pits arranged chequer-board fashion in longitudinal rows), and very short ± bifid stigma. All but one are annual. The seeds of *Sutera* are also bothrospermous, but are smaller than those of *Lyperia*, amber-coloured, pallid or bluish, and the stigma is lingulate, entire. The other species included by Bentham in *Lyperia*, as well as many more subsequently described in *Sutera*, are now transferred to *Jamesbrittenia* with reddish-brown seeds, the testa ± reticulate when seen under a lens, and a very shortly bifid stigma. It includes 83 species, nearly all perennial. *Camptoloma*, a small genus of three species (Namibia, Grand Canary, Somalia-Socotra-Yemen) has aulocospermous seeds (externally ribbed by 6–8 longitudinal rows of transversely elongated cells), and a short ± bifid stigma.

3

The genus was originally placed by Bentham in the tribe Sibthorpieae, but the aestivation of the corolla lobes and the synthecous anthers are characteristic of Manuleae.

Manulea, like *Sutera* and *Lyperia*, has bothrospermous seeds that are either pallid or bluish, and a lingulate stigma sometimes very shortly bifid (variable within a species); 74 species are recognized here, either annual or perennial.

Polycarena is a genus of 17 species of annuals confined to the western and south-western Cape. The stigma is lingulate, the seeds mostly broadly 3-winged, unique in the tribe.

Phyllopodium comprises 26 species, all but one annual, mostly confined to the western and south-western Cape. The stigma is lingulate, the seeds mostly ± colliculate, amber-coloured, pallid or mauvish, smooth in four species, and blue.

In the general affinity of *Phyllopodium*, and with somewhat similar seeds and stigma, are *Glekia* (monotypic shrublet, essentially montane, E Cape and Lesotho) and *Trieenea* (nine species, annual or perennial, endemic to the mountains of the western and southern Cape).

Melanospermum comprises six species, all but one are annual, with lingulate stigma and deeply reticulate black seeds. The genus is centred on the Transvaal and Swaziland, with one species reaching Namibia and Zimbabwe.

In floral morphology, *Strobilopsis* is not unlike *Melanospermum*, but the seeds are very different, being irregularly colliculate and pale blue-grey or pallid. The only species is a short-lived perennial confined to the Drakensberg in Natal and Lesotho.

Glumicalyx is another small genus (six species, all perennial), confined to the Eastern Mountain Region of southern Africa. It has a lingulate stigma and ± trigonous red-brown seeds. It stands apart from all the other genera by virtue of its inflorescence, a compact head nutant at anthesis, becoming erect as it passes into seed, and the rather leathery corolla lobes with the pigment confined to vertically elongated epidermal cells with dense contents (see Hilliard & Burtt, 1977, p.159 fig. 1). Similar pigmentation of the corolla lobes possibly occurs in some species of *Manulea*, *Jamesbrittenia* and in *Zaluzianskya diandra*; this needs anatomical investigation.

Despite the fact that several species of *Glumicalyx* were originally described in *Zaluzianskya*, its relationship to that genus is not close. *Zaluzianskya* comprises 55 species, roughly half of them annuals occurring mainly in the western half of southern Africa including Namibia, the rest perennials confined to the eastern half of the area, from Zimbabwe to the eastern Cape. The stigma is lingulate, the seeds pallid or mauvish and irregularly angled (by pressure?).

Reyemia, a ditypic genus found in the arid north western Cape, resembles *Zaluzianskya* in its bract, calyx and stigma; the seeds are irregularly colliculate, greyish when young, later pallid. The form of the corolla is highly distinctive: the anticous lobe is brought into the posticous position by curvature in the upper part of the tube. The plants are small annuals.

Tetraselago comprises four species, all perennial, found in the Transvaal, Swaziland and Natal: stigma lingulate, ovules only two in each loculus. As the generic name implies, the species were originally described in *Selago* and they provide a convincing link between Manuleae and Selagineae.

One other genus needs mention. When Urban described the monotypic *Tuerckheimocharis* from San Domingo, he placed it in the Manuleae (Urban, 1912, p.373–374). An isotype of *T. domingensis* Urb. in the Kew herbarium lacks flowers, but the ripe capsules are quite unlike those of members of Manuleae, being septicidal with entire valves.

MORPHOLOGICAL CHARACTERS

DURATION AND HABIT

The Manuleae is largely a tribe of annuals and herbaceous perennials; some species are more or less shrubby and two genera, *Antherothamnus* and *Manuleopsis* (both monotypic), are large shrubs c.3m tall.

Two genera are composed entirely of annuals: *Polycarena* (17 species), and *Reyemia* (2). Three more each include only one perennial, though a few of the annuals possibly persist for more than one season: *Phyllopodium* (26), *Lyperia* (6) and *Melanospermum* (6). *Zaluzianskya* (55) has only 12 species strongly perennial and a few more may persist for more than one season, but the rest are annual. *Manulea* (74) has about 43 species strongly perennial, *Trieenea* (9) five. The single species of *Strobilopsis* is a short-lived perennial. Of the 83 species of *Jamesbrittenia*, 11 are annuals, five others can behave as annuals, and four more possibly perennate; the rest are strongly perennial, often bushy or shrubby. *Sutera* (49) has only one annual and that the type species, but eight more can behave as annuals. *Glumicalyx* (6), *Tetraselago* (4) and *Glekia* (1) are all strongly perennial, while *Camptoloma* (3) is perennial but at least two of the species may behave as annuals.

Many of the perennials are adapted to the fire-swept grasslands, scrub and open woodland that cover much of southern Africa, and develop thick woody subterranean rootstocks producing, sometimes only one, but often tufts, of flowering stems. These stocks are of systematic value particularly in *Zaluzianskya* and *Manulea*. Several species of *Zaluzianskya* produce a stock the crown of which is covered by a dense mass of loose white somewhat fleshy buds (see Hilliard & Burtt, 1983, p.4); a somewhat similar stock occurs in tropical African species of *Thesium* e.g. *T. bangweolense* R.E. Fries and *T. unyikense* Engl. Usually only one bud at a time develops into a flowering stem, and nothing is known of their duration. Other species (notably *Z. pulvinata*) form a cushion of numerous leaf rosettes, while still other species may have several loosely arranged rosettes on the crown. A few species have a carrot-like stock. The annual species produce only a slender taproot, a fact of great value in sorting out the tangle of perennial species that had been confused with the annual *Z. maritima*.

In *Manulea*, several species in sections *Dolichoglossa*, *Manulea* and *Medifixae* develop a woody taproot as an organ of perennation. This character is useful in distinguishing at a glance the three perennials, *M. crassifolia*, *M. buchneroides* and *M. dregei*, from the annual *M. bellidifolia*, also *M. parviflora* (perennial) from *M. obovata* (annual). All these species have hitherto been much confused, though they differ significantly in floral characters as well as in duration. Only one species, *M. florifera*, is somewhat stoloniferous; however, the basal parts of a good many of the perennials are unknown.

In the annuals, the organ of perennation is of course the seed.

INDUMENTUM

Indumentum is of great systematic value at the level of both genera and species; the disposition of the hairs as well as the type of hair can be of significance. For example, *Phyllopodium* and *Polycarena* are readily distinguished by differences in the hairs on the vegetative parts: always patent and gland-tipped in *Polycarena*, patent eglandular hairs always present at least on the lower margins of the bracts in *Phyllopodium*, while retrorse eglandular hairs always occur at least on the lowermost internode of the stem. Balloon-tipped hairs (fig. 11C) are present in many species of *Manulea* and are unique to the genus. In *Jamesbrittenia*, the species of the informal group 1b, 2 (*J. glutinosa* and allies) can be recognized by their remarkably coarse hairs (figs 10B, D). On the other hand, group 1b, 1 (*J. fruticosa* and allies) lack glistening glands (term elaborated below) or scales on all parts, whereas one or the other is present in all other species (glistening glands are commonplace in the tribe). Differences in hairs are easily seen under a dissecting microscope, often with a good hand lens, and much use has been made of them in the keys to species as well as in discussion following the formal descriptions: for example, whole suites of species in both *Sutera* and *Jamesbrittenia* look much alike and have in the past been thoroughly confused, but differences in indumentum are a valuable aid to identification.

The hairs exhibit considerable diversity, and, although no attempt has been made to survey them exhaustively, a sample has been examined under the light microscope to gain some idea of their range of structure. (The hairs were cleared in lactic acid, mounted in glycerol, and drawn with the aid of a camera lucida). The term 'hair' is used very loosely and covers structures that could more precisely be termed scales; the following descriptions and the figures should bring clarity.

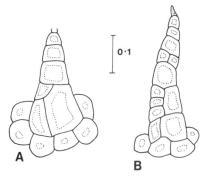

Fig. 2. Multiseriate heavily thickened hairs on stems and leaves of *Manulea ramulosa*. Scale bar = 0.1mm.

Both glandular and eglandular hairs occur throughout the tribe and range from unicellular to multicellular. They are nearly always uniseriate; multiseriate hairs (fig. 2) have been seen only in a few species of *Manulea* (e.g. *M. ramulosa*) and *Sutera* (e.g. *S. subnuda*); the glandular apical cell soon breaks off, while the walls of the stalk cells become very thick, the lumen sometimes almost disappearing. In the species descriptions, these hairs are described as 'almost scabrid'.

Unicellular hairs occur only on the upper surface of the corolla; they are frequently inside the tube and may extend out onto the lobes. These are the 'clavate hairs' of the text though they may be pointed rather than club-shaped (fig. 3). The cell wall is very thin and may be very finely striate. These hairs occur in nearly all the genera of Manuleae (as well as in some genera of Selagineae and other tribes), where they are frequently associated with an orange/yellow patch on the corolla tube and/or limb. Their disposition can be of importance particularly at generic level (see figs 48, 52 etc.) and their presence or absence can help to discriminate species. Clavate hairs never occur inside the corolla tube of *Zaluzianskya*, but they often form a circlet or partial circlet around the mouth, and are taxonomically useful. *Glumicalyx* is unique in Manuleae for its lack of clavate hairs (though they may be wanting in some species of the other genera). In *Antherothamnus*, clavate hairs occur only on the filaments; the hairs in the sinuses of the corolla lobes, though closely resembling clavate hairs when seen under a lens or dissecting microscope, consist of a big elliptical apical cell and a tiny foot cell (fig. 3E); the apical cell occasionally divides longitudinally into two cells.

Only comparatively rarely do hairs other than clavate ones occur on the upper surface of corolla lobes; both tiny glandular hairs and glistening glands have been recorded and can be of diagnostic value, for example in *Zaluzianskya* (gland-tipped hairs) and *Jamesbrittenia* (glistening glands).

In the text, frequent reference is made to glistening glands, so-called for obvious reasons. They are scales in the sense of Theobald *et al.* (1979, p.47 and fig. 5.6), but in this revision the term scale has been reserved for

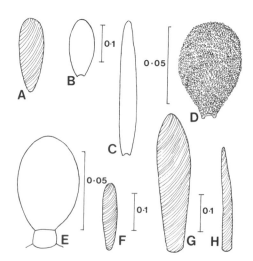

Fig. 3. Hairs on inside of corolla, mostly unicellular. A, *Manulea crassifolia*, clavate hair, delicately sculptured, inside tube; B, *M. cheiranthus*, clavate hair inside tube; C, *Zaluzianskya pumila*, somewhat pointed hair from circlet around mouth; D, *Manulea dubia*, heavily sculptured clavate hair from upper surface of corolla lobe; E, *Antherothamnus pearsonii*, two-celled hair from sinus of corolla lobes; F, *Manuleopsis dinteri*, delicately sculptured clavate hair, inside corolla tube; G, *Zaluzianskya capensis*, delicately sculptured clavate hair from circlet around mouth; H, *Manulea dubia*, sculptured hair from inside corolla tube. A, B, C, F, G, H: scale bar = 0.1mm; D, E: scale bar = 0.05mm.

peculiar trichomes that thickly cover all the external parts of several species of *Jamesbrittenia*.

Glistening glands show microscopic differences. The commonest sort comprises a very small foot cell and two much larger secretory cells forming the head; the head is slightly elliptic and measures roughly 0.05–1mm across the longest axis (fig. 4A). The smallest hairs are very common (seen in *Camptoloma, Glumicalyx, Lyperia, Manulea, Melanospermum, Phyllopodium, Polycarena, Reyemia, Strobilopsis, Sutera, Tetraselago, Trieenea, Zaluzianskya*; they are also common in some genera of Selagineae), but they are easily overlooked and are of little taxonomic value. It is this sort of hair that forms the 'dots' in the glandular-punctate leaves of, for example, *Glumicalyx* and *Tetraselago*; the head is usually 2-celled, sometimes 4-celled. Occasionally there is only one glandular secretory cell and a tiny foot cell; these have been noticed in some species of *Manulea* and *Zaluzianskya*. Glistening glands about 1mm in diameter are easily seen and can be useful taxonomically. They are commonplace in *Jamesbrittenia*, but group 1b, 1, as mentioned above, is characterized by their absence. They are also common in *Sutera, Lyperia* and *Manulea*.

In some species of *Jamesbrittenia* (for example, *J. accrescens, J. canescens, J. ramosissima*), *Manulea* (e.g. *M. altissima*), *Lyperia* (e.g. *L. tristis*), *Sutera* (e.g. *S. foetida*), *Zaluzianskya* (e.g. *Z. capensis, Z. diandra, Z. pumila*), the head may be 2-, 4- or 8-celled (fig. 4C), but this is of no practical classificatory value. In a few species of *Jamesbrittenia* (e.g. *J. grandiflora, J. huillana, J. pristisepala*) there may be a foot cell, a small neck cell, and a big 4-celled head (fig. 4G); again, this is of little practical value though the difference between stalked and sessile glands is visible under a dissecting microscope. Of some taxonomic value is the fact that in some species of *Jamesbrittenia* there is such copious secretion from these glands that they are rendered invisible and the stems and leaves look as though they have been varnished (see for example *J. aspalathoides, J. atropurpurea*).

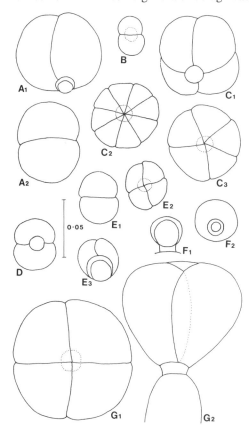

Fig. 4. Glistening glands on external parts. A1, A2, *Jamesbrittenia grandiflora*; B, *Manulea cheiranthus*; C1, 2, 3, *Jamesbrittenia ramosissima*; D, *Sutera aethiopica*; E1, 2, 3, *Glumicalyx montanus*; F1, 2, *Zaluzianskya capensis*; G1, 2, *Jamesbrittenia pristisepala*. Scale bar = 0.05mm.

An apparently unique type of glistening hair occurs on all the external epidermis of *Sutera halimifolia*, thereby distinguishing it from several species with which it is commonly confused. It comprises a tiny foot cell and two much larger almost oblong apical cells (figs 5B, 6A, and see also discussion under *S. halimifolia*).

The external epidermis of several species of *Jamesbrittenia* is, as mentioned above, covered in scales comprising a foot cell and a large head, strongly depressed and more or less scalloped in outline, composed of 4, 6 or 8 cells (figs 5A, 6B, C, D). My colleague Mrs Warwick scanned the scales of *J. pilgeriana* (fig. 6B) and was struck by their resemblance to those of some species of *Rhododendron*. She tells me that the scales of *Rhododendron* can be water-absorbent or water-repellant. Perhaps the scales on the stems of a species such as *Jamesbrittenia barbata*, which often has greatly reduced leaves, absorb water from the sea fogs that roll over its habitat, the Namib desert, and become translucent so that the chloroplasts in the underlying tissue can photosynthesize (cf. the velamen tissue of some epiphytic orchids).

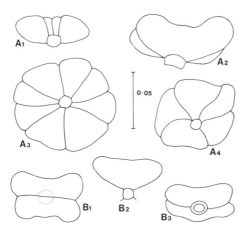

Fig. 5. Scales and hairs on external parts. A1,2,3,4, *James-brittenia chenopodioides*, scales; B1,2,3, *Sutera halimifolia*, glistening hairs. Scale bar = 0.05mm.

Throughout the tribe, hairs are usually simple; branched hairs have been seen in *Manulea altissima* and in some specimens of *Schlechter* 10442, thought to be of hybrid origin with *Sutera subspicata* as one parent (*S. subspicata* has only simple hairs). In both cases, the branches may terminate in either a neck cell supporting a globular secretory cell or an eglandular cell (fig. 7). The hairs are too infrequent and too difficult to see to be of diagnostic value.

Three- to many-celled gland-tipped hairs occur in many genera; under a compound microscope, the glandular head is seen to be composed of 1–36 cells; it is usually globose or depressed-globose, occasionally longer than broad; at the practical level, it is usually the length of the hairs or their texture, occasionally the diameter of the heads, that is of diagnostic value. A very common type of gland-tipped hair (also common in other tribes of Scrophulariaceae, see for example Grabias *et al.*, 1991, p.191, Raman, 1989–1990, as well as Gesneriaceae and other families including Cruciferae and Labiatae) comprises a stalk cell, a neck cell and a globose or depressed-globose 1- or 2-celled secretory head (figs 8C, G, I). There are many variants of this type of hair: the stalk cell can elongate and thus increase the length of the hair (fig. 8D) or the stalk can comprise up to ± 4 cells (fig. 8B, F) or many cells (fig. 8J), or the head may be composed of up to 8 cells (fig. 8H). Sometimes the stalk cells are very broad and the glandular head tiny (fig. 8E). All these types are common. Also common is the presence of foot cells, either one (fig. 8D) or several (fig. 8B).

A singular type of hair is found in all three species of *Camptoloma*. The hairs are up to 4mm long yet the stalk is but a single cell supporting a short neck cell and a subrotund glandular head composed of one or two cells (fig. 9A). The exudate is so copious that collectors have reported wetting of hands.

Hairs somewhat similar morphologically to those of *Camptoloma* are found in the species of *Jamesbrittenia* group 1b, 1 (*J. fruticosa* and its allies) a group notable within the genus for lack of glistening glands. The hairs are up to 2.5mm long, the stalk usually 1-celled, occasionally 2-celled, terminating in a short neck cell and a vertically elongated head composed of two cells (fig. 9B). No collector has noted copious exudate. The common type of glandular hair with an 8-celled head (fig. 8H) is mixed with the distinctive long hairs.

Multicellular glandular heads composed of more than 8 cells are uncommon. Those of *Phyllopodium lupuliforme* (fig. 10A) and *P.collinum* (fig. 10C) provide examples not seen elsewhere. *Jamesbrittenia* group 1b, 2 (*J. glutinosa* and allies) is characterized, as mentioned above, by its remarkably coarse hairs that are nevertheless thin-walled. The heads are 36-celled and can attain an astonishing size. The top of the head appears to be covered by a cuticle as no cell outlines are visible (fig. 10B).

In contrast, the stalk cells in a few species of *Jamesbrittenia* (*J. accrescens*, *J. grandiflora* and allies) are longitudinally striate probably due to a thickening of the cuticle (fig. 10E). The glandular head often breaks off and the rest of the hair persists as a coarse projection. In the descriptions of the species, the term 'glandular-hispid' is applied to these hairs.

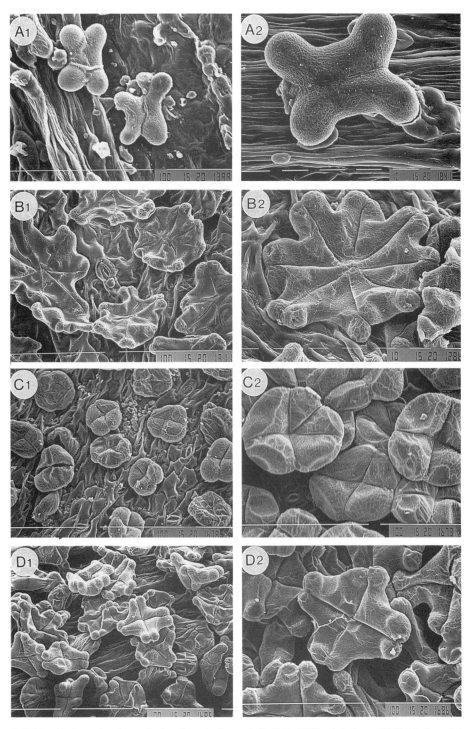

Fig. 6. SEMs of hairs and scales on leaf surface. A, *Sutera halimifolia* (*Hilliard & Burtt* 10712); B, *Jamesbrittenia pilgeriana* (*Giess* 10437); C, *J. barbata* (*Giess* 3046); D, *J. chenopodioides* (*Giess, Volk & Bleissner* 6233). Scale bars: A1, B1, C, D = 0.1mm; A2, B2 = 0.01mm.

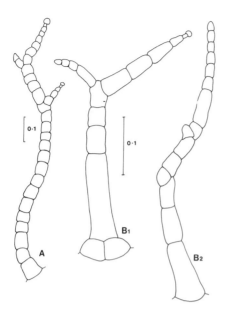

Fig. 7. Branched hairs. A, *Manulea altissima*; B1,2, *Sutera*, possibly hybrid (*Schlechter* 10442). Scale bar = 0.1mm.

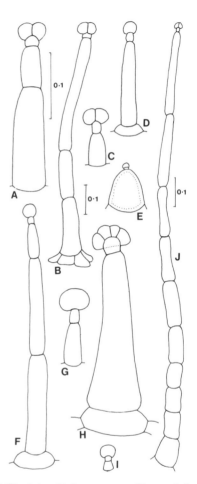

Fig. 8. Gland-tipped hairs on stems and leaves. A, *Lyperia tristis*; B, *Jamesbrittenia grandiflora*; C, *Trieenea laxiflora*; D, *T. taylorii*; E, *Sutera subspicata*; F, *Melanospermum rupestre*; G, *Trieenea laxiflora*; H, *Polycarena capensis*; I, *Glekia krebsiana*; J, *Jamesbrittenia tenella*. Scale bar = 0.1mm.

Fig. 9 (left). Gland-tipped hairs with a remarkably elongated stalk cell. A, *Camptoloma lyperiiflorum* (scale bar = 0.1mm); B, *Jamesbrittenia fruticosa* (scale bar = 0.05mm).

Eglandular hairs are also common throughout Manuleae and are often of diagnostic value; the indumentum may be wholly eglandular, or mixed glandular and eglandular, or glandular on some parts, eglandular or mixed on others (for example, glandular on inflorescence axes, eglandular or mixed on lower part of stem).

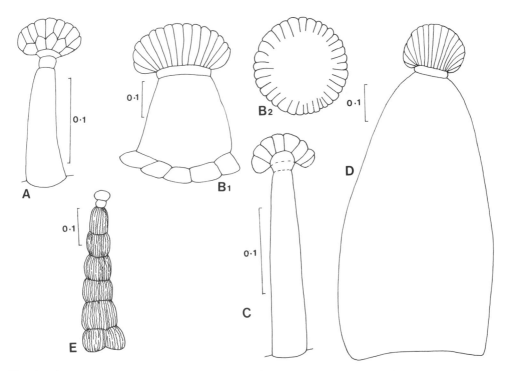

Fig. 10. A–D, hairs with multicellular glandular heads. A, *Phyllopodium lupuliforme*; B1, *Jamesbrittenia megadenia*; B2, head viewed from above, centre of cap presumably covered by a cuticle; C, *Phyllopodium collinum*; D, *Jamesbrittenia glutinosa*; E, *Jamesbrittenia grandiflora*, stalk cells presumably with a longitudinally striate cuticle, neck and head cells not sculptured. Scale bars = 0.1mm.

A remarkable variety of hairs, mixed glandular and eglandular, can occur in a single species, particularly species of *Manulea*. Balloon-tipped hairs (fig. 11C), so-called for obvious reasons, are unique to *Manulea* and occur in over half the species (see further in the introduction to *Manulea*). *Phyllopodium diffusum* is distinguished by hairs with a much inflated apical cell and a small stalk cell (fig. 11D) and somewhat similar hairs are found in other species, and in *Melanospermum* (figs 11J, L), and they often tend to be retrorse. Retrorse hairs, 2- to several-celled, the apical cell inflated or not, and then obtuse or acute, are found in several genera: *Glekia*, *Glumicalyx*, *Manulea*, *Melanospermum*, *Phyllopodium*, *Reyemia*, *Strobilopsis*, *Tetraselago*, *Zaluzianskya* (figs 11A, H, I, N). They can be of diagnostic value. Patent eglandular hairs are also common, and the apical cell may be acute to obtuse (figs 11E, F).

Sutera marifolia and *S. cinerea* are distinguished by their cobwebby-woolly indumentum. The hairs are so long that it was difficult to trace their full length in a microscope preparation, but fig. 11K indicates their structure: few-celled and acute.

Some species of *Sutera* have almost hispid hairs; *S. hispida* was an obvious choice for investigation. The hairs prove to have cell walls with reticulate thickenings (fig. 11G). The hairs of *S. decipiens*, hitherto confused with *S. hispida*, have thick but smooth walls and glandular tips (fig. 11M).

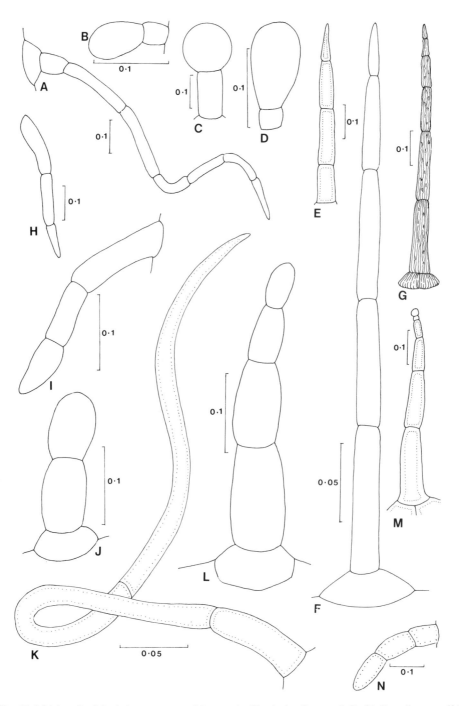

Fig. 11. Mainly eglandular hairs on stems and leaves. A, *Glumicalyx flanaganii*; B, *Phyllopodium cuneifolium*; C, *Manulea cheiranthus*; D, *Phyllopodium diffusum*; E, *Zaluzianskya diandra*; F, *Trieenea taylorii*; G, *Sutera hispida*, cells with sculptured walls; H, *Strobilopsis wrightii*; I, *Glekia krebsiana*; J, *Melanospermum transvaalense*; K, *Sutera cinerea*; L, *Melanospermum transvaalense*; M, *Sutera decipiens*; N, *Glumicalyx montanus*. Scale bar = 0.1mm except in F, K, where it = 0.05mm.

STEMS

The stems of most species are erect, decumbent or ascending. A few species have prostrate or trailing stems, sometimes rooting at the nodes, and one, *Manulea stellata*, is possibly more or less scandent. In both annuals and perennials, stems may be solitary or few to many tufted from the crown, simple to well-branched. Both the degree and the manner of branching may be of systematic value: for example, in both *Polycarena* and *Phyllopodium*, several species of annuals can be distinguished on divaricate versus corymbose branching. The point of branching may also be of taxonomic significance: from the base or higher. The flowering stem may be scapose (all the leaves radical) or sparsely to well leaved. Brachyblasts or axillary leaf tufts are characteristic of *Tetraselago*, *Antherothamnus*, and many species of *Jamesbrittenia*. The short lateral branches of *Antherothamnus* are ultimately leafless and subspinescent, as they are too in the intricately branched shrublets, *Jamesbrittenia ramosissima* and *J. merxmuelleri*. All three are plants of arid areas.

LEAVES

In many members of Manuleae, the leaves are opposite below, becoming alternate upwards, a common mode in Scrophulariaceae. In *Glekia*, *Lyperia* and *Manuleopsis*, the leaves are always opposite (or occasionally subopposite towards the tips of the inflorescences in *Glekia*), the bases (petioles) decurrent down the stem in two narrow ridges. In *Trieenea*, the leaves are opposite, bases connate, in three species, occasionally or always alternate above in the other six species. In *Manulea*, the leaves are usually opposite, bases connate, but in section *Medifixae* six of the species have nearly all the leaves alternate. In *Melanospermum*, the upper leaves rarely become alternate. On the other hand, the leaves are always alternate in *Tetraselago* and *Antherothamnus*, though in the latter genus the extreme base of the plant has not been seen; in *Camptoloma*, the leaves are opposite only at the extreme base.

The leaves are usually herbaceous; they become sclerophyllous in a number of species of *Jamesbrittenia* and one of *Sutera*. They are always simple with pinnate venation (camptodromus or reticulato-pinnatus); margins may be entire to variously toothed or deeply dissected in *Jamesbrittenia*. Differences in leaf-cutting can be of considerable taxonomic importance, particularly at species level, though the genera *Sutera* and *Jamesbrittenia* mostly have dissimilar leaves. Leaf-shape covers a wide range, from linear to suborbicular; the shapes of radical and cauline leaves may differ considerably.

The presence or absence of a petiole is of limited value because in several genera and many species the leaves are variously narrowed to the base and the distinction between sessile and petiolate is not clear-cut. This is particularly so among the annuals, though there are many exceptions.

Differences in indumentum are of systematic value particularly at species level.

INFLORESCENCE

In most genera of the Manuleae, the inflorescence is a spike, raceme or head, with one flower in the axil of each leaf or bract. The inflorescence may be lax or congested, sometimes congested in flower, lax in fruit, and simple or branched. A globose or oblong tightly congested head of ± sessile flowers is characteristic of *Glumicalyx* and *Strobilopsis*; in the former the head is nutant, straightening and elongating slightly as it passes into fruit. The form of the inflorescence, either capitate or elongate, is also of systematic importance in *Phyllopodium* (where it distinguishes two groups of species), *Polycarena*, *Melanospermum* and *Trieenea*. In these genera, at least the lowermost flowers are usually minutely pedicellate. In *Zaluzianskya*, the inflorescence is nearly always a spike, either elongate (but the flowers congested) or capitate (only one inadequately known and unnamed species from Namaqualand produces a raceme). *Tetraselago* is characterized by its corymbose inflorescence. *Reyemia* has a very lax, more or less panicled, raceme, *Lyperia* flowers crowded only initially, the raceme soon lax as it passes into fruit. In *Jamesbrittenia*, the inflorescence is essentially a raceme, often panicled, in *Sutera*, it is either racemose or cymose. This is also the case in *Manulea*, where the inflorescence is often a thyrse. In *Manuleopsis*, *Antherothamnus* and *Camptoloma* the inflorescence is cymose, though in the last two genera some flowers may be solitary, apparently by reduction, as a pair of bracteoles (?) are present on the pedicel. This is also seen in a few species of *Sutera*; in *Manulea*, cymes and solitary flowers may be mixed in an inflorescence.

The flowers of many species of Manuleae are small (often only c.2–5mm) and massing in heads is clearly of biological significance.

BRACTS

The flowers of Manuleae are always subtended by either a leaf or a bract, which may be leaflike or not. The shape of the bract and its indumentum may be useful in distinguishing between species. Adnation (recaulescence) of the bract to the pedicel or to the pedicel and calyx is of great systematic value at generic level. In *Zaluzianskya* and *Reyemia* the bracts are usually sharply differentiated from the leaves and are adnate to the calyx almost to its apex. In *Polycarena*, the bracts are mostly leaflike and adnate almost to the apex of the calyx. *Phyllopodium* shows similar adnation, and the bracts may be leaflike or not. They are leaflike in *Glekia*, and adnate halfway up the calyx tube; leaflike and adnate to ± the whole of the calyx tube in *Tetraselago*; leaflike or not in *Melanospermum* and *Trieenea* and adnate to the pedicel only or, rarely, to the extreme base of the calyx as well. In *Glumicalyx* and *Strobilopsis*, the bracts are sharply differentiated from the leaves and are adnate only to the minute pedicel. In *Sutera* and *Lyperia*, the flowers are in the axils of leaves, and occasionally there is slight adnation to the base of the pedicel. In *Lyperia lychnidea*, *L. tristis* and *Sutera violacea* (sometimes in *S. revoluta*) a pair of small bracts occurs immediately below the calyx, as they do in *Jamesbrittenia fodina*. In *Jamesbrittenia* and *Camptoloma*, the flowers are in the axils of the upper leaves frequently together with a vegetative bud (also seen in *Sutera*); there is no adnation, nor is there in *Manuleopsis* and *Antherothamnus*. In *Manulea*, the flowers are borne mainly in the axils of bracts, which are often very small and adnate to the base of the pedicel.

PEDICELS

The length of the pedicels (particularly the lowermost ones) may be useful in distinguishing allied species. Some species of *Jamesbrittenia* have extraordinarily long pedicels that probably play a part in pollination.

CALYX

The calyx is 5-lobed and may be distinctly to obscurely bilabiate. The measurements given in the formal descriptions were mostly made on rehydrated specimens, opened out flat. The length of the tube was measured from the base to the sinus between anticous and posticous lips when these were discernible, otherwise from the base to the sinus between the lobes. In a nearly regular calyx, a lobe was measured from base to apex and across its widest part. In a distinctly bilabiate calyx, the length of the two lips may be given, measured from the sinus between them to the tips of the lobes, or merely the size of an anticous lobe and a posticous one may be given. Flowers at anthesis were chosen whenever possible; the calyx may be accrescent, remarkably so in *Zaluzianskya* sect. *Macrocalyx*, where it is of diagnostic value. In some genera, Hiern (1904) made much use of the length of the corolla tube in relation to that of the calyx, but this is a difficult character to use, partly because the length of one or other organ may increase with age, partly because Hiern seems to have worked almost exclusively with dry specimens and ignored the shortening effect of withering.

Both the form of the calyx and individual characteristics are of systematic value. In *Sutera*, the calyx is regular to obscurely bilabiate, the lobing not reaching more than two thirds of the way to the base (fig. 53). On the other hand, in *Jamesbrittenia* the calyx is nearly regular, the lobes united briefly at the base (fig. 48) in all but *J. ramosissima* in which there is a distinct tube. Within the genus, differences in the shape of the lobes distinguish groups of allied species; the lobes are sometimes foliaceous. The calyx of *Manuleopsis*, *Camptoloma*, *Lyperia* and *Trieenea* is also divided nearly to the base; the lobes are never foliaceous. In *T. lasiocephala*, the lobing extends right to the base, and it may also do so in *Phyllopodium cephalophorum*.

In *Glumicalyx*, the calyx is unequally 5-lobed; in *G. apiculatus*, the two anticous lobes may be obsolete, and the calyx is always split to the base on the anticous side; it is too in *G. lesuticus*, *G. alpestris* and *G. goseloides*, (but not in *G. flanaganii*); in *G. montanus*, all five lobes are free, and membranous (glumaceous), whence the generic name. See fig. 66.

Zaluzianskya has a distinctive calyx, quite different from that of *Glumicalyx*, several species of which were originally described under *Zaluzianskya*. It is bilabiate, strongly 5-ribbed and plicate, the tissue between the ribs very delicate. The anticous lip is either 2-toothed or almost entire, the posticous lip 3-toothed (fig. 64E). The calyx is weakly to strongly accrescent and persistent. The calyx of *Reyemia* resembles that of *Zaluzianskya*. The calyx of *Polycarena* is also slightly accrescent and persistent, conspicuously ribbed and slightly plicate, the anticous lip only very briefly divided, but the sinus between the two lips is much deeper than in *Zaluzianskya*. The calyx of *Phyllopodium* differs from that of *Polycarena* in having the anticous lip often much more deeply divided; a briefly or deeply divided anticous lip is useful in distinguishing groups of species (fig. 63).

In *Manulea*, the calyx may be nearly regular to strongly bilabiate, the lobes much longer than the tube. Variation in development of two lips is helpful in distinguishing sections, subsections and species.

Particular types of hairs on the calyx characterize certain genera, for example *Phyllopodium*, and in many genera they are useful in distinguishing species or groups of allied species.

COROLLA

The corolla is gamopetalous, tubular and 5-lobed; the two posticous lobes are external in the bud, the anticous one included. Flowers were always soaked before measurement, the tube being measured from the base to the sinus between the anticous and posticous lips, and across the top immediately below the limb; the anticous lobe was measured from the sinus to the apex and across the broadest part. Sometimes the length and breadth of the whole posticous lip is given, more often that of the lobes, from the sinus between a posticous lobe and a lateral one to the apex, and across the broadest part.

The shape of the tube varies from narrowly to broadly cylindric and is often but not always more or less abruptly dilated to accommodate the included anthers, or it may be narrowly to broadly funnel-shaped; the shape of the tube is constant (or very nearly so) within a genus. The limb may be almost regular to distinctly bilabiate; usually the bilabiate limb is composed of a bilobed posticous lip, a trilobed anticous one, but in some species of *Manulea* and *Zaluzianskya* and in both species of *Reyemia*, the two lateral lobes approach the posticous ones to leave the anticous lobe isolated, sometimes strongly so. In *Reyemia*, unusual curvature of the upper part of the corolla tube permits the isolated anticous lobe to stand erect in what appears to be the posticous position. The shape of the lobes varies from orbicular to ovate, obovate, elliptic, spathulate, oblong or lanceolate. In *Manulea* in particular, but also in some species of *Jamesbrittenia* and *Zaluzianskya*, the margins of the lobes may be strongly revolute; should the lobes be lanceolate, they will then appear subulate to the eye, suborbicular lobes more or less oblong. The apex of the lobe is usually rounded or truncate, sometimes retuse or bifid; in a few species the lobes are divided again. All these characters are of systematic value. Bifid corolla lobes are characteristic of a good many species of *Zaluzianskya* and some of *Manulea*, retuse lobes of some species of *Jamesbrittenia*.

The mouth of the corolla tube is usually round. It is strongly compressed laterally in one species of *Manulea* and a few of *Jamesbrittenia*; it is dorsoventrally compressed in a group of allied species of *Jamesbrittenia*. In *Manuleopsis*, the mouth is elliptic due to slight lateral compression, and the tissue around the mouth is thickened. Thickening also occurs in some species of *Zaluzianskya* and *Lyperia*.

The corolla is often hairy outside, and differences in the types of hairs and in their lengths can be of systematic value especially in distinguishing species. Unicellular clavate hairs are often present inside the corolla tube and on the upper surface of the limb; differences in their disposition may characterize genera, and they can also be very useful in distinguishing species or groups of species. *Glumicalyx* is the only genus in which the corolla is completely glabrous.

In *Zaluzianskya*, the corolla tube is always glabrous inside, but many species have a circlet or partial circlet of clavate or somewhat acute hairs around the mouth, occasionally extending onto the bases of the lobes. The lobes may be minutely glandular as well, sometimes over the whole upper surface, sometimes in a band around the mouth.

In most species of *Manulea*, the posticous filaments are bearded with clavate hairs and there is either a posticous longitudinal band of hairs in the throat or up to five bands; in most species of section *Medifixae* there is in addition a roughly star-shaped pattern of clavate hairs on the upper surface of the corolla lobes.

In several genera, the base of the posticous lip is often bearded with clavate hairs, which may extend down the back of the throat and to the lateral or anticous lobes. This is so in many species of *Phyllopodium*, *Polycarena*, *Glekia*, *Melanospermum*, *Trieenea*, *Sutera* (where they are often broken up into discrete longitudinal bands reminiscent of *Manulea*, but the posticous filaments are rarely bearded) and *Strobilopsis*. The hairs are often associated with a yellow or orange patch in the back of the throat.

In *Jamesbrittenia,* the clavate hairs are nearly always confined to a band encircling the inflated part of the corolla tube, and they always extend, at least very briefly, onto the base of the anticous lip. All four filaments may be bearded with clavate hairs and minute glandular hairs are always present at the extreme apex. There is a somewhat similar arrangement in *Camptoloma* where a band of clavate hairs encircles the upper part of the corolla tube and there are minute glandular hairs at the apex of the anticous filaments; the posticous filaments may be bearded as well.

In *Lyperia*, clavate hairs are found inside the inflated part of the corolla tube and minute glands encircle the mouth in a broad band. In *Reyemia*, there is a broad band of clavate hairs in the back of the throat but no orange patch; in *Tetraselago* clavate hairs encircle the mouth, and in *Manuleopsis* a broad band runs down the whole length of the tube on the posticous side and a broad band of sessile glands encircles the mouth. In addition, the posticous filaments are heavily bearded with clavate hairs, with some glandular hairs at the apex, and only glandular hairs on the anticous filaments. *Antherothamnus* is unique in having little tufts of clavate hairs in the sinuses of the corolla lobes; these hairs are very broad in relation to their length and are most unusual in sometimes being divided longitudinally into two cells.

Some species of *Jamesbrittenia* are unusual in having glistening glands on the upper surface of the corolla lobes (they are commonplace on the lower surface).

The commonest flower colours in Manuleae are white or creamy white and shades of violet, lilac, mauve and pink, often associated with a contrasting yellow or orange patch in the posticous position, less often all round the mouth or the whole tube orange. Shades of yellow, including orange, are not infrequent in *Zaluzianskya*, *Jamesbrittenia* and *Manulea*, rare in *Phyllopodium* and *Polycarena*. The corolla limb in *Glumicalyx* is always pale to deep yellow or orange. The three largest genera, *Zaluzianskya*, *Manulea* and *Jamesbrittenia*, not unexpectedly show the widest range of flower colour, including shades of red. Contrasting colour patterns, other than the yellow/orange mentioned above, are common throughout the tribe on the upper surface of the corolla limb and in the throat, and sometimes on the backs of the lobes (particularly in *Zaluzianskya*, but also in *Glekia* and *Polycarena*). Apart from being useful systematically, flower colour and patterning is functionally significant, and is dealt with in greater detail under Floral Biology (p.53).

ANDROECIUM

There are usually four, didynamous, fertile stamens and no trace of a staminode in the truly posticous position (but in the following text the stamens are designated posticous and anticous pair). The only exception is *Antherothamnus*, where the fifth stamen is represented by a fully developed filament lacking an anther. In some genera, parts of genera, or in individual species, there are only two stamens by abortion of either the posticous or the anticous pair. (See further below, and also Table 1). The disposition of the stamens (anthers included or exserted) can also be of generic importance (see Table 2). The anthers are perfectly syntheceous, this being one of the salient characters of the whole tribe. They are dorsifixed and usually split wide open with the two valves rolling back; sometimes, particularly in very small anthers, they may open by a longitudinal slit, and they always do so in *Antherothamnus* and *Manuleopsis*. (For information on pollen, see p.61). The filaments are usually inserted about or above the middle of the corolla tube (near the base in *Antherothamnus* and some very short-tubed species of *Jamesbrittenia*). The posticous filaments are often decurrent in very narrow wings or ridges: this is a generic character, and must also be of biological significance (a channel to guide the tongue of a pollinator?); only in *Glekia*, *Manulea* and *Sutera* are the posticous filaments not decurrent.

There are misleading statements about the anticous pair of stamens being sterile in *Zaluzian-skya* (Hiern, 1904, p.123, 333; Dyer, 1975, p.542). In this genus, there are either four fertile stamens or two by abortion of either the anticous or the posticous pair; staminodes are rarely present and never consistently so. The posticous anthers are elongate, included, held vertically and often larger than the anticous anthers; the anticous anthers are held ± tranversely and are either shortly exserted or in the mouth.

As mentioned above, reduction in the number of stamens from four to two occurs in several genera, and is often useful in distinguishing between species; only the ditypic *Reyemia* is characterized by the possession of only two stamens. The following summary is expanded under the relevant species; in some species of *Phyllopodium* and *Polycarena* reduction is associated with autogamy.

Jamesbrittenia: 2 posticous stamens in mouth, 2 anticous stamens either wanting or greatly reduced (almost staminodes, but a few pollen grains often present).
   J. glutinosa.

Lyperia: 2 posticous stamens in mouth, 2 anticous stamens reduced to staminodes.
   L. antirrhinoides, L. formosa, L. tenuiflora, L. violacea.

Manulea: Either 2 stamens only (posticous pair) or anticous pair represented by 1 or 2 staminodes or greatly reduced stamens.
   M. diandra.

Phyllopodium: total or partial reduction of posticous stamens.
   P. anomalum: only 2 stamens.
   P. micranthum: normally 4 stamens, but individual plants show various stages of reduction to 2 stamens, 1 or 2 staminodes.
   P. mimetes: normally 4 stamens, rarely posticous anthers much reduced in size.

Polycarena: either the posticous or the anticous stamens aborted or reduced.
   P. aemulans: either 4 stamens, or anticous pair aborted or greatly reduced.
   P. comptonii: either 4 stamens, or posticous pair aborted or reduced to staminodes.
   P. tenella: either 4 stamens, or reduced to 2 or 3, sometimes a staminode present.

Reyemia: 2 anticous stamens reduced to staminodes.

Strobilopsis: either 4 stamens, or 1 or 2 anticous ones aborted.

Zaluzianskya
    Sect. *Zaluzianskya*: posticous pair retained, anticous pair aborted.
      Z. affinis, Z. gracilis, Z. parviflora, Z. pilosissima, Z. sanorum, Z. villosa.
    Sect. *Holomeria*: anticous pair retained, posticous pair aborted.
      Z. benthamiana, Z. diandra.

Table 2, showing the position of the anthers, indicates the diagnostic value of this character. It must also have biological significance.

Table 2. Position of anthers.

	All included	All exserted	Anticous only exserted	Anticous visible in mouth
Antherothamnus			+	
Camptoloma			+	
Glekia			+	
Glumicalyx		+	+	
Jamesbrittenia	+			+ rarely
Lyperia	+			+
Manulea	+			+
Manuleopsis			+	
Melanospermum		+	+	
Phyllopodium		+		
Polycarena		+		+
Reyemia		+		
		(only posticous pair present)		
Strobilopsis		+		
Sutera	+	+	+	
	(2 species only)			
Tetraselago		+		
Trieenea		+		+
Zaluzianskya	+		+	+

GYNOECIUM

Stigma

The stigma in most genera of Manuleae, including *Manulea* itself (see Table 1), is lingulate with two marginal bands of stigmatic papillae confluent at the tip; the tip is occasionally very minutely bifid, but this is not constant within a species. In most of these genera, the stigma is shorter than to roughly as long as the style, from which it may or may not be sharply differentiated, and it is usually exserted, though sometimes only briefly. In *Manulea* sect. *Dolichoglossa*, the stigma is much longer than the style and exserted; in sect. *Manulea* and sect. *Medifixae* it is usually longer than the style, and is included; in sect. *Thyrsiflorae*, it is either shorter than the style or about equals it, and is included. Post-fertilization enlargement of the ovary will of course often push the stigma out of a corolla in which it is included at anthesis.

    In all but two species of *Sutera*, the stigma is lingulate and exserted; in the close allies, *S. cooperi* and *S. griquensis*, the stigma is shortly bifid, the two arms united by a jelly-like receptive surface, and it is deeply included.

    In *Jamesbrittenia*, the stigma is minutely bifid with terminal papillae and is included; in *Camptoloma* too it is minutely bifid, the branches initially pressed together, separating presumably when the stigmatic papillae became receptive, and it reaches the mouth; the papillae are in a very short lateral groove that is continued over the tip, making the stigma virtually terminal. In *Lyperia lychnidea* and *L. tristis*, the stigma consists of two elliptic, eventually spreading, sticky cream-coloured lobes, in the other species of two short narrow lobes closely pressed together, the external faces covered in papillae; it is included in all species. The stigma of *Antherothamnus* is very short, almost capitate, with terminal papillae. *Manuleopsis* has a very shortly bifid, almost clavate, stigma, the two branches pressed together, the papillae mostly terminal. See fig. 12.

The position of the stigma in relation to the anthers is, in most genera, such that the pollen is shed onto it; see further under breeding systems, (p.53).

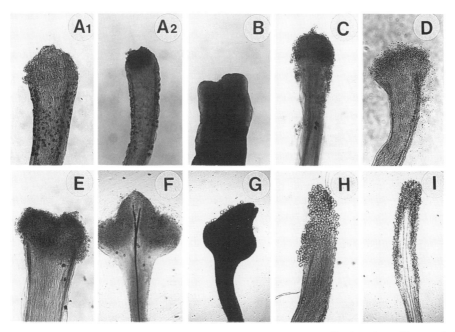

Fig. 12. Stigmatic form in genera of Manuleae. A1,2, *Antherothamnus pearsonii*, stigmatic papillae terminal; B, *Manuleopsis dinteri*, stigmatic papillae wanting; C, *Camptoloma lyperiiflorum*, stigmatic papillae terminal; D, *Jamesbrittenia foliolosa*, stigmatic papillae terminal; E, *J. jurassica*, minute bifurcation visible; F, *Lyperia tristis*, stigmatic papillae wanting; G, *L. lychnidea*, stigmatic papillae wanting, minute droplets (lipid?) visible; H, *L. antirrhinoides*, stigmatic papillae smothered in pollen grains; I. *Sutera foetida*, lingulate stigma with lateral and terminal stigmatic papillae; a lingulate stigma can be much longer than this and is sometimes acute. A–E, H × 33, F, G, I × 15.

Style

The style in all the genera is simple, ± cylindric, and may be glabrous or hairy, the hairs glandular or eglandular, characters sometimes useful in distinguishing between species.

Ovary

The ovary is bicarpellary, bilocular, with axile placentation; the posticous loculus is often slightly shorter than the anticous because a nectariferous gland intrudes into it at the base. In all but four genera, the nectary always comprises a small gland, often strongly adnate to the base of the posticous loculus; it does so too in all genera of Selagineae. Precisely this type of nectary has not yet been found in any other tribe of Scrophulariaceae.

In *Zaluzianskya*, the shape of this gland, peg-like or rounded, and its adnation or not to the ovary, helps to distinguish the four sections of the genus (fig. 64). In *Jamesbrittenia*, there is either a distinct gland or an annular disc that is sometimes more strongly developed on the posticous side. In *Sutera*, the nectary is semi-annular and often more strongly developed on the posticous side, while in *Camptoloma*, *Manuleopsis* and *Antherothamnus*, there is no clearly evident nectary.

Nine of the seventeen genera of Manuleae (in common with most genera of Scrophulariaceae) have many ovules in each loculus. The other eight genera show varying degrees of reduction in ovule-number (see Table 1); reduction may characterize species or genera. For example, *Polycarena batteniana* has ± 16 ovules in each loculus, *P. exigua* c.7, *P. filiformis* and *P. tenella* c.4; the other 13 species all have many ovules. *Phyllopodium alpinum*, which is without close allies, produces 5–8 seeds in each loculus; the other 25 species all have many-seeded capsules. *Antherothamnus* produces 6–12 ovules in each loculus, while the closely allied *Manuleopsis*, with much larger flowers, has many. *Glekia* has 9–16 ovules. All three genera are monotypic. *Trieenea* (9 species) always has few ovules in the loculus and consequently sets few seeds: *T. laxiflora* (up to 9), *T. lasiocephala* (c.10), *T. lanciloba* (c.6–7), *T. taylori* (c.6–10), *T. frigida* (c.6–10), *T. elsiae* (c.8), *T. schlechteri* (c.8–12), *T. longipedicellata* (up to 8), *T. glutinosa* (up to c.14).

A reduction in ovule-number that appears to be of phylogenetic significance is seen in the three genera, *Glumicalyx*, *Strobilopsis* and *Tetraselago*, which are placed in this sequence at the end of the taxonomic enumeration. They are allied, but not closely so. *Glumicalyx goseloides*, *G. nutans*, *G. flanaganii* and *G. lesuticus* all have many (at least 35) ovules in each loculus, in *G. montanus* there are mostly 20–25, but up to 35, and 7–12 in *G. apiculatus*. The monotypic *Strobilopsis* has 1–6 ovules in each loculus (see also Hilliard & Burtt, 1977, p.157), while *Tetraselago* (4 species) has only two, with the further peculiarity that they arise in the middle of the placenta and one ovule faces upwards, the other down (fig. 68, and see also Junell, 1961, p.188, fig. 7; note too the nectariferous gland in fig. 7a and the massive cushion funicle in fig. 7b).

The significance of this reduction in ovule-number, particularly that of *Tetraselago*, in a consideration of the relationship of Selagineae to Manuleae, is elaborated on p.67.

FRUIT

The fruit is always a dry capsule that dehisces septicidally; the valves are shortly split loculicidally. This type of capsule is characteristic of Manuleae, but within the tribe its features are mostly of minor taxonomic importance.

The capsule is either ovate or oblong-oval in outline. A very acute to beaked capsule is one of the discriminating features of *Manulea* section *Thyrsiflorae*, and the capsule of all three species of *Camptoloma* is also sharply beaked. The relatively large capsules of *Lyperia* are acute and very glandular with both glistening glands and gland-tipped hairs. In most genera, the capsules are glabrous; they may be glabrous, hairy or clad in glistening glands in *Jamesbrittenia*, glabrous or with glistening glands in *Sutera*, rarely minutely glandular-puberulous in *Manulea*, and this may be helpful in distinguishing species. In *Melanospermum*, *Phyllopodium*, *Polycarena* and *Trieenea*, the ovary is sometimes minutely glandular-puberulous, but the hairs rarely persist on the capsule except occasionally at the apex.

The young fruit in many genera is parasitized by a beetle (bristly weevil), the loculus then being occupied by a single beetle, its larva nourished by the ovules. Galling has also been seen in *Manulea* and *Sutera*. Under normal circumstances, seed-set is good, particularly among the annuals, where selfing seems to be common. The manner in which the seeds are dispersed has not been recorded; it is possibly a censer mechanism.

PLACENTAE

At maturity, the placentae are covered in pulvini, more or less round cushion-like structures, centrally depressed and there with a round scar left by the fallen seed. Each pulvinus is derived from a funicle. These funicular pulvini are not confined to Manuleae; they occur in other tribes as well (see Weber, 1989, figs 9, 13, 14, *Charadrophila*, *Sutera*, *Rehmannia*, *Anticharis*,

Alonsoa, and accompanying text). The pulvini are often discrete, as in all the genera illustrated by Weber as well as in most genera of Manuleae, (figs 13–18). In some genera, however, the pulvini are strongly confluent. They are always so in *Antherothamnus* (fig. 13B), *Camptoloma* (fig. 13C), *Lyperia* (fig. 14B), in *Zaluzianskya* section *Nycterinia* (fig. 17A), and in many species of *Jamesbrittenia* (fig. 13D). In *Polycarena*, they are discrete but tightly packed (figs 16B, C). In the few-seeded capsule of *Strobilopsis*, the pulvini are strongly deflexed (fig. 18B) and suggest a possible evolutionary pathway that could have led to the pair of funicles in each loculus of the 4-seeded capsule of *Tetraselago* (fig. 18D). A further small step along that pathway could have led to the similar but solitary massive funicle in the 1-seeded cocci of members of Selagineae.

SEEDS

Seed morphology

Seed morphology has long been used as a character in the classification of the Scrophulariaceae, at family, tribal, generic and species level. The advent of scanning electron microscopy has heightened its usefulness as evidenced by many recent publications including Sutton's revision of Antirrhineae (Sutton, 1988). The importance of seed morphology in the classification of the Manuleae has hitherto gone unrecognized, yet, at the practical level, many specimens can be assigned without hesitation to their genus on the basis of the seed alone, notably *Camptoloma* (longitudinally 6–8-ribbed, brown), all but two species of *Jamesbrittenia* (reticulate, brown), *Lyperia* (relatively large, transversely elongated pits, black), *Melanospermum* (longitudinal rows of relatively large deep round pits, black), *Polycarena* (3-winged or at least trigonous, buff), *Phyllopodium* subgen. *Leiospermum* (smooth, blue), *Zaluzianskya* sect. *Macrocalyx* (deeply furrowed and ridged or winged, pallid or mauve), *Glumicalyx* (irregularly colliculate, dark reddish-brown), *Strobilopsis* (somewhat compressed, sinuously colliculate and pitted in ± longitudinal bands, pallid becoming blackish-brown), *Tetraselago* (8 sinuous longitudinal bands, amber turning dull black). *Sutera* and *Manulea* have similar seeds with transversely elongated pits, mostly pallid to grey-blue or violet-blue, amber-coloured in some species of *Sutera*. The two genera differ in the ornamentation of the testa as seen under the SEM. See Table 1 and fig. 1 for a summary of seed characters.

The paragraph above immediately draws attention to the taxonomic importance of differences in surface sculpturing, and differences in colour, a much neglected but often useful discriminatory character. Seeds appear sometimes to change colour as they mature, and in *Sutera* and *Manulea* in particular I have not always been sure that seeds are consistently either pallid or violet-blue, which diminishes the value of the character; immature seeds can be deep violet-blue, mature ones pallid. Colour possibly results, not from pigmentation but from anatomical differences that develop as the seed matures. Hartl (1959, p.109) pointed out that differences in the structure of the inner endothelial wall result in differences in colour in the seeds of *Verbascum* (drab ochre to bright brown) and *Scrophularia* (black to brown). Sutton (1988, p.47), in discussing seed-colour in *Linaria*, states 'This effect of colour may be the result of optical interference patterns arising from the fine surface ornamentation of the cuticle.' In *Polycarena*, the testa is very thin, loose, and transparent; the buff colour of the seed possibly arises merely from the creamy-coloured endosperm being visible through it.

Scrophulariaceae as a whole is microspermous. The seeds of Manuleae range in length from 0.25–2mm. Seeds up to c.0.5mm long are characteristic of *Camptoloma* and *Phyllopodium* subgen. *Leiospermum*; the largest seeds occur in *Antherothamnus* (c.1.4mm), *Lyperia* (1–1.5mm), *Glekia* (1–1.25mm), *Zaluzianskya* sect. *Nycterinia* and sect. *Macrocalyx* (1–2mm), *Strobilopsis* (1.4mm). The commonest size-range is 0.5–1mm. See Table 1.

In all members of Manuleae, the hilum is approximately terminal. The commonest shape of the seed is ellipsoid or oblong-ellipsoid; the seeds of *Polycarena* are 3-winged or trigonous. In *Antherothamnus*, *Manuleopsis*, *Jamesbrittenia* and *Phyllopodium* subgen. *Leiospermum*, the

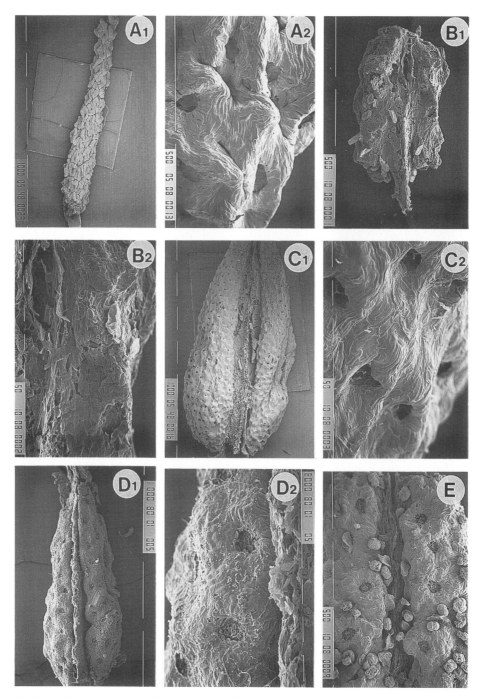

Fig. 13. Form of the placentae: A, *Manuleopsis dinteri* (*Giess & Robinson* 13262); B, *Antherothamnus pearsonii* (*Örtendahl* 533); C, *Camptoloma rotundifolium* (*Merxmüller & Giess* 28146); D, *Jamesbrittenia aurantiaca* (*Hilliard & Burtt* 3113); E, *J. grandiflora*, note glistening glands (*Compton* 25133). Scale bars: A1, C1, D1 = 1mm, A2, B1, E = 0.5mm, B2, C2, D2 = 0.05mm.

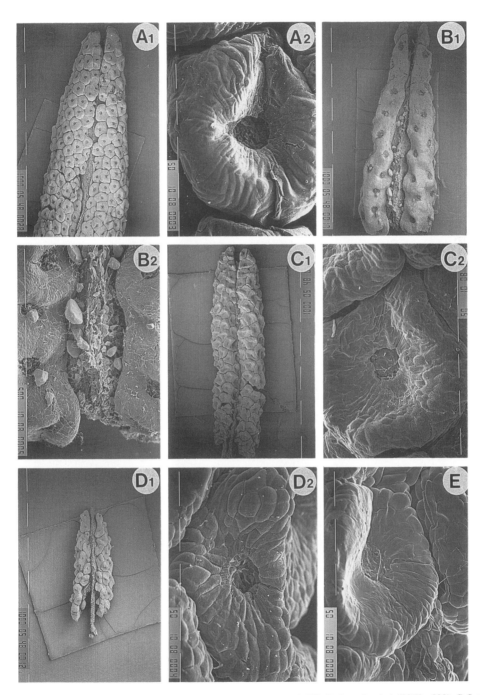

Fig. 14. Form of the placentae: A, *Jamesbrittenia fruticosa* (*Middlemost* 2130); B, *Lyperia tristis* (*Müller* 833); C, *Sutera foetida* (*Batten* 496); D, *Sutera archeri* (*Dean* 960); E, *Sutera floribunda* (*Alexander* 8). Scale bars = A1, B1, C1, D1 = 1mm, B2, C2, D2, E = 0.5mm, A2 = 0.05mm.

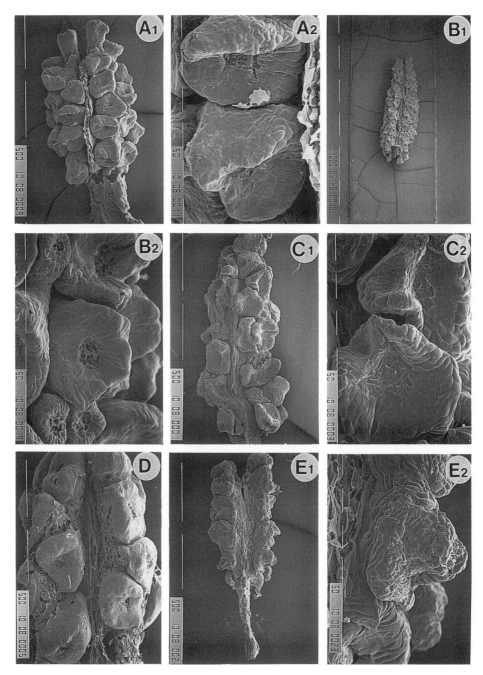

Fig. 15. Form of the placentae: A, *Manulea cheiranthus* (*Goldblatt* 5921); B, *Manulea robusta* (*Dinter* 6075); C, *Melanospermum italae* (*Hilliard & Burtt* 10025); D, *Glekia krebsiana* (*Hilliard & Burtt* 12247); E, *Trieenea laxiflora* (*Esterhuysen* 34828). Scale bars: B1 = 1mm, A1, C1, D, E1 = 0.5mm, A2, B2, C2, E2 = 0.05mm.

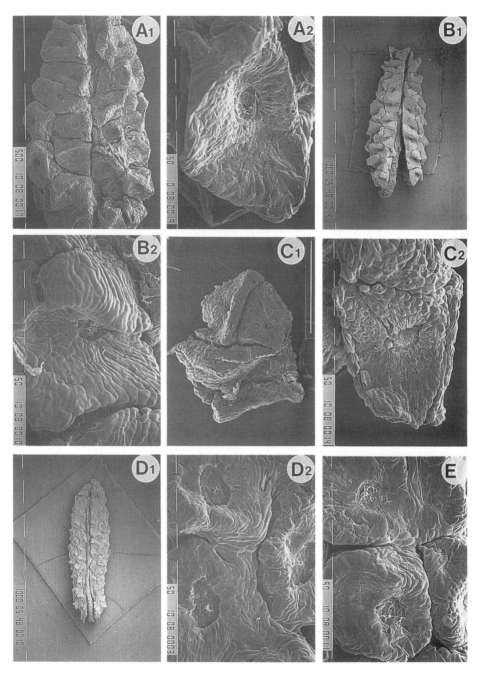

Fig. 16. Form of the placentae: A, *Trieenea glutinosa* (*Esterhuysen* 35915); B, *Polycarena capensis* (*Hilliard & Burtt* 13029); C, *Polycarena tenella* (*Esterhuysen* 20552); D, *Phyllopodium bracteatum* (*Hilliard & Burtt* 11008); E, *Phyllopodium pumilum* (*Batten* 1028). Scale bars: B1, D1 = 1mm, A1, C1 = 0.5mm, A2, B2, C2, D2, E = 0.05mm.

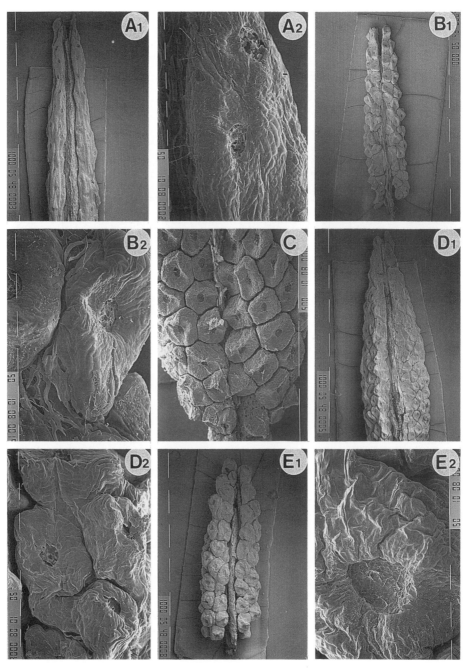

Fig. 17. Form of the placentae: A, *Zaluzianskya capensis*, section *Nycterinia* (*Batten* 1051); B, *Z. villosa*, section *Zaluzianskya* subsection *Zaluzianskya* (*Batten* 862); C, *Z. peduncularis*, section *Zaluzianskya* subsection *Noctiflora* (*Batten* 1037); D, *Z. divaricata*, section *Holomeria* (*Bokelmann & Paine* 10); E, *Z. pumila*, section *Macrocalyx* (*Batten* 1012). Scale bars: A1, B1, D1, E1 = 1mm, C = 0.5mm, A2, B2, D2, E2 = 0.05mm.

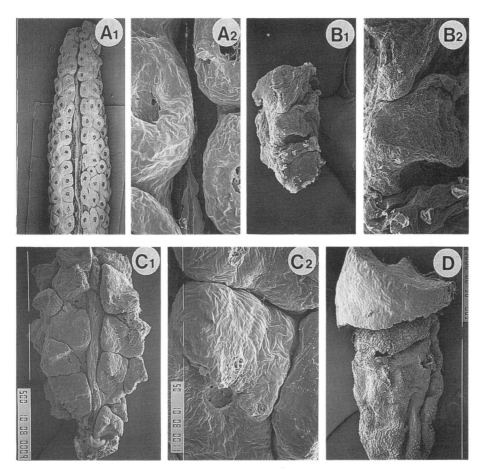

Fig. 18. Form of the placentae: A, *Reyemia chasmanthiflora* (*Batten* 1016); B, *Strobilopsis wrightii* (*Hilliard & Burtt* 17540); C, *Glumicalyx montanus* (*Hilliard* 8136); D, *Tetraselago natalensis* (*Hilliard & Burtt* 19168), massive funicle with seed attached, one of a pair in each loculus, one pendulous, one erect. Scale bars: A1 = 1mm, B1, C1, D = 0.5mm, A2, B2, C2 = 0.05mm.

seed is round in transverse section; in the rest of the genera (excluding *Zaluzianskya* sect. *Macrocalyx*, seeds strongly ridged and/or winged; *Strobilopsis* and *Tetraselago*, seeds somewhat compressed), the essential roundness is masked by the ridging, pitting or ruminations of the endosperm. All gross surface sculpturing, including wings, is modelled by the underlying endosperm. These surface patternings of the seed are visible at very low magnification, and are therefore of practical diagnostic value.

Further details of classificatory value reside in the cells of the testa and are visible under the SEM. In describing these, I have followed the sequence suggested by Barthlott (1984, p.95), namely, 1) cellular arrangement or pattern, which is visible under low magnification, 2) shape of the cells, 3) relief of the outer cell walls, 4) cuticular secretions.

Features of the seed-surface

1, *Antherothamnus* and 2, *Manuleopsis*: about 20 longitudinal rows of roughly isodiametric tetragonal to hexagonal cells forming a reticulum, anticlinal walls straight, raised, outer periclinal walls concave, irregularly verruculose (figs 19A, B).

3, *Camptoloma*: about 20 longitudinal rows of roughly isodiametric tetragonal to hexagonal cells forming a reticulum, the cells crowded over the tops of the 6–8 longitudinal ridges, anticlinal walls ± straight, raised, outer periclinal walls concave, irregularly verruculose and longitudinally striate (fig. 19C).

4, *Jamesbrittenia* Group A: about 15–30 longitudinal rows of roughly isodiametric tetragonal to hexagonal cells forming a reticulum, anticlinal walls straight, raised, outer periclinal walls concave, striate (seen in 12 species) or verruculose (seen in 8 species). Figs 19D, 20, 21, 22A, B.

Jamesbrittenia Group B: as group A with striate outer periclinal walls but anticlinal walls verrucose ('knotted'). Figs 22C, 23A (5 species examined). *Jamesbrittenia racemosa* is further distinguished by longitudinal rows of transverse pits in the endosperm, arranged like a chequer-board (fig. 22D).

Jamesbrittenia Group C: about 15–30 longitudinal rows of cells, the outline of individual cells not clear, but the rows demarcated by irregular thickenings ('knots') on the longitudinal anticlinal walls, which are linked by irregular raised striations (figs 23B, C, D, 24). At low magnifications the impression given is that of a reticulum.

5, *Lyperia*: 6–8 longitudinal rows of transversely elongated pits arranged like a chequer-board. In *L. lychnidea* and *L. tristis*, the cells of the testa are roughly isodiametric, pentagonal to hexagonal, anticlinal walls straight, their boundaries raised, outer periclinal walls plane, minutely and irregularly verruculose. The testa of the other species is probably similar, but good scans are lacking (fig. 25A).

6, *Sutera* sect. *Sutera*: 6–8 longitudinal rows of transversely elongated pits arranged like a chequer-board. The cells of the testa are slightly elongated longitudinally, anticlinal walls straight, their boundaries raised, outer periclinal walls plane, minutely verruculose (figs 25B, C).

Sutera sect. *Chaenostoma*: either 6–8 rows of transversely elongated pits as in sect. *Sutera* (fig. 25D), or the pits apparently weakly developed and creating 6–8 sinuous ridges (figs 26A, B, C). The individual cells of the testa are not clearly visible, the whole surface being verrucate.

7, *Manulea*: 6–8 rows of transversely elongated pits as in *Sutera* sect. *Sutera*. The individual cells of the testa are not clearly visible, the whole surface being verrucate and not unlike that of *Sutera* sect. *Chaenostoma* (fig. 26D).

8, *Melanospermum*: 6–8 longitudinal rows of 3–8 deep round pits. The cells of the testa line the pits, are isodiametric, the boundaries of the anticlinal walls channelled, the outer periclinal wall hemispherical, surface minutely verruculose (fig. 27A).

9, *Polycarena*: seed 3-winged or trigonous, individual cells of the testa not easily seen, isodiametric, outer periclinal walls with radial wrinkles culminating in a short, blunt central tubercle, probably cuticular in origin (figs 27B, C).

10, *Glekia*: seed irregularly wrinkled, cells of the testa isodiametric, anticlinal walls straight, their boundaries channelled, outer periclinal wall strongly domed with a somewhat flattened apex, whole surface minutely ruminate (fig. 27D).

11, *Trieenea*: seed irregularly wrinkled, cells of testa isodiametric, anticlinal walls straight, their boundaries channelled, outer periclinal walls convex, verrucate (figs 28A, B, C).

12, *Phyllopodium* subgen. *Phyllopodium*:
Groups 1 and 2: seed irregularly wrinkled, cells isodiametric, tetragonal or hexagonal, outlines not always easily seen.
Subgroup 1A: anticlinal walls straight, their boundaries slightly raised, outer periclinal walls convex, minutely verruculose (figs 28D, 29A).

Subgroup 1B: anticlinal walls straight, their boundaries slightly raised, outer periclinal walls with radial wrinkles culminating in a short blunt central pustule, probably cuticular in origin (fig. 29B).

Subgroup 1C: anticlinal walls straight, their boundaries slightly raised and verruculose, outer periclinal walls convex, minutely verruculose (fig. 29C).

Subgroup 2A: anticlinal walls straight, their boundaries channelled, outer periclinal walls convex, minutely verruculose (fig. 29D).

Subgroup 2B: anticlinal walls straight, their boundaries channelled, outer periclinal walls plane or slightly convex, minutely verruculose (fig. 30A).

Phyllopodium anomalum, P.alpinum (appendage of Group 2): anticlinal walls straight, their boundaries channelled, outer periclinal walls convex, verrucate (figs 30B, C).

Phyllopodium subgen. *Leiospermum*: seed perfectly smooth, cells isodiametric, tetragonal to hexagonal, anticlinal walls straight, their boundaries channelled, outer periclinal walls plane to slightly convex, minutely verruculose (fig. 30D).

13, *Zaluzianskya*: seeds either irregularly wrinkled or, in sect. *Macrocalyx*, deeply ridged or winged, cells isodiametric, pentagonal to heptagonal, anticlinal walls straight, their boundaries channelled, outer periclinal walls either almost plane to convex and verruculose (sect. *Nycterinia*, sect. *Zaluzianskya*, sect. *Holomeria*, figs 31A, D, 32A, B, C) or convex, radially wrinkled and culminating in an obscure to well-defined tubercle (sect. *Macrocalyx*, fig. 31B, C).

14, *Reyemia*: seed irregularly wrinkled, cells roughly isodiametric, anticlinal walls weakly to strongly undulate, their boundaries channelled, outer periclinal walls almost plane to convex, obscurely verruculose (figs 32D, 33A).

15, *Glumicalyx*: seed irregularly wrinkled, cells isodiametric, pentagonal to hexagonal, anticlinal walls straight, their boundaries channelled, outer periclinal walls convex, apex rounded or flattened, verruculose (figs 33B, C, D).

16, *Strobilopsis*: seed somewhat flattened with sinuous shallow longitudinal ridges and shallow pits, cells isodiametric, pentagonal to hexagonal, anticlinal walls straight, their boundaries channelled, outer periclinal walls convex, verruculose (fig. 34A).

17, *Tetraselago*: seed with eight sinuous flattened ridges, cells roughly isodiametric, pentagonal to hexagonal, anticlinal walls straight, their boundaries channelled, outer periclinal walls weakly to strongly convex, verruculose (fig. 34B).

Seed anatomy

The seeds of *Camptoloma*, *Lyperia*, *Manulea*, *Sutera* and two species of *Jamesbrittenia* have alveolate endosperm. The alveolations (pits) result from certain cells of the endothelium intruding deeply into the endosperm; such cells are termed bothroblasts. The endothelium consists of plane hexagonal cells arranged in 6–8 longitudinal rows running from micropyle to chalaza. The width of these cells is always much greater than their length so that 6–8 of them encompass the seed (they are easily seen if a whole young seed is cleared in lactic acid and the testa stripped off). Further development follows one of two patterns. In the first pattern, alternate cells (abothroblasts) in each longitudinal row divide transversely into four; these divisions are then followed by anticlinal divisions in the micropyle-chalaza direction. The inner periclinal wall of each cell (bothroblast) alternating with the abothroblasts intrudes deeply into the underlying endosperm thus forming a pit, elliptic in outline because it is so strongly stretched transversely. Only the abothroblasts and that part of each bothroblast external to the endosperm can stretch transversely; the inner part of each bothroblast is restrained by the endosperm in which it is embedded, and the cell becomes mushroom-shaped. In transverse section, the endosperm is star-shaped as a result of the 'stalks of the mushrooms' intruding into it (fig. 38). (Hartl, 1959, pp.95, 99, fig. 1,b,c). As the seed matures, the testa collapses into the

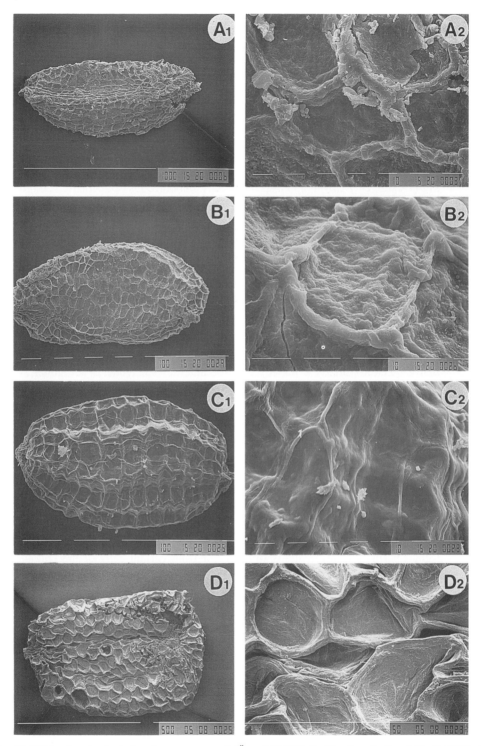

Fig. 19. SEMs of seeds. A, *Antherothamnus pearsonii* (*Örtendahl* 533); B, *Manuleopsis dinteri* (*Rodin* 2934); C, *Camptoloma lyperiiflorum* (*Gallagher* 6252/11); D, *Jamesbrittenia fodina* (*Wild* 5775). Scale bars: A1 = 1mm, D1 = 0.5mm, B1, C1 = 0.1mm, D2 = 0.05mm, A2, B2, C2 = 0.01mm.

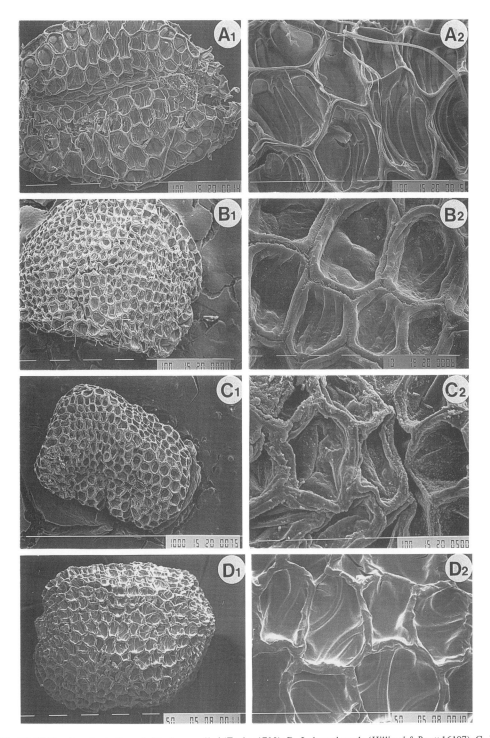

Fig. 20. SEMs of seeds. A, *Jamesbrittenia carvalhoi* (*Taylor* 1703); B, *J. dentatisepala* (*Hilliard & Burtt* 16197); C, *J. huillana* (*Hilliard & Burtt* 10318); D, *J. filicaulis* (*Hilliard & Burtt* 12237). Scale bars: C1 = 1mm, A1,2, B1, C2 = 0.1mm, D1,2 = 0.05mm, B2 = 0.01mm.

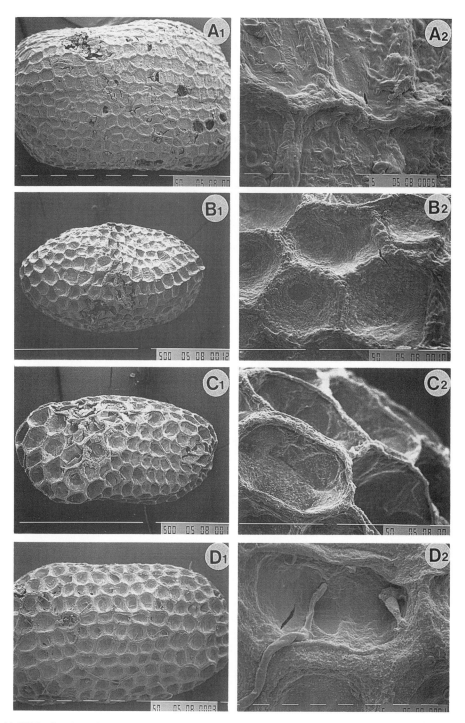

Fig. 21. SEMs of seeds. A, *Jamesbrittenia tortuosa* (*Vlok* 2305); B, *J. maritima* (*Joffe* 589); C, *J. argentea* (*Gillett* 1200); D, *J. tenuifolia* (*Middlemost* 2191). Scale bars: B₁, C₁ = 0.5mm, A₁, B₂, C₂, D₁ = 0.05mm, A₂, D₂ = 0.005mm.

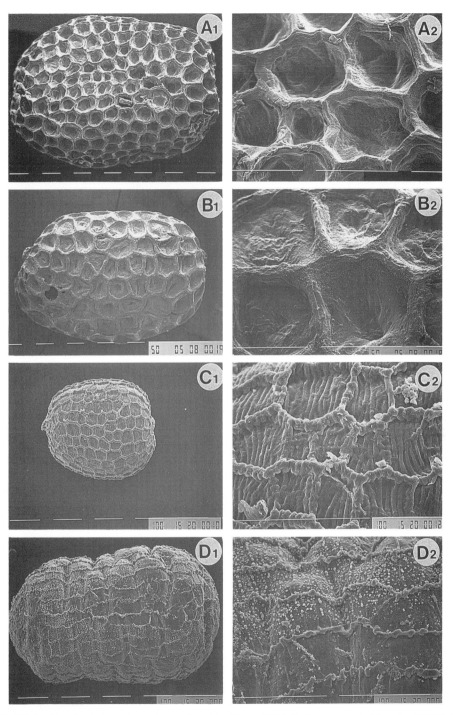

Fig. 22. SEMs of seeds. A, *Jamesbrittenia aspalathoides* (*Hugo* 1266); B, *J. microphylla* (*Compton* 13237); C, *J. amplexicaulis* (*Boucher* 3132); D, *J. racemosa* (*Batten* s.n.). Scale bars: C, D = 0.1mm, A, B = 0.05mm.

Fig. 23. SEMs of seeds. A, *Jamesbrittenia thunbergii* (*Thompson* 374); B, *J. pallida* (*Mendes* 147); C, *J. pilgeriana* (*Leippert* 4046); D, *J. crassicaulis* (*Hilliard* 3939). Scale bars = 0.1mm except C = 0.05mm.

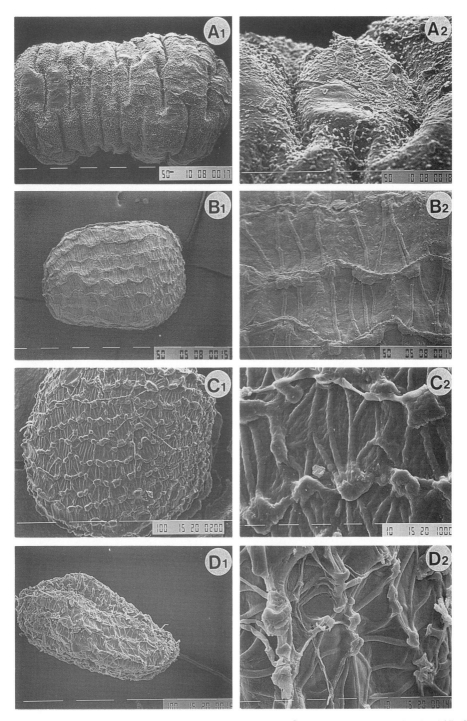

Fig. 24. SEMs of seeds. A, *Jamesbrittenia pedunculosa* (*Rosch & le Roux* 274); B, *J. tenella* (*Schelpe* 164); C, *J. aurantiaca* (*Hilliard & Burtt* 14417); D, *J. dissecta* (*Madden* 263). Scale bars: C1, D1 = 0.1mm, A1,2, B1,2 = 0.05mm, C2, D2 = 0.01mm.

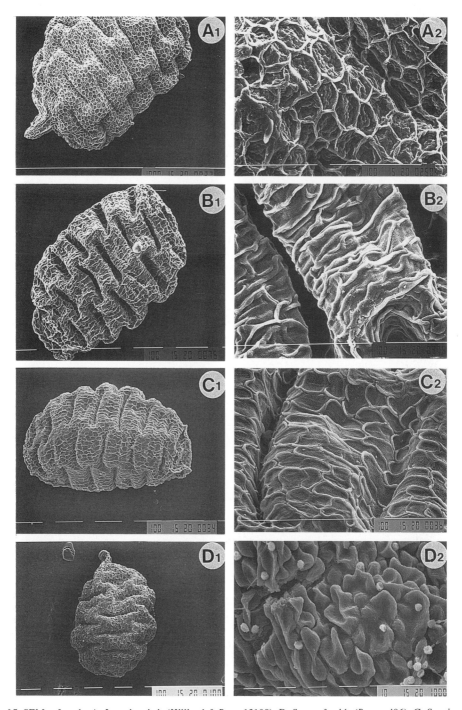

Fig. 25. SEMs of seeds. A, *Lyperia tristis* (*Hilliard & Burtt* 13108); B, *Sutera foetida* (*Batten* 496); C, *S. griquensis* (*Esterhuysen* 995); D, *S. caerulea* (*Schlechter* 8259). Scale bars: A1 = 1mm; A2–D1 = 0.1mm; D2 = 0.01mm.

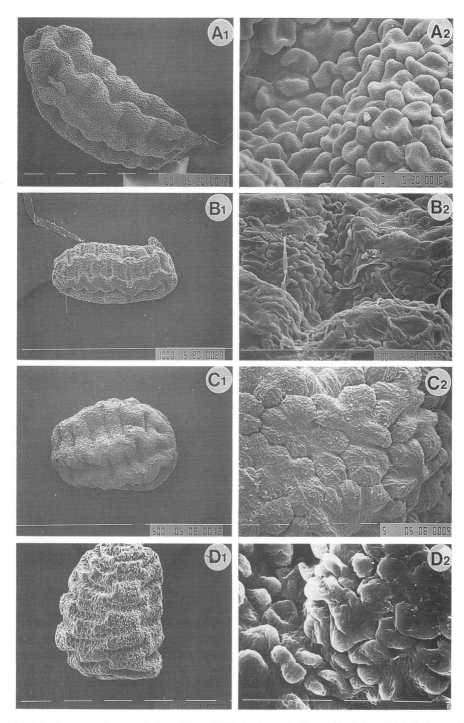

Fig. 26. SEMs of seeds. A, *Sutera subspicata* (*Oliver* 8519); B, *Sutera floribunda* (*Schlechter* 2834); C, *Sutera debilis* (*Maguire* 1417); D, *Manulea gariesiana* (*Batten* 739A). Scale bars: B1 = 1mm; C1 = 0.5mm; A1, B2, D1 = 0.1mm; A2, D2 = 0.01mm, C2 = 0.005mm.

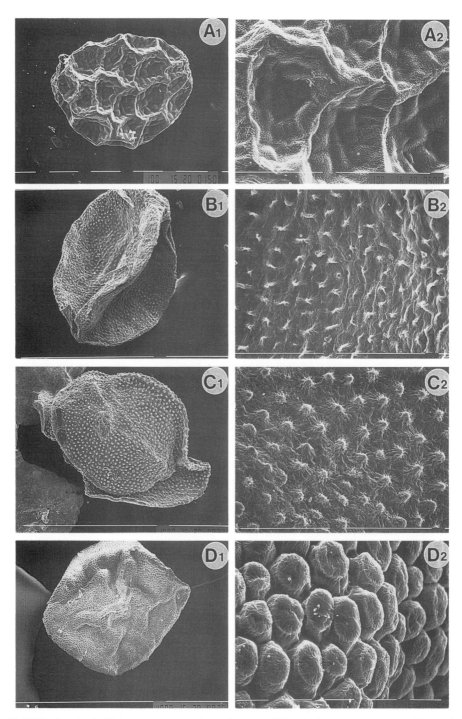

Fig. 27. SEMs of seeds. A, *Melanospermum swazicum* (*Compton* 25620); B, *Polycarena capensis* (*Hilliard & Burtt* 13029); C, *Polycarena filiformis* (*L.E. Taylor* 2798); D, *Glekia krebsiana* (*Hilliard & Burtt* 10652). Scale bars: B1, C1, D1 = 1mm; A, B2, C2 = 0.1mm; D2 = 0.01mm.

Fig. 28. SEMs of seeds. A, *Trieenea taylorii* (*Compton* 12733); B, *Trieenea lanciloba* (*Taylor* 11903); C, *Trieenea schlechteri* (*Esterhuysen* 22449); D, *Phyllopodium cuneifolium* (*Hilliard & Burtt* 10857). Scale bars: A1, B1, C1, D1, D2 = 0.1mm, A2, B2, C2 = 0.01mm.

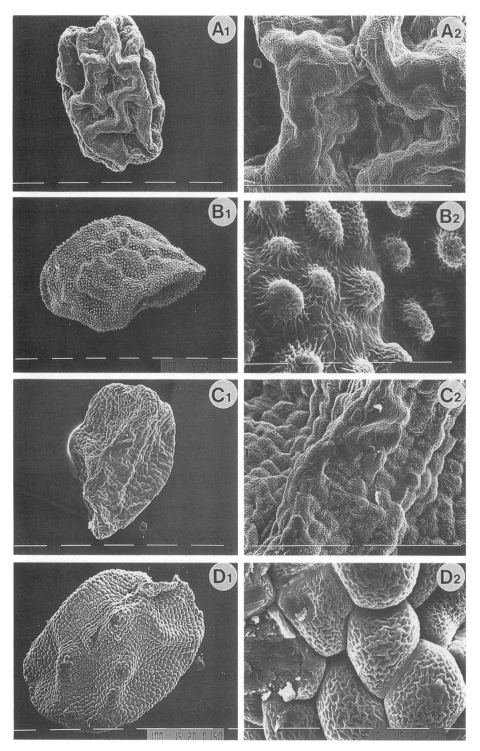

Fig. 29. SEMs of seeds. A, *Phyllopodium diffusum* (*Sim* 1322); B, *P. dolomiticum* (*Acocks* 20463); C, *P. micranthum* (*Compton* 16233); D, *P. heterophyllum* (*Hilliard & Burtt* 13069). Scale bars: A, B1, C, D1 = 0.1mm, B2, D2 = 0.01mm.

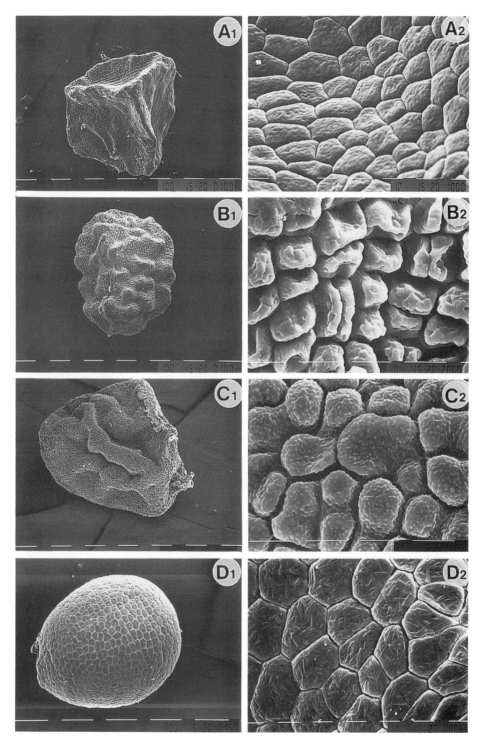

Fig. 30. SEMs of seeds. A, *Phyllopodium maxii* (*Max Schlechter* 102); B, *P. anomalum* (*Schlechter* 8195); C, *P. alpinum* (*Gibbs* 10035); D, *P. pumilum* (*Thompson* 1084). Scale bars: A1–D1 = 0.1mm, A2–D2 = 0.01mm.

41

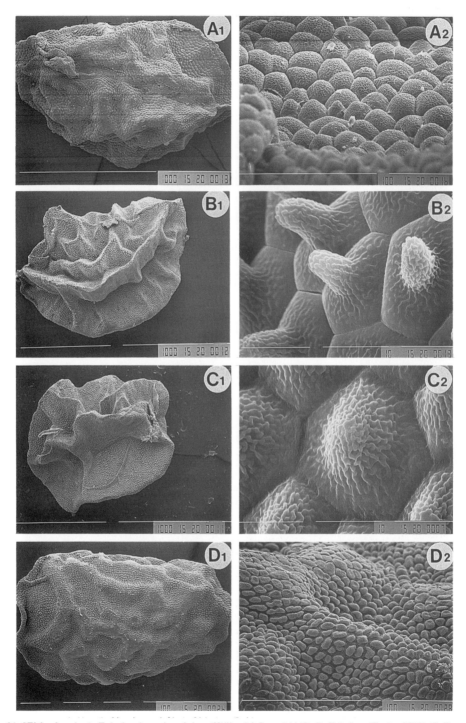

Fig. 31. SEMs of seeds. A, *Zaluzianskya microsiphon* (*Hilliard & Burtt* 14449); B, *Z. inflata* (*Taylor* 2819); C, *Z. pumila* (*Batten* 742); D, *Z. villosa* (*Lavranos* 11681). Scale bars: A1, B1, C1 = 1mm, A2, D = 0.1mm, B2, C2 = 0.01mm.

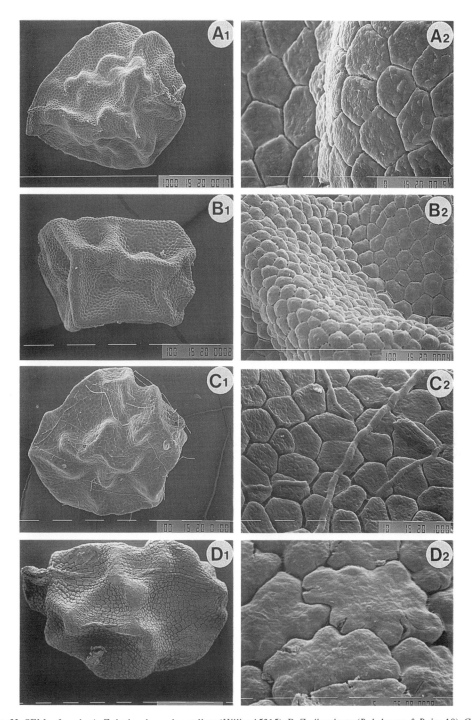

Fig. 32. SEMs of seeds. A, *Zaluzianskya rubrostellata* (*Hilliard* 5315); B, *Z. divaricata* (*Bokelmann & Paine* 10); C, *Z. diandra* (*Hardy* 748); D, *Reyemia chasmanthiflora* (*Batten* 1016). Scale bars: A1 = 1mm, B, C1 = 0.1mm, D1 = 0.05mm, A2, C2 = 0.01mm, D2 = 0.005mm.

Fig. 33. SEMs of seeds. A, *Reyemia nemesioides* (*Steiner* 800); B, *Glumicalyx montanus* (*Hilliard* 8136); C, *G. apiculatus* (*Hilliard & Burtt* 16666); D, *G. goseloides* (*Hilliard & Burtt* 6915). Scale bars: B1, D1 = 1mm; A1, C1 = 0.1mm, A2, B2, C2, D2 = 0.01mm.

Fig. 34. SEMs of seeds. A, *Strobilopsis wrightii* (*Hilliard & Burtt* 17540); B, *Tetraselago natalensis* (*Hilliard & Burtt* 14179). Scale bars: A1 = 1mm, B1 = 0.5mm, A2 = 0.01mm, B2 = 0.005mm.

pits. Such seeds are termed bothrospermous (bothro = pit): the surface of the seed has alternating rows of transversely elongated pits producing a chequer-board effect, as in *Lyperia*, *Manulea*, *Sutera*, *Jamesbrittenia racemosa* and *J. pedunculosa*.

The second pattern of development results in aulacospermous seeds, seeds with 6–8 ridges alternating with furrows (aulax = furrow), seen in *Camptoloma*. In aulacospermous seeds, all the endothelial cells remain the same size, their inner periclinal walls protrude into the endosperm (which is thus star-shaped in transverse section), and when the outer layers of the integument, including the testa, collapse into these invaginations of the endothelial cells, a furrow is formed. Weber (1989, pp.99, 109, fig. 10) describes and illustrates the aulacospermous seeds of *Charadrophila*, which closely resemble those of *Camptoloma*.

The seeds of *Melanospermum* are patterned with deep round pits, but their origin is obviously not due to the development of bothroblasts (fig. 39). Many of the other genera have irregularly wrinkled or deeply and irregularly furrowed or winged endosperm, as noted above. There is great scope for developmental studies, which might help to elucidate phylogeny.

Sectioning of seeds proved difficult; in nearly all cases, seed had to be taken from herbarium specimens and was often old; also, orientation of such small seeds often proved difficult. Although the anatomical diagrams offered are far from satisfactory, they make one point clear: the strong development of the cuticle surrounding the endosperm and the role it must play in maintaining the, often intricate, shape of the seed. In contrast, the testa is thin and unsupportive. Figs 35–42.

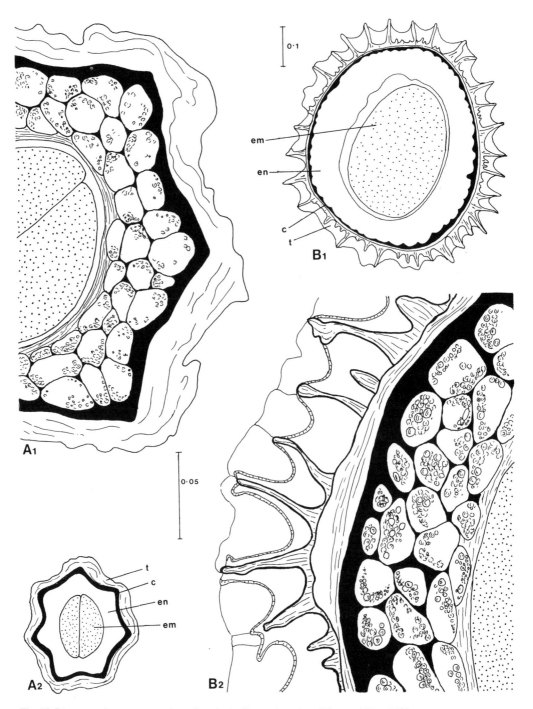

Fig. 35. Diagrams of transverse sections of seeds. A, *Camptoloma lyperiiflorum* (*Miller* 2755), mature aulacospermous seed, integument and testa collapsed into the invaginations of the endothelial cells that form the furrows; B, *Jamesbrittenia huillana* (*Hilliard & Burtt* 10318), outer periclinal walls of the cells of the testa collapsed inwards. Scale bars 0.1, 0.05mm. t = testa, c = cuticle, en = endosperm, em = embryo.

46

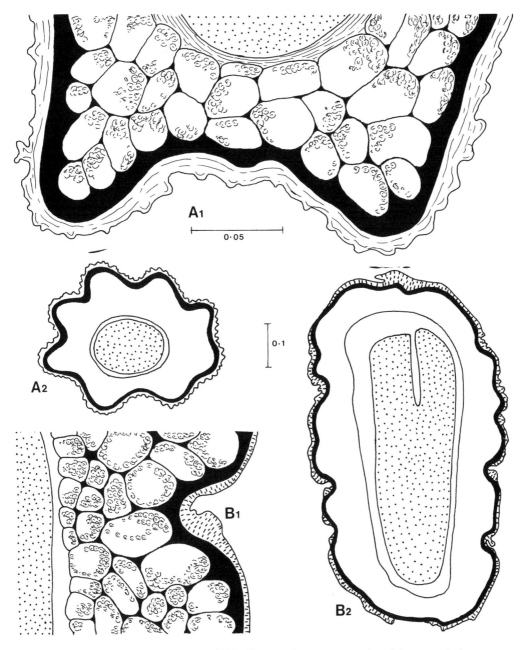

Fig. 36. A, *Lyperia tristis* (*Hilliard & Burtt* 13108): diagrams of a transverse section of the mature, bothrospermous, seed, the testa collapsed inwards into the invaginations of the endothelial cells; B, *Jamesbrittenia racemosa* (*Batten* s.n.): diagram of a longitudinal section of the mature, bothrospermous, seed, the testa collapsed inwards into the invaginations of the endothelial cells. Scale bars 0.1, 0.05mm.

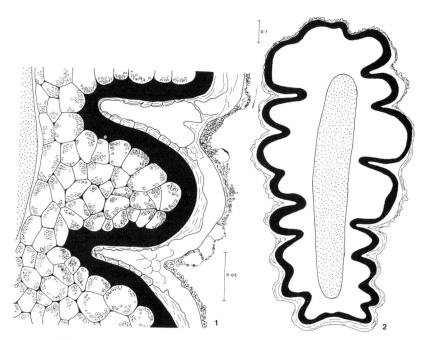

Fig. 37. *Lyperia tristis* (*Hilliard & Burtt* 13108), diagrams of a longitudinal section of the mature, bothrospermous, seed. Scale bars 0.1, 0.05mm.

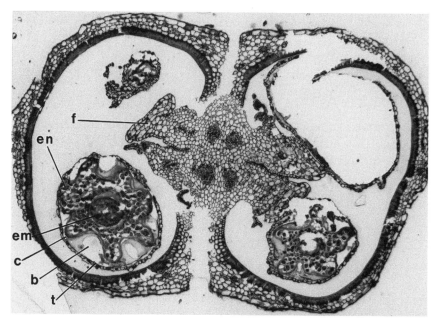

Fig. 38. *Sutera roseoflava* (*Batten* 1000), transverse section of nearly mature capsule showing two young seeds in transverse section: f = funicle; b = bothroblast (slightly distorted by pressure); c = cuticle surrounding endosperm; en = star-shaped endosperm; em = embryo; t = testa, not yet collapsed into the pits being formed by the bothroblasts. c. × 50.

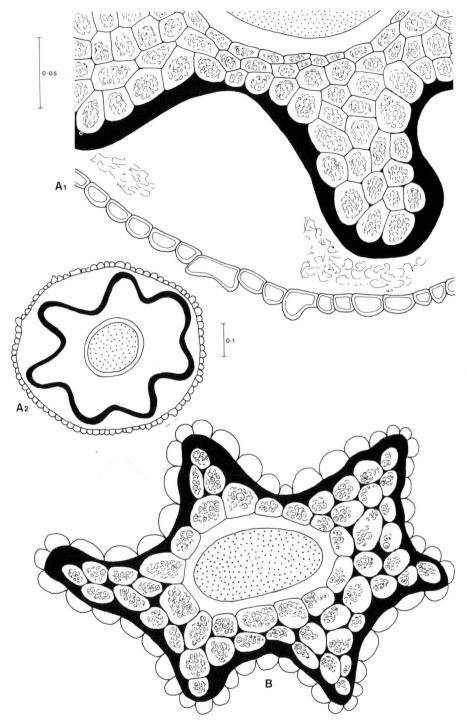

Fig. 39. Diagrams of transverse sections of seeds. A, *Sutera roseoflava* (*Batten* 1000), immature bothrospermous seed, testa not yet collapsed into the invaginations of the endothelial cells that form the pits; B, *Melanospermum swazicum* (*Compton* 25620), large cells of the testa lining the pits. Scale bars 0.1, 0.05mm.

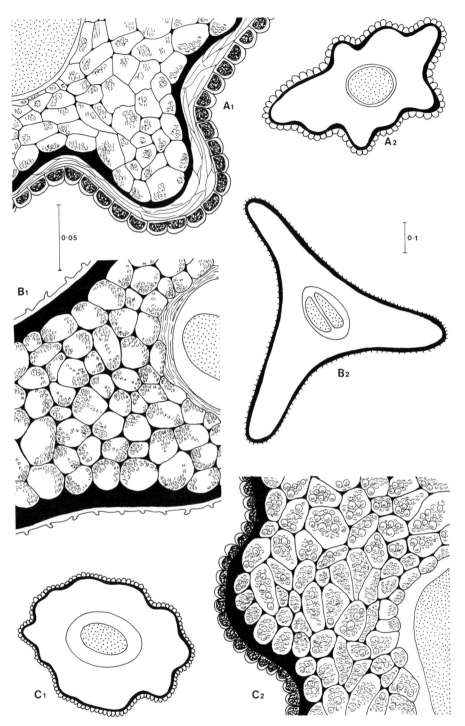

Fig. 40. Diagrams of transverse sections of seeds. A, *Glekia krebsiana* (*Hilliard & Burtt* 10652); B, *Polycarena capensis* (*Hilliard & Burtt* 13029); C, *Tetraselago natalensis* (*Hilliard & Burtt* 19168). Scale bars = 0.1, 0.05mm.

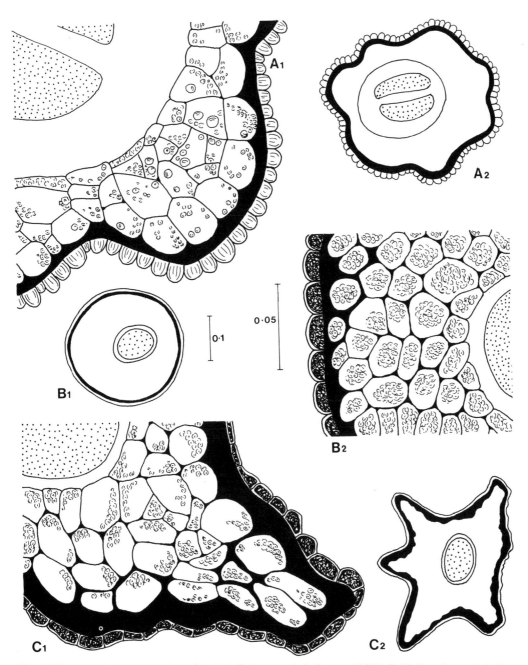

Fig. 41. Diagrams of transverse sections of seeds. A, *Trieenea taylorii* (*Compton* 12733); B, *Phyllopodium cephalophorum* (*Bokelmann* 35); C, *Phyllopodium multifolium* (*Maguire* 2596). Scale bars = 0.1, 0.05mm.

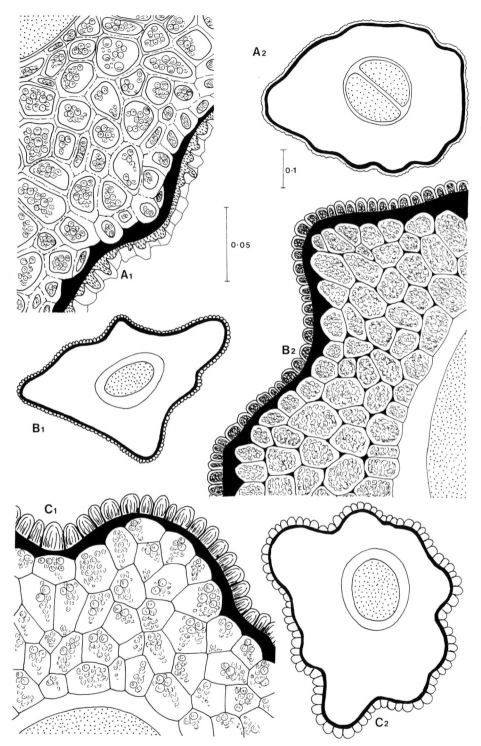

Fig. 42. Diagrams of transverse sections of seeds. A, *Zaluzianskya villosa* (*Lavranos* 11681); B, *Strobilopsis wrightii* (*Hilliard & Burtt* 17540); C, *Glumicalyx montanus* (*Hilliard* 8136). Scale bars = 0.1, 0.05mm.

CHEMICAL CHARACTERS

The only information on the chemical characteristics of members of Manuleae known to me is that published by Kooiman (1970). Kooiman wrote (p.333) 'The Manuleae are heterogeneous in respect of the presence of iridoid glycosides; even the genus *Sutera* is heterogeneous'. Kooiman was using the name *Sutera* in the broad sense then current. If one rearranges his results according to my classification, a more significant picture emerges:

Taxa devoid of glycosides: *Sutera caerulea* [*Sphenandra viscosa*], *S. campanulata*, *S. foetida*.
Taxa containing aucuboside and/or catalpol as well as other unidentified glucosides: *Lyperia antirrhinoides* [*Sutera antirrhinoides*], *Zaluzianskya capensis*.

FLORAL BIOLOGY

POLLINATION

This revision of Manuleae is almost entirely an herbarium study: my field knowledge is confined to a relatively small number of species in less than half the genera dealt with; I never had the luck to see pollinating agents at work and there is very little published information. However, even in the herbarium, patterns of diversification in certain floral features present themselves and suggest response to different pollinators, the selective force most likely to be operative on the functional unit, the flower. Throughout the tribe, the form of the corolla conforms to a single structural plan: a tubular flower with a 5-lobed limb. The features subject to diversification are the overall size of the flowers, their massing or not in heads or spikes, the length and breadth of the corolla tube, the shape of the limb, the shape and stance of the lobes, their colour and contrasting colour patterns, the presence or absence of almost sessile glands on the upper surface of the limb, the pattern of distribution of one-celled clavate hairs mainly inside the tube, the presence or absence inside the tube of a narrow channel formed by the decurrent posticous filaments, the degree of exsertion or inclusion of the anthers and stigma and their position in relation to each other, and the presence or absence of a nectary. Glandular hairs are common on the vegetative parts, and many plants are pleasantly or unpleasantly scented; the flowers also emit various odours and there may be strong periodicity in their emission. Unfortunately, collectors seldom record information on colour or odour, but at least contrasting colour patterns can often be detected on dried specimens that have not blackened (blackening is common in sections of *Zaluzianskya*, *Jamesbrittenia* and *Lyperia*).

Among the genera of Manuleae there are many instances of convergence in the shape and colouring of the flowers, probably in response to a set of pollinators held in common; convergence may extend to whole genera, to parts of genera, sometimes to species in other tribes or even families. Field studies would certainly extend examples of the latter beyond those I can offer. Ethologically based convergence is a well documented phenomenon: see for instance Procter & Yeo (1973, p.375), Pijl, van der (1961, p.47), Vogel (1954, p.45), Dafni *et al.* (1990). The colour plates, plates 1–5, give a sample, not comprehensive coverage of the range of corolla form and colour patterning in Manuleae.

Flowers pollinated by bees and bee-flies

A common and striking colour patterning among the small-flowered members of Manuleae is the presence of a yellow or orange patch associated with the posticous lip; this often extends down the back of the tube. Occasionally both posticous lobes are wholly orange or orange in the upper half in addition to the patch in the throat. Some examples are shown in plate 1. The yellow/orange patch is nearly always associated with a white or light violet/mauve limb, rarely with a creamy to clear yellow one. The corolla tube is either narrowly funnel-shaped or cylindric widening somewhat in the throat, mostly 1.5–7.5mm long, 1–2mm across the throat, the limb bilabiate with rounded or oblong lobes, stamens normally four, either all exserted or sometimes two in the throat, the posticous filaments decurrent down the tube forming a narrow channel (to guide an insect's tongue? to act as a nectar capillary?) and the lingulate stigma is exserted with the stamens. There is a small nectariferous gland at the base of the ovary on the posticous side (the same side as the filament-channel). This suite of characters occurs in *Melanospermum*, *Phyllopodium*, *Trieenea*, many (12) species of *Polycarena* and in *Strobilopsis*. Furthermore, the flowers are massed in rounded heads or crowded spikes or racemes, and they may sometimes simulate a single flower (e.g. *Melanospermum*, plate 1I, *Strobilopsis*, *Polycarena*, plate 1G).

An orange/yellow patch at the back of the throat also occurs in *Manulea* sect. *Dolichoglossa* subsect. *Isantherae* where the flowers resemble those of *Melanospermum* etc.: the limb may be mauve or white, the lobes are rounded, the clavate hairs spread laterally, and the anticous anthers are visible in the mouth (the posticous ones are included). *Manulea buchneroides* (sect. *Dolichoglossa* subsect. *Dolichoglossa*) also displays this colour pattern (plate 1A); the flowers are borne in corymbose heads and bear a striking resemblance to those of the white form of *Buchnera simplex* (tribe *Buchnerae*) with which it grows.

The colour patterns described above are not confined to members of Manuleae, but occur in the allied tribe Selagineae as well as in other families. Those species of *Selago* that fall into sect. *Spurieae* display the combination of a yellow/orange patch at the back of the throat and clavate hairs either there alone or all round the mouth, a white or mauve limb surmounting a narrowly cylindric corolla tube, decurrent posticous filaments, and exserted stamens and stigma. The flowers are massed.

A white or light violet/mauve/pink limb with either a yellow/orange tube or throat is characteristic of *Sutera*: tube 2.5–22mm long, funnel-shaped, 1.5–5mm across the mouth, limb 7–15mm across, clavate hairs usually present at least in the back of the throat, stamens four, mostly well exserted together with the lingulate stigma, posticous filaments not decurrent, a nectariferous gland at the base of the ovary on the posticous side. There is thus considerable similarity to the syndrome of characters evinced by *Melanospermum* etc.

Little has been published on the floral ecology of South African plants and the only records pertinent in the present context are those of Scott Elliot (1891), Marloth (1932) and Vogel (1954). Vogel recorded small bees as the pollinator of '*Sutera* cf. *caerulea*'. I have no means of checking Vogel's determination but his illustration (1954, p.243 fig. 133, 1) shows the typical corolla of *S. caerulea*, which has a very short yellow tube and rotate mauve limb not unlike that of *S. archeri* (plate 1H). Vogel also recorded small bees as the pollinators of *Selago quadrangularis* (sect. *Spurieae*) (Vogel, 1954, pp.247, 248, fig. 137, 1). Marloth (1932, plate 37A) shows a butterfly (*Pseudonymphe vigilans*) visiting an inflorescence of *Selago serrata* (sect *Spurieae*) and on the same plate (37B), *Stilbe ericoides* (Verbenaceae) (with mauve flowers and exserted yellow anthers in a crowded head, looking not at all unlike a species of *Phyllopodium* or *Polycarena*) being visited by a wasp (*Odynerus*). But Marloth points out in the text (p.144) that the wasp, although it may carry pollen from plant to plant, is really there to hunt its animal prey such as spiders, which are in turn waiting for the bees or small butterflies 'to which visitors the flowers are really adapted'.

Mrs Batten (in litt.) recorded long-tongued bee-flies on *Polycarena batteniana*, which has a white limb and yellow throat: 'they hovered, inserting the proboscis into flower after flower'. It would seem then that the flowers displaying this particular suite of features may be pollinated by bees, bee-flies, butterflies, or, accidentally, by wasps.

Flowers pollinated by hawkmoths and moths

A very different syndrome, presumably adapted to hawkmoths, obtains in nearly all the species of *Zaluzianskya* sect. *Nycterinia* and sect. *Zaluzianskya* subsect. *Noctiflora*, which are nocturnal. The flowers have a narrowly cylindric corolla tube up to 50mm long with the decurrent posticous filaments forming a channel down to the posticous nectariferous gland at the base of the ovary. The limb is regular, the backs of the lobes are mostly red to dark purplish brown and when folded over into a 'bud' during the day, render the flower very inconspicuous. When light intensity drops, the limb unfolds to reveal the pure white (sect. *Nycterinia*, plate 2A) or cream to bright yellow (subsect. *Noctiflora*, plate 2F) upper surface of the limb, clearly visible in dim light. The lobes reflex and the flowers, scentless during the day, emit a sharp sweet fragrance. The limb of each flower is carried more or less horizontally, the lobes in sect. *Nycterinia* being bifid, those in subsect. *Noctiflora* entire. The limb in some species of sect. *Nycterinia* has a bright orange 'eye' (and then looks remarkably like the flower of *Jamesbrittenia dentatisepala*, which may be growing nearby, plate 4A,D), while the limb in three species of subsect. *Noctiflora* has either a scarlet star-shaped patch around the mouth (plate 2F) or a scarlet band in the throat; the colour patch is probably an osmophore. These characters (long corolla tube, tongue groove, closure during day, strong periodicity of scent) indicate adaptation to visiting hawkmoths: see for example Faegri & Pijl (1971, p.136). The upper surface of the lobes may be glandular, particularly around the mouth, another character associated with hawkmoths. Vogel (1954, p.245, fig. 134, 2) records *Z. capensis* as being sphingophile and draws attention to the mimetic similarity between certain day-flowering as well as night-flowering Cape species of *Zaluzianskya* and species of *Silene*. Hilliard & Burtt (1987, pp.27, 28), writing about floral guilds in the Natal Drakensberg, also mentioned (independently of Vogel and with reference to different species) the mimicry between *Zaluzianskya* and *Silene* in that area. To this floral guild can now be added *Jamesbrittenia lesutica*, whose flowers are startlingly similar to those of the sympatric *Zaluzianskya ovata*.

Mrs Batten made observations on *Reyemia chasmanthiflora* (plate 2C), which is nocturnal, and noted not only changes in the position of the corolla lobes but also the emission of a strong scent of carnations. Further details are given under the species.

Vogel (1954, p.244) also notes (as species of *Sutera*) *Lyperia tristis*, *L. lychnidea* and *Jamesbrittenia atropurpurea* as being sphingophile and becoming sweetly scented at night. Flowers in both these genera may have the long narrow corolla tube, almost regular limb, included stamens and basal nectariferous gland characteristic of *Zaluzianskya*. The corolla of the two species of *Lyperia* ranges from greenish or pale drab yellow to bright yellow and is glandular round the mouth (plate 2I); it is whitish in *L. antirrhinoides* (plate 2G).

Jamesbrittenia atropurpurea has thick-textured corolla lobes ranging in colour from greenish yellow through dark oranges and browns to maroon or nearly black, always with a paler, thin-textured margin, and rendered narrower than they really are by the rolling under of the margins: a deeply dissected corolla limb is known to be attractive to hawkmoths. A corolla limb more or less resembling that of *J. atropurpurea* in form, colour and texture is found in *J. huillana*, *J. burkeana* (plate 4F), *J. silenoides* (plate 2D), *Zaluzianskya acutiloba* (plate 2J), *Jamesbrittenia albomarginata* (plate 2H), *J. merxmuelleri*, *Zaluzianskya diandra* (plate 2K) and a number of species of *Manulea* including *M. cheiranthus* (see further below). In *Jamesbrittenia albomarginata* (plate 2H) and *J. merxmuelleri* the narrowing of the lobes is an optical illusion stemming from broad white margins flanking a median brown patch. *Zaluzianskya acutiloba* is recorded as opening in daylight and emitting a strong scent. The thick-textured tissue of the corolla lobes is probably an osmophore, a tissue producing volatile secretions as attractants for pollinators (Vogel, 1990, defines and discusses osmophores). The dark patches in *Lyperia antirrhinoides* (plate 2G) and the red ones in *Zaluzianskya rubrostellata* (plate 2F) are also possibly osmophores. All six species of *Glumicalyx* have thick-textured, pale-margined corolla lobes, but they are almost round and the tightly congested inflorescence is nutant (plate 2E).

Vogel (1954, p.246, fig. 135, 2) suggested that *Manulea cheiranthus* is phalenophile (moth-pollinated); this species, and others including *M. adenodes*, *M. burchellii* and *M. gariepina*, have a relatively short cylindric corolla tube (up to c.12mm long) widening slightly in the throat to accommodate the included anthers, and a zygomorphic limb up to c.15mm in diameter (plate 2B) held more or less vertically in elongated racemes; the anticous lobe is isolated from the other four lobes, which form an upper lip above the oblique mouth; the corolla lobes are more or less subulate to the eye (revolute margins) and range in colour from greenish to cream, fawn, mustard-yellow, dull orange and reddish-brown, colour probably dependent on the age of the flower (see further below). Vogel (loc. cit.) thought *Holothrix* (Orchidaceae) and *Agathelpis* (Scrophulariaceae-Selagineae), among others, could form a floral guild with *Manulea* species.

Flowers pollinated by butterflies

Vogel (1954, pp.244, 245, figs 134, 1, 135, 1) gave several examples of members of Manuleae that are pollinated by butterflies (psychophily). These include *Zaluzianskya villosa* and *Z. divaricata*, both of which are diurnal. *Zaluzianskya villosa* (plate 3B) and its allies have corolla tubes mostly 8–30mm long, the limb regular, the lobes bifid or retuse, white to shades of pink or violet on the upper surface with yellow/orange/red nectar guides radiating from the mouth; *Z. divaricata* and its allies have corolla tubes similar in length to those of *Z. villosa* and allies, but the lobes are entire and creamy-white to bright yellow or orange above, patterned with orange/red streaks (see *Z. collina*, plate 3E). Four species of *Manulea* (*M. acutiloba*, *M. silenoides*, *M. gariesiana*, *M. praeterita*) always have a regular corolla limb with bifid white, blue or mauve lobes marked with yellow nectar guides; *M. fragrans* (plate 3M) sometimes has bifid lobes. The flowers of these manuleas bear a striking resemblance to those of *Zaluzianskya villosa* and its allies. Some species of *Polycarena* also share this syndrome, but the corolla lobes are rounded, not bifid (see *P. capensis*, plate 3J).

Change in colour of nectar guides and corolla limb

Nectar guides are present in species other than those mentioned above, including many in *Jamesbrittenia* (e.g. plate 3A,C,L, plate 4C), in some sections of *Zaluzianskya* (notably absent in section *Nycterinia*, where all flowers are nocturnal), in *Camptoloma* (plate 3I), *Lyperia* (plate 3G), *Manuleopsis* (plate 4E), *Antherothamnus* (plate 4H). Some at least exhibit striking changes in the colour of these guides as the flower passes from maturity to age. In other species, it is the whole limb, or limb and guides, that changes colour (plate 5). Dr Martha Weiss (1991) has pointed out that floral colour change was first noted almost 200 years ago but that the presence and significance of the phenomenon have gone largely unrecognized. She showed that 'flowers in at least 74 diverse angiosperm families undergo dramatic, often localized, colour changes which direct the movements of a variety of pollinators to the benefit of both participants'. She found that 'retention of older flowers increases a plant's attractiveness to pollinators from a distance, that pollinators discriminate between floral colour phases at close range, and that discrimination involves learning'. The dipteran, hymenopteran and lepidopteran pollinators she observed consistently visited pre-change flowers (that is, those that were ready for pollination and offered a reward to the pollinator) more frequently than the older, differently coloured, flowers. Vogel (1990, p.154) has discussed the relationship between change in colour and emission of fragrance.

In the course of field work, we had noted (and sometimes photographed) colour changes in *Zaluzianskya* and *Manulea*; Mrs Batten sent colour slides of several species of *Zaluzianskya* and *Jamesbrittenia* showing this feature, and in working through thousands of herbarium specimens, it became obvious that change in colour in relation to pollination is a common occurrence, though much will of course have escaped me. Some examples can be given. Two allied species of *Zaluzianskya*, *Z. villosa* and *Z. affinis*, common and widespread in the western and south-western Cape, bear abundant flowers in corymbose clusters; the corolla limb, initially white or pale mauve with a conspicuous yellow star-shaped patch around the mouth, gradually turns mauve, the star turning orange then crimson (plate 3B shows the initial phase). A similar colour change occurs in *Z. crocea* (plate 5) and *Z. vallispiscis* from the eastern Cape. In *Z. pumila*, the 'star' is formed by a thick-textured bar occupying the lower half of each corolla lobe; this changes from yellow to orange to bright scarlet, with the rest of the limb changing from white to mauve. At least the 'star' changes colour in other species of *Zaluzianskya*, including *Z. sanorum*, *Z. pilosissima*, *Z. karrooica*, *Z. minima* and *Z. divaricata*; the limb is yellow in the latter species. In *Z. acutiloba* the whole limb changes from yellow to dark brown (plate 2J) and a collector recorded 'strong scent'.

Several different patterns of colour-change seem to be common in *Jamesbrittenia*. From dried specimens alone, it is clear that the nectar guides in *J. stellata* change from yellow to orange to red, and the limb has been recorded as white, pink or mauve. In *J. fruticosa*, the limb opens rich violet and fades to white; the nectar guides probably darken with age: purple-maroon-blackish. In *J. racemosa*, the very strongly developed nectar guides appear to change from yellowish-brown to deep violet, while the limb passes from white to violet. *Jamesbrittenia atropurpurea* and its allies have been mentioned above; in these species, the limb possibly changes from greenish- or yellowish-brown to dark brown. The limb of *J. angolensis* has been recorded as 'at first pale yellow later white'; the red-brown throat probably changes colour too; *J. dolomitica* appears to change from creamy-white to yellow, the throat from yellow to orange; *J. fleckii* and *J. lyperioides* (plate 4C) may change from yellow to orange.

Jamesbrittenia ramosissima signals to its pollinators with a colour-change from mauve or blue to white (or possibly vice versa), as do *Manulea nervosa* and *M. schaeferi*. There is a remarkable degree of floral mimicry between these three species (further comment on p.196).

Many more species of *Manulea* may exhibit colour-change in the corolla limb, particularly from pale shades of yellow to darker yellow, orange and brown, often accompanied by a darkening of the colour-patch at the back of the throat, for example *M. burchellii*, *M. karrooica*, *M. thyrsiflora*, *M. platystigma*, *M. obovata*, *M. cheiranthus*, *M. decipiens*, *M. gariepina*, *M. pusilla*, *M. namibensis*, *M. virgata*. In species with a white limb and a yellow mouth, the mouth changes to orange, for example, *M. buchneroides* (plate 1A), *M. corymbosa*, *M. adenocalyx*, *M. psilostoma*, *M. plurirosulata*, while other species have a star-shaped patch that changes from yellow to orange or red.

Much more information can be culled from the formal descriptions of the species, particularly those of *Jamesbrittenia*, *Manulea* and *Zaluzianskya*.

Colour contrast between upper and lower surface of corolla limb

The biological significance of the undersurface of the corolla limb being differently and often more brilliantly coloured than the upper is not clear. As mentioned above, it is strikingly displayed in the nocturnal flowers of *Zaluzianskya* but coppery-coloured backs can be seen in some of the day-flowering species (plates 3, 5). In several species of *Polycarena*, the corolla lobes are flushed purple on the backs (this and other colour patterns are commented on in the introduction to the account of *Polycarena*) while in *Phyllopodium heterophyllum* the outside of the lobes usually has dark blotching or feathering. In *Glekia krebsiana*, the lobes always have strongly marked blue-purple patches on the backs, and these turn brownish as the flower ages (plate 4I).

Floral mimicry

Floral mimicry has been mentioned above in connection with a good many species of Manuleae. Another remarkable example is that of *Jamesbrittenia pedunculosa*, the flowers of which are unlike those of any other species of the genus (plate 4G), but closely resemble those of species of *Hemimeris* (Scrophulariaceae tribe Hemimerideae) with which *J. pedunculosa* may grow. The flowers are borne on remarkably long pedicels, which Vogel cites as characteristic of micromellitophile flowers, giving as an example *Sutera [Jamesbrittenia] kraussiana* (plate 4B and Vogel, 1954, pp.45, 46, fig. 4, 1). *Jamesbrittenia kraussiana* is just one of many quite unrelated species (members of 13 different families are quoted by Vogel) that have flowers on filiform pedicels borne singly in the axils of the upper leaves. There are several such species in *Jamesbrittenia* other than the two mentioned above (e.g. *J. argentea*, *J. phlogiflora*, *J. pinnatifida*), all with a white or mauve/violet strongly bilabiate limb that is yellow in the throat as in *J. kraussiana*.

Corolla hairs

The biological function of corolla hairs in the Manuleae is virtually unknown. In nearly all the species, clavate hairs are present either inside the corolla tube or, more rarely, on the upper surface of the limb; glandular hairs may also be present, but they are not nearly so common (however, they are frequently present on the outside of the corolla).

It is a striking fact that clavate hairs are often associated with a yellow/orange patch; the disposition of the hairs thus differs from genus to genus. In those species with one or two yellow patches on the posticous lip, often extending down the back of the throat, clavate hairs are usually confined to those coloured patches (*Melanospermum, Phyllopodium, Trieenea, Strobilopsis*, many species of *Polycarena*). In those species of *Polycarena* with yellow all round the mouth, the hairs follow that pattern as they do in species of *Manulea* resembling *Polycarena*. In a great many species of *Manulea*, there is a yellow and purple patch at the back of the throat and the hairs are found there and on the posticous filaments adjoining the patch, the hairs on the filaments being yellow, orange or purple. Scott Elliot (1891, p.368) suggested that in *M. cheiranthus* and its immediate allies, the infolding of the lines of union of the petals form two canals down which an insect's proboscis must pass; the clavate hairs arranged in longitudinal bands down these folds prevent any of the plant's own pollen being shaken down the tube, but brush pollen onto the proboscis as it is withdrawn. Unfortunately he failed to detect a pollinator.

In *Sutera*, it is mostly the top of the corolla tube that is yellow and the hairs are distributed accordingly. The flowers of *Glumicalyx* lack yellow patches, and are notable for lack of clavate hairs, though clavate hairs occur in *Tetraselago* where the flowers lack any contrasting pattern. Reduction in number of clavate hairs or their total disappearance in flowers that self (e.g. the autogamous form of *Sutera griquensis*) points to a role for the hairs in pollination. In *Manulea juncea* and its allies, clavate hairs are arranged in a star-shaped pattern on the upper surface of the corolla lobes, while in, for example, *Jamesbrittenia barbata, J. ramosissima, J. hereroensis, J. tenella*, they are distributed all over the upper surface; a circlet or partial circlet of clavate hairs around the mouth is commonplace in *Zaluzianskya*.

Glandular hairs around the mouth have already been mentioned in connection with pollination of *Zaluzianskya* flowers by hawkmoths, and further information is given in the introduction to the taxonomic account of that genus. A glandular mouth is characteristic of the flowers of *Lyperia* and also occurs in a number of species of *Jamesbrittenia* (*J. carvalhoi* and associated species, *J. fruticosa* and its allies). Plates 2–4, fig. 52.

Compression of the corolla mouth

Lateral compression of the corolla mouth occurs in *Manulea altissima*, in *Manuleopsis*, and in a number of species of *Jamesbrittenia* (e.g. *J. grandiflora*, plate 3K; *J. glutinosa* and allies) while dorsoventral compression is seen in *J. jurassica* (plate 1D) and other species in that general affinity.

This is by no means an exhaustive account of features that may be related to pollination in members of Manuleae, but it suffices to show the existence of a rich field for biological and evolutionary studies.

AUTOGAMY AND CLEISTOGAMY

In the Manuleae, the stigma is often exserted together with the anthers, or, when the anthers are included, it often pushes up through what is virtually an anther tube. Self-pollination is therefore a possibility, and, so far as I know, neither self-incompatibility nor dichogamy within the tribe has ever been investigated. Both cleistogamy and autogamy appear to be commonplace in some of the annual species of Manuleae that grow in the arid and semi-arid areas of the western, central and northern Cape, which is not surprising as 'autogamy is in fact a common condition in hot arid and semi-arid areas' (Procter & Yeo, 1973, p.379). In discussing annual plants of impermanent habitats, these two authors point out that 'the short generation time allows relatively rapid production of new gene combinations even with only occasional crossing, and favourable new combinations can rapidly become established in the populations. To the plant that suffers great fluctuation in numbers, self compatibility has the further advantage that a site can be colonized and a population built up by a single individual.' (Procter & Yeo, 1973, p.382).

Examples of autogamy can be offered from several genera of Manuleae. The flowers of *Phyllopodium anomalum* appear always to be selfed: there are only two stamens and the abnormally short stigma curls round the anthers to which it is often firmly attached; *P. micranthum* show reduction in flower-size associated with reduction in number of stamens and a curled back stigma. Similar indications of autogamy have been noticed in *Trieenea elsiae, T schlechteri* and *T. glutinosa*, where the corolla may also lack any colour patterning; *T. schlechteri* sometimes produces cleistogamous flowers. *Zaluzianskya affinis, Z. cohabitans* and *Z. parviflora* all exhibit reduction in flower-size and associated loss of colour patterning linked to inclusion of the stigma and, in *Z. parviflora*, reduction to only two stamens.

Sutera griquensis and *S. cooperi*, two coarse bushy herbs probably perennial, are unique in the genus in having deeply included anthers and a style so short that the stigma is positioned below or between the anthers; the stigma is invariably coated in pollen, and selfing seems inevitable. Indeed, the type of *S. burchellii* (now in synonymy under *S. griquensis*), has abnormally small flowers, corolla tube drab-coloured (not richly coloured as is normal) and scarcely any clavate hairs inside the tube. This plant is still present in the Asbestos Hills and is surely no more than an autogamous form of *S. griquensis*.

Lyperia comprises six species, five annual, one perennial. All set seed so prolifically that selfing seems certain (every flower of several plants of *L. antirrhinoides* grown under glass at the Royal Botanic Garden Edinburgh set seed). Similarly, in several of the annual species of *Jamesbrittenia* every flower sets a capsule full of seed (e.g. *J. dissecta, J. myriantha, J. micrantha, J. concinna*, all with very small white or yellow flowers). As in *Lyperia*, the stigma is included and lies between the anticous anthers; in herbarium specimens the stigma is so smothered in pollen that its form is often difficult to distinguish. Nevertheless, hybridization appears to be commonplace in *Jamesbrittenia* (and in *Sutera*), which indicates outcrossing. It is striking that species with large and often colourful flowers set far fewer capsules than species with small, insignificant, ones, suggesting that they rely on cross-pollination.

Cleistogamy is easy to detect because the small unopened corolla persists as a cap on the developing fruit. This is common in *Zaluzianskya peduncularis, Z. kareebergensis* and *Z. divaricata* and occasional in *Z. benthamiana* (every flower of several plants of *Z. benthamiana* grown under glass at the Royal Botanic Garden Edinburgh was cleistogamous and set seed). In *Z. peduncularis*, cleistogamous and chasmogamous flowers can be found in a single inflorescence (see further in the introduction to the taxonomic account of *Zaluzianskya*).

Polycarena rariflora produces cleistogamous flowers as well as small autogamous ones and larger allogamous ones. The autogamous flowers lack the orange posticous patch that presumably attracts insect pollinators; the plant with larger, patterned, corollas was originally described as a separate species (*P. leipoldtii*; more information under the species). Other small-flowered species, for example *P. tenella*, that lack the colour-patterning characteristic of their congeners, may also be autogamous.

Cleistogamy has also been noted in *Sutera platysepala* and *S. debilis,* as well as in *Camptoloma lyperiiflorum*. These three species all perennate, but they may also flower when very small. They grow in partial shade in the shelter of rock outcrops and cliffs; reduced light intensity may be a factor inducing cleistogamy.

Cleistogamy, autogamy and gynodioecism occur in *Cromidon* (tribe Selagineae), another genus of Cape annuals closely resembling *Polycarena* in floral morphology. Autogamy appears to occur in *Chenopodiopsis hirta* and *C. retrorsa*, also tribe Selagineae (Hilliard, 1990, pp.319, 325, 333, 334, 337, 339).

HYBRIDIZATION

The distinctions between a good many species of Manuleae appear to have been blurred by hybridization (introgression) in areas of sympatry. This is particularly so in *Jamesbrittenia*, *Manulea*, *Polycarena* and *Sutera*. Specimens intermediate in character between two species (sporadic hybrids) are also not uncommon.

A notable example of sporadic hybridization between *Jamesbrittenia pristisepala* and *J. breviflora* is known to me from field work in the Natal Drakensberg, but analysis of a large number of herbarium specimens indicates much back-crossing to produce a wide range of intermediates. *Jamesbrittenia pristisepala* also seems to cross and back-cross with *J. stricta*. A full account of both crosses is given under *J. pristisepala* and *J. stricta,* and at the end of the introduction to the genus there is a list of species under which hybridization is mentioned. Similar lists are given under *Manulea* and *Sutera*. All these crosses appear to be fully fertile, but a remarkable sterile hybrid (of garden origin) between *S. caerulea* and *S. hispida* is in cultivation in botanic gardens and has been given the cultivar name *S. × Longwood*. Possible hybridization between species of *Polycarena* is discussed in the introduction to the genus.

Hybridization seems to be rare in *Zaluzianskya*: *Z. violacea* and *Z. bella*, also *Z. pusilla* and *Z. acrobareia* may introgress. It is also rare in *Glumicalyx*; hybrids between *G. nutans* and *G. goseloides* were reported in some detail by Hilliard & Burtt (1977).

CHROMOSOMES

The chromosome studies of my colleague, Dr K. Jong, support my taxonomic decisions based on morphological characters. The four genera included under *Sutera* by Hiern (1904) prove to have disparate chromosome numbers as do *Sutera* and *Manulea*, whose technical distinction Hiern found 'very uncertain' (Hiern, 1904, p.222). *Phyllopodium* and *Polycarena* (united by Mrs Levyns, 1939) may also have consistently different numbers, but as yet too few species in each genus have been investigated. The following counts were taken from Jong (1993), and all except those otherwise attributed were made by him. A paper (Jong & Wright) in preparation will illustrate chromosome morphology.

	2n	
Camptoloma		
canariense	28 (Borgen)	*Sunding* s.n.
lyperiiflorum	56	*Miller et al* 8391
Jamesbrittenia		
breviflora	24	*Hilliard & Burtt* 12642
breviflora	24	*Hilliard & Burtt* 19142
filicaulis	24	*Compton, D'Arcy & Rix* 948
foliolosa	24	*Compton, D'Arcy & Rix* 936
fruticosa	24	*Oliver* 9468
grandiflora	24	*sine coll.*
jurassica	24	*Hilliard & Burtt* 19148
Lyperia		
antirrhinoides	16	*Bokelmann & Paine* 9
tristis	16	*Bokelmann & Paine* 43
Sutera sect. **Sutera**		
foetida	12	*Batten* 1107
Sutera sect. **Chaenostoma**		
archeri	14	*Dean* 960
x *Longwood*	14	RBG Edinb. C781071
roseoflava	14	*Batten* 1000

Manulea sect. **Manulea**		
chrysantha	16	*Vlok* 2514
obovata	18	*Bokelmann & Paine* 23
thyrsiflora	16	*Bokelmann & Paine* 25
Manulea sect. **Dolichoglossa**		
crassifolia	16	*Hilliard & Burtt* 16688
parviflora	16	*Balkwill* 6514
Manulea sect. **Thyrsiflora**		
tomentosa	16	*Bokelmann & Paine* 22
Polycarena		
pubescens	14	*Batten* 1087
Trieenea		
glutinosa	14	*Vlok* 2564
taylorii	12	*Taylor* 11894
Phyllopodium		
cephalophorum	12	*Bokelmann* 65
cuneifolium	12	*Batten* 1125
pumilum	12	*Bean & Viviers* 2551
Zaluzianskya		
affinis	12	*Batten* 867
benthamiana	12	*Batten* 1030
capensis	12	*Batten* 1051
collina	12	*Batten* 1025
elgonensis	12 (Hedberg)	
glareosa	12	*Hilliard & Burtt* 9787
maritima	12	*Batten* s.n.
peduncularis	12	*Batten* 1037
villosa	12	*Compton, D'Arcy & Rix* 732
villosa	12	*Batten* 862
villosa	12	*Batten* 865
Glumicalyx		
goseloides	14	*Hilliard & Burtt* 8757
montanus	14	*Hilliard* 4842A
Tetraselago		
natalensis	14	*Hilliard & Burtt* 19168

SYNOPSIS OF POLLEN

Charles L. Argue

Pollen grains of the Manuleae fall into three groups. Those of *Glekia*, *Glumicalyx*, *Melanospermum*, *Phyllopodium*, *Polycarena*, *Strobilopsis*, *Tetraselago*, *Trieenea*, and *Zaluzianskya* are single, isopolar, radially symmetrical, and tricolporate. Colpi very long to medium long (PAI, Table 3). Colpus membrane essentially psilate, often roughened adjacent to endoaperture. Endoaperture single, more or less distinct, equatorial, lalongate (H/W, Table 3) with more or less thickened and tapering to sometimes parallel polar margins and acute to sometimes rounded ends extending laterally under the mesocolpia (W/Wec, Table 3). Intersection of endo- and ectoaperture often more or less rectangular and sometimes covered with granules. Exine 1.2–2.2μm thick; sexine simplicolumellate, usually as thick as or thicker than the nexine, columellae distinct. Tectum usually microreticulate to reticulate, lumina variable in size and shape, mostly angular and more or less polygonal, reduced on apocolpia, abruptly reduced at margo. Margo distinct, imperforate to sparsely microperforate, psilate to slightly rugulate, narrowed toward poles.

Table 3. Generic summary of selected pollen data[a].

Taxon	P[b]	E[b]	PAI[b,c]	H[b]	W[b]	H/W[d]	Wec[b]	W/Wec[d]	Exine[d]	Max Lum	Sculpt[e,f]	Margo[f,g]	EndCos[f,h]
Antherothamnus [1]	15 (14–17)	14 (13–15)	20 (19–24)	4.2 (3–6.7)	3.1 (2.7–3.8)	1.36	3.1 (2.7–3.8)	1	1.2	0.7–1	m	+	–
Camptoloma [3]	14–21 (13–26)	15–18 (14–19)	13–20 (11–24)	6.7 (4.8–9.1)	5.9 (3.8–7.7)	1.14	5.2 (3.8–7.7)	1.1	1.2–1.4	0.6–1.2	m, (r), (m–g?)	+, +/–	–, (+)
Glekia [1]	20 (19–21)	20 (19–21)	15 (13–17)	4.6 (3.9–5.3)	11.1 (9.6–13.4)	0.41	4.2 (2.7–5.3)	2.6	2	0.9–1.2	m, (r)	+	+
Glumicalyx [6]	17–21 (16–22)	17–20 (15–21)	14–19 (13–24)	3.8–4.7 (2.9–5.8)	7.9–11.4 (5.8–14.4)	0.39–0.57	3.2–4.4 (2–5.8)	1.9–2.8	1.4–2.2	1.5–2.9	r	+	+
Jamesbrittenia [13]	14–21 (13–25)	14–22 (12–23)	13–33 (12–37)	3.8–6.4 (1–8.6)	2.2–5.8 (1.1–6.7)	0.89–2.27	2.2–5.8 (1.1–6.7)	1–1.1	1–1.3	0.5–2.1	m, (r), (im), (ru–r)	+/–, (+), (–)	–
Lyperia Type 1 [4]	21–28 (19–29)	24–27 (22–28)	11–20 (9–21)	4.5–6.5 (3.8–7.7)	4.8–6.9 (3.8–7.7)	0.89–1.15	4.8–6.7 (3.8–7.7)	1–1.1 (3–8.6)	1.6–1.9	1.2–2.1	r	+	–, (+/–)
Type 2 [2]	29–40 (27–44)	31–42 (29–47)	48–52 (41–58)	4.2–6.5 (3.8–6.9)	4.4–7.4 (3–11)	0.89–0.95	4.1–6.7 (1–6.7)	1.1–1.1	1.9–1.9	2.9–5.4	r	+?	–, (+/–)
Manulea [27]	8–16 (8–17)	8–15 (8–15)	16–32 (15–36)	1.9–3.8 (1–5.3)	1.6–4.1 (1–7.2)	0.73–1.88	1.4–4 (1–5.3)	1–1.2	0.9–1.3	0.3–1.9	m, (r)	–, +/–, +	–, +/–
Manuleopsis [1]	16 (13–16)	14 (13–16)	25 (16–32)	3.9 (2.1–5.8)	3 (2.4–3.8)	1.28	3 (2.4–3.8)	1	1.2	0.6–1.1	m	+	+/–
Melanospermum [4]	17–23 (16–24)	16–20 (15–21)	17–22 (15–28)	2.5–5 (1.9–6.7)	6.2–9.4 (4.8–11.5)	0.37–0.69	3–4.1 (1.9–5.3)	2.1–3.7	1.3–1.7	0.4–1.9	r, m, (r–ru)	+	+
Phyllopodium [17]	14–25 (13–30)	14–23 (12–25)	12–25 (10–30)	2.3–5.1 (1.2–6.7)	7.3–10.7 (5.8–13.9)	0.31–0.49	2.2–5.4 (1.3–7.1)	1.9–3.8	1.1–1.9	0.7–1.9	m, r, (ir), (ru)	+	+
Polycarena [10]	21–34	18–30	13–23	3.3–8	7–15	0.38–0.64	2.8–6	2–3.1	1.7–2	1–2.9	r, (m),	+	+

Reyemia [1]	24 (22–26)	29 (21–26)	21 (19–23)	5 (3.4–5.6)	7.5 (6.7–8.2)	0.67	4.4 (3.8–5.8)	1.7	1.1–1.9	1.9–2.9	r	+	+
Strobilopsis [1]	20 (18–21)	19 (17–21)	16 (14–18)	4.2 (3.7–4.8)	10.8 (9.1–12.5)	0.39	3.7 (2.7–4.8)	2.9	1.7	1–1.8	r	+	+
Sutera [6]	13–21 (12–25)	15–18 (13–19)	16–24 (13–29)	2.8–3.6 (2–5)	3.6–5.3 (2.2–6.7)	0.68–0.85	3.3–4.5 (2.2–6.7)	1–1.2	1.1–1.8	0.7–1.7	m, r	+, (+/–)	–
Tetraselago [4]	16–18 (15–20)	16–17 (14–18)	14–19 (11–21)	3–3.9 (2–4.8)	7.8–9.2 (5.8–10.5)	0.36–0.48	2.8–3.2 (1.7–4.8)	2.8–3	1.3–1.7	0.5–1.5	r, m	+	+
Trieenea [7]	15–26 (13–29)	15–25 (14–28)	14–27 (12–33)	3.4–5 (1.9–7.2)	5.1–9 (3.8–10.5)	0.41–0.89	2.6–4.8 (1.4–6.2)	1.4–2.6	1.4–1.9	0.8–3.8	r, (m), (ir), (ru)	+	+, +/–
Zaluzianskya [33]	16–33 (14–37)	16–32 (14–37)	10–35 (9–37)	3.9–6.6 (2.6–7.7)	7.1–18.3 (5.6–21.1)	0.28–0.67	2.3–6.6 (1.9–8.2)	1.5–3.7	1.3–2.2	0.9–6.2	r, (m), (ir)	+	+

a Measurements are in micrometers (µm). P, polar axis; E, equatorial axis; PAI, polar area index; W, height of endoaperture; Wec, width of endoaperture; W, width of ectoaperture at equator; Exine, exine thickness; Max Lum, range of maximum lumina diameters on mesocolpia; Sculpt, exine sculpturing; Margo, distinctness of margo; EndCos, endoaperture margin. Figure in brackets after each generic name signifies number of species examined.

b Mean values or range of species means followed by range in parentheses.

c Conventions on colpus length: short (PAI > 40), medium (25 < PAI ≤ 40), long (15 < PAI ≤ 25), very long (PAI ≤ 15).

d Mean values or range of species means.

e g, granulate; im, irregularly microreticulate; ir, irregularly reticulate; m, microreticulate; r, reticulate; ru, rugulate.

f ( ) = sometimes.

g +, margo distinct; +/–, margo less distinct or uncertain with LM; –, margo absent.

h +, endoaperture margin thickened; +/–, endoaperture margin slightly thickened; –, endoaperture margin unthickened.

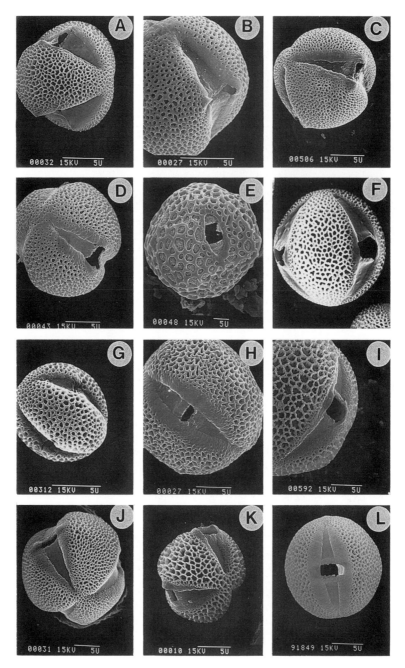

Fig. 43. SEMs of pollen of Manuleae. A, *Antherothamnus pearsonii* (*Werger* 341, PRE), oblique view; B, *Manuleopsis dinteri* (*Dinter* 333, E), oblique view; C, *Camptoloma canariense* (*Murray* s.n., E), oblique view; D, *Jamesbrittenia fruticosa* (*van der Merwe* 214, K), oblique view; E, *Lyperia tristis* (*Hilliard & Burtt* 13018, K), equatorial view; F, *Lyperia tenuiflora* (*Compton* 14921), equatorial view; G, *Manulea dregei* (*Hilliard & Burtt* 6671, E), subequatorial view; H, *Melanospermum rupestre* (*Burtt* 3112, E), subequatorial view; I, *Polycarena aurea* (*Batten* 973, E), subequatorial view; J, *Glekia krebsiana* (*Richardson* 255, E), oblique view; K, *Trieenea lasiocephala* (*Esterhuysen* 30016, BOL), oblique view; L, *Phyllopodium rustii* (*Viviers & Vlok* 478, E), equatorial view. Scale bars = 5μm. All micrographs by courtesy of Charles L. Argue.

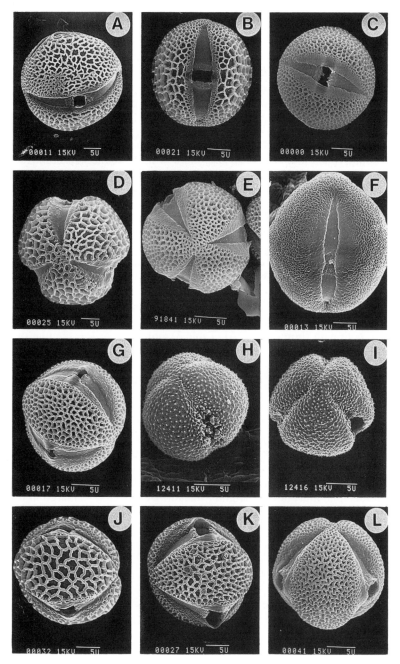

Fig. 44. SEMs of pollen of Manuleae, Selagineae and Globulariaceae. A, *Zaluzianskya ovata* (*Hilliard & Burtt* 9401, E), subequatorial view; B, *Glumicalyx lesuticus* (*Hilliard* 5397, E), equatorial view; C, *Strobilopsis wrightii* (*Hilliard & Burtt* 17031, E), equatorial view; D, *Gosela eckloniana* (*Pillans* 9142, BOL), oblique view; E, *Cromidon austerum* (*Batten* 749, E), subpolar view; F, *Chenopodiopsis retrorsa* (*Esterhuysen* 18942, BOL), subequatorial view; G, *Globulariopsis wittebergensis* (*Esterhuysen* 3767, BOL), oblique view; H, *Globularia sintenisii* (*Rechinger* 179, E), oblique view; I, *Globularia orientalis* (*Dudley* 36048, E), oblique view; J, *Agathelpis dubia* (*Sidey* 4181, MIN), subequatorial view; K, *Hebenstretia integrifolia* (*Giess* 11686), oblique view; L, *Dischisma squarrosa* (*Goldblatt* 3623), oblique view. Scale bars = 5μm. All micrographs by courtesy of Charles L. Argue.

Pollen grains of *Sutera, Jamesbrittenia, Camptoloma, Manulea, Manuleopsis, Antherotham-nus* and some grains of *Lyperia* (Table 3) are similar to grains of *Glekia* and associated genera (above). However, the endoapertures often lack thickened margins and are only slightly lalongate, more or less isodiametric, or lolongate (H/W, Table 3) with lateral ends little or not at all extended under the mesocolpia (W/Wec, Table 3). Margo present or absent. Grains in some species of *Trieenea* (e.g., *T. schlechteri, T. longipedicellata, T. lanciloba*), with endoaper-tures less extended under the mesocolpia and slightly thickened margins, resemble the pollen of this group.

Lyperia includes two distinct pollen types. The first type includes *L. antirrhinoides, L. formosa, L. tenuiflora,* and *L. violacea*. Grains here single, isopolar, radially symmetrical, suboblate to prolate spheroidal (Table 3), amb more or less circular, sometimes with slightly sunken colpi or convex-triangulaperturate in *L. formosa* and in some grains of *L. tenuiflora*. Tetracolporate in a few grains of *L. antirrhinoides*. Colpi very long to long (PAI, Table 3). Colpus membrane psilate. Endoaperture, slightly lalongate to lolongate (H/W, Table 3), usually little or not at all extended under mesocolpia (W/Wec, Table 3), shape elliptic to more or less rectangular, endoaperture margin unthickened or only slightly thickened. Exine 1.6–1.9μm thick; sexine simplicolumellate, as thick as or thicker than nexine, columellae distinct. Tectum reticulate; maximum diameter of lumina 1.2–2.1μm (mean maximum 1.6–1.9μm), lumina reduced in size on apocolpia; margo distinct, psilate (LM,) narrowed to poles; free-standing sculpturing elements occur in some grains of *L. formosa*.

The second type, unique in the Manuleae-Selagineae complex, includes *L. tristis* and *L. lychnidea*. Grains single, isopolar, radially symmetrical, 3 (4)-colporate in *L. tristis*, 4-colporate in *L. lychnidea*, both with short colpi (PAI, Table 3) amb more or less circular or sometimes square in tetracolporate grains with apertures on the planar surfaces. Colpus membrane psilate to minutely granular, granules sometimes also present on endoaperture. Endoapertures slightly lalongate (H/W, Table 3), little extended under the mesocolpia (W/Wec, Table 3), shape more or less rectangular to elliptic. Exine c.1.9μm thick; sexine simpli- or duplicolumellate, as thick as or thicker than nexine, columellae distinct. Tectum reticulate, maximum diameter of lumina 2.9–3.8μm (mean maximum 3.4μm) (*L. lychnidea*) to 4.8–5.4μm (mean maximum 5.1μm) (*L. tristis*), variable in size, usually more or less rounded in *L. tristis*, polygonal to rounded in *L. lychnidea*; unreduced on apocolpia. Nexine in *L. tristis* of granular, acetolysis-resistant com-ponent supported by a nexinous reticulum with dimensions approximating those of the tectum. Margo murus-like, psilate, encircling ectocolpi, not narrowed towards poles; width about equal to width of mesocolpial muri or only slightly thicker. Designation as margo in doubt ("?" in Table 3). See figs 43, 44.

TRIBAL AFFINITIES

There is currently no satisfactory tribal classification of Scrophulariaceae; indeed, doubt has been expressed that the family is monophyletic (Barringer, 1984). The relationships and placement of Manuleae within the family is a difficult problem for one with only limited knowledge of the family as a whole. The most recent tribal classification to be published is that of Hartl (1975, pp.17–19). Hartl there included in his circumscription of Scrophularieae the Eurasian genera *Scrophularia* and *Verbascum* as well as 'Chaenostoma Benth. und etwa 10 weitere fremdländische Gattungen (insbesondere die sog. Manuleen)'. The perfectly synthe-cous anthers of *Scrophularia* and *Verbascum* as well as their bothrospermous seeds do suggest some affinity with members of Manuleae (where three genera, *Chaenostoma* (now included in *Sutera*), as well as *Lyperia* and *Manulea*, have bothrospermous seeds). However, neither *Scrophularia* nor *Verbascum* have pulvinate placentae, and only *Verbascum* has capsules that dehisce as do those of Manuleae. Although the wider affinities of Manuleae may be uncertain, its immediate relationship is with Selagineae, a wholly Southern African tribe; this is more fully discussed below.

Hartl's published tribal classification is superseded by another in which he retains only *Scrophularia* and *Verbascum* in Scrophularieae and recognizes both Manuleae and Selagineae as separate tribes (Hartl, in litt.). This harks straight back to Wettstein's treatment of Scrophulariaceae (1891, pp.39–107), which follows Bentham (1876) very closely except that Selagineae, regarded as a separate order (that is, family) by Bentham, was included by Wettstein as a tribe of Scrophulariaceae, to follow Manuleae (comprising those genera included by Bentham). Baillon (1888, pp.371–374) was the first to include Selagineae in Scrophulariaceae, in two series: VI Série des *Selago* (including *Microdon, Agathelpis, Gosela*) and VII? Série des *Hebenstreitia* (including *Dischisma*). However, Selagineae was maintained as a separate order (family) by Rolfe (1901) and by Hutchinson (1926); these authors have been followed in most South African publications including that of Dyer (1975). Dyer intercalated his Selaginaceae in Scrophulariaceae in order to follow Dalla Torre & Harms (1904, p.456 et seq.) in their sequence of generic numbers (which follow Wettstein, 1891), and this device was also adopted by Gibbs Russell *et al.* (1987).

Southern African authors have virtually ignored Junell's study of ovarian morphology undertaken to test the reduction of Selaginaceae to a tribe of Scrophulariaceae (Junell, 1961). Junell investigated species of *Glumicalyx, Manulea, Phyllopodium* (as *Polycarena heterophylla*), *Sutera, Lyperia* (as *Sutera antirrhinoides*) and *Zaluzianskya* (Junell, 1961, pp.170–172, fig. 1), as well as a good many species of *Selagineae*, in the genera *Walafrida, Cromidon, Hebenstretia, Dischisma, Microdon, Gosela, Agathelpis, Globulariopsis*, and *Selago* itself, from which he segregated *Tetraselago* (Junell, 1961, figs 2–8 are very instructive). Junell concluded that 'it must be from such [a member of Manuleae with a reduced number of ovules] a scrophulariaceous type that the ovary of *Tetraselago*, with its four ovules, is to be derived' (Junell, 1961, p.190), and that 'the morphology of the ovary obviously confirms the opinion that Selagineae is a tribe within the Scrophulariaceae'. (Junell, 1961, p.191). It should be noted that Junell did not see a mature fruit of *Tetraselago* (p.189) and was therefore unaware that it is a capsule, not a coccus. The genus is a significant link between Manuleae and Selagineae. Reduction in ovule-number within the Manuleae is outlined above (p.20) and is summarized in Table 1 (see also Hilliard & Burtt, 1977). Only a small genetic switch would result in a single ovule in the loculus as in the genera of Selagineae.

Manuleae and Selagineae share the character of perfectly synthecous dorsifixed anthers. In addition, nearly all the genera of Manuleae, all of Selagineae, possess a lingulate stigma with marginal papillae, a dorsal nectariferous gland at the base of the ovary (a type of nectary that in its precise form appears unique to Manuleae and Selagineae), and there are marked similarities in the form of the funicle (see under placentae, p.20). The unicellular, clavate hairs that are so commonly present inside the corolla of members of Manuleae are also common in Selagineae. The flowers of several genera (for example *Selago, Walafrida, Cromidon, Polycarena, Phyllopodium, Sutera, Manulea*) can be so alike in general form that a good many species have been described in the wrong genera (Hilliard, 1990, pp.315–317, and this revision of Manuleae).

There is, however, currently still no agreement on the placement of Selagineae/Selaginaceae. Because the ovary in some genera of Selagineae may be reduced to one uniovulate loculus and thus resemble that of Globulariaceae, Cronquist (1981, followed by Mabberley, 1987), in his recent system of classification of the flowering plants, has been led to include Selagineae in Globulariaceae. Metcalfe & Chalk (1950) are quoted by Cronquist as having treated Selaginoideae and Globulariaceae sens. strict. collectively because of anatomical similarities. Cronquist claims that 'One of the unifying anatomical features of the Globulariaceae (sensu mei) is the structure of the glandular trichomes with a single stalk cell and a head of two or seldom four cells separated by vertical partitions'. (Cronquist, 1981, p.955). These hairs are commonplace in Scrophulariaceae-Manuleae and Selagineae and occur in other tribes as well (for example, Hemimerideae); they are also commonplace in Gesneriaceae. Being small (head c.0.03–0.06mm in diam.) and often masked by coarser, different, hairs, they are usually most easily seen by first clearing a piece of stem-epidermis in lactic acid and then examining it under a compound light microscope.

In three other recent systems of classification of the Angiospermae, Takhtajan (1980) included Globulariaceae and Selaginaceae in Scrophulariaceae, Dahlgren (1980) maintained them as three separate families, while Thorne (1983) included Selagineae in his subfamily Scrophularioideae, Globulareae in subfamily Rhinanthoideae.

Globulariaceae differs from Selagineae in a number of important characters: hairs inside the corolla tube uniseriate, acute, 2–3-celled in *Globularia*, 1-celled in *Poskea* (not clavate and always unicellular as in Selagineae); filaments tapered upwards, anthers versatile (not broadened apically, anthers dorsifixed); mature anthers with the four pollen sacs discernible by constrictions in the epidermis (*Poskea*), in both genera the sacs confluent but remnants of the internal dividing walls still visible (not completely synthecous); stigma shortly bifid (not lingulate); nectary cupular, persistent on the receptacle, the ovary disarticulating from it (not nectary a small persistent lateral gland).

Dr Charles L Argue, in the course of his work on the pollen of Scrophulariaceae, has found that 'The pollen data support a close relationship between the Manuleae and Selagineae, as well as the separation of the Selagineae from the Globulariaceae. Pollen morphology in *Glekia*, *Melanospermum*, *Phyllopodium*, *Polycarena*, *Zaluzianskya*, *Glumicalyx*, *Strobilopsis*, and *Tetraselago* is at once uncommon in the Scrophulariaceae and indistinguishable from that of the Selagineae. This pollen morphology is not found in the Globulariaceae. The pollen of all examined species of *Globularia* is entirely different from that of all taxa of the Selagineae and Manuleae, and while some similarities exist between the unspecialized and elsewhere very widely occuring pollen morphology found in *Poskea* and some species of, for example, *Manulea*, the pollen evidence reflects no affinity between the Selagineae and Globulariaceae.' (Argue, in litt. See also Argue, 1993). See fig. 44.

Carlquist studied the wood anatomy of *Globularia*, *Walafrida nitida* and *Selago thunbergii* and concluded that Globulariaceae is distinct from Selaginaceae (Carlquist, 1992, p.314). His further conclusion that Selaginaceae may be retained as a distinct family does not accord with the views arrived at in this revision of Manuleae. However, Professor Carlquist was working with a very inadequate sample of both Selagineae and Manuleae. He is now researching a wider sample that I have sent him and will in due course publish a reassessment.

The presence of a single carpel with an apical pendulous ovule in Globulariaceae and in some genera of Selagineae indicates convergence, not affinity. The many characters shared by Selagineae and Manuleae confirm their close relationship: posticous corolla lobes external in the bud, unicellular clavate hairs inside the corolla, synthecous, dorsifixed anthers, similar pollen, lingulate stigmas, a dorsal gland at the base of the ovary, pulvinate funicles.

HISTORICAL BACKGROUND

THE RECOGNITION OF MANULEAE

Bentham (June 1835) laid the foundations of a modern subdivision of Scrophulariaceae with the publication of his tribal classification of the family. The tribe Buchnereae Benth. was now recognized for the first time: *Corolla tubo tenui, limbo subplano, 4–5-fido, laciniis saepe bifidis. Stamina fertilia 4 didyma adscendentia. Antherae uniloculares vel loculis demum divaricatis confluentibus. Capsula 2-valvis valvulis integris bifidisve. Genera nonnulla Selagineis habitu affines*. It may be mentioned here that David Don also published, in July 1835, a tribal classification that included Buchnereae, though without reference to Bentham.

Bentham included five genera in his Buchnereae: *Buchnera* L., *Nycterinia* D. Don, *Erinus* L. (based on *E. alpinus* L.), *Manulea* L. [= *Nemia* Berg.], *Sutera* Roth (based on *S. glandulosa* Roth), and made an informal division into two groups: *Buchnera* in which the valves of the capsule remain entire, and the other four genera in which the valves become bifid.

A year later (July 1836), Bentham published a much fuller account of the Buchnereae (1836, pp.356–384). He had worked through two extensive collections of South African Scrophulariaceae made by Ecklon & Zeyher and by Drège, and much newly collected material from Australia, as well as the older collections in the herbaria of Linnaeus, the British Museum, Sir W.J. Hooker (the foundation of the Kew herbarium), Dr Lindley and himself. In the preamble to the paper, Bentham pointed out that 'three Linnaean genera, *Buchnera*, *Erinus* and *Manulea*, have been included in the Buchnereae, and appear to have been considered by many authors as so many common receptacles for all Scrophulariaceae with slender tubes to the corolla and plane lobes to the limb; the scabrous species, which dry black, being referred to *Buchnera*, and the remainder to *Erinus* or *Manulea*, according to whether the lobes of the corolla were supposed to be bifid or entire.' He goes on to show how Linnaeus himself (1759, p.1117) created a muddle by transposing the descriptions appropriate to *Buchnera* and *Erinus*, an error copied by Willdenow (1800, pp.333–334). It was left to Robert Brown (1810, p.437) to publish an unequivocal definition of *Buchnera* and its two sections.

Bentham further points out that Linnaeus based his concept of *Erinus* on *E. alpinus*, and was followed in this by subsequent authors who, however, also described various South African species of Buchnereae under the name *Erinus*. It was David Don (1835, t. 239) who confined the genus *Erinus* to *E. alpinus* and established a new genus, *Nycterinia*, to accommodate a plant that had wrongly been passing as *E. lychnidea* (L.) L.f. (now *Sutera lychnidea* (L.) Hiern = *Lyperia lychnidea* (L.) Druce). Don's plant is now *Zaluzianskya maritima* (L.f.) Walp. Don also mentioned as belonging to his new genus *E. fragrans* Ait., *E. tristis* L.f. and *E. africanus* L. (the last = *Zaluzianskya villosa* F.W. Schmidt). But, as Bentham noted, two of these species are at variance with the characters of *Nycterinia*, and belong elsewhere (*E. tristis* = *Lyperia tristis* (L.f.) Benth. and *E. fragrans* = *Lyperia fragrans* Benth. nom. illegit. = *Lyperia lychnidea* (L.) Druce). Bentham now (1836) excluded *Erinus* from his tribe Buchnereae (it is currently included in Digitaleae Benth.).

Linnaeus (Oct. 1767, p.12) established the genus *Manulea* for his species *M. cheiranthus*, which has subulate corolla lobes; this character was used by subsequent authors although they also included species with oblong, obovate or emarginate lobes. Bergius (Sept. 1767) had published a genus *Nemia* that included not only *Manulea cheiranthus* (L.) L. (basionym *Lobelia cheiranthus* L.), but a new species, *N. rubra* (= *Manulea rubra* (Berg.) L.f.), with obtuse corolla lobes, thus giving a broader generic concept, but Linnaeus's generic name was traditionally given precedence until it was finally conserved in modern times.

To return to Bentham (1836). He now published a revised definition of his Buchnereae: *Corollae limbus 5-fidus vel inaequaliter 4-fidus, interdum bilabiatus, laciniis omnibus planis. Stamina adscendentia, didyma, vel rarius 2 approximata. Antherae uniloculares. Capsula bivalvis, valvulis integris bifidisve, rarissime carnosa indehiscens. Stylus apice integer, stigmate simplici.*

This circumscription of the tribe entailed the removal of *Erinus* (confined to *E. alpinus*) and *Sutera* (confined to *S. glandulosa* Roth) to Gratioleae 'on account of their bilocular parallel-celled anthers'. However, Bentham was clearly under some misapprehension both as to the structure of the anthers in *Sutera* and the correct placement of *Erinus* because ten years later (1846) he included *Sutera* in Gratioleae subtribe Manuleieae (with unilocular anthers) and *Erinus* was relegated to Digitaleae.

But to return to the Synopsis of Buchnereae (1836). Bentham now included in this tribe *Striga* Lour., *Buchnera*, *Rhamphicarpa* Benth. and *Cycnium* [E. Mey. ex] Benth., segregated in an informal group on the basis of 'capsulae valvulae integrae'. A second group with 'capsulae *valvulae bifidae*' comprised *Nycterinia*, *Polycarena* Benth., *Phyllopodium* Benth., *Sphenandra* Benth., *Chaenostoma* Benth., *Lyperia* Benth. and *Manulea*. These genera are very nearly confined to southern Africa and the five new genera included Cape species that had originally been described, erroneously, in *Buchnera*, *Erinus* and *Manulea*.

In 1839, S.L. Endlicher gave formal recognition to Bentham's two informal groups by his creation of two subtribes, the Eubuchnereae (comprising *Striga*, *Buchnera*, *Rhamphicarpa*, *Cycnium* and *Doratanthera* Benth.) and the Manuleae (strictly Manuleinae) comprising *Zaluzianskya* F.W. Schmidt [*Nycterinia*], *Polycarena*, *Phyllopodium*, *Sphenandra*, *Chaenostoma*, *Lyperia* and *Manulea* (Endlicher 1839, p.685).

By 1846, Bentham, in his account of Scrophulariaceae in De Candolle's Prodomus 10, had revised his classification and now recognized three suborders, for the first time giving weight to differences in the aestivation of the corolla lobes. On this basis, the tribe Gratioleae (which included the Manuleae) is placed in the suborder Antirrhinideae Benth., with posticous lobes external, and the tribe Buchnereae (now comprising *Buchnera*, *Striga*, *Rhamphicarpa*, *Cycnium* and *Hyobanche* Thunb.) in the suborder Rhinanthideae Benth., with the posticous lobes always included (1846, p.189).

The tribe Gratioleae is further subdivided into four subtribes, only one of which concerns us: subtribe Manuleieae (Manuleinae) Benth.: *Folia saltem inferiora opposita. Antherae uniloculares*. It included *Nycterinia*, *Polycarena*, *Phyllopodium*, *Sphenandra*, *Chaenostoma*, *Lyperia*, *Sutera*, and *Manulea* (1846, p.341).

It was not until 1876 (in Bentham & Hooker f., 1876, pp.915, 919) that Bentham finally accorded full tribal rank to the Manuleae, and for the first time used the generic name *Zaluzianskya* in place of *Nycterinia*; the genera are otherwise as given in the Prodromus, 1846. Bentham gave as the chief distinguishing marks of the Manuleae their perfectly one-locular anthers (the loculi being divaricate and confluent), and the septicidal capsule, the two valves briefly split loculicidally.

Hiern (1904, p.123) used, without explanation or validation, his own tribal name, Nemieae, for the Manuleae, but the name has since disappeared into the oblivion where it belongs.

HISTORY OF CERTAIN GENERA OF MANULEAE

There has been much confusion over the circumscription of nearly all the genera of Manuleae. It was Bentham (1836) who first set our clearly the limits of the Linnaean genera *Buchnera*, *Erinus* and *Manulea* (see above); at this time he also described several new genera, which have subsequently been much misunderstood, partly because the characters upon which Bentham relied are not always tenable now that many more species are known and no attempt was made to seek others, and partly because full cognisance was not taken of Bentham's generic descriptions dating from 1876; notably the value of characters pertaining to stigmas and seeds was either ignored or dismissed.

Bentham (1846, pp.362, 363) confined his use of the generic name *Sutera* to *S. glandulosa* Roth (1821) and specifically excluded *S. foetida* Roth and *S. brachiata* Roth (dating from 1807), which became (1846, p.356) *Chaenostoma foetida* (Roth) Benth. and *C. hispidum* (Thunb.) Benth. [*Manulea hispida* Thunb., *M. oppositiflora* Vent., *Sutera oppositiflora* Roth; the last name is a slip: *S. brachiata* Roth was meant].

Roth's *Sutera* of 1807 does not accord with his *Sutera* of 1821, and Bentham's use of the name is nomenclaturally inadmissible, as O. Kuntze pointed out when he re-named *Sutera* Roth (1821) *Jamesbrittenia* O. Kuntze (1891, p.461). *Chaenostoma* Benth. became a synonym of *Sutera* Roth (1807) and Kuntze (loc. cit. p.467) made many of the necessary new combinations.

By 1898, Kuntze had changed his mind about the status of *Sutera* and, on very slender grounds, reduced it (including of course *Chaenostoma*) as well as *Lyperia* Benth. to synonymy under *Manulea* (Kuntze, 1898, p.235), and again made a great many new combinations.

In his reduction of *Sutera* to *Manulea*, Kuntze had been anticipated by Willdenow (1809, p.653), who gave *Buchnera foetida* Andrews as a synonym of *Manulea*, under the misapprehension that this is the basionym of *Sutera foetida* Roth. It is not. Roth clearly states that he had not seen Andrews plant and did not know if the specimen under his hand were the same thing or not; Roth's name is therefore independent of Andrews and dates from 1807; this error was perpetuated by Kuntze (1891, p.461) and by Bruce (1940, p.63).

Wettstein (1891, p.68) reduced *Lyperia* and *Urbania* Vatke to *Chaenostoma*, but maintained the rest of the genera of Manuleae enumerated by Bentham (1876), reserving *Sutera* for *S. glandulosa* Roth. *Camptoloma* Benth. was retained (p.88) in Digitaleae where Bentham had placed it.

Diels (1897, pp.492–496) went a step further than Wettstein and reduced *Sphenandra* as well to *Chaenostoma*, but his treatment does not otherwise differ from that of Wettstein (neither *Jamesbrittenia* nor *Camptoloma* was mentioned).

Hiern (1904) followed Kuntze's 1891 treatment of *Sutera* and, under *Sutera*, adopted the sections used by Diels (1897) for *Chaenostoma*, at the same time describing many more species. He prefaced his account of *Manulea* with the caveat 'The technical distinction between this genus and *Sutera* is very uncertain' and goes on to quote Kuntze, 1898 (Hiern 1904, p.222).

Hemsley & Skan (1906) maintained *Jamesbrittenia* as a distinct, monotypic, genus, and adopted a broad concept of *Sutera* for the 21 species they knew from tropical Africa, treating *Chaenostoma*, *Lyperia*, *Camptoloma* and *Urbania* as synonyms of *Sutera*.

Bruce (1941) sought to clear up the confusion over *Sutera*, but only compounded it. Hemsley & Skan had separated *Jamesbrittenia* from *Sutera* on the character of the shortly bifid stigma (cf. Bentham, 1876). Bruce examined *J. dissecta* (Del.) O. Kuntze, found it does indeed have a shortly bifid stigma, then examined *Sutera elliotensis* Hiern, in which she found the tip of the stigma to be minutely bifid. Claiming that a bifid stigma 'cannot therefore be taken as a separating generic character', she reduced *Jamesbrittenia* to *Sutera* (Bruce, 1941, p.64). However, *S. elliotensis* proves to be a species of *Manulea* (*M. paniculata* Benth.). In any case, Bruce overlooked the fundamental difference in the structure of the stigmas in the two genera. In *Jamesbrittenia dissecta*, the stigma is very short (c.0.2mm), almost capitate, and bifid; in *Manulea*, it is lingulate (as it is in *Sutera sens. strict.*); in *Manulea paniculata*, it is 3.5–5mm long, normally entire, very occasionally minutely bifid, as it may be in several other species of *Manulea*. *Jamesbrittenia* is easily distinguished from *Sutera* on several characters in addition to that of the stigma, but the reduction has been followed by all subsequent authors.

The descriptions of *Sutera* given by Hiern (1904), Phillips (1951) and Dyer (1975) are all misleading, particularly with regard to stigma and seed: only Phillips covered the range of stigmatic form in the broad concept of *Sutera* that they all adopted (the others give a description applicable only to *Sutera sens. strict.*) and all their descriptions of the testa are applicable only to *Sutera sens. strict.* and *Lyperia pro parte*.

Bentham's genera *Polycarena* and *Phyllopodium* have also undergone vicissitudes. Bentham wrote (1836, p.359) 'The one which I have called *Phyllopodium* is closely allied to the small-flowered *Polycarena* and indeed is scarcely to be distinguished, but by the equally 5-cleft calyx and more deciduous corolla'.

All the species that Bentham assigned to *Phyllopodium* have the calyx tube shorter than the anticous lobes and the anticous lip is split at least halfway to the base, mostly two thirds of the way or more. In contrast, the species assigned to *Polycarena* have the anticous lip only very shallowly notched. But this character alone is inadequate to discriminate the genera, and Bentham himself included some diverse elements under *Polycarena*. Despite his own warning that 'In appearance some species come so near to the Selagineae as to be known from them only by an inspection of the ovarium or fruit' (Bentham, 1836, p.360), the species described as *Polycarena intertexta* belongs in Selagineae (= *Cromidon decumbens* (Thunb.) Hilliard). Two other species are better placed in *Phyllopodium*.

Hiern (1904, p.123) did not look further than Bentham's one discriminating character (the degree of persistence of the corolla is untenable) and assembled species from an amazing diversity of genera under the name *Phyllopodium* in particular: his species of '*Phyllopodium*' included members of *Manulea*, *Selago* (Selagineae) and *Zaluzianskya*, as well as species now placed in *Glekia* Hilliard, *Melanospermum* Hilliard and *Trieenea* Hilliard. Under *Polycarena*, he included five species of *Phyllopodium* and three of *Melanospermum*.

Mrs Levyns, in preparing to write up Scrophulariaceae for Adamson & Salter, Flora of the Cape Peninsula, found she could not separate *Phyllopodium* from *Polycarena* on 'Bentham's distinguishing character, the calyx' (Levyns, 1939, p.36), reduced *Phyllopodium* to *Polycarena*, and made many either unnecessary or superfluous combinations.

Bentham's taxonomic judgement was far sounder than that of his successors. More than a hundred years have elapsed since Bentham's final account of Manuleae (1876), nearly a hundred since Hiern's account (1904), the most comprehensive since that of Bentham because Hiern was dealing with the rich southern African flora. Since this last account was published, botanical exploration has covered a much wider geographical area and a wealth of material has accumulated in herbaria. The present study has had the advantage not only of large numbers of specimens, but also that of greatly improved technical aids, which have permitted the detailed study of seeds in particular, but also hairs, chromosomes and pollen, all of which have yielded characters of taxonomic value. At the practical level, the characters of both seeds and hairs can be seen with a dissecting microscope or even a good hand lens.

TAXONOMY

Scrophulariaceae tribe Manuleae Benth.
in Benth. & Hook. f., Genera plantarum 2: 915, 919 (1876).

Herbs or shrubs, often glandular, often aromatic or foetid. *Leaves* simple, exstipulate, usually opposite at least at the base, often alternate upwards. *Flowers* sometimes in a raceme of cymes, often solitary in axils of leaves or bracts and then often forming terminal racemes, sometimes panicles, occasionally corymbose, bisexual, either subactinomorphic or zygomorphic. *Calyx* usually gamosepalous, persistent, obscurely to distinctly bilabiate, shallowly to deeply divided, occasionally right to base, lobes usually 5, rarely 6 to 9, or 3 by abortion. *Corolla* gamopetalous, sometimes persistent, tube cylindric or infundibuliform, limb bilabiate or nearly regular, posticous lip 2-lobed, exterior in bud, anticous lip 3-lobed (occasionally the posticous lip appearing 4-lobed by isolation of the anticous lobe, very rarely resupinate). *Stamens* usually 4, rarely 2 by abortion or reduction of either the anticous or posticous pair, in one genus the true posticous stamen represented by a staminode with filament equalling that of fertile stamens, fertile stamens didynamous, anticous pair exceeding posticous pair, often one or both pairs exserted, sometimes posticous pair or all 4 stamens included, filaments usually inserted at or above middle of corolla tube, occasionally near base, posticous filaments sometimes decurrent; anthers synthecous, posticous ones often larger than anticous. *Stigma* often lingulate with marginal bands of stigmatic papillae confluent over apex, apex usually entire, rarely bifid; or stigma minutely bifid to entire with terminal papillae; or rarely stigma bifid with the arms united by a jelly-like substance. *Style* solitary, simple, terminal. *Ovary* superior, bilocular, placentation axile, funicles persistent, often depressed pulvinate, the pulvini often discrete, sometimes weakly to strongly confluent, ovules two to many in each loculus; nectary often a small dorsal gland adnate to or ± free from base of ovary; sometimes annular and then more strongly developed on dorsal side; rarely obscure. *Fruit* a septicidal capsule, each valve with a short loculicidal split at the tip. *Seeds* small, usually ± oblong or ellipsoid, often angled by pressure, rarely winged, outer periclinal walls of testa often variously ornamented, endosperm copious, outer face smooth or variously invaginated or wrinkled.

Type genus: *Manulea* L.
Other genera: *Antherothamnus* N.E. Br., *Camptoloma* Benth., *Glekia* Hilliard, *Glumicalyx* Hiern, *Jamesbrittenia* O. Kuntze, *Lyperia* Benth., *Manuleopsis* Thell., *Melanospermum* Hilliard, *Phyllopodium* Benth., *Polycarena* Benth., *Reyemia* Hilliard, *Strobilopsis* Hilliard & Burtt, *Sutera* Roth, *Tetraselago* Junell, *Trieenea* Hilliard, *Zaluzianskya* F.W. Schmidt.

Key to the genera of Manuleae

1a. Posticous filaments not decurrent down corolla tube 2
1b. Posticous filaments decurrent down corolla tube, often to base 5

2a. Bract adnate halfway up calyx tube .... **10. Glekia**
2b. Bract or floral leaf not or scarcely adnate to calyx tube (at most adhering to part or all of pedicel and sometimes extreme base of calyx) 3

3a. Posticous stamens included, inserted halfway up tube or higher, anticous stamens either included or anthers just visible in mouth .... **7. Manulea**
3b. One or both pairs of stamens exserted at anthesis (caution: both pairs deeply included in 2 species of *Sutera*, inserted near base of tube), ... 4

4a. Inside of corolla tube either glabrous or with longitudinal bands of clavate hairs, these sometimes coalescing, sometimes extending from throat onto bases of corolla lobes; seeds patterned with transversely elongated pits arranged like a chequer board ...... **6. Sutera**
4b. Clavate hairs confined to orange/yellow patch at base of posticous lip; seeds sinuously wrinkled in longitudinal bands .... **11. Trieenea**

5a. Stigma very short (up to 2mm, but often much shorter), often minutely bifid, stigmatic surface ± terminal, either jelly-like or composed of minute papillae, these either terminal or terminal and extending very briefly downwards 6
5b. Stigma lingulate (tip occasionally minutely bifid), often ± equalling or longer than style, flattened, with 2 marginal bands of stigmatic papillae 10

6a. Leaves opposite, bases decurrent forming ridges or narrow wings down stem 7
6b. Leaves either alternate at least on upper part of stem or, if opposite, not decurrent 8

7a. Large shrub with whitish bark; corolla tube funnel-shaped, longitudinal bands of clavate hairs running down back of tube, seeds brown .... **2. Manuleopsis**
7b. Annual or perennial herbs; corolla tube cylindric, abruptly dilated in throat, V-shaped patch of clavate hairs in throat, seeds black .... **5. Lyperia**

8a. Corolla tube funnel-shaped; large staminode (a filament lacking an anther) near base of tube on posticous side .... **1. Antherothamnus**
8b. Corolla tube cylindric; staminodes usually wanting: sometimes present in one species but then minute and inserted in throat 9

9a. Corolla tube scarcely expanded at apex, anticous anthers shortly exserted, seeds with 6–8 longitudinal ribs, testa reticulate between ribs .... **3. Camptoloma**
9b. Corolla tube abruptly expanded at apex, at most tips of anticous anthers visible in mouth, seeds not ribbed, testa reticulate .... **4. Jamesbrittenia**

10a. Bract either adnate to pedicel (or part of pedicel) only or to pedicel and base of calyx tube (never more than lower third) 11
10b. Bract adnate at least halfway up calyx tube, rarely less 14

11a. Inflorescence capitate, nodding in flower, erect in fruit .... **15. Glumicalyx**
11b. Inflorescence either capitate or extended, always erect 12

12a. Floral bracts roughly as long as broad, sharply differentiated from leaves **16. Strobilopsis**
12b. Floral bracts longer than broad, often leaflike 13

13a. Corolla with an orange/yellow patch at base of posticous lip extending down back of tube, seeds pallid, greenish-blue or amber-coloured, sinuously wrinkled in vertical bands
11. Trieenea
13b. Orange/yellow patch on corolla confined to posticous lip, seeds black, coarsely reticulate
8. Melanospermum

14a. Ovules 2 in each loculus, one pointing upwards, one downwards; inflorescence a corymb
17. Tetraselago
14b. Ovules at least 4 in each loculus and never diametrically opposed; inflorescence a spike or raceme, sometimes condensed into a head, sometimes panicled 15

15a. Calyx distinctly bilabiate, anticous lip ± 2-toothed, posticous lip 3-toothed, strongly 5-ribbed, plicate in flower (expanded in fruit), stamens 4 or 2, no staminodes
13. Zaluzianskya
15b. Calyx distinctly bilabiate or not, rarely plicate and then 2 stamens, 2 staminodes, lobing various, often deep, sometimes right to base, stamens 4 or 2, staminodes present or absent 16

16a. Hairs on stems (and elsewhere) always patent, gland-tipped; testa translucent, loosely enveloping seed, seed 3-winged or 3-angled .... **9. Polycarena**
16b. Hairs on stems either wholly eglandular or mixed with glandular hairs, but some eglandular hairs always present, particularly near base of stem and there nearly always ± retrorse; patent eglandular hairs always present on margins of bracts; testa ± opaque, tightly investing seed; seed never winged   17

17a. Calyx mostly lobed halfway or more to base, upper lip of corolla 2-lobed, lower 3-lobed
12. Phyllopodium
17b. Calyx lobes much shorter than tube, one corolla lobe isolated, the other 4 closely associated .... **14. Reyemia**

1. ANTHEROTHAMNUS

N.E. Brown placed his new genus *Antherothamnus* in the tribe Cheloneae and diagnosed it against *Freylinia*; however, the two genera differ in the aestivation of the corolla lobes (anticous lobe inside, two posticous lobes outside in *Antherothamnus*, anticous lobe outside, two posticous lobes inside in *Freylinia*) and in their anthers, which are synthecous in *Antherotham-nus*, dithecous in *Freylinia*. Barringer's inclusion of *Antherothamnus, Manuleopsis, Phygelius* and *Freylinia* in a tribe Freylineae Barringer (Barringer, 1993) stems from inadequate examination of the plants, which diverge in aestivation of the corolla lobes, in anther thecae, staminodes and capsule.

Antherothamnus possesses a well-developed staminode; it was possibly this that influenced N.E. Brown's placement of *Antherothamnus* in Cheloneae, a tribe (now considered highly heterogeneous) characterized principally by the cymose disposition of the flowers: Brown described the inflorescence of *Antherothamnus* as racemose (he saw only Pearson's specimens). *Antherothamnus* has all the essential characters of Manuleae, in which cymose inflorescences may occur (e.g. *Sutera, Manulea*). Schinz & Thellung (1929) diagnosed their new genus *Selaginastrum* [= *Antherothamnus*] against *Chaenostoma* (including *Sutera* and *Sphenandra*) and thus by implication placed it in Manuleae. They did not realise that the inflorescences they saw were racemes of reduced cymes. In a footnote (1929, p.121) they pointed out that *Manuleopsis* Thellung has a cymose inflorescence and they therefore considered *Selaginastrum* to be generically distinct from *Manuleopsis*; had a range of material been available to them, they would have seen that reduced cymes occur in *Manuleopsis*. In other words, the inflorescences of *Manuleopsis* and *Selaginastrum* (that is, *Antherothamnus*) are essentially the same. They share several more characters: decurrent leaf-bases, the peculiar thickened and persistent base of the petiole (unknown elsewhere in the tribe) and similar seeds (figs 19A and 19B). I have reservations about their being generically distinct but as they are very easily distinguished on both floral and vegetative characters, I have maintained them. An examination of their chromosomes might tip the scales in favour of union. There is no doubt however that the two species belong in Manuleae, where Hartl (in litt.) too placed them.

Antherothamnus N.E. Br. in Hook., Ic. Pl. 31 t.3007 (Jan. 1915); Schweickerdt in Kew Bull. 1937: 447 (1937); Merxm. & Roessler in Prodr. Fl. S.W. Afr. 126: 9 (1967); Philcox in Fl. Zamb. 8(2): 20 (1990). **Figs 12A, 13B, 19A, 45, Plate 4H.**

Syn.: *Selaginastrum* Schinz & Thell. in Viert. Nat. Ges. Zürich 74: 119 (1929).

Shrub, foetid. *Branches* rigid, subspinescent with age, shallowly ribbed by decurrent leaf-bases when young, glandular-pubescent, glabrescent. *Leaves* alternate, fascicled on young wood, simple, entire to toothed, base of petiole persistent. *Inflorescence* composed of miniature thyrses arranged on the ultimately spinescent branchlets. *Bracts* free from pedicel and calyx. *Calyx* obscurely bilabiate, lobed nearly to base, ± glabrous. *Corolla* tube funnel-shaped, mouth ± round, limb bilabiate, lobes spreading, suborbicular, glabrous outside, inside with clavate unicellular or bicellular hairs in sinuses, otherwise glabrous except for glandular filaments. *Stamens* 4, staminode 1, filaments inserted near base of tube, decurrent, anthers synthecous, anticous ones shortly exserted, posticous included. *Stigma* very minutely bifid, arms not separating, papillae terminal, very shortly exserted. *Ovary* deltoid in outline, glabrous, nectary not conspicuous, possibly very shallowly annular, ovules c.6–12 in each loculus. *Fruit* a septicidal capsule, each valve with a short loculicidal split. *Seeds* seated on a round centrally depressed pulvinus (fig. 13B), seed ± ellipsoid in outline, brown, testa thin, reticulate (fig. 19A).

Type: *Antherothamnus pearsonii* N.E. Br.
Distribution: Zimbabwe, Botswana, N and W Transvaal, northern Cape, southern Namibia. Monotypic.

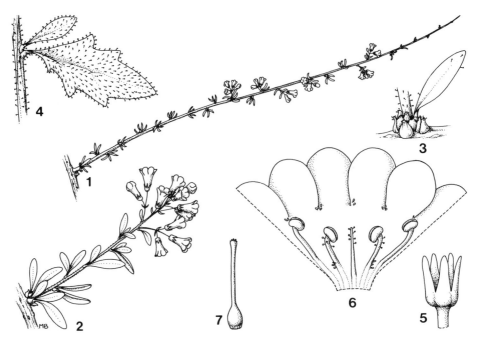

Fig. 45. *Antherothamnus pearsonii*: 1, twig to show fascicled leaves and subspinescent tip, × 0.6 (*Hansen* 3472); 2, inflorescence showing a 3-flowered cymule, × 1.5 (*Giess, Volk & Bleissner* 7180a); 3, swollen leaf bases, × 4; 4, coppice growth, × 3 (*Giess & Müller* 11993); 5, calyx, swollen at base of lobes, × 6; 6, corolla opened out to show large staminode and tufts of hairs in sinuses of lobes, × 6; 7, stigma, style and ovary lacking a lateral nectary, × 6.

Antherothamnus pearsonii N.E. Br. in Hook., Ic. Pl. 31 t.3007 (Jan. 1915); Merxm. & Roessler in Prodr. Fl. S.W. Afr. 126: 9 (1967); Philcox in Fl. Zamb. 8(2): 20, t.9 (1990). **Fig. 45.** Lectotype (chosen here): Cape, Bushmanland, [2920 BB] rocky places at foot of kopje near Groot Rozynbosch, 10 xii 1908, *Pearson* 3619 (K, isolecto. PRE).

Syn.: *Sutera rigida* L. Bolus in Ann. Bol. Herb. 1: 99 (1915) nomen & in Ann. S. Afr. Mus. 9: 267, t.6 fig. C (April 1915), nom. illegit.
  Manuleopsis karasmontana Dinter in Feddes Repert. (Beih.) 53: 29 (1928) nom. nud.
  Selaginastrum rigidum Schinz & Thell. in Viert. Nat. Ges. Zürich 74: 121 (1929).
  Selaginastrum rigidum var. (?) *karasmontanum* Schinz & Thell. loc. cit. 122 (1929). Type: [Namibia], Keetmanshoop distr., [2718 CA], Klein Karas, 11 xi 1923, *Dinter* 5088 (holo. Z, iso. K, PRE).
  Selaginastrum karasmontanum Schinz & Thell. loc. cit. 121 (1929) in obs.
  Antherothamnus rigida Phillips in Bothalia 3: 271 (1937) as *A. rigida* (L. Bol.) N.E. Br.

Shrub up to 1–3m tall, well branched, ultimate branchlets eventually rigid and subspinescent, shallowly ribbed by decurrent leaf bases, minutely glandular-puberulous (glandular hairs to 0.25mm on coppice shoots), older branches glabrescent, bark reddish to greyish, longitudinally striate, lenticels often prominent, all but oldest stems closely leafy, old stems rough with persistent bases of fascicles. *Leaves* alternate, fascicled except on coppice shoots, up to 4–15 × 1–2.5(–3.5)mm, oblanceolate, obtuse to acute and sometimes apiculate, base narrowed to a petiolar part, extreme base eventually swollen and suberized (?), persistent, margins plane to subrevolute, usually entire, occasionally with 1–4 very small teeth, midrib impressed above, slightly raised below, lateral veins invisible, both surfaces glandular-punctate, small glistening glands as well, sometimes confined to midrib, base of petiole glandular-puberulous, sometimes a few small hairs on blade and margins, *leaves on coppice shoots* up to 10–15 × 6–7mm, narrowly ovate in outline narrowed to a petiolar part, subacute, margins with 2–3 pairs of coarse teeth, midrib and lateral veins impressed above, raised below, both surfaces glandular-pubescent, hairs up to 0.25mm long. *Flowers* sweetly scented, racemosely arranged on short (up to c.35mm) lateral branchlets, these often panicled, individual flowers often solitary, sometimes in 3-flowered cymes, lowermost in axil of a leaf, these soon reduced to bracts. *Pedicels* c.1–5mm long, often with 2–3 tiny bracteoles. *Calyx* tube c.0.2mm long (measured inside), lobes 1–1.4 × 0.4–0.7mm, ± oblong, ± acute, rather fleshy and somewhat swollen on basal fused part forming tube, glabrous or with a few minute glandular hairs on margins. *Corolla* tube 3–4 × 2–3.4mm in mouth, funnel-shaped, limb bilabiate, c.4–8mm across lateral lobes, posticous lobes 1.2–3.5 × 1.7–3mm, anticous lobe 1.2–3.5 × 1.7–2.8mm, all subrotund, white, at least the anticous lobe marked with 1 or 3 reddish-purple lines (sometimes all 5 lobes), the lines running back down the tube, tube yellow (deduced from dried specimens) or rarely wholly flushed purplish, glabrous outside, inside with a small tuft of clavate hairs in each sinus. *Stamens* inserted near base of tube, posticous filaments 1.2–2mm long, clavate hairs at base, apex either glabrous or glandular-puberulous, anthers 0.5–0.8mm long, anticous filaments 1–2.5mm long, glabrous, anthers 0.4–0.7mm long, all filaments somewhat decurrent, staminode 1.7–2.5mm long, glandular-puberulous. *Stigma* c.0.1–0.2mm long, minutely papillose. *Style* 1.8–3mm long. *Ovary* 0.7–1 × 0.5–0.8mm, glabrous. *Capsules* 2.5–4 × 2–2.5mm, comparatively large ± clavate unicellular hairs on placentae. *Seeds* (few seen) 1–1.4 × 0.5–0.8mm.

Selected citations:

Botswana. 2227BB, Selebi-Phikwe, 30 ix 1978, *Hansen* 3472 (PRE). 2425BD, 5½ miles NW of Gaberones, 3400ft, 1 xii 1954, *Codd* 8909 (K, PRE). 2525AB, Kanye, near irrigation scheme, 4000ft, 19 xi 1948, *Hillary & Robertson* 628 (BM).
Transvaal. Zoutpansberg distr., 2229 AC, farm Breslau 2, 21 iv 1984, *Straub* 213 (K, PRE). Pietersburg distr. (?), 2328 DB, Swerwerskraal, 22 i 1929, *Hutchinson* 2629 (PRE).
Cape. Kenhardt div., 2820CB, Aughrabies National Park, c.700m, 7 v 1969, *Werger* 341 (K, PRE); 2820 DC, 40.3 miles from turnoff to Kakamas from Kenhardt, 9 ii 1972, *Ellis* 1046 (PRE).
Namibia. Lüderitz-Sud distr., 2716DD, farm Namuskluft (LU 88), 14 ix 1973, *Giess* 12949 (PRE, WIND). Keetmans-hoop distr., farm Kuchanas [2718BA] - Narubis [2719 AA], 19 vii 1931, *Örtendahl* 533 (K, S); 2718 BC, farm Noachabeb (KEE 97), 7 iv 1968, *Giess* 10333 (K, PRE, WIND). Warmbad distr., 2718BD, farm Sandmund (WAR 270), 21 v 1963, *Giess, Volk & Bleissner* 7180a (PRE, WIND).

Antherothamnus pearsonii has a relatively wide distribution, from the Matopas in S.W. Zimbabwe (fide Philcox in Fl. Zamb. 8, 2), then there are records down the south-eastern border of Botswana (Selebi-Phikwe to Kanye) and nearby parts of the Transvaal, the Cape around Kakamas, and the southernmost part of Namibia with records mainly from the Karasberge and west to the Namusberge. The plants favour rocky sites, among big boulders on mountain slopes and in ravines; granite, sandstone and quartzite have been mentioned by collectors. Flowering takes place between November and May.

Three collectors have recorded that the flowers are sweet smelling, one claiming the air filled with fragrance; Pearson (fide N.E. Brown in his original description of the genus) said that the flowers are delightfully scented at night. All four stamens and the style lie along the anticous side of the corolla tube, the anticous pair of anthers exserted; the tip of each filament is slightly curved to present the anthers with their line of dehiscence upwards, the stigma also presented between the anticous and posticous anthers by a slight upward curvature of the style-tip. Hooker's Icones t.3007 fig. 1 shows the anticous anthers in the posticous position; the illustration in Flora Zambesiaca is similar; perhaps the flowers are presented upside down, a feature not easily detected in dried material.

Brown remarked on the peculiarity that the dried flowers of *Antherothamnus* became subdeliquescent when softened in boiling water, but this is the result of overheating the material during the drying process: they are cooked!

2. MANULEOPSIS

Manuleopsis Thell. in Vierteljahrss. Nat. Ges. Zürich 60: 406 (Dec. 1915); Merxm. & Roessler in Prodr. Fl. S.W. Afr. 126: 33 (1967). **Figs 12B, 13A, 19B, 46, Plate 4E.**

Syn.: *Freyliniopsis* Engl. in Bot. Jahrb. 57: 609 (1922).

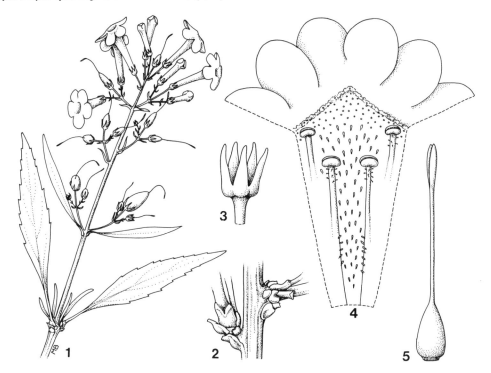

Fig. 46. *Manuleopsis dinteri* (*Giess* 13506): 1, cymose inflorescence, × 1; 2, swollen leaf bases, × 2; 3, calyx, swollen at base of lobes, × 4; 4, corolla opened out showing thickened tissue around laterally compressed mouth, minutely glandular there, clavate hairs down back of tube, × 4; 5, stigma, style and ovary lacking a lateral nectary, × 4.

Shrub, foetid. *Branches* brittle, ribbed by decurrent leaf bases when young, glandular-puberulous on uppermost parts. *Leaves* opposite and decussate, sometimes fascicled, simple, subentire to serrate, base of petiole persistent. *Inflorescence* a thyrse. *Calyx* obscurely bilabiate, lobed nearly to base, glandular-puberulous. *Corolla* tube cylindric, mouth laterally compressed, elliptic, limb bilabiate, lobes spreading, suborbicular, glandular-puberulous outside, inside minutely glandular around mouth, band of clavate hairs from mouth to base of tube on posticous side. *Stamens* 4, posticous ones included, filaments decurrent to base of tube, anticous ones inserted in throat, very shortly exserted, anthers synthecous. *Stigma* bifid, arms not separating, just reaching mouth. *Ovary* deltoid in outline, glabrous, nectary not conspicuous, possibly very shallowly annular, ovules many. *Fruit* a septicidal capsule, each valve with a short loculicidal split. *Seeds* seated on flattened, coalescing pulvini (fig. 13A), seed ± ellipsoid in outline, brown, testa thin, reticulate (fig. 19B).

Type: *Manuleopsis dinteri* Thellung
Distribution: central and northern Namibia. Monotypic.

Manuleopsis dinteri Thell. in Vierteljahrsschr. Nat. Ges. Zürich 60: 406 (1915); Merxm. & Roessler in Prodr. Fl. S.W. Afr. 126: 33 (1967). **Fig. 46, Plate 4E.**
Lectotype (chosen here): Namibia, Hereroland, Windhoek, ii 1899, *Dinter* 277 (Z).

Syn.: *Freyliniopsis trothae* Engl. in Bot. Jahrb. 57: 609 (1922). Lectotype (chosen here): Namibia, Okahandja, 1300m, 20 xii 1906, *Dinter* 333 (Z; isolecto. E, PRE).

Shrub up to 1.5–3(–5)m tall, branches brittle, twigs ribbed by strongly decurrent leaf bases, bark on older stems whitish, longitudinally striate, lenticels often prominent, rough with persistent leaf-bases, these often on short (c.2–10mm) brachyblasts closely beset with both leaf-bases and bud scales, often growing out into a flowering axis, minutely glandular-puberulous on the flowering axes, otherwise glabrous. *Leaves* opposite and decussate, sometimes fascicled, blade up to 35–90 × 6–60mm, lanceolate, apex acute, base cuneate and tapering into a petiolar part c.7–30mm long, extreme base of petiole eventually swollen and suberized (?), persistent, margins plane, subentire to serrate, penninerved, both surfaces glandular-punctate, base of petiole minutely glandular-puberulous. *Flowers* in terminal thyrses up to c.150mm long, each component often a 3-flowered cyme (or sometimes 7-flowered or reduced to 1 or 2 flowers) or occasionally up to c.11 flowers in an irregular cyme up to 30mm long with seemingly racemose branches, components in axils of uppermost leaves, soon reduced to bracts, bracteoles on pedicels up to c.2 × 0.2mm. *Calyx* tube c.0.1–0.2mm long (measured inside), lobes 2.5–3.6 × 1mm, lanceolate or oblong-lanceolate, acute, rather fleshy and swollen on basal fused part forming tube, glandular-puberulous. *Corolla* tube 10.5–12 × 3.5–4mm at apex, cylindric, mouth somewhat compressed laterally, elliptic, thickened around the orifice, limb bilabiate, 10–15mm across lateral lobes, posticous lobes 3–4 × 3–5mm, anticous lobe 4–5.5 × 4.2–6mm, all lobes subrotund, white (variously recorded as pure white, white with crimson centre, white with dark blue spots ringing throat, white with purple stripes on lower lip), glandular-puberulous outside, thickened rim around mouth glandular-puberulous, the hairs descending briefly down back of throat, with a band of clavate hairs running right down back of tube. *Stamens*: posticous filaments c.3–3.5mm long, decurrent to base of tube, bearded with clavate hairs, posticous anthers 1.1–1.5mm long, included, anticous filaments 1–1.6mm long, glandular, anthers 1–1.5mm long, very shortly exserted. *Stigma* c.1mm long, tip just visible in mouth between anticous anthers. *Style* 5.5–7.2mm long. *Ovary* 2.8–3.5 × 1.5–2mm, glabrous. *Capsules* 9–13 × 3.5–4mm. *Seeds* (few seen) c.0.8 × 0.4mm.

Selected citations:

Namibia. Kaokoveld, 1813DB, 16 miles N of Ombombo, 23 v 1957, *de Winter & Leistner* 5890 (WIND). Outjo distr., 1915CA, farm Cauas-Okawa, 6 iv 1955, *de Winter* 3072 (K, WIND). Omaruru distr., 2114 BA, Brandberg, Tsisab Valley, 31 v 1963, *Nordenstam* 2816 (M). Windhoek distr., 2217DA, Goreangab dam, 26 iii 1968, *Wanntorp* 359 (K); 2216DD, farm Claratal, 26 xii 1957, *Merxmüller & Giess* 875 (PRE, WIND). Maltahöhe distr., 2416AB, Bergzebrapark Naukluft (MAL 9), 24 i 1974, *Giess & Robinson* 13262 (WIND).

Manuleopsis dinteri appears to be confined to Namibia where it has a wide distribution from the Kaokoveld and the environs of Otavi and Tsumeb south to the Naukluft Mountains and Kalkrand, c.150km due west of those mountains, that is, from roughly the 25th parallel northwards and west of 18° longitude. The plants favour rocky sites, growing between boulders and in rock fissures on mountain slopes, on hills, and in gorges, sometimes near watercourses (dolomite, mica, granite and basalt all recorded as substrates).

The plants flower between November and June. No collector has mentioned scent, but the floral morphology suggests an insect pollinator. The mouth of the corolla tube is laterally compressed to form an elongated ellipse rimmed by thickened, glandular, tissue marked with dark blue or violet against a white background. The anticous anthers and the stigma are presented in the mouth of the tube just above the base of the anticous lobe, with the posticous anthers lying further back down the tube; all four anthers lie horizontally with the line of dehiscence upwards. This mode of presentation is similar to that in the much smaller flowers of the closely allied *Antherothamnus pearsonii*; see there for a discussion on the relationship of the two genera.

3. CAMPTOLOMA

Bentham (1876) placed his new genus *Camptoloma* in the Digitaleae with some reservation as he had not seen the aestivation of the corolla lobes. When Diels described *Chaenostoma heucheriifolium* (1897, p.475) he added a footnote about *Camptoloma*, noting the likeness between *Camptoloma rotundifolium*, *Chaenostoma canariense*, *C. lyperiaeflorum*, *C. socotranum* (in error for *Camptoloma villosa*, from Socotra) and *Chaenostoma heucheriifolium*, and concluding (correctly) that they should all be placed in Manuleae. The likeness between *Camptoloma* and *Chaenostoma heucheriifolium* is superficial; that species belongs in *Jamesbrittenia*.

The later inclusion of *Camptoloma rotundifolium* and *C. canariense* in *Sutera* obscured not only the close relationship of these two species with *C. lyperiiflorum* but also their remarkable disjunct geographical distribution: one species is confined to the Canary Islands, another to the western deserts of Angola and Namibia, the third to the southern coastal fringe of the Arabian peninsula, Socotra and Somalia. This is a classic example of the now well known pattern of disjunction between the arid areas of northern and south-western Africa (see Burtt, 1971, p.135–149; De Winter, 1971, p.424–437; Nordenstam, 1974, p.1–67).

The three species are remarkably similar in habit, foliage and long glandular hairs (fig. 9A) on the vegetative parts, and all favour the same sorts of habitat and appear to flower at any time of the year (after rain?). They differ only in small detail, though *C. canariense* stands slightly apart from the other two by virtue of its bracteate rather than leafy inflorescence and the strongly curved fruiting pedicels.

Camptoloma does not appear to be closely allied to *Sutera*. It differs in its calyx divided almost to the base, cylindric corolla tube, decurrent posticous filaments, different stigma, ill-defined nectary, strongly beaked capsule, placentae with strongly confluent pulvini, and 6–8-ribbed seeds. Although there is no easily discernible nectary, the base of the ovary is somewhat fleshy and is slightly convoluted or swollen on one side; this tissue may be secretory.

Camptoloma Benth. in DC., Prodr. 10: 430 (1846) and in Benth. & Hook. f., Gen. Pl. 2: 960 (1876). **Figs 12C, 13C, 19C, 35A, 47.**

Syn.: *Urbania* Vatke in Oesterr. Bot. Zeitschr. 25: 10 (1875). Type species: *U. lyperiiflora* Vatke.

Shrublets that may flower in the seedling stage, brittle-stemmed, remarkably glandular. *Leaves* opposite only at extreme base of seedlings, otherwise alternate, simple, petiolate, toothed. *Flowers* in the axils of the upper leaves, either solitary, paired or in cymules. *Bracts,* if present, very small, free from pedicels. *Calyx* 5-lobed nearly to base, very obscurely bilabiate. *Corolla* tube cylindric, scarcely swollen apically, throat bearded or not with clavate hairs, limb slightly zygomorphic, 2 posticous lobes joined for ± half their length. *Stamens* 4, didynamous, anticous pair inserted in throat, very shortly exserted, posticous pair inserted ± halfway up tube, included, filaments decurrent to base of tube forming a channel for the style, all filaments hairy or not, anthers synthecous, anticous ones often slightly larger than posticous (unusual in Manuleae). *Stigma* very short, stigmatic papillae in a lateral groove carried over the apex, the tip sometimes very shortly bifid, either held in the throat or very shortly exserted. *Ovary* cuneate in outline, tapering into the style, glabrous, ovules very many in each loculus; nectary not easily seen, but probably annular, either convoluted or swollen on one side. *Fruit* a sharply beaked septicidal capsule, tip of each valve with a short loculicidal split. *Seeds* seated on strongly confluent pulvini (fig. 13C), very small, red-brown, oblong in outline, longitudinally 6–8-ribbed (aulacospermous), under the SEM the furrows reticulate (fig. 19C).

Type species: *Camptoloma rotundifolium* Benth.
Distribution: Canary Islands, Angola, Namibia, South Yemen, Socotra, Kuria Muria Islands (Oman), Somalia.

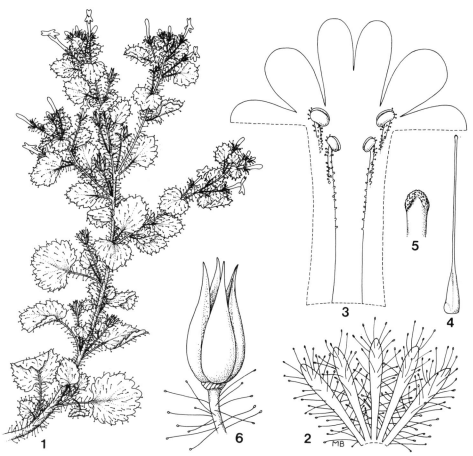

Fig. 47. *Camptoloma lyperiiflorum* (*Miller* 8391, together with slides and living plant): 1, twig to show habit, × 0.6; 2, calyx opened out, × 5; 3, corolla opened out to show posticous filaments decurrent down tube and all filaments with both clavate and glandular hairs, × 5; 4, stigma, style and ovary with ill-defined lateral swelling, probably nectariferous, × 5; 5, stigma to show apical stigmatic papillae, × 50; 6, capsule, sepals removed, × 5.

Key to species

1a. Inflorescence mainly bracteate, the bracts very small .... **1. C. canariense**
1b. Inflorescence decidedly leafy 2

2a. Flowers mostly arranged in cymules, only occasionally solitary in leaf-axils, upper part of corolla tube well bearded within .... **2. C. rotundifolium**
2b. Flowers usually solitary in the leaf-axils, occasionally two from one axil, but never cymose; corolla tube glabrous within except for hairs on the filaments   **3. C. lyperiiflorum**

1. Camptoloma canariense (Webb & Berth.) Hilliard in Edinb. J. Bot. 48: 341 (1991).

Type: Gran Canaria, near Las Palmas, Bandama Crater, *Webb* 1848 (K).

Syn.: *Lyperia canariensis* Webb & Berth., Hist. Nat. Isl. Canar. 3: 146 (1845) and Phytog. 3 t. 184 (no date given);
    Bramwell & Bramwell, Wild Flowers of the Canary Islands 198 cum ic. (1974).
  Sutera canariensis (Webb & Berth.) Sunding & Kunkel in Cuad. Bot. Canar. 14–15: 50 (1972); Henriquez et al.,
    Flora y vegetacion del archipielago canario 191 fig. 1 (1986).

Shrublet, branching from the base and above, stock at least 10mm in diam., stems up to c.700mm long, densely glandular-pubescent, hairs up to 1–2.5mm long, leafy. *Leaves* alternate, blade c.9–30 × 9–34mm, broadly elliptic to orbicular, apex obtuse, base cuneate passing into a petiole c.4–20mm long (often roughly half length of blade), margins coarsely toothed, both surfaces glandular as the stems. *Flowers* few to many, solitary in the axils of the upper leaves, which rapidly pass into small bracts, the whole forming a terminal raceme, axis often flexuous. *Bracts* (largest) c.2–6 × 0.2–1mm, often 1 or 2 on the pedicels as well as subtending them. *Pedicels* (longest) c.20–40mm, recurved in fruit, hairy as the stems. *Calyx* tube c. 0.2mm long, lobes c.6–6.5 × 1.3–1.8mm, narrowly elliptic, glandular-pubescent, hairs up to 1–2mm long. *Corolla* tube c.13–15 × 3–3.8mm, limb c.15–16mm across lateral lobes, posticous lobes c.5–6 × 5.4–6mm, anticous lobe 5–5.5 × 4.5–5mm, all lobes white, main veins purplish, they and the tube glandular-puberulous outside, glabrous inside except on the filaments. *Stamens* 4, posticous filaments c.3.5–4mm long, glabrous, posticous anthers 1.2–2mm long, included, anticous filaments 1–1.3mm long, minutely glandular-puberulous, anthers 1.2–1.8mm long, shortly exserted. *Stigma* 1–1.5mm long, scarcely exserted. *Style* c.6.5–9mm long. *Ovary* c.4 × 1.8mm, glabrous. *Capsule* c.8–10 × 4.5–6mm. *Seeds* c.0.3–0.5 × 0.15–0.25mm.

Selected citations:

Gran Canaria. SW side, Valle de Agaete, 500m, 26 vii 1960, *Andrews* R19 (K). Roque de Bentayga und du Fortalera nächst Mogan, 1890, *Simony* (W). Caldera de Bandama, iii 1846, *Bourgeau* 380 (E, K, LE, P, W, Z).

Camptoloma canariense is endemic to Gran Canaria, where it is known from several localities. It grows from crevices in volcanic rock faces and possibly flowers in any month. It is a peculiarity of the plant that the pedicels curl back as the capsules ripen; this is possibly a response to habitat: the branches are pendent and the curvature probably presents the capsules to the rock face; we have seen a similar response in species of *Streptocarpus* growing under rock overhangs. This type of dispersal is classified as autochory (see Pijl, L. van der, 1982).

2. Camptoloma rotundifolium Benth. in DC., Prodr. 10: 431 (1846).

Type: Angola, Benguela, Elephants Bay, 1842, *Curror* 13(K).

Syn.: *Chaenostoma corymbosum* Marloth & Engl. in Bot. Jahrb. 10: 253 (1888). Type: [Namibia] Hereroland, Davieb, 350m, vi 1886, *Marloth* 1461 (holo. B†; iso. BOL, K, M, PRE, SAM, W).
 Lyperia corymbosa (Marloth & Engler) N.E. Br. in Kew Bull. 1896: 163 (1896).
 Chaenostoma corymbosum Marloth & Engl. var. *huillanum* Diels in Bot. Jahrb. 23: 474 (1896). Type: Angola, Mossamedes, near Pedra do Rei and Pedra de Sal, by the river Maiombo, x 1859, *Welwitsch* 5806 (BM, COI, K).
 Sutera corymbosa (Marloth & Engl.) Hiern, Cat. Welw. Afr. Pl. 1: 757 (1898); Hemsley & Skan in Fl. Trop. Afr. 4(2): 304 (1906); Merxm. & Roessler in Prodr. Fl. S.W. Afr. 126: 50 (1967); Nordenstam in Dinteria 11: 38 (1974) cum ic.
 S. corymbosa (Marloth & Engl.) Hiern var. *huillana* (Diels) Hiern, Cat. Welw. Afr. Pl. 1: 757 (1898).
 S. welwitschii Skan in Fl. Trop. Afr. 4(2): 304 (1906). Type as for *Chaenostoma corymbosum* var. *huillanum*.

Plant herbaceous or woody, capable of flowering when only c.50mm tall or perennating to form a shrublet up to at least 700mm tall, with branches up to c.10mm in diam. springing from a stock at least 25mm in diam., densely glandular-pubescent, hairs up to 2–3mm long, leafy. *Leaves* alternate, blade c.13–75 × 13–70mm, broadly ovate, apex obtuse to subacute, base cuneate to cordate and passing into a petiolar part c.7–55mm long (about equalling to twice as long as blade), margins serrate, both surfaces glandular as the stems. *Flowers* sometimes solitary in the leaf axils, mostly in 2–11-flowered cymules, further arranged in leafy racemes or narrow panicles. *Bracts* up to c.5–10 × 0.5–1.2mm, narrowly elliptic. *Peduncles* up to 10–40mm long, *pedicels* up to 12–30mm long, hairy as the stems. *Calyx* tube c.0.2–0.5mm long, lobes 4.5–12 × 0.8–3.6mm, elliptic-spathulate, densely glandular, hairs up to 0.6–1.5mm long. *Corolla* tube 9–12 × 1.5–3.5mm, glabrous outside, inside throat bearded with broad band of clavate hairs, limb c.5.5–12mm across lateral lobes, posticous lobes 3–5 × 1.5–3.7mm, joined for ± half their length, anticous lobe 2–4.8 × 1.7–3mm, all lobes either glabrous or glandular-pubescent outside, hairs up to 0.1–0.6mm long, lobes varying in colour from white or yellowish to shades of pink or lilac, the three main veins in each lobe red or violet. *Stamens* 4, posticous filaments 2–3.5mm long, decurrent, bearded with long clavate hairs carried almost to base of tube, anthers 0.7–1.3mm long, included, anticous filaments 1–1.8mm long, minutely glandular at apex, anthers (0.4–)1.2–1.8mm long, shortly exserted. *Stigma* c.1–2mm long, very shortly exserted. *Style* 5.5–8mm long. *Ovary* 2–4 × 1.6–2.5mm, glabrous. *Capsules* 7–11 × 0.4–0.6mm. *Seeds* c.0.3–0.5 × 0.2–0.3mm.

Selected citations:

Namibia. 1712 AA, c.60km S of Cunene River and 30km inland, 21 vii 1973, *Robinson* R-K 25 (WIND). 1813 AC, Omutati, Huarasib river, 9 vi 1951, *Hall* 395 (NBG). 2014CB, farm Twyfelfontein OU534, 12 iv 1964, *Giess* 7911 (K, M, S, WIND). 2214DB, Haikamchab, 26 i 1907, *Galpin & Pearson* 7516 (BOL, K, PRE, SAM). 2114BA, Brandberg, The White Lady, 4 v 1963, *Nordenstam* 2532(M, S). 2416AB, farm Blässkranz (REH 7), 1 ix 1972, *Merxmüller & Giess* 28146(M).

Camptoloma rotundifolium was described from Baie dos elephantes, Benguela, Angola (13°13' S 12°44' E), while Welwitsch collected it further south, in Mossamedes district. These are the only collections I have seen from Angola, but the species appears to be common in western Namibia as far south as the Naukluft Mountains.

The plant is extraordinarily variable in stature and degree of woodiness. The material before me includes simple-stemmed plants 50–450mm tall with a simple taproot and leaves covering the whole size-range, as well as woody specimens with tufts of stems springing from a woody caudex. The plants favour shady places under overhanging rocks, in gullies, between boulders, and on cliff faces. Presumably if they get their roots deep into a moist crevice, they will perennate. Flowering and fruiting have been recorded in all months. One collector noted that the glandular stems and leaves wetted his hands; a similar comment was made of the very closely allied species *C. lyperiiflorum*, which differs in having the flowers always solitary or paired in the leaf axils (never in cymules) and the corolla lacks a broad band of clavate hairs in the throat, though these are present on the filaments.

3. Camptoloma lyperiiflorum (Vatke) Hilliard in Edinb. J. Bot. 48: 341 (1991). **Fig. 47, Plate 3I.**

Type: Somaliland, Ahl Mountains, 3300ft, March 1873, *Hildebrandt* 863 (holo. B†?; iso. W).

Syn.: *Urbania lyperiiflora* Vatke in Oest. Bot. Zeitschr. 25: 10 (1875).
   Camptoloma villosa Balf. f. in Proc. Roy. Soc. Edinb. 12: 84 (1883). Type: Socotra, ii–iii 1880, *Balfour* 237 (holo. K, iso. E).
   Chaenostoma lyperiiflorum (Vatke) Wettst. in Engl. & Prantl, Pflanzenfam. 4(3b): 69 (1891); Schwartz in Mitt. Inst. Alg. Bot. Hamb. 10: 243 (1939).
   C. oxypetalum Wagner & Vierh. in Oest. Bot. Zeitschr. 56: 258 (1906); Vierhapper in Denkschr. Akad. Wiss. Wien 71 t. 14, 2 (1907). Type: Socotra, Kustengebiet von Akarhi, 30–31 Jänner 1899, *Paulay* s.n. (n.v.).

Plant herbaceous or woody, capable of flowering when only c.70mm tall or perennating to form a shrublet up to c.600mm tall, stems up to c.6mm in diam, densely glandular-pubescent, hairs up to 2–4mm long, leafy. *Leaves* alternate, blade c.9–35 × 7–42mm, usually suborbicular, sometimes broadly ovate, apex obtuse to subacute, base cordate passing into a petiolar part c.6–40mm long (often about equalling, sometimes longer than the blade), margins shallowly lobed or toothed, lobes mostly obtuse, sometimes subacute, both surfaces glandular as the stems. *Flowers* mostly solitary in the leaf-axils, occasionally 2 pedicels from one axil, sometimes a branchlet as well, thus forming a narrow leafy panicle, otherwise the inflorescence racemose, sometimes tufts of very short shoots at ground level producing cleistogamous flowers. *Bracts* wanting. *Pedicels* up to c.10–17mm long, hairy as the stems. *Calyx* tube c.0.1–0.2mm long, lobes 5–5.5 × 0.8–1.2mm, spathulate, glandular-pubescent, hairs up to 1–2mm long. *Corolla* tube c.9.5–11 × 1.5–2.75mm, limb c.5–6mm across lateral lobes, posticous lobes c.3.2–4 × 1.6–2.5mm, joined for ± half their length, anticous lobe c.2.5–3.3 × 1.5–2.2mm, all lobes (and tube) glandular-puberulous outside, hairs up to 0.1–0.2mm long, glabrous inside except on filaments, limb white or creamy-white, main veins pale purple or red. *Stamens* 4, posticous filaments 1.5–2.5mm long, minutely glandular-puberulous at apex, short stout clavate hairs below them and carried ± halfway down tube by the decurrent filaments, anthers 1–1.5mm long, included, anticous filaments c.0.5mm long, hairy as the posticous ones, anthers 1–1.5mm long, shortly exserted. *Stigma* 0.2–0.5mm long, scarcely exserted. *Style* 6–7.5mm long. *Ovary* 2–2.5 × 1–1.5mm, glabrous. *Capsule* 5–8 × 3–4mm. *Seeds* 0.3–0.5 × 0.25–0.3mm.

Selected citations:

P.D.R. Yemen. Hadramaut, Kathiri, Raidat al Ma'arah, 5000ft, iv 1961, *Kerfoot* 3075 (K). Socotra, SE of Qalansiyah, 450m, 23 ii 1989, *Miller et al.* 8391 (E, UPS).
Sultanate of Oman. Kuria Muria Islands, Al Hallaniya, 1–10m, 16 x 1979, *Miller* 2755 (E).
Somalia. S side of Wadi Nogal, 2km from Eil towards the coast, 3 i 1973, *Bally & Melville* 15537 (K).

Camptoloma lyperiiflorum is known from several sites along the southern coast of the Arabian peninsula, from Mefa' to Al Makallã, from the off-shore islands of Kuria Muria (though not as yet from the mainland of Oman), Socotra, and Somalia, a compact distribution if one squeezes out the Gulf of Aden! The plants grow in the crevices and at the foot of cliffs, on scree, in rocky wadis and other similar sites from near sea level to c.1525m, and can possibly be found in flower in any month. One collector recorded the plants as 'sticky and wet with exudation' (cf. *C. rotundifolium*), another 'soapy-textured'.

Gallagher 6252/21 (E), collected on the Kuria Muria Islands, was growing in the silt of a steep rocky wadi under a south-facing cliff; it is simple-stemmed, 150mm tall, with a cluster of branchlets c.20mm long at the base, these full of old capsules produced by cleistogamous flowers; the fresh growth above is in first flower and fruit. *Gallagher* 6252/14 (E), from slopes of granite cliffs and scree facing north to the sea, shows exactly the same phenomenon as does *Bally & Melville* 15537 (cited above) from Somalia and the type material of *Camptoloma villosa* in the Kew and Edinburgh herbaria; this came from Socotra, and the plants are shrubby with clusters of old capsules from cleistogamous flowers at the base.

4. JAMESBRITTENIA

In 1807, Roth described a genus, *Sutera*, the lectotype of which is *S. foetida* Roth. In 1821, he inexplicably used the same generic name, but with a different diagnosis, for another plant, *Sutera glandulosa* Roth. Otto Kuntze, in 1891, judged these two plants to be generically distinct and therefore provided a new name for the latter, namely *Jamesbrittenia*, in honour of James Britten, at that time Keeper of Botany at the British Museum (Natural History) and obviously admired by Kuntze as a strong upholder of the nomenclatural Rule of Priority and editor of the Journal of Botany.

Kuntze's new name subsequently received scant attention and, if used at all, it has been confined to *J. dissecta* (Del.) O. Kuntze [= *Sutera glandulosa* Roth], an insignificant weed of Egypt, Sudan, India and Bangladesh. However, it proves to be a large genus (83 species recognized here), many species of which were originally described by Bentham under his generic name *Lyperia*. Bentham described the corolla tube of *Lyperia* as 'being more or less gibbous or incurved near the apex' while that of *Chaenostoma* 'has a corolla contracted at the base into a tube which is often elongated and always campanulate or infundibuliform at the orifice'. This sums up very neatly a basic distinction between *Jamesbrittenia* [*Lyperia* p.p.] and *Sutera* [*Chaenostoma*], though the very short corolla tube of *Jamesbrittenia pedunculosa* deceived Bentham into describing it under *Chaenostoma*.

The essential characters of *Jamesbrittenia* are the *calyx divided almost to the base* in all but one species; the *corolla tube always cylindric below, abruptly expanded above* to accommodate the *anthers*, which are *usually included, posticous filaments decurrent* down the tube at least as far as the top of the ovary, and at least the *posticous filaments pubescent* (the hairs may be minute and confined to the vicinity of the connective). There is *always a broad transverse band* (sometimes interrupted) *of clavate hairs in the throat*, and these usually extend only very briefly onto the anticous lip. The *stigma is always very short* (0.1–1.0mm long), *minutely bifid*, the two arms joined by a jelly-like substance and *crowned with minute stigmatic papillae*, and *included*; the style usually widens abruptly near the apex and carries the stigmatic surface at the tip of this broadened section, which is often slightly curved and lies between the anticous anthers. The *nectary* is *irregularly annular*, very shallow on one side, deeper on the other, there *sometimes forming a distinct lateral gland*. The *seeds* are ± oblong with a *reticulate testa*, easily seen with a hand lens in all but the smallest seeds; in only two species are the seeds further ornamented with transversely elongated furrows arranged like a chequer board.

The species fall into two major groups distinguished by the micromorphology of their seeds. In Group 1 (species 1–51) the polygonal cells of the testa are clearly visible (even under a hand lens or dissecting microscope), arranged in longitudinal rows. Two subgroups can be distinguished: in Group 1a (species nos. 1–37) the radial walls are smooth, in Group 1b (species 38–51) they are 'knotted' (figs 20–24). In both subgroups, the outer periclinal wall of each cell has collapsed inwards, forming a shallow pit. The inner periclinal wall appears to have scalariform thickenings, but these are not always clearly visible (age of the seed and degree of collapse of the outer periclinal wall may be contributory factors).

In Group 2 (species 52–83), only the longitudinal radial walls are clearly visible because they are irregularly thickened or 'knotted' (under a dissecting microscope, the seed looks more or less echinate). The longitudinal rows of 'knotted' walls are linked by smooth tranverse scalariform thickenings, and the transverse radial walls cannot be easily distinguished (figs 23, 24).

Within the two major groups, the species can be further broken down into assemblages of seemingly related species, but only two of these assemblages can be defined satisfactorily by mutually exclusive characters, namely, Group 1b, 1, which lacks glistening glands, and Group 1b, 2, which has extraordinarily coarse hairs (see pp.6–8 and figs 4, 5, 6, 8, 9, 10 for different types of hairs). Within the assemblages, vegetative characters and indumentum together, in some instances, with shape of the corolla lobes, give a classification that possibly reflects phylogenetic relationships. However, it is sometimes pollination systems that associate species on floral characters e.g. *J. silenoides* (Group 2,3) with corolla limb like that of *J. atropurpurea, J. huillana* and *J. namaquensis* (Group 1a, 2). These are floral guilds (not relationship groups) that stretch across species in unrelated families and genera: see further under pollination, p.55.

GROUP 1

Polygonal cells of testa clearly visible, arranged in longitudinal rows, outer periclinal walls collapsed inwards forming a shallow pit.

Group 1a Radial walls of testa smooth (figs 19A, 20, 21, 22A, B).

1a,1. Shrubs, shrublets or suffrutices, glandular-hairy, glistening glands as well, also often present on *upper* surface of corolla lobes and on placentae; leaves pseudofasciculate or not, margins mostly crenate to serrate, pinnatipartite only in *J. elegantissima*; corolla tube mostly 15–25mm long (c.5.5–7.5mm in *J. elegantissima*), corolla lobes rounded or slightly retuse, white, yellow, burnt orange to brown, or shades of violet to lavender, mouth round or laterally compressed, seeds mostly relatively large (c.0.8–1.5mm long). 1. *J. fodina*, 2. *giessii*, 3. *angolensis*, 4. *carvalhoi*, 5. *candida*, 6. *grandiflora*, 7. *macrantha*, 8. *albobadia*, 9. *burkeana*, 10. *accrescens*, 11. *dentatisepala*, 12. *elegant-*

issima.

Angola, Namibia, Zambia, Zimbabwe, Botswana, Malawi, Transvaal, Swaziland, High Drakensberg in Lesotho and Natal (one species).

1a,2. Shrublets or suffrutices, glandular-pubescent, glistening glands often predominant on leaves, leaves either pseudofasciculate or crowded on brachyblasts, margins entire or few-toothed; corolla tube c.15–27mm long, corolla lobes rounded in one species, white, ± truncate in others, shades of brown, mouth ± laterally compressed, seeds relatively large (c.0.8–1mm long).

13. *J. zambesica*, 14. *atropurpurea*, 15. *huillana*, 16. *namaquensis*.

Angola, Namibia, Malawi, Tanzania (Mbeya), Zimbabwe, Zambia, Botswana, Transvaal, Swaziland, Natal, Lesotho, O.F.S., Cape.

1a,3. Shrublets or suffrutices, glandular-pubescent, glistening glands sometimes predominant on leaves; leaves pseudofasciculate or crowded on brachyblasts, margins mostly coarsely toothed to deeply dissected, entire (and then very small) in a few species; corolla tube 7–15(--28)mm long, corolla lobes mostly oblong-cuneate, ± truncate to retuse, white, pink, mauve, violet, magenta, yellow in one species, shades of brown with yellowish margins in two species, mouth ± laterally compressed, seeds 0.4–1mm long.

17. *J. incisa*, 18. *tortuosa*, 19. *tysonii*, 20. *filicaulis*, 21. *albanensis*, 22. *phlogiflora*, 23. *maritima*, 24. *kraussiana*, 25. *pinnatifida*, 26. *argentea*, 27. *integerrima*, 28. *albiflora*, 29. *tenuifolia*, 30. *foliolosa*, 31. *zuurbergensis*, 32. *microphylla*, 33. *aspalathoides*, 34. *calciphila*, 35. *stellata*, 36. *albomarginata*, 37. *merxmuelleri*.

Mainly Cape, two extending into Namibia, two into Natal and Lesotho, one nearly confined to O.F.S.

Group 1,b Radial walls of testa 'knotted' (figs 22C, D, 23A).

1b,1. Shrublets or perennial herbs, one species doubtfully annual, glandular-hairy, glistening glands wanting on all parts; leaves entire or few-toothed; corolla tube mostly c.10–25mm long, shorter in one species, corolla lobes rounded, white, mauve, pink, rose, violet, creamy white to yellow in one species; mouth round; seeds c.0.25–0.6mm long.

38. *J. fruticosa*, 39. *maxii*, 40. *sessilifolia*, 41. *major*, 42. *megaphylla*, 43. *bicolor*, 44. *amplexicaulis*.

Mainly Namibia, Namaqualand and northern Cape, one species reaching Angola.

1b,2. Coarse annual herbs, three possibly perennating, glandular-hairy, some hairs (mainly on stems, inflorescence-axes and pedicels) extraordinarily coarse (either basal cell of some hairs 0.1–0.4mm across, or heads c.0.1–0.3mm across, or both), glistening glands mainly on leaves; leaves coarsely toothed; corolla tube c.7.5–25mm long, corolla lobes mostly ± cuneate, apex rounded to retuse or bifid, white or mauve, mouth laterally compressed or round; seeds 0.3–1mm long.

45. *J. racemosa*, 46. *thunbergii*, 47. *aridicola*, 48. *megadenia*, 49. *glutinosa*, 50. *primuliflora*, 51. *fimbriata*.

Namaqualand, northern Cape, Namibia.

GROUP 2

Only longitudinal radial walls of cells of testa clearly visible, irregularly thickened or 'knotted', linked transversely by smooth ± scalariform thickenings (on the outer periclinal walls?). Figs 23B, C, D, 24.

2, 1. Shrublets or suffrutices, glandular-hairy, glistening glands as well; leaves lobed and toothed; corolla tube 12–25mm long, corolla lobes cuneate-oblong, ± retuse, white, mauve, yellow, orange, mouth round; seeds 0.3–0.6mm long.

52. *J. acutiloba*, 53. *dolomitica*, 54. *heucherifolia*, 55. *pallida*, 56. *fleckii*, 57. *lyperioides*.

Namibia, one species extending into Angola.

2, 2. Shrublets or suffrutices, scales or glistening glands predominant, glandular-puberulous as well in three species; leaves entire to toothed (within a single species); corolla tube 6–11mm long, corolla lobes rounded, either yellow or varying from yellow to red, purple, violet or chestnut brown, mouth round; seeds 0.25–0.6mm long.

58. *J. pilgeriana*, 59. *barbata*, 60. *chenopodioides*, 61. *canescens*.

Three species confined to Namibia, *J. canescens* extending into Botswana and the arid northern Cape.

2, 3. Shrublets or suffrutices, glistening glands often predominant, glandular-pubescent as well in two species (but hairs possibly introduced into *J. pristisepala* by hybridization); leaves often pseudofasciculate, pinnatifid to pinnatisect; corolla tube c.5.5–20mm long, corolla lobes ± oblong to cuneate, often retuse, deeply bifid in one species, mostly white, creamy-white, yellowish, purplish or reddish, brown in one species, mouth laterally compressed; seeds 0.5–0.9mm long.

62. *J. crassicaulis*, 63. *stricta*, 64. *pristisepala*, 65. *lesutica*, 66. *silenoides*, 67. *beverlyana*.

Eastern Mountain Region, O.F.S., northern Natal, E Transvaal.

2, 4. One shrublet (*J. ramosissima*, an isolated species with well developed calyx tube), the rest annual but possibly sometimes perennating, glandular-hairy together with glistening glands; leaves coarsely toothed or lobed, corolla tube 4–10mm long, almost campanulate above, lobes suborbicular, white, lilac, rose, mauve, blue, but bright yellow in *J. pedunculosa*, mouth round; seeds 0.2–1mm long.

68. *J. ramosissima*, 69. *hereroensis*, 70. *tenella*, 71. *fragilis*, 72. *pedunculosa*.

Namibia, Namaqualand, arid northern Cape.

2, 5. Annual or perennial herbs (some perennials may flower young and look annual), glandular-hairy, glistening glands as well; leaves often pseudofasciculate, pinnatifid to pinnatisect; corolla tube 2.5–9mm long, often subcampanulate above, corolla lobes suborbicular to elliptic, white, pink, red, terracotta, orange, yellow, mouth often dorsoventrally compressed, seeds 0.2–0.7mm long.

73. *J. breviflora*, 74. *jurassica*, 75. *aspleniifolia*, 76. *multisecta*, 77. *aurantiaca*, 78. *montana*, 79. *concinna*, 80. *micrantha*, 81. *adpressa*, 82. *myriantha*, 83. *dissecta*.

Spread throughout the area of the genus.

Of the 83 species of *Jamesbrittenia*, only one, *J. dissecta* (2,5) has its area north of the equator. It is an annual of damp sandy places in Egypt, Sudan, India and Bangladesh, its nearest relatives being two more annual herbs of similar habitats, *J. adpressa* from Namibia (a typical north-south arid area connection, in the Manuleae paralleled in the genus *Camptoloma*), and *J. myriantha* from Zimbabwe.

Fourteen species are found in the Flora Zambesiaca area (Zambia, Malawi, Zimbabwe, Mozambique, Botswana), five of them endemic: *J. fodina* (1a,1), *J. myriantha* (2,5), *J. zambesica* (1a,2), all known only from Zimbabwe; *J. carvalhoi* (1a,1) from Zimbabwe and Mozambique; *J. albobadia* (1a,1) from Zimbabwe and Malawi. Nine of the 14 species are shrubs or suffrutices, three are perennial herbs that may behave as annuals, while two are annuals.

Thirty five species are found in Angola and Namibia, one endemic to Angola (*J. angolensis* 1a,1), one to Angola and Namibia (*J. heucherifolia*, 2,1), and 14 species and one variety to Namibia: *J. acutiloba* (2,1), *J. barbata* (2,2), *J. bicolor* (1b,1), *J. canescens* var. *laevior* (2,2), *J. chenopodioides* (2,2), *J. dolomitica* (2,1), *J. fimbriata* (1b,2), *J. fleckii* (2,1), *J. fragilis* (2,4), *J. giessii* (1a,1), *J. hereroensis* (2,4), *J. lyperioides* (2,1), *J. pallida* (2,1), *J. pilgeriana* (2,2), *J. primuliflora* (1b,2). Nearly all the endemics are shrubs, shrublets or suffrutices; only two are annuals (the allied species *J. hereroensis* and *J. fragilis*, both of which grow in moist shady places under overhanging rocks), and one (*J. primuliflora*) possibly an annual that may perennate. Indeed, nearly all the species are perennial, 24 clearly so, three possibly so (at least perennating when conditions prove favourable) and eight annual. The annual *J. concinna* (2,5) is confined to Namibia and Botswana.

There are 24 species in the Transvaal, Swaziland, O.F.S., Natal, Lesotho, Transkei and that part of the Cape falling within the Eastern Mountain Region. All are shrubs, suffrutices or perennial herbs, and 15 of them are confined, or very nearly confined, to the area. There are 7 endemics in the Eastern Mountain Region: *J. aspleniifolia* (2,5), *J. beverlyana* (2,3), *J. breviflora* (2,5), *J. dentatisepala* (1a,1), *J. jurassica* (2,5), *J. lesutica* (2,3), *J. pristisepala* (2,3). The five allied species, *J. accrescens*, *J. burkeana*, *J. candida*, *J. grandiflora* and *J. macrantha* (1a,1), are very nearly confined to the Transvaal: *J. burkeana* extends just across the western border into Botswana and south into northern Natal, *J. grandiflora* across the south eastern border into Swaziland. *Jamesbrittenia silenoides* (2,3) is confined to the eastern Transvaal and northern Natal, while *J. multisecta* (2,5) is very nearly confined to Transkei; its area extends briefly southwards over the Kei. *Jamesbrittenia albiflora* (1a,3) is very nearly confined to the O.F.S.

Forty seven species have been recorded from the Cape, 38 of them being shrubs or suffrutices, the rest annuals. Twenty one species are endemic, while 14 more are nearly confined to the arid Cape and Namibia.

Of the species strictly endemic to the Cape, 14 are closely allied shrublets (Group 1a,3) confined to the southern Cape, and include the only species (*J. stellata*) found on the Peninsula (*J. albanensis*, *J. albomarginata*, *J. argentea*, *J. aspalathoides*, *J. calciphila*, *J. foliolosa*, *J. maritima*, *J. microphylla*, *J. phlogiflora* (just enters Transkei), *J. pinnatifida*, *J. stellata*, *J. tenuifolia*, *J. tortuosa*, *J. zuurbergensis*). Five more endemics are confined to Namaqualand, two perennial (*J. amplexicaulis*, 1b,1, and *J. namaquensis*, 1a,2) while three are annuals (the allied *J. racemosa* and *J. thunbergii*, 1b,1, and *J. pedunculosa*, 2,4). The two remaining endemics are both shrublets, *J. incisa* (1a,3) in the west central Cape, *J. tysonii* (1a,3) in central and east central Cape.

There are 14 species nearly confined to Namibia and the Cape, either to Namibia and Namaqualand or to Namibia and the arid northern Cape; 'nearly' because *J. canescens* var. *seineri* (2,2) extends eastwards from Namibia into Botswana, while *J. maxii* (1b,1) extends northwards into Angola. The species are *J. canescens* var. *canescens* and var. *seineri* (perennial, 2,2), the five allied species (1b,1) *J. fruticosa*, *J. major*, *J. maxii*, *J. megaphylla*, *J. sessilifolia* (only *J. megaphylla* is doubtfully perennial; the others are shrublets), three allied annuals from Group 1b,2 (*J. aridicola*, *J. glutinosa*, *J. megadenia*), the annual *J. adpressa* (2,5) and *J. tenella* (2,4), and the shrublets *J. integerrima* (1a,3), *J. merxmuelleri* (1a,3) and *J. ramosissima* (2,4).

Jamesbrittenia is largely a genus of shrubs, shrublets, suffrutices and perennial herbs. The transition from woody plants to annuals seems to have taken place independently in several different assemblages, as follows:

Group 1a (37 species), all woody
Group 1b,1 (7 species), only one doubtfully annual, rest perennial
Group 1b,2 (7 species), coarse annuals of which three possibly perennate
Group 2,1 (6 species), all woody
Group 2,2 (4 species), all woody
Group 2,3 (6 species), all woody
Group 2,4 (5 species), one shrublet, not closely allied to the other four annuals
Group 2,5 (11 species), seven perennial (two of which may behave as annuals), four annual

Hybridization seems to be commonplace in *Jamesbrittenia*. Possible hybrids are mentioned in the text under the following species: *J. accrescens*, *J. argentea*, *J. aspleniifolia*, *J. atropurpurea*, *J. calciphila*, *J. filicaulis*, *J. fodina*, *J. foliolosa*, *J. fruticosa*, *J. huillana*, *J. kraussiana*, *J. maxii*, *J. pallida*, *J. phlogiflora*, *J. primuliflora*, *J. pristisepala*, *J. sessilifolia*, *J. stricta*, *J. tenuifolia*, *J. tysonii*. Hybridization does not appear to be confined to closely allied species:

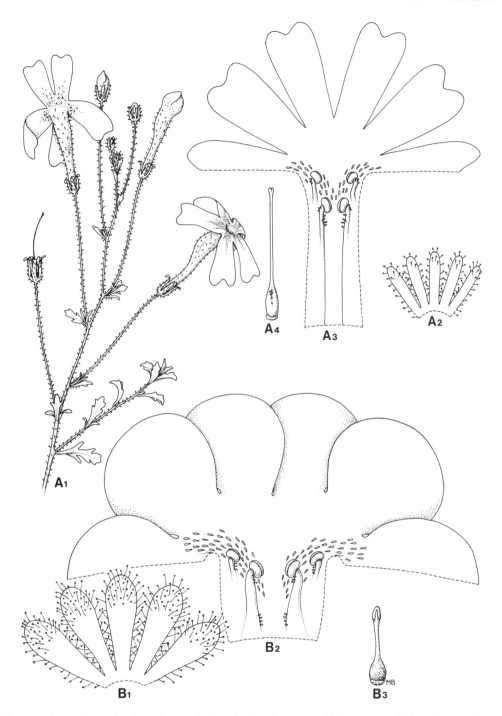

Fig. 48. A, *Jamesbrittenia foliolosa* (*Compton, D'Arcy & Rix* 936): A1, part of inflorescence, × 2; A2, calyx opened out, divided nearly to base, a generic character, × 5; A3, corolla opened out to show posticous filaments decurrent down tube, all anthers included, generic characters, × 5; A4, shortly bifid stigma (generic character), style, ovary with ± annular nectary slightly enlarged on one side, × 5; B, *J. breviflora* (*Hilliard & Burtt* 9825): B1, calyx opened out, × 5; B2, corolla opened out; note clavate hairs extending briefly onto lower lip, characteristic of genus, × 5; B3, stigma, style and ovary, × 5.

89

hybrids between Groups 1 and 2 are suspected (*J. huillana* × *lyperioides*, *J. canescens* × *primuliflora*), but most crossing takes place within one or the other major group, and then between species belonging to the same or different assemblages. Some hybrids are merely sporadic crosses, but sometimes complex backcrossing may be taking place, and it is possible that more than two species are involved.

Jamesbrittenia O. Kuntze, Rev. Gen. Pl. 2: 461 (1891); Hemsley & Skan in Fl. Trop. Afr. 4(2): 298 (1906); Andrews, Fl. Pl. Sudan 3: 137 (1956). **Figs 12D, E, 13D, E, 14A, 35B, 36B, 48.**

Syn.: *Sutera* Roth, Nov. Pl. Spec. 291 (1821), non 1807.
    Lyperia Benth. in Hook., Comp. Bot. Mag. 1: 377 (1836) p.p., excl. lectotype.

Shrubs, suffrutices, perennial or annual herbs, glandular or rarely clad in scales, aromatic, often unpleasantly so. *Stems* leafy. *Leaves* opposite, often alternate upwards, sometimes pseudofasciculate or crowded on very short brachyblasts, simple to deeply divided. *Inflorescence* racemose, flowers solitary in axils of leaves, these sometimes much reduced in size and bract-like. *Bracts* very rarely present, never adnate to pedicel nor calyx. *Calyx* 5-lobed almost to base (divided ± halfway in one species), almost regular, lobes elliptic, lanceolate, oblong or spathulate, usually entire, sometimes toothed, pubescent. *Corolla* tube cylindric, abruptly expanded near apex, mouth often round, sometimes laterally or dorsoventrally compressed, limb nearly regular to distinctly bilabiate, lobes spreading, orbicular to oblong or cuneate, often entire, sometimes retuse to bilobed, usually glandular outside, either glistening glands or glandular hairs or both, rarely scales, inside a broad transverse band of clavate hairs in throat, usually extending very briefly onto base of anticous lip, descending lower into tube on posticous than anticous side so that band is roughly V-shaped, rarely upper surface of limb minutely pubescent, more rarely clavate hairs or glistening glands present. *Stamens* 4 (in one species anticous pair wanting or much reduced), didynamous, filaments inserted in swollen part of tube, always puberulous, posticous filaments strongly decurrent, anthers synthecous, usually included, tips of anticous pair rarely just visible in throat or very shortly exserted. *Stigma* minutely bifid, included. *Ovary* ± deltoid in outline, usually glandular, nectary apparently annular, swollen to a greater or lesser extent on one side, sometimes resulting in a conspicuous lateral gland, ovules many in each loculus. *Fruit* a septicidal capsule with a short loculicidal split at tip of each valve. *Seeds* seated each on a round centrally depressed pulvinus, the pulvini discrete or confluent (figs 13D, E, 14A), more or less oblong, reddish- to greyish-brown, testa thin, tightly investing the endosperm, the cells arranged in longitudinal rows, forming a reticulum, under the SEM the walls seen to be smooth (figs 19D, 20, 21, 22A, B) or 'knotted' (figs 22C, D, 23, 24) (in two species the testa is furrowed with transversely elongated pits arranged in chequer-board fashion; fig. 24A).

Type: *Jamesbrittenia dissecta* (Del.) O. Kuntze.

Distribution: mainly Africa from Angola and Zambia southwards, one species (*J. dissecta*) ranging from Egypt and Sudan to the Indian subcontinent.

Key to species
(followed by regional keys)

Note: 1. All floral parts soaked before measurement; 2. Calyx tube measured on *inside*; 3. Corolla tube measured from base to point on anticous side where tube curves into limb; 4. Hybrids not catered for; 5. Consult discussion after species descriptions for further discriminatory information.

1a. Pair of sepal-like bracts c.4–8 × 0.8–1.7mm on pedicel immediately below calyx
1. J. fodina
1b. Pedicel very rarely with 1 or 2 tiny bracts but these not paired immediately below calyx   2

2a. Calyx lobed ± halfway, lobes triangular .... **68. J. ramosissima**
2b. Calyx lobed nearly to base, lobes variously shaped 3

3a. Stout perennial herb, glandular-pilose, hairs up to 1–2mm long, leaves pinnatipartite to bipinnatisect, flowers in upper leaf-axils, corolla tube 5.5–7.5mm long, limb 4–6mm across lateral lobes, yellow . . . . **12. J. elegantissima**
3b. Plant not as described above 4

4a. Corolla lobes shades of greenish-, yellowish-, orange-, red-, purple- to dark-brown, thick-textured (veins invisible on upper surface) with a very narrow to broad whitish to yellowish thin-textured margin (sometimes partly obscured because margins revolute) . . 5
4b. Corolla lobes either white or variously coloured, thin-textured (veins visible on upper surface); thick-textured tissue occasionally around mouth or radiating from mouth 12

5a. Corolla lobes deeply bifid . . . . **66. J. silenoides**
5b. Corolla lobes slightly retuse, truncate or rounded 6

6a. Thick tissue occupying nearly whole surface of corolla lobes, thin-textured margins very narrow 7
6b. Thick tissue forming a broadly cuneate median patch, thin-textured margin relatively broad 11

7a. Leaves elliptic to obovate in outline, glandular-hispid at least on lower surface; corolla lobes rounded 8
7b. Leaves ± oblong to spathulate in outline, often glabrous above, glistening glands only below (may ooze and then invisible but leaf looks varnished), glandular-pubescent in one species; corolla lobes ± truncate) 9

8a. *Upper* surface of leaves either glabrous or with glistening glands, very rarely a few hairs; at least primary leaves often folded and ± falcate, margins with mostly 1–3 pairs of teeth
9. J. burkeana
8b. Leaves glandular-hispid above and below, hairs up to 0.2–1mm long, glistening glands as well, leaves not folded and falcate, margins serrate . . . . **10. J. accrescens**

9a. Leaves glandular-puberulous on both surfaces, hairs up to 0.1–0.2mm long, glistening glands as well below . . . . **16. J. namaquensis**
9b. Leaves either glabrous above or with scattered glistening glands, thickly clad in glistening glands below (may ooze, then surface looks varnished), sometimes with hairs up to 0.1–0.2mm long on lower margins and midrib on undersurface 10

10a. Primary leaves mostly 1.5–10 × 0.4–3(–4)mm, mostly entire; flowers scattered irregularly along branches, which are often leafy at the tips . . . . **14. J. atropurpurea**
(see under species to distinguish subsp. **pubescens**)
10b. Primary leaves mostly (3–)10–30 × (1–)2–10mm, often with up to 5 pairs of teeth; some flowers often scattered along branches, nevertheless branch usually terminating in a distinct raceme . . . . **15. J. huillana**

11a. Primary leaves opposite with small axillary leaf-tufts, glabrous above, glistening glands below, sometimes oozing to give varnished effect . . . . **36. J. albomarginata**
11b. Primary leaves alternate, secondary leaves on axillary brachyblasts opposite and decussate, very closely packed, big glistening glands on both surfaces
37. J. merxmuelleri

12a. *Upper* surface of corolla lobes with glistening glands either confined to area around mouth or scattered all over, sometimes with tiny glandular hairs as well (not to be confused with clavate hairs extending outwards from throat), leaves variously toothed and lobed but never pinnately divided to midrib 13
12b. Upper surface of corolla lobes usually glabrous (except for clavate hairs extending outwards from throat onto base of anticous lip or rarely further); very rarely *either* with a

few glistening glands, and then leaves pinnatisect or bipinnatisect, *or* minutely glandular-puberulous, but without any glistening glands 20

13a. Upper surface of corolla lobes glandular-puberulous together with glistening glands . . 14
13b. Upper surface of corolla lobes with glistening glands only 16

14a. Corolla limb white, stems and leaves glandular-pubescent, hairs up to 0.6–1mm long, some glistening glands as well . . . . **5. J. candida**
14b. Corolla limb shades of lavender blue to deep violet 15

15a. Stems and leaves glandular-hispid, hairs up to 0.2–1mm long (glistening glands as well), leaf venation visible . . . . **6. J. grandiflora**
15b. Stems glandular-pubescent, hairs up to 0.1–0.2mm, leaves almost glabrous (a few glistening glands, sometimes a few minute glandular hairs as well on lower part), lateral veins in leaf invisible . . . . **7. J. macrantha**

16a. Leaves obovate in outline, often folded and ± falcate with 2–7 pairs of teeth (lower leaves sometimes lobed) . . . . **8. J. albobadia**
16b. Leaves plane, also differing from above in shape or in cutting of margins 17

17a. Upper leaf-surface either with glistening glands only or glabrous 18
17b. Upper leaf-surface with glandular hairs as well as glistening glands 19

18a. Leaves ovate to elliptic, margins crenate or doubly crenate, lower leaf-surface glandular-pubescent, hairs up to 0.5–1mm long, scattered glistening glands as well
3. J. angolensis
18b. Leaves oblanceolate, acute, 3–6 pairs of coarse teeth towards apex, both surfaces with sunken glistening glands only . . . . **13. J. zambesica**

19a. Shrub or suffrutex up to c.2.4m tall, calyx lobes oblong-lanceolate, acute, corolla limb 15–20mm across lateral lobes, throat brownish, the colour making an irregular ring around mouth . . . . **4. J. carvalhoi**
19b. Dwarf shrublet, calyx lobes spathulate, often toothed, corolla limb 10–15mm across lateral lobes, conspicuous orange star-shaped patch around mouth studded with glistening glands . . . . **11. J. dentatisepala**

20a. Plant completely lacking glistening glands 21
20b. Glistening glands present at least on lower leaf-surface (sometimes small and scattered, most easily seen on young leaves not yet fully expanded) 27

21a. Leaves usually consistently opposite, very rarely alternate at extreme tips of branches; calyx lobes relatively broad (c.1.1–2.1mm) . . . . **39. J. maxii**
21b. Leaves alternate at least towards tips of flowering branches; calyx lobes mostly narrow (c.0.6–1.4mm) 22

22a. Corolla tube 15–26mm long, limb white or shades of violet to purple 23
22b. Corolla tube 6.5–13mm long, limb white, creamy or yellow 25

23a. Cauline leaves usually elliptic tapering at both ends (rarely apex obtuse), petiolar part up to 1–2mm long (ratio length: breadth 2–4.8:1) . . . . **38. J. fruticosa**
23b. Cauline leaves elliptic to broadly ovate, petiolar part up to 0–7mm long (ratio length: breadth c.1–1.2:1) 24

24a. Corolla limb white (coloured markings around mouth); twiggy shrublet, branches usually remarkably flexuous . . . . **40. J. sessilifolia**
24b. Corolla limb lilac, rose or violet; perennial herb with long straight ascending branches
41. J. major

25a. Corolla tube 6.5–9mm long; stems weak, often decumbent or sprawling, leaves sharply ascending, often imbricate, margins entire to serrate in upper half .. **44. J. amplexicaulis**

25b. Corolla tube 9–13mm long; stems erect or nearly so, leaves patent or ascending, mostly entire 26

26a. Herb, possibly annual, largest leaves c.15–34mm broad, pedicels up to 7–16mm long
42. J. megaphylla

26b. Twiggy shrublet, largest leaves c.3–6mm broad, pedicels up to 2–6mm long **43. J. bicolor**

27a. Some hairs on stems (and sometimes pedicels and other parts) remarkably coarse: *either* basal cell of largest hairs up to 0.1–0.4mm across when flattened *or* the glandular heads up to c.0.1–0.3mm across (these on stems or pedicels), *or* hairs both stout-based and big-headed. Annual herbs, becoming coarse and possibly sometimes surviving a second season 28

27b. Hairs on stems and pedicels either neither coarse nor big-headed (caution: sand and dust stick to glandular heads and may give a false impression of size), or sometimes coarse but then remarkably small-headed 34

28a. Anticous stamens either aborted, reduced to staminodes or with tiny anthers 0.1–0.2mm long containing a few pollen grains; corolla tube 16–24mm long .... **49. J. glutinosa**

28b. Anticous stamens normal, anthers 0.4–1.2mm long 29

29a. Corolla lobes deeply notched (sinus c.0.8–3mm deep) 30

29b. Corolla lobes subentire to retuse (sinus up to c.0.3mm deep, or rarely up to 2mm, but then corolla limb not strongly patterned and seeds without transversely elongated depressions) 31

30a. Veins in corolla lobes dark and conspicuous; seeds patterned with transversely elongated depressions arranged like a chequer-board .... **45. J. racemosa**

30b. Veins in corolla lobes pale and inconspicuous; seeds without transversely elongated depressions .... **47. J. aridicola**

31a. Corolla limb white, relatively huge-headed glandular hairs scattered on stems and pedicels, heads c.0.2–0.3mm across (they look like miniature mushrooms under a lens)
48. J. megadenia

31b. Corolla limb mauve or white, glandular hairs without outstandingly large heads 32

32a. Anthers deeply included, not visible in mouth, mouth laterally compressed
46. J. thunbergii

32b. Anticous anthers either visible in mouth or very shortly exserted; mouth round 33

33a. Clavate hairs inside corolla tube on anticous side extending very briefly onto anticous lip and there c.0.25–0.4mm long, no band of hairs between posticous filaments
50. J. primuliflora

33b. Clavate hairs inside corolla tube extending in a broad fan to base of anticous lip and there up to 1.2mm long, band of hairs down back of tube between posticous filaments
51. J. fimbriata

34a. Corolla tube 18–28mm long 35

34b. Corolla tube up to 17mm long 40

35a. Mouth and top of throat encircled by a thick-textured red-brown to dark brown patch
8. J. albobadia

35b. Mouth and throat not encircled by thick brown tissue (throat often yellow or orange) 36

36a. Stems and leaves clad in glistening glands only .... **65. J. lesutica**
36b. Stems and leaves clad in glandular hairs, glistening glands also present 37

37a. Hairs on leaves up to 0.2–0.3mm long .... **17. J. incisa**
37b. Hairs on leaves up to 0.5–3mm long 38

38a. Leaves always opposite, pedicels up to 1–6mm long .... **52. J. acutiloba**
38b. Leaves opposite then alternate upwards, pedicels up to 9–30mm long 39

39a. Principal leaf blades ovate to subrotund (ratio length: breadth 1–1.2:1), shallowly lobed, rarely as much as halfway to midrib, petiolar part c.7–18mm long   **54. J. heucherifolia**
39b. Principal leaf blades oblong-elliptic to ovate (ratio length: breadth 1.2–2.3:1), usually lobed halfway or more to midrib, occasionally less, petiolar part up to 2–8(–10)mm long
    55. J. pallida

40a. Tips of corolla lobes ± truncate to retuse 41
40b. Tips of corolla lobes rounded 69

41a. Leaves with glandular hairs up to 0.1–3mm long as well as glistening glands on both surfaces 42
41b. Upper leaf surfaces mostly ± glabrous or with scattered glistening glands, occasionally with minute (less than 0.1mm) glandular hairs, lower surface thickly clad in glistening glands, sometimes a few minute glandular hairs as well 58

42a. Calyx lobes spathulate, tips 2–5-toothed (*J. pristisepala* hybridizes with *J. stricta* and *J. breviflora*: see under relevant species) .... **64. J. pristisepala**
42b. Calyx lobes variously shaped, entire 43

43a. Leaves without conspicuous tufts of axillary leaves 44
43b. Leaves with conspicuous tufts of axillary leaves or brachyblasts 49

44a. Hairs on stems mostly c.0.1mm long, sometimes scattered longer (0.2–0.3mm) hairs as well with much smaller heads than short hairs .... **57. J. lyperioides**
44b. Hairs on stems up to 0.5–4mm long 45

45a. Petiolar part of leaf up to 1–3mm long, blade finely crenate-serrate to doubly crenate, glandular hairs up to 0.5mm long mainly on petiolar part, shorter elsewhere . **2. J. giessii**
45b. Petiolar part of leaf 2–18mm long, blade lobed and toothed, glandular hairs up to 0.5–4mm long 46

46a. Hairs on vegetative parts very coarse, up to 1.2–4mm long on stems and leaves, stiffly patent on stems (see further under the species) .... **53. J. dolomitica**
46b. Hairs more delicate, not stiffly patent 47

47a. Principal leaf blades oblong-elliptic to ovate (ratio length: breadth 1.2–2.3:1), usually lobed and toothed halfway or more to midrib, rarely only c.⅓ of way, petiolar part usually 2–8mm long, rarely 10mm .... **55. J. pallida**
47b. Principal leaf blades ovate to subrotund (ratio length: breadth 1–1.4:1), ± shallowly lobed and toothed, rarely as deeply as halfway to midrib, petiolar part up to 5–18mm long .. 48

48a. Calyx lobes elliptic .... **54. J. heucherifolia**
48b. Calyx lobes linear-oblong to subspathulate .... **56. J. fleckii**

49a. Main primary leaves (ignore coppice shoots, old withered leaves at base of stem, and axillary leaves) entire or with 1–4 pairs of teeth near apex or coarsely toothed, sometimes with an occasional basal lobe cut ⅔ or more of way to midrib (at both leads consult notes after species descriptions to help discriminate these closely allied species) 50
49b. Main primary leaves cut ⅓–⅔ of way to midrib, lobes sometimes lobed again 54

50a. Leaves either entire or with 1–4 pairs of teeth near apex 51
50b. Leaf margins coarsely serrate 52

51a. Leaves mostly 1–3mm broad, often folded lengthwise, axillary leaf-tufts conspicuous; dark median streak at base of each corolla lobe; ovaries and capsules glandular-pubescent
18. J. tortuosa
51b. Leaves mostly 2.8–9mm broad, plane, axillary leaf-tufts often small or wanting; throat yellow to orange, colour sometimes radiating onto bases of lobes, but dark streaks very rarely present; ovaries and capsules with glistening glands all over, dominating any glandular hairs that may be present .... **27. J. integerrima**

52a. Flowers in leaf axils and not forming terminal racemes .... **24. J. kraussiana**
52b. Flowers running up into bracteate racemes 53

53a. Leaves lanceolate, coarsely and irregularly toothed throughout ...... **22. J. phlogiflora**
53b. Leaves mostly obovate, toothed in upper half .... **23. J. maritima**

54a. Hairs on stems up to 0.5–1.2mm long, on leaves up to 0.4–0.5mm ..... **25. J. pinnatifida**
54b. Hairs on stems and leaves up to 0.3mm long 55

55a. Corolla limb mauve to magenta; leaf-lobes obtuse .... **21. J. albanensis**
55b. Corolla limb white, or sometimes tinged with mauve but then leaf-lobes ± acute 56

56a. Leaf lobes ± acute, corolla limb white or tinged mauve .... **22. J. phlogiflora**
56b. Leaf lobes obtuse, corolla limb white 57

57a. Leaves oblanceolate in outline, lobes oblong, usually entire, occasionally larger ones with 1–2 teeth (see notes under species) .... **19. J. tysonii**
57b. Leaves ± ovate in outline, lobes ± spathulate or cuneate in outline, at least the largest lobes toothed with every gradation to bipinnatisect leaves .... **20. J. filicaulis**

58a. Corolla lobes cream-coloured to yellow .... **31. J. zuurbergensis**
58b. Corolla lobes either white or shades of pink, magenta, purple, mauve or blue 59

59a. Secondary leaves crowded on brachyblasts, all leaves seldom more than 3mm long and 1mm broad 60
59b. Secondary leaves pseudofasciculate, primary leaves mostly at least 3mm long (up to c.25mm) and mostly 1–10mm broad 62

60a. Most leaves 1–1.5mm long, ovate or cuneate, always acute, entire, margins revolute, large glistening glands in centre of lower surface, upper surface glabrous; calyx distinctly glandular-puberulous, hairs c.0.1–0.2mm long, glistening glands as well
32. J. microphylla
60b. Most leaves 0.8–3mm long, oblong to spathulate or subrotund, apex subacute to obtuse, entire or occasionally a pair of small teeth near apex, glistening glands all over lower surface, upper surface either glabrous, minutely glandular-pubescent or clad in glistening glands all over, very minute glandular hairs sometimes present as well 61

61a. Hairs on stems and pedicels rarely more than 0.1mm long, often oozing to give a varnished look (also leaves); leaves oblong to narrowly spathulate (ratio length: breadth 2.5–7:1) .... **33. J. aspalathoides**
61b. Hairs on stems and pedicels up to 0.2mm long, never oozing; leaves broadly spathulate to subrotund (ratio length: breadth 1–2.5(–3):1) .... **34. J. calciphila**

62a. Stems glandular-puberulous at least on upper parts (caution: hairs may be minute and are often mixed with glistening glands) 63
62b. Stems clad in glistening glands only 67

63a. Branches wiry, straggling, primary leaves mostly rhomboid to cuneate, coarsely few-toothed in upper half, filiform pedicels up to 10–25mm long in axil of each leaf tuft, not running up into distinct racemes . . . . **26. J. argentea**

63b. Plant not as described above (either differing in habit, or in shape and cutting of leaves, or in disposition of flowers) 64

64a. Stems, inflorescence axes and pedicels minutely glandular-puberulous (hairs less than 0.1mm long), principal leaves spathulate to oblanceolate, entire or few-toothed in upper half (teeth cut ½ way or occasionally ⅔ of way to midrib) . . . . **28. J. albiflora**

64b. Glandular hairs up to 0.1–0.4mm long present at least on upper parts of stem and inflorescence axes (as well as glistening glands, though these may be sparse) 65

65a. Leaves elliptic or spathulate in outline, at least primary leaves mostly with a pair of large teeth near apex (occasionally 2 or 3 pairs), pedicels up to 5–10(–15)mm long, flowers tending to run up into leafy racemes, no median bar on corolla lobes (see discussion under species) . . . . **29. J. tenuifolia**

65b. *Either* leaves with up to 4 pairs of lobes cut ½–¾ of way to midrib, no median bar on corolla lobes, *or* leaves entire or with up to 4 pairs of teeth, yellow to orange or red median patch at base of each corolla lobe (visible in dried specimens) 66

66a. Leaves with up to 4 pairs of lobes cut ½–¾ of way to midrib, nearly always conduplicate, flowers running up into distinct racemes, no median bar at base of corolla lobes
30. J. foliolosa

66b. Leaves entire or with 1 (rarely up to 4) pairs of teeth, usually plane, flowers scattered irregularly along branches, well marked yellow, orange or red median patch at base of each corolla lobe . . . . **35. J. stellata**

67a. Flowers borne in loose, few-flowered (up to c.14) bracteate racemes carried well above foliage . . . . **62. J. crassicaulis**

67b. Flowers many, crowded, the racemes sometimes leafy 68

68a. Calyx lobes linear-oblong, acute or subspathulate, entire (or very rarely a lobe with 1 or 2 teeth: see discussion after species), corolla tube 5.5–8mm long, limb 3–5mm across lateral lobes, shades of maroon or purple-red . . . . **63. J. stricta**

68b. Calyx lobes spathulate, tips 2–5-toothed, corolla tube 8–11.5mm long, limb 6.5–11mm across lateral lobes, creamy white, yellowish or lilac pink to mauve . . . **64. J. pristisepala**

69a. Tips of calyx lobes well dissected 70

69b. Tips of calyx lobes entire or rarely with a single tiny tooth 71

70a. Leaves opposite becoming alternate upwards, corolla tube 8–10mm long, limb pale yellow . . . . **67. J. beverlyana**

70b. Leaves always opposite, corolla tube 5–6mm long, limb dull terracotta, throat yellow
76. J. multisecta

71a. Stems and leaves thickly clad in glistening glands or scales (one species sometimes with glandular hairs less than 0.1mm long on stems, upper leaf-surface sometimes glabrous) 72

71b. Stems and leaves with a mixture of glandular hairs and glistening glands 77

72a. Flowers in leaf-axils, not running up into racemes, upper leaf-surface sometimes glabrous, stems with minute glandular hairs as well as glistening glands . . . . **31. J. zuurbergensis**

72b. Flowers running up into bracteate racemes, upper leaf-surface always clad in glistening glands or scales 73

73a. Clavate hairs extending out of mouth of corolla tube onto bases of lobes and at least to the sinus of the posticous lip; leaves often, but not always, so reduced that the branches seem bare .... **59. J. barbata**
73b. Clavate hairs extending at most to base of anticous lip and never onto posticous lip; leaves never strongly reduced 74

74a. Upper surface of corolla limb shades of red or purple-red .... **63. J. stricta**
74b. Upper surface of corolla limb yellow 75

75a. All external parts of plant as well as ovaries and capsules clad in scurfy scales; corolla tube 8.5–9.3 × 1.2–1.4mm at apex .... **58. J. pilgeriana**
75b. Stems and leaves clad in glistening glands (minute glandular hairs may occur on calyx); corolla tube 6.5–11.5 × 1.5–2mm at apex 76

76a. Plant usually single-stemmed at base, profusely branched above; largest leaves 10–55 × 3–30mm (ratio length: breadth 1–3(–5):1) .... **60. J. chenopodioides**
76b. Plant with stems tufted from base, largest leaves 10–34 × 2–6.5mm (ratio length: breadth (3–)4–9:1) .... **61. J. canescens** var. **laevior**

77a. Glandular hairs up to 0.4mm long on stems, to 0.1mm on leaves, leaves either entire or toothed up to halfway to midrib .... **61. J. canescens**
77b. Either glandular hairs on stems and/or leaves up to 0.5–3mm long, or, if shorter, leaves deeply dissected 78

78a. Corolla limb light to dark violet, lilac, rose or whitish; leaf margins coarsely and irregularly lobed or toothed 79
78b. *Either* corolla limb yellow, orange, terracotta or red, *or*, if violet, bright pink or whitish, leaves deeply dissected 81

79a. Leaf margins irregularly lobed; corolla tube 7.5–10 × 4mm in throat, limb glabrous above **71. J. fragilis**
79b. Leaf margins coarsely and irregularly toothed, corolla tube either 5–13 × 3–4mm or 4.8–7.2 × 1.7–2.5mm, limb either hairy or glabrous above 80

80a. Limb c.6.7–18mm across lateral lobes; clavate hairs in throat (and often extending out onto anticous lip) c.0.4–0.5mm long descending into tube on anticous side about as far as point of insertion of anticous filaments, on posticous side from that point almost to base of tube; filaments strongly bearded .... **69. J. hereroensis**
80b. Limb c.3–6mm across lateral lobes; inside of corolla tube sometimes glabrous, usually with 2 small patches of ± acute, retrorse unicellular hairs up to 0.5–0.8mm long on posticous side level with anticous anthers, narrow band of smaller hairs down back of tube between posticous filaments; filaments with tiny glandular hairs at apex ... **70. J. tenella**

81a. Pedicels up to 34–80mm long; corolla yellow, large dark patch at back of inflated part of tube often with 2 small dark outgrowths at its upper margin, sometimes a dark patch just above hooded part of posticous lip .... **72. J. pedunculosa**
81b. Pedicels up to 35mm long; corolla white or variously coloured, including yellow, always lacking a dark patch on the tube and posticous lip 82

82a. Leaf margins crenate; corolla limb red to rose pink .... **73. J. breviflora**
82b. Leaf margins cut at least halfway to midrib 83

83a. Leaves opposite, corolla tube 6.5–9mm long, limb violet or pink .... **74. J. jurassica**
83b. Leaves opposite below, alternate above, corolla tube mostly up to 7mm long, if longer, limb whitish, yellow, orange or red 84

84a. Leaves bipinnatisect, upper surface clad in glistening glands only .... **77. J. aurantiaca**
84b. *Either* leaves not bipinnatisect *or* upper surface with glandular hairs (sometimes sparse) as well as glistening glands, *or* both 85

85a. Corolla limb red; leaves cut once or twice nearly to midrib, lobes rounded, glandular-pubescent, hairs up to 0.2–0.5mm long, glistening glands few
75. J. aspleniifolia
85b. Corolla limb yellow or whitish, or occasionally pink; leaves not as described above ... 86

86a. Glandular hairs on stems and leaves less than 0.1mm long, pedicels up to 2–10mm long, corolla tube c.6.3–10mm long, limb yellow .... **81. J. adpressa**
86b. Glandular hairs on stems and leaves up to 0.1–0.5mm long, pedicels mostly up to 10–35mm long (if only c.3–12mm, corolla tube 2.5–4mm long, limb whitish), corolla tube 2.5–7mm long, limb mostly whitish to yellow, rarely pink or mauve 87

87a. Corolla limb c.2–4mm across lateral lobes, whitish .... **83. J. dissecta**
87b. Corolla limb c.5–10mm across lateral lobes, mostly yellow, white or pinkish to mauvish in one species 88

88a. Corolla limb white or pinkish to mauvish .... **82. J. myriantha**
88b. Corolla limb yellow 89

89a. Leaves divided nearly to midrib, 1 or 2 major lobes divided again, upper surface thickly clad in glistening glands, glandular hairs sparse .... **78. J. montana**
89b. Leaves divided from halfway to nearly to midrib, upper surface glandular-pubescent, glistening glands as well 90

90a. Annual, style usually 2.5–3.5mm long (shorter (c.2mm) only in tiny flowering seedlings that are probably selfing) .... **79. J. concinna**
90b. Perennial (but may flower when young and look annual), style 1.8–2.2mm long
80. J. micrantha

Regional Keys

EGYPT, SUDAN, INDIA

83. J. dissecta

ANGOLA, NAMIBIA

1a. Corolla limb with conspicuous band of glistening glands around mouth, white, 12–15mm across lateral lobes .... **3. J. angolensis**
1b. Corolla limb without glistening glands around mouth 2

2a. Stout perennial herb, glandular-pilose, hairs up to 1–2mm long, leaves pinnatipartite to bipinnatisect, flowers in upper leaf-axils, corolla tube 5.5–7.5mm long, limb 4–6mm across lateral lobes, yellow .... **12. J. elegantissima**
2b. Plant not as described above 3

3a. Upper surface of corolla limb shades of brown (yellow-, orange-, red-, purple-, chocolate-brown), usually with very narrow to broad whitish to yellowish margins; secondary leaves crowded on brachyblasts in axils of primary leaves; suffrutices 4
3b. Upper surface of corolla limb wholly white to shades of yellow, orange, violet or pink (very rarely brown in one species but then conspicuous axillary leaf-tufts wanting); brachyblasts or axillary leaf tufts present in few species; habit various 6

4a. Corolla lobes with a cuneate thick-textured median patch of colour with contrasting thin-textured broad pale margins .... **37. J. merxmuelleri**

4b. Corolla lobes almost entirely dark-coloured and thick-textured (veins invisible on upper surface) thin-textured pale margins very narrow and often not easily seen because margins revolute 5

5a. Primary leaves 1–6.5 × 0.5–1mm, entire or with up to 2 pairs of small teeth near apex, flowers scattered irregularly along branches, which often terminate in leaves, not in distinct racemes .... **14. J. atropurpurea** var. **pubescens**

5b. Primary leaves (3–)10–30 × (1–)2–10mm, margins with mostly up to 5 pairs of teeth, these sometimes coarse (axillary leaves often much smaller and entire), branches often with scattered flowers but usually terminating in distinct racemes, each leaf subtending a flower and a few, smaller, leaves .... **15. J. huillana**

6a. Calyx tube 1–1.2mm long, lobes triangular, 2.4–3 × 1.4–1.8mm; dense rounded shrublet, ultimate branchlets eventually leafless and subspinescent, corolla tube c.4–6 × 4mm in mouth, limb almost round, 10–16mm across, pale blue or mauve fading to white
68. J. ramosissima

6b. Calyx lobed almost to base, tube up to 0.1–0.4mm long, calyx lobes, habit and corolla not as described above 7

7a. Glistening glands or scales wanting on all parts 8

7b. Glistening glands or scales present at least on lower surface of leaves 13

8a. Leaves usually consistently opposite, very rarely alternate at extreme tips of branches, calyx lobes relatively broad (c.1.1–2.1mm) .... **39. J. maxii**

8b. Leaves alternate at least on flowering branches; calyx lobes mostly narrow (c.0.6–1.4mm) 9

9a. Corolla tube 15–26mm long, limb white or shades of violet to purple 10

9b. Corolla tube 9–13mm long, limb white, creamy or yellow 12

10a. Cauline leaves usually elliptic tapering at both ends (rarely apex obtuse), petiolar part up to 1–2mm long (ratio of length: breadth 2–4.8:1) .... **38. J. fruticosa**

10b. Cauline leaves elliptic to broadly ovate, petiolar part up to 1–7mm long (ratio of length: breadth c.1–2.6:1) 11

11a. Corolla limb white (coloured markings around mouth); twiggy shrublet, branches usually remarkably flexuous .... **40. J. sessilifolia**

11b. Corolla limb lilac, rose or violet; perennial herb with long straight ascending branches
41. J. major

12a. Herb, possibly annual, largest leaves c.15–34mm broad, pedicels up to 7–16mm long
42. J. megaphylla

12b. Twiggy shrublet, largest leaves c.3–16mm broad, pedicels up to 2–6mm long
43. J. bicolor

13a. Leaves (usually stems too) clad in glistening glands or scales only (or occasionally with minute (less than 0.1mm) hairs on pedicels, leaf margins and midrib on lower surface) 14

13b. Stems and leaves with mostly easily seen glandular hairs as well as glistening glands 17

14a. All external parts of plant as well as ovaries and capsules enveloped in scurfy scales; corolla tube 8.5–9.3 × 1.2–1.4mm at apex .... **58. J. pilgeriana**

14b. Both leaf-surfaces clad in glistening glands only (or sometimes with very minute (less than 0.1mm) glandular hairs as well on margins and midrib); corolla tube 6.5–11.5 × 1.5–2mm at apex 15

15a. Corolla limb with clavate hairs extending out of mouth onto bases of corolla lobes and reaching at least as far as sinus of posticous lip .... **59. J. barbata**
15b. Corolla limb glabrous above except for clavate hairs extending very briefly onto base of anticous lip 16

16a. Plant usually single-stemmed at base, branching profusely above; glandular hairs (as opposed to glistening glands) wanting on all parts; largest leaves 10–55 × 3–30mm (ratio length: breadth 1–3(–5):1) .... **60. J. chenopodioides**
16b. Plant with stems tufted from base, minute glandular hairs sometimes present on pedicels and often on leaf-margins and midrib on lower surface; largest leaves 10–34 × 2–6.5mm (ratio length: breadth (3–)4–9:1) .... **61. J. canescens** var. **laevior**

17a. Leaf blade cut nearly or quite to midrib at least in lower part 18
17b. Leaves entire to variously toothed or lobed but none divided to midrib 22

18a. Hairs on stems and leaves up to 0.5–2mm long 19
18b. Hairs on stems and leaves up to 0.3mm long 20

19a. Shrublet, corolla limb white or yellow .... **55. J. pallida**
19b. Annual herb, corolla limb lilac, rose or violet .... **71. J. fragilis**

20a. Leaves bi- or tripinnatisect, upper surface clad in glistening glands only **77. J. aurantiaca**
20b. Leaves pinnatisect, glistening glands and glandular hairs (less than 0.1–0.3mm) on upper surface 21

21a. Stems and leaves with glandular hairs up to 0.2–0.3mm long, pedicels up to 10–30mm long .... **79. J. concinna**
21b. Stems and leaves with glandular hairs less than 0.1mm long, pedicels up to 2–10mm long
81. J. adpressa

22a. Hairs on stems mostly up to 0.2mm long, rarely some 0.3–0.4mm 23
22b. Hairs on stems attaining 0.5–4mm 25

23a. Corolla limb white; largest leaves with up to 4 pairs of obtuse teeth .. **27. J. integerrima**
23b. Corolla limb yellow, orange, violet or purple-red; largest leaves either entire or shallowly to deeply lobed or toothed, teeth ± acute 24

24a. Blade of main leaves broadly ovate (roughly as long as broad), usually lobed and toothed, corolla tube 11–15mm long, limb 8–18mm across lateral lobes ...... **57. J. lyperioides**
24b. Blade of main leaves elliptic to oblanceolate (at least twice as long as broad, often much longer), entire or with up to 6 pairs of teeth; corolla tube 6–9.5mm long, limb c.4–9.5mm across lateral lobes .... **61. J. canescens**

25a. Some hairs on stems (and sometimes pedicels and other parts) remarkably coarse: *either* the basal cell of the largest hairs up to 0.15–0.4mm across when flattened, *or* the glandular heads up to c.0.1–0.3mm across (these on stems and pedicels), *or* hairs both stout-based and big-headed. Annual herbs, possibly sometimes surviving for a second season 26
25b. Hairs on stems and pedicels mostly neither coarse nor big-headed (caution: sand and dust stick to glandular heads and may give a false impression of size); hairs coarse (and very small-headed) in two species, both shrubs or subshrubs 30

26a. Anticous stamens either aborted, reduced to staminodes, or with tiny anthers 0.1–0.2mm long containing a few pollen grains; corolla tube 16–24mm long .... **49. J. glutinosa**
26b. Anticous stamens normal, anthers 0.4–1.3mm long 27

27a. Corolla lobes deeply notched, the sinus c.0.8–1.6mm deep; corolla limb white (orange and violet markings around mouth) .... **47. J. aridicola**

27b. Corolla lobes either entire or retuse, sinus up to c.0.3mm deep; corolla limb white or mauve 28

28a. Stems and pedicels sprinkled with hairs the glandular heads of which are 0.2–0.3mm in diam. (they look like miniature mushrooms); corolla mouth slightly compressed laterally, anthers included .... **48. J. megadenia**

28b. Glandular heads of some hairs up to 0.15mm in diam.; corolla mouth round, anticous anthers visible in mouth or very shortly exserted 29

29a. Clavate hairs inside corolla tube c.0.2–0.4mm long, extending only very briefly onto base of anticous lip .... **50. J. primuliflora**

29b. Clavate hairs up to 1.2mm long in a broad fan at base of anticous lip ... **51. J. fimbriata**

30a. Leaves always opposite (even on inflorescence axes), pedicels up to 1–6mm long
52. J. acutiloba

30b. Leaves always becoming alternate at least on upper stem and inflorescence axes; pedicels often, but not always, more than 6mm long 31

31a. Corolla tube 12–22mm long 32
31b. Corolla tube c.5–8mm long 38

32a. Corolla limb white, creamy, yellow or orange 33
32b. Corolla limb light to dark violet or blue 37

33a. Petiolar part of leaf up to 1–3mm long, blade finely crenate-serrate to doubly crenate, glandular hairs up to 0.5mm long mainly on petiolar part, shorter elsewhere .. **2. J. giessii**

33b. Petiolar part of leaf 2–18mm long, blade lobed and toothed, glandular hairs up to 0.5–4mm long 34

34a. Hairs on vegetative parts very coarse, up to 1.2–4mm long on stems and leaves, stiffly patent on stems (see further under the species) .... **53. J. dolomitica**

34b. Hairs more delicate, not stiffly patent 35

35a. Principal leaf-blades oblong-elliptic to ovate (ratio length: breadth 1.2–2.3: 1), usually lobed or toothed halfway or more to midrib, rarely only c.⅓ of way; petiolar part usually 2–8mm long, rarely 10mm .... **55. J. pallida**

35b. Principal leaf-blades ovate to subrotund (ratio length: breadth 1–1.4: 1), ± shallowly lobed and toothed, rarely as deeply as halfway to midrib; petiolar part up to 5–18mm long .. 36

36a. Calyx lobes elliptic .... **54. J. heucherifolia**
36b. Calyx lobes linear-oblong to subspathulate .... **56. J. fleckii**

37a. Hairs on stems and leaves up to 1.5–3mm long, calyx lobes c.5–9 × 0.8–1.2mm, hairs up to 0.6–3mm long, corolla tube 12–18 × 2.2–3mm at apex, lobes ± cuneate-oblong
54. J. heucherifolia

37b. Hairs on stems and leaves up to 0.5–1.2mm long, calyx lobes c.2.1–5.5 × 0.3–0.8mm, hairs up to 0.25–0.4mm long, corolla tube 9–13 × 3–4mm at apex, lobes subrotund
69. J. hereroensis

38a. Corolla limb c.6.7–12mm across lateral lobes, clavate hairs at base of anticous lip c.0.4–0.5mm long, on anticous side descending into tube about as far as point of insertion of anticous filaments, on posticous side almost to base of tube .... **69. J. hereroensis**

38b. Corolla limb 3–6mm across lateral lobes, inside of corolla tube sometimes glabrous, usually with 2 small patches of ± acute retrorse hairs up to 0.5–0.8mm long on posticous side level with anticous anthers, narrow band of smaller retrorse hairs down back of tube between posticous filaments .... **70. J. tenella**

BOTSWANA, MALAWI, MOZAMBIQUE, ZAMBIA, ZIMBABWE

1a. Pair of sepal-like bracts c.4–8 × 0.8–1.7mm on pedicel immediately below calyx
1. J. fodina
1b. Pedicels very rarely with 1 or 2 tiny bracts, but these not paired immediately below calyx   2

2a. Corolla tube 14–27mm long 3
2b. Corolla tube c.3–9mm long 9

3a. Corolla limb white or mauve 4
3b. Corolla limb shades of orange-brown, red-brown or dark brown, often with a contrasting
whitish or yellowish border 7

4a. Upper surface of corolla lobes glandular-pubescent, scattered glistening glands as well;
limb mauve . . . . **6. J. grandiflora** (garden escape)
4b. Upper surface of corolla lobes glabrous except for glistening glands around mouth; limb
white or mauve 5

5a. Leaves elliptic to ovate-elliptic, margins crenate to serrate, both surfaces glandular-
pubescent, hairs up to 0.3–0.8mm long . . . . **4. J. carvalhoi**
5b. Leaves obovate or oblanceolate with 2–7 pairs of teeth, leaves either glandular-pubescent at
least on lower surface, hairs up to 0.15–0.6mm long, or apparently glabrous (sunken
glands, sometimes oozing) 6

6a. Leaves glandular-pubescent at least on lower surface, calyx with glistening glands as well
as hairs up to 0.15–0.5mm long . . . . **8. J. albobadia**
6b. Leaves lacking stalked glandular hairs (sunken, often oozing, glands present), calyx with
oozing glistening glands only . . . . **13. J. zambesica**

7a. Primary leaves obovate in outline tapering to a petiolar part, often folded and ± falcate;
corolla tube glandular-pubescent outside, hairs up to 0.2–0.6mm long (glistening glands as
well) . . . . **9. J. burkeana**
7b. Primary leaves spathulate, oblong-spathulate, oblong or narrowly ovate, not folded and ±
falcate, (margins may be revolute or involute); corolla tube almost glabrous to very
minutely glandular-puberulous outside (glistening glands may be present on backs of
lobes) 8

8a. Pedicels and at least young parts of stems glandular-pubescent, hairs up to 0.15–0.5mm
long, glistening glands wanting . . . . **14. J. atropurpurea** subsp. **pubescens**
8b. Pedicels either glabrous or with scattered glistening glands, occasionally a few glandular
hairs less than 0.1mm long as well, at least young parts of stems with glistening glands and
hairs up to 0.1–0.15(–0.2)mm long . . . . **15. J. huillana**

9a. Corolla tube 6.3–9mm long, leaves elliptic to oblanceolate, entire or with up to 6 pairs of
teeth cut up to halfway to midrib . . . . **61. J. canescens** var. **seineri**
9b. Corolla tube c.3–7.5mm long, leaves lobed halfway or more to midrib 10

10a. Stems, leaves and calyx glandular-villous, hairs up to 0.5–2mm long   **12. J. elegantissima**
10b. Hairs on stems, leaves and calyx up to 0.3mm long 11

11a. Leaves bipinnatisect or occasionally less deeply dissected, upper surface clad in glistening
glands only; corolla limb pale mauve with orange throat (peculiar to Botswana)
77. J. aurantiaca
11b. Leaves divided halfway or more to midrib, glandular hairs (sometimes sparse)
c.0.1–0.3mm long as well as glistening glands on upper surface; corolla limb yellow or
white tinged pink or mauve 12

12a. Corolla limb white or tinged with pink or mauve, throat orange or chocolate-coloured .... **82. J. myriantha**
12b. Corolla limb yellow, throat orange 13

13a. Perennial, leaves divided nearly to midrib, 1 or 2 major lobes divided again, upper surface thickly clad in glistening glands, glandular hairs sparse .... **78. J. montana**
13b. Annual or perennial, leaves divided from halfway to nearly to midrib, upper surface glandular-puberulous, glistening glands as well 14

14a. Annual, style mostly 2.5–3.5mm long, shorter (c.2mm) only in tiny flowering seedlings that are probably selfing .... **79. J. concinna**
14b. Perennial (but may flower when young and look annual), style 1.8–2.2mm long .... **80. J. micrantha**

CAPE (EASTERN MOUNTAIN REGION ONLY), LESOTHO, NATAL, ORANGE FREE STATE, SWAZILAND, TRANSKEI, TRANSVAAL

1a. Corolla tube 15–27mm long 2
1b. Corolla tube c.3–12mm long 11

2a. Corolla limb white or various shades of mauve or blue (there may be whitish, yellow or orange markings around the mouth), thin-textured, veining visible 3
2b. Corolla limb greenish-, yellowish-, orange- or red-brown to shades of dark brown, thick-textured, veins invisible on upper surface (often with a very narrow whitish or yellowish thin-textured margin) 7

3a. Corolla limb white, stem and leaves clad in glistening glands only .... **65. J. lesutica**
3b. Corolla limb white or mauve, stem and leaves glandular-pubescent or glandular-hispid, often glistening glands as well 4

4a. Corolla limb white with conspicuous orange star-shaped patch around mouth, this clad in glistening glands .... **11. J. dentatisepala**
4b. Corolla limb white or mauve to blue, no orange star-shaped patch around mouth 5

5a. Corolla limb white, stems and leaves softly glandular-pubescent, hairs up to 0.6–2mm long, glistening glands few .... **5. J. candida**
5b. Corolla limb shades of mauve or blue, stems and leaves either glandular-hispid, hairs up to 0.2–1mm long, or stems glandular-pubescent, hairs 0.1–0.2mm long, leaves with a few glistening glands as well 6

6a. Leaves glandular-hispid (glistening glands as well), veins impressed above, raised below .... **6. J. grandiflora**
6b. Leaves with a few glistening glands only, lateral veins invisible .... **7. J. macrantha**

7a. Tips of corolla lobes deeply bifid .... **66. J. silenoides**
7b. Tips of corolla lobes rounded, retuse or almost truncate 8

8a. Corolla tube glandular-pubescent outside, hairs up to 0.2–0.6mm long (glistening glands may be present on backs of lobes) 9
8b. Corolla tube almost glabrous to very minutely glandular-puberulous outside (glistening glands may be present on backs of lobes) 10

9a. Upper surface of leaf either glabrous or with glistening glands, very rarely a few hairs; at least primary leaves often folded and ± falcate, mostly with 1–3 pairs of teeth .... **9. J. burkeana**
9b. Leaves glandular-hispid above and below, hairs up to 0.2–1mm long, leaves not folded and falcate, margins serrate .... **10. J. accrescens**

10a. Primary leaves 1.5–10 × 0.4–3(–4)mm, mostly entire, sometimes 1, rarely 2, pairs of teeth; flowers scattered irregularly along branches, which often terminate in leaves, not in distinct racemes .... **14. J. atropurpurea**
10b. Primary leaves (3–)10–30 × (1–)2–10mm, often with up to 5 pairs of teeth, sometimes entire (axillary leaves often much smaller and entire), branches often with scattered flowers but usually terminating in a distinct raceme, the leaves of which are very small
15. J. huillana

11a. Corolla limb white with a conspicuous orange star-shaped patch around mouth, this clad in glistening glands .... **11. J. dentatisepala**
11b. Corolla limb not described above 12

12a. Leaves always opposite 13
12b. Leaves becoming alternate upwards 16

13a. Calyx lobes deeply dissected .... **76. J. multisecta**
13b. Calyx lobes ± spathulate, entire 14

14a. Leaf margins crenate; corolla limb shades of red, rarely rose-pink ..... **73. J. breviflora**
14b. Leaf margins either entire or variously toothed or divided; corolla limb white, violet or lilac-pink 15

15a. Corolla limb violet or lilac-pink, stems and leaves glandular-pubescent, hairs up to 0.5–0.8mm long .... **74. J. jurassica**
15b. Corolla limb white, glandular hairs less than 0.1mm long, glistening glands predominant on undersurface of leaf .... **28. J. albiflora**

16a. Pedicels up to 1–5mm long, corolla limb creamy-coloured, yellow, pink, mauve or purple-red to dark red 17
16b. Pedicels up to 6–35mm long; corolla limb variously coloured or white, and always white when pedicels only 6mm long 19

17a. Flowers in leaf-axils, not running up into racemes .... **67. J. beverlyana**
17b. Flowers running up into long crowded bracteate racemes 18

18a. Calyx lobes 1.8–2.5 × 0.3–0.4mm, linear to subspathulate, acute to subobtuse, entire or rarely 1 lobe (out of 5) with 1 or 2 teeth, corolla tube 5.5–8mm long, limb 3–5mm across lateral lobes, shades of red or purple-red .... **63. J. stricta**
18b. Calyx lobes 2.3–4.5 × 0.3–2mm, tip 2–5-toothed, corolla tube 8–11.5mm long, limb 6.5–11mm across lateral lobes, creamy to yellowish, lilac or mauve. (*J. pristisepala* hybridizes with both *J. stricta* and *J. breviflora*: see under the relevant species)
64. J. pristisepala

19a. Leaf margins crenate, stems and leaves glandular-pubescent, hairs up to 1–3mm long
73. J. breviflora
19b. Leaf margins variously toothed or lobed (sometimes to midrib), entire to toothed in one species, glandular hairs, when present, up to 0.7mm long, mostly much shorter, leaves sometimes with glistening glands predominating 20

20a. Limb of corolla deep pink to red (may fade to mauve), leaves divided once or twice nearly to midrib, glandular-pubescent, hairs up to 0.2–0.5mm long, few glistening glands
75. J. aspleniifolia
20b. Limb of corolla mostly white, mauve or yellow, if purple-pink, reddish or orange, then glistening glands predominating on upper surface of leaves 21

21a. Corolla limb white or mauve, lobes ± cuneate-oblong, corolla tube (6.5–)7–10mm long 22

21b. Corolla limb yellow, orange, reddish or purple-pink, lobes suborbicular, corolla tube
3–7mm long 25

22a. Stems minutely glandular-puberulous (hairs less than 0.1mm long), leaves ± spathulate,
entire or bluntly few-toothed (1–4 pairs of teeth), upper surface glabrous or with a few
glistening glands, lower surface thickly clad in glistening glands . . . . **28. J. albiflora**
22b. Stems with glandular hairs 0.1–0.7mm long, glistening glands as well at least on young
parts, leaves ovate to elliptic in outline, coarsely toothed to deeply divided,
glandular-pubescent on both surfaces, often glistening glands as well particularly on lower
surface 23

23a. Leaves elliptic, margins coarsely serrate, flowers all along branches, pedicels filiform,
often spreading and curved in an elongated S . . . . **24. J. kraussiana**
23b. *Either* leaves ovate in outline and deeply divided (from c.⅓ of way to midrib) and flowers
not running up into definite racemes, *or* leaves elliptic to lanceolate in outline and
coarsely toothed (seldom more than ⅔ of way to midrib) and flowers in racemes 24

24a. Leaves deeply lobed, lobes ± spathulate or cuneate in outline, obtuse, flowers in leaf axils,
not forming definite racemes . . . . **20. J. filicaulis**
24b. Leaves coarsely toothed, teeth ± acute, flowers in racemes . . . . **22. J. phlogiflora**

25a. Leaves bipinnatisect, upper surface clad in glistening glands only . . . . . **77. J. aurantiaca**
25b. Leaves lobed ⅓ or more to midrib, upper surface clad in mixture of glistening glands and
glandular hairs (latter may be sparse) 26

26a. Leaves lobed nearly to midrib, upper surface thickly clad in glistening glands, sparse
glandular hairs as well . . . . **78. J. montana**
26b. Leaves divided roughly ⅓–½ of way to midrib, glandular hairs more plentiful than
glistening glands on upper surface . . . . **80. J. micrantha**

CAPE

1a. Corolla lobes shades of brown (greenish-, yellowish-, orange-, red-, purple-, chocolate-),
this tissue thick (veins invisible on upper surface) with very narrow to broad thin-textured
whitish to yellowish margins (sometimes visible only in sinuses) 2
1b. Corolla lobes white or variously coloured, lacking a contrasting border, thin-textured (veins
visible) 6

2a. Corolla lobes with only a very narrow thin-textured pale margin (often invisible except at
sinuses because margins revolute) 3
2b. Corolla lobes with a cuneate median patch of thick dark tissue, and broad pale margins   5

3a. Leaves glandular-puberulous on both surfaces, hairs up to 0.1–0.2mm long, glistening
glands as well below . . . . **16. J. namaquensis**
3b. Leaves either glabrous above or with scattered glistening glands, glistening glands below,
often oozing to give a varnished look, glandular hairs, if present, confined to lower parts of
margins and lower surface, particularly midrib 4

4a. Primary leaves mostly 1.5–10 × 0.4–3(–4)mm, mostly entire, flowers scattered irregularly
along the branches, many of which are vegetative at the tips . . . . **14. J. atropurpurea**
(see under the species to distinguish subsp. **pubescens**)
4b. Primary leaves mostly (3–)10–30 × (1–)3–10mm, often with up to 5 pairs of teeth, some
flowers often scattered on branches, nevertheless branches terminating in a distinct raceme
15. J. huillana

5a. Primary leaves opposite with small axillary leaf-tufts, glabrous above, glistening glands below, sometimes oozing to give a varnished look .... **36. J. albomarginata**
5b. Primary leaves alternate, secondary leaves very closely packed on axillary brachyblasts, opposite and decussate, big glistening glands on both surfaces ...... **37. J. merxmuelleri**

6a. Plant completely lacking glistening glands 7
6b. Glistening glands present at least on lower leaf-surface (sometimes small and scattered, most easily seen on young leaves not yet fully expanded) 13

7a. Leaves usually consistently opposite, very rarely alternate at extreme tips of branches; calyx lobes relatively broad (c.1.1–2.1mm) .... **39. J. maxii**
7b. Leaves alternate at least towards tips of flowering branches; calyx lobes mostly narrow (c.0.6–1.4mm) 8

8a. Corolla tube 15–26mm long, limb white or shades of violet to purple 9
8b. Corolla tube 6.5–13mm long, limb white, creamy or yellow 11

9a. Cauline leaves usually elliptic tapering at both ends (rarely apex obtuse), petiolar part up to 1–2mm long (ratio length: breadth 2–4.8:1) .... **38. J. fruticosa**
9b. Cauline leaves elliptic to broadly ovate, petiolar part up to 0–7mm long (ratio length: breadth c.1–1.2:1) 10

10a. Corolla limb white (coloured markings around mouth); twiggy shrublet, branches usually remarkably flexuous .... **40. J. sessilifolia**
10b. Corolla limb lilac, rose or violet; perennial herb with long straight ascending branches
41. J. major

11a. Corolla tube 6.5–9mm long; stems weak, often decumbent or sprawling, leaves sharply ascending, often imbricate, margins entire to serrate in upper half .. **44. J. amplexicaulis**
11b. Corolla tube 9–13mm long; stems erect or nearly so, leaves patent or ascending, mostly entire 12

12a. Herb, possibly annual, largest leaves c.15–34mm broad, pedicels up to 7–16mm long
42. J. megaphylla
12b. Twiggy shrublet, largest leaves c.3–6mm broad, pedicels up to 2–6mm long **43. J. bicolor**

13a. Some hairs on stems (and sometimes pedicels and other parts) remarkably coarse: *either* basal cell of largest hairs up to 0.1–0.4mm across when flattened *or* the glandular heads up to c.0.1–0.3mm across (these on stems or pedicels), *or* hairs both stout-based and big-headed. Annual herbs, becoming coarse and possibly sometimes surviving a second season 14
13b. Hairs on stems and pedicels neither coarse nor big-headed (caution: sand and dust stick to glandular heads and may give a false impression of size) 18

14a. Anticous stamens either aborted, reduced to staminodes or with tiny anthers 0.1–0.2mm long containing a few pollen grains; corolla tube 16–24mm long .... **49. J. glutinosa**
14b. Anticous stamens normal, anthers 0.4–1.2mm long 15

15a. Corolla lobes deeply notched (sinus c.0.8–3mm deep) 16
15b. Corolla lobes subentire to retuse (sinus up to 0.3mm deep, rarely to 2mm, but then limb not strongly patterned, seeds without transversely elongated depressions) 17

16a. Veins in corolla lobes dark and conspicuous; seeds patterned with transversely elongated depressions arranged like a chequer-board .... **45. J. racemosa**
16b. Veins in corolla lobes pale and inconspicuous; seeds without transversely elongated depressions .... **47. J. aridicola**

17a. Corolla limb mauve, glandular hairs without outstandingly large heads　**46. J. thunbergii**
17b. Corolla limb white, relatively huge-headed glandular hairs scattered on stems and
　　pedicels, heads c.0.2–0.3mm across (they look like miniature mushrooms under a lens)
　　　　　　　　　　　　　　　　　　　　　　　　　　　　　　48. J. megadenia

18a. Calyx tube 1–1.2mm long, lobes triangular, 2.4–3 × 1.4–1.8mm; dense rounded shrublet,
　　ultimate branchlets eventually leafless and subspinescent, corolla tube c.4–6 × 4mm in
　　mouth, limb almost round, 10–16mm across, pale blue or mauve fading to white
　　　　　　　　　　　　　　　　　　　　　　　　　　　　　　68. J. ramosissima
18b. Calyx lobed almost to base, tube up to 0.1–0.4mm long; calyx lobes, habit and corolla not
　　as described above 19

19a. Corolla tube c.18–28mm long, ovary minutely glandular-puberulous . . . . . . **17. J. incisa**
19b. Corolla tube up to 15mm long, rarely 17mm but then ovary with glistening glands
　　all over 20

20a. Corolla limb decidedly bilabiate, lobes ± cuneate or cuneate-oblong to oblong (lobes often
　　twice as long as broad), tips ± truncate to retuse, white or shades of blue, violet, magenta,
　　pink, rarely yellowish or maroon 21
20b. Corolla limb either bilabiate or almost regular, lobes suborbicular to broadly elliptic or
　　oblong-elliptic (roughly as long as broad), tips rounded, often yellow, orange, red,
　　terracotta, rarely shades of violet or rose 43

21a. Leaves with glandular hairs up to 0.1–0.6mm long as well as glistening glands on both
　　surfaces 22
21b. Upper leaf surface mostly ± glabrous or with scattered glistening glands, occasionally
　　with minute (less than 0.1mm) glandular hairs, lower surface thickly clad in glistening
　　glands, sometimes a few minute glandular hairs as well 32

22a. Calyx lobes spathulate, tips 2–5-toothed (*J. pristisepala* hybridizes with *J. stricta* and *J.
　　breviflora*: see under relevant species) . . . . **64. J. pristisepala**
22b. Calyx lobes variously shaped, entire 23

23a. Main primary leaves (ignore coppice shoots, old withered leaves at base of stem, and
　　axillary leaves) entire or with 1–4 pairs of teeth near apex or coarsely toothed, sometimes
　　with an occasional basal lobe cut ⅔ or more of way to midrib (consult notes after species
　　description to help discriminate these closely allied species) 24
23b. Main primary leaves cut ⅓–⅔ way to midrib, lobes sometimes lobed again 28

24a. Leaves either entire or with 1–4 pairs of teeth near apex 25
24b. Leaf margins coarsely serrate 26

25a. Leaves mostly 1–3mm broad, often folded lengthwise, axillary leaf-tufts conspicuous;
　　dark median streak at base of each corolla lobe; ovaries and capsules glandular-pubescent
　　　　　　　　　　　　　　　　　　　　　　　　　　　　　　18. J. tortuosa
25b. Leaves mostly 2.8–9mm broad, plane, axillary leaf-tufts often small or wanting; throat
　　yellow to orange, colour sometimes radiating onto bases of lobes, but dark streaks very
　　rarely present; ovaries and capsules with glistening glands all over, dominating any
　　glandular hairs that may be present . . . . **27. J. integerrima**

26a. Flowers in leaf axils and not forming terminal racemes . . . . **24. J. kraussiana**
26b. Flowers running up into bracteate racemes 27

27a. Leaves lanceolate, coarsely and irregularly toothed throughout . . . . . . **22. J. phlogiflora**
27b. Leaves mostly obovate, toothed in upper half . . . . **23. J. maritima**

28a. Hairs on stems up to 0.5–1.2mm long, on leaves up to 0.4–0.5mm .... **25. J. pinnatifida**
28b. Hairs on stems and leaves up to 0.3mm long 29

29a. Corolla limb mauve to magenta; leaf-lobes obtuse .... **21. J. albanensis**
29b. Corolla limb white, or sometimes tinged with mauve but then leaf-lobes ± acute 30

30a. Leaf lobes ± acute, corolla limb white or tinged mauve .... **22. J. phlogiflora**
30b. Leaf lobes obtuse, corolla limb white 31

31a. Leaves oblanceolate in outline, lobes oblong, usually entire, occasionally larger ones with
1–2 teeth (see notes under species) .... **19. J. tysonii**
31b. Leaves ± ovate in outline, lobes ± spathulate or cuneate in outline, at least the largest
lobes toothed with every gradation to bipinnatisect leaves .... **20. J. filicaulis**

32a. Corolla lobes cream-coloured to yellow, tips rounded .... **31. J. zuurbergensis**
32b. Corolla lobes either white or shades of pink, magenta, purple, mauve or blue, tips ±
truncate or retuse 33

33a. Secondary leaves opposite and decussate, crowded on brachyblasts, all leaves seldom
more than 3mm long and 1mm broad 34
33b. Secondary leaves pseudofasciculate, primary leaves mostly at least 3mm long (up to
c.25mm) and mostly 1–10mm broad 36

34a. Most leaves 1–1.5mm long, ovate or cuneate, always acute, entire, margins revolute, large
glistening glands in centre of lower surface, upper surface glabrous; calyx distinctly
glandular-puberulous, hairs c.0.1–0.2mm long, glistening glands as well
32. J. microphylla
34b. Most leaves 0.8–3mm long, oblong to spathulate or subrotund, apex subacute to obtuse,
entire or occasionally a pair of small teeth near apex, glistening glands all over lower
surface, upper surface either glabrous, minutely glandular-pubescent or clad in glistening
glands all over, very minute glandular hairs sometimes present as well 35

35a. Hairs on stems and pedicels rarely more than 0.1mm long, often oozing to give a
varnished look (also leaves); leaves oblong to narrowly spathulate (ratio length: breadth
2.5–7:1) .... **33. J. aspalathoides**
35b. Hairs on stems and pedicels up to 0.2mm long, never oozing; leaves broadly spathulate to
subrotund (ratio length: breadth 1–2.5(–3):1) .... **34. J. calciphila**

36a. Stems glandular-puberulous at least on upper parts; caution: hairs may be minute and are
often mixed with glistening glands 37
36b. Stems clad in glistening glands only 41

37a. Branches wiry, straggling, primary leaves mostly rhomboid to cuneate, coarsely
few-toothed in upper half, filiform pedicels up to 10–25mm long in axil of each leaf tuft,
not running up into distinct racemes .... **26. J. argentea**
37b. Plant not as described above (either differing in habit, or in shape and cutting of leaves, or
in disposition of flowers) 38

38a. Stems, inflorescence axes and pedicels minutely glandular-puberulous (hairs less than
0.1mm long), principal leaves spathulate to oblanceolate, entire or few-toothed in upper
half (teeth cut ½ way or occasionally ⅔ of way to midrib) .... **28. J. albiflora**
38b. Glandular hairs up to 0.1–0.4mm long present at least on upper parts of stem and
inflorescence axes (as well as glistening glands, though these may be sparse) 39

39a. Leaves elliptic or spathulate in outline, at least primary leaves mostly with a pair of large
teeth near apex (occasionally 2 or 3 pairs), pedicels up to 5–10(–15)mm long, flowers

tending to run up into leafy racemes, no median bar on corolla lobes (see discussion under species) .... **29. J. tenuifolia**

39b. *Either* leaves with up to 4 pairs of lobes cut ½–¾ of way to midrib, no median bar on corolla lobes, *or* leaves entire or with up to 4 pairs of teeth, yellow to orange or red median patch at base of each corolla lobe (visible in dried specimens) 40

40a. Leaves with up to 4 pairs of lobes cut ½–¾ of way to midrib, nearly always conduplicate, flowers running up into distinct racemes, no median bar at base of corolla lobes
30. J. foliolosa

40b. Leaves entire or with 1 (rarely up to 4) pairs of teeth, usually plane, flowers scattered irregularly along branches, well marked yellow, orange or red median patch at base of each corolla lobe .... **35. J. stellata**

41a. Flowers borne in loose, few-flowered (up to c.14) bracteate racemes carried well above foliage .... **62. J. crassicaulis**

41b. Flowers many, crowded, the racemes sometimes leafy 42

42a. Calyx lobes linear-oblong, acute or subspathulate, entire (or very rarely a lobe with 1 or 2 teeth: see discussion after species), corolla tube 5.5–8mm long, limb 3–5mm across lateral lobes, shades of maroon or purple-red .... **63. J. stricta**

42b. Calyx lobes spathulate, tips 2–5-toothed, corolla tube 8–11.5mm long, limb 6.5–11mm across lateral lobes, creamy white, yellowish or lilac pink to mauve .. **64. J. pristisepala**

43a. Stem clad in glistening glands only .... **63. J. stricta**

43b. Stem clad in glistening glands plus glandular hairs at least on upper part 44

44a. Leaves clad in glistening glands only (sometimes upper surface glabrous) 45

44b. Leaves clad in both glistening glands and glandular hairs though latter may be very small 46

45a. Twiggy shrublet, young branches thickly clad in glistening glands, racemes few-flowered, leafy, panicled .... **31. J. zuurbergensis**

45b. Stems tufted from base, simple to virgately branched, glandular hairs c.0.1–0.2mm long, conspicuous, sometimes predominating over glistening glands on upper parts of stem and inflorescence axes, racemes many-flowered, bracteate, only occasionally panicled
61. J. canescens

46a. Leaves entire to coarsely toothed, lobed as well in one species, margins crenate in another 47

46b. Leaves divided nearly or quite to midrib, the lobes sometimes divided again 50

47a. Leaf margins crenate, corolla tube c.3.5–6mm long, limb 7–20mm across lateral lobes, red to rose pink; branches sprawling .... **73. J. breviflora**

47b. Leaf margins entire to coarsely toothed or lobed; either corolla or habit not as described above 48

48a. Suffrutex, glandular hairs on stems and leaves up to 0.4mm long ...... **61. J. canescens**
(see under species to distinguish varieties)

48b. Annual herbs, glandular hairs on stems up to 0.5–3mm long (c.0.2–2mm on leaves) .. 49

49a. Pedicels up to 10–28mm long, corolla limb 3–6mm across lateral lobes, violet, lilac, rose or whitish .... **70. J. tenella**

49b. Pedicels up to 34–80mm long, corolla limb 8–11mm across lateral lobes, bright yellow
72. J. pedunculosa

50a. Calyx lobes deeply dissected .... **76. J. multisecta**

50b. Calyx lobes entire 51

51a. Leaves bipinnatisect, upper surface clad in glistening glands only, corolla limb yellow, orange or terracotta . . . . **77. J. aurantiaca**
51b. Leaves pinnatisect or bipinnatisect, upper surface with glandular hairs (minute in one species) as well as glistening glands, these sometimes few, corolla limb either red or bright yellow 52

52a. Twiggy shrublet, corolla tube 4–5 × 3mm in mouth, limb red (sometimes fading to mauve) . . . . **75. J. aspleniifolia**
52b. Annual herb, mat-forming, corolla tube 6.3–10 × 1.8–2.2mm in mouth, limb golden yellow . . . . **81. J. adpressa**

GROUP 1a, 1

1. Jamesbrittenia fodina (Wild) Hilliard in Edinb. J. Bot. 49: 229 (1992).
Type: Zimbabwe, Sipolilo [Guruve] distr., Great Dyke, Mpingi Pass, 4500ft, 17 v 1962, *Wild* 5775 (holo. SRGH; iso. K, M).

Syn.: *Sutera fodina* Wild in Kirkia 5: 79 (1965); Philcox in Fl. Zamb. 8(2): 31 (1990).

Shrublet c.0.3–1m tall, up to c.2m across, branches glandular-pubescent, hairs up to 0.4–2.3mm long, twigs very leafy. *Leaves* opposite, up to 4–23 × 2–8mm, oblong to narrowly elliptic, apex obtuse to subacute, base narrowed to petiolar part up to 1.5mm long, margins crenate, often revolute, texture thick, veins impressed above, raised below, both surfaces glandular-pubescent, hairs up to 0.3–2mm long, glistening glands as well particularly below. *Flowers* sweetly scented, solitary in upper leaf-axils forming short terminal racemes. *Bracts* 4–8 × 0.8–1.7mm, a pair immediately below the calyx, closely resembling calyx lobes. *Pedicels* up to 1–3mm long. *Calyx* tube c.0.2mm long, lobes 5–8 × 0.8–1.5mm, spathulate or oblong-spathulate, acute, glandular-pubescent, hairs up to 0.2–1.2mm long, a few glistening glands as well. *Corolla* tube 14–20 × 2–3mm in throat, cylindric, abruptly expanded near apex, limb nearly regular, 10–17mm across lateral lobes, posticous lobes 3–7 × 3–7mm, anticous lobe 3–6 × 3–6mm, broadly obovate to cuneate, apex rounded or retuse, glandular-pubescent outside, hairs up to 0.2–2mm long, glistening glands as well, broad band of clavate hairs in throat extending very briefly onto base of lower lip and there c.0.6mm long, slender, lobes white, creamy-white to shades of yellow, throat green to yellow, orange or brown, dark purplish radiating streaks as well (always?). *Stamens*: posticous filaments 1.5–2mm long, anthers 0.8–1.2mm, anticous filaments 0.4–0.8mm, anthers 0.5–1mm, all filaments bearded with clavate hairs. *Stigma* 0.2–1mm long. *Style* c.10.5–17mm long. *Ovary* 1.75–2.2 × 1–1.2mm. *Capsules* c.4–6 × 2.5–3mm (few seen), either minutely glandular-puberulous all over or clad in glistening glands. *Seeds* c.0.8–1 × 0.4–0.5mm.

Selected citations:

Zimbabwe. 1631AC, Mvuradonna Mountains, 4800ft, 12 v 1955, *Whellan* 919 (K, SRGH). 1730, Umvukwes, vicinity Imshi Mine, 4 ix 1960, *Leach & Bayliss* 10473 (K, SRGH). 1830CD, Mhlaba Hill near Windsor Chrome Mine, 16 i 1962, *Wild* 5603 (K, M, SRGH). 1930CA, Sebakwe National Park, 15 xi 1970, *Wild* 7797 (E, K, SRGH). 2029DD, 20 miles from West Nicholson on Belingwe road, 19 viii 1967, *Biegel* 2210 (K, SRGH).

Jamesbrittenia fodina is endemic to the ultramafic and mafic rocks of the Great Dyke of Zimbabwe, which runs south westwards from the edge of the Zambezi escarpment at Mvuradonna Mountain in the north to a point about 30km NE of West Nicholson in the south, between c.930 and 1700m above sea level. To quote Wild (in Kirkia 5: 80, 1965): 'This plant is presumably tolerant of very high chrome concentrations in the soil as it appears very rapidly in the spoil from chrome diggings. It occurs on both northern and southern serpentine'. Its natural habitat is among rock outcrops in grassland or woodland; flowering has been recorded in most months.

The plants display striking variation in indumentum, which has a geographical basis. For instance, specimens from around Sebakwe (*Wild* 7797 and 8010) have remarkably short hairs on all parts including the calyx and corolla, and the capsules are thickly clad in glistening glands. Elsewhere, the capsules are glandular-puberulous, and the hairs on the stems are up to c.1–2.3mm long, on the leaves c.0.5–2mm, on the calyx 1–1.2mm. On the northern part of the Great Dyke (from the Umvukwes northwards) the swollen upper part of the corolla tube is hirsute, the hairs up to c.1.5–2mm long; elsewhere they are only c.0.1–0.4mm long.

Another remarkable feature is the pair of bracts immediately below the calyx, the lobes of which they closely resemble. They are unmatched elsewhere in the genus, though a few species may have tiny bracts on the pedicels that are not a constant feature.

Wild 6412 (LISC, M), collected 20 miles north of West Nicholson on the Belingwe road, may be a hybrid between *J. fodina* and *J. huillana*. The plant has flowers very like those of *J. fodina*, including the pair of bracts below the calyx, but the leaves are spathulate, toothed at the apex, in form closely resembling those of *J. huillana* but differing in their short glandular-pubescence, also present on the stems and calyx. Wild recorded 'early colonist of chrome tip on serpentine slope', which seems to imply that the plant was not rare. If it is common on the Great Dyke, it could be given formal recognition.

2. Jamesbrittenia giessii Hilliard in Edinb. J. Bot. 49: 229 (1992). **Fig. 49.**
Type: Namibia, Otjiwarongo distr., 2016CD, [sic 2116 AB] farm Otjihaenamaparero (OTJ 92), 17 v 1978, *Giess* 15235 (holo. WIND, iso. M).

Subshrub up to c.450mm tall, stem c.3mm diam. (but no basal parts seen), possibly solitary and branching only above, branches long (up to c.300mm), ascending, glandular-puberulous with scattered longer hairs up to 0.5–1mm, leafy only on the upper parts. *Leaves* opposite, soon alternate upwards, up to c.4–13 × 3–7mm, ovate to elliptic, obtuse to subacute, base abruptly contracted to petiolar part c.1–3mm long, margins finely crenate-serrate to doubly crenate, texture thick, veins impressed above, raised below, upper surface minutely glandular-puberulous, glistening glands as well, lower surface similar but hairs slightly longer, rare on petiolar part, up to c.0.5mm. *Flowers* solitary in upper leaf axils forming long racemes. *Pedicels* up to 2–9mm long. *Calyx* tube c.0.2mm long, lobes c.3–5 × 0.7–0.8mm, subspathulate, sometimes with a pair of small teeth near apex, ± obtuse, glandular-pubescent, hairs up to c.0.4mm long, glistening glands as well. *Corolla* tube c.15–17 × 2mm in throat, cylindric, abruptly expanded near apex, limb nearly regular, c.8–14mm across lateral lobes, posticous lobes c.3–5.5 × 3–7mm, anticous lobe similar, broadly obovate, apex retuse, glandular-pubescent outside, hairs up to c.0.4mm long, glistening glands as well, broad band of clavate hairs in throat extending very briefly onto base of lower lip and there c.0.5mm long, slender, lobes white, throat yellow. *Stamens*: posticous filaments c.2mm long, anthers c.1.25mm, anticous filaments c.0.5mm, anthers 0.8mm, all filaments bearded with clavate hairs. *Stigma* 0.5–0.8mm. *Style* c.13mm. *Ovary* c.1.8 × 1mm. *Capsule* (immature) 5 × 3mm, glandular-puberulous all over, glistening glands on sutures. *Seeds* (immature) 0.6 × 0.4mm.

Specimen seen:

Namibia, Otjiwarongo distr., 2116AB, Etjo plateau, northern side, 3 ii 1979, *Craven* 1039 (WIND).

It gives me much pleasure to have this species bear the name of Mr W Giess, whose careful field notes, including details of flower colour and markings, have been invaluable to my study of the Manuleae. He found the plants in flower in May, in rather dry crevices of quartzite rocks, on a streambank near the dinosaur footprints ('nähe Quellbach bei den Dinosaurierspuren'). These are marked on the 1: 500000 map (1984 edition) about 25km east of Otjihaenamaparero (2116AB, not 2016CD as given on Giess's label and derived from Leistner & Morris, S. African place names). The Etjo plateau, whence came the only other specimen seen, lies about 7km

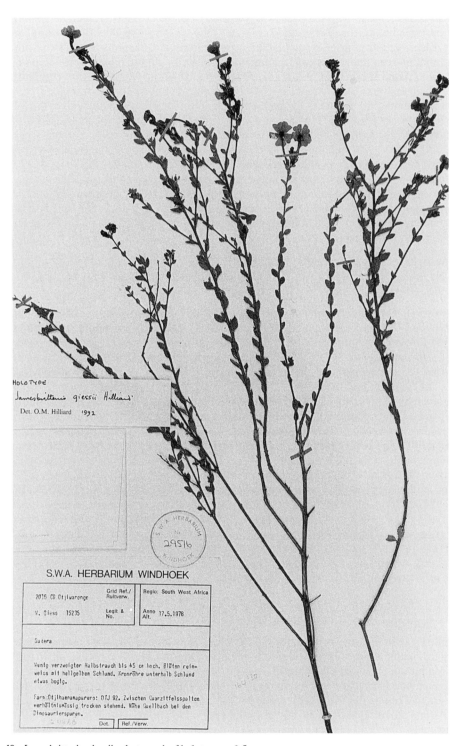

Fig. 49. *Jamesbrittenia giessii*: photograph of holotype, × 0.5.

south east of the footprints. This specimen is only a scrap with one flower, collected in February. The altitude is roughly 1600–2000m above sea level.

The affinity of *J. giessii* probably lies with *J. angolensis* to which it bears a superficial resemblance, but is easily distinguished by its leaves, glandular-puberulous on both surfaces (not glistening glands on the upper surface only), by its corolla limb, which has no glistening glands around the mouth and more delicate clavate hairs at the base of the anticous lip, and by its glandular-puberulous capsules.

3. Jamesbrittenia angolensis Hilliard in Edinb. J. Bot. 49: 226 (1992).
Type: Angola, Huila, Sá da Bandeira, Tundavala, 24 vii 1962, *Henriques* 71 (holo. LISC; iso. BM, COI, K).

Subshrub up to 0.8–1m tall, stem c.6mm diam. (but no basal parts seen), possibly solitary, well branched from low down, branches up to c.100mm long, ascending, glandular-pubescent, hairs up to c.1mm long, very leafy, often with small axillary leaf-tufts. *Leaves* alternate, up to 6–22 × 4.5–13mm, ovate to elliptic, acute, base cuneate and tapering into a short petiolar part up to 1–5mm long, margins finely crenate or doubly crenate, upper surface clad in glistening glands, lower surface glandular-pubescent, hairs up to 0.5–1mm long, scattered glistening glands as well. *Flowers* solitary in upper leaf-axils, the leaves greatly reduced, flowers crowded, forming short terminal racemes, often panicled by the branching of the main stem. *Pedicels* mostly up to 2–3mm long, up to c.16mm when flowers borne directly on main stem. *Calyx* tube c.0.2mm long, lobes c.6 × 1.3–1.5mm, oblong-spathulate, acute, glandular-pubescent, hairs up to c.0.8mm long, glistening glands as well. *Corolla* tube c.17–20 × 3mm in throat, cylindric, abruptly expanded near apex, limb nearly regular, c.12–15mm across lateral lobes, posticous lobes c.4–5 × 4–4.5mm, anticous lobe c.4–5 × 3–5mm, oblong-elliptic to subrotund, glandular-pubescent outside, hairs up to 0.4–0.6mm long, glistening glands as well, inside very stout clavate hairs at extreme base of anticous lip, descending into tube and diminishing in size towards the posticous side, a broad band of glistening glands around mouth, lobes 'at first pale yellow later white', throat red-brown. *Stamens*: posticous filaments c.2mm long, anthers 1.2–1.25mm, anticous filaments c.1mm long, anthers 0.8–1mm, all filaments bearded with clavate hairs. *Stigma* c.0.5mm long. *Style* c.15mm. *Ovary* c.2.3 × 1.3mm. *Capsules* c.8 × 4mm, covered in glistening glands, also present on placentae. *Seeds* c.1 × 0.6mm.

Specimens seen:

Angola. Huila distr., Sá da Bandeira, 1800m, 5 viii 1968, *Brito Teixeira et al* 12532 (LISC); environs de Humpata, Serra da Chella, viii 1937, *Humbert* 16559 (BM).

Jamesbrittenia angolensis is known to me from only a small area of Angola: Sà da Bandeira and Serra da Chela, roughly 14°S, 13°E. No ecological information is available other than the plants flower in July and August, but the species is the western counterpart of the eastern species, *J. carvalhoi*, so it probably occupies a similar habitat, montane grassland or scrub. The two species are remarkably similar in facies, in the colouring of the corolla limb and throat, very coarse clavate hairs at the base of the anticous lip, and the presence of glistening glands around the mouth, though they are much more conspicuous in *J. angolensis* than in *J. carvalhoi*. But *J. angolensis* is easily distinguished by the indumentum on its leaves: only glistening glands on the upper surface, not long hairs. The capsules too are clad in glistening glands, while those of *J. carvalhoi* are minutely puberulous.

4. Jamesbrittenia carvalhoi (Engl.) Hilliard in Edinb. J. Bot. 49: 228 (1992).
Type: Mozambique, Gorongoza, *Carvalho* s.n. (holo. COI, n.v.).

Syn.: *Cycnium carvalhoi* Engl., Planzenw. Ost-Afr. C 360 (1895) and in Bot. Jahrb. 23: 513 (1897).
   Sutera carvalhoi (Engl) Skan in Fl. Trop. Afr. 4(2): 307 (1906); Philcox in Fl. Zamb. 8(2): 32, tab. 13 (1990).

Shrub or subshrub up to c.2.4m tall, after fire stems regenerating from woody stock and then simple or subsimple, otherwise branching above and bushy, glandular-pubescent, hairs up to c.1mm long. *Leaves* opposite or ternate, sometimes alternate upwards, up to 10–55 × 6–25mm, elliptic to ovate-elliptic narrowed to a petiolar part c.2–6mm long, apex obtuse to acute or sometimes acuminate in uppermost leaves, margins finely crenate to serrate, upper surface drying blackish, lower surface brownish, both surfaces glandular-pubescent, hairs up to 0.3–0.8mm long, sometimes ± confined to veins on lower surface, glistening glands as well. *Flowers* solitary in upper leaf-axils producing crowded or lax terminal racemes sometimes branching into narrow panicles. *Pedicels* up to 4–14mm long. *Calyx* tube 0.1–0.2mm long, lobes 5.5–10 × 1.2–1.5mm, oblong-lanceolate, acute, glandular, hairs up to 0.5–1mm long, glistening glands as well. *Corolla* tube 14–22 × 3–3.2mm in throat, cylindric, abruptly expanded near apex, limb nearly regular, 15–20mm across lateral lobes, posticous lobes 5–7 × 5–6mm, anticous lobe 5–7 × 4–7mm, oblong-elliptic to subrotund, glandular-pubescent outside, hairs up to 0.4–0.7mm long, glistening glands as well, inside very stout clavate hairs at extreme base of anticous lip, descending into tube and diminishing in size towards the posticous side, a few glistening glands scattered around mouth, lobes white or pale mauve, throat yellowish-brown or orange-brown to purplish-brown. *Stamens*: posticous filaments 1.8–2.5mm, anthers 1.1–1.7mm, anticous filaments 0.8–1.2mm, anthers 0.7–1mm, all filaments puberulous. *Stigma* 0.4–0.5mm long. *Style* 11–19mm. *Ovary* 2.5–3 × 1–1.5mm. *Capsules* 7–8 × 4–5mm, minutely glandular-puberulous, glistening glands on placentae. *Seeds* c.1 × 0.5–0.7mm.

Selected citations:

Zimbabwe. Inyanga, Troutbeck, 5 viii 1957, *Hall* 1107 (NBG, STE). Pungwe hills, 6000ft, 8 viii 1950, *Chase* 2858 (SRGH). Umtali [Mutare], Vumba Mts., 2 vii 1957, *Schweickerdt* 2421 (M); Umtali, 4500ft, ix 1922, *Henkel* sub *Eyles* 3635 (BOL, SRGH). Chimanimani Mts, Mt. Peza, 6500ft, 15 x 1950, *Wild* 3628 (S, SRGH).
Mozambique. Gorongosa, Serra da Gorongosa, prox. do Pico Gogôgo, 26 ix 1943, *Torre* 5947 (LISC); ibidem, monte Nhandore, 1840m, 19 x 1965, *Torre & Pereira* 12412 (LISC).

Jamesbrittenia carvalhoi is known only from the eastern highlands of Zimbabwe, from Mt. Inyangani in the north to the Chimanimani Mountains in the south, and from the Gorongosa massif in Mozambique, from c.1370–2285m above sea level. Although I have seen no records from the Mozambique side of the Chimanimani Mountains, the plant will almost certainly occur there. It grows in grassland on open mountain slopes or in scrub above streams and on forest margins, flowering between June and October. The plants regenerate after fire: several herbarium sheets carry a burnt stem with new growth springing from its base. No underground parts have been seen, but *Rutherford-Smith* 90 (S, SRGH) recorded 'cluster of tubers at root' and *Torre & Pereira* 12412 noted 'com rizoma crasso'.

The species is so distinctive it is not easily confused with any other, except perhaps *J. grandiflora*, which occupies the same ecological niche in the eastern Transvaal and Swaziland. The figure in Flora Zambesiaca (cited above) gives an impression of the plant, but the drawing of the corolla laid open fails to show that the posticous filaments are decurrent (a generic character), nor does it show the characteristic stigma and nectary, and what purports to be a fruit is in fact an ovary.

5. Jamesbrittenia candida Hilliard in Edinb. J. Bot. 49: 228 (1992).
Type: Transvaal, Pietersburg distr., 2429BA, Chunies Poort, 14 x 1938, *Hafstrom & Acock* 1167 (holo. S, iso. PRE).

Suffrutex with a thick woody stock, stems several from the base, up to c.300–500mm long, spreading, ascending or erect, well-branched to twiggy at least in upper half, glandular-pubescent, hairs up to 1–2mm long, young parts densely leafy. *Leaves* alternate, pseudofasciculate, primary leaves up to 7–20 × 3–12mm (secondary leaves smaller), obovate to oblong, rather abruptly contracted into a petiolar part accounting for up to c.⅓ total leaf-length, apex obtuse, margins lobed (sometimes nearly to midrib), the lobes often toothed, or margins serrate, lobes obtuse, teeth obtuse to subacute, both surfaces glandular-pubescent, hairs up to 0.6–2mm long, a few scattered glistening glands as well, veins impressed above, raised below. *Flowers* solitary in upper leaf-axils, sometimes running up into short, ± bracteate racemes, sometimes the flowering part continuing as a vegetative twig. *Pedicels* up to 3–13mm long. *Calyx* tube c.0.1mm long, lobes c.4.5–6.5 × 1–1.7mm, oblong to elliptic, subacute, glandular-pubescent, hairs up to 0.3–0.6mm long. *Corolla* tube c.18–23 × 2.5–3mm in throat, cylindric, abruptly expanded near apex, mouth laterally compressed, limb slightly bilabiate, c.15–22mm across lateral lobes, posticous lobes 5.5–9 × 5.2–7mm, anticous lobe 6.5–9 × 5.5–7mm, cuneate to broadly ovate, apex sometimes shallowly retuse, glandular-pubescent outside, hairs up to 0.25–0.6mm long, glistening glands as well, inside delicate clavate hairs at extreme base of anticous lip descending into well-bearded throat, upper surface of lobes minutely glandular-puberulous, scattered glistening glands as well around mouth, lobes white. *Stamens*: posticous filaments c.2–3mm long, anthers 1–1.8mm, anticous filaments c.1mm long, anthers 0.8–1.2mm. *Stigma* 0.5–0.7mm long. *Style* 13–20mm. *Ovary* c.2 × 1.2mm. *Capsules* 5–8 × 2.8–3.5mm, glabrous or with a few tiny glandular hairs near apex, glistening glands on placentae. *Seeds* c.0.7–1.4 × 0.4–0.7mm.

Specimens seen:

Transvaal. Pietersburg distr., 2429BA, Chuniespoort, 4500ft, 4 x 1938, *Wall* s.n. (S); 2429BB, Bewaarkloof, 8 i 1986, *Raal & Raal* 693 (PRE); 2430AA, no locality given, 1000m, 22 xi 1985, *Raal* 672 (PRE); The Downs, 1100m, 20 xi 1986, *Raal* 950 (PRE). Lydenburg distr., (?), 2430AD, Kromellenboog, 860m, 17 viii 1989, *Glen* 1801 (PRE).

Jamesbrittenia candida is known from a small area of the NE Transvaal, from the Strydpoortberge east to the Transvaal Drakensberg between The Downs and Kromellenboog. The plants appear to favour rocky or cliffed wooded sites along rivers and streams between c.860 and 1375m above sea level; flowering has been recorded in October, November and January.

The flowers of *J. candida* resemble those of *J. grandiflora* and *J. macrantha*, but they are white, not mauve; the species differs further in its soft glandular-pubescence and paucity of glistening glands on stem, leaves and calyx. It appears to be a straggling plant, but twiggy, with a few flowers either terminating each branchlet or overtopped by new vegetative growth. In both *J. grandiflora* and *J. macrantha* the flowers run up into long bracteate terminal racemes.

6. Jamesbrittenia grandiflora (Galpin) Hilliard in Edinb. J. Bot. 49: 229 (1992). **Plate 3K.**
Lectotype (chosen here): Transvaal, Barberton, hillsides and valleys, 2200–3000ft, vii x 1889, *Galpin* 394 (K; isolecto. BOL, PRE).

Syn.: *Lyperia grandiflora* Galpin in Kew Bull. 1895: 151 (1895).
   Sutera grandiflora (Galpin) Hiern in Fl. Cap. 4(2): 304 (1904); De Wild., Pl. Nov. Herb. Hort. Then. 2 t.62 (1908); Pole Evans in Fl. Pl. S. Afr. 4: 131 (1924); Gard. Chron. ser. 3, 88: 277 (1930); Milne-Redhead in Curtis's Bot. Mag. 159 t.9452 (1936).

Subshrub up to c.0.5–1.2m tall, stem simple to sparingly branched, eventually well-branched above into a paniculate inflorescence, glandular-hispid, hairs up to 0.5–1mm long, glistening glands as well, closely leafy. *Leaves* alternate, pseudofasciculate, primary leaves up to 15–55 × 5–16mm (secondary leaves smaller), elliptic, obtuse to subacute, base tapering into a petiolar part, margins serrate to crenate-serrate, both surfaces glandular-hispid, hairs up to 0.2–0.6mm long, densely clad in glistening glands particularly on lower surface, veins impressed above, raised below. *Flowers* solitary in upper leaf-axils, the leaves greatly reduced in size and soon degenerating into bracts, forming a long terminal raceme, often panicled. *Pedicels* up to 7–25mm long. *Calyx* tube 0.1–0.2mm long, lobes c.5.5–8.5 × 1–1.4mm, oblong to oblong-spathulate, acute, glandular hispid, hairs up to c.0.2mm long, glistening glands as well. *Corolla* tube 20–23 × 3–4mm in throat, cylindric, abruptly expanded near apex, mouth laterally compressed, limb slightly bilabiate, 17–30mm across lateral lobes, posticous lobes 6–12 × 5–12mm, anticous lobe 6–10 × 5–10mm, cuneate to broadly ovate, apex ± rounded to retuse, glandular-pubescent outside, hairs up to 0.3–0.6mm long, glistening glands as well, inside clavate hairs at extreme base of anticous lip and descending into the throat, upper surface of lobes minutely glandular-puberulous, scattered glistening glands as well particularly around mouth, lobes lavender-blue or mauve, white, yellowish or greenish around mouth and in throat. *Stamens*: posticous filaments 1.8–2.5mm long, anthers 1.2–1.5mm long, anticous filaments 0.8–1mm, anthers 0.75–1.1mm, all filaments bearded with clavate hairs. *Stigma* 0.5–1mm long. *Style* 14–18mm. *Ovary* 2.5–3.5 × 1.2–1.75mm. *Capsules* 8–11 × 4–5mm, glandular-puberulous, glistening glands as well, both outside and on the placentae. *Seeds* c.1 × 0.6mm.

Selected citations:

Transvaal. Pilgrim's Rest distr., 2430DD, MacMac, 4500ft, vii 1894, *Goldie* in herb. aust. afr. 1639 (SAM, W, Z). Lydenburg distr., 2530BA, foot of Mt. Anderson, ± 4000ft, 18 iii 1933, *Galpin* 13719 (BOL). Nelspruit distr., 2530BD, west of Nelspruit, 25 x 1938, *Hafstrom & Acock* 1172 (S); 2530DB, Berlin State Forest, 1550m, 1 v 1987, *Balkwill et al* 3694 (E, J). Barberton distr., 2530DD, Nelshoogte, 17 iii 1962, *Strey* 4073 (M, SRGH); 2531 CC, Barberton, 3000ft, vi 1921, *Rogers* 24100 (S, SAM, Z).
Swaziland. Pigg's Peak distr., 2531CC, Havelock, c.4000ft, 21 iii 1961, *Compton* 30633 (NBG, PRE). Mbabane distr., 2631AC, Mbabane, 4400ft, xii 1905, *Bolus* 12186 (BOL). Hlatikulu distr., 2631CD, Sibowe river, c.2500ft, 22 vi 1959, *Dlamini* s.n. (NBG, PRE).

Jamesbrittenia grandiflora ranges from Mt Sheba and MacMac on the eastern escarpment of the Transvaal south to Waterval-Boven, Kaapschehoop and Barberton thence through the Barberton Mountains to Havelock, Mbabane and Hlatikulu in the western highlands of Swaziland, from c.600 to 2000m above sea level. It grows in rough grassland or scrub, often on the margins of forest or bush clumps; flowering has been recorded in all months, but the principal period is probably March to July. It is a handsome plant, often cultivated. It occurs in Zimbabwe as a garden escape (*Carter & Coates-Palgrave* 2102 (K), Mutare distr., Cecil Kop).

The flowers of *J. grandiflora* resemble those of *J. macrantha*, which is easily distinguished by its almost glabrous leaves with lateral veins invisible; its leaves are very like those of *J. accrescens* (which do, however, tend to blacken on drying; they never do so in *J. grandiflora*), which has flowers with a much smaller limb ranging in colour from greenish to yellowish to orange-brown or maroon.

7. **Jamesbrittenia macrantha** (Codd) Hilliard in Edinb. J. Bot. 49: 230 (1992).

Type: Transvaal, cultivated in Pretoria, *Pole-Evans* 4693A (holo. PRE; iso. K, S, SRGH).

Syn.: *Sutera macrantha* Codd in Fl. Pl. Afr. 30 t. 1162 (1954).

Subshrub up to c.0.5–1m tall, sparingly branched from base, branches ascending or decumbent, forking above into long rod-like flowering branches, glandular-pubescent, hairs c.0.1–0.2mm long, very leafy on upper parts. *Leaves* alternate, pseudofasciculate, primary leaves up to 10–20(–30) × 2–4(–10)mm (secondary leaves smaller), often folded along midrib and ± falcate, thick-textured, narrowly oblanceolate to elliptic tapering into a short petiolar part, apex acute, margins with 2–5(–8) pairs of teeth, both surfaces with a few scattered glistening glands, minute glandular hairs often present on petiolar part and scattered along lower part of midrib below, lateral veins invisible. *Flowers* solitary in upper leaf-axils, the leaves greatly reduced in size and quickly degenerating into small bracts, forming long (up to c.300mm in fruit) terminal racemes. *Pedicels* up to 10–28mm long. *Calyx* tube 0.1–0.2mm long, lobes c.4.5–5 × 1mm, oblong, acute, glandular-pubescent, hairs up to 0.1–0.25mm long, glistening glands as well. *Corolla* tube c.19–22 × 2.5–3mm in throat, cylindric, abruptly expanded near apex, mouth somewhat compressed laterally, limb slightly bilabiate, c.22–25mm across lateral lobes, posticous lobes c.10–11 × 5–9mm, anticous lobe c.10–10.5 × 5.7–10mm, cuneate, apex ± truncate or slightly retuse, glandular-pubescent outside, hairs up to 0.1–0.3mm long, glistening glands as well, inside clavate hairs at extreme base of anticous lip descending into well-bearded throat, upper surface of limb glandular-puberulous, scattered glistening glands as well mainly around mouth, lobes lilac to deep violet-mauve, throat yellowish-green. *Stamens*: posticous filaments c.2mm long, anthers 1–1.6mm, anticous filaments c.0.7–1mm long, anthers 0.7–1.1mm, all filaments bearded with clavate hairs. *Stigma* 0.6–1mm long. *Style* 14–17mm. *Ovary* 2.5–4 × 1.3–2mm. *Capsules* c.8–11 × 4mm, glabrous except for a few minute glandular hairs near apex, glistening glands on placentae. *Seeds* c.1.5 × 0.8mm.

Specimens seen:

Transvaal. Lydenburg distr., Sekoekoeniland, vii 1907, *Gray* 3768 (PRE); Lulu Mountains [2430CA, CC], farm Avontuur, 3800ft, 9 iii 1936, *Mogg* 16877 (PRE); 2430CC, 9 miles NW of P.O. Maartenshoop, 4000ft, 10 vi 1954, *Codd* 8795 (BM, K, PRE, S, SRGH); Steelpoort river valley, 30 v 1951, *Pole-Evans* 4693 (PRE); 2529BB, 9 miles N of Roossenekal, 4000ft, 14 x 1957, *Codd* 9776 (K, M, SRGH).

Jamesbrittenia macrantha is known from only a small area of the eastern Transvaal, along the valley of the Steelpoort river flanked on the west and north by the Sekukunis Mountains and the Lulu Mountains. The plants grow on grassy slopes with other scattered shrubs, flowering at least between June and October, but probably both earlier and later.

The flowers closely resemble those of *J. grandiflora*, but the almost glabrous folded leaves at once distinguish it. Its area lies immediately north west of that of *J. grandiflora*.

8. Jamesbrittenia albobadia Hilliard in Edinb. J. Bot. 49: 226 (1992). **Fig. 50.**
Type: Zimbabwe [Rhodesia], W Bulawayo-Essexvale, Hope Fountain Mission, c.1400m, 20 vi 1974, *Norrgrann* 565 (holo. S).

Syn.: *Sutera burkeana* auct. non (Benth.) Hiern; Hemsley & Skan in Fl. Trop. Afr. 4(2): 308 (1906); Philcox in Fl. Zamb. 8(2): 31 (1990) p.p. and excl. *Wild* 6412.

Suffrutex or shrubby, up to 0.6–1.5m tall, stems several from a thick underground stock, ascending to erect, becoming bare below, densely leafy above, simple below, above branching into long rod-like stems that eventually branch into narrow panicles, glandular-hispid, hairs up to 0.2–1mm long, scattered glistening glands as well. *Leaves* alternate, pseudofasciculate, primary leaves c.5–20 × 2–10mm, obovate in outline tapering to a petiolar part up to c.⅓ total leaf-length, blade often folded and somewhat falcate, thick-textured, apex acute to obtuse, margins sharply and boldly toothed (occasionally lobed in large lower leaves), the teeth in 2–7 pairs, themselves sometimes toothed, upper surface covered in glistening glands, sometimes with glandular-hispid hairs as well, very few to many, lower surface also clad in glistening glands and always glandular-hispid hairs mainly on midrib and main veins, hairs up to 0.1–0.6mm long, veins impressed above, raised below. *Flowers* solitary in upper leaf-axils, the leaves greatly reduced in size and quickly degenerating into bracts, forming long (up to c.150mm, 30 flowers) terminal racemes, lateral racemes with far fewer flowers. *Pedicels* up to 4–15mm long, sometimes with a pair of linear bracts c.2–6 × 0.1–0.7mm. *Calyx tube* 0.1–0.2mm long, lobes c.4.7–8 × 0.8–1.2mm, ± oblong, acute, glandular, hairs up to 0.15–0.5mm long, glistening glands as well. *Corolla* tube 18–21 × 2.6–3mm, cylindric, abruptly expanded near apex, mouth somewhat compressed laterally, limb slightly bilabiate, 12–17mm across lateral lobes, posticous lobes 4–6 × 4.5–6mm, anticous lobe 4.5–7 × 4–5.2mm, cuneate to broadly ovate, apex ± rounded, glandular-pubescent outside, hairs up to 0.2–1.3mm long, glistening glands as well, inside clavate hairs at extreme base of anticous lip descending into well-bearded throat, occasionally a few glistening glands at sinuses on upper surface of limb, lobes thin-textured, white (or mauve fide *Gilliland* 177, BM), mouth and top of throat red-brown to dark brown, the tissue thick. *Stamens*: posticous filaments 1.5–2mm long, anthers 1.1–1.8mm, anticous filaments 0.5–1mm, anthers 0.8–1.2mm, all filaments bearded with clavate hairs. *Stigma* 0.5–1mm long. *Style* 14–17mm. *Ovary* 2.5–3.2 × 1.2–1.8mm. *Capsules* c.7–9 × 4–4.5mm, glistening glands mainly down sutures, minutely glandular-puberulous as well or not, glistening glands on placentae. *Seeds* c.1–1.2 × 0.6–0.7mm.

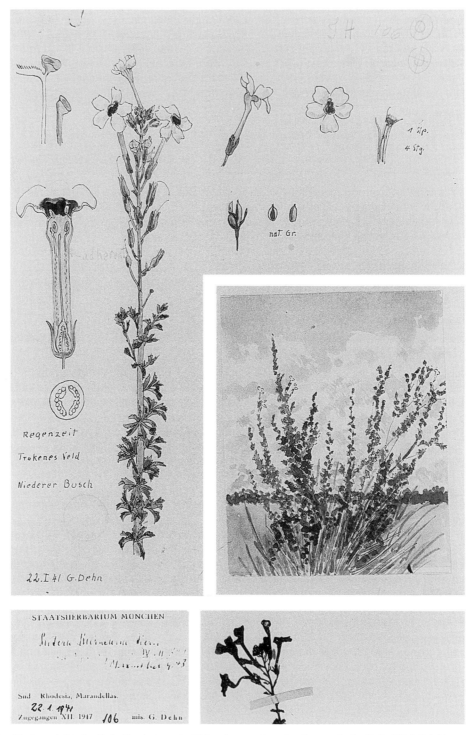

Fig. 50. *Jamesbrittenia albobadia*, photograph of ink and watercolour sketches attached to *Dehn* 106 (M); inflorescence × 1; half flower showing dark band in throat, decurrent posticous filaments, clavate hairs at base of lower lip, × 2; the enlarged portion of the half flower (above) shows the clavate hairs more clearly.

Selected citations:

Malawi. Dedza distr., Chongoni Forest Reserve, Chencherere Hill, 1675–1800m, 23 iv 1970, *Brummitt* 10077 (K, SRGH); Chiwan Hill, 17 vii 1969, *Salubeni* 1371 (K, SRGH).
Zimbabwe. Sipolilo distr., Nyamnyetsi Estate, on Great Dyke, 8 viii 1978, *Nyariri* 287 (SRGH). Harare, near Cleveland Dam, 13 viii 1916, *Eyles* 581 (SAM). Marandellas, 21 i 1941, *Dehn* 106 (M). Wankie distr., Victoria Falls road, 21 viii 1957, *Martin* 1069 (NBG, STE). Bubi distr., Inyati Mission, Ndumba Hill, 4500ft, 18 iv 1947, *Keay* sub FHI 21202 (K, SRGH). Matobo distr., Besner Kobila, 4700ft, vii 1953, *Miller* 1838 (SRGH). Inyanga village, c.1700m, 17 ii 1931, *Norlindh & Weimarck* 5097 (K, S, SRGH). Umtali distr., Inyamatshira Mt., 7 ii 1962, *Corner* s.n. (E). Idzabu River Valley, 1914, *Teague* 18 (BOL, K).

Jamesbrittenia albobadia has been known for a hundred years, but it has always been confused with *J. burkeana*. It differs strikingly from both *J. burkeana* and *J. accrescens* in its flowers: the corolla limb is thin-textured and white, with a brown, thick-textured band around the mouth; this band corresponds to the osmophore that occupies nearly the whole surface of the limb in the other two species, which is therefore shades of brown with a narrow thin-textured pale margin. The leaves of *J. albobadia* are more or less intermediate between those of the other two species: often folded and ± falcate as in *J. burkeana*, margins serrate as in *J. accrescens* but the toothing coarser, as in *J. burkeana*. The indumentum on the upper surface varies from glistening glands only to few to many glandular hairs as well, whereas in *J. burkeana* there are glistening glands only, in *J. accrescens* stalked hairs as well as glistening glands. *Dehn* 106, cited above, includes a water colour painting of the plant, showing both habit and floral details (fig. 50).

Jamesbrittenia albobadia is widespread on high ground in Zimbabwe, from roughly Sipolilo in the north south to the Matopos in the west and the border highlands in the east, between c.1200 and 1700m above sea level; it ranges north to Malawi, where it has been recorded only in Dedza and Lilongwe districts. Many records are from rocky, grassy, lightly wooded slopes, and flowers have been recorded between January and October.

9. Jamesbrittenia burkeana (Benth.) Hilliard in Edinb. J. Bot. 49: 227 (1992). **Plate 4F.**
Type: Transvaal, Macalisberg (sic; Magaliesberg), *Burke* 513 (holo. K; iso. PRE, SAM).

Syn.: *Lyperia burkeana* Benth. in DC., Prodr. 10: 361 (1846).
  Chaenostoma burkeanum (Benth.) Diels in Bot. Jahrb. 23: 491 (1897); Thonner, Blutenpfl. Afr. t. 137 (1908).
  Sutera burkeana (Benth.) Hiern in Fl. Cap. 4(2): 299 (1904) p.p., excl. some of the specimens cited; Phillips in Fl. Pl. S. Afr. 5 pl. 196 (1925); van Wyk & Malan, Field guide to the Wild Flowers of the Witwatersrand & Pretoria region 280 fig. 729 (1988).

Suffrutex, sometimes shrubby, stems c.0.3–1.5m long, erect or straggling, becoming twiggy with many lateral, often short, flowering branchlets towards the tips, glandular-hispid, hairs up to 0.15–0.4mm long, glistening glands as well, closely leafy. *Leaves* alternate, pseudofasciculate, primary leaves 4–12(–35) × 2–6(–12)mm, obovate or occasionally elliptic in outline tapering into a petiolar part often accounting for ± half total length, blade usually folded and ± falcate, thick-textured, apex often acute, margins mostly boldly and sharply toothed (1–3 pairs of teeth), larger, lower, leaves coarsely and irregularly lobed as well as sharply toothed, often almost to midrib, occasionally secondary leaves entire, both surfaces clad in glistening glands, glandular-hispid as well mainly on veins and margins of lower surface, hairs up to 0.1–0.2mm long but reduced to minute projections on upper leaves (the lower, lobed and toothed leaves may be glandular-hispid on both surfaces), veins impressed above, raised below. *Flowers* solitary in upper leaf-axils, the leaves greatly reduced in size and soon degenerating into bracts, forming mostly few-flowered racemes. *Pedicels* up to c.5–17mm long, the lower ones often with 1 or 2 tiny bracts. *Calyx* tube 0.1–0.2mm long, lobes 4–8 × 0.8–1.25mm, oblong, acute, glandular-pubescent, hairs up to 0.15–0.25mm long, glistening glands as well. *Corolla* tube 20–22 × 2.8–3mm, cylindric, abruptly expanded near apex, limb nearly regular, c.12–13mm across lateral lobes, posticous lobes 4–5 × 4–6.5mm, anticous lobe 4–5 × 3.5–6.5mm, cuneate to broadly ovate, lateral margins often revolute, apex rounded, glandular-pubescent outside, hairs up to 0.2–0.6mm long, glistening glands as well, inside clavate hairs at extreme base of anticous lip descending into well-bearded throat, very rarely clavate hairs all round mouth, rarely minutely glandular, rarely a few glistening glands at upper margins of lobe, seen only in buds, lobes thick-textured, varying in colour from olive green to rich shades of russet-orange, deep red, brownish purple or dark brown (colour probably deepening as flower ages), always with a paler thin-textured margin, greenish-yellow to saffron yellow. *Stamens*: posticous filaments 1.5–2.2mm long, anthers 1.1–1.2mm, anticous filaments 0.6–1mm, anthers 0.8–1mm, all filaments bearded with clavate hairs. *Stigma* 0.5–1mm long. *Style* 16–19mm. *Ovary* 2.7–3 × 1.2–1.5mm, minutely glandular-puberulous, glistening glands as well. *Capsule* 8–10 × 3–4mm, only glistening glands remaining, these also on placentae. *Seeds* c.1.2 × 0.7mm (few seen).

Selected citations:

Botswana. Pharing distr., 2525AB, Kanye, 4000ft, ii 1949, *Miller* B/838 (PRE); 2525BA, Ooste [sic; Ootsi] Mountain, 4600ft, 16 vi 1984, *Woollard* 1428 (SRGH).
Transvaal. Pietersburg distr., 2329AA, Blauwberg, 5200ft, 13 i 1955, *Codd & Dyer* 9130 (PRE). 2329AB, Zoutpansberg, top of Mt. Letsjume, 5510ft, 25 vii 1981, *Venter* 6210 (PRE). Potgietersrus distr., 2429AA, 11–12km SE Potgietersrus, 21 xii 1974, *Maguire* 8627 (J). 2429AA, Matopans [sic; Makapans] Poort, 1601m, 31 i 1894, *Schlechter* 4330 (BOL, E, PRE, S, STE, Z). Rustenburg, 2527CA, Hex river, viii 1903, *Pegler* 948 (BOL, PRE). Pretoria distr., 2528CB, Bynespoort, 7 i 1960, *Strey* s.n. (M, PRE, SRGH). Potchefstroom, 2627CA, 29 ix 1938, *Hafstrom & Acock* 1165 (S). Johannesburg, 2628AA, Bedford View Hills, 5700ft, 25 ii 1950, *Mogg* 18893 (J). Heidelberg distr., 14 miles SE Heidelberg, 5600–6000ft, Kuilfontein 289, 16 xii 1954, *Mogg* 24341 (J, S).
Swaziland. 2631AA, 16 miles NW of Mbabane, Londosi valley, 9 iv 1966, *Maguire* 7626/235 (J).
Natal. Louwsburg distr., 2731CA, Itala Nature Reserve, Craigadam Farm, 2450ft, 13 i 1978, *McDonald* 409 (E, PRE). Nkandla distr., 2830DB, Insuzi Valley, c.3500ft, 13 vi 1946, *Acocks* 12718 (PRE, SRGH).

Jamesbrittenia burkeana is widely distributed in the Transvaal, from the Zoutpansberg southwards almost to the border with the Orange Free State and eastwards to Ohrigstad and Lydenburg; I have seen one record from Swaziland (north of Mbabane) and there are a number of records from Natal: Itala near Louwsburg, Kranskop, the Nsuze valley, Melmoth and Eshowe. In the west, it enters Botswana: I have seen four records from the environs of Kanye, Lobatsi and Molepolole. The plants favour rocky grassy slopes in open woodland or scrub, and scrub on forest margins. The species appears to be particularly common along the Magaliesberg, Witwatersrand and Suikerbosrand (though this may simply reflect ease of collection), from c.900–1800m above sea level, though down to c.750m in the deep Nsuze valley in Natal. Quartzite, conglomerate, amygdaloidal lava and sandstone have been noted as substrate. Flowers can be found in any month.

Jamesbrittenia burkeana is characterized by its leaves, mostly with 1–3 pairs of bold teeth, the blade folded and falcate (leaves low on the stem may be jaggedly toothed and lobed, and plane), the upper surface gland-dotted, the lower with glandular-hispid hairs as well as glistening glands (only the lowermost leaves may be lobed and toothed with glistening glands on both surfaces); by its few-flowered racemes and by its flowers. The corolla lobes are dark-coloured and thick-textured except for the pale thin-textured margins. The thick tissue is probably an osmophore. No collector has mentioned scent (this is not unusual, but it may be because the flowers are scented at night). *Jamesbrittenia accrescens* has similar flowers, but there are nearly always glistening glands on the upper surface of the corolla limb at the sinuses; also, the leaves are always glandular-hispid above, more evenly serrate than in *J. burkeana* and mostly plane; the racemes are many-flowered. The two species are partially sympatric, and where they grow together, they may hybridize: see under *J. accrescens*.

The corolla of *J. albobadia* (hitherto confused with *J. burkeana*) is in sharp contrast to that of *J. burkeana*: the corolla lobes are white and thin-textured; the brown osmophore merely forms a band in the throat.

10. Jamesbrittenia accrescens (Hiern) Hilliard in Edinb. J. Bot. 49: 225 (1992).
Type: Transvaal, 2330CC, Houtboschberg, 2000m, 4 ii 1894, *Schlechter* 4382 (holo. K; iso. BOL, C, E, J, PRE, S, STE, Z).

Syn.: *Sutera accrescens* Hiern in Fl. Cap. 4(2): 305 (1904).
  S. brunnea Hiern var. *macrophylla* Hiern in Fl. Cap. 4(2): 306 (1904), p.p. quoad *Rehmann* 6008, Houtbosch, tantum.
  S.grandiflora auct. non (Galpin) Hiern; Hemsley & Skan in Fl. Trop. Afr. 4(2): 308 (1906).

Suffrutex, stems few to several from a woody stock, c.0.4–2m long, erect or straggling, simple or subsimple below, above with long ascending rod-like branches bearing flowers, glandular-hispid, hairs up to 0.25–1(–2)mm long, scattered glistening glands as well, closely leafy. *Leaves* alternate, pseudofasciculate, primary leaves c.10–40 × 5–14mm, elliptic to obovate tapering into a petiolar part up to c.⅓ total leaf-length, blade usually plane, thick-textured, apex acute to obtuse, margins serrate, both surfaces glandular-hispid, hairs up to 0.2–1mm long, glistening glands as well, veins impressed above, raised below. *Flowers* solitary in upper leaf-axils, the leaves greatly reduced in size and soon degenerating into bracts, forming long (up to c.350mm in fruit) racemes, often with further growth continuing beyond the fruiting inflorescence. *Pedicels* up to 14–22mm long. *Calyx* tube 0.1–0.2mm long, lobes 4–7 × 1–1.5mm, ± oblong, acute or subacute, glandular, hairs up to 0.15–0.3mm long, glistening glands as well. *Corolla* tube 18–25 × 2.5–3mm, cylindric, abruptly expanded near apex, limb nearly regular, c.12–14mm across lateral lobes, posticous lobes 4–6 × 4–6mm, anticous lobe 4.2–5 × 4–7mm, cuneate to broadly ovate, apex ± rounded, glandular-pubescent outside, hairs up to 0.2–0.3mm long, glistening glands as well, inside clavate hairs at extreme base of anticous lip descending into well-bearded throat, nearly always a few glistening glands at sinuses on upper surface of limb, lobes thick-textured, varying in colour from greenish to dull yellow, orange-brown, maroon and dark brown (colour probably changing as flower ages), always with a paler margin cream to greenish-yellow. *Stamens*: posticous filaments c.2mm long, anthers 1–1.5mm, anticous filaments 0.7–1mm, anthers 0.8–1mm, all filaments bearded with clavate hairs. *Stigma* 0.5–1mm long. *Style* 14–22mm. *Ovary* 2.5–3 × 1.2–1.5mm. *Capsules* 7–10 × 3.5–4.5mm, minutely glandular-puberulous, glistening glands as well, also on placentae. *Seeds* c.1.2 × 0.7mm (few seen).

Selected citations:

Transvaal. Zoutpansberg, 2329BB, Louis Trichardt, v 1921, *Breyer* sub TM 22105 (PRE). Letaba distr., 2330CA, Duiwelskloof, Westfalia Estate, c.4000 ft, 18 ix 1960, *Scheepers* 998 (M, PRE, S, SRGH, Z). Pietersrust distr., 2330CC, Houtbosch, ii 1904, *Bolus* 10953 (BOL); 2430AB, Shilouvane, *Junod* 716(Z). Potgietersrust distr., 2429AA, Makapan Valley, 16 i 1952, *Maguire* 1513 (NBG). Lydenburg distr. 2530AB, 10 miles N Lydenburg, 18 iii 1962, *Strey* 4112 (M, PRE). Belfast distr., 2530CB, Waterval Boven, old road near tunnel, 4 iii 1981, *Hilliard & Burtt* 14186(E).

Hiern (loc. cit.) wrote of *Sutera [Jamesbrittenia] accrescens*: 'This species is remarkable for having an accrescent corolla, the limb of which opens when the tube just equals the calyx in length, the tube subsequently elongating'. The type specimen certainly gives this impression but it is possibly an artefact of drying and is not displayed in the many other specimens seen.

Jamesbrittenia accrescens ranges from the Zoutpansberg down the eastern highlands of the Transvaal to Waterval Boven, Godwan river, and probably Barberton; it has not been collected further west than the environs of Potgietersrust, between c.1200 and 1600m above sea level. It favours rocky places in scrub, often on the margins of forest patches; flowering has been recorded between January and October.

The characteristic features of the species are its elliptic to obovate, serrate, leaves, glandular-hispid on both surfaces, and the thick-textured corolla limb, shades of orange-brown to dark brown above, with thin-textured pale margins. The flowers are remarkably similar to those of *J. burkeana*, whose area lies mostly west of that of *J. accrescens*, but where they are sympatric, in the Zoutpansberg, at Makapan near Potgietersrust, and near Ohrigstadt, they hybridize.

Normally, the species are easily distinguished by differences in the shape and indumentum of the leaves; also, in *J. accrescens*, there are almost always glistening glands on the upper surface of the corolla limb around the sinuses (*J. burkeana* rarely has glistening glands on the upper surface of the limb, and then they are along the upper margins). The putative hybrids have leaves more or less intermediate between those of the two species (± shape of those of *J. accrescens*, but margins more boldly and irregularly toothed, upper surface with few hairs), and there may or may not be glistening glands around the sinuses of the corolla limb. The specimens seen are: Zoutpansberg, 15 miles E of Louis Trichardt, *Schlieben* 7127(M). Maka-pan, *Maguire* s.n. (J). Ohrigstadt, farm Rustplaats, *Wilms* 1065a (E,Z). Both Wilms's sheets are a mixture of *J. burkeana* and the possible hybrid. *Codd* 5453 A (SRGH) is *J. accrescens* that flowered in Pretoria in May 1952 from seed collected on Schaagen Hill near Rosehaugh, but a specimen purportedly from the same seed that flowered in June 1961 (*Van Hoepen* 6, PRE) has leaves more like those of *J. burkeana*, and is possibly hybrid (one is left to wonder if *J. burkeana* was also in cultivation in the Pretoria garden; in any case, the species grows wild on the Magaliesberg).

11. Jamesbrittenia dentatisepala (Overkott) Hilliard in Edinb. J. Bot. 49: 228 (1992). **Plate 4A.**
Type: Natal, Bergville distr., Cathedral Peak Forest Reserve, boulder bed of Tseketseke river, 6700ft, *Killick* 1827 (holo. PRE, fragment M).

Syn.: *Sutera dentatisepala* Overkott in Bothalia 7: 491 cum ic. (1961); Trauseld, Wild Flow. Natal Drakensberg 169 cum ic. (1969).

Dwarf shrublet, many stems tufted from the base, c.150–260mm long, decumbent to erect, mostly sparingly branched near base, glandular-pubescent, hairs up to 0.2–0.8mm long, glistening glands as well, leafy. *Leaves* opposite, bases connate, up to 8–22 × 6–15mm, broadly ovate abruptly contracted to a petiolar part accounting for c.¼–⅓ total leaf length, apex obtuse, margins crenulate to serrulate and shallowly lobed, lobes ± oblong, ± truncate to obtuse, both surfaces glandular-pubescent, hairs up to 0.2–0.7mm long, many glistening glands as well. *Flowers* solitary in upper leaf axils. *Pedicels* up to 4–5mm long. *Calyx* tube 0.4–0.5mm long, lobes c.4.8–9 × 1.2–1.8mm, ± spathulate, entire to few-toothed at apex, glandular-pubescent, hairs up to 0.2–0.7mm long, glistening glands as well. *Corolla* tube 12–18 × 2–2.8mm in throat, cylindric, abruptly expanded near apex, limb bilabiate, c.10–15mm across lateral lobes, posticous lobes 4–7 × 2.2–4mm, anticous lobe 3.5–6.5 × 2.75–5mm, oblong to cuneate, apex ± truncate to retuse, glandular-pubescent outside, hairs up to c.0.15mm long, glistening glands as well, glistening glands on upper surface confined to orange zone around mouth, broad band of delicate clavate hairs in throat extending very briefly onto base of anticous lip and there c.1mm long, lobes white (see comment below), each with a conspicuous orange to orange-brown median patch forming a stellate pattern around mouth. *Stamens*: posticous filaments c.1.2–1.5mm long, anthers 1–1.2mm, anticous filaments c.0.4–0.7mm long, anthers 0.7–0.8mm, all filaments bearded with clavate hairs. *Stigma* c.0.4–0.5mm long. *Style* c.10–14mm. *Ovary* 2–2.5 × 0.8–1.5mm. *Capsules* c.4–8 × 2.5mm, thickly clad in glistening glands, a few tiny glandular hairs as well, few glistening glands on placentae. *Seeds* c.0.8–1 × 0.5mm.

Selected citations:

Lesotho. Mamalapi, 2928AC, 9000ft, 28 xii 1948, *Compton* 21343 (NBG, fragment M). 2929AA, Cleft Peak, c.10000ft, 21 i 1956, *Edwards* 1157 (PRE).
Natal. Bergville distr., 2828DB, Inner Tower gully, 6500ft, vii 1949, *Esterhuysen* 15541 (BOL). Estcourt distr., 2929AD, Langalebalela Pass, 9500ft, 16 xii 1965, *Trauseld* 497 (PRE). Mpendhle distr., 2929BC, Highmoor Forestry Reserve, ridge S of Giant's Castle, c.8100ft, 5 i 1983, *Hilliard & Burtt* 16197 (E, K).

Jamesbrittenia dentatisepala is endemic to the Eastern Mountain Region, where it has been recorded from the high Lesotho mountains and the Drakensberg in Natal, from Royal Natal

National Park south to the ridge running SE from Giant's Castle; the altitudinal range is c.1980–3000m. The plants grow in stream gullies, flowering between July and January. It is a most distinctive species, without close allies, and is easily recognized by its white corolla limb vividly marked with orange-brown. Compton (cited above) recorded 'mauve with yellow centre', but no trace of colour is discernible: the lobes are white.

12. Jamesbrittenia elegantissima (Schinz) Hilliard in Edinb. J. Bot. 49: 229 (1992).
Type citation: [Deutsch-Südwest-Afrika] Oshiheke bei Olukonda (see comments below).

Syn.: *Lyperia elegantissima* Schinz in Verh. Bot. Ver. Brandenb. 31: 192 (1890).
    Sutera elegantissima (Schinz) Skan in Fl. Trop. Afr. 4(2): 302 (1906): Merxm. & Roessler in Prodr. Fl. S.W. Afr.
    126: 50 (1967); Philcox in Fl. Zamb. 8(2): 27, tab. 12 A (1990).

Perennial herb with a woody taproot, stems several, tufted from the crown, c.150mm to more than 1m in length, up to c.6mm in diam., erect to prostrate, well branched but probably always spindly, glandular, hairs very delicate, up to 1–2mm long, very leafy. *Leaves* opposite, soon alternate upwards, pseudofasciculate, primary leaves up to 10–45 × 3–14mm, axillary leaves smaller but otherwise similar, oblong in outline, apex obtuse to subacute, base tapering into a short petiolar part, margins pinnatipartite to bipinnatisect, the lobes mostly obtuse, glandular-pubescent, hairs up to 0.5–1mm long. *Flowers* solitary in upper leaf axils. *Pedicels* up to 8–18mm long, hairy as stems. *Calyx* tube 0.1–0.25mm long, lobes 4.5–9 × 0.5–1mm, linear, subacute, glandular-pilose, hairs up to 1–2mm long. *Corolla* tube 5.5–7.5 × 1.5–2mm in throat, cylindric abruptly dilated near apex, limb nearly regular, 4–6mm across lateral lobes, posticous lobes 1.2–2.2 × 1.2–2.2mm, anticous lobe 1.2–2 × 1.2–2mm, all suborbicular, glandular-pubescent outside, hairs up to 0.4–1mm long, glistening glands as well, broad band of clavate hairs in throat extending very briefly onto anticous lip, lobes dull to bright yellow. *Stamens*: posticous filaments 1–1.2mm long, anthers 0.7–1mm, anticous filaments 0.3–0.5mm long, anthers 0.4–0.6mm, all filaments minutely puberulous. *Stigma* c.0.2mm long. *Style* 3.2–5mm long. *Ovary* 1.8–2 × 1mm. *Capsules* 4.5–6 × 2–3mm, glabrous except for a few glistening glands on sutures. *Seeds* c.0.25–0.3 × 0.25–0.3mm.

Selected citations:

Angola. Baixo Cunene, Pereira d'Eça, 1000m, 25 xi 1941, *Gossweiler* 13183 (LISC). Rio Cunene ..... próximo do terminus da picada das Quedas do Monte Negro, pr. Oncócua, 18 i 1955, *Mendes* 1340 (LISC).
Botswana. Northern div., 1923DC, 38km N.E. of Maun, Shorobe, 19 iii 1965, *Wild & Drummond* 7180 (K, M, SRGH); 1923AA, Okavango Swamp, Gwetshaa Island, 25 ii 1973, *Smith* 446 (SRGH).
Zambia. Barotseland, 1724AA, Masese, 20 vi 1960, *Fanshawe* 5759 (K, SRGH); 1622AB, 45 miles W. of Nangweshi, 3400ft, 6 viii 1952, *Codd* 7417 (K, SRGH).
Zimbabwe. Sekungwe distr., near Binga, 1470m, 6 xi 1958, *Phipps* 1364 (K, SRGH).
Namibia. Kaokoveld, 1712AB, Otjnungwa, 4 x 1960, *Giess & Wiss* 3286 (WIND). Grootfontein Distr., Unterer Omuramba ua Matako, v 1934, *Dinter* 7228 (BOL, K, S, WIND, Z). Grootfontein-Nord/Caprivizipfel, 1821BA, Bagani, 22 v 1939, *Volk* 2147 (M); Popa Falls near Andara, 20 vii 1952, *Maguire* 1703 (NBG, PRE).

Schinz did not cite a collection when he published *Lyperia elegantissima*: he merely gave a locality. I have failed to trace the specimen he saw; it was probably destroyed in the Berlin fire. Two sheets in the Zürich herbarium (Z) bear type labels; the specimens were collected by Rautanen at Olukonda on 10 February 1890; the information was written by a clerk and there is nothing to indicate that Schinz ever saw them (they bear a manuscript name *Chaenostoma elegantissima* (Schinz) Thell.). It is highly unlikely that a specimen collected on 10 February 1890 could have become the type of a name published on 30 May 1890. Two more sheets of Rautanen's collection (Z) bear Rautanen's printed labels with the details filled in by Rautanen himself: he underlined the relevant habitat, namely 'Oshiheke-Formationen' and filled in the locality Olukonda. It is 'Oshiheke bei Olukonda' that Schinz gave in his type citation. Under *Sutera elegantissima*, Skan (in Fl. Trop. Afr.) cited 'Oshiheke near Olukonda, *Schinz*', but he did not see the specimen; Merxmüller & Roessler must have followed Skan in assuming that it was a specimen collected by Schinz himself that is the type of the name; they too saw no specimen. *Schinz 136* (Z) from Okasima Ka Namutenya (not traced by me), collected in February 1886, is written up in Schinz's hand as *Lyperia elegans* [sic] Schinz. It could, if necessary, be adopted as a neotype.

Jamesbrittenia elegantissima, with its finely divided, almost fern-like, leaves, long delicate glandular hairs, and small yellow corolla scarcely exceeding the long calyx lobes, is so distinctive a species that it is never confused with any other. Its affinities are obscure. The line drawing in Flora Zambesiaca (reference above) gives a rough impression of the plant, but A3, 'corolla opened showing androecium', fails to show that the posticous filaments are strongly decurrent (a generic character) and A4 'fruit', is an ovary with style attached (no stigma); the calyx is divided nearly to the base (a generic character), not less than halfway as shown.

The species is confined to the Cunene-Cubango-Cuanavale-Cuando-Okavango-Zambezi drainage systems, the easternmost record being from Binga, near the western end of Kariba dam. Many records are from river banks, the margins of 'islands' in floodplains, islands in the Okavango swamps, around pans, and in dambos, but drier sites have also been noted: '*Terminalia* woodland' and 'dry grassland' (but in the Okavango delta). Flowers can be found in any month.

GROUP 1a, 2

13. Jamesbrittenia zambesica (R. E. Fries) Hilliard in Edinb. J. Bot. 49: 233 (1992). **Fig. 51.** Type: Zimbabwe, Victoria Falls, 27 vii 1911, *Fries 99* (holo. UPS, iso. Z, fruit only).

Syn.: *Lyperia zambesica* R. E. Fries, Wiss. Ergebn. Schwed. Rhod. Kongo Exped. 1911–1912, 1: 288 (1916).
    Sutera zambesica (R. E. Fries) R. E. Fries loc. cit.

Suffrutex, possibly several stems from base, up to 550mm long (possibly longer: no basal parts seen), c.2mm diam., probably sprawling, simple or subsimple, glandular-pubescent, hairs up to 0.1–0.2mm long, very leafy. *Leaves* alternate, pseudofasciculate, up to 30 × 9.5mm, oblanceolate, acute, base tapering to a long petiolar part, 3–6 pairs of coarse teeth in upper part, largest cut ± halfway to midrib, both surfaces with immersed glands oozing to give the blade a varnished appearance, midrib impressed above, raised below, lateral veins scarcely visible. *Flowers* solitary in upper leaf-axils. *Pedicels* up to 13–18mm long. *Calyx* tube c.0.1mm long, lobes c.5 × 1mm, linear-lanceolate, acute, clad in oozing glistening glands. *Corolla* tube c.17 × 2.8mm in throat, cylindric, abruptly expanded near apex, limb c.20mm across lateral lobes, lobes c.9 × 5.8mm, ± cuneate, glandular-pubescent outside, hairs c.0.1mm long, glistening glands as well, inside clavate hairs at extreme base of anticous lip descending into bearded throat, a few scattered glistening glands around mouth, lobes white. *Stamens*: posticous filaments c.1.5mm long, anthers c.1.3mm, anticous filaments c.1mm long, anthers c.0.9mm, all filaments bearded with clavate hairs. *Stigma* c.0.7mm long. *Style* c.12mm. *Ovary* c.3.4 × 1.7mm. *Capsules* c.9 × 4.5mm, clad in scattered glistening glands. *Seeds* c.0.8 × 0.4mm.

This interesting species is known to me only from the type collection, which came from the crevices of dry rocks along the edge of the gorge some distance below Victoria Falls ('in trockenen Felsenritzen der Canonbildungen ziemlich weit von dem Falle'). It is interesting because in habit, foliage and indumentum it strongly resembles *J. huillana* (Fries diagnosed it against *J. atropurpurea* and *J. huillana*) but its flowers are strikingly different, the corolla lobes being thin-textured (not thick, with veins invisible) and white (not shades of brown). Unfortunately, the few flowers present on the type specimen are now in poor condition, but there seems to be a dark median bar at the base of each corolla lobe, and a few glistening glands scattered around the mouth as in several other species in this group (though not in *J. huillana*). The plant should be sought again; in July, when Fries collected it, it was mostly in fruit; the one stem bearing three open flowers also has several side branches, in young bud.

The species is not mentioned in Flora Zambesiaca 8(2), 1990.

Fig. 51A. *Jamesbrittenia zambesica*: photograph of holotype, by courtesy of Dr Moberg (UPS).

Fig. 51B. *Jamesbrittenia zambesica*: photograph of part of holotype to show flowers, c. × 2.

14. Jamesbrittenia atropurpurea (Benth.) Hilliard in Edinb. J. Bot. 49: 227 (1992).

Two subspecies are recognized:

1a. Pedicels glabrous or with scattered glistening glands (often oozing, then ± invisible and the pedicels 'varnished'), rarely a few minute stalked glands also present .... subsp. **atropurpurea**
1b. Pedicels pubescent, hairs 0.15–0.5mm long, glistening glands wanting .. subsp. **pubescens**

Subsp. **atropurpurea**
Lectotype (chosen here): Cape of Good Hope, without precise locality, *Masson* s.n. (BM).

Syn.: *Lyperia atropurpurea* Benth. in Hook., Comp. Bot. Mag. 1: 380 (1836).
  Lyperia crocea Benth. in DC., Prodr. 10: 361 (1846), nom. illegit.
  Chaenostoma croceum Diels in Bot. Jahrb. 23: 491 (1897), nom. illegit.
  Manulea atropurpurea (Benth.) O. Kuntze, Rev. Gen. Pl. 3(2): 235 (1898).
  Sutera atropurpurea (Benth.) Hiern in Fl. Cap. 4(2): 306 (1904).

Wiry well branched shrublet up to c.1m tall but often much dwarfed (browsed), stems many from a stout caudex up to c.25mm in diam., young parts clad in glistening glands, often oozing to give a varnished look, sometimes minute (mostly less than 0.1mm long, rarely 0.15mm) stalked glands as well, twigs crowded with closely leafy brachyblasts. *Leaves* opposite, often alternate upwards, primary leaves 1.5–10 × 0.4–3(–4)mm (secondary leaves mostly much smaller and sometimes elliptic), spathulate or spathulate-oblong, thick-textured, obtuse to subacute, mostly entire, sometimes with a pair of teeth near apex, very rarely two pairs, margins revolute, involute or plane, lower surface beset with glistening glands, often oozing and then invisible, upper surface usually glabrous, sometimes with a few glistening glands. *Flowers* scattered in axils of leaves subtending brachyblasts with short to long vegetative sections intervening. *Pedicels* up to 7–35mm long, glabrous or with scattered glistening glands, often oozing, rarely a few minute stalked glands as well. *Calyx* tube 0.1–0.3mm long, lobes 2.5–3.8 × 0.8–1.2mm, linear-lanceolate, acute to subobtuse, thickly clad in glistening glands, often oozing and lobes looking varnished. *Corolla* tube 15–24 × 2–2.7mm in throat, cylindric abruptly expanded near apex, limb bilabiate, c.9–17mm across lateral lobes, posticous lobes 3.2–7.5 × 2–3.2mm, anticous lobe 3.5–8 × 2.2–3.7mm, all lobes oblong to cuneate, apex very obtuse to truncate, lateral margins strongly revolute making lobes look very narrow, tube minutely glandular-puberulous to almost glabrous outside, glistening glands on backs of lobes but not always easily seen, broad transverse bands of clavate hairs in throat extending very briefly onto base of anticous lip, lobes thick-textured, upper surface shades of yellowish-, orange-, red- or chocolate-brown, often very dark, venation invisible, margins thin-textured and paler (creamy or yellowish). *Stamens:* posticous filaments 1–1.8mm long, anthers 1–1.4mm, anticous filaments 0.5–1mm, anthers 0.6–1.1mm, all filaments shortly bearded. *Stigma* 0.3–0.8mm long. *Style* 11–19.5mm. *Ovary* 1.8–2.8 × 0.8–1.2mm. *Capsules* 6–10 × 3–4.5mm, glistening glands all over. *Seeds* 0.8–1.1 × 0.4–0.6mm.

Selected citations:

Cape. Barkly West div., 2824AB, Pienaarsfontein, 19 iii 1936, *Acock* 162 (PRE). 2824BB, Warrenton, 11 iii 1903, *Adams* 138 (BOL, PRE, Z). Kenhardt div., 25 miles SE of Pofadder, Gannapoort, 3400ft, 21 v 1961, *Leistner* 2462 (M, PRE). Aliwal North div., 3026AC, Bethulie, Tussen-die-Rivieren Nature Reserve, 16 xi 1969, *Edwards* 4172 (PRE). Middelburg div., 3125AC, Middelburg, 28 iv 1947, *Theron* 277 (E, PRE). Calvinia div., 3119BC, 31km from Calvinia on Loeriesfontein road, 700m, 17 iii 1988, *Powrie* 626 (STE). Beaufort West div., 3223AA, Nelspoort, 22 iii 1953, *Martin* 948 (NBG). Cradock div., 3225AB, near Post Chalmers, 22 i 1976, *Brenan* 14086 (K, NBG, PRE). Oudtshoorn div., 3322CA, Cango Valley, Boomplaas, 25 vi 1974, *Moffett* 102 (PRE, STE). Willowmore div., 3323AC, flats between Hotsprings and Toorwater, 2500ft, 5 x 1971, *Thompson* 1396 (STE). Uniondale div., 3323DD, Joubertina, iii 1955, *Esterhuysen* 24223 (BOL).
Transvaal. Marico distr., 2525DC, 5 miles from Mafeking border, iv 1899, *F. Bolus* sub BOL 6438 (BOL). 2526AB, Rooderand farm, 3500ft, 8 iv 1970, *Carter* 957 (PRE). Pretoria distr., 2528CA, 8 miles S of Pretoria, Swartkop area, 10 iv 1944, *Repton* 1859 (PRE).

Orange Free State. Bethlehem distr., 2828CB, farm Dunblane, c.2000m, 9 iii 1972, *Scheepers* 1834 (PRE, SRGH). Bloemfontein distr., 2926AA, O.F.S. Botanic Garden, 25 ii 1967, *Müller* 96 (NBG). Trompsburg distr., 3025BA, 12 miles beyond Trompsburg, near Blaauwheuvel, 25 iii 1969, *Werger* 206 (SRGH).
Lesotho. Leribe, 2828CC, slopes of Qoqolosi Peak, 24 i 1913, *Dieterlen* 938 (NBG, P, PRE, SAM). 3028AB, White Hill, i 1912, *Jacottet* 343 – B445 (Z).
Transkei. 3028AD, Ongeluks Nek, hills south of York, c.5700ft, 6 xii 1985, *Hilliard & Burtt* 18705 (E, S).

Under his new name *Lyperia atropurpurea*, Bentham cited the manuscript name *Manulea atropurpurea* Herb. Banks. The specimen bearing this name is in the BM and was collected at the Cape by Masson; I have chosen it as lectotype. Mounted on the same sheet is a specimen cultivated at Kew in 1792; the leaves are larger than those of the plant from the wild. Bentham also cited specimens collected in 'Karro and Karroid districts' by Ecklon and by Drège, which are now housed in the Kew herbarium. *Drège* 827a, from Klipplaatrivier, and *Drège* 827b, from the Winterveld (Richmond division) are *J. atropurpurea*, but *Ecklon*, collected at the Gauritz river, proves to be *J. tortuosa*.

The corolla lobes of *J. atropurpurea* are thick-textured (veins invisible on upper surface), dark-coloured (shades of brown, maroon and ochre), more or less oblong with strongly revolute margins making them appear much narrower than they really are. The closely allied *J. huillana* has similar distinctive flowers; the two species are distinguished mainly by differences in their leaves and to some extent by a difference in the disposition of their flowers, but these distinctions are not always clearly apparent in parts of the very extensive geographical areas of the two species, especially where their areas meet (for example, eastern Cape and Transvaal, mentioned below).

Typical *J. atropurpurea* has short leaves, seldom exceeding 10 × 3mm (examine primary leaves), mostly entire, sometimes with a pair of small teeth near the apex, rarely two pairs of teeth, and glistening glands on the lower surface; the upper surface is usually glabrous; the stems are clad in scattered glistening glands sometimes mixed with glandular hairs seldom more than 0.1mm long; the pedicels often look glabrous, but there are usually glistening glands present and occasionally a few minute hairs. The flowers are scattered along the stems, each in the axil of a primary leaf at the base of a brachyblast. Not every axil produces a flower; often there are one to a few flowers separated from each other by one to a few sterile brachyblasts then a longer sterile gap before a few more flowers are produced; it is therefore usual to find capsules, flowers, and buds on a single stem, which may terminate in a sterile tip. In typical *J. huillana* on the other hand flowers are mostly produced in consecutive axils often interspersed with flowerless leafy spans, but the branch terminates in a distinct raceme with few and very small secondary leaves in the axil of each small primary leaf. The primary leaves of *J. huillana* are mostly larger than those of *J. atropurpurea* (c.10–30 × 2–10mm), often more toothed and with glistening glands on both surfaces, but these distinctions are by no means always absolute. The reasons for this are not clear; growing conditions may play a significant role in leaf-size (see further under *J. huillana*), but other factors are surely involved.

Typical *J. atropurpurea* ranges widely in the Cape, from the environs of Pofadder, Calvinia, the Tanqua Karoo and the environs of Montagu eastwards to the Transkeian border, from c.300 to over 2000m above sea level. There are few records from Transkei (only Cala and Ongeluks Nek known to me), but the plant appears to be common in the Orange Free State south of Bloemfontein and eastwards to Bethlehem and Harrismith, and in western Lesotho. In the western half of the Transvaal (about as far north as Pietersburg) and around Vryburg in the NE Cape, young stems and sometimes pedicels are often hairier than normal; they thus begin to approach subsp. *pubescens*, whose area they adjoin. The hairy plants are frequently recorded as growing on limestone, dolomite and calcrete, and are possibly confined to calcareous soils. Typical *J. atropurpurea* is not so confined, but it is always on stony or rocky sites. Flowers can possibly be found in any month.

Esterhuysen 2389 (BOL, NBG, PRE) from the Kuruman Hills in the NE Cape (2723CD) may be a hybrid between *J. atropurpurea* and *J. integerrima*. The plant has the aspect of *J.*

atropurpurea but is glandular-puberulous on all parts including the calyx and the corolla limb is thin-textured, described by Miss Esterhuysen as 'petals creamy towards edges, reddish-brown towards centre'.

Where the areas of *J. atropurpurea* and *J. huillana* meet in the Transvaal and in the eastern Cape, it has not always been easy to assign specimens to one species or the other. Hybridization may be taking place; indeed, it would be surprising if it did not between two species whose flowers are identical, in a genus where hybridization appears to be commonplace.

Subsp. **pubescens** Hilliard in Edinb. J. Bot. 49: 227 (1992).

Type: Botswana, northern district, 20°49.75'S, 21°28.7'E, near small limestone pan, 18 iii 1980, *Smith* 3247 (holo. SRGH).

Syn.: *Sutera atropurpurea* (Benth.) Hiern; Philcox in Fl. Zamb. 8(2): 30 (1990) quoad spec. Botswana and excl. *Lyperia aspalathoides.*

Spindly or rounded shrublet up to c.600mm tall, young parts of stems glandular-pubescent, hairs 0.15–0.5mm long. *Primary leaves* 1–6.5 × 0.5–1.5mm, mostly ± oblong to subspathulate, thick-textured, acute, entire or with up to 2 pairs of small teeth near apex, upper surface glabrous or nearly so, lower surface thickly clad in glistening glands, a few glandular hairs c.0.1–0.2mm long often present on lower margins. *Pedicels* up to 4–18mm long, glandular-pubescent, hairs 0.15–0.5mm long.

Selected citations:

Botswana. Northern distr., 2022BC, western end of Lake Ngami, roadside Phathane to Xara, 18 ix 1974, *Smith* 1088 (SRGH). Ghanzi & Kgalagadi districts, 23°57'S 22°01'E, Tshawe Pan, 5 xii 1979, *Skarpe* 372 (K). South-west distr., Takatshwane Pan, 20 ii 1960, *Wild* 5085 (M, PRE, SRGH). South-East distr., [2425 AA], Letlaking valley, 16 ii 1960, *Wild* 4966 (K, PRE, SRGH). Parla Camp, ix 1896, *Klingberg* s.n. (S).
Namibia. Gobabis distr., [c.2218BD], Gobabis-Oas, 1 ii 1913, *Dinter* 2710 (SAM). 2118DB, farm Sturmfeld (GO 252), beim Bushman Vley, 6 i 1962, *Tölken* s.n. (WIND).
Cape. 2520AA, Kalahari-gemsbok National Park, Bayip, 6 vi 1986, *van Rooyen* 3720 (PRE). Postmasburg div., 2722CC, Pearson's Hunt, 8 xii 1960, *Leistner* 2050 (M). Kuruman div., 2723AB, Cotton End, *Esterhuysen* 2871 (BOL).

Subsp. *pubescens* differs from typical *J. atropurpurea* in its conspicuously hairy upper stems and pedicels, which lack glistening glands. It is widespread in Botswana, and extends westwards into Namibia (the only two specimens seen are cited above) and southwards into the northernmost Cape from roughly Upington north east to Vryburg. Most records are in and around pans, probably always on calcareous soils. The plants can become very woody and gnarled with short spreading branchlets subspinescent with age. These tough weatherbeaten and browsed bushes bear very small leaves and the specimens have frequently been misidentified as *Sutera aspalathoides* and *S. microphylla*.

15. Jamesbrittenia huillana (Diels) Hilliard in Edinb. J. Bot. 49: 230 (1992).

Lectotype (chosen here): Angola, Serra da Huilla, Cunene, vi 1860, *Welwitsch* 5834 (BM; isolecto. COI, K).

Syn.: *Chaenostoma huillanum* Diels in Bot. Jahrb. 23: 477 (1897).
    Sutera huillana (Diels) Hiern, Cat. Welw. Afr. Pl. 1(3): 757 (1898); Hemsley & Skan in Fl. Trop. Afr. 4(2): 309 (1906).
    S. brunnea Hiern in Fl. Cap. 4(2): 305 (1904); Philcox in Fl. Zamb. 8(2): 30 (1990). Lectotype (chosen here): Transvaal, Barberton, Queen's River Valley, 2400ft, x 1889, *Galpin* 645 (K; isolecto. BOL, PRE, SAM, Z).
    S. brunnea var. *macrophylla* Hiern in Fl. Cap. 4(2): 306 (1904). Lectotype: Mozambique, between Lebombo Mts and Komati River, c.500ft, viii 1886, *Bolus* 7609 (K, isolecto. BOL).
    Lyperia dinteri Pilger in Bot. Jahrb. 48: 440 (1912). Type: Namibia, Windhoek, 30 i 1889, *Dinter* 259 (holo. B†; iso. Z).
    L. longituba Dinter in Feddes Repert. 19: 94 (1923). Type: Namibia, Aus, 1400m, 10 i 1910, *Dinter* 1075 (holo. B†, iso. SAM).

Sutera atropurpurea auct. non (Benth.) Hiern; Hemsley & Skan in Fl. Trop. Afr. 4(2): 309 (1906); Merxmüller & Roessler, Prodr. Fl. S. W. Afr. 126: 49 (1967); Philcox in Fl. Zamb. 8(2): 30 (1990) quoad spec. Zambia et Zimbabwe tantum.

S. longituba (Dinter) Range in Feddes Repert. 38: 265 (1935).

Shrublet c.150mm–1.2m tall, stems many from a stout caudex up to c.30mm in diam., at least young parts glandular-pubescent, hairs up to 0.1–0.15(–0.2)mm long, glistening glands as well, branches crowded with very leafy brachyblasts, these often less leafy in well-grown, large-leaved specimens. *Leaves* opposite, often alternate upwards, primary leaves (3–)10–30 × (1–)2–10mm (secondary leaves often much smaller), spathulate or sometimes oblong-spathulate, often thick-textured, large leaves thinner with lateral veins sometimes visible, obtuse to subacute, margins mostly toothed at least near apex, up to c.5 pairs of teeth, these sometimes coarse, secondary leaves in particular often entire, lower surface thickly beset with glistening glands, these sometimes oozing, also often present on upper surface but more thinly dispersed, usually tiny (less than 0.1mm) glandular hairs as well on margins of petiolar part and lower part of midrib on undersurface. *Flowers* scattered in axils of primary leaves, often in a few consecutive axils, then a sterile portion, followed by more flowers, the twigs normally terminating in a raceme, each flower subtended by few very small leaves. *Pedicels* up to c.5–18mm long, glabrous or with scattered glistening glands, occasionally tiny (less than 0.1mm) glandular hairs as well. *Calyx* tube c.0.2mm long, lobes 3–4.5 × 0.8–1mm, linear-lanceolate, acute to subobtuse, thickly clad in glistening glands, oozing or not. *Corolla* tube 18–27 × 2.2–3mm in throat, cylindric, abruptly expanded near apex, limb bilabiate c.9–16mm across lateral lobes, posticous lobes 4–8 × 3–4.5mm, anticous lobe 4–8 × 3–4mm, all lobes oblong to cuneate, apex very obtuse to truncate, lateral margins strongly revolute making lobes look very narrow, minutely glandular-puberulous outside, glistening glands on backs of lobes, broad transverse band of clavate hairs in throat extending very briefly onto base of anticous lip, lobes thick-textured, upper surface shades of yellow-, orange-, red- or chocolate-brown, the colour often rich and dark, venation invisible, margins thin-textured and pale. *Stamens*: posticous filaments c.1.4–1.8mm long, anthers 1–1.5mm, anticous filaments c.0.5–0.8mm long, anthers 0.8–1mm long, all filaments shortly bearded. *Stigma* 0.5–1mm long. *Style* 14–23mm. *Ovary* 2.5–3 × 1.2–1.6mm. *Capsules* c.8–10 × 4mm, glistening glands all over. *Seeds* c.1 × 0.5–0.75mm.

Selected citations:

Angola. Huila, Serra da Chela, Tchivinguiro, Humpata, 1750m, 2 x 1941, *Gossweiler* 13182 (LISC); Sá da Bandeira, Tchivinguiro, 9 xi 1961, *Santos* 437 (LISC); Leba, 2000m, 30 vii 1954, *Pritchard* 338 (BM, LISC).

Namibia. Kaokoveld, 1813BC, flats round Kaoko Otavi fountain, 20 iv 1957, *De Winter & Leistner* 5551 (M, PRE). Grootfontein distr., 1917BA, Tsumeb, 19 iii 1934, *Dinter* 7506 (B, BM, BOL, K, PRE, S, WIND, Z); 2017BB, farm Ravenna, 15 iii 1960, *Seydel* 2144 (B, WIND). Omaruru distr., 2114BA, Brandberg, summit of Königstein, 2580m, 31 v 1963, *Nordenstam* 2797 (M, S). Windhoek distr., 2217CA, Auasberge am Auaspass, 19 xii 1957, *Merxmüller & Giess* 779 (M, PRE, WIND). Rehoboth distr., 2316BA, farm Friedendal, c.1700m, 9 v 1963, *Kers* 180 (S, WIND). Bethanien distr., 2516DD, farm Helmeringhausen, 20 iv 1949, *Kinges* 2171 (M, PRE). Keetmanshoop distr., 2718AD, farm Carolina (KEE 99), 18 v 1972, *Giess & Müller* 12053 (WIND).

Zambia. Kafue distr., King Edward VII Copper Mine [15°30'S 27°55'E], 11 ix 1929, *Sandwith* 75 (K, SRGH); 5 miles E of Lusaka, Kabulonga, 4100ft, 17 vi 1955, *Robinson* 1307 (K, SRGH); ibidem, 27 xii 1957, *Noak* 295 (K, SRGH).

Zimbabwe. Charter distr., Mhlaba Hills, 10 miles S of Ngesi, Great Dyke, 16 i 1962, *Wild* 5595 (K, M, PRE, SRGH). Bulawayo, 18 ii 1912, *Rogers* 5749 (BOL, Z mixed with *J. albobadia*). Selukwe, v 1875, *Klingberg* s.n. (S). Belingwe distr., southern end Great Dyke near Otto Mine, 17 iii 1964, *Wild* 6390 (K, M, SRGH). Umtali [Mutare] distr., Inyamatshira Mts, 4500ft, 22 vi 1958, *Chase* 6937 (K, SRGH).

Transvaal. Zoutpansberg, [2230CD?], 31 x 1937, *Van den Berg* 12 (W). Potgietersrust distr., 2429CD, between Grass Valley and Stavoren tin mine, 19 i 1955, *De Winter* 2299 (M, Z). Middelburg distr., 2529CD, Buffelsvlei, 4500ft, 18 xii 1933, *Rudatis* 25 (STE). Barberton distr., 2531CA, North Kaap, 2560ft, i 1924, *Thorncroft* 2038 (PRE).

Swaziland. 2631CD, Hlatikulu, xi 1910, *Stewart* sub TM 9902 (PRE).

Natal. Ingwavuma distr., 2632CD, Ndumu Game Reserve, 7 xii 1968, *Pooley* 227a (E). Umfolozi distr. [2832 AC?], Umfolozi river, vii 1922, *Wager* sub TM 22390 (PRE). Ixopo distr., 2930 AC, valley of Umkomaas river above Hela-hela, 29 iv 1977, *Hilliard & Burtt* 10318 (E). Alexandra distr., [now Umzinto], Ellesmere, 600m, 18 vii 1911, *Rudatis* 1427 (E, M, S, W, Z).

Cape. Komgha div., 3227DB, Kei Bridge, i 1890, *Flanagan* 436 (PRE, SAM); Kei River, 11 i 1895, *Krook* in herb. Penther 3091 (W). Ceded territory, Gegend um Fort Beaufort, Oct., [distributed under code 31.10], *Ecklon & Zeyher* (E, M, P, S, W). Not all collections distributed as 31.10 are *J. huillana*. Albany div., 3325BB, near Kommadagga, Anne's Villa, 1200ft, 1 ii 1978, *Bayliss* 8504 (M, Z). Uitenhage div.?, 3325BC, near Slagboom below the Zuurberg, 500–1000ft, 29 ix 1932, *Rennie* 463 (BOL).

Jamesbrittenia huillana is distributed in a broad arc stretching across southern Africa from southernmost Namibia northwards to the highlands of Huila in Angola, thence a distributional gap (real?) to Kafue district in Zambia, and so south through western Zimbabwe to the eastern half of the Transvaal (and apparently just entering Mozambique near Komatipoort), western Swaziland, and Natal, from Ndumu (on the southern Mozambique border) southwards (scattered records mainly from the deep hot river valleys that carry coastal vegetation well inland) to the Kei River (on the Transkei-E Cape border), with a few scattered records in the eastern Cape. (A similar distribution pattern, namely highlands of Angola, then Zimbabwe and southwards to Natal and the E Cape is known in other species: Composites such as *Helichrysum aureum* (Houtt.) Merrill var. *monocephalum* (DC.) Hilliard, *Kleinia fulgens* Hook. f., *Senecio lygodes* Hiern spring readily to mind.) The plants favour rocky or stony sites; in Namibia, flowering is distributed over the whole year; records from Zambia and Zimbabwe are too few to give a clear picture; in Transvaal, Swaziland and Natal, the main season appears to be from October to April.

The type specimen of *J. huillana* and all other Angolan material came from the high well-watered plateau (c.2000m) lying between the source of the Cunene (near Nova Lisboa) and Serra da Chela near Huila. All the specimens have relatively large leaves and distinct racemes terminating the branches, and are indistinguishable from the type specimen of *Sutera brunnea* collected on the other side of southern Africa at the foot of the mountains at Barberton in the south-eastern Transvaal. Over the great range of the species, there is considerable variation in size of leaves, and plants with small narrow leaves look very different from those with large broad ones; the former are apt to be misdetermined as *J. atropurpurea*. The two species are very closely allied; they differ in their leaves (in *J. huillana*, primary leaves mostly 10–30 × 2–10mm and mostly toothed, in *J. atropurpurea* mostly 1.5–10 × 0.4–3mm, mostly entire) and to some degree in the disposition of their flowers: in *J. huillana*, the branches terminate in definite racemes, the flowers subtended by few small leaves, whereas in *J. atropurpurea*, the flowers are scattered irregularly along the branches, which often terminate in a leafy shoot. But these distinctions may be difficult to see, especially in stunted, browsed, or otherwise atypical specimens. How much variation in leaf size is attributable to genetic factors, how much to environmental ones is impossible to judge accurately from herbarium specimens alone, but a few significant examples may be quoted:

In Zimbabwe, *Wild* 6390, recorded as growing on serpentine, has much smaller leaves than most Zimbabwean specimens, though *Wild* 5595, also on serpentine, has leaves more typical of those of *J. huillana*. In the northern districts of Namibia (roughly north of the 21st parallel), all the substrates recorded are calcareous. *De Winter & Leistner* 5551 came from flats around a spring and is large-leaved, exactly like the Angolan material; so are some of the specimens from the Great Waterberg and its environs (others are small-leaved, but no ecological information is given). There are a great many collections from the environs of Windhoek (whence came the type of *Lyperia dinteri*); most have relatively small leaves (primary ones c.7–10 × 1.5–2mm), but others may have much larger ones, for example *Müller* 2 (WIND), collected 12km from Windhoek on the road to the airport (no ecological information given), while *Nordenstam* 2370 (M, S), found among rocks near the windmill on the farm Gamsberg, SW of Windhoek, has mostly small leaves (c.7 × 1.5mm), but on the sheet in S, larger ones (8 × 4, 12 × 3.2, 12 × 3mm) are present near the base of the plant: clearly the potential for large leaves is present. Similarly, around Aus (whence came the type of *Lyperia longituba*, with leaves up to c.7 × 2mm), leaves can be relatively large (10 × 4mm in *Wendt* 98, WIND; 15 × 4mm in *Dinter* 6075, B) or tiny (4 × 2mm in *Dinter* 6179, B, but sometimes larger on sheets in other herbaria).

In the Transvaal, *J. huillana* stretches down the whole of the eastern side of the province and is typically large-leaved (sometimes reaching 30 × 6mm) with well developed terminal racemes, but in *Balkwill* 1503 (E, PRE) 'growing on a ridge of serpentine' between Kaapmuiden and Barberton, the largest leaves measure only c.8 × 1.8mm (cf. *Wild* 6390, Great Dyke). The area Chuniespoort-Bewaarkloof (2429BB, BC) needs investigation: I have seen only poor specimens from there, which are possibly *J. huillana* (*Hafstrom & Acocks* 1169, S and *Stalmans* 1874, J).

In Natal, large-leaved plants have come from the well-watered Zululand coastal plain, smaller-leaved ones from the hot dry river valleys. The sheets of *Rudatis* 1427 (cited above) in M, W and Z carry separate thin twigs bearing large leaves, while the more woody flowering branches are small-leaved; this is also so in *Rudatis* 1091 (STE), from the same locality. Unfortunately, *Rudatis* gave no written information. I have seen no records from Transkei, but a few from the eastern Cape, from hot dry scrub-filled valleys as in much of Natal, and again leaves are often much smaller than from well-watered sites. With similar evidence over the whole area of the species, one is led inescapably to the conclusion that edaphic conditions and water regimes contribute very largely to variation in leaf-size (it led Dinter to create five manuscript names, in addition to the one he published, for specimens he collected in Namibia).

Jamesbrittenia huillana may also hybridize with other species; this has already been mentioned under *J. fodina* (no. 1) on the Great Dyke in Zimbabwe. *Dinter* 849 (SAM), from the farm Hoffnung (Windhoek distr., 2217AC) may represent *J. huillana* × *lyperioides*; Dinter himself thought the plant to be hybrid. In facies, it closely resembles *J. huillana*, but all parts, including leaves and calyces, are glandular-puberulous; unfortunately, the only two flowers on the sheets are now in very poor condition, and one cannot judge the texture of the corolla limb. *Meyer* 96 (WIND) came from Moltkeblick (Windhoek distr., 2217CA) and comprises a twig of *J. huillana* and a second twig that matches *Dinter* 849; the flowers are in good condition and the corolla limb is like that of *J. huillana* (the calyx is not). *Meyer* 107 (M, WIND), from the same locality as *Meyer* 96, is the putative hybrid. *Volk* 12861 (M, WIND), farm Erichsvelde (Okahandja distr., 2116DB) has leaves not unlike these of Meyer's specimens, but the corolla limb is thin-textured (clearly visible on the sheet in WIND); it too could be *J. huillana* × *lyperioides*. *Giess* 11247 (WIND), from the farm Blyerust (Grootfontein distr., 1916DB) has the foliage and other characters of *J. huillana* but the corolla limb is thin-textured and quite unlike that of *J. huillana*; Giess described the flower as 'blau mit gelblichweissen Rändern und Spitzen'. It too appears to be hybrid and *J. lyperioides* could be the second parent. *Giess* 11247 in PRE is normal *J. huillana*, and so is *Giess* 11246 (WIND), collected two days earlier than 11247. The PRE sheet numbered 11247 matches 11246 precisely; it is probable that it simply bears the wrong label.

16. Jamesbrittenia namaquensis Hilliard in Edinb. J. Bot. 49: 231 (1992).
Type: Cape, Namaqualand, Richtersveld, 2817 AA, Kodaspiek, 900m, 2 ix 1977, *Oliver, Tölken & Venter* 448 (holo. STE, iso. PRE).

Twiggy dwarf shrublet c.80–600mm tall, branches erect or decumbent, young parts glandular-pubescent, hairs up to 0.1–0.2mm long, twigs crowded with leaf-fascicles. *Leaves* opposite, alternate upwards, primary leaves 2–12 × 1–7mm (secondary leaves often much smaller), spathulate, larger ones tapering into a distinct petiolar part, apex obtuse, margins entire to coarsely toothed (usually 1 or 2, occasionally 3, pairs of teeth), both surfaces glandular-puberulous, hairs up to 0.1–0.2mm long, lower surface with crowded glistening glands as well. *Flowers* scattered irregularly in axils of primary leaves. *Pedicels* up to 6–17mm long, glandular-puberulous, hairs up to 0.1–0.15mm long. *Calyx* tube c.0.2mm long, lobes 2.8–4.1 × 1mm, oblong to linear-lanceolate, acute, glandular-puberulous, hairs up to 0.1–0.15mm long, scattered glistening glands as well. *Corolla* tube 17–22 × 2–2.2mm in throat, cylindric, abruptly expanded near apex, limb bilabiate, 9–14mm across lateral lobes, posticous lobes 4–5 ×

2.2–5.5mm, anticous lobe 4.5–6 × 2.7–3.5mm, all lobes ± oblong, apex ± truncate, lateral margins strongly revolute making lobes look very narrow, tube minutely glandular-puberulous outside, glistening glands on backs of lobes, broad transverse band of clavate hairs in throat extending very briefly onto base of anticous lip, lobes thick-textured, venation invisible above and there chocolate-brown, sometimes with very narrow pale thin-textured margins. *Stamens*: posticous filaments 1–1.8mm long, anthers 1–1.2mm, anticous filaments 0.3–0.5mm, anthers 0.8–1mm, filaments glandular-puberulous. *Stigma* c.0.3–1mm long. *Style* 13–17mm. *Ovary* 2–3 × 1.2–1.4mm. *Capsules* c.6.5–8 × 3.5–4mm, glistening glands all over, sometimes a few minute glandular hairs as well. *Seeds* c.1 × 0.6mm (few seen).

Selected citations:

Cape. Namaqualand, Richtersveld, 2816 BD, ridge north of Numees camp, 460m, 21 ix 1981, *McDonald* 707 (PRE, STE). 2917BD, 22 miles N by W of Springbok, Paddagat, c.3100ft, 28 v 1961, *Leistner* 2549 (M, PRE). 2917DB, Springbok, c.3200ft, 25 ix 1952, *Acocks* 16538 (PRE). 3017 BB, Kamieskroon, 24 vii 1941, *Esterhuysen* 5456 (BOL); ibidem, 24 vii 1941, *Bond* 1145 (NBG). Nama'land Minor, viii–ix 1883, *Bolus* 9439 (BOL).

The flowers of *J. namaquensis* resemble those of *J. atropurpurea*, with which it has hitherto been confused. It is, however, a much hairier plant; not only are the stems, pedicels and leaves distinctly pubescent, so too is the calyx, where the hairs are mingled with glistening glands (*J. atropurpurea* has only glistening glands, often oozing so that the calyx looks varnished). The leaves are minutely pubescent on both surfaces (only in *J. atropurpurea* subsp. *pubescens* are there sometimes a few minute hairs on the lower leaf-margins: glistening glands predominate) and the leaves tend to be broader in relation to their length than those of *J. atropurpurea* (ratio of length to breadth 2–4:1, not 3–9:1 and 3–6:1 in subsp. *pubescens*).

Jamesbrittenia namaquensis appears to be confined to Namaqualand, from the Richtersveld south to Kamieskroon and Leliefontein. It favours rocky or stony sites, sometimes growing in clefts between rocks, c.460–1600m above sea level, and flowering between May and September.

GROUP 1a, 3

17. Jamesbrittenia incisa (Thunb.) Hilliard in Edinb. J. Bot. 49: 230 (1992).
Type: Cape, without precise locality, *Thunberg* (sheet no. 14406 in herb. Thunb., UPS).

Syn.: *Erinus incisus* Thunb., Prodr. 103 (1800) & in Fl. Cap. ed. Schultes 476 (1823).
  Lyperia incisa (Thunb.) Benth. in Hook., Comp. Bot. Mag. 1: 379 (1836) & in DC., Prodr. 10: 359 (1846).
  Chaenostoma stenopetalum Diels in Bot. Jahrb. 23: 477 (1897). Type: Cape, Calvinia div., Hantam Mts., *Meyer* (B†).
  Sutera stenopetala (Diels) Hiern in Fl. Cap. 4(2): 305 (1904).
  S. incisa (Thunb.) Hiern in Fl. Cap. 4(2): 303 (1904).
  S. esculenta Bond in J.S. Afr. Bot. 40: 311 (1974) cum ic. Type: Cape, Calvinia div., 3120 CA, farm Langfontein, 55 km E of Calvinia, 26 ix 1973, *Hall* 4448 (holo. NBG; iso. K, MO, PRE, STE).

Dwarf twiggy shrublet, branches up to 300mm long but often heavily browsed and stunted, up to 6–10mm in diam., erect, ascending or spreading, densely glandular-pubescent, hairs up to 0.25–0.5mm long, leafy. *Leaves* opposite or sometimes alternate towards tips of branchlets, pseudofasciculate, largest primary leaves 6–25 × 4–13mm, ovate to elliptic tapering to a petiolar part accounting for up to ⅓ total leaf length, often much less, apex obtuse or occasionally subacute, blade divided up to halfway to midrib into 2–4 pairs of broad obtuse to subacute lobes, these nearly always entire, or some leaves ± entire, both surfaces densely glandular-pubescent, hairs up to 0.2–0.3mm long, glistening glands as well mostly on lower surface. *Flowers* up to 10 in upper leaf axils forming small lax terminal racemes, sometimes panicled. *Pedicels* up to 5–22mm long. *Calyx* tube c.0.1–0.15mm long, lobes c.4.5–5 × 0.6–1mm, ± spathulate, densely glandular-pubescent, hairs up to 0.25–0.4mm long, glistening glands as well. *Corolla* tube 18–28 × 2.2–2.5mm in throat, cylindric, abruptly expanded near apex, limb oblique, 12–15mm across lateral lobes, posticous lobes 7.5–10 × 2.1–3mm, anticous lobe 7–9 × 2.1–3.2mm, all lobes oblong, ± truncate, glandular-pubescent outside, hairs up to 0.25–0.4mm long, glistening glands as well mainly on backs of lobes, transverse band of clavate hairs in throat extending briefly onto lower lip, lobes white, each marked at the base with a dark red-brown median bar. *Stamens*: posticous filaments 1.8–2.2mm long, anthers 1.1–1.2mm long; anticous filaments 0.7–1mm long, anthers 0.8–0.9mm long, all filaments minutely puberulous. *Stigma* 0.25mm long. *Style* c.14–20mm. *Ovary* 2–2.5 × 1mm. *Capsules* 4.5–7 × 2.5–3mm, minutely glandular-puberulous. *Seeds* not seen.

Selected citations:

Cape. Calvinia div., 3020CC, Soutputs – Breekbeenkolk, ± 6km from Soutputs, 2 ix 1986, *Burger & Louw* 158 (PRE, STE). 3119BD, 24 miles N of Downes Station, 12 v 1934, *Salter* 4460 (BOL). 3120AC, farm Zandbult 723, 1079m, 1 ix 1986, *Cloete & Haselau* 150 (STE).

The one flower on Thunberg's specimen of *Erinus incisus* is in poor condition, but a sheet collected by Masson (who was travelling with Thunberg) and now housed in the Natural History Museum (BM) bears several complete flowers and shows the long corolla tube that is characteristic of the species. Although Meyer's specimen from the Hantam Mountains was destroyed in the Berlin fire, Diels's full description of the plant and his comparison of it with *Chaenostoma incisum* (said to differ in its more richly branched inflorescence and longer corolla lobes) leaves no doubt as to its identity and I have therefore reduced *C. stenopetalum* to synonymy under *Jamesbrittenia incisa*. Thunberg and Masson travelled past the Hantam Mountains on their journey to the Roggeveld and their specimens probably came from this area. Indeed, *J. incisa* seems to have a very limited distribution from the environs of Calvinia north east to Swartputs and Breekbeenkolk, a linear N-S distance of c.100km, and c.150km to the south east, to the farm Aarfontein, between Sutherland and Fraserburg, and the environs of Sutherland itself.

The plants are often found among boulders (possibly the only place where they survive browsing!) in Karoo vegetation, and flowering has been recorded in September and May. The species is easily recognized by its dense glandular indumentum, few-toothed, thick-textured leaves, and particularly by its long corolla tube. The midrib and lateral veins are deeply incised into the upper leaf surface, which possibly suggested the trivial name.

18. Jamesbrittenia tortuosa (Benth.) Hilliard in Edinb. J. Bot. 49: 233 (1992).
Type: Cape, Beaufort West div., [c.3221DD] Gamka river, May, *Burke* 144 (holo. K; iso. BM, SAM; BOL, BM, S, W sub *Zeyher* 1307).

Syn.: *Lyperia tortuosa* Benth. in DC., Prodr. 10: 362 (1846).
 Sutera tortuosa (Benth.) Hiern in Fl. Cap. 4(2): 309 (1904).

Twiggy dwarf shrublet, caudex up to c.20mm in diam., stems up to c.150–500mm long, often heavily browsed, erect or decumbent, glandular-pubescent, hairs from less than 0.1–4mm long, leafy. *Leaves* opposite, alternate upwards, pseudofasciculate, largest primary leaves 3–11(–18) × 1–3(–6)mm, spathulate, often conduplicate, often entire, sometimes with 1–2 pairs of small teeth near apex, rarely with 4 deeply cut lobes, the largest toothed (the latter seen only in coppice shoots and sometimes persisting at base of older stems), both surfaces glandular-pubescent, hairs from less than 0.1–3mm long, many glistening glands as well. *Flowers* solitary in axils of upper reduced leaves forming small terminal racemes, or solitary in axils of primary leaf of pseudofascicles and then sometimes overtopped by new growth. *Pedicels* up to 5–22mm long, hairy as stems. *Calyx* tube 0.2–0.4mm long, lobes 3.2–5 × 0.7–0.8mm, ± oblong-spathulate, glandular-pubescent, hairs up to 0.2–0.3mm long, few glistening glands as well on lower part. *Corolla* tube 9.5–12 × 1.5–2mm in throat, cylindric, abruptly expanded near apex, limb oblique, 10–15mm across lateral lobes, posticous lobes 5–7.5 × 2–4mm, anticous lobe 4.5–7 × 2–4mm, all lobes cuneate-oblong, glandular-pubescent outside, hairs up to 0.15–0.2mm long, glistening glands as well on backs of lobes, transverse band of clavate hairs in throat extending very briefly onto base of lower lip, lobes often white, sometimes pale pink or mauve, dark median streak at base of each lobe. *Stamens*: posticous filaments 1–1.4mm long, anthers 0.8–1.3mm, anticous filaments 0.3–0.9mm long, anthers 0.5–1mm, all filaments puberulous. *Stigma* 0.2mm long. *Style* 5.5–10mm. *Ovary* 1.5–2 × 0.8–1mm. *Capsules* c.5 × 3mm, glandular-puberulous, glistening glands sometimes present down sutures. *Seeds* c.0.7–0.8 × 0.4mm.

Selected citations:

Cape. Prince Albert div., 3322BC, near Klaarstroom, x 1952, *Zinn* sub SAM 66455 (SAM). Oudtshoorn div., 3322DA, De Rust, near wall of Stompdrift dam, 1500ft, 31 v 1990, *Vlok* 2305 (E, K, S); ibidem, farm Otse Kloof, 28 ix 1971, *Dahlstrand* 2101 (J, STE). Willowmore div., 3323AC, flats between Hotsprings and Toorwater, 2500ft, 5 x 1971, *Thompson* 1397 (STE). Uniondale div., 3323CA, near Uniondale, x 1952, *Zinn* sub SAM 66453 (SAM). Jansenville div., 3324AB, Mount Stewart, 5 xii 1947, *Compton* 20322 (NBG); 3324BD, Saltpansnek, c.2800ft, 1 ix 1951, *Acocks* 16001 (PRE).

Burke collected the type of *Jamesbrittenia tortuosa* near the Gamka river when he was travelling with Zeyher back to the Cape after their epic journey to what is now the Transvaal. Material distributed as *Zeyher* 1307 is so palpably part of *Burke* 144 that the whole should be regarded as a single collection and I have labelled the isotypes accordingly. The specimens are dated 'May', when the travellers must have been about to turn SW towards the Dwyka river, and I judge the collection to have been made somewhere near present day Kruidfontein. I have seen three collections from the Little Karoo on the heights above the Gauritz river; the others have all come from further east: Klaarstroom and De Rust (north and south respectively of Meiring's Poort through the Great Swartberg) east to the environs of Willowmore, Uniondale, Steytlerville, Mount Stewart in the gap through the Groot River Heights, and Saltpansnek through the Klein Winterhoekberge. The plants are a constituent of karroid scrub on stony or shaley slopes and flats, between c.300 and 750m above sea level. Flowering has been recorded in most months.

The characteristic form of the plant is a dwarf gnarled shrublet (often heavily browsed, as in the type collection, which probably suggested the epithet *tortuosa*), the branchlets crowded with tiny leaves in dense tufts. These leaves are entire or 2–4-toothed near the apex and often folded lengthwise. *Dahlstrand* 986 (J) has simple stems tufted from a stock: these are regeneration shoots and they bear relatively enormous leaves (up to 18 × 6mm, the measurement in parentheses in the formal description) that are plane and deeply lobed (similar leaves can sometimes be found, withered and dry, at the base of old stems). Dahlstrand made many collections on farms in the neighbourhood of De Rust and most have tiny leaves.

Jamesbrittenia tortuosa is allied to *J. tysonii*, which has similar indumentum on stem, leaves and calyx; the leaves of *J. tysonii* are plane and mostly well cut: differences in leaf form will therefore distinguish the species, except in exceptional cases such as that of *Dahlstrand* 986. They may also differ in the colour-patterning of the corolla limb, but unfortunately collectors seldom record colour, and even less often a pattern. In *J. tortuosa*, there many always be a dark median streak at the base of each lobe (not necessarily visible in dried material). On the other hand, *J. tysonii* appears only rarely to have dark median streaks. The area of *J. tortuosa* lies south of that of *J. tysonii*, but they possibly meet near Beaufort West and Aberdeen.

19. Jamesbrittenia tysonii (Hiern) Hilliard in Edinb. J. Bot. 49: 233 (1992).
Type: Cape, Murraysburg div., 3123DD, in rocky places near Murraysburg, 4100ft, x 1879, *Tyson* 302 (holo. BM).

Syn.: *Lyperia pinnatifida* (L.f.) Benth. var. *macrophylla* Benth. in DC., Prodr. 10: 361 (1846) nomen, as to *Drège*, Camdeboosberg, only (*Burke* from the Magaliesberg is *Jamesbrittenia aurantiaca*).
   Sutera tysonii Hiern in Fl. Cap. 4(2): 296 (1904).
   S. altoplana Hiern in Fl. Cap. 4(2): 302 (1904). Type: Cape, Fraserburg div., 3121DC., elevated plains near Fraserburg, 4200ft, iv 1886, *Bolus* 9181 (holo. K, iso. BOL.).

Dwarf shrublet, stems up to c.10mm in diam., up to c.300mm long but often heavily browsed, possibly always ± erect unless affected by browsing, glandular-pubescent, hairs up to 0.1–0.3mm long, leafy. *Leaves* opposite becoming alternate towards tips of branchlets, pseudofasciculate, largest primary leaves c.5–20 × 2–8mm, oblanceolate in outline tapering into a petiolar part up to c.⅓ total leaf-length, margins (except in some smaller leaves) with 1–4 pairs of teeth or lobes cut c.⅓ to ⅔ of way to midrib, rarely more deeply, lobes often ± oblong, occasionally cuneate, obtuse, mostly entire, occasionally the largest lobes with 1 or 2 teeth, both surfaces glandular-pubescent, hairs up to 0.1–0.3mm long, glistening glands as well particularly on lower surface. *Flowers* solitary in the axils of the leaves, sometimes running up into nearly nude racemes. *Pedicels* up to 5–16mm long. *Calyx* tube 0.1–0.2mm long, lobes 2–4 × 0.5–1mm, ± oblong, glandular-pubescent, hairs up to 0.2–0.3mm long, glistening glands as well. *Corolla* tube 8–14.5 × 1.5–2mm in throat, cylindric, abruptly expanded near apex, limb oblique, 10–15mm across lateral lobes, posticous lobes 5–8 × 1.5–3mm, anticous lobe 4.5–8 × 1.8–3.8mm, all lobes cuneate-oblong, ± truncate or retuse, glandular-pubescent outside, hairs up to 0.25–0.3mm long, glistening glands as well mainly on backs of lobes, transverse band of clavate hairs in throat extending very briefly onto base of lower lip, lobes white or cream, throat yellow, sometimes a dark violet median streak at base of each lobe. *Stamens*: posticous filaments 1–1.2mm long, anthers 0.7–1.1mm, anticous filaments 0.3–0.6mm long, anthers 0.4–0.8mm, all filaments minutely puberulous. *Stigma* c.0.2mm long. *Style* 5.5–12mm. *Ovary* 1.4–1.8 × 0.8–1mm. *Capsules* 3.5–5 × 2–2.5mm, minutely glandular-puberulous, a few glistening glands on sutures. *Seeds* c.0.6 × 0.3mm (few seen).

Selected citations:

Cape. Barkly West div., 2824AD, Newlands, 18 iii 1939, *Esterhuysen* 988 (BOL, NBG); Herbert div., 2923BA, Mazelsfontein, x 1923, *Anderson* sub Wilman 1607 (BOL). Hay div., between Abrams Dam [2922BA] and Springfield [2922 BB], x 1936, *Acocks* 1200 (BOL). Victoria West div., 3123AC, NW of Victoria West near dam, c.1370m, 14 v 1976, *Thompson* 3082 (STE); ibidem, *Martin* 953 (NBG). Fraserburg div., 3121CD, 12 miles SW of Fraserburg on farm Grootfontein, 1290m, 10 iv 1971, *Coetzer* 72 (PRE, STE). Cradock div., 3225BB, Rasfonteinpoort, 22 xi 1977, *Hilliard & Burtt* 10564 (E, K).

The type of *J. tysonii* has deeply lobed leaves, the lobes ± oblong, obtuse, mostly entire, some of the larger lobes with a small tooth; both stems and leaves are glandular-puberulous, the hairs up to 0.1mm long, with glistening glands as well on the leaves, mainly on the lower surface. In the type of *Sutera altoplana*, the indumentum is longer (hairs up to 0.2mm), the leaves are less deeply divided, lobes all entire. There seems to be some geographical patterning to variation in length of indumentum, but not to degree of lobing of the leaves. Specimens with short indumentum are widespread in the eastern half of the central Cape, from roughly Williston and Carnarvon east to Cradock and Burghersdorp (specimens seen from the following quarter degree squares: 3021BD, 3022AB, CC, CD, 3024AD, CA, 3025CA, 3026CD, 3121DB, 3122BC, CD, 3123AC, BC, CC, DD, 3125AC, AD, 3222BA, BC, 3224AC, 3225BB). Specimens with longer indumentum (as in *Sutera altoplana*) have been seen from the area around Fraserburg, which is the type locality for *S. altoplana* (3121CD, DC, 3221BA, BB) then a great disjunction to the Barkly West - Hay area (2724CD, 2823BA, DB, DD, 2824AD, CA, 2922BA, BB, 2923BA, CB, 2924CA). There are thus two areas of relatively long-haired plants, one north of the area of short-haired plants, one more or less west of it. It must be understood that differences in hair-length are not absolute; nevertheless, I do not understand the pattern of variation, which is further complicated by suites of specimens, from several different areas, that are intermediate in character between *J. tysonii*, in the broad sense adopted here, and other species. This is elaborated below.

The character that holds together the specimens I have grouped under the name *J. tysonii* is the shape of the leaves (oblanceolate in outline) and their lobing: lobes cut more than ⅔ of the way to the midrib, more or less oblong, obtuse, usually entire, occasionally the larger ones with one or two teeth. These leaf-characters distinguish *J. tysonii* from the closely allied *J. filicaulis*, which has leaves more or less ovate in outline, divided from ⅓ of the way to right to the midrib, the lobes more or less spathulate or cuneate in outline, at least the largest lobes toothed and with every gradation to bipinnatisect leaves. Specimens of *J. tysonii* tend to dry black, but this is not constant; specimens of *J. filicaulis* never blacken. Also, the inflorescence of *J. filicaulis* is leafy; that of *J. tysonii* may run up into nearly nude racemes, but the heavy browsing to which the plants are subjected often disguises or disturbs this feature.

One of the 'problem' areas mentioned above is in the environs of De Aar (3024AD, CA). Some specimens can be assigned to *J. tysonii* but others may be of hybrid origin: the geographical areas of *J. tysonii*, *J. albiflora* and *J. filicaulis* meet hereabouts. *Esterhuysen* 1000 (BOL, NBG), *Moss* 11250 (J), *Markotter* s.n. (STE 9939), *Rogers* 17987 (BOL) are possibly of hybrid origin.

A second 'problem' area is Colesberg (3025CA, CB), at the interface of the ranges of *J. tysonii*, *J. albiflora* and *J. filicaulis*: *Bayliss* 1211 (Z) is very close to *J. albiflora*, differing in its more glandular-pubescent calyx; *Thorne* s.n. (SAM 51879) is *J. filicaulis*; *Thorne* s.n. (SAM 51880), *Louw* s.n. (STE 31959) and *Nutt* s.n. (E) appear to be of hybrid origin.

A third area where studies are needed is the environs of Cradock (3225AB, AD, BA). Unequivocal *J. tysonii* has not been seen from Cradock, though *Brynard* 11 (PRE) and *Zietsman* 1277 come very close to it. *Hilliard & Burtt* 10587 (E) and *Penzhorn* 5922 (PRE), collected on the Bankberg in the Mountain Zebra Park, can be assigned to *J. filicaulis*, though they are possibly not 'pure' *J. filicaulis* (possibly *J. filicaulis* × *foliolosa*). *Muller* 538 (PRE, SRGH) from the Park, and *Acocks* 16324 (PRE), collected 5¼ miles north of Cradock (which matches it exactly) is an odd-looking plant with many of the leaves ± entire, but some with two basal lobes so deeply cut that the leaf looks trifoliolate. They have the indumentum of *J. tysonii*. A specimen from Rasfonteinpoort (*Hilliard & Burtt* 10564, E),*Retief & Germishuizen* 383, PRE (3125DC, farm Grootfontein) and *Retief & Germishuizen* 323, PRE (3125DD, farm Echodale), all assigned to *J. tysonii*, have leaves only slightly more lobed. *Jamesbrittenia foliolosa* also occurs in the environs of Cradock.

There are also problems in Herbert and Hay divisions (2823CC, 2923BA, BB, 2924CA) and near Kimberley (2824DB), where the area of *J. tysonii* meets those of *J. albiflora* and *J. integerrima*. At Mazelsfontein (2923BA), Mrs Anderson collected both *J. tysonii*, with glandular hairs up to 0.3mm long (sub *Wilman* 1607, BOL) and a plant that may be a hybrid between *J. tysonii* and *J. albiflora* or possibly *J. integerrima* (*Anderson* sub SAM 51176). *Orpen* 114 (SAM) from St. Clair (2923BB) resembles *Anderson* sub SAM 51176. *Wilman* sub BOL 16199 from Kimberley (2824DB) may also represent *J. tysonii* × *albiflora*. *Acock & Hafström* 1028 (PRE) from Doringbult (2924CA) and *Orpen* sub BOL 6483 from Griquatown (2823 CC) are very close to *J. integerrima*; they differ in their glandular-pubescent capsules, which resemble those of *J. tysonii*.

The area around Victoria West dam (3123AC) should also be investigated for possible hybridization. Typical *J. tysonii* occurs there, but *Hardy* 848 (M, SRGH, Z) is very hairy and the larger leaves in particular are more toothed than is usual in *J. tysonii*.

Jamesbrittenia tysonii favours rocky places; both dolerite and limestone have been recorded several times. Flowers can possibly be found in any month.

20. Jamesbrittenia filicaulis (Benth.) Hilliard in Edinb. J. Bot. 49: 229 (1992).
Type: Cape, Aliwal North div., in rocky places on the Witteberg, 6000ft, 15 i 1833, *Drège* (holo. K, iso. P).

Syn.: *Lyperia filicaulis* Benth. in Hook., Comp. Bot. Mag. 1: 380 (1836) and in DC., Prodr. 10: 361 (1846).

    L. pinnatifida (L. f.) Benth. var. *subbipinnatisecta* Benth. in Hook., Comp. Bot. Mag. 1: 380 (1836) and in DC., Prodr. 10: 361 (1846). Lectotype (chosen here): Cape, Graaff Reinet div., 3224AB, Oudeberg, 3500ft, 6 ix 1829, *Drège* (K; iso. LE, P).

    L. pinnatifida (L.f.) Benth. var. *canescens* Benth. in Hook., Comp. Bot. Mag. 1: 380 (1836). Type: no specimen annotated by Bentham found, lectotype (chosen here): Cape, near Graaff Reinet, *Drège* (K; isolecto. P, S).

L. pinnatifida (L.f.) Benth. var. *visco-pubescens* Benth. in Hook., Comp. Bot. Mag. 1: 380 (1836). Lectotype (chosen here): Cape, [c.3226BC], Philipstown on the Kat River, *Ecklon* (K; isolecto. LE, M, P, S, W).

Sutera filicaulis (Benth.) Hiern in Fl. Cap. 4(2): 298 (1904).

S. pinnatifida Hiern in Fl. Cap. 4(2): 296 (1904), p.p. min. Lectotype (chosen here): Cape, Cathcart, 1400m, 25 ii 1894, *Kuntze* s.n. (NY).

S. henrici Hiern in Fl. Cap. 4(2): 300 (1904). Lectotype (chosen here): Tembuland, near Emgwali River, 2900ft, i 1896, *Bolus* 8758 (BOL; isolecto. K, PRE).

S. virgulosa Hiern in Fl. Cap. 4(2): 302 (1904). Type: Natal, East Griqualand, near Matatiele, 5300ft, i 1884, *Tyson* 1634 (holo. BOL; iso. NBG, PRE, SAM).

Dwarf shrublet with stems tufted from a woody caudex up to c.20mm in diam., stems up to c.450mm long, often diffuse and then sometimes rooting, occasionally ± erect, simple to loosely branched and forming tangled masses, glandular-pubescent, hairs up to 0.2–0.3mm long, some scattered glistening glands as well, very leafy. *Leaves* opposite, alternate towards tips of branchlets, pseudofasciculate, largest primary leaves 8–25 × 4–13mm, up to c.⅓ of the length petiolar, blade ovate in outline, ± deeply divided (from c.⅓ of the way to the midrib virtually to the midrib), the lobes ± spathulate or cuneate in outline, obtuse, at least the largest lobes toothed, with all gradations to every lobe divided ± to the midrib, all segments obtuse, both surfaces glandular-pubescent, hairs up to 0.15–0.2mm long, glistening glands as well particularly on lower surface. *Flowers* solitary in the axils of the leaves, the inflorescence so-formed usually conspicuously leafy. *Pedicels* up to 6–20mm long. *Calyx* tube 0.1–0.3mm long, lobes 2.5–3.7 × 0.4–1mm, oblong to subspathulate, densely glandular-pubescent, hairs up to 0.15–0.2mm long, glistening glands as well. *Corolla* tube 7–10 × 1.3–2mm in throat, cylindric, abruptly expanded near apex, limb oblique, 8–12mm across lateral lobes, posticous lobes 3–6.5 × 1.5–3.7mm, anticous lobe 3–6.5 × 1.5–3.2mm, all lobes cuneate-oblong, ± truncate or retuse, glandular-pubescent outside, hairs up to 0.1–0.2mm long, glistening glands as well particularly on backs of lobes and swollen part of tube, transverse band of clavate hairs in throat extending very briefly onto base of lower lip, lobes white or creamy, occasionally with a dark median streak at base, throat yellow to orange. *Stamens*: posticous filaments 0.8–1.4mm long, anthers 0.5–1mm, anticous filaments 0.3–1mm, anthers 0.4–0.6mm, all filaments minutely puberulous. *Stigma* 0.15–0.3mm long. *Style* 5–7mm. *Ovary* 1.5–2 × 0.7–1mm. *Capsules* 3–5 × 2–2.5mm, minutely glandular-puberulous, glistening glands as well, these sometimes confined to sutures. *Seeds* c.0.4–0.6 × 0.3mm.

Selected citations:

Orange Free State. Harrismith distr., 2828DB, Qwa Qwa Mt. above 'Bluegumbosch', 8 i 1979, *Hilliard & Burtt* 11977 (E, K, M, S). Ficksburg distr., 2827DD, Westbury, 5900ft, 14 x 1939, *Galpin* 13935 (BOL, PRE, W).

Lesotho. Leribe, 2828CC, *Dieterlen* 260 (P, Z). Thaba Bosiu, 2927BC, i–ii 1903, *Junod* 1807 (M, Z).

Cape. Lady Grey div., 3027CB, lower slopes Witteberg, Joubert's Pass, c.6800ft, 19 i 1979, *Hilliard & Burtt* 12237 (E, S). Aliwal North div., 3126BB, Jamestown, Vogelfontein Farm, 16 xii 1942, *Barker* 2217 (NBG). Graaff Reinet div., 3224AA, Koudeveldberge SE of Doornbosch, 6000ft, 6 xi 1974, *Oliver* 5206 (PRE, STE). Stockenstrom div., 3226DB, Elandspost [= Seymour], *Cooper* 332 (E, PRE, W, Z).

Natal. Underberg distr., Coleford Nature Reserve, Sunnyside, 5500ft, 25 xii 1976, *Hilliard & Burtt* 9545 (E, S).

Transkei. Between Engcobo [3127DB] and Nquamakwe [3227BB], 3600ft, i 1896, *Bolus* 8757 (BOL, K). Cofimvaba div., 3127CD, Qamata Poort, 30 iv 1955, *Lewis* 5103 (PRE, SAM). Tsolo div., 3128BC, 1 mile E of Tsolo on Umtata road, 29 i 1966, *Hilliard & Burtt* 3729 (E).

Many specimens of *J. filicaulis* will be found in herbaria under the name *Sutera pinnatifida*. Hiern's entry, *Sutera pinnatifida* O. Kuntze, is indefensible (Hiern, 1904, p.296). In 1898, O. Kuntze did not accept the genus *Sutera* (which he placed under *Manulea*). Under *Manulea multifida* (Benth.) O. Kuntze (Kuntze, 1898, p.236), Kuntze refers to 'der sehr ähnlichen *S. pinnatifida* mit milchweissen basal gelben Corollen und linealen Kelchzipfeln'. This is what Hiern quotes as the original publication of *S. pinnatifida*. There is nothing to show what the *S.* stands for: on p.235 we find '*Manulea* L. incl. *Sutera* Roth', and on p.236, *M. pinnatifida* L. (*Lyperia pinnatifida* Benth.). '*S. pinnatifida*', in the discussion a few lines above, can only have been a slip for *M. pinnatifida*, not a new species.

'*Manulea pinnatifida* L.' (of Otto Kuntze) is *Manulea pinnatifida* L.f.; *Lyperia pinnatifida* Benth. is clearly a new combination for *Manulea pinnatifida* L.f., which Hiern reduced to synonymy under *Sutera kraussiana*. This was a mistake; the species is enumerated here as *Jamesbrittenia pinnatifida* (L.f.) Hilliard.

The references that Hiern gives under *Sutera pinnatifida* lead nomenclaturally to the plant described by Otto Kuntze. *Sutera pinnatifida* must be attributed to Hiern, and it is now lectotypified by one of the specimens quoted by O. Kuntze. It proves to be synonymous with *Lyperia filicaulis* Benth., that is, *Jamesbrittenia filicaulis*. I have ignored the varieties that Hiern lists under *Sutera pinnatifida*; the varietal names are those that Bentham gave under *Lyperia pinnatifida* (L.f.) Benth., and these I have lectotypified and placed in synonymy under the appropriate species. Hiern's use of the names in *Sutera* covers a jumble of species.

Jamesbrittenia filicaulis ranges from Qwa Qwa Mountain near Witzieshoek in the NE corner of the Orange Free State through the sandstone areas of the eastern OFS and western Lesotho to the Witteberg and Cape Drakensberg, thence west on the high ground to the environs of the Oudeberg, Koudeveldberg and other mountains around Graaff Reinet; the southernmost records are from the Elandsberg and the Amatole Mountains, the easternmost from the high ground in the Transkei, with two records from southernmost Natal, at Matatiele, and Coleford on the Ngangwane river, which here forms the boundary with Transkei. The plants favour rocky sites on hill and mountain slopes, often growing under rock outcrops and overhangs, partially shaded, between c.800 and 2100m above sea level. Flowering is mainly between October and May.

The distinguishing features of the species are the leafy flowering branches (not running up into distinct racemes), all parts glandular-pubescent, the white corolla limb (creamy tints, even palest yellow, have also been recorded; the throat is always yellow to orange), and the decidedly obtuse leaf lobes, which are themselves toothed. The degree of lobing varies tremendously, and there is some geographical patterning to this. In the OFS, Lesotho, Natal and most of the Transkei, leaves are mostly pinnatifid to pinnatipartite (divided roughly ½ to ⅔ of way to midrib), as in the type specimens of *Sutera henrici* and *S. virgulosa*. At least the lower, largest, lobes are toothed. On the Witteberg (type locality of *J. filicaulis*), the lobing is often deeper and some leaves are pinnatisect (lobed to the midrib). Around Queenstown, Cathcart (type locality of *Sutera pinnatifida*) and Engcobo leaves may be bipinnatisect. Pinnatipartite and pinnatisect leaves dominate over the western part of the geographical range of the species. The cause of such diversity in leaf-lobing is problematical. It is known that edaphic factors can be responsible: it may be significant that least lobing is shown by plants from the Cave Sandstone (Clarens Formation) areas of the OFS, Lesotho, Natal and Transkei. Hilliard & Burtt specimens from the Witteberg were collected on basalt, but most other collections lack information on substrate. Field and laboratory investigations are needed. Hybridization may also be a factor to be considered: see under *J. tysonii*.

21. Jamesbrittenia albanensis Hilliard in Edinb. J. Bot. 49: 225 (1992).
Type: Cape, Albany div., Queen's Pass, 2000ft, 16 viii 1964, *Bayliss* 2268 (holo. NBG, iso. Z).

Syn.: *Lyperia pinnatifida* (L.f.) Benth. var. *microphylla* Benth. in Hook., Comp. Bot. Mag. 1: 380 (1836) and in DC., Prodr. 10: 361 (1846). Lectotype: Cape, Zwischen Zondagrivier und Ado, 400–600', 4 xii 1829, *Drège* (K; isolecto. K, P, S, W).

Twiggy shrublet, branches up to c.450mm long, erect, decumbent or spreading, glandular-pubescent, hairs up to 0.15–0.3mm long, very leafy. *Leaves* opposite becoming alternate on the flowering twiglets, pseudofasciculate, largest primary leaves 4–15 × 1.5–8mm, spathulate to ovate in outline tapering into a petiolar part up to nearly half total leaf length, lobed from ⅔ to whole way to midrib, lobes oblong to spathulate, obtuse, sometimes entire, the larger ones often few-toothed to almost bipinnatisect, both surfaces glandular-pubescent, hairs up to 0.15–0.3mm long, glistening glands as well, mainly on lower surface. *Flowers* solitary in the axils of the upper leaves, these much reduced in size, the inflorescence thus racemose. *Pedicels* up to c.10–23mm long. *Calyx* tube 0.1–0.2mm long, lobes 2.3–3.4 × 0.5–0.7mm, ± oblong, densely glandular-pubescent, hairs up to 0.15–0.4mm long, glistening glands as well. *Corolla* tube 7,5–10 × 1.8–2.2mm in throat, cylindric, abruptly expanded near apex, limb oblique, 9–14mm across lateral lobes, posticous lobes 5–8 × 2.8–3.7mm, anticous lobe 4.5–6.5 × 2.8–3.2mm, all lobes oblong-cuneate, truncate to retuse, glandular-pubescent outside, hairs up to 0.2–0.25mm long, glistening glands as well mainly on backs of lobes, transverse band of clavate hairs in throat extending very briefly onto base of lower lip, lobes shades of purple-pink and mauve, throat probably yellow/orange. *Stamens*: posticous filaments 1–2mm long, anthers 0.6–1.1mm, anticous filaments 0.5–1mm, anthers 0.4–0.7mm, all filaments puberulous. *Stigma* c.0.2mm long. *Style* 4.8–7.4mm. *Ovary* 1.8–2 × 0.8–1mm. *Capsule* 5–7 × 3mm, glistening glands on sutures, minute glandular hairs sometimes present on rest of capsule. *Seeds* c.0.6–0.7 × 0.4mm.

Selected citations:

Cape. Albany div., 3326BD, near Fraser's Camp, 30 xi 1950, *Martin* 688 (NBG); 3326BB, near Breakfast Vlei, 8 xi 1938, *Hafstrom & Acocks* 1177 (PRE, S). Victoria East div., 3226DD, near Alice on Grahamstown road, 2 i 1943, *Barker* 2171 (NBG). Alexandria div., 3326CB, Nanaga, 10 viii 1953, *Archibald* 5945 (BOL). Uitenhage div., 3325DC, Aloes Station, c.154ft, 2 ix 1912, *Burtt Davy* sub BOL 51642 (BOL). Uniondale div., c.3323CB, Bo-Kouga, off Long Kloof, 1000ft, 9 x 1975, *Bayliss* 5159 (M, Z).

Jamesbrittenia albanensis is characterized by its bluntly lobed leaves, glandular-pubescent on both surfaces, flowers borne in pseudoracemes, and mauve to magenta corolla limb. In foliage, it resembles forms of *J. filicaulis*, which differs in its leafy inflorescences and white corolla limb. In leaf cutting, inflorescence and flower colour, it resembles *J. foliolosa*, but that species is distinguished by the indumentum on the leaves: lower surface thickly clad in glistening glands, minute glandular hairs present or not, upper surface either glabrous or minutely glandular-pubescent. *Jamesbrittenia foliolosa* possibly hybridizes with *J. albanensis* in the Grahamstown area (see under *J. foliolosa*); they may also hybridize around Alice: *Barker* 2171 (cited above) is typical *J. albanensis*, but *Bokelmann* 4 Pl. 42 (NBG), collected on a roadside near Alice, comprises several twigs with the indumentum of *J. foliolosa* but one twig is much more glandular and not unlike *J. albanensis*.

    Jamesbrittenia albanensis appears to be common around Grahamstown, and its area extends thence north-east to Alice, south-east to Kowie, and west to the environs of Ado and Aloes, with one record from much further west, in the Long Kloof. The earliest records of the species go back to Ecklon & Zeyher and to Drège, who collected it between the Sundays river and Ado. Bentham enumerated these collections as a variety of *Lyperia pinnatifida*, and it is under this epithet that most specimens will be found in herbaria.

    Only Bayliss has given any ecological information: 'Karroid' and 'Euphorbia bush'. Flowering has been recorded in most months.

22. **Jamesbrittenia phlogiflora** (Benth.) Hilliard in Edinb. J. Bot. 49: 232 (1992).
Type: Cape [c.3327AB], on the Keiskamma, 500–1000ft, 17 vi 1832, *Drège* (holo. K; iso. E, K, P, W, fragment in S, mixed with *J. foliolosa*).

Syn.: *Lyperia phlogiflora* Benth. in Hook., Comp. Bot. Mag. 1: 379 (1836) and in DC., Prodr. 10: 360 (1846).
    Chaenostoma phlogiflorum (Benth.) Diels in Bot. Jahrb. 23: 491 (1897).
    Sutera phlogiflora (Benth.) Hiern in Fl. Cap. 4(2): 298 (1904) p.p. and excluding many of the specimens cited.

Shrublet up to c.500mm tall, stems up to 5–8mm diam. at base, erect or decumbent, virgately branched, glandular-pubescent, hairs up to 0.2–0.25mm long, often coarse, scattered glistening glands as well on young parts, densely leafy except at the flowering tips. *Leaves* opposite becoming alternate upwards, pseudofasciculate, largest primary leaves 7–25 × 3–10mm, elliptic to lanceolate in outline gradually tapering into a petiolar part up to ¼–⅓ total leaf-length, apex acute, margins coarsely serrate or occasionally more deeply cut but seldom more than ⅔ of way to midrib and then only in lowermost pair of lobes, lobes usually ± acute, both surfaces glandular-puberulous, hairs often rather coarse, up to 0.15–0.25mm long, below thickly beset with glistening glands, these few above. *Flowers* solitary in the axils of the uppermost leaves, these rapidly smaller and passing into bracts, thus forming long racemes, sometimes panicled. *Pedicels* up to 12–20mm long. *Calyx* tube 0.15–0.3mm long, lobes 2.5–3.7 × 0.5–0.75mm, oblong, acute, glandular-puberulous, hairs up to 0.2–0.3mm long, scattered glistening glands as well. *Corolla* tube 7.5–9 × 1.6–2mm in throat, cylindric, abruptly expanded near apex, limb oblique, 10–16mm across lateral lobes, posticous lobes 3.3–7 × 1.8–3mm, anticous lobe 4–8 × 2.5–3.8mm, all lobes ± cuneate-oblong, truncate to retuse, glandular-puberulous outside, hairs up to 0.15–0.25mm long, glistening glands as well mainly on backs of lobes and swollen part of tube, transverse band of clavate hairs in throat extending briefly onto base of lower lip, lobes white, white tinged mauve at tips, or wholly mauve, throat orange. *Stamens*: posticous filaments 1–1.5mm long, anthers 0.7–1mm, anticous filaments 0.4–1mm long, anthers 0.4–0.8mm, all filaments minutely puberulous. *Stigma* c.0.2mm long. *Style* 5.4–6.5mm long. *Ovary* 1.7–2 × 0.8–1mm. *Capsules* c.5 × 3mm, minutely glandular-puberulous, glistening glands as well, sometimes confined to sutures.

Selected citations:

Transkei. Butterworth distr., 3228AC, 2 miles SW of Ndabakazi, 1700ft., 10 iii 1955, *Codd* 9244 (PRE).
Cape. Stutterheim div., 3227CB, Stutterheim, xii 1913, *Rogers* 12719 (BOL, J, Z). King William's Town div., 3227CC, slopes round King William's Town, 1500ft, i 1887, *Tyson* 849 (BOL, SAM, W). 3227CD, SE of King William's Town, Mt. Coke, 14 xii 1977, *Hilliard & Burtt* 11017 (E, K, PRE, S). East London div., 3327BB, East London, 17 xii 1943, *Barker* 2793 (NBG). Peddie div., 3327AB, 15 miles from Peddie on East London road, 1 iii 1946, *Barker* 3978 (NBG).

The characteristic features of *J. phlogiflora* as the name is now used are the lanceolate coarsely serrate leaves (the lobes all ± acute) and the flowers in the axils of the leaves but these rapidly reduced in size so that the impression gained is that of bracteate rather than leafy racemes. The glandular hairs on stems and leaves are often rather coarse, a quality impossible to quantify, but more delicate indumentum seems to occur indiscriminately over the geographical range of the species, which appears to be confined to a limited area: the environs of Stutterheim, King William's Town, Komgha, East London and Peddie, and possibly the heights above the Great Fish river with one record from southernmost Transkei. The plants favour rough grassland near patches of forest or bush clumps; flowers can be found in any month.

The type material came from somewhere near the Keiskamma river, east of Peddie. The holotype (a large branch in excellent condition) has coarsely toothed leaves (at most cut ± halfway to midrib), somewhat coarse hairs up to 0.2mm long on the stems, and greatly reduced leaves on the inflorescence axes. The twigs on the six isotypes seen are not all uniform with the holotype: the lowest leaf lobes are often more deeply incised (though the lobing of the holotype can be matched in, for example, specimens from East London and Stutterheim), and the indumentum may be coarse or fine, the inflorescence leafy or not. One isotype in P (bearing Drège's own label) has fine indumentum and leafy inflorescence, the other coarse indumentum and 'nude' racemes, that in W fine indumentum and 'nude' racemes, in E fine indumentum and leafy inflorescence. Some of these specimens may have been introgressed by *J. albanensis*. The specimen collected from a site geographically closest to that of Drège (*Barker* 3978, cited above) has foliage similar to that of Drège's own specimen in P, but the inflorescence is less leafy and the indumentum coarse.

Jamesbrittenia phlogiflora is closely allied to *J. filicaulis* from which it may be distinguished by its ± acute leaf lobes and less leafy inflorescence.

23. Jamesbrittenia maritima (Hiern) Hilliard in Edinb. J. Bot. 49: 230 (1992).
Type: Cape, Bathurst div., Port Alfred, 50ft, 21 xii 1898, *Galpin* 2933 (holo. K).

Syn.: *Sutera maritima* Hiern in Fl. Cap. 4(2): 277 (1904).

Perennial herb, sometimes shrubby, basal parts not seen, up to c.700mm tall, stems erect or ascending, occasionally ± scrambling, glandular-pubescent, hairs up to 0.2–0.5mm long, very few glistening glands, very leafy. *Leaves* opposite, alternate upwards, pseudofasciculate, largest primary leaves 12–30 × 5.5–14mm, obovate or sometimes oblanceolate, rarely ovate, often very obtuse, sometimes subacute, base cuneate tapering into a short petiolar part, margins coarsely toothed in upper half, the larger teeth occasionally 1-toothed, both surfaces glandular-pubescent, hairs up to 0.2–0.25mm long, glistening glands as well, scattered mainly on lower surface. *Flowers* solitary in axils of upper leaves, these rapidly smaller upwards, often producing terminal racemes. *Pedicels* up to 10–35mm long, mostly ascending at an angle of c.45°. *Calyx* tube 0.3–0.4mm long, lobes 3–4.5 × 0.8–1mm, oblong-lanceolate, subacute, glandular-puberulous, hairs up to 0.2–0.25mm long, a few glistening glands as well. *Corolla* tube 7–10 × 2–2.1mm in throat, cylindric, abruptly expanded near apex, limb oblique, c.10–15mm across lateral lobes, posticous lobes 4.2–8 × 2.4–4mm, anticous lobe 4.5–8.8 × 2.3–4.4mm, all lobes oblong to cuneate-oblong, tips often ± truncate or retuse, glandular-puberulous outside, hairs up to c.0.2mm long, glistening glands as well mainly on backs of lobes, broad transverse band of clavate hairs in throat extending briefly onto base of lower lip, all lobes white or mauve, yellow in throat. *Stamens*: posticous filaments 1–1.2mm long, anthers 0.7–1mm, anticous filaments 0.6–0.8mm long, anthers 0.5–0.8mm, all filaments glandular-puberulous. *Stigma* c.0.2mm long. *Style* 5.2–6mm. *Ovary* c.2 × 0.8mm. *Capsules* 3.5–7 × 2.5–3.5mm, minutely glandular-puberulous, glistening glands as well, often ± confined to the sutures. *Seeds* c.0.7–0.9 × 0.5–0.7mm.

Selected citations:

Cape. East London div., 3327BA, Kayser's Beach, 24 ii 1929, *Galpin* 10434 (PRE). Bathurst div., c.10km from Port Alfred on East London road, 6 xii 1977, *Hilliard & Burtt* 10895 (E, K, PRE). Alexandria div., 3326DA, Bushman's River mouth, 5 xii 1941, *Barker* 1510 (NBG). 3326CB, Olifantshoek, *Pappe* s.n. (S, SAM).

Jamesbrittenia maritima ranges along the eastern Cape coast from the environs of East London to Alexandria, in grassy places on the margins of forest patches, usually not far from the sea, but recorded up to c.180m above sea level. It can be found in flower in any month.

Its distinguishing features are the mostly obovate leaves toothed in the upper half and coarsely pubescent on both surfaces, and flowers arranged in the axils of the upper leaves often forming terminal racemes though occasionally the flowers remain axillary rather than racemose. In herbaria, it is sometimes confused with *J. kraussiana*, but that species is easily distingushed by its elliptic leaves and flowers arranged all along the leafy stems on delicate filiform pedicels. The calyx is larger in *J. maritima* than in *J. kraussiana* and the corolla tube mostly longer.

24. Jamesbrittenia kraussiana (Bernh.) Hilliard in Edinb. J. Bot. 49: 230 (1992). **Plate 4B.** Type: Natal, in graminosis prope Natal Bay, Aug. 1828, *Krauss* 119 (holo. MO; iso. K, fragment S).

Syn.: *Chaenostoma kraussianum* Bernh. apud Krauss in Flora 27: 835 (1844).
    Lyperia kraussiana (Bernh.) Benth. in DC., Prodr. 10: 360 (1846), excl. var. *latifolia* Benth.
    Manulea kraussiana (Bernh.) O. Kuntze, Rev. Gen. Pl. 3(2): 235 (1898).
    Sutera kraussiana (Bernh.) Hiern in Fl. Cap. 4(2): 301 (1904) p.p. and excl. *Manulea pinnatifida* L.f. and *S. kraussiana* var. *latifolia* (Benth.) Hiern.

Perennial herb; stems c.150–600mm long, eventually several tufted from a woody caudex up to c.20mm in diam., but will flower when young and then single-stemmed, erect or diffuse, eventually well-branched, glandular-puberulous, hairs up to 0.1–0.7mm long, glistening glands as well, very leafy. *Leaves* opposite becoming alternate upwards, pseudofasciculate, largest primary leaves 15–45 × 3–20mm, narrowly to broadly elliptic, apex obtuse to subacute, base cuneate tapering into a short petiolar part, margins coarsely serrate, the teeth sometimes few-toothed, glandular-puberulous, hairs up to 0.1–0.5mm long all over upper surface, hairy mainly on veins below, glistening glands as well mainly on lower surface. *Flowers* solitary in the leaf-axils. *Pedicels* mostly 12–27mm long, filiform, often spreading and curved in an elongated S. *Calyx* tube 0.1–0.2mm long, lobes 2–2.25 × 0.4–0.75mm, oblong, acute or subacute, glandular-puberulous, hairs up to 0.15–0.4mm long, glistening glands as well. *Corolla* tube 6.5–7 × 1.2–1.8mm in throat, cylindric, abruptly expanded near apex, limb oblique, c.8–9mm across lateral lobes, posticous lobes 4–5 × 1.7–2.3mm, anticous lobe 4.2–5 × 2–3mm, all lobes oblong to cuneate-oblong, tips often ± truncate or retuse, glandular-puberulous outside, hairs up to 0.2–0.3mm long, glistening glands as well mainly on backs of lobes, transverse band of clavate hairs in throat extending briefly onto lower lip, all lobes white or pale mauve, orange in throat. *Stamens*: posticous filaments 0.7–1mm long, anthers 0.7–0.8mm, anticous filaments 0.6–0.7mm long, anthers 0.4–0.5mm, all filaments glandular-puberulous. *Stigma* c.0.2mm long. *Style* 4.3–4.8mm long. *Ovary* 1–1.4 × 0.7–0.8mm. *Capsules* c.2.5–5 × 2–2.5mm, glistening glands and minute glandular hairs all over. *Seeds* c.0.7–1 × 0.6–0.75mm.

Selected citations:

Natal. Durban distr., 2931CC, Avoca, 250ft, 1 viii 1893, *Schlechter* 3010 (BOL, W, Z). [Umzinto distr.,] 3030BC, Umgaye Flat, Ifafa valley, 15 v 1910, *Rudatis* 991 (E, S, STE, Z, W). Port Shepstone distr., 3030CD, Uvongo Beach, 4 viii 1967, *Strey* 7605 (M, PRE, S).
Transkei. Pondoland, July 1915, *Burtt Davy* 15308 (BOL). Mazeppa Bay, 3228BC, 61m, 17 v 1981, *Phillipson* 314 (PRE).
Cape. Komgha div., 3228CB, 10 miles W of Kei Mouth, 24 xi 1945, *Compton* 17634(NBG); near Kei Mouth, 100ft, xi 1892, *Flanagan* 1346 (BOL, SAM, Z).

Jamesbrittenia kraussiana ranges from roughly the Umfolozi river in Natal (2831AD, 2832AC) south to the environs of East London in the Cape (3227DD), along the coast and ranging inland to c.800m or a little higher. It favours grassy or scrubby places, often on the margins of forest patches; flowering has been recorded in all months.

It is easily recognized by its elliptic serrate leaves and flowery branches with a long filiform spreading pedicel in each leaf-axil. In foliage it much resembles *J. phlogiflora* (with flowers in terminal racemes) and the two species may hybridize where their areas meet somewhere north of East London; *Winkler* 57 (NBG, STE), collected between the Quinera (Kwerera) and Gonubie rivers appears to be just such a hybrid.

25. Jamesbrittenia pinnatifida (L.f.) Hilliard in Edinb. J. Bot. 49: 232 (1992).
Type: Cape, between the Sundays and Fish rivers, *Thunberg* (sheet no. 14376 in herb. Thunb., UPS). See note below.

Syn.: *Manulea pinnatifida* L.f., Suppl. 286 (1782); Thunb., Prodr. 102 (1800) & Fl. Cap. ed. Schultes 473 (1823).
   Lyperia pinnatifida (L.f.) Benth. in Hook., Comp. Bot. Mag. 1: 380 (1836) & in DC., Prodr. 10: 361 (1846), excl. vars.
   L. mollis Benth. in Hook., Comp. Bot. Mag. 1: 380 (1836) & in DC., Prodr. 10: 360 (1846). Lectotype (chosen here): Cape, Colesberg, 4500ft, 25 ii 1833, *Drège* (K; isolecto. P, S, W).
   Chaenostoma pinnatifidum (L.f.) Diels in Bot. Jahrb. 23: 491 (1897).
   C.molle (Benth.) Diels in Bot. Jahrb. 23: 491 (1897).
   Sutera mollis (Benth.) Hiern in Fl. Cap. 4(2): 295 (1904) p.p. and excluding many of the specimens cited.

Perennial herb, loosely branched, stems straggling, c.300–450mm long (no basal parts seen), glandular-pubescent, hairs up to 0.5–1.2mm long, few scattered glistening glands as well, very leafy at least on the younger parts. *Leaves* opposite becoming alternate upwards, pseudofasciculate, blades of largest primary leaves 5–22 × 4–19mm, ovate in outline abruptly contracted into a petiolar part 3–10mm long (roughly ⅓–½ length of blade), apex obtuse, often 3-toothed, margins divided roughly ½ way to midrib or deeper into 4 major pairs of broad oblong lobes, lobes mostly 1–3-toothed, both surfaces glandular-pubescent, hairs up to 0.4–0.8mm long, glistening glands as well particularly on lower surface. *Flowers* solitary in the leaf-axils. *Pedicels* c.9–20mm long, filiform, sparsely to densely glandular-pubescent, hairs up to 0.2–0.8mm long. *Calyx* tube 0.1–0.25mm long, lobes 3–4 × 0.7–1mm, spathulate, ± acute, glandular-pubescent, hairs c.0.3–1mm long, glistening glands as well. *Corolla* tube c.7.3–8.5 × 2mm in throat, cylindric, abruptly inflated near apex, limb oblique, c.7–11mm across lateral lobes, posticous lobes 3–5.5 × 2.3–3mm, anticous lobe 3–7 × 2.3–3.5mm, all lobes cuneate-oblong, tips ± retuse, glandular-pubescent outside, hairs up to 0.25–0.5mm long, glistening glands as well mainly on backs of lobes, broad transverse band of clavate hairs in throat extending very briefly onto base of lower lip, all lobes white, throat yellow. *Stamens*: posticous filaments 0.7–1.5mm long, anthers 0.8–1mm; anticous filaments 0.3–0.7mm long, anthers 0.5–0.8mm, all filaments minutely puberulous. *Stigma* c.0.2mm long. *Style* 5–6.5mm. *Ovary* 1.5–1.8 × 0.7–0.8mm. *Capsules* 3.5–5 × 2–3mm, clad all over in glistening glands and minute glandular hairs. *Seeds* c.0.9 × 0.8mm.

Selected citations:

Cape. Albany div., 3326AD, 10 miles W of Grahamstown, 12 iv 1952, *Barker* 7834 (NBG); 3326BC, near Grahamstown, 2500ft, v 1893, *Schlechter* 2668 (BOL, E, J, PRE, S, Z). Uniondale div., 3323CB, Haarlem, 2 miles beyond Stone's Hill, iv 1919, *Schonland* 3150 (PRE).

The name *Manulea pinnatifida* was based on a specimen purportedly collected by Thunberg; a 'precise' locality is given only in the 1823 edition of Thunberg's *Flora capensis*: 'inter *Sondags et Vischrivier*'. Thunberg travelled no further than the west bank of the Sundays river; it was his friend Sparrman who, in 1775, crossed the Sundays river and on 14th December was at Assegaibosch, between modern Alicedale and Sidbury, about 18 miles (30km) WSW of Grahamstown, and roughly halfway between the Sundays and Great Fish rivers (see Karsten in J. S. Afr. Bot. 23: 136, 1957). It seems possible that it was Sparrman who collected what was to become the type of the name and gave the specimen to Thunberg. See also under *J. microphylla*.

Since the publication of Hiern's account of *Sutera*, the epithet *pinnatifida* has been grossly misused. The muddle originated with Bentham who was mistaken in his concept of *Manulea pinnatifida* Thunb. and re-described the species as *Lyperia mollis*. However, under his new combination *L. pinnatifida* (L.f.) Benth. he erected a number of varietal names, which were taken up by Hiern in *Sutera*. Hiern muddled things still further, both nomenclaturally and taxonomically; this is dealt with under *Jamesbrittenia filicaulis* (no. 20), with which *Sutera pinnatifida* Hiern is synonymous.

Bentham cited three collections under *Lyperia mollis*. Ecklon's specimen from the Zuurberg was possibly collected on the northern flank of the range not far from Assegaibosch, Sparrman's site, which lies at the eastern end of the Zuurberg. There is no Ecklon collection from Albany in the Kew herbarium, but I have seen material in W and S, and it is typical *J. pinnatifida*. Bentham's third syntype is a specimen collected by Drège at Colesberg. It is rather more hairy than the plant from Albany division and I have chosen it as the most appropriate specimen to lectotypify the name *Lyperia mollis*. Its leaves are also more deeply dissected than those of most other material of *Jamesbrittenia pinnatifida*. Colesberg lies roughly 300km NNE of Grahamstown; *J. pinnatifida* is otherwise known only from a small area of the eastern Cape, most material seen having come from the environs of Assegaaibosch and Grahamstown, but one collection was from further west, near Haarlem in the Langekloof. No ecological information has been given by collectors, but the plants probably grow in scrubby grassland and flower in any month.

Neither Thunberg nor Sparrman reached the area of *J. kraussiana*, with which Hiern equated *Manulea pinnatifida*; the two species are closely allied, but may be distinguished by differences in the shape and toothing of the leaves; also, the flowers of *J. pinnatifida* are possibly always slightly larger than those of *J. kraussiana*, but I have seen too little material to be sure (calyx lobes 3.5–4mm versus 2–2.5mm, corolla tube 7.3–8.5mm versus 6.5–7mm).

26. Jamesbrittenia argentea (L.f.) Hilliard in Edinb. J. Bot. 49: 226 (1992).
Type: Cape, without precise locality, *Thunberg* 404 (holo. LINN 787.15; iso. prob. sheet no. 14343 in herb. Thunb., UPS, and an unnumbered sheet, S).

Syn.: *Manulea argentea* L.f., Suppl. 386 (1782); Thunb., Prodr. 102 (1800) & Fl. Cap. ed. *Schultes* 472 (1823).

Buchnera pedunculata Andr., Bot. Rep. 2: t. 84 (1800). Type: no specimen found, iconotype t. 84.

Manulea pedunculata (Andr.) Pers., Syn. 2: 148 (1806).

Lyperia argentea (L.f.) Benth. in Hook., Comp. Bot. Mag. 1: 379 (1836) & in DC., Prodr. 10: 359 (1846) excl. var. β.

L. pedunculata (Andr.) Benth. in Hook., Comp. Bot. Mag. 1: 379 (1836) & in DC., Prodr. 10: 360 (1846).

L. cuneata Benth. in Hook., Comp. Bot. Mag. 1: 380 (1836) & in DC., Prodr. 10: 361 (1846). Type: Cape [3421 BC], near the Gauritz river, *Drège* b (holo. K; iso. E, P, S, W).

Lyperia kraussiana (Bernh.) Benth. var. *latifolia* Benth. in DC., Prodr. 10: 360 (1846). Type: Cape, Addo and Bushmans River, *Ecklon* (holo. K; iso. LE, FI, M, S, W). See note below.

Chaenostoma cuneatum (Benth.) Diels in Bot. Jahrb. 23: 491 (1897), non *C. cuneatum* Benth. (1836), which is *Sutera hispida*.

Sutera argentea (L.f.) Hiern in Fl. Cap. 4(2): 301 (1904), p.p., excl. many of the specimens cited.

S. pedunculata (Andr.) Hiern in Fl. Cap. 4(2): 307 (1904), p.p., excl. some of the specimens cited.

S. rhombifolia Schinz in Vierteljahrsschr. Nat. Ges. Zürich 74: 117 (1929). Type: Cape, George div. [3322 DC], Zwarte rivier, 23 iii 1893, *Schlechter* 2380 (holo. Z; iso. BOL, C, E, J, PRE, S).

Shrub up to c.750mm to 1.5m tall, stems ± erect to straggling, well-branched, the ultimate branchlets often short (less than 100mm), patent or wide-spreading, glandular-puberulous, hairs up to 0.1–0.2mm long, scattered glistening glands as well, closely leafy. *Leaves* alternate, pseudofasciculate, largest primary leaves 3–25 × 1–10mm, rhomboid, cuneate or sometimes oblong in outline, apex obtuse (often almost truncate but toothed), base cuneate tapering into a short petiolar part up to c.⅓ total leaf-length, margins sometimes entire in small secondary leaves, usually with up to 5 pairs of ± evenly-sized, acute teeth in upper half, the largest teeth occasionally 1-toothed, glabrous above or occasionally a few glistening glands, thickly clad in glistening glands below. *Flowers* solitary in the axils of leaf tufts scattered along the branches. *Pedicels* up to 10–25mm long, filiform, often wide-spreading. *Calyx* tube 0.2–0.3mm long, lobes 2–2.6 × 0.5–0.8mm, oblong-lanceolate, ± acute, clad in glistening glands only. *Corolla* tube 7–9.5 × 1.8–2mm in throat, cylindric, abruptly expanded near apex, limb oblique, c.8–14mm across lateral lobes, posticous lobes 4–6.5 × 1.8–2.8mm, anticous lobe 4–6 × 2–3.5mm, all lobes cuneate-oblong, tips ± truncate or retuse, glandular-puberulous outside, hairs up to 0.2–0.25mm long, glistening glands as well mainly on backs of lobes and swollen part of tube, transverse band of clavate hairs in throat extending briefly onto base of lower lip, all lobes white, throat yellow. *Stamens*: posticous filaments 1–1.2mm long, anthers 0.7–0.8mm, anticous filaments 0.6–0.7mm long, anthers 0.4–0.5mm, all filaments minutely glandular. *Stigma* c.0.2mm long. *Style* 6–6.8mm. *Ovary* 1.5–2 × 0.8–1mm. *Capsules* 3.5–6 × 2.2–3mm, glistening glands all over. *Seeds* c.0.7 × 0.4mm.

Selected citations:

Cape. Swellendam div., 3320DC, Tradouws Pass, 1100ft, 29 v 1982, *Viviers* 364 (STE). Montagu div., 3320CC, Cogman's Kloof, 5 v 1940, *Compton* 8758 (NBG). Mossel Bay div., 3421BD, Gouritz river bridge, 15 iv 1952, *Lewis* 3892 (SAM). George div., 3322DC, valley of Klein Zwarte river, 8 x 1928, *Gillett* 1200 (BOL, STE). Humansdorp div., 3424BA, Kromme river, 250ft, viii 1942, *Fourcade* 5613 (BOL, NBG, STE). Port Elizabeth div., 3325DC, Baakens river valley, xi 1909, *Paterson* 851 (Z).

Jamesbrittenia argentea shows considerable variation in the size and toothing of the leaves, but the plant is unmistakable: wiry, straggling habit, ultimate branchlets often short and almost patent, densely leafy, flowers borne on spreading filiform pedicels in the axils of the leaf-tufts and not running up into discrete racemes, corolla limb white. The leaves are often cuneate in outline and 3-toothed at the apex, but large leaves are often rhomboid to oblong with several pairs of coarse teeth. They are glabrous above, densely clad in glistening glands below (which possibly suggested the epithet *argentea*, that is, silvery); the calyx lobes too are clad only in glistening glands.

Thunberg's original collection is unlocalized, but its relatively large well-toothed leaves can be precisely matched by specimens from around Humansdorp and Uitenhage, areas traversed by Thunberg. The type of *Sutera rhombifolia* (from the Zwarte river between George and Knysna) is also large-leaved, but the toothing is less acute than in Thunberg's collection. The painting of *Buchnera pedunculata* is much like Thunberg's plant. Drège's plant from the Gauritz river (type of *Lyperia cuneata*) is small-leaved, and Bentham (1836, p.380) shrewdly suggested it might be the wild form of *Buchnera pedunculata*, though it is now clear that large leaves appear in the wild as readily as small ones.

The species is widely distributed in the southern Cape, from Cogman's Kloof (Montagu) and Tradouw's Pass (near Barrydale) in the west to Port Elizabeth in the east, from sea level to c.600m, and flowering in any month. The plants appear to favour partly shaded, and probably damp, spots on forest margins and, in more exposed situations, among rocks: both sandstone and limestone have been recorded.

Hiern misdetermined specimens of *J. integerrima* and *J. foliolosa* as *Sutera argentea*, and subsequently the name has scarcely ever been correctly applied; much material will be found in herbaria under *Sutera pedunculata*. The habit of *Jamesbrittenia argentea* is similar to that of *J. kraussiana*, but they are easily distinguished by their leaves, elliptic in *J. kraussiana* (not rhomboid, cuneate nor oblong), glandular-puberulous above (not glabrous or with a very few glistening glands) and glandular-pubescent calyx (not glistening glands only). The type specimen of *Lyperia kraussiana* var. *latifolia* Benth. is possibly a hybrid with *J. argentea* as one parent (several other species occur in the area including *J. albanensis* and *J. aspalathoides*). It differs from *J. argentea* in its glandular-pubescent leaves and calyx, from *J. kraussiana* (which is only known from much further east) by its hairier leaves. Bentham originally placed the plant as an unnamed variety (var. δ) of *Lyperia argentea*. It is not matched by any subsequent collection.

27. Jamesbrittenia integerrima (Benth.) Hilliard in Edinb. J. Bot. 49: 230 (1992).

Lectotype: Cape, [Griqualand West 2623AC] Chue Spring [Heuningvlei], 18 x 1812, *Burchell* 2393 (G-DC, isolecto K).

Syn.: *Lyperia integerrima* Benth. in DC., Prodr. 10: 359 (1846).
   Manulea linifolia var. *integerrima* (Benth.) O. Kuntze, Rev. Gen. Pl. 3(2): 235 (1898).
   Sutera integerrima (Benth.) Hiern in Fl. Cap. 4(2): 284 (1904) p.p.
   S. asbestina Hiern in Fl. Cap. 4(2): 285 (1904). Lectotype (chosen here): Cape [Griqualand West 2922 BB] Asbestos Mountains near the Kloof Village, 25 ix 1811, *Burchell* 1665 (K).
   S. batlapina Hiern in Fl. Cap. 4(2): 265 (1904). Type: Cape, Bechuanaland, Batlapin country, *Nelson* 37 (holo. K, iso. PRE).
   Lyperia squarrosa Pilger in Bot. Jahrb. 48: 442 (1912). Lectotype (chosen here) Namibia, Aus, 1400m, 11 i 1910, *Dinter* 1097 (SAM).
   Sutera squarrosa (Pilger) Range in Feddes Repert. 38: 265 (1935); Merxm. & Roessler in Prodr. Fl. S.W. Afr. 126: 54 (1967).
   S. ausana [Dinter ex] Range, nomen, in Feddes Repert 38: 264 (1935). Specimen cited: Aus, *Dinter* 6088 (B, BOL, E, K, M, S, SAM, STE).

Dwarf shrublet or perennial herb, stems 100–600mm long, up to 8–15mm in diam. at base of plant, erect, ascending or spreading, fastigiate or more laxly branched, glandular-puberulous, hairs up to 0.1–0.2mm long, sometimes less than 0.1mm, glistening glands as well, these most conspicuous when hairs less than 0.1mm, leafy mainly on the lower parts. *Leaves* opposite becoming alternate upwards, axillary leaf tufts often small or wanting, largest primary leaves 7–25 × 2.5–9mm, elliptic to oblanceolate, apex obtuse to subacute (the broader the leaf the more obtuse the tip), base tapering to form a petiolar part accounting for up to c.⅓ total leaf length, margins entire or with up to 4 pairs of small teeth, lowermost pair occasionally reaching c.halfway to midrib, both surfaces glandular-puberulous, hairs up to 0.1–0.2mm long, sometimes less than 0.1mm, glistening glands as well. *Flowers* solitary in the axils of the upper leaves, these rapidly becoming smaller and passing into bracts, forming long racemes or panicles. *Pedicels* up to 7–26mm long. *Calyx* tube 0.2–0.3mm long, lobes 2.1–5 × 0.5–1mm, lanceolate-oblong to subspathulate, glandular-puberulous, hairs up to 0.1–0.2mm long, glistening glands as well. *Corolla* tube (7–)9.5–12(–17) × 1.8–2.2mm in throat, cylindric, abruptly expanded near apex, limb oblique, 8–18mm across lateral lobes, posticous lobes 3.5–8 ×

2.7–4.5mm, anticous lobe 3–8.5 × 2.4–5mm, all lobes cuneate-oblong, ± truncate to retuse, glandular-puberulous outside, hairs up to 0.15–0.2mm long, glistening glands as well, transverse band of clavate hairs in throat extending briefly onto lower lip, lobes usually white, very rarely cream or yellow, throat yellow to orange, the colour sometimes radiating onto base of lobes, these sometimes also marked with a light blueish streak. *Stamens*: posticous filaments c.1–2mm long, anthers 0.8–1.3mm; anticous filaments 0.5–1mm long, anthers 0.5–0.8mm, all filaments glandular-puberulous. *Stigma* 0.2–0.3mm long. *Style* (5–)7–10(–14)mm long. *Ovary* c.2–3 × 0.8–1.2mm. *Capsules* 4–9 × 3–3.4mm, glistening glands all over, often minute glandular hairs as well. *Seeds* c.0.6–0.7 × 0.3–0.5mm.

Selected citations:

Namibia. Windhoek distr., 2217CA, 10 miles S of Windhoek, iv 1958, *de Winter* 6054 (WIND). Lüderitz-Süd distr., 2616CB, farm Augustfelde: LUS 42, 7 ix 1973, *Giess* 12814 (PRE, WIND); Aus, 1400m, 4 vi 1922, *Dinter* 3608 (B, BOL, K, NBG, SAM, STE, Z).
Cape. Namaqualand, Richtersveld, 2817CA, Cornellsberg gorge, 17 viii 1986, *Williamson* 3565 (BOL, NBG). Kenhardt div., 2919 AB, 2 miles N of Pofadder, 24 viii 1954, *Barker* 8356 (NBG, STE). Prieska, 2922DA, iii 1935, *Bryant* 52 (K), 12 ii 1920, *Bryant* J20 (PRE), 1933, *Bryant* sub STE 18264 (STE). Hay div., 2822BD, Floradale, iv 1940, *Esterhuysen* 2311 (BOL, NBG). Barkly West div., 2823BA, Daniel's Kuil, 21 iii 1939, *Esterhuysen* 993 (BOL, NBG, PRE). Vryburg div., 2624DC, Vryburg, Tiger Kloof, 4 iv 1945, *Brueckner* 401 (PRE, SRGH).

When Bentham described *Lyperia integerrima* he cited two specimens, *Burchell* 1665 and 2393; the former has broad leaves, the latter narrow, and Bentham allowed for this in his description: '... *foliis ... obovatis oblongisve* ...' Hiern chose to re-describe the broad-leaved plant as a new species, *Sutera asbestina*, and quoted two specimens, *Burchell* 1665 and 2056. I take the specimen that Hiern left under *Lyperia integerrima*, namely *Burchell* 2393, as the lectotype of that name. The three specimens collected by Burchell represent part of the variation range of but one species. Burchell recorded the corolla of his 2393 as *aurantiaca*. This is very unusual in *J. integerrima* where the limb is normally white, the throat yellow to orange or orange-brown. However, two other collectors have noted 'corolla cream with orange-brown throat' (*Esterhuysen* 2311, cited above) and 'flowers yellow and orange' (*Wilman* 2319 (BOL) from Dunmurray in Hay division).

Jamesbrittenia integerrima is widely distributed from the southern half of Namibia and the Richtersveld in northern Namaqualand, east through the dry northern Cape as far as Vryburg, Taungs and Hopetown. The plants grow among rocks and boulders on hill- and mountain-slopes, between c.760 and 1675m above sea level; flowering has been recorded in all months. They are not confined to any particular rock formation: banded jasper, dolerite, granite, ironstone, limestone, mica schist, quartzite and sandstone have all been recorded.

The leaf blade is narrowly to broadly elliptic, sometimes oblanceolate, and entire to toothed, but it is always pubescent on both surfaces with a mixture of glistening glands and tiny stalked glandular hairs. Another characteristic feature of the species is the dense covering of glistening glands on the capsules (and ovaries); tiny glandular hairs may also be present, but they are dominated by the glistening glands. In Namibia, these glistening glands are smaller than in specimens from the Cape (including the Richtersveld) and very shortly stalked; also the leaves are always toothed. These are perhaps the marks of a distinct race, but scarcely of a distinct species, and *Sutera squarrosa* is here reduced to synonymy. Three specimens from 'Namaqualand' (*Meyer* sub STE 9146) and from Koeboes in the Richtersveld (*Herre* sub STE 122743, *Verdoorn & Dyer* 1828, K), are close to *J. integerrima* but have hairs up to c.0.2mm long on stems, leaves, pedicels and calyces.

The overall hairiness of the leaves will at once distinguish *J. integerrima* from *J. albiflora*, in which the leaf blade is either glabrous or has a few scattered glistening glands above, thickly clad in glistening glands below with only a few tiny stalked hairs confined to the midrib. *Jamesbrittenia integerrima* has been confused with *J. argentea* ever since Hiern misdetermined a specimen (*Mrs Orpen* sub Bolus 6483) as that species. Mrs Orpen's plant is in point of fact

J. tysonii, which sufficiently resembles *J. integerrima* to have perpetuated the error. *Jamesbrittenia integerrima* and *J. argentea* can be distinguished at a glance: the inflorescence mostly bracteate in *J. integerrima*, decidedly leafy in *J. argentea*, which is confined to the southern Cape coastal districts.

28. Jamesbrittenia albiflora (Verdoorn) Hilliard in Edinb. J. Bot.49: 226 (1992).

Type: Orange Free State, Fauresmith distr., Veld Reserve, *Verdoorn* 1053 (holo. PRE).

Syn.: *Sutera albiflora* Verdoorn in Fl. Pl. S. Afr. pl. 485 (1933).
    Manulea linifolia Thunb. var. β *heterophylla* O. Kuntze, Rev. Gen. Pl. 3(2): 235 (1898). Type: Orange Free State, Bloemfontein, 1560m, v 1894, *Kuntze* s.n. (holo. NY, iso. K).
    Sutera linifolia (Thunb.) O. Kuntze var. *heterophylla* (O. Kuntze) Hiern in Fl. Cap. 4(2): 272 (1904).

Dwarf twiggy shrublet, caudex up to 15mm diam., branches c.100–450mm long, often heavily browsed and stunted, up to 6–8mm in diam., erect, ascending or spreading, minutely glandular-puberulous, hairs less than 0.1mm long, leafy. *Leaves* opposite or sometimes becoming alternate on the inflorescence axes, pseudofasciculate, largest primary leaves 4–16(–20) × 1–6(–10)mm, spathulate or oblanceolate, apex obtuse, base cuneate and tapering into a petiolar part, blade entire or with 1–4 pairs of teeth in upper part, these reaching up to halfway, occasionally ⅔ of the way or more, to the midrib, broad, obtuse, entire, lower surface thickly clad in glistening glands, a few minute glandular hairs on the midrib below, upper surface glabrous or with a few scattered glistening glands. *Flowers* solitary in the upper leaf-axils forming many-flowered terminal racemes, sometimes panicled. *Pedicels* up to 6–16mm long. *Calyx* tube 0.1–0.2mm long, lobes 2.2–2.6 × 0.5–0.6mm, ± spathulate, thickly covered in glistening glands, a few minute glandular hairs as well. *Corolla* tube 7–9 × 1.3–1.5mm in throat, cylindric, abruptly expanded near apex, limb oblique, c.7–10mm across lateral lobes, posticous lobes 3–5 × 1.7–3mm, anticous lobe 3–5 × 1.5–2.6mm, all lobes oblong, ± truncate or retuse, glandular-puberulous outside, hairs up to 0.1–0.2mm long, glistening glands as well on swollen part of tube and backs of lobes, transverse band of clavate hairs in throat extending briefly onto lower lip, lobes white, each marked at the base with a dark brown or crimson median bar. *Stamens*: posticous filaments 1–1.3mm long, anthers 0.7–1.2mm, anticous filaments 0.4–0.7mm long, anthers 0.3–0.8mm long, all filaments minutely glandular at apex. *Stigma* 0.15–0.2mm long. *Style* 4.5–6.4mm long. *Ovary* 1.5–1.8 × 0.8–1mm. *Capsules* 4–5 × 2.5mm, glistening glands all over. *Seeds* 0.4–0.5 × 0.25–0.3mm.

Selected citations:

Transvaal. Bloemhof distr., 2725CC, Christiana, Kameelpan, 9 i 1934, *Theron* 504 (PRE).
Cape. Kimberley div., 2824DB, Mauretzfontein Koppie, 16 iii 1939, *Esterhuysen* 989 (BOL, NBG, PRE); zwischen Kimberley und Vaalriver, iv 1886, *Schenk* 765 (Z). Colesberg, 3025CA, 11 iv 1963, *Bayliss* 1211 (Z).
Orange Free State. Bloemfontein, 2926AA, Naval Hill, 16 vii 1943, *Wasserfall* 224 (NBG). 3025BA, 12 miles beyond Trompsburg on way to Philippolis, near Blaauheuvel, c.5000ft, 25 iii 1969, *Werger* 223 (K, SRGH).

Jamesbrittenia albiflora appears to be nearly confined to the southern half of the Orange Free State, extending slightly northwards to Christiana and Makwassie in the extreme SW Transvaal and westwards to the environs of Kimberley in the Cape and southwards just across the Orange River to Colesberg. It grows between c.1370 and 1525m above sea level in stony ground, may flower in any month, and provides good browse. The salient features of the species are the very minute glandular hairs on the stems, inflorescence axes and pedicels, the ± spathulate leaves, either entire or bluntly few-toothed, the lower surface thickly covered in glistening glands, the upper glabrous or with a few scattered glands, and the white corolla limb.

The possible involvement of *J. albiflora* in hybridization is mentioned under *J. tysonii* (no. 19).

29. Jamesbrittenia tenuifolia (Bernh.) Hilliard in Edinb. J. Bot. 49: 232 (1992).
Type: Cape, George div., [3322DC] sandy places near Zwarte Valley, ii 1839, *Krauss* 1611 (holo. n.v.; iso. M, W, Z).

Syn.: *Lyperia tenuifolia* Bernh. apud Krauss in Flora 27: 835 (1844); Hiern in Fl. Cap. 4(2): 310 (1904) as an imperfectly
known species.
  Sutera tenuifolia (Bernh.) Fourcade, Mem. Bot. Surv. S. Afr. 20: 74 (1940).
  S. atrocaerulea Fourcade in Trans. Roy. Soc. S. Afr. 21: 100 (1932). Type: Cape, Humansdorp div., hills 5 miles
N of Humansdorp, 1500ft, vii 1922, *Fourcade* 2224 (holo. BOL, iso. NBG, STE).
  S. atrocaerulea var. *latifolia* Fourcade in Trans. Roy. Soc. S. Afr. 21: 100 (1932). Type: Cape, Humansdorp div.,
Groot Hoek, 600ft, ix 1911, *Fourcade* 739 (holo. BOL, iso. Z).

Dwarf shrub up to 200–600mm tall, stems spreading, decumbent or erect, well branched, branches mostly ascending at an angle of c.45°, glandular-pubescent, hairs up to 0.1–0.4mm long, a few glistening glands as well, glands sometimes oozing so that stems (also leaves and calyx) look varnished, closely leafy. *Leaves* alternate, pseudofasciculate, largest primary leaves 1.5–10(–25) × 0.4–3(–11)mm, elliptic to spathulate, often conduplicate, obtuse or subacute, base narrowed, becoming petiolar only in exceptionally large leaves, usually with a pair of prominent teeth near the apex, occasionally 2 pairs, exceptionally 3 pairs, secondary leaves in particular sometimes entire, upper surface glabrous or with a few glistening glands, thickly clad in glistening glands below, minute big-headed glandular hairs as well (glands not easy to see when oozing). *Flowers* solitary in the axils of the leaf tufts with a strong tendency to run up into leafy racemes. *Pedicels* up to 5–15mm long, glandular-puberulous, hairs less than 0.1–0.15mm. *Calyx* tube 0.2–0.3mm long, lobes 2.2–2.8(–3.4) × 0.5–0.7mm, oblong-lanceolate, ± acute, thickly clad in glistening glands and minute big-headed glandular hairs. *Corolla* tube 8–10(–13.5) × 1.5–2.2mm in throat, cylindric, abruptly expanded near apex, limb oblique, c.8–18mm across lateral lobes, posticous lobes 4–9 × 3–7mm, anticous lobe 3.8–7.5 × 2.6–6.8mm, all lobes cuneate, ± retuse, glandular-puberulous outside, hairs up to 0.15–0.2mm long, glistening glands as well mainly on backs of lobes, broad transverse band of clavate hairs in throat extending briefly onto base of lower lip, lobes purple, deep purple-blue or deep blue (exceptionally white, mauve or pink; see below). *Stamens*: posticous filaments 0.7–1.5mm long, anthers 0.7–1.2mm; anticous filaments 0.3–0.6mm long, anthers 0.4–0.6mm, all filaments glandular-puberulous. *Stigma* c.0.2mm long. *Style* 5.6–6.8(–8.5)mm. *Ovary* 1.7–2 × 0.8–1.2mm. *Capsules* c.3–6.5 × 2–3mm, glistening glands all over. *Seeds* c.0.8–1 × 0.4–0.6mm.

Selected citations:

Cape. Mossel Bay div., 3422AA, Great Brak river, ii 1930, *Ryder* 50 (BOL). George div., 3422AB, Mount Pleasant, 27
xii 1949, *Martin* 131 (NBG). Knysna div., 3422BB, Sedgefield, 29 x 1961, *Middlemost* 2191 (NBG, STE); 3423AB,
Ganse Vlei, 300ft, x 1921, *Keet* 835 (STE). Uniondale div., 3323CC, Prince Alfred's Pass, 1000ft, xi 1927, *Fourcade*
3488 (BOL, STE). Humansdorp div., [3324CB?] Kouga hills, 12 xi 1941, *Esterhuysen* 6696 (BOL, PRE); 3324CD, Zuur
Anys, along road to Kouga, 1500ft, ix 1925, *Fourcade* 3037 (STE).

Lyperia tenuifolia was unknown to Hiern and Fourcade re-described the species as *Sutera atrocaerulea* (Fourcade's variety is untenable: some leaves on the type of his new name are as broad as those mentioned on the variety). The type of *J. tenuifolia* came from Zwarte Valley (now Swartvlei) between George and Knysna, and the plant seems still to be common in that area: records stretch along the coast from Mossel Bay to Plettenberg Bay and inland to the Long Kloof between Avontuur and Assegaibosch, thence north up the valley of the Gamtoos river, and so into Baviaanskloof lying parallel to and north of the Long Kloof, reaching up to c.800m above sea level. Fourcade's original material was found about 5 miles north of Humansdorp on the road to Patensie in the Gamtoos valley. Scant ecological information is available; the plants seem to grow on sandy or rocky scrub-covered hills and stabilized dunes near the coast, probably on hill and mountain slopes inland. Flowering has been recorded in most months.

The characteristic features of *J. tenuifolia* are its leaves, elliptic or spathulate in outline, at least the primary leaves mostly with a pair of large teeth near the apex (occasionally two or

149

even three pairs), its shortish pedicels, mostly up to 5–10mm long but sometimes reaching 15mm, a tendency for the flowers to run up into leafy racemes, and the purple-blue corolla limb. It is often confused with *J. argentea* (as *Sutera pedunculata*), which is distinguished by its flowers scattered all along the branches (thus giving the plant a different aspect), the pedicels often longer than those of *J. tenuifolia*, and the corolla limb is white. Confusion with *J. aspalathoides* is also common, but in *J. aspalathoides* the leaves are often smaller (rarely more than 3mm long), oblong or narrowly spathulate and mostly entire; the flowers tend to be scattered along the branches.

I do not understand the complexities of plants collected in the upper Gamtoos valley, roughly between Loerie Forest Station, Otterford, Patensie and Cambria, thence into Baviaanskloof. The corolla limb of typical *J. tenuifolia* is always rich shades of blue or purple, but the puzzling plants have flowers variously described as mauve, pale mauve, pale bluish-mauve, light lavender, pink or white, and the corolla tube tends to be longer than in typical *J. tenuifolia* (10–13.5mm, the measurement given in parentheses in the formal description). The greatest variation in colour has been recorded around Cambria (but many of the problematical sheets give no information whatsoever) and here too some of the specimens bear relatively very large and coarsely toothed or lobed leaves (measurements in parentheses in description). I would accept these large leaves as part of the normal variation range were they not associated with an exceptionally long corolla tube and pale limb. Obviously a thorough field investigation is needed: one collector recorded 'common shrubs'. The relevant herbarium material comprises *Acocks* 13696 (PRE), *Barker* 7856, 7860, 7889 (NBG), *Batten* 1146, 1147, 1159, 1160, 1162 (E), *Bayliss* 6013 (Z), *Bond* 917 (NBG), *Compton* 24086 (NBG, STE), *Dix* 214 (BOL), *Ferreira* 9 (STE), *Fourcade* 3659 (STE), 5146, 5193 (BOL), *Hugo* 1421 (STE), *Lewis* 3891, 3893 (SAM), *Long* 1024 (PRE). Whether or not the typical plant occurs with plants bearing pale flowers is impossible to judge because so many collectors give no information.

It is possible that *J. tenuifolia* hybridizes with other species. *Bond* 912 (NBG) from Kouga near Joubertina may represent *J. tenuifolia × aspalathoides*: the leaves are mostly entire or with one pair of teeth, but they are longer than is normal in *J. aspalathoides* (up to c.6mm), the pedicels are long (up to 16–20mm) and the flowers tend to run up into racemes.

30. Jamesbrittenia foliolosa (Benth.) Hilliard in Edinb. J. Bot. 49: 229 (1992). **Fig. 48A, Plate 3C.**
Lectotype (chosen here): Cape, Uitenhage distr., near the Zwartkops River, *Ecklon* [& *Zeyher*] (K, left-hand piece marked 'b' by Bentham; isolecto. E, FI, LE, M, NY, P, S, SAM, W, Z. Some of the duplicates bear the number 3513 or 3513b or the coded locality 2.1, but not all specimens so marked are duplicates of the specimen Bentham saw).

Syn.: *Lyperia foliolosa* Benth. in Hook., Comp. Bot. Mag. 1: 380 (1836) & in DC., Prodr. 10: 361 (1846).
    L. pinnatifida Benth. var. *subcanescens* Benth. in Hook., Comp. Bot. Mag. 1: 380 (1836). Lectotype (chosen here):
        Uitenhage div., Enon, 11 xi 1829, *Drège* (K; isolecto. E, LE, P, S, W).
    Manulea foliolosa (Benth.) O. Kuntze, Rev. Gen. Pl. 3(2): 235 (1898).
    Sutera foliolosa (Benth.) Hiern in Fl. Cap. 4(2): 298 (1904), p.p., excluding *Pappe*, Zwart Valley, and *Bolus* 2419,
        Zwart Bergen (*Bolus* 1519 not seen).

Well-branched dwarf shrublet up to 150–400mm tall, branches erect, decumbent or ascending, young parts thickly clad in glistening glands, glandular hairs up to c.0.1–0.2mm long as well on uppermost parts, sometimes more conspicuous than the glistening glands, very leafy except on the floriferous parts. *Leaves* opposite becoming alternate on the inflorescence branches, pseudofasciculate, largest primary leaves c.3–11 × 1–7mm, spathulate in outline, thick-textured and sometimes conduplicate, apex obtuse to subacute, base narrowing into a broad petiolar part, upper ⅓–⅔ with up to 4 pairs of lobes mostly cut ½–¾ of the way to the midrib, sometimes the lowermost one or two pairs nearly or quite to the midrib, oblong or narrowly spathulate, often entire, occasionally the largest with 1–3 small teeth, lower surface thickly clad in glistening glands, a few very minute glandular hairs as well mainly on petiolar part, upper surface glabrous or with a few minute glandular hairs. *Flowers* borne singly in the axils of the uppermost leaves, which decrease rapidly in size upwards. *Pedicels* up to 12–24mm long, glandular-pubescent, hairs up to 0.1–0.25mm long, very few glistening glands. *Calyx* tube 0.2–0.3mm long, lobes 2.2–3.2 × 0.6–0.7mm, oblong, obtuse to subacute, glandular-puberulous, hairs up to 0.1–0.2mm long, glistening glands as well. *Corolla* tube 7–9 × 1.8–2mm in throat, cylindric, abruptly expanded near apex, limb oblique, c.7–13mm across lateral lobes, posticous lobes 3–6 × 2–4mm, anticous lobe 2.8–7 × 1.8–4.2mm, all lobes oblong to cuneate, apex more or less truncate, retuse, glandular-puberulous outside, hairs up to 0.1–0.2mm long, glistening glands as well particularly on backs of lobes, transverse band of clavate hairs in throat extending briefly onto lower lip, lobes often purple to violet blue, sometimes white, throat yellow/orange. *Stamens*: posticous filaments 0.8–1.3mm long, anthers 0.7–1.1mm, anticous filaments 0.4–0.8mm, anthers 0.4–0.7mm, all filaments minutely puberulous. *Stigma* 0.2mm long. *Style* 4–6.3mm. *Ovary* 2 × 0.7–1mm. *Capsules* 5–6.5 × 2.5–3mm, often streaked with purple, glistening glands ± confined to sutures, minutely glandular-puberulous or glabrous elsewhere. *Seeds* c.0.5–0.7 × 0.4–0.5mm.

Selected citations:

Cape. Uniondale div., 3322DB, Mannetjieberg, 5 xi 1941, *Esterhuysen* 6424 (BOL). 3323CA, hills NE of Avontuur, 3000ft, xii 1930, *Fourcade* 4542 (STE). 3323DD, N of Joubertina, 2000ft, i 1924, *Fourcade* 2938 (BOL). Humansdorp div., 3424BB, 4.1 miles from Humansdorp to Cape St. Francis, 25 ix 1969, *Marsh* 1089 (STE). Uitenhage div., 3325CD, near Uitenhage, 22 iv 1893, *Schlechter* 2589 (E, MO, S, SAM, Z). Port Elizabeth div., 3325DC, Redhouse, iv 1913, *Paterson* 13 (BOL). Alexandria div., 3326CB, Debega, 16 v 1931, *Galpin* sub BOL 51690 (BOL). Albany div., 3326AC, W of Grahamstown, 12 iv 1952, *Barker* 7829 (NBG). Somerset East div., 3325BB, Komadagga, 14 i 1934, *Adamson* 468 (BOL). Graaff Reinet div., 3224BD, Graaff Reinet - Somerset East road near Melk River, 30 xi 1977, *Hilliard & Burtt* 10779 (E, K, PRE, S).

Specimens of *J. foliolosa* are commonly misdetermined as *Sutera pinnatifida*, a legacy of the confusion generated by Hiern in Flora Capensis. The plant is a dwarf shrublet, the many branches densely leafy, but the terminal floriferous parts nearly nude because of the sudden gross reduction in leaf-size. The leaves are borne in closely set pseudofascicles. They are small (up to c.11mm long), spathulate in outline, nearly always conduplicate, the margins in the upper part deeply cut into 1–4 pairs of ± oblong lobes; only the largest lobes are occasionally toothed; the smallest leaves may be entire. The backs of the leaves are thickly beset with large glistening glands; these are absent from the upper surface, which may or may not produce scattered minute glandular hairs. Similar minute hairs are often present on the petiolar part of the leaves, but they are usually very inconspicuous. The corolla limb is usually shades of purple, blue or mauve, rarely white; flowers can be found in any month. The capsules have glistening glands confined or very nearly confined to the sutures, with tiny glandular hairs elsewhere, though these may disappear with age.

The species is found in the south eastern Cape, from Cloete's Pass at the eastern end of the Langeberg, Mannetjiesberg at the eastern end of the Kammanassie Mountains through the Long Kloof and its flanking mountains to Uitenhage, Port Elizabeth and Alexandria; hereabouts it may come down almost to sea level, but on the mountains it may reach 900m. From Alexandria, the species ranges north and west through Albany division to the environs of Adelaide, Cradock, and the Karoo near Graaff Reinet. Typical *J. foliolosa* occurs around Grahamstown, but some specimens from that area appear to be of hybrid origin. For example, *Blackbeard* sub NBG 1151/2 is more glandular-hairy than is normal for *J. foliolosa* but a sheet under the same number in SAM has the more usual indumentum; similarly, *Bayliss* 8250 (Z) is more glandular-pubescent than normal, whereas *Hilliard & Burtt* 10837 (E, K, PRE, S) is normal; both collections came from Coombs Valley, east of Grahamstown, on the same day. These abnormally glandular plants could have been introgressed by *J. albanensis*.

31. Jamesbrittenia zuurbergensis Hilliard in Edinb. J. Bot. 49: 233 (1992).
Type: Cape, Uitenhage div., 3325AD, Rietberg, the mountain above Kirkwood, Zuurberg range, fynbos on summit, 10 vi 1962, *Nordenstam* 298 (holo. E, iso. S).

Well-branched shrub up to c.500mm tall, young parts of branches thickly clad in glistening glands, minute glandular hairs as well towards tips of twigs, leafy. *Leaves* opposite, alternate upwards, pseudofasciculate, largest primary leaves c.7–20 × 2–7mm, elliptic in outline, apex acute, base tapering, distinctly petiolar in larger leaves, sharply toothed or lobed (usually 4 pairs, sometimes more) in upper half ± ½ way or less to midrib, lower surface thickly clad in glistening glands, upper surface similar or glands few or wanting. *Flowers* borne singly in the axils of the primary leaves towards the tips of all the twiglets. *Pedicels* up to 6–11mm long, clad in glistening glands and glandular hairs less than ± 0.1mm. *Calyx* tube c.0.15mm long, lobes 2.6–3.4 × 0.6–0.7mm, oblong, acute, glistening glands all over, glandular hairs as well less than ± 0.1mm. *Corolla* tube c.7–8 × 1.8–2mm in throat, cylindric, abruptly expanded near apex, limb c.7–8mm across lateral lobes, posticous lobes 3–4 × 1.8–2mm, anticous lobe 3–4 × 1.8–2mm, all lobes ± oblong, tips rounded, glandular-puberulous outside, hairs up to 0.1–0.15mm long, glistening glands as well, transverse band of clavate hairs in throat, lobes pale yellow to cream. *Stamens*: posticous filaments c.1mm long, anthers 0.7–1mm, anticous filaments c.0.5mm long, anthers 0.4–0.5mm, all filaments minutely puberulous. *Stigma* 0.2mm long. *Style* 4.5–5mm. *Ovary* 2 × 1mm. *Capsules* c.5–5.5 × 2.5mm, thickly clad in glistening glands all over. *Seeds* c.1 × 0.4mm.

Citations:

Cape. Uitenhage div., Zuurbergen, nordlichster Bergabhang, 2500–3000ft, 30 x 1829, *Drège* distributed as *Lyperia argentea* Benth. d (E, LE, P, S, W); Zuurberg Forest Reserve, 2300ft, 28 viii 1932, *Long* 719 (PRE); Zuurberg Inn, 2000ft, 24 ix 1953, *Johnson* 769 (PRE).

Jamesbrittenia zuurbergensis is possibly confined to the Zuurberg, where it forms part of the scrub on the range. Drège first collected it there in 1829, in October, Long in August 1932, Johnson in September 1953, then Nordenstam found it in 1962, in June; in those months the plants were in flower and fruit. I have seen no other collections.

It is distinguished from *J. foliolosa*, which it somewhat resembles in its indumentum and leaves, by its leafy inflorescences, mostly shorter pedicels, yellow corolla limb, and ovary and capsules thickly covered in glistening glands.

32. Jamesbrittenia microphylla (L.f.) Hilliard in Edinb. J. Bot. 49: 231 (1992).
Type: Cape of Good Hope, *Sparrman* 328 (lecto. LINN 787.8, iso. centre bottom specimen sheet. 14375 herb. Thunb., UPS). See note below.

Syn.: *Manulea microphylla* L.f., Suppl. 285 (1782); Thunb., Prodr. 100 (1800) & Fl. Cap.ed. Schultes 466 (1823).
    Lyperia microphylla (L.f.) Benth. in Hook., Comp.Bot. Mag. 1: 381 (1836) & in DC., Prodr. 10: 362 (1846).
    Chaenostoma microphyllum (L.f.) Wettst. in Engl. & Prantl, Pflanzenfam. 4 3B: 68, fig. 31 G (1891); Diels in Bot. Jahrb. 23: 489 (1897).
    Sutera microphylla (L.f.) Hiern in Fl. Cap.4(2): 309 (1904); E. P.Phillips in Fl. Pl. S. Afr. 22: t. 873 (1942); Batten & Bokelmann, Wild Flow. E Cape Prov. 130 pl. 103, 5 (1966).
    S. densifolia Hiern in Fl. Cap.4(2): 309 (1904). Type: Cape, Bathurst div., near Port Alfred, 21 xii 1898, *Galpin* 2932 (holo. K, iso. PRE).

Stout dwarf twiggy shrublet up to 150–500mm tall, branches crowded with closely leafy brachyblasts, young parts and pedicels glandular-pubescent, hairs up to 0.2–0.25mm long, scattered glistening glands as well. *Leaves* opposite and decussate, very close-set, mostly up to 1–1.5 × 0.8–1mm, ovate or cuneate or sometimes the lowermost on each brachyblast up to 1.8–3 × 0.5–0.8mm, oblong, all thick-textured, apex acute, base broad, margins entire, revolute, upper surface glabrous, lower with big glistening-glands in centre of leaf, margins glabrous. *Flowers* solitary in the axils of the primary leaves on all the branchlets. *Pedicels* c.6–13mm long. *Calyx* tube 0.2–0.3mm long, lobes 2–3.3 × 0.5–0.8mm, ± oblong, acute, tips often recurved, glandu-

lar-puberulous, hairs up to 0.1–0.2mm long, glistening glands as well. *Corolla* tube 7–10 ×
1.8–2mm in throat, cylindric, abruptly expanded near apex, limb c.7.5–18mm across lateral
lobes, posticous lobes 3–8 × 2–5.5mm, anticous lobe 3–8 × 2–5mm, cuneate-oblong, often
retuse, glandular-puberulous outside, hairs up to 0.1–0.2mm long, glistening glands as well
mainly on backs of lobes and swollen part of tube, broad transverse band of clavate hairs in
throat, lobes bright purple-blue or deep mauve, yellow in throat, red median streak at base of
each lobe. *Stamens*: posticous filaments 1–1.4mm long, anthers 0.7–1.1mm, anticous filaments
0.6–1mm long, anthers 0.4–0.8mm, all filaments minutely puberulous. *Stigma* 0.2–0.3mm long.
Style 4–7mm long. *Ovary* 1.3–1.8 × 0.75–1mm. *Capsules* 4.5–7 × 2.5–3mm, scattered glistening glands, sometimes ± confined to sutures, minute glandular hairs as well near apex. *Seeds*
c.0.25–0.5 × 0.2–0.3mm.

Selected citations:

Cape. Knysna div., 3423AB, Robbeberg Nature Reserve, c.250ft, 30 v 1970, *Taylor* 7716 (STE). Humansdorp div.,
3424BB, Jeffreys Bay, 31 xii 1939, *Barker* 605 (NBG). Port Elizabeth div., 3325DC, Coega river, 200ft, i 1872, *Bolus*
1891 (BOL, SAM); Uitenhage to Grahamstown, 28 xi 1894, *Penther* 1780 (W).

When Linnaeus the younger described *Manulea microphylla* he cited Thunberg as the collector
of the specimen he saw; however, the specimen in the Linnean herbarium (LINN 787.8) was
collected by Sparrman (Sp.328 inscribed on the sheet probably by Linnaeus himself). There is
a specimen in Thunberg's herbarium (in UPS) that may be a duplicate of it (there are also 3
sheets attributed to Thunberg in S). The pragmatic course seems to dictate the acceptance of
LINN 787.8 as lectotype and the specimen in UPS as an isotype; not all the specimens on the
sheet (no. 14375) are *J. microphylla*, as already noted by Hiern. Smith (as noted in Savage's
catalogue of the Linnean herbarium) gave a reference to Plukenet (Pluk., Phytogeog. t. 272 fig.
7); the figure, though crude, is unmistakably *J. microphylla*.

The distinctive habit of *J. microphylla* (a twiggy shrublet, the branches closely set with very
short brachyblasts clad in tiny, neatly 4-ranked, superimposed leaves) is shared with *J.
merxmuelleri*, from the coastal areas of southern Namimbia and northern Namaqualand.
Jamesbrittenia microphylla is confined to the coastal areas of the southern Cape, from the
environs of Knysna in the west to Port Alfred in the east, being particularly common around
Port Elizabeth. It grows in coastal scrub or grassland, from sea level to c.120m, flowering in
all months. It is distinguished from *J. merxmuelleri* by its differently shaped leaves and
differently coloured flowers (see under *J. merxmuelleri*).

33. Jamesbrittenia aspalathoides (Benth.) Hilliard in Edinb. J. Bot. 49: 227 (1992).
Lectotype (chosen here): Cape, Uitenhage div., Grassrug [3325DA Grassridge], March, *Ecklon*
(K; isolecto. FI, L, E, M, P, S, W).

Syn.: *Lyperia aspalathoides* Benth. in Hook., Comp.Bot. Mag. 1: 381 (1836) & in DC., Prodr. 10: 362 (1846).
    Chaenostoma aspalathoides (Benth.) Diels in Bot. Jahrb. 23: 491 (1891).
    Sutera aspalathoides (Benth.) Hiern in Fl. Cap.4(2): 308 (1904) p.p.

Spreading dwarf shrublet up to c.200–500mm tall, branches prostrate, ascending or erect, very
twiggy, often with short stiff lateral branchlets, crowded with closely leafy brachyblasts,
minutely puberulous with a mixture of large glistening glands and minute big-headed hairs
mostly less than 0.1mm long, occasionally up to 0.15mm, rarely 0.2mm, the glands sometimes
oozing so that stems and leaves look varnished. *Leaves* opposite and decussate, crowded,
sometimes more distant and alternate upwards, primary leaves 1.5–3(–5.5) × 0.3–1(–2)mm
(ratio of length to breadth 2.5–7: 1), oblong to narrowly spathulate, mostly entire, occasionally
a primary leaf with a pair of small teeth near the apex, thick-textured, often conduplicate, apex
obtuse to subacute, upper surface glabrous or with a very few glistening glands or minute
stalked hairs especially when young, lower surface thickly beset with glistening glands
(sometimes oozing, then leaf looks varnished). *Flowers* solitary in the axils of the primary

leaves, irregularly scattered all along the branchlets. *Pedicels* up to (8–)10–22mm long, sparsely and very minutely puberulous, glistening glands as well, sometimes looking varnished. *Calyx* tube c.0.2mm long, lobes 1.7–2.8 × 0.6–0.8mm, ± oblong, subacute, glistening glands all over, minute glandular hairs on margins. *Corolla* tube 9.5–11 × 1.8–2mm in throat, cylindric, abruptly expanded near apex, limb oblique, 10–18mm across lateral lobes, posticous lobes 3.8–7 × 2.3–4mm, anticous lobe 4–7.5 × 2.5–5.3mm, all lobes cuneate-oblong, often retuse, glandular-pubescent outside, hairs up to 0.15–0.2mm long, glistening glands as well mainly on backs of lobes, broad band of clavate hairs in throat extending very briefly onto base of lower lip, lobes shades of pink, purple, mauve and blue, occasionally white, probably always yellow in throat, once recorded as having a red spot at base of each lobe (invisible in dried material). *Stamens*: posticous filaments 1.2–1.5mm long, anthers 0.8–1.2mm, anticous filaments 0.5–1mm, anthers 0.5–0.8mm, all filaments minutely puberulous. *Stigma* c.0.2mm long. *Style* 6–8mm. *Ovary* 1.7–2 × 0.8–1mm. *Capsules* 4–8 × 2.5–5mm, glistening glands all over. *Seeds* c.0.7–0.8 × 0.4mm.

Selected citations:

Cape. Bredasdorp div., 3420CA, Bontebok Park, 17 ii 1951, *Compton* 22620 (NBG). Riversdale div., 3421BD, 3km north of The Fisheries, c.500ft, 25 ix 1978, *Hugo* 1266 (STE). Prince Albert div., 3321AD, Seven Weeks Poort, 2500ft, 27 ix 1932, *Compton* 4019 (BOL, NBG). Ladismith div., 3321DA, Rooiberg, just west of Teeboskop, 3000ft, 9 xi 1974, *Oliver* 5327 (PRE, STE). Oudtshoorn div., 3321CD, Gamka Mountain Reserve, east of Tierkloof, 850m, 17 v 1982, *Cattell & Cattell* 23 (STE). George div., 3322DC, Ezeljachtpoort, c.1800ft, 19 v 1966, *Acocks* 23843 (PRE, STE). Willowmore div., 3323DB, Kouga Mountains, Vleikloof NE of Smutsberg, 3100ft, 19 ix 1973, *Oliver* 4623 (PRE, STE). Uniondale div., 3323DD, near Joubertina, 2000ft, 2 v 1935, *Compton* 5212 (BOL).

Bentham cited two specimens under his new name *Lyperia aspalathoides*: 'Grasrugg and Krakakamma in Uitenhage, *Ecklon*'. The specimen from Grasrug best fits Bentham's description of the leaves as being '*oblongis linearibusve*' and it is therefore chosen as lectotype. The specimen from Krakakamma in Bentham's own herbarium (now in K) proves to be *J. microphylla*, which has ovate or cuneate leaves and calyx with conspicuously stalked glands. Specimens distributed to various herbaria under the code 3.2 (Krakakamma, February) prove to be a mixture of *J. aspalathoides*, *J. microphylla* and *J. atropurpurea*.

Jamesbrittenia aspalathoides is widely distributed in the southern Cape from the environs of Koude River near Elim in Bredasdorp division east to the Baviaanskloof Mountains south of Steytlerville in the east; the lectotype came from Grassridge (Grasrug) about 30km south of Addo, and I have seen a somewhat hairier specimen from Addo National Park (*Archibald* 3730, PRE). The plants are a constituent of scrub, from near sea level to c.1070m, and flowers have been recorded in all months.

The distinguishing features of the species are its small narrowly oblong to spathulate leaves (ratio of length to breadth 2.5–7:1), mostly entire, some of the primary leaves occasionally with a pair of small teeth near the apex (rarely a plant will produce an occasional leaf as much as 7 × 3mm with two pairs of teeth), the normally very short indumentum on stems, leaves and pedicels (a mixture of glistening glands and minute gland-tipped hairs that often ooze and give these parts as well as the calyx a varnished look, the individual hairs then often invisible), often longish pedicels, the calyx clad predominantly in large glistening glands, and the corolla limb usually shades of pink, mauve or blue, possibly always yellow in the throat (collector's notes often meagre or wanting) and, once, a red spot was recorded at the base of each lobe; this is not visible in the dried specimen.

Jamesbrittenia aspalathoides is often confused with a variety of other microphyllous species; Philcox (in Fl. Zam. 8(2): 3, 1990) went so far as to reduce it to synonymy under *Sutera atropurpurea*, but *Jamesbrittenia atropurpurea* is at once distinguished by its flowers: corolla tube c.15–24mm long, either glabrous or very minutely puberulous, limb shades of brown, the margins of the lobes revolute giving the lobes a peculiarly narrow appearance, the texture thick, venation invisible on the upper surface.

34. Jamesbrittenia calciphila Hilliard in Edinb. J. Bot. 49: 227 (1992).
Type: Cape, Riversdale div., 3421BC, Canca se Laagte, south of Albertinia, 160m, 20 iii 1975, *Oliver* 5719 (holo. STE).

Dwarf shrublet, obten gnarled, prostrate to erect, branches up to 100–450mm long, very twiggy with many short stiff lateral branchlets, crowded with closely leafy brachyblasts, glandular-pubescent, hairs up to 0.15–0.2mm long, scattered glistening glands as well. *Leaves* opposite and decussate, sometimes alternate towards the tips of the branchlets, primary leaves 0.8–2.2(–3.2) × 0.6–1.2mm (ratio of length to breadth 1–2.5(–3): 1), subrotund, broadly elliptic or broadly spathulate, mostly entire, occasionally a few primary leaves with a pair of teeth near apex, thick-textured, upper surface with a few glistening glands and sometimes minute glandular hairs as well, often glabrescent, lower surface thickly clad in glistening glands. *Flowers* solitary in the axils of the primary leaves, irregularly scattered all along the branchlets. *Pedicels* up to 6–14mm long, glandular-pubescent, hairs up to 0.1–0.2mm long, a few glistening glands as well. *Calyx* tube 0.15–0.2mm long, lobes 2–3 × 0.5–0.8mm, ± oblong, obtuse to subacute, glistening glands all over, minute glandular hairs as well. *Corolla* tube 9–11 × 2mm in throat, cylindric, abruptly expanded near apex, limb oblique, c.10–14mm across lateral lobes, posticous lobes 4–7 × 2.2–4.5mm, anticous lobe 4–5.5 × 2.2–4mm, all lobes cuneate-oblong, often retuse, glandular pubescent outside, hairs up to 0.15–0.2mm long, glistening glands as well mainly on backs of lobes and swollen part of tube, broad transverse band of clavate hairs in throat extending very briefly onto base of lower lip, lobes shades of pink, mauve and blue, rarely white, throat probably always yellow. *Stamens*: posticous filaments 1–1.5mm long, anthers 0.8–1mm, anticous filaments 0.5–0.8mm long, anthers 0.5–0.8mm, all filaments minutely puberulous. *Stigma* c.0.2mm long. *Style* 6.5–8.3mm. *Ovary* 1.2–2 × 0.8–1mm. *Capsules* 3–4 × 2.5mm, glistening glands all over. *Seeds* not seen.

Selected citations:

Cape. Bredasdorp div., 3419DA, Hagelkraal, 26 xii 1946, *Compton* 19041 (NBG); 3420BC, Potteberg, 19 ix 1954, *Esterhuysen* 23303 (BOL). Riversdale div., 3421AD, Still Bay Hills, 9 viii 1949, *Morris* 261 (NBG); 3421BA, near Albertinia, 30 i 1951, *Compton* 22582 (NBG).

Jamesbrittenia calciphila is confined to limestone outcrops in Bredasdorp and Riversdale divisions, from roughly Hagelkraal in the west (just east of Pearly Beach) east to the environs of Albertinia. It grows on steep rocky slopes, ridges and cliffs, sometimes in rock crevices, possibly never more than c.200m above sea level, and often forms a dense gnarled shrublet lying flat against the rock or with the branches spreading to erect.

It is allied to *J. aspalathoides* from which it may be distinguished by the longer hairs on stems and pedicels (up to 0.2mm long, rarely more than 0.1mm in *J. aspalathoides*) and these never ooze to produce the varnished effect so common in *J. aspalathoides*. The leaves mostly differ in shape, being broader in relation to their length than in *J. aspalathoides* (ratio of leaf length to breadth mostly 1–2.5:1 versus 2.5–7:1).

Bohnen 5813 (STE) may be a hybrid between *J. calciphila* and *J. aspalathoides*; it came from Still Bay, Olive Grove Farm (3421AD).

35. Jamesbrittenia stellata Hilliard in Edinb. J. Bot. 49: 232 (1992).
Type: Cape, Riversdale div., 3421AD, Still Bay Hills, 9 viii 1949, *Morris* 257 (holo. NBG).

Syn.: *Sutera pedunculata* auct., non (Andr.) Hiern; Levyns in Adamson & Salter, Fl. Cape Penins. 715 (1950).

Dwarf shrublet c.200–400mm tall, branches erect, spreading or ascending, well branched, glandular-pubescent, hairs up to 0.1–0.25mm long, scattered glistening glands as well, all but the oldest stems closely leafy. *Leaves* pseudofasciculate, opposite, sometimes alternate upwards, primary leaves 2–7(–22) × 0.8–3(–12)mm (ratio of length to breadth 2–5: 1), spathulate, either entire or with a pair (rarely up to 4 pairs) of teeth in upper part, usually plane, either glabrous above or with a few minute glandular hairs, glabrescent, lower surface densely clad in glistening glands, these sometimes oozing to give a varnished appearance (sometimes seen on young stems and calyx as well). *Flowers* solitary in the axils of the primary leaves, irregularly scattered all along the branchlets. *Pedicels* up to (7.5–)10–21mm long, glandular pubescent, hairs up to 0.1–0.15mm long. *Calyx* tube 0.1–0.3mm long, lobes 2.5–3.5 × 0.7–1mm, lanceolate-oblong, subacute, clad in a mixture of glistening glands and minute glandular hairs. *Corolla* tube 10–15 × 1.8–2.3mm in throat, cylindric, abruptly expanded near apex, limb oblique, c.12–15mm across lateral lobes, posticous lobes 4.8–6.2 × 3–4.8mm, anticous lobe 5–6 × 3–4.2mm, all lobes cuneate-oblong, often retuse, glandular-pubescent outside, hairs up to 0.1–0.2mm long, glistening glands as well mainly on backs of lobes, broad band of clavate hairs in throat extending very briefly onto base of lower lip, lobes varying from white to pink or mauve with a wedge-shaped, slightly raised, yellow to orange or red median patch at the base of each lobe (usually clearly visible in the dried state). *Stamens*: posticous filaments 1.2–2mm long, anthers 0.8–1.2mm, anticous filaments 0.6–1mm long, anthers 0.5–0.8mm, all filaments minutely glandular. *Stigma* c.0.2mm long. *Style* 7.5–12mm. *Ovary* 1.7–2 × 0.8–1.2mm. *Capsules* 5–6.5 × 2.8–3.5mm, glistening glands and minute glandular hairs all over. *Seeds* c.0.6–0.8 × 0.4mm.

Selected citations:

Cape. Cape Peninsula, 3418AD, Buffels Bay, 6 i 1933, *Salter* 2931 (BM, BOL); ibidem, 30 x 1949, *Middlemost* 1687 (NBG, STE). Bredasdorp div., 3419DB, Poontrivier, 25 ix 1953, *Leistner* 1122 (STE); 3420CA, Bredasdorp Poort, 2 ix 1943, *Barker* 2477 (NBG). Riversdale div., 3421AD, Still Bay, ridge below reservoir, 60m, 9 viii 1978, *Bohnen* 3978 (STE); 3421BC, Ystervarkpunt, 110m, 3 vii 1987, *Willemse* 397 (STE).

Jamesbrittenia stellata is known from Buffels Bay near the tip of the Cape Peninsula south of Simonstown, then from Bredasdorp and Riversdale divisions, between Poontrivier in the west and Ystervarkpunt in the east. Four collectors recorded it on limestone, another on 'Sweet sand dunes', so it may be another limestone endemic. It grows in shrub communities, not much above sea level, and flowers mainly between July and January.

The distinguishing marks of the species are its leaves, relatively large in relation to those of *J. aspalathoides* to which it is allied, and the star-shaped patch around the mouth. At Buffels Bay, the leaves may be extraordinarily large (up to c.22 × 12mm), lobed rather than toothed, and with a well developed petiole. Salter recorded the plants as growing in mossy tufts on a limestone cliff, so one assumes the site is damp and sheltered. The star-shaped patch around the mouth is not as conspicuous as it is in plants from Bredasdorp and Riversdale. The population has probably been isolated from the eastern sites since the Middle Pleistocene. *Jamesbrittenia stellata* is the only species of *Jamesbrittenia* to occur on the Peninsula and is there known to me only from Buffels Bay.

The star-shaped patch around the mouth appears to be an osmophore, the tissue therefore different from that of the rest of the limb, and visible in the dried state. Indeed, it survives in the earliest collection seen, made by Mund (in SAM) in c.1825 near the Kars river, which rises south of Bredasdorp and flows in a northerly direction to join the Sout river.

The type specimen, *Morris* 257, came from Still Bay. It is interesting to note that Morris also collected *J. calciphila* on the same day and in the same locality (*Morris* 261, NBG). He noted the colour of *J. calciphila* as pinkish-mauve, that of *J. stellata* as pink, ignoring the prominent star around the mouth. Clearly Still Bay carries a wealth of species, several of them hitherto unrecognized. The area deserves a thorough investigation, with *Sutera* as well as *Jamesbrittenia*, and the possibility of interspecific hybridization, in mind.

36. Jamesbrittenia albomarginata Hilliard in Edinb. J. Bot. 49: 226 (1992). **Plate 2H.**
Type: Cape, Bredasdorp div., [3420CA], near Struys Bay, x 1940, *Esterhuysen* 5112 (holo. BOL, iso. NBG).

Well-branched dwarf shrublet up to 100–400mm tall, branches erect or decumbent and then sometimes rooting, glandular-puberulous, hairs up to 0.1–0.2mm long, few scattered glistening glands as well, very leafy. *Leaves* opposite, pseudofasciculate, largest primary leaves 1.5–9(–11) × 0.5–2.5(–6)mm (the largest leaves on coppice shoots after fire), oblong-spathulate, spathulate or cuneate, obtuse, entire or with a pair of teeth near apex, glabrous above, glistening glands below, sometimes oozing and then invisible but leaves look varnished. *Flowers* borne singly in axils of each leaf tuft on ultimate twiglets, sometimes overtopped by new growth. *Pedicels* up to 9–30mm long, glandular-puberulous, hairs up to 0.1–0.2mm long, scattered glistening glands as well. *Calyx* tube 0.2mm long, lobes 2.5–4.4 × 0.8mm, oblong to oblong-spathulate, subacute, glistening glands all over, sometimes minute glandular hairs as well on margins. *Corolla* tube 12–18 × 2–2.5mm in throat, cylindric abruptly expanded near apex, limb almost regular, c.10–15mm across lateral lobes, posticous lobes 3.3–5.4 × 3.2–4.5mm, anticous lobe 3.5–5.4 × 3.5–4.5mm, all lobes cuneate, ± truncate, glandular-puberulous outside, hairs up to 0.15–0.2mm long, glistening glands as well mainly on backs of lobes, broad transverse band of clavate hairs in throat extending briefly onto lower lip, lobes with a broadly cuneate median patch usually brown in colour, rarely orange or maroon, lateral margins white or cream. *Stamens*: posticous filaments 1.2–2mm long, anthers 0.8–1.1mm, anticous filaments 0.6–1mm long, anthers 0.5–0.8mm, all filaments puberulous. *Stigma* 0.3–0.4mm long. *Style* 9.5–14mm long. *Ovary* 2–3 × 1.2–1.7mm. *Capsules* 6–8 × 2.5–3.5mm, glistening glands all over. *Seeds* c.0.7–0.8 × 0.4–0.5mm.

Selected citations:

Cape. Caledon div., 3419CB, Danger Point, 10 vi 1950, *Maguire* 13 (NBG). Bredasdorp div., 3419DA, Ratel Rivier, 50ft, 12 xii 1896, *Schlechter* 9724 (BOL, PRE, Z); 3419DB, Mierkraal, 200ft, 24 iv 1897, *Schlechter* 10509 (BOL, E, PRE, S, SAM, W, Z). 3420AD, De Hoop Provincial Farm, 10 iv 1957, *Lewis* 5161 (NBG, STE). 3420BD, NE of Port Beaufort, 300ft, 6 iv 1984, *Oliver* 8423 (PRE, STE). Riversdale div., 3421AD, 16 miles S of Riversdale, Blombos road, 14 iv 1962, *Lewis* 5937 (NBG, STE).

Jamesbrittenia albomarginata is endemic to the limestone areas of Caledon, Bredasdorp and Riversdale divisions, from the environs of Stanford in the west to near Still Bay in the east. The plants grow on coastal sand dunes or sandy flats and hillocks from sea level to c.230m, in shrub communites or restioid vegetation. Flowers can be found in any month. The corolla limb is very distinctive: the broadly cuneate lobes are shades of brown (orange and maroon have also been recorded) with narrow white or cream-coloured lateral margins, the overall effect being that of a 5-rayed Maltese cross (only *J. merxmuelleri* has similar flowers). The leaves are usually very small and densely tufted, but clearly size is related to growing conditions because *Oliver* 4292 (STE) comprises very short coppice shoots (after fire) that have produced relatively enormous leaves c.11 × 6mm; c.2–3 × 0.5–1.5mm is the common size. Most of the specimens seen have been mistaken for *J. argentea* [*Sutera pedunculata*, *S. cuneata*] from which they are readily distinguished not only by the corolla limb (white with a yellow throat in *J. argentea*) but also by the longer corolla tube (12–18mm versus 7–9.5mm), the much less toothed leaves, and the flowers produced only towards the tips of the twiglets.

The earliest collection I have seen was made by Ecklon in October 1828 in 'Zoutendaalsvalley', Caledon (specimen in S). The peculiar colour pattern on the limb is still clearly visible.

37. Jamesbrittenia merxmuelleri (Roessler) Hilliard in Edinb. J. Bot. 49: 231 (1992).
Type: Namibia, Klinghardtberge, 18 ix 1977, *Merxmüller & Giess* 32154 (holo. M, iso. WIND).

Syn.: *Sutera merxmuelleri* Roessler in Mit. Bot. Staatss. München 16 (Beih.): 41 (1980).
S. microphylla auct. non (L.f.) Hiern; Hiern in Fl. Cap.4(2): 309 (1904) p.p., quoad *Drège* 3116 b tantum; Dinter in Feddes Repert. 23: 368 (1927); Range in Feddes Repert 38: 265 (1935); Merxm. & Roessler in Prodr. Fl. S.W. Afr. 126: 52 (1967).

Stout dwarf twiggy shrublet up to 150–600mm tall, many short lateral divaricate branchlets crowded with closely leafy brachyblasts, becoming bare and subspinescent with age, young parts and pedicels glandular-puberulous, hairs up to 0.25mm long. *Leaves*: primary leaves spirally arranged, up to 5 × 2.5mm, those on the brachyblasts opposite and decussate, very close-set, largest ones 1.5–3.5 × 1–2.5mm, subrotund, broadly obovate or broadly spathulate, thick-textured, very obtuse, base somewhat narrowed, entire, both surfaces with big glistening glands in centre of leaf, margins bare. *Flowers* solitary in the axils of the primary leaves on all the branchlets. *Pedicels* c.3–8mm long. *Calyx* tube 0.2–0.3mm long, lobes 2.5–4 × 0.8–1.3mm, oblong-lanceolate, ± obtuse, clad in glistening glands only or sometimes a few very short-stalked, big-headed glandular hairs as well near base. *Corolla* tube 9.5–14 × 2–2.5mm in throat, cylindric, abruptly expanded near apex, limb nearly regular, c.7–14mm across lateral lobes, posticous lobes 2.3–5 × 2.5–5.5mm, anticous lobe 2.3–5 × 2.3–5mm, all lobes broadly ovate to broadly oblong, tips rounded or retuse, glandular-puberulous outside, hairs less than 0.1mm long, glistening glands as well mainly on backs of lobes, broad transverse band of clavate hairs in throat, lobes bicoloured, with a broad cuneate brown or purple-brown thick-textured patch bordered in lower part with thin-textured white to creamy yellow. *Stamens*: posticous filaments 1.5–2mm long, anthers 0.8–1.3mm, anticous filaments 0.6–1mm long, anthers 0.7–1mm, all filaments bearded with clavate or glandular hairs. *Stigma* 0.3–0.5mm long. *Style* 7.5–11mm. *Ovary* 1.7–2.2 × 1.2–1.6mm. *Capsules* 5.5–9 × 4–5mm, glistening glands all over. *Seeds* c.1 × 0.7mm.

Selected citations:

Namibia Lüderitz-Süd distr. Lüderitzbucht, 2615CA, vii 1922, *Dinter* 4063 (BOL, PRE, SAM, Z); 2715AB, Pomona, 11 v 1929, *Dinter* 6340 (BOL, E, M, S, PRE, SAM, STE, Z); 2715DC, 15 miles S of Chameis, 11 ix 1977, *Williamson* 2656 (BOL). 2715 BC, Klinghardtberge, 17 ix 1977, *Merxmüller & Giess* 32092 (M).
Cape. Namaqualand. 2816DA, mouth of the Gariep, 4 x 1830, *Drège* (distributed as *Lyperia microphylla* Benth. a) (E,P,S,W); 2816DA, between Holgat and mouth of Orange River, Witbank, x 1926, *Pillans* 5109 (BOL). 2917CA, 3km south of Kleinzee, 23 vii 1983, *Van Wyk* s.n. (BOL).

Drège collected *J. merxmuelleri* in 1830 but Hiern equated the specimen with *J. microphylla* and 150 years passed before it was recognized as an undescribed species. The two species resemble one another in habit: short branchlets thickly beset with short brachyblasts, the tiny leaves of which are 4-ranked and close-set (as in several species of *Crassula*!); the leaves of *J. merxmuelleri* are very obtuse, those of *J. microphylla* acute. The species are also strikingly different in their flowers: the corolla lobes shades of purple in *J. microphylla*, brown margined white to pale yellow in the lower half in *J. merxmuelleri* (precisely as in *J. albomarginata*). The margins of the lobes tend to roll back and the nearly regular limb must then resemble that of certain species of *Manulea*, e.g. *M. robusta*.

Jamesbrittenia merxmuelleri has been recorded from the environs of Lüderitz in Namibia south to Kleinsee in Namaqualand, never far from the sea and growing from crevices in rocks: limestone, granite and quartzite have all been noted. Flowering has been recorded between May and October; old capsules are nearly always present as well.

GROUP 1b, 1

38. Jamesbrittenia fruticosa (Benth.) Hilliard in Edinb. J. Bot. 49: 229 (1992). **Plate 3L.**
Lectotype (chosen here): Cape, Namaqualand, [3017DD] Zwartdoorn-rivier, 500–1000ft, 11 viii 1830, *Drège* (K; isolecto. E, LE, P, S, SAM, W); distributed as *Lyperia fruticosa* Benth. b.

Syn.: *Lyperia fruticosa* Benth. in Hook., Comp. Bot. Mag. 1: 377 (1836) & in DC., Prodr. 10: 358 (1846).
    L. litoralis Schinz in Verh. Bot. Ver. Brandenb. 31: 192 (1890). Lectotype (chosen here): Namibia, Angra Pequena [Lüderitz], viii 1885, *Schenk* 15 (Z).
    Chaenostoma fruticosum (Benth.) Wettst. in Engl. & Prantl, Natürlich. Pflanzenfam. 4 (3B): 69 (1891).
    Peliostomum oppositifolium Engl. in Bot. Jahrb. 19: 149 (1894). Type: Namibia, Angra Pequena, *Herman* 12 (holo. B†, iso. Z).
    Chaenostoma litorale (Schinz) Diels in Bot. Jahrb. 23: 477 (1897).
    Sutera litoralis (Schinz) Hiern in Fl. Cap. 4(2): 291 (1904).

S. *fruticosa* (Benth.) Hiern in Fl. Cap: 4(2): 288 (1904); Merxm. & Roessler in Prodr. Fl. S.W. Afr. 126: 51 (1967); le Roux & Schelpe, Namaqualand ed. 2: 150, 151 cum ic. (1988).

S. *lilacina* Dinter in Feddes Repert. 30: 185 (1932). Type: Namibia, Klinghardtgebirge, 25 ix 1922, *Dinter* 3992 (holo. B†; iso. M, PRE, STE, Z).

Aptosimum oppositifolium (Engl.) Phillips in J.S. Afr. Bot. 16: 22 (1950).

Shrublet up to c.1.3m tall and as much across, well branched, main branches up to c.10mm in diam., young parts glandular-pubescent, hairs up to 0.5–2mm long, very delicate. *Leaves* opposite becoming alternate towards tips of flowering branchlets, up to 7–40 × 2–15mm (ratio 2–4.8:1), cauline ones ± patent, narrowly to broadly elliptic (upper leaves subtending flowers often ovate), apex acute to rarely obtuse, base tapered, often into a short petiolar part 1–2mm long, margins entire or very rarely with 1–3 tiny teeth, both surfaces thinly glandular-pubescent, hairs up to 0.5–1mm long. *Flowers* solitary in upper leaf axils, usually crowded. *Pedicels* up to 2–10mm long. *Calyx* tube 0.1–0.2mm long, lobes 6–12 × 0.7–1 (–1.5)mm, oblong to oblong-elliptic, subacute, glandular-pubescent, hairs up to 1–1.8mm long. *Corolla* tube (16–) 20–26 × 2.2–3mm in mouth, cylindric, abruptly expanded near apex, limb nearly regular, 13–20mm across lateral lobes, posticous lobes 5–8 × 4–6.8mm, anticous lobe 5–8 × 3.7–5.5mm, broadly ovate to suborbicular, glandular-puberulous outside, hairs up to 0.15–0.4mm long, very minutely glandular around mouth on upper surface of limb, lightly bearded inside with clavate hairs up to 0.1–0.25mm long scarcely extending onto base of anticous lip, lobes ranging from white to shades of blue, purple or mauve, dark median bar at base of each lobe (maroon, purple, black recorded), throat yellow (always?). *Stamens*: posticous filaments 1.7–2.5mm long, anthers 1.2–2mm, anticous filaments 0.8–1.5mm, anthers 1–1.4mm, visible in mouth, filaments bearded with tiny clavate hairs. *Stigma* 0.5–1mm long. *Style* 13–22mm. *Ovary* 2–3 × 1–1.2mm. *Capsules* 6–10 × 3–5mm, glandular-puberulous at apex. *Seeds* 0.3–0.5 × 0.3–0.4mm.

Selected citations:

Namibia. Lüderitz-Süd distr., 2615CA, Angra Pequena [= Lüderitz], 16 i 1907, *Galpin & Pearson* 7521 (BOL, SAM); ibidem, Diamantberg, 5 viii 1959, *Giess & Van Vuuren* 1959 (M, WIND); 2716DD, farm Zebrafontein (LU 87), 1200m, 11 ix 1973, *Giess* 12880 (K, S, WIND). 2816BB, Kahanstal, 3 xii 1934, *Dinter* 8157 (BM, BOL, K, S, WIND, Z). **Cape**. Namaqualand, 2816BD, Richtersveld, Numees camping site, 26 ix 1981, *Hugo* 2790 (PRE, STE); 2817CD, Richtersveld, Stinkfontein, 1000ft, 10 vii 1961, *Middlemost* 2130 (NBG, STE). 2916BB, Doornpoort, 2010ft, *Pearson* 6137 (BOL, SAM). 2917DC, Wildeperdehoek Pass SW of Springbok, 1 ix 1991, *Batten* 1104 (E). 2917BD, 20 miles N of Steinkopf, viii 1949, *Hall* 1018 (M, NBG,). 3018CA, Loerkop 4 miles N.E. of Garies, c.1000ft, 4 x 1956, *Leistner* 737 (K, NBG, PRE). Van Rhynsdorp distr., 3118DA, 12 km S of Van Rhynsdorp, N of Wiedouw river, 9 ix 1974, *Goldblatt* 2543 (M, S).

Bentham based his name *Lyperia fruticosa* on three specimens collected by Drège in Namaqualand. I have chosen the specimen from the Zwartdoorn river as lectotype because the material is good and is represented in several herbaria. One of the syntypes was collected near the mouth of the Gariep (Orange river). It has very small leaves (c.9 × 3mm) and is precisely matched by the type of *Lyperia litoralis*, collected c.250km to the north at Lüderitzbucht. Not all small-leaved specimens are littoral, and small-leaved plants are also found inland.

Jamesbrittenia fruticosa is distinguished from its allies primarily by its leaves: the cauline leaves are nearly always elliptic, tapering at both ends (ratio of length: breadth 2–4.8:1) and spread nearly at right angles to the stem; only the uppermost leaves subtending flowers are sometimes ovate and broad-based. The pubescence on the leaves is often (but by no means always) shorter than in allied species; it is useful to examine the margins, where the hairs are easily seen in profile. The pedicels are short (up to 10mm, but 4–5mm is more usual), the calyx lobes mostly less than 1mm broad (occasionally up to 1.5mm), the corolla tube long (mostly 20–26mm, occasionally as short as 16mm) and only lightly bearded with short clavate hairs that scarcely extend onto the base of the lower lip. The whole plant often blackens on drying.

Jamesbrittenia fruticosa possibly hybridizes with *J. maxii.* They are sympatric around Lüderitz, and there some specimens appear to be more or less intermediate. For example, *Giess & Van Vuuren* 684 (K, M, NBG, PRE, W, WIND), from the south-facing slope of Nautilus, north of Lüderitz, has leaves, indumentum and calyx lobes ± intermediate between those of the two species, while many more specimens (which I have assigned to *J. fruticosa*) have leaves blunter than is usual (this could be induced by edaphic factors?). Another possible hybrid is mentioned under *J. maxii.*

Jamesbrittenia fruticosa occurs in southern Namibia and the western Cape, from Lüderitz southwards along the coast and inland to Witpütz, Namuskluft and Namusberg, thence into Namaqualand and south to about the Doring river, roughly 30km south of Van Rhynsdorp. There is one record from Langebaan, but its occurrence there needs confirmation. The plants grow on sandy flats and slopes, among rocks, in gullies or scattered with other plants in the open, from littoral dunes to about 1200m above sea level. Flowering has been recorded in all months, but most frequently in August and September.

39. Jamesbrittenia maxii (Hiern) Hilliard in Edinb. J. Bot. 49: 230 (1992).
Lectotype (chosen here): Cape, Little Bushmanland, 2917BB, Schakalswater, 27 xi 1897, *Max Schlechter* 6 (BOL; isolecto. E, K, PRE, W).

Syn.: *Sutera maxii* Hiern in Fl. Cap. 4(2): 288 (1904); Hemsley & Skan, Fl. Trop. Afr. 4(2): 307 (1906); Merxm. & Roessler in Prodr. Fl. S.W. Afr. 126: 52 (1967), p.p.
 S. amplexicaulis auct. non (Benth.) Hiern; Hemsley & Skan in Fl. Trop. Afr. 4(2): 206 (1906) quoad *Marloth* 1225 (*Guerich* 141 not seen: B†?).

Shrublet, 150mm–1m tall, sometimes broader than tall, several-stemmed from base, well-branched above, often with new growth developing below old inflorescences, main stems up to 12mm diam., young parts glandular-pubescent, hairs up to 1–2mm long, very delicate. *Leaves* opposite, very rarely becoming alternate at extreme tips of inflorescences, 7–28 × 4–16mm (ratio 1.1–2.1:1), mostly broadly ovate to ovate-elliptic, lowermost sometimes elliptic, obtuse to subacute or acute, base cordate to rounded and clasping in most leaves, tapered to a petiolar part 1–2mm long in lower leaves, margins entire or with up to 4 pairs of teeth, both surfaces densely glandular-pubescent, hairs up to 0.8–2mm long. *Flowers* solitary in upper leaf axils forming lax to crowded racemose infloresences up to c.250mm long but often much shorter. *Pedicels* up to 2–6 (–18)mm long. *Calyx* tube 0.1–0.2mm long, lobes 5.5–13 × 1.1–2.1mm, subspathulate, ± obtuse, glandular-pubescent, hairs up to 1–2mm long. *Corolla* tube (8–) 10–25 × 2.5–3mm in mouth, cylindric, abruptly expanded at apex, limb nearly regular, (5–) 7–14mm across lateral lobes, posticous lobes (1.3–) 3–6 × (1.3–) 3–6mm, anticous lobes (1.7–) 3–5 × (1.2–) 3–6mm, suborbicular, minutely glandular to glabrous outside, very minutely glandular around mouth on upper surface of limb, well bearded inside with clavate hairs c.0.2–0.4mm long where they extend onto extreme base of anticous lip, lobes white or occasionally creamy-white, dark violet streaks at base of each lobe, throat yellow. *Stamens:* posticous filaments 1.4–2.8mm long, posticous anthers 1.2–1.8mm, anticous filaments 0.4–1mm, anthers 1–1.3mm, all filaments puberulous. *Stigma* 0.3–1mm long. *Style* (5–) 10–20mm. *Ovary* 2–3 × 1.2–1.5mm. *Capsules* 6–13 × 3.5–5mm, glandular-pubescent near apex. *Seeds* 0.3–0.6 × 0.25–0.3mm.

Selected citations:

Angola. Mossamedes distr., 8km from Mossamedes, 1 vi 1937, *Carrisso & Sousa* 225 (BM, COI); vallée du Rio Giraul, 50–150m, 1937, *Humbert* 16433 (BM).
Namibia . Kaokoveld, 1812BA, 20 meilen Südlich Oropembe, 9 vi 1963, *Giess & Leippert* 7429 (NBG, PRE, WIND). Omaruru distr., 2114DC, Nordufer des Omaruru, 35km Östlich der Salzstrasse, 10 iv 1964, *Giess* 7858 (PRE, WIND). Swakopmund distr., 2214DA, in der Swakopmündung am Südufer, 14 i 1957, *Seydel* 810 (S, Z); 2315DB, gorge of Kuiseb river on Gamsberg road to Walvis Bay, 13 vii 1967, *Hilliard* 4691 (E, WIND). Lüderitz-Süd distr., 2615CA, Kovis Mountain, 6 iii 1963, *De Winter & Hardy* 7900 (K, PRE, WIND). 2616CA, farm Klein Aus, 12 v 1949, *Kinges* 2231 (M, PRE). Bethanien distr., 2616BA, farm Gamochas (BET 31), 14 ix 1977, *Merxmüller & Giess* 32005 (WIND). Warmbad distr., 2718CA, Klein Karas, 18 viii 1923, *Dinter* 4911 (BOL, K, PRE, SAM, Z); 2818BC, bank of Hom River, 11 xi 1937, *Galpin* 14160 (BOL, PRE, W).

Cape. Namaqualand, 2817DC, just south of Vyfmylspoort, *Fellingham* 1100 (STE). Kenhardt distr., 25 miles SE of Pofadder, Gannapoort, 3400ft, 18 v 1961, *Leistner* 2405 (K, M, PRE, SRGH); 5 miles W of Pofadder, 23 v 1961, *Schlieben* 8969 (K, M, NBG, PRE, S, SRGH, Z).

Hiern cited four collections under his new name *Sutera maxii*: *Schinz* 39 and 50, from the Tsirub Mountains and Tiras in Namibia, *Schlechter* 11430 from I'us in Namaqualand, and *Max Schlechter* 6, from Jakhalswater in Little Bushmanland. The latter was chosen as lectotype, not only because it is the eponymous specimen, but also because the material is good and widely distributed. *Schlechter* 11430 is also good material. Schinz's specimens are in K, but duplicates under the numbers 39 and 50 have not been traced; however, *Schinz* 114, 115 and 116 (all in Z) came from the appropriate localities and are probably part of the same collections; all are *J. maxii*, but only 115 is a good specimen.

Schinz 50 (K) comprises two twigs of *J. maxii*, but a third twig differs in having alternate leaves.

Jamesbrittenia maxii is characterized by its shrubby habit, broad sessile or nearly sessile leaves that are nearly always opposite right to the tips of the flowering branches, relatively broad calyx lobes and its small white (or sometimes creamy or yellow) corolla limb. The plants may or may not blacken on drying. The pedicels are usually short (up to c.6mm), but one of the syntypes (*Schlechter* 11430, the sheet in S) has pedicels up to 18mm long; it came from I'us, not traced precisely, but somewhere between Steinkopf and Goodhouse in Namaqualand (2918AC). Otherwise, plants with long pedicels have been seen only from the gorge of the Kuiseb river and its environs (2315BD, CA, DB, DD; 2314BB, DB; 2215AB); plants with short pedicels also occur in the area. The corolla tube is usually at least 18mm long, but it can be as short as 8mm. Only two collections have been seen from Angola, and both are short-tubed (c.14mm). Other short-tubed plants have been seen from north of Sesfontein (1913AB. 14mm), near the Vogelfederberg (2314BB. 11.5mm), Natab on the Kuiseb river (2315CA. 8mm), Gobabeb (2315CA. c.15mm), Kovisberge (2615CD. 10.5, 15.5mm), and the farm Plateau near Aus (2616CB. 13mm).

All these plants with short-tubed flowers have the general facies of *J. maxii* and are probably no more than part of the normal variation range of the species. One other collection may, however, represent a hybrid: *de Winter & Giess* 6306 (WIND), collected on a dolomite bank 2 miles north of Witpütz police station (2716DA), has corollas only 9mm long, and the leaves look intermediate between those of *J. maxii* and *J. fruticosa*. *Kraeusel & Wiss* 2036 b (WIND) from Pockenbank (2716BA) may represent *J. maxii* × *J. bicolor*; the facies of the plant is that of *J. maxii* and the corolla tube is c.17mm long, but the calyx lobes are narrow (c.0.6mm) and very acute, as in *J. bicolor*. The four closely allied species, *J. bicolor*, *J. fruticosa*, *J. maxii* and *J. sessilifolia*, all occur around Pockenbank and Witpütz, where some complicated interbreeding may be taking place. Other possible hybridization around Witpütz is mentioned under *J. sessilifolia*.

The overall range of *J. maxii* is from the environs of Mossamedes in Angola south through the western part of Namibia then trending slightly east into Bethanien and Warmbad districts, and across the Orange into the northernmost Cape, roughly between Vioolsdrift and Pofadder. Many records are from sandy river beds and watercourses, but sandy pockets among rocks, and gravel plains have also been mentioned, from about sea level to just over 1000m. Flowers can be found in any month.

40. **Jamesbrittenia sessilifolia** (Diels) Hilliard in Edinb. J. Bot. 49: 232 (1992).
Type: Namibia, Great Namaqualand, lower Orange River, *Steingröver* 17 (holo. Z).

Syn.: *Chaenostoma sessilifolium* Diels in Bot. Jahrb. 23: 476 (1897).
    Sutera sessilifolia (Diels) Hiern in Fl. Cap. 4 (2): 291 (1904); Merxm. & Roessler in Prodr. Fl. S.W. Afr. 126: 54 (1967), excl. *Lyperia major* Pilger.
    S. gariusana Dinter in Feddes Repert. 30: 186 (1932). Type: Namibia, Garius [2717DB, Karios], 9 x 1923, *Dinter* 5011 (holo. B†, iso. Z).

Twiggy shrublet but will flower when young and then look annual, woody caudex up to c.10mm diam. at crown, stems 250mm–1m tall, erect, often remarkably flexuous, glandular-pubescent, hairs up to 0.6–2mm long, very delicate. *Leaves* opposite, alternate upwards, ± patent, 11–30 × 5–22mm (ratio 1–2.6:1), lower leaves elliptic, subacute, base tapering to petiolar part 1–7mm long, upper leaves often ovate, base subcordate, amplexicaul, margins entire or with 1–5 pairs of teeth in upper part, both surfaces glandular-pubescent, hairs up to 0.6–1.5mm long. *Flowers* solitary in upper leaf axils forming short rather lax racemose inflorescences. *Pedicels* up to 12–31mm long. *Calyx* tube 0.1–0.2mm long, lobes 5.5–7 × 0.8–1.2mm, oblong-spathulate, subacute, glandular-pubescent, hairs up to 0.8–1.2mm long. *Corolla* tube 16–21 × 2.2–2.5mm in mouth, cylindric, abruptly expanded near apex, limb nearly regular, 12–20mm across lateral lobes, posticous lobes 4–6.5 × 4–6.5mm, anticous lobes 3.8–6.5 × 4–6mm, oblong-elliptic to suborbicular, glandular-puberulous outside or nearly glabrous, very minutely glandular around mouth on upper surface of limb, well bearded inside with clavate hairs c.0.3mm long where they extend onto extreme base of anticous lip, lobes white with 3 dark purplish streaks at base of each lobe, throat yellow. *Stamens*: posticous filaments 1.8–2.5mm long, anthers 1.2–1.6mm, anticous filaments 0.5–1.2mm, anthers 1–1.25mm, filaments very minutely puberulous. *Stigma* 0.7–1mm long. *Style* 12.5–16mm. *Ovary* 2.2–3 × 1–1.5mm. *Capsules* 7–9 × 3.5–4mm, glandular-puberulous at apex. *Seeds* 0.3–0.5 × 0.25–0.3mm.

Selected citations:

Namibia. Lüderitz-Süd distr., 2616DB, Kuibis Ravine, 30 i 1912, *Pearson* 8003 (BOL, K); Bethanien distr., 2717AC, farm Huns (BET 106), 8 vi 1976, *Giess & Müller* 14282 (WIND); 2717AD, Fish River Canyon, *le Roux* 7 (WIND). Warmbad distr., 2717DA, farm Hobas (WAR 374), 30 v 1972, *Giess & Müller* 12295 (WIND). Lüderitz-Süd distr., 2816BB, östl. Sendlingsdrift, 21 ix 1972, *Merxmüller & Giess* 28666 (M).

Jamesbrittenia sessilifolia appears to be confined to southernmost Namibia from about Kuibis (2616DB) south to the Orange river, east to Klein Karas and west to Witpütz. It has been recorded on the floodplain of the Orange (so may yet be found south of the river in the Sendlingsdrift area) and in the beds or ravines of several other rivers, and may flower in any month.

The characteristic features of *J. sessilifolia* are its stems, often remarkably flexuous, leaves alternate at least on the flowering twigs, and distinctly petiolate (only the uppermost leaves subtending flowers may become broad-based and clasping and are possibly responsible for the misleading epithet), long pedicels (up to 12–31mm) and white corolla lobes marked with dark lines, the throat yellow. The type is a very poor specimen from the 'lower Orange river'. The isotype of *Sutera gariusana* in Z (the only one traced) is also now in poor condition (of the flowers, only the imprints of two remain, and a loose corolla is caught up in the leaves), but the habit of the plant, its zigzag twigs, foliage, long pedicels and big capsules are all well displayed.

Giess 13397 (M, PRE, WIND) from the farm Genot (2616BB) is scarcely separable from *J. sessilifolia*: it differs chiefly in its mostly shorter pedicels (longest ones 5–16mm); the leaves are also remarkably petiolate (petiolar part up to c.12mm long).

Far more puzzling specimens have been collected in the environs of Witpütz (2716DA). *De Winter & Giess* 6302 (M, WIND), collected two miles north of Witpütz police station on the road to Aus, has mostly opposite flowers on straight twigs (though the old stems are zigzag) and the longest pedicels measure only 5–7mm. *Dinter* 8086 (B), from 'Südlich Wittpüts' has pedicels up to 6mm long, while *Dinter* 8244 (B), from 'Schwarzkalk, 20 km nördl. Wittpüts' has pedicels up to 15mm long; both specimens have the facies of *De Winter & Giess* 6302, and all three specimens have leaves rather smaller and narrower than is normal for *J. sessilifolia* (c.10–15 × 4–5mm) and in aspect reminiscent of those of *J. fruticosa*. *Dinter* 308 (M), 'Wittpütz-Sendlingsdrift', can also be associated here, as well as *Merxmüller & Giess* 28850 (K, PRE, WIND), collected '10km N. von Witpütz-Polizeistation'; this last specimen differs from typical *J. sessilifolia* only in its short pedicels (up to 4mm). *Jamesbrittenia fruticosa, J. bicolor* and *J. maxii* also occur in the area (and clearly grow mixed with the problematical plants) so hybridization is a distinct possibility (see also under *J. maxii*).

41. Jamesbrittenia major (Pilger) Hilliard in Edinb. J. Bot. 49: 230 (1992).
Type: Namibia, Keidorus [=Karious of Range's map = Karios, 2717DB], 220m, viii 1909, *Range* 721 (holo. B†, iso. BM). See comments below.

Syn.: *Lyperia major* Pilger in Bot. Jahrb. 48: 441 (1912).
 Sutera major (Pilger) Range in Feddes Repert. 38: 265 (1935).

Perennial herb, taproot lignified, up to 10mm diam. at crown, stem c.200–400mm tall, main stem branching from base and above, erect, glandular-pubescent, hairs up to 1–2.5mm long, very delicate. *Leaves* opposite, alternate above but often only near tips of inflorescences, ± patent, 15–65 × 6–32mm (ratio 1.3–2.6:1), broadly elliptic to broadly ovate, lower ones tapering to a petiolar part 1–7mm long, upper leaves subtending flowers often broad-based, clasping, apex acute to obtuse, margins mostly entire, occasionally with 1–2 pairs of teeth in upper part, both surfaces glandular-pubescent, hairs up to 1–2mm long. *Flowers* solitary in upper leaf axils forming long (c.150–300mm) often crowded racemose inflorescences. *Pedicels* up to 6–30mm long. *Calyx* tube 0.1–0.2mm long, lobes 5.6–12 × 0.8–1.4mm, narrowly oblong-spathulate, subacute, glandular-pubescent, hairs up to 1–2.5mm long. *Corolla* tube 15–22 × 2.5–2.8mm in mouth, cylindric, abruptly expanded near apex, limb nearly regular, 15–20mm across lateral lobes, posticous lobes 4.5–7.5 × 4–6.5mm, anticous lobe 5–7.5 × 3.5–7mm, oblong-elliptic, glandular-puberulous or almost glabrous outside, very minutely glandular around mouth on upper surface of limb, well bearded inside with clavate hairs, hairs c.0.25–0.5mm long where they extend onto extreme base of anticous lip, lobes shades of lilac, rose and violet, 3 purple streaks at base of each lobe, throat yellow. *Stamens*: posticous filaments 1.8–2.2mm long, anthers 1.5–1.8mm, anticous filaments 0.8–1mm long, anthers 1.2–1.4mm, all filaments puberulous. *Stigma* 0.5–1mm long. *Style* 10–17.5mm. *Ovary* 2.2–3 × 1.2–1.5mm. *Capsules* 4–9 × 2.5–4mm, glandular-puberulous at apex. *Seeds* 0.25–0.6 × 0.2–0.3mm.

Selected citations:

Namibia. Warmbad distr., Klein Karas - Aiais [2718CA - 2717 CB], 29 vii 1931, *Örtendahl* 585 (K, PRE, S); 2717DA, farm Altdorn (WAR 3), 17 vi 1976, *Giess & Müller* 14448 (K, M, PRE, WIND). 2817AA, Nuob rivier, 3km nördlich Einmündung in den Oranje, 2 x 1975, *Giess* 13841 (M, PRE, WIND). Lüderitz-Süd distr., 2816BB, östl. Sendlingsdrift, 21 ix 1972, *Merxmüller & Giess* 28667 (M).
Cape. Namaqualand, Richtersveld, 2817AD, Tatasberg, 2 x 1930, *Herre* sub STE 12275 (STE).

The name *Lyperia* major was based on *Range* 721; the holotype was destroyed in the Berlin fire, but there is a fragment in BM that fits Pilger's detailed description perfectly. A specimen under the same number in SAM does not fit Pilger's description (inter alia 'folia ..... subsessilia vel sessilia') and is *J. glutinosa*. Pilger diagnosed his new species against *Lyperia amplexicaulis* Benth., and Roessler (in Prodr. Fl. S.W. Afr.), presumably without seeing a type, suggested the plant might be *Sutera sessilifolia*. Range (in Feddes Repert., cited above) attributed the combination *Sutera major* to Hiern; I have failed to trace publication of this, which is in any case improbable given the date of publication of the basionym. It is curious that Range commented under *Sutera major* 'Bestimmung später in *Lyperia glutinosa* geändert.'

Jamesbrittenia major is closely allied to *J. sessilifolia*, but differs in habit (long straight ascending branches, not short twiggy often markedly flexuous ones) and in its lilac pink to violet (not white) corolla limb. They are partially sympatric; Merxmüller & Giess collected them growing together on the floodplain of the Orange river east of Sendlingsdrift and it is instructive to look at their specimens side by side (nos. 28666 and 28667, both cited in this revision). Although the specimen of *J. sessilifolia* is a very young plant (the collectors judged it to be annual), it clearly shows the zigzag axes characteristic of the species.

Range 721 came from Karios, and the species occurs from thereabouts south to the Orange river and across into the Richtersveld (mainly 2717 and 2817 degree squares). But there are three collections from northwest Namibia that I am unable to distinguish from *J. major*: *Wiss* 1942 (M, WIND) from the environs of Twyfelfontein (2014CB), *Craven* 1384 (WIND) from north of Gaias (2013DB), and *Craven* 1506 (WIND) from near the watershed between the Ugab and Huab rivers (2014CC). It is an astonishing disjunction, and this northern plant needs investigation, including information on the habit and basal parts of the plant.

Like *J. sessilifolia*, *J. major* grows in river beds or on mountain slopes (in shelter of rocks?); flowering records range from June to October, but two of the northern collections were in flower in March and May.

42. Jamesbrittenia megaphylla Hilliard in Edinb. J. Bot. 49: 231 (1992).
Type: Namibia, Warmbad distr., 2817DA, Schwarzkalkhang am Oranje, 12km westlich Nordufer, 7 viii 1976, *Giess* 14524 (holo. WIND; iso. K, M, PRE).

Herb, possibly annual, taproot woody, c.8mm diam. at crown, one main stem c.350–600mm long, sparingly branched at base and above, erect or possibly decumbent, glandular-pubescent, hairs up to 0.5–1mm long, very delicate. *Leaves* opposite only at extreme base, soon alternate, more or less patent, up to 16–45 × 15–34mm (ratio 1–1.3:1), lowermost leaves broadly elliptic, base slightly narrowed to petiolar part up to 2mm long, most leaves broadly ovate (but narrowing upwards) base cordate, amplexicaul, apex obtuse then abruptly apiculate becoming acute upwards, margins entire or very occasionally with a tiny tooth, both surfaces glandular-pubescent, hairs up to 0.5–1mm long. *Flowers* solitary in leaf axils almost from base of plant, often crowded. *Pedicels* up to 7–16mm long. *Calyx* tube c.0.2mm long, lobes 5.5–7.3 × 0.7–0.8mm, linear-lanceolate or very narrowly subspathulate, acute, glandular-pubescent, hairs up to 0.8–1mm long. *Corolla* tube 9.5–10 × 2.2–2.8mm in mouth, cylindric, scarcely broadened at apex, limb ± bilabiate, 8–9mm across lateral lobes, posticous lobes 3–3.4 × 2.4–2.8mm, anticous lobe 3–3.8 × 2.4–3.2mm, ± oblong-elliptic, glandular-puberulous outside, very minutely glandular around mouth on upper surface of limb, well bearded inside with clavate hairs c.0.2mm long where they extend onto extreme base of anticous lip, lobes white with dark red to blackish streaks at base of each lobe, throat yellow. *Stamens*: posticous filaments 1.8–2mm long, anthers 1.2–1.6mm, anticous filaments c.0.6mm long, anthers 1–1.3mm, all filaments puberulous. *Stigma* 0.4–0.5mm long. *Style* 5–7mm. *Ovary* 2.2–5 × 1–1.2mm. *Capsules* 5–7 × 3–4mm, glandular-puberulous near apex. *Seeds* 0.3–0.4 × 0.2–0.25mm.

Specimens seen:

Cape. Namaqualand, 2817DC, about 1 mile S.E. of Viools Drift and sides of Khusies river, ix 1931, *Pillans* 6383 (BOL, K); 6 miles from Viools Drift towards Steinkopf, 900ft, 22 vii 1954, *A.S.L. Schelpe* 232 (BOL); 4km SE of Viools Drift towards Steinkopf, 16 ix 1976, *van Jaarsveld* 1487 (NBG).

Jamesbrittenia megaphylla is known to me only from a small area of Namibia and Namaqualand around Viools Drift on the Orange river. The plants grow among rocks or at the foot of cliffs on mountain slopes; Giess recorded 'Schwarzkalk' on the Namibian side of the river, but south of the river the rock is andesite. Flowers appear in winter and spring, between July and September.

The species is characterized by its relatively large alternate leaves, long pedicels, narrow acute calyx lobes and short corolla tube. It is distinguished from *J. bicolor*, which also has alternate leaves, narrow acute calyx lobes and short corolla tube, by its habit (annual or perennial herb, not twiggy shrublet), larger (especially broader) leaves and longer pedicels (up to 7–16mm, not 2–6mm). *Jamesbrittenia major* also has large leaves and long pedicels, but the leaves are mostly opposite, the hairs on stem, leaves and calyx are longer (1–2.5mm, not 0.5–1mm), the corolla tube longer (15–22mm, versus 9.5–10mm), the limb larger (13–20mm across the lateral lobes, not 8–9mm) and shades of rose or violet (not white).

43. Jamesbrittenia bicolor (Dinter) Hilliard in Edinb. J. Bot. 49: 227 (1992).
Type: Namibia, Klinghardtgebirge, 16 ix 1922, *Dinter* 3909 (holo. B; iso. BM, BOL, C, PRE, SAM, STE, Z).

Syn.: *Sutera bicolor* Dinter in Feddes Repert. 30: 186 (1932).
    S. amplexicaulis auct. non (Benth.) Hiern; Merxm. & Roessler, Prodr. Fl. S.W. Afr. 126: 48 (1967).

Twiggy shrublet but will flower when young and only just beginning to branch, woody caudex up to c.10mm diam. at crown, stems c.100–500mm tall, erect, glandular-pubescent, hairs up to 0.5–1.5mm long, very delicate. *Leaves* opposite, alternate upwards especially on the flowering parts and there usually imbricate, 8–27 × 3–16mm (ratio c.1.1–2.8:1), lower leaves elliptic, subacute to acute, tapering to a petiolar part 1–4 (–7)mm long, upper leaves elliptic to ovate, base ± rounded to cordate, often amplexicaul, margins entire or very rarely with a few tiny teeth on some lower leaves, both surfaces glandular-pubescent, hairs up to 0.5–1.5mm long. *Flowers* solitary in upper leaf-axils forming long (up to c.150mm) crowded racemose inflorescences. *Pedicels* up to 2–6mm long. *Calyx* tube 0.1–0.2mm long, lobes 4.2–7 × 0.6–1mm, linear-lanceolate, acute to very acute, glandular-pubescent, hairs up to 0.5–1.3mm long. *Corolla* tube 9–13 × 2–2.6mm in mouth, cylindric, abruptly expanded near apex, limb bilabiate, 7–13mm across lateral lobes, posticous lobes 3–5.5 × 1.8–4mm, anticous lobe 2.7–5 × 2.2–4mm, ± oblong-elliptic, sparsely glandular-puberulous to glabrous outside, very minutely glandular around mouth on upper surface of limb, well bearded inside with clavate hairs c.0.15–0.2mm long where they extend onto extreme base of anticous lip, lobes ranging from whitish to yellow, throat orange/yellow, 3 dark streaks at base of each lobe (see further in discussion). *Stamens*: posticous filaments 1.5–2.5mm long, anthers 1–1.5mm, anticous filaments 0.3–1mm long, anthers 0.7–1mm, all filaments puberulous. *Stigma* 0.25–0.5mm long. *Style* 6.5–10mm. *Ovary* 2–3 × 1–1.5mm. *Capsules* 4–7 × 2.5–3.5mm, glandular-puberulous near apex. *Seeds* 0.25–0.4 × 0.25–0.3mm.

Selected citations:

Namibia. Lüderitz-Süd distr., 2715BC/BD, Klinghardtberge, umgebung des 'Sargdeckels', 16 ix 1977, *Merxmüller & Giess* 32060 (WIND); 2716BA, farm Pockenbank, 26 viii 1963, *Merxmüller & Giess* 3151 (M, PRE, WIND); 2716DA, Witpüts, 29 xi 1934, *Dinter* 8080 (B, K, PRE, S, WIND, Z); farm Witpütz Nord (LU 22), 30 ix 1975, *Giess* 13783 (K, PRE, W).

Jamesbrittenia bicolor is known from a small area of Namibia: the Klinghardtberge, Tsausberge, and the high ground from Pockenbank (almost due east of the Tsausberge) south to Witpütz. No ecological information is available other than 'Berghang' [mountain slopes] and two records of the plants being on limestone (Schwarzkalk). Most flowering records are between August and November.

Dinter's epithet '*bicolor*' refers to the corolla, which he recorded as yellowish to white with yellow throat and orange-yellow veins (a reference to the nectar guides). *Merxmüller & Giess* 3167 records the corolla as yellowish (gelblich) fading to white above, yellow below. The nectar guides have been variously described as dark, black, and carmine red. *Giess* 13750 records 'Blüten hellblau mit dunklem Schlund'. The flowers have dried exactly the same colour as those in all the rest of the material, with no trace of blue visible; possibly this was a slip of the pen (typewriter!).

In the Prodromus Flora von Süd-West Afrika, the species was equated with *Sutera [Jamesbrittenia] amplexicaulis*, from which it differs not only in its twiggy habit, but also in its petiolate lower leaves, linear-lanceolate acute calyx lobes, and corolla with mostly longer tube and larger lobes.

For possible hybridization involving *J. bicolor* see under *J. maxii*.

44. Jamesbrittenia amplexicaulis (Benth.) Hilliard in Edinb. J. Bot. 49: 226 (1992).
Lectotype (chosen here): Cape, Namaqualand [c.3017DB], zwischen Groenrivier und Kweek-
rivier, 500–1000 Fuss, 13 viii 1830, *Drège* (K; isolecto. E, LE, P, S, W).

Syn.: *Lyperia amplexicaulis* Benth. in Hook., Comp. Bot. Mag. 1: 377 (1836) & in DC., Prodr. 10: 358 (1846).
 Sutera amplexicaulis (Benth.) Hiern in Fl. Cap. 4(2): 287 (1904).

Perennial herb, stems several tufted from a woody caudex, c.200–450mm long, up to 5mm
diam. at base, simple to laxly branched from base, erect, decumbent or sprawling, glandular-
villous, hairs up to 1.5–2.5mm long, very delicate. *Leaves* opposite only at extreme base, soon
alternate, imbricate except in very lax specimens, up to 5–23 × 3–18mm (ratio c.1–1.5:1), often
small near base of twig, suddenly increasing in size, then decreasing again on flowering part,
ovate to broadly elliptic or suborbicular, base ± rounded, sessile, apex obtuse becoming acute
in uppermost leaves, margins entire to serrate in upper half, both surfaces glandular-villous,
hairs up to 1–2mm long. *Flowers* solitary in upper leaf-axils, forming long (up to c.300mm)
crowded racemose inflorescences. *Pedicels* up to 3–8mm long. *Calyx* tube 0.1–0.2mm long,
lobes 5.5–7.5 × 0.7–1.25mm, oblong to oblong-elliptic, subacute, glandular-pubescent, hairs
up to 1–1.6mm long. *Corolla* tube 6.5–9 × 1.8–2.6mm in mouth, cylindric abruptly expanded
near apex, limb bilabiate, 4.2–6.2mm across lateral lobes, posticous lobes 2–3 × 1.2–2.5mm,
anticous lobe 1.8–2 × 1–2.5mm, broadly ovate to suborbicular, sparsely and minutely glandular-puberulous outside or glabrous, very minutely glandular around mouth on upper surface of
limb, well bearded inside with clavate hairs up to 0.3–0.5mm long extending onto extreme base
of anticous lip, lobes ranging from creamy white to pale yellow with a median brown streak at
base of each lobe. *Stamens*: posticous filaments 1–1.6mm long, anthers 0.7–1.2mm, anticous
filaments 0.4–0.8mm, anthers 0.6–1mm, all filaments puberulous. *Stigma* 0.25–0.4mm long.
Style 3.7–5.7mm. *Ovary* 2–2.5 × 0.8–1.1mm. *Capsules* 6–8 × 3–3.5mm, glabrous. *Seeds* c.0.25
× 0.2–0.25mm.

Selected citations:

Cape. Namaqualand, 2817DC, about 1 mile SE of Viools Drift, Khusies [Viools] river, ix 1931, *Pillans* 6384 (BOL,
K); 2917 DB, 13½ miles N by W of O'okiep, c.3000ft, 21 vii 1957, *Acocks* 19347 (K, M, PRE, SRGH); 3018DA,
Kamabies, 24 xii 1908, *Pearson* 3466 (BOL, K, PRE, SAM, Z); 3017DB, Groenrivier just south of Garies, 7 ix 1980,
Goldblatt 5685 (NBG).

The distinguishing features of *J. amplexicaulis* are its habit (weak stems tufted from a woody
caudex), the stems terminating in long crowded inflorescences, leaves sharply ascending,
imbricate, the lowermost much reduced in size, then more or less abruptly larger, again
diminishing in size on the floriferous part of the stem (this is of course not always visible, not
only because only the upper part of a stem may be present on the sheet, but also because the
lower leaves may have fallen), the short corolla tube and small bilabiate limb.

It appears to be confined to Namaqualand, from the Richtersveld south to Eenkokerboom
(3018CC) and Loeriesfontein (3019CD). It favours streambeds, often in gorges or ravines. It
may grow rooted in sand or silt, or in the crevices of bare rock; Bolus recorded it as 'weakly
dependent from clefts of rocks' (*Bolus* 6567, BOL). Flowering occurs between May and
September.

The plant in Namibia mistakenly referred to *J. amplexicaulis* in the Prodromus Flora is *J.
bicolor;* see there to distinguish them.

GROUP 1b, 2

45. Jamesbrittenia racemosa (Benth.) Hilliard in Edinb. J. Bot. 49: 232 (1992). **Fig. 22D, Plate 3A.**

Lectotype (chosen here): Cape, Namaqualand, zwischen Mierenkasteel und Zwartdoorn rivier, 1000–2000 Fuss, 9 viii 1830, *Drège* (K; isolecto. E, LE, P, PRE, S, SAM, W, distributed as *Lyperia racemosa* Benth. a).

Syn.: *Lyperia racemosa* Benth. in Hook., Comp. Bot. Mag. 1: 378 (1836) & in DC., Prodr. 10: 359 (1846).
  Chaenostoma racemosum (Benth.) Diels in Bot. Jahrb. 23: 489 (1897), nom. illegit., non *C. racemosum* Benth. (= *Sutera racemosa* (Benth.) O. Kuntze).
  Sutera dielsiana Hiern in Fl. Cap. 4(2): 284 (1904). Type as above.
  Chaenostoma dielsianum (Hiern) Thell. in Vierteljahrss. Nat. Ges. Zürich 60: 410 (1915).
  Sutera tomentosa auct., non Hiern; le Roux & Schelpe, Namaqualand ed. 2: 150 cum ic. 151 (1988).

Annual herb, stems 50–600mm tall, up to c.4mm diam. at base, erect, simple but soon sparingly branched from the base and above, glandular pubescent, hairs up to 0.4–1mm long, some of them stout (up to 0.1–0.2mm across at base when flattened), leafy. *Leaves* opposite becoming alternate upwards, up to 20–55 × 5–30mm, elliptic tapering to a petiolar part rapidly shorter upwards, ± acute, margins serrate, teeth up to 1–2 (–3.5)mm deep, the larger ones mostly entire or 1-toothed, or occasionally several-toothed, both surfaces glandular pubescent, hairs up to 0.2–0.6mm long, small scattered glistening glands as well. *Flowers* solitary in upper leaf axils eventually forming long pseudoracemes. *Pedicels* up to 17–55mm long, glandular pubescent, hairs 0.2–0.4mm long. *Calyx* tube c.0.2mm long, lobes 3.5–6 × 0.8–1.3mm, lanceolate-oblong, glandular pubescent, hairs up to 0.2–0.6mm long. *Corolla* tube 15.5–18 × 2–3mm in throat, cylindric, abruptly expanded near apex, limb nearly regular, 12–24mm across lateral lobes, posticous lobes 5–10 × 3–6.3mm, anticous lobe 5–9 × 3.5–7mm, all lobes cuneate, conspicuously notched, the sinuses c.1–3mm deep, glandular puberulous outside, hairs up to 0.1–0.15mm long, transverse band of clavate hairs in throat extending briefly onto base of anticous lip, lobes often white, sometimes suffused with mauve or wholly shades of mauve, dark veins conspicuous, base of each lobe with a prominent 3-pronged mark running back down the tube, seemingly changing colour as the flower matures, from yellowish brown to deep violet, often drying orange, throat yellow/orange. *Stamens:* posticous filaments c.1–1.2mm long, anthers 1.2–1.8mm, anticous filaments c.0.8–1mm long, anthers 0.8–1.2mm, all filaments minutely puberulous. *Stigma* c.0.5–1mm long. *Style* 10–11mm. *Ovary* 2.5–3.5 × 1–1.8mm. *Capsules* 6–10 × 4–5mm, glandular-puberulous, glistening glands as well, confined to sutures. *Seeds* c.0.8–1 × 0.3–0.5mm, minutely reticulate but also patterned with transversely elongated pits arranged in chequer-board fashion.

Selected citations:

Cape. Namaqualand, 2917DA, Spektakel Pass, 17 viii 1974, *Goldblatt* 2373 (NBG, M). 3017BB, 15 miles N of Kamieskroon, 25 vii 1950, *Barker* 6228 (NBG, STE). 3017BD, Brakdam, 4 ix 1945, *Leighton* 1145 (BOL, PRE). 3017DB, 5 miles SW of Garies, near Groenerivier, 19 viii 1970, *Hall* 3768 (NBG, PRE, STE). Van Rhynsdorp div., 3118BA, 8 miles SE of Bitterfontein, 14 ix 1948, *Acocks* 14783 (K, PRE); 3118AA, Kliphuis se Kop, 7 ix 1974, *Nordenstam & Lundgren* 1713 (E,S). Between Van Rhynsdorp (3318DA) and Bitterfontein (3118AB), 21 ix 1929, *Grant* 4754 (E, M, PRE, S).

Jamesbrittenia racemosa is very nearly confined to Namaqualand: it ranges from about Henkries, just south of the Orange river, south and then south east to the environs of Van Rhynsdorp, between c.300 and 900m above sea level, growing in sandy places between shrubs and boulders. It flowers in spring, between July and October.

Its distinguishing marks are the deeply notched corolla lobes, the veins dark and conspicuous, the mouth boldly marked with trident-shaped purplish patches (which seem to change colour as the flower ages, and dry orange), capsules with small glistening glands down the sutures and seeds patterned with transversely elongated depressions arranged in chequer-board fashion, one of only two species of *Jamesbrittenia* to have such seeds (see fig. 22D).

It is often confused with *J. thunbergii*, which usually has more shallowly notched corolla lobes with inconspicuous veins, a less pronounced and differently coloured pattern around the mouth, capsules lacking glistening glands, and the reticulate seeds normal for the genus. The area of *J. thunbergii* lies mainly south and east of that of *J. racemosa*, but their areas meet in the 3118 degree square near Van Rhynsdorp.

46. Jamesbrittenia thunbergii (G. Don) Hilliard in Edinb. J. Bot. 49: 233 (1992).
Type: Cape, [Calvinia div., 3119], crescit in Carro infra Bockland, prope rivulos, Dec., *Thunberg* (sheet no. 14412 in herb. Thunb., UPS).

Syn.: *Erinus tomentosus* Thunb., Prodr. 103 (1800) & Fl. Cap. ed. Schultes 476 (1823), non Miller (1768). Type as above.
Manulea thunbergii G. Don, Gen. Syst. 4: 596 (1837–1838; precise date unknown).
? *Chaenostoma gracile* Diels in Bot. Jahrb. 23: 476 (1896). Type: Cape, Calvinia div., Hantam Mts., *Meyer* (B†).
Sutera fraterna Hiern in Fl. Cap. 4(2): 274 (1904). Type: Cape, Calvinia div., 3220CD, Brand Vley, *Johanssen* 3 (holo. K, iso. SAM).
? *S. gracilis* (Diels) Hiern in Fl. Cap. 4(2): 283 (1904).
Lyperia tomentosa Pilger in Bot. Jahrb. 48: 442 (1912). Type as for *Erinus tomentosus* Thunb.
Chaenostoma tomentosum Thell. in Vierteljahrss. Nat. Ges. Zürich 60: 410 (1915). Type as for *Erinus tomentosus* Thunb.
C. fraternum (Hiern) Thell. in Vierteljahrss. Nat. Ges. Zürich 60: 410 (1915).

Annual herb, stems 20–600mm tall, up to 5mm diam. at base, erect, simple but soon sparingly branched from the base and above, glandular pubescent, hairs up to 0.2–0.7mm long, up to c.0.1mm across flattened base, leafy. *Leaves* opposite becoming alternate upwards, up to 13–60 × 5–28mm, blade elliptic to ovate tapering to a petiolar part accounting for up to half total length, rapidly shorter upwards, ± acute, margins serrate, teeth up to 3mm long, themselves sometimes sparingly toothed, both surfaces glandular-pubescent, hairs up to 0.2–0.5mm long, small scattered glistening glands as well. *Flowers* solitary in upper leaf axils, eventually forming long pseudoracemes. *Pedicels* up to 12–30 (–54)mm long, glandular pubescent, hairs 0.2–0.5mm long. *Calyx* tube 0.2–0.5mm long, lobes 3.4–5.4 × 0.6–1mm, lanceolate-oblong, glandular-pubescent, hairs up to 0.1–0.3mm long. *Corolla* tube 7.5–17 × 1.5–2.3mm in throat, cylindric, abruptly expanded near apex, mouth laterally compressed, limb nearly regular, 10–18mm across lateral lobes, posticous lobes 4–8 × 2–5mm, anticous lobe 3.8–8 × 1.8–5mm, all lobes cuneate-oblong, shallowly notched (but sinus occasionally up to 2mm deep), glandular-puberulous outside, hairs up to c.0.1mm long, transverse band of clavate hairs in throat extending briefly onto base of anticous lip, lobes pale to deep mauve, throat yellow/orange, the colour extending onto the base of each lobe in a shortly 3-fid patch outlined with purple. *Stamens* deeply included, posticous filaments 1–2mm long, anthers 1–1.5mm, anticous filaments 0.5–1mm long, anthers 0.7–1.1mm, all filaments minutely puberulous. *Stigma* 0.2–0.5mm long. *Style* 4–10.4mm. *Ovary* 2–3 × 1–1.8mm. *Capsules* 5–10 × 3–5mm, very sparingly and minutely glandular-puberulous. *Seeds* 0.5–0.8 × 0.2–0.4mm.

Selected citations:

Cape. Van Rhynsdorp div., 3118DB, 15.8 miles from Van Rhynsdorp to Calvinia, c.900ft, 25 viii 1967, *Thompson* 374 (K, PRE, STE); 3119AC, Van Rhyn's Pass, 28 viii 1941, *Esterhuysen* 6009 (BOL, PRE). Calvinia div., 3119AB, Kareeboom, 550m, 4 ix 1986, *Burger & Louw* 259 (STE); 3119AC, near Naresie, 23 viii 1990, *Batten* 1024 (E); 3119BD, ± 28 miles N of Calvinia, 25 ix 1952, *Maguire* 1949 (NBG, STE). Ceres div., 3219DA, Gansfontein, 26 viii 1935, *Compton* 5523 (NBG). Sutherland div., 3220CA, Houthoek, 13 viii 1968, *Hanekom* 1062 (K, PRE). Laingsburg div., 3320BA, Whitehill Ridge, 18 viii 1942, *Compton* 11242 (NBG). Oudtshoorn div., Gamka Mountain Reserve, Tierkloof, 1100ft, 10 vii 1982, *Cattell & Cattell* 85 (STE).

Erinus tomentosus Thunb. is a later homonym of *E. tomentosus* Miller, which was clear to George Don, who re-named Thunberg's plant *Manulea thunbergii*. *Sutera tomentosa* Hiern

(1904, p.283) is illegitimate: he cited *Lyperia glutinosa* Benth., and the type of that name is automatically the type of *Sutera tomentosa*. *Sutera fraterna* Hiern is conspecific with Thunberg's plant. The type (sheet 14412 in herb. Thunberg, UPS) is precisely matched by *Masson* (BM), obviously part of one collection made when Thunberg and Masson were travelling together. Sheet 14370 (in herb. Thunberg, UPS) bears the manuscript name *Manulea incisa;* on the back of the sheet is written 'cult. in hort. *Uppsaliensis, Thunberg*'. The plant is probably *Jamesbrittenia thunbergii* grown from seed collected by Thunberg. (I have only been able to refer to the microfiche of the Thunberg herbarium; my knowledge was inadequate when I saw the sheet in UPS in 1989, and I could not then make any decision.)

The type of *Chaenostoma gracile* was destroyed in the Berlin fire. I have reduced the name to synonymy under *Jamesbrittenia thunbergii* with some doubt: Diels diagnosed his new species against '*Chaenostoma racemosum* Benth.' (which is *Sutera racemosa* (Benth.) O. Kuntze; Diels clearly meant *Lyperia racemosa* Benth.) saying, *inter alia*, it was much more glabrous; also, the capsules were said to be 4–5mm long. These are the two sticking points: *J. thunbergii* is decidedly glandular, and the capsules are normally never as short as 4mm. Of course, the plant may have been described in the wrong genus, but I have not been able to place it satisfactorily elsewhere.

Jamesbrittenia thunbergii is widely distributed in the western part of the Cape: most collections have come from Calvinia division, thence south to Ceres division and east to the Great Swartberg, but there is one record from Kenhardt division, on the road to Putsies (in which direction?). The plants often grow in karroid scrub, in open sandy or shaly patches between the bushes and rocks, between c.250 and 1000m above sea level. They appear after spring rain, flowering between June and October.

The species is often confused with *J. racemosa*; see there to distinguish them.

47. Jamesbrittenia aridicola Hilliard in Edinb. J. Bot. 49: 226 (1992).
Type: Cape, Namaqualand, 2917DB, 20 miles NE.of Springbok, 8 ix 1950, *Maguire* 340 (holo. NBG, iso. STE).

Annual herb, stem c.25–600mm tall, up to c.5–7mm diam. at base, erect, simple at first, soon branched from the base and above, glandular-pubescent, hairs up to 0.8–2.0mm long, some up to 0.2mm diam. at base, heads less than 0.1mm diam. *Leaves* opposite at base, soon alternate upwards, up to 25–75 × 10–40mm, blade narrowly to broadly ovate, apex acute, base ± abruptly contracted to a petiolar part up to roughly half total leaf-length, margins coarsely serrate, teeth entire to few-toothed, both surfaces glandular-pubescent, hairs up to 0.3–1.5mm long, longest on petiolar part, glistening glands as well particularly on lower surface. *Flowers* solitary in leaf axils almost from base of plant. *Pedicels* up to 15–56mm long, glandular-pubescent, hairs up to 0.2–0.6mm long, heads less than 0.1mm across (though basal stalk cell can be very stout). *Calyx* tube 0.2–0.4mm long, lobes 3.7–9 × 0.5–1mm, slightly accrescent in fruit, narrowly oblong, ± acute, glandular-pubescent, hairs up to 0.2–0.6mm long. *Corolla* tube 16–21.5 × 1.8–2.5mm in throat, cylindric, abruptly expanded near apex, mouth laterally compressed, limb nearly regular, c.10–18mm across lateral lobes, posticous lobes 4.2–7 × 3.2–6.5mm, anticous lobe 4.5–6.5 × 3–6.5mm, all lobes cuneate in outline, apex distinctly notched, sinus c.0.8–1.6mm deep, glandular-puberulous outside, hairs less than 0.1mm long, broad band of clavate hairs in throat extending very briefly onto base of anticous lip, lobes white, throat orange/yellow, the colour radiating into 3 points at base of each lobe, the median point tipped with a violet streak (always?). *Stamens*: posticous filaments 1.5–2.1mm long, anthers 1.2–1.7mm, anticous filaments 0.5–1mm, anthers 0.4–0.8mm, all filaments minutely puberulous. *Stigma* 0.6–1mm long. *Style* 12–19mm. *Ovary* 2–4 × 0.8–1.4mm. *Capsules* 7–10 × 3–4mm, glandular-puberulous. *Seeds* 0.4–0.7 × 0.2–0.25mm.

Selected citations:

Namibia. Lüderitz-Süd distr., 2716BA, farm Pockenbank, 26 viii 1963, *Merxmüller & Giess* 3147 (WIND). Warmbad distr., 2818DA, farm Auros (WAR 127), 7 x 1977, *Merxmüller & Giess* 32522 (WIND); 2817DA, slopes between Modder Drift [2817 DA] and Sjambok, ix 1931, *Pillans* 6445 (BOL, K).
Cape. Namaqualand, 2917DB, 25 miles N of O'okiep, 25 viii 1959, *Barker* 9047 (NBG, STE). Kenhardt distr., 2921AC, 6 miles W of Kenhardt, 15 vi 1961, *Schlieben* 8831 (K, M, PRE, S, SRGH, Z); 2919AA, Groot Pellaberg, 10 viii 1982, *Van Jaarsveld & Patterson* 6695 (NBG); 2820DC, 15 miles WSW of Alheit, c.2900ft, 28 iii 1948, *Acocks* 14266 (PRE).

Jamesbrittenia aridicola has a wide west-east distribution in the northernmost Cape and southernmost Namibia, from about Vioolsdrif and Springbok to Upington and Kenhardt, but there are also several records from Pockenbank (2716BA) in Lüderitz-Süd district. Its area thus lies mainly south of that of its ally *J. megadenia*, but they are known to be sympatric between roughly Warmbad and Upington; Pockenbank lies west of the area of *J. megadenia*. Both species have a white corolla limb patterned round the mouth in orange and violet, but the corolla lobes are much more deeply notched in *J. aridicola* (c.0.8–1.6mm) than in *J. megadenia* (up to c.0.3mm), and *J. aridicola* lacks the relatively huge glands that are a notable feature of *J. megadenia* (care must be taken because sand and dust stick to glandular heads and may give a false impression of size).

The plants grow in sand or gravel associated with rocks, on hills as well as in river valleys; flowering has been recorded between March and October. Seed must germinate after a shower of rain and plants will flower when only a few centimetres tall, but can persist long enough to become tall and stout-stemmed.

48. Jamesbrittenia megadenia Hilliard in Edinb. J. Bot. 49: 230 (1992).
Type: Namibia, Bethanien distr., 2617CC, farm Irene (BET 161), 8 vi 1976, *Giess & Müller* 14278 (holo. WIND, iso. PRE).

Annual herb, stems 60–500mm tall, up to c.8mm diam. at base, erect, simple at first, soon branched from the base and above, glandular-pubescent, hairs up to 0.5–1.8mm long, some up to 0.1mm diam. at base, very coarse scattered glands as well, c.0.2–0.4mm long, head 0.2–0.3mm in diam., light brown. *Leaves* opposite at base, soon alternate upwards, up to c.25–50 × 10–20mm, blade narrowly ovate, apex acute, base cuneate tapering into a petiolar part, margins coarsely serrate, teeth mostly entire, both surfaces glandular-pubescent, hairs up to 0.4–1.8mm long, longest on petiolar part, scattered small glistening glands as well, mostly on lower surface. *Flowers* solitary in leaf axils almost from base of plant. *Pedicels* up to 13–55mm long, glandular-pubescent, hairs up to 0.25–1mm long, some short coarse glands as well, heads c.0.2mm across. *Calyx* tube 0.2–0.3mm long, lobes 4.5–6.5 × 0.7–1mm, elongating somewhat in fruit, narrowly oblong, subacute, glandular-pubescent, hairs up to c.0.15–0.6mm long. *Corolla* tube 12.6–20 × 2.2–3mm in throat, cylindric, abruptly expanded near apex, mouth laterally compressed, limb nearly regular, c.12–17mm across lateral lobes, posticous lobes 4–8 × 3.4–7mm, anticous lobe 4–7 × 3.4–6mm, all lobes obovate to cuneate, apex ± truncate or slightly emarginate, glandular-puberulous outside, hairs mostly less than 0.1mm long, broad band of clavate hairs in throat extending very briefly onto base of anticous lip, lobes white, throat orange or golden yellow, the colour radiating into 3 points at base of each lobe, each point tipped with a dark violet streak. *Stamens*: posticous filaments 1–2mm long, anthers 1–1.5mm long, anticous filaments 0.4–0.8mm, anthers 0.6–0.8mm, all filaments minutely puberulous. *Stigma* 0.6–1.5mm long. *Style* 9.5–14.5mm. *Ovary* 2.5–3.8 × 1–1.8mm. *Capsules* 5.5–10 × 3.2–5mm, glandular-puberulous. *Seeds* c.0.6–0.8 × 0.3–0.4mm.

Selected citations:

Namibia. Keetmanshoop distr., 2617DD, Seeheim, iv 1913, *Dinter* 2951 (SAM); 2628AA, farm Itzawisis, 13 v 1963, *Giess, Volk & Bleissner* 6873 (K, PRE, WIND); 2718CA, Klein Karas, 31 vii 1923, *Dinter* 4746 (BOL, K, PRE, SAM, STE, Z); ibidem, 800m, 19 iv 1931, *Örtendahl* 691 (S, UPS).
Cape. Gordonia, 49 miles N of Upington, Grondneus, 21 iv 1928, *Pole-Evans* 2135 (PRE).

Jamesbrittenia megadenia derives its trivial name from the relatively huge glands that dot the pedicels and stems: under a lens they look like miniature mushrooms (fig. 10B). Although this whole group of species is notable for coarse glandular hairs, those of *J. megadenia* are outstanding and at once distinguish the species from its allies; *J. glutinosa* may have hairs almost as gross, but it is easily distinguished by its greatly reduced anticous anthers. *Jamesbrittenia aridicola* differs not only in its lack of huge glands but also in its decidedly notched corolla lobes.

Jamesbrittenia megadenia appears to be common in the south-eastern part of Namibia, mainly in Keetmanshoop and Warmbad districts, and there is one record from across the border in South Africa, north of Upington. The plants have been recorded on various stony sites (shale koppie, limestone and shale cliffs, stony watercourse), as well as in sandy river beds, flowering between April and June.

49. Jamesbrittenia glutinosa (Benth.) Hilliard in Edinb. J. Bot. 49: 229 (1992).

Type: Cape [Namaqualand], near the Gariep, 500ft, 11 viii 1830, *Drège* 3119 (holo. K; iso. LE, P, S, W).

Syn: *Lyperia glutinosa* Benth. in Hook., Comp. Bot. Mag. 1: 378 (1836) & in DC., Prodr. 10: 359 (1846).
   Sutera tomentosa Hiern in Fl. Cap. 4(2): 283 (1904), nom. illegit; Merxm. & Roessler in Prodr. Fl. S.W. Afr. p.p.
   Type as for *Lyperia glutinosa*.

Annual (or possibly perennating) herb, stems 20–450mm tall, up to 8mm diam. at base, erect, simple but soon sparingly branched from the base and above, glandular-pubescent, hairs up to 0.8–2mm long, up to 0.2–0.4mm across flattened base, heads c.0.1–0.2mm in diam., leafy. *Leaves* opposite but quickly alternate upwards, up to 18–45 × 8–30mm, blade of largest leaves ovate abruptly contracted to a petiolar part up to roughly ⅓ total leaf length, uppermost leaves elliptic or narrowly ovate and more gradually tapered, apex acute, margins serrate, teeth up to 3mm long, both surfaces glandular-pubescent, hairs up to 0.5–1mm long, small scattered glistening glands as well. *Flowers* solitary in each leaf axil, almost from base of plant, capable of forming very stout and crowded pseudoracemes. *Pedicels* up to 25–60mm long, glandular-pubescent, hairs up to 0.5–1mm long, some very stout, up to 0.25mm across flattened base, heads up to 0.15mm across. *Calyx* tube 0.25–0.4mm long, lobes 5.5–8.5 × 0.7–0.8mm, linear-oblong, glandular-pubescent, hairs up to 0.25–0.8mm long. *Corolla* tube 16–24 × 2.5–2.8mm in throat, cylindric, abruptly expanded near apex, mouth somewhat compressed laterally, limb nearly regular, 14–17mm across lateral lobes, posticous lobes 5–7.4 × 3.5–6mm, anticous lobe 5–7.5 × 4–6mm, all lobes cuneate, ± truncate or emarginate, glandular-puberulous outside, hairs c.0.1mm long, transverse band of clavate hairs in throat extending very briefly onto base of anticous lip, lobes mauve, lilac, rose or white, throat orange/yellow, the colour extending out around the mouth, 3 median deep violet bars near base of each lobe. *Stamens*: posticous filaments 2–3mm long, glandular-puberulous, anthers 1.4–2mm, anticous filaments 0.1–0.2mm long, anthers greatly reduced, 0.1–0.2mm long, or rarely anticous stamens aborted. *Stigma* 1–1.2mm long. *Style* 13.5–21mm. *Ovary* 3–4 × 1.2–1.8mm. *Capsules* 7.5–11 × 3–4.5mm, glandular-puberulous. *Seeds* c.0.3–0.7 × 0.2–0.3mm.

Selected citations:

Namibia. Lüderitz-Süd distr., 2716DC, farm Spitskop (LU 111), 14 vi 1976, *Giess & Müller* 14382 (PRE, WIND). 2816BB, Kahanstal, 5 xii 1934, *Dinter* 8126 (K, M, Z); 2816BB, Östlich Sendlingsdrif, 13 viii 1976, *Giess* 14607 (K, PRE, WIND). Warmbad distr., 2718CA, Klein Karas, 800m, 19 xi 1931, *Örtendahl* 74 (S). 2817BA, Ai-Ais, 26 vi 1974, *Nordenstam & Lundgren* 132 (E, S). 2817DD, Haib rivier, 11 ix 1963, *Merxmüller & Giess* 3647 (WIND). 2818CA, farm Sperlingspütz (WAR 259), 27 v 1972, *Giess & Müller* 12250 (PRE, WIND).
Cape. Namaqualand, 2816BD, Richtersveld, Numees Mine, 27 viii 1987, *Germishuizen* 4543 (PRE). 2817AD, Tatasberg, 21 ix 1986, *Williamson* 3623 (NBG). 2817AC, Hakiesdoringhoek in the Ganakouriep at the base of the Rosyntjiesberg, 400m, 7 vii 1987, *McDonald* 1313 (STE). 2818CC, 9 miles S of Goodhouse, 27 vii 1950, *Barker* 6272 (NBG, STE).

Jamesbrittenia glutinosa is confined to a small area of southernmost Namibia and northernmost Namaqualand, from the Namusberge in the west to the Klein Karas Berge in the east, thence south to the Orange River and the mountains of the Richtersveld, the westernmost record near Sendlingsdrif, the easternmost near Goodhouse. It has frequently been recorded from mountain slopes and gorges, often in granite, growing in the shelter of rocks or in their fissures, but there are also records from dry sandy and rocky watercourses. Flowering takes place principally between July and September, but there are scattered records between April and December.

The salient features of the species are its extraordinarily stout glandular hairs and the reduction of the anticous stamens: these are occasionally aborted, but mostly there is a minute anther containing a few pollen grains. The mouth is somewhat compressed laterally; in this, and in the colour patterning around the mouth, it resembles *J. megadenia*, with which it is partially sympatric.

50. Jamesbrittenia primuliflora (Thellung) Hilliard in Edinb. J. Bot. 49: 232 (1992). Type: Namibia, 2716AB, Tsaus südlich von Aus, 19 vii 1885, *Schenk* 142 (holo. Z).

Syn.: *Chaenostoma primuliflorum* Thellung in Vierteljahrsschr. Nat. Ges. Zürich 60: 409 (1915).
   Sutera primuliflora (sphalm. *primulina*) (Thell.) Range in Feddes Repert. 38: 265 (1935).

Herb, possibly sometimes perennating, stem c.25–600mm tall, up to 10mm diam. at base, erect, simple at first, soon branching from base and above, glandular-pubescent, hairs up to c.0.8–1.6mm long, sometimes very stout and up to 0.15mm diam. at base. *Leaves* opposite at base, soon alternate upwards, up to c.15–75 × 10–40mm, blade ovate, apex acute, base ± abruptly contracted to petiolar part up to roughly total leaf-length, margins coarsely serrate, teeth entire to few-toothed, both surfaces glandular-pubescent, hairs up to 0.5–1.2mm long, longest on petiolar part, scattered glistening glands as well. *Flowers* solitary in leaf axils almost from base of plant. *Pedicels* up to 6.5–50mm long, glandular-pubescent, hairs up to 0.2–0.6mm long, sometimes stout, head up 0.1mm across, base of stalk cell up to 0.15mm across. *Calyx* tube 0.2–0.4mm long, lobes 3.2–6.5 × 0.5–0.8mm, slightly accrescent in fruit, narrowly oblong, ± acute, glandular-pubescent, hairs up to 0.2–0.4mm long, glistening glands as well. *Corolla* tube 9–20 × 2–2.5mm in throat, cylindric, abruptly expanded near apex, mouth round, limb nearly regular, c.9–15.5mm across lateral lobes, posticous lobes 3.5–6.8 × 3.1–5.7mm, anticous lobes 3–6 × 2.7–5mm, all lobes obovate to cuneate, apex rounded, truncate or slightly emarginate, glandular-puberulous outside, hairs mostly less than 0.1mm long, often glistening glands as well, patch of clavate hairs inside throat on anticous side extending very briefly onto base of lip, lobes usually shades of light violet to lilac, sometimes white, throat orange/yellow, the colour radiating into 3 points at base of each lobe, the median point sometimes tipped with a dark violet streak. *Stamens*: posticous filaments 1–2.5mm long, anthers 1–1.8mm, anticous filaments 0.5–1.5mm long, anthers 0.6–1.2mm, very shortly exserted or at least visible in mouth, all filaments minutely puberulous. *Stigma* 0.2–0.8mm long. *Style* 7.5–18mm. *Ovary* 1.5–3 × 0.8–1.5mm. *Capsules* 4–8.5 × 2.5–3.2mm, glandular-puberulous. *Seeds* 0.4–0.7 × 0.25–0.3mm.

Selected citations:

Namibia. Rehoboth distr., 2316CA, farm Ubib (REH 396), Spreetshoogte Pass, 31 viii 1972, *Merxmüller & Giess* 28117 (S, WIND). Maltahöhe distr., 2416AB, Bergzebrapark, Naukluft (MAL 9), 1 ix 1972, *Merxmüller & Giess* 28162 (WIND). Gibeon distr., 2417AD, Fischfluss bei Kub, 15 viii 1963, *Merxmüller & Giess* 2801 (PRE, WIND); 2417DA, Haribes, 13 iv 1956, *Volk* 12371 (M); 2519AC, 17 miles SE of Eindpaal, c.3000ft, 10 iv 1960, *Leistner* 1801 (K, PRE, WIND). Lüderitz-Nord, 2516CB, farm Sinclair Mine, 29 ix 1959, *Giess* 2311 (PRE, WIND). Bethanien distr., 2516DC, Kunjas, 24 xi 1934, *Dinter* 8025 (B, BOL, K, M, S, Z); 2616BA, farm Gamochas (BET 31), 5 ix 1966, *Giess* 9444 (PRE, WIND); 2717CD, 14 miles W of Konkiep, 14 iv 1963, *Nordenstam* 2201 (M). Keetmanshoop distr., 2518CA, farm Mukorob (KEE 14), 13 v 1963, *Giess, Volk & Bleissner* 6839 (M, PRE, WIND). Lüderitz-Süd distr., 2616CA, farm Klein Aus, plateau of mountains, 11 viii 1959, *Giess & van Vuuren* 763 (M, PRE, WIND); Zwartaus, 31 iii 1929, *Dinter* 6232 (B, BM, BOL, E, K, M, PRE, SAM, STE, Z); 2716DD, farm Namuskluft (LU 88), 14 ix 1973, *Giess* 12958 (WIND).

The type of *J. primuliflora* is a seedling c.80mm tall, but the plant can become astonishingly robust, with the basal part of the stem c.10mm in diameter; like many desert annuals, the plants are opportunists that will flower when very young, set seed quickly, then persist if conditions are favourable. The species is easily distinguished from its allies by the round corolla mouth in which the anticous anthers are visible or even shortly exserted.

Jamesbrittenia primuliflora appears to be confined to Namibia, where it is widely distributed from the Auas Mountains near Windhoek south almost to the Orange river, on the high ground, possibly always above c.900m. The plants favour rocky sites, on mountain tops or slopes, or in river beds, growing in sand between and under rocks. Flowering has been recorded in all months.

Merxmüller & Giess 3609 (M, PRE, WIND), from the side of the road at Wasser railway station in Keetmanshoop district, is either an undescribed species or a hybrid between *J. primuliflora* and *J. canescens*. It has the habit, foliage and small capsules of *J. canescens*, but the indumentum is far coarser than in that species and resembles that of *J. primuliflora*; the corolla too is similar to that of *J. primuliflora*.

51. Jamesbrittenia fimbriata Hilliard in Edinb. J. Bot. 49: 229 (1992).
Type: Namibia, Lüderitz-Nord distr., 2415DD, Vreemdelingspoort, 10 v 1976, *Oliver, Müller & Steenkamp* 6528 (holo. PRE, iso. K).

Possibly perennial, height unknown, branches c.200mm long, 3mm diam. at base, very twiggy, glandular-pubescent, hairs up to 0.8–1.5mm long, glistening glands as well. *Leaves* alternate, up to c.14–47 × 8–30mm, blade ovate in outline, abruptly or more gradually contracted to a petiolar part up to c.⅓ total leaf length, margins serrate, both surfaces glandular-pubescent, hairs up to c.0.5mm long, glistening glands as well mainly on lower surface. *Flowers* solitary in all the leaf axils. *Pedicels* up to 17–24mm long, glandular pubescent with delicate hairs up to c.0.6–0.8mm long, stout coarse hairs as well, up to c.0.25mm long, head c.0.1mm in diam. *Calyx* tube 0.2mm long, lobes 3.7–4 × 0.4–0.6mm, narrowly oblong, ± acute, glandular-pubescent, hairs up to 0.4–0.5mm long, glistening glands as well. *Corolla* tube 12.5–13 × 2–2.6mm in throat, cylindric, abruptly expanded near apex, mouth round, limb nearly regular, 10–11mm across lateral lobes, posticous lobes 4–4.2 × 3.1–3.2mm, anticous lobe 4 × 2.8–3mm, all lobes cuneate-oblong, apex ± truncate, glandular-puberulous outside, hairs up to 0.1-0.15mm long, inside a band of clavate hairs down back of tube between posticous filaments, extending in a broad fan to base of anticous lip and there up to 1.2mm long, lobes mauve, throat yellow. *Stamens*: posticous filaments 2–2.2mm long, anticous filaments 1mm long, anthers 0.7–0.8mm long, very shortly exserted. *Stigma* 0.2–0.5mm long. *Style* c.11.5mm. *Ovary* 2 × 1mm. *Capsules* 2.5 × 2mm (probably ill-developed), glandular-puberulous. *Seeds* c.0.6 × 0.4mm (only 2 seen).

Jamesbrittenia fimbriata is known only from the type collection. The trivial name refers to the remarkable fringe of long hairs at the base of the lower lip; these extend down the throat and end in a longitudinal band about 4mm long between the posticous filaments. The hairs inside the tube are c.0.4mm long, those on the anticous lip 1mm or more. The flowers of *J. fimbriata* and *J. primuliflora* are similar in the general form of the corolla, but they differ markedly in the size and disposition of the hairs inside the tube; in *J. primuliflora* the hairs are much smaller (c.0.25–0.4mm long) and fewer: there is no posticous band between the filaments. In *J. fimbriata* only the throat is yellow; there is no radiation of the colour onto the limb as in *J. primuliflora*, but this needs confirmation on a range of specimens. The plants were in full flower in May. Only two capsules are present (on the holotype) and they are surely abnormally small, though one contained two normal seeds.

The collectors recorded neither the habit nor habitat of the plant. It is probably bushier than *J. primuliflora* and the leaves have dried grey-green; also, the leaves are probably less deeply cut and less sharply toothed than those of *J. primuliflora*: only one large leaf is represented (in a capsule on the K sheet); the largest of the other leaves are all damaged.

GROUP 2, 1

52. Jamesbrittenia acutiloba (Pilger) Hilliard in Edinb. J. Bot. 49: 225 (1992).
Lectotype (chosen here): Namibia, Waterberg, an der Quelle in Halbschatten, 6 ii 1911, *Dinter* 1797 (SAM).

Syn.: *Lyperia acutiloba* Pilger in Bot. Jahrb. 48: 440 (1912).
    Chaenostoma acutilobum (Pilger) Thellung in Vierteljahrsschr. Nat. Ges. Zürich 60: 411 (1915).
    Sutera acutiloba (Pilger) Roessler in Mitt. Bot. München 6: 16 (1966); Merxmüller & Roessler, Prodr. Fl. S.W.
        Afr. 126: 48 (1967).

Shrublet up to at least 600mm tall, drying fuscous, well branched, branches up to at least 5mm in diam. at base, divaricate, glandular, villous, hairs up to 2–3mm long. *Leaves* always opposite, blade up to 8–30 × 8–30mm, ovate, abruptly contracted to a petiolar part up to 2–4mm long, apex obtuse to acute, base often ± truncate, margins shallowly lobed and toothed, lobes often broadly oblong, teeth obtuse to acute, some of the reduced upper leaves with linear-oblong lobes cut up to halfway to midrib, both surfaces glandular, hairs up to 1.5–2mm long, scattered glistening glands as well mainly on lower surface. *Flowers* solitary in upper leaf-axils. *Pedicels* up to 1–6mm long. *Calyx* tube c.0.2mm long, lobes 6.5–9 × 0.5–0.8mm, linear-oblong, glandular, hairs up to 0.8–1mm long. *Corolla* tube 19–25 × 2.5–3mm in mouth, cylindric, abruptly expanded near apex, limb nearly regular, c.13–14mm across lateral lobes, posticous lobes c.5–5.5 × 4–5mm, anticous lobe c.5.5–7 × 4–5mm, cuneate-oblong, apex ± truncate to emarginate, glandular-pubescent outside, hairs up to 0.3–0.4mm long, broad band of clavate hairs in throat extending very briefly onto base of anticous lip, lobes white, throat yellowish, dark streaks in throat. *Stamens*: posticous filaments 2.3–3mm long, anthers 1.3–1.8mm; anticous filaments 0.5–0.8mm, anthers 1–1.2mm, all filaments weakly bearded. *Stigma* 0.3–1mm long. *Style* 14–21mm. *Ovary* 2–2.5 × 1mm. *Capsules* 5–6 × 3–3.5mm, minutely glandular puberulous, sometimes a few glistening glands at apex. *Seeds* c.0.4–0.6 × 0.3mm.

Selected citations:

Namibia. Otjiwarongo distr., 2017AC, Waterberg, 5500ft, 14 vii 1954, *A.S.L. Schelpe* 181 (BOL, K); ibidem, Great Waterberg, just below the plateau, 3 iv 1968, *Kers* 2975 (S); 2017CA, Klein Waterberg, 2 vi 1940, *Volk* 3001 (M). Waterberg, farm Hohensee, 25 v 1968, *Meyer* 1201 (WIND).

Pilger cited two specimens, *Dinter* 775 *and Dinter* 1797, both of which were destroyed in the Berlin fire. A duplicate of *Dinter* 1797 is lodged in SAM, and this has been designated lectotype.

    Jamesbrittenia acutiloba appears to be confined to limestone rocks in the Waterberg, where it grows in the crevices of cliffs or in the shelter of outcropping rocks, damp and partially shaded. It is in full flower between April and July.

    The distinguishing features of the species are the long, glandular hairs on all parts, constantly opposite leaves with very short petioles, very short pedicels, long narrowly oblong calyx lobes, long corolla tube and white limb.

53. Jamesbrittenia dolomitica Hilliard in Edinb. J. Bot. 49: 228 (1992).
Type: Namibia, Grootfontein distr., 1917BB, near Meteor, 5000ft, 19 vii 1965, *Leach & Bayliss* 13007 (holo. WIND, iso. K, S).

Subshrub between c.450mm and 1m tall, probably spindly though well branched, main stem up to c.6mm in diam. (no basal parts seen), very glandular with minute hairs overlaid by stout coarse ones up to 1.2–4mm long, rather stiffly patent. *Leaves* opposite becoming alternate upwards, blade up to 10–27 (–45) × 10–26(–45)mm, broadly ovate, obtuse to acute, abruptly contracted to a petiolar part up to 2.5–10(–18)mm long (ratio of petiole to blade 1: 2.5–4.4), margins lobed and toothed roughly ⅓–½ of way to midrib, uppermost leaves sometimes very sharply and irregularly toothed, glandular hairs as on stems, scattered glistening glands as well on lower surface. *Flowers* solitary in upper leaf-axils. *Pedicels* up to 6–18mm long, glandular as stems. *Calyx* tube c.0.2–0.3mm long, lobes 5.5–9 × 1–1.8mm, narrowly elliptic, subacute, very glandular, hairs up to 0.4–1mm long. *Corolla* tube 12–14.5 × 2.5–3mm in mouth, cylindric, abruptly expanded near apex, limb 8–11mm across lateral lobes, posticous lobes 2.7–5 × 2.3–4.2mm, anticous lobe 2.8–5.2 × 2.8–4.3mm, cuneate-oblong, apex ± truncate to retuse, glandular puberulous outside, hairs up to c.0.2mm long, broad band of clavate hairs in throat extending briefly onto base of anticous lip, lobes ranging from creamy white to yellow, throat yellow/orange. *Stamens*: posticous filaments 2.1–3.8mm long, anthers 1.2–1.4mm, anticous filaments 0.6–1.8mm long, anthers 0.8–1mm, visible in mouth. *Stigma* 0.5–0.8mm long. *Style* 10.2–13.5mm. *Ovary* 1.8–2.3 × 1.1–2mm. *Capsules* 3.5–5 × 2.5–3mm, glandular puberulous. *Seeds* c.0.5 × 0.25mm (few seen).

Selected citations:

Namibia. Grootfontein distr., 1917DA, Auros (Otavi), 14 ii 1925, *Dinter* 5641 (B, BOL, M, PRE, SAM, Z). 1917DB, Guchab, 18 iv 1939, *Volk* 50 (M); ibidem, 12 iii 1974, *Merxmüller & Giess* 30260 (WIND). West of Grootfontein, c.1917DB, *Schoenfelder* 340 (K).

Jamesbrittenia dolomitica is known only from limestone rocks between Auros in the west and Grootfontein in the east, and the epithet *dolomitica*, which Dinter wrote on a sheet of *Dinter* 5641(B), has therefore been adopted. On that sheet, Dinter diagnosed against *J. [Sutera] acutiloba* what he recognised as an undescribed species. It differs from *J. acutiloba* by its leaves, alternate upwards (not always opposite), pedicels up to 6–18mm long (not 1–6mm), corolla tube 12–14.5mm long (not 19–25mm). Like *J. acutiloba*, *J. dolomitica* turns fuscous on drying; the allied *J. heucherifolia* on the other hand remains green. *Jamesbrittenia dolomitica* and *J. heucherifolia* look different, but the differences are difficult to express in precise terms. *Jamesbrittenia dolomitica* has coarser hairs on the vegetative parts; they are clearly patent, while the remarkably delicate hairs of *J. heucherifolia* tend to mat together; the petioles are mostly shorter (up to 2.5–10mm long, not 7–18mm) as well as being shorter in relation to the length of the blade (blade: petiole 2.5–4.4: 1, not 1.1–2.4: 1), and the inflorescences are more congested (length of second internode from base of inflorescence 2–8 (–10)mm long, not (7–) 10–16(–26)mm). The hairs on the calyx of *J. dolomitica* are up to 0.4–1mm long (not 0.6–3mm).

All three species occupy geographically discrete areas: the area of *J. acutiloba* lies SW of that of *J. dolomitica*, on the Waterberg, where it too is a limestone endemic. The area of *J. heucherifolia* is away to the N.W., a linear distance of roughly 200km to the nearest known site, in the Kaokoveld. Plants of *J. dolomitica* grow in the partial shade of cliffs and boulders about 1500m above sea level, and possibly flower in any month.

54. Jamesbrittenia heucherifolia (Diels) Hilliard in Edinb. J. Bot. 49: 230 (1992).
Type: Angola, Mossamedes distr., banks of the river Bero, near Cavalheiros, vii & viii 1859, *Welwitsch* 5805 (holo. B†; iso. BM, COI, K, PRE). See comments below.

Syn.: *Chaenostoma heucherifolium* Diels in Bot. Jahrb. 23: 475 (1897).
    Sutera heucherifolia (Diels) Hiern in Cat. Welw. Afr. Pl. 1: 757 (1898); Hemsley & Skan in Fl. Trop. Afr. 4(2): 306 (1906).
    S. gossweileri Skan in Fl. Trop. Afr. 4(2): 307 (1906). Type: Angola, Mossamedes, Fazenda Boa Vista, 200ft above sea level, xi 1900, *Gossweiler* 62 (holo. K, iso. C).

Dwarf shrublet, but will flower young and look annual, several stems tufted from the base, c.150–700mm long, erect or straggling, well branched, main branches up to 6mm in diam., very glandular, some hairs minute, many up to 1.5–3mm long, remarkably delicate, often matted together in dried specimens. *Leaves* opposite becoming alternate upwards, blade up to 11–34 × 10–30mm, ovate to broadly ovate or subrotund, obtuse to subacute, base abruptly contracted to a petiolar part up to 7–18mm long (ratio of petiole to blade 1: 1.1–2.4), margins lobed and toothed, lobing usually shallow, occasionally up to ½ way to midrib, glandular as stems, but glistening glands more conspicuous. *Flowers* solitary in leaf-axils, the plant very floriferous.

Pedicels up to 13–25mm long, glandular as stems. *Calyx* tube 0.1–0.3mm long, lobes 5–9 × 1–2.5mm, elliptic or ± oblong to elliptic-oblong, acute to subacute, very glandular, hairs up to 0.6–3mm long. *Corolla* tube 12–18 × 2.2–3mm in mouth, cylindric, abruptly expanded near apex, limb 6.5–13mm across lateral lobes, posticous lobes 2–5 × 2–5mm, anticous lobe 2–4.5 × 2–5mm, ± cuneate-oblong, apex ± truncate, glandular puberulous outside, hairs up to c.0.1–0.2mm long, occasionally longer, broad band of clavate hairs in throat extending briefly onto base of anticous lip, lobes white, mauve or yellow, throat shades of yellow and orange, once recorded as red. *Stamens*: posticous filaments 1.5–2mm long, anthers 1–1.8mm, anticous filaments 0.6–1mm, bearded or glabrous, anthers 0.8–1.5mm, visible in mouth or very shortly exserted. *Stigma* 0.3–0.8mm long. *Style* 10–16mm. *Ovary* 1.8–2 × 0.8–1.3mm. *Capsules* 3–5 × 2.5–3mm, glandular-puberulous. *Seeds* c.0.4–0.6 × 0.25–0.3mm.

Selected citations:

Angola. Mossamedes distr., Pediva, 1 vi 1965, *Henriques* 486 (K); Rio Giraul, 21 ii 1961, *Barbosa* 9493 (COI, K, PRE, SRGH). Iona, 13 i 1956, *Torre* 8433 (LISC); ibidem, 825m, 23 i 1975, *Ward* 103 (WIND).
Namibia. Kaokoveld, 1712BA, Otjimborombonga, 700m, 14 vii 1976, *Leistner et al* 137 (PRE, SRGH, Z); 1812BA, Sanitatas, 5000ft, 15 vi 1961, *Hall* 434 (NBG, PRE). 1712BC, Kapupa valley, 18 viii 1956, *Story* 5871 (M, PRE, WIND).

The holotype of *J. heucherifolia* was destroyed in the Berlin fire. Diels described his plant as suffruticose, his specimen comprising a well branched twig about 30cm long. The right hand specimen on the sheet of *Welwitsch* 5805 in BM is precisely that; there is also another, smaller, twig, and two whole young plants, together with a young plant of *Camptoloma rotundifolia*. The specimen in K is a whole young plant that looks annual. Skan would have taken this specimen as typical *J. heucherifolia* when he described *Sutera gossweileri*, distinguishing it from *J.* [*Sutera*] *heucherifolia* by its perennial habit.

The types of both names came from the environs of Mossamedes in Angola; Welwitsch described the flowers as white, throat yellow ('*flores albi tubo et fauce obiter flavescentes*'); Gossweiler wrote 'flowers yellow'. All subsequent collections from Angola (all from Mossamedes district, 1412 DD, 1512 AA, 1612 AA, DC) have flowers described either as white, or white with a yellow throat. The species extends southwards into the Kaokoveld of northernmost Namibia. There, only two collections (*Leistner* et al 137, Otjimborombonga, 1712 BA, and *Davies et al* 56, Otjihipaberg, 1712 BC) have white flowers with a yellow throat; *Davies et al* 57 had 'flowers blue with darker stripes in centre'; both specimens came from the same place, about nine miles up the Kapupa river on the south side of the Otjihipaberg. *Giess* 10529, from the Otjihipaberg near Okango, also had flowers with a coloured limb: 'violett mit dunkel-violetten Malen, in Schlund gelb'. Plants with similarly coloured flowers have been collected near Otjinungua (1712 AB), in the Cunene gorge (1712 AB), in the Kapupa valley (1712 BC), Hartmans valley (1712 CB) and at Sanitatas (1812 BA). In blue or violet flowers, the calyx lobes are often oblong or oblong-elliptic, not elliptic as in white or yellow flowers. Plants with flowers variously described as 'yellow with red throats' (*Leistner et al* 237), 'yellow with orange centre' (*Leistner et al* 170), 'leuchtend gelb' (*Giess* 8988), and 'dottergelb' (*Meyer* 1301) have come from the vicinity of the Epupa Falls (1713AA), Otjimborombonga (1712BA), Baynes Mountains at Okonbambi (1712BB) and at Otjipemba (1712BB) and near Orukawe (1712BC).

In Angola, the plants have been found in damp ground around springs, and along sandy and rocky river and stream banks, from near sea level to c.825m; in Namibia, the collections are all montane, often from gorges and ravines, growing in rock fissures around springs and other damp sites, or in the beds of sandy watercourses, between c.700 and 1000m above sea level. Only field work can establish whether or not yellow-flowered plants are confined to dolomite, white and mauve to granite; collectors' notes are scanty. Plants can possibly be found in flower in any month.

In Prodr. Fl. S.W. Afr., the Namibian collections were included under *J. [Sutera] pallida*, from which *J. heucherifolia* is distinguished by its much more delicate and often longer indumentum (for example, hairs on calyx c.0.6–3mm long, not 0.2–1mm) and leaves with mostly longer petioles and more shallowly lobed margins. The species can also be confused with *J. dolomitica*: see there.

55. Jamesbrittenia pallida (Pilger) Hilliard in Edinb. J. Bot. 49: 232 (1992).
Type: Namibia [2116 AD], Giftkopje, 12 ii 1900, *Dinter* 1433 (holo. B†, iso. Z).

Syn.: *Lyperia pallida* Pilger in Bot. Jahrb. 48: 441 (1912).
  Chaenostoma ambleophyllum Thell. in Vierteljahrsschr. Nat. Ges. Zürich 60: 408 (1915). Type: Namibia, Gam-Koichas, 19 ii 1900, *Dinter* 1465 (holo. Z).
  C. schinzianum Thell. in Vierteljahrsschr. Nat. Ges. Zürich 60: 410 (1915). Type: Namibia, Quaaiputs, i 1899, *Dinter* 203 (holo. Z).
  Sutera pallida (Pilger) Roessler in Mitt. Bot. Staats. München 6: 16 (1966); Merxm. & Roessler in Prodr. Fl. S.W. Afr. 126: 53 (1967) pp., excl. *Chaenostoma fleckii* and many of the specimens cited under *S. pallida* and in the discussion.
  S. tomentosa auct., non Hiern; Hemsley & Skan in Fl. Trop. Afr. 4(2): 305 (1906).

Dwarf shrublet, woody caudex c.10–20mm in diam., stems several tufted from the base, c.150–450mm tall, well branched, branches up to c.6mm diam., very glandular, most hairs minute but few to many up to 0.5–2mm long, these sometimes only on lower part of stem. *Leaves* opposite becoming alternate upwards, blade up to 6–20(–40) × 3–14(–36)mm, oblong-elliptic to ovate in outline, base cuneate or more abruptly contracted to a petiolar part up to 2–8(–10)mm long (roughly ¼–½ as long as blade), apex obtuse to subacute, margins typically lobed halfway or more to midrib, the lowermost sometimes to midrib, sometimes lobing shallower (only ⅓ of way to midrib), lobes ± oblong, toothed, both surfaces glandular as stems, some scattered glistening glands as well. *Flowers* solitary in upper leaf-axils. *Pedicels* up to 9–20(–30)mm long, glandular. *Calyx* tube c.0.2mm long, lobes 4–8 × 0.7–1.8mm, oblong to spathulate, subacute, very glandular, some hairs up to 0.2–1mm long. *Corolla* tube (13–) 16–22 × 2–2.8mm in mouth, cylindric, abruptly expanded near apex, limb c.12–14mm across lateral lobes, posticous lobes 3.5–6 × 3.7–5.2mm, anticous lobe 4–6.5 × 3.2–4.8mm, oblong to cuneate-oblong, apex ± truncate to retuse, glandular puberulous outside, hairs mostly up to 0.15–0.3mm long, occasionally longer, broad band of clavate hair inside tube extending very briefly onto base of anticous lip, lobes either pale yellow with throat orange, or white with (always?) violet streaks at base of lobes, and throat yellow (always?). *Stamens*: posticous filaments 1.5–2.5mm long, anthers 1–1.6mm, anticous filaments 0.5–1mm long, anthers 0.8–1mm, just visible in mouth, filaments bearded with clavate hairs or glabrous. *Stigma* 0.4–1mm long. *Style* 10–17mm. *Ovary* 2–2.5 × 1–2mm. *Capsules* 4–7 × 2.5–3.5mm, glandular puberulous. *Seeds* 0.3–0.6 × 0.2–0.3mm.

Selected citations:

Namibia. Karibib distr., 2115DC, Ameib, 18 i 1934, *Dinter* 6855 (BOL, BM, S, WIND, Z): farm Ameib (KAR 60), Erongo Südseite, 31 v 1978, *Giess* 15268 (WIND); farm Otjosondu (KAR 36), Rote Berge, 6 vi 1961, *Giess* 3448 (K, NBG, W, WIND). 2216AC, Otjimbingwe, 900m, vi 1886, *Marloth* 1407 (M, SAM, W). Swakopmund distr., 2115CC, Zwischen der Kleinen Spitzkoppe und der Strasse Usakos - Hentiesbaai, 23 xii 1967, *Meyer* 1034 (WIND).

Typical *J. pallida* has leaves more or less oblong in outline, lobed halfway or more to the midrib, and abruptly contracted to a petiolar part roughly a quarter the length of the blade, both stems and leaves densely glandular puberulous together with some hairs up to c.1mm long, and corolla tube c.20mm long. Pilger described the flowers as pale yellow (*pallide flavidi*), though no colour is mentioned on the label of the only isotype now extant. Dinter wrote on the label the manuscript name *Lyperia pallida* and Pilger obviously adopted his epithet. The collectors of three specimens that match the type closely recorded yellow flowers (*Wanntorp* 843, S, *Puff* 790729 - 1/1, J and *Ihlenfeldt* 1959, PRE); the first two came, one in 1961, one in 1979, from a site (2115DB) c.20km south of Omaruru, roughly 40km SW of the Giftkuppe (2116AD), provenance of the type specimen; the third was found on the farm Otjimbojo-West, 2116CC. The colour is still visible in *Wanntorp* 843, and the throat is orange. The rest of the material that I have assigned to *J. pallida* has flowers variously described as white (weiss), pure white (reinweiss), or white with violet streaks (reinweiss mit violetten Strichmalen), though many collectors fail to mention colour; neither colour nor patterning is visible in these specimens because the flowers have darkened on drying. They came from the Erongo Mountains, 2115DA, DC, and the Klein and Gross Spitzkoppe, the Rote Berge and other granite mountains, 2115CC, DD. One specimen (*Dinter* 7033, B, BM, BOL, K, M, S, WIND, Z) from Unduasbank (2115DD) bears a manuscript name (mentioned in Prodr. Fl. S.W. Afr. 126 p.54). It was found under granite overhangs and is a particularly lax specimen with large leaves, both petioles and pedicels long for the species (measurements in parentheses in the formal description). Specimens from the Erongo Mountains most closely resemble the type of *Chaenostoma ambleophyllum*, which came from Gam-Koichas, and has leaves mostly less deeply cut than is typical. I have failed to trace Gam-Koichas, but the specimen was collected a week after the type of *Jamesbrittenia pallida*, so obviously during Dinter's expedition from Windhoek to the twin Omatako peaks (2116BA), thence to the massif Etjo (2216AB), Giftkoppe (2116AD) and Omburo (2116CB), and back to Windhoek via Karibib (2115DD) and Swakopmund (see Dinter, Botanische Reisen in Deutsch-Südwest-Afrika, in Feddes, Repertorium Beiheft 3: 15, 1918). Then in 1934, Dinter collected the species again at Ameib (2115DC) on the southern flank of the Erongo Mountains, and gave the specimen yet another manuscript name (*Dinter* 6855, BM, BOL, S, WIND, Z); the foliage is that of typical *Jamesbrittenia pallida* and the flowers may have been white. Other specimens with typical foliage but no mention of flower-colour have come from Orumbo (2217BD) and the farm Otjozondu (2216AA) in Okahandja district, and from gneiss hills on the road from Tinkas to Walvis Bay (2215CD). The type of *Chaenostoma schinzianum* also has foliage typical of *Jamesbrittenia pallida*, and the flowers were probably white. It came from Quaaiputs, in January 1899, when Dinter botanised from his base at Salem towards Windhoek (see Feddes Repertorium Beiheft 3: 11, 1918), so, although I have failed to trace the locality, it is possibly somewhere in the Khomas Hochland (2216C).

Even with the adoption of a broad concept of *J. pallida*, some specimens that are clearly close to *J. pallida* remain problematical. For instance, *Dinter* 6999 (B, K), which in most characters agrees well with typical *J. pallida*, had pale blue flowers ('fl. pall. - coer.' recorded by Dinter) and the sheet in B bears a manuscript name reflecting the colour. The plant came from 'Grosser Granitberg S.Ö. Karibib' (c.2115DD). *Giess* 12766 (WIND), from a valley at the western foot of the Pontokberge (2115CC) records yellowish-white flowers (gelblich-weiss); the leaves are relatively shallowly cut. *Seydel* 4037 (M, WIND), from the farm Onjossa (2216AB) has the typical foliage of *J. pallida*, but the flowers were variable in colour, often on the same shrublet ('Blüten verschiedenfarbig, weiss, hellgelb, dunkelgelb oft am selben Zwergstrauch').

Two orange-flowered specimens are before me: *Schwerdfeger* 2/337 (WIND) from the farm Westfalenhof (2216AB, the same ¼ degree square as *Seydel* 1407), and *Ihlenfeldt* 1959 (M) from the farm Otjimbojo-West (2116CC). It was suggested (in Prodr. Fl. S.W. Afr. 126: 53, 1967) that *Seydel* 4037 might be a hybrid between *J. pallida* and the orange-flowered *J. lyperioides*. This cannot be ruled out, but on the other hand, *J. pallida* may be displaying an inherent tendency to marked variation in flower-colour (cf. *J. heucherifolia*, no. 54), or the colour may change with age. Field work is needed to investigate both flower-colour and leaf-cutting. Two collections are a mixture of twigs with typical deeply cut leaves, others with only shallow lobing: *Seydel* 181 (M, SRGH), from Nudis (2215BC), *De Winter & Hardy* 8005 (K, M, PRE, WIND) from the farm Anschluss (2215DB?).

The plants grow in the crevices and crannies of granite (or occasionally gneiss) rocks on mountains and hillocks, at least 1000m above sea level, flowering mainly between December and July.

It may be helpful to comment on some of the specimens cited under *Sutera pallida* in Prodr. Fl. S.W. Afr. 126: 53, 1967. I have maintained *Chaenostoma fleckii* as a distinct species: *Fleck* 758, type, *Merxmüller & Giess* 884, which the authors of the Prodromus thought might be a hybrid. *De Winter* 6054, also cited as a possible hybrid, is *J. integerrima*. *Dinter* 5641, 7652, *Schoenfelder* 340, *Volk* 50 are *J. dolomitica*. *Story* 5871, 5872, *De Winter & Leistner* 5795, *Hall* 434, *Giess* 8925 are *J. heucherifolia*.

56. Jamesbrittenia fleckii (Thell.) Hilliard in Edinb. J. Bot. 49: 229 (1992).
Type: Namibia, 'Gansberg & Tabgebiet' [see below], v 1889, *Fleck* 758 (holo. Z).

Syn.: *Chaenostoma fleckii* Thell. in Vierteljahrsschr. Nat. Ges. Zürich 60: 408 (1915).

Perennial herb, stock c.5mm in diam., stems several tufted from the base, c.200–400mm tall, well branched, branches up to c.3.5mm in diam., very glandular, hairs up to 1–2.5mm long. *Leaves* opposite becoming alternate upwards, blade up to 7–30 × 7–27mm, broadly ovate abruptly contracted to a petiolar part up to 5–15mm long (roughly ⅓–½ length of blade), apex obtuse to subacute, margins typically shallowly lobed (up to ⅓ of way to midrib), rarely lobing somewhat deeper, lobes typically oblong, ± obtuse, toothed, both surfaces glandular as stems, dense glistening glands as well. *Flowers* solitary in upper leaf-axils. *Pedicels* up to 9–35mm long, glandular. *Calyx* tube 0.2–0.4mm long, lobes 5.6–9 × 1–1.8mm, linear-oblong to subspathulate, ± acute, very glandular, hairs up to 0.5–2mm long, dense glistening glands as well. *Corolla* tube 13–16 × 2.2–3mm in mouth, cylindric, abruptly expanded near apex, limb 10–16mm across lateral lobes, posticous lobes 4.5–6.4 × 2.5–5mm, anticous lobe 3.5–6.7 × 2.5–4.6mm, oblong to cuneate-oblong, apex ± truncate to retuse, glandular-pubescent outside, hairs up to 0.3–1mm long, usually glistening glands as well, broad band of clavate hairs inside tube extending briefly onto base of anticous lip, lobes yellow to orange. *Stamens*: posticous filaments 1.3mm–3mm long, anthers 1–2mm, anticous filaments 1–1.5mm, anthers 0.8–1mm, just visible in mouth, all filaments minutely glandular at apex, occasionally a few clavate hairs on lower part of posticous ones. *Stigma* 0.3–1mm long. *Style* 9–14.5mm. *Ovary* 2–3 × 1–1.5mm. *Capsules* 6–8 × 2.5–3mm, minutely glandular, usually glistening glands as well. *Seeds* c.0.5–0.6 × 0.3mm.

Selected citations:

Namibia. Swakopmund distr., 2315BD, Kuiseb-durchfahrt bei Brücke, 13 x 1961, *Giess* 3821 A (PRE, WIND). 2315CA, Kuiseb river, east of Gobabeb, vii 1977, *Theron* 3740 (PRE). Rehoboth distr., 2316AD, almost at top of Gamsberg Pass, 1 viii 1982, *Craven* 1434 (WIND); Gamsberg Pass, 28 ii 1978, *Müller* 946 (M, PRE).

Fleck's label on the type specimen is written in an indecipherable scrawl, which Thellung interpreted as 'Gansberg & Tabgebiet'. I in turn interpret this as Gamsberg, whence has come much of the material of the species available to me. The type specimen has broadly ovate shallowly lobed leaves and delicate glandular pubescence, the hairs up to 2.5mm long on the stems, 1mm on the outside of the corolla, corolla tube c.13.5mm long.

In Prodromus Flora S.W. Afrika, *Chaenostoma fleckii* was reduced to synonymy under *Sutera pallida*, that is, *Jamesbrittenia pallida*, but it differs in its less woody habit (perennial herb, not shrublet), the leaves typically broader in relation to their length, less deeply lobed, and often with longer petioles, mostly shorter corolla tube, and longer hairs on the outside of the corolla.

Jamesbrittenia fleckii appears to have a limited distribution from the farms Mahonda and Göllschau west to the Kuiseb river, but the collections mainly reflect the route of the Swakopmund-Windhoek road via the Gamsberg Pass (specimens seen from 2315BD, CA, 2316AB, AD, BA, BC). The plant has been collected in the sandy bed of the Kuiseb river, on the rocky roadside of the Gamsberg Pass, and, at Göllschau, in the shelter of granite rocks. Flowering records are scattered throughout the year.

57. Jamesbrittenia lyperioides (Engl.) Hilliard in Edinb. J. Bot. 49: 230 (1992). **Plate 4C.**
Type: Namibia, 2116DD, near Okahandja, Kaiser Wilhelmsberg, 1400m, vi 1886, *Marloth* 1351 (holo. B†; iso. PRE, SAM, W).

Syn.: *Chaenostoma lyperioides* Engl. in Bot. Jahrb. 10: 253 (1888); Hemsley & Skan in Fl. Trop. Afr. 4(2); 310 (1906).
     Sutera lyperioides (Engl.) Range in Feddes Repert. 38: 265 (1935); Merxm. & Roessler in Prodr. Fl. S.W. Afr. 126: 52 (1967).

Subshrub, woody caudex up to c.25mm in diam., stems several tufted from the base, subsimple to well branched, c.150–600mm tall, very glandular, hairs mostly c.0.1mm long, heads resembling glistening glands, sometimes 0.2–0.3mm long with smaller heads and then often scattered among shorter ones. *Leaves* opposite becoming alternate upwards, blade up to 4.5–40 × 3.5–36mm, ovate to broadly ovate, occasionally subrotund, abruptly contracted to a petiolar part up to 2–11mm long (roughly ½–¼ length of blade), apex obtuse to subacute, margins usually shallowly lobed and toothed or coarsely toothed, rarely uppermost subentire, both surfaces glandular as stems. *Flowers* solitary in upper leaf-axils, the leaves rapidly diminishing in size, resulting in long pseudoracemes, often panicled. *Pedicels* up to 10–22mm long, glandular. *Calyx* tube 0.2–0.4mm long, lobes 4–6.5 × 0.5–1mm, linear-oblong to narrowly spathulate, subacute, glandular, hairs up to 0.1–0.25mm long. *Corolla* tube 11–15 × 2–2.5mm in mouth, cylindric, abruptly expanded near apex, limb 8–18mm across lateral lobes, posticous lobes 4–7.8 × 2–4.8mm, anticous lobe 4–7 × 2–5.2mm, oblong to cuneate-oblong, apex ± truncate or retuse, glandular-pubescent outside, hairs c.0.1mm long, broad band of clavate hairs in throat extending very briefly onto base of anticous lip, lobes yellow to orange (or rarely a white sport), the colour deepening into the throat and the 3 main veins at base of each lobe darkening as they descend into the throat. *Stamens*: posticous filaments 1.8–2.5mm long, anthers 1–1.5mm, anticous filaments 0.5–1mm long, anthers 0.6–1mm, all filaments minutely glandular. *Stigma* 0.4–1mm long. *Style* 9–12.5mm. *Ovary* 2–2.5 × 1–1.5mm. *Capsules* 5–7 × 3–3.4mm, minutely glandular. *Seeds* c.0.3 × 0.2–0.25mm.

Selected citations:

Namibia. Okahandja distr., 2116DD, Okahandja, 1300m, iii 1906, *Dinter* 78 (B, E, PRE, SAM, Z). Windhoek distr., 2217CA, Grossherzog-Friedrichs-Berg, 22 viii 1972, *Merxmüller & Giess* 28032 (PRE, WIND); 10 miles NE of Windhoek, 23 vii 1949, *Steyn* 176 (NBG, STE); 2217AC, road Windhoek-Okahandja, Brakwater, 26 iii 1968, *Wanntorp* 372 (PRE, S). Maltahöhe distr., 2416AB, Bergzebra Park Naukluft (MAL 9), 1 vi 1968, *Giess* 10415 (WIND). Lüderitz-Süd distr., 2616AB, S slopes of Tiras Mountains, farm Excelsior, 4 vii 1974, *Nordenstam & Lundgren* 575 (S).

I have seen two collections from Ovamboland, one unlocalized (*Höpfner* 57, Z), the other from the southern shore of Etosha Pan (*Kers* 678, WIND); the rest of the considerable material has come from Okahandja, Karibib, Windhoek, Swakopmund, Maltahöhe and Rehoboth districts, as well as the Great Tiras Mountains in Lüderitz-Süd district, above c.1000m. At Etosha, the plants were found under mopane trees, near the pan; elsewhere, stony or gravelly soils, sandy river banks and beds, partially shaded cliffs, and crevices of rock outcrops have all been recorded. Flowers may be found in any month.

The distinguishing features of the species are its very short indumentum, relatively short corolla tube, and yellow/orange limb.

GROUP 2, 2

58. Jamesbrittenia pilgeriana (Dinter) Hilliard in Edinb. J. Bot. 49: 232 (1992).
Type: Namibia, [2416AB] Kalkberg, Büllsport, 4 iv 1911, *Dinter* 2108 (holo. B†, iso. SAM).

Syn.: *Lyperia pilgeriana* Dinter in Feddes Repert. 19: 93 (1923).
     Sutera pilgeriana (Dinter) Range in Feddes Repert. 38: 265 (1935).

Twiggy dwarf shrublet up to 150–300mm tall and as much across, caudex up to at least 15mm in diam., branches many from the crown, erect or ascending, leafy, all parts (stems, leaves, pedicels, calyx lobes, outside of corolla, ovary, capsule) enveloped in glistening felted scurfy scales, apparently silvery-grey when fresh, often drying rufous. *Leaves* opposite becoming alternate upwards, up to 8–17 × 3.5–7mm, elliptic to narrowly obovate in outline, apex acute to subobtuse, base narrowed to a distinct petiolar part, margins sometimes entire, mostly with up to 3 pairs of teeth or lobes cut ½–⅔ of way to midrib. *Flowers* solitary in the axils of the upper leaves, these rapidly decreasing in size upwards, producing a short 'bracteate' raceme terminating each twiglet. *Pedicels* up to 5–13mm long. *Calyx* tube 0.15–0.2mm long, lobes 2.3–3 × 0.4mm, linear, acute, minute glandular hairs up to c.0.15mm long on margins, these often obscured by the scales. *Corolla* tube 8.5–9.3 × 1.2–1.4mm in throat, cylindric, abruptly slightly widened near the apex, limb nearly regular, 2.7–5.5mm across lateral lobes, posticous lobes 1.1–2.1 × 1–1.5mm, anticous lobe 0.8–2 × 0.8–1.5mm, all lobes suborbicular, scurfy outside, broad band of clavate hairs in throat extending briefly onto base of lower lip, lobes lemon yellow, throat orange. *Stamens*: posticous filaments 1–1.4mm long, anthers 0.8–1mm, anticous filaments 0.4–0.6mm, anthers 0.3–0.6mm, all filaments minutely puberulous. *Stigma* 0.2–0.3mm long. *Style* 6.5–7mm long, sometimes scurfy at the base. *Ovary* 1.5–2 × 0.8–1mm. *Capsules* 3–4 × 2–2.4mm. *Seeds* 0.25–0.6 × 0.25–0.3mm.

Selected citations:

Namibia, Maltahöhe distr., 2416AB, Pad Büllsport – Schlangenpoort, 31 xii 1934, *Dinter* 8316 (B); farm Buellsport, Büllenkopf, 10 iv 1947, *Strey* 2131 (BOL, K, PRE, SRGH); Blässkranz, 1 ix 1972, *Merxmüller & Giess* 28142 (M, PRE, WIND).

Jamesbrittenia pilgeriana is known only from the Naukluft Mountains where it grows in the cracks of large fallen rocks, on cliffs, and in other rocky sites. It possibly flowers in any month.

In the Prodromus Fl. S.W. Afr. it was equated with *Sutera canescens* (*Jamesbrittenia canescens*), but it is easily distinguished by its indumentum of scurfy scales, which envelop all the external parts of the plant as well as the ovaries and capsules. These scales will rub off and one collector recorded 'covered with fine white powder-like stuff'. However, the scales usually dry rufous, whereas in allied species they dry white. The corolla tube is extraordinarily narrow in relation to its length; for example, the longest tube seen (9.3mm) measured only 1.2mm across the flattened throat; in *J. canescens* (and other allied species) a corolla tube of comparable length will measure c.2–2.3mm across the throat.

59. Jamesbrittenia barbata Hilliard in Edinb. J. Bot. 49: 227 (1992).
Type: Namibia, Swakopmund distr., 2214DB, Welwitschia-Fläche bei Haigamkab, 1 v 1965, *Giess* 8738 (holo. WIND).

Very twiggy dwarf shrublet (but will flower when little more than a seedling) up to 150–350mm tall and as much, or more, across, branches erect or spreading, often thickly clad in white glistening scales, sometimes with minute (up to 0.1mm) glandular hairs as well, these hairs occasionally predominating, either leafy or leaves so reduced in size that the branches look bare. *Leaves* opposite becoming alternate upwards, axillary leaves inconspicuous, primary leaves up to 2–18 × 1–6mm, narrowly elliptic to subrotund, apex more or less obtuse, base tapering into a short petiolar part, margins entire or with 1–3(–6) pairs of small teeth, both surfaces thickly clad in white glistening scales, very minute glandular hairs sometimes present as well on margins, and midrib on lower surface. *Flowers* solitary in axils of upper leaves to produce innumerable terminal racemes, these often panicled. *Pedicels* up to 3–15mm long, either entirely clad in white scales or these sparse and glandular hairs up to 0.1mm long plentiful. *Calyx* tube 0.1–0.25mm long, lobes 1.5–3 × 0.5–0.8mm, linear, acute, ± thickly clad in white glistening scales, glandular hairs up to 0.1mm long present as well though sometimes very inconspicuous. *Corolla* tube 6.2–8.5 × 1.5–2.1mm in throat, cylindric, abruptly expanded near apex, limb nearly regular, 4.5–7mm across lateral lobes, posticous lobes 1.2–3.2 × 1–2.5mm, anticous lobe 1.3–2.8 × 1–2mm, all lobes suborbicular to broadly elliptic, ± thickly clad outside in glistening glands, glandular hairs up to 0.15mm long as well, broad band of clavate hairs in throat extending out onto bases of lobes and up to at least the sinus on the posticous lip, lobes variously coloured shades of yellow, red or chestnut brown, throat (always?) red when lobes yellow, the colour extending into a median streak at base of each lobe. *Stamens*: posticous filaments 0.8–1.4mm long, anthers 0.7–1.3mm, anticous filaments 0.5–1mm long, anthers 0.5–0.8mm, all filaments minutely puberulous. *Stigma* c.0.2mm long. *Style* 4–6mm. *Ovary* 1.5–2 × 0.8–1mm. *Capsules* 3–4 × 2–2.5mm, thickly clad in glistening glands, sometimes very minute glandular hairs as well. *Seeds* 0.3–0.4 × 0.2–0.25mm.

Selected citations:

Namibia. Swakopmund distr., 2114DC., Nordufer des Omaruru, 35 km östlich der Salzstrasse, 10 iv 1964, *Giess* 7859 (WIND); 2214DB, near Swakop river, along track to Guanicontes [Goanikontes], c.250m, 29 v 1963, *Kers* 1305 (S, WIND); 2215 CA, Flächen nördl. und südl. des Swakop bei den Husabbergen (Tsawichab), 19 v 1959, *Giess* 3046 (M, WIND); 2315CA, zwischen Natab und Homeb, 11 x 1961, *Giess* 3766 (M, NBG, W, WIND). Maltahöhe distr., 2416DD, farm Grootplaas, 8 ix 1972, *Merxmüller & Giess* 28231 (M, PRE, WIND). Bethanien distr., 2617AC, Nördlich Bethanien, 8 vi 1976, *Giess & Müller* 14276 (WIND).

Jamesbrittenia barbata is known to range from about Cape Cross south through the Namib desert, thence inland to Maltahöhe and Bethanien districts. The plants grow on open gravel or rubble flats, often in association with *Welwitschia*, sometimes in sand or gravel in river beds or in small watercourses, and flower in any month.

The species has hitherto been confused with *J. canescens*, from which it is distinguished by its twiggy habit and the remarkably bearded corolla limb: the hairs are relatively large and extend out onto the bases of all the lobes and at least as far as the sinus of the posticous lip. Specimens from Maltahöhe and Bethanien districts have leafy stems and sometimes conspicuous glandular hairs on stems, pedicels and calyx; in specimens from the Namib, the leaves are so reduced that the branches look bare; also, glandular hairs (as opposed to large sessile scales) are almost entirely wanting.

60. Jamesbrittenia chenopodioides Hilliard in Edinb. J. Bot.49: 228 (1992).
Type: Namibia, Omaruru distr., 2114AB, Brandberg, Tsisab valley, 20 ii 1963, *Kers* 1010 (holo. WIND, iso. S).

Perennial herb but will flower when very young and then looks annual, probably always only one main stem up to 300mm to 1m tall, up to c.10mm in diam. at the base, becoming woody in the lower part, erect, virgately branched, enveloped in glistening glands as are pedicels, leaves, calyx and capsules, very leafy except sometimes on the flowering twigs. *Leaves* opposite becoming alternate upwards, without axillary leaf-tufts, leaves up to c.10–55 × 3–30mm, ovate to elliptic in outline, apex acute, base abruptly or more gradually narrowed to a short petiolar part, margins entire, subentire or with up to 5 pairs of teeth or lobes cut up to ½ way to midrib. *Flowers* solitary in the axils of the upper leaves, these rapidly smaller upwards and passing into bracts, eventually forming long terminal racemes, often panicled. *Pedicels* up to 4–25mm long. *Calyx* tube c.0.2mm long, lobes 1.8–4.5 × 0.5–0.7mm, linear, acute, thickly clad in glistening glands, minute (less than 0.1mm) glandular hairs on margins. *Corolla* tube 8.5–11.5 × 1.8–2mm in throat, cylindric, abruptly expanded near apex, limb nearly regular, 4–6mm across lateral lobes, posticous lobes 1.7–2.5 × 1.6–2.2mm, anticous lobe 1.6–2.3 × 1.1–2.2mm, all lobes suborbicular to broadly elliptic, thickly clad in glistening glands outside, minute glandular hairs as well mainly on lower half of tube, broad transverse band of clavate hairs in throat extending very briefly onto base of anticous lip, lobes shades of yellow (lemon, egg, sulphur). *Stamens*: posticous filaments 1–1.5mm long, anthers 0.8–1.2mm, anticous filaments 0.5–1mm, anthers

0.5–0.8mm, all filaments minutely puberulous. *Stigma* c.0.2mm long. *Style* 5–8.5mm. *Ovary* 1.5–2 × 0.8–1mm. *Capsules* 4–4.5 × 2.5–2.8mm, thickly clad all over in glistening glands. *Seeds* c.0.3 × 0.2mm.

Selected citations:

Namibia. Kaokoveld distr., 1913BD, W entrance to Khowarib gorge, 3 iv 1977, *Lavranos & Barad* 15354 (E, WIND); 1913BA, Sesfontein, 4 vii 1951, *Hall* 481 (NBG). Outjo distr., 2014DD, Südufer des Ugab, 2 mls südwestlich der Brücke auf der pad Fransfontein - Omaruru, 6 iv 1963, *Ihlenfeldt et al* 3253 (M, PRE, WIND); 2014AC, Farm Vrede, Berge 13km nördlich des Huab rivier, 14 iv 1964, *Giess* 7945 (PRE, WIND). Omaruru distr., 2114BA, Brandberg, Tsisab valley mouth, 3 v 1963, *Nordenstam* 2543 (M, S); ibidem, 16 v 1976, *Oliver et al* 6683 (SRGH, Z).

The likeness of *Jamesbrittenia chenopodioides* to a chenopod such as *Chenopodium album* L. suggested the trivial name. It is a perennial herb that becomes woody at the base and seems usually to be single-stemmed there though it branches profusely above; one collector (*Giess* 7969) described it as a *Rutenstaude*, that is, a foxtail bush. (However, *Nordenstam* 3724 (S) is tufted from the base; the plant appears to have been browsed.) In this it contrasts sharply with *J. canescens*, with which it has hitherto been confused: that species is several-stemmed from the base. *Jamesbrittenia canescens* differs further in its possession of conspicuous glandular hairs on at least the pedicels and calyx, its often shorter corolla tube, and corolla limb ranging in colour from yellow through shades of red to brown, often with a contrasting median streak at the base of each lobe. Whereas *J. canescens* favours sandy river beds, *J. chenopodioides* commonly grows on granite inselberge or among rock outcrops, on screes, and in rubble, sometimes shaded by overhanging rocks, but there are a few records from river beds. It flowers in any month.

It has been recorded from the Kaokoveld (1913BA, BD) south to the Brandberg.

61. Jamesbrittenia canescens (Benth.) Hilliard in Edinb. J. Bot. 49: 228 (1992).

Three varieties are recognized:

1a. Stems, leaves and pedicels glandular-pubescent, hairs up to 0.15–0.4mm long, glistening glands very few or wanting on stems, present on leaves; capsules (and ovaries) minutely glandular-puberulous, glistening glands either scattered or confined to sutures **var. seineri**
1b. Glistening glands more conspicuous than glandular hairs at least on lower part of stem; capsules thickly clad in glistening glands 2

2a. Glistening glands predominating on stems and inflorescence axes, glandular hairs, if present, mostly less than 0.1mm long .... **var. laevior**
2b. Glandular hairs up to 0.1–0.2mm long, conspicuous and sometimes predominating on inflorescence axes .... **var. canescens**

var. canescens
Type: Cape, Namaqualand, zwischen Verleptpram und der Mündung des Garip, Ufer des Garip, 300ft, 16 ix 1830, *Drège* (holo. K; iso. E, LE, P, S, W, Z).

Syn.: *Lyperia canescens* Benth. in Hook., Comp. Bot. Mag. 1: 379 (1836) & in DC., Prodr. 10: 359 (1846).
  Chaenostoma canescens (Benth.) Diels in Bot. Jahrb. 23: 490 (1897).
  Sutera canescens (Benth.) Hiern in Fl. Cap. 4(2): 303 (1904); Merxm. & Roessler, Prodr. Fl. S.W. Afr. 126: 49 (1967) p.p. excl. *Lyperia seineri, L. confusa, L. pilgeriana* and *Sutera canescens* var. *laevior* Dinter.
  S. dioritica Dinter in Feddes Repert 30: 187 (1932). Type: Namibia [2818 BA], Kalkfontein - Süd, *Dinter* 4738 (holo. B†; iso. BOL, SAM, Z).

Subshrub or perennial herb, caudex up to at least 15mm in diam., stems several tufted from the base, c.150–750mm long, erect or decumbent, simple to virgately branched, thickly clad in glistening glands especially on lower parts, glandular hairs up to c.0.1–0.2mm long often present as well but becoming conspicuous, and sometimes predominating, only on the inflorescence axes, leafy mainly on the lower parts. *Leaves* opposite becoming alternate upwards, axillary leaves inconspicuous, primary leaves up to 8–45 × 3–12mm, elliptic in outline tapering into a petiolar part, apex ± acute, margins entire or with up to 6 pairs of acute teeth cut up to ½–⅓ of way to midrib, both surfaces thickly clad in glistening glands, glandular hairs up to 0.1–0.15mm long sometimes present as well, but generally inconspicuous. *Flowers* solitary in axils of upper leaves these rapidly becoming smaller and passing into bracts, forming long racemes or occasionally panicles. *Pedicels* up to 6–20mm long, glandular-pubescent, hairs up to 0.1–0.2mm long, scattered glistening glands sometimes present, often wanting. *Calyx* tube 0.1–0.2mm long, lobes 2.8–3.4 × 0.7–0.8mm, linear, acute, glandular-pubescent, hairs up to 0.1–0.15mm long, glistening glands as well. *Corolla* tube 7–9.5 × 1.5–2.2mm in throat, cylindric, abruptly expanded near apex, limb nearly regular, 3.8–9.5mm across lateral lobes, posticous lobes 1.2–3.6 × 1.2–2.8mm, anticous lobe 1.5–4 × 1–2.8mm, all lobes suborbicular to broadly elliptic, glandular-puberulous outside, hairs up to 0.1mm long, many glistening glands as well, broad transverse band of clavate hairs in throat extending briefly onto base of lower lip, lobes variously coloured pale or lemon yellow with a dark median streak at base of each lobe, light to dark violet or purple with yellow throat, wine-red or dusky-red with yellow throat and yellow median streaks (always?) at base of lobes. *Stamens*: posticous filaments 0.8–1.8mm long, anthers 0.7–1.2mm, anticous filaments 0.6–1mm, anthers 0.6–1mm, all filaments glandular-puberulous. *Stigma* 0.2–0.3mm long. *Style* 4.5–7mm long. *Ovary* 1.8–2 × 1–1.2mm. *Capsules* 4–6 × 2.5–3mm, glistening glands all over, occasionally a few minute glandular hairs near tips of valves. *Seeds* 0.3–0.5 × 0.2–0.4mm.

Selected citations:

Namibia. Maltahöhe distr., [c.2416AB], pad Büllsport-Schlangenpoort, 31 xii 1934, *Dinter* 8317 (M, Z). Bethanien distr., 2617DA, farm Naiams, 11 viii 1976, *Giess* 14589 (WIND). Lüderitz - Süd distr., 2616AB, farm Weissenborn, Rietrivier, 9 vii 1949, *Kinges* 2444 (M). Keetmanshoop distr., 2718BA, foothills of Groot Karasberge, farm Dassiefontein, 1300m, 2 ii 1974, *Davidse* 6260 (M, PRE, S). Lüderitz-Süd distr., 2816BB, am Oranje bei Fähre östl. Sendlingsdrift, 21 ix 1972, *Merxmüller & Giess* 28653 (M, S, WIND). Warmbad distr., 2818BC, farm Iris, Teil von farm Ortmansbaum, 22 v 1952, *Giess & Müller* 12113 (WIND).
Cape. Prieska div., 2922DA, banks of Orange River, iii 1937, *Bryant* K21 (BOL). Kenhardt div., 2820CB, Aughrabies, 28 ix 1938, *Middlemost* 1916 (NBG, STE).

The original material of *J. canescens* was collected on the banks of the Orange River (Gariep), probably near Verleptpram, on the south bank of the river opposite its confluence with the Oub (Fish). Subsequent collections in South Africa have all come from the banks of the Orange as far east as Prieska; a specimen collected by Zeyher and now in SAM bears Zeyher's own label with the locality 'Bitterfontein' (between Van Rhynsdorp and Springbok), but it is far more likely to have come from the banks of the Orange whither Zeyher travelled about the same time as did Drège. The species is widespread in the southern half of Namibia as far north as Windhoek. It commonly grows in sandy watercourses and on the banks of rivers and streams; there is one record of its growing from cracks in rocks in an intermittently wet streambed. Flowers can be found in any month; they vary greatly in colour, from yellow to shades of red and purple even in a single colony. Thus it is impossible to uphold Dinter's *Lyperia dioritica*, which he sought to distinguish by its deep purple flowers (there is of course no record of the colour of the flowers in the type material of *J. canescens*!).

The characteristic feature of the species is its indumentum, a dense covering of glistening glands at least on the lower part of the stems and on the leaves, sometimes with glandular hairs as well up to 0.1–0.2mm long, but these usually subordinate to the glistening glands; the pedicels are always glandular-pubescent with few or no glistening glands. The var. *seineri* has markedly pubescent stems and leaves; although glistening glands are often present as well, they are subordinate to the hairs. The area of var. *seineri* lies mostly south and east of that of var. *canescens*; where they are sympatric around Windhoek, above the Swakop river, and south to about Remhoogte, the distinctions may blur and some specimens are difficult to place.

Jamesbrittenia canescens var. *laevior* differs from both var. *canescens* and var. *seineri* in the predominance of glistening glands on the stem, including the inflorescence axes. It is also subtly different in facies because the leaves are often narrower in relation to their length than in the other two varieties. The ratio of length to breadth in the main primary leaves of var. *laevior* is mostly 4–9: 1, whereas in the other two varieties it is mostly 2.5–4:1. Its area lies well to the north of that of var. *seineri* (between the 17th and 20th parallels) though I have seen one collection from 2115DC, on the margin of the NW limits of var. *seineri*.

Jamesbrittenia canescens may hybridize with *J. primuliflora*: see under that species.

var. **seineri** (Pilger) Hilliard in Edinb. J. Bot. 49: 228 (1992).
Lectotype: Namibia, Epata, Omaheke [c.1919BA], *Seiner* 405 (BM).

Syn.: *Lyperia seineri* Pilger in Bot. Jahrb. 48: 441 (1912).
    L. confusa Dinter in Feddes Repert. 19: 93 (1923). Type: Namibia, Okahandja [2116 DD], 16 iv 1906, *Dinter* 130 (holo. B†, iso. SAM).
    Sutera canescens (Benth.) Hiern sensu Hemsley & Skan in Fl. Trop. Afr. 4(2): 303 (1906), quoad *Lugard* 268 (K).
    Sutera micrantha (Klotzsch) Hiern sensu Philcox in Fl. Zam. 8(2): 29 (1990) p.p., quoad *Brown* 6737 tantum.

Stems glandular-pubescent, hairs up to 0.1–0.4mm long, glistening glands few or wanting. *Leaves* (primary) up to 7–47 × 2.8–18mm, both surfaces glandular-pubescent, hairs up to 0.1–0.2mm long, glistening glands as well, these sometimes sparse. *Pedicels* up to 7–32mm long, glandular-pubescent, hairs up to 0.2mm long, often scattered glistening glands as well. *Calyx* tube 0.15–0.2mm long, lobes 1.8–3 × 0.5–0.7mm, glandular-pubescent, hairs up to 0.15–0.2mm long, scattered glistening glands as well. *Corolla* tube 6.3–9 × 1.5–2mm in throat, limb 4.5–8mm across lateral lobes, variously coloured as in *J. canescens*. *Capsules* c.3.5–4 × 2–2.2mm, minutely glandular-puberulous, glistening glands as well, sometimes confined to sutures.

Selected citations:

Namibia. Okahandja distr., 2116DD, Okahandja, 1200m, 14 viii 1906, *Dinter* 263 (B, E, SAM, Z). Karibib distr., 2215BA, farm Tsabichab, 10 v 1973, *Giess* 13485 (WIND). Windhoek distr., 2216DA, farm Karanab, 13 v 1973, *Giess* 13500 (NBG, PRE, WIND); 2217CA, municipal area, Windhoek, 5 xi 1962, *Hanekom* 210 (WIND); 2218CC, farm Renette, 8 i 1958, *Merxmüller & Giess* 1060 (M, PRE, SRGH, WIND). Swakopmund distr., 2315BD, Namib desert, gorge of Kuiseb river on Gamsberg road to Walvis Bay, 13 vii 1967, *Hilliard* 4692 (E, WIND). Gobabis distr., 2318BA, farm Donnersberg, iii 1956, *Volk* 11710 (WIND). Gibeon distr., 2418AB, Rohrbeck, 17 iv 1960, *Freyer* 111 (WIND). **Botswana**. Ghanzi distr., 2220AA, Okwa valley, 13 xii 1976, *Skarpe* 110 (K). [2124AD] 21°07.8'S, 24°59'E, 19 ii 1980, *Smith* 3106 (K). 2121DA, 15 miles W of Ghanzi, 3500ft, 20 x 1969, *Brown* 6737 (K, SRGH). **Cape**. Gordonia, 2620BC, Kalahari Gemsbok National Park, 13 miles NW of Twee Rivieren, c.2900ft, 20 x 1961, *Leistner* 2900 (M, PRE, SRGH, Z).

Jamesbrittenia canescens var. *seineri* is widely distributed across central Namibia (between the 21st and 24th parallels) and extends eastwards into Botswana (records from around Ghanzi and the Boteti river floodplain); there are also records from around Twee Rivieren in the Kalahari Gemsbok National Park in the Cape. The plants favour river beds and the margins of pans and springs, sometimes on calcareous soils, and may flower in any month.

Both Pilger, in describing *Lyperia seineri*, and Dinter, in describing *L. confusa* (he was clearly unaware of Pilger's earlier name), distinguished their new species from *L. canescens* by its glandular pubescence. I am hesitant about accepting this as a specific difference because there is so much variation in the length of the hairs and some specimens, from Windhoek district in particular, are more or less intermediate in indumentum; infraspecific rank seems more appropriate.

Pilger cited *Seiner* 189, 303 and 405, but none of these specimens survived the Berlin fire; fortunately there is a duplicate of *Seiner* 405 in the BM. The holotype of *Lyperia confusa* was also destroyed, but there are two good sheets in SAM.

var. laevior (Dinter) Hilliard in Edinb. J. Bot. 49: 228 (1992).
Type: Namibia, Otavi, 2 iii 1925, *Dinter* 5720 (holo. B†).

Syn.: *Sutera canescens* (Benth.) Hiern var. *laevior* Dinter in Feddes Repert. 30: 188 (1932).

Stems usually thickly clad in glistening glands, often with minute (less than 0.1–0.1mm) glandular hairs as well (exceptionally to 0.2mm and glistening glands sparse in some specimens from Kaoko Otavi, 1813BC). *Leaves* (primary) up to 10–34 × 2–6.5mm, both surfaces thickly clad in glistening glands, usually glandular hairs up to 0.1mm long also present. *Pedicels* up to 7–25mm long, ± thickly clad in glistening glands, often minute glandular hairs as well. *Calyx* tube 0.15–0.2mm long, lobes 1.6–3 × 0.5–0.7mm, glandular pubescent, hairs less than 0.1–0.15mm long, glistening glands as well. *Corolla* tube 6.5–9 × 1.5–2mm in throat, limb 5–7mm across anticous lobes, probably always yellow, with or without a dark median streak at base of each lobe. *Capsules* 3.5–4.5 × 2–3mm, thickly clad in glistening glands.

Selected citations:

Angola. Huila, próximo Oncócua, Rio Cunene, próximo do terminus da picada das Quedas do Monte Negro, 18 i 1956, *Mendes* 1341 (LISC).
Namibia. Kaokoveld distr., 1713CD, banks of Ososou river at Otjiwero, 3 iv 1957, *De Winter & Leistner* 5366 (M, WIND). 1812DD, W of Purros, Hoarusib river, 26 xi 1976, *Viljoen* 388 (WIND). 1813BC, Kaoko Otavi, 18 vi 1951, *Hall* 455 (NBG). Outjo distr., 1914 CB, farm Grootberg, *Giess* 7760 (WIND). Grootfontein distr., 1917BA, farm Auros, 11 iii 1973, *Giess* 12575 (WIND). 1917BC, Otavifontein, 3 ii 1959, *De Winter & Giess* 6789 (M, PRE, WIND). Karibib distr., 2115DC, Ameib, *Volk* 558 (M).

There is one record of *J. canescens* var. *laevior* from Angola, otherwise it appears to be confined to the northern part of Namibia; records stretch from the Kaokoveld to Grootfontein district, but there is one collection from much further south, in Karibib district. The plants grow in riverbeds and around pans and springs, often in calcareous soil or even in the cracks in limestone pavements, and may flower in any month.

The type of the name was destroyed in the Berlin fire and I have seen no duplicate. However, Dinter's description is explicit: he distinguished his plant from *Lyperia canescens* 'durch völlige Unbehaarheit der Stengel, Blatter, Blütenstiele, Kelche, Corollen und Kapseln; die ganze Pflanze ist nur dicht mit nicht Klebenden glitzernden (unter der Lupe) Drüschen bedeckt.' *Dinter* 649 from Otavi (the type locality) is clad only in glistening glands, but other specimens from the area have minute glandular hairs as well. Some specimens from Kaoko Otavi (well to the west of Otavi) are aberrant in their possession of glandular hairs up to 0.2mm long.

GROUP 2, 3

62. Jamesbrittenia crassicaulis (Benth.) Hilliard in Edinb. J. Bot. 49: 228 (1992).
Lectotype: Cape, [3126CA] Wildschutsberg, 5500–6000ft, 13 xii 1832, *Drège* (K; isolecto. E, K, LE, P, S, W).

Syn.: *Lyperia crassicaulis* Benth. in Hook., Comp. Bot. Mag. 1: 379 (1836) & in DC., Prodr. 10: 360 (1846) p.p. excluding *Drège*, Wittebergen.
  Chaenostoma crassicaule (Benth.) Thell. in Viertelj. Nat. Ges. Zürich 74: 116 (1929) in obs.
  Sutera crassicaulis (Benth.) Hiern in Fl. Cap. 4(2): 294 (1904), p.p. & excluding var. *purpurea* Hiern.

Rigid dwarf shrublet, stock stout, woody, up to at least 20mm diam., stems c.60–300mm long, erect, decumbent or spreading, virgately well branched from the base and above, thickly clad in large glistening glands, leafy. *Leaves* opposite becoming alternate on the inflorescence-axes, pseudofasciculate, crowded well below the inflorescences, largest 10–25 × 4–9mm, spathulate or oblong in outline with a broad petiolar part, blade cut mostly ⅔–¾ of way to midrib, sometimes a little deeper, lobes oblong, obtuse to subacute, entire or the largest 2–3-toothed, both surfaces clad in large glistening glands. *Flowers* up to c.14, distant, solitary in the axils of bracts, forming racemes carried well above the foliage. *Bracts* (lowermost) 3–5.5 × 0.8–2.5mm, spathulate, entire or rarely the lowermost 2-toothed. *Pedicels* up to 3–5mm long. *Calyx* tube 0.2–0.3mm long, lobes 2.4–3.5 × 0.5–1mm, oblong to spathulate, thickly clad in glistening glands and minute hairs up to c.0.15mm long. *Corolla* tube 9–12 × 2–2.8mm in throat, cylindric, abruptly expanded near apex, mouth laterally compressed, limb oblique, 6–7.5mm across the lateral lobes, posticous lobes 2.2–3.3 × 1.5–2mm, anticous lobe 1.8–3 × 1.8–2.3mm, all lobes oblong, retuse, glandular pubescent outside, hairs up to 0.15mm long, glistening glands as well, broad transverse band of clavate hairs in throat extending briefly onto lower lip, lobes yellow, posticous lobes sometimes flushed purple at tips, tube purplish. *Stamens*: posticous filaments 1.3–1.5mm long, anthers 0.9–1.1mm, anticous filaments 0.8–1mm long, anthers 0.5–0.8mm, all filaments minutely glandular at apex. *Stigma* 0.2mm long. *Style* 5.5–7.8mm. *Ovary* 1.5–2 × 0.8–1mm. *Capsules* 5–7 × 2–3mm, glistening glands all over. *Seeds* (few seen) c.0.7 × 0.5mm.

Citations:

Cape. Graaff Reinet div., 3224AA, Koudeveldberge SE of Doornbosch, 6000ft, 6 xi 1974, *Oliver* 5208 (STE); 3224AA, Toorberg, c.2100m, *Nordenstam* 1953 (S). 3224AB, St Olive's, c.5500ft, 29 xi 1977, *Hilliard & Burtt* 10771 (E, K); 3124DC, Compassberg, iii 1869, *Bolus* 1975 (BOL, S); ibidem, 8000ft, 25 xii 1951, *Esterhuysen* 19721 (BOL). Cradock div., 3225AB, Mountain Zebra National Park, c.5000ft, 22 xi 1977, *Hilliard & Burtt* 10570 (E, K); 3225AB, Wilgeboom River valley, 7 iv 1981, *Phillipson* 276 (K, PRE). Queenstown div., 3126DA, summit Andriesberg, 6800ft, ii 1896, *Galpin* 2030 (BOL, K).

Bentham's syntypes of *Lyperia crassicaulis* are not the same. His brief description applies equally well to both plants; I have chosen the Drège specimen from the Wildeschutsberg to lectotypify the name because the second syntype, a Drège collection from the Witteberg, near Aliwal North, appears to be a hybrid between *J. pristisepala* and *J. stricta* (see under *J. stricta*).

On the type sheet from Bentham's herbarium (K), Bentham has transposed Drège's localities. This was noted by N. E. Brown, substantiated by a reference to Drège, Zwei Pflanzengeographische Documente. Further verification is given by a specimen in the Paris herbarium (P) bearing a ticket in Drège's own hand, giving the Wildeschutsberg as the provenance of his collection, together with a precise date. *Jamesbrittenia crassicaulis* has never been recorded on the Witteberg; its area lies to the south and west of the Witteberg, on the mountains north of the Great Karoo, the Koudeveldberge, Oudeberg and Sneeuwberg in the Graaff Reinet-Middelburg area, thence east to the Bankberg near Cradock, the Bamboesberge near Sterkstroom and the Wildeschutsberg to the south east, and the Andriesberg due east of the Bamboesberge. The plants favour rocky sites including cliff faces, between c.1500 and 2400m above sea level, flowering between November and March. They blacken on drying.

The flowers are borne in loose, few-flowered racemes well above the foliage, in contrast to the crowded, many-flowered and often more leafy racemes of *J. pristisepala*, which differs further in its more deeply cut primary leaves and toothed, not entire, calyx lobes.

63. Jamesbrittenia stricta (Benth.) Hilliard in Edinb. J. Bot. 49: 232 (1992).
Type: Orange Free State, between the Vet and Sand rivers, Jan., *Burke* 337 (holo. G-DC, seen on microfiche; iso. BM, K, PRE, SAM).

Syn.: *Lyperia stricta* Benth. in DC., Prodr. 10: 360 (1846).

Dwarf shrublet, stock stout, woody, up to at least 30mm in diam., stems up to 450mm long, virgately branched, erect or decumbent and then forming a rounded bush, clad in glistening glands, making the plant look greyish, leafy. *Leaves* opposite becoming alternate on the inflorescence axes, pseudofasciculate, largest 14–37 × 6–20mm, lanceolate to ovate in outline narrowed to a broad petiolar part, blade cut to midrib, lobes linear-oblong or oblong, entire to

1–2-toothed or occasionally the largest cut to midrib, acute to subacute, both surfaces clad in glistening glands. *Flowers* many, solitary in the axils of the leaves forming long crowded racemes or panicles, largest reduced leaves subtending flowers c.4–8 × 0.8–1mm, spathulate, either entire or 2–8-toothed. *Pedicels* up to 1–5mm long. *Calyx* tube 0.1–0.2mm long, lobes c.1.8–2.5 × 0.3–0.4mm, linear to subspathulate, acute or subobtuse, entire or rarely one lobe with 1 or 2 teeth, clad in glistening glands and minute hairs. *Corolla* tube c.5.5–8 × 1–1.8mm in throat, cylindric, abruptly expanded near apex, limb oblique, 3–5mm across lateral lobes, posticous lobes 1–2 × 1–1.8mm, anticous lobe 1.1–2 × 1–2mm, all lobes oblong, retuse, glandular pubescent outside mainly on tube, hairs up to 0.1–0.15mm long, glistening glands as well mainly on backs of lobes, rarely a few on upper surface as well, broad transverse band of clavate hairs in throat extending briefly onto base of lower lip, lobes maroon to crimson, magenta pink or deep purple. *Stamens*: posticous filaments 0.5–1.3mm long, anthers 0.5–0.8mm, anticous filaments 0.3–0.8mm long, anthers 0.2–0.5mm. *Stigma* 0.1–0.2mm long. *Style* 3.8–5.5mm. *Ovary* 1.4–1.6 × 0.6–0.8mm. *Capsules* 3–4 × 1.5–2.2mm, glistening glands all over. *Seeds* c.0.5 × 0.25mm.

Selected citations:

Transvaal. Between Trigardsfontein [Trichardt, 2629AC] and Standerton [2629CD], *Rehmann* 6763 (BM, BOL, K, Z); 2629DB, Ermelo, *Burtt Davy* 9060 (PRE); 2629CD, between Standerton and Morgenzon, 13 ii 1979, *Balsinhas* 3372 (K, PRE).
Orange Free State. 2727BD, Heilbron, 20 i 1931, *Goossens* 463 (PRE). Winburg distr., 2827AC, Willem Pretorius Game Reserve, 10 iv 1962, *Leistner* 2995 (K, M, PRE, SRGH, Z). Between the Vet and Sand Rivers, *Zeyher* 1297 (BM, BOL, S, SAM, W). Groot Vet river, *Zeyher* 1308 (SAM, W). Bloemfontein, 2926AA, Hillandale, xi 1916, *Potts* sub SAM 98537 (SAM). Zastron, 3027AC, Aasvoëlberg, 27 iii 1980, *Smook & Gibbs Russell* 2298 (K).
Lesotho. 2927BB, near Mateka, c.6000ft, 22 iii 1951, *Bruce* 348 (PRE). 2927BD, Mountain road, between Rual and Mpao, iii 1977, *Schmitz* 7463 (PRE).
Cape. 3026DA, Aliwal North, 15 xi 1933, *Gerstner* 135 (PRE).

Hiern relegated *Lyperia stricta* to synonymy under *Sutera crassicaulis*, but it is an independent species, the most important characters of which are the linear-oblong, acute or subspathulate, entire calyx lobes and the small flowers (tube 5.5–7.5mm, limb 3–5mm across the lateral lobes), the limbs of which are mostly dark shades of red though once recorded as 'almost port wine, varying from pink to magenta'. (Change of colour with age may be involved.) The leaves are typically pinnatisect, the lobes linear or oblong, either entire or toothed, but sometimes the secondary lobing is more pronounced and then the leaves are almost bipinnatisect. Both stems and leaves are clad in glistening glands only.

The type material came from somewhere near modern Winburg, in the east central Orange Free State, and is precisely matched by specimens collected in the Willem Pretorius Game Reserve near that town. The overall range of the species is from the south eastern Transvaal (around Trichardt, Standerton, Ermelo, Amersfoort, Volksrust, and just entering Natal near the latter town) south and west through the eastern half of the Orange Free State (no records seen further west than Bloemfontein), the western part of Lesotho, and just entering the Cape (records from Aliwal North, the Witteberg and Indwe). The plants favour rocky sites, often on scrub-covered slopes, between c.1300 and 2200m above sea level, flowering between November and April.

Jamesbrittenia stricta is allied to *J. pristisepala* and where their areas meet, in Lesotho, around Fouriesburg, Golden Gate and Witzieshoek in the Orange Free State, the low Drakensberg on the Transvaal-Natal border, and the Cape Witteberg, hybridization may occur. *Jamesbrittenia pristisepala* differs from *J. stricta* in its more finely dissected leaves, strongly toothed calyx lobes, and larger flowers, particularly the larger corolla limb, which ranges in colour from yellowish to shades of mauve. The varying degrees of toothing of the calyx lobes encountered in areas of sympatry are indicative of possible hybridity, and so is variation in leaf-lobing and in the size and colour of the corolla limb. Field observations are needed, and the following information indicates possibly lucrative areas of study.

LESOTHO AND ORANGE FREE STATE

On the Maseru to Thaba-Tseka road (the so-called mountain road, in south central Lesotho), unequivocal *J. stricta* has not been collected east of 28° longitude. As one travels east, so *J. stricta* appears to grade towards *J. pristisepala*. The following series is from the 2927BD quarter degree square: *Schmitz* 7463 (PRE), between Rual and Mpao, is unequivocal *J. stricta*; less than 5km eastwards, *Richardson* 211 (E), from Ha Matjoka woodlot, c.6km south of Ha Ntsi, and *Thode* 42 (BOL, K, STE, a syntype of *Sutera crassicaulis* var. *purpurea*), from Matela's Peak, are very close to *J. stricta* (Richardson's specimen differs merely in corolla tube 9mm, limb 6mm). In *Thode* 42 (BOL, STE) corolla tube c.7.5mm, limb 5mm, calyx lobes with an occasional tooth, *Thode* 42 (K) calyx lobes well-toothed; flower colour was recorded as 'lilac or whitish' on the STE sheet, 'purple or lilac' in BOL, K. On Bushman's Pass, about 7km further east, *Hilliard & Burtt* 12121 (E) has bipinnatisect primary leaves (as in *J. pristisepala*), the corolla tube 8mm long, limb 5.5mm, reddish (*J. stricta*), calyx lobes entire or few-toothed. *Whellan* 1451 (SRGH) and *Werdermann & Oberdieck* 1556 (PRE), from the same pass, are similar to *Hilliard & Burtt* 12121. Another c.7km eastwards, on Molimo Nthuse Pass, *Hilliard & Burtt* 12084 (E, S) resembles *J. pristisepala* in foliage and size of flowers (corolla tube 9mm, limb 8mm) but the calyx lobes are mostly entire and the corolla limb very pale lilac or more rarely light reddish-purple. About 3km further east, on Blue Mountain Pass, *Hilliard & Burtt* 12005 (E) is the glandular-pubescent form of *J. pristisepala*; c.13km east again, beyond Likalaneng (2928AC), *Halliwell* 5028 (PRE) is glandular-pubescent *J. pristisepala* while *de Kruif* 1164 (PRE), about 45km east of Blue Mountain Pass, near Auray (2928CA) is typical *J. pristisepala*.

Other possible hybrids have been seen from the Semonkong area (2928CC), and see under *J. pristisepala* for hybrids between that species and *J. breviflora*, *J. jurassica* and *J. aspleniifolia*.

The material distributed under *Dieterlen* 424, from Qoqolosi Peak (10km due east of Leribe, in northern Lesotho) is not uniform. The sheets in P, SAM and Z have the foliage and indumentum of *J. stricta*; the calyx lobes are either entire or 1–2-toothed (often in a single calyx), the corolla tube is c.8.5mm long, limb 5mm, colour cream. *Dieterlen* 424a (PRE) differs from the foregoing sheets in having a few minute glandular hairs on stems and leaves, but 424 (PRE) is decidedly glandular-puberulous (hairs up to 0.2mm long), glistening glands very sparse on the stems, and leaves bipinnatisect. The material is probably mostly *J. stricta* × *pristisepala*, the specimens in SAM and Z approaching *J. stricta* closely, while 424 (PRE) inclines more towards *J. pristisepala*. A seventh sheet, in STE, is glandular-puberulous with well toothed calyx lobes and can be assigned to *J. pristisepala*. *Phillips* 951 (SAM) from Leribe inclines very strongly to *J. stricta* (and the flowers were described as 'dark coloured') but the calyx lobes are well toothed in the three twigs I have marked A (in the other four twigs they are entire or very nearly so).

Hybridization similar to that around Leribe seems to have taken place in the nearby Fouriesburg, Clarens, Witzieshoek area of the Orange Free State. The type of *J. pristisepala* came from a site roughly 18km almost due south of Witzieshoek, at c.2000m above sea level. *Thode* 5478 (M, STE) from Witzieshoek (2828DB), at c.1800m, is almost pure *J. stricta* (differing in some calyx lobes bearing an occasional tooth, flowers lilac); *Junod* sub TM 17504 (PRE), also from Witzieshoek, has larger flowers and calyx lobes more toothed than in Thode's specimens, but the leaves are similar. *Zietsman* 537 (PRE), from Golden Gate (2828DA) at 2200m, has the leaves of *J. stricta*, calyx lobes entire or with 1 or 2 teeth, and flowers more like those of *J. pristisepala* in size and with a 'white' limb. *Acocks* 11203, from the same general locality, has more deeply dissected leaves (as in *J. pristisepala*), the calyx lobes mostly few-toothed, corolla small (tube 7mm, limb 4.5mm), described as 'purplish-white'. *Scheepers* 1835 came from the farm Dunblane (2828CB) near Clarens, altitude c.2000m. The sheet in PRE is *J. pristisepala*, that in K has smaller flowers and less dissected leaves and may be of hybrid origin. *Gemmell*

5824 (PRE) from Fouriesburg (2828CA) has most of the characters of *J. pristisepala*, but the corolla limb is only 3mm across (as in *J. stricta*) and has dried dark red (the label records 'flowers white', but in other collections with flowers recorded as white they have dried white).

TRANSVAAL AND NATAL

Plants along the Transvaal-Natal border between Volksrust and Wakkerstroom are problematical: they seem to be more or less intermediate between *J. stricta* and *J. pristisepala*. *Turner* 909 (PRE) from the environs of Amersfoort (2629DD) about 40km due north of Volksrust, is unequivocal *J. stricta*. No collections between Amersfoort and Volksrust have come into my hands. *Mogg* 7509 (PRE, SRGH) from Volksrust (2729BD) unfortunately is in fruit only; the calyx lobes are entire or 1–2-toothed (in any one calyx); the leaves on the PRE sheet are dissected as in *J. pristisepala*, those on the sheet in SRGH incline more towards those of *J. stricta*. *Wood* 5740 (E) came from the valley of the Buffalo river, almost certainly (judging from other specimens collected by Wood) near Charlestown, a few kilometres south of Volksrust, in Natal. It has the leaves, calyx and floral dimensions of *J. stricta*, but Wood described the flowers as '*lutei*'. *Devenish* 1684 (E) from the farm Nauwhoek (2730AD), about 30km due east of Volksrust, on the Natal side of the border, matches Mogg's specimens from Volksrust: the corolla tube is 7mm long, limb 3.5mm, 'dark yellow'. *Devenish* 1306 (M), collected on the same farm eleven years earlier, is more lush, the leaves more dissected, calyx lobes entire or 1–2-toothed, flowers few, corolla tube c.7.5mm, limb 5mm, 'yellow'. *Devenish* 464 (PRE) from Oshoek, adjoining Nauwhoek but on the Transvaal side of the border, comes very close to *J. pristisepala*, differing in its calyx lobes, either entire or with a pair of tiny teeth (corolla tube 7.5mm, limb 6mm and therefore very near the lowest part of the range in *J. pristisepala*). I have seen no unequivocal *J. pristisepala* from the area, the nearest known locality being on Rensburg's Kop, roughly 130km south west of Volksrust. All these localities (Rensburg's Kop to Nauwhoek) lie along the escarpment of the low Drakensberg, which is not well explored botanically.

E CAPE

Hilliard & Burtt 12212, from Joubert's Pass across the Witteberg near Lady Grey, appears to be a mixture of *J. pristisepala*, *J. stricta* and hybrids between them, though it is not improbable that all the material is more or less hybrid (the corolla limb was recorded as yellow). On the sheet in E, the twigs marked A are *J. stricta*, that marked B, *J. pristisepala*. The material distributed to S comprises three twigs, one with calyx lobes entire, one with lobes mostly entire, another with toothed lobes; those in M and PRE are wholly *J. pristisepala*. Specimens collected by Drège at the same locality show a somewhat similar range of variation: they were distributed as *Lyperia crassicaulis* Benth. b, and are syntypes (*not* lectotype) of that name. *Cooper* 597 (BM, BOL, W), also from the Witteberg, is all intermediate; the sheet in BOL was inscribed by Cooper as 'whitish yellow'. *Hilliard & Burtt* 16538 (E) from the foot of Ben McDhui, roughly 65km almost due east of Joubert's Pass, is unequivocal *J. pristisepala*, and *J. stricta* has not been recorded from this area.

Jamesbrittenia aspleniifolia (glandular-pubescent, calyx lobes entire) is sympatric with *J. pristisepala* in the Cape Drakensberg and thereabouts, as well as in south-eastern Lesotho, and could also be involved in complex hybridization. See further under *J. pristisepala* for hybridization with *J. breviflora* and other species.

64. Jamesbrittenia pristisepala (Hiern) Hilliard in Edinb. J. Bot. 49: 232 (1992).
Type: Orange Free State, near cave at foot of Mont aux Sources, 6800ft, i 1894, *Flanagan* 2085 (holo. K, iso. BOL, PRE, SAM).

Syn.: *Sutera pristisepala* Hiern in Fl. Cap. 4(2): 293 (1904).

Lyperia crassicaulis Benth. in Hook., Comp. Bot. Mag. 1: 379 (1836) & in DC., Prodr. 10: 360 (1846) as to *Drège*,
   Wittebergen, only, excluding lectotype.
Sutera crassicaulis var. *purpurea* Hiern in Fl. Cap. 4(2): 294 (1904). Lectotype (chosen here): Top of Mont aux
   Sources, 11000ft, iii 1898, *Evans* 760 (K).

Hybrids between *J. pristisepala* × *breviflora*?:
Chaenostoma saccharatum Thell. in Vierteljahrss. Nat. Ges. Zürich 74: 115 (1929). Type: Basutoland [Lesotho],
   White Hill [3028AB], i 1912, *Jacottet* 245-359 (holo. Z). See comments below.
C. jacottetianum Thell. in Vierteljahrss. Nat. Ges. Zürich 74: 113 (1929). Type: Basutoland [Lesotho], Qacha's
   Nek [3028 BA], xi 1911, *Jacottet* 118-247 (holo. Z). See comments below.

Dwarf shrublet, stock stout, woody, up to c.30mm in diam., stems up to at least 350mm long, erect or decumbent, well branched, either thickly clad in glistening glands or with varying degrees of development of glandular hairs up to c.0.5mm long, these sometimes predominating (see comments below), leafy. *Leaves* opposite becoming alternate on the inflorescence axes, pseudofasciculate, largest 10–30 × 5–17mm, ovate, elliptic or oblong in outline with a broad petiolar part, blade divided to midrib, lobes elliptic or cuneate in outline, deeply cut, segments acute, subacute or obtuse, both surfaces clad in glistening glands, often with the addition of glandular hairs c.0.1–0.6mm long, their presence very variable in density. *Flowers* many, solitary in the axils of the upper leaves forming long crowded racemes. *Pedicels* up to 2–4mm long (to c.8mm in some hybrids). *Calyx* tube (0.15–)0.3–0.5mm long, lobes 2.3–4.5 × 0.3–2mm at tips, spathulate, the tip broad and 2–5-toothed in the typical plant, narrowly spathulate to oblong in some specimens (possibly of hybrid origin), the lobes entire or with varying degrees of toothing, either glistening glands predominating or various degrees of development of glandular hairs up to 0.2–0.7mm long, then glistening glands sometimes few. *Corolla* tube 8–11.5 × 1.5–2mm in throat, cylindric, abruptly expanded near apex, mouth laterally compressed, limb oblique, bilabiate, c.6.5–11mm across the lateral lobes, posticous lobes 1.8–6 × 1.2–3.2mm, anticous lobe 2.2–5.5 × 1.5–4.5mm, all oblong, retuse, glandular-pubescent outside, hairs up to 0.25–0.6mm long, glistening glands as well, broad transverse band of clavate hairs in throat extending briefly onto base of lower lip, lobes varying in colour from creamy white or yellowish to lilac-pink and shades of mauve, throat whitish to pale yellow often with dark purple radiating lines at base of lobes, tube often purplish outside. *Stamens*: posticous filaments 1–1.3mm long, anthers 0.7–1.1mm long, anticous filaments 0.3–1mm long, anthers 0.4–0.8mm long, all filaments minutely glandular at apex. *Stigma* 0.2–0.3mm. *Style* 5.5–7mm long. *Ovary* 1.5–2 × 0.7–1mm. *Capsules* 3.5–5 × 2–2.5mm, glistening glands all over. *Seeds* 0.5–0.9 × 0.25–0.6mm, sometimes as few as 6 in each loculus.

Selected citations (many possible hybrids are mentioned in the discussion below and under *J. stricta*):

Orange Free State. Harrismith distr., 2829AD, Swinburne, Rensburg's Kop, c.7300ft, 23 ii 1970, *Hilliard* 4985 (E, K, PRE). 2828DB, c.33km SW of Witzieshoek, Sentinel Peak, 25 ii 1974, *Davidse* 6961 (E, M).
Lesotho. 2928CA, about 45km E of Blue Mountain Pass, near Auray, c.2400m, 17 xi 1983, *de Kruif* 1164 (PRE).
Natal. Bergville distr., 2828DB, Royal Natal National Park, bed of Tugela river, 30 x 1938, *Häfstrom & Acocks* 1174 (PRE, S); 2829CC, Cathedral Peak Forestry Reserve, Ndedema river valley, Schoongezicht Cave, 25 x 1973, *Hilliard & Burtt* 6924 (E, K); Estcourt distr., 2929AB, valley below Ship's Prow Pass, 2100m, 7 xii 1983, *Balkwill et al.* 1047 (E).
Transkei. Thaba Chitja distr., 3028AD, Marshall Clarke, 1800m, 9 xii 1982, *Granger* 3662 (PRE).
Cape. Barkly East distr., 3027DB, Ben McDhui, Bell River Gorge, 8000ft, 17 ii 1983, *Hilliard & Burtt* 16538 (E, K). 3127BB, Mfecani Pass, 4km from Barkly Pass, 2000m, 18 xii 1982, *Phillipson* 613 (K, PRE).

Jamesbrittenia pristisepala is endemic to the Eastern Mountain Region of southern Africa, from the low Drakensberg on the Natal-Orange Free State border, southwards over the high Drakensberg of Natal and Transkei, and the mountains of Lesotho, to the Cape Drakensberg near Barkly East and the neighbouring Witteberg, between c.1500 and 3000m above sea level, growing on cliffs and other rocky places, and flowering principally between November and April. It is very variable in indumentum, degree of toothing of the calyx lobes and colour of the corolla limb. A good deal of variation can probably be attributed to hybridization and subsequent backcrossing.

The name *J. pristisepala* was based on a specimen collected in the Elands River valley below Mont aux Sources on the Orange Free State side. This plant is characterized by the presence of big glistening glands on the stems and leaves almost to the exclusion of glandular hairs (a few tiny ones are present mainly on the petiolar part of the larger primary leaves), the leaves divided to the midrib, the lobes elliptic or cuneate in outline and deeply cut, and the calyx lobes, broadened at the tips and there deeply toothed (the epithet *pristisepala* is derived from the latin name for the sawfish and refers to the ragged sepals). *Evans* 760, lectotype of *Sutera crassicaulis* var. *purpurea*, came from the top of Mont aux Sources and precisely matches the type of *J. pristisepala*, from the foot of Mont aux Sources. These type specimens can be matched precisely by material collected in the same general area (below The Sentinel and Mont aux Sources) and ranging thence down the face of the high Drakensberg in Natal as far south as Ship's Prow Pass (2929AB), thence a distributional gap to a site in the Transkei Drakensberg (3127BB) and again a gap to the Lady Grey-Barkly East area (3027CB, DB, 3127BB). The gap between Ship's Prow Pass and Qacha's Nek (3028BA) is filled by plants in which glandular hairs, ranging in length from c.0.2–0.6mm long, are intermingled with glistening glands on the stems and leaves, and sometimes predominate. There is also great variation in the calyx lobes, which range from narrow and entire (though broader, toothed tips will be found in any one specimen, or one calyx) to variously broadened and toothed at the tips. The corolla limb also varies greatly in size and colouring. Similar plants also range northwards over the area of the typical plant, and westwards into Lesotho, coinciding with the geographical area of *J. breviflora*, which is glandular-pubescent, with entire calyx lobes and large red corolla limb. In the southern part of its range, in the Cape Drakensberg, the nearby Witteberg, as well as in its northernmost part and in western Lesotho, *J. pristisepala*, appears to hybridize with *J. stricta* (see discussion under that species). The Drakensberg in Transkei, and most of the mountainous interior of Lesotho, is ill-explored botanically, and much work needs to be done there.

During the course of field work in the southern Natal Drakensberg, Hilliard & Burtt became aware that obvious hybrids between *J. pristisepala* and *J. breviflora* are not infrequent; they reported this in 1982, and gave a full description of the hybrid (Hilliard & Burtt 1982, p.291); further information and colour photographs appeared in 1987 (Hilliard & Burtt, 1987 pp.35, 200, plate 28 A, B, C). Hilliard & Burtt remarked (loc. cit. p.35) that '..... they always occur as scattered individuals and there is no evidence yet that they reproduce or backcross to either parent'. But now that a wide range of herbarium material has been studied, it seems clear that backcrossing does indeed take place and a great many specimens of *J. pristisepala* seen from the southern Drakensberg are introgressed: glandular hairs up to c.0.3–0.6mm long predominate on stems and leaves, and the calyx lobes can be well-toothed to few-toothed or entire. However, some specimens approach the typical plant more closely. The type specimen of *Chaenostoma jacottetianum* (in synonymy above) is surely of hybrid origin: glandular hairs up to c.0.3mm long predominate over glistening glands, the leaves are less deeply dissected than those of *J. pristisepala*, the pedicels are up to 8mm long, the calyx lobes entire, the corolla limb small. The most likely parentage is *J. pristisepala* × *J. breviflora*, but it is possible that *J. aspleniifolia* is involved: both species have been recorded from Qacha's Nek, whence came *Jacottet* 118-247, type of *Chaenostoma jacottetianum*. *Killick* 4273 (PRE) was collected near Taung in eastern Lesotho (2928DB); it may represent *J. pristisepala* introgressed by *J. aspleniifolia*, which was collected at the same locality (*Killick* 4276, K).

The type of *Chaenostoma saccharatum* (in synonymy above), from White Hill near Qacha's Nek, also appears to be *J. pristisepala* × *breviflora*: it closely resembles typical *J. pristisepala* in leaves, calyx and corolla, but tiny glandular hairs are scattered on the stems and backs of the leaves.

In central and eastern Lesotho, *J. pristisepala* appears to have hybridized mainly with *J. breviflora*. However, *Hilliard* 5438 (E), from Sani Top (2929CB), at c.2900m above sea level, may represent *J. pristisepala × jurassica*: the specimen was collected near *J. jurassica* (*Hilliard* 5434, type); other material from Lesotho could also be this hybrid, as the species has a wide distribution in easternmost Lesotho. *Phillipson* 1403 (PRE) may provide an example of such a cross; it came from Mahlasela Pass (2828DC) near Oxbow, at 3100m.

A selection of specimens possibly of hybrid origin follows:

J. pristisepala × breviflora (sporadic plants obviously intermediate between the two species):
Natal. Underberg distr., 2929CB, Sani Pass, 6500ft, 15 xii 1984, *Hilliard & Burtt* 17981 (E, K, PRE); ibidem, 7500ft, 24 iii 1977, *Hilliard & Burtt* 9826 (E, K, PRE), 9827 (E, PRE); ibidem, 8400ft, 6 i 1984, *Hilliard & Burtt* 17293 (E); Chameleon Cave area, c.5 miles N of Castle View Farm, 7250ft, 1 xii 1984, *Hilliard & Burtt* 17751 (E); ibidem, 7800ft, 4 xii 1984, *Hilliard & Burtt* 17858 (E); Upper Polela Cave area, c.7600ft, 13 ii 1979, *Hilliard & Burtt* 12504 (E).

J. pristisepala introgressed by *J. breviflora* (showing varying degrees of intermediacy):
Orange Free State. 2828DB, northern slopes of The Sentinel, near chain ladder to Mont aux Sources, 9000ft, 3 iii 1976, *Rourke* 1527 (NBG); Mont aux Sources, 8000ft, iv 1913, *Dyke* in herb *Marloth* 5413 (PRE p.p., SAM).
Lesotho. 2928AC, Mamalapi, 10,000ft, 26 xii 1948, *Compton* 21301 (NBG). 2929AC, Mokhotlong, 7000ft, 25 ii 1949, *Compton* 21501 (NBG). 2929CC, Sehlabathebe Reserve, 9000ft, 24 i 1975, *Bayliss* Lesotho 57 (Z). 2929AD, c.1½ miles downstream from summit of Langalibalele Pass, [c.10,000ft], 11 ii 1972, *Wright* 1253 (E). 3028BA, near Qacha's Nek, Rapase, c.5000ft, 15 iii 1936, *Galpin* 14038 (BOL, K, PRE).
Natal. Impendhle distr., 2929AD, Hlatimba Pass, c.8700ft, 2 iii 1971, *Wright* 1133 (E). Underberg distr., 2929CB, Sani Pass, c.8000ft, 22 iii 1977, *Hilliard & Burtt* 9763 (E, K, S); 5–7 miles NNW Castle View Farm, headwaters Mlahlangubo river, 8500ft, 23 i 1982, *Hilliard & Burtt* 15362 (E, K, S).

65. Jamesbrittenia lesutica Hilliard in Edinb. J. Bot. 49: 230 (1992).

Type: Lesotho, 2929AC, Phutha, 8000ft, 28 ii 1949, *Compton* 21604 (holo. NBG).

Rigid dwarf shrublet up to c.1m tall, virgately branched, stems thickly clad in large glistening glands, leafy. *Leaves* opposite becoming alternate upwards, pseudofasciculate, largest primary leaves 10–15 × 4–7mm, ± oblong in outline tapering to a broad petiolar part, blade cut almost to midrib into c.4–7 pairs of lobes, lobes oblong to narrowly elliptic, acute, entire or few-toothed, both surfaces clad in large glistening glands. *Flowers* solitary in the upper leaf-axils, forming few-flowered (up to c.15) terminal racemes. *Pedicels* c.3mm long. *Calyx* tube 0.2–0.3mm long, lobes c.3.5–3.7 × 0.5–0.7mm, oblong to narrowly spathulate, glandular-puberulous, hairs up to 0.15–0.2mm long, glistening glands as well. *Corolla* tube c.18–21 × 2–2.2mm in throat, cylindric, abruptly expanded near apex, limb oblique, c.15mm across lateral lobes, posticous lobes c.7 × 3.5–4.3mm, anticous lobe c.6–6.5 × 4.2–5.2mm, all lobes cuneate, deeply retuse, glandular-pubescent outside, hairs up to 0.15–0.2mm long, glistening glands as well on backs of lobes and swollen part of tube, broad transverse band of clavate hairs in throat extending very briefly onto base of lower lip, lobes white. *Stamens:* posticous filaments 1.5–1.7mm long, anthers 1–1.1mm, anticous filaments 0.8–1.1mm long, anthers 0.7–0.8mm, all filaments puberulous. *Stigma* 0.4–0.8mm long. *Style* 14–15.5mm. *Ovary* c.2.5 × 1mm, thickly clad in glistening glands. *Capsules* not seen.

Citations:

Lesotho. Mokhotlong distr., 2929AC, Mokhotlong, 7000ft, 25 ii 1949, *Compton* 21492 (NBG); ibidem, 8000ft, 6 iii 1938, *Brooke* 39 (BM); ibidem, c.8000ft, 6 xii 1956, *Dohse* 313 (M, PRE); 2929AA, Merareng on the Senqubetu river, ± 8000ft, i 1953, *Liebenberg* 5691 (PRE).

Jamesbrittenia lesutica is known only from the environs of Mokhotlong in eastern Lesotho, an area little known botanically. Miss Dohse recorded it on 'rocky ledges', Miss Brooke on 'sunny slopes; makes a brilliant white cushiony mass'. It flowers between December and March. The long corolla tube and white limb with deeply retuse lobes resembles that of, for example, *Zaluzianskya ovata*, which is common in the mountains of Lesotho.

The large white flowers at once distinguish *J. lesutica* from its ally, *J. pristisepala*, with which it is sympatric, but they differ further in their leaves and calyx lobes: leaves pinnatisect in *J. lesutica*, bipinnatisect in *J. pristisepala*, and calyx lobes entire, not toothed.

Jamesbrittenia lesutica hybridizes with *J. aspleniifolia*: see under that species, no. 75.

66. Jamesbrittenia silenoides (Hilliard) Hilliard in Edinb. J. Bot. 49: 232 (1992). **Plate 2D.**
Type: Natal, 2730DB, Vryheid distr., Hlobane, 8 x 1950, *Johnstone* 474 (holo. NU, iso. E).

Syn.: *Sutera silenoides* Hilliard in Notes RBG Edinb. 43: 216 (1986).
   S. burkeana auct. non (Benth.) Hiern; Fabian & Germishuizen, Transvaal Wild Flowers 238 pl. 114 j (1982).

Herbaceous perennial, woody caudex up to c.25mm in diam., stems several from the caudex, c.200–500mm long, erect, ascending or decumbent, well branched, the whole plant eventually bushy, clad in glistening glands, leafy on lower part. *Leaves* opposite, bases connate, becoming alternate upwards and then passing rapidly into small floral bracts, somewhat pseudofasciculate, primary leaves up to c.25–32 × 6–20mm, blade oblanceolate to ovate in outline, ± ⅓–½ total length petiolar, pinnatisect to bipinnatisect, lobes narrowly oblong to narrowly oblanceolate, entire to few-toothed, both surfaces clad in glistening glands. *Flowers* many, solitary in the axils of uppermost leaves or bracts, producing lax racemes up to c.150–200mm long. *Bracts* up to c.3–7 × 0.4–0.5mm, linear, acute. *Pedicels* up to 5–7(–18)mm long. *Calyx* tube c.0.2mm long, lobes c.2.7–4.6 × 0.7–0.9mm, linear, acute to subobtuse, clad in glistening glands and glandular hairs c.0.1mm long. *Corolla* tube c.20–21 × 2–2.5mm in throat, cylindric, abruptly expanded near apex, limb c.10–15mm across lateral lobes, oblique, bilabiate, mouth somewhat compressed laterally, posticous lobes 5–6.7 × 2–3mm, anticous lobe 4.5–6.5 × 2–4mm, all lobes ± oblong, apex deeply bifid, the tips very acute, margins strongly revolute, glandular-puberulous outside, hairs c.0.1mm long, glistening glands as well especially on backs of lobes and upper half of inflated part of tube, occasionally a few glistening glands on upper surface of lobes, otherwise glabrous there, broad transverse band of clavate hairs in throat extending very briefly onto anticous lip, lobes brown or reddish-brown, thick-textured (no venation visible on upper surface). *Stamens*: posticous filaments 1.2–2mm long, anthers 0.8–1.3mm, anticous filaments 0.7–1mm long, anthers 0.8–1.1mm, all filaments bearded. *Stigma* c.0.8–1mm long. *Style* c.16mm. *Ovary* 2.2–2.5 × 1–1.2mm. *Capsules* 7.5–8 × 2.5–3mm, glistening glands all over. *Seeds* 0.5–0.6 × 0.3–0.5mm.

Citations:

Transvaal. Potgietersrus distr., 2429AA, farm Portugal, 1800m, 16 i 1967, *Maguire* 7658 (E). Belfast distr., 2530AA, near Draaikraal, farm Saaihoek, 17 xi 1953, *Codd* 8056A (K, M, PRE, SRGH); 2530AC, farm Kurunitzi, ii 1979, *Fabian* 13 (PRE); 2530CD, Doornkop, c.5500ft, i 1912, *Thode* 5081 (STE). Middelburg distr., Tantesberg, 8 xi 1933, *Young* A139 (PRE).
Natal. Vryheid distr., 2730DB, Hlobane, c.5000ft, 12 x 1950, *Johnstone* 554 (K).

Jamesbrittenia silenoides is known from the highlands of the eastern Transvaal, from the environs of Potgietersrus south to Belfast and Dullstroom, with two records from northern Natal, both from the top of Hlobane Mountain. It favours rocky sites, and flowers between October and February.

Jamesbrittenia silenoides is allied to *J. crassicaulis* and *J. pristisepala*, from which it differs strikingly in its flowers with brown, forked corolla lobes. The corolla limb, with its thick brown lobes, rendered narrower than they really are by the strongly revolute margins, is similar to those of *J. atropurpurea* and *J. huillana*, no doubt in response to a common pollinator, because *J. silenoides* is not at all closely allied to these two species. *Jamesbrittenia silenoides* and *J. huillana* (as well as *J. burkeana*, also with brown flowers) are sympatric in the eastern Transvaal.

67. Jamesbrittenia beverlyana (Hilliard & Burtt) Hilliard in Edinb. J. Bot. 49: 227 (1992).
Type: Lesotho, 2929CC, Sehlabathebe National Park, c.300m downstream from Phororong, c.2325m, 16 ii 1976, *Beverly* 510 (holo. PRE).

Syn.: *Sutera beverlyana* Hilliard & Burtt in Notes RBG Edinb. 43: 216 (1986).

Perennial herb with a woody taproot c.5mm diam. at crown, stems several from crown, up to 350mm long (but material young), probably prostrate, well branched, glandular-pilose, hairs up to c.1mm long, leafy throughout. *Leaves* opposite becoming alternate upwards, somewhat pseudofasciculate, primary ones c.20–45 × 12–24mm including a petiolar part 5–15mm long, blade ovate in outline, pinnatisect, lobes more or less oblong, deeply lobed to toothed, lobes obtuse, upper surface clad in glistening glands together with short (c.0.4mm) scattered glandular hairs, lower surface with glistening glands and glandular hairs up to 1mm long. *Flowers* solitary in upper leaf-axils, much shorter than the leaves. *Pedicels* c.3mm long, glandular-pilose. *Calyx* tube c.2mm long, lobes c.5 × 3mm, foliaceous, glandular-pilose, hairs c.0.8mm long. *Corolla* tube c.8–10 × 2.5mm in throat, cylindric, abruptly expanded near apex, limb bilabiate, c.6mm across lateral lobes, posticous lobes c.2.5–3 × 1.5–1.75mm, anticous lobe c.2.5–3.5 × 2mm, all lobes oblong, apex rounded, glandular-puberulous outside, hairs up to c.0.3mm long, glistening glands as well mainly on backs of lobes, broad transverse band of clavate hairs in throat, lobes pale yellow, the 3 lower ones each with a dark median streak. *Stamens*: posticous filaments c.1.5–2mm long, anthers c.1.2mm long, anticous filaments c.1.3mm long, anthers c.1mm. *Stigma* c.0.2mm. *Style* c.4.5–6mm long. *Ovary* 1.8–2 × 1–1.25mm, glandular-puberulous at apex, glistening glands on sutures. *Capsules* not seen.

Jamesbrittenia beverlyana is still known only from the type collection. The seeds remain unknown, and I am not certain that I have placed the species correctly, that is, with *J. pristisepala* and its allies rather than with *J. dentatisepala*. However, the floral morphology is that of *J. pristisepala*, from which *J. beverlyana* is easily distinguished by flowers in the axils of the leaves, not in distinct racemes.

Beverly recorded the plant as growing in 'rocky soil in the shade of an overhanging outcrop, southern exposure; mountainous grassveld region', and it was in first flower in the middle of February.

GROUP 2, 4

68. Jamesbrittenia ramosissima (Hiern) Hilliard in Edinb. J. Bot. 49: 232 (1992).
Lectotype (chosen here): Cape, Great Bushmanland, [2919AA], Pella, 14 viii 1898, *Max Schlechter* 131 (K; isolecto. BOL, E, PRE, S, W, Z).

Syn.: *Sutera ramosissima* Hiern in Fl. Cap.4(2): 265 (1904); Merxm. & Roessler in Prodr. Fl. S.W. Afr. 126: 54 (1967).

Dense rounded shrublet c.300mm–1.5m tall and as much or more across, intricately branched, branches divaricate, ultimate branchlets rather short and eventually leafless and subspinescent, densely clad in short-stalked glistening glands together with delicate glandular hairs up to 0.3–1mm long, these few to many, sometimes wanting on older parts. *Leaves* alternate, blade c.3–10 × 2.5–10mm, broadly ovate to rhomboid in outline, base cuneate to ± truncate passing abruptly into a petiolar part 1.5–5mm long (roughly half as long as blade), apex very acute to subacute, margins usually with 2–4 pairs of coarse acute teeth, teeth entire or occasionally 1-toothed, occasionally teeth obscure, both surfaces thickly clad in short-stalked glistening glands, delicate glandular hairs as well up to 0.2–0.8mm long. *Flowers* solitary in upper leaf axils, whole plant very floriferous. *Pedicels* up to 3–8mm long, glandular. *Calyx* tube 1–1.2mm long, lobes 2.4–3 × 1.4–1.8mm, triangular, thickly clad in stalked glistening glands and delicate glandular hairs up to 0.3–0.8mm long. *Corolla* tube c.4–6 × 4mm in mouth, cylindric below, upper half abruptly expanded, there obliquely campanulate, pouched on posticous side, limb nearly regular, 10–16mm across lateral lobes, posticous lobes 3.2–5 × 4.3–7.8mm, anticous lobe 3.3–5 × 4–8mm, all lobes subrotund, glandular-puberulous outside, hairs up to 0.1–0.2mm long, glistening glands as well, upper surface of limb minutely glandular puberulous, broad band of tiny clavate hairs all round mouth, a patch of longer ones inside tube on posticous side, lobes pale blue, mauve or lilac on opening, fading to white, throat yellow, posticous pouch dark violet. *Stamens*: posticous filaments 0.8–1mm long, inserted level with top of ovary and not decurrent, anthers 1.1–1.25mm, anticous filaments 0.4–0.6mm long, anthers 0.8–1.2mm long, visible in mouth, all filaments glabrous. *Stigma* 0.2–0.3mm long. *Style* 2–3mm. *Ovary* 1.3–1.7 × 1–1.3mm. *Capsules* c.4–5 × 3mm, covered in glistening glands and minute glandular hairs. *Seeds* c.0.5–0.7 × 0.25–0.4mm.

Selected citations:

Namibia. Lüderitz-Süd distr., 2716DC, farm Spitskop III, 23 ix 1972, *Merxmüller & Giess* 28718 (PRE, WIND). Warmbad distr., 2819CB, Velloorsdrift, 10 vi 1924, *Dinter* 5173 (BOL, PRE, SAM, W, Z).
Cape. Namaqualand, 2817AC, Lelieshoek, 760m, 29 viii 1977, *Oliver et al* 320 (PRE, STE); 2817DA, 6 miles from Modderdrift, road to Stinkfontein, 16 ix 1961, *Hardy* 674 (M, NBG, S, SRGH, Z); 2918BB, Aggeneys, c.3000ft, 25 v 1961, *Leistner* 2507 (M, PRE, SRGH, Z). Kenhardt distr., 2820CB, Aughrabies Falls National Park, c.2000ft, 7 v 1969, *Leistner* 3337 (K, PRE, WIND). Gordonia, 2820AD, Riemvasmaak, 3 x 1988, *Balkwill* 4181 (E, J, M).

Jamesbrittenia ramosissima is confined to the drainage system of the Orange (Gariep) river from the mountains about Spitskop, Namuskluft and Rosh Pinah in southernmost Namibia and the mountains of the Richtersveld in the northernmost Cape east to the environs of Aughrabies Falls. It grows in the shelter of rocks, mostly on steep slopes or on cliffs, but it has also been recorded from a stony dry river bed. It flowers between March and October.

It is so distinctive a species that it is never mistaken for any other. The corolla limb is nearly regular, and is hairy above (which is not unusual in this group of species, nos. 68–70); it probably changes colour as the bud opens, from shades of blue or mauve to white; the throat is yellow, and there is a dark patch on the posticous side. The corolla bears a remarkable resemblance to that of *Manulea nervosa* and *M. schaeferi*, in both form and colour-patterning, including colour-change. In *Jamesbrittenia*, it is very unusual for there to be a dark patch at the back of the corolla-throat, whereas it is commonplace in *Manulea*. Furthermore, even the calyx of *Jamesbrittenia ramosissima* resembles that of *Manulea nervosa*, being lobed only about two thirds of the way to the base, whereas in all other species of *Jamesbrittenia* it is lobed nearly to the base. The three species are at least partially sympatric.

69. Jamesbrittenia hereroensis (Engl.) Hilliard in Edinb. J. Bot. 49: 230 (1992).
Type: Namibia, [2214DB], Hykamkab, 250m, April 1886, *Marloth* 1207 (holo. B†; iso. BOL, PRE, SAM).

Syn.: *Chaenostoma hereroense* Engl. in Bot. Jahrb. 19: 150 (1894).
  Sutera hereroensis (Engl.) Skan in Fl. Trop.Afr. 4(2): 301 (1906); Merxm. & Roessler, Prodr. Fl. S.W. Afr. 126: 51 (1967) p.p.
  S. pedunculosa auctt., non (Benth.) O. Kuntze; Hemsley & Skan in Fl. Trop.Afr. 4(2): 301 (1906) p.p. quoad *Marloth* 1207 tantum.

Annual herb, but possibly perennating, stem 30–600mm tall, up to 6mm diam. at base, usually solitary, soon well branched from the base and above, glandular-pubescent, hairs up to 0.5–1.2mm long, leafy. *Leaves* opposite near base, alternate upwards, up to c.10–70 × 5–35mm, blade ovate in outline abruptly contracted to a petiolar part accounting for at least ⅓–½ total leaf length, sometimes longer than blade, apex ± acute, margins coarsely and irregularly toothed (lower seedling leaves may be ± entire), teeth very acute to subacute, both surfaces glandular-pubescent, hairs up to 0.2–1mm long, longest on petiolar part, glistening glands as well particularly on lower surface. *Flowers* solitary in nearly all the leaf axils. *Pedicels* up to 15–40(50)mm long, glandular-pubescent, hairs up to 0.2–0.8mm long. *Calyx* tube c.0.2(0.2–0.4)mm long, lobes 2.4–3.5 × 0.4–0.6mm (2.1–5.5 × 0.3–0.8mm), linear-lanceolate, subacute, glandular-pubescent, hairs up to 0.2–0.3 (0.25–0.4)mm long, glistening glands as well. *Corolla* tube 5–8 × 3–4mm in mouth (9–13 × 3–4mm), cylindric, abruptly expanded and almost campanulate above, limb oblique, 6.7–12mm (9.5–18mm) across lateral lobes, posticous lobes 2–3.6 × 2–4.5mm (3.2–5.8 × 3.3–6mm), anticous lobe 2–4 × 2.2–4.8mm (2.7–6 × 3–6mm), all lobes subrotund, glandular-puberulous outside, hairs up to 0.1–0.25mm long, glistening glands as well, upper surface of lobes clad in small clavate hairs (glabrous), base of anticous lip bearded with clavate hairs c.0.4–0.5mm long, these also within the tube, particularly on the posticous side between the filaments, all lobes light to dark violet, throat orange/yellow, the colour extending out onto base of anticous lip, sometimes outlined deep violet. *Stamens*: posticous filaments 1–1.6mm long (1.5–1.8mm), anthers 0.7–1mm (0.8–1.3mm), anticous filaments 0.3–0.8mm (1–1.5mm), anthers 0.6–0.7mm (0.7mm–1.2mm), all filaments bearded with clavate hairs c.0.2mm long. *Stigma* 0.1–0.2mm long (0.2–0.5mm). *Style* 2.7–4.3mm (6.5–8.5mm). *Ovary* 1.2–1.5 × 0.7–1mm (1.3–2 × 0.7–1mm). *Capsules* 3–4 × 1.8–2.2mm (3–6 × 2–4mm), glandular-puberulous, hairs c.0.2mm long, glistening glands as well. *Seeds* 0.25–0.5 × 0.2–0.3mm.

Selected citations:

Namibia. Karibib distr, 2115DD, Karibib, 28 i 1934, *Dinter* 6905 (BOL,K, M, PRE, S, WIND, Z); 2215BA, farm Tsabichab (KAR 58), 10 v 1973, *Giess* 12741 (K, WIND). Swakopmund distr., 2214DB, Khan river, near farmhouse Palmenhorst, c.250m, 5 iii 1963, *Kers* 1110a (S); 2315BA, Blutkuppe, 2 v 1965, *Giess* 8754 (W, WIND).

The most notable features of typical *J. hereroensis* are the shape of the corolla tube and the indumentum inside the tube and on the limb. The corolla tube is almost campanulate above (as it is in several of its allies); the upper surface of the corolla limb is covered in clavate hairs, which become longer on the palate (c.0.4–0.5mm) and descend into the tube, on the anticous side about as far as the point of insertion of the anticous filaments, on the posticous side from that same point almost to the base of the tube; the filaments too are strongly bearded with clavate hairs. In the allied *J. tenella* (confused with *J. hereroensis* in Prodr. Fl. S.W. Afr.), the limb may also be hairy above, but the clavate hairs inside the tube are much stouter than those of *J. hereroensis*, are strongly retrorse, and are confined to two patches on the posticous side about level with the anticous anthers, and a band running almost to the base of the tube between the posticous filaments (occasionally the hairs are much reduced in number or are completely wanting). The filaments in *J. tenella* are merely minutely glandular or glabrous.

I have included within my concept of *J. hereroensis* a plant that differs from it in its larger flowers with the limb glabrous above (measurements in parentheses in the formal description). In all other respects, including the remarkable development of hairs within the corolla tube, it seems indistinguishable from typical *J. hereroensis*. However, field work is needed to see if the glabrous-limbed plant differs from *J. hereroensis* in habit: it forms a rounded clump (Dinter gave it the manuscript epithet *pulvinata*, meaning cushion-forming), whereas *J. hereroensis* appears to have but one main stem, though that stem will branch profusely. The pulvinate plant is known only from the Naukluft Mountains (2316CD, 2416AA, AB). I have seen a number of specimens, including *Dinter* 8291 (BM, BOL, K, PRE, S, WIND, Z); *Giess* 10462 (WIND), *Giess* 10417 (K, PRE, WIND); *Merxmüller & Giess* 28150 (M); *Volk* 839 (M). In Prodr. Fl. S.W. Afr. 126: 55, *Dinter* 8291 and *Volk* 839 were mentioned under *Sutera tomentosa*.

The area of typical *J. hereroensis* lies north of the Naukluft Mountains, ranging from the Namib Desert Park north to the Brandberg and east to Karibib (2115DD) and Okongawa (2215BB). The plants grow in moist shady places under overhanging rocks and cliffs and in suitable rock crevices, often along watercourses and at springs; one record is from coarse gravel on the floor of a cave. Flowering has been recorded in all months, but the peak period is possibly March to May.

70. Jamesbrittenia tenella (Hiern) Hilliard in Edinb. J. Bot. 49: 232 (1992).
Type: Cape, Hay div., between the Kloof village [2922BB], in the Asbestos Mountains, and Wittewater (or Gattakamma) [2823 CC], 17 ii 1812, *Burchell* 2071 (holo. K).

Syn.: *Sutera tenella* Hiern in Fl. Cap.4(2): 278 (1904).
    S. flexuosa Hiern in Fl. Cap.4(2): 252 (1904). Type: Cape, Hay div., in the Asbestos Mountains at the Kloof village [2922 BB], 25 ix 1811, *Burchell* 1655 (holo. K).
    S. hereroensis auctt., non (Engl.) Skan; Merxmüller & Roessler, Prodr. Fl. S.W. Afr. 126: 51 (1967), p.p.
    S. pedunculosa auctt., non (Benth.) O. Kuntze; Hemsley & Skan in Fl. Trop.Afr. 4(2): 301 (1906) p.p. quoad *Marloth* 1380 tantum.

Annual herb, stem c.40–400mm tall, up to c.4–5mm diam. at base, solitary but soon well branched from the base and above, glandular-pubescent, hairs up to 1–3mm long, scattered glistening glands as well, leafy. *Leaves* opposite near base, alternate upwards, 15–80 × 6–45mm, blade ovate in outline, abruptly contracted to a petiolar part accounting for roughly half total leaf-length, apex ± acute, margins coarsely and irregularly toothed (lower seedling leaves may be ± entire), teeth very acute to subacute, both surfaces glandular-pubescent, hairs up to 0.5–2mm long, longest on petiolar part, glistening glands as well particularly on lower surface. *Flowers* solitary in nearly all the leaf axils, sometimes running up into pseudoracemes. *Pedicels* up to 10–28mm long, glandular-pubescent, hairs up to 0.3–1mm long. *Calyx* tube 0.1–0.2mm long, lobes 2.4–4 × 0.3–0.8mm, linear-lanceolate, glandular-pubescent, hairs up to 0.2–1mm long, glistening glands as well. *Corolla* tube 4.8–7.2 × 1.7–2.5mm in mouth, cylindric below, abruptly expanded above and narrowly campanulate, limb 3–6mm across lateral lobes, posticous lobes 1–2 × 0.9–1.7mm, anticous lobe 1–2 × 0.8–2mm, all lobes subrotund, glandular-puberulous outside, hairs up to 0.1–0.3mm long, glistening glands as well, upper surface of lobes sometimes glabrous, often clad in small clavate hairs, inside sometimes glabrous, usually with 2 small patches of ± acute, retrorse unicellular hairs up to 0.5–0.8mm long on posticous side level with anticous anthers, a narrow band of smaller retrorse hairs down back of tube between posticous filaments, all lobes shades of violet, lilac or rose, occasionally very pale or whitish, tube and throat yellow, the colour extending onto base of anticous lip, sometimes margined with a darker band of rose or violet. *Stamens*: posticous filaments 0.5–1mm long, anthers 0.2–0.5mm, anticous filaments 0.2–0.5mm, anticous anthers 0.15–0.4mm long, all filaments either glabrous or with a few minute glandular hairs at apex. *Stigma* c.0.1–0.2mm long. *Style* 2.7–4.8mm. *Ovary* 1.1–1.6 × 0.5–1mm. *Capsules* 3–4.5 × 1.7–3mm, glandular-puberulous, hairs up to 0.15–0.2mm long, glistening glands, if present, very few, near apex. *Seeds* 0.2–0.6 × 0.2–0.3mm.

Selected citations:

Cape. Hay div., c.2823CC, Asbestos Mountains, 1200m, vii 1894, *Marloth* 2024 (K, PRE). Prieska div., 2922DA, Buisvlei, 3500–4000ft, 22 ix 1938, *Wall* s.n. (S).
Namibia. Kaokoveld, 1712AB, 12 meilen südwestlich Otjinungua, 11 vi 1965, *Giess* 8915 (PRE, WIND). Okahandja distr., 2116DD, Okahandja, 5 iv 1907, *Dinter* 481 (E, SAM, Z); ibidem, 4500ft, 12 vii 1954, *A.S.L. Schelpe* 164 (BOL, K, M, PRE). Omaruru distr., 2114BA, Brandberg, Tsisab Valley, 5 v 1963, *Nordenstam* 2540 (M, S). Windhoek distr., c.2216DB, Daan Viljoen Park, Okaikasdamm, 20 viii 1972, *Merxmüller & Giess* 28001 (PRE, WIND). Rehoboth distr., 2316AD, Farm Gamsberg, 25 iv 1963, *Nordenstam* 2368 (M). Maltahöhe distr., 2416AA/AB, Bergzebrapark, Naukluft (MAL 9), 2 ix 1972, *Merxmüller & Giess* 28153 (WIND). Bethanien distr., 2616BA, Tiras, on road between Aus and Helmeringhausen, 8 v 1976, *Oliver & Müller* 6453 (K).

Both *Sutera tenella* and *S. flexuosa* were described from specimens collected by Burchell in the Asbestos Mountains, NE Cape, early in the 19th century. The type of the name *S. tenella* comprises three flowering seedlings, that of *S. flexuosa* a number of fruiting twigs in which the axes are markedly flexuous; the specimens represent the two extremes of the life cycle of the plant. Neither name has ever been taken into use. The plant has scarcely been re-collected in South Africa (the only specimens seen are cited above), but its area stretches north-westwards through Namibia; the southernmost record seen came from Sandmodder (2618DD) near the

Gross Karasberge, thence to the high ground stretching north from Aus to Rehoboth, Windhoek, Okahandja, the Waterberg and the Brandberg, with one record from much further north, near the Cunene, on a small tributary of the Marienfluss (1712AB). The plants grow in damp shady places under overhanging cliffs and rocks, or in rock crevices, but always in shade; granite is commonly mentioned by collectors, but basalt, gneiss, pegmatite and slate have also been recorded. Flowering possibly occurs in any month, but most records are between March and September.

In the Prodr. Fl. S.W. Afr., most of the specimens cited under *Sutera hereroensis* are *Jamesbrittenia tenella*; *Dinter* 6905 (which is *J. hereroensis*) is quoted as having somewhat larger flowers than normal, which suggests that the authors did not see the type of the name. The flowers of *J. hereroensis* are not only larger in the limb than those of *J. tenella*, they differ in the indumentum inside the corolla tube (see under *J. hereroensis*), and the ovary and capsule of *J. tenella* mostly lack glistening glands.

The area of *J. tenella* lies mostly east of that of *J. hereroensis* and possibly mostly at higher altitudes.

71. Jamesbrittenia fragilis (Pilger) Hilliard in Edinb. J. Bot. 49: 229 (1992).
Type: Namibia, 1917CB, Otaviberge, 14 i 1909, *Dinter* 740 (holo. B, iso. SAM).

Syn.: *Sutera fragilis* Pilger in Bot. Jahrb. 48: 439 (1912); Merxm. & Roessler, Prodr. Fl. S.W. Afr. 126: 50 (1967).

Annual herb, stems up to c.55–650mm long, up to c.5mm diam. at base, erect to prostrate, simple at first, soon well branched from base and above, glandular-pubescent, hairs up to c.0.6–1mm long, glistening glands as well, leafy. *Leaves* opposite near base, alternate upwards, up to 15–50 × 8–30mm, blade ovate in outline abruptly contracted to a petiolar part accounting for up to half total leaf-length, apex obtuse to subacute, margins of lowermost 1 or 2 pairs of leaves shallowly serrate, other leaves irregularly lobed, sometimes to midrib, lobes elliptic or oblong-elliptic, few-toothed, both surfaces glandular-pubescent, hairs up to c.0.5–0.8mm long, glistening glands as well particularly on lower surface. *Flowers* solitary in upper leaf axils. *Pedicels* up to 19–45mm long, glandular-pubescent, hairs up to 0.2–0.6mm long. *Calyx* tube 0.2–0.25mm long, lobes 2.3–4 × 0.3–0.7mm, oblong-lanceolate, subacute, glandular-pubescent, hairs up to 0.2–0.3mm long, glistening glands as well. *Corolla* tube c.7.5–10 × 4mm in mouth, cylindric, abruptly expanded and almost campanulate above, limb oblique, c.8.5–12mm across lateral lobes, posticous lobes 2.3–4.5 × 2–4.1mm, anticous lip porrect, anticous lobe 2.5–5 × 2–4.2mm, all lobes subrotund to oblong-elliptic, glandular-puberulous outside, hairs up to 0.15–0.2mm long, glistening glands as well, broad band of clavate hairs in throat extending onto base of anticous lip, lobes lilac, rose or light violet, yellow in throat and at base of anticous lip. *Stamens*: posticous filaments 1.8–2.5mm long, minutely glandular above point of insertion, glabrous or puberulous above, anthers 0.8–1.2mm long, anticous filaments 0.4–0.8mm long, glandular, anthers 0.6–1mm long. *Stigma* 0.3–0.5mm long. *Style* 3.4–4.5mm. *Ovary* 1.5–2 × 0.7–1mm. *Capsules* 3.5–4.5 × 2–3mm, glandular-puberulous, glistening glands as well. *Seeds* c.0.5–0.6 × 0.25–0.3mm.

Selected citations:

Namibia. Grootfontein distr., 1917BC, farm Elandshoek (GR 771), 23 iv 1978, *Giess* 15133 (PRE, WIND); 1917CA, farm Goab-Pforte (GR 56), 23 iv 1963, *Giess, Volk & Bleissner* 6387 (WIND); 1918CA, farm Karstberg, 1 iv 1968, *Wanntorp* 570 (M, S); 1917BC, Bobos, 2 iv 1934, *Dinter* 7573 (BOL, K, PRE, S, WIND, Z).

Jamesbrittenia fragilis appears to be confined to the limestone of the Otavi Mountains: records stretch from the farm Nimitz (OU 353) in the west to the farm Karstberg in the east. The plants grow in shade under rocks and in the crevices of cliffs; many records are from forest patches or woodland; flowering has been recorded between December and July, mostly in April.

It is easily distinguished from its allies by its deeply lobed leaves.

72. Jamesbrittenia pedunculosa (Benth.) Hilliard in Edinb. J. Bot.49: 232 (1992). **Fig. 24A, plate 4G.**

Type: Cape, Namaqualand, [2918 CC], Zilverfontein, 2000ft, 30 viii 1830, *Drège* (holo. K; iso. E, LE, M, MO, P, S, W).

Syn.: *Chaenostoma pedunculosum* Benth. in Hook., Comp.Bot. Mag. 1: 377 (1836) & in DC., Prodr. 10: 357 (1846).
    Sutera pedunculosa (Benth.) O. Kuntze, Rev. Gen. Pl. 2: 467 (1891); Hiern in Fl. Cap.4(2): 267 (1904); le Roux
    & Schelpe, Namaqualand ed. 2: 150, 151 cum ic.(1988).

Annual herb, stems up to c.45–400mm long, up to 0.25mm diam. at base, well branched from base and above, sprawling, glandular-pubescent, hairs up to 0.5–1.4mm long, some coarse and up to 0.1–0.2mm across flattened base, leafy. *Leaves* opposite near base, soon alternate upwards, up to 10–50 × 10–25mm, ovate in outline abruptly contracted to a petiolar part accounting for up to half total leaf-length, apex ± acute, margins irregularly and coarsely lobed and toothed (cut up to ½–¾ of way to midrib), both surfaces glandular-pubescent, hairs up to 0.2–1.4mm long, longest on petiolar part, scattered glistening glands as well. *Flowers* solitary in all the leaf axils. *Pedicels* up to 34–80mm long, almost filiform, glandular-pubescent, hairs up to 0.2–0.25mm long, some stout. *Calyx* tube 0.1–0.25mm long, lobes 2.3–4 × 0.6–1.1mm, narrowly rhomboid, ± acute, glandular-pubescent, hairs up to 0.1–0.2mm long, glistening glands as well. *Corolla* tube c.4–7.5×2.5–3.5mm in throat, cylindric then abruptly campanulate in upper half, limb distinctly bilabiate, c.8–11mm across lateral lobes, posticous lobes 3–6 × 2.5–5mm, base of posticous lip shallowly cucullate, anticous lobe 2–4 × 2.4–3.6mm, all lobes subrotund, minutely glandular-puberulous outside, band of clavate hairs inside throat, extending briefly onto base of anticous lip and there c.0.75–1mm long, ± acute, lobes bright yellow, sometimes with a dark purplish patch at base of posticous lip, always with a big dark patch on posticous side of tube, often with 2 small dark outgrowths on upper margin of patch about level with top of anticous anthers. *Stamens*: posticous filaments c.1–1.2mm long, minutely glandular above point of insertion, anthers 0.5–0.9mm long, anticous filaments 0.4–0.6mm long, anthers 0.3–0.7mm long, visible in mouth. *Stigma* c.0.2mm long. *Style* 3–5.8mm. *Ovary* 1.2–1.5 × 0.7–0.8mm. *Capsule* 3–6.5 × 2.5–4mm, glandular-puberulous, glistening glands on sutures. *Seeds* 0.7–1 × 0.3–0.4mm, reticulate, eventually patterned with transversely elongated pits arranged in chequer-board fashion.

Selected citations:

Cape. Namaqualand, 2917DB, O'okiep, 3200ft, ix 1883, *Bolus* in herb. norm. austr. Afr. 661 (BOL, E, K, PRE, SAM, W, Z); 2917DD, Mesklip, 24 viii 1941, *Esterhuysen* 5845 (BOL, K, PRE); 2917DB, 15 miles N of Springbok, 27 v 1961, *Schlieben* 9033 (K, M, NBG, S, SRGH, Z); 3018CB, southern Khamiesberg, Klippoort SE of Hoedberg, 520m, 6 vii 1990, *Oliver* 9535 (E, STE).

Jamesbrittenia pedunculosa appears to be confined to a small area of Namaqualand, from about Steinkopf and O'okiep south to the Khamiesberg, between c.500 and 1000m above sea level. The plants favour shady moist places under boulders and rock outcrops, flowering between May and December.

  The small yellow flowers have a unique corolla, the tube campanulate in the upper half, the upper lip larger than the lower and slightly hooded at its base, the lower lip with a median tuft of long hairs at its base immediately in front of the two anticous anthers, visible in the mouth. There is a large dark patch at the back of the inflated part of the corolla tube, often with two small dark outgrowths on its upper margin, and sometimes a dark patch just above the hooded part of the upper lip.The flowers are borne on remarkably long slender pedicels, which suggests pollination by small bees. They bear a strong resemblance to flowers of some species of *Hemimeris*, which occupy similar habitats in Namaqualand.

GROUP 2, 5

73. Jamesbrittenia breviflora (Schltr.) Hilliard in Edinb. J. Bot. 49: 227 (1992). **Fig. 48B.**
Type: Natal, foot of Drakensberg, Polela, 5000–6000ft, ii 1896, *Evans* 631 (holo. n.v., iso. BOL).

Syn.: *Lyperia breviflora* Schltr. in J. Bot. 34: 393 (Sept. 1896).
    L. punicea N.E. Br. in Kew Bull. 1896: 163 (Oct. 1896). Lectotype (chosen here): Natal, Griqualand East, 6000ft, 1883, *Tyson* 1363 (K; isolecto. BOL, E, PRE, SAM, Z).
    Chaenostoma woodianum Diels in Bot. Jahrb. 23: 474 (May 1897). Lectotype (chosen here): Natal, Maritzburg County, 18 xii 1885, *Wood* 3572 (Z).
    C. breviflorum (Schltr.) Diels in Bot. Jahrb. 23: 493 (1897).
    Sutera breviflora (Schltr.) Hiern in Fl. Cap.4(2): 263 (1904); Hilliard & Burtt, Botany of southern Natal Drakensberg pl. 28b (1987).

Perennial herb developing a stout caudex up to c.10mm diam. at the crown, stems c.150–600mm long, simple to laxly branched, sprawling, glandular-villous, hairs up to 1–3mm long, leafy. *Leaves* opposite or becoming alternate upwards, pseudofasciculate, largest blades 10–35 × 8–35mm, ovate to subrotund tapering to a petiolar part 3–10mm long (roughly ½–⅕ length of blade), obtuse, margins crenate, the longest lobes occasionally with a tooth, both surfaces glandular-villous, hairs up to 1–3mm long (longest hairs on petiolar part), glistening glands also present particularly on lower surface. *Flowers* solitary in the leaf axils forming long leafy racemes. *Pedicels* up to 9–35mm long. *Calyx* tube 0.2–0.8mm long, lobes c.3.8–5.5 × 1–3mm, spathulate, glandular-villous, hairs up to 0.5–2mm long, glistening glands as well. *Corolla* tube 3.5–6 × 2–4mm, abruptly expanded into the limb, mouth dorsoventrally compressed, limb oblique, c.7–20mm across lateral lobes, posticous lobes 3–7 × 3–7mm, anticous lip porrect, anticous lobe 2–8 × 2.5–8.5mm, all lobes suborbicular, glandular-pubescent outside, hairs up to c.0.5–1mm long, glistening glands as well, broad transverse band of clavate hairs in throat extending briefly onto base of lower lip, lobes dark red to terracotta or rose-pink, rarely dark lilac, main veins dark, palate bright yellow. *Stamens*: posticous filaments 1.5–2.5mm long, glandular-puberulous above, minute sessile glands at base, anthers 0.8–1.4mm long, anticous filaments 0.5mm–1mm long, glandular-puberulous, anthers 0.6–1mm long. *Stigma* 0.3–0.5mm long, included. *Style* 2.2–3mm long. *Ovary* 1–1.5 × 0.8–1.2mm. *Capsules* c.4–6 × 2.5–3mm, glistening glands all over, a few minute hairs at apex. *Seeds* c.0.3–0.6 × 0.3–0.4mm.

Selected citations:

Natal. Bergville distr., 2828DB, Royal Natal National Park, 5000ft, v 1946, *Lewis* 1944 (SAM). Estcourt Distr., 2929BC, Kamberg Mt., c.6300ft, 28 xii 1968, *Hilliard & Burtt* 5721 (E,K, Z). Underberg distr., Bushman's Nek, ridge near Thamathu Cave, c.7500ft, 6 ii 1976, *Hilliard & Burtt* 8993 (E, K, PRE, S).
Transkei. Mt. Ayliff distr., 3029DC, Fort Donald, 4000ft, 21 xi 1945, *Story* 570 (PRE, Z).
Cape. Elliot-Maclear distr. boundary, 3127BB, Bastervoetpad, c.7200ft, 15 ii 1983, *Hilliard & Burtt* 16696 (E).
Lesotho. 2929AC, Putha, 8500ft, 28 ii 1949, *Compton* 21612 (NBG). 2929CC, Sehlabathebe, 8200ft, 7 i 1973, *Bayliss* 5425 (NBG, Z).

Jamesbrittenia breviflora is a plant of the Eastern Mountain Region, with records stretching from Royal Natal National Park near Bergville in Natal along the Drakensberg, its foothills and outliers (Karkloof range, Swartkop above Pietermaritzburg, Ngeli Mountain) and the eastern fringe of Lesotho to the Cape Drakensberg near Maclear, between c.1370 and 3000m above sea level.

It favours grassy and often rocky slopes and will colonize bare soil, flowering mainly between October and March, and is easily recognized; sprawling habit, glandular-villous stems and leaves, large red to rose-pink flowers. In the Natal Drakensberg, *J. breviflora* and *J. pristisepala* frequently hybridize: see further under *J. pristisepala*.

74. Jamesbrittenia jurassica (Hilliard & Burtt) Hilliard in Edinb. J. Bot.49: 230 (1992). **Plate 1D.**

Type: Lesotho, Sani Top, 2900m, 2 i 1974, *Hilliard* 5434 (holo. E; iso. K, NU, PRE).

Syn.: *Sutera jurassica* Hilliard & Burtt in Notes RBG Edinb. 40: 292 (1982).

Perennial herb, stout caudex up to c.15mm in diam. at crown, stems many from the crown, up to c.60–120mm long, simple to laxly branched, prostrate, glandular-pubescent, hairs up to c.0.5–0.8mm long, glistening glands as well, leafy. *Leaves* opposite, largest blades 10–20 × 7–12mm, oblong to elliptic in outline tapering to a broad petiolar part, divided ⅔ or more to the midrib, the lobes sometimes divided again, lobes oblong to elliptic, obtuse, both surfaces glandular-pubescent, hairs up to 0.4–0.8mm long, glistening glands as well. *Flowers* solitary in the leaf axils. *Pedicels* up to 4–10mm long. *Calyx* tube c.0.8mm long, lobes 3–3.8 × 0.7–1.5mm, spathulate, glandular-pubescent, hairs up to 0.5–1mm long, glistening glands as well. *Corolla* tube 6.5–9 × 2.5–2.8mm, abruptly expanded into the limb, mouth dorsoventrally compressed, limb oblique, c.14mm across, posticous lobes 5–6 × 5–6.3mm, anticous lip porrect, anticous lobe 5–6 × 5–6.5mm, all lobes suborbicular, glandular-pubescent outside, hairs up to 0.3–0.6mm long, glistening glands as well, broad band of clavate hairs in throat extending onto bases of lobes, lobes light violet or occasionally pink, dark violet streaks at base, throat and base of anticous lip yellow. *Stamens*: posticous filaments c.2mm long, well bearded, anthers 0.8–1.2mm long, anticous filaments 1–1.2mm long, minutely glandular, anthers 0.5–0.8mm long. *Stigma* 0.2–0.3mm long, included. *Style* 5.25–6.4mm long. *Ovary* 1.5–2 × 1–1.2mm. *Capsules* c.3–3.5 × 2.5mm, glistening glands all over, tiny glandular hairs as well scattered on upper half.

Selected citations:

Lesotho. 2929CA, Black Mountains, 10400–10600ft, 13 i 1976, *Hilliard & Burtt* 8760 (E, K, NU, PRE). Above confluence of Mangamay and Sani streams, c.8000ft, 1935, *Milford* 730 (K). 2928BD, S of Mokhotlong, Menoaneng Pass, 3100m, 30 iii 1986, *Phillipson* 1431 (PRE).

Jamesbrittenia jurassica is easily distinguished from its ally, *J. breviflora*, by its more deeply divided leaves, shorter pedicels, and violet-coloured flowers with a longer corolla tube.

The plant forms a small flowery mat on bare gravelly ground between c.2500 and 3230m above sea level. Specimens have been seen only from the Sani Top – Black Mountains-Mokhotlong-Menoaneng Pass area, but there is a colour slide in the Edinburgh herbarium (E) of the plant growing near Oxbow (taken by the late Col. Anderson) so it is clearly widely distributed over the high mountains of Lesotho. It is in full flower in January and we collected seed from dry withered stems in April.

75. Jamesbrittenia aspleniifolia Hilliard in Edinb. J. Bot. 49: 227 (1992).

Type: Cape, Barkly East div., 3028CC, Rhodes to Naude's Nek, Dunley, c.7500ft, 13 ii 1983, *Hilliard & Burtt* 16615 (holo. E; iso. K, NU, PRE, S).

Twiggy shrublet up to c.350mm tall, stems erect or decumbent, glandular-pubescent, hairs up to 0.25–0.5mm long, leafy. *Leaves* opposite, alternate upwards, pseudofasciculate, largest 10–22 × 2.5–6mm, blade oblong in outline with a conspicuous petiolar part, divided nearly to midrib, the largest lobes often divided again, lobes broadly elliptic to subrotund, apex rounded, both surfaces glandular-pubescent, hairs up to 0.2–0.5mm long (longest on petiolar part), glistening glands as well, mainly on lower surface. *Flowers* solitary in axils of upper leaves, these rapidly very small upwards, forming long racemes sometimes panicled. *Pedicels* up to 8–15(–23)mm long. *Calyx* tube c.0.2–0.25mm long, lobes 2.3–3.2 × 0.6–1mm, spathulate, glandular-pubescent, hairs up to 0.2–0.6mm long, glistening glands as well. *Corolla* tube c.4–5 × 3mm in throat, cylindric, abruptly expanded into the limb, mouth dorsoventrally compressed, limb oblique, c.6–8mm across lateral lobes, posticous lobes 2.2–3 × 2.5–3mm, anticous lip porrect, anticous lobe 2.2–3 × 2.2–3.3mm, all lobes suborbicular, glandular-pubescent outside, hairs up to 0.25–0.5mm long, glistening glands as well, broad transverse band of clavate hairs in throat extending briefly onto base of anticous lip on either side of slightly raised 3-lobed palate, lobes brick-red, bright red or dark red, perhaps rarely white (see discussion), palate yellow. *Stamens*: posticous filaments 1.2–1.6mm long, either glabrous or minutely glandular above, minutely pubescent on decurrent part, anthers 0.5–0.8mm long, anticous filaments 0.4–1mm long, minutely puberulous, anthers 0.4–0.7mm long. *Stigma* 0.2–0.4mm long, included. *Style* 1.8–2.2mm long. *Ovary* 1–1.4 × 0.7–1mm. *Capsules* c.3.5–6 × 2–2.5mm, glistening glands all over, a few minute hairs at apex. *Seeds* c.0.6–0.75 × 0.3mm.

Citations:

Cape. Barkly East div., 3028CC, ascent to Naude's Nek from Rhodes, 30 i 1966, *Hilliard & Burtt* 3774 (E). 3027DC, Barkly East - Clifford road through 'Fonteins Kloof', c.5600ft, 14 x 1980, *Hilliard & Burtt* 13127 (E, K, PRE, S). 3027DA, Witteberg, Avoca Farm, 3 v 1928, *Hutchinson* sub Moss 16843 (BM). Barkly East div., 10 xii 1933, *Gerstner* 178(K).
Lesotho. 3028BA, near Qacha's Nek, Rapase, 5500ft, 10 iii 1936, *Galpin* 14048 (PRE and BOL sub BOL 51387). 3028CA, Buffalo River waterfall, c.7500ft, 11 iii 1904, *Galpin* 6804 (BOL, K, NBG, PRE, SAM); without precise locality, *Staples* 133 (K). 2928DB, 3km from Taung on road to Matabeng, 1980m, 1 xii 1977, *Killick* 4276 (K). 2929AC, Sehonghong river crossing, 13km from Mokhotlong, 2150m, 10 ii 1987, *Killick* 4549 (PRE). 3027BC, 9 miles from Quthing on road to Mt. Moorosi, 1520m, 10 xii 1977, *Killick* 4378 (K). 2929AC, Mokhotlong, 8000ft, 6 iii 1938, *Brooke* 40 (BM). Without locality, 1894, *Glass* 856 (NBG).

Jamesbrittenia aspleniifolia is known only from the southern half of Lesotho and the neighbouring mountainous parts of the Cape around Barkly East. It favours rocky sites such as banks and cliffs, possibly always on basalt, between c.1675 and 2300m above sea level, and flowers between October and March. The plant dries black, but alive, it is pretty, with narrow finely cut leaves clothing all the twiglets, the flowers carried aloft in nearly nude racemes (they are in the axils of mostly very small leaves). The colour of the corolla limb ranges from deep pink through terracotta to bright red, fading to shades of mauve, the main veins darker in colour, and the 3-pronged slightly raised palate yellow.

In herbaria it has been confused with *J. aurantiaca*, from which it is easily distinguished by its narrower and less finely cut leaves, pubescent on both surfaces: in *J. aurantiaca* the upper surface of the leaf is clad in glistening glands to the almost total exclusion of stalked hairs.

Specimens collected by Compton at Mokhotlong (*Compton* 2159, NBG, STE) were recorded as having white flowers, though *Brooke* 40 from the same locality had red flowers. The sheet of *Compton* 2159 in STE is wholly *J. aspleniifolia*; that in NBG comprises five branchlets, one of which is surely a hybrid between *J. aspleniifolia* and the white-flowered *J. lesutica*, with which it is there sympatric. The hybrid has flowers intermediate in size between those of *J. aspleniifolia* and *J. lesutica* and carries far more large glistening glands, which are characteristic of *J. lesutica*.

76. Jamesbrittenia multisecta Hilliard in Edinb. J. Bot. 49: 231 (1992).
Type: Transkei, 3128DA, Umtata-Engcobo road near Umtata, 21 xi 1977, *Hilliard & Burtt* 10552 (holo. E).

Perennial herb, woody caudex up to c.15mm diam. at crown, stems many from the crown, c.70–200mm long, prostrate, profusely branched, glandular-pubescent, hairs up to 0.6–1.3mm long, scattered glistening glands as well, closely leafy. *Leaves* opposite, pseudofasciculate, largest c.5–22 × 5–15mm, broadly ovate in outline with a short petiolar part, bi- to tripinnatisect, lobes linear, acute, thinly glandular-pubescent below, hairs c.0.2–1mm long, few glistening glands as well, crowded glistening glands above. *Flowers* solitary in the axils of the uppermost leaves. *Pedicels* up to c.3–5mm long. *Calyx* tube 0.2mm long, lobes c.4.5–6 × 2.5–3mm, accrescent in fruit and enfolding the capsule, spathulate in outline, the expanded upper part deeply dissected, glandular hairs up to c.0.5–0.8mm long, glistening glands as well, as in leaves.

Corolla tube 5–6 × 2.5–3mm, cylindric, expanded in throat, mouth dorsoventrally compressed, limb oblique, c.11mm across lateral lobes, posticous lobes 3.8–4 × 3–3.5mm, anticous lip porrect, anticous lobe 4–4.8 × 3.2–4mm, all lobes suborbicular, glandular-puberulous outside, hairs c.0.25mm long, glistening glands as well, broad transverse band of clavate hairs in throat extending onto base of anticous lip, lobes dull terracotta, throat and base of anticous lip yellow. *Stamens*: posticous filaments c.2mm long, anthers 0.8mm, anticous filaments c.0.5mm long, anthers 0.5–0.6mm long, filaments glabrous. *Stigma* c.0.2–0.4mm long, included. *Style* 2.75–3mm long. *Ovary* c.1.5 × 1.5mm. *Capsules* c.3–5 × 2–3mm, glistening glands all over. *Seeds* 0.6–0.7 × 0.4–0.5mm.

Citations:

Transkei. Butterworth distr., 3228AC, 2 miles SW of Ndabakazi, 1700ft, 10 iii 1955, *Codd* 9252 (K, M, PRE, S, SRGH); between Idutywa and Butterworth, 2000ft, 8 xi 1938, *Wall* s.n. (S). Port St Johns distr., 3129CB, Mt. Thesiger, 7 xi 1938, *Wall* s.n. (BOL, S); between Cala and Elliot, 4500ft, i 1896, *Flanagan* 2636 (PRE).
Cape. Komgha, 8 xi 1938, *Hafstrom & Acocks* 1178 (S); grassy hill near Komgha, 2000 ft, xi 1890, *Flanagan* 79 (PRE).

Jamesbrittenia multisecta is a most distinctive plant, with finely cut leaves, the segments very narrow and acute, the calyx lobes similarly dissected. In herbaria, it has been confused with *J. aurantiaca*, from which it is distinguished by its constantly opposite and more finely dissected leaves, dissected calyx lobes, flowers only in the upper leaf-axils, and much shorter pedicels. It seems to be confined to the Transkei and neighbouring Komgha, and thus its area lies well outside that of *J. aurantiaca*. It grows in short grassland, c.300 to 600m above sea level, flowering between November and March, when it is well on into fruit.

77. Jamesbrittenia aurantiaca (Burchell) Hilliard in Edinb. J. Bot. 49: 227 (1992).
Type: Cape [Herbert div., 2823CD], Spuigslang Fontein, 25 x 1811, *Burchell* 1727 (holo. K; iso. M, NY, PRE, W).

Syn.: *Buchnera aurantiaca* Burchell, Trav. S. Afr. 1: 270 (1822).
   Lyperia multifida Benth. in Hook., Comp.Bot. Mag. 1: 38 (1836) & in DC., Prodr. 10: 361 (1846). Lectotype (chosen here): Cape, [Queenstown div., 3126AB], Sternbergspruit, *Drège* 797a (K; isolecto. E, LE, P, S, W).
   Manulea multifida (Benth.) O. Kuntze, Rev. Gen. Pl. 3 (2): 236 (1898).
   Sutera aurantiaca (Burch.) Hiern in Cat. Afr. Pl. Welw. 1: 757 (1898) & in Fl. Cap.4 (2): 292 (1904) p.p.; Pole Evans in Fl. Pl. S. Afr. 13: pl. 484 (1933); van Wyk & Malan, Field Guide to the Wild Flowers of the Witwatersrand and Pretoria region 280 fig. 728 (1988); Philcox in Fl. Zamb. 8(2): 29 (1990).
   Chaenostoma aurantiacum (Burch.) Thell. in Viertel. Nat. Ges. Zürich 74: 116 (1929) in obs.

Perennial herb (can flower when very young and look annual) with a woody rootstock up to c.20mm diam. at the crown, stems many from the crown, c.60–300(–750)mm long, prostrate or ascending, sometimes rooting, well branched from the base and above, glandular-puberulous, hairs up to 0.1–0.25mm long, leafy throughout. *Leaves* opposite at extreme base, soon alternate, pseudofasciculate, up to 5–30 × 5–16mm, elliptic to broadly ovate in outline contracted into a petiolar part, bipinnatisect, lobes oblong or narrowly elliptic, subacute, thinly glandular-puberulous below, hairs up to 0.1–0.2mm long, crowded glistening glands above, very rarely a few stalked hairs as well. *Flowers* solitary in the axils of the leaves, these smaller upwards but nearly always dissected. *Pedicels* up to 7–25mm long. *Calyx* tube 0.1–0.3mm long, lobes 2.4–5 × 0.6–1(–2)mm, oblong to spathulate, usually entire, very rarely with a few teeth, glandular-pubescent, hairs up to 0.1–0.25mm long, glistening glands as well. *Corolla* tube 5–7 × 1.7–2mm, cylindric, throat slightly expanded, mouth dorsoventrally compressed, limb oblique, 6–14mm across lateral lobes, posticous lobes 1.7–4 × 1.5–4mm, anticous lip porrect, anticous lobe 1.7–4.5 × 1.7–5mm, all lobes suborbicular, glandular-puberulous outside, hairs up to 0.15mm long, glistening glands as well, broad transverse band of clavate hairs in throat extending to base of anticous lip, lobes varying in colour from bright orange to shades of red and purple-pink, throat yellow, the colour extending onto base of lower lip, the main veins coloured dark brown to purple. *Stamens*: posticous filaments 1–1.5mm long, posticous anthers

0.6–1mm, anticous filaments 0.4–0.8mm, anticous anthers 0.4–0.7mm, filaments minutely puberulous. *Stigma* 0.2–0.25mm long, included. *Style* 3.2–4.5mm long. *Ovary* 1.2–1.5 × 0.8–1mm. *Capsules* 3.5–5 × 2–3mm, scattered glistening glands and a few minute hairs. *Seeds* 0.3–0.4 × 0.3mm.

Selected citations:

Namibia. Oranjemund, 2816CB, 22 iii 1957, *Merxmüller & Giess* 2277 (M).
Botswana. 2525BA, S of Ootze, 17 viii 1978, *Hansen* 3458 (PRE, SRGH). 2425DB, 6 miles SW of Gaborone, 3200ft, 2 ix 1974, *Mott* 361 (K, PRE, SRGH).
Transvaal. Marico distr., 2525BD, Linokana, v 1887, *Holub* s.n. (Z). Pretoria distr., 2528CA, Pretoria, 1660m, 3 x 1893, *Schlechter* 3599 (BM, BOL, E,K, S, STE, Z). Belfast distr., 2530CA, near Belfast, 7 xii 1965, *Burtt* 3113 (E, NBG, S).
Orange Free State. Dewetsdorp, 2926DA, 15 iv 1950, *Steyn* 934 (NBG). Ficksburg, 15 xi 1937, *Fawkes* 220 (NBG).
Natal. Without precise locality, *Gerrard* 1517 (W).
Lesotho. 2927BC, Thaba Bosiu, i 1903, *Junod* 1899 (Z). Leribe, 2828CC, *Dieterlen* 198 (NBG, P, PRE).
Cape. Barkly East div., 3027BC, Three Drifts Stream below Pitlochrie, c.5800ft, 6 xii 1981, *Hilliard & Burtt* 14732 (E, K). Albert div., 1861, *Cooper* 573 (BOL, E, K, W, Z). Vryburg div., 2624, Armoedsvlakte, 19 ii 1921, *Mogg* 8691 (PRE, STE). Herbert div., 2923BB, St Clair, xii 1896, *Orpen* 139 (NBG, SAM).

Jamesbrittenia aurantiaca is generally perennial, and develops a stout caudex, but Burchell's original specimens (particularly those in M and W) show that the plant can flower when only simple-stemmed with a thin taproot, and thus look annual. The colour of the corolla limb is variable and there appears to be some geographical patterning to this. In the Cape, whence came the original material of both names, *aurantiaca* and *multifida*, the limb varies from orange to brick red or salmon pink, the throat and base of the anticous lip yellow, the principal veins brown to blackish.

Similar colouring is found in Lesotho, the Orange Free State, the western and southernmost parts of the Transvaal; from roughly the Johannesburg area eastwards the limb has been recorded as purple-red, magenta or shades of pink, yellow or 'dark' in the throat and on the palate, the main veins purple. In Botswana the limb is pale mauve with an orange throat. The two specimens from Botswana cited above have the leaves typical of *J. aurantiaca*; those of *Smith* 4252 (SRGH, from the bed of the Sibuyu river) and *Harbor* from Mochudi (sub 16381 and 16382 in SAM) have leaves less deeply dissected than is normal in *J. aurantiaca*, but are mauve-flowered as in the other material from Botswana.

The distinguishing marks of the species are its bipinnatisect leaves (cut very nearly to the midrib), thinly glandular-puberulous below, the upper surface clad in crowded glistening glands with very rarely a few glandular hairs as well. The leaves of *J. micrantha*, with which *J. aurantiaca* is often confused, are less deeply dissected and with glandular hairs much more plentiful on the upper surface than glistening glands; the flowers are also smaller than those of *J. aurantiaca*.

Jamesbrittenia aurantiaca is widely distributed in the interior of southern Africa, from the Transvaal south of the 29th parallel (excluding the Lowveld) and neighbouring Botswana through the Orange Free State and the western part of Lesotho to the eastern Cape in the mountains above Lady Grey, Barkly East, Queenstown and Molteno, thence through Colesberg to Herbert div., Kuruman, Vryburg and Mafeking. There is one record as far west as Aughrabies falls and another from the mouth of the Orange River, which suggests stray plants from seeds washed down the river. There is one old, unlocalized, record from Natal, made by Gerrard about 150 years ago, but as the plant occurs at Harrismith it is possible that it grows just within Natal's border with the Orange Free State. The plants favour sandy or stony places that are nevertheless damp or even marshy, between c.1370 and 1900m above sea level. Flowering has been recorded in nearly all months of the year, but the chief season is October to March.

78. Jamesbrittenia montana (Diels) Hilliard in Edinb. J. Bot. 49: 231 (1992).
Type: Natal, Biggarsberg range, bei de Taagen, x 1888, *Wilms* 1051 (B†, iso. BM).

Syn.: *Chaenostoma montanum* Diels in Bot. Jahrb. 26: 121 (1899).
   Sutera montana (Diels) S. Moore in J. Bot. 38: 467 (1900); Hiern in Fl. Cap. 4(2): 262 (1904); Hilliard & Burtt
      in Notes RBG Edinb. 40: 293 (1982) excl. syn. *S. bolusii* Hiern.
   S. luteiflora Hiern in Fl. Cap. 4(2): 294 (1904). Lectotype: Natal, Colenso, 3300ft, 23 x 1888, *Wood* 4044 (K;
      isolecto. BM, E, SAM).

Perennial herb but can flower young and look annual, woody rootstock eventually up to c.10mm in diam., stems solitary at first, soon several from the crown, c.70–400mm long, erect, ascending or decumbent and then sometimes rooting, soon branching from the base and above, glandular-puberulous, hairs up to 0.1–0.3mm long, few scattered glistening glands as well, leafy. *Leaves* opposite becoming alternate upwards, pseudofasciculate, up to 8–30 × 6–22mm, blade elliptic to broadly ovate in outline tapering into a petiolar part, blade cut nearly to midrib, one or two of the largest pairs of lobes cut again, other lobes entire or 1-toothed, plentiful glistening glands on both surfaces, scattered glandular hairs as well up to 0.1–0.3mm long particularly on lower surface. *Flowers* solitary in the axils of the leaves, these becoming very small upwards, the whole forming long racemes or panicles. *Pedicels* up to 9–27mm long. *Calyx* tube 0.2–0.25mm long, lobes 2–3.8 × 0.6–1.1mm, oblong to spathulate, glandular-pubescent, hairs up to 0.15–0.25mm long, glistening glands as well. *Corolla* tube 2.8–4.2 × 1.8–2mm, cylindric, throat slightly expanded, mouth dorsoventrally compressed, limb oblique, 5–10mm across lateral lobes, posticous lobes 1.5–2.7 × 1.7–2.3mm, anticous lip porrect, anticous lobe 1.5–3.6 × 1.8–3.6mm, all lobes suborbicular, yellow, deeper yellow in throat and at base of anticous lip, glandular-puberulous outside, hairs c.0.15mm long, glistening glands as well, a broad band of clavate hairs in throat extending briefly onto base of lower lip. *Stamens*: posticous filaments 0.8–1.5mm long, anthers 0.4–1mm, anticous filaments 0.5–0.7mm long, anthers 0.5–1mm, filaments minutely puberulous. *Stigma* 0.2–0.4mm long, included. *Style* 1.7–2.7mm long. *Ovary* 1.2–1.5 × 0.8–1mm. *Capsules* 3–5 × 2–3mm, glistening glands all over.

Selected citations:

Zimbabwe. Beit Bridge distr., c.2128CB, Shashi river, near the District Commissioner's Rest Hut, 1 ii 1973, *Ngoni* 183 (K, SRGH).
Transvaal. Waterberg distr., 2428DA, Naboomspruit, Mosdene, 9 v 1921, *Galpin* 531 (E, PRE, SAM). Klerksdorp distr., 2626CD, Wolwerand, 9 x 1972, *Hanekom* 1874 (PRE, SRGH).
Natal. Babanango distr., 2831AC, west slopes of Babanango plateau, c.4000ft, 8 x 1946, *Acocks* 12929 (PRE). Klip River distr., 2829DB, Colenso, 1500m, 27 ix 1893, *Schlechter* 3374 (BM, BOL, E, K, S, Z).

Jamesbrittenia montana is allied to *J. micrantha*, which has similar yellow flowers. They differ in their leaves: those of *J. montana* are cut nearly to the midrib and one or two of the major lobes are cut again, while in *J. micrantha* they are divided roughly ½ to ⅓ of the way to the midrib, and the lobes are either entire or toothed. Also, in *J. montana* the upper surface of the leaf is conspicuously gland-dotted and there are few stalked glands; in *J. micrantha*, stalked glands are far more plentiful than glistening glands. It has also been confused with *J. aurantiaca*, which has indumentum similar to that of *J. montana* but the leaves are more dissected, the flowers larger (longer corolla tube, longer style) and usually differently coloured limb; the limb of *J. montana* is normally yellow, but Thode recorded it as vermilion-coloured or dull yellow (*Thode* 4524, STE, from Elandslaagte in Natal).

Jamesbrittenia montana has been recorded from one site in south western Zimbabwe, then from the western part of the Transvaal and northern Natal. Throughout its range, its area lies west of that of *J. micrantha*, but it is partly sympatric with *J. aurantiaca*. It favours damp places, in grassland (sometimes among rocks), around marshes and pans and along rivers and streams, from c.1070 to 1500m above sea level, flowering from about September to February.

79. Jamesbrittenia concinna (Hiern) Hilliard in Edinb. J. Bot. 49: 228 (1992).
Type: Bechuanaland, *Passarge* 37 (B†); Botswana, Kangwa (Xanwe), 27km NE of Aha Hills, 12 iii 1965, *Wild & Drummond* 6943 (neotype SRGH; isoneotype K, PRE).

Syn.: *Sutera concinna* Hiern in Fl. Cap.4(2): 293 (1904).
    S. tenuis Pilger in Bot. Jahrb. 48: 438 (1912); Merxm. & Roessler in Prodr. Fl. S.W. Afr. 126: 54 (1967). Lectotype: Namibia, Grootfontein, viii 1911, *Dinter* 3038 (B†, iso. SAM).

Annual herb, stems c.20–400mm long, erect, decumbent or prostrate, solitary at first, later several tufted from crown of woody taproot up to c.7mm in diam, simple at first, soon well branched from base and above, glandular-puberulous, hairs up to 0.2–0.3mm long, scattered glistening glands as well, leafy throughout. *Leaves* opposite becoming alternate upwards, often pseudofasciculate, largest 10–35 × 5–18mm, blade elliptic in outline tapering into a petiolar part, blade divided roughly halfway or more to midrib, lobes oblong to elliptic, entire or the largest with up to 4 teeth, both surfaces hairy as the stems. *Flowers* solitary in the axils of the leaves, these becoming very small and entire upwards, forming racemes or narrow panicles. *Pedicels* up to 10–30mm long. *Calyx* tube 0.2–0.3mm long, lobes 2–3 × 0.5–1mm, oblong to spathulate, glandular-pubescent, hairs up to 0.1–0.25mm long, glistening glands as well. *Corolla* tube 4–7 × 1.5–2mm, cylindric, throat slightly expanded, (mouth dorsoventrally compressed?), limb (4–)6–11mm across lateral lobes, posticous lobes (1.2–)2–4 × (1.4–)2–4.7mm, anticous lip porrect, anticous lobe (1.4–)2.3–4 × (1.5–) 2.3–4.25mm, all lobes suborbicular, bright yellow or very rarely pink, orange patch in throat extending onto base of lower lip, glandular-puberulous outside, glistening glands as well, broad transverse band of clavate hairs in throat and at base of anticous lip.*Stamens:* posticous filaments (0.8–)1–1.5mm long, anthers (0.5–)0.7–1.1mm, anticous filaments 0.5–1mm, anticous anthers (0.4–)0.5–0.8mm. *Stigma* 0.1–0.25mm long, included. *Style* (2–)2.5–3.5mm long. *Ovary* 1–1.5 × 0.8–1mm. *Capsules* c.3.5–4 × 2–2.5mm, few minute stalked glands at apex, few glistening glands on sutures. *Seeds* c.0.25–3 × 0.15–0.25mm.

Selected citations:

Botswana. 2123BD, Central Kalahari Game Reserve, pan north of Deception Pan, 13 v 1986, *Chadwick* 241 (PRE). 1921CA, Qangwa stream, 28 iv 1982, *Mavi* 1718 (SRGH).
Namibia. Kaokoveld, 1712DA?, Kunene river, 24 vi 1951, *Hall* 471 (NBG). 1815CC/CD, Etosha National Park, Elandsdraai, 27 v 1974, *le Roux* 1160 (WIND). 1915BD, Etoschapfanne, nähe Ombika, 17 iv 1978, *Giess* 15081 (S, WIND). 1918CA, Grootfontein, 11 vii 1934, *Dinter* 7537 (BOL, K, PRE, S, WIND). 2116DD, North of Okahandja, 1 iv 1963, *Kers* 317 (S, WIND).

Although the type of *Sutera concinna* was lost in the Berlin fire, Hiern's detailed description in Flora capensis (1904) enables me confidently to equate his plant with that subsequently described as *S. tenuis*. The species is widely distributed across the northern half of Namibia with records as far east as central Botswana (Deception Pan area). It probably also occurs across the Cunene in Angola. The southernmost records in Namibia are from the environs of Okahandja. It was hereabouts that Dinter collected some tiny flowering seedlings (*Dinter* 7843, B) as well as better-grown specimens (*Dinter* 600, SAM, as well as unnumbered specimens under SAM 74453, 74454, 74455) that he misdetermined as *Sutera aurantiaca*. *Jamesbrittenia concinna* is easily distinguished from *J. aurantiaca* by its 1-fid leaves. Its leaves are more like those of *J. micrantha*, but whereas *J. micrantha* is perennial, *J. concinna* is an annual; the style of *J. micrantha* is up to 2.2mm long, that of *J. concinna* is usually 2.5–3.5mm long; it is shorter only in tiny flowering seedlings (the flowers will self). The species is not included in *Flora zambesiaca* 8(2) because Philcox confounded it with *J. micrantha*.

Jamesbrittenia concinna favours damp ground around marshes, pans and along river banks; several collectors recorded its growing in calcareous soils, even growing from the cracks in limestone pavements. It is normally yellow-flowered, but *Kers* 317 (cited above) records pink flowers, the plants growing in red sand, and in *Volk* 5173 (M), from the farm Ongombeanavita, in Okahandja district, the corolla has dried mauve with a golden throat. Flowering has been recorded between January and September, most records in April.

80. Jamesbrittenia micrantha (Klotzsch) Hilliard in Edinb. J. Bot. 49: 231 (1992).
Type: Mozambique, Rios de Sena, *Peters* s.n. (B†); neotype: near Lupata, Oct. 1858, *Kirk* s.n. (K).

Syn.: *Lyperia micrantha* Klotzsch in Peters, Reise Mossamb. Bot. 222 (1861).
  L. pedicellata Klotzsch in Peters, Reise Mossamb. Bot. 223 (1861); Hemsley & Skan in Fl. Trop.Afr. 4(2): 309 (1906). Type: Mozambique, Rios de Sena, banks of the Zambezi, *Peters* (B†).
  Chaenostoma micranthum (Klotzsch) Engl., Pflanzenw. Ost. Afr. C.356 (1895); Diels in Engl. Bot. Jahrb. 23: 489 (1897).
  C.pedicellatum (Klotzsch) Engl., Pflanzenw. Ost. Afr. C.356 (1895).
  Sutera fissifolia S. Moore in J. Bot. 38: 467 (1900). Type: Zimbabwe, Bulawayo, i 1898, *Rand* 155 (holo. BM; iso. BR n.v., C).
  S. micrantha Hiern in Fl. Cap.4(2): 263 (1904); Philcox in Fl. Zam. 8(2): 29 (1990), p.p. Lectotype (chosen here): South Africa, Transvaal, between Spitzkop and Komati River, *Wilms* 1075 (K).
  S. bolusii Hiern in Fl. Cap.4(2): 300 (1904). Type: South Africa, Transvaal, near the Crocodile River on the road between Delagoa Bay and Barberton, c.750ft, ix 1886, *Bolus* 7670 (holo. K; iso. BOL, PRE).
  S. luteiflora Hiern in Fl. Cap.4(2): 294 (1904) p.p., quoad *Galpin* 1069 and 1137.
  S. blantyrensis Skan in Fl. Trop.Afr. 4(2): 304 (1906). Lectotype (chosen here): Malawi, *Buchanan* in herb Wood 6630 (K, isolecto. PRE).

Perennial herb but will flower when very young and then looks annual, woody rootstock eventually up to c.15mm in diam. at crown, stems solitary at first, ultimately several from the crown, erect, ascending or prostrate, c.80–600mm long, soon branching from the base and above, glandular-puberulous, hairs up to 0.2–0.3mm long, scattered glistening glands as well, leafy throughout. *Leaves* opposite becoming alternate upwards, often pseudofasciculate, largest 12–45 × 5–28mm, blade elliptic or ovate-elliptic in outline tapering into a petiolar part, blade divided roughly ½–⅓ to midrib into oblong-elliptic obtuse or subobtuse lobes, lobes either entire or the largest with 1–3 teeth, both surfaces glandular-puberulous, hairs up to 0.1–0.25mm long, glistening glands as well particularly on lower surface. *Flowers* solitary in the axils of the leaves, these becoming very small and entire upwards, the whole forming long racemes. *Pedicels* up to 10–35mm long. *Calyx* tube 0.25–0.4mm long, lobes 2.5–3 × 0.5–1.2mm, oblong to spathulate, glandular-pubescent, hairs up to 0.15–2mm long, glistening glands as well. *Corolla* tube 3.8–5 × 2mm, cylindric, throat slightly expanded, mouth dorsoventrally compressed, limb oblique, 5–6mm across lateral lobes, posticous lobes 1.6–2.2 × 1.6–2mm, anticous lip porrect, anticous lobe 2–3 × 2.3mm, all lobes suborbicular, lemon-yellow, orange patch in throat extending onto base of lower lip, glandular-puberulous outside, glistening glands as well, a broad transverse band of clavate hairs in throat and at base of anticous lip. *Stamens*: posticous filaments 1–1.4mm long, anthers 0.7–1.2mm, anticous filaments 0.3–0.7mm, anthers 0.4–0.6mm, all filaments minutely puberulous. *Stigma* 0.2–0.5mm long, included. *Style* 1.8–2.2mm long. *Ovary* 1–1.2 × 0.8–1mm. *Capsules* c.3.5–5 × 2.5–3mm, minutely glandular-puberulous, glistening glands as well.

Selected citations:

Tanzania. Uyole, 11km E of Mbeya, 1800m, 25 xii 1974, *Aleljung* 172 (K).
Malawi. Dedza distr., Chongoni Forestry School, 12 x 1962, *Banda* 461 (SRGH).
Zambia. Mazabuka distr., Mapanza, Choma, 3500ft, 28 ix 1958, *Robinson* 2894 (K, M, PRE, SRGH).
Zimbabwe. Harare distr., Highlands near junction of Glenara and Grosvenor roads, 1520m, 14 x 1988, *Adams* 679 (SRGH). Belingwe [Mberengwa] distr., Emberengwa Mountain, 7 vi 1965, *Leach & Bullock* 12880 (SRGH). Wankie [Hwange] distr., Kazuma range, 1000m, 9 v 1972, *Russell* 1909 (SRGH).
Botswana. Tati Concession, 2027BD, Ramaquabane River, 4 vii 1962, *Wild* 5851 (K, SRGH). 2127CB, Macloutsi, ix 1876, *Klingberg* s.n. (S).
Transvaal. Zoutpansberg, 2229DC, Waterpoort, 2000ft, ix 1918, *Rogers* 21634 (PRE, Z). Pilgrim's Rest distr., 2431CB, Hermitage, Manyeleti Game Reserve, 400m, 22 ii 1977, *Bredenkamp* 1698 (PRE). Nelspruit distr., 2531AB, 14 miles N of Pretorius Kop, 1500ft, 4 ii 1949, *Codd & de Winter* 4978 (PRE).

Swaziland. Manzini distr., 2631AD, St. Josephs, c.3000ft, 12 xii 1963, *Compton* 31830 (NBG, PRE).
Mozambique. Sul do Save, Chibuto, Maniquenique, 17 vii 1947, *Pedro & Pedrogão* 1485 (K, PRE, SRGH).
Natal. Ingwavuma distr., 2732AC, Otobotini area, c.250ft, *Scott-Smith & Ward* 44 (PRE).

The types of *Lyperia micrantha* and *L. pedicellata* were destroyed in the Berlin fire, but Klotzsch's descriptions are so precise that I judge the type of *L. pedicellata* to have been no more than a young plant of *L. micrantha*. Philcox (loc.cit.) chose a neotype from much the same area as that from which the type of *L. micrantha* came, and I have accepted this as a neotype for the name *Jamesbrittenia micrantha*. When Hiern adopted the epithet *micrantha* in *Sutera*, he wrote (in Fl. Cap.4(2): 263, 1904) 'It is doubtful whether this is the same species as *Lyperia micrantha* Klotzsch ... '. I have therefore treated *Sutera micrantha* Hiern as an independent name and lectotypified it.

Jamesbrittenia micrantha is very variable in stature and robustness. Although it perennates, it will also flower when it is very young and therefore very small and annual-like. This has undoubtedly contributed to the lengthy synonymy. The natural habitat of the plant is damp sandy or clayey soils along riverbanks, around pans and dams and at the margins of marshes, but it readily becomes a weed in fallow fields and similar habitats. Flowers can be found in any month, but the principal season is June to November. It is obvious that the flowers normally self, and in consequence the seed-set is enormous. The species is characterized by its once-lobed leaves, small flowers (tube 3.8–5mm long), short style (1.8–2.2mm), and yellow limb with a porrect lower lip orange at the base.

It is widely distributed from Malawi and neighbouring Mbeya in Tanzania, Mozambique, Zambia, Zimbabwe and the fringe of north eastern Botswana south through eastern parts of the Transvaal and Swaziland, just reaching the coastal plain of NE Natal, which is a direct continuation of the coastal plain in Mozambique. The plant ranges up to c.1200m above sea level in the Transvaal and Swaziland, to c.1800m in tropical Africa.

81. Jamesbrittenia adpressa (Dinter) Hilliard in Edinb. J. Bot. 49: 225 (1992).
Type: Namibia, near Keetmanshoop, Azab, 29 vii 1923, *Dinter* 4737 (holo. B; iso. BOL, K, SAM).

Syn.: *Sutera adpressa* Dinter in Feddes Repert. Beih. 53: 9 (1928); Merxm. & Roessler, Prodr. Fl. S.W. Afr. 126: 48 (1967).

Annual herb well branched from the base and above, stems c.20–300mm long, prostrate, mat-forming, minutely glandular-puberulous, hairs mostly less than 0.1mm long, leafy throughout. *Leaves* opposite becoming alternate upwards, often pseudofasciculate, largest c.6–20 × 3–7mm, blade elliptic in outline tapering into a petiolar part, first 1 or 2 pairs of leaves entire or subentire, others divided nearly or quite to midrib into entire or 1–few-toothed lobes, lobes oblong to spathulate, obtuse, minutely glandular-puberulous on both surfaces, glistening glands plentiful. *Flowers* solitary in the axils of the leaves, the whole plant very floriferous. *Pedicels* up to 2–10mm long. *Calyx* tube 0.2–0.3mm long, lobes c.2.5–4 × 0.6–0.8mm, oblong to spathulate, glandular-pubescent, hairs up to 0.1–0.15mm long, glistening glands as well. *Corolla* tube c.6.3–10 × 1.8–2.2mm in throat, cylindric below, expanded in throat, limb c.6–7.5mm across lateral lobes, posticous lobes 2–3 × 1.5–2.5mm, anticous lobe 2–3.2 × 1.8–2.5mm, all lobes golden-yellow, oblong-elliptic, glandular-puberulous outside, glistening glands as well, broad transverse band of clavate hairs inside throat. *Stamens*: posticous filaments c.1–1.4mm long, posticous anthers c.1mm long, included, anticous filaments c.0.4–0.6mm long, anticous anthers c.0.4–0.7mm long, tips just visible in mouth or included, all filaments minutely puberulous. *Stigma* c.0.2–0.3mm long, included. *Style* 5–7mm long. *Ovary* c.1.1–1.6 × 0.7–1mm. *Capsules* c.3–4 × 2–2.5mm, very minutely puberulous, a few glistening glands as well mainly down sutures. *Seeds* c.0.2–0.5 × 0.15–0.2mm.

Selected citations:

Namibia. Swakopmund distr., 2315BC, Zebrapfanne, 12 x 1961, *Giess* 3775 (WIND). Maltahöhe distr., 2416DD, farm Daweb (MAL 43), 5 ix 1972, *Merxmüller & Giess* 28237 (K, S, WIND). Gibeon distr., 2417BA, 11km SE of Kalkrand, 22 vii 1974, *Nordenstam & Lundgren* 977 (E, S). Bethanien distr., 2617DD, Seeheim, 30 v 1974, *Goldblatt* 2012 (NBG, PRE, WIND).
Cape. Namaqualand, 3018BC, Platbakkies, c.3300ft, 10 ix 1976, *Hugo* 489 (PRE, STE). Upington 2821, Lang Klip [not traced by me], 11 ix 1968, *Mostert* 1677 (PRE).

Although the name *Sutera adpressa* is attributed to Dinter, he merely mentioned the plant in a travelogue, saying that it was prostrate with golden yellow flowers. Merxmüller & Roessler (in Prodr. Fl. S.W. Afr. 126: 48, 1967) claimed that this, together with their fuller description (in German), is sufficient to validate the name. As I am doubtful about that, I take the precaution of giving a brief latin diagnosis: *herba annua prostrata, foliis ad costam semel divisis, lobis plerumque integris; corolla tubo c.6–10 × 2mm, limbo 6–7.5mm diam. aureo-flavo.*

Jamesbrittenia adpressa ranges from south of the 23rd parallel in Namibia to Namaqualand; I have seen only two collections from south of the Orange river. It grows in sandy places, often in river beds or around pans and seasonal marshes, flowering mainly between May and September.

Its distinguishing marks are its leaves, only once divided into lobes that are mostly entire, its short pedicels up to 10mm long, its relatively long corolla tube and style (6.3–10mm, style 5–7mm, the longest in this group) and its golden-yellow limb, the lower lip not porrect.

82. Jamesbrittenia myriantha Hilliard in Edinb. J. Bot. 49: 231 (1992).
Type: Zimbabwe, Sebungwe distr., Simchembai camp, 1500ft, ix 1955, *Davies* 1433 (holo. SRGH; iso. K, PRE).

Annual herb well branched from the base and above, stems c.100–400mm long, prostrate to erect, glandular-pubescent, hairs up to 0.1–0.3mm long, scattered glistening glands as well, leafy. *Leaves* opposite, soon alternate, pseudofasciculate, up to 15–30 × 4–10mm, blade elliptic in outline tapering into a petiolar part, blade divided roughly ⅔ or more of way to midrib, lobes oblong-elliptic, entire or with up to 4 teeth, both surfaces glandular-pubescent, hairs up to 0.25mm long, glistening glands as well. *Flowers* solitary in the axils of the leaves, these becoming very small and entire upwards, the whole forming very well branched panicles. *Pedicels* up to 10–23mm long. *Calyx* tube c.0.2mm long, lobes 2–2.7 × 0.6–0.9mm, oblong or oblong-spathulate, glandular-pubescent, hairs up to 0.1–0.15mm long, glistening glands as well. *Corolla* tube c.4–5 × 2.5mm, cylindric, throat slightly expanded, (mouth dorsoventrally compressed?), limb oblique, c.5–7mm across lateral lobes, posticous lobes 1.5–2.5 × 1.5–2mm, anticous lip porrect, anticous lobe 1.5–3 × 1.7–2.5mm, all lobes suborbicular, white or sometimes tinged with pink or palest mauve, orange or chocolate-brown patch in throat and extending onto base of lower lip, glandular-puberulous outside, glistening glands as well, transverse band of clavate hairs in throat and at base of anticous lip.*Stamens*: posticous filaments 0.7–1.2mm long, posticous anthers 0.4–0.8mm long; anticous filaments 0.5–1mm long, anticous anthers 0.3–0.5mm long. *Stigma* c.0.2mm long. *Style* 2–2.8mm long. *Ovary* c.1–1.3 × 0.8–1mm. *Capsules* c.3.5–4 × 2–2.5mm, glistening glands all over together with a few minute hairs at the apex. *Seeds* c.0.25 × 0.2mm.

Selected citations:

Zimbabwe. Sebungwe distr., Zambezi river, 1500ft , ix 1955, *Davies* 1496 (K, SRGH); ibidem, Sebungwe camp, ix 1955, *Davies* 1463 (K); near Binga, 1470ft, 6 xi 1958, *Phipps* 1360 (K, PRE, SRGH). Gokwe distr., Sengwa Research Station, 25 viii 1976, *Guy* 2450 (SRGH); ibidem, 26 vii 1976, *Guy* 2357 (K, SRGH). Gwelo [Gweru] distr., Gwelo, 4600ft, 28 ix 1966, *Biegel* 1314 (PRE).

Jamesbrittenia myriantha is allied to *J. dissecta*, from which it may be distinguished by its flowers in panicles (not simple racemes), longer pedicels and larger flowers (limb 4–7mm across lateral lobes versus 2–4mm, anticous lobe 1.5–3mm long versus 0.8–1.5mm, style 2–2.8mm versus 1.2–1.5mm). The species has hitherto been confounded with *J. micrantha*, from which it differs in its annual habit, more deeply dissected leaves, white or whitish flowers, and mostly longer style.

It is known to me only from north western and western Zimbabwe, in drying mud along the banks of the Zambezi, in river sand at Sengwa Research Station (where it is also a weed in the Institute's lawn!), as a weed in a tree nursery at Gwelo, and in a marsh near Bulawayo. Flowering has been recorded in July, August and September, but in July the plants were already well on into fruit. As in several of its close allies, the flowers usually self.

83. Jamesbrittenia dissecta (Del.) O. Kuntze, Rev. Gen. Pl. 2: 461 (1891); Hemsley & Skan in Fl. Trop.Afr. 4(2): 298 (1906); Andrews, Fl. Pl. Sudan 3: 137 (1956). Type: Egypt, *Delile* (P).

Syn.: *Capraria dissecta* Del., Fl. d'Egypte 95 t. 32 f. 3 (1813).
    Sutera glandulosa Roth, Nov. Pl. Spec.291 (1821); Wight, Ic.pl. Ind. orient. 3 t. 856 (1844–1845); Hook. f., Fl. Brit. India 4: 258 (1884); Duthie, Fl. Upper Gangetic plain 2: 141 (1911); Gamble, Fl. Madras 2: 946 (1922); Haines, Botany of Bihar & Orissa 621 (1922); Broun & Massey, Fl. Sudan 326 (1927); Täckholm, Students Fl. Egypt ed. 2: 494 (1974). Type: India, *Heyne* (n.v.).
    Sutera dissecta (Del.) Walp., Repert. 3: 271 (1844); Verma et al, Fl. Raipur, Durg & Rajnandgaon 263 (1985); Ugemuge, Fl. Nagpur distr. 271 (1986); Singh, Fl. E. Karnataka 472 (1988).
    Chaenostoma dissectum (Del.) Thell. in Vierteljahrss. Nat. Ges. Zürich 74: 116 (1929), in obs.

Annual herb branching from the base and above, stems c.50–350mm long, sometimes erect, often prostrate or ascending, glandular-pubescent, hairs up to 0.2–0.5mm long, glistening glands as well, leafy throughout. *Leaves* opposite becoming alternate upwards, often pseudo-fasciculate, blade elliptic in outline, c.7–40 × 5–20mm tapering into a petiolar part c.2–20mm long (up to ½–⅓ as long as blade), apex obtuse to subacute, margins deeply divided c.⅔–¾ of way to midrib, lobes ± oblong, usually toothed, both surfaces hairy as stems. *Flowers* few to many, solitary in the axils of the upper leaves, these passing into filiform bracts, forming crowded racemes. *Pedicels* up to c.3–12mm long. *Calyx* tube 0.1–0.3mm long, lobes 1.7–3.5 × 0.4–1mm, oblong-spathulate, glandular-pubescent, hairs c.0.15–0.25mm long, glistening glands as well. *Corolla* tube c.2.5–4 × 1mm, cylindric expanding to c.1.5mm across throat, limb c.2–4mm across lateral lobes, posticous lobes 0.7–1.5 × 0.5–0.9mm, anticous lobe 0.8–1.5 × 0.7–1mm, all lobes oblong-elliptic, minutely glandular-puberulous outside, glistening glands as well, inside throat a broad transverse band of minute clavate hairs, lobes whitish. *Stamens*: posticous filaments c.0.9–1.5mm long, anthers 0.3–0.5mm, included, anticous filaments c.0.4–0.5mm long, anthers 0.2–0.3mm long, included or just visible in throat. *Stigma* c.0.1–0.2mm long, often attached to anthers. *Style* c.1–1.5mm long. *Ovary* c.1.3–1.5 × 1mm, densely clad in glistening glands, minute glandular hairs as well. *Capsules* c.3.5–5 × 2.2–3mm, hairy as ovaries. *Seeds* c.0.2–0.4 × 0.2–0.3mm.

Selected citations:

Egypt. Thebes, on the banks of the Nile, 22–23 ii 1881, *Letourneaux* (FI, P, W).
Sudan. Sennar Province, near Chartum [Khartoum], 9 iii 1840, *Kotschy* 330 (LE, M, W).
India. Utta Pradesh, from below Mirzapur to junction of Gogra with Ganges, *Madden* 263 (E).

Jamesbrittenia dissecta is a weed of moist places, often recorded along river banks, in gullies or around reservoirs, flowering between December and May. It has been recorded throughout the Nile valley, in Sudan, and in India and Bangladesh, from the Gangetic plain south to the environs of Madras.

This is the most northerly representative of the genus and the only one to range outside Africa. Its closest allies are *J. adpressa* from Namibia and *J. myriantha* from Zimbabwe.

Doubtful Species

Erinus patens Thunb., Prodr. Pl. Cap. 102 (1800). The type of this name is sheet no. 14409 in herb. Thunberg (UPS). Hiern (1904, p.274) referred it to *Sutera antirrhinoides* (that is, *Lyperia*), but it is not a species of *Lyperia*. It is possibly a *Jamesbrittenia*, but I failed to place it when I saw the specimen in 1989, prior to revising *Jamesbrittenia*, and still cannot place it.

5. Lyperia

The genus *Lyperia* has been lectotypified by *L. tristis* (Pfeiffer, 1874, p.186). The generic name is derived from the Greek word for grief and Bentham wrote '... I have given the name *Lyperia*, partly because it contains *Erinus tristis* [*tristis* is the Latin for sad] and other species with that peculiar - coloured flower, and partly because the corolla almost constantly, and often the whole plant dry black ...' (Bentham, 1836, p.360). I have adopted a generic concept narrower than Bentham's and restrict the name to species agreeing in essential characters with *L. tristis*. Those characters are: *stems* narrowly winged by decurrent leaf bases, *corolla tube* cylindric, ± abruptly dilated in the throat, stamens either 4 or 2 plus 2 staminodes, included, posticous filaments decurrent to base of corolla tube, *stigma* either broadly hastate without papillae and sticky, or narrowly hastate and clad throughout in stigmatic papillae, *ovary and capsule* glandular-pubescent and with glistening glands as well, capsule relatively large. *Placentae* composed of strongly confluent pulvini. *Seeds* bothrospermous, patterned with transversely elongated pits, relatively large, black.

It is a peculiarity of the genus that the mouth is surrounded by a broad band of minute glandular hairs. These hairs sometimes occur just inside the throat as well. There is always a broad V-shaped band of upward-pointing clavate hairs in the throat. Both *L. tristis* and *L. lychnidea* have 'sad-coloured' flowers, that is, shades of yellow, greenish-yellow, greenish-cream or dirty white; the lobes reflex strongly at night, the lobes narrow by reflexion of the margins, and a strong scent of cloves is emitted, all characters indicative of moth-pollination. The flowers of *L. antirrhinoides* may also be moth-pollinated as the limb is white or creamy white and the lobes reflex strongly, but the mouth is surrounded by a strongly marked 'eye', lacking in both *L. tristis* and *L. lychnidea*. The other three species, *L. violacea, L. tenuiflora* and *L. formosa*, all have a prettily coloured limb with a contrasting 'eye' and probably attract day-flying insects. However, seed-set is so heavy in all the species that is seems likely selfing is commonplace.

Blackening upon drying is an almost constant feature of *L. lychnidea* and *L. violacea*, but it is not strongly marked in the other species. In herbaria, the species are repeatedly confused with *Zaluzianskya,* but a glance at the calyx alone will immediately distinguish the genera: deeply 5-lobed in *Lyperia*, shortly 5-toothed in *Zaluzianskya*.

Lyperia Benth. in Hook., Comp. Bot. Mag 1: 377 (1836) and in DC., Prodr. 10: 357 (1846) p.p. **Figs 12 F, G, H, 14B, 25, 36A, 37, 52.**

Mostly annual herbs, one species perennial, markedly glandular. *Stems* leafy throughout, glandular-pubescent, narrowly winged by decurrent leaf bases. *Leaves* opposite, simple, entire to toothed. *Inflorescence* a terminal raceme, flowers solitary in axils of leaves or leaflike bracts. *Bracts* alternate, leaflike, adnate to extreme base of pedicel. *Calyx* obscurely bilabiate, 5-lobed almost to base. *Corolla* tube narrowly cylindric, ± abruptly dilated in throat, pubescent outside, throat bearded inside with clavate hairs, limb nearly regular, lobes rotate or reflexed, minutely glandular around mouth.*Stamens* 4 or 2 by reduction of anticous pair to staminodes (sometimes with a minute head containing a few pollen grains), didynamous, inserted in throat, included or just visible in throat, posticous filaments decurrent to base of tube forming a channel for the style, anthers synthecous. *Stigma* very short, either broadly hastate with 2 lateral sticky receptive surfaces, or narrowly hastate clad all round in stigmatic papillae, either included or

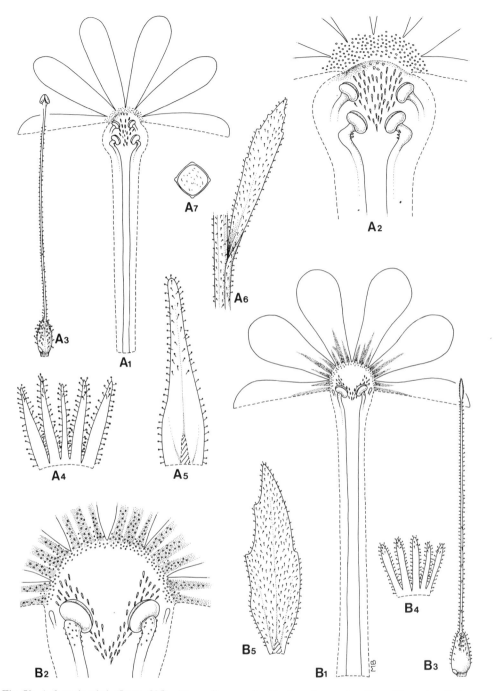

Fig. 52. A. *Lyperia tristis* (*Batten* 825): A1, corolla opened out to show decurrent posticous filaments, × 3; A2, upper part of corolla to show glands around mouth, glandular hairs and clavate hairs in throat, × 8; A3, stigma, style and ovary, × 3; A4, slightly bilabiate calyx, opened out, × 6; A5, bract showing adnation scar, × 6; A6, leaf decurrent down stem, × 3; A7, TS stem to show narrow wings, × 8. B. *Lyperia tenuiflora* (*Levyns* 11606): B1, corolla opened out to show decurrent posticous filaments and colour-pattern around mouth, × 3; B2, upper part of corolla to show anticous stamens reduced to staminodes, glands around mouth, clavate hairs in throat, × 8; B3, stigma, style and ovary, × 3; B4, slightly bilabiate calyx opened out, × 3; B5, bract showing adnation scar, × 3.

very shortly exserted. *Ovary* cuneate in outline, tapering into the style, glandular-pubescent, often densely clad in glistening glands as well, nectariferous gland lateral, very small, completely adnate to base of ovary, ovules many in each loculus. *Fruit* a somewhat beaked, relatively large, septicidal capsule with a short longitudinal split at tip of each valve, conspicuously glandular. *Seeds* arising from a flattened placenta, the funicles coalescing into wavy lines (fig. 14B), seed roughly elliptic in outline, black, pale terminal wings in one species, testa thin, tightly investing the endosperm, which has 6–8 longitudinal rows of transversely elongated pits arranged in chequer-board fashion (bothrospermous), under the SEM the cells of the testa seen to be polygonal, plane (fig. 25A).

Lectotype species: *Lyperia tristis* (L.f.) Benth.
Distribution: Namibia, western and southern Cape.

Key to species

1a. Stamens 4, apex of corolla tube almost globose, no contrasting colour-pattern on limb . . . 2
1b. Stamens 2 + 2 staminodes, apex of corolla tube ± funnel-shaped, contrasting pattern of
    radiating streaks or patches around mouth 3

2a. Perennial herb usually drying black, leaves ± pseudofasciculate, capsules c.13–23mm
    long, seeds with a broad pale wing at each end . . . . **1. L. lychnidea**
2b. Annual herb not or scarcely drying black, leaves not fascicled, capsules c.4–15mm long,
    seeds usually wingless, occasionally with an incipient lateral or terminal wing   **2. L. tristis**

3a. Corolla tube less than 20mm long 4
3b. Corolla tube at least 22mm long 5

4a. Corolla limb white to creamy white, orange to black rounded patches at base of each
    lobe . . . . **3. L. antirrhinoides**
4b. Corolla limb pink, mauve or blue patterned with dark radiating lines around the mouth
                                                                                   4. L. violacea

5a. Corolla limb pink, rose-mauve or lavender, lowermost pedicels 4–11mm long
                                                                                   5. L. tenuiflora
5b. Corolla limb white, lowermost pedicels 18–23mm long . . . . **6. L. formosa**

1. Lyperia lychnidea (L.) Druce in Rep. Bot. Exch. Club Brit. Isles 1913, 3: 420 (1913).
Type: Burm., Afr. 138, t.49, f.4.

Syn.: *Lychnidea villosa, foliis oblongis, dentatis, floribus spicatis* Burm., Afr. 138 (1738).
  Selago lychnidea L., Pl. Rar. Afr. 89 (1760) & Sp. Pl. ed. 3, 877 (1764).
  Erinus lychnideus (L.) L.f., Suppl. 287 (1782) as to basionym only; Thunb., Prodr. 102 (1800) & Fl. Cap. ed.
    Schultes 474 (1823), excl. *Erinus capensis* L. [= *Zaluzianskya capensis*].
  Erinus fragrans Ait., Hort. Kew ed.1, 2: 357 (1789) p.p., nom. illegit.
  Manulea lichnidea (L.) Desrouss. in Lam., Encycl. 3: 707 (1792).
  Lyperia fragrans Benth. in Hook., Comp. Bot. Mag. 1: 378 (1836) & in DC., Prodr. 10: 358 (1846), nom. illegit.
  Lyperia macrocarpa Benth. in Hook., Comp. Bot. Mag. 1: 378 (1836) & in DC., Prodr. 10: 358 (1846). Type:
    Cape, without precise locality, *Masson* (holo. BM).
  Nycterinia fragrans of Index Kewensis 320 (1894), nom. inval.
  Sutera lychnidea (L.) Hiern in Fl. Cap. 4(2): 290 (1904); Levyns in Adams. & Salter, Fl. Cape Penins. 715 (1950).

Perennial herb branching from the base and above, branches ± erect, decumbent or sprawling, eventually forming a bushy clump up to 0.5–1 m tall and as much across, main stems c.4–8mm in diam., glandular pubescent, hairs c.0.2–0.3mm long, glistening glands as well, leafy, whole plant blackening on drying. *Leaves* opposite to alternate, pseudofasciculate, the leaf tufts often elongating into short branchlets, primary leaves c.30–65 × 4–13mm, smaller upwards and passing into floral leaves, oblanceolate, obtuse, margins entire to shortly toothed in upper half, both surfaces hairy as the stems. *Flowers* many, corymbose at first, soon elongating into racemes up to c.300mm long, often panicled. *Bracts* (floral leaves) c.20–35 × 2–5mm at base of inflorescence; pair of true bracts ± immediately below calyx, lowermost c.2.5–4.5 × 0.2–0.7mm, hairy as leaves. *Pedicels* (lowermost) 4–7mm long. *Calyx* tube c.0.2–0.5mm long, lobes 4.5–7 × 0.5–1.25mm, glandular-pubescent, hairs 0.25–0.5mm long, glistening glands as well. *Corolla* tube 23–28 × 1.5–1.8mm, cylindric, abruptly expanded in throat and there c.3–4mm in diam., contracted abruptly to mouth and there c.1.5mm in diam., limb nearly regular, rotate, c.15–21mm across, lobes 6–9 × 5–7mm, ± broadly cuneate, margins revolute and therefore lobes look oblong to the eye, tube and backs of lobes glandular-pubescent, hairs up to 0.4–0.6mm long, glistening glands as well, upper surface of lobes glabrous except for a broad band of minute stalked glands around mouth, these also within upper part of throat above 2 patches of clavate hairs, limb greenish-cream to lime-yellow or primrose yellow, darkening with age, tube sometimes tinged with mauve, sweetly clove-scented at night. *Stamens* 4, anticous filaments 1–1.5mm long, anthers 1.2–1.5mm long, posticous filaments similar but almost knee-jointed, anthers 1.5–2mm long. *Stigma* c.1mm long, hastate. *Style* 20–23mm long. *Ovary* c.3 × 1.25mm, thickly clad in glistening glands, glandular hairs as well c.0.2–0.3mm long. *Capsules* c.13–23 × 5–7mm, oblong, valves acute, prominently hairy as ovary. *Seeds* c.3.5–4 × 1mm, elongate, with broad pale wings at both ends, these accounting for c.⅔ total length of seed.

Selected citations:

Cape. Malmesbury div., 3318AB, near Hopefield, x 1885, *Bachmann* 1274 (Z). Cape Town div., 3318CD, Blouberg-strand, 14 viii 1988, *Bokelmann* 46(E). Cape Peninsula, 3418AB, Karbonkel Berg, 26 x 1940, *Kies* 191 (NBG). 3418BA, Cape Flats, Swartklip, 22 ix 1972, *Taylor* 8206 (STE). Caledon div., 3419AC, Vermont, 19 ix 1982, *Burman* 959 (BOL).

Lyperia lychnidea has a limited distribution range from about Saldanha Bay south to the Cape Peninsula and east to the environs of Still Bay, a constituent of scrub on sandy flats and dunes, never far from the sea, and up to c.80m above sea level. It flowers mainly between September and November. Both Burmann and Aiton, followed by Hiern, claimed that the flowers are purple in colour; possibly neither author saw living material and was deceived by the change in colour upon drying. Specimens from the southern Cape coast roughly between Gansbaai and Stilbaai look somewhat atypical, having broader and more deeply toothed leaves than usual. The possibility of hybridization with *L. tristis* should not be overlooked.

Lyperia lychnidea is often confused with the closely allied species, *L. tristis*, which has similar flowers but differs in its annual habit, leaves not fascicled, mostly smaller capsules and seeds not or scarcely winged. The annual habit of *L. tristis* is easy to detect in herbaria because most specimens have been pulled up entire, whereas specimens of *L. lychnidea* are usually just branches of the plant.

2. Lyperia tristis (L. f.) Benth. in Hook., Comp. Bot. Mag. 1: 376 (1836) & in DC., Prodr. 10: 358 (1846). **Fig. 52A, Plate 2I**.

Type: Cape of Good Hope, without precise locality, *Thunberg* sheet no. 14413 (UPS, probable isotype in herb. Montin, S).

Syn.: *Erinus tristis* L. f., Suppl. 287 (1782); Thunb., Prodr. 103 (1800) & Fl. Cap. ed. Schultes 476 (1823).
    E. fragrans Ait., Hort. Kew. ed. 1, 2: 357 (1789), p.p., nom. illegit.
    Lyperia simplex Benth. in Hook., Comp. Bot. Mag. 1: 378 (1836) & in DC., Prodr. 10: 358 (1846), non *Erinus simplex* Thunb. which is *Buchnera simplex* (Thunb.) Druce. Lectotype: Graaff Reinet, Karroo desert, 1836, *Ecklon* (K, isolecto. SAM).
    Chaenostoma triste (L. f.) Wettst. in Engl. & Prantl, Pflanzenfam. 4(3B): 69 (1891).
    C. triste var. *montanum* Diels in Bot. Jahrb. 23: 477 (1897). Type: Namaland, Hantam-Gebirge, 1869, *Meyer* (B†).
    Sutera tristis (L. f.) Hiern in Fl. Cap. 4(2): 289 (1904) including var. *montana* (Diels) Hiern; Levyns in Adamson & Salter, Fl. Cape Penins. 715 (1950); Merxm. & Roessler in Prodr. Fl. S.W. Afr. 126: 55 (1967); le Roux & Schelpe, Namaqualand ed. 2: 150, 151 cum ic. (1988).

Annual herb, very variable in stature and robustness, c.30–600mm tall, stem simple to sparingly branched from the base and above, main stem up to 5–6mm in diam., erect, glandular-pubescent, hairs up to 0.3–0.4mm long, glistening-glands as well, leafy throughout, plant not or scarcely blackening on drying. *Leaves* opposite to alternate, often crowded at base of stem, more distant upwards, c.17–75 × 3–25mm, smaller upwards and passing quickly into floral leaves, elliptic to oblong-elliptic, apex obtuse, base of lower leaves tapering into a petiole accounting for up to ± half the total leaf length, becoming sessile upwards, margins entire to deeply toothed (teeth up to c.3–5mm long), both surfaces hairy as the stems. *Flowers* many, corymbose at first, soon elongating into racemes up to 300–500mm long in robust plants. *Bracts* (floral leaves) c.10–40 × 3–10mm at base of inflorescences; pair of true bracts ± immediately below calyx, lowermost c.1–5 × 0.1–0.6mm, hairy as leaves. *Pedicels* (lowermost) 2–4mm long. *Calyx* tube 0.25–0.5mm long, lobes 3.5–4.5 × 0.6–1mm, glandular-pubescent, hairs c.0.25–0.5mm long, glistening glands as well. *Corolla* tube 20–29 × 1.2–2mm, cylindric, abruptly expanded in throat and there c.2.5–3mm in diam., contracted abruptly to mouth and there c.1.5mm in diam., limb nearly regular, c.9–23mm across, lobes 3.4–10 × 1.8–6mm, oblong-elliptic, margins often revolute, tube and backs of lobes glandular-pubescent, hairs up to c.0.25mm long, glistening glands as well, upper surface of lobes glabrous except for a broad band of minute stalked glands around mouth, these also often within upper part of throat above two patches of clavate hairs, limb whitish to greenish-yellow or clear yellow, darkening with age, sweetly clove-scented at night. *Stamens* 4, anticous filaments 0.5–1mm long, posticous ones c.1mm long, knee-jointed, sometimes bearded with a few clavate hairs, anthers 0.8–1.4mm long. *Stigma* c.0.5–1mm long, hastate. *Style* 17–23mm long. *Ovary* c.2–3 × 1.5mm, thickly clad in glistening glands, glandular hairs as well c.0.2–0.3mm long. *Capsules* 4–15 × 4–6mm, ± ovate, prominently hairy as ovary. *Seeds* 0.8–1.5 × 0.4–0.7mm, usually wingless, occasionally with an incipient lateral or terminal wing.

Selected citations:

Namibia. 2215DC, Ganab area towards Tinkas in Namib Desert Park, 13 v 1976, *Oliver, Müller & Steenkamp* 6623 (SRGH). 2616CB, 6mls. westlich Aus, am Wege nach Lüderitz, 24 ii 1963, *Giess, Volk & Bleissner* 5470 (M, WIND). 2715BC, Klinghardtgebirge, 27 ix 1922, *Dinter* 3891 (BOL, PRE, SAM, Z).
Cape. Namaqualand, 2917DB, 4 miles W of Springbok, 8 ix 1950, *Barker* 6639 (NBG, STE). Kenhardt div., 2918BC, farm Gannapoort, 21 v 1951, *Schlieben* 8948 (K, NBG, PRE, S, SRGH, Z). Calvinia div., 3119AC, Naresie, 23 viii 1990, *Batten* 1022 (E). Clanwilliam div., 3219AA, 5 miles from turnoff to Wupperthal into Bidouw Valley, 12 ix 1974, *Mauve & Oliver* 98 (PRE, STE). Malmesbury div., 3318CB, near Darling, Buck Bay, 16 ix 1980, *Hilliard & Burtt* 13018 (E, S). Riversdale div., 3321CC, Sopieshoogte, N entrance Garcia's Pass, 15 ix 1981, *Fellingham* 128 (K, STE). Willowmore div., 3323AC, flats between Hotsprings and Toorwater, 5 x 1971, *Oliver* 3647 (STE). Cape Peninsula, 3418AB, Karbonkelberg, 12 ix 1942, *Compton* 13616 (NBG). Riversdale div., 3420AD, 2.5km SW of Ouplaas, 16 x 1984, *Fellingham* 785 (STE).

The type of the name *Erinus tristis* comprises two sterile plants and a flowering seedling, but there is little doubt that the name is being used correctly and that *L. simplex* is synonymous with it; the types of the latter epithet are precisely matched by several specimens from the dry interior of the Cape, including *Schlieben* 8948, cited above. The type of Diels's variety *montana* was destroyed in the Berlin fire; the brief description *could* apply to a small plant of *L. tristis*.

Lyperia tristis is widely distributed from the western part of Namibia (Klein Spitskuppe, west of Usakos, southwards), south through the western and central Cape, reaching the Cape Peninsula in the west, Graaff Reinet at the eastern end of the Great Karoo, Toorwater and Dysselsberg at the eastern end of the Little Karoo. The altitudinal range is also great, from near sea level to c.1000m. The plants favour sandy, gravelly or stony ground, often growing in scrub among boulders or under overhanging rocks. Flowers have been recorded in all months, but the principal season is July to October.

3. Lyperia antirrhinoides (L. f.) Hilliard in Edinb. J. Bot.48: 341 (1991). **Plate 2G.**
Lectotype: Cape of Good Hope, without precise locality, *Thunberg* (sheet no. 14339 in herb. Thunb., UPS).

Syn.: *Manulea antirrhinoides* L. f., Suppl. 286 (1782); Thunb., Prodr. 101: 1800 & Fl. Cap. ed. Schultes 469 (1823); Murray, Syst. Veg. 14: 569 (1784); Benth. in Hook., Comp. Bot. Mag. 1: 384 (1836).
    Sutera antirrhinoides (L. f.) Hiern in Fl. Cap. 4(2): 274 (1904) as to type only.
    S. ochracea Hiern in Fl. Cap. 4(2): 282 (1904). Type: Cape, Little Namaqualand, near Garies, Spektakel Mt., 3600 ft, *Bolus* in *herb. Norm. Austr. Afr.* 664 (holo. K; iso. BM, BOL, SAM, W).

Annual herb more or less blackening on drying, main stem c.25–300mm tall, simple at first, later loosely branched from the base and above, lower branches decumbent or ascending, glandular-puberulous, hairs up to c.0.25–0.3mm long, scattered glistening glands also present, distantly leafy. *Leaves* opposite, c.10–35 × 3–15mm, elliptic tapering to a petiolar part accounting for up to c.⅓ total leaf length, becoming nearly sessile upwards, apex obtuse to subacute, margins coarsely toothed, both surfaces hairy as stems. *Flowers* few to many in lax terminal racemes. *Bracts* alternate, leaflike, lowermost c.7–25 × 2–14mm. *Pedicels* (lowermost) c.3–13mm long. *Calyx* tube c.0.2–0.3mm long, lobes 3.5–5 × 0.8–1.2mm, densely glandular-pubescent, hairs up to c.0.6–1.2mm long, scattered glistening glands as well. *Corolla* tube 14–18 × 1.2–1.4mm, cylindric, expanded in throat to c.2mm, limb nearly regular, 6–10mm across, lobes 2–4 × 1.3–2.5mm, oblong-elliptic, tube and backs of lobes glandular-pubescent, hairs c.0.4–0.5mm long, glistening glands as well particularly on upper parts, upper surface of lobes glabrous except for broad band of minute glandular hairs around mouth, band of upward-pointing clavate hairs in throat, limb white to creamy-white with an orange-yellow, brown, purplish or black rounded patch at the base of each lobe. *Stamens* 2, filaments 0.6–1mm long, anthers 0.8–1.5mm long, either included or just visible in mouth, staminodes 2, 0.2–0.3mm long. *Stigma* c.1–1.5mm long, included. *Style* 10–15mm long, glandular-puberulous. *Ovary* 2.5–3.5 × 1.4–2mm, densely clad in glistening glands, hirsute as well with glandular hairs c.0.2–0.5mm long. *Capsules* 10–13 × 5.5–7mm, hairy as ovaries. *Seeds* c.0.8–1 × 0.6–0.75mm.

Selected citations:

Cape. Piquetberg div., 3218DD, Piquetberg Mountain, 6 ix 1894, *Schlechter* 5200 (BOL, PRE, Z). Ceres div., 3220CC, farm Thyskraal 80, 7 ix 1986, *Cloete & Haselau* 278 (STE); 3319AD, Gydo Pass, 9 ix 1938, *Hafstrom & Acocks* 1162 (K PRE, S). Malmesbury div., 3318BD, Bothmaskloof Pass, 25 viii 1970, *Acocks* 24321 (K, PRE, SRGH, Z); 3318BB, Porterville, 5 viii 1897, *Schlechter* 10737 (BM, BOL, E, K, PRE, S, Z). Montagu div., 3319DB, Eendracht, ix 1946, *Lewis* SAM 60020 (SAM). 3320CA, NE slopes of Wagensboombergen, 11 vii 1954, *Middlemost* 1870 (NBG). Caledon div., 3419AB, Caledon, 1 x 1846, *Prior* s.n. (PRE). Ladismith div., 3321AC, Noukloof Nature Reserve, 6 vii 1982, *Laidler* 56 (STE). Oudtshoorn div., 3322BC, De Rust, 7 xii 1990, *Vlok* 2334 (E).

Sheet no. 14339 in Thunberg's herbarium has been chosen as the lectotype of *L. antirrhinoides* because the detailed description given in the 1823 edition of Thunberg, Flora capensis, could have been based only on it (sheet no. 14338 is a small seedling). See further comments under *L. violacea*, with which *L. antirrhinoides* has been much confused.

    Lyperia antirrhinoides has been recorded mainly on the high ground of the SW Cape, from around Piquetberg, Tulbagh, Worcester, Swellendam and Montagu east to De Rust near Oudtshoorn, between c.180 and 1065m above sea level, but the type of *Sutera ochracea* (in synonymy above) came from the Spektakelberg in Namaqualand; the species seems never to have been re-collected in Namaqualand. It appears to favour stony ground and flowers between July and October.

4. Lyperia violacea (Jarosz) Benth. in Hook., Comp. Bot. Mag. 1: 379 (1836) & in DC., Prodr. 10: 359 (1846).
Type: cult. in Hort. Bot. Berol. from Cape seeds (no specimens extant in B, but a specimen in Bentham's herbarium, K, from same source, dated 4 ix 30, is here taken as neotype).

Syn.: *Manulea violacea* [Link ex] Jarosz, Pl. Nov. Cap. 16 (1821); Link, Enum. Pl. Hort. Bot. Berol. 2: 142 (1822).
    M. crystallina Weinm.in Hornschuch, Syll. Pl. Nov. Soc. Reg. Bot. Ratisb. 1: 226 (1822-24). Type: not found (see
      below).
    ?*Lyperia diandra* E. Mey., Hort. Regimont. Seminif. 5 (1848) & in Ann. Sci. Nat. sér. 3, 11: 254 (1849). Type:
      not found (see below).
    Sutera antirrhinoides auct. non (L.f.) Hiern; Levyns in Adamson & Salter, Fl. Cape Penins. 715 (1950).

Annual herb blackening on drying, main stem c.50–450mm tall, simple at first, later loosely branched from the base and above, lower branches decumbent or ascending, sparsely glandular-puberulous to almost glabrous, hairs c.0.1–0.2mm long, scattered glistening glands as well, distantly leafy. *Leaves* opposite, c.10–60 × 5–20mm, elliptic tapering to a petiolar part accounting for c.⅓ total length of lower leaves, becoming nearly sessile upwards, apex ± obtuse, margins coarsely toothed, lower surface in particular sparsely hairy as the stems. *Flowers* few to many in lax terminal racemes. *Bracts* alternate, leaflike, lowermost c.6–22 × 2–11mm. *Pedicels* (lowermost) c.3–10mm long. *Calyx* tube c.0.1–0.25mm long, lobes c.3–5 × 0.7–1mm, thinly glandular-puberulous, hairs up to 0.15–0.25mm long, glistening glands as well. *Corolla* tube 9–15 × 1–1.2mm, cylindric, expanded in throat to c.1.8–2.25mm, limb nearly regular, c.6–14mm across, lobes 2.3–4 × 1.4–3.5mm, oblong-elliptic to subspathulate, tube and backs of lobes glandular-pubescent, hairs 0.15–0.25mm long, glistening glands as well particularly on upper parts, upper surface of lobes glabrous except for a broad band of minute glandular hairs around mouth, band of upward-pointing clavate hairs in throat, limb pink, mauve or blue marked with 3 dark streaks at base of each lobe. *Stamens* 2, filaments 1mm long, anthers 1–1.4mm long, just visible in throat or included, staminodes 2, 0.2–0.4mm long. *Stigma* 0.7–1mm long, shortly exserted or included. *Style* c.5–13mm long, glandular-puberulous. *Ovary* 2.2–2.5 × 1.3–1.7mm, densely clad in glistening glands, minute glandular hairs (c.0.1mm long) particularly on upper part. *Capsules* 7–12 × 4–6mm, hairy as ovaries, the glandular hairs very inconspicuous. *Seeds* 0.8–1.3 × 0.7–1mm.

Selected citations:

Cape. Cape Peninsula, 3318CD, Signal Hill above Green Point, 800ft, vii 1908, *Dümmer* 1677 (E). Swellendam div., 3320CD, below Leeuwriviersberg, 26 vi 1952, *Wurts* 199 (NBG); 3420AB, Bontebok Park, 24 viii 1965, *Grobler* 480 (STE). Bredasdorp div., 3420BC, Potteberg, 19 ix 1954, *Esterhuysen* 23276 (BOL). Riversdale div., 3421AB, farm Windsor, 2 viii 1983, *Bohnen* 8225 (STE); 3421BA, near Albertinia, Botteliersfontein, 100ft, ix 1914, *Muir* 1822 (PRE, Z). George div., 3422AA, Brak River, 8 x 1928, *Gillett* 1247 (STE). Uniondale div., 3323CA–3322DB, between Uniondale and Laudina, ix 1944, *Fourcade* 6458 (BOL, STE); 3323CC, Prince Alfred's Pass, near foot, iv 1928, *Fourcade* 3784 (BOL).

Hiern did not distinguish *L. violacea* from *L. antirrhinoides* and compounded the confusion by re-describing *L. antirrhinoides* as *Sutera ochracea*. The confusion has persisted to this day. *Lyperia violacea* is a far less hairy plant than *L. antirrhinoides* (stems, leaves, calyx, ovary and capsules all ± hirsute in *L. antirrhinoides*), and they also differ strikingly in the corolla limb, which is shades of pink or violet in *L. violacea* marked with short dark radiating lines around the mouth, white or creamy-white in *L. antirrhinoides* with a rounded patch at the base of each lobe varying in colour from orange to black. *Zeyher* 3512, cited by Hiern as *Sutera antirrhinoides*, is in many herbaria a mixture of *L. antirrhinoides* and *L. violacea*.

The type specimen of *Manulea violacea*, if one ever existed, was destroyed in the Berlin fire, but Bentham received a specimen from Berlin in 1830, and this has been chosen as a neotype. No type of *M. crystallina* has been found, but it was described only a year later than *M. violacea* and was almost certainly grown from the same seed: two old specimens of *L. violacea* in the Vienna herbarium (W), three in Munich (M), bear the name *Manulea crystallina*.

Lyperia diandra is a much more difficult problem. The plant was grown at Konigsberg in 1848 from seed sent to Meyer by Drège. Meyer (1849, p.254) wrote that Drège gave no locality, but described the plant as being '*Nemesiae affinis*'. Hiern saw no specimen and suggested (1904, p.353) that Meyer's plant might be compared with *Zaluzianskya nemesioides* (now *Reyemia nemesioides*), which has only two stamens and a bearded throat and therefore fits

Meyer's meagre description. However, the description applies equally well to *Lyperia violacea*. There are two specimens in Vienna (W) named *L. diandra* E. Mey.; one came from the Berlin garden, and the other is inscribed 'Konigsberg 24 Juli 1869' and could well have been a descendant of Meyer's plant as the flowers are often autogamous. The big capsules of *L. violacea* could have occasioned Drège's likening his plant to a *Nemesia*.

Lyperia violacea has been recorded on the Cape Peninsula (Lion's Head, Signal Hill, Devil's Peak), then from the environs of Swellendam east to the Cango valley and the foot of Prince Alfred's Pass in Uniondale division, from near sea level to c.300m. There is also a record purportedly from Brakdam in Namaqualand (3017BD, *Pearson* s.n., Z), but its presence there needs confirmation. The plants grow in sandy or stony ground, sometimes over limestone, flowering principally between June and September.

5. Lyperia tenuiflora Benth. in Hook., Comp. Bot. Mag. 1: 378 (1836) & in DC., Prodr. 10: 358 (1846). **Fig. 52B, Plate 3G.**
Type: Cape, without precise locality, *Drège*, (holo. K, iso. P).

Syn.: *Sutera tenuiflora* (Benth.) Hiern in Fl. Cap. 4(2): 291 (1904).

Annual herb not or scarcely drying black, main stem c.30–200mm tall, simple at first, later loosely branched near the base, branches ± sharply ascending, thinly glandular-puberulous, hairs up to 0.2–0.4mm long, scattered glistening glands as well, leafy. *Leaves* opposite, few (stems often floriferous nearly from base), c.7–33 × 3–9mm, narrowly elliptic tapering to a petiolar part up to c.⅓ total leaf length, rapidly shorter upwards, apex acute to subobtuse, margins entire to few-toothed, both surfaces hairy as stems. *Flowers* few to many in terminal racemes. *Bracts* alternate, leaflike, lowermost c.8–24 × 2–7mm. *Pedicels* (lowermost) 4–11mm long. *Calyx* tube 0.2–0.5mm long, lobes 4–6.5 × 0.6–1mm, thinly glandular-pubescent, hairs up to 0.25–0.6mm long, sparse glistening glands as well. *Corolla* tube c.24–32 × 1.4–1.7mm, cylindric, expanded in throat to c.2.8mm, limb nearly regular, c.16–18mm across, lobes c.5–8 × 3–8mm, broadly and bluntly spathulate, tube and backs of lobes glandular-puberulous, hairs c.0.25mm long, few glistening glands as well on upper part, upper surface of lobes glabrous except for a broad band of minute glands around mouth, band of upward-pointing clavate hairs in throat, limb shades of pink, rose-mauve and lavender, base of each lobe marked with 3 orange/yellow streaks coalescing into a band around the throat, tips of streaks sometimes dark violet. *Stamens* 2, filaments 1–2mm long, anthers 1.6–2mm long, just visible in mouth, staminodes 2, c.0.3mm long. *Stigma* 1–1.5mm long, just exserted. *Style* 21–27mm long, glandular-puberulous. *Ovary* 2.5–3 × 1.3–1.5mm, thinly glandular-puberulous, hairs c.0.2–0.25mm long, glistening glands nearly confined to a broad band down the sutures. *Capsules* c.7–10 × 3.5–4.5mm, hairy as the ovaries. *Seeds* c.0.8–1 × 0.5mm.

Selected citations:

Cape. Ceres div., 3220CC, Yuk Riviers Hoogte, 19 vii 1811, *Burchell* 1246 (K); 3220CC, Bantamsfontein Kop, *Acocks* 23666 (PRE). Beaufort West div., 3222 CC, near road between Seekoeigat and Kruidfontein, farm Klein-Waterval, 900m, 27 x 1991, *Vlok* 2516 (E). Laingsburg div., 3320AA, Whitehill, 20 ix 1943, *Compton* 14921 (NBG). Prince Albert div., 3321BC, Bosluis Kloof, 17 vii 1967, *Levyns* 11606 (BOL).

Lyperia tenuiflora is known from a limited area of the southern Cape, from hills north of Ceres east to Bosluis Kloof on the northern flank of the Great Zwartberg and on high ground to the north of it. It grows on mountain slopes, in sandy or gravelly places, between c.670 and 900m above sea level, flowering between June and October. Only twelve collections have been seen, five of them made by Compton at Whitehill near Laingsburg between 1929 and 1943, while Miss Thompson found it on nearby Ngaap Kop in 1971, so it is possibly under-collected rather than rare.

It is easily recognized by its large colourful flowers with radiating dark lines around the mouth of the corolla tube, relatively narrow leaves, and rather sparse indumentum, particularly on the ovaries and capsules. This last character will at once distinguish it from the closely allied *L. formosa*; see under that species.

6. Lyperia formosa Hilliard in Edinb. J. Bot. 48: 341 (1991).
Type: Cape, 3319DB, southern foothills of Voetpadsberg, farm Doringkloof, 1500ft, 23 viii 1985, *Morley* 420 (holo. STE, iso. PRE).

Annual herb scarcely blackening on drying, main stem c.200–400mm tall, simple at first, later loosely branched from the base and above, branches ascending, glandular-puberulous, hairs up to c.0.3mm long, very sparse glistening glands as well, leafy. *Leaves* opposite, c.15–47 × 6–20mm, elliptic tapering to a petiolar part up to c.⅓ total leaf length, rapidly shorter upwards, apex ± obtuse, margins serrate, both surfaces hairy as stem. *Flowers* few to many in lax terminal racemes. *Bracts* alternate, leaflike, lowermost c.10–23 × 5–10mm. *Pedicels* (lowermost) c.18–23mm. *Calyx* tube c.0.2mm long, lobes c.4.2–5 × 1mm, glandular-pubescent, hairs up to 0.5–1mm long, sparse glistening glands as well. *Corolla* tube c.24 × 1.4mm, cylindric, expanded in throat to c.2.8mm, limb nearly regular, c.14mm across, lobes c.5.2–6 × 4.2–5mm, broadly and bluntly spathulate, tube and backs of lobes glandular-puberulous, hairs up to 0.4–0.5mm long, very few glistening glands on upper part, upper surface of lobes glabrous except for a broad band of minute glandular hairs around mouth, band of upward-pointing clavate hairs in throat, limb white, base of each lobe with a broad orange patch 3-toothed at apex. *Stamens* 2, filaments 1.2mm long, anthers c.1.7mm long, included, staminodes 2, 0.3mm long. *Stigma* c.1–1.3mm long, included. *Style* c.20mm long, glandular-puberulous. *Ovary* c.3.5 × 2mm, densely clad in glistening glands and glandular hairs c.0.3–0.6mm long. *Capsule* (young) c.14 × 6mm, hairy as ovary. *Seeds* immature.

Lyperia formosa is known only from the type collection. The plants were growing on a river bank in shallow soil over rock, in full flower and young fruit near the end of August. The species is closely allied to *L. tenuiflora*, but it is a more robust plant, mostly broader in the leaf than *L. tenuiflora* (ratio of length: breadth c.2–3: 1 versus 4–5: 1) with longer pedicels (lowermost c.18–23mm, not 4–11mm), and larger, hairier ovaries and capsules. The corolla limb is white (shades of pink and mauve in *L. tenuiflora*) and the colour-patterning around the mouth more pronounced.

6. SUTERA

Sutera is characterized by a funnel-shaped corolla tube (cylindric in two species) always broad in the mouth, posticous filaments not decurrent down the corolla tube, semi-annular nectariferous gland, and pallid white, grey-blue or amber-coloured seeds ornamented with longitudinal rows of transversely elongated pits arranged in chequer-board fashion. The limb of the corolla is white or shades of pink, mauve or blue and lacks nectar-guides; the tube is yellow or orange at least in the upper part. In all but two species one or both pairs of stamens are exserted; the stigma is also exserted. The stigma is nearly always lingulate with two marginal rows of papillae; in *S. platysepala* it is very short and bifid. In the two very closely allied species, *S. cooperi* and *S. griquensis*, the stamens and stigma are deeply included; the stigma is shortly bifid and lacks papillae; instead, the two branches are firmly joined by a jelly-like receptive zone. The bracts are at most adnate to the extreme base of the pedicel.

Forty nine species are recognized here. They are confined to Africa south of the Cunene and Zambezi rivers. Only one, *S. patriotica*, occurs in Namibia, but its main area lies further east, principally in the NE Cape, W Transvaal, Orange Free State and Lesotho. There are three species in Zimbabwe, one endemic (*S. septentrionalis*), one, *S. debilis*, occurring also in the Transvaal, one, *S. floribunda*, the most widespread species in the genus, extending from Inyangani to the southern part of Transkei. Twenty species in all occur in the eastern half of southern Africa, 33 in the southern and western Cape, three sharing the E–W distribution.

The species fall into two seemingly natural groups:

Section Sutera. Hairs present on upper surface of corolla lobes near mouth. Testa ornamented with ± oblong reticulations visible under the SEM. Always some 3–c.11-flowered cymules in each inflorescence, mixed with solitary flowers.

Section Chaenostoma. Upper surface of corolla lobes glabrous; hairs, if present, confined to inner surface of upper part of corolla tube. Testa ornamented with irregular pustules visible under the SEM. Of 46 species, only one always has some flowers in cymules; they are sometimes present in two more species.

Section *Sutera* consists of three species, *S. foetida*, *S. cooperi* and *S. griquensis*.

My section *Chaenostoma* corresponds closely to Bentham's genus *Chaenostoma*, which he described (1836, p.360) as having 'a corolla contracted at the base into a tube which is often elongated and always campanulate or infundibuliform at the orifice.' (*Chaenostoma* means gaping mouth). Bentham did not become aware of *Sutera* Roth until 1846 and then (in DC., Prodr. 10: 362) used it exclusively for *S. glandulosa* Roth (1821), that is, *Jamesbrittenia dissecta* (Del.) O. Kuntze. I have included Bentham's monotypic genus *Sphenandra* in section *Chaenostoma*; Bentham's distinguishing characters, a very short corolla tube and rotate limb, do not hold now that many more species are known. The only species of *Chaenostoma* now excluded is *C.pedunculosum* Benth. which is transferred to *Jamesbrittenia*; Bentham himself queried its placement in *Chaenostoma*.

Bentham (in DC., Prodr. 10, 1846) divided *Chaenostoma* into two unranked groups, *Breviflora* and *Longiflora*, but the groups are untenable as they cut across affinities: this is most strikingly demonstrated in four closely allied species with distinctive leaf-tips: *S. subnuda* (tube 3–5.5mm), *S. affinis* (tube 5–8mm), *S. denudata* (tube 9–12.5mm), *S. tenuicaulis* (tube 10.5–16.5mm).

Diels (in Bot. Jahrb. 23: 492, 1897) followed Bentham in having *Sutera* (that is, *Jamesbrittenia*) as a monotypic genus, but included *Lyperia* Benth. and *Sphenandra* in *Chaenostoma*, and then recognized four unranked groups: 1, *Breviflorae* Benth. [including *Sphenandra*]; 2, *Intermediae* Diels [*Chaenostoma* § *Longiflorae* Benth. and *Lyperia* § *Racemosae* Benth.]; 3, *Spicatae* (Benth.) Diels [*Lyperia* § *Spicatae* Benth.]; 4, *Foliolosae* (Benth.) Diels [*Lyperia* § *Foliolosae* Benth.]. Hiern (in Fl. Cap. 4(2), 1904) used these as sectional names, adding many more species: his section *Breviflorae* includes one species of *Manulea*, three of *Jamesbrittenia*; section *Intermediae* is a mixture of *Manulea, Jamesbrittenia, Lyperia* and *Sutera*; sect. *Spicatae* is a mixture of *Lyperia* and *Jamesbrittenia*; sect. *Foliolosae* is wholly *Jamesbrittenia*. Hiern did not deal with *J. dissecta*, its geographical area being outside that of Flora Capensis.

The species of *Sutera* are extraordinarily difficult to classify. There are a few very distinctive ones, sometimes without close allies, then there are a number of suites of seemingly allied species. Relationships between the groups are not readily apparent and the linear sequence in the enumeration is somewhat arbitrary. Major breaks are marked by a line.

Within suites, the limits of individual species are often clear-cut, but in others the limits are blurred. Hybridization, either occasional crossing or persistent back-crossing, is very probably a major cause of blurring: it is striking that geographically isolated species present no difficulty. Cognisance should be taken of all species growing in the area when a specimen cannot be determined satisfactorily. To make the geographical distributions clearer, the quarter degree squares from which specimens have been seen are sometimes listed under the citations of specimens. Possible hybrids are mentioned under the following species: *S. aethiopica, affinis, archeri, caerulea, calciphila, decipiens, hispida, langebergensis, levis, longipedicellata, paniculata, pauciflora, polyantha, revoluta, subspicata, uncinata,violacea*.

Hybridization has clearly taken place between species in cultivation; I have given a cultivar name, 'Longwood', to a plant that passes in Europe and America as *S. hispida* but is a sterile hybrid devoid of pollen.

Many species are superficially similar; meticulous attention to details of indumentum and floral characters is necessary to distinguish them. Casual matching is virtually impossible. Even so great a botanist as Bentham based names on a mixture of two or even three different species when he was working with very limited material. Hiern misdetermined dozens of specimens. There are several names in herbaria proposed by Miss Overkott; these were never published and have no validity. Mixed collections on herbarium sheets are not uncommon. I am sure I have contributed my own quota of mistakes. But I have sought, first, to clarify the application of names, lectotypifying them where necessary, and secondly to build up a taxonomic framework that will enable others to carry out the field and laboratory investigations that are patently needed.

The key provided is highly artificial and allied species do not necessarily come down together. Also, the key is often only a guide; the descriptions and particularly the notes following the descriptions need to be read in conjunction with it. Most hybrids are not catered for as this would often have led to the absurdity of keying individual specimens.

In the descriptions and key, all floral measurements are based on rehydrated material. The calyx was opened out and the tube measured from the base of the sinus between the two lips to the junction of torus (receptacle) and tube; this is shorter than an external measurement. Measure both calyx and corolla at anthesis; the calyx tube may elongate as the fruit develops. The corolla tube was measured from the sinus between the lobes to the base, throat measured across the flattened corolla at the junction between lobes and tube. Filaments were measured from the point of insertion to the middle of the connective.

Sutera Roth, Bot. Bemerk. 172 (1807), non Roth, Nov. Pl. Spec.291 (1821); Hiern in Fl. Cap. 4(2) p.p. min. and excl. Sect. Spicatae and Sect. Foliolosae; also excl. all subsequent authors. **Figs 14, 25, 26, 38, 39A, 53.**

Syn.: *Chaenostoma* Benth. in Hook., Comp. Bot. Mag. 1: 374 (1836), nom. cons. Lectotype: *C.aethiopicum* (L.) Benth., typ. cons.

Sphenandra Benth. in Hook., Comp. Bot. Mag. 1: 373 (1836). Type: *S. viscosa* (Ait.) Benth. [that is, *Sutera caerulea* (L.f.) Hiern].

Palmstruckia Retz., Observ. Bot. Pugill 15 (1810), nom. rej. Type: *Manulea foetida* (Andr.) Pers.

Shrublets, suffrutices or perennial herbs, rarely annual, mostly glandular, and sometimes aromatic or foetid. *Stems* leafy throughout. *Leaves* usually opposite, sometimes alternate upwards, bases either ± connate or decurrent in narrow wings or ridges, simple, entire to toothed, rarely more deeply lobed. *Inflorescence* ± racemose, flowers mostly solitary in axils of leaves or bracts, sometimes in cymules or cymose racemes, sometimes panicled. *Bracts* at most adnate to extreme base of pedicel. *Calyx* bilabiate, sometimes obscurely so, anticous lip 2-lobed, posticous lip 3-lobed, or rarely regularly divided into 6–9 lobes, lobes ± linear-lanceolate, usually pubescent. *Corolla* tube usually funnel-shaped, cylindric in 2 species, mouth round, limb nearly regular, lobes spreading, suborbicular to oblong, entire, usually glandular-pubescent outside, often with glistening glands as well, inside usually with either 1–5 longitudinal bands of clavate hairs in throat, or glabrous (Sect. *Chaenostoma*), rarely with clavate hairs extending from throat out onto lower part of lobes (Sect. *Sutera*). *Stamens* 4 (a fifth occasionally developed), didynamous, filaments usually inserted in upper part of corolla tube, not decurrent, anticous pair exserted, posticous pair either included or exserted; in 2 species inserted in lower part of tube, all included; all anthers synthecous. *Stigma* usually lingulate with 2 marginal bands of stigmatic papillae (short and minutely bifid in one species), exserted, in 2 species shortly bifid, the two branches united by the jelly-like (not papillose) stigmatic surfaces, deeply included. *Ovary* ± elliptic in outline, often with glistening glands at least on the sutures, rarely glandular-pubescent as well, nectariferous gland semi-annular, ovules many in each loculus. *Fruit* a septicidal capsule with a short loculicidal split at tip of each valve, glabrous or with glistening glands. *Seeds* seated on a round centrally depressed pulvinus (fig. 14 C, D, E), seed roughly elliptic in outline, sometimes angled by pressure, amber-coloured, pallid or grey-

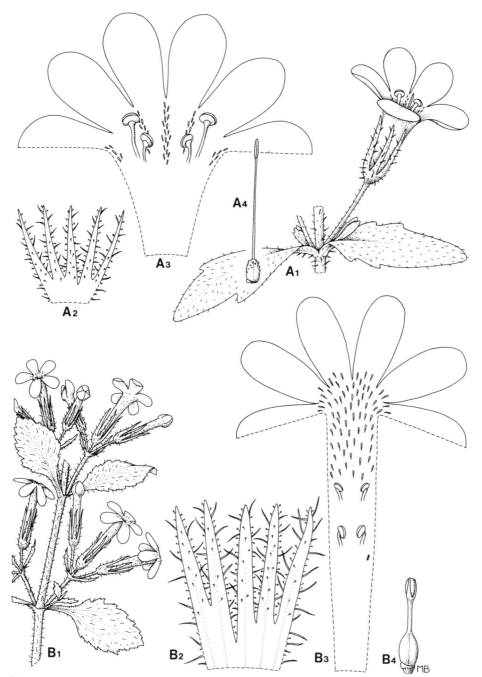

Fig. 53. A. *Sutera roseoflava* (*Hilliard & Burtt* 10886): A1, part of inflorescence, flower solitary in leaf axil, stamens exserted (typical of section *Chaenostoma*), × 3; A2, calyx opened out, slightly bilabiate, tube well developed, a generic character, × 5; A3, corolla opened out to show arrangement of clavate hairs in section *Chaenostoma*, and non-decurrent filaments, a generic character, × 5; A4, gynoecium, stigma lingulate with marginal papillae, nectary lateral, × 5. B. *Sutera cooperi* (*Acocks* 23476): B1, part of inflorescence, showing flowers in axillary cymules (typical of section *Sutera*), stamens deeply included, × 1; B2, calyx opened out, distinctly bilabiate, × 5; B3, corolla opened out, showing clavate hairs extending well down tube and onto base of limb, stamens deeply included, × 5; B4, gynoecium, showing stigma peculiar to *S. cooperi* and *S. griquensis*, × 5.

to violet-blue, testa thin, tightly investing the endosperm, which is alveolate, with several longitudinal rows of transversely elongated pits arranged in chequer-board fashion, under the SEM seen to be ornamented with ± oblong reticulations (sect. *Sutera*, fig. 25C) or irregular pustules (sect. *Chaenostoma*, figs 25D, 26A, B, C).

Lectotype: *Sutera foetida* Roth
Distribution: Africa south of the Cunene and Zambezi rivers, mainly Cape, Natal, Transvaal.

Key to species

Note: read last two paragraphs of introduction, p.222.

1a. At least some of the flowers in 3–several-flowered cymules; clavate hairs present around mouth and on bases of corolla lobes (rarely clavate hairs wanting, but then glistening glands present). **Section Sutera** 2
1b. Flowers usually solitary in the axils of leaves or bracts, in cymules in a few species; upper surface of corolla lobes always glabrous (clavate hairs often present inside throat). **Section Chaenostoma** 4

2a. Nearly glabrous (or rarely glandular-pubescent) annual herb; anticous stamens and stigma shortly exserted . . . . **3. S. foetida**
2b. Coarse bushy perennial herbs, hairs up to 0.5–1.5mm long on stems and leaves; stamens and stigma all deeply included 3

3a. Leaves always opposite, bases connate (the uppermost bracts may be alternate)
1. S. cooperi
3b. Only the lowermost leaves sometimes opposite and then bases not connate   **2. S. griquensis**

4a. Hairs on stems and leaves cobwebby-woolly 5
4b. Hairs on stems and leaves not cobwebby-woolly 6

5a. Principal leaves elliptic, c.5–22 × 2.5–13mm; corolla tube 9–13mm long . **25. S. marifolia**
5b. Principal leaves linear to narrowly elliptic, c.15–20 × 2–3mm; corolla tube c.5–
5.5mm long . . . . **26. S. cinerea**

6a. Blades of principal leaves about as long as broad to roughly twice as long as broad 7
6b. Blades of leaves more than twice as long as broad, often considerably longer 29

7a. One pair of stamens exserted, one pair included 8
7b. All four stamens exserted 25

8a. Stigma very short (0.2–0.5mm) and distinctly bifid . . . . **9. S. platysepala**
8b. Stigma c.0.5–2mm long, lingulate, entire 9

9a. Hairs on stems (ignore inflorescence axes) nearly or quite confined to 2 bands running down the internodes from the point at which the bases of the pair of leaves meet 10
9b. Stems hairy all round 11

10a. Leaves spreading, thin-textured, either with scattered hairs up to 0.2mm long or nearly glabrous . . . . **23. S. langebergensis**
10b. Leaves usually sharply ascending, ± appressed, imbricate, thick-textured, hairs either wanting or few on blade, always present on lower margins and carried down the internode by the decurrent bases . . . . **27. S. subspicata**

11a. Leaves on floriferous part of stem becoming alternate upwards 12
11b. Leaves always opposite though bracts may sometimes become alternate towards tips of inflorescence axes 13

12a. Largest teeth on leaves up to c.2mm along upper margins, inflorescence panicled in well-grown specimens, corolla tube lightly bearded within .... **7. S. septentrionalis**
12b. Largest teeth on leaves c.3–5mm along upper margin, inflorescence always a raceme, corolla tube glabrous within .... **8. S. debilis**

13a. Branches sprawling, flowers always solitary in axils of leaves 14
13b. Branches normally erect or decumbent, flowers in the axils of either leaves or bracts .. 17

14a. Hairs on stems up to c.0.4–2mm long 15
14b. Hairs on stems c.0.1mm long .... See under **45. S. impedita**

15a. Petiolar part of principal leaves roughly half as long as blade; ovary either glabrous or with a few glistening glands on the sutures .... **43. S. cordata**
15b. Petiolar part of principal leaves roughly ⅕–⅓ as long as blade; ovary ± wholly clad in glistening glands 16

16a. Hairs on stems up to 0.8–2mm long; calyx lobes always 5 .... **44. S. glandulifera**
16b. Hairs on stems up to 0.4–1mm long; calyx lobes 5–7 .... **45. S. impedita**

17a. Hairs on calyx up to 0.25mm long 18
17b. At least some hairs on calyx 0.3–2mm long 20

18a. Leaves entire or very nearly so .... See under **42. S. pauciflora**
18b. Leaves coarsely toothed 19

19a. Petiolar part of main foliage leaves 0–6mm long, leaves and bracts on inflorescence axes becoming alternate towards the tips. Perennial .... **5. S. polelensis subsp. polelensis**
19b. Petiolar part of main foliage leaves c.5–8mm long, leaves and bracts on inflorescence axis always opposite. Probably annual .... **6. S. racemosa**

20a. Flowers c.2–6 in each raceme, sometimes overtopped by further vegetative growth
28. S. titanophila
20b. Flowers many in each raceme 21

21a. Leaves entire or very nearly so .... See under **42. S. pauciflora**
21b. Most leaves distinctly toothed 22

22a. Hairs on stem, leaves and calyx almost hispid .... **24. S. hispida**
22b. Stems, leaves and calyx softly glandular-pubescent or with some coarse obtuse hairs on the stems mixed with glandular ones 23

23a. Leaf bases decurrent forming narrow wings or ridges down to the node below (careful observation needed) .... **22. S. decipiens**
23b. Leaf bases not decurrent 24

24a. Flowers mostly in a mixture of 3-flowered cymules, cymose-racemes and racemes, only the lowermost sometimes solitary in the axils of leaves; longest pedicels 4–10mm; petioles c.4–9mm .... **4. S. floribunda**
24b. Flowers mostly solitary in the axils of the uppermost leaves and bracts; longest pedicels (6–)10–30mm; petioles c.0–6mm long .... **5. S. polelensis**

25a. Annual or perennial herb 26
25b. Twiggy shrublet 28

26a. Hairs on stem all glandular .... **47. S. polyantha**
26b. Eglandular hairs (often mixed with glandular ones) at least on lower part of stem 27

27a. Stems sprawling, flowers in upper leaf-axils forming leafy racemes; ovaries and capsules clad all over in glistening glands .... **48. S. roseoflava**
27b. Stems usually ± erect, sometimes sprawling; ovaries with glistening glands on sutures only; only the lowermost flowers in the axils of leaves, most in the axils of bracts
49. S. campanulata

28a. (Note: 3 leads) Leaf blade suborbicular to broadly rhomboid abruptly contracted to a petiolar part c.2–10mm long (roughly half as long as blade); hairs on both leaves and stems c.0.1–0.25mm long .... **41. S. rotundifolia**
28b. Leaf blade suborbicular to broadly elliptic abruptly contracted to a petiolar part c.3–5mm long (roughly ⅓–½ as long as blade); hairs on stem up to c.0.5–0.8mm long, on leaves c.0.3–0.4mm .... **42. S. pauciflora**
28c.Leaf blade broadly elliptic tapering to a short petiolar part; hairs on stem and leaves c.0.1–0.15mm long .... See under **42. S. pauciflora** (Gamtoos river)

29a. Stem, leaves and calyx clad in vesicular hairs (which look like tiny round scales), very rarely minute glandular hairs present as well .... **36. S. halimifolia**
29b. Stem etc.either glabrous or clad in manifestly stalked hairs 30

30a. All four stamens exserted; corolla tube up to 6(–8.5)mm long 31
30b. One pair of stamens exserted, one pair either included or just visible in throat; corolla tube (5–)8–28mm long 44

31a. Calyx lobes 6–9 .... **46. S. calycina**
31b. Calyx lobes 5 32

32a. Style and ovary either glandular-pubescent or with glistening glands, or both 33
32b. Style glabrous, ovary either glabrous or with a few glistening glands on the sutures ... 37

33a. Style and top of ovary with glandular hairs up to 0.5–1.5mm long, glistening glands as well .... **40. S. caerulea**
33b. Style and ovary either with glistening glands only or a few minute glandular hairs as well 34

34a. Corolla tube up to 3.5mm long, limb 5–7mm across .... **38. S. levis**
34b. Corolla tube (3–)4–9mm long, limb c.8.5–15mm across 35

35a. Corolla tube 3–6mm long, hairs on stem glandular, up to 0.25(–0.3)mm long 36
35b. Corolla tube 6–9mm long, hairs on upper part of stem glandular or sometimes eglandular as well, up to 0.2–0.6mm long, mainly eglandular on lower part of stem, c.0.3–0.7mm long .... **49. S. campanulata**

36a. Rootstock a woody caudex of irregular shape; vegetative buds in leaf axils rarely developing (see further under S. patriotica) .... **37. S. neglecta**
36b. Rootstock a simple woody taproot; leaves pseudofasciculate .... **39. S. patriotica**

37a. Leaves glabrous to minutely puberulous, linear to narrowly elliptic or oblanceolate, acute to mucronate 38
37b. Leaves either distinctly pubescent or, if glabrous or only very minutely puberulous, then obtuse 39

38a. Corolla tube 3–5.5mm long; flowers usually in well-branched twiggy panicles, bracts sharply differentiated from leaves .... **33. S. subnuda**
38b. Corolla tube 5–8mm long; flowers in racemes or very loose panicles (see further under S. affinis) .... **34. S. affinis**

39a. Stem (ignore inflorescence axis) very narrowly winged by decurrent leaf bases, hairs confined to the grooves so formed; leaves either glabrous or with a few scattered hairs, scattered glistening glands on lower surface .... **31. S. placida**

39b. Stem terete, pubescent all round; leaves pubescent as stems 40

40a. Stems with glandular and some eglandular hairs up to 0.2–0.6mm long on upper part, mainly eglandular hairs to 0.3–0.7mm on lower part .... **49. S. campanulata**

40b. Stems glandular-puberulous, hairs up to 0.1–0.25mm long 41

41a. Leaves linear to narrowly elliptic, obtuse, primary ones up to 10–30(–40) × 0.6–3(4–5)mm 42

41b. Leaves more or less oblanceolate to elliptic, acute to subacute, primary ones c.9–27 × 3–10mm 43

42a. Longest pedicels 1–9mm; anticous filaments glabrous .... **13. S. revoluta**

42b. Longest pedicels (8–)10–25mm; anticous filaments weakly to strongly bearded
15. S. archeri

43a. Lowermost bracts c.1–2mm broad, flowers mauve or violet(see notes under *S. calciphila*)
30. S. calciphila

43b. Lowermost bracts (equivalent to uppermost leaves) c.2.5–7mm broad; flowers white
47. S. polyantha

44a. Principal leaves c.9–26mm broad; ovary with a broad band of glistening glands on sutures and scattered on top of ovary and up style .... **5. S. polelensis subsp. fraterna**

44b. Principal leaves mostly up to 5mm broad; if broader then ovaries either glabrous or with a few glistening glands on the sutures 45

45a. Calyx nearly glabrous, a few hairs up to 0.1–0.15mm long mainly on margins; leaves glabrous or minutely glandular-puberulous, glistening glands also present particularly on lower surface .... **17. S. integrifolia**

45b. Calyx variously pubescent; leaves glabrous to hairy 46

46a. Leaves narrowly ovate, cuneate or elliptic, bases connate; stems, leaves and calyx clad in almost hispid hairs c.0.5–1.5mm long; stems terete; flowers usually in upper leaf axils, sometimes uppermost in axils of bracts, always in opposite pairs .... **24. S. hispida**

46b. Characters not as combined above (leaf shape different; leaf bases decurrent and stem thereby winged or ridged; indumentum short or not hispid; inflorescence different) ... 47

47a. Hairs on stem (ignore inflorescence axis) nearly or quite confined to 2 bands running down the internodes from the point at which the bases of the pairs of leaves meet
23. S. langebergensis

47b. Stems either glabrous or hairy all round (but confined to two bands in specimens from Mossel Bay area, discussed under 29. *S. aethiopica*) 48

48a. Stems narrowly winged or ridged by decurrent leaf bases (this sometimes needs very careful observation, and must not be confused with ridges formed by raised vascular strands running into the midrib) 49

48b. Leaf bases not decurrent 55

49a. Hairs on stem up to 0.5–1.8mm long 50

49b. Stems either glabrous or with hairs up to 0.4mm long 52

50a. Only the lowermost flowers in the axils of leaves, most in the axils of small bracts; leaves linear to elliptic 51

50b. Flowers mostly in the axils of leaves, forming leafy racemes; when the inflorescence is panicled, the flowers on the short side branches may be in the axil of bracts; leaves lanceolate, entire or nearly so .... **21. S. subsessilis**

51a. Main primary leaves linear to narrowly elliptic, 7–35 × 0.8–5(–7)mm, margins entire or with up to 4–6 pairs of tiny callose teeth; calyx tube (1–)1.5–2mm long   . **18. S. uncinata**

51b. Main primary leaves elliptic narrowed to a short petiolar part, 10–32 × 3–15mm, margins with 2–4 pairs of coarse teeth; calyx tube 0.5–0.8mm long .... **22. S. decipiens**

52a. Brachyblasts present in leaf-axils; tips of main leaves obtuse to subacute, stems often puberulous to pubescent 53

52b. No brachyblasts in leaf-axils; tips of leaves acute to mucronate; stems glabrous

35. S. tenuicaulis

53a. Main leaves 9–20mm long, calyx tube 0.6–0.8mm long, glandular-puberulous, without coarse broad-based hairs .... **20. S. multiramosa**

53b. Leaves often more than 20mm long (but they may be as short as 7mm), calyx tube 0.8–2mm long, glandular-puberulous plus coarse broad-based hairs 54

54a. Longest pedicels 5–10(–15)mm, calyx tube (1–)1.5–2mm long, anticous filaments 1.8–3mm long .... **18. S. linifolia**

54b. Longest pedicels 11–21mm, calyx tube 0.8–1mm, anticous filaments 3.2–4mm long (see notes following description) .... **19. S. longipedicellata**

55a. Stems glabrous to very minutely puberulous or scabridulous, leaves linear (rarely narrowly elliptic), apex acute to mucronate .... **32. S. denudata**

55b. Stems variously pubescent, leaves, if linear, usually obtuse (linear acute, but not mucronate, only in *S. macrosiphon*) 56

56a. Stem glandular-puberulous, sometimes with scattered ± hispid hairs as well, these up to c.0.5–1mm long, contrasting with the leaf blade, either completely glabrous (though gland-dotted on lower surface) or with scattered ± hispid hairs; leaves oblanceolate to narrowly cuneate, acute to subobtuse, usually with up to 3 pairs of prominent teeth (see also *Burgers* 1127 mentioned at end of discussion under 13. *S. revoluta*); calyx glandular-puberulous plus scattered coarse almost hispid hairs up to c.0.4–0.6mm long

29. S. aethiopica

56b. Indumentum on stems and leaves similar; calyx glandular-puberulous to glandular-pubescent (0.1–1mm) without any additional coarse hairs 57

57a. Calyx tube 2–3.5mm long, corolla tube 14–28mm; inflorescence a 1–few-flowered raceme, sometimes panicled .... **18. S. macrosiphon**

57b. Calyx tube up to 2mm long; corolla tube 5–17mm; flowers many in leafy racemes or panicles 58

58a. No conspicuous brachyblasts (leaf tufts) in axils of leaves; stems, leaves and calyx glandular-puberulous (hairs up to 0.1–0.2mm long), no tiny bracts immediately below calyx; corolla tube 12.5–17mm long .... **10. S. glabrata**

58b. Brachyblasts usually conspicuous; hairs on stems, leaves and calyx c.0.1–1mm long, tiny bracts (0.2–7mm long) sometimes present below calyx, corolla tube 5–15mm long (nos. 11–14 are closely allied and species limits may be blurred by hybridization; read notes following descriptions) 59

59a. Tiny bracts always present below calyx; flowers either solitary in the leaf-axils or in 3-flowered cymules; corolla tube 11–15mm long .... **12. S. violacea**

59b. Tiny bracts rarely present below calyx; flowers always solitary in the leaf-axils; corolla
tube 5–13mm long 60

60a. Hairs on stem and leaves usually c.0.1mm long, rarely up to 0.2mm (see also *Thompson*
2101 under 11. *S. paniculata* with stem-hairs c.0.2mm long) .... **13. S. revoluta**

60b. Hairs on stem up to 0.3–1mm long 61

61a. Main leaves 20–30mm long, hairs on calyx up to 0.25–0.4mm, corolla tube 6.5–9.5mm
11. S. paniculata

61b. Main leaves 5–18mm long, hairs on calyx up to 0.4–1.2mm long, corolla tube
10.5–12.5mm .... **14. S. comptonii**

Section Sutera

1. Sutera cooperi Hiern in Fl. Cap. 4(2): 282 (1904). **Fig. 53B.**
Type: Lesotho, [3027BA] near Bethesda, *Cooper* 732 (holo. K; iso. BM, E, PRE, W, Z).

Syn.: *Chaenostoma cordatum* var. β *hirsutior* Benth., nomen, in Hook., Comp. Bot. Mag. 1: 377 (1836) & in DC., Prodr.
10: 357 (1846). Specimen cited: Cape, Aliwal North div., [3027CA] on the Wittebergen, 5500–6000ft, 15 i
1833, *Drège* (K, P).
Sutera cordata var. *hirsutior* Hiern in Fl. Cap. 4(2): 280 (1904).
S. latifolia Hiern in Fl. Cap. 4(2): 281 (1904). Type: Orange Free State, [3027AD] Kornet Spruit, between the
Orange and Caledon rivers, 5000–6000ft, Nov., *Zeyher* (P, PRE, S, W, Z).
Chaenostoma cooperi (Hiern) Thell. in Viert. Nat. Ges. Zürich 60: 409 (1915) in obs.
Sutera bracteolata Hiern in Fl. Cap. 4(2): 276 (1904). Type: 'Natal', 1862, *Cooper* 2857 (holo. K, iso. Z).
S. cymulosa Hiern in Fl. Cap. 4(2): 267 (1904). Type: Orange River Colony, Leeuwspruit and Vredefort,
Barrett-Hamilton s.n. (holo. BM).

Coarse bushy perennial herb, basal part seen on only one specimen (woody tap-root c.8mm
diam. at crown) and height not recorded but possibly reaches 1m, upper stems c.3mm diam.,
vegetative parts (including calyx) glandular-puberulous together with much coarser patent hairs
c.1–3mm long, minutely gland-tipped, leafy throughout. *Leaves* opposite, bases connate, blade
c.12–40 × 10–40mm, sometimes smaller on the inflorescence, ovate, abruptly contracted to a
petiolar part c.5–30mm long (nearly equalling to half as long as blade), apex obtuse to subacute,
base cordate-cuneate, margins coarsely toothed, the larger teeth denticulate. *Flowers* either
solitary or in 3-flowered cymules in the upper leaf axils, in racemes or panicles. *Bracts* (as
opposed to leaves) present or wanting, if present, a pair immediately below the calyx, c.2–7 ×
0.1–1mm, linear. *Peduncles* and pedicels (lowermost) 5–20mm long. *Calyx* obscurely bilabiate,
tube 1–3mm long, lobes 4.5–9 × 1–2mm, the anticous lobes slightly broader than the posticous
ones, linear-lanceolate. *Corolla* tube 8–14 × 1.5–2.5mm, cylindric, scarcely broadened upwards, limb almost regular, 6–15mm across, posticous lobes 2.2–5.5 × 1.8–3mm, anticous lobe
2.4–5 × 1.8–3mm, all oblong or oblong-elliptic, white, tube orange, glandular-pubescent
outside, hairs up to 0.25–0.4mm long, densely pubescent inside on lower part of lobes and upper
part of tube, hairs clavate, up to c.0.5mm long. *Stamens* 4, all deeply included, anticous anthers
0.8–1.1mm long, posticous anthers 1.1–1.3mm long, filaments short (c.1mm) not decurrent,
glabrous. *Stigma* 0.6–1.5mm long, gelatinous, deeply included. *Style* 1.2–2.5mm long. *Ovary*
1.5–2 × 1–1.2mm. *Capsule* 4–5 × 2.5–3mm, glabrous except for a few glistening glands on the
sutures. *Seeds* c.0.7–0.8 × 0.3mm, amber-coloured.

Citations:

Orange Free State. Bethlehem distr., 2828AB, Bethlehem, 5100ft, i 1894, *Bolus* 8227 (BOL, K). Senekal distr.,
2827CB, Marquard, Klein Leeukop, c.5800ft, 21 viii 1964, *Acocks* 23476 (K, PRE). Ficksburg distr., 2827DD, farm
Braamhoek, 6500ft, 10 x 1934, *Galpin* 13889 (PRE, mixed with *S. polelensis*).
Lesotho. Mafeteng distr., 2927CD, Maboloka Mt, 6 iii 1915, *Dieterlen* 325a (NBG, P, PRE). Sine loc. *Wahlberg* (S).

When Bentham published *Chaenostoma cordata* var. *hirsutior* he noted that it 'may perhaps be a distinct species, but the specimens are past flower, and imperfect.' It is indeed a distinct species, and proves to be the plant to which Hiern gave four different names, of which *S. cooperi* has been chosen to stand because the holotype is in the Kew herbarium and there are several good isotypes.

Sutera cooperi has been much confused with *S. floribunda*, by Hiern and by those who followed him. However, it bears scant resemblance to that species. It is closely allied to *S. griquensis*, the two species being unique in their possession of deeply included stamens and stigma, the style being remarkably short; the stigma itself is unique: it consists of two short branches firmly held together by a jelly-like receptive surface (there are no stigmatic papillae). Furthermore, there are long clavate hairs on the lower part of the corolla lobes and down the tube to the point of insertion of the anticous stamens; *S. foetida* too has clavate hairs around the mouth and on the lower part of the corolla lobes, but these do not extend down the tube.

Sutera cooperi is easily distinguished from *S. griquensis* by its constantly opposite, connate, leaves (the upper bracts may be alternate). It ranges from the environs of Bethlehem in the eastern Orange Free State along the Lesotho border to the foot of the Witteberg near Lady Grey in the Cape, so it probably also occurs in the Herschel district of Transkei. The type of *S. bracteolata* (in synonymy) was said to have come from Natal, but that is unlikely. Similarly, the type of *S. cymulosa* bears Barrett-Hamilton's printed ticket 'Leeuwspruit and Vredefort, Orange River Colony', which are sited at 2627DD and 2727AB, and are an unlikely provenance for the plant; Hiern recorded the specimen as having come from Roodeval, which I have failed to trace satisfactorily.

The plants grow at the foot of cliffs and under overhanging rocks between c.1500 and 1800m above sea level, flowering between August and March.

2. Sutera griquensis Hiern in Fl. Cap. 4(2): 279 (1904).
Type: Cape, Griqualand West, Hay div., near Griquatown, *Orpen in herb. Bolus* 5738 (holo. K; iso. BOL, SAM).

Syn.: *S. burchellii* Hiern in Fl. Cap. 4(2): 279 (1904). Type: Cape, Griqualand West div., Klip Fontein, 26–29 xii 1812, *Burchell* 2622 (holo. K).

A coarse bushy spreading herb up to 200–300mm tall (possibly taller), flowering when young and probably perennating, stems becoming woody and up to 4mm in diam., vegetative parts (including calyx) glandular-puberulous, hairs up to c.0.2mm long, coarser hairs up to 0.8–1.5mm long present or not, leafy throughout. *Leaves* often alternate, sometimes opposite but bases not connate, blade c.5–32 × 5–30mm, ovate, abruptly contracted to a petiolar part c.4–30mm long (often nearly equalling blade, sometimes up to c.one third as long), apex acute to ± obtuse, base cordate-cuneate, margins coarsely toothed, the larger teeth sometimes sparingly denticulate. *Flowers* either solitary or in 3-flowered cymules in the upper leaf axils, arranged in racemes or loose panicles, inflorescence axis often flexuous. *Bracts* (as opposed to leaves) present or wanting, if present, a pair not far below the calyx, c.3.5–7 × 0.6–1mm, linear-oblong or elliptic-lanceolate. *Peduncles* and pedicels (lowermost) c.7–30mm long. *Calyx* obscurely bilabiate, tube 0.4–1mm long, lobes 5.5–7 × 1–1.2mm, linear-lanceolate. *Corolla* tube c.6.6–11 × 1.3–1.8mm, cylindric, scarcely broadened upwards, limb almost regular, 4.3–10mm across, posticous lobes 2–3 × 1–1.8mm, anticous lobe 2–4 × 1–2.5mm, all oblong or oblong-elliptic, white or very rarely mauve, tube yellow or orange, glandular-puberulous outside, hairs 0.1–0.4mm long, densely pubescent inside on lower part of lobes and usually down the tube to the point of insertion of the anticous stamens, hairs very few in flowers with very small limb and pale tube, hairs clavate, up to c.0.5mm long. *Stamens* 4, all deeply included, anticous anthers 0.7–1mm long, posticous anthers 1–1.4mm long, filaments short (c.1mm), glabrous, not decurrent. *Stigma* 1–1.2mm long, gelatinous, deeply included. *Style* 1.2–2.3mm long. *Ovary* c.2 × 1.2–1.5mm. *Capsules* 5–6.5 × 3mm, glabrous except for a few glistening glands on the sutures. *Seeds* c.0.8 × 0.4–0.5mm, amber-coloured.

Citations:

Transvaal. Zeerust distr., 2525BD, Linokana [Dinokana], 10 vi 1887, *Holub* s.n. (Z). 2526 DA, between Swartruggens and Lichtenburg, 15 viii 1979, *Botha* 2579 (PRE). Rustenburg distr., 2527CA, Rustenburg Kloof, 2 ix 1930, *Hutchinson* sub *Moss* 18880a (J).
Cape. Hay div., 2823AB, between Daniels Kuil and Postmasburg, Malansrust, 25 iii 1936, *Acock* 324 (BOL, K, PRE); 2823AA, Beeshoek, 5 miles WNW of Postmasburg, c.4500ft, 28 vi 1961, *Leistner & Joynt* 2705 (K, SRGH, PRE); 2823AC, Klipfontein, iii 1938, *Wilman* 5358 (PRE); Postmasburg, 29 v 1936, *Bester* s.n. (BOL). 2823DC, Griquatown Commonage, iv 1937, *Acocks* 2184 (BOL, PRE). 2822BC, slopes of the Langebergen, Dunmurry, 27 ix 1939, *Vigne* 5669 (BOL, K); c.2822DD, Asbestos Hills, Klipvlei, 22 iii 1939, *Esterhuysen* 995 (BOL, NBG). Prieska div., 2922DB, Spitzkop (N of Orange River), 30 iii 1920, *Bryant* J52 (NBG, PRE).

Hiern distinguished *S. burchellii* (in syn. above) from *S. griquensis* by its smaller flowers. The type, *Burchell* 2622, came from Klipfontein, north of present-day Postmasburg, where Miss Wilman (cited above) collected it again over a hundred years later. The two specimens are identical: the flowers are abnormally small for *S. griquensis* (corolla tube c.6.6mm long, limb 4.3mm across), the tube is not richly coloured, and there are scarcely any clavate hairs inside the corolla tube, all indicative of selfing rather than outcrossing, and this appears to be the case, as the stigma is smothered in pollen from the contiguous anthers.

Most collections of *S. griquensis* have come from the Langeberge and Asbestos Hills in Griqualand West, but it is also on the high ground north of Prieska and there is one old record from near Zeerust, roughly 250km to the NE of the Asbestos Hills, one from near Zwartruggens, and another from Rustenburg, about 120km almost due east of Zeerust. It probably always grows at altitudes above c.1300m, and favours the shelter of cliffs and rock outcrops, as does its ally, *S. cooperi*, away to the east along the mountainous border between the Orange Free State and Lesotho. From the material available, it is difficult to judge its flowering season; possibly it is any time after rain, as old twigs with decaying capsules may have fresh flowering growth above them.

3. Sutera foetida Roth, Bot. Bemerk. 172 (1807); Hiern in Fl. Cap. 4(2): 275 (1904).
Type: a garden plant, *Roth* (lectotype M).

Syn.: *Buchnera foetida* Andr., Bot. Rep. 2 t. 80 (1800); Spreng. in Schrad., Journ. 2: 196 (1801); Jacq., Hort. Schoenbr. 4: 23, t. 448 (1804). Type: the plate in Bot. Rep., cited above.
Manulea alternifolia Pers., Syn. 2: 148 (1806). No specimen found.
M. foetida (Andr.) Pers., Syn. 2: 148 (1806).
M. aequipetala Sternberg in Bot. Zeit., Regensburg 6: 340 (1807). Type: a garden plant; there is an old specimen in W ex hort. Gerh.
Palmstruckia foetida (Andr.) Retz., Obs. Bot. Pugill. 15 (1810).
Chaenostoma foetidum (Andr.) Benth. in Hook., Comp. Bot. Mag. 1: 377 (1836) & in DC., Prodr. 10: 357 (1846).

Annual herb, foetid, root simple, main stem erect, c.150–600mm tall, often flexuous, branching from the base and higher, branches decumbent or divaricate, minutely glandular particularly on the upper parts, a few scattered glistening glands as well, very coarse scattered hairs present or not, usually present on lower parts, up to 0.5–1mm long, glandular tip minute, stalk cells relatively enormous, leafy throughout. *Leaves* opposite below becoming alternate upwards, blade c.10–45 × 5–25mm, broadly lanceolate to ovate, apex acute, base cuneate tapering into a petiolar part c.7–45mm long (about equalling to half as long as blade), margins coarsely and irregularly serrate, both surfaces minutely glandular when young, often ± glabrescent except for a few coarse hairs mainly on margins and petiolar part. *Flowers* in 1–11-flowered irregular cymules racemosely arranged in the upper leaf axils, often loosely panicled. *Bracts* (as opposed to leaves) wanting or present and then c.3–4 × 0.2–0.4mm, linear. *Peduncles* up to 40mm long, usually short, pedicels up to c.10mm long, often recaulescent. *Calyx* ± bilabiate, tube 0.5–1mm long, lobes 2.5–4 × 0.6–0.8mm, linear-lanceolate, glabrous except for minute glands on lower margins. *Corolla* tube 10–11.5 × 1.8–2.3mm across the mouth, narrowly funnel-shaped, limb almost regular, 7–8mm across, posticous lobes 2.3–3 × 1.6–2.4mm, anticous lobe 2.1–3 × 1.7–2.2mm, all lobes oblong-elliptic, white to pale pink or pale violet, tube yellow or orange, glabrous outside, bearded with clavate hairs all round mouth, these rarely wanting, glistening glands also present. *Stamens* 4, anticous anthers 0.5–1.1mm long, very shortly exserted, filaments c.1–2mm long, posticous anthers 0.8–1.5mm long, included, filaments c.1mm, glabrous. *Stigma* 1.5–2mm long, just included or shortly exserted. *Style* 6–10mm long. *Ovary* 1.5–2.5 × 1–1.5mm. *Capsules* 5–8 × 3–4mm, glabrous except for a few glistening glands on the sutures. *Seeds* 0.4–0.75 × 0.25–0.4mm, amber-coloured.

Selected citations:

Cape. Namaqualand, 3017BB, Bowesdorp, south slopes of Sneeuwkop, viii 1929, *Pillans* 7020 (BOL). Clanwilliam div., 3219AA, N Cedarberg, Pakhuis, 3000ft, 7 ix 1953, *Esterhuysen* 21746 (BOL, PRE). Piquetberg div., 3218DB, Piquenierskloof, 360m, 22 viii 1894, *Schlechter* 4953 (BOL, E, K, PRE, S, STE, Z). Worcester div., 3319AD, near Mitchell's Pass, Mostertsberg, 1800ft, xi 1897, *Bolus* 5214 (BOL, Z). Montagu div., 3320DC, Barrydale, Joubert's farm Weltevreden, 9 ix 1979, *Batten* 496 (E). Oudtshoorn div., 3322AC, Cango Caves, x 1932, *Barker* 95 (BOL, K); 3322BC, Meirings Poort, 16 x 1955, *Esterhuysen* 24887 (BOL). Uniondale div., 3323CA, Avontuur Poort, 2600ft, xii 1928, *Fourcade* 4275 (BOL, STE). Bredasdorp div., 3420AA, 15km ESE of Stormsvlei, Witkop, 300m, 1 x 1983, *Burgers* 3183 (STE).

The name *Sutera foetida* is attributable to Roth alone; he stated clearly that he had not seen Andrews's plant and did not know if his was the same as *Buchnera foetida* or not. The specimen in M, quoted above as a lectotype, bears a label in Roth's hand: *'Büchneriae foetidae* nomine semina accepi, ad neque ad Buchnera neque ad Erinum referri potest ob corollam aequalem. An eadem cum illa in Andrews Bot. Repersory [sic] Vol. 2 Tab 81?' The species has long been in cultivation in botanical gardens in Europe and was described several times from garden material. It is a distinctive plant, not easily confused with any other. The glistening glands around the mouth of the corolla tube are remarkable; they are usually mixed with clavate hairs.

It is widely distributed in the Cape, from about Kamieskroon in Namaqualand south to Malmesbury and Worcester and east to the Great Swartberg and Kougaberg. The plants favour damp shady spots, often in the shelter of rocks, mostly c.300–900m above sea level, but down to c.75m on the Potteberg in Bredasdorp division, flowering between July and December.

Section Chaenostoma

Section Chaenostoma (Benth.)Hilliard, comb. nov.
Lectotype: *C. aethiopicum* (L.) Benth., typ. cons.
Syn.: *Chaenostoma* Benth. in Hook., Comp. Bot. Mag. 1: 374 (1836), nom. cons.

4. Sutera floribunda (Benth.) O. Kuntze, Rev. Gen. Pl. 2: 467 (1891); Hiern in Fl. Cap. 4(2): 277 (1904) p.p.; Philcox in Fl. Zam. 8(2): 32 (1990), p.p., excl. most of the specimens cited. Type: Natal, Port Natal [Durban] under 500ft, 1 iv 1832, *Drège* (holo. K; iso. E, LE, MO, P, S, W).

Syn.: *Chaenostoma floribundum* Benth. in Hook., Comp. Bot. Mag. 1: 376 (1836) & in DC., Prodr. 10: 356 (1846).
  C. natalense [Bernh. ex] Krauss in Flora 27: 835 (1844); Benth. in DC., Prodr. 10: 357 (1846). Type: in sylvis circa Natalbai, Aug., *Krauss* 119 (holo. MO).
  Sutera natalensis (Krauss) O. Kuntze, Rev. Gen. Pl. 2: 467 (1891); Hiern in Fl. Cap. 4(2): 310 (1904).
  Manulea floribunda (Benth.) O. Kuntze, Rev. Gen. Pl. 3(2): 235 (1898).
  Manulea thyrsiflora var. *versicolor* O. Kuntze, Rev. Gen. Pl. 3(2): 236 (1898). Type: Natal, Clairmont, 50ft, 10 iii 1894, *Kuntze* (NY).
  Sutera arcuata Hiern in Fl. Cap. 4(2): 266 (1904). Lectotype (chosen here): Transvaal, Steenkampsbergen, Klip Spruit, *Nelson* 385 (K; isolecto. PRE).
  S. compta Hiern in Fl. Cap. 4(2): 268 (1904). Type: Natal, hills above Greytown, 4000–5000ft, *Wood* 4333 (holo. K, iso. BOL).

S. macleana Hiern in Fl. Cap. 4(2): 276 (1904). Type: Transvaal, on mountain declivities near Pilgrim's Rest Gold Fields, Drakensberg, 1874, *McLea* in herb. Bolus 3032 (holo. K; iso. BOL, SAM). See note below.

S. noodsbergensis Hiern in Fl. Cap. 4(2): 265 (1904). Type: Natal, Noodsberg, 2500ft, *Wood* 105 (holo. K).

S. pulchra Norlindh in Bot. Not. 1951: 107 (1951). Type: Zimbabwe, Inyanga distr., Mt Inyangani, c.2450m, 14 ii 1931, *Norlindh & Weimarck* 4994 (holo. S; iso. BOL, K, M, PRE).

Perennial herb up to 1m tall, main stems up to 3–4mm in diam., sometimes simple below, usually branched or several-stemmed from the base, branching above, erect, decumbent or straggling, and then sometimes rooting at the nodes, pubescent, hairs up to 0.3–1mm long, either all gland-tipped or mixed with eglandular hairs, leafy. *Leaves* opposite (bracts occasionally alternate on ultimate twiglets), largest blades 10–40 × 10–30mm, ovate or occasionally elliptic abruptly contracted into a petiolar part c.4–9mm long (proportion of blade to petiole c.3–9:1), apex acute to obtuse, margins with c.5–16 pairs of teeth, both surfaces glandular pubescent as stems, glistening glands as well. *Flowers* ultimately innumerable, arranged in large panicles, occasionally the lowermost flowers solitary in the axil of a leaf, mostly in 3-flowered cymules, or mixtures of racemes and cymes, the main branches of the panicle always opposite, the flowers on the ultimate twiglets sometimes alternate and subtended by small bracts rather than leaves. *Pedicels* (longest) c.4–10mm. *Calyx* obscurely bilabiate, tube 0.7–2.5mm long, posticous lobes 1.5–3.5 × 0.4–1mm, anticous lobes slightly broader, glandular-puberulous together with longer hairs up to 0.3–0.8mm long. *Corolla* tube 6–11 × 2.2–3.8mm in throat, funnel-shaped, limb nearly regular, 8–12mm across, posticous lobes 2.8–5 × 1.7–3.2mm, anticous lobe 2.5–5 × 1.5–3mm, all elliptic, white or creamy-white to shades of mauve, tube yellow, glandular-puberulous outside, hairs up to 0.15–0.3mm long, glistening glands as well, inside either bearded with clavate hairs or subglabrous to glabrous. *Stamens* 4, anticous pair exserted, filaments 2.2–3.5mm long, posticous pair included, filaments 1–2mm long, anthers 0.8–1.2mm long. *Stigma* 0.5–2mm long, exserted. *Style* 5–9mm long. *Ovary* 1.2–1.8 × 0.6–1mm, glistening glands either confined to sutures or sometimes a few on top of ovary and on base of style. *Capsules* 2.6–5 × 1.5–3mm. *Seeds* 0.5–1 × 0.25–0.5mm, pallid.

Selected citations:

Zimbabwe. Nyanga distr., ± 5km east of Mt Chera, 6100ft, 16 ii 1974, *Burrows* 343 (NBG, SRGH). Stapleford, 22 ii 1946, *Wild* 845 (K, SRGH).
Transvaal. Messina distr., 2230BD, Tate Vondo, *Netshungani* 1220 (J). Louis Trichardt distr., 2329BB, Zoutpansberg, 3100ft, vi 1918, *Rogers* 21144 (SAM, Z). Letaba distr., 2330CA, Piesangskop, 3750ft, 7 iv 1960, *Scheepers* 939 (M, PRE, SRGH, Z). Waterberg distr., 2428AD, W extremity Waterberg, farm Groothoek 1246, 6000ft, 7 iv 1948, *Codd* 3960 (K, PRE). Pilgrim's Rest distr., 2430DD, Black Hill, 6700ft, 1 iii 1937, *Galpin* 14306 (BOL, PRE, SRGH). Barberton distr., 2530DB, 7km from Kaapsche Hoop towards Ngodwana, 23 v 1975, *van Jaarsveld* 497 (K, NBG, PRE).
Swaziland. Mbabane distr., 2631AC, Ukutula, 10 ii 1955, *Compton* 24901 (NBG, PRE). Hlatikulu, 2631CD, ii 1911, *Stewart* 94 (K, PRE, SAM).
Orange Free State. Harrismith distr., 2829AC, Platberg, 2300m, 2 v 1974, *Jacobsz* 2504 (NBG).
Natal. Ngotshe distr., 2731CD, Ngome, c.4000ft, 31 iii 1977, *Hilliard & Burtt* 9851 (E). Bergville distr., 2929AB, Injasuti area, 5000ft, vii 1952, *Esterhuysen* 20249 (BOL, NBG, PRE). Durban distr., 2931CC, Durban, 26 vi 1893, *Schlechter* 2834 (BOL, E, J, PRE, S, STE, Z). Lions River distr., 2930AC, Curry's Post, Methley's Bush, 27 iv 1969, *Hilliard* 4850 (E, SRGH). Underberg distr., 2929CB, Sani Pass, c.8600ft, 23 iii 1977, *Hilliard & Burtt* 9796 (E, S). Alexandra distr. [Umzinto distr.], 3030BC, Ifafa, c.500m, 31 v 1908, *Rudatis* 395 (E, PRE, S, STE, W, Z).
Transkei. Umzimkulu distr., 3029BD, Clydesdale, 2500ft, xii 1884, *Tyson* 2111 (BOL, NBG, Z). Engcobo distr., 3127DB, Engcobo Mountain, 3000ft, 15 i 1896, *Flanagan* 2788 (NBG, PRE).
Cape. Maclear distr., 3128AA, Ugie, Pomona, ii 1928, *Gill* 157 (STE).

Sutera floribunda is the mostly widely distributed species, ranging as it does over 15° of latitude, from eastern Zimbabwe (records along the eastern escarpment and from Bukwa Mountain in the south) over the high ground of the Transvaal and Swaziland (c.1200–2100m), through Natal (sea level to c.2600m) to the Transkei, and Ugie in that part of the eastern Cape lying between southern Lesotho and Transkei. The plants grow among boulders in grassland, or along streambanks and on forest margins, flowering mainly between January and May, though there are records from all months.

Typically, the plant has more or less ovate shortly petiolate leaves with shortly toothed margins and big panicles of flowers mostly arranged in cymes or cymose-racemes. The main inflorescence branches are nearly always opposite; they rarely become alternate towards the tips. Naturally, plants just coming into flower look very different from plants that have had time to open innumerable flowers, and specimens straggling up through other vegetation or from otherwise shaded sites may never develop a big panicle. Variation in leaf-toothing, petiole-length, and length of indumentum may have a genetic basis, but there is no strict geographical patterning to this variation and plants from Zimbabwe or the Transvaal can be matched precisely with plants from Natal and Transkei.

A critical character in distinguishing *S. floribunda* from its close allies is the inflorescence: most flowers are arranged in cymes, racemes, or cymose-racemes, though the lowermost flowers may be solitary in the axil of a leaf. Mis-observation of this character has contributed to the synonymy: Norlindh sought to distinguish *S. pulchra* from *S. floribunda* on flowers solitary in the leaf-axil and the corolla limb mauve; however, the flowers in the holotype as well as in the isotypes seen by me are in cymes, and although the corolla appears always to be mauve in Zimbabwe, mauve-flowered specimens are as common as white or parti-coloured ones over the rest of the geographical range. Similarly, Hiern (Fl. Cap. 4(2): 245) keyed out *S. noodsbergensis* and *S. arcuata* under 'Flowers axillary or arranged in simple terminal racemes or open panicles', but the type specimens all have the usual panicle with a mixture of some solitary flowers but mostly axillary racemes and some cymes. The type of the name *S. macleana* is *McLea* 43 distributed as no. 3032 in herb. Bolus. Hiern's description of *S. macleana* is clearly based on the specimen in the Kew herbarium; there is a duplicate in BOL (bearing McLea's original ticket) but a second sheet under the same Bolus number is *Selago elata* Rolfe, nom. illegit. This sheet also bears McLea's original label; it is *McLea* 148 from MacMac. A sheet in SAM is a mixture of *S. macleana* and *Selago elata*.

In Zimbabwe, *S. floribunda* is most likely to be confused with *S. septentrionalis*: see there. South of the Limpopo, confusion is likely with *S. polelensis*, which is distinguished by differences in inflorescence as well as in foliage (see under that species). In the high Drakensberg on the Natal-Lesotho border, *S. polelensis* and *S. floribunda* are often sympatric. Hilliard & Burtt (Botany of the southern Natal Drakensberg p.199) listed *S. floribunda* as *Sutera* aff. *polelensis*. However, now that the genus has been revised and a great deal of material examined it is clear that a broad species concept must be adopted for *S. floribunda*.

5. Sutera polelensis Hiern in Fl. Cap. 4(2): 252 (1904).
Lectotype (chosen here): Natal, Drakensberg Range, 2929CC, Polela, 6000–7000ft, vii 1895, *Evans* 518 (K).

Two subspecies are recognized:

1a. Leaf blade usually about as long as broad, or broader than long; glistening glands (most easily seen on backs of corolla lobes under a good dissecting microscope) ± round:
subsp. **polelensis**
1b. Leaf blade usually longer than broad; glistening glands decidedly oblong:　　subsp. **fraterna**

subsp. **polelensis**

Perennial herb well branched from the base and above, main stems up to c.600mm long, up to c.3–5mm in diam., erect, decumbent or sprawling, glandular-pubescent, hairs rarely only c.0.2mm long, usually up to 0.4–2mm, leafy. *Leaves* opposite, becoming alternate only near the tips of the inflorescence branches, these often flexuous, largest blades 10–35 × 11–40mm, rhomboid or ovate abruptly contracted into a petiolar part 1–5mm long, or virtually sessile (proportion of blade to petiole (3–)7–35:1), apex acute to obtuse, margins with mostly 4–10 pairs of coarse irregular teeth or lobes, up to 2–8mm measured along top edge, often denticulate, both surfaces glandular-pubescent as stems, glistening glands as well. *Flowers* ultimately innumerable, mostly solitary, occasionally paired, rarely a 3-flowered cymule, in the axils of the upper leaves, passing into bracteate racemes, sometimes loosely panicled. *Pedicels* (longest) c.10–23mm. *Calyx* obscurely bilabiate, tube 1–2mm long, posticous lobes 2.7–6 × 0.6–1.3mm, anticous lobes scarcely broader, glandular-pubescent, hairs (0.2–)0.5–2mm long. *Corolla* tube 7.5–12 × 3–4mm in throat, funnel-shaped, limb nearly regular, 8–13mm across, posticous lobes 3–4.8 × 1.5–3.5mm, anticous lobe 3–4.8 × 2–4mm, all elliptic, usually white, occasionally palest mauve, tube yellow/orange, glandular-puberulous outside, hairs up to 0.25–0.6mm long, ± round glistening glands as well, inside either bearded with clavate hairs or sometimes glabrous. *Stamens* 4, anticous pair exserted, filaments 2–3mm long, posticous pair included, filaments 1.3–1.6mm long, anthers 1–1.5mm long. *Stigma* 1–2.5mm long, exserted. *Style* 5.5–9.5mm. *Ovary* 1.2–2 × 0.7–1.1mm, usually only a few glistening glands on sutures and occasionally on base of style, rarely a broad band on sutures. *Capsule* c.4–5 × 2.8–3mm. *Seeds* c.0.8 × 0.4–0.6mm, pallid.

Selected citations:

Orange Free State. Bethlehem distr., 2828DA, Golden Gate National Park, ± 2300m, i 1963, *Liebenberg* 6809 (PRE, Z). Harrismith distr., 2829AC, Platberg, One Man Pass, 3 xii 1976, *Hilliard & Burtt* 9514 (E).
Lesotho. Leribe, 2828CC, 5–6000ft, *Dieterlen* 325 (K, NBG, P, PRE, SAM). 2927BC, Roma to Mafeteng, 5800ft, 16 ii 1976, *Schmitz* 6512 (PRE).
Natal. Bergville distr., 2829CC, Mnweni area, 6000ft, vii 1949, *Esterhuysen* 15545 (BOL, PRE). Underberg distr., 2929CB, Gxalingenwa valley, 6400ft, 10 xii 1983, *Hilliard & Burtt* 17150 (E, K, PRE).
Cape. [Lady Grey div.] Albert, top of the Wittebergen, 1861, *Cooper* 2872 (K, Z); ibidem, 3027DA, farm Beddgelert, c.6200ft, 1 xii 1981, *Hilliard & Burtt* 14575 (E). Maclear distr., 3028CD, summit Pot River Berg, c.6400ft, 20 iii 1904, *Galpin* 6802 (BOL, K, PRE, SAM).

Sutera polelensis subsp. *polelensis* is endemic to the Eastern Mountain Region; records are very nearly confined to Cave Sandstone (Clarens Formation) outcrops from Harrismith and Bethlehem in the Orange Free State along the Orange Free State–Lesotho border to the Witteberg and Cape Drakensberg around Lady Grey, Barkly East and Maclear, then along the high Drakensberg in Natal from Royal Natal National Park to Bushman's Nek and neighbouring Sehlabathebe in Lesotho. It will surely also be in the Drakensberg on the Lesotho–Transkei border. The plants grow under damp overhangs and in rock tumbles between c.1525 and 2300m above sea level; flowering has been recorded in all months. Drège was the first to collect the species, in January 1833, on the Witteberg near Lady Grey, but the specimen was misdetermined as *Chaenostoma pauciflorum*.

Sutera polelensis is most frequently confused with *S. floribunda*, but may be distinguished by its inflorescence: the flowers are nearly always solitary in the leaf-axils and the lowermost pedicels are mostly longer; also, the petioles are mostly shorter, in absolute terms as well as in the proportion of blade to petiole.

subsp. **fraterna** Hilliard in Edinb. J. Bot. 48: 344 (1991).
Type: Swaziland, between Mbabane and Forbes Reef, c.5000ft, xii 1905, *Bolus* 12183 (holo. BOL, iso. K).

Perennial herb well branched from the base and above, main stems up to c.450mm long, up to c.3–4mm in diam., erect or decumbent, glandular-pubescent, hairs c.0.5–1mm long, leafy. *Leaves* opposite, becoming alternate only near the tips of the inflorescence branches, these often flexuous, largest blades 13–35 × 9–26mm, elliptic or ovate, cuneately contracted into a petiolar part 0–6mm long (proportion of blade to petiole (4–)7–25:1), apex usually acute, sometimes obtuse, margins coarsely and irregularly toothed, largest teeth c.1.5–3.5mm along top edge, denticulate or not, both surfaces glandular-pubescent as stems, glistening glands as well. *Flowers* ultimately innumerable, solitary in the leaf-axils, passing into bracteate racemes, loosely panicled. *Pedicels* (longest) 6–30mm. *Calyx* obscurely bilabiate, tube c.1mm long, posticous lobes 3–5 × 0.7–1mm, anticous lobes scarcely broader, glandular-pubescent, hairs up to c.0.4–0.8mm long. *Corolla* tube 6.5–9.5 × 2.5–3.6mm in throat, funnel-shaped, limb nearly regular, c.6–11mm across, posticous lobes 2.1–4.5 × 1.6–3.5mm, anticous lobe 2.4–4.2 × 1.7–3.6mm, all elliptic, usually white though recorded as 'yellowish', tube yellow/orange, glandular-puberulous outside, hairs up to 0.25–0.4mm long, oblong glistening glands as well, well bearded inside, hairs clavate. *Stamens* 4, anticous pair exserted, filaments 2–3mm long, posticous pair included, filaments 0.8–2mm long, anthers 0.8–1.5mm long. *Stigma* 0.5–2mm, exserted. *Style* 5.5–8mm, occasional glistening glands. *Ovary* 1.4–1.7 × 0.8–1mm, broad bands of glistening glands on sutures, also scattered on top of ovary and up style. *Capsules* c.3–4.5 × 2.5–3mm. *Seeds* c.0.8 × 0.4mm, pallid.

Selected citations:

Transvaal. Pilgrim's Rest distr., 2430DD, Graskop Peak, 5900ft, 7 iii 1937, *Galpin* 14356 (BOL, K, PRE). 2530CA, Belfast, 6500ft, xii 1912, *Bolus* 12182 (BOL, K, PRE). 2630AC, near Lake Chrissie, Blaaupan A, 27 ii 1928, *Moss* 16410 (E, J).
Swaziland. Mbabane distr., Miller's Falls, c.4500ft, 6 ii 1957, *Compton* 26552 (NBG). Mukusini Hills, c.4000ft, 6 iii 1964, *Karsten* s.n. (NBG, PRE).

Sutera polelensis subsp. *fraterna* is confined to the south eastern Transvaal and neighbouring parts of Swaziland, from about Graskop, Dullstroom and Belfast south to Lake Chrissie and east to Kaapsche Hoop, Mbabane and Forbes Reef. It grows among rock outcrops, between c.1050 and 2000m above sea level, flowering between December and March.

The two subspecies appear to occupy discrete areas, no specimen of either having been recorded between Lake Chrissie and Harrismith. The leaves of subsp. *fraterna* are often decidedly longer than broad, in contrast to leaves roughly as long as broad in subsp. *polelensis*, but there are exceptions. The difference in the shapes of the glistening glands on the two entities is remarkably reliable, but a minimum magnification of ×25 and good lighting is needed to see the glands successfully (round in subsp. *polelensis*, oblong in subsp. *fraterna*). They are present on the leaves and calyx as well as on the corolla, but on the corolla (backs of the lobes in particular) they are less obscured by other hairs. Also, there are far more glistening glands on the sutures and top of the ovary in subsp. *fraterna* than in most specimens of subsp. *polelensis*.

Subsp. *fraterna* has been confused mainly with *S. floribunda*, which differs in its complex inflorescence. They are sympatric over the whole area of subsp. *fraterna*: comparison of two specimens from nearby localities will show how strikingly different they look.

6. Sutera racemosa (Benth.) O. Kuntze, Rev. Gen. Pl. 2: 467 (1891); Hiern in Fl. Cap. 4(2): 276 (1904), excl. *Flanagan* 892.
Type: Cape, Zuurebergen, between Hoffman's Kloof (3326AC) and Driefontein (3325AC), precisely localized as Rietfontein on sheet in P, 2000', 18 xi 1829, *Drège* (holo. K; iso. E, LE, MO, P, S, W).

Syn.: *Chaenostoma racemosum* Benth. in Hook., Comp. Bot. Mag. 1: 377 (1836) & in DC., Prodr. 10: 357 (1846), non Diels (1897).

Herbaceous, possibly annual, sparingly branched from base, stems c.200–350mm long, terete, weak, glandular-puberulous, the hairs longest on lower part of stem (up to c.0.4–0.6mm), distantly leafy. *Leaves* opposite, largest c.20–30 × 10–20mm, ovate, apex obtuse to subacute, base cuneate and tapering into a short petiolar part (c.5–8mm) then abruptly expanded and connate, margins with 5–6 pairs of obscure to coarse and irregular teeth, both surfaces glandular-puberulous, some longer hairs particularly on lower margins, these up to 0.6mm. *Flowers* in terminal racemes sometimes with 1 or 2 branches at base, the lowermost flowers occasionally in the axils of the uppermost leaves, usually opposite, rarely some ternate. *Bracts* (lowermost) c.4–7 × 2–5mm, leaf-like (largest leaf subtending a flower c.21 × 14mm). *Pedicels* (lowermost) 12–30mm long, glandular-puberulous. *Calyx* bilabiate, tube c.1mm long, posticous lobes c.4 × 0.8mm, anticous lobes slightly broader, glandular-puberulous, occasional longer hairs c.0.2–0.25mm. *Corolla* tube c.8–9.5 × 3mm in throat, narrowly funnel-shaped, limb c.8.5mm across, posticous lobes c.2.5–2.8 × 2mm, anticous lobe c.2.5–3 × 2mm, all oblong-elliptic, possibly white, tube yellow, glandular-puberulous outside, hairs up to c.0.15–2mm, glabrous inside. *Stamens* 4, anticous pair exserted, filaments 2–2.3mm long, posticous pair included, filaments 1–1.5mm long, anthers c.1.1–1.2mm long. *Stigma* 1mm long, exserted. *Style* c.5.5–6.2mm long. *Ovary* c.1.5–1.8 × 1mm, a few glistening glands on sutures. *Capsules* 4.5–6 × 3mm. *Seeds* c.0.7 × 0.4mm, pallid.

Sutera racemosa is known only from the type collection made by Drège on the south face of the Zuurberg north of Port Elizabeth. It is difficult to relate Drège's route to a modern map, but the plant can probably be found on the Zuurberg between Enon and Alicedale. Hereabouts, the Zuurberg is traversed by three passes, and there are other, secondary, roads giving access to the range. When Drège collected it in November, the plants were already well on into fruit. The stems look so weak that the plants probably grow in the shelter of rocks, partly shaded. An isotype in P has a thin taproot, and two branches from the base of the primary stem, indicating that the plant is either annual or at least flowers in the first year of growth.

7. Sutera septentrionalis Hilliard in Edinb. J. Bot. 48: 345 (1991).
Type: Zimbabwe, Marandellas [Marondera] distr., ± 12 miles S of Marandellas [Marondera] on Wedza road, 22 v 1968, *Mavi* 761 (holo. K).

Perennial herb, branched from the base, stems erect or procumbent, up to c.200–700mm long, c.1–3mm in diam. at base, initially simple, soon laxly branched, branching often alternate in upper parts, branchlets often flexuous, glandular-pubescent, hairs up to 0.25–0.4mm long, leafy. *Leaves* opposite becoming alternate upwards, largest blades c.8–19 × 5–14mm, elliptic or ovate abruptly contracted into a petiolar part c.3–7mm long (proportion of blade to petiole c.2–4:1), apex ± acute, margins with 4–5 pairs of teeth (largest teeth up to c.2mm measured along top edge), both surfaces glandular-pubescent, hairs up to c.0.2–0.3mm long, a few glistening glands as well. *Flowers* solitary in axils of upper leaves, panicled in well-grown specimens. *Pedicels* (longest) c.8–18mm. *Calyx* bilabiate, tube 1–1.7mm long, posticous lobes 2.5–3.7 × 0.5–0.7mm, anticous lobes a little broader, glandular-pubescent, hairs up to 0.2–0.25mm long. *Corolla* tube 7–9 × 1.5–2.5mm in throat, funnel-shaped, limb nearly regular, 7–8mm across, posticous lobes 2.3–3 × 1.6–2mm, anticous lobe 2.2–3 × 1.5–2mm, all elliptic, white or mauve, throat yellow, glandular-puberulous outside, hairs up to 0.15–0.2mm long, scattered glistening glands as well, lightly bearded inside. *Stamens* 4, anticous pair exserted, filaments 1.6–2.3mm long, posticous pair included, filaments c.1mm long, anthers 0.8–1mm long. *Stigma* 0.5–1.6mm long, exserted. *Style* 4.8–6.5mm long. *Ovary* 1.1–1.2 × 0.6–1mm, a few glistening glands on sutures. *Capsules* 2–4 × 1.5–2mm. *Seeds* c.0.8 × 0.4mm, pallid.

Selected citations:

Zimbabwe. Makoni distr., Rusapi, *Eyles* 7563 (K, SRGH); Makoni Reserve, Timarm, 5000ft, v 1961, *Plowes* 2181 (SRGH). Mazowe distr., Iron Mask summit, 5100ft, v 1907, *Eyles* 549 (K, SRGH, Z).

Sutera septentrionalis is so named because it is the most northerly *Sutera*, known from a few sites in Zimbabwe, on Iron Crown Mountain and near Marondera and Rusape, in the north east of the country. It grows around large granite boulders or on cliffs; flowering collections have been made in April and May, when fruits are often present as well.

Sutera septentrionalis is most often confused with *S. floribunda*, but in that species the leaves are always opposite, while in *S. septentrionalis* they, and of course the branches, often become alternate upwards, and the flowers are seldom solitary in the leaf axils (always so in *S. septentrionalis*). The characters of alternate leaves and solitary flowers are shared with *S. debilis*, to which *S. septentrionalis* seems to be very closely allied; it differs in its less strongly toothed leaves (upper margin of largest teeth up to c.2mm versus 3–5mm), in its inflorescence (panicles in well grown specimens, not simple racemes), and in the corolla tube, lightly bearded within, not glabrous. Field investigation and further collecting might show that *S. septentrionalis* is no more than a form of *S. debilis*, but the two entities appear to be wholly allopatric.

8. Sutera debilis Hutch., Botanist in S Africa 353 (1946).
Type: Transvaal, 2329BC, Matoks, 24 viii 1930, *Hutchinson & Gillett* 4464 (holo. K).

Syn.: *S. hereroensis* auct., non Skan; Philcox in Fl. Zam. 8(2): 27 (1990), excl. *Greatrex* 13245.

Herb, either annual or, presumably, a short-lived perennial, stems simple to loosely branched from the base and above, c.15–450mm long, up to c.4mm diam. at base, semi-erect or decumbent, often flexuous in upper part, glandular-pubescent, hairs up to c.0.15–1.5mm long, leafy. *Leaves* opposite becoming alternate upwards, largest blades c.4–33 × 3–25mm, rhomboid to broadly ovate abruptly contracted into a petiolar part c.3–11mm long (proportion of blade to petiole c.1–3:1), apex acute, margins with 3–6 pairs of coarse teeth or lobes themselves sometimes sparingly toothed (largest teeth c.3–5mm measured along the top edge), both surfaces glandular-pubescent, hairs as stems, scattered glistening glands as well. *Flowers* solitary in axils of upper leaves. *Pedicels* (longest) c.6–23mm. *Calyx* bilabiate, tube 0.7–1.2mm long, posticous lobes 2–5 × 0.5–1mm, anticous lobes a little broader, glandular-pubescent, hairs up to 0.1–1.4mm long. *Corolla* tube 6–11 × 2.2–3.5mm in throat, funnel-shaped, limb nearly regular, 6–10mm across (or smaller if flowers are selfing), posticous lobes 2–4 × 1.5–2.6mm, anticous lobe 1.8–4 × 1.5–2.5mm, all elliptic, white or pale pink or mauve, throat yellow, glandular-puberulous outside, hairs up to 0.2–0.5mm long, scattered glistening glands as well, glabrous within. *Stamens* 4, anticous pair exserted, filaments 1.7–3.2mm long, posticous pair included, filaments 0.8–1.7mm long, anthers 0.5–1.1mm long. *Stigma* 1–2.5mm long, usually exserted. *Style* 3.5–9mm long. *Ovary* 1.1–1.7 × 0.7–1mm, a few glistening glands on sutures and sometimes scattered up style. *Capsules* 2.5–5 × 2–5mm. *Seeds* c.0.5–0.8 × 0.3–0.4mm, pallid.

Selected citations:

Zimbabwe. Matobo distr., farm Chesterfield, 4800ft, iii 1958, *Miller* 5172 (K, SRGH). Victoria distr. (Masvingo distr.), 10 miles N of Fort Victoria on Chilimanzi road, 4 v 1962, *Drummond* 7940 (K, PRE, SRGH). Chipinga distr., Vermont Farm, 17 v 1974, *Cannell* 605 (K, SRGH).
Transvaal. Pietersburg distr., 2329CD, campus U.C.N. 18 miles E of Pietersburg, road to Tzaneen, 14 iii 1962, *van Vuuren* 1416 (K). Waterberg distr., 2428BA, Hanglip, 6 i 1952, *Maguire* 1417 (BOL, NBG); 2428BC, Bokpoort, 27 xii 1931, *Galpin* 11655 (BOL, K, PRE). Lebowa, 2429CD, Arabie, on northern slope of Motsephiri stream, 2760ft, 1 ii 1982, *Ellery* 386 (PRE).

Sutera debilis has been recorded from southern Zimbabwe and the northern Transvaal, between c.1450 and 1980m above sea level. In Zimbabwe, it appears to be confined to granite outcrops, where it grows deep in under the rock in moist shade; in the Transvaal, it has been recorded on shady cliff faces as well as under outcropping rocks, possibly not always granitic. Plants may be found in flower and fruit in any month, but most records are between November and May. Plants will flower when minute, and may then bear tiny flowers that probably self (e.g. *Miller* 5791 (SRGH) growing at the entrance to a cave in Matoba district) while *Drummond* 7940 (cited above, and found in cracks in a shaded granite outcrop) is mostly in fruit but bears traces of cleistogamous flowers.

   Though the plants vary in habit from tiny annuals to bushy tangled herbs with stems about 450mm long, and the hairs on stem, leaf and calyx vary greatly in length, the species is easily recognized by its stems, often flexuous above, ovate jaggedly toothed leaves, flowers always solitary in the axils of the leaves, and corolla tube glabrous inside. The species with which it is most likely to be confused is *S. septentrionalis*, which differs in its less deeply toothed leaves, flowers eventually borne in panicles and corolla tube lightly bearded within; see further under *S. septentrionalis*.

9. Sutera platysepala Hiern in Fl. Cap. 4(2): 249 (1904).
Type: Natal, Zululand [2831CD] Entumeni, 2000ft, 18 iv 1888, *Wood* 3982 (holo. K, iso. BOL).

Syn.: *S. pallescens* Hiern in Fl. Cap. 4(2): 280 (1904). Type: Natal, [Port Shepstone distr., 3030CA], near Murchison, 2 v 1883, *Wood* 3035 (holo. K, iso. BOL).
S. humifusa Hiern in Fl. Cap. 4(2): 281 (1904). Type: Natal, Noodsberg, v, *Wood* 124 (holo. K).

Perennial herb but will flower when young and then looks annual, loosely branched from the base and above, stems up to c.200–600mm long, up to 5mm in diam. at base, trailing, glandular-pubescent, hairs c.0.1–1.2mm long, leafy. *Leaves* opposite but often alternate upwards, largest blades c.15–30 × 10–25mm, ovate abruptly contracted into a petiolar part c.5–18mm long (roughly ⅓–⅔ as long as blade), apex acute to obtuse, margins with 3–5 pairs of coarse teeth or lobes themselves sometimes sparingly toothed, both surfaces glandular-pubescent, hairs as on stems. *Flowers* solitary in axils of upper leaves, in well-grown plants forming very long (c.450mm) inflorescences. *Pedicels* (longest) c.10–25mm. *Calyx* bilabiate, tube at anthesis 1–2.5mm long, about doubling in length in fruit, posticous lobes 3–5 × 1–2mm, anticous lobes slightly broader, glandular-pubescent, hairs up to 0.2–2mm long. *Corolla* tube 8–14.5 × 4–5mm in throat, funnel-shaped, limb nearly regular, 8–14mm across, posticous lobes 3–4.5 × 2–3mm, anticous lobe 3–6.5 × 2–4.5mm, all elliptic, usually white, one record 'purple', tube yellow/orange, sparsely glandular-puberulous outside, hairs 0.2–0.3mm long, scattered glistening glands as well, lightly bearded with clavate hairs inside. *Stamens* 4, anticous pair exserted, filaments 3–5mm long, posticous pair in throat, filaments c.2mm long, anthers 1–1.3mm long. *Stigma* c.0.2–0.5mm long, shortly bifid, exserted. *Style* 8–12.5mm long. *Ovary* 1.5–2.5 × 0.8–1.5mm, a few glistening glands on sutures and sometimes on base of style. *Capsules* 3.5–6.5 × 2.5–3mm. *Seeds* 0.5–0.75 × 0.3–0.5mm, pallid.

Selected citations:

Natal. Stanger distr., 2931AC, Upper Tongaat, 23 vii 1944, *Hillary* 44 (E). New Hanover distr., 2930DA, about 8km south of Wartburg, farm Windy Hill, 850m, 10 x 1989, *Balkwill & Balkwill* 4714 (E, J). Pinetown distr., 2930DD, Hillcrest, Umhlatuzane valley, c.2000ft, ix 1917, *Thode* sub STE 5078 (STE). Alexandra [Umzinto] distr., 3030BC, Umgaye, 19 vi 1910, *Rudatis* 1036 (BM, E, K, PRE, STE, W, Z). Port Shepstone distr., 3030CD, Excelsior Farm, above Vungu river, 1600ft, 21 xii 1965, *Hilliard & Burtt* 3381 (E).

Sutera platysepala is a well-marked species, easily recognized by its broad-lobed accrescent calyx and very short, minutely bifid stigma, not matched elsewhere in the genus (the stigma is papillose, not jelly-like as in the bifid stigma of *S. ccoperi* and *S. griquensis*). There is considerable variation in the length of the hairs on stem, leaves and calyx; this is possibly induced, at least in part, by habitat, as similar variation is also shown by *S. debilis*, which grows under outcropping rocks and in shallow caves.

Sutera platysepala appears to be endemic to the Table Mountain Sandstone rocks of coastal Natal, with a distribution range from Entumeni in the north to the Vungu (Uvongo) river in the south, from c.450–850m above sea level. The plants grow from crevices in cliffs at the upper margins of forest patches, often in river gorges, or under overhanging rocks on steep grassy slopes where the valleys are wider. Flowering has been recorded in most months. *Balkwill & Balkwill* 4714 (cited above) is of particular interest as the sheet in J carries a normal lax flowering branch as well as two twiggy branches with crowded flowers, some of which are cleistogamous, with the unopened bud capping the capsule. The only other species of *Sutera* in which cleistogamy has been detected is *S. debilis* (though flowers of *S. griquensis* may self). Low light intensity or limited periods of illumination may be factors inducing cleistogamy.

---

10. Sutera glabrata (Benth.) O. Kuntze, Rev. Gen. Pl. 2: 467 (1891); Hiern in Fl. Cap. 4(2): 271 (1904) p.p. excluding many of the specimens cited.
Lectotype (chosen here): Cape, Swellendam div., c.3320CC, Kannaland, *Ecklon* (K).

Syn.: *Chaenostoma glabratum* Benth. in Hook., Comp. Bot. Mag. 1: 375 (1836) & in DC., Prodr. 10: 355 (1846), p.p., as to lectotype only.

Coarse herb or shrublet, c.150–450mm tall, many-stemmed from the base, erect or decumbent, virgately branched, main branches c.2–4mm in diam., either terete or almost imperceptibly ridged by lines decurrent from the leaf bases and midrib, minutely glandular-puberulous, hairs up to 0.1–0.2mm long, leafy, axillary vegetative buds usually small and inconspicuous, or wanting. *Leaves* opposite, bases connate, largest c.10–35 × 0.8–3mm, linear to oblong, apex obtuse, margins strongly revolute to plane (these leaves the broadest), entire, indumentum as on stems. *Flowers* solitary in axils of upper leaves thus forming leafy racemes or sometimes branching into narrow panicles. *Bracts* (the upper leaves) c.3–18 × 0.5–1.8mm at base of inflorescence, smaller upwards. *Pedicels* (lowermost) 5–25mm long, stout. *Calyx* bilabiate, tube 1–2mm long, posticous lobes 3–4.7 × 0.4–0.8mm, anticous lobes a little broader, glandular-puberulous, some hairs up to 0.1–0.3mm long but coarse hairs wanting. *Corolla* tube 12.5–17 × 2.5–3.5mm in throat, funnel-shaped, limb almost regular, 10–15mm across, posticous lobes 3–7 × 2–4mm, anticous lobe 3–7 × 2–4.5mm, all oblong-elliptic, white to pink, lilac-mauve or reddish, tube yellow, glandular-puberulous, hairs up to 0.1–0.2mm long, glistening glands as well, 5 longitudinal bands of clavate hairs inside. *Stamens* 4, anticous pair exserted, posticous pair in throat, anthers 1–1.3mm long, posticous filaments 1.4–2.2mm long, anticous ones 2.2–3mm. *Stigma* 1.6–3.5mm long, exserted. *Style* 9.5–13mm long. *Ovary* 1.4–2 × 1mm, a few glistening glands on sutures or glabrous. *Capsules* 5–8 × 2.5–3mm. *Seeds* c.1 × 0.4mm (few seen), pallid.

Selected citations:

Cape. Worcester div., 3319BC, De Doorns, iv 1907, *Bolus* 13166 (BOL, PRE). 3319DB, Concordia east of Koo, c.900m, 10 vii 1980, *Hugo* 2412 (K, PRE, STE). 3319CC, Tafelberg, 3000ft, x 1921, *Pillans* sub BOL 14163 (BOL). Montagu div., 3320CA, Baden, ix 1946, *Lewis* 1943 (SAM). 3320CB, Dobbelaarskloof, 25 ix 1946, *Compton* 18445 (NBG). Swellendam div., 3320DC, 15 miles W of Barrydale, 31 x 1928, *Gillett* 1895 (PRE, STE).
Seen from the following quarter-degree squares: 3319BC, BD, CC, DA, DB; 3320CA, CB, CC, CD, DC.

Under his new name, *Chaenostoma glabratum*, Bentham cited four collections. That from the Kei river, collected by Ecklon, is not in Bentham's herbarium (K) and no duplicate has presented itself. The other three specimens, glued to one sheet, belong to three different species. The matter is further complicated by Bentham's having mistakenly marked the 'c' specimen as Ecklon's from Swellendam. This collection is Drège's from Kendo (it matches the sheet in P bearing Drège's own label); the top centre specimen marked 'b' is also part of Drège's collection from Kendo (Kendo's or Kendous Mountain). The left-hand specimen marked 'b' is Ecklon's collection from Swellendam division, and it has been chosen as the lectotype because it is so finely puberulous that it appears to be glabrous (*Drège* from Kendo has narrowly winged stems and is distinctly hispid; it is *S. uncinata*). The second Drège specimen ('a' from Zwanepoelspoortberg) is very finely puberulous and is *S. denudata*.

Sutera glabrata has been recorded on the southern slopes of Tafelberg (Worcester div.) east to Montagu and Barrydale and north to De Doorns, south of the Hex river mountains. Scant ecological information is available, but it seems to occur on slopes, sometimes on cliffs, and is possibly a constituent of scrub, between c.500 and 900m above sea level, flowering principally between July and December, one record in April.

The plant is characterized by its stiff virgate branching, obtuse linear leaves, lack of conspicuous brachyblasts, minute pubescence on all parts (no coarse hairs on calyx) and flowers borne in leaf-axils. When the inflorescence axis branches into a panicle, the branches are often only 30–50mm long, and, together with the stout pedicels, give the plant a peculiarly stiff, twiggy aspect.

It is almost wholly sympatric with *S. uncinata* in the eastern part of the range of that species. It can be distinguished by its terete or very indistinctly ribbed stems, lack of conspicuous brachyblasts, and the lack of coarse hairs on the calyx.

11. Sutera paniculata Hilliard in Edinb. J. Bot. 48: 344 (1991).
Type: Cape, Clanwilliam div., 3219AC, Brakfontein, 800ft, 1 viii 1896, *Schlechter* 7975 (holo. E; iso. BOL, K, MO, PRE, S, W, Z).

Syn.: *Chaenostoma revolutum* Benth. var. β *pubescens*, nomen, in Hook., Comp. Bot. Mag. 1: 375 (1836) p.p. Specimen cited: near Boschkloof (Zwischen Clanwilliam und Boschkloof in the Zwei Pflanzeng. Doc. and Boschkloof, 1000–1500', 26 xi 1828 on a specimen in P), *Drège* (E, K, LE, P, S, W).
Sutera revoluta var. β *pubescens* Hiern, nomen, in Fl. Cap. 4(2): 271 (1904), p.p.

Suffrutex up to c.750mm tall, spindly or bushy, branches terete, glandular-pubescent, hairs up to 0.3–1mm long, brachyblasts present, leafy at least on the upper parts. *Leaves* opposite or sometimes becoming alternate upwards, largest primary leaves 30–40 × 1–4.3mm, linear to narrowly elliptic, obtuse, base slightly narrowed then abruptly expanded and clasping, connate or not, margins usually strongly revolute, sometimes weakly so, both surfaces hairy as stems. *Flowers* solitary in the axils of the leaves forming long (often 150–300mm) narrow panicles. *Bracts* usually wanting, occasionally present in specimens from Namaqualand and then up to c.2.5mm long. *Pedicels* (lowermost) 2–9mm long. *Calyx* distinctly bilabiate, tube 0.7–1mm long, posticous lobes 1.8–2.6 × 0.3–0.8mm, anticous lobes slightly broader, glandular-pubescent with some hairs up to 0.25–0.4mm long. *Corolla* tube 6.5–9.5 × 3–4mm in throat, funnel-shaped, limb nearly regular, 8–11mm across, posticous lobes 2–4 × 2–3.1mm, anticous lobes 2.5–4 × 2.5–3.2mm, all elliptic-oblong, shades of pink or mauve, throat yellow/orange, glandular-pubescent outside, hairs up to 0.1–0.3mm long, inside bearded with 5 longitudinal bands of clavate hairs. *Stamens* 4, anthers 0.8–1.4mm long, anticous pair exserted, filaments 2–2.8mm long, posticous pair in throat, filaments 1.4–1.8mm long. *Stigma* 1.3–3mm long, exserted. *Style* 5.5–8mm. *Ovary* 1–1.5 × 0.6–1mm. *Capsules* c.3 × 2mm. *Seeds* c.0.8 × 0.4mm, pallid.

Selected citations:

Cape. Namaqualand, 3017BB, southern slopes of Sneeuwkop, 12 xii 1910, *Pearson & Pillans* 5880 (BOL). Van Rhynsdorp div., 3118DA, Gifberg, 11 v 1971, *van Breda* 4177 (PRE). 3118DD, top of Nardouwsberg, farm Vondeling, 11 vi 1980, *Snijman* 290 (NBG). Calvinia div., 3119CA, Nieuwoudtville, Oorlogskloof Nature Reserve, 680m, 3 viii 1988, *Pretorius* 56 (PRE, STE); Lokenburg, c.2200ft., 29 vii 1956, *Acocks* 18913 (K, M). Clanwilliam div., 3218BB, 5 miles from Clanwilliam on road to Lambert's Bay, 8 ix 1938, *Gillett* 4048 (BOL, K). 3219AA, Cedarberg, Dwarsrivier, Boskloof, c.320m, 21 vi 1984, *Taylor* 10952 (PRE, STE).
Seen from the following quarter-degree squares: 3017BB; 3018AC; 3118AB, DA, DB, DC, DD; 3119CA; 3218BB; 3219AA, AC, CB and possibly CA (see below).

The two earliest collections of *S. paniculata* were made by Zeyher in the Khamiesberg and by Drège near Clanwilliam, but both were confused with *S. revoluta* as were most subsequent collections. It is distinguished from *S. revoluta* by the longer pubescence on stem and calyx and the mostly longer leaves (largest primary leaves 20–40mm long versus 3.5–25mm). In indumentum and leaf-length it resembles *S. violacea* but the flowers arranged in panicles (not simple racemes), the shorter calyx (tube 0.7–1mm, not 1–2mm, and posticous lobes 1.8–2.6mm, not 3.2–5mm) and shorter corolla tube (6.5–9.5mm versus 11–15mm) will distinguish it; furthermore, there are seldom any bracts immediately below the calyx (they have been seen only in specimens from Namaqualand) whereas in *S. violacea* they are always present.
Sutera paniculata has been recorded from Sneeuwkop (just north of Kamieskroon) and the Kamiesberg, then from the Bokkeveld mountains near Nieuwoudtville, the Gifberg, Matsikammaberg and Nardouwsberg south east of Van Rhynsdorp, the Cedarberg, Baliesgatberg and near Clanwilliam. It grows on sandy streambanks or among boulders on hillslopes, a constituent

of scrub, c.250–1600m above sea level. Flowering seems to occur chiefly between May and September, but the specimens from Namaqualand were collected in December and January.

Hybridization may account for the difficulty in placing some specimens. For example, *Esterhuysen* 12732 (BOL) from Hondverbrand ridge in the S Cedarberg (3219CA) appears to be intermediate between *S. paniculata* and *S. violacea*: the short calyx lobes (2.5mm), short corolla tube (8mm) and short anticous filaments (1.5mm) are characteristic of *S. paniculata*, but only one of the four twigs on the sheet has the flowers in a panicle; the other three are racemes, characteristic of *S. violacea*, as are the bracts immediately below the calyx. Hondverbrand ridge would be the southernmost record for either *S. paniculata* or *S. violacea*; unequivocal *S. paniculata* has been collected at Pakhuis (3219AC) and Dwarsrivier (3219AA), while *S. violacea* has been collected at Bidouw (3219AB), near Wupperthal (3219AC), and Matjesrivier (3219AD).

The Drège collection from Boschkloof cited as *S. revoluta* var. *pubescens* (in synonymy above) is also problematical. The specimens in Bentham's herbarium (K) and in LE and S are unequivocal *S. paniculata*. The two twigs in Hooker's herbarium (K) appear to have come from different plants: the one on the left has longer hairs on the stem and shorter calyx lobes than the one on the right; the sheet in E comprises a similar mixed pair; P also has both, on separate sheets, while that in W comprises one twig of the plant with short calyx lobes. Field investigations at Boschkloof (3218BB) are needed to explore the possibility of hybridization there between *S. paniculata* and *S. violacea*.

A problem of a different nature exists at Matsikamma. *Compton* 7228 (NBG) from Matsikamma is *S. paniculata*. A plant collected by Miss Thompson (*Thompson* 2101, STE) on the Matsikammaberg, SE end between Waterval and Vaalsyfer at 365m, is possibly an undescribed species allied to *S. paniculata*. The plants were 'occasional soft shrublets to 60cm' with mauve/pink flowers (in full flower but no fruit on 15 vii 1974), growing 'in sandy ground against large rocks': all parts glandular-pubescent, hairs up to 0.2mm long, *stem* terete, *leaves* opposite becoming alternate upwards, largest primary leaves 36 × 3.5mm, narrowly elliptic, margins weakly revolute, *flowers* in leaf-axils, *bracts* wanting, *pedicels* up to 12mm long, *calyx* tube 0.6mm, lobes 3 × 0.6mm, hairs up to 0.2–0.3mm long, *corolla* tube 9.5 × 4mm, limb 11.5mm, posticous lobes 4.2 × 3.2mm, anticous lobe 4.5 × 4.2mm, *anthers* 1mm, anticous *filaments* 3mm, posticous filaments 1.5mm, glabrous, *stigma* 2mm, *style* 8mm, *ovary* 1.2 × 1mm. The plant should be sought again to establish its status.

12. Sutera violacea (Schltr.) Hiern in Fl. Cap. 4(2): 286 (1904).
Type: Cape, Clanwilliam div., Koudeberg, 3219AB, near Hoeck, c.1500ft, 27 viii 1896, *Schlechter* 8708 (holo. B†; iso. BOL, K, PRE, Z).

Syn.: *Chaenostoma violaceum* Schltr. in Bot. Jahrb. 27: 180 (1899).

Suffrutex between 500mm and 1.5m tall, probably spindly but regenerating from base, branching loose, virgate, branches terete, glandular-pubescent, hairs up to 0.1–0.4mm long, brachyblasts present, leafy. *Leaves* opposite or sometimes becoming alternate upwards, largest primary leaves 20–40 × 1–2.5mm, linear, obtuse, base slightly narrowed then abruptly expanded and clasping, connate or not, margins revolute, both surfaces hairy as stems. *Flowers* usually solitary in the leaf-axils forming long (up to c.300mm) simple racemes, or occasionally in 3-flowered cymules particularly in the lower part of the inflorescence. *Bracts* 1–7 × 0.1–0.8mm, in pairs immediately below calyx. *Pedicels* (lowermost) c.2–8mm long (peduncles of cymules up to c.15mm). *Calyx* distinctly bilabiate, tube 1–2mm long, posticous lobes 3.2–5 × 0.4–1mm, anticous lobes slightly broader, glandular-pubescent with some hairs 0.25–0.7mm long. *Corolla* tube 11–15 × 3–4mm in throat, limb nearly regular, 10–16mm across, posticous lobes 3.2–5 × 2.5–4mm, anticous lobe 4–6 × 2.5–3.8mm, all elliptic-oblong, shades of mauve or purple, throat yellow/orange, glandular-pubescent outside, hairs up to c.0.2mm long, inside weakly bearded with clavate hairs. *Stamens* 4, anthers 1–1.6mm long, anticous pair exserted, filaments 2.5–3mm long, posticous pair in throat, filaments 1.5–3mm long. *Stigma* 2–4mm long, exserted. *Style* 10.5–13mm long. *Ovary* 1.6–2 × 1–1.3mm, either glabrous or few glistening glands on sutures. *Capsules* 4.5–6 × 3–3.5mm. *Seeds* 0.8–1 × 0.3–0.4mm, pallid.

Selected citations:

Cape. Calvinia div., 3119AC, Nieuwoudtville, Menzieskraal, 29 iii 1933, *Markotter* sub STE 18705 (STE); 3119CD, top of Botterkloof Pass, 21 viii 1986, *Bean & Viviers* 1732 (BOL); 3219AC, Wupperthal, 20 viii 1959, *Middlemost* 2032 (NBG, STE). Sutherland div., 3220DB, Komsberg Pass, c.4500ft., 20 x 1954, *Acocks* 17812 (K). Prince Albert–Oudtshoorn div., 3322BC, Meiringspoort, 12 xi 1941, *Thorns* s.n. (NBG 46766).
Seen from the following quarter-degree squares: 3119AC, DB; 3219AB, AC, AD, DC; 3319BC; 3322BC.

Schlechter allied his new species to *Chaenostoma glabratum* and *C. revolutum*. However, he had confused *Sutera subspicata* with *S. glabrata* and the species here described as *S. paniculata* with *S. revoluta*. *Sutera violacea* differs from *S. paniculata* in the mostly shorter hairs on the stem (0.1–0.4mm versus 0.25–1mm), its flowers in racemes (not panicles), the presence of bracts below the calyx (very rarely present in *S. paniculata* and then minute), longer calyx (tube 1–2mm, posticous lobes 3.2–5mm versus 0.7–1mm and 1.8–2.6mm), and possibly always longer corolla tube (11–15mm versus 6.5–9.5mm). It is closely allied to *S. revoluta*, which differs in its mostly shorter primary leaves (7–20mm versus 20–40mm) and very short almost velvety glandular pubescence on stem, leaves and calyx.

The salient features of *S. violacea* are the presence of rather coarse hairs scattered in the underlying velvety pubescence on stems, leaves and calyx, its long narrow leaves, flowers in the axils of the leaves forming long racemes, a pair of small bracts immediately below the calyx with a flower sometimes produced in each axil and thus giving rise to a 3-flowered cymule. The species is known from the Nieuwoudtville, Botterkloof, Wupperthal, Cedarberg area, thence south to Zwart Ruggens and Karoopoort, and east to the Komsberg Pass in the Roggeveld, and Meiringspoort, a route through the Great Zwartberg between De Rust and Klaarstroom. It is a constituent of scrub, flowering between April and November. When Dr Markotter found the species at Nieuwoudtville, she made two separate collections, recording the flowers of one as blue, the other as yellow; yellow flowers are quite exceptional in the genus.

It seems likely that *S. violacea* and *S. revoluta* hybridize. Schlechter collected *S. violacea* on Botterkloof Pass in 1897 (*Schlechter* 10885, E, PRE, W, Z) and there have been several subsequent collections: *Barker* 6500 (NBG, STE), *Lewis* 3175 and 3176 (SAM), *Esterhuysen* 5770 (BOL) and *Bean & Viviers* 1732 (BOL). All these specimens accord well with the type of *S. violacea*, differing only in the weak development of coarse hairs on stem, leaves and calyx. However, *Barker* 1944 (NBG, PRE), also from Botterkloof, lacks coarse hairs, has a short calyx tube (0.6mm), short corolla tube (8mm) and lacks a pair of bracts below the calyx, all characters pointing to *S. revoluta* sens. lat. Field studies are desirable, Botterkloof being the only known northern locality where these two species are sympatric.

Sutera violacea may also hybridize with *S. paniculata*: see under that species. Field studies are also needed between Zwartruggens and Karoopoort. The specimen from Zwartruggens (*Levyns* 1811, BOL) deviates from *S. violacea* only in its possession of exceptionally long hairs on the stems and calyx, c.0.7–1mm. One specimen from Karoopoort (*Compton* 16057, NBG) deviates in its short anticous filaments (2.2mm) and rather short leaves (15–25mm); the hairs on the stem are c.0.4mm long. A second specimen, (*Marloth* 9001, STE, PRE), which is *S. comptonii*, has similar indumentum and leaves but lacks bracts below the calyx. Hybridization may have taken place, possibly with *S. comptonii*; see there.

13. Sutera revoluta (Thunb.) O. Kuntze, Rev. Gen. Pl. 2: 467 (1891); Hiern in Fl. Cap. 4(2): 270 (1904), p.p.
Type: Cape, without precise locality, *Thunberg* (sheet no. 14378, UPS).

Syn.: *Manulea revoluta* Thunb., Prodr. 100 (1800) & Fl. Cap. ed. Schultes 467 (1823).

Chaenostoma revolutum (Thunb.) Benth. in Hook., Comp. Bot. Mag. 1: 375 (1836) & in DC., Prodr. 10: 355 (1846), excl. vars.

Suffrutex up to c.600mm tall, but often much shorter (well grazed), developing a stout woody caudex up to at least 20mm in diam., well branched, branching lax or congested, branches often erect, sometimes decumbent or ascending, terete to very weakly ribbed, glandular-puberulous, hairs normally up to c.0.1mm long, exceptionally 0.2mm (see discussion), very leafy at least on upper parts, brachyblasts present. *Leaves* opposite, very rarely alternate towards the tips of the twigs, largest primary leaves 3.5–25 × 0.6–4.5mm, linear to narrowly elliptic, apex obtuse, base slightly narrowed then abruptly expanded and clasping, sometimes connate, often decurrent to produce barely discernible ridges, margins strongly to weakly revolute, both surfaces glandular-puberulous as stems. *Flowers* solitary in the axils of the leaves, forming racemes often branching into narrow panicles. *Bracts* rarely present except in northern part of range, c.0.2–4 × 0.1–1mm, in pairs or sometimes solitary, immediately below calyx. *Pedicels* (lowermost) 1–9mm long. *Calyx* often distinctly bilabiate, tube 0.3–1.8mm long, posticous lobes 1.5–4 × 0.4–1mm, anticous lobes slightly broader, glandular-puberulous, hairs up to 0.1mm long or occasionally some up to 0.2–0.4mm (see discussion). *Corolla* tube 5–13 × 3–4.5mm across throat, funnel-shaped, limb nearly regular, 8–15mm across, posticous lobes 2.4–5 × 1.8–5mm, anticous lobe 3–6 × 2–4.5mm, all oblong-elliptic, shades of pink or mauve but often white along the southern Cape coast, throat yellow/orange, glandular-puberulous outside, hairs up to 0.15–0.25mm long, inside weakly to strongly bearded with longitudinal bands of clavate hairs. *Stamens* 4, anthers 0.8–1.5mm long, anticous pair exserted, filaments 2.5–5mm long, posticous pair in throat or sometimes shortly exserted, filaments 1.2–2.8mm long, either glabrous or with a few tiny hairs. *Stigma* 1.2–4mm long, exserted. *Style* 3–11mm long. *Ovary* 1–1.8 × 0.6–1.2mm, glabrous or with a few glistening glands on sutures. *Capsules* 3–5 × 2–3mm. *Seeds* 0.5–1 × 0.3–0.5mm, pallid or greyish-blue.

Selected citations:

Cape. Sutherland div., 3220DA, Roggeveld escarpment, farm Jakhals Valley 99, 1300m, 6 ix 1986, *Cloete & Haselau* 231 (STE). Ladismith div., 3320DB, near Plathuis station, farm Comae, 22 vi 1976, *Van Breda* 4404 (K, PRE); 3321BA, 3.4 miles from Ladismith to Barrydale, 28 ix 1969, *Marsh* 1426 (PRE, STE). 3321BC, Huis River Mountains, 3000ft, 28 ix 1932, *Compton* 4047 (BOL, NBG). Oudtshoorn div., 3321CB, Gamka Mountain Reserve, 1100ft, 27 vi 1982, *Cattell & Cattell* 69 (STE). 3322DA, De Rust, farm Doornkraal, 28 v 1971, *Dahlstrand* 2019 (J, PRE, STE). 3322CC, Cango, 7 v 1938, *Compton* 7186 (NBG). Willowmore div., 3323AD, about 20km from Willowmore in Baviaanskloof, 1 xi 1980, *van Wyk* 380 (STE). Caledon div., 3419AA, near Bot River, *Lewis* 133 (SAM). Swellendam div., 3420AB, Bontebok Park, c.250ft, 20 vi 1962, *Acocks* 22242 (K, PRE, STE). Riversdale div., 3421AD, Still Bay, Botterkloof farm, 50m, 30 vii 1981, *Bohnen* 7953 (PRE, STE). Mossel Bay div., 3421BB, eastern side of Gauritz river, 5 xi 1814, *Burchell* 6416 (K, PRE).
Seen from the following quarter-degree squares: 3119CD; 3220DA, DC; 3319BC, DB; 3320AC, BC, DB, DC, DD; 3321AD, BA, BC, CA, CB, CC, DA, DB; 3322AC, BC, CA, CB, CC, CD, DA, DB; 3323AD, BA, CA; 3419AA; 3420AA, AB, AD, BD, CA; 3421AC, AD, BB.

Sutera revoluta is wide-ranging on the high ground flanking the karoos, from Calvinia and Sutherland divisions south to Touws River and east across the Cape fold mountains to the eastern end of the Swartberg and the mountains near Uniondale. There the altitudinal range is c.300–1500m, but along the southern coast in Caledon, Bredasdorp, Riversdale and Mossel Bay divisions the plant comes down almost to sea level. It is in the coastal area that plants with the broadest leaves occur (margins only very weakly revolute); this could be a response to less harsh growing conditions, but of course genetic factors cannot be ruled out. There too flowers may be white, though mauve-flowered plants seem just as common, and two collectors noted 'white to light blue' and 'mauve or white.' Over the whole range, the plants seem to favour stony or rocky sites and form part of a range of shrub communities: coastal renosterveld, succulent karoo and Spekboom veld have been mentioned by collectors. Plants can possibly be found in flower in any month, but the peak period appears to be May to September.

A broad species concept has been adopted for *S. revoluta*. Among its close allies, the distinguishing marks of the species are its very short (hairs c.0.1mm long) almost velvety glandular pubescence on stems, leaves and calyx, short leaves seldom exceeding 20mm, and short pedicels. The type of the name is an unlocalized specimen collected by Thunberg, and it has proved difficult to match precisely: the calyx tube (measured on the *outside*) is about 2.5mm long, there are two small bracts (c.2mm long) immediately below the calyx, and the corolla tube is at the short end of the variation range. Specimens from 'Calvinia, mountain slopes in karoo' (*Schmidt* 263, E, PRE, and 354, E) approach the type closely, as do specimens from the Roggeveld escarpment (3220DA), Klein Roggeveld (3320AB), and the Touws River area (3319DB), though there the corolla tubes are c.10–12mm long. Elsewhere over the area of the species, bracts are seldom present below the calyx, the calyx tube is short, and the full range of variation in the length of the corolla tube is covered. Then there are a good many specimens that differ from *S. revoluta* mainly in the possession of longer than usual hairs on stems, leaves and calyx. It seems possible that these plants are hybrids between *S. revoluta* and species sympatric with it. That *S. revoluta* may hybridize with *S. pauciflora, S. violacea* and *S. decipiens* is discussed under those species; however, there is scope here for further comment, and undoubtedly scope for field observations.

Plants with more or less the facies of *S. revoluta* but with slightly longer hairs than is normal for that species on stems, leaves, pedicels and calyx, and sometimes with abnormally short anticous filaments as well, can be found in three different areas; *S. revoluta* occurs in the same general areas and so do other species that could have contributed the aberrant characters. One of these areas is Seven Weeks Poort (3321AD), another is on the mountains from about Meirings Poort (3322BC) east to the Willowmore area (3323AD, BA), and south to Uniondale (3323CA). Drège collected *S. revoluta* at Klaarstroom, just north of Meirings Poort; the specimen in Bentham's herbarium (K) is normal *S. revoluta* while the specimen in Hooker's herbarium (K) and that in Paris (P) is the plant with abnormally long hairs. The third area lies in Robertson, Swellendam and Bredasdorp divisions, with specimens recorded from De Hoop Nature Reserve and nearby (3420AD), Stormsvlei (3420AA), Nachtwacht (3420CA), Bonnievale (3320CC) and Riviersonderend (3419BB). Some very odd plants have been collected in the Vrolijkheid Nature Reserve (3319DD): *Van der Merwe* 2438 (PRE) has four twigs on the sheet from, I judge, three different plants, one with hairs on the stem up to c.0.5mm (two twigs), 0.15mm and 0.2mm on the other two twigs. On the twig with the shortest indumentum, the anticous filaments are c.3–3.5mm long, and thus approach *S. revoluta*. Anticous filaments 2.5mm long are associated with stem hairs 0.5mm. *Jooste* 131 (STE) has hairs c.0.1mm long on the vegetative parts and the calyx and anticous filaments 3.5mm long. In *Hahndiek* 7 (STE), the hairs are 0.3–0.5mm long, anticous filaments 2.2mm. Three other species are known to occur in this area, namely *S. caerulea, S. decipiens* and *S. uncinata*. It is the last species that is possibly involved in hybridization with *S. revoluta*: *S. uncinata* in this area (the only one where it is sympatric with *S. revoluta*) has abnormally short hairs on the stem. Possible hybrids between *S. revoluta* and *S. decipiens* are mentioned under that species.

Manulea satureoides Desrouss. (in Lam., Encycl. Meth. 3: 705, 1792. Type: C.B.S., *Vaillant* in P-LAM) is possibly a hybrid between *S. revoluta* and *S. aethiopica*. The specimens have the facies of *S. revoluta* including the very short pedicels (2mm), but in addition to the usual velvety pubescence, there are coarse hairs on both stems and pedicels as well as on the calyx, particularly the calyx lobes, a character that could have come from *S. aethiopica*.

14. Sutera comptonii Hilliard in Edinb. J. Bot. 48: 342 (1992).
Type: Cape, Laingsburg div., Witteberg (Whitehill section), 5000ft, 7 xi 1948, *Compton* 21097 (holo. NBG).

Dwarf shrublet or suffrutex, c.150–250mm tall or sometimes straggling to c.310mm, branches erect or decumbent, terete or very weakly ribbed, glandular-pubescent with delicate hairs up to 0.4–1mm long, leafy at least on upper parts, brachyblasts present. *Leaves* opposite, occasionally alternate upwards, largest primary leaves c.5–18 × 1–1.5mm, linear, obtuse, base clasping, connate, margins strongly revolute, both surfaces glandular-pubescent as stems. *Flowers* solitary in the upper leaf-axils forming short racemes or occasionally narrow panicles, old inflorescences sometimes overtopped by fresh growth. *Bracts*, when present, very small, c.1–2 × 0.1–0.3mm, immediately below calyx. *Pedicels* (lowermost) 2–14mm long. *Calyx* bilabiate, tube 1–1.5mm long, posticous lobes 3–5 × 0.7–0.8mm, anticous lobes slightly broader, glandular-pubescent, hairs up to 0.4–1.2mm long. *Corolla* tube 10.5–12.5 × 2.5–3.5mm across throat, funnel-shaped, limb nearly regular, 8–11mm across, posticous lobes 3–3.2 × 1.6–2.8mm, anticous lobe 3–4 × 2.2–2.8mm, all oblong-elliptic, mauve, throat yellow, glandular-puberulous outside, hairs up to 0.2–0.3mm long, inside bearded with longitudinal bands of clavate hairs. *Stamens* 4, anthers 1–1.4mm long, anticous pair exserted, filaments 2.2–3mm long, posticous pair included, filaments 1.4–1.5mm long. *Stigma* 2–4mm long, exserted. *Style* 7.5–10.5mm long. *Ovary* 1.5–2 × 0.8–1.2mm, few glistening glands on sutures. *Capsules* 3–5 × 1.8–2.8mm. *Seeds* not seen.

Citations:

Cape. Laingsburg div., 3320BA, Witteberg, Whitehill, 4500ft, 23 x 1939, *Compton* 7990 (NBG); Witteberg (Bantams), 4000ft, 27 x 1941, *Compton* 12232 (NBG); Witteberg, north side, 4500ft, 24 x 1943, *Compton* 15197 (NBG); Witteberg, 3320BC, south slopes, 4000ft, 12 x 1929, *Compton* 3561 (BOL). Worcester div., 3320AA, Pienaarskloof, 29 x 1950, *Barker* 6822 (NBG, STE). 3319 BC, Karroopoort, 2250ft, viii 1919, *Marloth* 9001 (PRE, STE).

Sutera comptonii is known from the Witteberg, west of Laingsburg, Pienaarskloof, a few kilometres to the west of the Witteberg, and from Karoopoort, again to the west. *Compton* 16057 (NBG), from Karroopoort, has longer leaves than typical *S. comptonii*, conspicuous bracts immediately below the calyx, and some flowers in cymules, all characters pointing to *S. violacea*, from which it differs in its mostly shorter leaves and anticous filaments; see under *S. violacea*, where the possibility of hybridization is mooted.

The long hairs on stems, leaves and calyx distinguish *S. comptonii* from *S. revoluta*, and must account for the misuse of the name *S. revoluta* var. *pubescens* for some specimens in herbaria. However, this varietal name is a *nomen nudum*: the Drège specimen from Boschkloof quoted by Bentham as *Chaenostoma revolutum* var. *pubescens* is *S. paniculata*, while that from Ezelsbank in the Cedarberg is *S. subsessilis*.

Scant ecological information is available, but the species is possibly confined to quartzitic rocks. Flowering has been recorded in October and November.

15. Sutera archeri Compton in Trans. Roy. Soc.S. Afr. 19: 307 (1931). **Plate 1H.**
Lectotype (chosen here): Cape, Laingsburg div., 3320BA, Whitehill, 2700ft, 15 vii 1923, *Compton* 2849 (BOL).

Dwarf twiggy shrublet up to c.350mm tall, but often heavily browsed and regenerating shoots initially subsimple, stout woody caudex up to c.20mm in diam., branches erect or decumbent, terete or very weakly longitudinally ridged, minutely glandular-puberulous, hairs up to c.0.1mm long, leafy at least on upper parts, brachyblasts present. *Leaves* opposite, largest primary ones c.10–30(–40) × 1–3(–4.5)mm, linear to narrowly elliptic, apex obtuse, base tapering to a petiolar part then abruptly expanded and clasping, not or slightly connate, sometimes decurrent in scarcely perceptible ridges, margins usually entire, rarely with 1–6 pairs of tiny teeth, strongly to weakly revolute, both surfaces glandular-puberulous. *Flowers* solitary in axils of leaves forming leafy racemes, sometimes overtopped or interrupted by sterile growth. *Pedicels* (lowermost) (8–)10–25mm long. *Calyx* distinctly bilabiate, tube 1–1.8mm long, posticous lobes 2.4–4.5 × 0.7–1.2mm, anticous lobes slightly broader, minutely glandular-puberulous with scattered longer hairs up to 0.15–0.2mm. *Corolla* tube 5–8 × 3–5mm in throat, broadly funnel-shaped, limb nearly regular, posticous lobes 2.6–7 × 2.2–5.2mm, anticous lobe

3.5–6.5 × 2.8–5mm, all suborbicular, usually shades of mauve, rarely white, throat yel-
low/orange, glandular-puberulous outside, hairs c.0.1–0.2mm long, well-bearded inside with
5 longitudinal bands of clavate hairs. *Stamens* 4, all exserted, posticous filaments 2–3.8mm
long, usually bearded, anticous filaments 4–5mm long. *Stigma* 2–5mm long, exserted. *Style*
3.5–7mm long. *Ovary* 1.4–1.8 × 1–1.3mm, few glistening glands on sutures. *Capsules* 4.5–7 ×
2.5–3mm. *Seeds* (few seen) c.0.6 × 0.4mm, amber-coloured, but young.

Selected citations:

Cape. Fraserburg div., 3221BB, Layton, 1 v 1965, *Shearing* 22 (PRE). Beaufort West div., 3222BA, Nieuweveld,
Leeuwkloof, 4500–5000ft, x 1935, *Thorne* s.n. sub SAM 51882 (SAM). Laingsburg div., 3320BA, Whitehill ridge, 20
ix 1943, *Compton* 14878 (NBG). Prince Albert div., 3322AB, Tierberg, 900m, 18 vi 1990, *Dean* 960 (E). 3323AC, 30
miles W of Vondeling Station, c.2500ft, 13 ix 1955, *Leistner* 208 (K, PRE).
Specimens seen from following quarter-degree squares: 3221BB; 3222BA; 3320BA, BB; 3321AA; 3322AB; 3323AC.

Sutera archeri is confined to the Karoo, and has been recorded from Layton and Beaufort West
south to the Witteberg and Moordenaars Karoo near Laingsburg and east to Prince Albert. The
plants grow on rocky hillsides, between c.750 and 1500m above sea level, flowering between
May and October. One collector noted that the plants were browsed by sheep and sprout again
quickly after rain; several other collections have clearly been heavily browsed.

   Sutera archeri is easily recognized by its very short glandular pubescence on stems and
leaves, calyx similarly pubescent but with the addition of some coarser, longer hairs, long
pedicels, short corolla tube, and all four stamens well exserted; the posticous filaments are often
bearded. It has been confused in herbaria with both *S. halimifolia* and *S. caerulea*, both species
with short corolla tubes and far-exserted stamens, but the flowers are arranged in bracteate
racemes. Furthermore, *S. caerulea* is a much hairier plant with a glandular-pubescent style, and
S. halimifolia is clad in peculiar vesicular hairs. It is possible that *S. archeri* and *S. caerulea*
hybridize. The specimens collected by Leistner (cited above) have the leaves toothed (they are
normally entire in *S. archeri*), while some specimens of *S. caerulea* from the precincts of
Oudtshoorn are far less hairy than normal (e.g. *Dahlstrand* 2254, STE, from the side of a dam).

16. Sutera macrosiphon (Schltr.) Hiern in Fl. Cap. 4(2): 286 (1904).
Type: Cape, Queenstown div., near Bailey, Andriesberg, c.6400ft, ii 1896, *Galpin* 2004 (holo.
K; iso. BOL, PRE).

Syn.: *Chaenostoma macrosiphon* Schltr. in J. Bot. 34: 502 (1896).

Perennial herb eventually forming a small woody caudex, stems many from the base, c.150–
450mm long, loosely to more closely branched, sprawling or decumbent and then rooting at
the nodes, possibly sometimes erect, terete, glandular-pubescent, hairs up to c.0.1–0.25mm
long, brachyblasts present, closely leafy. *Leaves* opposite, bases connate forming a transverse
ridge, largest primary leaves 6–30 × 1–5mm, linear to narrowly elliptic, acute, margins often
entire, rarely with up to 3 pairs of small callose teeth, plane to weakly revolute, glandular-puberulous as the stems. *Flowers* solitary in the axils of the uppermost leaves, the longest of these
c.3–15 × 0.7–3.5mm, often becoming alternate, each terminal inflorescence 1–few-flowered,
forming either racemes or narrow panicles. *Pedicels* (lowermost) 7–28mm long. *Calyx* distinctly bilabiate, tube 2–3.5mm long, posticous lobes 3–10 × 0.5–1mm, anticous lobes slightly
broader, glandular-pubescent, hairs up to 0.15–0.3mm long, some glistening glands as well.
Corolla tube 14–28 × 2–3mm in throat, limb slightly bilabiate, 9–14mm across, posticous lobes
3.2–6 × 2–4mm, anticous lobe 3.2–6 × 2–4mm, all broadly elliptic to oblong-elliptic, often
mauve, sometimes white or tinged with pink, throat yellow, glandular-pubescent outside, hairs
0.15–0.3mm long, inside well-bearded with clavate hairs in throat below the posticous lip.
Stamens 4, anthers 1–1.5mm long, anticous pair exserted, filaments 2–3mm long, posticous
pair included, filaments 1–2mm long. *Stigma* 1.3–4mm long, exserted. *Style* 12.5–25mm long.

Ovary 1.4–2 × 0.8–1mm, glabrous. *Capsules* usually c.5–9 × 2–2.5mm, but an occasional tiny capsule present. *Seeds* c.1 × 0.4mm, pallid.

Selected citations:

Orange Free State. Fauresmith, 2925CB, iv–v 1935, *Pont* 1469 (BM, K). Philippolis distr., 3025BC, 18 miles E of Springfontein, 5300ft, 24 iii 1947, *Acocks* 13154 (PRE).
Cape. Richmond div., 3124CA, Rhenosterfontein, c.6000ft, 26 iv 1950, *Acocks* 15812 (K). Steynsburg div., 3125BB, Zuurbergen, 4500ft, iv 1944, *Thorns* s.n. (NBG). Middelburg div., 3125AC, Sherbourne, 13 iv 1948, *Theron* 479 (K, PRE). Queenstown div., 3126BD, Stormberg, Gretna, c.5000ft, 17 iv 1907, *Galpin* 7724 (BOL, PRE). Graaff Reinet div., 3224AA, Sneeuwbergen, c.4000ft, iv, *Bolus* 1976 (BOL, PRE, S, SAM, Z). Beaufort West div., 3222AD, Karoo National Park, 1780m, 22 iii 1988, *Steiner* 1685 (NBG, PRE).
Seen from the following quarter-degree squares: 2925CB; 3025AD, BC; 3026CA, DA; 3124CA, DC; 3125AC, AD, BB, BD; 3126BD, DA; 3221BB; 3222AD; 3224AA.

Sutera macrosiphon is on the hills and mountains from about Fauresmith in the Orange Free State south to Andriesberg near Queenstown, thence west over the mountains north of the Great Karoo to Beaufort West and Fraserburg, between c.1200 and 2200m above sea level. It favours rocky sites and flowers mainly from March to May.

The plants can be very lax and sprawling to much more compact and more or less upright; they appear sometimes to have been browsed down. The species is easily recognized by its remarkably long calyx and corolla tube. Over the whole of its geographical range the indumentum remains uniformly short and glandular.

---

17. Sutera integrifolia (L.f.) O. Kuntze, Rev. Gen. Pl. 2: 467 (1891); Hiern in Fl. Cap. 4(2): 269 (1904), excluding var. *parvifolia*.
Type: Cape, without precise locality, *Thunberg* 329 (LINN 787.9).

Syn.: *Manulea integrifolia* L.f., Suppl. 285 (1782); Thunb., Prodr. 100 (1800) & Fl. Cap. ed. Schultes 467 (1823).
    Chaenostoma integrifolium (L.f.) Benth. in Hook., Comp. Bot. Mag. 1: 376 (1836), excl. var. *parvifolium*, & in DC., Prodr. 10: 356 (1846).

Twiggy shrublet, branches spreading, up to 150–600mm long, terete, glandular-puberulous, hairs up to c.0.1mm long, closely leafy, often with axillary leaf tufts. *Leaves* opposite, bases connate, primary leaves 5–28 × 1.8–9mm, elliptic, oblanceolate or lanceolate, shortly petiolate, apex obtuse to subacute, base cuneate, margins entire or with up to 5 pairs of small teeth, sometimes revolute, glistening glands particularly on lower surface, otherwise both surfaces glabrous or minutely and sparsely glandular-puberulous. *Flowers* solitary in the upper leaf axils of every twiglet, forming leafy racemes or narrow panicles. *Pedicels* up to 6–12mm long, filiform. *Calyx* obscurely bilabiate, tube 0.5–1mm long, lobes 2.2–4 × 0.2–0.8mm, linear-lanceolate, glandular-puberulous mainly on the margins, hairs up to 0.1–0.15mm long. *Corolla* tube 8–11 × 2.5–3mm in throat, narrowly funnel-shaped, limb almost regular, 7–14mm across, posticous lobes 2.5–5 × 1.5–3mm, anticous lobe 2.8–6 × 1.8–3mm, all oblong, white or rarely pale mauve, tube yellow/orange at least in the throat, glandular-puberulous outside, hairs up to 0.1mm long, a few glistening glands as well, clavate hairs inside throat arranged in 5 longitudinal bands. *Stamens* 4, anticous pair shortly exserted, filaments 2–3mm long, posticous pair included, filaments 1–1.75mm long, anthers 0.75–1mm long. *Stigma* 1–1.5mm long, shortly exserted. *Style* 6–9mm long. *Ovary* 1–1.5 × 0.6–1mm, glabrous or with a few glistening glands on the sutures. *Capsules* 3.5–5 × 2–2.5mm. *Seeds* c.1 × 0.4mm, pallid or dull blue-grey.

Selected citations:

Cape. Riversdale div., 3421BA, hills behind Albertinia, 100ft, vi 1914, *Muir* 1367 (BOL, PRE). Mossel Bay div., 3422AB, Gwaing river mouth, ± 10m, 25 v 1984, *O'Callaghan et al.* 163a (STE). George div., 3422BA, Victoria Bay, 63m, 23 iii 1893, *Schlechter* 2399 (BOL, E, J, K, PRE, S, Z). George div., 3322DC, Wilderness, 28 iv 1941, *Compton* 10691 (NBG). Knysna div., 3423AA, Touw river, Knysna Lakes road, 13 ix 1953, *Taylor* 4097 (STE). Humansdorp div., 3324DA, Kouga River Poort, 1100ft, ix 1925, *Fourcade* 3063 (BOL, K).

Sheet no. 787.9 (LINN) can be taken as the type: it bears the name *Manulea integrifolia* in the hand of Linnaeus the younger. There are four sheets in Thunberg's herbarium under the name, but only 14372 is *S. integrifolia*.

Sutera integrifolia grows along the southern Cape coast from about Albertinia in the west to Humansdorp in the east; most collections have been made not much above sea level, but Fourcade twice found it in the passes through the mountains, at Kouga River Poort (335m) and Prince Alfred's Pass (300m). It appears to be a common constituent of scrub, near the seashore and on forest margins or in the forest understorey, sometimes straggling up through other vegetation, sometimes forming small compact bushes. Flowering has been recorded in all months.

The name has been much misused, but the species is easily recognized. Its salient features are the terete stems clad in very short glandular hairs, leaves that seem glabrous on casual inspection, and the calyx nearly glabrous, with short glandular hairs almost confined to the margins of the lobes. No other species of *Sutera* has a similar calyx.

18. Sutera uncinata (Desrouss.) Hilliard in Edinb. J. Bot. 48: 346 (1991).
Type: C.B.S., *Sonnerat* (P-LAM).

Syn.: *Manulea uncinata* Desrouss. in Lam., Encycl. 3: 706 (1792).
    M. linifolia Thunb., Prodr. 100 (1800) & Fl. Cap. ed. Schultes 466 (1823). Type: Cape of Good Hope, *Thunberg* (sheet no. 14374 in herb. Thunberg, UPS).
    Chaenostoma linifolium (Thunb.) Benth. in Hook., Comp. Bot. Mag. 1:375 (1836) & in DC., Prodr. 10: 355 (1846).
    Chaenostoma fasciculatum Hort. ex Vilmorin's Blumeng. ed. 3, Sieb. & Voss 1: 772 (1895) in syn.
    Chaenostoma linifolium (Thunb.) Benth. var. *hispidum* Krauss in Flora 27(2): 834 (1844), *nomen nudum*. Specimen quoted: Ad lat. m. Tigerberg, *Krauss* 1637 (MO, W sine num.).
    C. glabratum Benth. p.p. (excl. lecto.), *Drège*, Kendo [Kendousberg or Kandos Mts] (K, P).
    Sutera linifolia (Thunb.) O. Kuntze, Rev. Gen. Pl. 2: 467 (1891); Hiern in Fl. Cap. 4(2): 272 (1904) p.p. and excluding var. *heterophylla* (O. Kuntze) Hiern.

Perennial herb soon forming a well-branched shrublet up to 600mm tall, base eventually woody and gnarled, branches erect, decumbent or somewhat spreading, narrowly winged or ridged, glandular-puberulous throughout including pedicels, lower parts sometimes glabrescent, few to many coarse patent glandular hairs also present, up to 1mm long, rather broad-based, the basal part sometimes forming ± scabrid projections, or these entirely wanting, leafy throughout, brachyblasts present. *Leaves* opposite or sometimes ternate, largest primary leaves 7–35 × 0.8–5(–7)mm, axillary leaves smaller, all linear to narrowly elliptic, apex ± obtuse, base narrowed then abruptly expanded and decurrent in a narrow wing or ridge, margins plane to strongly revolute, entire to denticulate, hairy as the stems. *Flowers* often opposite, sometimes subopposite or alternate, in terminal racemes, sometimes branching into narrow panicles, very rarely some of the lowermost flowers in 3-flowered cymules. *Bracts* (lowermost) c.3–12 × 0.8–2mm, linear to narrowly lanceolate, or occasionally the lowermost flowers in the axils of the uppermost leaves c.3–17 × 0.8–3mm (vegetative bud visible in axil). *Pedicels* (lowermost) 5–10(–15)mm long. *Calyx* bilabiate, tube (1–)1.5–2mm long, posticous lobes (2.5–)3–4.8 × 0.5–1mm, anticous lobes somewhat broader, glandular-puberulous, few to many coarser hairs also present, 0.2–0.6mm long. *Corolla* tube 9.5–19 × 2.5–4mm in throat, narrowly funnel-shaped, limb almost regular, c.9–15mm across, posticous lobes 2.7–6 × 2–3.5mm, anticous lobes 3–6 × 2–4mm, all oblong-elliptic, usually shades of pink, mauve and purple, occasionally white, tube yellow or orange, glandular-puberulous outside, hairs up to 0.25mm long, no glistening glands, either glabrous within or weakly, rarely strongly, developed longitudinal bands of clavate hairs. *Stamens* 4, anthers 1–1.4mm long, posticous filaments 1–2mm long, anthers in mouth, anticous filaments 1.8–3mm long, anthers exserted. *Stigma* 1.5–4mm long, exserted. *Style* 7–14.5mm long. *Ovary* 1.5–2 × 0.8–1.2mm. *Capsule* c.5–7.5 × 2.2–3mm. *Seeds* c.0.7–0.8 × 0.3mm, pallid.

Selected citations:

Cape. Piquetberg div., 3218DC, Piquetberg Mt., 10 ix 1950, *Barker* 6724 (NBG, STE). Tulbagh div., 3319AA, Saron, 800ft, 18 viii 1894, *Schlechter* 4866 (BOL, E, J, K, PRE, S, STE, Z). Malmesbury div., 3318BD, Riebeck Kasteel, 14 ix 1941, *Compton* 14626 (NBG). Cape Town div., 3318DC, Sewefontein above Kuils River, 1000ft, 2 x 1973, *Oliver* 4730 (PRE, STE). Stellenbosch div., 3418BB, 27 v 1943, *Parker* 3804 (BOL, NBG). Worcester div., 3319CB, Worcester Karoo Garden, 12 vii 1948, *Compton* 20528 (NBG). Robertson div., 3319DD, Robertson, 6 viii 1951, *Middlemost* 1710 (NBG). Montagu div., 3320CC, Bonnievale, 1600ft, ix 1933, *Levyns* 4576 (BOL). 3320CC, Olifantsberg, 600ft, 11 vi 1982, *Viviers* 424 (STE).
Seen from the following quarter-degree squares: 3118DC; 3218BD, CC, DA, DB, DC, DD; 3219CA, CC; 3317BB; 3318AA, AD, BA, BC, BD, CB, CD, DA, DB, DC, DD; 3319AA, AB, AC, BC, CB, CC, CD, DA, DC, DD; 3320CC; 3322BD; 3418BB. See also penultimate paragraph of discussion below.

Sutera uncinata is variable in aspect. Frequently, the branching is virgate with the flowers carried in long terminal racemes, but under some conditions the plants become twiggy and then narrow panicles may develop. The leaves are always narrow in relation to their length, but the margins may be plane (the leaves then at their broadest) to strongly revolute (then the width much reduced), and they may be entire to denticulate. An essential character is the presence of brachyblasts and no less important are the narrowly winged or ridged stems: the leaves are opposite or occasionally ternate, the bases decurrent, each pair forming a shallow groove running the full length of the internode. The bracts are nearly always distinctly smaller than the leaves, but this character falls away in lush specimens where the lowermost flowers may be produced in the axils of the uppermost leaves.

Coarse broad-based hairs are nearly always present on the calyx (few or wanting in some specimens from around Saldanha Bay) in addition to a dense covering of tiny glandular hairs, and the calyx tube is well-developed ((1–)1.5–2mm long). The inflorescence axis and the uppermost parts of the stem are always glandular-puberulous; in addition there is a weak to strong development of much coarser hairs up to 1mm long (as in the type specimen); indumentum on the leaves parallels that on the stems, and this ranges from pubescent to almost glabrous. From De Doorns south to Worcester and east to Robertson and Montagu (3320CC) development of coarse hairs on stem and leaves is very weak and may amount to no more than ± scabrid projections, or they may be entirely wanting, though the minute glandular pubescence often persists (see under *S. revoluta* for possible hybridization). This is the most easterly range of the species, except for a collection made by Drège in the Kendous Mountains (3322BD) east of Prince Albert, which has well developed coarse hairs. Specimens from around Saldanha Bay may also show scant development of coarse hairs.

To sum up: *S. uncinata* is characterized by its narrowly winged stems, brachyblasts in the leaf-axils, coarse almost scabrid hairs at least on the calyx (very rarely wanting there) and well developed calyx tube. Also, the capsules are relatively long and narrow; care is needed here because short, half-formed capsules will split open on herbarium specimens and look mature: they of course contain only dried-up ovules. It can be confused with *S. glabrata*; see there to distinguish them.

The overall range of the species is roughly from the Gifberg, near Klaver, and Grey's Pass (Pikenierskloof) south to Somerset West, bounded on the east by the Cold Bokkeveld Mountains, Winterberge and Hex River Mountains, thence east to Robertson and Montagu, bounded on the south by the Riviersonderend Mountains. There are three records from the Peninsula: at Orange Kloof and Lion's Head made by Schlechter in 1892 (*Schlechter* 716 (Z) and 1043 (Z) respectively) and one by Krauss, unnumbered (M, W) made in July 1838 on 'Duvelsberg' (Devils Peak). As the species has never been re-collected on the Peninsula, one is left to wonder if it is now extinct there. The plant should also be sought again in the Kendous Mountains where Drège is said to have found it, roughly 250km ENE of any modern collection.

The plants are often a constituent of scrub, often on hillslopes, sometimes on flats, the sites rocky or not, from near sea level in the west to c.500m on the mountains. They flower chiefly between June and October, sometimes as early as April or May.

19. Sutera longipedicellata Hilliard in Edinb. J. Bot. 48: 343 (1991).
Type: Cape, Calvinia div., 3119AC, Nieuwoudtville, ix 1930, *Lavis* sub BOL 20679 (BOL).

Small shrub, probably between c.300 and 800mm tall, branches erect or decumbent, narrowly winged, glandular-puberulous, glabrescent, leaving scattered ± scabrid projections up to c.0.2mm long, leafy throughout, brachyblasts present. *Leaves* opposite, largest primary leaves c.25–27 × 2.2–6mm, those subtending flowers c.5–25 × 1–4mm, all linear to narrowly elliptic, ± obtuse, base narrowed then abruptly expanded and decurrent in a narrow wing, margins plane or revolute (leaves then appearing narrower than they really are), entire or rarely with a pair of minute teeth, hairy as the stems. *Flowers* opposite becoming alternate upwards, solitary in the leaf-axils forming long terminal racemes. *Pedicels* (lowermost) 11–21mm long. *Calyx* obscurely bilabiate, tube 0.8–1mm long, posticous lobes 4–4.5 × 0.6–0.9mm, anticous slightly broader, glandular-puberulous with scattered coarser glandular hairs up to 0.25–0.4mm long. *Corolla* tube 8–11 × 4–5mm in throat, funnel-shaped, limb nearly regular, 12–15mm across, posticous lobes 4–5 × 3–5mm, anticous lobe 4–6 × 3.3–5mm, all broadly elliptic, mauve, tube yellow, glandular-puberulous outside, hairs up to 0.2mm long, 5 longitudinal bands of clavate hairs inside. *Stamens* 4, anthers 1–1.3mm long, posticous pair included, filaments 2–2.5mm long, anticous pair exserted, filaments 3.2–4mm long. *Stigma* 1.8–2.5mm long, exserted. *Style* 6–9mm long. *Ovary* 1.7–2 × 1.2–1.3mm. *Capsules* (few seen) 5 × 2.5mm. *Seeds* c.0.7 × 0.3mm, pallid.

Citation:

Cape. Van Rhynsdorp div., 3119AC, Vanrhyn's Pass, 21 vi 1931, *Herre* sub STE 11440 (STE).

Sutera longipedicellata is known from only two collections, one from Nieuwoudtville, the other from the nearby Van Rhyn's Pass. It appears to be closely allied to *S. uncinata*, from which it differs in its leafier inflorescences, mostly longer pedicels, mostly shorter calyx, and longer anticous filaments. Its clear circumscription is complicated by the occurrence on Van Rhyn's Pass of what seem to be hybrids between *S. longipedicellata* and another species. *Sutera caerulea*, *S. violacea* and *S. subsessilis* have been recorded from this area (there is a doubtful record of *S. uncinata* from Klaver, but the main area of the species appears to lie much further south, from Piquetberg southwards, while the area of *S. decipiens* lies further to the west, from Spitsberg (near Nuwerus) and Gifberg (near Klaver) southwards). Clearly this whole area needs careful investigation, and Van Rhyn's Pass seems an excellent place to begin. The possible hybrids from Van Rhyn's Pass are: *Batten* 873 (E), *Compton* 11118 (NBG), *Salter* 2486 (BOL), together with *Stirton* 11051 (NBG, PRE, STE) from Arendskraal in the same quarter-degree square. The plants probably grow among rock outcrops, flowering mainly between June and September.

20. Sutera multiramosa Hilliard in Edinb. J. Bot. 48: 344 (1991).
Type: Cape, Vredendal div., 3118CD, north of Lambert's Bay, road to Vredendal, Witwatersbult, 31 v 1976, *van Jaarsveld* 1289 (holo. NBG).

Dwarf shrublet becoming very twiggy and gnarled, up to c.130–300mm tall, branches erect or decumbent, narrowly winged or ridged, glandular-puberulous throughout including pedicels, scattered coarser glandular hairs as well up to c.0.2–0.25mm long, leafy throughout, brachyblasts present. *Leaves* opposite, largest primary leaves 9–20 × 2–4mm, upper ones subtending flowers c.6–18 × 1.5–4mm, axillary leaves smaller, all narrowly elliptic, apex obtuse to subacute, base narrowed into a short petiolar part then abruptly expanded and decurrent in a narrow wing or ridge, margins plane, entire or with a pair of obscure teeth, sparsely glandular-puberulous mainly on upper surface. *Flowers* solitary in upper leaf axils, either opposite or becoming alternate upwards, forming leafy terminal racemes. *Pedicels* (lowermost) 7–20mm long. *Calyx* obscurely bilabiate, tube 0.6–0.8mm long, posticous lobes 2.5–6 × 0.4–0.7mm, anticous pair slightly broader, glandular-puberulous, some scattered hairs up to 0.15–0.3mm long. *Corolla* tube 8.5–13 × 3–4mm in throat, narrowly funnel-shaped, limb almost regular, 9–14mm across, posticous lobes 2.5–5.2 × 2–3.8mm, anticous lobe 3.8–5 × 2.8–4mm, all elliptic-oblong, pink, mauve or blue, tube yellow or orange, glandular-puberulous outside, hairs up to 0.15–0.2mm long, no glistening glands, weakly bearded within, hairs clavate, in three longitudinal bands. *Stamens* 4, anthers 1–1.3mm long, posticous filaments 1–1.2mm long, anthers in mouth, anticous filaments 1.8–2mm long, anthers exserted. *Stigma* 1–2.4mm long, eventually exserted. *Style* 6.2–10.5mm long. *Ovary* 1.2–1.3 × 0.8–1mm, glabrous. *Capsules* c.4.5 × 2.4mm. *Seeds* c.0.8 × 0.4mm (few seen), pallid.

Citations:

Cape. Vredendal div., 3118CA, Strandfontein, 16m, 11 iv 1971, *Boucher* 1491 (STE, PRE). Van Rhynsdorp div., [but modern Vredendal div., 3118CB?], Viswater-Ebenezer, 12 vii 1964, *Booysen* 147 (NBG, S).

Sutera multiramosa is known to me only from a small area of the western Cape coast, very roughly between the Olifants river and Lamberts Bay, in Vredendal division. The provenance of *Booysen* 147 (quoted above) is in some doubt: there is a village Ebenhaeser just south of the Olifants river in Vredendal division (not modern Van Rhynsdorp division) and I have failed to trace Viswater. The plants grow in sand among rock outcrops, and were recorded in flower between April and July.

The stems are longitudinally ribbed by the decurrent leaf bases, as they are in both *S. uncinata* and *S. decipiens*, but *S. multiramosa* can be distinguished from *S. uncinata* by its shorter calyx tube and lack of the coarse almost scabrid hairs that are normally present on the calyx of *S. uncinata*, and by its shorter ovaries (1.2–1.3mm long versus 1.4–1.8mm), which give rise to shorter capsules. The flowers of *S. multiramosa* arise in the upper leaf-axils; in *S. uncinata* they are usually in the axils of bracts sharply distinguishable from the leaves. The area of *S. multiramosa* probably lies outside that of *S. uncinata*, whose main area is from Clanwilliam southwards. It lies immediately west of the northern part of the range of *S. decipiens*, from which it differs in the much shorter pubescence on stem and leaves, and in its entire or subentire leaves.

21. Sutera subsessilis Hilliard in Edinb. J. Bot. 48: 345 (1991). **Figs 54, 55.**
Type: Cape, Ceres div., Elandskloof, 3000ft, 3 x 1940, *Levyns* 7243 (holo. BOL).

Syn.: *Chaenostoma revolutum* Benth. var. β *pubescens*, nomen, in Hook., Comp.Bot. Mag. 1: 375 (1836) p.p. Specimen cited: on the Cedarbergen, Ezelsbank, [3219AC], 3000ft, 23 xii 1830, *Drège* (K, P).
Sutera revoluta (Benth.) Hiern var. β *pubescens* Hiern, nomen, in Fl. Cap.4(2): 271 (1904), p.p.

Perennial herb, stems several tufted from the base, c.150 to at least 350mm long, simple or very laxly branched, branches up to 1.5mm diam., sharply ascending, very weakly ribbed, glandular-pubescent with delicate hairs up to c.1.2–1.5mm long, leafy, without brachyblasts. *Leaves* opposite or occasionally alternate upwards, largest c.10–17 × 3–8mm, smaller upwards, thin-textured, lanceolate, apex obtuse, base broad or narrowed, half-clasping, margins usually entire, rarely with 1 or 2 very obscure teeth, decurrent in very narrow ridges running to node below, both surfaces glandular-pubescent as stems. *Flowers* solitary in upper leaf axils forming leafy racemes or narrow panicles. *Pedicels*: lower on main racemes c.12–15mm, on branches of panicles c.5–9mm. *Calyx* distinctly bilabiate, tube 0.7–1.2mm long, posticous lobes c.3.5–5 × 0.7–0.9mm, anticous lobes broader (θ.9–1.3mm), glandular-pubescent, hairs up to 0.3–1mm long. *Corolla* tube c.8–10 × 4mm across throat, funnel-shaped, limb nearly regular, c.10mm across, posticous lobes c.3.5 × 2.5–3mm, anticous lobe c.3.5–4 × 3–3.5mm, all suborbicular to broadly elliptic, pale mauve, throat yellow/orange, glandular-puberulous outside, hairs up to c.0.2mm long, inside either very weakly bearded with longitudinal bands of clavate hairs or glabrous. *Stamens* 4, anthers 0.7–1mm long, anticous pair very shortly exserted, filaments 2–2.5mm long, posticous pair included, filaments c.0.8–1mm long. *Stigma* shortly exserted, in

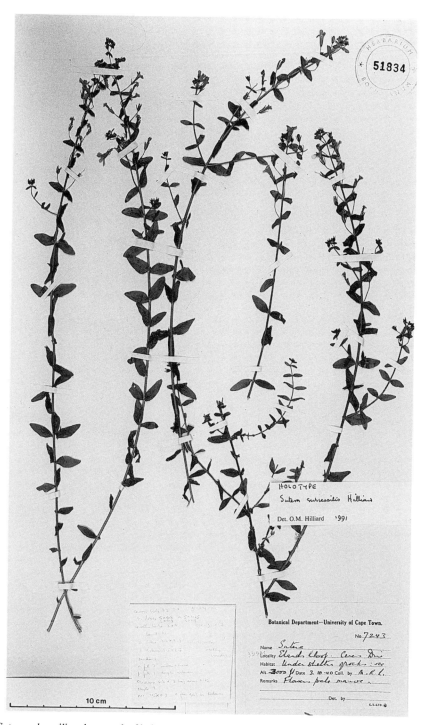

Fig. 54. *Sutera subsessilis*: photograph of holotype.

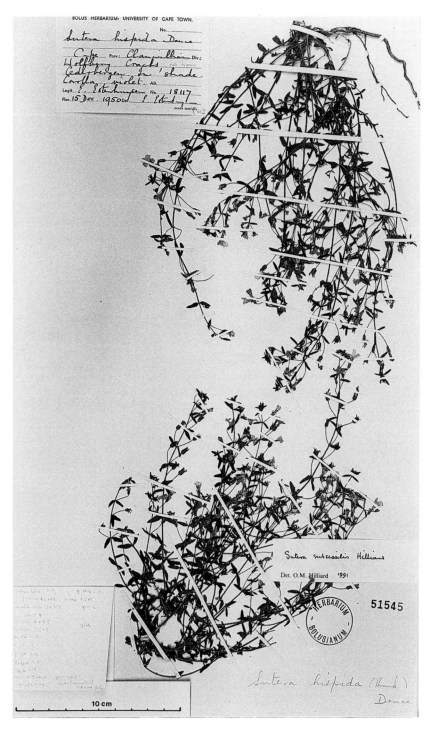

Fig. 55. *Sutera subsessilis*: photograph of specimens bushier than the type.

the type specimen c.0.8mm long with 2 tiny arms c.0.25mm long, in the other specimens c.3.8–4.2mm long, normal. *Style* c.3.3–7mm long. *Ovary* c.1.2–1.5 × 0.8–0.9mm, a few glistening glands on sutures. *Capsules*: c.3–4 × 2mm. *Seeds* (immature) c.0.8 × 0.4mm, pallid.

Citation:

Cape. Clanwilliam div., Cedarbergen, Wolfberg Cracks, 15 xii 1950, *Esterhuysen* 18117 (BOL).

Drège collected this plant in the Cedarbergen in 1830, Miss Esterhuysen in 1950, while Mrs Levyns had it in Elandskloof, Ceres div., in 1940, so it probably ranges from the Cedarbergen south through the Kouebokkeveldberge and Skurweberg to the Elandskloofberg. The exact location of Mrs Levyns's Elandskloof is in some doubt, but it is probably the one c.15km SE of Citrusdal, 3219CA (H.C.Taylor, in litt.). Mrs Levyns found her plant under the shelter of rocks, flowering in October, Miss Esterhuysen in shade, in flower and young fruit in December.

The relationship of *S. subsessilis* is obscure. Bentham included it as a variant of *S. revoluta* from which it is easily distinguished by the long hairs on stems, leaves and calyx, its thin-textured lanceolate leaves with plane margins, lack of brachyblasts, and longer pedicels. The combination of thin-textured lanceolate leaves scarcely narrowed to the base, long delicate glandular pubescence, and rather short-tubed flowers on long filiform pedicels make recognition easy. Drège's and Mrs Levyns's collections are identical: long (to 310mm) slender subsimple branches with the largest leaves c.17 × 6.5–8mm; Miss Esterhuysen's collection is a whole plant, tufted and loosely branched from the base, with stems reaching c.210mm, and much smaller leaves, the largest c.10–15 × 3–4mm.

22. Sutera decipiens Hilliard in Edinb. J. Bot. 48:342 (1991).Type: Cape, Clanwilliam div., 3219AA, N Cedarberg, Pakhuis, 3000ft, 7 ix 1953, *Esterhuysen* 21745 (holo. BOL, iso. PRE).

Perennial, eventually forming a woody caudex up to c.10mm diam. at crown, stems eventually many from the crown, more or less virgately branched, often lax, sometimes densely twiggy, erect to sprawling, c.220–500mm long, distinctly to obscurely longitudinally ridged, glandular-puberulous particularly on upper parts including pedicels, with many long delicate glandular hairs as well, up to 0.5–1.8mm long, leafy, brachyblasts often present though generally inconspicuous. *Leaves* opposite, bases not connate, decurrent down to the next node forming two very narrow parallel ridges, these sometimes very obscure, largest primary leaves 10–32 × 3–15mm (measured across spread of teeth) elliptic in outline tapering into a short clasping petiolar part, apex obtuse to subacute, margins usually with 2–4 pairs of coarse teeth sometimes reaching ± halfway to midrib, very rarely teeth obscure, glandular-pubescent, hairs up to 0.5–0.8mm long. *Flowers* often but not always solitary in axils of upper leaves but quickly running out into bracteate racemes sometimes branching into narrow panicles, upper flowers often alternate, the rhachis straight or flexuous. *Pedicels* (lowermost) 6–20(–23)mm long. *Calyx* obscurely bilabiate, tube 0.5–0.8mm long, posticous lobes 2.8–6 × 0.4–0.8mm, anticous lobes slightly broader, glandular-pubescent, hairs delicate, up to 0.4–1mm long, minutely glandular-puberulous as well. *Corolla* tube 8.5–12 × 2.5–3.8mm in throat, narrowly funnel-shaped, limb nearly regular, c.8–15mm across, posticous lobes 2.5–5 × 2.2–3mm, anticous lobe 3–5.3 × 2–4mm, all oblong elliptic, white to pale pink or mauve, or saxe-blue (type), tube orange or yellow, glandular-pubescent outside, hairs up to 0.25–0.6mm long, either glabrous inside or weakly bearded with longitudinal bands of clavate hairs. *Stamens* 4, anticous pair exserted, filaments 1.5–2.5mm long, posticous pair in throat, filaments 0.8–1.5mm long, anthers 0.8–1.25mm long. *Stigma* 1.5–2mm long, exserted. *Style* 7–10mm long. *Ovary* 1.3–1.6 × 0.8–1mm, glabrous or with a few glistening glands on the sutures. *Capsules* 3.5–5 × 2.5–3mm. *Seeds* c.0.5–0.8 × 0.3mm, pallid.

Selected citations:

Cape. Van Rhynsdorp div., 3118DC, Gift Berg Range, 1–2000ft, ix 1911, *Phillips* 7354 (SAM). Clanwilliam div., 3218BA, Graafwater, 1000ft, 18 viii 1896, *Schlechter* 8569 (BOL, E, K, PRE, S, W, Z). Piquetberg div., 3218DC, Piquetberg, 2500ft, 3 xi 1951, *Johnson* 294 (NBG). Ceres div., 3319AD, Mitchell's Pass, 10 v 1941, *Compton* 10827 (NBG). Worcester div., 3319BC, Hex River Mountains, Prospect Peak, 2000-3000ft, 2 x 1949, *Esterhuysen* 15942 (BOL). Laingsburg div., 3320BC, Witteberg, 3500ft, 30 xi 1924, *Compton* 2803 (BOL). Ladismith div., 3320DB, Touwsberg, W end, 3000ft, 19 viii 1956, *Wurts* 1437 (NBG). 3321AD, Klein Swartberg near Hoeko, 2000ft, 10 x1956, *Wurts* 1479 (NBG). Montagu div., 3320CA, Baden, ix 1946, *Lewis* 1945 (SAM). George div., 3322CD, Camfer, 2 xii 1941, *Esterhuysen* 7103 (BOL, K, PRE).

Seen from the following quarter-degree squares: 3118AB, CA, CB, CD, DA, DC; 3218AB, BA, BB, DA, DC; 3219AA, AC; 3319AC, AD, BC, BD, CA, CB, CC, CD, DD; 3320BC, CA, CC, DA, DB; 3321AD, BD, CC; 3322BC, CD; 3323CA.

The geographical range of *S. decipiens* is from the Spitsberg east of Nuwerus and the Gifberg east of Klaver, in Van Rhynsdorp division, south over the higher ground to the mountains around Ceres, Worcester and Fransch Hoek, thence east over the Cape fold mountains, the most easterly records being around Uniondale and Avontuur. The plants have frequently been recorded as growing in kloofs, sometimes in shady spots there and also under rocks out on open mountain slopes, between c.300 and 1200m above sea level. The chief flowering period is August to October, but there are records as early as March and as late as December.

Sutera decipiens has hitherto been confused with *S. hispida*; it differs in its long soft glandular pubescence, impossible to quantify, but soon appreciated when a series of specimens of both species is examined (see last paragraph, p.11 and fig. 11G, M), by its decurrent leaf bases forming two longitudinal grooves down each internode (very careful observation may be needed here), by a strong tendency for the leaves to be longer in relation to breadth than in *S. hispida* ((2–)2.5–3.5(–4):1 versus 1–2(–3):1), and for the flowers to be in the axils of bracts rather than leaves and to become alternate on the upper part of the raceme. The calyx tube is usually shorter in *S. decipiens* than in *S. hispida* (0.5–0.8mm versus mostly 1–2mm). The two species are virtually allopatric; their areas appear to meet near Tulbagh, Worcester and Fransch Hoek and hybridization may occur thereabouts. For example, material collected by Rehmann on the mountains at Worcester and distributed under his number 2486 is a mixture of *S. hispida* (Z) and specimens that appear to be *S. decipiens* × *hispida* (BM, PRE, Z). There are many more examples.

Hybridization also seems to occur where *S. decipiens* is sympatric with *S. revoluta*, and possibly with *S. caerulea* in the Van Rhynsdorp area, and almost certainly in Avontuur Poort. This of course needs investigation in the field.

The following specimens may be *S. decipiens* × *revoluta*: 3321AD, Zwartberg near Ladismith, *Esterhuysen* 13956 (BOL); Sevenweekspoort, *Bohnen* 6931 (STE); spur of Elandsberg (Torenberg), *Wurts* 1143 and 1085 (NBG); Elandsberg, *Wurts* 1181 (NBG); 3321BD, Toverkop, *Wurts* 1199 (NBG). The stems are mostly terete (as in *S. revoluta*), the indumentum may be shorter than in *S. decipiens* (short in *S. revoluta*), and the toothing of the leaves is less pronounced. Some specimens have longer anticous filaments than is normal for *S. decipiens* (long anticous filaments are characteristic of both *S. revoluta* and *S. caerulea*).

The two specimens from the Van Rhynsdorp area that may be of hybrid origin are: 3118DC [no precise locality], sandy stony slopes, 2400ft, *Oliver* 4932 (STE) and 3118DA, Snorkfontein, *Compton* 7221 (NBG).

Hybridization also seems a factor to be considered in Uniondale Poort-Avontuur Poort (alternative names for the same place?), involving crossing between *S. decipiens* and *S. caerulea* (though *S. revoluta* and *S. denudata* are also in the area). *Fourcade* 4274 (BOL, K, NBG) came from Avontuur Poort at 2600ft in December 1928; it is almost certainly a hybrid between *S. decipiens* and *S. caerulea*. The length of the calyx tube (1.2mm) and the anticous filaments (2.6–2.7mm) as well as the glandular-pubescent style and ovary indicate *S. caerulea*, while the length of the corolla tube (8.5–9.5mm) indicates *S. decipiens*. *Fourcade* 4273 (BOL,

K, NBG), collected at the same time, has the short calyx tube (0.7mm), short anticous filaments (1.6mm), corolla tube (9mm) and glabrous style and ovary of *S. decipiens*, but the stem is terete (as in *S. caerulea*). The facies of the plants differs because the leaves of 4723 are broader than those of 4724. Unfortunately, Fourcade gave no field information, though he recorded the flowers of 4724 as lavender. His no. 2918 (BOL, K, STE) came from 'north side of old pass between Uniondale and Avontuur, 2300ft, Jan. 1924. Flowers white, throat yellow.' The material comprises almost leafless upper parts with spent inflorescences; floral details are: calyx tube 0.7mm, corolla tube 8mm, anticous filaments 1.8mm, style and ovary glabrous. It thus accords pretty well with *S. decipiens*, but the stems are terete. *Acocks* 1997 (PRE) from 'Uniondale Poort, 2700ft, steep rocky fynbos, occasional under bushes, scrambling, leaves grey, flowers mauve. 15 xi 1958' approaches *S. decipiens* very closely, differing slightly in its nearly terete stems, less distinctly toothed leaves, slightly shorter pubescence, and blue seeds.

23. Sutera langebergensis Hilliard in Edinb. J. Bot. 48: 343 (1991).
Type: Cape, Swellendam div., 3320DC, Tradouw Pass, 6 viii 1950, *Martin* 402 (holo. NBG, iso. STE).

Perennial herb or shrublet, stems many from the base, up to c.300mm long, 2mm in diam., profusely branched, branches probably ± sprawling, often tangled, very weakly ridged from leaf bases and the midrib to the node below, sparse glandular hairs up to 0.2–0.3mm long nearly confined to narrow bands from the leaf bases down to the node below but inflorescence axes glandular-puberulous all round, stems leafy, brachyblasts wanting. *Leaves* opposite, sometimes alternate on the inflorescence, largest leaves 10–20 × 3–9mm, oblanceolate, apex subacute, base narrowed to a short petiolar part then abruptly expanded and half-clasping, margins weakly revolute, 1–3 pairs of prominent teeth or smallest leaves subentire, both surfaces with scattered glandular hairs up to 0.2mm long or nearly glabrous, lower surface gland-dotted. *Flowers* in c.4–20-flowered lax racemes very loosely panicled, solitary in axils of uppermost leaves, these rapidly smaller upwards. *Pedicels* up to 5–15mm long. *Calyx* obscurely bilabiate, tube c.0.5–0.7mm long, lobes c.2.5–3 × 0.4–0.5mm, glandular-puberulous with coarser glandular hairs as well up to 0.2–0.5mm long. *Corolla* tube c.8.5–11.5 × 2.5–3mm in throat, narrowly funnel-shaped, limb almost regular, c.9–10mm across, posticous lobes 3–3.8 × 1.8–2.1mm, anticous lobe 3.7–4.1 × 1.8–2.4mm, all oblong, white or pale mauve, throat yellow, glandular-puberulous outside, hairs up to c.0.15mm long, scattered glistening glands as well, either glabrous inside or weakly bearded. *Stamens* 4, anticous pair shortly exserted, filaments 1.5–2mm long, posticous pair included, filaments 0.8–1.3mm long, anthers 0.7–1mm long. *Stigma* 1–1.2mm long, exserted. *Style* c.7–11mm long. *Ovary* c.1 × 0.8mm, glabrous or with a few glistening glands on the sutures. *Capsules* c.3.5–4 × 2–2.5mm. *Seeds* c.0.7–0.8 × 0.3mm, pallid.

Citations:

Cape. Riversdale div., 3321CC, kloof at Garcia Forest Station, 25 v 1950, *Esterhuysen* 17239 (BOL, NBG); Garcia's Pass, c.1000ft, 29 ix 1897, *Galpin* 4379 (K, PRE); south entrance to Gysmanshoek Pass, 16 ix 1981, *van Wyk* 700 (PRE, STE). 3421AA, Korente River dam, 200m, 14 v 1979, *Bohnen* 5702 p.p. (STE).

Sutera langebergensis is known from four sites in the Langeberg, from Tradouw Pass in the west to Garcia Forest Station in the east. Van Wyk recorded the plants as growing on a 'roadside along steep bouldery cliff-hang, west aspect, full sun', while Bohnen recorded 'steep slope ... growing in light shade.' The plants were collected in full flower in May, August and September.

In facies, the species much resembles *S. decipiens*, which may be distinguished by the profusion of long delicate glandular hairs on stems, leaves and calyx. This is probably where the relationship of *S. langebergensis* lies, and the two are known to be sympatric on or near Garcia's Pass. Four of the five specimens seen had been misidentified as *S. integrifolia*; that species differs in its terete stems very minutely glandular-puberulous all round, leaves either entire or with very small teeth, and the calyx with hairs very nearly confined to the margins of the lobes (no trace of the coarse hairs seen in *S. langebergensis*). *Sutera integrifolia* has not been recorded on the Langeberg; its area extends eastwards from Albertinia and is markedly coastal.

Bohnen 5702, cited above, comprises two twigs of *S. langebergensis* and three more twigs that are possibly of hybrid origin. Bohnen made several more collections from various sites around the Korente river dam (5963, PRE, STE; 7898, PRE, STE; 8321, STE); all these seem to be hybrid, with *S. langebergensis* a likely parent. *Sutera caerulea* and *S. decipiens* are both in the area. See also under *S. aethiopica*.

24. Sutera hispida (Thunb.) Druce, Rep.Bot. Exch. Club Brit. Isles, 1916: 649 (1917); Levyns in Adamson & Salter, Fl. Cap.Penins. 715 (1950); Kidd, Cape Peninsula. S. Afr. Wild Flow. Guide 3: 36, 9 (1983).
Lectotype (chosen here): C.B.S., sheet no. 14368 in herb. Thunberg (UPS).

Syn.: *Manulea hispida* Thunb., Prodr. 102 (1800) & Fl. Cap.ed. Schultes 473 (1823).
 M. oppositiflora Vent., Jard. Malm. 15, t. 15 (1803); O. Kuntze, Rev. Gen. Pl. 3(2): 236 (1898). Type: cult. Malmaison (iso. P ex herb. Bonpland).
 Sutera brachiata Roth, Bot. Bemerk. 173 (1807); Hiern in Fl. Cap.4(2): 272 (1904) p.p. Type not known.
 Chaenostoma hispidum (Thunb.) Benth. in Hook., Comp.Bot. Mag. 1: 376 (1836) & in DC., Prodr. 10: 356 (1846) excl. var.
 C. cuneatum Benth. in Hook., Comp.Bot. Mag. 1: 376 (1836) & in DC., Prodr. 10: 356 (1846). Lectotype (chosen here): Cape, Stellenbosch div., Hottentotshollandberge, 1000–3000 Fuss, June, *Ecklon* (K; isolecto. E, P, PRE, W, Z).
 C. integrifolium var. *parvifolium* Benth., in Hook., Comp.Bot. Mag. 1: 376 (1836), nomen (see comments below).
 Buchnera oppositifolia Steud., Nomencl. Bot. ed. 2, 1: 234, 338 (1841).
 Sutera oppositiflora (sphalm. *oppositifolia*) (Vent.) O. Kuntze, Rev. Gen. Pl. 2: 467 (1891).
 S. cuneata (Benth.) O. Kuntze, Rev. Gen. Pl. 3(2): 236 (1898).
 ?*S. integrifolia* var. *parvifolia* Hiern in Fl. Cap.4(2): 269 (1904). Type: Cape, Tygerberg, *Drège* 291 (holo. K, iso. P). See comments below.

Bushy shrublet c.100–500mm tall, branches erect, decumbent or sprawling, terete, almost hispid, hairs either very acute or minutely gland-tipped, up to 0.5–1.5mm long, glandular-puberulous as well particularly on upper parts, including pedicels, small glandular hairs often wanting or very sparse on lower parts, leafy, brachyblasts present. *Leaves* opposite, bases connate forming a small ridge across the stem, largest primary leaves 7–25(–30) × 3–18(–25)mm, ovate, cuneate or elliptic tapering or more abruptly contracted into a short clasping petiolar part, apex subacute to obtuse, margins often slightly revolute, usually with 1–5 pairs of prominent teeth, teeth occasionally obscure in upper leaves, both surfaces usually sparsely to densely hairy, hairs up to 0.5–1.25mm long, very acute, much shorter glandular hairs often present as well particularly on lower surface, rarely leaves almost glabrous, a few acute hairs present, more or less confined to margins and veins. *Flowers* solitary in the axils of the upper leaves or leaf-like bracts, sometimes running up into distinct racemes or narrow panicles, normally always in opposite pairs. *Pedicels* (lowermost) 5–20mm long. *Calyx* ± distinctly bilabiate, tube (0.6–)1–2mm long, posticous lobes 3–4.75 × 0.5–0.9mm, anticous lobes slightly broader, glandular-puberulous together with coarse acute hairs up to c.0.5–1mm long. *Corolla* tube 8–12 × 2.3–3.5mm in throat, narrowly funnel-shaped, limb almost regular, 9–18mm across, posticous lobes 3–7 × 2–5mm, anticous lobe 3.5–8 × 2.2–5mm, all oblong-elliptic, often white, sometimes pale shades of pink, lilac or mauve, tube orange or yellow, pubescent outside, hairs up to 0.8mm long, some gland-tipped, usually with 5 longitudinal bands of clavate hairs inside throat, hairs sometimes few. *Stamens* 4, anticous pair exserted, filaments 2–2.6mm long, posticous pair in throat, filaments 1–2mm long, anthers 0.8–1.2mm long. *Stigma* 1.2–2.5mm long, exserted. *Style* 6–10mm. *Ovary* 1–1.5 × 0.8–1mm, glabrous. *Capsules* c.3.5–5 × 2.2–3mm. *Seeds* c.0.7–1 × 0.4–0.6mm, pallid.

Selected citations:

Cape. Stellenbosch div., 3318DB, Paarl Mt., c.1500ft, 26 ix 1961, *Jordaan* 1277 (PRE, STE). 3318DD, Jonkershoek, 13 xi 1941, *Levyns* 8479 (BOL). Cape Peninsula, 3318CD, N slopes Devils Peak, 16 xii 1952, *Esterhuysen* 20827 (BOL, PRE). 3418AB, lower slopes of Kalk Bay mountain, 6 iv 1974, *Goldblatt* 1396 (PRE). Caledon div., 3418BD, Hangklip, 10 x 1951, *Compton* 22936 (NBG). 3419AB, Caledon, Zwartberg, 30 ix 1980, *Hilliard & Burtt* 13065 (E, PRE, S). 3419AC, Hermanus, 1 ii 1933, *Gillett* 640 (STE). Bredasdorp div., 3419CB, Fransche Kraal, 14 ix 1954, *Maguire* 2625 (NBG). 3419DC, De Dam, c.10ft, 25 iv 1962, *Acocks* 22293 (K, PRE).
Seen from the following quarter-degree squares: 3318CD, DB, DD; 3319CC; 3418AB, BA, BB, BC; 3419AA, AB, AC, AD, BA, BC, CB, DA.

In Thunberg's herbarium there are four sheets labelled *Manulea hispida*: α 14365, β 14366, γ14367, and δ 14368; the 1ast has been chosen as lectotype because it best fits Thunberg's (1800) description: *foliis ovatis serratis villosis, caule decumbente*. The varieties date from 1823 and were not given epithets, only brief diagnoses: Var. δ 14368 was described as having '*ramis secundis*'. Var. β 14366 is also *S. hispida*, and γ 14367 possibly is; var. α 14365 is very glandular and *could* be of hybrid origin (see comments below on hybrids). Unfortunately, I saw Thunberg's herbarium before I had studied the species closely and was not at that time able to place 14365 and 14367 satisfactorily.

Although there appears to be no type for the name *S. brachiata*, Roth's diagnosis against *S. foetida*, the only other species he knew, is sufficiently full to be reasonably confident that the reduction to synonymy is justified: Roth commented, *inter alia*, on the subwoody branches, opposite leaves, opposite simple hispid (*setaceus*) peduncles in the axils of the leaves, corolla tube about half an inch long (*semiuncialis*), and the hispid (*setaceus*) calyx.

Sutera hispida is confined to a relatively small area of the SW Cape, from Paarl, Jonkershoek and the Peninsula south east to the mouth of the Ratel river and the Poort in Bredasdorp division, from near sea level to c.650m. It is a constituent of scrub on mountain slopes, often growing in the shelter of rocks, or in scrub on fixed dunes near the sea, and can be found in flower in any month. The much wider distribution accorded it by, for example, Hiern in Flora Capensis and Bond & Goldblatt, Plants of the Cape Flora, is the result of persistent confusion with other species, notably *S. decipiens*.

The salient features of *S. hispida* are its terete, almost hispid, stems, opposite leaves with flowers in the axils, the hairs on stems, leaves and calyx long and very acute (often eglandular but sometimes minutely gland-tipped); shorter glandular hairs may be present as well but these are usually insignificant except on the pedicels, inflorescence rhachis and calyx. Leaf shape is variable, but the proportion of length to breadth is mostly of the order 1–2:1, very rarely 3:1.

Some variation may be due to environmental conditions or to genetic factors; for example, Wolley-Dod collected remarkably lax specimens with relatively huge leaves on 'bushy slopes towards Smitswinkel Bay' (*Wolley-Dod* 3025, K, PRE), while J. B. Gillett (789, BOL, STE) found very glandular (not hispid) specimens with tiny leaves at Cape Point, about 10km due south of Smitswinkel Bay. *Phillips* 4084 (SAM), from a valley above Smitswinkel Bay, is as glandular as Gillett's specimen, while a second sheet of the same collection (Z) bears two twigs, one glandular, one hispid. Then there are two glandular twigs in SAM, mounted with material of *S. hispida*. One twig was purportedly collected by Pappe on Devil's Peak, the other by Ecklon, but they are clearly part of a single collection. The label on the 'Ecklon' specimen is in Zeyher's hand; the locality is illegible. Another glandular specimen, with narrower leaves than the foregoing, was collected by Rehmann on Table Mountain (*Rehmann* 821, Z). This specimen is the closest match found for the type of the varietal name *S. integrifolia* var. *parvifolia*, which came from Tigerberg. Work is needed on Tigerberg and on the Cape Peninsula to investigate variation in *S. hispida* and to determine the status of the glandular plant.

Other puzzling specimens appear to be hybrids, either sporadic hybrids or the products of introgression. For instance, *S. hispida* and *S. decipiens* may hybridize where their areas meet near Tulbagh, Worcester and French Hoek. Similarly, *S. hispida* and *S. subspicata* appear to introgress where their areas meet in Bredasdorp district: *Bond* 760 (NBG) from Stanford

(3419AD), *Thompson* 3972 (STE) from Bredasdorp forest reserve (3419DC) and *Thompson* 3809 (STE) from Kok's River (3419DA) have much longer hairs on the stem than is normal in *S. subspicata*. At Papiesvlei (3419BC) Schlechter, under his no. 10442, collected a whole series of twigs that show varying degrees of intermediacy between *S. hispida* and *S. subspicata* (sheets in BOL, E, K, MO, PRE, W, Z). There must be a long history of hybridization in this area (see also under *S. calciphila*).

Bentham mentioned a plant cultivated in England together with *S. hispida* and *S. caerulea*, which was intermediate between them and possibly a garden hybrid. He gave this a tentative varietal epithet (*Chaenostoma hispidum* β ? *breviflorum* Benth. in DC., Prodr. 10: 356, 1846), tentative because the specimen he saw in De Candolle's herbarium (G-DC) had come from Paris labelled *Manulea oppositiflora* Vent. But Ventenat's plant is *Sutera hispida* and can be matched perfectly with plants from the wild. I have not had the opportunity to examine the specimen Bentham saw but what appears (from the microfiche of De Candolle's herbarium) to be the same plant was cultivated in European gardens as *Manulea oppositifolia* (sphalm.) and I have seen specimens in E, LE and W, the latter herbarium having several sheets from a number of different garden sources. This plant is very glandular, the flowers mostly alternate, the corolla tube short and the style glandular-pubescent (characters of *S. caerulea*) and the pollen is mostly malformed, so it is surely as Bentham suspected, a hybrid between *S. caerulea* and *S. hispida*. No wild hybrids have been seen but as *S. hispida* and *S. caerulea* appear to be wholly allopatric, this is scarcely surprising.

The cross may have taken place more than once in cultivation because a plant grown in the Royal Botanic Garden Edinburgh as *S. hispida*, and which came from the Longwood garden in Philadelphia, is also a hybrid, with anthers completely devoid of pollen. Exactly the same plant was in cultivation in Boissier's garden at Vaud, Switzerland, before 1888 (herb. J. Vetter, 27 viii 1888, Z). It is a compact shrublet with opposite leaves, the flowers in the upper leaf axils, the limb white, throat yellow, and thus has the aspect of *S. hispida*. But it differs in its soft glandular pubescence and sterile anthers. The glandular pubescence and the presence of a few tiny glandular hairs on the style hint at the possibility that *S. caerulea* was the other parent. As this plant is still in cultivation, it is to be known by the cultivar name 'Longwood'.

---

25. Sutera marifolia (Benth.) O. Kuntze, Rev. Gen. Pl. 2: 467 (1891); Hiern in Fl. Cap. 4(2): 270 (1904).
Lectotype (chosen here): Cape, Uitenhage div., Vanstaadensriviersberge, xii, *Ecklon* [& *Zeyher*] (K; isolecto. BOL, E, LE, M, MO, P, PRE, S, SAM, W, Z).

Syn.: *Chaenostoma marifolium* Benth. in Hook., Comp. Bot. Mag. 1: 376 (1836) & in DC., Prodr. 10: 356 (1846), excl. syn.

Perennial herb, virgately branched from base, eventually developing into a twiggy shrublet, stems up to c.150–400mm long, erect, decumbent or possibly sometimes prostrate, terete, glandular-puberulous, but these hairs masked by long (up to 2mm) delicate, cobwebby-woolly, greyish-white hairs, leafy throughout. *Leaves* opposite, largest c.5–22 × 2.5–13mm, elliptic, acute to obtuse, base narrowed into a short petiolar part then abruptly expanded and connate, margins distinctly toothed with c.4–10 pairs of teeth to subentire, slightly revolute, hairs as on stem, sometimes thinner on upper surface. *Flowers* solitary in axils of upper leaves, these sometimes alternate upwards, the lowermost c.2–12 × 1.5–8mm, eventually forming well-branched leafy panicles, simple terminal racemes if branching very lax. *Pedicels* (lowermost) 2–9mm long. *Calyx* distinctly bilabiate, tube 0.7–1mm long, posticous lobes 2.4–4.2 × 0.6–1mm, anticous lobes slightly broader, glandular-puberulous, but masked by long cobwebby-woolly hairs. *Corolla* tube 9–13 × 2–3mm in throat, narrowly funnel-shaped, limb nearly regular, 8–10mm across, posticous lobes 2.6–4.6 × 1.8–3mm, anticous lobe 3–4.6 × 1.8–2.8mm, all oblong elliptic, often white, sometimes pale pink or mauve, tube yellow or orange, ±

cobwebby outside, hairs up to c.0.5–0.6mm long, glandular-puberulous as well, weakly bearded inside with clavate hairs on posticous side. *Stamens* 4, anticous pair exserted, filaments 1.4–2.2mm long, posticous pair included, filaments 1–1.2mm long, anthers 0.7–1mm long. *Stigma* 1–2.5mm long, exserted. *Style* 7–9.5mm long. *Ovary* c.1.2 × 0.8mm, a few glistening glands on sutures. *Capsules* 2.5–3.5 × 2.5–3mm. *Seeds* 0.8–1 × 0.4mm, grey-blue.

Selected citations:

Cape. Uniondale div., 3323DD, Twee Rivieren, 29 xi 1941, *Esterhuysen* 7054 (K, BOL); Joubertina, 25 xi 1941, *Esterhuysen* 6915 (BOL, K, PRE). Uitenhage div., 3325CA, Groendal Wilderness Reserve, Chase's Kloof, 1850m, 9 iv 1974, *Scharf* 1352 (PRE). 3325CC, Van Staadens, v 1914, *Paterson* 2260 (BOL, Z).

Sutera marifolia has a restricted distribution in the southern Cape fold mountains, from about Prince Alfred's Pass in the west to Vanstadensberg in the east, having been recorded on the mountains at the western end of the Lange Kloof, the northern flanks of the Tsitsikamma Mountains, the southern flanks of the Kougaberge and the Elandsberge, which terminate in the Vanstadensberg. The plants favour rocky sites, along rivers and on mountain slopes, and can possibly be found in flower in any month.

The greyish-white cobwebby indumentum on all parts including the calyx and corolla (though the hairs may be rather scanty there) makes this a very distinctive species; similar indumentum is found only in *S. cinerea*, which has narrow leaves and a much shorter corolla tube.

26. Sutera cinerea Hilliard in Edinb. J. Bot. 48: 342 (1991).
Type: Cape, Willowmore div., 3324CB, Klein Rivier, along Enkeldoorn track, 28 ix 1980, *Snijman* 341 (holo. NBG, iso. PRE).

Perennial, virgately branched from base, forming a small shrublet, stems up to 150–400mm long, erect, decumbent or semiprostrate, terete, glandular-puberulous but these hairs masked by whitish woolly-cobwebby hairs, leafy. *Leaves* opposite, largest c.15–20 × 2–3mm, narrowly elliptic or linear, acute, base slightly narrowed, clasping, connate, margins entire or with a few tiny teeth, both surfaces woolly-cobwebby. *Flowers* in short terminal racemes, mostly alternate. *Bracts* (lowermost) c.2.5–6 × 1–2mm, lanceolate, glandular-puberulous, often with cobwebby hairs as well. *Pedicels* (lowermost) 4–12mm long, glandular-puberulous. *Calyx* distinctly bilabiate, tube 0.5–0.7mm long, posticous lobes c.2.7–3 × 0.4–0.8mm, anticous lobes slightly broader, glandular-puberulous, some longer hairs as well, up to 0.25–0.6mm. *Corolla* tube c.5–5.5 × 3.2–3.7mm in throat, broadly funnel-shaped, limb c.7–8mm across, posticous lobes 2.5–3 × 1.8–2.1mm, anticous lobe 3 × 2–2.5mm, all elliptic, pale mauve, tube yellow, pubescent outside, hairs up to c.0.2–0.3mm long, some glistening glands as well, inside well bearded with clavate hairs all round throat. *Stamens* 4, all exserted, anticous filaments c.2.2–3.7mm long, posticous filaments 1.6–2mm, anthers 0.8–1.1mm long. *Stigma* 1.1–2mm long, exserted. *Style* 4–5mm long. *Ovary* 1–1.3 × 0.7–1mm, a few glistening glands on sutures. *Capsules* 2.5–3 × 2–2.5mm. *Seeds* c.0.8–1 × 0.4mm, grey-blue.

Specimens seen:

Cape. Willowmore div., 3324CB, Baviaanskloof-Enkeldoorn track, 515m, ii 1977, *Bond* 916 (NBG); Goedehoop Farm, 1 x 1981, *Zantovska* 126 (PRE). Humansdorp div., S.W. aspect above Combrink, c.1800ft, 29 iv 1947, *Acocks* 13697 (PRE).

Sutera cinerea is known from only four collections, three of them made on the south-facing lower slopes of the Baviaanskloof Mountains, south of Steytlerville; the fourth came from Combrink, which I failed to trace, but it possibly lies below these mountains further to the east. The plants were found growing in scrub, flowering and fruiting in September, January and February. The species is closely allied to *S. marifolia*: it has similar cobwebby indumentum, very small capsules, and grey-blue seeds, but it is at once distinguished by its differently shaped leaves (linear to narrowly elliptic, not broadly elliptic) and shorter, broader corolla tube with all four stamens exserted. *Sutera marifolia* has not been recorded from the Baviaanskloof Mountains, but only on the mountains to the south and east of them.

---

27. Sutera subspicata (Benth.) O. Kuntze, Rev. Gen. Pl. 2: 467 (1891); Hiern in Fl. Cap. 4(2): 287 (1904) p.p. excluding *Schlechter* 10442 p.p.
Type: Cape, 3419DA, zwischen Honigklip und Hagelkraal, 500–1000Fuss, 25 v 1833, *Drège* 7908b (holo. K; iso. LE, MO, P, W).

Syn.: *Chaenostoma subspicatum* Benth. in Hook., Comp. Bot. Mag. 1: 376 (1836) & in DC., Prodr. 10: 356 (1846).

Stiff gnarled dwarf shrublet c.50–300mm tall, branches stout, erect, decumbent or ± prostrate, pubescent in two narrow bands down length of each internode, hairs up to 0.2–0.3mm long, gland-tipped or not, old parts of branches bare, upper parts closely leafy, often with small brachyblasts. *Leaves* opposite, thick-textured, usually sharply ascending, ± appressed, imbricate, 3–15 × 1.5–7mm, obovate, elliptic or cuneate in outline, apex acute to subobtuse, margins with 1–3(–4) pairs of prominent teeth, revolute, base clasping, decurrent, lower surface glandular-punctate, hairs either wanting or few on blade, always present on lower part of margins and carried down the internode in a narrow band. *Flowers* solitary in the upper leaf axils forming short crowded (rarely lax) racemes sometimes branching into narrow panicles, vegetative growth continuing above floriferous part then flowering again. *Pedicels* 1–6(– 13)mm long, stout, glandular-puberulous. *Calyx* bilabiate, tube 0.8–2mm long, posticous lobes 3–6 × 0.8–1.7mm, anticous somewhat broader, always puberulous on the margins otherwise glabrous to thickly beset with coarse hairs 0.2–0.7mm long. *Corolla* tube 9–15 × 2.3–4mm in throat, narrowly funnel-shaped, limb nearly regular, 9–15mm across, posticous lobes 3–7 × 1.8–3.8mm, anticous lobe 3.5–7 × 2.2–4.2mm, all oblong, usually shades of pink, occasionally white, tube sometimes purplish, throat yellow or orange, glandular-puberulous outside, hairs 0.15–0.25mm long, a few glistening glands as well, 5 longitudinal bands of clavate hairs inside, sometimes coalescing. *Stamens* 4, anticous pair shortly exserted, filaments 2–2.7mm long, posticous pair included, filaments 1–1.5mm long, rarely with a few clavate hairs near base, anthers 0.8–1.3mm long. *Stigma* 1.4–2.5mm long, exserted. *Style* 7–13.5mm long. *Ovary* 1.2–1.8 × 0.8–1.2mm, glabrous. *Capsules* 4–7 × 2–3mm. *Seeds* (few seen) c.0.8–1 × 0.3– 0.5mm, pallid.

Selected citations:

Cape. Bredasdorp div., 3419CB, near Strandkloof, 12 vi 1951, *Martin* 350 (NBG). 3419DA, Klein Hagelkraal just east of Pearly Beach, c.500ft, 10 iv 1979, *Hugo* 1719 (PRE, STE). 3420AC, Van der Stel's Kraal, c.400ft, 1 viii 1968, *Acocks* 24043 (BOL, PRE). 3420AD, De Hoop area, NE of Hardevlakte, 200ft, 20 vi 1984, *Oliver* 8519 (E, STE). 3420CA, The Poort, 4 viii 1940, *Compton* 9024 (NBG).

Sutera subspicata is endemic to limestone substrates in Bredasdorp division, roughly from Stanford and Pearly Beach to Cape Infanta, from near sea level on consolidated dunes up to c.200m on hills, as a constituent of low scrub, and flowering mainly between April and October. It is a most distinctive species, easily recognized by its stems with hairs confined to two narrow bands running down each internode from the bases of the leaf pairs, thick and coarsely toothed imbricate leaves, and terminal racemes of crowded flowers eventually overtopped by further growth, resulting in knots of old capsules below the young flowering shoots. *Schlechter* 10442 (from Papiesvlei, ESE of Stanford), cited by Hiern as both *S. subspicata* and *S. aethiopica*, is a mixture of *S. subspicata, S. hispida* and possible hybrids between them. Sheets in various herbaria may carry only *S. subspicata*, or a mixture. See further under *S. hispida*.

28. Sutera titanophila Hilliard in Edinb. J. Bot. 48: 345 (1991).
Type: Cape, Bredasdorp div., 3420AD, De Hoop-Potberg Nature Reserve, 15m, 16 x 1978, *Burgers* 1356 (holo. STE).

Twiggy gnarled shrublet up to c.200mm tall, main branches up to c.6mm in diam., bark pale, twigs terete, glandular-puberulous, hairs c.0.1–0.3mm long, closely leafy. *Leaves* opposite, largest c.6–10 × 3–5.5mm, ± oblanceolate, narrowed to a short petiolar part, base abruptly expanded, half-clasping, connate, apex subacute, margins weakly revolute, usually with 1–3 pairs of teeth, occasionally entire, both surfaces glandular-puberulous, hairs up to 0.1–0.3mm long. *Flowers* 2–6 in terminal racemes sometimes overtopped by new growth, panicled by the close branching of the stems. *Bracts* (lowermost) c.2.5–5 × 1–3mm, leaflike. *Pedicels* (lowermost) c.3–7mm long. *Calyx* bilabiate, tube 0.7–1mm long, posticous lobes 2.5–3.1 × 0.8mm, anticous lobes slightly broader, minutely glandular-puberulous with scattered coarse hairs as well up to c.0.4–0.5mm long. *Corolla* tube c.7–10 × 3mm in throat, narrowly funnel-shaped, limb almost regular, c.9–11mm across, posticous lobes c.3–3.5 × 2.5mm, anticous lobe 3.5–4 × 2.3–2.8mm, all oblong-elliptic, white to pale pink or mauve (probably changing colour as the flower ages), throat yellow/orange, glandular-pubescent outside, hairs c.0.25mm long, clavate hairs inside throat arranged in 5 longitudinal bands. *Stamens* 4, anticous pair shortly exserted, filaments 2–2.5mm long, posticous pair included, filaments 1.2–1.3mm long, anthers 0.8–1.2mm long. *Stigma* c.0.5–0.8mm long, exserted. *Style* 6.5–9mm long. *Ovary* 1–1.5 × 0.8–1mm, glabrous. *Capsule* (only 1 seen) 4.5 × 2.8mm. *Seeds* shed.

Citation:

Cape. Bredasdorp div., De Hoop, 1971, *van der Merwe* 1894 (STE).

Sutera titanophila appears to be yet another species confined to the limestone of Bredasdorp division. It is known from only two collections made on the margins of De Hoop vlei, where the plants grow in cracks in the limestone cliffs. The type material was in full bloom in mid-October; the second collection has only one flower and a few old pedicels, and is dated to year only. Burgers recorded the flowers as 'white or light purple', van der Merwe 'off white or pale pink', which suggests that they change colour with age.

The relationship of the species possibly lies with *S. aethiopica*, from which it differs in its glandular-pubescent leaves and shorter calyx tube and calyx lobes; the stigma is also remarkably short (less than 1mm long), but the constancy of this character needs checking on further material. The type sheet was confused with *S. hispida*, which differs in the hispid hairs up to 0.5–1.5mm long on stems and leaves and more flowers in the inflorescence, these being mostly in the axils of the uppermost leaves. A glandular variant of *S. hispida* is mentioned under that species, but even those specimens have longer coarser hairs than *S. titanophila*.

29. Sutera aethiopica (L.) O. Kuntze, Rev. Gen. Pl. 2: 467 (1891); Hiern in Fl. Cap. 4(2): 269 (1904), p.p.
Type: C.B.S., campis arenosis, *Tulbagh* (LINN 790.6).

Syn.: *Buchnera aethiopica* L., Mant. alt. 251 (1771).
　　Manulea aethiopica (L.) Thunb., Prodr. Pl. Cap. 101 (1800).
　　Chaenostoma aethiopicum (L.) Benth. in Hook., Comp. Bot. Mag. 1: 375 (1836) & in DC., Prodr. 10: 355 (1846), p.p., quoad typus tantum.
　　C. fastigiatum Benth. in Hook., Comp. Bot. Mag. 1: 376 (1836) & in DC., Prodr. 10: 356 (1846) excl. syn. Type: Cape, Caledon div., 3419AD, Babylonstorensbergen, August, *Ecklon* (holo. K; iso. LE, MO, P, S, W).
　　C. fastigiatum var. *glabratum* Benth. loc. cit., nomen. Specimens cited: Caledon div., 3419AD, Klynriviersberge, *Ecklon* (K, SAM, W) and near Caledon, *Drège* (K, P).
　　Sutera cephalotes Hiern in Fl. Cap. 4(2): 268 (1904), nom. illegit., non *S. cephalotes* (Thunb.) O. Kuntze.
　　S. cephalotes var. *glabrata* Hiern loc. cit. Lectotype (chosen here): Caledon div., Klein River Mountains, *Ecklon* (K; isolecto. SAM, W).
　　S. fastigiata (Benth.) Druce in Rep. Bot. Exch. Club Brit. Isles 1916: 649 (1916).
　　Polycarena aethiopica (L.) Druce in Rep. Bot. Exch. Cl. Brit. Isles 1916: 641 (1917).

Twiggy shrublet, many-stemmed from crown of thick woody stock, branches erect or decumbent, up to 150–300mm long, main ones c.2–4mm in diam. at base, either weakly ridged by prominent vascular strands running up into the leaf bases or ± terete, glandular-puberulous, often with scattered ± hispid hairs as well, these up to c.0.5–1mm long, closely leafy on upper parts, brachyblasts present. *Leaves* opposite, largest primary leaves 6–20 × 1.5–5mm, oblanceolate or narrowly cuneate, acute to subobtuse, base narrowed to a short petiolar part then abruptly expanded, half-clasping, ± connate, margins strongly to weakly revolute, usually with 1–2(–3) pairs of prominent teeth, smaller leaves sometimes entire, or teeth hidden by revolute margins, sometimes minutely glandular-puberulous on petiolar part, lower surface glanddotted, otherwise leaf glabrous or with scattered ± hispid hairs up to c.0.2–1mm long, these hairs sometimes reduced to tiny scabrid projections. *Flowers* up to c.16 in short terminal racemes, solitary in the axils of bracts or the lowermost pair in the axils of the uppermost pair of leaves. *Bracts* (lowermost) or uppermost pair of leaves c.3–10 × 1.5–3.5mm, lanceolate, otherwise leaf-like. *Pedicels* (lowermost) 4–20mm long. *Calyx* bilabiate, tube 1.2–2.2mm long, posticous lobes 3.6–5 × 0.6–1.1mm, anticous lobes slightly broader, minutely glandular-puberulous with scattered coarse hairs as well up to 0.5–1mm long. *Corolla* tube 10–19.5 × 3mm in throat, narrowly funnel-shaped, limb almost regular, 10–14mm across, posticous lobes 3.5–6.5 × 2–3mm, anticous lobe 4.2–6.5 × 2.2–3.2mm, all oblong-elliptic, shades of pink or violet, rarely white, throat yellow/orange, glandular-pubescent outside, hairs up to 0.2–0.3mm long, a few glistening glands as well, clavate hairs inside throat arranged in 5 longitudinal bands. *Stamens* 4, anticous pair shortly exserted, filaments 1.7–2.5mm long, posticous pair included, filaments 1–1.5mm long, anthers 0.8–1.6mm long. *Stigma* 2.5–4mm long, exserted. *Style* 8–15mm long. *Ovary* 1.4–1.8 × 0.8–1mm, either glabrous or with a few glistening glands on the sutures. *Capsules* c.4–6 × 2–2.5mm. *Seeds* c.0.5–0.8 × 0.3–0.4mm, grey-blue or pallid.

Selected citations:

Cape. Caledon div., 3419AB–BD, between Caledon and Napier, 3 viii 1940, *Esterhuysen* 3046 (BOL). 3419BB, east of Rivierzonderend, 4 viii 1951, *Johnson* 100 (NBG). Swellendam div., 3420BB, near Heidelberg, 10 viii 1949, *Steyn* 349 (NBG). Bredasdorp div., 3419BD, 11 miles SW of Bredasdorp, 19 ix 1934, *Salter* 4844 (BOL, K).
Seen from following quarter-degree squares: 3319DD; 3419AD, BB, BD, DB; 3420AA, AB, AD, BA, BB, BD, CA.

Sutera aethiopica is characterized by its minutely glandular-puberulous stems often with long (up to 1mm) scattered almost hispid hairs as well, by its leaves, the largest ones narrowly to broadly wedge-shaped with mostly 1–2, or occasionally 3, pairs of coarse teeth in the upper half (teeth often obscured by the revolute margins), the blade either glabrous or with scattered almost hispid hairs up to 1mm long, but sometimes represented by no more than tiny projections, and by its few-flowered terminal racemes, the calyx with prominent ± hispid hairs as well as minute glandular ones. The corolla tube is long (c.10–20mm), the anticous filaments short (1.7–2.5mm). The type specimen of *S. aethiopica* has glandular-puberulous stems, the hairs on the young parts up to c.0.25mm long, and leaves with relatively few scattered hairs up to 0.2mm long. In contrast, the type specimen of *S. fastigiata* has long (c.0.5–0.8mm) scattered hairs on stems and leaves (in addition to minute glandular hairs on the stem), and so do two of the three specimens singled out by Bentham (who was followed by Hiern) as var. *glabratum*: only the left-hand 'b' specimen in Bentham's herbarium has very few hairs on the leaves. However, the shape of the leaves, their toothing, and lack of any short glandular pubescence, contrasting with the presence of such hairs on the stem, is so characteristic that I have no hesitation in uniting *S. fastigiata* with *S. aethiopica*.

Typical *S. aethiopica* is almost confined to Caledon, Swellendam, Bredasdorp and Riversdale divisions, from the Babylonstoringberge and Kleinriviersberge, north and east of Hermanus in the west, to Riversdale in the east, and north to Riviersonderend, Stormsvlei, Swellendam and Heidelberg, with one record from Robertson; there are doubtful records east to the environs of Mossel Bay; these are mentioned below.

Barker 8748 (NBG), from De Hoop Provincial Farm, is atypical: the stems and leaves are those of *S. aethiopica*, but the inflorescence is panicled, many-flowered, and the corolla tube is 9mm long, the anticous filaments 3mm; these last four characters point to *S. calciphila* with which *S. aethiopica* is here sympatric.

Specimens collected by Ecklon & Zeyher and widely distributed under 3503 and 3503b (K, P, PRE, S, SAM, W, Z) are a mixture of *S. aethiopica* and what may be hybrids with *S. uncinata*; the putative hybrids have weakly to quite strongly ridged stems and entire leaves with strongly revolute margins. The locality is given on a sheet in Z as 89.9, that is, 'Rivierzondereinde bei Stormsvalei, Hassaquaskloof u.s.w. bei Breederivier', but a label written by Zeyher himself (3503, with b added possibly by another hand, in W) gives 'Hassaquaskloof und Malabarshoogde; und Stormsvalley.' A sheet of 3503b in P (written in a clerk's hand) gives 'Swellendam. Secus Rivierzondereinde, ad Appelskraal vicinosque montes.' It is notoriously difficult to localize Ecklon & Zeyher collections, especially as collections from more than one locality were distributed under one number, but these particular specimens almost certainly came from somewhere near Stormsvlei, as they are matched precisely by a modern collection from that area (*Oliver* 6011, STE, from Korlands, just west of Stormsvlei, 7 ix 1975). If hybridization is a factor to be considered, then it is of very long standing. It may also be of significance that it is in this area, from Riviersonderend east to Heidelberg, a linear distance of c.75km, occur nearly all the specimens of *S. aethiopica* lacking the long-hair component on stems and leaves.

There are several collections of *S. aethiopica* from the Bontebok National Park at Swellendam, roughly midway between Riviersonderend and Heidelberg. One of them, *Liebenberg* 6454 (PRE), consists of one twig of *S. aethiopica*, while a second twig is a possible hybrid with *S. aethiopica* as one parent: the stems are clad in a short (c.0.2mm) coarse almost hispid pubescence, unlike the normal short fine glandular pubescence of *S. aethiopica*.

Four specimens from the vicinity of Riversdale come close to typical *S. aethiopica*: *Schlechter* 1954 (Z) from Riversdale, *Winkler* 70 (NBG), collected between Riversdale and Albertinia, *Levyns* 3547 (BOL) from Oakdale just outside Riversdale, and *Bolus* 11354 (BOL) 'prope Riversdale', but differ in the rather short calyx tube. They may be of hybrid origin; under *S. caerulea* I have mentioned possible hybrids between that species and *S. calciphila* or *S. integrifolia* in the Riversdale-Albertinia area, and possible *S. langebergensis* × *caerulea* or *decipiens* occur around the Korente River dam, north west of Riversdale.

But the problems extend east to the environs of Mossel Bay. Five specimens from that area may be *S. aethiopica* introgressed by another species. They are: *Lewis* s.n. (BOL) from Little Brak River on the shores of Mossel Bay; *Van Niekerk* 104 (BOL) from Ruytersbosch, about 20km in a straight line NNE of Mossel Bay on the road to Robinson Pass at the western end of the Outeniqua Mountains; *Ryder* 54 (K), near Great Brak river; *Burchell* 6337 (K), between Mossel Bay and Zoute River; *Wall* s.n. (S), Blanco. They have the inflorescence and floral characters of *S. aethiopica*, but differ in having the pubescence on the stems (glandular hairs up to c.0.2mm long) in two bands running from the bases of each pair of leaves down to the node below; the leaves are virtually glabrous and either entire or with up to three pairs of teeth. Five more collections can be associated with these two: they have similar pubescence on the stems (though two of the sheets include twigs that are hairy all round), somewhat similar leaves (glabrous, but toothing different), and the calyx tube is shorter (1–1.2mm versus 1.5–1.7mm). These specimens are: *Compton* 23503 p.p. (NBG), 9 miles W of Mossel Bay; *Muir* 1627 (BOL, PRE), Botteliersfontein near Albertinia; *Bohnen* 8540 (STE), from the upper reaches of the Soetmelk river; *Compton* 7184 (NBG), near the Gouritz river, Albertinia; *Barker* 7939 (NBG), Gouritz Bridge, Riversdale to Mossel Bay.

A thorough field study is needed in the area from Riversdale to Mossel Bay and up into the foothills of the Langeberge. The need for similar studies in Riversdale and Bredasdorp divisions is mentioned under *S. calciphila*. These problems are intractable without such a study.

30. Sutera calciphila Hilliard in Edinb. J. Bot. 48: 342 (1991). **Fig. 56.**
Type: Cape, Caledon div., [?] [3420CA/CC], near Struys Bay, x 1940, *Esterhuysen* 5119 (holo. BOL).

Perennial herb, taproot becoming woody, up to c.8mm diam. at crown, stems many from crown, loosely branched, branches c.130–310mm long, either prostrate and rooting from lower nodes, or decumbent to erect eventually forming a twiggy shrublet, terete to weakly ridged by prominent vascular strands, glandular-puberulous, hairs up to 0.1–0.2mm long, leafy, often with axillary leaf tufts. *Leaves* opposite, largest primary ones c.10–20 × 3–8mm, more or less oblanceolate, occasionally upper ± linear, tapering to a short petiolar part, apex acute or subacute, margins either entire or with 1–4 pairs of small teeth, usually revolute, both surfaces hairy as stems, glistening glands as well particularly on lower surface. *Flowers* in small (up to c.20 flowers) racemes terminating the twiglets and producing ± corymbose panicles. *Bracts* (lowermost) c.2–10 × 1–2mm, linear or lanceolate, acute or subacute, entire or with 1 or 2 pairs of teeth. *Pedicels* (lowermost) 6–15mm long. *Calyx* obscurely bilabiate, tube 0.7–1(–2)mm long, lobes 1.8–5 × 0.7–1mm, glandular-puberulous with coarser hairs c.0.2–0.8mm long, glistening glands as well. *Corolla* tube 4.7–8.5 × 3.5–5mm in throat, broadly funnel-shaped, limb almost regular, 8–16mm across, posticous lobes 2.7–5.5 × 1.6–5mm, anticous lobe 2.8–6.3 × 1.5–6.2mm, all oblong-orbicular, mauve-pink to violet-blue, tube orange, glandular-puberulous outside, glistening glands as well, clavate hairs inside in 5 longitudinal bands. *Stamens* 4, all exserted, anthers 0.8–1.3mm long, posticous filaments 1.3–2.5mm long, anticous ones 2.3–4mm. *Stigma* 1–2.5mm long, exserted. *Style* 3–7mm long. *Ovary* 1.2–1.8 × 0.8–1mm, glistening glands on sutures, sometimes very few. *Capsules* 3–5.5 × 2–3mm. *Seeds* (few seen) c.0.5 × 0.4mm, pallid.

Selected citations:

Cape. Bredasdorp div., 3419BD, Mierkraal, 250ft, 24 iv 1897, *Schlechter* 10499 (BOL, E, K, PRE, S, W, Z, sometimes mixed with *S. hispida*). 3420AD, De Hoop, 9 iv 1957, *Barker* 8695 (NBG, STE); farm Melkkamer near De Hoop, 60m, 15 v 1984, *van Wyk* 1508 (PRE, STE). Riversdale div., 3421AC, Puntjie, 18 v 1950, *Esterhuysen* 16968 (BOL).

Sutera calciphila is confined to the limestone of Bredasdorp and Riversdale divisions from about Bredasdorp and Struys Bay east to Still Bay, a constituent of scrub up to c.80m above sea level. Flowers can be found in any month.

It is not impossible that most of the specimens included under *S. calciphila* are of hybrid origin between *S. calciphila* and *S. campanulata*, or perhaps *S. placida* or *S. aethiopica*, but they form a recognizable entity and it seems best to treat them as a species. I have chosen as the type of the name the most distinctive specimen, which came from somewhere near Struys Bay (Struisbaai). Struys Bay lies south of Bredasdorp and well west of the nearest known locality for both *S. campanulata* and *S. placida* (Still Bay), but within the area of *S. aethiopica*. The type specimen has long simple to sparingly branched stems rooting at the lower nodes, oblanceolate toothed leaves, almost velvety glandular pubescence on stems, leaves and pedicels (hairs up to 0.2mm long) and up to ten flowers in a short terminal raceme; in October the plants were in first flower. I have associated with it plants with similar indumentum and relatively few-flowered inflorescences (up to c.20 flowers), but differing in habit (stems decumbent to erect) and considerable variation in leaf-width and toothing. The three specimens most strikingly different in facies came from Puntje and nearby (mouth of the Duivenhoks river, 3421AC), namely *Esterhuysen* 16968 (BOL) and *O'Callaghan et al.* 444 (STE), and *Marloth* 3558 (PRE) from 'Riversdale flats, 60m', which is the same general area. They have precisely the indumentum of the plant from Struys Bay, but the stems are erect and the leaves narrow and entire to sparingly toothed, the narrowness of the leaves being exaggerated by their strongly revolute margins, which also tend to hide any toothing. These three specimens have the facies of *S. placida*, known from a number of collections made at or near Still Bay; *S. placida* has linear leaves with strongly revolute margins, the margins either entire or with a pair of teeth,

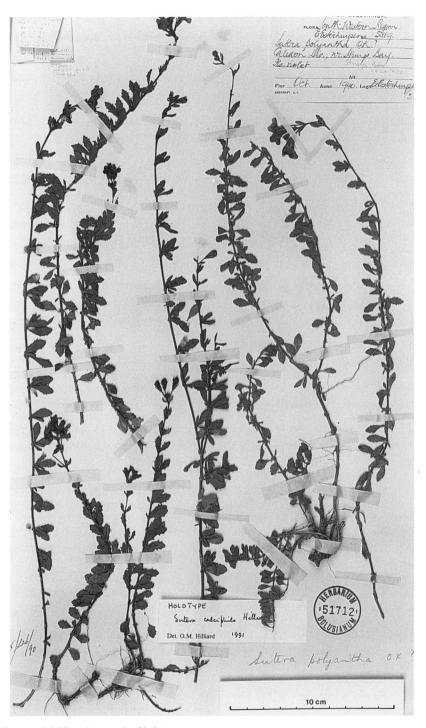

Fig. 56. *Sutera calciphila*: photograph of holotype.

and the blade glabrous or with a few coarse hairs; the stems are narrowly winged by decurrent leaf bases, the pubescence being confined to the grooves so formed. All the other specimens differ from the type mainly in habit: stems decumbent to erect. They have somewhat the aspect of *S. campanulata*, which differs in the long (0.3–0.7mm) acute hairs at least on the lower part of the stem: *Jordaan* 31827 (STE) from 'Stilbaai' is *S. campanulata*, *Bohnen* 7611 (PRE, STE) from 'Still Bay Hoogte' is *S. calciphila*.

The problem is compounded by a further set of specimens with much the same range of variation in leaf-shape as the specimens included under *S. calciphila*, but differing in indumentum: they all have a long-hair component, which could have been derived from either *S. aethiopica* or *S. campanulata* (stem hairs 0.4–0.8mm, leaf hairs 0.3–0.6mm). *Thompson* 703 (K, PRE, STE), collected '18 miles from Still Bay to Riversdale' approaches *S. aethiopica* in indumentum, but the corolla tube is short (7mm) and the anticous filaments long (3.8mm), as in *S. calciphila* (also *S. placida* and *S. campanulata*). *Oliver* 5984 (STE), from Wankoe se Rante, between Puntje and Still Bay, has the corolla tube 8mm long, anticous filaments 4mm, and could have had the same parentage. Other putative hybrids are *Esterhuysen* 16966 (BOL) and *Walgate* 825 (BOL), both from Whitesands (3420BD), *Brynard* 237 (PRE), from the old Bontebok Park at Bredasdorp (3420CA), *Snijman* 829 (NBG), farm Kleinheuwel south of Bredasdorp (3420CA), and *Blum* 123 (E) from Cape Infanta (3420BD).

Obviously the whole area of calcareous soils from Struys Bay to Still Bay and inland to Bredasdorp and Riversdale needs careful scrutiny to test the hypothesis that complex hybridization is taking place. See also under *S. aethiopica* and *S. caerulea*.

31. Sutera placida Hilliard in Edinb. J. Bot. 48: 344 (1991).
Type: Cape, Riversdale div., 3421AD, Still Bay Hills, 4 viii 1951, *Johnson* 116 (holo. NBG, iso. STE).

Perennial herb or shrublet, basal parts not seen, stems c.100–250mm long, c.1–2mm in diam., erect, decumbent or ascending, either simple or loosely branched, narrowly winged by decurrent leaf bases, glandular-pubescent with hairs c.0.2–0.3mm long confined to the groove between the pairs of wings, inflorescence axes glandular-puberulous all over, stems leafy, axillary leaf tufts either wanting or very small. *Leaves* opposite, sometimes becoming alternate upwards, largest leaves c.7–15 × 1–2mm, linear to oblanceolate, uppermost sometimes lanceolate, apex obtuse, base not narrowed, half-clasping, closely approaching opposite leaf then decurrent to node below forming a narrow groove, margins mostly strongly revolute (contributing to narrowness of leaves), entire or with 1–2 pairs of teeth near apex, either glabrous or with a few scattered hairs up to c.0.25mm long, scattered glistening glands on lower surface. *Flowers* in terminal racemes (up to c.20 flowers) often loosely panicled. *Bracts* (lowermost) c.2.5–10 × 1–3mm, lanceolate, leaf-like. *Pedicels* (lowermost) up to c.30mm long, but often only c.2–15mm. *Calyx* obscurely bilabiate, tube c.0.7–1mm long, posticous lobes c.3.5–4.3 × 0.7–0.8mm, anticous lobes slightly broader, minutely glandular-puberulous, thickly beset with much coarser hairs as well, these c.0.4–0.6mm long. *Corolla* tube c.6–8 × 5–6mm in throat, broadly funnel-shaped, limb almost regular, c.11–16mm across, posticous lobes c.4.5–6 × 3.3–5mm, anticous lobe c.4.5–6 × 3.3–5.2mm, all oblong-orbicular, mauve, throat yellow, glandular-puberulous outside, a few glistening glands as well, clavate hairs inside in 5 longitudinal bands. *Stamens* 4, all exserted, anticous filaments 3–4.2mm long, posticous filaments 1.8–2.2mm, anthers 1–1.1mm. *Stigma* c.3–4mm long, exserted. *Style* 4–5.2mm long. *Ovary* c.1.2 × 1mm, a few glistening glands on sutures. *Capsules* 3–5 × 2–3mm. *Seeds* c.0.7–0.8 × 0.3–0.4mm, pallid.

Citations:

Cape. Riversdale div., 3421AD, Still Bay Forest Reserve, viii 1934, *Laver* 20 (STE); near sea at Melkhoutfontein, vi 1924, *Muir* 3125 (SAM, PRE); Botterkloof north of Still Bay, 500ft, 21 iii 1975, *Oliver* 5763 (STE).

Sutera placida is so-named because it is known only from the environs of Still Bay. Oliver recorded 'limestone hills, open short vegetation' and flowering takes place between March and August. All the specimens seen had been determined as *S. linifolia* (that is, *S. uncinata*), with which *S. placida* shares the character of narrowly winged stems, but the leaves of *S. uncinata*, even the narrowest ones, are narrowed towards the base then abruptly expanded (not narrowed in *S. placida*), the calyx tube is mostly longer (1.5–2mm, not 0.7–1mm), the corolla tube longer and narrower (9.5–19 × 2.5–4mm versus 6–8 × 5–6mm) and the anticous filaments 1.8–3mm long (versus 3–4.2mm). The two species are allopatric.

The relationship of *S. placida* seems to lie with *S. calciphila*, with which it is sympatric, and some forms of which it much resembles in general facies and floral details. The distinguishing marks of *S. placida* are its stems, narrowly winged by the decurrent leaf bases and with pubescence confined to the grooves so formed (the inflorescence axes are glandular-puberulous), and its leaves, glabrous or very nearly so. *Sutera placida* and *S. calciphila* may be involved in complex hybridization; this is elaborated under *S. calciphila*.

---

32. Sutera denudata (Benth.) O. Kuntze, Rev. Gen. Pl. 2: 467 (1891); Hiern in Fl. Cap. 4(2): 261 (1904) p.p. (excl. all except lectotype).
Lectotype (chosen here): Cape, Uniondale div., 3323DC, Langekloof, Onzer, 2000ft, 12 i 1830, *Drège* 7916 (K, isolecto. P).

Syn.: *Chaenostoma denudatum* Benth. in Hook., Comp. Bot. Mag. 1: 375 (1836) & in DC., Prodr. 10: 355 (1846), excl.
    Ecklon, Langekloof.
  C. glabratum Benth. p.p. (excl. lecto.), *Drège* a, Zwanepoelspoortberge (K, MO, P, S, W) tantum.

Perennial herb, taproot eventually forming a woody stock, stems many from the crown, up to 150–450mm tall, simple or subsimple to loosely virgately branched, weakly ridged, the lines running from the leaf-bases and the base of the midrib, glabrous to very minutely puberulous or scabridulous, leafy on the lower parts or right up to the inflorescence, without brachyblasts. *Leaves* opposite, bases connate, largest c.10–25 × 1–2(–3)mm, linear or rarely narrowly elliptic, acute to mucronate, margins usually strongly revolute, rarely ± plane, entire or sometimes with a few callose teeth, indumentum as on stems. *Flowers* in terminal racemes or very loose panicles. *Bracts* sharply differentiated from the leaves, lowermost c.2–2.5 × 1–1.4mm, lanceolate, acute. *Pedicels* (lowermost) c.6–10mm long. *Calyx* bilabiate, tube 1–1.5mm long, posticous lobes (1.8–)2.5–4.5 × 0.6–1mm, anticous ones a little broader, glandular-puberulous together with much coarser glandular hairs up to 0.3–0.8mm long. *Corolla* tube 9.2–12.5 × 2.7–3mm in throat, funnel-shaped, limb almost regular, 7–12mm across, posticous lobes 2.2–4.2 × 1.6–3.2mm, anticous lobe 2.7–5 × 1.8–3mm, all oblong-elliptic, mauve or pink to purple, tube yellow/orange, glandular-puberulous outside, hairs up to 0.15–0.2mm long, glistening glands as well, 5 longitudinal bands of clavate hairs inside. *Stamens* 4, anticous pair exserted, filaments 1.8–2.5mm, posticous pair in throat, filaments 0.8–1.6mm long, anthers 0.7–1.2mm long. *Stigma* 1–2.3mm long, exserted. *Style* c.8–11mm long. *Ovary* 1.2–1.8 × 1mm. *Capsules* c.3.5–5 × 2–3mm. *Seeds* 0.8–1 × 0.4mm, violet-blue or grey-blue.

Selected citations:

Cape. Willowmore div., 3323AD, Buispoort, south of Willowmore, 19 xii 1950, *Theron* 1010 (K, PRE). Uniondale div., 3323CA, 14½ miles E by S of Uniondale, c.3500ft, 16 xi 1958, *Acocks* 19993 (K, PRE, NBG). 3323CB, Long Kloof near Haarlem, 2700ft, v 1921, *Fourcade* 1331 (BOL, K, SAM, STE). Humansdorp div., 3323CD, Kouga Hills, 12 xi 1941, *Esterhuysen* 6670 (BOL, K). 3324CD, hills N of Assegai Bosch, 750ft, 10 ii 1943, *Fourcade* 5883 (STE).
Seen from following quarter-degree squares: 3323AD, CA, CB, CC, CD, DB, DC, 3324CA, CD, DB.

Bentham said of his new species *Chaenostoma denudatum* 'calyx of *C. campanulatum*' (which implies that it is hispid), 'corolla of *C. pumilum*' (which implies that it is short) and 'capsule oblong, longer than calyx.' He quoted Ecklon and Drège collections, both from the Langekloof and both of which are in Bentham's own herbarium (K). The Drège specimen has terete very minutely puberulous stems (the longitudinal wrinkling is an artefact of drying), leaf bases connate and not decurrent, flowers in racemes very loosely panicled, and a hispid calyx; the corollas are now broken off, but as this is the specimen bearing a capsule slightly longer than the calyx, it has been chosen as lectotype. A duplicate in P still bears a withered flower (corolla tube c.10mm long) and a good capsule. The Ecklon specimen has similar stems, leaves and calyx, flowers in racemes, corolla tube c.7–8mm long, and no capsules. The duplicates in K, M, S, SAM and W have the corolla tube only c.5.5mm long. These specimens match the type of *S. affinis*.

Sutera denudata, S. subnuda, S. affinis and *S. tenuicaulis* form a close-knit group distinguished by their leaves with acute to mucronate tips, and the lack of brachyblasts. *Sutera denudata* and *S. subnuda* are distinguished by differences in stem and inflorescence, the length and breadth of the corolla tube (9.2–12.5 × 2.7–3mm in throat versus 3–5.5 × 3–4.5mm) and the length of the filaments (anticous 1.8–2.5mm versus 2.5–4mm, posticous 0.8–1.6mm versus 1.5–2.6mm), which results in only the anticous stamens being exserted in *S. denudata*, all four stamens in *S. subnuda*. Moreover, their areas appear to be mutually exclusive (but see comments under *S. affinis*). The area of *S. denudata* lies immediately east of that of *S. subnuda*, on the mountains about Willowmore and Uniondale east over the mountains and through the Langekloof to Melkhoutboom on the Great Winterhoek Mountains and the hills north of Assegaibosch in Humansdorp division (the Suuranysberge), between c.230 and 1000m above sea level. It has been recorded growing in scrub on rocky slopes, but also on river banks. Flowering records are scattered throughout the year.

33. Sutera subnuda (N. E. Br.) Hiern in Fl. Cap. 4(2): 258 (1904).
Type: Cape, Riversdale div., 3321CC, near Garcia's Pass, Muiskraal, c.1500ft, 3 x 1897, *Galpin* 4375 (holo. K, iso. PRE).

Syn.: *Chaenostoma subnudum* N.E. Br. in Kew Bull. 1901: 128 (1901).
    Sutera stenophylla Hiern in Fl. Cap. 4(2): 258 (1904); De Wildeman, Pl. Nov. Herb. Hort. Then. 2 t. 61 (1908). Type: Cape, Oudtshoorn div., 3322CD, in gravelly places at Klipdrift in the Great Karroo, 2000ft, 7 iii 1893, *Schlechter* 2250 (holo. K; iso. BOL, E, J, PRE, S, STE, Z).

Perennial herb ultimately forming a shrublet arising from a thick woody stock, stems many from the crown, up to 80–450mm tall, simple or subsimple, soon branching virgately into a lax much-branched panicle, narrowly ribbed, glabrous to minutely puberulous and then sometimes sparsely scabridulous from scattered broad-based hairs, both stems and leaves often shining as if from an exudate, leafy on the lower parts or up to the inflorescence, without brachyblasts. *Leaves* opposite, bases ± connate or not quite meeting, decurrent down the stem in narrow ridges, another ridge running into the midrib, largest c.12–35 × 1–2(–2.5–3mm), often linear, sometimes narrowly elliptic, acute to mucronulate, margins usually strongly revolute, entire, or rarely plane and then sometimes with a few callose teeth, glabrous to minutely puberulous or scabridulous as the stems, gland-dotted on lower surface. *Flowers* rarely in simple terminal racemes, usually in well-branched twiggy panicles, the branchlets often flexuous. *Bracts* sharply differentiated from the leaves, lowermost c.2–2.5 × 1.4mm, lanceolate, acute. *Pedicels* (lowermost) c.3–15mm long, minutely glandular-puberulous. *Calyx* bilabiate, tube 1–1.4mm long, posticous lobes 1.5–3.2 × 0.5–0.9mm, anticous ones a little broader, glandular-puberulous together with much coarser glandular hairs up to 0.2–0.8mm long. *Corolla* tube 3–5.5 × 3–4.5mm in throat, broadly funnel-shaped, limb almost regular, 7–12mm across, posticous lobes 2.5–4 × 2–3.2mm, anticous lobe 3–4.5 × 2.5–4.5mm, all oblong-elliptic, usually shades of mauve or pink, occasionally a white sport, tube yellow or orange, glandular-puberulous outside, hairs up to 0.15–0.25mm long, glistening glands as well, 5 longitudinal bands of clavate hairs inside. *Stamens* 4, all exserted, anthers 1–1.3mm long, posticous filaments 1.5–2.6mm long, anticous ones 2.5–4mm. *Stigma* c.1–2mm long, exserted. *Style* 4–5.5mm long. *Ovary*

1–1.5 × 0.8–1.1mm, glabrous or with a few glistening glands on sutures. *Capsules* 3–4 × 2–3mm. *Seeds* c.0.7–0.8 × 0.3–0.4mm, violet-blue.

Selected citations:

Cape. Ladismith div., 3321AD, ± 10km from Sevenweeks Pass to Vleiland, 13 iii 1981, *Van Wyk* 527 (PRE, STE). Oudtshoorn div., 3321CB, Gamka Mountain Reserve, 2000ft, 30 x 1982, *Cattell & Cattell* 207 (STE); 3321DA, Rooiberg State Forest, c.2050ft, 27 iv 1977, *Taylor* 9679 (K, STE); Roodeberg Pass, 8 viii 1949, *Steyn* 293 (NBG, STE). Mossel Bay div., 3321DD, Cloete's Pass, 1000ft, 14 v 1915, *Muir* 2196 (BOL, PRE). 3322CA, between Outdtshoorn and Moerass River, c.1600ft, xii 1905, *Bolus* 12184 (BOL, K, PRE). Oudtshoorn div., 3322CB, 6 miles W of de Rust, farm Die Krans, 400m, 29 v 1971, *Dahlstrand* 2037 (J, PRE, STE).
Seen from following quarter-degree squares: 3320DD; 3321AD, BD, CB, CC, CD, DA, DB, DD; 3322AC, CB, CC, CD, DA.

N. E. Brown's name *S. subnuda* has never been taken into use because Hiern re-described the plant as *S. stenophylla*. It is closely allied to, and often confused with, *S. denudata*; see under that species.

Sutera subnuda is found along the Klein and Groot Zwartberg (both flanks) south across the mountains to the northern flank of the Langeberg and the western part of the Outeniqua Mountains, a constituent of scrub on rocky or stony ground between c.300 and 1200m above sea level. It can possibly be found in flower in any month, but most records are between November and May.

The characteristics of typical *S. subnuda* are the stems, distinctly ribbed by the decurrent leaf bases, often glabrous and shining, sometimes almost scabridulous, the more or less mucronate leaves, the twiggy panicles, short corolla tube, and relatively long filaments with all four anthers well exserted. The leaves are usually linear, their narrowness partly attributable to the strongly reflexed margins; occasionally the margins are ± plane, then the leaf is seen to be broader (measurements in parentheses in description). The leaf margin is nearly always entire; however, there appears to be a local form around De Rust near Oudtshoorn (*Dahlstrand* 2037 cited above) in which the leaves are elliptic with callose-toothed margins.

34. Sutera affinis (Bernh.) O. Kuntze, Rev. Gen. Pl. 2: 467 (1891); Hiern in Fl. Cap. 4(2): 261 (1904), as to type only, excluding description.
Type: Cape, 'in collibus prope Knysna', 29 ii 1839, *Krauss* 1615 (holo. MO) (see note below about locality).

Syn.: *Chaenostoma affine* Bernh. in Flora 27: 834 (1844); Benth. in DC., Prodr. 10: 355 (1846).

Perennial herb developing a stout woody stock, stems many from the crown, up to 150–300mm tall, simple to loosely virgately branched, very weakly ridged, the lines running from the leaf bases and the base of the midrib, minutely puberulous, the hairs less than 0.1mm long, leafy throughout, brachyblasts wanting. *Leaves* opposite, bases connate, largest c.12–20 × 1–3mm, linear to narrowly oblanceolate, acute to mucronate, margins strongly to weakly revolute, entire or with 1 or 2 pairs of small callose teeth, very minutely scabridulous, lower surface gland-dotted as well. *Flowers* in terminal racemes or very loose panicles. *Bracts* (lowermost) c.2–5 × 1–1.2mm, lanceolate to linear-lanceolate, otherwise as leaves. *Pedicels* (lowermost) c.8–14mm long. *Calyx* bilabiate, tube 1–1.5mm long, posticous lobes 2–3.2 × 0.4–1mm, anticous ones a little broader, glandular-puberulous together with much coarser glandular hairs up to 0.3–1mm long. *Corolla tube* 5–8 × 3–4mm in throat, broadly funnel-shaped, limb nearly regular, 9–12mm across, posticous lobes 3–3.5 × 2.2–3.2mm, anticous lobe 3.2–4.5 × 2.5–4mm, all lobes broadly elliptic, mauve, tube yellow/orange, glandular-puberulous outside, glistening glands as well, 5 longitudinal bands of clavate hairs inside. *Stamens* 4, all exserted, anthers 0.8–1.1mm long, posticous filaments 1.6–2mm long, anticous filaments 2.5–3.7mm. *Stigma* 1.6–2mm long, exserted. *Style* 4.8–6mm long. *Ovary* 1.3–1.6 × 0.8–1.1mm, a few glistening glands on sutures. *Capsules* c.3.5–4 × 2.5mm. *Seeds* c.0.7 × 0.3mm, blue-grey.

Citations:

Cape. George div., 3323CC, Keurbooms river, Long Kloof, between main road and De Vlugt, 1800ft, xi 1927, *Fourcade* 3376 (BOL, STE); in valle 'Vlugt', xi 1870, *Bolus* 2413 p.p. (BOL, K). Uniondale div., 3323CA, Long Kloof near Avontuur, 2900–3000ft, *Fourcade* 1318 (BOL); road to Montagu Pass from Avontuur, 15 x 1928, *Gillett* 1601 (STE). Langekloof, 2000–4000ft, xii, *Ecklon* (syntype of *Chaenostoma denudatum* Benth., K, M, SAM, W).

The name *S. affinis* has never been correctly applied in herbaria, probably because Hiern misused it for *Schlechter* 10499, which is a mixture of *S. calciphila* and *S. hispida*. N. E. Brown probably had a Schlechter sheet before him when he made an annotation in the Kew herbarium equating *S. affinis* with *S. brachiata*, that is, *S. hispida*.

Krauss collected the type of *S. affinis* 'on hills near Knysna.' All subsequent collections have come from the environs of Avontuur (in the Long Kloof) and Prince Alfred's Pass across the Outeniqua Mountains on the road from Knysna to Avontuur; De Vlugt lies immediately south of the pass at the crossing of the Keurbooms River. The label on Krauss's specimen is in his own hand and is dated 29 February; on 28 February he arrived at the farm Ongelegen in the Langekloof, roughly 25km east of Avontuur. He stayed at Ongelegen until 2 March, so it is not improbable that his specimen actually came from present-day Avontuur! (O. H. Spohr, ed., Ferdinand Krauss Travel Journal Cape to Zululand pp. 43–44).

Before I saw Krauss's specimen, I had set aside the specimens cited above as possible hybrids between *S. denudata* and *S. subnuda*: Schonland collected *S. denudata* 'near top of Prince Alfred's Pass' (*Schonland* 3413, PRE) and *Bolus* 2413 (cited above) is a mixture of *S. denudata* and *S. affinis*; the geographically nearest collection of *S. subnuda* known to me came from Klipdrift in the Little Karoo about 20km SSE of Oudtshoorn, and c.75km W of Avontuur.

Specimens of *S. affinis* have the stem, leaf and inflorescence characters of *S. denudata*, but the length of the corolla tube is intermediate between that of *S. subnuda* (3–5.5mm) and *S. denudata* (9.2–12.5mm); the anticous filaments are 2.5–3.7mm long in *S. affinis* (2.5–4mm in *S. subnuda*, 1.8–2.5mm in *S. denudata*). The possible hybrid origin of *S. affinis* needs investigation; on the other hand, it could be merely a short-tubed form of *S. denudata*, but the length of the anticous filaments argues against that.

Gillett collected his specimen in waste ground; no other ecological information is available. The plants flower between October and February.

35. Sutera tenuicaulis Hilliard in Edinb. J. Bot. 48: 345 (1991).
Type: Cape, Prince Albert div., 3322AC, Swartberg Pass, 4000–4900ft, xii 1904, *Bolus* 11612 (holo. BOL; iso. K, PRE).

Perennial herb, stems several from the crown, loosely branched, branches slender, up to c.150–450mm long, c.1–1.5mm in diam., straggling, narrowly winged, glabrous but sometimes the inflorescence-axis and always the pedicels glandular-puberulous, laxly leafy, brachyblasts wanting. *Leaves* opposite, the largest c.10–30 × 0.8–3mm, smaller upwards, linear, linear-lanceolate or narrowly oblanceolate, apex acute to mucronate, base narrowed, half-clasping, decurrent in a narrow wing to the node below, margins revolute to ± plane, often with 1–3 pairs of small teeth (not always readily seen when margins revolute), entire in small leaves, both surfaces very minutely and sparsely scabridulous, glistening glands on lower surface. *Flowers* in lax terminal racemes sometimes loosely panicled. *Bracts* not always sharply differentiated from uppermost leaves, largest c.1.5–5 × 0.3–0.8mm, linear to linear-lanceolate, acute. *Pedicels* (lowermost) 5–20mm long, glandular-puberulous. *Calyx* bilabiate, tube 0.8–2mm long, posticous lobes 2.2–4 × 0.4–0.8mm, anticous lobes a little broader, glandular-puberulous together with much coarser hairs up to 0.2–0.7mm long. *Corolla* tube 10.5–16.5 × 2.5–2.7mm in throat, narrowly funnel-shaped, limb almost regular, 8–11mm across, posticous lobes 2.8–4 × 1.8–3mm, anticous lobe 2.5–4.5 × 1.5–3.2mm, all oblong-elliptic, pink, throat yellow, glandular-puberulous outside, hairs up to 0.2–0.25mm long, glistening glands as well, inside weakly bearded with longitudinal bands of clavate hairs. *Stamens* 4, anticous pair shortly exserted, filaments 1.5–2mm long, posticous pair included, filaments 1–1.3mm long, anthers 0.8–1mm long. *Stigma* 1–2.7mm long, exserted. *Style* 9–13mm long. *Ovary* 1–1.5 × 0.8–1mm, glabrous or a few glistening glands on sutures. *Capsule* 3–5 × 2–3mm. *Seeds* c.0.8–1 × 0.5mm, pallid.

Selected citations:

Cape. Prince Albert div., 3322AC, Swartberg Pass, 5300ft, v 1926, *Pocock* S118 (BOL, PRE, STE); ibidem, 3000ft, 8 v 1938, *Compton* 7187 (NBG).

Sutera tenuicaulis is known only from the Swartberg Pass and its immediate environs; *Moffett* 483 (STE) from Waenskloof in the Cango valley is a poor specimen, but seems to be this species. It grows on both the north and the south sides of the pass, and straggles among boulders in scrub, between c.900 and 1500m above sea level. It has been collected in both flower and fruit in May, November and December.

Its relationship lies with *S. denudata* and *S. subnuda*. It is sympatric with *S. subnuda* on the pass, and can be distinguished by its straggling habit, lax racemes, and much longer corolla tube. Its habit as well as its narrowly winged stems will distinguish it from *S. denudata*, which has similar flowers. The seeds of both *S. subnuda* and *S. denudata* are violet-blue or grey-blue, those of *S. tenuicaulis* are pallid, but this is not always a reliable character.

---

36. Sutera halimifolia (Benth.) O. Kuntze, Rev. Gen. Pl. 2: 467 (1891); Hiern in Fl. Cap. 4(2): 257 (1904).
Lectotype (chosen here): Cape, Aberdeen div., 3223AA, Hamerkuil [Hamelkuil], 3000ft, 31 i 1827, *Drège* (K; isolecto. E, LE, P).

Syn.: *Chaenostoma halimifolium* Benth. in Hook., Comp. Bot. Mag. 1: 375 (1836) & in DC., Prodr. 10: 354 (1846).
    C. pumilum Benth. in Hook., Comp. Bot. Mag. 1: 375 (1836) & in DC., Prodr. 10: 355 (1846). Lectotype (chosen here): Cape, Willowmore div., 3323CC, Zwanepoelspoortbergen, 3000–4000 Fuss, 14 viii 1829, *Drège* 7917b (K; isolecto. E, LE, MO, P).
    C. laxiflorum Benth. in Hook., Comp. Bot. Mag. 1: 374 (1836) & in DC., Prodr. 10: 354 (1846). Type: Cape, c.3326BB, Höhe von Keiskamma, 500–1000 Fuss, 17 vi 1832, *Drège* 7911a (holo. K; iso. E, LE, MO, P, W).
    Sphenandra cinerea Engl., Bot. Jahrb. 10: 253 (1889). Type: Cape, 2724CD, near Kuruman, Grootfontein, Boetsap, 1200m, ii 1886, *Marloth* 1126 (iso. BOL, E, K, M, NBG, SAM, W).
    Sutera pumila (Benth.) O. Kuntze, Rev. Gen. Pl. 2: 467 (1891).
    S. laxiflora (Benth.) O. Kuntze, Rev. Gen. Pl. 2: 467 (1891); Hiern in Fl. Cap. 4(2): 259 (1904).
    Manulea halimifolia (Benth.) O. Kuntze, Rev. Gen. Pl. 3(2): 235 (1895).
    M. oppositiflora var. *angustifolia* O. Kuntze, Rev. Gen. Pl. 3(2): 236 (1895). Type: Cape, Cathcart, 3227AC, 1400m, 26 ii 1894, *Kuntze* (holo. NY).

Ultimately a twiggy gnarled dwarf shrublet, but will flower when young and then looks like an annual or an herbaceous perennial, taproot eventually forming a woody caudex up to c.10mm diam. at the crown, sometimes becoming several-crowned, branches up to 50–350mm long, erect or decumbent, leafy, thickly to thinly clad in almost sessile glistening vesicular hairs that also clothe leaves, bracts, pedicels and calyces, minute glandular hairs sometimes present as well. *Leaves* opposite, bases connate, largest c.7–25(–30) × 1.5–5(–12)mm, elliptic tapering to a short petiolar part, subacute or acute, margins entire or with 1–8 pairs of teeth, teeth minute to pronounced, texture often leathery, herbaceous in some larger leaves. *Flowers* in long racemes terminating every branch, thus loosely panicled. *Bracts* usually sharply differentiated from the leaves, occasionally the lowermost one or two pairs of flowers in axils of uppermost leaves, lowermost bracts (as opposed to leaves) c.1.5–6 × 0.8–1.5mm, lanceolate to ovate, acute. *Pedicels* (lowermost) 5–20mm long. *Calyx* obscurely bilabiate, tube 0.7–1mm long, lobes 1.6–2.5 × 0.5–1mm. *Corolla* tube 3–5(5.5–7.5) × 3–4mm in throat, broadly funnel-shaped, limb almost regular, 7–10mm across, posticous lobes 2–3.2 × 2–3mm, anticous lobe 3–4 × 2.3–3.5mm, all oblong-orbicular, shades of pink or mauve, occasionally white, tube yellow, glandular-puberulous outside, hairs up to 0.15mm long, these hairs sometimes wanting, glistening glands always present, clavate hairs inside in 5 longitudinal bands. *Stamens* 4, all exserted, anthers 0.8–1.3mm long, posticous filaments 1.7–2.2(–3)mm long, anticous filaments (2.2–2.8)2.7–5mm long. *Stigma* 1.5–2.5mm long, exserted. *Style* 2.5–4mm long. *Ovary* 1–1.5 × 0.6–1.1mm, with glistening glands on sutures and often on upper half as well. *Capsules* c.3.5–4 × 2–2.5mm, glistening glands persistent. *Seeds* c.0.5–0.8 × 0.3–0.4mm, pallid or blue-grey.

Selected citations:

Namibia. Warmbad distr., 2718CA, Klein Karas, 7 viii 1923, *Dinter* 4868 (PRE).
Cape. Barkly West div., 2823BA, Asbestos Hills, Daniels Kuil, 21 iii 1939, *Lewis* sub NBG 110096 (NBG). Postmasburg div., 2823AC, 17 miles ENE of Postmasburg, c.4800ft, 27 iii 1959, *Leistner* 1398 (K, PRE). Kimberley div., 2824DB, Kimberley, 23 iii 1953, *Compton* 23943 (NBG, STE). Herbert div., 2923BA, Mazelsfontein, 14 iv 1937, *Acock* 2221 (PRE). Barkly East div., 3027DD, 5 miles E of Belmore on road from Barkly East to Rhodes, 30 i 1966, *Hilliard & Burtt* 3770 (E, K). Victoria West div., 3123CC, N of Three Sisters, 27 iii 1953, *Lewis* 3886 (SAM). Middelburg div., 3125AC, Bangor Farm, 4000ft, ix–x 1917, *Bolus* 14065 (BOL). Queenstown div., 3126DD, 3600–5300ft, 1893, *Galpin* 1650 (K, PRE). Graaff Reinet div., 3224AB, St. Olive's, c.5500ft., 29 xi 1977, *Hilliard & Burtt* 10756 (E). Prince Albert div. c.3322AA, Sandriver Mts, v 1907, *Marloth* 4483 (STE). Humansdorp div., 3324DD, between Hankey and Gamtoos river, 40ft, 12 ix 1942, *Fourcade* 5754 (BOL, STE). Peddie, 3327AA, 11 iv 1952, *Barker* 7814 (NBG).
Transkei. 3127CD, St. Marks, *Barker* 362 (K, PRE).
Orange Free State. Boshof distr., 2825CA, Kromrant, c.20km SW Boshof, 1300m, 12 ii 1985, *Zietsman* 195 (PRE). Fauresmith, 2925CB, v 1946, *Brueckner* 720 (PRE). Dewetsdorp, 2926DA, 15 iv 1950, *Steyn* 931 (NBG). Zastron distr., 3027AC, Kranskop, 6200ft, 6 xii 1973, *Ferreira* F098 (PRE).

The type of *S. laxiflora* is no more than a lush specimen of *S. halimifolia*; equally lush specimens occur sporadically throughout the range of the species, even in the driest parts of the interior of the Cape. For example, one of the most northerly records, *Rogers* sub BOL 12577 collected between Korannaberg and Tsenin (c.2723AA), not far from Kuruman, is a good match of Drège's type specimen of *S. laxiflora* from the heights above the Keiskamma river at the south-easterly limits of the range of *S. halimifolia*.

The corolla tube is normally short (up to 5mm long); the longer measurements given in parentheses in the description (also the shorter anticous filaments) refer to a plant seemingly common in Humansdorp division. Fourcade collected it several times between Zuurbron and the Gamtoos River drift, in Zuurbron Kloof on the way to Hankey, between Jeffrey's Bay rail and Kabeljauws, near Kabeljauws Station, and on hills above Humansdorp (*Fourcade* 2125, BOL; 3990, BOL, STE; 4915, STE; 6032, BOL, STE; 6349, BOL). Rogers also had it at Humansdorp (*Rogers* 2924, BOL, SAM). It differs from *S. halimifolia* solely in its longer corolla tube; short-tubed *S. halimifolia* is also present in the area: *Fourcade* 5754 (PRE) between Hankey and the Gamtoos River, *Fourcade* 2744 (BOL) from Groot River Poort; *Barker* 7887, 7888 (NBG) and *Lewis* 3882 (BOL) from Mistkraal. Population studies hereabouts might prove interesting.

Bentham's epithet *halimifolium* may be derived from the resemblance the leaves of the specimens he saw bear to those of *Halimione* (Chenopodiaceae), but it is also possible that he had in mind the Greek word *hals*, meaning salt, because the stems, leaves and calyx are clothed in peculiar hairs that at low magnification resemble grains of salt (see figs 5 and 6 for the true nature of the hairs). This is the characteristic feature of *S. halimifolia* and in combination with the lack of any coarse hairs on the calyx will distinguish it from *S. caerulea* and *S. campanulata* with which it is often confused in herbaria. The 'grains of salt' hairs are often so closely set that the plant looks grey, but in plants growing under very favourable conditions that lead to lax stems and broad leaves, the hairs are more thinly spread and this coloration is lost.

I have seen one record from Namibia (cited above, the specimens distributed as *Diascia karasmontana* Dinter MSS), otherwise the species is widely distributed in the dry interior of the Cape from Griqualand West south to Fraserburg and Prince Albert and east through the southern part of the Orange Free State to the borders of Lesotho, and through the eastern Cape to the borders of the Transkei at altitudes between c.1000 to 1900m above sea level. Only in Humansdorp division on the southern Cape coast does it descend almost to sea level. Collections from the southern and eastern Cape are meagre. There is one old record from 'plains near Potchefstroom' (*McLea* 117, BOL), but its presence in the Transvaal needs confirmation.

The plants favour rocky sites; flowering occurs principally between September and April.

37. Sutera neglecta (Wood & Evans) Hiern in Fl. Cap. 4(2): 260 (1904).
Lectotype (chosen here): Natal, 2729BD, Newcastle distr., near Charlestown, summit Drakensberg mountains, 5–6000ft, 10 i 1894, *Wood* 5241 (K; isolecto. BOL, E, PRE, Z, W).

Syn.: *Chaenostoma neglectum* Wood & Evans in J. Bot. 35: 352 (1897), Diels in Bot. Jahrb. 26: 121 (1898).

Perennial herb with an irregularly-shaped caudex up to c.30mm in diam. or possibly larger, stems few to several from the crown, erect or decumbent, c.150–600mm long, simple to branched mainly in the upper half to form a corymbose panicle, glandular-pubescent, hairs up to 0.15–0.25(–0.3)mm long, glistening glands as well, closely leafy. *Leaves* opposite, bases connate, largest ones c.15–40 × 3.5–17mm, smaller upwards and passing into bracts, oblong to elliptic, apex acute or subacute, base slightly narrowed or not, margins subentire or with up to 6 pairs of small teeth, both surfaces glandular-pubescent, hairs up to 0.3mm long, densely covered in glistening glands as well. *Flowers* in long terminal racemes, sometimes branching into corymbose panicles. *Bracts* (lowermost) c.2.5–14 × 2–4.5mm, usually at least twice as long as broad, ± lanceolate, hairy as the leaves. *Pedicels* (lowermost) 10–40mm long. *Calyx* obscurely bilabiate, tube 1.3–2mm long, lobes 2–3.5 × 1–1.3mm, lanceolate, glandular-pubescent, hairs up to 0.3–0.6mm long, glistening glands as well. *Corolla* tube 4–6 × 3.8–5mm in throat, broadly funnel-shaped, limb almost regular, 11–15mm across, posticous lobes 3.3–5 × 3.1–4.7mm, anticous lobe 3–6 × 3–5.5mm, all suborbicular, mauve to bright purple-pink, an occasional white sport, tube yellow/orange, glandular pubescent outside, hairs up to c.0.5mm long, glistening glands as well, clavate hairs inside all round throat. *Stamens* 4, all exserted, anthers 1.2–2.5mm long, posticous filaments 2–3mm long, anticous ones 3–4mm. *Stigma* 1.5–3mm long, exserted. *Style* 2.2–4mm long, clad in sessile glistening glands, glandular hairs up to 0.15–0.2mm long occasionally present as well. *Ovary* 1.5–2 × 1–1.25mm, with glistening glands on upper part and down sutures. *Capsules* 4–6 × 3–5mm, glistening glands persistent. *Seeds* c.0.6–0.8 × 0.3–0.4mm, pallid.

Selected citations:

Transvaal. Pilgrim's Rest distr., 2430DD, Graskop Peak, 5900ft, 7 iii 1937, *Galpin* 14512 (PRE). Lydenburg distr., 2530BA, top of Long Tom Pass, 7050ft, 20 i 1969, *Hilliard & Burtt* 6018 (E, K, PRE, Z). Carolina distr., 2530CD, 9.8 miles SE of Machadodorp on road to Slaaihoek, 20 i 1954, *Marais* 299 (PRE). Middelburg distr., 2529CD, Klein Olifants Rivier, 6000ft, 21 xii 1893, *Schlechter* 4018 (BOL, E, K, PRE, S, STE, Z). Barberton distr., 2531CC, Barberton, Saddleback Mt., 4800ft, iii 1890, *Galpin* 875 (BOL, K, PRE, Z). Ermelo distr., 2630AD, Knock Dhu farm near Lothair, 9 i 1984, *Welman* 368 (PRE). Wakkerstroom distr., 2730AD, farm Oshoek, c.1900m, 30 iii 1976, *Devenish* 1632 (E).
Swaziland. Mbabane distr., 2631AA, Mbabane to Pigg's Peak, c.16th mile from Ermelo road, 19 i 1966, *Hilliard & Burtt* 3570 (E).
Natal. Utrecht distr., 2730AD, farm Altemooi, c.6000ft, i 1920, *Thode* 4514 (STE). Klip River distr., 2829AD, Van Reenen, 5000ft, 2 iii 1895, *Schlechter* 6922 (BOL, NBG, W, Z). Bergville distr., 2828DD, Mont aux Sources, 2500m, iv 1913, *Dyke* in herb. *Marloth* 5412 (PRE, STE).
Orange Free State. Harrismith distr., 2828DB, Qwa Qwa Mt. above Bluegumbosch, c.6200–6600ft, 8 i 1979, *Hilliard & Burtt* 12002 (E, K, M, S). 2829AC, Manyanyeza Mt., 5 i 1979, *Hilliard & Burtt* 11953 (E). Bethlehem distr., 2828DA, Golden Gate, 12 v 1956, *Strey* 2892 (PRE). Fouriesburg distr., 2828CA, Dunelm Farm, 9 i 1918, *Potts* 3067 (PRE).
Lesotho. Leribe, 2828CC, *Dieterlen* 421 (K, P, PRE, SAM, Z). Marrakabei distr., 2928CC, Semonkong, ± 7000ft, 28 xi 1976, *Forbes* 687 (J).

Sutera neglecta is found on the high ground in the south-eastern Transvaal from Graskop and Mount Anderson south to the Natal border, reaching Middelburg in the west and, to the east, just entering Swaziland; it ranges thence along the Drakensberg on the Natal-Transvaal and Natal-Orange Free State borders to the environs of Fouriesburg in the Orange Free State and Leribe in Lesotho; there is also a record from central Lesotho. It favours rocky grassland, between c.1675 and 3300m above sea level, coming into flower in late November or December, peaking between January and March, and tailing off in May.

Both *S. neglecta* and its close ally *S. patriotica* have been much confused with *S. caerulea* and *S. campanulata*. *Sutera caerulea* is easily distinguished by the long glandular hairs on the style and ovary; *S. campanulata* has larger flowers and different indumentum on the lower part of the stems. The distinctions between *S. neglecta* and *S. patriotica* are given under that species, but not all specimens are easy to place.

38. Sutera levis Hiern in J. Bot. 51: 364 (1903) & in Fl. Cap. 4(2): 262 (1904).
Type: Transvaal, near Johannesburg, Bezuidenhout Valley, *Rand* 1156 (holo. BM).

Syn.: *S. palustris* Hiern in Fl. Cap. 4(2): 256 (1904); van Wyk & Malan, Field guide to Wild Flowers of Witwatersrand & Pretoria region 262 fig. 674 (1988); Philcox in Fl. Zamb. 8(2): 29 (1990). Lectotype (chosen here): Transvaal, Johannesburg, ridge above Jeppe's Town, c.5900ft, ii–iii 1894, *Galpin* 1387 (K; isolecto. BOL, PRE, SAM).

Bushy perennial herb that will flower in the seedling stage, up to c.600mm tall, developing a woody stock (taproot) up to c.10mm in diam. at crown, branching from base, branches erect or decumbent, glandular-puberulous, hairs up to 0.1–0.2mm long, densely leafy, often with axillary leaf tufts. *Leaves* opposite, bases connate, primary leaves up to c.20–40 × 1–3.5(–9)mm, smaller upwards (axillary leaves also smaller), linear to narrowly elliptic, apex acute, base somewhat narrowed, margins entire to minutely toothed (2–6 pairs of teeth), both surfaces glandular-puberulous, glistening glands also present. *Flowers* innumerable in much-branched panicles terminating every branchlet, ultimate axes sometimes ± flexuous. *Bracts* up to c.2.5–4.2 × 0.4–0.6mm, leaflike but much smaller than the leaves. *Pedicels* (lowermost) c.6–10(–16)mm long, filiform. *Calyx* obscurely bilabiate, tube 0.5–1.5mm long, lobes 1.4–2.6 × 0.4–0.8mm, lanceolate, glandular-puberulous with coarser hairs as well up to 0.25–0.7mm long. *Corolla* tube 2.5–3.5 × 2–2.5mm in throat, broadly funnel-shaped, limb almost regular, 5–7mm across, posticous lobes 1.5–2.5 × 1.5–2mm, anticous lobe 1.5–2.5 × 1.5–2.4mm, all suborbicular, shades of mauve or pink, white sports not uncommon, tube yellow or orange, glandular-puberulous outside, glistening glands as well, clavate hairs inside all round throat. *Stamens* 4, all exserted, anthers 0.5–1mm long, posticous filaments 1–2mm long, anticous ones 2–2.8mm. *Stigma* 1–1.8mm long, exserted. *Style* 1.6–3.5mm long, clad in glistening glands. *Ovary* 1–1.5 × 0.7–1mm, glistening glands at apex and down sutures. *Capsules* 2–3.5 × 1.5–2.5mm, glistening glands persistent. *Seeds* c.0.5–0.7 × 0.25–0.3mm, pallid.

Selected citations:

Transvaal. Thabazimbi distr., 2427BC, Kransberg, 1625m, 28 iii 1980, *Westfall* 01022 (PRE). Marico distr., 2525BD, Linokana [Dinokana], iv 1887, *Holub* s.n. (Z). Rustenburg Nature Reserve, 2527CB, c.5000ft, *Jacobsen* 978 (PRE). Pretoria distr., 2528CB, Leeuwhoek, 23 v 1968, *Strey & Leistner* 8280 (K, PRE). Middelburg distr., 2529CB, Botsabelo, 1260m, 27 xii 1893, *Schlechter* 4072 (BOL, E, J, K, PRE, S, STE, Z). Potchefstroom distr., 2627CD, 11 miles from Parys on Potchefstroom road, 5400ft, 8 iv 1967, *Vahrmeijer* 1573 (PRE).
Botswana. 2525BA, south east distr., Ootse Mountains, 1350–1450m, 9 x 1977, *Hansen* 3218 (K). 2525BB, W of Lobatse, 9 ii 1977, *Giess* 64 (M, WIND).

Sutera levis is endemic to the Transvaal and neighbouring parts of Botswana; it appears to be commonest along the Magaliesberg and the Witwatersrand and on the Highveld as far south as the Vaal and east to Middelburg; there is one record from Waterval Boven but its presence there needs confirmation. Holub collected it near Zeerust at the end of the last century and there is another old collection from Schweizer-Reneke and one from the western extremity of the Waterberg. Apart from these, there are no collections west of Rustenburg except those from Botswana; Ootse in Botswana is about 65km NNW of Zeerust. From the Magaliesberg, it ranges over the Springbok Flats about as far north as the 23rd parallel. The better-known epithet *palustris* must give precedence to the earlier name, *S. levis*, which is also more apt, as the species favours, not marshes, but stony or rocky sites on grassy hillslopes, often among boulders, between 1375 and 1850m above sea level, flowering principally between December and May.

It is a bushy and very floriferous herb bearing masses of small flowers on filiform pedicels; the leaves are often very narrow, but appear to broaden in response to shade and moisture. The small size of the flowers (corolla tube up to 3.5mm long, limb 5–7mm across) will at once distinguish it from *S. caerulea* and *S. campanulata*, with which it is most often confused; both these species are confined to the Cape. *Sutera levis* is partly sympatric with *S. neglecta*, and one specimen (*Turner* 123, PRE) appears to be a hybrid between them; it came from a roadside south east of Middelburg and has the facies of *S. levis* but the corolla tube is c.4.3mm long, the limb 9mm across, and staining with cotton blue showed a high proportion of sterile pollen grains (though in *Sutera* this does not always seem to be a reliable indication of possible hybridization).

39. Sutera patriotica Hiern in Fl. Cap. 4(2): 250 (1904).
Type: Cape, Wodehouse div., on the Stormberg Range near Patriot's Klip [3126BA], 5000ft, i 1897, *Wood* in herb. Galpin 2302 (holo. PRE, iso. K).

Syn.: *Chaenostoma calycinum* Benth. var. β *laxiflorum* Benth. in Hook., Comp. Bot. Mag. 1: 374 (1836), nomen, quoad Windvogelberg, *Drège* 548b tantum.
  Diascia avasmontana Dinter in Feddes Repert. 30: 189 (1932). Lectotype (chosen here): Namibia, Auasgebirge, farm Lichtenstein, *Dinter* 4701 (Z, isolecto. S).
  [*Sutera caerulea* auct. non (L.f.) Hiern; Merxmüller & Roessler, Prodr. Fl. S.W. Afr. fam. 126: 49 (1967)].

An erect herb, often annual, but capable of perennating, the taproot eventually woody and up to c.8mm in diam. at the crown, main stem erect to c.70–450mm, branching from the base, branches erect or decumbent, simple or branching from the base or higher into a corymbose panicle, glandular-pubescent, hairs up to 0.2–0.25mm long, glistening glands as well, densely leafy, with axillary leaf-tufts. *Leaves* opposite, bases connate, primary leaves up to 17–30(–45) × 2–8(–15)mm, smaller upwards (axillary leaves also smaller) and passing into bracts, oblong to elliptic, apex ± acute, base somewhat narrowed to distinctly petiolar, margins subentire or with up to 6 pairs of small teeth, hairy as the stems. *Flowers* in long terminal racemes eventually forming corymbose panicles, sometimes well-branched. *Bracts* (lowermost) leaflike, c.7–15 × 1.2–5mm, more than twice as long as broad. *Pedicels* (lowermost) 10–30mm long. *Calyx* obscurely bilabiate, tube 1–1.8mm long, lobes 1.8–3.1 × 0.8–1mm, lanceolate, densely glandular-pubescent, hairs up to 0.25–0.5mm long. *Corolla* tube 3–4.5 × 2.5–4.5mm in throat, broadly funnel-shaped, limb almost regular, c.8.5–14mm across, posticous lobes 3–5 × 2.5–4mm, anticous lobe 3–5 × 2.5–4mm, all suborbicular, shades of pink, mauve or blue-violet, tube yellow/orange, glandular-pubescent outside, hairs up to c.0.25mm long, glistening glands as well, clavate hairs inside all round throat. *Stamens* 4, all exserted, anthers 1.3–2mm long, posticous filaments 1.5–2.2mm long, anticous filaments 2.5–4mm. *Stigma* 2–3.5mm long, exserted. *Style* 1.5–3.5mm long, clad in glistening glands, occasionally minute glandular hairs as well. *Ovary* 1.3–2 × 1–1.5mm, with glistening glands at apex and down sutures. *Capsules* 3–5 × 2.5–3.5mm, glistening glands persisting. *Seeds* c.0.5 × 0.3mm, pallid.

Selected citations:

Namibia. Rehoboth distr., 2316BA, 80km from Windhoek to Gamsberg, 12 iii 1974, *Müller* 49 (PRE, WIND). Windhoek distr., 2217CA, farm Regenstein, WIN 32, 19 iii 1972, *Giess* 11685 (K, M, NBG, PRE, WIND).
Cape. Barkly West div., 2624CD, Schietpan, ii 1937, *Acock* 1644 (BOL, PRE), Vryburg div., 2723AD, near Kuruman, Batlharo, *Silk* 222 (BOL, SAM). Barkly West div., 2823BA, 1 mile N of Danielskuil, c.4400ft, 20 ii 1956, *Leistner* 610 (K, NBG, PRE). Kimberley div., 2824DB, Doornlaagte, ii 1937, *Wilman* sub BOL 51423 (BOL). Prieska div., 2922DA, banks of Orange River near Prieska, v 1935, *Bryant* 1114 (K, PRE). Albert div., 3026CD, Burghersdorp, 5000ft, xii 1892, *Flanagan* 1544 (PRE). Murraysburg, 3123DD, *Tyson* 133 (BOL). Cathcart div., 3227AC, Windvogelberg, 4000ft, 23 xi 1832, *Drège* (P). Barkly East div., 3027DB, Ben Mcdhui, c.8400ft, 3 ii 1983, *Hilliard & Burtt* 16381 (BOL, E, K, PRE, S).
Transvaal. Waterberg distr., 2328AC, Alexandersfontein, iv 1920, *Power* sub PRE 39907 (PRE). Pietersburg distr., 2329DC, 18 miles E of Pietersburg, road to Tzaneen, 27 xi 1963, *Van Vuuren* 1574 (K, PRE). Zeerust distr., 2525BD, Linokana [Dinokana], v 1887, *Holub* s.n. (Z). Lichtenburg distr., 2626AB, Witstinkhoutboom, 28 i 1926, *Liebenberg*

51 (PRE). Schweizer Reneke, 2725AB, ii 1919, *Rogers* 22713 (Z). Potchefstroom distr., 2627CC, road to Schoemans-drift, 21 xi 1984, *Bredenkamp* 187 (PRE). Standerton, 2629CD, *Rehmann* 6810 (Z).

Orange Free State. Kroonstadt, 2727CA, *Chennells* 107 (BOL). Boshof, 2825CA, xii 1920, *Power* sub BOL 16837 (BOL). Glen, 2826CD, School of Agriculture, ii 1921, *Potgieter* 11 (PRE). Near Winburg, 2827CA, 20 i 1936, *Gillett* 1117 (STE). Petrusburg, 2925AD, 2 xi 1921, *Marais* 139 (PRE). Bloemfontein, 2926AA, Onzerust Farm, 22 i 1936, *Gillett* 1127 (STE); Bloemfontein, *Rehmann* 3904 (BOL, Z). Fauresmith, 2925CB, Brandewynskuil, i 1928, *Smith* 5280 (PRE).

Lesotho. Near Mateka, 2927BB, 6000ft, 22 iii 1951, *Bruce* 345 (PRE). Thaba Bosiu, 2928BA, i 1903, *Junod* 1883 (Z). Sehlabathebe National Park, 2929CC, 2425m, 18 iv 1979, *Hoener* 2210 (PRE). Qomoqomong, 3027BD, iv 1923, *Ashton* 67 (BOL). White Hill, 3028AB, i 1912, *Jacottet* 259-B372 (Z).

Transkei. 3028AD, Ongeluks Nek Pass, 6900ft, 7 xii 1985, *Hilliard & Burtt* 18724 (E, K).

Sutera patriotica has a very wide distribution, from Windhoek, Rehoboth and Keetmanshoop districts in Namibia, with an apparent distributional gap to Hay, Barkly West and Prieska divisions in the Cape, thence east into the western Transvaal, with a few records from the northern Transvaal, common in the Orange Free State and Lesotho, thence south into the eastern Cape, reaching Murraysburg and the Windvogelberg at Cathcart, just entering the Transkei at the summit of the passes into Lesotho. In the more arid parts of its range, it is found on sandy flats or in open woodland or scrub on sand; in the wetter parts, it grows in grassland; it will also behave as a weed and appear along roadsides, cattle tracks and similar disturbed areas. Its altitudinal range is also great: 1200–2950m, and flowering occurs between November and May, peaking in February.

Although *S. patriotica* has been much confused with *S. caerulea* (easily distinguished by longer hairs on all parts, including style and ovary), it is more closely allied to *S. neglecta*. The two species have similar indumentum, but differ in foliage and rootstock. The leaves of *S. patriotica* are thinner than those of *S. neglecta* (a character difficult to quantify) and they are pseudofasciculate; although there are buds in the axils of the leaves of *S. neglecta*, they rarely develop into a tuft of leaves. The rootstock of *S. neglecta* is a woody caudex of irregular shape; that of *S. patriotica* is a simple woody taproot. Over the more arid part of its range, *S. patriotica* appears nearly always to be annual; in the eastern and better-watered part it seems often to perennate. *Sutera patriotica* and *S. neglecta* are nearly allopatric; their areas meet in the south eastern Transvaal (east of Bethal and Standerton) and in Lesotho, where it may be difficult to place some specimens.

40. Sutera caerulea (L.f.) Hiern in Fl. Cap. 4(2): 255 (1904) p.p.
Lectotype (chosen here): C.B.S., *Thunberg* (sheet 14354 UPS, left-hand specimen).

Syn.: *Manulea caerulea* L.f., Suppl. 285 (1782); Thunb., Prodr. 101 (1800) & Fl. Cap., ed. Schultes 467 (1823).
  Buchnera viscosa Ait., Hort. Kew. ed. 1(2): 357 (1789) & in Bot. Mag. t. 217 (1793). Type: a cultivated specimen
   'from the Greenhouse at Goodwood' (K).
  Erinus viscosus (Ait.) Salisb., Prodr. 94 (1796).
  Manulea rotata Desrouss. in Lam., Encycl. 3: 706 (1792). Type: garden plant, no specimen preserved.
  M. viscosa (Ait.) Willd., Sp. Pl. 3(1): 336 (1800) & in Enum. Pl. Hort. Berol. 653 (1809).
  Sphenandra viscosa (Ait.) Benth. in Lindl., Nat. Syst. Bot. 445 (1830), in Hook., Comp. Bot. Mag. 1: 373 (1836)
   & in DC., Prodr. 10: 353 (1846).
  S. caerulea (L.f.) O. Kuntze, Rev. Gen. Pl. 3(2): 239 (1898).

An erect herb flowering in the first year but probably perennating, taproot eventually woody and up to c.10mm diam. at the crown, main stem erect to c.1m, eventually branching from the base, branches erect or decumbent, simple or branching, usually above the middle, into a corymbose panicle, glandular-pubescent, hairs up to 0.25–1mm long, densely leafy, often with axillary leaf tufts. *Leaves* opposite, bases connate, primary leaves up to 6–30 × 1.5–11mm, smaller upwards (axillary leaves also smaller) and passing into bracts, ± oblong, apex acute, base narrowed or not, margins usually with 1–5 pairs of small teeth, sometimes incise-dentate, both surfaces glandular-pubescent. *Flowers* opposite or subopposite to alternate, arranged in long terminal racemes eventually forming corymbose panicles. *Bracts* (lowermost) c.2–8 × 1.5–4.5mm, roughly as long as broad to twice as long as broad, ± ovate, acute, glandular-pubescent. *Pedicels* (lowermost) 7–16mm long. *Calyx* obscurely bilabiate, tube 0.6–1.5mm long, lobes 2–4.3 × 0.8–1.6mm, lanceolate, densely glandular-pubescent, hairs up to 0.5–2mm long. *Corolla* tube 3–4.5 × 2.5–3.5mm in throat, broadly funnel-shaped, limb almost regular, 9–15mm across, posticous lobes 2.5–5.5 × 2.5–5mm, anticous lobe 3.2–6 × 2.5–5mm, all suborbicular, shades of mauve or violet, rarely a white sport, tube yellow or orange, glandular-pubescent outside, hairs up to 0.5–1mm long, clavate hairs inside all round throat. *Stamens* 4, all exserted, anthers 1–2mm long, posticous filaments 2–3.8mm long, anticous ones 3–4.8mm. *Stigma* 2–3.7mm long, exserted. *Style* 2–4mm long, glandular-pubescent, hairs up to 0.5–1.5mm long. *Ovary* 1.5–2.7 × 1–1.8mm, hairy as the style and with glistening glands as well. *Capsules* 2–5.5 × 2–4mm, hairs persisting near the apex. *Seeds* c.0.5–0.75 × 0.3–0.5mm, pallid or blue.

Selected citations:

Cape. Van Rhynsdorp div., 3118AB, Klein [Kleyn] Fontein, 1500ft, 21 vii 1896, *Schlechter* 8259 (BOL, E, PRE, S, W, Z). Clanwilliam div., 3219AA, near Pakhuis, 3 ix 1933, *Leipoldt* BOL 20218 (BOL). Worcester div., 3319BD, Triangle, 3000ft, 27 x 1931, *Compton* 3904 (BOL). Caledon div., 3419BD, east of Riviersonderend, 4 viii 1951, *Middlemost* 1694 (NBG, STE). Montagu div., 3320CC, Baden, 10 vii 1946, *Walgate* BOL 23355 (BOL, PRE). Laingsburg div., 3320AB, near Tweedside, 20 ix 1931, *Compton* 3779 (BOL, NBG). Riversdale div., 3321CD, Garcia Pass, 1900–2300ft, 12 v 1982, *Viviers* 262 (STE); 3421AA, Corente River, xi 1908, *Muir* 578 (PRE). Oudtshoorn div., 3322AC, Cango Valley, Waenskloof, 6 xi 1974, *Moffett* 542 (PRE, STE). Mossel Bay div., 3422AB, hills overlooking Great Brak River, c.500ft, 20 x 1971, *Taylor* 8005 (PRE, STE). Willowmore div., 3323AC, flats between Hotsprings and Toorwater, 2500ft, 5 x 1971, *Thompson* 1394 (PRE, STE). Uniondale div., 3323CA, 3.5 miles from Uniondale on road to Willowmore, 2700ft, ix 1932, *Fourcade* 4669 (STE). Knysna div., 3423AA, Knysna, 13 ix 1953, *Taylor* 4135 (NBG, STE). Humansdorp div., 3324DA, 8.4 miles from Patensie to Andrieskraal, 26 ix 1969, *Marsh* 1376 (K, PRE, SRGH, STE).

Sutera caerulea ranges from the environs of the Hol river (west of Van Rhynsdorp) and Calvinia south to Worcester and Robertson, thence east as far as Humansdorp division, from c.60–1400m above sea level; although it was recorded at only 80m a.s.l. near the Hol river, it is absent from the sandy flats of the SW Cape and approaches the sea only along the southern Cape coast, though there it also ranges inland over the mountains to Laingsburg and Willowmore divisions. It is often a constituent of scrub on flats and hillslopes, but has also been recorded in salt meadows (near the Great Brak River) and as a weed in sandy fields near Knysna. Mrs Levyns (in Adamson & Salter, Fl. Cape Penins. 715, 1950) claims that it was recorded once on sand dunes near the Black River, Cape Peninsula, but I have not seen the specimen. The record seems improbable. Flowering has been recorded in all months, but peaks between July and October.

The species is easily recognized by its very short corolla tube with rotate limb and the long glandular hairs on the style and ovary. The hairy style and ovary will at once distinguish it from *S. campanulata* (which also has bigger flowers), *S. patriotica* and *S. neglecta*, with which it is often confused. Its relationship lies with these last two species, both of which have sessile glistening glands on the style and ovary; occasionally there are also tiny glandular hairs on the style, but they are never as long as those of *S. caerulea*. *Sutera neglecta* is further distinguished by its lack of conspicuous leaf tufts, in *S. patriotica* the lowermost bracts are usually more than twice as long as broad, in *S. caerulea* they are roughly as long as broad or up to twice as long as broad. The area of *S. caerulea* lies well outside that of both its allies.

Sutera caerulea has also been confused with *S. archeri*, and under that species possible hybrids with *S. caerulea* are mentioned. *Sutera caerulea* also appears to hybridize with *S. calciphila*: *Bohnen* 626.1 (STE) from Riversdale division, 2km east of Reisiesbaan Siding, comprises five twigs of *S. caerulea* and a sixth that may be *S. caerulea* × *calciphila*, while *McDonald* 1285 (STE), collected on the roadside between Albertinia and Bergfontein, is a hybrid between *S. caerulea* and possibly *S. integrifolia* or *S. calciphila*. Possible complex hybridization involving *S. caerulea* is mentioned under *S. aethiopica*; see also under *S. hispida*.

41. Sutera rotundifolia (Benth.) O. Kuntze, Rev. Gen. Pl. 2: 467 (1891); Hiern in Fl. Cap. 4(2): 250 (1904).

Type: Cape, without precise locality, *Drège* 4842 (holo. K, iso. P).

Syn.: *Chaenostoma rotundifolium* Benth. in Hook., Comp. Bot. Mag. 1: 374 (1836) & in DC., Prodr. 10: 354 (1846).

Twiggy shrublet c.200–450mm tall, main branches up to c.6mm diam., all parts glandular-puberulous, hairs up to 0.1–0.25mm long, leafy. *Leaves* opposite, largest blades 4–20 × 4–20mm, suborbicular or broadly rhomboid, abruptly contracted to a petiolar part 2–10mm long (roughly half as long as blade), apex rounded to subacute, margins usually serrulate, occasionally subentire or few-toothed, sometimes slightly revolute, both surfaces glandular-puberulous (hairs up to 0.1–0.25mm), twice recorded as glaucous. *Flowers* solitary in axils of uppermost leaves, panicled. *Pedicels* (lowermost) 2–10mm long. *Calyx* obscurely bilabiate, tube 0.5–1mm long, lobes 1.8–3.5 × 0.5–1mm, glandular-puberulous, hairs up to 0.1–0.25mm long. *Corolla* tube 4–6.5 × 3–5mm in throat, broadly funnel-shaped, limb almost regular, 8–10mm across, posticous lobes 2.5–3.5 × 2.5–3.5mm, anticous lobe 3–4 × 2.6–3.5mm, all oblong-orbicular, dark mauve or rosy-mauve, tube yellow or orange, glandular-puberulous outside, hairs up to 0.2mm long, either glabrous inside or occasionally with 1–3 weak longitudinal bands of clavate hairs. *Stamens* 4, all exserted, anthers 1.2–1.5mm long, posticous filaments 2–3.2mm long, anticous ones 3–4.5mm. *Stigma* 1.3–3.5mm long, exserted. *Style* 4–5.5mm. *Ovary* 1.3–1.8 × 0.8–1mm, a few glistening glands on sutures. *Capsules* 3–4.5 × 2–2.5mm. *Seeds* c.0.5–0.7 × 0.3mm, amber-coloured.

Citations:

Cape. Griquatown div., c.2823CB, Asbestos Hills, ii 1911, *Marloth* 4915b (PRE, STE). Britstown div., 3023DD, Damfontein, 17 v 1938, *Acocks* 8722 (PRE). Victoria West div., 3122DB, farm Wolwefontein west of Leeukop rock, c.1524m, 13 v 1976, *Hugo* 302 (K, STE); 3123AC, farm Uitzigt, 12km NE of Victoria West, c.4000ft, 14 v 1976, *Thompson* 3072 (PRE, STE). Murraysburg div., 3123DD, Murraysburg, 4000ft, iv 1879, *Tyson* 173 (SAM, Z); 22 miles E of Murraysburg, c.4800ft, 12 v 1950, *Acocks* 15870 (K). Beaufort West div., 3222BB, Hartebeesfontein c.18km N.W. of Nelspoort, 3500ft., 23 iii 1985, *Shearing* 1009 (PRE). Graaff Reinet div., 3224AA, Sneeuwberg, Lunsklip, 4000ft, iv 1868, *Bolus* 466 (BOL, PRE, S, SAM, Z). Journey from Colesberg (3025CA) to Hopetown (2924CA), comm. 1873, *Shaw* s.n. (W).

Sutera rotundifolia has been recorded from the mountains of the central Karoo, from about Victoria West east to the Sneeuwberg, thence north to the Asbestos Mountains, c.1200 to 1525m above sea level, growing in crevices on rocky cliff faces and among rock outcrops. It flowers between February and May. The leaf blade seldom exceeds 10mm in length; a specimen with blades up to 20mm long came from a young plant collected by Acocks (no. 15870), who particularly noted the difference in leaf-size between young and old plants.

Sutera rotundifolia is very closely allied to *S. pauciflora*; see under that species, below.

42. Sutera pauciflora (Benth.) O. Kuntze, Rev. Gen. Pl. 2: 467 (1891); Hiern in Fl. Cap. 4(2): 250 (1904) p.p.

Lectotype (chosen here): Cape, Beaufort West div., Nieuweveld, between Rhinoster Kop (3222BB) and Ganzefontein [not traced], 3500–4000ft, 31 x 1826, *Drège* 745a (K; isolecto. MO, P).

Syn.: *Chaenostoma pauciflorum* Benth. in Hook., Comp. Bot. Mag. 1: 374 (1836) p.p. excluding *Ecklon* [& *Zeyher*], Krakakamma.

Twiggy shrublet c.200–450mm tall, main branches up to c.7mm diam., erect, decumbent or ascending, all parts glandular-pubescent, hairs up to 0.5–0.8mm (0.2–0.8mm) long, leafy. *Leaves* opposite, largest blades c.5–12 × 5–9mm (7–16 × 3.5–9mm) suborbicular to broadly elliptic abruptly contracted or tapering to a broad petiolar part c.3–5mm long (roughly ⅓ to ½ as long as blade), apex rounded to subacute, margins often serrulate, sometimes obscurely few-toothed, repand or entire (up to 5 pairs of coarse teeth or obscurely toothed to entire), often

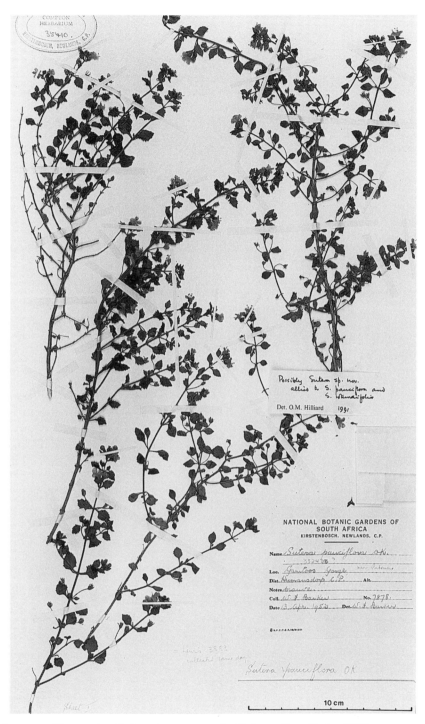

Fig. 57. *Sutera*, possibly an undescribed species; see text under 42. *S. pauciflora* (*Barker* 7878).

Table 4. Characters indicating possible hybrid influence in specimens from the Swartberg. Nos 1–9 are close to *S. pauciflora*; nos 10–11 are close to *S. revoluta*.

Collector	Locality	Hairs on stem and calyx (mm)	Leaf shape	Leaf margin	Length of corolla tube (mm)	Length of anticous filaments (mm)
Typical *S. pauciflora*	–	0.5–0.8	suborbicular to broadly elliptic	usually serrulate	5.5–7.5	3.8–4.5
1. *Markotter* s.n. sub STE 9962 (STE)	Swartberg Pass 3322AC	0.4	broadly elliptic	coarsely toothed	9.5	3
2. *Bolus* s.n. (BOL)	Swartberg Pass 3322AC	two twigs: 0.8	broadly elliptic	obscurely toothed to entire	5.5	c.3
		0.3	broadly elliptic	coarsely toothed to entire	7.5	c.3
3. *A. M. Krige* s.n. (BOL)	Prince Albert 3322AA	0.3–0.4	broadly elliptic	coarsely toothed	9	4
4. *D. Krige* s.n. (PRE)	Prince Albert 3322AA	0.3–0.7	broadly elliptic	coarsely toothed	7.2	3
5. *Marloth* 4486 (PRE, STE)	Sand River Mountains 3322AB	0.25	broadly elliptic	coarsely toothed	7.5	4
6. *Lewis* 4878 (PRE, STE, SAM) *Rycroft* 1615 (NBG, STE) is similar	Boschluiskloof 3321BC	0.2	elliptic	toothed to entire	9	4
7. *Acocks* 18431 (K)	Boschluiskloof Pass 3321BC	0.2	elliptic	toothed	8	3
8. *Thorne* s.n. sub SAM 51877 (SAM)	Gamka Poort 3321BC	0.2	elliptic to broadly elliptic	entire	4.5	3
9. *Acocks* 14324 (PRE)	Foothills of Swartberg 3322AC?	0.8	elliptic	entire	5	4
10. *Erasmus* 205 (E)	Gamka Poort 3321BC	0.2	elliptic	entire	7.5	3.5
11. *Lewis* 5100 (SAM)	Boschluiskloof to Prince Albert 3321BC–BD	0.2	narrowly elliptic	entire	8.4	3.3
Typical *S. revoluta*	–	0.1	linear	entire, strongly revolute	6.5–13	2.5–5

slightly revolute, both surfaces glandular-pubescent, hairs up to 0.3–0.4mm long (as on stem). *Flowers* solitary in axils of uppermost leaves, panicled. *Pedicels* (lowermost) 4–10mm long (6–15mm). *Calyx* obscurely bilabiate, tube 0.7–1.4mm long (0.7–1mm), lobes 2.8–4 × 0.8–1mm (2.4–4.8 × 0.5–0.9mm) glandular-pubescent, hairs up to 0.5–0.8mm long (0.2–0.8mm). *Corolla* tube 5.5–7.5 × 4–4.2mm in throat (4.5–9.5 × 3–5.5mm), broadly funnel-shaped, limb almost regular, 10–11mm across (8–15mm), posticous lobes 3.5–4 × 3–3.5mm (2–4.5 × 1.7–5mm), anticous lobe 3.5–5 × 3.3–5mm (2.5–6 × 2.2–4.3mm), all oblong-orbicular, mauve, tube yellow, glandular-puberulous outside, hairs up to 0.2–0.25mm long, either glabrous inside or with 1–3 weak longitudinal bands of clavate hairs (strongly to weakly bearded). *Stamens* 4, all exserted, anthers 1.3–1.4mm long (1–1.7mm), posticous filaments 2.5–3.5mm long (1.2–2.8mm), anticous ones 3.8–4.5mm (3–4mm). *Stigma* 2–3mm long, exserted. *Style* 3.5–6mm (3.6–6.8mm). *Ovary* 1.4–1.5 × 1mm (1–1.4 × 0.8–1mm), a few glistening glands on sutures (or glabrous). *Capsules* c.4 × 3mm. *Seeds* c.0.7 × 0.3mm, amber-coloured (pallid).

Citations:

Cape. Sutherland div., c.3121DC, Roggeveld, between Salt and Brakriver [virtually illegible], *Rehmann* 3211 (Z). Beaufort West div., 3222BA, Bleak House Farm, 16 iv 1978, *Gibbs Russell et al.* 320 (PRE); 3223AA, Nelspoort, Courland's Kloof, vii 1907, *Pearson* 1304 (SAM). Victoria West div., 3123CC, Three Sisters, 4200ft, 31 iii 1956, *Schelpe* 5858 (BOL, PRE). c.3319AB, Ceres Bokkeveld, *Rehmann* 3098 (Z). Specimens from the Great Swartberg are cited in Table 4.

Bentham (1836) cited two different species under his new name *Chaenostoma pauciflorum*; the description fits them both. Drège's specimen from the Nieuweveldberge has been chosen as lectotype; Ecklon's specimen (from Krakakamma near Uitenhage, no. 360, SAM, W) is *S. cordata*.

Sutera pauciflora is closely allied to *S. rotundifolia*; typical specimens of *S. pauciflora* differ from *S. rotundifolia* in having longer hairs on the vegetative parts and the calyx, and the leaves may be elliptic. Such plants occupy an area west and south of that of *S. rotundifolia*, from the Roggeveld east to Nelspoort and Three Sisters and possibly west to the Ceres Bokkeveld (said by Rehmann to have been collected there, but his localities are not always reliable). These sites all lie north and west of the Great Karoo, and the plants show no variation other than that attributable to growing conditions. But on the Great Swartberg on the southern flank of the Great Karoo, different circumstances obtain: here *S. pauciflora* shows considerable variation consistent with hybridization; in the formal description, data in parentheses refer to these plants. In the northern part of its range, *S. pauciflora* is sympatric only with *S. halimifolia* and there is no apparent interaction between them. But on the Swartberg, *S. pauciflora* is sympatric with several species (*S. archeri, caerulea, halimifolia, revoluta, violacea*), and appears to have hybridized (probably with *S. revoluta*) to such an extent that a 'pure' strain is not easy to find. The most significant variation is in the length of the indumentum, shape and toothing of leaves, length of corolla tube, length of anticous filaments, and colour of the seeds (amber in the typical plant, pallid white in the Swartberg specimens). Pertinent data are presented in Table 4; at the head of the table, data are given for typical *S. pauciflora*, at its base, typical *S. revoluta*.

The plants grow in karroid vegetation on rocky mountain slopes, between c.700 and 1200m above sea level, flowering principally between March and September.

Miss Barker and Dr G. J. Lewis were clearly together in April 1952 when they collected a plant in the kloof of the Gamtoos River between Patensie and Mistkraal (3324DB): *Barker* 7878, NBG, and *Lewis* 3883, SAM. It is allied to *S. rotundifolia* and *S. pauciflora*, but differs from the former in the shape of its leaves (broadly elliptic tapering into a short petiole) and from the latter in its very short indumentum. The flowers have a very short corolla tube (c.4.5 × 3.5mm), well exserted stamens (anticous filaments 4.8mm, posticous ones 3mm), and the seeds are pallid. It is possibly an undescribed species, and should be sought in the Elandsberg and Great Winterhoek Mountains; the Baviaanskloof and the Kouga Mountains should also be explored. **Fig. 57.**

43. Sutera cordata (Thunb.) O. Kuntze, Rev. Gen. Pl. 2: 467 (1891); Hiern in Fl. Cap. 4(2): 279 (1904) p.p. and excl. var. *hirsutior*.

Type: Cape, without precise locality, *Thunberg* (sheet no. 14357 herb. Thunberg, UPS).

Syn.: *Manulea cordata* Thunb., Prodr. 102 (1800) & Fl. Cap. ed. Schultes 473 (1823).
    Chaenostoma cordatum (Thunb.) Benth. in Hook., Comp. Bot. Mag. 1: 377 (1836) & in DC., Prodr. 10: 357 (1846), excluding β *hirsutior*.
    C. pauciflorum Benth. in Hook., Comp. Bot. Mag. 1: 374 (1836) p.p., excluding type, as to *Ecklon* [& *Zeyher*], Krakakamma, only.

Sprawling perennial herb, rootstock woody, up to c.10mm diam., branches many from the crown, of indeterminate length, laxly branched, often rooting at the nodes, softly villous, hairs up to 0.5–1.8mm long, both acute and gland-tipped, leafy. *Leaves* opposite, largest blades 6–25 × 5–23mm, ovate, broadly ovate, or sometimes broadly elliptic, abruptly contracted to a petiolar part c.4–13mm long, very roughly half length of blade, apex subacute to obtuse, margins crenate or serrate, both surfaces pubescent to villous, hairs up to 0.5–1mm long, longest often on petiolar part, either with scattered glistening glands as well or these wanting. *Flowers* solitary in axils of leaves. *Pedicels* (lowermost) c.9–25mm long. *Calyx* obscurely bilabiate, tube (1–)1.5–2.25mm long, lobes 2.5–6 × 0.7–1mm, villous, hairs up to 0.6–1.5mm long, some gland-tipped, glistening glands either wanting or a few on tube. *Corolla* tube 8–11 × 3.2–4mm in throat, funnel-shaped, limb almost regular, 8–12mm across, posticous lobes 2.5–4.3 × 1.8–3mm, anticous lobe 3–5 × 1.8–4mm, all oblong-orbicular, white, tube yellow or orange, glandular-pubescent outside, hairs up to 0.1–0.6mm long, some scattered glistening glands as well, 3–5 longitudinal bands of clavate hairs inside. *Stamens* 4, anthers 1–1.3mm long, anticous pair exserted, filaments 2.5–3.2mm long, posticous pair in mouth, filaments 1–1.6mm long. *Stigma* 0.7–1.4mm long, exserted. *Style* 5.5–8mm long. *Ovary* 1.2–1.4 × 0.8–1mm, glabrous or a few glistening glands on sutures. *Capsule* 3–5 × 2–3mm. *Seeds* c.0.4–0.8 × 0.3–0.6mm, amber-coloured.

Selected citations:

Cape. East London div., 3327BA, Igoda mouth, 17 xii 1977, *Hilliard & Burtt* 11121 (E, K). Alexandria div., 3326CB, near Alexandria, 6 vii 1947, *Compton* 19665 (NBG). Albany div., 3326BC, Amos Kloof, 2000ft, x 1888, *Galpin* 182 (K, PRE). Alexandria div., 3325BC, top of Zuurberg Pass, 3500ft, 13 viii 1973, *Bayliss* 6048 (NBG, Z). Humansdorp div., 3424AA, Eerste river, 500ft, viii 1924, *Fourcade* 3001 (BOL, STE). Knysna div., 3323DC, Nature's Valley, c.100ft, 7 vii 1960, *Acocks* 21162 (K, PRE, SRGH, STE); 3423AA, Noetzie, 27 xii 1956, *Middlemost* 1949 (PRE, STE). George div., 3322DC, Wilderness, 28 iv 1941, *Compton* 10685 (NBG); 3322CD, Outeniqua Mts., Montagu Pass, *Rehmann* 230 (Z).

There are no flowers on the type of *S. cordata* (just one old calyx) and Thunberg did not describe the flowers, but there is no doubt that the name is here used correctly. *Sutera cordata* has been much confused with *S. pauciflora* (one of the syntypes of that name being a specimen of *S. cordata*) and with *S. racemosa*; it is easily distinguished from *S. pauciflora* by its sprawling rooting branches, longer hairs on stems, leaves and calyces, longer corolla tube and white (not mauve) corolla limb. *Sutera racemosa* is a completely different plant, allied to *S. polelensis*; the name has never been correctly used in herbaria. The closest allies of *S. cordata* are *S. glandulifera* and *S. impedita*: see there.

    Sutera cordata is a common plant along stretches of the Cape coast, from George in the west to East London in the east, though there appear to be no records from Algoa Bay and its immediate environs. It favours partially shaded places, in forest and at their margins, or in scrub, including littoral scrub, and may flower in any month. Most records are from the coast and nearby, but the species penetrates inland to forested kloofs around Grahamstown, in the Zuurberg Pass (c.1000m) and on Montagu Pass across the Outeniqua Mountains.

44. Sutera glandulifera Hilliard in Edinb. J. Bot. 48: 343 (1991).
Type: Cape, Stockenstrom div., 3226DA, Katberg Pass, c.5800ft, 26 i 1979, *Hilliard & Burtt* 12390 (holo. E; iso. K, M, S).

Sprawling perennial herb, rootstock woody, up to c.10mm diam., branches many from the crown, of indeterminate length, laxly branched, sometimes rooting at the nodes, softly villous with gland-tipped hairs up to 0.8–2mm long, leafy. *Leaves* opposite, largest blades 6–21 × 8–24mm, ovate abruptly contracted to a petiolar part c.2–6mm long, roughly ¼ to ⅓ length of blade, apex acute, margins serrate to irregularly shallowly lobed and toothed, both surfaces pubescent to villous with glandular hairs up to 0.8–2mm long, densely clad in glistening glands as well. *Flowers* solitary in axils of leaves. *Pedicels* (lowermost) c.6–30mm long. *Calyx* obscurely bilabiate, tube 0.5–1mm long, lobes 3–4.7 × 0.4–0.7mm, villous with glandular hairs up to 0.4–2mm long, densely clad in glistening glands as well. *Corolla* tube 4.5–10 × 2.5–4.5mm in throat, funnel-shaped, limb almost regular, 7–9.5mm across, posticous lobes 2.5–3.5 × 1.8–3mm, anticous lobe 2.6–4 × 1.8–3.4mm, all oblong-elliptic to oblong-orbicular, white, occasionally faintly tinged mauve, tube yellow or orange, glandular-pubescent outside, hairs up to 0.2–0.6mm long, many glistening glands as well, inside well bearded all round with clavate hairs. *Stamens* 4, anticous pair exserted, filaments 2.5–3.5mm long, posticous pair included, filaments 1.5–2.5mm long, anthers 0.8–1.2mm long. *Stigma* 0.7–1.3mm long, exserted. *Style* 4–7mm long. *Ovary* 1–1.4 × 0.75–1.2mm long, ± wholly clad in glistening glands. *Capsule* 3.5–4 × 2.5–3mm, glistening glands persistent. *Seeds* c.0.4–0.7 × 0.3mm, amber-coloured.

Citations:

Cape. Wodehouse div., 3126BC, Penhoek Pass, c.6000ft, 24 i 1956, *Acocks* 18670 (K, M, PRE, SRGH). Queenstown div., 3126DD, Hangklip Mt., 31 xii 1962, *Roberts* 2029 (PRE); Queenstown Mt. Range, 4000ft, iii 1896, *Galpin* 2087 (PRE); Bongola Dam, c.4000ft., 19 ii 1955, *Acocks* 17930 (K, M). Stockenstrom div., 3226DA, Katberg Pass, c.5500ft, 25 i 1979, *Hilliard & Burtt* 12404 and 12371 (E); ibidem, 10 xii 1977, *Hilliard & Burtt* 10960 (E); ibidem, 5 i 1927, *A. Lewis Grant* 3098 (PRE, S); ibidem, 6 i 1927, *Young* sub Moss 14743 (BM, J) and 15285 (BM); ibidem, Balfourside, 3500ft, vii, *Baur* 863 (K, PRE). Amatole Mts., 3226DB, Menziesberg, 5200ft, 1 xii 1986, *Phillipson* 1542 (K); 3226DD, Wolf Ridge, 3600ft, 2 v 1943, *Giffen* 1465 (PRE). Stutterheim div., 3227CB, Dohne Hill, 4000–5000ft, 1897, *Sim* 20349, 20367, 124 (PRE); Perie Mts., Evelyn Valley, 3000ft, *Scott Elliot* 933 (BM, E); near King William's Town, 1200ft., xii 1891, *Sim* 233 (PRE).
Transkei. Tembuland, 3227AB, near St. Marks, Clark's Drift, Qumagneu River, 3000ft, i 1896, *Bolus* 8763 (BOL, K).

Sutera glandulifera is closely allied to *S. cordata* from which it is distinguished by its more glandular indumentum and greater development of glistening glands on leaves, calyx, and outside of corolla (the glands are all over the outside of the calyx; in *S. cordata* they are only at the base of the tube), leaves with a shorter petiolar part (roughly ¼–⅓ as long as the blade, not half as long), generally shorter calyx tube (up to 1mm long, not mostly 1.5–2.25mm) and ovary (also capsule) with glistening glands all over (not confined to the sutures).

Sutera glandulifera is known only from the mountains of the eastern Cape and neighbouring parts of the Transkei, from the Stormberg in the north south to the Amatole Mountains and the mountains around Queenstown and Stutterheim, between c.900 and 2000m above sea level. Its area thus lies north and east of that of *S. cordata* and generally at higher altitudes. It grows in the shelter of rock overhangs and rock outcrops, flowering mainly between December and March, though there is one record in July.

Sutera glandulifera can also be confused with *S. polelensis*, but that species has leaves with proportionately shorter petioles (ratio of blade-length to petiole mostly 7–35:1 versus 3–5:1), flowers running up into racemes or panicles, glistening glands on the ovaries confined to the sutures, and pallid seeds.

45. Sutera impedita Hilliard in Edinb. J. Bot. 48: 343 (1991).
Type: Cape, Komgha div., margins of woods near Komgha, 2000ft, viii 1891, *Flanagan* 892 (holo. SAM; iso. BOL, K, PRE).

Sprawling perennial herb, rootstock woody, up to c.10mm diam., branches eventually many from the crown, c.200–350mm long, soon laxly branched, glandular-pubescent, hairs up to 0.4–1mm long, leafy. *Leaves* opposite, largest blades c.16–25 × 14–22mm, ovate abruptly contracted to a petiolar part c.6–9mm long (c.⅓ to ¼ as long as blade), apex acute, margins serrate, both surfaces glandular-pubescent, hairs up to 0.5–0.6mm long, many glistening glands as well. *Flowers* solitary in axils of leaves. *Pedicels* (lowermost) c.8–30mm long. *Calyx* obscurely bilabiate, tube c.0.7mm long, lobes 5–7, 2.4–5 × 0.4–0.7mm, villous, hairs up to 0.25–0.6mm long, many gland-tipped, glistening glands all over as well. *Corolla* tube 6–10 × 3–4.2mm in throat, funnel-shaped, limb almost regular, 7–10mm across, posticous lobes 2.3–3.3 × 2–3.3mm, anticous lobe 2.3–4 × 2–4mm, all oblong-orbicular, mauve, tube orange, glandular-pubescent, hairs 0.1–0.4mm long, glistening glands as well, bearded inside with 5 longitudinal bands of clavate hairs. *Stamens* 4, anticous pair exserted, filaments 1.8–3mm long, posticous pair in throat, filaments 0.8–1.5mm long, anthers 0.8–1mm long. *Stigma* 0.8–1.5mm long, exserted. *Style* c.4–7mm. *Ovary* 1–1.2 × 0.7–1mm, glistening glands all over. *Capsules* c.3–5 × 2.5–3mm, glistening glands persistent. *Seeds* c.0.8 × 0.6–0.8mm, amber-coloured.

Citations:

Transkei. 3228CB, Kentani, 1200ft, iii 1905, *Pegler* 896 (BOL, PRE); Goluwen, 13 vii 1966, *Strey* 6702 (K, PRE).
Cape. [Near Stutterheim] 3227DA, Amabele, 24 i 1919, *Moss* 4848 (J).

The sprawling habit and axillary flowers give *S. impedita* the aspect of *S. cordata* and *S. glandulifera*, from which it may be distinguished by its calyx often with 6 or 7 lobes (not always 5) and its mauve (not white) corolla limb. It is further distinguished from *S. cordata* by its leaves with a relatively longer petiolar part, leaves, calyx and ovary with a much denser covering of glistening glands, and shorter calyx tube; from *S. glandulifera* by the generally shorter hairs on stems, leaves and calyx. Its area lies roughly east of that of both these species, around Komgha and Kentani, botanically ill-known areas. There, it is sympatric with *S. calycina*, with which it shares the peculiar character of a multi-lobed calyx. It differs from *S. calycina* in its longer indumentum, leaves with a longer petiolar part, flowers in the axils of the leaves, lack of coarse hairs on the calyx, and narrower corolla tube.

Flanagan recorded *S. impedita* as growing 'on margins of woods', Moss 'forests', Pegler 'valleys in sheltered places', Strey 'mountain slopes', between c.300 and 600m above sea level, flowering and fruiting in January, March and July.

A nearly sterile specimen from the south face of Boschberg at Somerset East (*van der Walt* 50, PRE) appears to be in the affinity of *S. impedita*, but the plant should be sought again to determine its status. It differs from both *S. impedita* and *S. racemosa* in its much shorter indumentum (c.0.1mm).

46. Sutera calycina (Benth.) O. Kuntze, Rev. Gen. Pl. 2: 467 (1891); Hiern in Fl. Cap. 4(2): 254 (1904).
Type: Transkei, zwischen Gekau und Basche [Bashee River], 1500–2000ft, 23 i 1832, *Drège* 4860 (holo. K; iso. P, S).

Syn.: *Chaenostoma calycinum* Benth. in Hook., Comp. Bot. Mag. 1: 374 (1836) & in DC., Prodr. 10: 354 (1846).
   C. calycinum var. *laxiflorum* Benth., loc. cit., nomen, quoad *Drège* 4732, tantum.
   Sutera calycina var. *laxiflora* Hiern in Fl. Cap. 4(2): 254 (1904). Lectotype: Transkei, near the Bashee river, 500ft, 30 i 1832, *Drège* 4732 (K, isolecto. P).
   Sutera polysepala Hiern in Fl. Cap. 4(2): 254 (1904). Type: Cape, Komgha div., grassy flat near Kei Mouth, 200ft, i 1892, *Flanagan* 1061 (holo. BOL, iso. PRE).

Perennial herb, initially with a woody taproot, eventually developing a small woody caudex, stems few to several from the crown, erect or decumbent, sometimes straggling, 100–450mm long, well-branched, glandular-puberulous, hairs up to 0.1–0.15mm long, glistening glands as well particularly on upper parts, leafy. *Leaves* opposite, bases connate, largest ones 12–30 × 4–15mm, smaller upwards and passing into bracts, elliptic, apex acute or subacute, base of lower leaves narrowed to a short petiolar part, upper ones sessile, broad-based, half-clasping, margins with c.2–10 pairs of small teeth, both surfaces glandular-puberulous, hairs up to c.0.15mm long, many glistening glands as well. *Flowers* in long racemes, often further arranged in panicles. *Bracts* (lowermost) c.3.5–15 × 2–4mm, leaflike, the smaller ones sometimes entire. *Pedicels* (lowermost) c.5–25mm long, filiform. *Calyx* regular, tube 1–1.6mm long, lobes (6–)8(–9), 1.4–3.8 × 0.5–1mm, lanceolate, glandular–puberulous, tube in particular with coarse glandular hairs as well c.0.15–0.4mm long, many glistening glands also present. *Corolla* tube 4–7.5 × 3–6mm in throat, broadly funnel-shaped, limb almost regular, 8–14mm across, posticous lobes 3–5 × 3–4mm, anticous lobe 3.2–5 × 3–4mm, all oblong-orbicular, mauve-pink, tube yellow or orange, glandular-puberulous outside, hairs up to c.0.2mm long, glistening glands as well, clavate hairs inside, more or less in 5 longitudinal bands around throat. *Stamens* 4, all exserted, anthers 1–1.7mm long, posticous filaments 1–2.5mm long, anticous filaments 2.5–4mm long. *Stigma* 1.2–2.5mm long, exserted. *Style* 2.5–6mm long, clad in glistening glands, glandular hairs up to 0.15mm long sometimes present as well. *Ovary* 1.4–1.8 × 1–1.25mm, glistening glands all over. *Capsules* 3–6 × 2–4mm, glistening glands persistent. *Seeds* 0.4–0.75 × 0.25–0.5mm, amber-coloured.

Selected citations:

Cape. Elliot distr., 3127BB, Barkly Pass, c.6700ft, 30 i 1966, *Hilliard & Burtt* 3767 (E). Komgha distr., 3227DB, sandy spots along the Kei River near Komgha, 1800ft, xii 1891, *Flanagan* 1082 (BOL, PRE, SAM).
Transkei. Cala, 3127DA, 4000ft, ii 1910, *Kolbe & Pegler* 1683 (BOL). 3128DB, 2 miies SW of Umtata, 2600ft, 11 iii 1955, *Codd* 9278 (PRE). Kentani distr., 3228CB, near Qolora Mouth, c.20ft, vii 1906, *Pegler* 1350 (BOL, PRE, SAM). Butterworth distr., 3228BC, Mazeppa Bay, 11 iii 1952, *Theron* 1233 (K). Elliotdale distr., 3228BB, near the Bashee, 830m, 16 i 1895, *Schlechter* 6297 (BOL, E, PRE, S, Z).

Sutera calycina has a limited distribution; most records are from the southern half of the Transkei, between sea level and c.1200m, but there is one record from Barkly Pass (Cape) at c.2000m, and the species also extends across the Kei into the eastern Cape at Komgha (550m) and down to sea level at Haga-Haga. The plants have been recorded from sandy flood plains, sandy grass slopes and in scrub on the seaward side of dunes; on Barkly Pass it was on earth banks, so it appears to be a colonizer of bare areas. Its habit depends, at least in part, on habitat: it has been recorded as sprawling in short grass, scrambling in tall; on open sites it is erect. It can possibly be found in flower in any month, but most records are in January.

In herbaria, it has been confused with *S. campanulata*, *S. caerulea* and *S. halimifolia*. It can at once be distinguished from all three by its 6–9-lobed calyx. Its relationship lies with *S. campanulata*, which differs further in the indumentum on the stems and ovaries.

47. Sutera polyantha (Benth.) O. Kuntze, Rev. Gen. Pl. 2: 467 (1891); Hiern in Fl. Cap. 4(2): 259 (1904), p.p.
Lectotype (chosen here): Cape, Uitenhage div., on the Zwartkops river, *Ecklon* (K, left hand piece; isolecto. M, S, W).

Syn.: *Chaenostoma polyanthum* Benth. in Hook., Comp. Bot. Mag. 1: 375 (1836) & in DC., Prodr. 10: 354 (1846).
    Manulea polyantha (Benth.) O. Kuntze, Rev. Gen. Pl. 3(2): 236 (1898).
    Chaenostoma procumbens Benth. in Hook., Comp. Bot. Mag. 1: 374 (1836) & in DC., Prodr. 10: 354 (1846).
      Type: Cape, an Vischrivier, bei Kookhuis, 2000ft, 10 x 1829, *Drège* 7914 (holo. K, iso. P).
    Sutera procumbens (Benth.) O. Kuntze, Rev. Gen. Pl. 2: 467 (1891); Hiern in Fl. Cap. 4(2): 254 (1904).
    S. infundibuliformis Schinz in Vierteljahrsschr. Nat. Ges. Zürich 74: 117 (1929). Type: Cape, 3325CD, Uitenhage, 80m, 18 iv 1893, *Schlechter* 2538 (holo. Z; iso. BOL, E, J, K, PRE, S).

S. campanulata auct. non (Benth.) O. Kuntze; Batten & Bokelmann, Wild Flow. E Cape Prov. 132 pl. 104, 5 (1966).

Herb, possibly perennating, with a simple, woody taproot up to c.8mm diam. at the crown, soon branching from the base and above, eventually bushy, branches erect or decumbent, c.60–300mm long, up to 0.2–0.3mm diam. at base, glandular-puberulous, hairs at base of stem up to 0.2–0.25mm long, shorter upwards, c.0.1mm on inflorescence branches and there mixed with glistening glands, leafy. *Leaves* opposite, largest c.9–27 × 3–10mm, elliptic tapering into a short petiolar part, apex acute or subacute, margins rarely subentire, usually obscurely to distinctly toothed, minutely glandular-puberulous (hairs c.0.1mm), glistening glands as well. *Flowers* in leafy racemes paniculately arranged. *Bracts* (lowermost) equivalent to uppermost leaves, c.5–15 × 2.5–7mm, rapidly smaller upwards. *Pedicels* (lowermost) c.5–20mm long, filiform. *Calyx* obscurely bilabiate, tube 0.5–1.5mm long, lobes 1.4–4 × 0.5–0.8mm, clad in glistening glands and glandular hairs, the coarser hairs up to 0.1–0.4mm long. *Corolla* tube 4.5–7 × 3.5–5mm in throat, broadly funnel-shaped, limb almost regular, 7–12mm across, posticous lobes 2–4 × 2–3.7mm, anticous lobe 2.5–5 × 2.8–4mm, all oblong-orbicular, white sometimes faintly tinged mauve with age, or rarely wholly pale mauve, tube yellow or orange, glandular puberulous outside, glistening glands as well, clavate hairs inside ± in 5 longitudinal bands around throat. *Stamens* 4, all exserted, anthers 1–1.3mm long, posticous filaments 1.5–2.8mm, anticous filaments 2.8–4mm. *Stigma* 1.2–2.2mm long, exserted. *Style* 3–5mm long. *Ovary* 1.3–1.8 × 0.8–1mm, with glistening glands on sutures only or on upper part as well. *Capsules* 2.5–4.5 × 2–2.5mm, glistening glands persistent. *Seeds* c.0.3–0.6 × 0.25–0.5mm, amber-coloured, sometimes with a whitish tinge.

Selected citations:

Cape. Knysna div., 3424AA, Groot Rivier, 20ft, iv 1928, *Fourcade* 3691 (BOL, STE). Victoria East div., 3226DD, Alice, 1700ft, 14 v 1943, *Giffen* 1500 (PRE). Humansdorp div., 3325CC, Loerie River, 29 xi 1899, *Penther* 3025 (W). Port Elizabeth div., 3325BD, Addo Elephant National Park, 15 v 1954, *Brynard* 354 (PRE). Peddie div., 3327AA, Peddie, 11 iv 1952, *Compton* 23365 (M, NBG). Albany div., 3326BB, Trompeters Drift, 16 i 1947, *Compton* 19341 (NBG); 3326BC, Grahamstown-East London road at Fort Beaufort turnoff, 8 xii 1977, *Hilliard & Burtt* 10902 (E, K).

Bentham's two specimens of *Chaenostoma polyanthum* appear to have come from two different collections; the left-hand specimen has been chosen as the lectotype because it alone has the paniculate inflorescence that probably suggested the trivial name. The specimen is decidedly time-worn, but there are two excellent sheets in Vienna (W) as well as a further twig on a third sheet. A note by N. E. Brown on the type sheet (K) of *C. procumbens* suggests that the plant is *C. halimifolium*, but this is not so: it has the glandular pubescence of *Sutera polyantha* (vesicular hairs only in *S. halimifolia*); Bentham erroneously described the plant as glabrous except for a few hairs on the upper parts and the calyx.

There has been a good deal of confusion in herbaria over the use of the name *S. polyantha* and Hiern cited a number of specimens belonging to different species. For example, *Rehmann* 6846 (Transvaal) is *S. neglecta, Drège* 5486 (Windvogelberg at Cathcart) is *S. patriotica. Sutera polyantha* appears to be confined to the drier parts of the eastern Cape, from Peddie in the east to Patensie in the west and inland to the environs of Cookhouse (near Somerset East) and Alice (near Fort Beaufort) at c.600m, then along the southern Cape coast, on littoral dunes, as far west as George. A Zeyher collection, purportedly from the Langekloof, and distributed as no. 3505 (S, W) is *S. polyantha*; it was cited by Hiern as *S. linifolia*.

Zeyher 1032 'on the fields near the Zwartkops River, district of Uitenhage', widely distributed as *Chaenostoma pumilum* Benth., is also *S. polyantha*.

Sutera polyantha can be recognized by the very short glandular pubescence on all parts; this distinguishes it from *S. campanulata* where there are long acute hairs at least near the base of the stem and the coarse hairs on the calyx are longer. However, some specimens are difficult to place, especially those from the southern Cape coast, and hybridization is a factor to be considered. *Scharf* 1295 (PRE) is possibly *S. polyantha* × *halimifolia*; it has the facies of *S. polyantha* and its short glandular pubescence and more or less hispid calyx, but the blue-grey seeds (and possibly the blue flowers) indicate crossing with *S. halimifolia*. The specimen came from Uitenhage division, in the catchment of the Elands River, at the roadside in dry valley bushveld.

48. Sutera roseoflava Hiern in Fl. Cap. 4(2): 251 (1904). **Fig. 53A.**

Lectotype (chosen here): East London div., slopes near sea coast, near East London, vi 1894, *Galpin* 1878 (K; isolecto. BOL, PRE).

Perennial herb capable of flowering when young, taproot woody eventually developing into a knotty rootstock c.15mm in diam. at crown, stem soon branching from base and above, branches decumbent or sprawling, erect in early stages of growth, up to c.600mm long, pubescent with acute hairs on lower parts up to 0.5–1mm long, soon mixed with glandular hairs, ultimate inflorescence axes and pedicels with glandular hairs only, glistening glands also present, leafy throughout. *Leaves* opposite, largest c.15–30 × 10–17mm, ovate or elliptic, either broad-based and sessile or tapering to a short petiolar part, apex ± acute, margins serrate, both surfaces pubescent, hairs up to 0.25–0.4mm long, glistening glands as well. *Flowers* solitary in the leaf axils, often from very low down on the stem, then branching towards the tips into very leafy panicles, the flowers then crowded (seen only in well-grown specimens). *Bracts*, that is leaves, at base of inflorescence c.9–20 × 4–16mm. *Pedicels* (lowermost) c.5–20mm long. *Calyx* obscurely bilabiate, tube 1mm long, lobes 3.5–4.8 × 0.8–1mm, glandular-pubescent, the coarser hairs up to 0.5–1.2mm long, glistening glands as well. *Corolla* tube 5–8 × 4–5mm in throat, broadly funnel-shaped, limb almost regular, 8–14mm across, posticous lobes 2.5–4 × 2.1–3.8mm, anticous lobe 3–5 × 3–4.2mm, all oblong-orbicular, pale mauve, tube yellow or orange, glandular-puberulous outside, glistening glands as well, 5 bands of clavate hairs inside. *Stamens* 4, all exserted, anthers 1–1.5mm long, posticous filaments 1.5–2mm long, anticous ones 2.5–4mm. *Stigma* 1–2.2mm long, exserted. *Style* 3.2–6mm long. *Ovary* 1–1.5 × 0.8–1.2mm, glistening glands all over, sometimes tiny glandular hairs as well. *Capsules* c.4–5.5 × 2.5–4mm, glistening glands persistent. *Seeds* 0.5–1 × 0.3–0.6mm, amber-coloured.

Selected citations:

Cape. King William's Town distr., 3227CD, King William's Town, 1300ft, xii 1886, *Tyson* 2883 (855 in herb. norm. austr. afr.) (BOL, E, PRE, SAM, W, Z); ibidem, 1 xi 1953, *Taylor* 4171 (NBG, STE). East London distr., 3228CC, Gonubie Beach, 19 xii 1985, *Hilliard & Burtt* 18902 (E, K, PRE, S); 3327BA, Kidd's Beach, 18 xii 1985, *Hilliard & Burtt* 18890 (E). Bathurst distr., Port Alfred, 20 x 1948, *Compton* 21080 (NBG). Alexandria distr., 3326DA, W side of Bushman's River mouth, 6 xii 1977, *Hilliard & Burtt* 10886 (E, PRE).

Hiern cited two Galpin collections under *S. roseoflava*; I have lectotypified the name by the collection from East London, where the plant is still common though its habitat is disappearing under urban sprawl; *Galpin* 4377 (PRE) proves to be *S. campanulata*.

Sutera roseoflava is very closely allied to *S. campanulata*. It differs from typical *S. campanulata* in its sprawling habit, leafy inflorescence, and ovaries (and capsules) with glistening glands all over. However, within the broad species concept here adopted for *S. campanulata*, these characters can in part be matched. For example, a form of *S. campanulata* common in Albany division (see under *S. campanulata*) is prostrate and has leafy inflorescences (though the facies is not that of *S. roseoflava*), but the ovaries are glandular only on the sutures. The whole *S. campanulata*, *S. roseoflava*, *S. polyantha*, *S. calycina* complex needs field and laboratory investigation.

Sutera roseoflava ranges along the Cape coast from about Port Alfred to East London and inland to King William's Town and Stutterheim; there is a single record from Somerset East. The plants favour sandy grassland, and can possibly be found in flower in any month.

49. Sutera campanulata (Benth.) O. Kuntze, Rev. Gen. Pl. 2: 467 (1891); Hiern in Fl. Cap. 4(2): 253 (1904), p.p.

Lectotype (chosen here): Cape [3325BD], Zuurbergen, 3500ft, 1 xi 1829, *Drège* (K; isolecto. E, LE, MO, P, S, W).

Syn.: *Chaenostoma campanulatum* Benth. in Hook., Comp. Bot. Mag. 1: 374 (1836) & in DC., Prodr. 10: 354 (1846); Krauss in Flora 27: 834 (1844).
  C. polyanthum auctt., non Benth.; Loudon, Ladies Fl. Gard. & Ornamental Greenh. Pl. t. 40 (1843); Floricult. Cab. 13: 241 (1843); Edwards Bot. Reg. 33 t. 32 (1847); Paxton, Mag. 13: 31 (1847).
  Manulea campanulata (Benth.) O. Kuntze, Rev. Gen. Pl. 3(2): 235 (1898).
  Sutera intertexta Hiern in Fl. Cap. 4(2): 255 (1904). Type: Cape, Port Elizabeth div., Algoa Bay, 1860, *Cooper* 1452 (holo. K, iso. W).
  S. roseoflava Hiern in Fl. Cap. 4(2): 251 (1904) p.p. as to *Galpin* 4377 only.

Perennial herb but flowering when young and then looking like an annual, taproot eventually thick and woody, up to 10mm diam. at crown, stem soon branching from the base and above, branches erect or decumbent, sometimes sprawling, c.50–450(–600)mm long, pubescent with glandular hairs up to 0.2–0.6mm long on upper parts, sometimes mixed with acute hairs, mainly acute hairs on lower part of stem, c.0.3–0.7mm long, leafy. *Leaves* opposite, largest c.10–40 × 5–15(–30)mm, ovate or elliptic tapering to a short petiolar part, apex acute, margins with few to several pairs of small teeth, both surfaces pubescent, hairs up to 0.25mm long, glistening glands as well. *Flowers* in racemes paniculately arranged, the lowermost flowers in the axils of the uppermost leaves. *Bracts* (lowermost, and often a leaf with an axillary bud) c.6–18 × 1.5–7mm, rapidly smaller upwards and then often entire. *Pedicels* (lowermost) c.5–28mm long. *Calyx* obscurely bilabiate, tube 0.7–1.5mm long, lobes 2.8–6 × 0.7–1mm, glandular-pubescent, the coarser hairs 0.5–1.2mm long, glistening glands also present. *Corolla* tube 6–9 × 4–6mm in throat, broadly funnel-shaped, limb almost regular, 9–14mm across, posticous lobes 3.3–4.5 × 3–4.8mm, anticous lobe 3.3–6 × 3.3–5mm, all oblong-orbicular, shades of mauve, tube yellow or orange, glandular puberulous outside, glistening glands as well, clavate hairs inside in 5 longitudinal bands. *Stamens* 4, all exserted, anthers 1–1.6mm long, posticous filaments 1.5–3mm long, anticous ones 3–4mm. *Stigma* 1–2.4mm, exserted. *Style* 4.8–7mm long, occasionally with a few glistening glands. *Ovary* 1–1.8 × 0.8–1.1mm, glistening glands on sutures and very rarely a few on lower part of style. *Capsules* c.4.3–6.5 × 2.5–3mm, glistening glands persistent. *Seeds* c.0.5–0.8 × 0.3mm, amber-coloured.

Selected citations:

Cape. Albany div., 3326AD, Grahamstown Nature Reserve, c.2500ft, 2 xii 1977, *Hilliard & Burtt* 10831 (E, K). Bathurst div., 3326BD, Trappes Valley, 11 i 1947, *Compton* 19119 (NBG). Alexandria div., 3326CB, Alexandria, 4 v 1931, *Galpin* 10649 (PRE). Port Elizabeth div., 3325DC, near Port Elizabeth, 100ft, 1871, *Bolus* 2236 (BOL). Humansdorp div., 3324DD, west bank of Gamtoos river near Hankey, 150ft, vii 1922, *Fourcade* 2260 (BOL, K, NBG, STE). George div., 3322DC, Wilderness, 15ft, 8 i 1929, *Mogg* 11550 (PRE). Riversdale div., 3421BA, Albertinia, near Eland's Kop, ix 1914, *Muir* 1766 (BOL, PRE).

I have adopted a broad species concept for *S. campanulata* notwithstanding the exclusion of the very closely allied *S. roseoflava*. The name is lectotypified by the Drège collection from the foot of the Zuurberg, of which a number of excellent sheets exist. The specimens collected by Ecklon and said by Bentham to be imperfect have not been seen by me; however a fragment of *S. campanulata* collected by Ecklon and written up in an old hand as *Chaenostoma campanulatum* Benth. is now in MO and came from Bernhardi's herbarium.

The type specimen is a robust plant with the flowers arranged in long lax racemes, the lowermost flowers being in the axils of the uppermost leaves (vegetative buds present in the axils). The inflorescence axes and the uppermost part of the stem are glandular pubescent with some hairs up to 0.5mm long (long hairs may be very few) while the lower stem is clad in mostly acute hairs up to 0.5–0.7mm long. It is these long hairs and the long coarse hairs on the calyx (as well as the short broad corolla tube) that comprise the distinguishing features of *S. campanulata* and which have played a major part in formulating my concept of the species.

The type of *S. intertexta*, which came from Algoa Bay, consists of a few small twigs with withered leaves and a few withered flowers. It is possibly the plant with congested inflorescences that is particularly common around Algoa Bay, though it also occurs elsewhere along the coast. Only this congestion distinguishes the plant from typical *S. campanulata* and it is not clearcut: all degrees of congestion occur. Plants with congested or relatively congested inflorescences seem to be common around Alexandria and Port Alfred, east of Algoa Bay; in facies they may either resemble the plant from Algoa Bay or they may begin to approach *S. roseoflava* (leafy inflorescences) and some have the ovaries (and capsules) glandular all over, which is characteristic of *S. roseoflava*; others have the glands confined to the sutures or sometimes spreading over the top of the ovary (as in *S. campanulata*).

Another variant is common around Grahamstown and it has also been collected at Manley's Flat, Martindale and Trappe's Valley, along the road to Bathurst and Port Alfred, and west of Grahamstown at Alicedale. This plant is prostrate with long leafy crowded racemes much panicled, and thus looks very different from typical *S. campanulata* but with which it accords in details of indumentum and flowers. It could perhaps be recognized as a variety, but infraspecific rank for this and other variants must wait upon a thorough field investigation; hybridization with *S. roseoflava* and *S. polyantha* with which it is partially sympatric could contribute to taxonomic difficulties.

Sutera campanulata ranges along the southern Cape coast from Still Bay and Albertinia in the west to Port Alfred in the east and inland to the foothills of the Zuurberg north of Algoa Bay and adjacent Albany district, from near sea level to c.700m. It favours sandy places, from stabilized littoral dunes to scrub and grassland; flowers can be found in any month.

It is adventive in Australia: Victoria, Coode Island, *Tovey* 3 (K), originally determined as *S. floribunda*.

EXCLUDED NAMES

Chaenostoma cymbalariifolium (Chiov.) Cufod. = *Stemodiopsis* sp.
Sutera cymbalariaefolia Chiov. = *Stemodiopsis* sp.
Sutera serrata Hochst., nomen = *Stemodia serrata* Benth.
Sutera multifida (Michx.) Walp. = *Leucospora multifida* (Michx.) Nutt.

7. MANULEA

Manulea is characterized by having the *bract always adnate to the base of the pedicel*; *stamens 4* (in one species often reduced to 2), the *anticous pair* either *included or just visible in the mouth*, the *posticous pair* often larger, *always included, filaments not decurrent down the tube*; *stigma lingulate*, with marginal papillae; *seeds* white to bright violet blue, *patterned with several longitudinal rows of transversely elongated pits* arranged in chequer-board fashion, under SEM testa seen to be ornamented with irregular pustules; the *disposition of the clavate hairs inside the corolla tube*: in 60 (out of 73) species, the posticous filaments are always bearded and in five more species a few hairs may be present; 53 species always have a posticous longitudinal band of hairs in the throat (there may be up to five bands round the throat) and in three more species a band may or may not be present (fig. 58).

In addition, more than half the species (47) have balloon-tipped hairs on stem, leaves, calyx or outside of corolla tube, or on several of these organs: 16 (out of 25) species in Section *Dolichoglossa*, eight (out of 11) in Section *Thyrsiflorae*, 23 (out of 28) in Section *Manulea*, none in Section *Medifixae* (10 species). These hairs are peculiar to *Manulea* (in a few species of *Phyllopodium* and in two species of *Melanospermum* the apical cell of a hair may be larger than the stalk cells, but they differ in detail from those of *Manulea*; see fig. 11).

Several different sorts of hairs are often present in any one species: balloon-tipped pairs, several-celled uniseriate hairs tipped with a neck cell and a globular unicellular gland, and relatively large glistening glands are a common combination. In *Manulea*, glistening glands often occur on the backs of the corolla lobes and on the ovary, but not exclusively so. Glistening glands should not be confused with balloon-tipped hairs, which do not glisten.

In Section *Medifixae*, many of the species possess a peculiar type of multiseriate hair terminating in a very small globular 1-celled gland (fig. 2). Only the stalk cells persist as a scabrid projection on stem, leaf and calyx. In this Section, species 66–70 proved difficult to delimit satisfactorily, lack of basal parts in most herbarium specimens being a contributory factor. Careful field work and good specimens are needed.

Differences in indumentum are very important taxonomically and meticulous attention to detail is frequently necessary to distinguish between species.

Many species of *Manulea* are aromatic, sometimes unpleasantly so, but information on herbarium sheets is so scanty that I have largely ignored this character.

Hybridization appears to be an important factor in blurring species limits and this needs to be confirmed by field and laboratory studies. Many instances are mentioned in the text, but it may be useful to draw the information together here. The following seem to be sporadic crosses: *M. arabidea × praeterita, M. buchneroides × parviflora* var. *limonioides, M. decipiens × psilostigma, M. deserticola × plurirosulata* (enumerated as sp. no. 54), *M. fragrans × schaeferi, M. glandulosa × altissima*. Introgression may blur the limits between *M. acutiloba* and *M. silenoides, M. caledonica* and *M. tomentosa, M. cheiranthus* and *M. pusilla, M. cheiranthus* and *M. decipiens, M. crassifolia* subsp. *thodeana* and *M. florifera, M. decipiens* and *M. pusilla, M. exigua* and *M. cheiranthus, M. fragrans* and *M. praeterita, M. gariesiana* and *M. silenoides, M. leiostachys* and *M. rubra, M. rubra* and *M. tomentosa*.

Manulea L., Mant. 1: 12 (Oct. 1767), nom. cons.; Benth. in Hook., Comp. Bot. Mag. 1: 381 (1836), in DC., Prodr. 10: 363 (1846) & in Benth. & Hook. f., Gen. Pl. 2(2): 946 (1876); Hiern in Thiselton-Dyer, Fl. Cap. 4(2): 321 (1904), p.p. **Figs 15, 26, 58.**

Syn.: *Nemia* Berg., Pl. Cap. 160 (Sept. 1767), nom. rejic.
     Lychnidea Moench, Suppl Meth. Pl. 160 (1802).

Shrubs, or, more often, annual or perennial herbs. *Stems* either leafy throughout or mostly at base, usually pubescent, often with a mixture of 2 or more different sorts of glandular and eglandular hairs, the eglandular ones often with a much inflated apical cell (referred to as balloon-tipped: fig. 11C). *Leaves* usually opposite, bases connate, or sometimes alternate, all radical or radical and cauline, simple, entire to variously toothed or lobed. *Inflorescence* a raceme or thyrse, sometimes panicled. *Bract* adnate to base of pedicel, often small, sometimes leaflike. *Calyx* bilabiate, sometimes obscurely so, anticous lip 2-lobed, posticous lip 3-lobed, lobes short to long, linear, lanceolate, oblong or subspathulate, usually pubescent at least on margins. *Corolla* tube usually narrowly cylindric and abruptly expanded in upper part, often slightly bent at point of inflation, rarely funnel-shaped, mouth usually round, sometimes oblique, laterally compressed in one species, limb bilabiate or nearly regular, lobes rotate or reflexed, suborbicular to oblong or lanceolate, obtuse to subulate, usually entire, sometimes retuse or once or twice bifid, margins often revolute, variously pubescent outside, rarely glabrous, inside usually with 1–5 longitudinal, or 1 transverse, bands of unicellular clavate hairs in throat, these hairs sometimes also present around mouth or on upper surface of lobes. *Stamens* usually 4, 2 or 2 + 1 or 2 staminodes in one species, didynamous, inserted in upper half of corolla tube, posticous pair always included, filaments often bearded with unicellular clavate hairs, anthers vertical, often cohering, anticous pair inserted just below mouth, anthers usually smaller, ± oblique, either included or very shortly and often only partially exserted, all anthers synthecous. *Stigma* lingulate with 2 marginal bands of stigmatic papillae, much longer than, about equalling or, rarely, shorter than style, exserted or included. *Ovary* elliptic or cuneate in outline, base slightly oblique with a nectariferous gland on the shorter side, glabrous (except for glistening glands sometimes present on sutures), ovules many in each loculus. *Fruit* a septicidal capsule with a short loculicidal split at the tip of each valve, usually glabrous, rarely minutely glandular-puberulous. *Seeds* seated on a funicle in the form of a round centrally depressed cushion (fig. 15A, B), seed roughly pyriform to elliptic in outline, testa thin, pallid or violet-blue, tightly investing the endosperm which is bothrospermous and patterned with several longitudinal rows of transversely elongated pits arranged in chequer-board fashion, seen under the SEM to be ornamented with irregular pustules (fig. 26D).

Type species: *Manulea cheiranthus* (L.) L.

Distribution: Southern Africa, mainly western and south western Cape, progressively fewer species northwards and eastwards, one species ranging from Transkei to Mozambique, Zimbabwe and Angola.

Bentham (1846) divided *Manulea* into three unranked groups, *Racemosae*, *Thyrsiflorae* and *Acutiflorae*. At that date, he knew only 25 species (73 are now recognized) and he based his subdivisions on differences in habit, inflorescence, and shape of the corolla lobes. A more natural subdivision can be achieved if differences in the form of the stigma as well as other morphological details are taken into account. The genus is accordingly now divided into four sections, one of which comprises three subsections.

   Bentham's *Acutiflorae* includes the type of the genus and thus becomes Section *Manulea*; all five species enumerated by him belong here. His *Thyrsiflorae* and *Racemosae* are heterogeneous. I have retained the name *Thyrsiflorae* at sectional level and designated *M. tomentosa* as the type species (of the species included by Bentham in his group *Thyrsiflorae*, only four, *M. leiostachys*, *M. rubra*, *M. tomentosa*, *M. turritis*, belong in Section *Thyrsiflorae*). My Section *Dolichoglossa* includes three of the species enumerated by Bentham under his group *Racemosae*, and two of these, *M. nervosa* and *M. corymbosa*, are now included in Section *Dolichoglossa* subsect. *Racemosae*. The rest of Bentham's group *Racemosae* are now distributed between Sections *Dolichoglossa* subsection *Dolichoglossa*, Section *Manulea* and Section *Medifixae*.

Key to sections and subsections

1a. Stigma much longer than style, usually shortly exserted, sometimes just reaching mouth   2
1b. Stigma shorter than to about equalling or longer than style, always included (tip seldom reaching more than halfway up tube) 3

2a. (*Note*: 3 leads) Annual or perennial herbs, throat bearded inside in longitudinal bands or rarely glabrous, possibly always a yellow/orange patch on posticous side, anticous anthers much smaller than posticous . . . . sect. **Dolichoglossa** subsect. **Dolichoglossa**
2b. Annual herbs, throat bearded inside with a longitudinal posticous band, the hairs usually fanning out transversely, sometimes extending to base of posticous lip or right round mouth in a ± star-shaped pattern, limb with a white to yellow ± star-shaped patch round mouth, anticous anthers much smaller than posticous
                                                sect. **Dolichoglossa** subsect. **Racemosae**
2c. Annual herbs, throat bearded inside with a longitudinal posticous band fanning out laterally to encircle mouth (hairs weakly developed in one species), yellow patch in back of throat, anthers often subequal, or anticous ones sometimes slightly smaller than posticous ones
                                                sect. **Dolichoglossa** subsect. **Isantherae**

3a. Perennial herbs or shrublets, stigma shorter than style or ± equalling it, ovary strongly tapered upwards, capsule very acute, almost beaked . . . . sect. **Thyrsiflorae**
3b. Plants either annual or perennial, stigma usually longer than style, rarely shorter than or equalling it, ovary ± obtuse, capsule not beaked 4

4a. Stamens either inserted above the middle (often in upper third) of corolla tube or, if inserted ± halfway, then corolla tube up to 6mm long . . . . sect. **Manulea**

4b. Stamens inserted ± halfway up corolla tube, tube 6.5–15mm long . . . . sect. **Medifixae**

Manulea sect. **Dolichoglossa** Hilliard, sect. nova

Herbae annuae vel perennes, plerumque caulibus pluribus e basi caespitosis. Folia opposita, basi connata, interdum sursum redacta et alterna. Corollae tubus vel anguste cylindricus prope apicem inflatus vel plus minusve infundibuliformis et brevissimus; faux saepe barbatus, pilis in vittas longitudinales vel interdum in fasciam transversalem dispositis, interdum in limbum extensis, interdum glaber. Antherae anticae in ore corollae plerumque visibiles (vel in speciebus duabus inclusae), posticis dimidio vel triente minores vel subaequales. Stigma stylo multo longius, plerumque breviter exsertum, interdum vix ad corollae orem attingens.

Type species: *M. crassifolia* Benth.

Annual or perennial herbs mostly with one to several stems tufted from the base. *Leaves* opposite, bases connate, sometimes reduced and alternate upwards. *Inflorescence* either a raceme or a thyrse. *Calyx* obscurely to distinctly bilabiate. *Corolla* tube either narrowly cylindric and inflated near apex or ± funnel-shaped and then very short, limb bilabiate or nearly regular, mouth sometimes slightly oblique, lobes oblong to elliptic or sometimes subrotund, obtuse, margins plane or revolute, often white, sometimes shades of yellow to rose-red or brown-red, rarely violet or blue, throat often bearded within, hairs usually in longitudinal bands (or sometimes a transverse band), sometimes extending on to the limb, or rarely glabrous. *Stamens*: anticous anthers usually visible in mouth (included in 2 species), either ½–⅓ the size of the posticous anthers or about equalling them, posticous anthers included, filaments often bearded or sometimes glabrous. *Stigma* much longer than style, usually shortly exserted, sometimes just reaching mouth.

Section *Dolichoglossa* comprises species 1–25. It can be divided into three subsections: subsect. *Dolichoglossa* (species 1–12), subsect. *Racemosae* (species 13–17) and subsect. *Isantherae* (species 18–25). Most of the species occurring in the summer rainfall area of southern Africa are included in this section.

subsect. **Dolichoglossa** Hilliard, subsect. nova. **Fig. 58F.**

Herbae annuae vel perennes. Corollae tubus anguste cylindricus, prope apicem inflatus, fauce vittis longitudinalibus barbato raro glabro; limbus bilabiatus vel fere regularis, albus, malvinus, vel varie flavus vel aurantiacus vel ruber, fortasse semper postice flavo- vel aurantiaco-notatus; anthere anticae posticis minores.

Type species: *M. crassifolia* Benth.

Annual or perennial herbs. *Corolla* tube narrowly cylindric, inflated near apex, throat bearded in longitudinal bands, rarely glabrous, limb bilabiate or nearly regular, white, mauve, livid, or shades of yellow to orange or red, possibly always with a yellow or orange patch on posticous side of throat. *Anthers*: anticous pair much smaller than posticous.

Fig. 58 (Opposite). A, *Manulea* sect. *Medifixae*, *M. rigida* (*van Jaarsveld* 6636): A1, corolla opened out to show posticous stamens inserted halfway up tube, tiny clavate hairs at base of lobes and on posticous filaments, glandular hairs in lower half of tube; A2, gynoecium with stigma slightly longer than style, small adnate nectariferous gland at base of ovary. B, *Manulea* sect. *Dolichoglossa* subsect. *Isantherae*, *M. calciphila* (*Esterhuysen* 19131): B1, corolla opened out to show subequal anthers, clavate hairs around mouth on posticous side of tube and on posticous filaments; B2, gynoecium, stigma longer than style. C, *Manulea* sect. *Thyrsiflorae*, *M. leiostachys* (*Compton* 4832): C1, corolla opened out to show clavate hairs confined to posticous filaments; C2, gynoecium, stigma shorter than style. D, *Manulea* sect. *Dolichoglossa* subsect. *Racemosae*, *M. corymbosa* (*Hilliard & Burtt* 13014): D1, corolla opened out to show posticous band of clavate hairs fanning out around mouth, clavate hairs on posticous filaments; D2, gynoecium, stigma much longer than style. E, *Manulea* sect. *Manulea*, *M. cheiranthus* (*Goldblatt* 5921): E1, corolla opened out to show longitudinal bands of clavate hairs, also on posticous filaments; E2, gynoecium, stigma longer than style. F, *Manulea* sect. *Dolichoglossa* subsect. *Dolichoglossa*, *M. crassifolia* (*Hilliard & Burtt* 12036): F1, corolla opened out to show longitudinal bands of clavate hairs, also on posticous filaments; F2, gynoecium, stigma much longer than style. All drawings × 7.

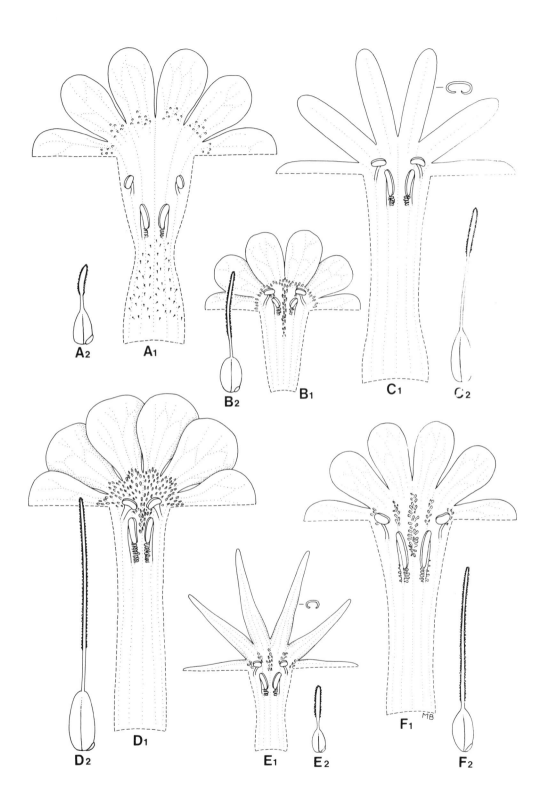

Three of the species (1. *M. leptosiphon*, 2. *M. burchellii*, 3. *M. conferta*), are confined to Namibia and/or northern Namaqualand and the semidesert areas of the northern Cape and Botswana. The rest occur in the summer rainfall area but *M. crassifolia* ranges westwards across the mountains to Victoria West, Oudtshoorn and Willowmore divisions, *M. bellidifolia* ranges into the area of all-year-round rainfall (Alexandria and Bathurst divisions), while *M. parviflora* reaches Zimbabwe, Mozambique and Angola and is the most widely distributed of all the species.

subsect. **Racemosae** Hilliard, subsect. nova. **Fig. 58D.**

Herbae annuae. Corollae tubus anguste cylindricus prope apicem inflatus, fauce intus vitta pilorum longitudinali postica barbato, pilis plerumque transverse extensis et interdum ad basin labii postici attingentibus, vel pilis circum orem ordinationem stellatam formantibus; limbus fere regularis, albus (vel in duobus speciebus albus vel azureus), area stelliformi alba vel lutea circum orem praeditus. Antherae anticae posticis minores.

Type species: *M. nervosa* Benth.

Annual herbs. *Corolla* tube narrowly cylindric, inflated near apex, throat bearded inside with a longitudinal posticous band of hairs that usually fan out transversely, so reaching the base of the posticous lip or encircling the mouth in a ± star-shaped pattern, limb nearly regular, white (or white to blue in two species) with a white to yellow, often star-shaped, patch around mouth. *Anticous anthers* much smaller than posticous.

Manulea schaeferi is the most widely distributed species, ranging from southern Namibia into the north central Cape; its close ally, *M. nervosa*, replaces it in northern Namaqualand. The other three species (*M. corymbosa*, *M. adenocalyx* and *M. psilostoma*) are all narrowly endemic in the western and south-western Cape, where they occupy mutually exclusive areas.

subsect. **Isantherae** Hilliard, subsect. nova. **Fig. 58B.**

Herbae annuae. Corolla vel infundibuliformis (et brevissimus), vel inferne cylindrica superne ampliata, intus vitta postica longitudinali pilorum praedita et pilis lateraliter circum orem dispersis (in specie una infirme evoluti); limbus fere regularis vel distincte bilabiatus, saepe albus, cremeus vel in una specie sulphureus, area flava postice in fauce praedita.

Type species: *M. latiloba* Hilliard

Annual herbs. *Corolla* either funnel-shaped and then very short, or cylindric below, expanded above, inside with a posticous longitudinal band of hairs fanning out laterally to encircle mouth (weakly developed in one species), limb nearly regular to distinctly bilabiate, often white, cream to sulphur yellow in one species, yellow patch in back of throat. *Anthers* usually subequal, anticous ones sometimes slightly smaller than posticous.

Manulea latiloba, *M. karrooica* and *M. derustiana* are confined to the Karroo, the former in the western part, the latter two in the eastern; *M. calciphila* is endemic to the limestone formations in Bredasdorp and Riversdale divisions; the other four species are narrowly endemic in the western Cape.

sect. **Thyrsiflorae** Hilliard, sect. nova. **Fig. 58C.**

Herbae perennes vel suffrutices. Folia opposita, basi connata, interdum sursum redacta et alterna. Corollae tubus anguste cylindricus, prope apicem inflatus; faux vel valide vel tenuiter intus barbatus, pilis in vittas longitudinales dispositis, vel glaber. Antherae inclusae, anticae posticis dimidio minoribus. Stigma plerumque stylo brevius, raro plus minusve stylum aequans, inclusum (apice plus minusve medium tubum attingente). Ovarium sursum breviter angustatum; capsula fere rostrata.

Type species: *M. tomentosa* (L.) L.

Perennial herbs or subshrubs. *Leaves* opposite, bases connate, sometimes reduced and alternate upwards. *Inflorescence* usually a thyrse, rarely a raceme and then cymules often mixed with solitary flowers. *Calyx* obscurely bilabiate. *Corolla* tube narrowly cylindric, inflated near apex, limb bilabiate or nearly regular, lobes oblong, elliptic or subrotund, obtuse, margins plane or revolute, shades of yellow, orange, red or brown, throat either strongly to weakly bearded within, the hairs in longitudinal bands, or glabrous. *Stamens* included, anticous anthers roughly half size of posticous ones, posticous filaments often bearded, sometimes glabrous. *Stigma* usually shorter than style, rarely ± equalling it, included (tip reaching ± halfway up tube). *Ovary* strongly tapered upwards, capsule almost beaked.

Section *Thyrsiflorae* comprises species 26–36, which are confined to the western and south western Cape, from Van Rhynsdorp in the north to Mossel Bay in the east, and inland to Laingsburg and Ladismith divisions.

sect. **Manulea. Fig. 58E.**

Type species: *M. cheiranthus* (L.) L.

Annual or perennial herbs. *Leaves* opposite, bases connate, sometimes reduced and alternate upwards. *Inflorescence* either a raceme or a thyrse, cymules sometimes mixed with solitary flowers. *Calyx* often obscurely, sometimes distinctly, bilabiate. *Corolla* tube narrowly cylindric, inflated in upper part, limb usually bilabiate, nearly regular in four species, *either* lobes mostly elliptic, oblong or lanceolate, entire, margins plane or revolute and then apex sometimes subulate to eye, mostly shades of yellow to orange, red or brown, white in one species, always with an orange patch in back of throat, *or* lobes sometimes subrotund, margins plane, rarely entire, usually once or twice bifid or at least retuse (oblong lobes also sometimes retuse), white, mauve or blue (wholly yellow in one species) with a yellow star-shaped patch around mouth, possibly also a dark patch in back of throat; usually 3–4 longitudinal bands of hairs in back of throat, glabrous or nearly so in three species, rarely a transverse band of hairs also present in throat. *Stamens* usually 4, 2 or 2+ 1 or 2 staminodes in one species, anticous anthers often visible in mouth, about ½–⅓ size of posticous anthers, posticous anthers included, posticous filaments usually bearded, sometimes glabrous. *Stigma* usually much longer than style, about equalling it or a little shorter in two species, always included (tip reaching roughly halfway up tube).

Section *Manulea* comprises species 37–64. Twenty of the 28 species are found in Namibia, Namaqualand and the western, south-western and central Cape; two (41. *M. platystigma*, 42. *M. dregei*) are endemic to the Eastern Mountain Region, one (44. *M. obovata*) is confined to the area of all-year-round rainfall between Knysna and Port Alfred. Of the species favouring arid areas, twelve occur in Namibia, Namaqualand and the north central Cape, with two of them just reaching the south western part of the Orange Free State, two more are confined to the Roggeveld and nearby (43. *M. incana*, 59. *M. diandra*) and one (41. *M. chrysantha*) to the Karroo. The remaining ten are concentrated in the western and south western Cape, with two of them extending into Namaqualand, one (48. *M. cheiranthus*) eastwards along the coast as far as Knysna.

sect. **Medifixae** Hilliard, sect. nova. **Fig. 58A.**

Herbae perennes vel fruticuli. Folia vel alterna basi decurrentia, vel opposita basi connata sursum redacta et alterna. Corollae tubus anguste cylindricus, supra medium abrupte flexus et sursum paulo ampliatus; faux intus plerumque glaber (in speciebus tribus vittis posticis longitudinalibus adsentibus), tubi dimidium inferius intus saepe minutissime glanduloso-puberulum et limbi pagina superior pilis minutis clavatis ordinationem stellatam circum orem formantibus. Antherae anticae ad dimidium longitudinis posticarum, omnes profunde inclusae. Stigma stylo plus minusve aequale vel interdum paulo brevius, profunde inclusum (apice plus minusve trientem tubi attingente).

Type species: *M. juncea* Benth.

Perennial herbs or shrublets, one species annual. *Leaves* either alternate, bases decurrent, or opposite, bases connate, reduced and alternate (but bases not decurrent) upwards. *Inflorescence* either a raceme or a thyrse. *Calyx* obscurely bilabiate. *Corolla* tube narrowly cylindric, abruptly bent about the middle and slightly widened upwards, mouth laterally compressed in one species, limb almost regular, lobes broadly elliptic to subrotund, obtuse, margins plane or revolute, *either* white with a greenish to yellow or orange star-shaped patch around mouth and throat either glabrous within or with a longitudinal band of hairs (two species), *or* white to shades of yellow, orange or red-brown with a ± star-shaped patch of tiny clavate hairs around mouth and throat usually glabrous within (posticous longitudinal band of hairs in one species) but tube very minutely glandular-puberulous below insertion of posticous filaments. *Stamens* deeply included, anticous anthers up to half as long as posticous ones, posticous filaments bearded. *Stigma* roughly equalling style or sometimes a little shorter, deeply included (tip reaching ± ⅓ of way up tube).

Section *Medifixae* comprises species 65–74, which are easily recognized by the distinctive form of the corolla tube: scarcely widened upwards but bent about the middle at the point of insertion of the posticous filaments; the style and stigma are remarkably short in relation to the length of the corolla tube. The balloon-tipped hairs so common in *Manulea* are wanting in this section. Two species stand apart because they are annual or perennial herbs with mainly radical leaves; one (*M. androsacea*) occurs in southern Namibia and Namaqualand, while the other (*M. altissima*) ranges from Namaqualand south to Saldanha Bay. This species has the corolla mouth laterally compressed. The other ten species are shrublets or suffrutices. There are often peculiar glandular hairs on the stems, leaves and calyces: the basal cell is very broad and stout and persists as a ± scabrid projection, and the upper surface of the limb is patterned with tiny clavate hairs. Frequently there are minute glandular hairs *inside* the lower half of the corolla tube. Two of these species (*M. robusta*, *M. dubia*) are nearly confined to Namibia: *M. robusta* just reaches the Richtersveld in northern Namaqualand. They, like *M. androsacea* and *M. altissima*, have opposite leaves connate at the base. The remaining six species are unique in having alternate leaves with bases decurrent on the stem; the lowermost leaves are possibly opposite with connate bases, but this needs confirmation (most herbarium specimens lack basal parts). These six species are confined to the western and south western Cape.

Key to species

Note: calyx tube and lobes measured *inside*, between the posticous and anticous lobes, after rehydration, corolla tube measured an anticous side, using a fully mature flower.

1a. Stigma either exserted or just included (**Section Dolichoglossa**) 2
1b. Stigma included, often deeply so 31

Section Dolichoglossa

2a. Corolla tube glabrous or very nearly so inside (ignore any hairs on posticous filaments) . . 3
2b. Corolla tube with either 3–5 longitudinal bands or a broad transverse band of clavate hairs
    inside upper, swollen, part 8

3a. Inflorescence a raceme, with flowers solitary or occasionally paired in the axil of each
    bract, or in 2–3-flowered cymules near the base of the raceme 4
3b. Inflorescence thyrsoid with a few- to many-flowered cymule in the axil of each bract . . . 5

4a. Flowering axis leafless above the base, lowermost bracts c.3–3.5mm long, calyx lobes
    c.2.5mm long, hairs all gland-tipped, up to 0.2mm long . . . . **1. M. leptosiphon**
4b. Flowering axis usually laxly leafy, lowermost bracts c.4–10mm long, calyx lobes
    2.75–5.5mm long, hairs both gland- and balloon-tipped, up to 0.4–0.8mm long
                                                                        2. M. burchellii

5a. Calyx distinctly bilabiate, lobes 3.5–6mm long, strongly pouched in lower half, abruptly
    narrowed to the linear ± folded tips, corolla limb c.7–11.5mm across . . . . . **3. M. conferta**
5b. Calyx obscurely bilabiate, lobes 1–3.3mm long, linear to oblong, flat, corolla limb
    3–7.2mm across 6

6a. Annual, calyx lobes glandular-pubescent all over, balloon-tipped hairs sparse (and inconspicuous) or wanting .... **7. M. flanaganii**
6b. Perennial, calyx lobes pubescent mainly on margins, balloon-tipped hairs conspicuous .. 7

7a. Both stem and inflorescence axis puberulous .... **6. M. parviflora** var. **parviflora**
7b. Stem and inflorescence axis glabrous .... **6. M. parviflora** var. **limonioides**

8a. Inflorescence thyrsoid with a few- to many-flowered cymule in the axil of each bract (the axis can be simple or branched, capitate or elongate) 9
8b. Inflorescence an elongate or capitate raceme with flowers solitary or occasionally paired in the axil of each bract 17

9a. Hairs inside swollen part of corolla tube in ± 5 longitudinal bands 10
9b. Hairs inside swollen part of corolla tube in a broad fan extending to base of posticous lip
17. M. psilostoma

10a. Coarse perennial herb with stems branching from the base and above into large spreading panicles .... **9. M. paniculata**
10b. Stems ± simple below the inflorescence, 1 to several tufted from the base 11

11a. Inflorescence a well-branched panicle 12
11b. Inflorescence essentially simple, at most one or two branches 13

12a. Flowers yellow to orange .... **4. M. crassifolia**
12b. Flowers mauve or whitish .... **8. M. florifera**

13a. Corolla white, the flowers always apical, in a ± flat-topped corymb even when the inflorescence is much elongated and passing into fruit .... **10. M. buchneroides**
13b. Corolla variously coloured, flowers never in a flat-topped corymb, usually scattered down the axis in various stages of maturity 14

14a. Calyx distinctly bilabiate, posticous anthers 0.8–1.5mm long 15
14b. Calyx lobes subequal, posticous anthers 1.6–1.8mm long 16

15a. Hairs on inflorescence axis and calyx c.0.2–0.25mm long
4. M. crassifolia subsp. **crassifolia**
15b. Hairs on inflorescence axis and calyx c.0.3–1mm .... **4. M. crassifolia** subsp. **thodeana**

16a. Corolla limb deep red to rosy pink, seeds pallid .... **5. M. rhodantha** subsp. **rhodantha**
16b. Corolla limb shades of yellow and brownish yellow, seeds usually violet-blue
5. M. rhodantha subsp. **aurantiaca**

17a. Calyx markedly bilabiate, lobes pouched at base 18
17b. Calyx obscurely bilabiate, lobes not pouched at base 20

18a. Corolla limb glabrous around mouth (it may be yellow there), hairs confined to anticous and posticous bands inside throat .... **14. M. nervosa**
18b. Corolla limb with a star-shaped yellow patch around mouth and there with clavate hairs running back into throat 19

19a. Anticous calyx lobes 3.8–4.5 (–5.5)mm long, posticous lobes 3.75–4.8mm long, all lobes glabrous on the backs except for a few minute glands, margins heavily bearded with eglandular hairs up to 0.5–1mm long .... **15. M. corymbosa**
19b. Anticous calyx lobes 4.3–9.5mm long, posticous lobes 4.5–9.5mm long, all lobes densely glandular-puberulous all over the backs, few or no long eglandular hairs on margins
16. M. adenocalyx

20a. Corolla limb glabrous around mouth 21
20b. Corolla limb bearded around mouth, hairs sometimes best developed on posticous side   23

21a. Throat of corolla bearded all round inside . . . . **12. M. plurirosulata**
21b. Hairs in corolla throat in 1–5 discrete longitudinal bands 22

22a. Balloon-tipped hairs present on both stem and calyx, corolla limb distinctly bilabiate
11. M. bellidifolia
22b. Only gland-tipped hairs present on stem and calyx, corolla limb nearly regular
13. M. schaeferi

23a. Corolla tube at least 5.5mm long 24
23b. Corolla tube up to 4mm long 26

24a. Posticous filaments glabrous . . . . **19. M. paucibarbata**
24b. Posticous filaments bearded 25

25a. Hairs on stem less than 0.1mm long, pedicels 2–4mm long, anticous anthers 0.6–0.8mm
long, posticous anthers 1mm . . . . **18. M. augei**
25b. Hairs on stem up to 0.25mm long, pedicels 4–20mm long, anticous anthers 0.3–0.5mm
long, posticous anthers 0.5–0.8mm . . . . **20. M. annua**

26a. Posticous filaments bearded 27
26b. Posticous filaments glabrous 28

27a. Pedicels 4–20mm long, acute eglandular hairs on calyx margins and outside of corolla
tube . . . . **20. M. annua**
27b. Pedicels 1.5–6mm long, balloon-tipped hairs mixed with gland-tipped ones on calyx
margins and outside of corolla tube . . . . **22. M. calciphila**

28a. Lowermost bracts 4–11 × 1–3 mm, often leaf-like, balloon-tipped hairs to 0.4mm long on
margins of both bracts and calyx lobes (tiny glandular hairs also present) **21. M. arabidea**
28b. Lowermost bracts 1.5–4 × 0.2–1 mm, not leaf-like, mainly minute gland-tipped hairs on
margins of bracts and calyx lobes (any balloon-tipped hairs ± sessile or up to 0.25mm
long) 29

29a. Lateral corolla lobes ascending and lying close to the two posticous lobes leaving the
anticous lobe isolated . . . . **23. M. karrooica**
29b. Lateral corolla lobes lying at right angles to the anticous lobe 30

30a. Anticous calyx lobes 1.5–2 × 0.6–1 mm, balloon-tipped hairs ± sessile, corolla tube
2–2.5mm long . . . . **24. M. latiloba**
30b. Anticous calyx lobes c.2.5–3 × 0.8 mm, balloon-tipped hairs up to 0.25mm long, corolla
tube c.4mm long . . . . **25. M. derustiana**

31a. Perennial herbs or shrublets, stigma shorter than style or occasionally ± equalling it, ovary
strongly tapered upwards, capsule very acute, almost beaked (**Section Thyrsiflorae**) . . 32
31b. Plants either annual or perennial, stigma usually longer than style (or rarely shorter than or
about equalling it), ovary ± obtuse, capsule not beaked 42

Section Thyrsiflorae

32a. At least lower part of stem clad in long (0.4–2mm) ± acute hairs; short gland-tipped hairs
may also be present (Note: specimens ± intermediate between two species occur and are
not catered for in the key; see discussion under individual species) 33
32b. At least lower part of stem clad in either balloon-tipped or gland-tipped hairs, or a mixture
of both (length varies from minute to c.1mm) 37

33a. Three to five well developed longitudinal bands of clavate hairs inside throat (for plants from Piquetberg and Pikeniers Kloof, see under *M. leiostachys*) 34
33b. Throat either glabrous or with a few posticous hairs (ignore bearded filaments) 36

34a. Stems leafy only at or near the base, leafless or remotely bracteate above, simple or very sparingly branched at base, upper stem and inflorescence axis glabrous or nearly so, hairs on outside of corolla tube confined to upper, swollen, portion .... **27. M. rubra**
34b. Stems more or less leafy throughout, plant often shrubby, both stem and inflorescence axis persistently hairy, corolla tube hairy all over outside 35

35a. Leaf margins serrate to crenate, both surfaces villous, hairs up to 0.5–1.5mm long, ± acute, gland-tipped hairs either wanting or few, much smaller than acute hairs
26. M. tomentosa
35b. Leaf margins entire or with a very few obscure teeth, both surfaces pubescent with short gland-tipped hairs and longer (to 0.5mm) ± acute ones .... **28. M. obtusa**

36a. Leaves elliptic in outline: blade c.18–60 × 7–20mm .... **30. M. turritis**
36b. Leaves ovate or occasionally oblong in outline: blade c.12–23 × 7–20mm
31. M. ovatifolia

37a. Three or four well developed longitudinal bands of clavate hairs inside throat
29. M. caledonica
37b. Throat either glabrous or with a few posticous hairs (ignore bearded filaments) 38

38a. Young parts of stems, also leaves, very minutely puberulous (hairs scarcely visible with × 10 lens), hairs balloon-tipped, leaf margins either entire or with a few obscure teeth
34. M. pillansii
38b. Hairs on stem 0.1–1mm long, either balloon- or gland-tipped, or a mixture of both, leaf margins entire to serrate 39

39a. Stems villous near the base, hairs up to 0.6–1mm long, both glandular and eglandular
32. M. leiostachys
39b. Hairs on lower part of stem up to 0.3mm long 40

40a. Hairs all gland-tipped, up to 0.2–0.3mm long .... **35. M. montana**
40b. Hairs balloon-tipped, up to 0.1–0.2mm long, sometimes mixed with smaller gland-tipped hairs 41

41a. Leaves grey (at least in dried state), margins entire or with a few tiny remote teeth, corolla tube 6–9mm long, limb 5–8mm across lateral lobes, anticous lobe 2–3.5mm long, posticous lobes 2.2–4mm .... **33. M. laxa**
41b. Leaves green, margins sharply serrate, corolla tube c.8–10.25mm long, limb c.7–11mm across lateral lobes, anticous lobe (3.4–) 3.75–4.2mm long, posticous lobes 3.75–5mm long .... **36. M. glandulosa**

42a. Stamens either inserted above the middle (often in upper third) of corolla tube, or, if inserted ± halfway, then corolla tube up to 6mm long (**Section Manulea**) 43
42b. Stamens inserted ± halfway up corolla tube, tube 6.5–15mm long (**Section Medifixae**) 74

Section Manulea

43a. Perennial herbs (either well branched to shrubby or ± simple stems from a carrot-like rootstock) 44
43b. Annual herbs 51

44a. Shrubby or at least well branched 45
44b. Simple or subsimple stems from a perennial stock 48

45a. Greyish shrub with entire leaves .... **37. M. cinerea**
45b. Green, shrubby or subshrubby, leaves variously toothed 46

46a. Longest hairs on calyx and outside of corolla either balloon-tipped or gland-tipped; these
    may also be present on stems and leaves .... **38. M. thyrsiflora**
46b. Longest hairs on calyx and outside of corolla ± acute (small gland-tipped hairs may also
    be present) 47

47a. Corolla tube 4.25–5.5mm long, posticous corolla lobes 3–3.5mm long, anticous lobe
    2.75–3.2 mm; branches erect, ascending or decumbent .... **39. M. virgata**
47b. Corolla tube 5.5–7mm long, posticous corolla lobes 3.1–5mm long, anticous lobe
    3.2–5mm long; branches sprawling or almost scandent .... **40. M. stellata**

48a. Inflorescence thyrsoid 49
48b. Inflorescence a raceme (lowermost flowers occasionally paired or in threes) 50

49a. Hairs on lower part of stem up to 0.1mm long, mixed gland- and balloon-tipped, corolla
    tube 6.5–8mm long, hairy all over outside .... **41. M. platystigma**
49b. Hairs on lower part of stem 0.3–0.5mm long, mostly glandular, occasionally mixed with
    balloon-tipped hairs; corolla tube 4.5–5.75mm long, pubescent mainly on backs of lobes
    and inflated part of tube .... **44. M. obovata**

50a. Plant green, hairs on stem up to 0.1mm long, mixed gland- and balloon-tipped; calyx tube
    1mm long .... **42. M. dregei**
50b. Plant hoary, hairs on stem up to 0.2–0.4mm long, balloon-tipped (tiny glandular hairs also
    present), calyx tube 1.5–2mm long .... **43. M. incana**

51a. Corolla limb mostly shades of yellow (including greenish), orange or brown, white or
    cream in one species but then (as in all this group) with an orange/yellow patch in tube but
    no contrasting 'eye'; stamens 4 52
51b. Corolla limb usually white, blue or mauve with a contrasting yellow to white 'eye',
    wholly bright yellow in one species but then stamens either reduced to 2 or anticous pair
    of anthers very small or in various stages of reduction to staminodes 67

52a. Stem clad in a mixture of balloon- and gland-tipped hairs, balloon-tipped hairs often
    prominent on inflorescence axis 53
52b. Stem clad only in gland-tipped hairs, these often very small and diminishing upwards,
    inflorescence axis often glabrous (sessile glistening glands may be present) 63

53a. Corolla limb white or occasionally flushed pink .... **46. M. exigua**
53b. Corolla limb shades of yellow, orange or brown 54

54a. Corolla lobes elliptic or oblong, tips obtuse to ± acute 55
54b. Corolla lobes lanceolate or long-lanceolate, tips very acute to subulate, the extreme tip
    occasionally retuse or forked 57

55a. Inflorescence thyrsoid, lowermost bracts 4–15 × 0.5–3mm .... **44. M. obovata**
55b. Inflorescence a raceme, lowermost bracts 1.5–3 × 0.4–1mm 56

56a. Stem and inflorescence axis with conspicuous balloon-tipped hairs less than 0.1mm long;
    corolla tube 3–4.5mm long .... **45. M. chrysantha**
56b. Lower part of stem glandular-puberulous only, hairs up to 0.3mm long, balloon-tipped
    hairs on inflorescence axis only; corolla tube 5.5–6mm long .... **55. Manulea sp.**

57a. Posticous filaments bearded (but consult notes under *M. pusilla* and *M. decipiens*) 58
57b. Posticous filaments glabrous 61

58a. Gland-tipped hairs at base of stem minute 59
58b. Gland-tipped hairs at base of stem up to 0.3–0.5mm long 60

59a. Lowermost pedicels up to 5mm long, corolla tube 3–4.5mm long, posticous lobes up to
 c.4mm long . . . . **45. M. chrysantha**
59b. Lowermost pedicels up to 1.5mm long, corolla tube 4.5–5.2mm long, posticous lobes
 4–4.8mm long . . . . **54. M. deserticola**

60a. Stem (including inflorescence axis) and leaves clad in glandular hairs up to 0.2–0.6mm
 long, balloon-tipped hairs in upper parts with apical cell up to c.0.05mm in diameter;
 inflorescence a raceme . . . . **47. M. adenodes**
60b. Glandular hairs prominent only at base of stem, balloon-tipped hairs dominant on
 inflorescence axis, apical cell c.0.1mm in diameter; inflorescence thyrsoid in all but
 flowering seedlings . . . . **48. M. cheiranthus**

61a. Inflorescence thyrsoid, sometimes panicled, lowermost flower occasionally solitary,
 otherwise in 2–11(–23)-flowered cymules 62
61b. Inflorescence a raceme, individual flowers mostly solitary in the axil of each bract, very
 rarely paired . . . . **50. M. pusilla**

62a. Calyx lobes 2.5–4.5mm long, corolla tube 3.5–5.5mm long, limb 9–15mm across lateral
 lobes . . . . **49. M. decipiens**
62b. Calyx lobes 2–2.4mm long, corolla tube 2.4–2.8mm long, limb 5–8mm across lateral
 lobes . . . . **51. M. aridicola**

63a. Corolla tube 2.5–3.5mm long; inflorescence a thyrse (raceme in flowering seedlings)
 . . . . **52. M. minuscula**
63b. Corolla tube (3.5–) 4.5–8mm long; inflorescence a raceme, panicle or thyrse 64

64a. Inflorescence a raceme or panicle, corolla lobes lanceolate, acute, stigma longer than style
 or about equalling it 65
64b. Inflorescence a thyrse, corolla lobes oblong-elliptic, stigma shorter than style
 57. M. namibensis

65a. Gland-tipped hairs on stem and calyx up to 0.2mm long, racemes unbranched
 56. M. tenella
65b. Gland-tipped hairs on stem and calyx less than 0.1mm long 66

66a. Inflorescence often paniculate, calyx lobes mostly 1.2–2mm long, corolla tube 5–7mm
 long . . . . **53. M. gariepina**
66b. Inflorescence always simple, calyx lobes mostly 2.5–4.5mm long, corolla tube 3.5–4.5mm
 long . . . . **58. M. fragrans**

67a. Flowers blue, mauve or white with a yellow to white 'eye', stamens 4; inflorescence lax,
 much elongated in fruit 68
67b. Flowers wholly yellow, stamens rarely 4, usually 2 plus 2 staminodes or staminodes
 wanting; inflorescence compact, up to 20mm long in fruit . . . . **59. M. diandra**

68a. Corolla lobes entire 69
68b. Tips of corolla lobes bifid or twice bifid 70

69a. Stem minutely glandular-puberulous, often glabrescent (but sessile glistening glands often
 present); corolla blue to mauve or white, lobes oblong . . . . **58. M. fragrans**
69b. Lower part of stem pubescent with hairs up to 0.2–0.3mm long, mostly gland-tipped,
 minutely glandular-puberulous above on inflorescence axis; corolla limb white, lobes
 subrotund . . . . **60. M. minor**

70a. At least base of stem with glandular (sometimes mixed with balloon-tipped) hairs up to 0.1–0.3mm long, calyx (margins only or all over backs) also with glandular hairs 0.1–0.5mm long 71

70b. Stem minutely glandular-puberulous, often glabrescent, minute glandular hairs on margins of calyx lobes (sessile glistening glands may also be present) . . . . **58. M. fragrans**

71a. Flowers mostly solitary in the axil of each bract or sometimes a few flowers in an inflorescence grouped in twos or threes 72

71b. Inflorescence thyrsoid, flowers in 2–7-flowered cymules mixed with an occasional solitary flower . . . . **61. M. praeterita**

72a. Hairs on calyx lobes nearly confined to margins, mostly glandular, lowermost pedicels mostly short, c.1.5–9(–13)mm long 73

72b. Calyx lobes hairy on backs and margins, mixed gland- and balloon-tipped, lowermost pedicels mostly long, c.(6–)10–20mm . . . . **63. M. silenoides**

73a. Lowermost bracts c.2.5–10mm long or leaflike and then larger; corolla limb 5.5–7mm across lateral lobes . . . . **62. M. gariesiana**

73b. Lowermost bracts c.1.6–2mm long, corolla limb 8–14mm across lateral lobes
64. M. acutiloba

Section Medifixae

74a. Corolla limb white or creamy with a greenish, yellow or orange star around mouth, glabrous above 75

74b. Corolla limb wholly pale to deep yellow to orange or red-brown, white in one species, usually with a ± star-shaped pattern of clavate hairs encircling mouth, often well out on the limb 77

75a. Mouth of corolla round, 3 longitudinal bands of clavate hairs in back of throat, calyx glandular-puberulous, corolla pubescent to almost glabrous outside . . **65. M. androsacea**

75b. Mouth of corolla compressed laterally, throat glabrous or *very* rarely with a few posticous hairs, calyx and corolla with hairs up to 0.3–1mm long outside 76

76a. (*Note*: 3 leads) stem and inflorescence axis glandular-puberulous overlaid with long (0.3–1.5mm) glandular hairs, varying in density but always present
66. M. altissima subsp. **altissima**

76b. Stem glandular-puberulous at least at base, glandular hairs to 0.3–1mm always present (though sometimes sparse) immediately below inflorescence, sometimes extending further back down stem . . . . **66. M. altissima** subsp. **longifolia**

76c. Stem either glabrous or minutely glandular-puberulous near base, glabrous above (though inflorescence axis may be glandular-puberulous) . . . **66. M. altissima** subsp. **glabricaulis**

77a. Leaves mostly alternate (except at extreme base), bases decurrent on stem in two very narrow ridges; outside of corolla tube pubescent with mainly eglandular ± retrorse hairs 78

77b. Most leaves opposite, bases connate, alternate only on upper part of stem as they pass into bracts; outside of corolla tube pubescent with glandular ± patent hairs 83

78a. Twiggy shrublet scabrid all over, leaves small (5–25 × 1–6mm), oblanceolate to oblong, sparingly toothed to entire, flowers in short (up to 20mm) racemes . . . . **70. M. ramulosa**

78b. Plants not as described above (differing either in habit, indumentum, leaf size, shape and toothing, or inflorescence, or a combination of these characters) 79

79a. Plant minutely glandular-puberulous and totally lacking in scabrid projections; branches often widely divaricate, conspicuously leafy or bracteate, leaves broad-based
69. M. rigida

79b. Plants either with few to many scabrid projections particularly on stem and calyx (these
projections the persistent broad bases of glandular hairs) or branches not divaricate, or
leaves mostly contracted at the base or petiolate (or a combination of these characters)   80

80a. Inflorescence essentially a narrow thyrse (often panicled) with cymules crowded, solitary
flowers occasionally present especially at base of inflorescence 81
80b. Inflorescence essentially a raceme (often panicled) with flowers distant, mostly solitary
in the axil of each bract, sometimes 2–3-flowered cymules interspersed with solitary
flowers 82

81a. Stem (including inflorescence axis) and leaves pubescent, some hairs ± scabrid, main
leaves oblong to lanceolate, thin-textured, up to 25–60 × 4–10mm with c.2–4 pairs of
teeth in upper half . . . . **71. M. cephalotes**
81b. Stem (including inflorescence axis) and leaves glabrous, main leaves linear,
thick-textured, up to 12–22 × 1–1.5 mm, entire or very rarely with a single stubby tooth
72. M. linearifolia

82a. Corolla tube c.(8.5–)10–15mm long, limb c.7–10mm across lateral lobes, capsules densely
glandular-puberulous . . . . **67. M. juncea**
82b. Corolla tube up to 8.5mm long, limb 5.5–7.5mm across lateral lobes, capsules glabrous
68. M. multispicata

83a. Primary leaves oblanceolate to elliptic, c.8–40 × 2–9 mm, narrowed to a very short
petiolar part, lowermost floral bracts always linear (c.3.5–10 × 0.5–2mm), calyx lobes
c.3–3.5mm long . . . . **73. M. robusta**
83b. Primary leaves ovate to broadly elliptic, c.12–57 × 5–30 mm, roughly ¼–½ the length
petiolar, lowermost floral bracts usually leaflike (c.5–24 × 1–9mm), calyx lobes
(3.2–)4–5.5mm long . . . . **74. M. dubia**

Section Dolichoglossa subsection Dolichoglossa

1. Manulea leptosiphon Thellung in Vierteljahrss. Nat. Ges. Zürich 57: 559 (1912); Roessler
in Mitt. Bot. München 6: 12 (1966).
Lectotype: Cape, Rietfontein am Rande der Kalachari, 1889, *Fenchel* 445 (Z).

Herb, possibly persisting for more than one season, taproot woody, c.2.5mm diam. at crown,
stems several from the crown, c.190mm tall in flower (will elongate in fruit), simple, glandular-puberulous, hairs sparse on inflorescence axis, occasional hairs up to 0.2mm long, nude.
Leaves all radical, crowded in a rosette, opposite, bases connate, blade up to 25 × 8.5mm, elliptic
tapering to a petiolar part 14mm long, apex subacute, margins irregularly serrate, both surfaces
glandular-puberulous. *Flowers* many in long simple terminal racemes, distant. *Bracts* (lowermost) c.3–3.5 × 0.8mm, linear-lanceolate. *Pedicels* (lowermost) 3–4mm long. *Calyx* obscurely
bilabiate, tube c.0.3mm long, lobes subequal, c.2.5 × 0.6mm, lanceolate, glandular-puberulous
all over, some hairs, mainly on margins, to 0.2mm long, glistening glands as well. *Corolla* tube
c.7.5 × 0.8mm in mouth, cylindric, inflated in upper part, limb bilabiate, c.9mm across the
lateral lobes, posticous lobes c.5 × 1.2mm, anticous lobe c.5 × 1mm, all oblong-lanceolate,
margins revolute, tips acute to eye, colour unknown, tube almost glabrous outside (a few minute
glandular hairs), glistening glands on backs of lobes, glabrous inside. *Stamens* 4, anticous
anthers 0.6mm long, tips just visible in mouth, posticous anthers 1.2mm long, included,
posticous filaments either glabrous or with 1 or 2 hairs. *Stigma* c.5mm long, reaching mouth.
Style c.1.2mm long. *Ovary* 1.2 × 0.8mm. *Capsules* unknown. (Thellung's original description
referred to capsules of a different species).

Thellung based his name *M. leptosiphon* on three different elements as Roessler pointed out (in Mitt. Bot. München 6: 12, 1966) when he chose *Fenchel* 445 as lectotype of the name. But Roessler was mistaken in equating *M. leptosiphon* with *M. gariepina* subsp. *campestris* (that is, *M. gariepina* of this revision): *M. leptosiphon* differs fundamentally from *M. gariepina* in its long stigma that reaches the mouth of the corolla tube, and they differ further in details of bracts and calyx.

The relationship of *M. leptosiphon* lies with *M. burchellii*, from which it may be distinguished by its scapose (not ± leafy) flowering stems, shorter bracts, and calyx with subequal lobes, mostly shorter than those of *M. burchellii* and clad in shorter hairs, all of which are gland-tipped (not mixed with balloon-tipped hairs). *Manulea leptosiphon* is known only from Fenchel's original collection, just one plant in the early stages of flowering, gathered by Fenchel together with the type material of *M. simpliciflora* Thellung, a synonym of *M. burchellii*. In the original description, Thellung stated that the type of *M. leptosiphon* came from Rietfontein, which lies just on the Cape side of the Cape - Namibia border; the label states 'zwischen Keetmanshoop und Rietfontein'. However, the precise locality is really immaterial; what is important is that the species grows in the Kalahari, is sympatric with *M. burchellii* at least on the western edge of the area of that species, and it should be sought again.

2. Manulea burchellii Hiern in Fl. Cap. 4(2): 228 (1904).
Type: Bechuanaland [i.e. Northern Cape], between Moshowa river and Chue Vley, x 1812, *Burchell* 2403 (holo. K).

Syn.: *M. simpliciflora* Thell. in Vierteljahrsschr. Nat. Ges. Zürich 57: 560(1913). Type: Namibia, Sanddünen zwischen Keetmanshoop und Rietfontein an der Grenze der Kalachari, *Fenchel* 446 (holo. Z).

Herb, flowering in seedling stage but clearly persisting for more than one season, taproot woody, to 5mm diam. at crown, stems (100–)150–350mm tall, rarely solitary (seedlings), usually several from the crown, simple to laxly branched, glandular-puberulous, nearly nude to laxly leafy. *Leaves* mainly radical, crowded in a rosette, opposite, bases connate, only the upper stem leaves alternate, blade (of radical leaves) 12–55 × 3–13mm, narrowly to broadly elliptic or oblanceolate tapering to a petiolar part c.3–20mm long (up to roughly half as long as the blade, but often much shorter), margins entire to distinctly but distantly serrulate to serrate, both surfaces glandular-puberulous; stem leaves similar but much smaller and sessile. *Flowers* many in long simple terminal racemes, crowded at first, soon distant, very rarely in few-flowered monochasial cymes near the base of the inflorescence, solitary in the upper part. *Bracts* (lowermost) 4–10(–28) × 0.3–1(–1.2)mm, occasionally ± leaflike (the longest ones), otherwise linear. *Pedicels* (lowermost) 1.4–3mm long. *Calyx* bilabiate, tube 0.4–1.2mm long, anticous lobes 2.75–5.5 × 0.7–1.7mm, anticous lip split 2.5–5.4mm, posticous lobes 2.5–5 × 0.5–1.2mm, posticous lip split 2.5–4.5mm, lobes lanceolate, basal part very slightly pouched, pubescent all over backs and margins, hairs to 0.4–0.8mm, both balloon- and gland-tipped. *Corolla* tube 7–14 × 1–1.5mm in throat, cylindric, inflated near apex, limb bilabiate, 6–13mm across lateral lobes, posticous lobes 2–4.5 × 1.4–3mm, anticous lobe 2.2–6 × 1.1–2.2mm, all oblong, margins revolute, creamy-yellow to orange, tube and lobes puberulous outside with both gland- and balloon-tipped hairs, glabrous inside. *Stamens* 4, anticous anthers 0.5–0.8mm long, visible in mouth, posticous anthers 1.2–1.8mm long, included, posticous filaments very sparsely bearded or occasionally glabrous. *Stigma* 4–11mm long, shortly exserted. *Style* 1.25–2mm long. *Ovary* 1.5–2 × 0.7–1.1mm. *Capsules* c.4.5 × 3mm, very minutely glandular-puberulous. *Seeds* c.0.5 × 0.3mm, blue-violet.

Selected citations:

Cape. Gordonia, 2520AD, Sewepanne, 17 ix 1959, *Barnard* 722 (PRE); 2822AD, farm England, c.4000ft, 7 xii 1960, *Leistner* 2028 (K, PRE); 2622CC, 36 miles ENE of Van Zylsrus, c.3350ft, 17 x 1961, *Leistner* 2884 (PRE). Postmasburg div., 2722CC, Pearson's Hunt, c.3400ft, 8 xii 1960, *Leistner* 2049 (K, PRE).
Namibia. Keetmanshoop div., 2719DB, Dawignab, 900 m, 28 x 1913, *Range* 1931 (SAM).

Manulea burchellii has been much confused with *M. conferta* to which it is closely related and under which name Roessler mentioned it (Prodr. Fl. S.W. Afr. 126: 31, 1967), but the two species can usually be distinguished at a glance: flowers in monochasial cymules in *M. conferta*, solitary in the axils of the bracts in *M. burchellii*, where only very rarely are some of the lowermost flowers arranged in few-flowered cymules. Thellung's epithet *simpliciflora* (in synonymy above) draws attention to the solitary flowers. The two species differ further in the calyx: in *M. burchellii* the calyx lobes are lanceolate and flat, scarcely pouched at the base, whereas in *M. conferta* the lobes are strongly folded and the bases are deeply pouched. This is to some extent reflected in absolute measurements, e.g. anticous lobes 0.7–1.7mm broad at base, posticous lobes 0.5–1.2mm (*M. burchellii*) versus 1–2.8mm and 1–2.2mm (*M. conferta*). In both species, the calyx is accrescent in fruit and this difference is then very clearly apparent, but it can be seen even in buds: The capsule is hidden by the calyx in *M. conferta*, but not so in *M. burchellii*. *Manulea burchellii* is possibly never as robust as *M. conferta*, which has a stout woody stock and strong branches.

The area of *M. burchellii* lies south of that of *M. conferta*, in the arid parts of the northern Cape, from the Kalahari Gemsbok National Park south to Pofadder in the west and Heuningsvlei (the Chue Vley of Burchell's collection) and Seekoeibaardsnek in the east. It has also been recorded just across the border in Namibia, from about Aroab south to Dawignab. It will certainly also be found in southernmost Botswana. The plants grow in loose red Kalahari sand, often on the slopes of dunes, c.900–1200 m above sea level, and have been found in flower in most months of the year.

3. Manulea conferta Pilger in Bot. Jahrb. 48: 437 (1912); Merxm. & Roessler, Prodr. Fl. S.W. Afr. 126: 31 (1967); Philcox in Fl. Zam. 8(2): 34 (1990).
Lectotype (chosen here): Namibia, Damaraland, Okahandjarivier auf Sand, März 1906, *Dinter II 8* (E; isolecto. SAM, WU, Z).

Herb, flowering in seedling stage but clearly persisting for more than one season, taproot woody, to 10–15mm in diam. at crown in old plants, stems (150–)300–600mm tall, rarely solitary (seedlings), usually several from the crown, simple to laxly branched from the base, eventually stout (c.3–5mm diam.) and woody, puberulous, hairs both gland- and balloon-tipped, leafy. *Leaves* opposite, bases connate, in young plants a tuft of radical leaves, blade c.35–85 × 7–12mm tapering to a petiolar part 5–15mm long (much shorter than blade), cauline leaves only in older plants, mostly 20–50 × 2–10mm, sessile or nearly so, elliptic to linear, tapering at both ends, margins mostly distantly but sharply serrate, both surfaces puberulous with gland- and balloon-tipped hairs. *Flowers* in long narrow rather lax terminal thyrses, cymules monochasial, to 30mm long, somewhat spreading, flowers crowded. *Bracts* (lowermost subtending cymules) c.7–21 × 1–1.5mm, sometimes leaflike, usually linear, bracteoles shorter than calyx. *Peduncles* (lowermost cymules) c.1.5–6mm long, flowers sessile or very nearly so. *Calyx* markedly bilabiate, tube 0.2–0.5mm long, anticous lobes 3.5–5 × 1–2.8mm, anticous lip split 2.3–4.5mm, posticous lobes 3.5–6 × 1–2.2mm, posticous lip split 2–3.5mm, all lobes remarkably pouched in lower half and bulging outwards, abruptly narrowed to the linear, folded, tips, accrescent and hiding the capsules, pubescent all over backs and margins, hairs to 0.3–0.8mm, both gland- and balloon-tipped. *Corolla* tube 8–14 × 1.3–1.6mm in throat, cylindric, inflated near apex, limb bilabiate, 7–11.2mm across lateral lobes, posticous lobes 2.5–4.5 × 1–2.5mm, anticous lobe 3–5 × 1–2.2mm, all oblong, margins revolute, yellow, tube and lobes puberulous outside with both balloon- and gland-tipped hairs, glabrous inside. *Stamens* 4, anticous anthers 0.5–0.8mm long, visible in mouth, posticous anthers 1.2–1.8mm long, included, posticous filaments very sparsely bearded. *Stigma* 6.5–11mm long, shortly exserted. *Style* 1–2.6mm long. *Ovary* 1.1–1.6 × 0.75–1mm. *Capsules* c.3–4 × 2–3mm, very minutely glandular-puberulous. *Seeds* c.0.6–0.75 × 0.3–0.5mm, violet-blue to pallid.

Selected citations:

Namibia. Swakopmund distr., 2214DB, Haikamchab, 26 i 1907, *Galpin & Pearson* 7513 (PRE, SAM). Okahandja distr., 2116DD, 35 km S of Okahandja, 23 v 1974, *Goldblatt* 1900 (NBG, PRE). Windhoek distr., 2217CA, farm Regenstein, 9 vi 1973, *Giess* 12758 (PRE). Gobabis distr., c.2219BD, between Karolinenhof and Babi-Babi, xii 1921, *Wilman* s.n. (BOL, SAM).
Botswana. 2220AC, SWA border on Ghanzi-Gobabis road, 2 viii 1955, *Story* 5082 (K).

Manulea conferta has been recorded in Namibia along the route from the mouth of the Swakop river to the Klein Spitzkoppen, Okahandja, Windhoek and Gobabis, then just across the Botswana border en route Ghanzi, an altitudinal range from sea level to c.1400 m. It grows in sand, on flats and in river beds, a constituent of scrub, and can be found in flower in any month. The plant is aromatic.

Its relationship to *M. burchellii*, with which it has been much confused, is discussed under that species.

4. Manulea crassifolia Benth. in Hook., Comp. Bot. Mag. 1: 382 (1836) excl. spec. *witbergensi* & in DC., Prodr. 10: 364 (1846) partly; Hiern in Fl. Cap. 4(2): 231 (1904) partly; Hilliard & Burtt in Notes RBG Edinb. 40: 288 (1982).

Two subspecies are recognized:

subsp. **crassifolia**. **Fig. 58F.**
Lectotype: Cape, Albert div., Mooyplaats ('Moogplats'), *Drège* 7919b (K; isolecto.? E; TCD).*

Perennial herb, stock narrowly obconical, up to 5–10mm diam. at the crown, stems (100–)300–600(–1000)mm tall, solitary to several, often simple, sometimes very sparingly branched in upper half, or occasionally paniculate, puberulous, hairs to 0.2–0.25mm, balloon-tipped, minute gland-tipped hairs as well, these predominating on lower part of stem, nude to very sparsely bracteate or sometimes leafy throughout. *Leaves* often mostly radical, crowded in rosettes, opposite, bases connate, (30–)60–170 × 7–30mm, elliptic to oblanceolate, obtuse or acute in smaller cauline leaves, base narrowed to a short petiolar part, margins entire to serrulate or serrate, both surfaces glandular-puberulous. *Flowers* many in a long narrow terminal thyrse, sometimes branching into a panicle, cymules several to many-flowered, dichasial, becoming distant as the thyrse elongates. *Bracts* (of lowermost cymules) 5–20 × 1–2mm, linear or oblong. *Peduncles* (of lowermost cymules) 1–5mm long, pedicels very short or flowers sessile. *Calyx* distinctly bilabiate, tube 0.5–1.5mm long, anticous lobes 2–3.8 × 0.5–1mm, lip split 1.2–2.5mm, posticous lobes 1.8–3.8 × 0.4–1mm, posticous lip split 1.5–3mm, all lobes lanceolate or oblong-lanceolate, puberulous, hairs to 0.2–0.25mm, balloon-tipped, smaller gland-tipped hairs as well. *Corolla* tube (4.5–)5–8 × 1–1.5mm in throat, cylindric, expanded in upper part, limb distinctly bilabiate, 5–9.5mm across lateral lobes, posticous lobes 2–3.2 × 1.1–2.3mm, anticous lobe 2.4–4 × 0.8–1.6mm, all oblong-elliptic, margins often ± revolute, shades of dull yellow (*fulvus* or *ochraceus*) or orange-yellow, sometimes throat tinged orange, tube and backs of lobes puberulous, hairs to 0.1mm, balloon-tipped, inside a longitudinal posticous band of clavate hairs in throat, usually dark-coloured in lower part, lesser bands in lateral and anticous positions. *Stamens* 4, anticous anthers 0.4–0.7mm long, visible in mouth, posticous anthers 0.8–1.5mm long, included, posticous filaments bearded, hairs usually dark. *Stigma* 3–5.5mm

---

* Bentham gave the locality 'Moogplats' [i.e. Mooyplaats] for Drège 7919b; but this is possibly an error: Drège's distributed specimens of *M. crassifolia* are localized as Stormbergen, Witbergen and Nieuweveldbergen; his specimens from the Stormbergen exactly match Bentham's specimen and are cited above as isolectotypes. 'Mooyplaats' (I, a, 38) is the station adjacent to 'Stormbergen' (I, a, 39) in Drège's printed list.

long, shortly exserted to just included. *Style* 0.4–1mm long. *Ovary* 1.2–1.5 × 0.6–1.1mm. *Capsule* 3–5 × 2–2.5mm. *Seeds* c.0.8 × 0.5mm, pallid.

Selected citations:

Lesotho. Berea distr., 2828CC, Leribe, *Phillips* 960 (SAM). Maseru distr., 2927BD, Blue Mt Pass, c.9660ft, 13 i 1979, *Hilliard & Burtt* 12097 (E). 2928AA, Meniaming Pass, 9000ft, 9 i 1955, *Coetzee* 506 (PRE). 2928AC, Mamalapi, 10000ft, 31 xii 1948, *Compton* 21401 (NBG); 32.5 km east of Thaba Putsoa, 2500m, 19 iii 1983, *Halliwell* 5036 (PRE).
Cape. Barkly East div., 3027DB, Ben McDhui, c.8400ft, 4 ii 1983, *Hilliard & Burtt* 16398 (E); 3028CC, Rhodes to Naude's Nek, 8600ft, 22 ii 1971, *Hilliard & Burtt* 6704 (E, K). Elliot-Maclear div. boundary, 3127BB, Bastervoetpad, c.7200ft, 15 ii 1983, *Hilliard & Burtt* 16688 (E). Murraysburg div., 3123DD, near Murraysburg, c.3500ft, *Tyson* sub BOL 6490 (BOL). Oudtshoorn div., 3322AC, Upper Cango Valley, Bassonsrus, 29 iii 1975, *Moffett* 675 (PRE, STE). Uniondale div., 3323CA, Avontuur Poort, 2600ft, i 1928, *Fourcade* 3571 (BOL, K, NBG, PRE, STE).
Transkei. 3028AD, Ongeluks Nek Pass, 6800ft, 7 xii 1985, *Hilliard & Burtt* 18734 (E).

Bentham included two different species in his original description of *M. crassifolia*, Hiern half a dozen (*inter alia M. bellidifolia*); subsequent confusion over the use of the name was inevitable. *Manulea crassifolia* is characterized by its perennial habit, rosulate leaves, thyrsoid inflorescence, distinctly bilabiate calyx and yellow corolla. *Manulea thodeana*, here treated as a subspecies of *M. crassifolia*, differs principally in its longer indumentum; in addition, the flowers of *M. thodeana* are often livid or pure mauve. Commonly, in the species as a whole, the flowering stem is nearly nude and terminates in a long narrow thyrse; in the Cape in particular, but also elsewhere, the flowering stem is often leafy, and, almost exclusively in the Cape, it may branch into a panicle.

Manulea rhodantha is a close ally, often mistaken for *M. crassifolia* and distinguished principally by its calyx, deeply divided into five subequal lobes (see further under *M. rhodantha*).

Manulea crassifolia subsp. *crassifolia* is known mainly from Lesotho and the north-eastern Cape, with a range from about Harrismith (Platberg) and Clarens in the Orange Free State, just north of Lesotho's northern border, through the mountainous parts of Lesotho (and just entering Transkei on the upper slopes of the Drakensberg escarpment but apparently absent from the Natal Drakensberg where it is replaced by subsp. *thodeana*) to the mountains of the north-eastern Cape, from the Witteberg and Cape Drakensberg west over the Stormberg, Bankberg, Sneeuwberg and Koudeveldberg to the environs of Murraysburg, and south to the Kouga Mountains and the Groote Zwartberg, as low as c.1050 m above sea level on the flanks of the Cape mountains in the west and south, but reaching 3000 m in Lesotho. It favours damp grass slopes, flowering between December and March, principally January.

subsp. **thodeana** (Diels) Hilliard in Notes RBG Edinb. 46: 50 (1989).
Type: Orange Free State, Mont aux Sources, 7000–8000ft, ii 1891, *Thode* 72 (holo. B†, prob. iso. STE 8205).

Syn.: *M. thodeana* Diels in Bot. Jahrb. 23: 479 (1896).

Calyx distinctly bilabiate, tube (0.75–)1–2mm long, anticous lobes 2.2–5 × 0.6–1mm, anticous lip split 1.6–3.4mm, posticous lobes 2.2–5 × 0.6–1.3mm, posticous lip split 1.6–3.5mm, all lobes lanceolate, hairs to 0.3–1mm long, balloon-tipped, smaller gland-tipped hairs as well. *Corolla* tube 6.5–9 × 1.5–1.8mm in throat, limb 6.5–11mm across lateral lobes, posticous lobes 1.8–4.5 × 1.3–2.4mm, anticous lobe 2.2–5 × 1–1.8mm, yellow (creamy to mustard), livid or vivid violet, tube pubescent outside, hairs to c.0.1–0.25mm long.

Selected citations:

Lesotho-Orange Free State border. 2828DD, slopes of Mont aux Sources, 7000ft, i 1894, *Flanagan* 2077 (PRE); foot of The Sentinel, c.9000ft, 22 i 1977, *Stewart* 1966 (E).
Natal. Bergville distr., 2829CC, Cathedral Peak Forest Reserve near the Met. Station, c.6700ft, 18 i 1983, *Hilliard & Burtt* 16294 (E); ibidem, below Organ Pipes Pass, c.7000ft, 22 i 1956, *Edwards* 1192 (PRE); ibidem, above Tseketseke river, c.6900ft, 18 i 1983, *Hilliard & Burtt* 16297 (E). Estcourt distr., 2929AD, Giant's Castle, 8500ft, xii 1914, *Symons*

269 (PRE). Mpendhle distr, 2929 BC, Highmoor, ridge SE of Giant's Castle, 7800ft, 5 i 1983, *Hilliard & Burtt* 16243 (E, S). Underberg distr., 2929CB, Sani Pass, c.7900ft, 5 i 1984, *Hilliard & Burtt* 17269 (E). 2929CC, vicinity of Tarn Cave above Bushman's Nek, c.7800ft, 18 i 1984, *Hilliard & Burtt* 17317 (E).

Manulea crassifolia subsp. *thodeana* is distributed along the face of the Drakensberg escarpment from the vicinity of Mont aux Sources on the Natal-Lesotho-Orange Free State border south to Bushman's Nek in southernmost Natal. Like typical *M. crassifolia*, it favours damp grassy places and flowers between December and March. It is distinguished from subsp. *crassifolia* chiefly by its longer indumentum (hairs on inflorescence axis and calyx 0.3–1mm long), but the flowers may sometimes be distinctly larger, as shown by the calyx and corolla measurements given above. Thode described the flowers as yellow and this colour predominates in the northernmost part of the range of subsp. *thodeana*. It is here too that its area abuts on that of typical *M. crassifolia*, which is always yellow-flowered. A little to the north east of Mont aux Sources on Mota Pass in Lesotho (*Coetzee* 433, NBG) and in Golden Gate National Park, Orange Free State (*Roberts* 3336, PRE), plants may have the calyx tube rather short for subsp. *thodeana* (0.75–1mm) and the indumentum short (0.3mm). Travelling south down the face of the Drakensberg, in the Cathedral Peak Reserve, flowers may be 'dull light violet' (*Hilliard & Burtt* 16294, E), 'light purple' (*Killick* 1441, E), 'cream corolla, yellow inside mouth, outside of tube dull violet' (*Hilliard & Burtt* 16297, E); in the vicinity of Giant's Castle 'blue-purple' (*Goldblatt & Manning* 8442, E), 'vivid violet' (*Hilliard & Burtt* 16243, E); Sani Pass 'livid mauve changing to pale mustard' (*Hilliard & Burtt* 17269, E); Cobham Forest Station 'corolla tube dull livid mauve outside, limb dull pale purple' (*Hilliard & Burtt* 9707, E); then at the southernmost localities, near Tarn Cave above Bushman's Nek, 'livid mauve; a few plants in the colony of 17317' (*Hilliard & Burtt* 17448, E), 'flowers opening and remaining pale yellow, slightly deeper around throat' (*Hilliard & Burtt* 17317, E). A short distance further south, at Ramatseliso's Gate and Ongeluks Nek, in Transkei, only typical yellow-flowered *M. crassifolia* has been recorded.

It can thus be seen that plants with flowers wholly light violet have been recorded only from the Cathedral Peak and Giant's Castle areas. *Manulea florifera* with light violet flowers occurs at Highmoor, below Giant's Castle, and some specimens thereabouts (e.g. *McKeown* 106, E) may be difficult to place. Field work is needed along the face of the Drakensberg in the ill-collected area between Cathedral Peak and Giant's Castle, and also at lower altitudes to trace the full distribution range and colour-patterning of *M. crassifolia*.

5. Manulea rhodantha Hilliard in Notes RBG Edinb. 46: 52(1989).

Two supspecies are recognized:

subsp. rhodantha
Type: Cape, Barkly East distr., 3027DC, Witteberg, river below Pitlochrie, c.5800ft, 4 xii 1981, *Hilliard & Burtt* 14706 (holo. E; iso. K, NU).

Perennial herb, stock narrowly obconical, up to c.10mm diam. at crown, stems 300–750mm tall, solitary to several, often simple, sometimes very sparingly branched in upper half, puberulous, hairs mostly gland-tipped, balloon-tipped hairs most frequent on inflorescence axis, nude to very sparsely bracteate. *Leaves* radical, crowded in rosettes, opposite, bases connate, 30–95 × 7–18mm, elliptic to oblanceolate, ± obtuse, base narrowed to a short petiolar part, margins entire to serrulate, or obscurely serrate, both surfaces minutely glandular-puberulous. *Flowers* many in a long narrow terminal thyrse, cymules few-flowered (mostly 3), dichasial, eventually distant. *Bracts* (subtending lowermost cymules) 4–7 × 0.8–1mm, linear to linear-lanceolate, *Peduncles* (of lowermost cymules) 2.5–7mm long, pedicels very short. *Calyx* tube 0.3–0.5mm long, lobes subequal, 2.2–3.75mm long, oblong-lanceolate, puberulous, hairs to 0.1–0.25mm long, balloon-tipped, tiny gland-tipped hairs as well. *Corolla* tube 6.75–8.5 × 1.2–1.75mm in throat, cylindric, expanded in upper part, limb bilabiate, 5–8mm across lateral lobes, posticous lobes 2–3.5 × 1.8–2.6mm, anticous lobe 2–4.5 × 1.3–2.75mm, all oblong to oblong-elliptic, margins ± flat, deep red to rosy pink, tube and backs of lobes puberulous, hairs to 0.1mm, balloon-tipped, also ± sessile glistening glands on backs of lobes, inside a posticous band of clavate hairs in throat, dark or reddish in lower part, lesser bands in lateral and anticous positions. *Stamens* 4, anticous anthers 0.3–0.6mm long, tips visible in throat, posticous anthers 1.6–1.8mm long, included, posticous filaments bearded, hairs often dark or reddish. *Stigma* 4–5.8mm long, shortly exserted to just included. *Style* 0.5–1.5mm long. *Ovary* 1.5–2 × 0.8–1.4mm. *Capsules* 5–7 × 3–4mm. *Seeds* c.0.6 × 0.3mm, pallid.

Selected citations:

Lesotho. Berea distr., 2828CC, Cana, xi 1911, *Dieterlen* 309 (PRE, SAM, STE, caution needed, may be mixed with *M. buchneroides*); 2927BA, Teyateyaneng, 5500ft, 29 xii 1964, *Jacot Guillarmod* 4748 (PRE). Maseru distr., 2928AC, 77 km from Maseru, Likalaneng, 2450m, 29 xi 1977, *Killick* 4231 (PRE). 2927CD or 3027AB, Thaba Tsueu, 5500ft, xii 1920, *Page* sub BOL 16820 (BOL).
Cape. Molteno distr., 3126AD, Broughton, 6300ft, xii 1892, *Flanagan* 1622 (BOL, PRE, SAM). Barkly East distr., 3127BA, 14 miles SE by S of Barkly East, c.6000ft, 13 i 1959, *Acocks* 20233 (BOL).

Manulea rhodantha is allied to *M. crassifolia*, which it resembles in habit and inflorescence; it differs in the calyx (deeply divided into five subequal lobes, not distinctly bilabiate with a 2-lobed anticous lip and a 3-lobed posticous one) and longer posticous anthers (1.6–1.8mm versus 0.8–1.5mm).

Manulea rhodantha subsp. *rhodantha* ranges from western Lesotho south to the environs of Barkly East and Molteno in the mountains of the NE Cape. There is also a record from 'Tsomo river, British Kaffraria' (*Bowker* 785, TCD); this river rises in the Cape Drakenberg due south of Barkly East. The area of typical *M. rhodantha* therefore lies wholly within that of *M. crassifolia* subsp. *crassifolia* and over the same altitudinal range, but *M. rhodantha* flowers earlier, mainly in November and December, passing into fruit in January. In addition to the discriminating characters given above, the two species may also differ consistently in flower colour: shades of red in *M. rhodantha*, yellow in *M. crassifolia*. An element of doubt is raised by the Bowker specimen quoted above, where the flowers are recorded as 'buff-coloured'. However, the label was written by Mrs Barber (sister of Mr Bowker) and she noted 'this plant grows also on the Zwart Kei, Queenstown' a reference to her own collection (no. 315, TCD) of *M. bellidifolia*, which she failed to distinguish as a separate species and which she described as having buff-coloured flowers (they are usually white). Mrs Barber also collected the white-flowered *M. plurirosulata* and described the flowers as 'pale buff'. White flowers may fade to buff, and as she thought her brother's plant to be the same as her own collections, she may have extrapolated. Unfortunately, there are no other collections from the upper Kei and Tsomo rivers, an area much neglected botanically.

subsp. **aurantiaca** Hilliard in Notes RBG Edinb. 46: 52(1989).
Type: Transvaal, Belfast distr., 2530CA, 8 miles from Belfast on Stoffberg road, 20 xii 1964, *Burtt & Scheepers* 2948 (holo. E, iso. PRE).

Selected citations:

Transvaal. Pietersburg distr., 2430AA, The Downs, 1460 m, 11 ix 1985, *Stalmans* 683 (PRE). Middelburg distr., 2529CD, near Middelburg, c.4800ft, ix 1886, *Bolus* 7669 (BOL). Belfast distr., 2530CA, c.5 miles NW of Belfast, c.5800ft, 12 xii 1955, *Leistner* 522 (NBG). Carolina distr., 2630AA, near Carolina, c.5600ft, xii 1905, *Bolus* 12178 (BOL, PRE). Standerton distr., 2628DD, Waterval Rivier, 4600ft, 17 x 1899, *Schlechter* 3477 (BOL).
Natal. Utrecht distr., 2730AC, 7 miles NW of Groenvlei, 6000ft, 11 i 1947, *Codd* 2524 (E, PRE); 2730AD, Altemooi, c.6000ft, x–xi 1920, *Thode* 4563 (STE, mixed with *M. buchneroides*).

Subsp. *aurantiaca* differs from the typical plant chiefly in the colour of the corolla (shades of yellow and brownish-yellow: *aureus, ochraceus, aurantiacus*) and the seeds (blue-violet, not whitish). One collection from the southernmost part of the range has white seeds (*Devenish* 190, PRE, from the farm Oshoek near Wakkerstroom) while a second collection from the same farm has blue seeds (*Devenish* 1655, E). There is also a tendency for the cymules to be several-to many-flowered, rather than few-flowered as in the typical plant.

Subsp. *aurantiaca* is almost confined to the eastern and south-eastern Transvaal, where it has been recorded from Haenertsburg south to Val (NW of Standerton) and Wakkerstroom, just entering Natal near Groenvlei. It grows in moist grassland, sometimes among rocks, between c.1400 and 2100 m above sea level, flowering mainly between October and December (specimens can be found in herbaria under the names *bellidifolia*, *crassifolia* and *parviflora*).

There appear to be no records from the north eastern Orange Free State, and there is thus a gap about 200 km wide between the northernmost record of *M. rhodantha* sens. strict. and the southernmost record of subsp. *aurantiaca*.

A plant that is common in damp sandy places between the rock outcrops at Kaapschehoop needs special mention. It differs from subsp. *aurantiaca* in its small capsules (3mm long) that are minutely glandular-puberulous all over, though the ovaries are glabrous except for the glistening sessile glands down the sutures, which are commonplace in *Manulea*. Whether or not this is a purely local variant of subsp. *aurantiaca* needs further investigation (*Gilmore* 2220, PRE; *Hilliard & Burtt* 18463, E, S; *Onderstall* 1264, UPS; *Pole Evans* 982, PRE; *Prosser* 1482, J, PRE; *Thode* A 1629, PRE).

6. Manulea parviflora Benth. in Hook., Comp. Bot. Mag. 1: 383 (1836) & in DC., Prodr. 10: 365 (1846); Hiern in Fl. Cap. 4(2): 237 (1904), p.p.
Type: Near the Omtata in the Amaponda country, *Drège* (holo. K).

Syn.: *Manulea natalensis* [Bernh. ex] Krauss in Flora 27: 835 (1844). Type: Port Natal, near the Umlaas river, Oct., *Krauss* 408 (holo. W; iso. BM, K, TCD).
M. angolensis Diels in Bot. Jahrb. 23: 478 (1896); Hemsley & Skan in Fl. Trop. Afr. 4(2): 299 (1906). Lectotype (chosen here): Angola, Huilla, in moist bushy places near a cataract, x 1859, *Welwitsch* 5836 (K, isolecto. BM).
Nemia parviflora (Benth.) Hiern in Cat. Afr. Pl. Welw. 1: 758 (1898).
N. angolensis (Diels) Hiern in Cat. Afr. Pl. Welw. 1: 757 (1898).
Manulea rhodesiana Sp. Moore in J. Bot. 49: 156 (1911); Philcox in Fl. Zamb. 8(2): 36(1990). Type: [Zimbabwe] Rhodesia, Victoria [Masvingo], *Monro* 924 (holo. BM, iso. S).
M. crassifolia auct. non Benth.; Philcox in Fl. Zamb. 8(2): 34 (1990).

Two varieties are recognized:

var. **parviflora**

Perennial herb, stock narrowly obconical, up to c.15mm diam. at crown, stems up to 1 m tall, solitary to several, simple to laxly branched in upper half, puberulous, hairs to 0.1–0.3mm long, both balloon- and gland-tipped, distantly leafy to bracteate. *Leaves* mostly crowded at base of stem becoming distant and smaller upwards, opposite, bases connate, becoming alternate upwards, lowermost 30–120 × 4–22mm, elliptic to oblanceolate, obtuse to subacute, base narrowed to a short petiolar part, margins entire to serrate, both surfaces glandular-puberulous. *Flowers* many in a long narrow terminal thyrse, paniculate when stem branched, the branches ascending, cymules normally many-flowered, up to c.40mm long, mostly once-dichasial (occasionally twice-dichasial), thereafter monochasial, often becoming distant in fruit. *Bracts* (subtending lowermost cymules) c.4–10×0.5–1mm, linear. *Peduncles* (of lowermost cymules) c.2–7mm long, pedicels very short. *Calyx* obscurely bilabiate, tube 0.2–0.4(–0.5)mm long, lobes subequal, 1–2.8 × 0.3–0.6(–0.8)mm, linear, acute to ± spathulate and thickened, hairy mainly on margins, hairs up to 0.2–0.5mm long, balloon-tipped, smaller gland-tipped hairs also present. *Corolla* tube 4–9 × 0.8–1.4mm in throat, cylindric, expanded in upper part, limb bilabiate, 3–7.2mm across lateral lobes, posticous lobes 1.3–3 × 0.8–2.6mm, anticous lobe 1.3–3.3 × 0.6–2mm, all oblong-elliptic to subrotund, margins often revolute, dull yellow to orange or brownish-red, tube occasionally purple-tinged, tube and backs of lobes puberulous, hairs to 0.1(–0.2)mm, balloon-tipped, inside rarely with a few clavate hairs on posticous side, usually glabrous there, usually weakly bearded on anticous side. *Stamens* 4, anticous anthers

0.2–0.4mm long, visible in mouth, posticous anthers 0.6–1.4mm long, included, posticous filaments sparsely bearded, hairs often very few. *Stigma* c.3–8mm long, shortly exserted to just included. *Style* 0.5–1.5mm long. *Ovary* 0.6–1.5 × 0.4–0.8mm. *Capsules* 1.5–3.4 × 1.5–2.5mm, (subject to galling, then larger), often glabrous, sometimes minutely glandular-puberulous (in addition to sessile shining glands often present along sutures). *Seeds* c.0.5–0.75 × 0.3–0.4mm, usually violet-blue, very rarely pallid.

Selected citations:

Angola. Huila, Humpata, Palanca, c.2000m, 4 vi 1937, *Exell & Mendonça* 2554 (BM); Lubando, Tundavala, 27 iv 1971, *Borges* 47 (BM).

Zimbabwe. Marandellas distr., 8 iv 1948, *Corby* 67 (K); Beatrice, 26 xii 1924, *Eyles* 4428 (K). Inyanga distr., c.9km from Inyanga village, 28 i 1931, *Norlindh & Weimarck* 4623 (BOL, K). Chipinga distr., Chipinga Experiment Farm, 15 ii 1962, *Chase* 7614 (BM, K). Victoria distr., Makohali Experiment Station, 17 iii 1978, *Senderayi* 285 (K).

Transvaal. Messina distr., 2230DA, Venda, Thengwe Bashasha, 3 i 1980, *Khorommbi* 1107 (PRE). Nylstroom distr., 2428CD, Warmbaths, 31 ix 1908, *Leendertz* 5552 (PRE). Pretoria distr., 2528CA, Hornsnek, 29 i 1956, *Schlieben* 7824 (K); Pretoria, x 1895, *McLea* 5743 (BOL, K, SAM). Johannesburg, 2628AA, 3 x 1960, *Macnae* 1264 (J, NBG). Barberton distr., 2530BD, Nelspruit, xi 1915, *Rogers* 18400 (K, PRE).

Swaziland. Mbabane distr., 2631AC, Black Mbuluzi Valley, 9 xii 1964, *Dlamini* s.n. (NBG, PRE). Stegi distr., 2631BB, Mbuluzi Poort, 27 ix 1960, *Compton* 30144 (NBG, PRE).

Mozambique. Sul do Save, Marracuene, 24 iii 1954, *Barbosa & Balsinhas* 5455 (BM).

Natal. Ingwavuma distr., 2632CD, Ndumu Game Reserve, 13 vi 1969, *Pooley* 596 (E); 2632DD, Kosi Estuary, 10 v 1965, *Vahrmeijer & Tölken* 925 (PRE). Hlabisa distr., 2832AD, St. Lucia Estuary, 15 vi 1978, *Pooley* 2092 (E). Vryheid distr., 2730DB, Hlobane, *Johnstone* 481 (E). Durban distr., 2930DD, Northdene, 19 iii 1900, *Wood* 7934 (E). Pietermaritzburg distr., 2930DA, Umgeni Dam, 5 i 1948, *Barker* 5223 (NBG). Camperdown distr., 2930DA, Cato Ridge, 16 ix 1893, *Schlechter* 3257 (BOL, E, STE, UPS); Harrison Flats near Inchanga, 11 x 1973, *Hilliard & Burtt* 6823 (E, NU). Ixopo distr., Umtwalume waterfall, 23 x 1962, *Strey* 4409 (PRE).

Manulea parviflora has a remarkably wide geographical range, having been recorded from the highlands of Angola, then an apparent disjunction to eastern Zimbabwe, and south to the Transvaal, Swaziland, Natal, Transkei, and coastal Mozambique at least as far north as the Komati river. Over that enormous area, it shows relatively little morphological variation. It is characterized by its small flowers (often small in the limb though the tube may be long) arranged in many-flowered thyrses, the individual cymules of which are normally well-spaced and almost invariably produce two lateral flowers at the first branching, and thereafter the branching is monochasial. Each monochasium tends to produce many, crowded, flowers, giving well-grown specimens a highly characteristic facies. In coastal Natal, the cymules may remain crowded and corymb-like particularly towards the tip of each rhachis, but there is no other significant difference from the typical plant, which also occurs in the area. The holotype of *M. natalensis* in Vienna (W) as well as the isotypes in the British Museum (Natural History) and Trinity College, Dublin, show a tendency towards corymb-like cymules, while that in the Kew herbarium does not. Other examples of these crowded cymules can be seen in *Pooley* 2092 (E), *Prosser* 1979 (PRE) and *Wood* 17 and 7934 (E); in the Pretoria herbarium this form of *M. parviflora* is frequently confused with *M. crassifolia*.

Spencer Moore described *M. rhodesiana* as differing from *M. parviflora* in 'the narrow leaves, the smaller calyx with spathulate segments and the longer corollas', characters that do not hold when a wide range of material is examined, and no others present themselves. Indeed, the calyx lobes in the type of *M. parviflora* have spathulate tips and the type of *M. rhodesiana* consists only of inflorescence branches without leaves. Some specimens from Inyanga have remarkably long hairs all over the calyx, and may warrant formal recognition (e.g. *Norlindh & Weimarck* 4623, cited above).

Diels mentioned no affinity for *M. angolensis*. He cited two Welwitsch collections from Huila province, nos. 5835 and 5836; these have the corolla glabrous or nearly so outside, whereas in typical *M. parviflora* it is densely puberulous. I have seen two other specimens from Angola (cited above): *Exell & Mendonça* 2554 closely resembles the Welwitsch collections, while *Borges* 47 is typical *M. parviflora*. Even in those specimens with the corolla glabrous, the stems and inflorescence axes are puberulous as in the typical plant. It is the general lack of indumentum that distinguishes var. *limonioides* from Transvaal (see below).

Manulea parviflora favours damp sandy places and may grow along streamsides, in open grassland, or in open woodland. It clearly has weedy propensities, and may be found in flower in any month. In Transkei and Natal, it occurs from sea level to c.1200m, in Swaziland from c.150–1200m, Transvaal 700–1600m, Zimbabwe 1500–1950m. There is a single record from the Cape, at Plettenberg Bay (*Rogers* 15497) but its presence there needs confirmation.

var. **limonioides** (Conrath) Hilliard in Notes RBG Edinb. 46:51(1989).
Type: Transvaal, near Johannesburg, Rietfontein, 7 xi 1897, *Conrath* 979 (holo. K).

Syn.: *M. limonioides* Conrath in Kew Bull. 1908: 225 (1908).

Citations:

Transvaal. Johannesburg, 2628AA, 25 xii 1903, *Pegler* 1040 (BOL, K, SAM); Jeppestown Ridge, x 1898, *Gilfillan* 76a (PRE); Elsburg, x 1913, *Rogers* 12130 p.p. (PRE). Heidelberg distr., 2628CB, Suikerbosrand, 19 xii 1972, *Bredenkamp* 957 (PRE); 2628AD, Heidelberg, 21 x 1893, *Schlechter* 3530 (BOL, K, PRE, UPS). Piet Retief distr., 2730BB, Assegai river between Piet Retief and Pongola, 17 xi 1976, *Balsinhas* 03062 (K, PRE).

Conrath diagnosed his new name against *M. obovata* but, although that species has a similar inflorescence, it is only distantly related to *M. parviflora* (the true affinity of *M. limonioides*) from which Conrath's plant differs only in the total lack of pubescence on the stem and inflorescence axes (calyx and corolla as well as the leaves are more or less puberulous). This nearly glabrous plant has been recorded on the Witwatersrand and south and east to the vicinity of Parys, Vereeniging, Heidelberg and Piet Retief, and seems worthy of recognition at varietal level. It has been found together with typical *M. parviflora* both on the Rand and in the Piet Retief area: both varieties were collected by F.A. Rogers at Elsburg (near Johannesburg) under the one number, 12130, while a collection from Piet Retief (*Galpin* 9619, PRE) comprises one nearly glabrous plant, one puberulous, as does another from Krugersdorp (*Gilliland* 26878, J); a duplicate in the Pretoria herbarium is typical *M. parviflora*. Over the rest of its great range, *M. parviflora* is always densely puberulous.

Manulea parviflora var. *limonioides* appears to hybridize with *M. buchneroides*. Dr L.E. Codd collected three sets of specimens in disturbed sandy soil near the banks of the Vaal at Visgat, east of Vereeniging, thinking at the time that hybridization might be taking place. His no. 4466 (E, PRE) has the criteria of *M. parviflora* var. *limonioides* except for the unusual colour of the flowers (strawberry-pink); no. 4467 (E) is almost pure *M. buchneroides* (deviating slightly in that some of the lower cymules carry open flowers at the same time as the terminal, corymbose, cluster); no. 4468 (E, PRE) seems to be intermediate between the two (calyx tube and anthers too large for true *M. parviflora*, floriferous, pink-flowered, thyrse alien to *M. buchneroides*). True *M. buchneroides* has been recorded on the Vaal at Schoemansdrif, west of Vereeniging (e.g. *Botha & Ubbink* 1555, PRE) and at Sasolburg, to the south-east (e.g. *Kroon* 26, PRE). Red-flowered specimens of *M. parviflora* have been recorded near Potchefstroom, further to the west (e.g. *Vahrmeyer* 1547, K). Clearly there is potential for a cross between red-flowered *M. parviflora* and white-flowered *M. buchneroides* to produce the pink-flowered specimens collected by Codd.

Gilfillan 412 (PRE), which was collected at Buffelshoek, Magaliesberg (Buffelshoek not traced, but Magaliesberg roughly 100 km north of the Vaal), is a good match of *Codd* 4468 in floral measurements and general facies but there are no records of *M. buchneroides* in that area; *M. parviflora* on the other hand is common around Pretoria.

7. Manulea flanaganii Hilliard in Notes RBG Edinb. 46:339(1990).
Type: Orange Free State, 3025BD, near Bethulie, Orange River, 4000ft, xii 1892, *Flanagan* 1505 (holo. SAM, iso. BOL).

Annual herb, taproot woody, up to 4mm diam. at crown, stems c.300–380mm long, erect or decumbent, well branched from the base and above, glandular-puberulous throughout, hairs up to 0.1mm long, occasional balloon-tipped hairs also present, laxly leafy. *Leaves* opposite, bases connate, becoming alternate on the inflorescence axes, main leaves c.30–55 × 4–6mm, smaller upwards and passing imperceptibly into bracts, narrowly elliptic, apex ± acute, base narrowed to a short petiolar part, lowermost leaves with ± serrate margins, upper entire, both surfaces glandular-puberulous. *Flowers* many in long spreading paniculate thyrses, cymules many-flowered, monochasial or dichasial, the lowermost in the axils of the uppermost leaves. *Bracts* (lowermost) either leaflike or, on small side branches, c.5–7 × 1mm, oblong. *Peduncles* (lowermost) c.1–5mm long. *Calyx* obscurely bilabiate, tube c.0.2mm long, lobes subequal, c.2.5–3.3 × 0.3–0.4mm, linear-lanceolate, acute, glandular-puberulous all over, hairs up to 0.2mm long, balloon-tipped hairs very occasional. *Corolla* tube c.5–6.5 × 0.8mm in throat, cylindric, inflated near apex, limb bilabiate, c.4.5–5.8mm across lateral lobes, posticous lobes 2–2.5 × 1.1–1.6mm, anticous lobe 2.2–2.8 × 1mm, all oblong-elliptic, colour unknown, tube and backs of lobes puberulous with gland- and balloon-tipped hairs up to 0.1mm long, glabrous within. *Stamens* 4, anticous anthers 0.5–0.6mm long, visible in mouth, posticous anthers 1mm long, included, posticous filaments either glabrous or with a few hairs. *Stigma* c.3–3.6mm long, just included. *Style* 0.3–0.75mm long. *Ovary* c.1–1.2 × 0.6mm. *Capsule* 3–3.5 × 2.25–2.5mm, glandular-puberulous all over. *Seeds* immature (or shed), pallid.

Manulea flanaganii is known only from the type collection, made in an area that is scarcely known botanically. Hiern cited the specimen (in Fl. Cap. 4(2): 238, 1904) as *M. parviflora*, to which it is certainly allied, but from which it differs in habit (annual, not perennial), in the stem and calyx being predominantly glandular-puberulous (not with balloon-tipped hairs predominant), the calyx lobes mostly longer (c.2.5–3.3mm versus 1–2.8mm) and glandular-puberulous all over (not with balloon-tipped hairs nearly confined to the margins), and longer anticous anthers (0.5–0.6mm versus 0.2–0.4mm).

Bethulie is remote from the area of *M. parviflora*. When Flanagan collected his specimens in December, they had both ripe capsules and flowers.

8. Manulea florifera Hilliard & Burtt in Notes RBG Edinb. 40: 290 (1982).
Type: Natal, Lion's River distr., Kamberg area, farm Allendale, 24 i 1978, *Hilliard & Burtt 11251* (holo. E; iso. K, NU, PRE).

Perennial herb somewhat stoloniferous at base and there to c.10mm diam., stems 1 or few from crown, erect to c.1m, simple below the inflorescence, puberulous, hairs balloon-tipped, to 0.3mm long, distantly leafy. *Leaves* opposite, bases connate, lowermost c.80–200 × 13–25mm, diminishing upwards and passing into bracts, narrowly elliptic, tapering at both ends, margins entire to serrulate or serrate, both surfaces glandular-puberulous, a few balloon-tipped hairs as well on midrib below. *Flowers* many in narrow thyrses composed of dichasial cymules further arranged in a large spreading panicle. *Bracts* subtending individual cymules c.2.5–10 × 0.5–1mm, linear. *Pedicels* very short or flowers sessile, peduncles of individual cymules c.1–10mm long. *Calyx* obscurely bilabiate, tube 0.3–0.75mm long, anticous lobes 1–1.8 × 0.4–0.6mm, anticous lip split 1–1.5mm, posticous lobes 0.8–2 × 0.3–0.7mm, posticous lip split 1–1.6mm, all lobes linear-oblong, tips ± thickened or subspathulate, puberulous on backs and margins, balloon-tipped hairs to 0.2–0.3mm long, smaller gland-tipped ones as well. *Corolla* tube 4.5–6.5 × 1–1.6mm in throat, cylindric, dilated near apex, limb bilabiate, 5.5–7mm across the lateral lobes, posticous lobes 2–2.8 × 1–1.8mm, anticous lobe 2–3 × 0.8–1.4mm, all oblong-elliptic, mauve or whitish, tube purplish or yellowish respectively at base, puberulous outside, hairs to 0.1mm, balloon-tipped, inside a short posticous band of clavate hairs, dark in lower part, 4 lesser bands round rest of throat. *Stamens* 4, anticous anthers 0.4–0.6mm long, very shortly exserted, posticous anthers 0.6–1mm long, included, posticous filaments bearded, hairs usually dark. *Stigma* 2.75–4.5mm long, shortly exserted or just included. *Style* 0.4–0.8mm long. *Ovary* 1–1.2 × 0.4–0.6mm. *Capsules* (immature) c.4.5 × 2.5mm. *Seeds* unknown.

Citations:

Natal. Estcourt distr., 2929AD, Highmoor Forest Reserve, c.7000ft, 18 ii 1968, *Hilliard* 4820 (E). Lion's River distr., 2930AC, near Nottingham Road, 4–5000ft, 31 xii 1890, *Wood* 4397 (BOL, SAM). 2929BC, Kamberg, ii 1953, *Smuts* 1059 (PRE). Mooi River distr., 2929BC, Mooi River, Bray Hill, 15 i 1902, *Johnston* 799 (E). 2930CA, Liddesdale [now Gowan Brae, Merrivale] 4000ft, 18 ii 1888, *Wood* 3935 (BOL).
Transkei. Umzimkulu distr., 3029BC, Zuurbergen, 5500ft, 1 ii 1895, *Schlechter* 6588 (BOL); Newmarket–Umzimkulu, 2 ii 1895, *Krook* in herb. Penther 3019 (W).

Manulea florifera is a poorly known species apparently confined to a small area of south central Natal and neighbouring Transkei, from Highmoor Forest Reserve, the Kamberg area, Mooi River, Nottingham Road and Merrivale south to the Zuurberg and environs in Umzimkulu district, Transkei. It must surely occur between Merrivale and the Zuurberg, though its habitat (damp or marshy grassland between c.1500 and 2000m above sea level) is rapidly being destroyed by afforestation. It flowers between December and February.

Manulea florifera and *M. paniculata* appear to be vicariants, similar in inflorescence and floral morphology, but very different in habit, possibly a reflection of different habitat preferences: *M. florifera* in open grassland, subject to fire, *M. paniculata* in the shelter of rock outcrops. Schlechter collected both species on the same day on the Zuurberg while Krook collected *M. paniculata* at Newmarket on 1 ii 1895, and *M. florifera* the following day when he was journeying on to Umzimkulu; it is there that the areas of the two species meet, *M. florifera* replacing *M. paniculata* north of that point.

See also under *M. crassifolia* subsp. *thodeana*.

9. Manulea paniculata Benth. in Hook., Comp. Bot. Mag. 1: 383 (1836) & in DC., Prodr. 10: 365 (1846); Hiern in Fl. Cap. 4(2): 237 (1904); Hilliard & Burtt in Notes RBG Edinb. 37: 315 (1979).
Lectotype (chosen here): Cape, Stormbergen, auf steinigen und felsigen Höhen, 5000-6000 Fuss, December, *Drège* (K; isolecto., E, LE, S, TCD, W).

Syn.: *Manulea thyrsiflora* var. *albiflora* O. Kuntze, Rev. Gen. Pl. 3(2): 236 (1898), nom. nud. Specimen cited: Cape, Molteno, *O. Kuntze* (NY).
   Sutera elliotensis Hiern in Fl. Cap. 4(2): 251 (1904). Type: Tembuland, by the Slang river, near Elliot, *Bolus* 8762 (K).
   [*Manulea elliotensis* [Overkott ex] Guillarmod, Fl. Lesotho 245 (1971) - nomen invalidum].

Coarse perennial herb branching from the base, stems to 1m tall, 5–7mm diam. near base, soon branching into a large leafy panicle, pubescent with patent eglandular hairs to 0.4–0.6mm long, much shorter gland-tipped ones as well. *Leaves* opposite, bases connate, becoming alternate only on the upper inflorescence branches, c.20–85 × 8–43mm, smaller on the inflorescence, elliptic, subacute, base narrowed but scarcely petiolar, margins closely serrate or crenate-serrate, both surfaces pubescent, hairs mostly gland-tipped. *Flowers* many, in narrow thyrses composed of dichasial cymules further arranged in a large spreading panicle terminating each stem, leafy in its lower part. *Bracts* c.3–7 × 0.7–1.2mm, linear, but the thyrses in the axils of the upper leaves, which pass progressively into bracts. *Pedicels* (and peduncles of cymules) very short. *Calyx* obscurely bilabiate, tube 0.3–0.5mm long, anticous lobes 2–3.7 × 0.4–0.5mm, anticous lip split 1.75–3.2mm, posticous lobes 2.4–3.4 × 0.4–0.6mm, posticous lip split 2–3.2mm, all lobes lanceolate, pubescent on backs and margins, hairs to 0.3–0.5mm, both glandular and eglandular. *Corolla* tube 6–7 × 1.25–1.5mm in throat, cylindric, inflated near apex, limb bilabiate, 5–6.5mm across the lateral lobes, posticous lobes 1.8–2.5 × 1.3–1.75mm, anticous lobe 2–2.75 × 1.1–1.6mm, all ± elliptic, white or creamy, yellow at base of posticous lip and down back of tube, tube and lobes pubescent outside, hairs to 0.2mm, mainly eglandular, inside a short posticous band of clavate hairs, 4 lesser bands in anticous position. *Stamens* 4, anticous anthers 0.4–0.6mm long, very shortly exserted, posticous anthers 0.8–1.2mm long, included, posticous filaments bearded. *Stigma* 3.5–5mm long, very shortly exserted or just included. *Style* 0.5–1.3mm long. *Ovary* 1–1.5 × 0.6–1mm. *Capsules* c.3.5–5.5 × 2–3mm. *Seeds* c.0.75–0.8 × 0.4mm, pallid.

Selected citations:

Transvaal. Potchefstroom distr., 2627CA, Buffeldooms, 6 iv 1945, *Louw* 1414 (J, PRE).
Orange Free State. Bethlehem distr., 2828CB, 24 miles NE of Clarens, c.6000ft, 26 iii 1951, *Bruce* 398 (PRE). Ficksburg distr., farm 'North End' near Wellington, 5700ft, 5 iii 1970, *van der Zeyde* s.n. (NBG).
Cape. Albert div., 3027CB, Witteberg, 1861, *Cooper* 1374 (E, K, NY, TCD, W). Elliot div. (?), c.3127BD, Slang river between Cala and Elliot, 5000ft, 1896, *Flanagan* 2624 (BOL, PRE).
Transkei. Umzimkulu distr., 3029BC, Zuurbergen, 5200ft, 1 ii 1892, *Schlechter* 6586 (BOL).
Natal. Mt. Currie distr., 3029AD, Kokstad, 5000ft, ii 1883, *Tyson* 1420 (BOL, PRE, SAM); Newmarket, 1 ii 1895, *Krook* in herb. Penther 1768 (W).
Lesotho. c.2927CB, Roma to Mafeteng, 5900ft, 16 ii 1976, *Schmitz* 6514 (PRE). 2828CC, Leribe, 5–6000ft, *Dieterlen* 436 (PRE, SAM).

Manulea paniculata ranges from the Witwatersrand south to Potchefstroom and Standerton, thence into the eastern Orange Free State and western Lesotho and so to the Witteberg, Cape Drakensberg and Stormberg in the north-eastern Cape. Then there are two isolated records from further east on the hills at Kokstad in southernmost Natal and the nearby Zuurberg in Transkei: the gap should be closed by field work in Mt Fletcher and Mt Frere districts of Transkei. It is absent from most of Natal, where it is replaced by its ally, *M. florifera*: see comments under that species.

Manulea paniculata favours rocky sites, and is often recorded as growing among boulders, between 1500 and 1980m above sea level, flowering principally in February and March.

10. Manulea buchneroides Hilliard & Burtt in Notes RBG Edinb. 40: 286 (1982). **Plate 1A.** Type: Orange Free State, Harrismith distr., Platberg, One Man Pass, c.1950–2100m, 13 xii 1976, *Hilliard & Burtt* 9512 (holo NU; iso. E, K, PRE).

Perennial herb, stock conical, slender, up to 10mm diam. at crown, stems (150–)300–450mm tall, solitary to several, simple or very sparingly branched in upper part, puberulous, hairs both balloon- and gland-tipped, the former commonest on upper parts, leafless or with a few much reduced leaves. *Leaves* mainly radical, in 1 to several rosettes on the crown, opposite, bases connate, blade c.15–70 × 4–25mm, narrowly to broadly elliptic tapering to a petiolar part c.2–20mm long (up to c.⅓ length of blade but often much shorter), margins entire to somewhat obscurely toothed, both surfaces glandular-puberulous. *Flowers* many in a long very narrow thyrse, the cymules dichasial, few-flowered, eventually distant, the apical, flowering, ones characteristically in a congested corymbose head. *Bracts* (lowermost) 3.5–10 × 0.5–1mm. *Peduncles* of lowermost cymules (or pedicels when lowermost flower occasionally solitary) c.4–7mm long, pedicels otherwise very short, the cymules congested. *Calyx* very obscurely bilabiate, tube 0.4–0.5mm long, lobes subequal, 2.6–3.75 × 0.6–0.8mm, linear-oblong, tips somewhat spathulate, pubescent outside and on margins, hairs to 0.2–0.4(–0.5)mm, both balloon- and gland-tipped. *Corolla* tube 6.5–9 × 1–1.6mm in throat, cylindric, expanded in ± upper third, limb slightly bilabiate, 6–8.5mm across lateral lobes, posticous lobes 1.6–3 × 1.5–3mm, anticous lobe 2–3.5 × 1.6–3mm, all broadly obovate to subrotund, white, sometimes yellowish-green to orange/yellow at base of posticous lip and down back of throat, tube puberulous, hairs to 0.2mm, both balloon- and gland-tipped, ± sessile glistening glands on backs of lobes, inside a posticous longitudinal band of clavate hairs, dark-coloured in lower part, 4 smaller bands in lateral and anticous positions. *Stamens* 4, anticous anthers 0.4–0.8mm long, visible in mouth, posticous anthers 1.3–1.8mm long, included, posticous filaments bearded with dark hairs. *Stigma* 4.1–6.5mm long, shortly exserted or just included. *Style* 0.4–1.2mm long. *Ovary* 1.2–1.75 × 0.8–1mm. *Capsules* c.4–5 × 3mm. *Seeds* c.0.5–0.75 × 0.3–0.4mm, pallid.

Selected citations:

Natal. Utrecht distr., 2730AD, farm Altemooi, c.6000ft, x–xi 1920, *Thode* 4563 (STE, mixed with *M. rhodantha*).
Orange Free State. Harrismith distr., 2829AD, Rensburg's Kop, 10 xii 1962, *Jacobsz* 331 (PRE). 2828DB, Besters
Vlei near Witzieshoek, 5400ft, 1893, *Bolus* 8229 (BOL); Qwa Qwa Mt above 'Bluegumbosch', 6200–6600ft, 8 i 1979,
Hilliard & Burtt 11978 (E). Bethlehem distr., 2828DA, Golden Gate, 6000–7000ft, 2 iii 1947, *Story* 1966 (PRE). Senekal
distr., 2827AD, Willem Pretorius Game Reserve, 30 xi 1971, *Muller* 788 (PRE).
Cape. Barkly East distr., 3027DA, Witteberg, Beddgelert, c.6200ft, i xii 1981, *Hilliard & Burtt* 14584 (E).
Lesotho. 2828CC, Leribe, *Phillips* 869 (SAM).

Manulea buchneroides has been much confused with several species, particularly *M. crassifolia*, *M. bellidifolia* and *M. benthamiana* (that is, *M. corymbosa*), yet it is a well-marked species
easily recognized by the combination of perennial habit and white flowers in cymules, these
always in a flat-topped terminal corymb even when the inflorescence is much elongated and
passing into fruit.

It is commonest in damp grassland, over Cave Sandstone, in an arc from the low Drakenberg
on the Natal-Transvaal-Orange Free State border (as far east as Groenvlei in Natal), through
the north-eastern Orange Free State near Vrede and Memel thence through Harrismith,
Bethlehem and Ficksburg districts and western Lesotho to the north-eastern Cape, on the
Witteberg near Lady Grey and Barkly East, between c.1500 and 2000m above sea level. But
there are also records from further north and west as far as the valley of the Vaal river between
Potchefstroom and Vereeniging, particularly along the river banks. Here, *M. buchneroides* may
hybridize with *M. parviflora*: see under that species.

It flowers principally between October and January, thereafter there may be a few flowers as
late as April as the inflorescences pass into fruit.

11. Manulea bellidifolia Benth. in Hook., Comp. Bot. Mag. 1: 382 (1836) & in DC., Prodr.
10: 364 (1846); Hiern in Fl. Cap. 4(2): 227 (1904) p.p.
Lectotype (chosen here): Cape, Uitenhage [sic; Alexandria div.] between the Sundays and
Bushmans rivers, Quaggavlakte, *Ecklon & Zeyher* (K; isolecto. LE, SAM, TCD, W, WU).

Annual herb, stems 150–450mm tall, solitary or up to c.12 loosely tufted from base, usually
simple, sometimes with a pair of branches low down, puberulous, hairs to 0.25mm, balloon-
tipped, glandular-puberulous as well, either nude or remotely leafy in lower part. *Leaves* in 2–4
pairs near base of each stem, often crowded, sometimes straggling to halfway up stem, bases
connate, blade 25–90 × 7–30mm, narrowly to broadly elliptic or oblanceolate tapering to a
petiolar part 5–20mm long (less than ¼ length of blade), margins usually finely to coarsely
serrate, occasionally subentire, both surfaces glandular-puberulous. *Flowers* many in simple
terminal racemes, crowded at first, often distant in fruit. *Bracts* (lowermost) 2.5–4 × 0.5–1mm,
lanceolate, acute. *Pedicels* (lowermost) 3–6mm long, shorter upwards. *Calyx* obscurely bilabiate, tube 0.4–0.75mm long, anticous lobes 2.5–5 × 0.6–1mm, anticous lip split 2–4mm,
posticous lobes 2.4–4.2 × 0.6–0.7mm, posticous lip split 2.5–4.2mm, all lobes oblong-lanceolate, puberulous all over outside and margins, both glandular and balloon-tipped hairs to
0.25mm long. *Corolla* tube 7–10 × 1.2–1.6mm in throat, cylindric, expanded in ± upper third,
limb distinctly bilabiate, 7.5–10mm across lateral lobes, posticous lobes 3–4 × 2–2.5mm,
anticous lobe 3–4.5 × 1.4–2.5mm, all ± elliptic, margins usually revolute, white or creamy-
white to dull yellow, orange/yellow around mouth and in throat, tube puberulous, hairs to
c.0.1mm, balloon-tipped, very occasional gland-tipped hairs as well, ± sessile glistening glands
on backs of lobes, inside a longitudinal posticous band of clavate hairs plus 4 lesser bands,
posticous hairs often dark. *Stamens* 4, anticous anthers 0.5–0.8mm long, tips just visible in
mouth, posticous anthers 1.4–1.8mm long, included, posticous filaments bearded, hairs dark.
Stigma 5–6.5mm long, shortly exserted or just included. *Style* 0.6–1.7mm long. *Ovary* 1.5–2 ×
0.8–1.4mm. *Capsules* c.4.5–6 × 3–4mm. *Seeds* c.0.7 × 0.4mm, pallid.

Selected citations:

Cape. Queenstown div., 3326BB, Shiloh, 3500ft, *Drège* (K). Stutterheim div., 3227AD, Turpins Bridge, c.2700ft, 5 xii 1942, *Acocks* 9410 (PRE). Komgha div., 3227DB, along Kabousie river near Komgha, 2000ft, 1891, *Flanagan* 792 (PRE, SAM). Bathurst div., 3326DB, Kasouga mouth, 29 ix 1920, *Britten* 2321 (PRE). Alexandria div., 3326DA, Bushman's river mouth, v 1950, *Leighton* 3129 (BOL, PRE); ibidem, 6 vii 1947, *Compton* 19684 (NBG). 3326CB, Alexandria, 2 v 1931, *Galpin* 10682 (BOL, PRE).

When Bentham described *M. bellidifolia*, he quoted Ecklon [& Zeyher] specimens from Katriviersberg, Addo and Quaggasvlakte, as well as a Drège specimen from the Klipplaat river at Shiloh. Both the Drège specimen and the Ecklon & Zeyher collection from Quaggasvlakte are in the Kew herbarium, and as the latter is a good specimen it has been chosen as lectotype. Hiern confounded the species with *M. crassifolia*, *M. dregei* and *M. plurirosulata*, so both the description and distribution of *M. bellidifolia* as given in Flora Capensis are misleading, and the name has since been much misused.

Manulea bellidifolia is an annual and may be recognized by its woody taproot crowned with several loosely arranged stems with leaves mostly crowded near the base, simple racemes, and distinctly balloon-tipped hairs on the corolla tube and also on the calyx and stems but there mixed with gland-tipped hairs. These characters (except that of the raceme) help to distinguish it from its ally *M. plurirosulata* (see further under that species). The annual habit, loosely arranged leaves and simple racemes at once distinguish it from *M. crassifolia*.

Manulea bellidifolia grows in sandy places, often along river banks, but also on hill slopes, from near sea level to c.900m, flowering mainly between May and October. It has been recorded along the eastern Cape coast from about Alexandria to the Kowie river, then inland to the valleys of the Kabusie, Toise, Zwart Kei and Klipplaat rivers, roughly between Shiloh (near Whittlesea) and Komgha.

12. Manulea plurirosulata Hilliard in Notes RBG Edinb. 46:52 (1989).
Type: Orange Free State, Bloemfontein distr., 2926AC, 45 miles S of Bloemfontein along railway line, c.4800ft, 1 ix 1925, *Smith* 376 (holo. PRE).

Syn.: *M. bellidifolia* auct. non Benth.; Harv., Thes. Cap. 2: 62 t. 197 (1863).

Annual herb, stems 20–180mm tall, solitary to many and then in discrete units crowded together into a small leafy rosette, usually simple, sometimes with a pair of branches low down, glandular-puberulous, plus very occasional tiny balloon-tipped hairs, usually nude, occasionally with a pair of leaves ± halfway. *Leaves* in 2–3 pairs at base of each stem, crowded, spreading, bases connate, blade 12–35 × 5–13mm, oblanceolate to narrowly elliptic tapering to a petiolar part 6–30mm long (roughly half as long as to nearly equalling blade), margins entire to serrate, both surfaces glandular-puberulous. *Flowers* many in simple terminal racemes, initially capitate, elongating a little in fruit, lowermost flowers sometimes distant. *Bracts* (lowermost) 2–4 × 0.6–1.2mm, lanceolate, acute. *Pedicels* (lowermost) 2–4mm long, shorter upwards. *Calyx* obscurely bilabiate, tube 0.4–0.8mm long, anticous lobes 3–5 × 0.8–1mm, anticous lip split 3–5mm, posticous lobes 2.8–4.5 × 0.6–1mm, posticous lip split 3–5mm, all lobes lanceolate, glandular-puberulous all over outside and on margins, hairs to 0.1mm, very occasional ± sessile balloon-tipped hairs as well, sometimes wanting. *Corolla* tube 6–9.5 × 1.2–2mm in throat, cylindric, expanded in ± upper third or half, limb distinctly bilabiate, 7–12mm across lateral lobes, posticous lobes 2.8–4.5 × 2–3.6mm, anticous lobe 3–5.4 × 1.7–3mm, all oblong to oblong-elliptic, margins often revolute, creamy-white, orange/yellow around mouth and in throat, tube puberulous with tiny glandular and eglandular hairs, ± sessile glistening glands on backs of lobes, inside well bearded with clavate hairs all round throat, darkish in posticous position, pale elsewhere. *Stamens* 4, anticous anthers 0.5–1mm long, tips just visible in mouth, posticous anthers 1.4–1.8mm long, included, posticous filaments bearded with dark hairs. *Stigma* 4–7.5mm long, shortly exserted or just included. *Style* 0.5–1.3mm long. *Ovary* 1.5 × 1mm. *Capsules* 3.5–6 × 2.5–4mm. *Seeds* c.0.5 × 0.3mm, pallid.

Selected citations:

Orange Free State. Bloemfontein distr., 2926AA, Bloemfontein, O.F.S. Botanic Garden, 27 iv 1968, *Muller* 274 (NBG). Fauresmith distr., c.2925CB, Dassiespoort, 24 ix 1935, *Pont* sub Henrici 2801 (K, PRE). 3025BB, Trompsburg, iv 1925, *Pole Evans* s.n. (PRE).
Cape. Colesberg div., 3025CA, Colesberg, x , *Botha* R120 (PRE). Middelburg div., c.3125AC, Culmstock, 3300 ft, x 1899, *Southey* sub *Galpin* 5593 (PRE); Bangor Farm, ix 1917, *Bolus* 14064 (BOL). Steynsburg div., 3125BD, Reed's Farm, ix 1925, *Pole Evans* 1740 (PRE). 3125AA, 11km from Noupoort on road to Colesberg via Oorlogspoort, 3 iv 1981, *Herman* 526 (PRE).

Manulea plurirosulata has been much confused with both *M. benthamiana* (that is, *M. corymbosa*) and *M. bellidifolia*. It can at once be distinguished from the former by its glabrous corolla mouth and very different calyx. Its relationship lies with *M. bellidifolia*, which differs in habit (stems loosely tufted at the base in *M. bellidifolia*, not crowded, as in *M. plurirosulata*, which suggested the trivial name), greater stature (150–450mm versus 20–180mm, mostly longer leaves with shorter petiolar part (blade 25–90 × 7–30mm, petiolar part less than ¼ length of blade versus 12–35 × 5–13mm, petiolar part roughly ½ as long as to nearly equalling blade) and coarser indumentum on stems and calyx, composed principally of balloon-tipped hairs to 0.25mm long (not principally gland-tipped hairs to 0.1mm with very occasional nearly sessile balloon-tipped hairs), and more scantily bearded throat (hairs in discrete bands, not one continuous band right round throat). Plants flowering in the seedling stage naturally do not show the crowded clump of leaf rosettes that are characteristic of *M. plurirosulata*.

Harvey's illustration under the name *M. bellidifolia* depicts a specimen collected by M.E. Bowker (no. 231, TCD) in 'Queenstown district on sandy flats'. The flower colour was recorded as pale buff: *M. plurirosulata* has creamy-white flowers but these may well fade to buff. The illustration is not very good, and there is no indication of hairs on the posticous filaments nor on the inside of the corolla tube.

Manulea plurirosulata ranges from the southern part of the Orange Free State to the Colesberg, Middelburg, Steynsburg area of the north-eastern Cape, and therefore north of the area of *M. bellidifolia*. Their areas probably meet somewhere in Queenstown division, but unfortunately the material available is not precisely localized. However, it seems likely that *M. plurirosulata* grows at higher altitudes than *M. bellidifolia*, c.1000–1500 m above sea level. Ecological information is scanty, but it has been recorded from 'open veld', 'stony ridges', 'moist flats' and flowers between April and October.

Section Dolichoglossa subsect. Racemosae

13. Manulea schaeferi Pilger in Bot. Jahrb. 48: 436 (1912); Merxm. & Roessler, Prodr. Fl. S.W. Afr. 126: 32 (1967).
Lectotype (chosen here): Namibia, 2718CA, Klein Karas, bei 1300m auf Sandboden, *Schäfer* 41, sub *Dinter* 1327 (SAM).

Annual herb, stems 20–180mm tall, few to many tufted from the base, in small discrete groups crowded together, mostly simple, sometimes with 1 or 2 branches very low down, glandular-puberulous, nude or with a few greatly reduced leaves near base. *Leaves* all radical or very nearly so, opposite and decussate, bases connate, blade 6–40 × 2–15mm, oblanceolate to elliptic tapering to a petiolar part 4–20mm long (roughly half to almost equalling length of blade), margins entire to obscurely or more sharply but remotely denticulate, both surfaces glandular-puberulous. *Flowers* many in simple terminal racemes, crowded at first, distant in fruit. *Bracts* (lowermost) c.1–4 × 0.2–0.75mm, linear or linear-lanceolate, acute. *Pedicels* (lowermost) 3.5–14mm long, shorter upwards. *Calyx* obscurely bilabiate, tube 0.3–1mm long, anticous lobes 3.5–4.8 × 0.6–1.1mm, anticous lip split 3–4.8mm, posticous lobes 3.1–5 × 0.6–1mm, posticous lip split 2.8–4.5mm, lobes all narrowly oblong, tips somewhat spathulate, glandular-puberulous all over outside, hairs to 0.3–0.5mm long. *Corolla* tube 7.5–10.5 × 1.2–2mm in throat, cylindric, expanded in upper third, limb 10–14mm in diam., nearly regular, lobes 3–5.5 × 2.5–6mm, broadly ovate to subrotund, reflexed, white to mauve-blue (changing colour on one plant), white or yellow around mouth and in throat, tube densely puberulous all over outside, hairs c.0.1–0.4mm, both glandular and eglandular, ± sessile glistening glands on backs of lobes, inside a longitudinal posticous band of clavate hairs (often dark-coloured), throat otherwise almost glabrous. *Stamens* 4, anticous anthers 0.4–0.6mm long, tips just visible in mouth, posticous anthers 1.1–2mm long, included, posticous filaments bearded, hairs mostly dark. *Stigma* 5.4–7.5mm long, shortly exserted or just included. *Style* 0.6–1.4mm long. *Ovary* 1.4–2 × 0.6–1mm. *Capsules* 4.5–7 × 3–4mm. *Seeds* 0.5–0.6 × 0.3–0.4mm, violet-blue (young?) or pallid.

Selected citations:

Namibia. Keetmanshoop div., Klein Karas, 2718CA, 4 viii 1923, *Dinter* 4791 (BOL, PRE, SAM, Z); 2717DB, farm Holoogberg, 30 v 1972, *Giess & Müller* 12300 (PRE). Warmbaths div., 2828DA, farm Aluriesfontein, *Giess* 14495 (PRE).
Cape. Kenhardt div., 2820CB, Aughrabies, 21 viii 1954, *Compton* 24473 (NBG); 2820DC, Kakamas, Letterkop Bot. Reserve, 15 vii 1946, *Wasserfall* 1112 (PRE); 2821AC, near Upington, *Glover* BOL 10440 (BOL). Prieska div., 2922DA, near Prieska, vi 1916, *Pearson* 9869 (BOL); 2921DA, Angelierspan, c.3300ft, 13 v 1946, *Acocks* 12673 (PRE).

Manulea schaeferi has been much confused with *M. nervosa*, which it resembles in habit and in its almost regular corolla limb, but it is readily distinguished by the calyx, which is lobed almost to the base whereas in the markedly bilabiate calyx of *M. nervosa* the two lips are each lobed less than halfway. The flowers of *M. schaeferi* are mostly smaller too (e.g. anticous lip of calyx 3.5–4.8mm long versus 4.6–8.5mm, and corolla limb 10–14mm in diameter versus 14–20mm) and the calyx lobes are narrower in relation to their length.

The area of *M. schaeferi* lies north and east of that of *M. nervosa*, from Aus and Keetmanshoop in southern Namibia south to Warmbaths and east across the Orange river to Kakamas, Upington, Kenhardt and Prieska in the northern Cape. The two species appear to be wholly allopatric. *Manulea schaeferi* flowers mainly between May and September.

14. Manulea nervosa [E. Mey. ex] Benth. in Hook., Comp. Bot. Mag. 1: 381(1836) & in DC., Prodr. 10: 363 (1846); Hiern in Fl. Cap. 4(2): 224 (1904).
Type: Namaqualand, Zilverfontein, *Drège* (holo. K; iso. E, LE, S, TCD, W).

Annual herb, stems 30–210mm tall, few to many tufted from the base, in small discrete groups crowded together, mostly simple, sometimes with 1 or 2 short branches very low down, glandular-puberulous, nude or with a few greatly reduced leaves near base. *Leaves* all radical or very nearly so, opposite and decussate, bases connate, blade 7–55 × 4–20mm, oblanceolate to elliptic tapering to a petiolar part c.5–20mm long (roughly ½–⅓ length of blade), margins entire, both surfaces glandular-puberulous. *Flowers* many in simple terminal racemes, crowded at first, distant in fruit. *Bracts* (lowermost) 1.5–4(–10) × 1–4mm, mostly ovate, acute. *Pedicels* (lowermost) 7–23mm long, shorter upwards. *Calyx* distinctly bilabiate, tube 0.1–1mm long, anticous lobes 4.6–8.5 × 1–3mm, anticous lip split 2.25–5mm, posticous lobes 4.4–8.5 × 1–2.5mm, posticous lip split 2–4.75mm, lobes all lanceolate, subacute, flat, glandular-puberulous all over, hairs to c.0.2–0.5mm long. *Corolla* tube 8–13 × 1.2–2mm in throat, cylindric, expanded in upper third, limb 14–20mm in diam., nearly regular, lobes 4.5–7.5 × 4–5mm in upper part, broadly obovate to subrotund with a broad claw, reflexed, white to blue (opening white and turning mauve-blue or vice versa), yellow round mouth and in throat, tube densely puberulous all over outside, hairs 0.1–0.4mm, both glandular and eglandular, ± sessile glistening glands on backs of lobes, inside a longitudinal posticous band of clavate hairs (some dark-coloured), a transverse anticous band in throat. *Stamens* 4, anticous anthers 0.5–1mm long, tips just visible in mouth, posticous anthers 1.8–2.2mm long, included, posticous filaments bearded. *Stigma* 6–10mm long, shortly exserted or just included. *Style* 0.5–2mm long. *Ovary*

2–3 × 1–1.5mm. *Capsules* 4–6.5 × 2.5–4mm. *Seeds* c.0.75 × 0.5mm, pallid or violet-blue (only when young?).

Selected citations:

Cape. Kenhardt div., 2919AA, 20 miles out of Pofadder on road to Springbok, 4 ix 1971, *Strauss* 122 (NBG). Namaqualand, 2918BC, c.100km west of Pofadder towards Springbok, c.700m, 13 ix 1973, *Coetzee & Werger* 1750 (PRE); 2918CA, near Ratelkraal, 7 ix 1950, *Maguire* 317 (NBG, STE); 2917DB, near Springbok, ix 1939, *Lewis* 1551 (SAM). 2917DD, Buffel rivier, 1600ft, 15 ix 1897, *Schlechter* 11263 (BOL, PRE); 3018BC, Platbakkies, c.3300ft, 10 ix 1976, *Hugo* 488 (PRE, STE).

When Bentham (1836) published Meyer's manuscript name, *M. nervosa*, he had doubts about its generic placement because the deeply bilabiate calyx was at variance with the calyces of the other species known to him. But it is now clear that varying degrees of development of a bilabiate calyx are characteristic of *Manulea*, and *M. nervosa* conforms to a whole group of allied species.

Manulea nervosa appears to be confined to Namaqualand where it has been recorded from about Pella, just south of the Orange river, in the east, to Springbok and the Buffels river in the west, south to Platbakkies, east of the Kamiesberg. It flowers between June and September, and is not easily mistaken for any other species except *M. schaeferi*, known from north and east of the area of *M. nervosa*. See further under *M. schaeferi*.

15. Manulea corymbosa Linn. f., Suppl. 286 (1782); Thunb., Prodr. 102 (1800) & Fl. Cap. ed. Schultes 472 (1823); Benth. in Hook., Comp. Bot. Mag. 1: 381 (1836) & in DC., Prodr. 10: 363 (1846); Hiern in Fl. Cap. 4(2): 231 (1904). **Fig. 58D.**
Type: Cape of Good Hope, *Thunberg* (holo. sheet no. 14358, UPS, iso. S).

Syn.: *M. benthamiana* Hiern in Fl. Cap. 4(2): 226 (1904). Lectotype (chosen here): [Malmesbury div.] between Groenekloof and Saldanha Bay, *Drège* (K; isolecto. E, LE, TCD, W).

Annual herb, stems 30–450mm tall, solitary or up to 7 tufted from the base, often simple, sometimes with 1 or 2 branches near the base, glandular-puberulous with occasional eglandular hairs to 1mm as well, nude or distantly leafy. *Leaves* opposite and decussate, bases connate, crowded at base of stem, greatly reduced and subopposite to mostly alternate upwards; blade of radical leaves 9–60 × 5–25mm, narrowly to broadly elliptic, tapering to a petiolar part 3–20mm long (roughly ⅓ length of blade), margins shallowly to deeply and sharply toothed, both surfaces glandular-puberulous. *Flowers* many crowded in simple terminal racemes, capitate or subcapitate though a few distant flowers often straggle down the stem. *Bracts* (lowermost) 3–35 × 0.5–7mm, linear to leaflike. *Pedicels* (lowermost) 1–10mm long, shorter upwards. *Calyx* distinctly bilabiate, tube 0.2–0.5(–1.1)mm long, anticous lobes 3.8–4.5(–5.5) × 0.8–1.5mm, anticous lip split 3.5–4mm, posticous lobes 3.75–4.8 × 0.8–1mm, posticous lip split c.4mm, all lobes remarkably pouched in lower half and bulging outwards, tips narrow, strongly folded, curving out then in, margins densely bearded with eglandular hairs 0.5–1mm long, tiny glandular hairs as well, otherwise very nearly glabrous outside except for sparse ± sessile glands, hairy inside. *Corolla* tube 8.5–12 × 1–1.5mm in throat, cylindric, dilated near apex, limb 7–12mm across lateral lobes, nearly regular, posticous lobes 2.5–4 × 2.5–4.3mm, anticous lobe 3–4.5 × 2.5–4.5mm, all broadly obovate to subrotund, creamy-white with yellow/orange star-shaped patch around mouth, tube densely pubescent outside, hairs c.0.1–0.3mm in upper part, glistening glands on backs of lobes, inside a longitudinal posticous band of clavate hairs (usually dark-coloured between the posticous anthers) running up and extending round mouth and out onto the yellow patch. *Stamens* 4, anticous anthers 0.25–0.4mm long, just included, posticous anthers 0.8–1.25mm long, deeply included, posticous filaments bearded with dark hairs. *Stigma* 4–7mm long, shortly exserted or just included. *Style* 0.5–3mm long. *Ovary* 1.5–2 × 0.6–1.2mm, elliptic. *Capsules* c.4 × 2.5–3mm. *Seeds* c.0.5 × 0.4mm, violet-blue.

Selected citations:

Cape. Piquetberg div., 3218CD, between Berg River and Sauer, 15 ix 1953, *Lewis* 3880 (SAM). Malmesbury div., 3318CB, Mamre Hills, 31 viii 1944, *Compton* 15849 (NBG); Buck Bay near Darling, 16 ix 1980, *Hilliard & Burtt* 13014 (E). 3318AD, near Darling, 25 ix 1926, *Grant* 2568 (PRE). 3317BB, Hopefield and Hoetje's Bay, 100–200ft, ix 1905, *Bolus* 12789 (BOL). 3318AB, Hopefield, 150ft, 12 ix 1894, *Schlechter* 5302 (BOL, E, PRE, STE, UPS). 3318AC, Yzerfontein, 21 viii 1938, *Compton* 7323 (NBG). 3318AB, Baarhuis-Zonquasfontein boundary, c.300ft, 4 x 1977, *Thompson* 3503 (PRE, STE).

Why Hiern should have thought Bentham was mistaken in his concept of *M. corymbosa* is not clear: the specimens that Bentham saw do not differ from *M. corymbosa* in inflorescence (as implied by Hiern's separation of the names *M. benthamiana* and *M. corymbosa* in his key to the species of *Manulea*) nor is the throat of the corolla naked in *M. corymbosa* (so described by Hiern) as opposed to papillose - pubescent in *M. benthamiana*. Hiern's description of *M. benthamiana* embraces five different species: *M. corymbosa* (two collections made by Drège, and cited by Bentham), *M. bellidifolia* (*MacOwan* 1445), *M. altissima* subsp. *longifolia* (*Schlechter* 8070), *M. adenocalyx* (*Schlechter* 10750, *Bolus* 6230), *M. exigua* (*Zeyher* 3499). In view of this very considerable confusion in the use of the name *M. benthamiana*, it seems best to lectotypify it by one of the Drège specimens cited by Bentham as *M. corymbosa* and so reduce it to synonymy under *M. corymbosa*. The Drège collection from Groenekloof- Saldanha Bay has been chosen; although Thunberg's collection (the type of *M. corymbosa*) is not localized, it too probably came from near Saldanha Bay, as that is where Masson found it, when the two men were travelling together.

The salient feature of *M. corymbosa* is the calyx: not only is it remarkably bilabiate but each lobe is deeply pouched in the lower half then abruptly contracted into the long narrow strongly folded free upper portion, which curves out then slightly in again at the tip. The margins of the lobes are fringed with long eglandular hairs (which dry white) while the backs are glabrous or very minutely glandular (apart from any sessile glistening glands that may be present) though occasionally one or two long hairs may occur on the midline. This is in marked contrast to *M. adenocalyx*, in which the calyx lobes are only shallowly pouched and are glandular-pubescent all over the backs, with only a few long eglandular hairs sometimes present on the margins (see further under *M. adenocalyx*).

The name *M. benthamiana* has been much misused for *M. altissima* and *M. androsacea*, which are similar in facies to *M. corymbosa* and *M. adenocalyx* (all have mainly radical leaves, nude or nearly nude stems, subcapitate inflorescence) but differ in details of indumentum and floral structure: see further under those names.

Manulea corymbosa is confined to the low-lying sandy scrubby area from just north of the Great Berg river south to Camps Bay and the Cape Flats, flowering between July and November.

16. Manulea adenocalyx Hilliard in Notes RBG Edinburgh 46: 337 (1990).
Type: Cape, Piquetberg div., 3218DB, Piquenierskloof, 700ft, 6 viii 1897, *Schlechter* 10750 (holo. E; iso. BOL, W).

Annual herb, stems 100–450mm tall, often solitary or up to 4 from base, simple or 1–4 short side shoots, glandular-puberulous, sometimes with very sparse eglandular hairs as well to 0.5mm, nude or with a few distant much reduced leaves. *Leaves* opposite and decussate, bases connate, crowded at base of stem, any reduced stem leaves alternate; blade of radical leaves 16–50×5–18mm, narrowly to broadly elliptic, tapering to a petiolar part 3–15mm long (roughly up to ⅓ length of blade), margins shallowly to deeply and sharply toothed, both surfaces glandular-puberulous. *Flowers* many, crowded in simple terminal racemes, capitate or subcapitate though a few distant flowers often straggle down the stem. *Bracts* (lowermost) 6–27 × 0.5–2.5mm, mostly linear, occasionally leaflike. *Pedicels* (lowermost) 1–6(–15)mm long, shorter upwards. *Calyx* distinctly bilabiate, tube 0.2–0.75mm long, anticous lobes 4.3–9.5 ×

0.8–1.75mm, anticous lip split 4.5–7.5mm, posticous lobes 4.5–9.5 × 0.6–1.25mm, posticous lip split 3.5–6mm, all lobes somewhat pouched in lower half, tips narrow, almost straight, glandular-puberulous all over, hairs up to 0.3–0.4mm long, sometimes a few eglandular hairs 0.5–1mm long present as well, particularly on margins, hairy inside. *Corolla* tube 7–12 × 0.9–1.4mm in throat, cylindric, dilated near apex, limb 7–11mm across lateral lobes, nearly regular, posticous lobes 2–4 × 2.1–4.5mm, anticous lobe 2.5–4 × 2.5–4mm, all broadly obovate to subrotund, creamy white with yellow/orange star-shaped patch around mouth, tube densely pubescent outside, hairs c.0.1mm long, glistening glands on backs of lobes, inside a longitudinal posticous band of clavate hairs (usually dark-coloured between the posticous anthers) running up and extending round mouth and out onto the yellow patch. *Stamens* 4, anticous anthers 0.25–0.5mm long, just included, posticous anthers 0.75–1mm long, deeply included, posticous filaments bearded with dark hairs. *Stigma* 3.5–8.5mm long, shortly exserted or just included. *Style* 1.5–3.8mm long. *Ovary* 1.5–2.4 × 0.8–2mm. *Capsules* 4–5 × 2.5–3mm. *Seeds* 0.5 × 0.4mm, violet-blue.

Selected citations:

Cape. Clanwilliam div., 3219AB, Modderfontein, 700ft, 6 ix 1933, *Compton* 4284 (BOL, NBG); 3219CA, Citrusdal, Theerivier, 26 viii 1968, *Hanekom* 1158 (PRE); 3218BC, Lange Vallei, 31 viii 1935, *Compton* 5530 (NBG). Piquetberg div., 3218DA, Het Kruis, 29 ix 1943, *Compton* 15013 (NBG); Kapitein's Kloof, viii 1948, *Stokoe* sub SAM 63909 (SAM); 3218DD, near Piquetberg, 300ft, x 1895, *Bolus* 13624 (BOL). Ceres div., 3319BC, Karroo Poort, 26 viii 1935, *Compton* 5531 (NBG).

Manulea adenocalyx differs from *M. corymbosa* principally in its calyx: lower part of the calyx lobes not so markedly pouched, the lobes mostly longer (anticous lobes in *M. adenocalyx* 4.3–9.5mm long versus 3.8–4.5(–5.5)mm and densely glandular-puberulous all over the backs), long eglandular hairs on margins wanting or few (not glabrous on the backs except for minute glands, contrasting strongly with the heavily bearded margins).

Manulea adenocalyx ranges from Modderfontein near Wupperthal south and west to the environs of Clanwilliam, Citrusdal, Het Kruis and Karroo Poort. Its area thus lies north and east of that of *M. corymbosa*, possibly mostly at higher altitudes, but in similar habitats and flowering at the same time.

Schlechter 5123 (BOL, E, PRE, S, UPS), collected at Alexander's Hoek (between Clanwilliam and Piquetberg) on 1 ix 1894, and *Penther* 1781 (W) collected at 'Olifantsrivier', viii 1894 (he was surely travelling with *Schlechter*?) need special mention. The calyx is glandular all over as in *M. adenocalyx* but in addition it may be shaggy with eglandular hairs c.1mm long: there is variation in the degree of development of these hairs and they may be confined to the margins. The corolla tube may also be hirsute with hairs to 0.6mm long, or the hairs may be shorter. The indumentum on the calyx points to *M. adenocalyx*, but the calyx is small for that species (anticous lobes c.4 × 0.7mm) and so is the corolla tube (6–7.5mm). All the specimens are young. Field investigations are needed.

17. Manulea psilostoma Hilliard in Notes RBG Edinburgh 46:341(1990).
Type: Cape, Clanwilliam div., 3218BA, Graafwater, 20 vii 1941, *Bond* 1059 (holo. NBG).

Annual herb, stems 45–230mm tall, solitary or up to 6 tufted from the base, simple, glandular-puberulous, either nude or with 1 or 2 leaves. *Leaves* opposite and decussate, bases connate, crowded at base of stem, greatly reduced and alternate upwards; blade of radical leaves c.14–50 × 5–20mm, elliptic tapering to a petiolar part 2–22mm long (very roughly up to half length of blade, usually much shorter), margins obscurely to sharply toothed, both surfaces glandular-puberulous. *Flowers* many in a thyrse 20–40mm in diam., crowded and subcapitate at apex, in robust stems individual cymules straggling down stem almost to base, cymules dichasial. *Bracts* (lowermost, as distinct from reduced leaves) c.3–15 × 0.25–1.3mm, linear, glandular-puberulous. *Pedicels* (lowermost) c.2.5–5mm long, shorter upwards. *Calyx* distinctly bilabiate, tube

0.3–0.4mm long, anticous lobes 2.4–3.5 × 0.6–0.75mm, anticous lip split 1.5–2.3mm, posticous lobes 2.3–3.2 × 0.5–0.8mm, posticous lip split 2.2–3.2mm, tips of lobes subspathulate, minutely glandular-puberulous particularly on margins outside, occasional eglandular hairs to c.0.25mm on margins, very sparsely hairy inside. *Corolla* tube c.4.8–6 × 1mm in throat, cylindric, dilated near apex, limb c.8.5–10mm across the lateral lobes, nearly regular, posticous lobes 3–4.3 × 2.5–4.3mm, anticous lobe 3.5–4.5 × 2.5–3.5mm, all broadly elliptic to subrotund, white with yellow star-shaped patch around mouth, tube glandular-puberulous with occasional eglandular hairs to c.0.1mm, backs of lobes with very few glistening glands, inside a broad posticous band of clavate hairs running from ± point of insertion of posticous filaments and fanning out towards base of posticous lip, mouth glabrous. *Stamens* 4, anticous anthers 0.25–0.3mm long, tips just visible in mouth, posticous anthers 0.6mm long, included, posticous filaments bearded with dark hairs. *Stigma* 3.5–5.5mm long, well exserted. *Style* 1–1.75mm long. *Ovary* 1–1.5 × 0.5–0.8mm. *Capsules* (young) c.4 × 3mm. *Seeds* (young) c.0.5 × 0.4mm, violet-blue.

Citation:

Cape. Clanwilliam div., 3218BA, Graafwater, 20 vii 1941, *Esterhuysen* 5606 (BOL).

Manulea psilostoma is known from only two collections made simultaneously at Graafwater near Clanwilliam. The showy thyrsoid inflorescence at once distinguishes it from all other species in this group (viz. *M. adenocalyx, M. corymbosa, M. nervosa, M. schaeferi*) while the glabrous mouth sets it even further apart from *M. adenocalyx* and *M. corymbosa* to which it seems most closely allied. It is possibly partly sympatric with *M. adenocalyx* (their areas at least adjoin). The calyx of *M. psilostoma* is smaller than that of both *M. adenocalyx* and *M. corymbosa* (e.g. anticous lobes 2.4–3.5mm versus 4.3–9.5 and 3.8–5.5mm respectively) and the corolla tube shorter (c.4.8–6mm versus 7–12mm and 8.5–12mm respectively) and there are differences too in details of indumentum, particularly that of the calyx.

Two more specimens must be mentioned here. The first is *Compton* 10943 (NBG) also collected at Graafwater on 20 vii 1941, clearly on a joint outing with the Misses Esterhuysen and Bond. It bears some resemblance to *M. psilostoma* (thyrsoid inflorescence with reduced leaves) but the corolla lobes are ± oblong, the margins plane or somewhat revolute, the tips ± acute to obtuse, and there are four longitudinal bands of hairs in the throat. In these characters (and also in the thyrsoid inflorescence) it is reminiscent of *M. decipiens*, which not only occurs in the area, but was also collected by Compton (10937, NBG) on the same day, but at Alpha (precise locality unknown to me). *Compton* 10943 may well be a hybrid between *M. psilostoma* and *M. decipiens* though *M. adenocalyx, M. praeterita* and *M. annua* (as well as a number of other species) also occur in the environs of Clanwilliam. (*Compton* 11017, NBG, is *M. annua* from Uitkomst, again collected on 20 vii 1941). Clearly, field studies are needed, and as *M. psilostoma* itself is known only from Graafwater, its pedigree too needs investigation.

Section Dolichoglossa subsection Isantherae

18. Manulea augei (Hiern) Hilliard in Notes RBG Edinb. 46:49(1989).
Type: Cape, without locality, *Auge* (Nelson) (holo. BM).

Syn.: *Phyllopodium augei* Hiern in Fl. Cap. 4(2): 313 (1904).
    ?*Phyllopodium sordidum* Hiern in Fl. Cap. 4(2): 314 (1904). Type: Cape, Malmesbury div., neighbourhood of
        Hopefield, *Bachmann* 52 (holo. B†).
    Polycarena augei (Hiern) Levyns in J.S. Afr. Bot. 5: 36 (1939).
    P. sordida (Hiern) Levyns in J.S. Afr. Bot. 5: 36 (1939).

Annual herb, stems one to several from base, 100–150mm long, erect or decumbent, sparingly branched, minutely glandular-puberulous, ± distantly leafy. *Leaves* opposite and decussate at base, bases connate, soon alternate and passing imperceptibly into bracts, blade up to 17–37 × 6–12mm, smaller upwards, elliptic, tapering to a flat petiolar part 3–10mm long, margins sharply and narrowly toothed, both surfaces densely glandular-puberulous. *Flowers* many in simple terminal racemes, subcapitate at first, soon elongate, flowers distant in fruit. *Bracts* (lowermost) c.12–19 × 3–6mm, leaflike, diminishing somewhat in size upwards but raceme decidedly leafy. *Pedicels* (lowermost) 2–4mm long, shorter upwards. *Calyx* obscurely bilabiate, tube 0.75–2mm long, anticous lobes 3.2–6.8 × 0.8–1.8mm, glandular-puberulous all over, sparse coarse eglandular hairs to c.1mm as well. *Corolla* tube 7–9.5 × 1.2–1.5mm in throat, cylindric, widening near apex, limb 6–12mm across lateral lobes, nearly regular, posticous lobes 2–3.5 × 2–3.5mm, anticous lobe 2.4–5 × 1.8–4mm, all broadly elliptic, very obtuse, white (?) or pale mauve, yellow patch down back of tube, tube glandular-puberulous outside plus occasional eglandular hairs to 0.25–0.4mm, glistening glands on backs of lobes, inside a longitudinal posticous band of dark clavate hairs fanning out into pale hairs encircling mouth. *Stamens* 4, anticous anthers 0.6–0.8mm long, visible in mouth, posticous anthers 1mm long, included, posticous filaments bearded with dark hairs. *Stigma* 7–9mm long, well exserted. *Style* 1–3mm long. *Ovary* 1.5–2.5 × 0.75–1.3mm, elliptic. *Capsules* 4–5 × 3mm. *Seeds* 0.5 × 0.3–0.4mm, pallid.

Citations:

Cape. Malmesbury div., 3317BB, Hoetjies Bay, viii 1924, *Mathews* sub NBG 1579/23 (BOL), ibidem, Hoedjies Point, x 1938, *Walgate* s.n. (BOL); Coenvadenberg, xi, *Bachmann* sub Bolus 9202 (BOL); peninsula west of Langebaan, 11 x 1933, *Pillans* 6964 (BOL).

Hiern described *M. augei* as a species of *Phyllopodium*, but the bract adnate to the base of the pedicel, the structure of the eglandular hairs, and the disposition of the clavate hairs inside the corolla tube are all characteristic of *Manulea* and are foreign to *Phyllopodium*.

From Hiern's description (bract ... adhering to the lower part of the pedicel) it is clear that *Phyllopodium sordidum* is also a *Manulea*. The holotype was destroyed in the Berlin fire and no isotype bearing Bachmann's collection number has been traced. However, Bachmann presented to Bolus a specimen that is almost certainly the plant described by Hiern as *P. sordidum* and it came from 'Coenvadenberg' which Mrs Bean at Bolus Herbarium traced for me: there is a Coeratenberg, ± 2 miles south of Hopefield. There are two discrepancies between Hiern's description of *P. sordidum* and Bolus's specimen. Hiern described the inflorescence of *P. sordidum* as 'dense terminal racemes, racemes subhemispherical or globose'. In Bolus's specimen, only the young inflorescences fit this description: the racemes going into fruit are elongate. Then Hiern described the corolla tube as 'glabrous inside, a little dilated at the naked throat', which is at variance with Bolus's plant. This description can probably be discounted because Hiern also described the mouth in *M. augei* as glabrous, which is simply not true. There is little doubt in my mind that *P. augei* and *P. sordidum* are the same species: Hiern's discriminatory character in the key 'corolla lobes about ⅙ inch [4mm] long': *augei*, versus 'corolla lobes ¹⁄₂₄–¹⁄₁₂ inch [1–2mm]', leading to *sordidum*, fits the material perfectly, the corolla lobes being c.3.25–4mm long in Auge's specimen, 2–2.4mm in Bachmann's.

The only other specimens of *M. augei* seen by me came from Hoetjes Bay near Saldanha and the peninsula west of Langebaan; the species is possibly confined to the sandy lowlands between Saldanha Bay and Hopefield. It is allied to *M. annua*, a more laxly leafy species, glandular-pubescent rather than glandular-puberulous, with longer pedicels (4–20mm versus 2–4mm at base of inflorescence) and smaller flowers (e.g. calyx tube 0.2–0.5mm long versus 0.75–2mm, corolla tube 4–5.5mm long versus 7–9.5mm).

19. Manulea paucibarbata Hilliard in Notes RBG Edinb. 46:51(1989).
Type: Cape, Clanwilliam div., 3218DB, Boontjes Rivier, 2200ft, 25 viii 1896, *Schlechter* 8667 (holo. K; iso. BOL, E, MO, PRE, S, STE, W, Z).

Syn.: *Sutera annua* var. *laxa* Hiern in Fl. Cap. 4(2): 264 (1904).

Annual herb, stems erect or decumbent, 60–400mm long, solitary and simple at first, soon several- to many-stemmed and loosely branched from the base, glandular-pubescent, hairs to c.0.25mm, moderately leafy. *Leaves* at base of plant opposite and decussate, bases connate, soon alternate upwards and passing into bracts, lowermost blades c.17–42 × 5–22mm, lanceolate to elliptic in outline, tapering to a petiolar part 17–25mm long (roughly ⅓ to ½ length of blade), smaller upwards and petiolar part shorter, margins mostly coarsely and irregularly toothed, sometimes almost pinnatipartite, whole leaf glandular-pubescent. *Flowers* many in simple terminal racemes, usually solitary in each axil, sometimes paired, very lax, the axis usually markedly flexuous. *Bracts* (lowermost, and the lowermost flowers are often in the axils of the uppermost leaves) c.10–25 × 2–10mm, leaflike, rapidly smaller, linear to filiform, glandular-pubescent. *Pedicels* (lowermost) 8–40mm, shorter upwards. *Calyx* obscurely bilabiate, tube 0.2–1mm long, anticous lobes 2.5–4.5 × 0.5–0.7mm, posticous ones slightly smaller, glandular-pubescent with several-celled eglandular hairs to 0.3–0.4mm as well. *Corolla* tube 5.5–9 × 1–2mm in throat, ± narrowly funnel-shaped, limb 6–13mm across lateral lobes, posticous lobes 2.4–5 × 1.5–3.5mm, anticous lobes 2.6–5.5 × 1.6–4mm, all oblong to elliptic, obtuse, white or rarely flushed with pinkish-mauve, yellow patch down back of tube, tube glandular-puberulous outside plus eglandular hairs to 0.1–0.25mm long, glistening glands on backs of lobes, inside a short posticous band of clavate hairs, often dark-coloured in lower half, and three short lateral bands, the hairs fanning out round mouth, these hairs pale, c.0.1–0.2mm long. *Stamens* 4, anticous anthers 0.5–1mm long, shortly exserted, posticous anthers 0.7–1.4mm long, included, all filaments glabrous. *Stigma* 3.5–8.2mm long, well exserted. *Style* 0.5–2.5mm long. *Ovary* 1–1.5 × 0.75–1mm. *Capsules* 4–5 × 2.5–3mm. *Seeds* 0.5–0.6 × 0.4mm, pallid.

Selected citations:

Cape. Clanwilliam div., 3118DB, Boterkloof, 29 viii 1941, *Compton* 11499 (NBG, PRE); 3118DC, Nardouw Road, 2 ix 1945, *Barker* 3626 (NBG); 3218BB, Pakhuis Pass, c.1500ft, 23 viii 1967, *Thompson* 323 (K, PRE); 3218BB, near Clanwilliam, 600ft, 2 x 1897, *Bolus* 9068 (BOL, K, NBG, PRE, STE); 3219 AA, N Cedarbergen, Krakadouw Poort, 27 x 1945, *Esterhuysen* 12262 (BOL).

This plant was originally described as *Sutera annua* var. *laxa*, but, the epithet *laxa* already being in use in *Manulea*, a new epithet, *paucibarbata*, was chosen: one of the most easily observed differences between *M. annua* and *M. paucibarbata* is the weak bearding around the corolla mouth of the latter species (hairs sparse, up to 0.2mm long versus plentiful, 0.4–0.6mm long). The epithet *laxa* was also apt, because the terminal flowers in the inflorescence are never crowded as they are in *M. annua* and the racemes are very lax with the axis usually markedly flexuous (straight or weakly flexuous in *M. annua*). These characters, together with the mostly longer pedicels and larger flowers, as well as the glabrous filaments, combine to separate *M. paucibarbata* from *M. annua* at species, not varietal, level. It should be pointed out that the corolla tube in the type material of *M. paucibarbata* is short, c.4.75mm long, but the only flowers are either depauperate or young ones at the tips of fruiting axes.

Manulea paucibarbata is also allied to *M. arabidea*. They share the characters of short hairs bearding the mouth and glabrous filaments, but *M. arabidea* has the flowers initially crowded in the inflorescence and the axis is never flexuous; also, the flowers are much smaller (e.g. corolla tube 2.75–4mm long versus 5.5–9mm).

Manulea paucibarbata occupies a relatively small area, from Boterkloof and Nardouw Pass, south of Van Rhynsdorp, to Clanwilliam and the Cedarberg, thence south to Boontjesrivier. Ecological notes are scanty, but the plants probably spring up in sandy or stony terrain after rain, between August and October. The altitudinal range lies between c.180 and 900m above sea level. Whether or not *M. paucibarbata* is wholly sympatric with *M. annua* is not clear: the specimens seen indicate an area between 3118DB and 3218DB for *M. paucibarbata*, between

3218BB and 3318BB for *M. annua*. *Manulea arabidea* is known only from 3219AA, and is sympatric with *M. paucibarbata*.

20. Manulea annua (Hiern) Hilliard in Notes RBG Edinb. 46:49(1989).
Lectotype: Cape, Clanwilliam div., 3218BA, Graafwater, 1000ft, 18 viii 1896, *Schlechter* 8567 (K; isolecto. BOL, E, MO, PRE, S, W, Z).

Syn.: ?*Chaenostoma divaricatum* Diels in Bot. Jahrb. 23: 476 (1897). Type: Cape, Calvinia div., Hantam Mts, *Meyer* s.n. (B†).
　Sutera annua Hiern in Fl. Cap. 4(2): 264 (1904) p.p. and excluding var. *laxa* Hiern.
　?*S. divaricata* (Diels) Hiern in Fl. Cap. 4(2): 281 (1904).

Annual herb, stems erect or ascending, 30–300mm long, solitary and simple at first, soon several- to many-stemmed and loosely branched from base, glandular-pubescent, hairs to c.0.25mm, moderately leafy at base of plant. *Leaves* opposite and decussate, bases connate, soon alternate upwards and passing imperceptibly into bracts, lowermost blades to 12–45 × 6–18mm, lanceolate to elliptic in outline, tapering to a petiolar part 5–35mm long (roughly ½–⅓ length of blade), smaller upwards and petioles relatively shorter, margins mostly deeply and irregularly toothed, sometimes almost pinnatipartite, occasionally only obscurely toothed or subentire, whole leaf glandular-pubescent. *Flowers* many in simple terminal racemes, usually solitary in each axil, occasionally paired, often ± crowded at first, soon distant. *Bracts* (lowermost) 4–22 × 0.5–4mm, mostly leaflike, rapidly smaller upwards, linear to filiform, glandular-pubescent. *Pedicels* (lowermost) 4–20mm long, shorter upwards. *Calyx* obscurely bilabiate, tube 0.2–0.5mm long, anticous lobes 2–4 × 0.4–0.8mm, posticous ones slightly smaller, oblong-lanceolate, glandular-pubescent with several-celled acute hairs to 0.4–0.6mm as well. *Corolla* tube 4–5.5(–7) × 1–1.5mm in throat, ± narrowly funnel-shaped, limb 4–8mm across lateral lobes, posticous lobes 1.5–3 × 1.1–2.5mm, anticous lobes 1.6–3.5 × 1–2.75mm, all oblong-elliptic, obtuse, white, yellow patch down back of tube, tube glandular-puberulous outside plus eglandular hairs to 0.1–0.3mm, glistening glands on backs of lobes, inside a short longitudinal posticous band of clavate hairs, often dark-coloured, running out onto base of posticous lip and encircling mouth, these hairs pale, c.0.4–0.6mm long. *Stamens* 4, anticous anthers 0.35–0.5mm, visible in mouth, posticous anthers 0.5–0.8mm, posticous filaments dark-bearded. *Stigma* 3–6mm long, well exserted. *Style* 0.75–1.3mm. *Ovary* 1–1.7 × 0.6–1mm. *Capsules* 2.5–4 × 2–3mm. *Seeds* 0.4–0.5 × 0.3–0.4mm, pallid.

Selected citations:

Cape. Clanwilliam div., 3218BA, 3½ miles E by N of Graafwater, c.1100ft, 22 viii 1958, *Acocks* 19644 (K, M, PRE). 3218DB, Oliphants River valley near Warmbaths, 21 ix 1911, *Stephens* 6869 (BOL). 3219AC, top of Nieuwoudt Pass, 27 ix 1942, *Stokoe* sub Esterhuysen 8134 (BOL). 3219CA, Elandskloof bridge 10 miles SE of Citrusdal, *Maguire* 1811 (NBG). Piquetberg div., 3218CB, kloof at foot of Grey's Pass, 1 x 1952, *Esterhuysen* 20447 (BOL, PRE). 3218DA, Berg River near Piquetberg, 200ft, 10 ix 1894, *Schlechter* 5269 (BOL, E, K, PRE, S, STE, UPS).

Hiern described this plant as a species of *Sutera*, but the combination of bract adnate to the base of the pedicel, posticous band of hairs, stamens nearly included, and long lingulate stigma are characteristic of species of *Manulea*, not of *Sutera*, even in the broad circumscription Hiern accorded the genus.

　Hiern cited five collections under the name *Sutera annua*, namely *Schlechter* 5060, 8022, 8567 and 10768, and *Bolus* 9068. All but *Schlechter* 8567 and 10768 prove to be the same plant as *Schlechter* 8667, type of *Sutera annua* var. *laxa*, that is, *Manulea paucibarbata*. *Schlechter* 8567 has been chosen as the lectotype of *M. annua*: several sheets of this number have been written up by Schlechter himself with the epithet 'annuum' but under the genus *Chaenostoma*.

　The name *Chaenostoma divaricatum* is included in the synonymy with reservation. The type was destroyed in the Berlin fire. Diels's description seems applicable to both *Manulea annua* and *M. paucibarbata*; because certainty is impossible, I am adopting the pragmatic course of maintaining the two younger names.

The relationship of *M. annua* clearly lies with *M. arabidea* and its allies. It can be distinguished from *M. arabidea* not only by its bearded posticous filaments but also by its longer pubescence and mostly larger flowers (anticous calyx lobes 2–4mm long versus 1.25–2mm, corolla tube 4–7mm long versus 2.75–4mm). *Manulea paucibarbata* can at once be distinguished by the scant short hairs in the throat and glabrous filaments (see further under *M. paucibarbata*).

Manulea annua ranges from about Pakhuis Pass and the northern Cedarberg to Citrusdal, the Oliphants River valley and Piquetberg, with its southernmost station at Porterville. It seems to be equally at home on hillslopes and valley bottoms, probably in damp sandy places, flowering between July and October. Its altitudinal range is not clear, but it possibly lies between c.60 and 600 m.

21. Manulea arabidea [Schltr. ex] Hiern in Fl. Cap. 4(2): 227 (1904).
Type: Cape, Clanwilliam div., 3219AA, Lamm Kraal, 1000ft, 14 viii 1897, *Schlechter* 10848 (holo. BOL; iso. E, K, PRE, S, W).

Annual herb, stems several tufted from base or solitary in young plants, 30–210mm tall, simple or sparingly and distantly branched, minutely glandular-puberulous, occasional eglandular hairs to 1mm as well, distantly leafy. *Leaves* opposite and decussate, bases connate, 2 pairs at base of plant then distant upwards at forking of stems, blade up to 7–27 × 3–8mm, elliptic, tapering to a petiolar part 4–15mm long (very roughly half to equalling length of blade), margins entire to shallowly toothed, both surfaces minutely glandular-puberulous. *Flowers* many in simple terminal racemes, crowded at first, distant in fruit. *Bracts* (lowermost) c.4–11 × 1–3mm, leaflike or lanceolate, smaller and linear-lanceolate to linear upwards, glandular-puberulous, a few several-celled, balloon-tipped hairs to 0.4mm as well. *Pedicels* (lowermost) 4–11mm long, shorter upwards. *Calyx* obscurely bilabiate, tube 0.4–0.5mm long, anticous lobes 1.25–2 × 0.4–0.5mm, posticous ones slightly smaller, subspathulate, tips often incurved, pubescent all over with tiny gland-tipped hairs and several-celled, balloon- tipped ones to 0.4mm. *Corolla* tube 2.75–4 × 1.1–1.3mm in throat, ± funnel-shaped, limb 5–6mm across lateral lobes, posticous lobes 1.5–2 × 1.25–1.6mm, anticous lobe 2–2.5 × 1.3–1.75mm, all oblong, very obtuse, white, yellow patch down back of tube, tube minutely glandular-puberulous outside, ± sessile balloon-tipped hairs on backs of lobes, inside a longitudinal posticous band of clavate hairs running out onto lower part of posticous lip and extending around mouth. *Stamens* 4, anticous anthers 0.3–0.4mm long, visible in mouth, posticous anthers 0.4–0.6mm, included, all filaments glabrous. *Stigma* 2–3.4mm long, well exserted. *Style* 0.5mm. *Ovary* 1–1.5 × 0.6– 1mm, subrotund. *Capsules* c.3.5 × 3mm. *Seeds* (immature) 0.4 × 0.3mm, pallid or violet-blue (possibly only when very young).

Citation:

Cape. Clanwilliam div., [3218BB?], Jan Dissel's Rivier, 12 viii 1897, *Leipoldt* 588 (NBG, SAM).

Schlechter collected the type material at Lamkraal, about 10km due east of the Brandewyn river, on the Pakhuis to Calvinia road, while Leipoldt's specimen came from Jan Dissel's river, possibly not far from Clanwilliam, collected on a sandy riverbank in the same year and almost on the same day as Schlechter's. I have seen no other material.

Manulea arabidea is allied to *M. paucibarbata* and *M. annua* with which it is sympatric: see further under these two species. For a possible hybrid between *M. arabidea* and *M. praeterita* see under the latter species.

22. Manulea calciphila Hilliard in Notes RBG Edinb. 46:50(1989). **Fig. 58B.**
Type: Cape, Bredasdorp div., 3420CA, near Bredasdorp, 15 x 1951, *Esterhuysen* 19131 (holo. BOL).

Annual herb, stems several tufted from base, 20–150mm long, erect or ascending, simple or sparingly and distantly branched, glandular-pubescent, hairs to 0.2(–0.3)mm, distantly leafy or sometimes leafless. *Leaves* opposite and decussate, bases connate, mostly ± radical, distant upwards at forking of stems, blade up to 7–35 × 4–15mm, narrowly to broadly elliptic tapering into a flat petiolar part 4.5–22mm long (roughly half to equalling length of blade), margins entire to distinctly toothed, both surfaces glandular-puberulous. *Flowers* many in simple terminal racemes, crowded at first, distant in fruit. *Bracts* (lowermost) 2–10 × 0.4–3.5mm, sometimes leaflike, mostly lanceolate to linear, smaller upwards, glandular-pubescent, a few several-celled, balloon-tipped hairs to 0.2mm as well. *Pedicels* (lowermost) 1.5–4(–6)mm, shorter upwards. *Calyx* obscurely bilabiate, tube 0.4–0.6mm long, anticous lobes 2–3 × 0.6–0.8mm, posticous ones slightly smaller, subspathulate, glandular-pubescent all over, several-celled, balloon-tipped hairs to 0.4–0.5mm as well mainly on margins. *Corolla* tube 3.2–4 × 1–1.8mm in throat, ± funnel-shaped, limb 4–6mm across lateral lobes, posticous lobes 1.4–2 × 1.3–2.1mm, anticous lobe 1.5–2.4 × 1–2mm, all oblong-obovate, very obtuse, white, yellow patch down back of tube, tube glandular-puberulous, several-celled balloon-tipped hairs to 0.1mm as well, ± sessile balloon-tipped hairs on backs of lobes, inside a longitudinal posticous band of clavate hairs running out onto base of posticous lip and extending around mouth. *Stamens* 4, anticous anthers 0.4–0.6mm long, visible in mouth, posticous anthers 0.5–0.75mm, included, posticous filaments bearded. *Stigma* 2–3.5mm long, well exserted. *Style* 0.5–1mm. *Ovary* 0.8–1.1 × 0.6–0.8mm, subrotund. *Capsules* 2.5–3 × 1.5–2mm. *Seeds* 0.4–0.5 × 0.3mm, blue.

Citations:

Cape. Namaqualand, 3017BC, c.8km from Wallekraal on road to Kamieskroon, 17 ix 1991, *Goldblatt* 9232 (E). Bredasdorp div., 3419BD, Zoetendalsvlei, c.100ft, ix 1926, *Smith* 3096 (PRE); 3419DB, The Poort between Bredasdorp and Elim, 24 ix 1933, *Leighton* BOL 21105 (BOL); ibidem, c.100ft, 11 xi 1962, *Acocks* 22996 (PRE); ibidem, 25 ix 1933, *Acock* 1523 (S); 3420CA, De Hoop, 80ft, 21 ix 1984, *Van Wyk* 1972 (PRE); ibidem, 1971, *Dixon* 135 (STE); De Hoop - Potberg Nature Reserve, 45m, 26 vii 1979, *Burgers* 2037 (PRE, STE); 3420BD, De Hoop-Witwater, 6 viii 1984, *Van Wyk* 1698 (STE); 3420CA, Die Mond, 8 x 1950, *Compton* 22204 (NBG); Arniston, c.100ft, 24 viii 1962, *Acocks* 22614 (PRE). Riversdale div., 3421AD, Still Bay hills, 4 viii 1951, *Johnson* 119 (NBG). Swellendam div., 3320DC?, Bushman's River, 25 ix 1935, *Compton* 8194 (NBG). Without precise locality, *Ecklon & Zeyher* sub SAM 48293 (SAM).

Manulea calciphila is endemic to limestone outcrops (whence the trivial name) and has been recorded from The Poort, between Elim and Bredasdorp in the west, to Still Bay in the east, with a disjunction to Namaqualand. It grows in damp sandy soil not much above sea level and flowers between July and November, when it is passing into fruit. The species is frequently misdetermined as *M. minor*, *M. fragrans* or *M. corymbosa* (also as *Polycarena*). The well exserted stigma and pubescent corolla mouth distinguish it at a glance from *M. minor* and *M. fragrans*, while the calyx of *M. corymbosa* is very different from that of *M. calciphila*, and the corolla tube is much longer.

The relationship of *M. calciphila* lies with *M. arabidea*, to which it bears a strong resemblance. It can however at once be distinguished by its bearded posticous filaments. Also, the pedicels are mostly shorter (1.5–4(–6)mm versus 4–11mm), the anticous calyx lobes are mostly larger (2–3 × 0.6–0.8mm versus 1.25–2 × 0.4–0.5mm), and the bracts and corolla tube in particular are more markedly glandular.

23. Manulea karrooica Hilliard in Notes RBG Edinb. 46: 50 (1989).
Type: Cape, 3221DD, Prince Albert div., 2 miles S of Kruidfontein station, bed of Gamka river, c.1800ft, 24 viii 1963, *Acocks* 23330 (holo. PRE).

Annual herb, stems several tufted from base, erect to 40–130mm, simple, minutely glandular-puberulous with almost sessile balloon-tipped hairs as well. *Leaves* opposite and decussate, bases connate, crowded at base of plant, blade up to 7–15 × 3–10mm, elliptic or oblanceolate tapering into a petiolar part about equalling the blade, margins subentire to distinctly toothed, both surfaces glandular-puberulous. *Flowers* many in simple terminal racemes, distant. *Bracts* (lowermost) 2 × 0.5mm, smaller upwards, lanceolate, glandular-puberulous on margins only, occasional ± sessile balloon-tipped hairs on back. *Pedicels* (lowermost) 2.5–5mm long, shorter upwards. *Calyx* obscurely bilabiate, tube 0.2–0.3mm long, anticous lobes 2–3 × 0.4–0.8mm, posticous ones slightly smaller, oblong, tips subspathulate or not, margins minutely glandular-puberulous, occasional ± sessile balloon-tipped hairs on backs. *Corolla* tube 2.3–3.5 × 0.8–1.3mm in throat, ± funnel-shaped, limb 4–6mm across lateral lobes, these ascending, the anticous lobe thus isolated from the other four, posticous lobes 1.2–2.5 × 0.6–1.3mm, anticous lobe 1.5–2.8 × 0.8–1.7mm, all lanceolate-oblong, subacute, cream to pale sulphur yellow, tube with very few ± sessile balloon-tipped hairs, these more plentiful on backs of lobes, inside a longitudinal posticous band of clavate hairs running out onto lower part of posticous lip and extending to lateral lobes or right round mouth. *Stamens* 4, anticous anthers 0.25–0.5mm long, visible in mouth or very shortly exserted, posticous anthers 0.4–0.6mm long, included or tips just visible in mouth, posticous filaments glabrous. *Stigma* 2–3mm long, well exserted. *Style* 0.4–1.5mm long. *Ovary* 1.2–1.5 × 0.8–1mm, subrotund. *Capsules* 3–4 × 2mm. *Seeds* c.0.5 × 0.3mm, white.

Citations:

Cape. Graaff Reinet div., 3224BC, Graaff Reinet, 2000–3000ft, *Ecklon* (SAM). Prince Albert div., 3221 DC, 3½ miles N of Prince Albert Road station, c.2200ft, 19 ix 1953, *Acocks* 17110 (PRE); Prince Albert, 3322AA, 6 viii 1929, *Erasmus* sub Marloth 13696 (PRE). Beaufort West div., 3222BC, Beaufort West, townlands west of Walker Dam, 14 ix 1979, *Bohnen* 6511 (STE). Fraserburg div., 3221BB, Layton, 3000ft, 20 viii 1965, *Shearing* 1075 (PRE); ibidem, *Shearing* 1074 (PRE); ibidem, 3500ft, 24 vii 1967, *Shearing* 92 (PRE).

Manulea karrooica is known only from the environs of Graaff Reinet, Beaufort West, Fraserburg, Prince Albert Road and Prince Albert; clearly it must be widely distributed in the Great Karroo but, being so small, it is easily overlooked. Acocks recorded 'stony ridges, rare' and 'bed of Gamka river', Shearing stated 'garden' and 'on top of a rant' while Bohnen noted 'karroo vegetation on sand'.

The relationship of *M. karrooica* lies with *M. latiloba* from which it can be distinguished by its longer and narrower calyx lobes (anticous lobes 2.2–3 × 0.4–0.8mm versus 1.5–2 × 0.6–1mm), mostly longer corolla tube (2.3–3.5mm versus 2–2.5mm), and the differently shaped limb: in *M. karrooica* the lateral lobes ascend and lie close to the two posticous lobes, leaving the anticous lobe isolated; the mouth is thus very oblique. In *M. latiloba* the lateral lobes lie at right angles to the anticous lobe (as is usual in this whole subsection) and the mouth is round. Both species inhabit karroid terrain, but the area of *M. karrooica* lies further east than that of *M. latiloba*.

Manulea karrooica bears a strong superficial resemblance to *M. pusilla*, with which it shares the character of a '4 + 1' limb with oblique mouth, but *M. karrooica* is easily distinguished by its well exserted stigma (deeply included in *M. pusilla*), anthers nearly equal in size, and prominently bearded upper lip and mouth (three discrete posticous longitudinal bands of hairs just visible in mouth of *M. pusilla*). It has also been much confused with *M. fragrans*, but again, characters of stigma, anthers and disposition of clavate hairs will distinguish it; also, *M. fragrans* has the limb nearly regular.

24. Manulea latiloba Hilliard in Notes RBG Edinb. 46:50(1989).
Type: Cape, 3219DD, Ceres div., Kareekolk in southern Tanqua Karroo, 1800ft, 1 ix 1973, *Oliver* 4376 (holo. STE, iso. PRE).

Annual herb, stems several tufted from base, 10–120mm long, simple, erect, minutely glandular-puberulous, leafless. *Leaves* opposite and decussate, bases connate, crowded at base and appearing rosulate, blade up to 6–24 × 3–15mm, oblanceolate to obovate, tapering into a flat petiolar part 4–18mm long (very roughly half length of blade), margins entire to shallowly toothed, both surfaces minutely glandular-puberulous. *Flowers* many in simple terminal racemes, crowded at first, distant in fruit. *Bracts* (lowermost) 1.5–3 × 1mm, shorter upwards, lanceolate, minutely glandular-puberulous. *Pedicels* (lowermost) 1.5–6.5mm long, shorter upwards. *Calyx* obscurely bilabiate, tube 0.2–0.4mm long, anticous lobes 1.5–2 × 0.6–1mm, posticous ones slightly smaller, oblong-lanceolate, minutely glandular-puberulous on margins only, sparse ± sessile balloon-tipped hairs on tube and backs of lobes. *Corolla* tube 2–2.5 × 1.2–2mm in throat, ± funnel-shaped, limb 4–5mm across lateral lobes, posticous lobes 1–2 × 1–1.1mm, anticous lobe 1.4–2.5 × 0.8–1.5mm, all ovate, obtuse to subacute, white, yellow patch down back of tube, tube minutely glandular-puberulous to almost glabrous outside, always with ± sessile balloon-tipped hairs on backs of lobes, inside a longitudinal posticous band of clavate hairs running out onto lower half of posticous lip and often right round mouth. *Stamens* 4, anticous anthers 0.3–0.5mm long, shortly exserted, posticous anthers 0.3–0.5mm long, visible in mouth, all filaments glabrous. *Stigma* 1.1–2mm long, well exserted. *Style* 0.4–0.8mm. *Ovary* 1–1.2 × 0.6–1mm, subrotund. *Capsules* 2.5–3 × 1.8–2.25mm. *Seeds* c.0.5 × 0.4mm, violet-blue, grey-blue, or creamy-coloured.

Citations:

Cape. 3219DD, Ongeluks River, 18 vii 1811, *Burchell* 1229 (K). 3319AD, Ceres Karroo, 3000ft, ix 1931, *Compton* 3804 (BOL), ibidem, 1 x 1932, *Compton* 4075 (BOL); 3319BB, Flats west of Inverdoorn, 700m, 16 ix 1971, *Thompson* 1264 (STE); 3319BC, 10 miles from Karroo Poort, 1800ft, 24 ix 1962, *Levyns* 11350 (BOL); 3320DC?, Swellendam, Bushman's River, 25 ix 1935, *Compton* 5707 (NBG); c.3321CA, Ladismith road to Riversdale, 1800ft, vii 1967, *Levyns* 11616 (BOL).

Manulea latiloba is distinguished from *M. arabidea* by its leafless flowering stems, somewhat broader anticous calyx lobes (c.0.6–1mm versus 0.4–0.5mm) clad in minute hairs only (not with some hairs to 0.4mm), shorter corolla tube (2–2.5mm versus 2.75–4mm), subacute corolla lobes (versus very obtuse), and all four anthers visible in the mouth (not just the anticous pair). The trivial name refers to the calyx lobes.

The area of *M. latiloba* lies south east of that of *M. arabidea*, ranging from Kareekolk in the southern Tanqua Karroo, south west to the Ceres Karroo and south east to karroid areas abutting on the Langeberge, between c.550 and 900m above sea level. The plants grow in damp sand or shaly ground, flowering between July and early October. Burchell was the first to collect the species, in 1811, but Hiern mistook it for *M. fragrans*; over a hundred years elapsed before a second collection was made, by Compton.

25. Manulea derustiana Hilliard in Edinb. J. Bot. 49: 297 (1992).
Type: Cape, Oudtshoorn distr., 3322CB, De Rust, north of Le Roux station, 1800ft, 5 viii 1990, *Viviers & Vlok* 479 (holo. E; iso. K, NBG, S).

Annual herb, stems several tufted from base, c.100–200mm long, simple, erect, glandular-puberulous, sometimes a very few balloon-tipped hairs as well, up to 0.25mm long, leafless. *Leaves* opposite and decussate, bases connate, crowded at base and appearing rosulate, blade up to 12–30 × 3.5–12mm, elliptic tapering into a flat petiolar part up to 5–25mm long (very roughly half length to equalling length of blade), margins coarsely and irregularly toothed, both surfaces minutely glandular-puberulous. *Flowers* many in simple terminal racemes, crowded at first, distant in fruit. *Bracts* (lowermost) c.1.5–4 × 0.2–0.4mm, shorter upwards, linear-lanceolate, glandular-puberulous. *Pedicels* (lowermost) c.3–6mm long, shorter upwards. *Calyx* obscurely bilabiate, tube c.0.4–0.5mm long, anticous lobes 2.5–3 × 0.8mm, posticous ones slightly smaller, oblong-lanceolate, glandular-puberulous, hairs up to c.0.1–0.15mm long, occasional balloon-tipped hairs up to 0.25mm long on backs and margins. *Corolla* tube c.4 × 2mm in throat, ± funnel-shaped, limb c.7mm across lateral lobes, posticous lobes c.2 × 2mm, anticous lobe c.2.5–2.7 × 1.8–2mm, all suborbicular to broadly elliptic, obtuse, white, yellow patch down back of tube, tube minutely glandular-puberulous, sessile glands on backs of lobes, inside a longitudinal posticous band of clavate hairs fanning out right round throat and onto bases of all 5 lobes. *Stamens* 4, anticous anthers 0.7mm long, shortly exserted, posticous anthers 0.8mm long, visible in mouth, all filaments glabrous. *Stigma* c.1.7mm long. *Style* c.2.8mm. *Ovary* 1.3 × 1mm, subrotund. *Capsules* c.4 × 2.5mm. *Seeds* c.0.7 × 0.4 (young), bright violet.

Manulea derustiana is known only from the type collection. It was found on a south-facing slope, in deep sandy soil in disturbed Little Karoo vegetation, in full flower early in August.

It is allied to *M. latiloba*, which differs in its relatively broader calyx lobes and shorter corolla tube, as well as in the lack of relatively long balloon-tipped hairs on the calyx. De Rust lies well east of the area of *M. latiloba*, whose nearest known station was recorded as 'Ladismith road to Riversdale'.

Section Thyrsiflorae

26. Manulea tomentosa (Linn.) Linn., Mant. alt. 420 (1771); Thunb., Prodr. Pl. Cap. 101 (1800) & Fl. Cap. ed. Schultes 470 (1823); Link & Otto, Ic. Pl. Sel. t. 19 (1821); Benth. in Hook., Comp. Bot. Mag. 1: 383 (1836) & in DC., Prodr. 10: 365 (1846); Hiern in Fl. Cap. 4(2): 238 (1904).
Type: Herb. Sloane vol. 102 folio 72, top right specimen (BM).

Syn.: *Selago tomentosa* Linn., Pl. Afr. Rar. 13 (1760) & Amoen. Acad. 6: 90 (1764).
Lychnidea tomentosa (L.) Moench, Suppl. Meth. Pl. 160 (1802).

Perennial, but capable of flowering in seedling stage, branching from the base and above, old plants becoming bushy shrublets, base woody and there up to c.8mm in diam., stems c.30–600mm long, erect or decumbent, villous with spreading white acute hairs up to 1–2mm long, the inflorescence axes also usually villous, but the hairs occasionally sparse, normally leafy throughout. *Leaves* opposite, bases connate, often crowded, sometimes becoming more distant upwards, uppermost often alternate and passing into inflorescence bracts, c.7–45 × 5–25mm (12–75 × 5–15mm), slightly smaller upwards, broadly spathulate to obovate, apex rounded (oblanceolate, apex rounded to subacute), base tapering to a broad petiolar part, margins serrate to crenate, both surfaces villous, hairs acute, up to 0.5–1.5mm long, gland-tipped hairs wanting or ± sessile (stalked gland-tipped hairs occasional, much smaller than acute hairs). *Flowers* many in a dense (sometimes ± lax) terminal thyrse c.30–140 (–300)mm long, panicled in well-branched plants, cymules dichasial, many-flowered, crowded. *Bracts* (lowermost) c.3–20 × 0.5–7mm, either ± lanceolate, or the larger ones somewhat leaflike, or the lowermost few cymules in the axils of the uppermost leaves. *Peduncles* (of lowermost cymules) 2–5(–8)mm long, pedicels very short or flowers sessile. *Calyx* very obscurely bilabiate, tube 0.25–0.6mm long, lobes 1.75–4.5 × 0.4–0.8mm, subspathulate, the tips somewhat thickened, villous on backs and margins or hairs sometimes nearly or quite confined to margins, hairs to 0.6–0.75mm long, acute, much smaller glandular hairs also present. *Corolla* tube 6.5–10 × 1–1.5mm in throat, cylindric, inflated at apex, limb obscurely bilabiate, 5.5–8mm across lateral lobes, posticous lobes 2.5–4 × 2–3.5mm, anticous lobes 2–3.5 × 1.5–3mm, all broadly elliptic to subrotund, margins often revolute, dull brownish-yellow to deep dull orange, or sometimes red-brown, pubescent outside, hairs up to 0.25–0.3mm long, occasionally sparse in lower part, inside 4–5 strong longitudinal bands of clavate hairs in throat. *Stamens* 4, anticous anthers 0.4–0.5mm long, just included, posticous anthers 0.8–1mm long, included, posticous filaments bearded. *Stigma* 0.75–1.75mm long, shorter than style, included. *Style* 1.5–2.6mm long. *Ovary* 1–1.75 × 0.6–1mm, conical. *Capsules* c.3.5–7 × 2–3.5mm. *Seeds* c.0.5 × 0.3mm, violet-blue.

Selected citations:

Cape. Malmesbury div., 3217DB, Stompneus Point, 4 ix 1955, *Taylor* 1530 (SAM). 3317BB, Hoetjes Bay, ix 1905, *Bolus* 12788 (BOL). Cape Town div., 3318CD, Blaauwberg Beach, 12 x 1978, *Boucher* 3978 (PRE, STE). 3418AD,

Olifant's Bos, 25 x 1953, *Compton* 24443 (NBG). 3418AB, Fish Hoek, x 1888, *MacOwan* 2974 (E); Nordhoek, 3 x 1980, *Hilliard & Burtt* 13095 (E). Caledon div., 3418BD, Betty's Bay, 16 x 1949, *Martin* 773 (NBG).

The essential features of typical *M. tomentosa* are its villous (hairs acute) stems and inflorescence axes, obtuse leaves running up to the base of the dense many-flowered inflorescence, corolla tube hairy all over outside and with strong longitudinal bands of clavate hairs inside the throat, and corolla lobes almost or quite as long as broad with rounded tips (the true breadth often obscured by the revolute margins).

The species is common on sand dunes along the seashore from about Stompneus Point north of Saldanha Bay to Pearly Beach in Bredasdorp division, between Danger Point and Quoin Point. The principal flowering season is August to December. When growing close to the sea, the plants are compact, closely leafy, the leaves thick-textured and merging into inflorescence bracts, and the thyrses are very dense. Such was the plant illustrated by Plukenet in 1691 and described by Linnaeus in 1760. In presumably more sheltered positions either near the sea or on sandy flats and slopes away from the shore, the plants are more lax with leaves thinner in texture and becoming somewhat distant upwards and often longer, narrower and more acute; the thyrse too may be less compact and the axis only very thinly hairy or nearly glabrous, and the hairs on the outside of the corolla tube may be confined to the inflated part. In the formal description, the measurements and characters in parenthesis refer to such plants, which retain the characters of bushy habit, stems leafy to the top, crowded thyrse with many-flowered cymules, and corolla lobes almost as broad as long. That Linnaeus himself recognized this plant as *M. tomentosa* is borne out by a specimen from Van Royen's herbarium (now in S) bearing the name *Selago tomentosa* in Linnaeus' hand.

Hiern quoted a number of illustrations, including Meerburgh (Pl. Rar. tab. 8, 1789), Lamarck (Ill. t. 520 fig. 1, 1794) and Wettstein (in Engl., Pflanzenfam. 4(3B): 68 fig. 31 A–D, 1891), but they are not good. The first two are straggly greenhouse plants; Wettstein's drawing of a flowering branch is accurate and depicts typical *M. tomentosa*, but the characteristic clavate hairs in the throat have been omitted from the drawing of a flower opened out to display the anthers. Plate 322 in Curtis, Botanical Magazine (1796) depicts a lax plant that Link & Otto described as *M. angustifolia* (cited below in synonymy under *M. rubra*) after receiving seed from England. Hiern confused the lax form of *M. tomentosa* with *M. rubra*, and possibly Mrs Levyns (in Fl. Cape Penins. 716, 1950) did too, at least in part (she quotes no specimens). The two species are certainly very closely related, but typical *M. rubra* can be distinguished from *M. tomentosa* by its simple stems, leafy near the base, nude or very nearly so above and becoming glabrous there and on the inflorescence axis, lax thyrses with few-flowered cymules, corolla tube hairy outside only on the upper part, and often narrower corolla lobes.

Some specimens are difficult to place; this may indicate incomplete speciation but as the two species are partly sympatric, hybridization is also a possibility. This is discussed further under *M. rubra* (see also *M. caledonica*).

27. Manulea rubra (Berg.) L.f., Suppl. 286 (1782); Thunb., Prodr. 102 (1800) & Fl. Cap. ed. Schultes 472 (1823); Benth in Hook., Comp. Bot. Mag. 1: 383 (1836) & in DC., Prodr. 10: 365 (1846); Hiern in Fl. Cap. 4(2): 234 (1904) p.p. and excluding var. *turritis*.
Type: C.B.S., *Grubb* (STB).

Syn.: *Nemia rubra* Berg., Pl. Cap. 161 (1767).
    Manulea angustifolia Link & Otto, Ic. Pl. Sel. 47, t. 20 (1844). Type: a plant raised from seed received from Kew (?) in 1819 (iso. LE).

Perennial herb but can flower in seedling stage, taproot eventually woody and up to 10–20mm diam. at the crown, stems c.250–700mm long, erect or decumbent, solitary in young plants, up to c.16 in old ones, tufted from the base, simple or very sparingly branched at the base and there villous with spreading white acute hairs up to c.0.4–1mm long, rapidly diminishing upwards, the inflorescence axis glabrous or very nearly so, leafy only at or near the base, naked or very remotely bracteate upwards. *Leaves* opposite, bases connate, c.25–100 × 3–12(–20)mm, oblan-ceolate, base petiole-like, apex subacute to ± obtuse, margins serrate, both surfaces villous, hairs mainly acute, up to 0.4–0.75mm long, sparse and much shorter gland-tipped hairs also present. *Flowers* many in a long narrow lax terminal thyrse, up to 300(–450)mm long in fruit, often simple, sometimes sparingly to quite richly branched, the branches sharply ascending, cymules dichasial, up to c.11-flowered, seldom more. *Bracts* (lowermost) c.2.5–11 × 1–2mm, lanceolate or occasionally ± oblong with 1 or 2 teeth. *Peduncles* (of lowermost cymules) 2.5–11mm long, pedicels very short or flowers sessile. *Calyx* very obscurely bilabiate, tube 0.25–1mm long, lobes c.1.1–2 × 0.4–0.8mm, subspathulate, the tips somewhat thickened, acute hairs up to 0.2–0.4mm long on margins together with minute gland-tipped hairs. *Corolla* tube 7.5–9 × 1–1.25mm in throat, cylindric, inflated at apex, limb bilabiate, 6–8mm across the lateral lobes, posticous lobes 2.1–3.8 × 1.2–2.5(–3)mm, anticous lobe 2.2–3.5 × 1.3–2(–2.6)mm, all elliptic, margins usually revolute, dark reddish-brown, opening paler, pubescent outside only on the backs of the lobes and on the inflated part of the tube or occasionally a few hairs near the base, inside 4–5 strong longitudinal bands of clavate hairs in throat. *Stamens* 4, anticous anthers 0.4–0.5mm long, just included, posticous anthers 0.8–1mm long, included, posticous filaments bearded. *Stigma* 0.7–1.75mm long, shorter than style, included. *Style* 1.5–2.5mm long. *Ovary* 1.25–1.5 × 0.5–0.8mm, conical. *Capsules* c.4–7 × 2.5–3mm. *Seeds* c.0.5 × 0.3mm, violet-blue.

Selected citations:

Cape. Piquetberg div., 3318BB, Bridgetown, 29 v 1952, *Esterhuysen* 20157 (BOL). Malmesbury div., 3318AB, near Hopefield, 4 x 1920, *Garside* 1637 (K). 3318 AD, 2½ miles NW of Darling, c.600ft, 15 x 1959, *Acocks* 20704 (PRE). 3318CB, Buck Bay, 16 ix 1980, *Hilliard & Burtt* 13027 (E). Cape Town div., 3318CD, Milnerton, 17 ix 1913, *Phillips* s.n. (NBG); Camp Ground, 26 vii 1895, *Wolley Dod* 135 (BOL). 3318DC, Kuils River, x 1912, *Worsdell* s.n. (K). 3418AB, Constantia, 1 xi 1916, *Purcell* sub SAM 90621 (SAM). Stellenbosch div., French Hoek, 900ft, xi 1913, *Phillips* 1256 (SAM). 3418BB, Somerset West, viii, *Ecklon & Zeyher* (E, PRE).

Manulea rubra has been much confused with the lax form of *M. tomentosa* and, indeed, it can be difficult to place some specimens. Although the type specimen comprises only the upper part of a flowering stem, it is clear that typical *M. rubra* has the leaves crowded near the base of the stem, which is therefore scapose below the inflorescence, and entirely nude or with one or two very reduced leaves; the apex of the stem and the inflorescence axis are glabrous or very nearly so and the hairs on the corolla are almost entirely confined to the swollen part of the corolla tube and the backs of the lobes (a few hairs may also be present near the base); the corolla lobes are somewhat narrower than in *M. tomentosa*. The stems never become twiggy as they frequently do in *M. tomentosa*; they may branch once or even twice at the base, but typically they are simple. The inflorescence may branch, and then there is usually a cymule in the angle between the main axis and the branch; in specimens of *M. tomentosa* with a paniculate inflorescence, the branches arise in the usual way, in the axil of a leaf.

Manulea rubra ranges from about the Great Berg River north of Hopefield (and possibly Piquetberg; see under *M. leiostachys*) south to Somerset West and inland to the environs of Paarl, Stellenbosch and Fransch Hoek, on sandy flats up to c.350m above sea level. It seems never to occur on the foreshore as *M. tomentosa* often does, but may grow very close to the sea as at Saldanha Bay and Buck Bay near Darling. When it does, it is sympatric with *M. tomentosa* and then more or less intermediate specimens may occur. For example, *Boucher* 4182 (PRE, STE) from Buck Bay has the bushy habit of *M. tomentosa* but the lax inflorescence composed of few-flowered cymules, glabrous axis, only partially hairy corolla tube and narrow corolla lobes that are typical of *M. rubra*. True *M. rubra* (*Hilliard & Burtt* 13027, cited above) also occurs at Buck Bay. Bolus collected typical *M. tomentosa* at Hoetje's Bay in Saldanha Bay (*Bolus* 12788, BOL) together with no. 12790 (BOL), which has the general facies of the lax form of *M. tomentosa* but the thyrse is lax and the corolla tube hairy only in the upper part. There are other examples, including some from the Cape Peninsula, but from the herbarium material available, it is clear that the area from Saldanha Bay to Buck Bay might be a profitable one for field studies.

Manulea rubra is also closely allied to *M. leiostachys*, and they are easily confused; see further under *M. leiostachys*.

28. Manulea obtusa Hiern in Fl. Cap. 4(2): 235 (1904).
Type: Cape, without locality, *Masson* (holo. BM).

Probably a bushy perennial herb (only twigs seen), stems (including inflorescence axis) villous with spreading white acute hairs up to 1mm long, probably leafy throughout though sometimes leaves distant below the inflorescence. *Leaves* opposite, bases connate, becoming alternate on the inflorescence branches, up to 20 × 3.5mm, oblanceolate, apex obtuse, base narrowed to a petiole-like part, margins entire or with a few very obscure teeth, both surfaces densely pubescent with short gland-tipped hairs and longer (to 0.5mm) acute ones. *Flowers* many in a ± dense terminal thyrse, sometimes branched and ± paniculate, cymules monochasial, many-flowered, crowded. *Bracts* (lowermost) c.3–4 × 1mm. *Peduncle* (lowermost) c.2–3mm long, pedicels very short or flowers sessile. *Calyx* very obscurely bilabiate, tube c.0.6mm long, lobes c.1.6 × 0.5mm or a little longer, subspathulate, the tips somewhat thickened, margins with a few eglandular hairs up to c.0.4–0.5mm, minutely glandular as well. *Corolla* tube c.7.25 × 1mm in throat, cylindric, inflated at apex, limb obscurely bilabiate, c.5.5mm across lateral lobes, posticous lobes c.3 × 2mm, anticous lobe 2.2 × 1.7mm, all broadly elliptic, margins revolute, presumably dull orange, pubescent outside, hairs sometimes sparse on anticous side of slender part of tube, inside strong longitudinal bands of clavate hairs in throat. *Stamens* 4, anticous anthers c.0.25mm long, just included, posticous anthers c.0.7mm long, included, posticous filaments bearded. *Stigma* c.1.5mm long. *Style* c.2mm long. *Ovary* c.1.5 × 0.7mm. *Capsules* c.4–5 × 2.5mm. *Seeds* c.0.5 × 0.3mm, violet-blue.

I have seen nothing to match this plant precisely. It is allied to *M. tomentosa*, and resembles the lax form of that plant, but it is at once distinguished by the entire or subentire leaves clad in a mixture of glandular and eglandular hairs, and by the monochasial cymes. In facies and in size of flowers it most closely resembles some of the plants mentioned under *M. caledonica* (viz. *Maguire* 29 and *Martin* 340) as possibly having been influenced by *M. tomentosa*, but the long hairs on the stems and leaves of these specimens are distinctly obtuse (almost balloon-tipped), whereas in *M. obtusa* they are acute, as in *M. tomentosa*. Masson's plant could well have come from the coastal parts of Caledon or Bredasdorp divisions, and it should be sought again; that it was of hybrid origin should not be discounted.

29. Manulea caledonica Hilliard in Notes RBG Edinb. 46:338(1990).
Type: Cape, Caledon div., 3419CD, West of Franskraal, ± 75ft, 9 iv 1979, *Hugo* 1699 (holo. STE, iso. PRE).

Perennial herb, taproot stout, woody, up to c.10mm diam. at the crown, stems c.150–750mm long, erect or decumbent, solitary and simple at first, soon several from the base and often branching there and above, pubescent at least in lower part, hairs often becoming much sparser on the inflorescence axis and sometimes glabrous there, hairs mostly up to 0.1–0.4mm long (rarely to c.0.75mm, see discussion below), mixed balloon- and gland-tipped, balloon-tipped sometimes dominant on upper parts, either leafy throughout or sometimes only near base. *Leaves* opposite, bases connate, sometimes becoming alternate above and passing into bracts, ± 15–70 × 2–10mm, narrowly oblanceolate, apex subacute, base narrowed and petiole-like, margins usually toothed, occasionally subentire in small uppermost leaves, both surfaces pubescent with both balloon-tipped and gland-tipped hairs up to 0.2–0.4mm long (rarely to c.0.75mm, see discussion below). *Flowers* many in a long narrow thyrse, this often becoming paniculate through branching of the upper stem or the main inflorescence axis, or both, lower flowers often solitary in the bract axils, thereafter in either c.2–25-flowered cymules, usually monochasial, sometimes dichasial and then often with only one branch well developed. *Bracts* (subtending lowermost flower or cymule) c.2–14 × 0.5–1mm, or cymules in axils of uppermost, reduced, leaves. *Peduncles* (of lowermost cymule or pedicel of flower) c.2–10mm long (rarely longer), pedicels very short or flowers sessile. *Calyx* obscurely bilabiate, tube 0.2–0.5mm long, lobes 1.4–2.7 × 0.4–0.8mm, subspathulate, tips somewhat thickened, margins glandular-puberulous, a few small balloon-tipped and gland-tipped hairs usually present as well, 0.1–0.3mm long. *Corolla* tube c.7–10 × 0.8–1.3mm in throat, cylindric, inflated near apex, limb bilabiate, 5–7mm across lateral lobes, posticous lobes 2.2–3.8 × 1–2.5mm, anticous lobes 2.1–3.5 × 1–2mm, all elliptic, margins more or less revolute, reddish brown to orange or orange-yellow, pubescent outside with balloon-tipped hairs, these sometimes confined to inflated part of tube, inside with 3 or 4 well developed longitudinal bands of clavate hairs in throat. *Stamens* 4, anticous anthers 0.3–0.5mm long, just included, posticous anthers 0.6–1mm long, included, posticous filaments well-bearded. *Stigma* 0.8–1.8mm long, shorter than style. *Style* 1.5–2.4mm long. *Ovary* 1.2–1.8 × 0.5–1mm, conical. *Capsules* c.3–5 × 2–3mm. *Seeds* c.0.5–0.6 × 0.3mm, violet-blue or pallid.

Selected citations:

Cape. Caledon div., 3419CB, near Die Kelders, Baviaansfontein, 243m, 22 v 1971, *Oliver* 3419 (PRE, STE); Danger Point, i 1941, *Pillans* sub BOL 47503 (BOL); ibidem, 30 xii 1940, *Compton* 10196 (NBG). 3419AD, Hermanus, Kleinrivier Lagoon, 5m, 5 ii 1983, *Williams* 912 (K, NBG). Bredasdorp div., 3420AD, 10 iv 1957, *Barker* 8710 (NBG). 3420CA, Nachtwacht, c.150ft, vi 1927, *Smith* 4254 (K, PRE). Riversdale div., 3421AD, Still Bay, 60 m, 23 xi 1978, *Bohnen* 4701 (PRE); ibidem, c.200ft, 11 viii 1960, *Acocks* 21371 (PRE). 3421BA, near Albertinia, 800ft, 20 x 1966, *Bayliss* 3652 (NBG). 3421BC, Roggeland Farm SE of Albertinia, c.500ft, 26 ix 1978, *Hugo* 1273 (PRE).

Manulea caledonica is frequently misidentified as *M. rubra* or *M. tomentosa*, but it can be distinguished from both by the balloon-tipped (not acute) hairs on stem and leaves mixed with conspicuous gland-tipped hairs, which may be dominant on the leaves and lowermost part of the stem.

   Manulea caledonica has been recorded from Caledon, Bredasdorp and Riversdale divisions of the southern Cape, from Kleinrivier lagoon near Hermanus in the west to Albertinia and Still Bay in the east, from sea level to c.240m, on flats and slopes with sandy calcareous soils. Flowers can probably be found in any month.

   Typical *M. caledonica* is a bushy plant with stems leafy up to the inflorescence, leaves seldom more than 5mm broad, and corolla tube pubescent all over outside. But I have broadened the circumscription of the species to include plants in which the leaves are up to 10mm broad and crowded at the base of the stem, and the corolla tube is pubescent mainly on the upper, inflated, part (this plant has the facies of *M. leiostachys*, which differs in its longer indumentum and glabrous throat). The typical plant ranges along the coast from Hermanus to about De Hoop; the ± rosulate plant occurs around Albertinia and Still Bay. Between De Hoop, the Potteberg, Albertinia and Still Bay can be found both bushy and rosulate plants together with many intermediate forms. The two extremes look so different that initially I took them to be two different species. Perhaps they were, but they now seem so thoroughly introgressed that the recognition of but one species is perhaps appropriate.

   It is probable that in the western part of its range, between Hermanus and Pearly Beach (midway between Danger Point and Quoin Point, both west of Cape Agulhas), *M. caledonica* hybridizes with *M. tomentosa*. It is here that the two species are sympatric, and it is here that some specimens show characters of both species. Several specimens from Gansbaai and thereabouts demonstrate this very well: for example, *Germishuizen* 4137 and *Taylor* 4905 (both NBG) are unequivocally *M. caledonica* but *Leighton* 1861 (BOL) has hairs up to 0.75mm long (thus resembling *M. tomentosa*) but they are balloon-tipped and intermixed with gland-tipped hairs, which is typical of *M. caledonica*, and the cymules tend to be dichasial and crowded. *Compton* 18235 (NBG) strongly resembles *Leighton* 1861 but the hairs are shorter; the specimens were collected at the same place within a day of each other. There is a similar range of specimens from Pearly Beach. For example, *Taylor* 4044 (STE, PRE) is unequivocally *M. tomentosa* whereas *Maguire* 29 and *Martin* 340 (both NBG, and the collectors were almost certainly together) have the habit and the balloon-tipped hairs of *M. caledonica* but the hairs are as long as those of *M. tomentosa* (c.1mm on the stem). *Martin* 348 and 349 (both NBG) and *Lewis* 3881 (SAM), collected inland from Pearly Beach, deviate from typical *M. caledonica* only in the length of the hairs (up to 0.75mm on stems).

30. Manulea turritis Benth. in Hook., Comp. Bot. Mag. 1: 383 (1836) & in DC., Prodr. 10: 366 (1846), as *M. turrita*.
Type: Cape, Tulbagh div., near Tulbagh, Nieuwekloof, *Drège* 7922 (holo. K; iso. LE, W).

Syn.: *M. hirta* Thunb., Prodr. Pl. Cap. 101 (1800) & Fl. Cap. ed. Schultes 471 (1823) non Gaertn. 1788. Specimen cited: Cape, Roodezand, *Thunberg* (sheet no. 14364, UPS; S, STB).
   M. rubra (Berg.) L.f. var. *turritis* (Benth.) Hiern in Fl. Cap. 4(2): 235 (1904).

Perennial herb but can flower quickly and then appears annual, stems c.200mm - 1 m, solitary to several from the base, simple or virgately branched, branches woody at base and there up to c.4mm diam., villous with spreading whitish acute hairs up to 1–2mm long, together with shorter, gland-tipped hairs, becoming sparser upwards, the inflorescence axes often glabrous or very nearly so, sometimes sparsely villous, more or less closely leafy throughout. *Leaves* opposite, bases connate, becoming alternate upwards and passing into bracts, blade c.18–60 × 7–20mm, smaller upwards, elliptic, apex acute, base narrowed to a short petiolar part c.6–14mm long (roughly ¼–⅓ length of blade), margins doubly serrate, both surfaces villous, hairs acute, up to c.1mm long, shorter gland-tipped hairs as well. *Flowers* many in a long narrow lax terminal thyrse composed of few-flowered (up to c.11) dichasial cymules. *Bracts* (lowermost) sometimes leaf-like, often linear-lanceolate, rapidly smaller upwards. *Peduncles* (of lowermost cymules) c.3–12mm long, pedicels very short or flowers sessile. *Calyx* obscurely bilabiate, tube 0.2–0.4mm long, anticous lobes 1.5–2.3 × 0.4–0.6mm, anticous lip split 1.1–2.3mm, posticous lobes 1.5–2.2 × 0.4–0.6mm, posticous lip split 1.2–2mm, all lobes linear-lanceolate, tips subspathulate, somewhat thickened, hairy usually on margins only, sometimes also thinly so on backs, some hairs acute, 0.3–0.75mm long, others gland-tipped and much smaller. *Corolla* tube 6.5–8.5 × 0.8–1.2mm in throat, cylindric, inflated near apex, limb bilabiate, 6–7mm across lateral lobes, posticous lobes 2.25–3.5 × 0.75–1.5mm, anticous lobes 1.75–3.1 × 0.75–1.3mm, all lobes oblong, tips sometimes retuse, margins revolute, dull brownish-yellow to brownish-orange, or occasionally reddish-brown, pubescent outside, hairs often confined to swollen part of tube and backs of lobes, hairs acute, 0.2–0.3mm long, inside either glabrous or a few clavate hairs at back of throat, occasionally a few in anticous position. *Stamens* 4, anticous anthers 0.4mm long, just included, posticous anthers 0.8mm long, more deeply included, posticous filaments thinly bearded. *Stigma* 0.5–2mm long, shorter than or about equalling style, included. *Style* 1–2.25mm long. *Ovary* c.1.5 × 0.6–0.8mm, conical. *Capsules* c.4–4.5 × 2–2.5mm. *Seeds* c.0.5 × 0.3mm, white turning violet-blue.

Citations:

Cape. Piquetberg div., 3218DC, Piquetberg Mt Plateau, 2500ft, 12 ix 1954, *Esterhuysen* 23111 (BOL); ibidem, between Mouton's Vley and Gruys Kop, 7 xi 1934, *Pillans* 7253 (BOL). Tulbagh div., 3319AA, [Great] Winterhoek, c.1700ft, xi, *Bolus* s.n. (BOL). 3319AC, Tulbagh, Steendaal, x, *Ecklon* (SAM, TCD). Paarl div., 3319AC, foot of Elands Kloof Mtns north of Wellington, 6 x 1956, *Stokoe* sub SAM 70149 (SAM). Ceres div., 3319AD, Mitchell's (sic, Michell's) Pass, 1200ft, xi 1879, *Bolus* 214 (BOL); ibidem, 11 ix 1896, 1200ft, *Schlechter* 8959 (BOL); near Ceres, 1600ft, i 1888, *Bolus* 9180 (BOL). Worcester div., Worcester, Breede River, x, *Bolus* 2796a (BOL); Sebastians Kloof, 14 ix 1941, *Compton* 11643 (NBG); ibidem, 14 ix 1941, *Esterhuysen* 6117 (BOL). Without locality, anno 1831, *Verreaux* s.n. (TCD); anno 1772, *Oldenburg* 976 (BM); C.B.S., *Auge* (Nelson) (BM). Ceres Wild Flower Show, 4 x 1937, *Compton* 7106 (NBG).

The name *M. turritis* is antedated by *M. hirta* Thunb., but this is a later homonym of *M. hirta* Gaertner (1788). Hiern (1904) attributed the name to Gaertner only (and reduced it to synonymy under *M. rubra*) but as Thunberg does not cite Gaertner, his name must be accepted as independent. Gaertner's plant proves to be *M. cheiranthus*.

The stems of *M. turritis* are leafy throughout, which character, as well as the leaves themselves (elliptic in outline and with distinctly doubly serrate margins), will at once distinguish the species from both *M. rubra* and *M. leiostachys*, with which it has been confused. Furthermore, in *M. rubra* there are strong bands of clavate hairs in the throat; in *M. turritis*, the throat is glabrous or nearly so, as it is in *M. leiostachys*.

Collectors have not described the habit of *M. turritis* and although it is clear that it will flower when little more than a seedling, the plants possibly become at least subshrubby with age, the stems sometimes being decidedly woody and brittle at the base and laxly branched there and above, the branches sharply ascending and leafy. It has been recorded from Piquetberg Mountain, then from the Great Winterhoek Mountains, south east of Piquetberg, south to Tulbagh, Ceres and Worcester between c.300 and 760 m above sea level, probably always on rocky sites and flowering between September and January.

Manulea turritis and *M. rubra* appear to be sympatric in the neighbourhood of Tulbagh, but *M. rubra* possibly always occurs at lower altitudes than *M. turritis*; *M. turritis* and *M. leiostachys* on the other hand appear to be truly sympatric at least around Ceres and Worcester. See also *M. ovatifolia*.

31. Manulea ovatifolia Hilliard in Notes RBG Edinb. 46:51(1989).
Type: Cape, Piquetberg div., 3218DA, Kapitein's Kloof Mt., 21 x 1935, *Pillans* 7894 (holo. BOL).

Shrubby, stems up to 300–750mm long (only upper parts seen), 4mm in diam., woody, decumbent, ascending or erect (?), branches virgate, pubescent with spreading acute hairs up to 0.6–1.2mm long, long gland-tipped hairs also present, leafy throughout, dwarf axillary shoots often prominent. *Leaves* opposite, bases connate becoming alternate upwards and passing imperceptibly into bracts, blade of primary leaves c.12–23 × 7–20mm, ovate or occasionally oblong, apex acute to obtuse, base abruptly contracted to a petiolar part 3–15mm long, roughly ½–⅓ as long as, or occasionally almost equalling, blade, margins coarsely, sharply, and sometimes doubly, serrate, both surfaces villous, hairs up to c.0.8mm long, mainly gland-tipped. *Flowers* many in a long narrow lax terminal thyrse composed of few-flowered (up to c.11) dichasial cymules. *Bracts* (lowermost) either leaflike and then c.10–20 × 5–11mm including a short petiole, or narrowly elliptic and c.6 × 0.6mm. *Peduncles* (lowermost) c.2–8mm long, pedicels c.1mm. *Calyx* obscurely bilabiate, tube c.0.2–0.4mm long, lobes c.2.5–3 × 0.5–0.6mm, lanceolate, villous all over, hairs 0.5–0.6mm long, acute, smaller glandular hairs also present. *Corolla* tube c.6–7.2 × 1–1.2mm in throat, cylindric, inflated near apex, limb bilabiate, c.7–9.5mm across lateral lobes, posticous lobes c.3.5–4.8 × 1.5mm, anticous lobe c.3.2–3.5 × 1.1–1.5mm, all lobes oblong, tips often retuse, margins revolute, 'yellow-brown', pubescent all over outside, hairs acute, up to 0.4mm long, inside a few clavate hairs in back of throat, sometimes also on anticous side. *Stamens* 4, anticous anthers c.0.5mm long, just included, posticous anthers 0.7–1mm long, more deeply included, posticous filaments thinly bearded. *Stigma* 0.9–1.5mm long, shorter than or about equalling style, included. *Style* 1.3–2mm long. *Ovary* 1–1.5 × 0.6–0.8mm. *Capsules* not seen.

Citations:

Cape. Piquetberg div., 3218DA, hills NW of Mouton's Vlei, 6 xi 1934, *Pillans* 7408 (BOL). 3218DC, between Zebrakop and Die Toring, 3700ft, 13 x 1963, *Esterhuysen* 30383 (BOL).

Manulea ovatifolia is known only from rocky sites on the summit of Piquetberg Mountain, where it flowers in October and November. It is closely allied to *M. turritis*, but may be distinquished by its differently shaped leaves (usually ovate or occasionally oblong, not elliptic), this reflected in the ratio of leaf length to breadth: blade c.12–23 × 7–20mm in *M.*

ovatifolia, c.18–60 × 7–20mm in *M. turritis*. The flowers of *M. ovatifolia* are somewhat larger than those of *M. turritis*: calyx lobes c.2.5–3mm long (versus 1.5–2.3mm), corolla limb c.7–9.5mm across the lateral lobes (versus 6–7mm), posticous lobes c.3.5–4.8 × 1.5mm (versus 2.25–3.5 × 0.75–1.5mm). *Manulea ovatifolia* is also usually hairier on the inflorescence axes and calyx than *M. turritis*, and it is possibly shrubbier in habit. This needs field investigation, and Piquetberg Mountain is the obvious starting point: Pillans collected both species there within a day of each other. His gathering of *M. turritis* made 'on hills between Mouton's Vley and Gruys Kop' contrasts strikingly in habit with his collection of *M. ovatifolia* from hills NW of Mouton's Vley. The difference in indumentum is also marked.

32. Manulea leiostachys Benth. in Hook., Comp. Bot. Mag. 1: 383 (1836) & in DC., Prodr. 10: 365 (1846); Hiern in Fl. Cap. 4(2): 233 (1904) p.p. **Fig. 58C.**
Lectotype (chosen here): Cape, Clanwilliam div., 3219AA–AC, Cedarbergen, 'bei Honigvalei und auf Koudeberg, 3000–4000 Fuss, Dec.', *Drège* (K).

Syn.: *M. leiostachys* (sphalm. *leucostachys*), nomen; Drège, Zweipflanzengeogr. Doc. 73 (1843).

Perennial herb, taproot stout, woody, up to 10mm diam. at the crown, stems c.550–1200mm tall, erect, solitary to several tufted from the crown, usually simple below, rarely forking once or twice, often branching above to form a lax panicle, villous at the base with spreading white obtuse hairs up to 0.6–1mm long, gland-tipped hairs of similar length frequent, rapidly becoming glabrous upwards, leafy in the lower half, remotely bracteate upwards. *Leaves* opposite, bases connate, lower ones c.60–120 × 3–13mm, upper c.35–40 × 2–4mm thereafter passing into a very few distant ± alternate bracts, narrowly to broadly oblanceolate, apex ± acute, base narrowed to a petiolar part, margins serrate (coarsely so in larger leaves), both surfaces pubescent, hairs glandular and eglandular, up to c.0.4–0.75mm long, apical cell of eglandular hairs globose or at least obtuse. *Flowers* many in a long narrow lax terminal thyrse c.300–600mm long, simple, or with a few sharply ascending branches, the lowermost often in the axil of a bract, the uppermost with a cymule interposed between branch and main axis, cymules dichasial, up to 15-flowered. *Bracts* (subtending lowermost cymule on main axis) c.4–13 × 0.8–1.2mm, oblong. *Peduncles* (of lowermost cymules on main axis) c.4–15mm long, pedicels very short or flowers sessile. *Calyx* obscurely bilabiate, tube 0.1–0.5mm long, lobes 1–1.8 × 0.3–0.6mm, subspathulate, the tips somewhat thickened, obtuse eglandular hairs up to 0.2–0.3mm long on margins only, together with smaller gland-tipped hairs. *Corolla* tube 7.5–10 × 0.8–1.2mm in throat, cylindric, inflated at apex, limb bilabiate, 5–8mm across the lateral lobes, posticous lobes 2.3–4 × 1–2mm, anticous lobe 2.3–3.75 × 1–2mm, all narrowly elliptic, margins revolute, dull orange-brown or yellow-brown, pubescent outside only on backs of lobes and swollen part of tube, or occasionally a few hairs near the base, either glabrous inside or with a few clavate hairs in a longitudinal posticous line. *Stamens* 4, anticous anthers 0.4–0.5mm long, just included, posticous anthers 0.8–1mm long, included, posticous filaments weakly bearded. *Stigma* 0.8–1.6mm long, shorter than style, included. *Style* 1.4–3mm long. *Ovary* 1.3–1.6 × 0.6–0.8mm, conical. *Capsules* c.4–6 × 2–3mm. *Seeds* c.0.5 × 0.3mm, pallid or violet-blue.

Selected citations:

Cape. Clanwilliam div., 3219AC, Cedarbergen, xi 1929, *Thode* A2173 (K, PRE). 3219CA, Thee rivier, Citrusdal, 9 x 1969, *Hanekom* 1318 (PRE). Ceres div., 3319AC, Elands Kloof, 30 ix 1944, *Compton* 16148 (NBG). Worcester div., Du Toit's Kloof, 1000ft, 5 x 1947, *Barker* 4832 (NBG). Mountain slopes E of Fonteintjiesberg, 3000–4000ft, 2 i 1950, *Esterhuysen* 16699 (BOL).

Bentham based his name *M. leiostachys* on two different species. As the description applies equally well to both plants, the specimen first cited, that collected by Drège in the Cedarbergen, has been chosen as lectotype. The other collection, made by Drège at Baviaanskloof near

Genadendal, is *M. laxa*. The specimen collected by Kuntze at Cathcart and cited by Hiern as *M. leiostachys* is *M. bellidifolia*.

Manulea leiostachys appears to be confined to the mountainous tract from the Cedarbergen to Ceres and Worcester, but there are also puzzling specimens from Piquetberg and Piqueniers-kloof, discussed below. Its area therefore lies east of that of *M. rubra* (to which it is allied and with which it is very readily confused) and they may always be spatially separated because of different altitudinal preferences, but this needs confirmation. They look much alike and care is needed to separate them as they differ principally in details of indumentum: there are well-de-veloped longitudinal bands of clavate hairs in the throat of *M. rubra* whereas the throat in *M. leiostachys* is glabrous or very nearly so (apart from hairs on the posticous filaments). Also, in *M. rubra*, the eglandular hairs on stems and leaves are acute, whereas in *M. leiostachys* they tend to be obtuse, with some hairs almost balloon-tipped, and many more long gland-tipped hairs are present than in *M. rubra*, where they are very short and sparse.

Field investigations are needed at Piquetberg. There, on Piquetberg Mountain at 2500ft (760 m), Compton collected *M. leiostachys* (*Compton* 3640, BOL, NBG). Another specimen from the same altitude (*Johnson* 278, NBG, STE), is very close to true *M. rubra*, having acute eglandular hairs, very sparse glandular ones, but only one weak posticous band of hairs in the throat; it thus approaches *M. leiostachys*. A similar-looking specimen from Bosch Kloof (no altitude given) has four weak bands of hairs in the throat (*Bond* 537, NBG). Another, from Bokloof (*Pillans* 7818 (BOL) and again without altitude, appears to be true *M. rubra*, while *Pillans* 7512 (BOL) from Mouton's Vlei (no altitude given) has the external indumentum of *M. leiostachys*, but there are three strong bands of hairs in the throat. *Taylor* 10457 (STE) from Piekenierskloof at c.500 m has the indumentum of *M. rubra* but the hairs in the throat are very weakly developed.

33. Manulea laxa Schltr. in Bot. Jahrb. 27: 179 (1899); Hiern in Fl. Cap. 4(2): 233 (1904). Lectotype (chosen here): Van Rhynsdorp div., 3118DC, Windhoek, c.600 ped., 1 viii 1896, *Schlechter* 8363 (K; isolecto. BOL, PRE).

Syn.: *M. leiostachys* Benth. in Hook., Comp. Bot. Mag. 1: 383 (1836) excluding lectotype.

Perennial herb, taproot stout, woody, up to c.10mm diam. at crown, stems (150–) 300–800mm tall, erect, several to many tufted from the crown, simple or laxly branched low down, pubescent in lower part, hairs mostly up to 0.1–0.2mm long, all balloon-tipped or sometimes mixed with few to many gland-tipped hairs, glabrous in upper part, leafy at the base and scapose upwards in weak plants, ± leafy throughout in robust specimens. *Leaves* opposite, bases connate, sometimes becoming alternate as they pass upwards into bracts, 10–70 × 1.5–8mm, oblanceo-late, apex obtuse to subacute, base narrowed to a petiole-like part, margins entire to shortly and remotely toothed, both surfaces greyish, pubescent, hairs as on stems. *Flowers* many in a long lax terminal raceme or very narrow thyrse, this often branching into a cymose panicle, flowers often solitary, or 2–4 in monochasial cymules, rarely in 3-flowered dichasia. *Bracts* (subtending lowermost flowers or cymules) c.1.5–3 × 0.5–0.6mm. *Peduncle* or *pedicel* (lowermost) c.1.5–9mm long, flowers in cymules ± sessile. *Calyx* obscurely bilabiate, tube 0.1–0.5mm long, lobes 1.2–2 × 0.4–0.6mm, subspathulate, the tips somewhat thickened, minutely puberulous on margins, the hairs mostly balloon-tipped. *Corolla* tube 6–9 × 0.8–1.2mm in throat, cylindric, inflated near apex, limb bilabiate, 5–8mm across the lateral lobes, posticous lobes 2.2–4 × 1–2mm, anticous lobe 2–3.5 × 0.8–1.8mm, all narrowly elliptic, margins revolute, reddish-brown to orange-brown, pubescent outside on backs of lobes and swollen part of tube, occasionally a few hairs near the base as well, or extending throughout in a posticous band, either glabrous inside or a few clavate hairs in a posticous longitudinal line. *Stamens* 4, anticous anthers 0.4–0.5mm long, just included, posticous anthers 0.8–1mm long, included, posticous filaments either glabrous or with a few clavate hairs. *Stigma* 1–2mm long, a little shorter than or about equalling the style. *Style* 1.5–2.5mm long. *Ovary* 1–1.5 × 0.5–0.8mm, conical. *Capsules* c.3–5 × 1.5–2.5mm. *Seeds* c.0.5–0.8 × 0.3–0.5mm, violet-blue or pallid.

Selected citations:

Cape. Van Rhynsdorp div., 3118DC, Klaver, 29 x 1944, *Leipoldt* 4124 (BOL, PRE). Calvinia div., 3119CA, Lokenburg, 21 miles S of Niewoudtville, 2500ft, 11 x 1953, *Story* 4301 (PRE). Ceres div., 3319AD, Ceres, c.1500ft, i 1903, *Bolus* 8339 (BOL, SAM). Worcester div., 3319DA, Hex River valley near De Wet, 17 xii 1951, *Esterhuysen* 19671 (BOL, UPS). Robertson div., 3319DD, Robertson Karroo, 24 ix 1935, *Compton* 5708 (NBG). Laingsburg div., 3320BB, Cabidu, 25 x 1950, *Compton* 22211 (NBG). Ladismith div., 3321AD, Ladismith, near Waterkloof, 2000ft, 1 xi 1928, *Hutchinson* 1126 (K). George div., 3322CD, between Doorn River and Klipdrift, 2000ft, xi 1927, *Fourcade* 3431 (BOL, STE). Caledon div., 3419BA, Genadendal, *Ecklon* s.n. (SAM); ibidem, Baviaanskloof, 1000–2000ft, *Drège* (syntype of *M. leiostachys*)(K). Swellendam div., 3420AB, Swellendam, Bontebok Park, c.200ft, 20 vi 1962, *Acocks* 22246 (PRE).

Manulea laxa is characterized by its grey stems and leaves, these more or less obtuse with mostly entire to subentire margins (occasionally sparingly toothed), short indumentum, the hairs mostly balloon-tipped (though in the southern part of its range gland-tipped hairs may be intermixed) and delicate inflorescences in which the flowers are either solitary or in very few-flowered monochasial cymules. On the other hand, *M. leiostachys* (with which *M. laxa* is often confused) is green often drying brown with mostly longer indumentum (hairs to 0.6–1mm though they may be much shorter), serrate leaves, and coarser inflorescences with up to 15 flowers in each dichasial cymule. See also *M. pillansii* and *M. glandulosa*, which are also often confused with *M. laxa*.

The type material of *M. laxa* came from Windhoek near Klaver (Van Rhynsdorp div.,) and the plant appears still to be common thereabouts, probably always on the mountain slopes above c.180m. Thence south and east to the southern end of the Cold Bokkeveld Mountains and the mountains around Tulbagh, Ceres, Worcester, Robertson, Montagu and Genadendal and east to the environs of Laingsburg, Ladismith, the Langeberge, and the southern flank of the Outeniqua Mountains. There is one collection from Swellendam at only 60m above sea level: the others are all above c.180m and reach c.1370m in the Cold Bokkeveld Mountains and Klein Swartberg.

The typical plant is more or less leafy up to the inflorescences; many of the plants from higher altitudes have the leaves crowded near the base of stem, possibly in response to harsher conditions: for example, on the flanks of the Klein Swartberg at 1300m (*Moffett & Steensma* 3841, E, PRE, STE) in 'sandy rocky talus below lowest rock band' the plants have scanty radical leaves and long nude scarcely branched stems, while on 'dunes of Breede River' at 60m, the lowest elevation recorded, the plants are richly branched and leafy (*Acocks* 22246, PRE).

Flowers can possibly be found in any month although the principal season seems to be September to December.

34. Manulea pillansii Hilliard in Notes RBG Edinb. 46:51(1989).
Type: Cape, Piquetberg div., 3218BC, sandy slope between Verloren Vlei and Rooikransberg, 18 x 1935, *Pillans* 7973 (holo. BOL).

Subshrubby, stems up to 1m long, probably several from the base, more or less sprawling to erect, well-branched, at least main stems dark purple-red, minutely puberulous with almost sessile balloon-tipped hairs, glabrescent, distantly leafy. *Leaves* opposite, bases connate, or sometimes alternate upwards and then subtending branches, c.10–70 × 1.5–5.5mm, oblanceolate, apex ± obtuse, base narrowed and petiole-like, margins entire or with a few obscure teeth, both surfaces very minutely puberulous, probably glaucous and tinged with dark purple-red. *Flowers* many in long lax racemes, often thyrsoid, paniculately arranged through the branching of both stems and inflorescence axes, flowers either solitary or up to 7 in dichasial cymules. *Bracts* (lowermost) c.1–3 × 0.5–1mm, oblong. *Peduncles* and *pedicels* (lowermost) c.2–3mm long, shorter upwards, most flowers ± sessile. *Calyx* very obscurely bilabiate, tube c.0.3mm long, lobes 0.7–1.3 × 0.3–0.5mm, subspathulate, tips somewhat thickened, very minutely puberulous on margins. *Corolla* tube 6.75–7.5 × 0.75–1mm, cylindric, inflated near apex, limb bilabiate, 6–8mm across the lateral lobes, posticous lobes 2.8–4 × 1–1.5mm, anticous lobe 2.5–3.6 × 1–1.2mm, all narrowly elliptic, margins revolute, orange, almost glabrous outside or very minutely puberulous, a few glistening glands on backs of lobes, inside glabrous or with a few clavate hairs in posticous and anticous positions. *Stamens* 4, anticous anthers 0.4–0.6mm long, just included, posticous anthers 0.8–1mm long, posticous filaments weakly bearded. *Stigma* 1–1.8mm long, slightly shorter than style. *Style* 1.5–2mm long. Ovary 1.5–1.6 × 0.5–0.7mm, conical. *Capsules* c.4–6.5 × 1.5–2mm. *Seeds* c.1 × 0.5mm, pallid.

Citations:

Cape. Clanwilliam div., 3218AB, Lambert's Bay, Nortier Reserve, c.200ft, 29 x 1948, *Acocks* 15190 (PRE); ibidem, 14 iv 1968, *Van Breda* 4110 (PRE); ibidem, iv 1977, *Walters* 605 (NBG). 3218BB, near Clanwilliam, v 1946, *Leipoldt* 4241 (BOL).

Manulea pillansii is allied to *M. laxa*, but it is a stouter plant with main stems c.4mm in diameter (not mostly 2mm, and up to 3mm), and so minutely puberulous that the hairs on stems, leaves and corolla tube are scarcely visible with a ×10 lens, but in *M. laxa* they show up clearly as white dots or a short white fuzz.

I have seen only five collections of this plant that I could determine with certainty, three from the Nortier Reserve near Lambert's Bay, one from further south in the environs of Verloren Vlei, and one from further east, near Clanwilliam. But two more collections from Clanwilliam div. probably belong here, namely *Leipoldt* 3429 (BOL, PRE) and *Leipoldt* 3430 (BOL), both from Grootkliphuis, about 12km north of Clanwilliam; they comprise inflorescences only.

Collector's notes are scanty. Only Acocks recorded flower colour and described the plant as a 'rather sprawling virgate shrub'; Van Breda contributed 'very well eaten'. The plants were in both flower and fruit in October, December, April and May. It is clear from Leipoldt's collections that *M. laxa*, *M. leiostachys* and *M. pillansii* all grow within a 12km radius of Clanwilliam and this would undoubtedly be a good place in which to begin a study of these three closely allied species.

35. Manulea montana Hilliard in Notes RBG Edinb. 46:50(1989).

Type: Cape, Clanwilliam div., 3219AC–AD, top of Cedarberg, 4500ft, 12 x 1947, *L.E. Taylor* 2964 (holo. NBG).

Perennial herb, taproot becoming stout, woody, up to c.7mm diam. at crown, stems 270–455mm tall, erect, several tufted from the crown, either simple or the inflorescence sparingly branched, puberulous in lower part, hairs up to 0.2–0.3mm long, gland-tipped, glabrous above, leafy only near base, scapose above. *Leaves* opposite, bases connate, c.14–70 × 2–6mm, narrowly elliptic or oblanceolate, apex ± acute, base narrowed to a petiolar part, margins sharply serrate, both surfaces puberulous, hairs up to 0.2–0.3mm long, both gland-tipped and balloon-tipped. *Flowers* many in a lax very narrow terminal thyrse, inflorescence sometimes branched and then paniculate, lowermost flowers occasionally solitary, mostly in 3–5-flowered dichasial cymules. *Bracts* (subtending lowermost peduncle or pedicel) c.1.5–3 × 0.5–1mm, oblong. *Peduncles* or *pedicel* (lowermost) c.1–2mm long, flowers otherwise sessile or nearly so. *Calyx* obscurely bilabiate, tube 0.3–0.5mm long, lobes 1.4–1.8 × 0.4–0.6mm long, subspathulate, tips somewhat thickened, margins minutely glandular-puberulous, very occasional small balloon-tipped hairs also present on backs of lobes. *Corolla* tube 7.5–9 × 0.8–1.1mm in throat, cylindric, inflated near apex, limb bilabiate, c.7mm across the lateral lobes, posticous lobes c.3.1–4 × 1.1–1.4mm, anticous lobe 2.5–3.2 × 0.9–1mm, all narrowly elliptic, margins revolute, brown, pubescent outside on backs of lobes and tube, hairs sometimes scanty or wanting on anticous side of tube, glabrous inside or a very few clavate hairs at back of throat. *Stamens* 4, anticous anthers 0.4–0.5mm long, just included, posticous anthers 0.8–1mm long, included, posticous filaments weakly bearded. *Stigma* 0.6–1mm long, much shorter than style. *Style* 2–2.5mm long. *Ovary* 1.2–1.5 × 0.5–0.7mm, conical. *Capsules* c.4–5 × 2mm. *Seeds* unknown.

Citations:

Cape. Clanwilliam div., 3219AC, Cedarberg, mountain slopes near Algeria Forest Station, 24 x 1930, *Galpin* 10576 (PRE). 3219AA, Middelberg, SW slopes, xii 1939, *Esterhuysen* 2504 (BOL). 3218BB, Langberg, 15 xii 1941, *Bond* 1381 (NBG).

Manulea montana has hitherto been confused with *M. laxa*, from which it can be distinguished by its green, sharply serrate leaves (not grey and entire or almost so), both stems and leaves glandular-puberulous with sparse balloon-tipped hairs as well on leaves only (not clad in tiny balloon-tipped hairs that in the dried state at least make stems and leaves look lepidote). The margins of the calyx lobes in *M. montana* are glandular-puberulous (balloon-tipped hairs predominate in *M. laxa*), the corolla lobes are often narrower, and the stigma is not only mostly shorter in absolute terms (0.6–1mm versus 1–2mm) it is only ½–⅓ as long as the style, not about equalling it or just a little shorter.

 Manulea montana has been recorded from three mountain ranges near Clanwilliam: the Langeberg, Middelberg and Cedarberg; *M. laxa* has not yet been recorded there. Collectors have not given any ecological information, but the plants flower between October and December.

36. Manulea glandulosa E.P. Phillips in Ann. S. Afr. Mus. 9: 123 (1913).
Lectotype (chosen here): Cape, Van Rhynsdorp div., 3118DC, Giftberg Range, 1–2000ft, 17 xi 1911, *Phillips* 7357 (SAM, isolecto. PRE).

Perennial herb, taproot stout, woody, up to 10mm diam. at crown, stems (180–)400–550mm tall, erect, several tufted from the crown, often simple, occasionally branching above once or twice to bear subsidiary inflorescences, puberulous in lower part, hairs up to 0.2mm long, balloon-tipped, glabrous above, leafy only near the base, scapose above. *Leaves* opposite, bases connate, c.35–70 × 2–8mm, narrowly elliptic, apex ± acute, base narrowed to a petiolar part, margins usually sharply serrate, sometimes subentire, both surfaces puberulous with balloon-tipped hairs up to 0.2mm long. *Flowers* many in long narrow lax terminal thyrses, lowermost flowers sometimes solitary, mostly flowers up to 9 in dichasial cymules, occasionally mono-chasial when one branch fails to develop. *Bracts* (subtending lowermost flower or cymule) c.2.5 × 0.75mm, oblong. *Peduncle* or *pedicel* (lowermost) c.2–6mm long, flowers otherwise sessile or almost so. *Calyx* obscurely bilabiate, tube 0.4–0.8mm long, lobes 1.5–2 × 0.4–0.8mm, subspathulate, tips somewhat thickened, tiny balloon-tipped hairs on margins only. *Corolla* tube 7.5–10.25 × 1–1.4mm in throat, cylindric, inflated near apex, limb bilabiate, c.9–11mm across lateral lobes, posticous lobes 3.75–5 × 1.4–2.1mm, anticous lobe 3.4–4.2 × 1–2mm, all narrowly elliptic, margins revolute, orange to orange-brown, tube sometimes purplish, puberulous outside on backs of lobes and inflated part of tube, hairs often running down tube, particularly on posticous side or thinly puberulous all over, either glabrous inside or with a few posticous clavate hairs in throat. *Stamens* 4, anticous anthers 0.5–0.8mm long, just included, posticous anthers 0.8–1.2mm long, included, posticous filaments weakly bearded. *Stigma* 0.8–1.7mm long, shorter than style. *Style* 2–2.7mm long. *Ovary* 1–1.6 × 0.5–0.8mm, conical. *Capsules* c.5–6.5 × 2.5mm. *Seeds* c.0.8 × 0.3mm, violet-blue.

Citations:

Cape. Van Rhynsdorp div., 3118DC, Giftberg Range, 1–2000ft, ix 1911, *Phillips* 7366 (syntype BOL); ibidem, 2500ft, 15 x 1953, *Esterhuysen* 22115 (BOL); Giftberg Pass, 15 x 1976, *Hugo* 711 (PRE, STE). Clanwilliam div., 3118DC, Nardouw, 1000ft, 22 ix 1937, *Compton* 6990 (NBG); Nardouw Kloof, ix 1947, *Stokoe* sub SAM 63907 (SAM); 3118DD, Brandewyn river, 1100ft, 13 viii 1897, *Schlechter* 10820 (BOL); 3219AA, Pakhuisvlakte, 1.5km along jeeptrack from Pakhuis to Heuningvlei, c.970m, 8 x 1985, *Taylor* 11379 (STE).

Phillip's choice of epithet is unfortunate as the plant is eglandular: it has the balloon-tipped hairs characteristic of the genus as a whole on the lower stem, leaves and corolla tube. These hairs will distinguish *M. glandulosa* from *M. leiostachys*, as Phillips pointed out, but the species is probably more closely allied to *M. laxa*, which also has balloon-tipped hairs, but *M. laxa* is grey, well grown plants are more branched than those of *M. glandulosa*, the leaves are often entire, the flowers often solitary and tending to be smaller than those of *M. glandulosa*, with corolla tube up to 9mm long, limb up to 8mm in diameter and anticous lobe 2–3mm long.

Manulea glandulosa has been recorded from the Giftberg and neighbouring Nardouws Mountain, the northern Cedarberg and 'Worcester'; the type locality of *M. laxa* is just north of the Giftberg, so the two species are sympatric though it is not possible to judge if they are spatially isolated. The ecological information on *M. glandulosa* given by collectors is 'sandy old agricultural fields' (near top of Giftberg Pass), 'in sand on [Giftberg] plateau', and 'level sandy restionaceous flats' at Pakhuisvlakte, northern Cedarberg. That the species occurs at Worcester needs confirmation, the record being based on an undated collection in STE: *Naude* sub STE 31810, 'among river stones'. The plants flower in September and October.

A unique collection from Klaver may possibly represent a hybrid between *M. glandulosa* and *M. altissima*: it has the facies of the former but the indumentum is more like that of the latter. (*Lavis* sub BOL 20284 (BOL) collected viii 1932, flowers and young fruits; *Lavis* sub BOL 20283 is *M. altissima* subsp. *longifolia*).

Section Manulea

37. Manulea cinerea Hilliard in Notes RBG Edinb. 46:50(1989).
Type: Cape, Namaqualand, 2816DC, Witbank, x 1926, *Pillans* 5211 (holo. BOL).

Dwarf shrub up to c.600mm tall, whole plant ashy-grey, bark pale, main stems c.10mm diam., well branched, branches erect, young parts densely puberulous with very small shining balloon-tipped hairs, occasional gland-tipped hairs to 0.2mm long also present, closely leafy, dwarf axillary shoots prominent. *Leaves* opposite, bases connate, c.12–35 × 4–8mm, elliptic, apex subacute, base tapered to a short petiolar part, margins entire, both surfaces densely puberulous, hairs as on stem. *Flowers* many in long narrow lax terminal thyrses, cymules dichasial, up to c.11-flowered. *Bracts* (lowermost) leaflike, c.5–13 × 1–3mm, rapidly smaller upwards. *Peduncles* (lowermost) c.2–4mm long, flowers sessile or nearly so. *Calyx* very obscurely bilabiate, tube 0.2–0.3mm long, lobes c.1.3–2.5 × 0.3–0.6mm, spathulate, puberulous all over, hairs very short, balloon-tipped, shining. *Corolla* tube c.6–7 × 0.8–1.2mm in mouth, cylindric, inflated at apex, limb distinctly bilabiate, c.6–11mm across lateral lobes, posticous lobes c.2.2–5 × 0.8–1.5mm, anticous lobe c.2–4.5 × 0.6–1mm, all lobes elliptic-oblong, margins strongly revolute, tips acute to the eye, ochre yellow turning red-brown or purple-brown, densely puberulous outside with shining balloon-tipped hairs, inside with 4 very weak longitudinal bands of clavate hairs in throat. *Stamens* 4, anticous anthers 0.3–0.5mm long, just included, posticous anthers c.0.5–0.6mm long, included, posticous filaments weakly bearded. *Stigma* c.0.8–1mm long, either shorter than or about equalling style, included. *Style* c.0.8–1.5mm long. *Ovary* 0.8–1.5 × 0.5–0.8mm. *Capsules* c.2.5–3.5 × 2mm, glandular-puberulous. *Seeds* c.0.6–0.8 × 0.4mm, pallid to violet blue.

Citations:

Cape. Namaqualand, 3017AD, ¼ mile N of Hondeklip Bay, x 1924, *Pillans* sub BOL 18177 (BOL, STE); ibidem, N side Hondeklip Bay, x 1924, *Pillans* sub BOL 18178 (BOL); 3017DC, Groen River Mouth, 20 x 1980, *Le Roux & Parsons* 76 (STE).

Manulea cinerea is without close allies. Its shrubby habit, entire leaves and ashy-grey appearance (derived from the felt-like indumentum of tiny balloon-tipped hairs) render it very distinctive. The shape of the corolla tube and the form of the stigma, style and ovary suggest a distant relationship with *M. thyrsiflora*.

It grows on seashore sand dunes in Namaqualand; all four specimens were collected in October when both flowers and old capsules were present.

38. Manulea thyrsiflora Linn. f., Suppl. 286 (1782); Thunberg, Prodr. 102 (1800) & Fl. Cap. ed. Schultes 471 (1823); Benth. in Hook., Comp. Bot. Mag. 1: 383 (1836) & in DC., Prodr. 10: 365 (1846); Hiern in Fl. Cap. 4(2): 236 (1904), excl. vars.
Type: CBS, *Thunberg* (sheet no. 14380, UPS)

Shrublet c.350mm – 1 m tall, stems erect, ascending or decumbent, up to c.4mm diam. at base, woody, laxly branched, densely pubescent, hairs either gland-tipped or balloon-tipped, often mixed, sometimes glandular only, up to 0.25–0.7mm long, leafy throughout. *Leaves* opposite, bases connate, becoming alternate only on the inflorescence axes where they pass imperceptibly into bracts, often with dwarf axillary shoots, primary leaves c.12–50 × 5–30mm, broadly elliptic to subrotund or ovate, apex rounded to subacute, base cuneate and narrowed with a distinct petiolar part, margins serrate or sometimes doubly serrate, both surfaces pubescent, often mixed glandular and eglandular hairs, sometimes entirely glandular, up to 0.2–0.3(–0.6)mm long. *Flowers* in stout leafy or bracteate thyrses terminating each branchlet, axis sometimes flexuous, cymules dichasial, many-flowered. *Bracts* (lowermost) frequently leaflike or cymules in axils of uppermost leaves, occasionally ± elliptic, entire, c.10–35 × 1.5–12mm. *Peduncles* (lowermost) 2–40mm long, flowers sessile. *Calyx* distinctly bilabiate, anticous lip less deeply divided than posticous, tube 0.2–0.5mm long, lobes c.1.5–2.75 × 0.4–0.8mm, oblong-lanceolate, tips slightly thickened, pubescent all over, hairs 0.2–0.5mm long, balloon-tipped, tiny gland-tipped hairs also present. *Corolla* tube 5.5–7.5 × 1–1.5mm in throat, cylindric, inflated at apex, limb distinctly bilabiate, 5.5–8mm across lateral lobes, posticous lobes 2–3.5 × 1–2mm, anticous lobe 2–3.5 × 0.8–2mm, all elliptic, margins strongly revolute so that tips look pointed, greenish-yellow to pale- or golden-yellow turning orange or brownish, yellow/orange patch in back of throat, sometimes marked with purple-black, pubescent outside, hairs c.0.1mm long, balloon-tipped, inside 4 longitudinal bands of clavate hairs in throat. *Stamens* 4, anticous anthers 0.3–0.5mm long, just included, posticous anthers 0.6–0.8mm long, included, posticous filaments bearded. *Stigma* 1.25–2.25mm long, longer than style, included. *Style* 0.25–1mm long. *Ovary* 0.8–1.4 × 0.5–1mm. *Capsules* c.3.5 × 2mm. *Seeds* c.0.6–0.8 × 0.3mm, dark violet blue.

Selected citations:

Cape. Malmesbury div., 3217DB, Vredenburg, Stompneus Point, 4 ix 1955, *Taylor* 1524 (NBG, SAM); near Veldrif, Shell Bay, 14 viii 1969, *Marsh* 1279 (K, NBG, STE). 3218CC, 4½ miles NNE of Laaiplek, c.10ft, 25 viii 1958, *Acocks* 19688 (BOL). 3318AA, Langebaan, farm Panorama, 50m, x 1976, *Hugo* 611 (PRE, STE). 3318CB–3317BB, zwischen Groenekloof und Saldanha Baai, unter 500 Fuss, ix–x, *Drège* (E, LE, TCD). 3318CD, Bloubergstrand, ± 20m, 3 ix 1988, *Bokelmann & Paine* 21 (E). Bredasdorp div., 3420AC, De Hoop - Potberg Nature Reserve, Dronkvlei, 60m, 11 ix 1979, *Burgers* 2189 (PRE, STE). Riversdale div., 3421AD, Still Bay, viii 1924, *Muir* 3332 (BOL, PRE).

Manulea thyrsiflora grows in scrub on seashore dunes or other sandy places near the sea, flowering between August and October. It exhibits a very odd distribution pattern, having been recorded from the mouth of the Great Berg river along the coast to Blaauwberg Strand, then a great gap until it re-appears on the coast of Bredasdorp and Riversdale divisions, roughly between De Hoop in the west and Still Bay in the east. The two areas are presumably relicts of a once continuous distribution along the Cape coast during the Middle Pleistocene, when the coastline lay much further out than it does today and the great bays between Cape Point and Cape Agulhas were sandy wastes.

There is some variation over the present-day range. In the northernmost localities (Great Berg river to just north of Saldanha Bay), the indumentum on stems and leaves is almost entirely glandular; around Langebaan, balloon-tipped hairs appear on the leaves, especially on the lower surface; from Buck Bay to Blaauwberg Strand, balloon-tipped hairs predominate on the stems and the lower surfaces of the leaves, and this is also so in Bredasdorp and Riversdale divisions. The type specimen is not precisely localized, but as it has predominantly balloon-tipped hairs c.0.2mm long on both stem and leaves, and the lowermost peduncles are c.15mm long, it almost certainly came from the Buck Bay - Blaauwberg Strand area.

Along the southern Cape coast, the cymules tend to have shorter peduncles than specimens from the west coast and the corolla lobes may be broader in relation to their length than in western specimens, but these differences seem too trivial to warrant infraspecific division.

Manulea thyrsiflora is frequently confused with *M. tomentosa* but is is a greener plant (*M. tomentosa* is decidedly grey) with strongly developed axillary leaf tufts, the eglandular hairs on stem, leaf and calyx are distinctly balloon-tipped (± acute in *M. tomentosa*) and shorter, the corolla limb more distinctly bilabiate with narrower lobes, which are more acute to the eye. There is also a fundamental difference in that the stigma is much longer than the style in *M. thyrsiflora*; the reverse is true in *M. tomentosa*.

The relationship of *M. thyrsiflora* lies with *M. virgata*, which is montane in distribution, and easily distinguished by its delicate thyrses and acute hairs on stems, leaves and calyx.

39. Manulea virgata Thunb., Prodr. 101 (1800) & Fl. Cap. ed. Schultes 470 (1823); Hiern in Fl. Cap. 4(2): 240 (1904), p.p.
Type: Cape of Good Hope, without precise locality, *Thunberg* (holo., sheet no. 14385, UPS).

Syn.: *M. exaltata* Benth. in Hook., Comp. Bot. Mag. 1: 384 (1836). Lectotype (chosen here): Cape, Clanwilliam div., Olifants river and Brakfontein, *Ecklon* [& *Zeyher*] (K; isolecto, S, SAM, TCD, W).

Shrublet, stems c.250–450mm long, erect, ascending or decumbent, woody at base and there up to c.4mm diam., herbaceous above, well branched, densely pubescent, hairs up to 0.3–0.6mm long, either all glandular or mixed with acute hairs, leafy throughout. *Leaves* opposite, bases connate, becoming alternate above as they pass imperceptibly into floral bracts, blade c.7–27 × 5–16mm, broadly elliptic or sometimes ovate, apex ± acute, base cuneate and passing into a petiolar part c.3–16mm long (roughly ⅓–½ length of blade), margins coarsely serrate, doubly so in larger leaves, both surfaces pubescent, hairs up to 0.3–0.6mm long, either all gland-tipped or mixed with few to many acute hairs. *Flowers* many in long lax delicate terminal thyrses, either solitary or in few-flowered (up to 11) dichasial cymules. *Bracts* (lowermost) always leaflike, c.6–21 × 3–10mm including the short petiole. *Peduncle* or pedicel (lowermost) c.0.5–11mm long. *Calyx* obscurely bilabiate, tube 0.1–0.3mm long, anticous lobes 2–2.6 × 0.4–0.5mm, posticous lobes 2–2.5 × 0.3–0.4mm, lanceolate, pubescent all over, hairs up to 0.3–0.4mm long, acute, small glandular hairs also present. *Corolla* tube 4.25–5.5 × 0.8–1mm in throat, cylindric, inflated at apex, limb distinctly bilabiate, 6.8–7mm across lateral lobes, posticous lobes 3–3.5 × 0.8–1.1mm, anticous lobe 2.75–3.2 × 0.75–1mm, all lanceolate, acute, margins revolute, yellow turning brown, yellow/orange patch in back of throat, pubescent outside, hairs up to 0.2mm long, inside 4 longitudinal bands of clavate hairs in throat. *Stamens* 4, anticous anthers 0.3–0.5mm long, tips just visible in mouth, posticous anthers 0.6–0.8mm long, included, posticous filaments bearded. *Stigma* 1–1.5mm long, longer than style, included. *Style* 0.2–0.8mm long. *Ovary* 1–1.2 × 0.6–0.8mm. *Capsules* 2.5–3.5 × 1.5–2.5mm, glandular-puberulous. *Seeds* c.0.5–0.6 × 0.3–0.4mm, pallid.

Selected citations:

Cape. Clanwilliam div., 3118DC, Nardouw Kloof, ix 1947, *Stokoe* sub SAM 63922 (SAM). 3218BA, 3½ miles E by N of Graafwater, c.1100ft, 22 viii 1958, *Acocks* 19649 (BOL). 3218BB, N Cedarberg, Pakhuis, 3000ft, 7 ix 1953, *Esterhuysen* 21742 (BOL, PRE, S): 3218DA, Muishoek Berg, 25 x 1938, *Pillans* 8604 (BOL). 3219AA, Pakhuis Pass, 3000ft, 30 ix 1940, *Compton* 9609 (NBG). Without precise locality (but see below), *Masson* β (BM).

Thunberg's specimen is not localized, but it is precisely matched by the Masson specimen cited above. Hiern (loc. cit.) states that Masson's specimen came from Heerenlogement (3118DC) and as Masson and Thunberg travelled together, it is likely that Heerenlogement is also the provenance of Thunberg's specimen. Bentham did not see Thunberg's collection, and he re-described the species as *M. exaltata*, his type material having come from the Olifants River and Brakfontein, south-east of Heerenlogement, and from Bergvalei, south of Heerenlogement. *Manulea virgata* seems to be common in the mountains thereabouts and ranges as far east as the northern Cedarberg and south to Muis Hoek Berg (Olifants River Mountains?). It is probably a constituent of scrub on rocky sites, flowering mainly between August and October.

The plant bears a strong superficial resemblance to *M. ovatifolia* (itself confounded with *M. turritis*) but that species has longer hairs on stem and leaf (up to 0.6–1.2mm), longer corolla tube (6–7.2mm), mostly broader lobes (1.1–1.5mm), only a few hairs in the throat, and the stigma shorter than or about equalling the style.

Manulea virgata is very closely allied to *M. stellata* (see under that species).

40. Manulea stellata Benth. in Hook., Comp. Bot. Mag. 1: 384 (1836) & in DC., Prodr. 10: 366 (1846); Hiern in Fl. Cap. 4(2): 239 (1904).
Type: 'Mountains of Cape and Worcester districts', *Ecklon* (holo. K; iso. SAM, W).

Shrubby, stems to at least 1 m long (only upper parts seen), woody below and there up to 4mm in diam., herbaceous above, laxly well branched, probably sprawling or even scandent (upper part of branches often flexuous), densely pubescent, hairs up to 0.4–0.8mm long, both acute and gland-tipped, leafy throughout. *Leaves* opposite, bases connate, becoming alternate upwards and passing imperceptibly into floral bracts, blade c.8–26 × 5–19mm, ovate, apex ± acute to obtuse, base cuneate and passing into a petiolar part c.2.5–10mm long (roughly ⅓ length of blade), margins coarsely and irregularly lobed and toothed, both surfaces pubescent, hairs up to 0.4–0.6mm long, mixed acute and gland-tipped. *Flowers* many in long lax somewhat flexuous terminal thyrses, cymules 3–9-flowered, dichasial. *Bracts* (lowermost) leaflike, c.9–25 × 5–14mm including short petiole, rapidly smaller upwards. *Peduncles* (lowermost) c.2–20mm long, rapidly shorter upwards. *Calyx* obscurely bilabiate, tube 0.2–0.5mm long, anticous lobes c.2.1–3.2 × 0.4–0.6mm, posticous ones slightly narrower, lanceolate, pubescent all over, hairs up to 0.4–0.5mm long, acute, small glandular hairs also present. *Corolla* tube 5.5–7mm long, cylindric, inflated at apex, limb distinctly bilabiate, 7–10mm across the lateral lobes, posticous lobes c.3.1–5 × 0.8–1mm, anticous lobe c.3.2–5 × 0.6–1mm, all lanceolate, subulate, margins revolute, colour not stated but certainly orange or yellow turning brownish, orange patch in back of throat, pubescent outside, hairs up to 0.2mm long, inside 4 longitudinal bands of clavate hairs in throat. *Stamens* 4, anticous anthers 0.4–0.5mm long, tips visible in mouth, posticous anthers 0.7–0.8mm long, included, posticous filaments bearded. *Stigma* c.1.5mm long, longer than style, included. *Style* c.0.5–1.2mm long. *Ovary* 1–1.25 × 0.6–0.8mm. *Capsules* c.3.5 × 2mm, glandular-puberulous. *Seeds* not seen.

Citations:

Cape. Clanwilliam div., 3218BA, Vogelfontein, 1300ft, 13 viii 1896, *Schlechter* 8512 (BOL, E, PRE, S, W). [3218BC, Lange Valei fide Hiern, loc. cit.] *Masson* α (BM).

Bentham did not give a precise locality for his type material of *M. stellata*, saying merely 'mountains of Cape and Worcester district'. A specimen in SAM is so clearly part of the type collection that I have unhesitatingly designated it an isotype. This sheet is inscribed 'Tulbagh's Kloof, October' in Pappe's neat (but often misleading!) hand. However, the presence of *M. stellata* in the mountains around Tulbagh and Worcester needs confirmation; the other two collections known to me both came from sites west of Clanwilliam, roughly 120km north of Tulbagh, and in the middle of the area of *M. virgata*. That *M. stellata* is really distinct from *M. virgata* needs investigation: I cannot make a sound decision on the material available to me. The specimens have flowers somewhat larger than those of *M. virgata* (as Bentham said) but the distinction is not absolute. There also appears to be a difference in habit, *M. stellata* being a much laxer plant than *M. virgata* with almost scandent branches. However, field investigation is needed to see if this is merely a response to variation in growing conditions.

41. Manulea platystigma Hilliard & Burtt in Notes RBG Edinb. 37: 315 (1979).
Type: Lesotho, Sani Top, rocky ridge across Sani River, 2895m, 2 i 1974, *Hilliard* 5432 (holo. NU; iso. E, K, PRE).

Perennial herb, stock up to c.15mm diam. at crown and there briefly branching in old plants to bear several discrete but closely associated leaf rosettes at ground level, flowering stems several to many, simple, scapose, c.40–300mm long, erect or decumbent, puberulous, hairs up to 0.1mm long, both gland- and balloon-tipped. *Leaves* opposite, bases connate, crowded in radical rosettes, c.15–80 × 4–15mm, oblanceolate to spathulate, apex obtuse, base narrowed into a long flat petiolar part, margins entire or repand to crenate-serrate, thick-textured, both surfaces minutely glandular-puberulous. *Flowers* many in long lax very narrow terminal thyrses, mostly in 2(–3)-flowered cymules, occasionally solitary. *Bracts* (lowermost) 3.5–7 × 1–2mm, lanceolate. *Peduncles* (lowermost) 3–6mm long, flowers shortly pedicellate. *Calyx* bilabiate, tube 1–1.2mm long, anticous lobes 3–4.5 × 0.8–1.2mm, posticous lobes 2.75–4 × 0.6–1.1mm, oblong-lanceolate, puberulous all over outside, hairs up to 0.2–0.3mm long, usually both balloon- and gland-tipped, rarely glandular only. *Corolla* tube 6.5–8mm long, cylindric, inflated in upper part, limb bilabiate, c.9–12mm across lateral lobes, posticous lobes 4–5.5 × 1.5–2.5mm, anticous lobe 4–5.5 × 1.5–2mm, all lobes oblong, margins revolute, dull to bright yellow, sometimes turning deep red-brown, back of throat yellow/orange, puberulous outside, hairs both balloon- and gland-tipped, inside with 4 longitudinal bands of pale clavate hairs in throat. *Stamens* 4, anticous anthers 0.4–0.75mm long, tips visible in mouth, posticous anthers 1.1–1.5mm long, included, posticous filaments bearded. *Stigma* 1.4–2mm long, included. *Style* 0.4–0.9mm long. *Ovary* 1.3–2 × 0.8–1.2mm. *Capsules* c.7 × 3–4mm. *Seeds* (immature) c.8 × 4mm, amber-coloured.

Citations:

Lesotho. 2929AC, Mokhotlong, iii 1949, *Guillarmod* 1146 (PRE). 2929AD, Thabana Ntlenyana, 11000ft, 21 i 1955, *Coetzee* 600 (PRE); ibidem, 18 i 1955, *Guillarmod* 2328 (E). 2929CB, Sani Top, ii 1985, *Manning* 561 (E, K); ibidem, c.2740m, 14 i 1977, *Killick* 4107 (K, PRE). 2929CA, Black Mountains, 10400–10600ft, 13 i 1976, *Hilliard & Burtt* 8762 (E, K).

Manulea platystigma is known only from a small area of eastern Lesotho, from the top of Sani Pass over the Black Mountains to Mokhotlong, and a little further north on Ntabana Ntlenyana. (The record from Naude's Nek in the Cape Drakensberg, published with the original description is an error; the specimen, *Stewart* 1893, proves to be *M. dregei*.) It grows in damp turf, silt patches on rock sheets and on scree, between c.2900 and 3300 m above sea level, flowering mainly in January and February. Specimens from Ntabana Ntlenyana lack balloon hairs on the calyx and Dr Coetzee recorded that the corolla limb turns 'deep red to blackish brown' with an 'unpleasant heavy odour'.

42. Manulea dregei Hilliard & Burtt in Notes RBG Edinb. 40: 289 (1982).
Type: E Cape, 3027CB, Lady Grey dist., Witteberg, Joubert's Pass, c.7700ft, 18 i 1979, *Hilliard & Burtt* 12175 (holo. E; iso. NU, S).

Syn.: [*M. crassifolia* Benth. in Hook., Comp. Bot. Mag. 1: 332 (1836) quoad spec. witbergense, excl. lecto.].
    [*M. bellidifolia* auct. non Benth.; Hiern in Fl. Cap. 4(2): 227 (1904) quoad *Drège* 7919d tantum].

Perennial herb, stock carrot-like, up to c.10mm diam. at apex, crowned with one to several leaf rosettes, stems several to many, scapose, c.70–300mm tall, erect, simple, puberulous, hairs up to 0.1mm long, both balloon- and gland-tipped. *Leaves* opposite, bases connate, crowded in radical rosettes, c.15–75 × 5–13mm, oblanceolate, apex obtuse, base narrowed to a long flat petiolar part, margins obscurely to distinctly serrate, occasionally almost lobulate, both surfaces minutely glandular-puberulous, occasional balloon-tipped hairs also present. *Flowers* many in long lax terminal racemes, usually solitary, occasionally paired. *Bracts* (lowermost) 2–7 × 1–1.6mm, lanceolate. *Pedicels* (lowermost) 2–5mm long. *Calyx* bilabiate, tube 1mm long, anticous lobes c.1.8–3.1 × 0.8–1mm, posticous lobes slightly smaller, oblong-lanceolate, minutely glandular-puberulous all over, balloon-tipped hairs up to 0.1–0.2mm long also present. *Corolla* tube 6–7.5 × 1.2–1.5mm in throat, cylindric, inflated in upper part, limb bilabiate, 6–7mm across lateral lobes, posticous lobes 2.2–4 × 1.2–1.8mm, anticous lobes 2.5–4.5 × 1–1.5mm, all lobes elliptic-oblong, margins ± revolute, dull mustard yellow, back of throat yellow/orange with a dark purplish patch, puberulous outside, hairs balloon-tipped, very small, inside with 4 longitudinal bands of hairs in throat, these often dark in colour. *Stamens* 4, anticous anthers 0.5–0.75mm long, tips visible in throat, posticous anthers 1.4–1.75mm long, included, posticous filaments bearded. *Stigma* 2.4–3.1mm long, much longer than style, included. *Style* 0.5–0.8mm long. *Ovary* 1–2 × 0.6–1mm. *Capsules* c.5–6 × 2.5–3mm. *Seeds* c.0.7–0.8 × 0.3–0.4mm, pallid.

Selected citations:

E Cape. Lady Grey distr., 3027CB, Wittebergen, Joubert, i 1928, *Thode* A504 (PRE). Barkly East distr., 3027DB, Ben McDhui, c.8400ft, 3 ii 1983, *Hilliard & Burtt* 16371 (BOL, E); ibidem, c.9300ft, 11 iii 1904, *Galpin* 6805 (BOL, K); ibidem, 12 i 1984, *Dove* 106 (NBG, PRE). 3028CA, Rhodes to Naude's Nek, 7800ft, 21 ii 1971, *Hilliard & Burtt* 6671 (E, K); near top of Naude's Nek, c.7500ft, 13 xii 1976, *Stewart* 1893 (E); 3km SE of Cairntoul police hut, 2550m, *Phillipson* 716 (K, UPS).

Manulea dregei has been recorded from only a small area of the Witteberg and Cape Drakensberg, between Joubert's Pass above Lady Grey in the west (where Drège first found it about 160 years ago) and Naude's Nek in the east, a linear distance of about 80km. It grows in damp basaltic silt or gravel, between c.2285 and 2835m above sea level, flowering between December and March.

It has been much confused with *M. bellidifolia*, from which it is easily distinguished by its perennial habit, less deeply divided calyx, yellow corolla limb, and much shorter, included, stigma. However, its relationship lies with *M. platystigma* whose area lies roughly 180km north east of that of *M. dregei*, in the mountains of eastern Lesotho. *Manulea platystigma* is distinguished by its flowers being mostly in 2–3-flowered cymules, and having longer calyx lobes, larger corolla limb and shorter, broader and less tapered stigma.

43. Manulea incana Thunb., Prodr. 101 (1800) & Fl. Cap. ed. Schultes 468 (1823); Benth. in Hook., Comp. Bot. Mag. 1: 382 (1836) & in DC., Prodr. 10: 364 (1846); Hiern in Fl. Cap. 4(2): 230 (1904).
Type: C.B.S., *Thunberg* (sheet 14369 in UPS).

Perennial herb, hoary, caespitose, stock woody, up to 15mm in diam. at crown and there producing many dwarf leafy branchlets, flowering stems many, simple, scapose, up to c.60–130mm long, erect or decumbent, puberulous, hairs up to 0.2–0.4mm long, balloon-tipped, smaller glandular hairs also present. *Leaves* opposite, bases connate, crowded at the base of the branchlets, c.10–35 × 2.5–6mm, elliptic tapering to a flat petiolar part accounting for ½–⅔ the total leaf length, apex of blade ± acute, margin obscurely to distinctly toothed, both surfaces puberulous, hairs as on stem. *Flowers* few to many in short terminal racemes, always opposite, usually solitary in each bract-axil, rarely lowermost in 3-flowered dichasial cymules. *Bracts* (lowermost) c.2–5 × 0.8–1.75mm, lanceolate. *Pedicels* (lowermost) 1.5–13mm, shorter upwards. *Calyx* bilabiate, tube c.1.5–2mm long, anticous lobes c.2.5–4 × 1–1.5mm, posticous lobes c.2.5–4 × 0.8–1.2mm, lanceolate, puberulous all over, hairs up to 0.3–0.4mm long, balloon-tipped, tiny glandular hairs also present. *Corolla* tube c.5.5–7 × 1.2–1.4mm in throat, cylindric, inflated in upper part, limb bilabiate, c.6–7.2mm across lateral lobes, posticous lobes c.2.75–3.2 × 1.5–1.8mm, anticous lobe 2.8–3.5 × 1.4–1.8mm, all lobes oblong, margins revolute, yellow, dark reddish patch in back of throat, puberulous outside, hairs balloon-tipped, inside back of throat pubescent with clavate hairs. *Stamens* 4, anticous anthers 0.6–0.8mm long, tips just visible in mouth, posticous anthers 1.1–1.3mm long, included, posticous filaments bearded. *Stigma* 2.4–3.8mm long, included. *Style* 0.4–0.5mm long. *Ovary* 1.1–1.25 × 0.8–1mm. *Capsules* c.5–6 × 2.5mm. *Seeds* not seen.

Citations:

Cape. Sutherland div., 3220AD, farm Uitkyk, top of Sneeuwkrans, 1700m, x 1920, *Marloth* 9732 (PRE); ibidem, 1600m, x 1920, **Marloth** 9873 (PRE); 3220DA/DB, Komsberg escarpment, Besemgoedberg, c.5500ft, 22 ix 1953, *Acocks* 17196 (K). Without precise locality, *Masson* (BM).

Thunberg's and Masson's specimens of *M. incana* are not localized, but they will surely have collected them when they travelled together over the Roggeveld escarpment in 1774. It is there that the species has since been re-collected, first by Marloth in 1920 and then by Acocks in 1953. Acocks recorded 'sandstone slopes, rare' and his specimen was in full flower in September, as was Marloth's in October.

Acock's specimen was misdetermined as *M. altissima*, Marloth's as *M. virgata*, to neither of which does *M. incana* bear any resemblance. It is a dwarf tufted perennial herb, hoary all over (which clearly suggested the trivial name), and must look very pretty in flower, with the yellow corollas contrasting with the grey indumentum. It is allied to *M. dregei* and *M. platystigma*, also montane in their distribution, but occurring much further east in the Drakensberg massif. The grey indumentum will at once distinguish *M. incana*, but it differs further in floral detail, particularly in its longer calyx tube and the disposition of the clavate hairs in the throat.

44. Manulea obovata Benth. in Hook., Comp. Bot. Mag. 1: 383 (1836) & in DC., Prodr. 10: 365 (1846); Hiern in Fl. Cap. 4(2): 234 (1904), p.p.
Lectotype: Cape, Port Elizabeth div., Port Elizabeth, 1836, *Ecklon* 587 (K; isolecto. SAM, W).

Herb, either annual or a short-lived perennial, taproot becoming woody and up to 15mm in diam. at crown, stems c.150–600(–900)mm tall, erect, solitary to several from the crown, simple to laxly branched, pubescent at base, hairs there often entirely gland-tipped, up to 0.3–0.5mm long, mixed with balloon-tipped hairs upwards, upper stem and inflorescence axis either glabrous or with scattered balloon-tipped hairs, leafy throughout but in small plants leaves crowded at base. *Leaves* opposite, bases connate, c.17–90 × 7–25mm, narrowly to broadly elliptic, apex subacute to obtuse, base narrowed to a short petiolar part, margins serrate to crenate, sometimes ± lobulate, both surfaces pubescent with gland-tipped and balloon-tipped hairs up to 0.25–0.5mm long, balloon-tipped hairs particularly common on lower surface. *Flowers* many in long (up to c.300mm) terminal thyrses, the axis occasionally branched, cymules dichasial, many-flowered. *Bracts* (lowermost) c.4–15 × 0.5–3mm, usually elliptic-oblong, rarely leaf-like. *Peduncles* (lowermost) c.3–14mm long, all flowers sessile or nearly so. *Calyx* very obscurely bilabiate, tube 0.1–0.4mm long, lobes subequal, 1.5–2.5 × 0.4–0.6mm, spathulate, thick, rounded on backs, hairs on margins only, up to 0.2–0.25mm long, balloon-tipped, tiny glandular hairs also present. *Corolla* tube 4.5–5.75(–7.5–8) × 0.75–1(–1.2)mm in throat, cylindric, inflated in upper part, limb bilabiate, 4.8–6(–7–8)mm across lateral lobes, posticous lobes 1.6–2.75(–3–3.8) × 1–1.5(–2–2.5)mm, anticous lobe 1.6–2.5(–3–4) × 0.75–1.8 (–2)mm, all lobes elliptic, margins revolute, pale yellow to orange, later brownish, yellow/orange patch in back of throat, backs of lobes and inflated part of tube puberulous, hairs balloon-tipped (hairy all over; see discussion below), sometimes extending in a band down back of tube, inside with 4 longitudinal bands of clavate hairs in throat. *Stamens* 4, anticous anthers 0.25–0.5mm long, tips just visible in throat, posticous anthers 0.6–1mm long, included, posticous filaments bearded. *Stigma* 0.8–1.5(–1.8)mm long, about equalling or slightly longer than style, included. *Style* 0.7–1.25(–1.8–2)mm long. *Ovary* 0.75–1.3(–1.5–1.8) × 0.5–0.75mm. *Capsules* c.2.5–4 × 1.5–2.5mm. *Seeds* c.0.6–0.8 × 0.3–0.4mm, pallid.

Selected citations:

Cape. Humansdorp div., 3424BA, Slang River, xi 1921, *Fourcade* 1789 (BOL, NBG); 3424BB, Kabeljouws, viii 1928, *Fourcade* 3965 (BOL, SAM). Port Elizabeth div., 3325DC, Walmer, 8 xi 1896, *Kensit* sub Bolus 6486 (BOL); 5 miles from Port Elizabeth, 23 viii 1947, *Rodin* 1005 (BOL, E, PRE). Bathurst div., 3326DB, c.10 miles from Port Alfred on East London road, 6 xii 1977, *Hilliard & Burtt* 10893 (E, K). 3327CA, 20km from Port Alfred, Kleinemond, 26 viii 1988, *Bokelmann* 23 (E); ibidem, *Taylor* 4295 (NBG).

Manulea obovata has been recorded along the southern Cape coast at Plettenberg Bay then a disjunction to the Slang River (west of Cape St. Francis) and so along the coast to Kleinemond, east of Port Alfred, among sand dunes or in open sandy places in coastal scrub, never far from the sea, and flowering mainly between August and December.

The measurements in parenthesis refer to specimens from Plettenberg Bay. These not only have larger flowers than typical *M. obovata* but also the leaves are narrower in relation to their length and the corolla tube is hairy all over. The status of this plant remains uncertain, and good collections are needed. The two collections seen by me are *Rogers* 26682 (PRE), Plettenberg Bay, xi 1921, with a second sheet sub *Smart* (PRE ex Transvaal Museum Herbarium), and *Fourcade* 6524 (BOL), sand hills between Plettenberg Bay and Keurboomsriver, 25 xi 1944.

Manulea obovata is a distinctive species without close allies but it is often confused with *M. parviflora*, which not only differs fundamentally in its stigma (c.3–8mm long and usually exserted) but also in its decidedly perennial habit, leaves often longer in relation to their length than those of *M. obovata* and less markedly toothed, and in details of indumentum, on both the vegetative parts and the corolla. The two species appear to be allopatric, *M. parviflora* not having been recorded further south along the south-east African coast than the Umtata river.

45. Manulea chrysantha Hilliard in Notes RBG Edinb. 46:338(1990).
Type: Cape, 3224AC, Aberdeen, 4 v 1935, *L.E. Taylor* 451 (holo. BOL).

Annual herb, stems 1–50 tufted from the base, 15–150mm tall, erect, simple, minutely glandular puberulous plus conspicuous balloon-tipped hairs less than 0.1mm long, leafless. *Leaves* opposite, bases connate, crowded at base, blade c.7–25 × 3–12mm, elliptic to broadly ovate tapering to a broad flat petiolar part c.5–26mm long, (roughly equalling to twice as long as blade), apex subacute to obtuse, margins ± entire to serrate, both surfaces minutely glandular-puberulous with occasional balloon-tipped hairs as well particularly over main veins on lower surface. *Flowers* many in terminal racemes, crowded initially but axis soon elongating, eventually up to 100mm long. *Bracts* (lowermost) 1.5–3 × 0.6–1mm, lanceolate. *Pedicels* (lowermost) 1.5–5mm long, shorter upwards. *Calyx* obscurely bilabiate, tube 0.3–0.5mm long, lobes 2–4 × 0.4–0.7mm, oblong-lanceolate, margins minutely glandular-puberulous, conspicuous balloon-tipped hairs less than 0.1mm long particularly on backs of lobes. *Corolla* tube 3–4.5 × 0.75–1.1mm in throat, cylindric, inflated in upper part, limb obscurely bilabiate, rarely distinctly so, 4–7.5(–9)mm across lateral lobes, posticous lobes 1.6–2.8(–4.1) × 1–1.8mm, anticous lobe 1.6–3(–3.8) × 0.8–1.8mm, all lobes elliptic-oblong, tips ± obtuse, entire, bright yellow (or rarely cream), tube paler with orange patch down back of throat, outside conspicuous balloon-tipped hairs on backs of lobes and inflated part of tube, scattered downwards or wanting there, inside a posticous longitudinal band of clavate hairs together with 3 lesser lateral bands. *Stamens* 4, anticous anthers 0.2–0.4mm long, just included, posticous anthers 0.6–0.8mm long, included, posticous filaments bearded. *Stigma* 1–1.8mm long, deeply included. *Style* 0.2–0.6mm long, much shorter than stigma. *Ovary* 0.8–1.2 × 0.6–0.8mm. *Capsules* 3–4.5 × 2–3.5mm. *Seeds* c.0.5–0.6 × 0.3mm, pallid.

Citations:

Cape. Prince Albert div., 3221DD, 2 miles S of Kruidfontein Station, 1800ft, 24 viii 1963, *Acocks* 23329 (PRE); 3322BC, Meiring's Poort, 2300ft, 28 ix 1938, *Levyns* 6603 (BOL). Beaufort West div., 3222AD, Beaufort West, Karroo National Park, 4000ft, 17 vii 1983, *Braack* 26 (PRE). 3222CC, between Seekoeigat and Kruidfontein, farm Waterval, 850m, 27 x 1991, *Vlok* 2514 (E). 3222DB, farm Rystkuil, 8 x 1983, *Retief & Reid* 222 (PRE). 3223AA, Nelspoort, Courland's

Kloof, vii 1907, *Pearson* 1334 (SAM). Jansenville div., 3224DC, Jansenville, c.2000ft, 24 iv 1954, *Comins* 819 (PRE). Pearston div., 3225CA, between Vogel and Plat rivers on road between Graaff Reinet and Somerset East, 2500ft, v 1868, *Bolus* 1861 (BOL, SAM). Oudtshoorn div., 3322CB, farm Die Krans, 21 vii 1973, *Dahlstrand* 2409 (J). Willowmore div., 3323AD, Willowmore, *West* 218 (BOL). 3323AC, NW of Toorwater, 2500ft, 5 x 1971, *Thompson* 1403 (PRE, STE); near Toorwater, 30 vii 1955, *Van Niekerk* 501 (BOL). Uniondale div., 3323CA, Vet Vlei, vii 1928, *Markotter* sub STE 8807 (STE). 3324BB, 13½ miles ESE of Jansenville, c.1300ft, 26 v 1963, *Acocks* 23273 (K). Uitenhage Carroo, xi 1867, *Prior* s.n. (K).

Manulea chrysantha has been much confused with *M. fragrans* but it is distinguished by its conspicuous balloon-tipped hairs on stems, pedicels, calyces and outside of corolla lobes and tube, as well as by its bright yellow flowers. Its area lies south of that of *M. fragrans*, in the Karroo from the Gamka river in the west to Pearston in the east. It has several times been recorded in river beds, but also in gravel patches on rock outcrops, flowering between April and October.

Thompson 1403 from Toorwater is unusual in that the corolla limb is distinctly bilabiate and larger, with more acute lobes (the measurements in parentheses in the formal description refer to this plant). *Van Niekerk* 501, also from Toorwater, is typical.

46. Manulea exigua Hilliard in Notes RBG Edinb. 46:339(1990).

Type: Cape, Caledon div., Mossel River shore, 24 ix 1952, *Compton* 23617 (holo. NBG, iso. STE).

Annual herb, stems 1 to c.12 tufted from the base, 25–370mm long, main stem usually erect, laterals often decumbent or ascending, simple to branched, lowermost part glandular-puberulous, hairs up to 0.3–0.4mm long, soon mixed with balloon-tipped hairs up to 0.3mm long, apical cell round or elliptic and up to 0.2mm in diam., these hairs rapidly dominant upwards, any glandular hairs up to 0.1mm long and inconspicuous, leafy particularly near the base but a few distant leaves often straggling upwards, subtending any branches and passing into inflorescence bracts. *Leaves* (lower) opposite, bases connate, either opposite or alternate upwards, blade 5–60 × 3–35mm, broadly to narrowly elliptic tapering to a flat petiolar part 4–25mm long (roughly half length of blade), apex obtuse to subacute, margins coarsely serrate or doubly serrate, sometimes lobulate or nearly lyrate, both surfaces puberulous with balloon- and gland-tipped hairs up to 0.2–0.5mm long. *Flowers* many in long terminal thyrses or sometimes racemes in flowering seedlings, the lowermost few flowers in thyrses often solitary, thyrses simple or ± panicled, cymules 2–11-flowered, monochasial or dichasial. *Bracts* (lowermost) often leaflike albeit reduced in size, in small plants in particular commonly oblong-lanceolate, c.3–13 × 0.2–1.5mm. *Peduncles* or pedicels (lowermost) c.(1–)3–12mm long. *Calyx* obscurely bilabiate, tube 0.2–0.75mm long, lobes 1.5–3.6 × 0.3–0.8mm, oblong, ± obtuse, margins glandular-puberulous, hairs up to 0.1mm long, backs of lobes with balloon-tipped hairs up to 0.2mm long, apical cell c.0.1mm in diameter, small glistening glands particularly on tube. *Corolla* tube 2.75–4 × 0.7–0.8mm in throat, cylindric, expanded near apex, limb bilabiate, 3.5–6.5mm across lateral lobes, posticous lobes 1.8–3.2 × 1–1.75mm, anticous lobe 1.6–3 × 0.7–1.5mm, all lobes broadly oblong-elliptic to lanceolate, apex obtuse to acute, margins flat, white, back of tube orange or yellow often marked with a dark violet patch, tube and backs of lobes puberulous with balloon-tipped hairs up to 0.1mm long, inside with 3 or 4 longitudinal bands of clavate hairs. *Stamens* 4, anticous anthers 0.25–0.3mm long, just included, posticous anthers 0.5–0.75mm long, included, posticous filaments bearded. *Stigma* 0.6–1mm long, included, tip occasionally bifid. *Style* 0.2–0.5mm long, shorter than stigma. *Ovary* 0.8–1 × 0.5–0.8mm. *Capsules* c.2.5–3 × 1.5–2.2mm. *Seeds* 0.5–0.8 × 0.3–0.5mm, pallid or blue-violet.

Selected citations:

Cape. Caledon div., 3418BD, Palmiet River, 200ft, 8 xi 1942, *Compton* 14136 (NBG). 3419AA, Houw Hoek, vii 1885, *Bolus* 9920 (BOL). 3419AC, Hermanus, 200ft, xi 1921, *Rogers* 26589 (J, PRE); ibidem, [as Hermanus Pietersfontein], 200ft, vii 1896, *Guthrie* 4142 (NBG); ibidem, 4 x 1983, *Goldblatt* 7035 (NBG, PRE). 3419AD, Mündung der Kleynrivier, viii 1847, *Zeyher* 3499 (BOL, S, SAM, W). 3419CB, Gansbaai, 22 vii 1962, *Walters* 13 (NBG).

Hiern (in Fl. Cap 4(2), 1904) confused *M. exigua* with *M. minor*, and the confusion has persisted until now. Both species have small white flowers, but *M. exigua* is easily distinguished by its more or less leafy stems and leaf-like bracts, much more deeply lobed and toothed leaves, conspicuous balloon-tipped hairs on the inflorescence axis, pedicels, calyx and corolla, and differently shaped corolla lobes. The relationship of *M. exigua* lies with *M. cheiranthus*, which it much resembles in its vegetative parts and in indumentum, but *M. exigua* has white (not yellow to orange or red-brown) flowers with a smaller limb (3.5–6.5mm across the lateral lobes versus 5–16mm), the lateral lobes set at right angles to the posticous lobes (not ascending), all lobes oblong-elliptic to shortly lanceolate, the apex obtuse to acute, margins plane (never long-lanceolate, subulate, margins revolute), the difference in shape reflected in the proportions of breadth to length (posticous lobes in *M. exigua* 1.8–3.2 × 1–1.75mm versus 2.75–10 × 0.5–1.2mm in *M. cheiranthus*).

It seems likely that the two species hybridize. *Manulea exigua* appears to occupy a restricted area mainly in Caledon division, from sea level to c.200 (?)m and wholly within the range of *M. cheiranthus*. Within the area of sympatry (and beyond it), the flowers of *M. cheiranthus* are smaller than in the typical plant with markedly less pointed corolla lobes, as though the two species had introgressed yet retained their separate identities. On the other hand, occasional specimens have most of the characteristics of *M. exigua* but seem to have been contaminated by *M. cheiranthus*. For example, *L. Bolus* 19342 (BOL) from Hermanus comprises 10 mounted specimens plus others in a capsule; most are typical *M. exigua*, but two clearly have orange flowers, one with short obtuse corolla lobes, the other with short acute lobes (the collector noted flowers orange to yellow). Similarly, *McMurtry* 205 (PRE), from Betty's Bay, seems to be a mixture of *M. cheiranthus* and possible hybrids. *Compton* 13533 (NBG) from Hangklip has the aspect of *M. exigua* but the corolla lobes are very acute and the flowers were recorded as 'ochre'. Obviously, field investigations are needed along the coast between Cape Hangklip and Hermanus. But investigations are also needed on the Zwartberg at Caledon. Here we (*Hilliard & Burtt* 13070, E) collected a small (up to c.120mm tall) annual with flowers within the size range of those of *M. exigua* but with very acute lobes (the lateral ones ascending), pink outside in the bud and opening into a pinkish-cream limb. Schlechter (no. 5540, BOL, E, PRE, S, UPS) collected exactly the same plant on the Zwartberg in October 1894, nearly 100 years before us (September 1980). A distinctive entity that has survived for at least a hundred years deserves formal recognition, but I am reluctant to accord it without a local field study.

Manulea exigua seems to grow mainly on sandy flats and slopes and flowers in August and September. Our collection on the Zwartberg came from bare sandy stony ground at an elevation of c.1000m.

47. Manulea adenodes Hilliard in Notes RBG Edinb. 46:337(1990).
Type: Cape, Clanwilliam div., 3219AA, N Cedarberg, Pakhuis, 17 x 1954, *Esterhuysen* 23760 (holo. BOL).

Annual herb, stems 1 to several tufted from the base, 70–150mm long (possibly up to 200mm in fruit), erect or decumbent, simple to sparingly branched at the base, lower part glandular-puberulous, hairs up to 0.4–0.5mm long, mixed with balloon-tipped hairs upwards, these up to 0.3mm long, apical cell c.0.05mm in diameter, gland-tipped hairs up to 0.2mm long, leafy particularly near the base, branches or incipient branches in the axil of a reduced leaf. *Leaves* opposite, bases connate, blade (of principal leaves) c.15–26 × 6–13mm, broadly to narrowly elliptic in outline tapering to a long flat petiolar part c.7–15mm long (about equalling to ⅓ as long as the blade), apex ± obtuse, margins serrate to lyrate, these lobes toothed again, both surfaces glandular-puberulous, hairs up to 0.4–0.6mm long. *Flowers* many in a long lax terminal raceme, usually solitary or sometimes paired in the axil of each bract. *Bracts* (lowermost) 1.5–6 × 0.4–0.7mm, lanceolate or oblong-lanceolate, or occasionally lowermost flower subtended by a much reduced leaf. *Pedicels* (lowermost) 1–5mm long. *Calyx* very obscurely bilabiate, tube 0.3–0.4mm long, lobes 2.2–2.7 × 0.4–0.6mm, lanceolate, acute, puberulous all over with mainly balloon-tipped hairs up to 0.25–0.5mm long (apical cell up to c.0.05mm diam.) mixed with glandular hairs up to 0.3mm long. *Corolla* tube 3.5–3.75 × 0.8–0.9mm in throat, cylindric, expanded near apex, limb bilabiate, 6.5–8.5mm across the lateral lobes, which ascend, leaving the anticous lobe somewhat isolated, posticous lobes 4–5 × 0.6–0.8mm, anticous lobe 2.6–3.5 × 0.5–0.6mm, all lobes long-lanceolate, margins revolute, lobes subulate to eye, orange, back of tube also orange, tube and backs of lobes puberulous with balloon-tipped hairs up to 0.1–0.2mm long, inside with 3 longitudinal bands of clavate hairs on posticous side, lesser band on anticous side. *Stamens* 4, anticous anthers 0.25–0.3mm long, tips just visible in mouth, posticous anthers 0.6–0.7mm long, included; posticous filaments bearded. *Stigma* 0.7–0.9mm long, included, tip sometimes bifid. *Style* 0.3–0.5mm long. *Ovary* 0.9–1 × 0.7–0.8mm. *Capsules* not seen.

Citation:

Cape. Clanwilliam div., 3219AA, N. Cedarberg, Pakhuis, 3000ft, 7 ix 1953, *Esterhuysen* 21807 (BOL).

Manulea adenodes is a delicate-looking plant much like flowering seedlings of *M. cheiranthus* in facies but at once distinguished by the very glandular stems (including inflorescence axes) and leaves, which suggested the trivial name. The balloon-tipped hairs (present chiefly on the inflorescence axes, pedicels and calyces) have the apical cell roughly half the size of that in *M. cheiranthus*; in *M. cheiranthus*, these balloon-tipped hairs predominate on the upper part of the stem and gland-tipped hairs are mostly wanting, or minute; in *M. adenodes*, the upper stem is clad in a mixture of long glandular and balloon-tipped hairs.

Manulea adenodes is known only from the two collections made by Miss Esterhuysen at Pakhuis; in September and October they were still in flower only, and were found growing in sand. The area of typical *M. cheiranthus* lies well south of Pakhuis. In the Pakhuis area, *M. adenodes* may be sympatric with *M. decipiens*, which shares the character of balloon-tipped hairs with a small apical cell, but differs in its shorter indumentum with glandular hairs either wanting or minute on the upper part of the axis, flowers mostly in cymules, corolla limb mostly larger, and posticous filaments glabrous.

48. Manulea cheiranthus (Linn.) Linn., Mant. 88 (Oct. 1767); Thunb., Prodr. 101 (1800) & Fl. Cap. ed. Schultes 471 (1823); Bentham in Hook., Comp. Bot. Mag. 1: 384 (1836) & in DC., Prodr. 10: 366 (1846); Hiern in Fl. Cap. 4(2): 241 (1904) p.p.; Levyns in Adamson & Salter, Fl. Cape Penins. 716 (1950). **Fig. 58E, Plate 2B.**
Type: Commelin, Hort. Amst. 2: 83, t. 42 (1701).

Syn.: *Lobelia cheiranthus* Linn., Sp. Pl. ed. 1: 933 (1753).
   Nemia cheiranthus (Linn.) Berg., Pl. Cap. 160 (Sept. 1767).
   Manulea hirta Gaertn., Fruct. 1: 258, t. 55 f. 1 (1788). Type: C.B.S., *Masson* (holo. BM).
   Nemia capensis J.F. Gmelin, Syst. Nat. 2: 936 (1792), nom. illegit.
   Manulea rhynchantha Link, Enum. Hort. Berol. 2: 142 (1822), fide Bentham (1836). Type: not traced.

Annual herb, stems 1 to many tufted from the base, 35–300(–600)mm long, main stem erect, lateral ones often decumbent or ascending, simple to well branched, lowermost part glandular-pubescent, hairs up to 0.3–0.4mm long, sometimes mixed with balloon-tipped hairs up to 0.2mm long, apical cell c.0.1mm in diam., these hairs rapidly becoming dominant upwards, any glandular hairs minute and very inconspicuous, leafy particularly at the base but leaves often straggling upwards, distant, passing into inflorescence bracts. *Leaves* (lower) opposite, bases connate, either opposite or alternate upwards, blade (of principal leaves) (5–)15–60 × (3–)6–35mm, ovate to elliptic tapering to a flat petiolar part (3–)10–15mm long (about equalling to ½–¼ as long as blade), apex subacute to obtuse, margins coarsely toothed to almost lobed, very rarely less markedly toothed in flowering seedlings, both gland- and balloon-tipped hairs up to 0.2–0.3mm long. *Flowers* many in terminal racemes or thyrses, ± panicled in well branched plants, stems often floriferous almost to the base, flowers either solitary in the axil of each bract (commonplace in seedlings and at the base of a thyrse) or in 2–9-flowered lax or congested monochasial or dichasial cymules (the latter sometimes compounded). *Bracts* (lowermost) often leaflike though usually much reduced in size, or oblong-lanceolate and then c.1.5–8 × 0.4–2mm, the smallest bracts on seedlings. *Pedicels* or peduncles (lowermost) c.1–20mm long. *Calyx* obscurely bilabiate, tube 0.3–1mm long, lobes 1.8–5 × 0.3–0.8mm, oblong-lanceolate, ± acute, puberulous all over with both gland- and balloon-tipped hairs up to 0.2mm long, apical cell of latter up to 0.1mm in diam. *Corolla* tube 2.75–5 × 0.7–1mm in throat, cylindric, expanded near apex, limb bilabiate, 5–16mm across lateral lobes, which ascend, leaving the anticous lobe somewhat isolated, posticous lobes 2.75–10 × 0.5–1.2mm, anticous lobe 2–7.3 × 0.5–1.2mm, all lobes long-lanceolate, margins usually revolute and lobes subulate to eye, sometimes merely acute (see discussion), varying in colour from ochre-yellow to orange or shades of red-brown, back of tube yellow/orange with a dark violet patch between the posticous filaments (possibly not always present), tube and backs of lobes puberulous outside with balloon-tipped hairs up to 0.1mm long, inside with 3 longitudinal bands of clavate hairs on posticous side, sometimes a band on anticous side as well. *Stamens* 4, anticous anthers 0.25–0.3mm long, just visible in mouth, posticous anthers 0.5–0.8mm long, included, posticous filaments bearded. *Stigma* 0.75–1.5mm long, included. *Style* 0.2–0.6mm long, shorter than stigma. *Ovary* 0.8–1.2 × 0.5–0.8mm. *Capsules* c.3.5–5 × 2–2.5mm. *Seeds* c.0.5–0.6 × 0.4mm, white or blue-violet.

Selected citations:

Cape. Malmesbury div., 3318CB, Mamre Hills, 22 ix 1943, *Compton* 14953 (NBG). Cape Town div., 3318CD, Table Mt, 800ft, viii 1881, *Tyson* 827 (SAM); 3418AB, Kalk Bay Mountain, 15 ix 1974, *Goldblatt* 2679 (E, NBG). Paarl div., W foot of Vogelvlei Mts near Gouda, 9 ix 1951, *Esterhuysen* 18838 (BOL, PRE). Caledon div., 3419AD, Stanford, Klein River foothillls, x 1931, *Stokoe* 6485 (BOL). Bredasdorp div., near Klooster, Nachtwacht, c.150ft, ix 1926, *Smith* 3069 (PRE). Mossel Bay div., 12 miles Mossel Bay, George road, near sea level, 7 ix 1947, *Story* 3103 (K, PRE). Knysna div., 3422BB, Ruigte Vlei, 10ft, x 1921, *Fourcade* 1551 (BOL, PRE, SAM, STE).

Gaertner wrote of *M. hirta* (in synonymy above) '*an Selago hirta* .... Linn. ?' and also cited a specimen '*ex herbario Banksiano*'. As Gaertner was primarily interested in fruits and seeds, neither his description nor drawings are adequate for positive identification. However, the Masson specimen quoted above not only carries the annotation '*hirta* Gaertn. Sem. 1 p.258 t. 55' in an old hand (not identified by me) but also bears a strong superficial resemblance to *Selago hirta* (that is, *Chenopodiopsis hirta* (L.f.) Hilliard), which seems sufficient reason for accepting it as the holotype of the name.

Manulea cheiranthus is very variable in aspect; not only do single-stemmed seedlings a few centimetres tall look very different from robust well branched plants, but there is also considerable variation in the inflorescence (which ranges from simple racemes to lax or very congested thyrses), and in the size of the corolla limb, this dictated by the shape of the corolla lobes, which may be subulate or acute. But the species is easily recognized by certain salient features: hairs at base of stem gland-tipped, up to 0.3–0.4mm long, soon replaced upwards by balloon-tipped hairs with an extraordinarily large apical cell (c.0.1mm in diameter or even larger, and present also on pedicels, calyx, and outside of corolla); leaves very coarsely toothed, straggling upwards and passing into leaf-like bracts in all but the smallest plants; corolla lobes subulate to acute, shades of yellow, orange and reddish-brown; posticous filaments bearded.

The type of the name is the plate in Commelin's Hortus Medicus Amstelodamensis, which depicts a tufted plant with remotely leafy stems, a lax inflorescence, and subulate corolla lobes. Commelin received his specimen from Nicolaus Oortmans in 1697 and it almost certainly came from the Peninsula or nearby: despite the fact that no indumentum is indicated, the plate clearly accords with many specimens from the western part of the range of the species, roughly from Piquetberg and the Swartland south to the Peninsula and False Bay and inland to Tulbagh, Stellenbosch and French Hoek. Specimens there, apart from stature and branching, vary chiefly in the degree of congestion of the inflorescence: the extremes are a simple raceme and a very congested thyrse with cymules c.20mm long. It is just such a plant with long cymules borne in congested thyrses that Gaertner described as *Manulea hirta*. However, Masson's specimen has rather small flowers (limb c.6–7.5mm across) with acute rather than subulate corolla lobes and is best matched by specimens from Caledon and Bredasdorp divisions (where he is known to have travelled). *Manulea cheiranthus* ranges eastwards through Caledon and Bredasdorp divisions to Albertinia, Mossel Bay and Knysna, and here many, but not all, specimens tend to have a corolla limb smaller than is usual in the western part of the range, because the corolla lobes are acute rather than subulate. In this they somewhat resemble the allied species, *M. exigua*, with which *M. cheiranthus* is sympatric between roughly Cape Hangklip and Quoin Point, and introgression may take place; see further under *M. exigua*. It also seems possible that *M. cheiranthus* introgresses with *M. decipiens* and *M. pusilla*: see further under those species.

   Manulea cheiranthus has been recorded mainly in sandy soil, sometimes noted as 'damp', on flats and on dunes, also on rocky slopes, sometimes among shrubs in scrub, from sea level to c.600 m, but apparently commonest at low altitudes. It flowers between July and December.

49. Manulea decipiens Hilliard in Notes RBG Edinb. 46:338(1990).
Type: Cape, Van Rhynsdorp div., 3118CD, ± 10 miles east of Doringbaai, farm Kliphoek, 300ft, 29 viii 1970, *Hall* 3795 (holo. STE, iso. PRE).

Annual herb, stems 1–c.15 tufted from the base, 100–410mm tall, erect, simple but occasionally cymosely branched in lower part of inflorescence, puberulous with a mixture of gland-tipped and balloon-tipped hairs up to 0.1–0.25mm long, apical cell of latter c.0.05mm in diam., glandular hairs sparser and shorter upwards, leafy only near base. *Leaves* opposite, bases connate, crowded near base or with 1 or 2 pairs of small leaves subtending the lowermost branches or rarely a much reduced leaf subtending lowermost flower or cymule, blade c.9–35 × 3–17mm, ovate to elliptic tapering into a flat petiolar part c.3.5–20mm long (about equalling to roughly half as long as blade), apex obtuse to subacute, margins subentire to serrate, both surfaces puberulous with gland-tipped hairs up to 0.1–0.3mm long and and occasional balloon-tipped hairs of similar size. *Flowers* many in long lax terminal thyrses, lowermost flower occasionally solitary, otherwise in 2–11(–23)-flowered lax monochasial or dichasial cymules, occasionally the axis cymosely branched into a ± paniculate inflorescence. *Bracts* (lowermost) mostly small, 1.5–5(–10) × 0.6–0.8(–1.5)mm, linear to lanceolate, very rarely with 1–2 pairs of teeth, even more rarely larger and leaf-like. *Peduncles* or *pedicels* (lowermost) 3–15mm. *Calyx* obscurely bilabiate, tube 0.2–0.6mm long, lobes 2.5–4.5 × 0.5–0.6mm, oblong-lanceolate, ± acute, puberulous with glandular hairs up to c.0.1mm long (mainly margins) and balloon-tipped hairs (mainly on backs) up to 0.1–0.2mm long, the apical cell c.0.05mm in diam. *Corolla* tube 3.5–5.5 × 0.8–0.9mm in throat, cylindric, expanded near apex, limb bilabiate, 9–15mm across lateral lobes, which ascend, leaving the anticous lobe somewhat isolated, posticous lobes 5–10 × 0.7–1.2mm, anticous lobe 3–7 × 0.5–0.9mm, all lobes long-lanceolate, subulate to eye, margins revolute, varying in colour from pale to deep yellow to greenish, orange, brown or maroon, back of tube yellow/orange often with a dark violet (or brownish?) patch between the posticous filaments, and probably streaking the base of the posticous lobes, tube and backs of lobes puberulous outside with balloon-tipped hairs less than 0.1mm long, inside with 3 longitudinal bands of clavate hairs on posticous side. *Stamens* 4, anticous anthers 0.2–0.3mm long, just visible in mouth, posticous anthers 0.6–0.7mm long, posticous filaments normally glabrous, sometimes with a few hairs (but see discussion). *Stigma* 0.7–1.6mm long, included. *Style* 0.3–0.75mm long, shorter than stigma. *Ovary* 0.8–1.5 × 0.6–1mm. *Capsules* c.3–4.5 × 2mm. *Seeds* c.0.5 × 0.4mm, violet-blue or pallid.

Selected citations:

Cape. Namaqualand, 2917BA, Klipfontein, viii 1883, *Bolus* 9456 (BOL); 2917DA, Spektakel, 25 viii 1941, *Compton* 11510 (NBG); 2917DD, 10km W of Wildeperdehoek turnoff from Springbok-Garies road, 26 ix 1974, *Goldblatt* 2816

(PRE); 3017 BB, 7 miles N by W of Kamieskroon, c.2500ft, 25 ix 1952, *Acocks* 16486 (PRE). Van Rhynsdorp div., 3118CD, 4 miles from Vredendal/Lambert's Bay on road to Doornbaai, 500ft, 2 viii 1970, *Oliver* 3164 (PRE, STE). Clanwilliam div., 3218BA, 9 miles W of Clanwilliam on Lambert's Bay road, vii 1948, *Lewis* 3038 (SAM).

Manulea decipiens has hitherto been confused with *M. pusilla* and *M. cheiranthus*, which suggested the trivial name. It is allied to both and is more or less intermediate between them. Its area stretches from northern Namaqualand to roughly Verloren Vlei and Clanwilliam; from Clanwilliam to Piquetberg its distinction from *M. cheiranthus* may become blurred: this is discussed further below.

Throughout its range, it is sympatric with *M. pusilla*, with which it shares the characters of leaves crowded at the base and not straggling upwards, the bracts normally very small, the apical cell of the balloon-tipped hairs small (c.0.05mm in diam.), leaves ± entire to serrate and filaments glabrous; they differ in the inflorescence, essentially a lax thyrse in *M. decipiens*, a crowded raceme in *M. pusilla* (this difference best appreciated in specimens passing into fruit), the corolla tube of *M. decipiens* is mostly longer than that of *M. pusilla* (3.5–5.5mm versus 2.8–4mm) and the limb larger (9–15mm across the lateral lobes versus 4.8–8mm).

There are a number of specimens that I cannot place satisfactorily. These are: 2917DB, 2 miles NE of Springbok, *Martin* 508 (NBG; STE), Hester Malan Nature Reserve, *Rösch & le Roux* 751 (PRE); 3017DC, Groen River mouth, *le Roux & Ramsey* 317 (STE), ibidem, *Linder* 4264 (E); 3118AB, Bitterfontein, *Schlechter* 11025 (BOL, E, PRE, S), 6 miles NW Bitterfontein, farm 'Rietputs', *Hall* 4140 (NBG); 6km from Nuwerus, Qaggaskop farm, *le Roux* 2226 (STE), 7km N of Nuwerus, 22 viii 1974, *Nordenstam & Lundgren* 1470 (E, S).

The inflorescence may be a raceme but often with some cymules mixed with the solitary flowers, or it may be a definite thyrse with crowded cymules, different in aspect from the lax thyrse of *M. decipiens* (and more like that of *M. aridicola* from the Richtersveld), and the flowers may be small or large (calyx lobes 2.2–3.8mm long, corolla tube 3–4.25mm, limb 5.5–13mm across the lateral lobes), and in two of the specimens (*Schlechter* 11025 and *Linder* 4264) the tips of the corolla lobes may be subulate, retuse or forked (cf. *M. pusilla*). Clearly, a field investigation is needed in Namaqualand to see if *M. decipiens* hybridizes with *M. pusilla*.

As noted above, *M. decipiens* may be difficult to distinguish from *M. cheiranthus* in their area of sympatry. *Manulea decipiens* differs from *M. cheiranthus* in its normally leafless flowering axes with very small bracts (seldom leaflike as they usually are in *M. cheiranthus*), less deeply toothed leaves, the glandular hairs at the base of the stem shorter (up to 0.25mm not 0.3–0.4mm), the apical cell of the balloon-tipped hairs roughly half the size of that in *M. cheiranthus* (c.0.05mm versus c.0.1mm or even larger) and the filaments glabrous. But some specimens show varying degrees of intermediacy. For example, *Compton* 10937 (NBG; from Alpha, Clanwilliam div.) has the small balloon-tipped hairs of *M. decipiens* and the general facies of that species, but the posticous filaments are lightly bearded. *Compton* 24337 (NBG; from Cardouw, Clanwilliam div.) also has small balloon-tipped hairs, but reduced leaves straggle up the stem and the filaments are bearded. *Barker* 5735 (NBG; from Olifants River dam) is not unlike *Compton* 24337, while *Stirton* 6115 A (PRE; from Verloren Vlei) resembles *Compton* 10937. Whether these, and other, intermediate specimens are the result of introgressive hybridization or of incomplete speciation is a matter for field and laboratory investigation.

50. Manulea pusilla Benth. in Hook., Comp. Bot. Mag. 1: 384 (1836) & in DC., Prodr. 10: 366 (1846); Hiern in Fl. Cap. 4(2): 229 (1904) excluding var. *insigniflora* Diels.
Type: Namaqualand, 2918CC, Zilverfontein, an felsigen Oertern, 2000–3000 Fuss, Sept.–Oct. 1836, *Drège* (holo. K; iso. E, LE, S, TCD, W).

Annual herb, stems one to many tufted from the base, 5–120mm tall, erect, simple or cymosely branched at the base in larger plants, puberulous with a mixture of minute gland-tipped hairs and balloon-tipped ones up to 0.1–0.2mm long, apical cell of latter c.0.05mm in diam., leafy only below. *Leaves* opposite, bases connate, crowded near base or one or two pairs of much reduced leaves subtending a pair of branches, blade c.3–20 × 2–10mm, suborbicular or ovate to elliptic tapering into a flat petiolar part 3–15mm long (about equalling to half as long as blade), apex obtuse to subacute, margins entire to obscurely serrate, rarely distinctly serrulate, both surfaces puberulous with gland-tipped hairs up to 0.2–0.25mm long and occasional balloon-tipped hairs to 0.2–0.5mm. *Flowers* many in terminal racemes accounting for nearly the whole of the length of the stem, flowers solitary in the axil of each bract. *Bracts* (lowermost) 1.5–5 × 0.5–1mm, lanceolate. *Pedicels* (lowermost) c.1–5mm long. *Calyx* obscurely bilabiate, tube 0.3–0.5mm long, lobes 2–3.8 × 0.5–1mm, oblong-lanceolate, ± acute, puberulous with glandular hairs up to c.0.1mm long (mainly margins) and balloon-tipped hairs (all over) up to 0.15–0.25mm long, the apical cell c.0.05mm in diam. *Corolla* tube 2.8–4 × 0.7–1mm in throat, cylindric, expanded near apex, limb bilabiate, 4.8–8(–9.5)mm across the lateral lobes, which ascend leaving the anticous lobe somewhat isolated, posticous lobes 2.5–5(–7) × 0.7–1mm, anticous lobes 2–4.5 × 0.6–0.8(–1)mm, all lobes usually lanceolate with revolute margins and subulate to the eye, occasionally at least some of the lobes retuse to distinctly bifid, varying in colour from greenish or ochre-yellow to orange, sometimes livid, back of tube yellow/orange with a dark violet patch between the posticous filaments, tube and backs of lobes puberulous with mainly balloon-tipped hairs less than 0.1mm long, inside with 3 longitudinal bands of clavate hairs on posticous side. *Stamens* 4, anticous anthers 0.2–0.4mm long, just visible in mouth, posticous anthers 0.5–0.8mm long, included, posticous filaments usually glabrous or sometimes with a few hairs. *Stigma* 1–1.7mm long, included. *Style* 0.1–0.5mm long, shorter than stigma. *Ovary* 0.8–1.25 × 0.5–1mm. *Capsules* c.3–4 × 2–3mm. *Seeds* c.0.4–0.7 × 0.3–0.4mm, pallid or violet-blue.

Selected citations:

Cape. Van Rhynsdorp div., 3118BC, Zout Rivier, 450ft, 12 vii 1896, *Schlechter* 8108 (E, PRE, S); ibidem, ix 1941, *Stokoe* sub SAM 57677 (SAM). 3119AC, Van Rhyns Pass, 28 viii 1941, *Compton* 11433 (NBG). Calvinia div., 3119BD, Akkerdam Nature Reserve below Hantam Peak, 3300ft, 19 viii 1975, *Thompson* 2424 (STE). Ceres div., 3319BD, Bonteberg, 3000ft, 20 ix 1931, *Compton* 3785 (BOL, NGB). Laingsburg div., 3320BA, Whitehill, 20 ix 1943, *Compton* 14916 (NBG).

In the northern part of its range, in Namaqualand and in Van Rhynsdorp and Calvinia divisions, *M. pusilla* always has the posticous filaments glabrous, as they are in the type material, which came from Zilverfontein, south east of Springbok. Then there is a distributional gap across the Tanqua Karroo (possibly not real; the area is ill-served with roads) to Ceres division (Katbakkies Pass, the Bonteberg and Karroo Poort), the southern flank of the Klein Roggeveld and the environs of the Witteberge near Laingsburg and the Warmwatersberg (3320DC). Here, the posticous filaments may be glabrous or furnished with a few hairs, while a few specimens may have the filaments well bearded (eg. *Viviers* 1522 (BOL) from Katbakkies Pass, 3219DC, and *Fellingham* 1219 (PRE, STE) from the farm De Plaat, 3220DC). These southern specimens may also have the corolla limb larger than northern specimens (7.5–9.5mm across the lateral lobes).

Several collections of *M. pusilla* from the northern part of its range show remarkable variation in the tips of the corolla lobes: these are normally subulate and mostly appear so to the unaided eye, but in *Leistner* 469 (PRE) collected at Kareekom, c.25km north by west of Calvinia, close inspection shows that all the specimens on the sheet have the tips minutely retuse to shallowly or more deeply bifid or even trifid. *Schlechter* 8108 (cited above) also has the lobes subulate, retuse or bifid. *Manulea fragrans* shows similar variation. It is suggested that variation in *M. fragrans* may stem from introgression with species that normally have bifid lobes, and *Leistner* 469 (above) may have been contaminated by *M. fragrans*.

Manulea pusilla is characterized by its simple stems running up into racemes, stems, calyx and outside of corolla tube clad in a mixture of gland-tipped and balloon-tipped hairs but the latter usually the most conspicuous and the apical cell small (c.0.05mm diam.), and leaf margins mostly entire or very obscurely serrate; if distinctly serrate, then teeth small. In contrast, *M. cheiranthus* (with which *M. pusilla* is often confused) frequently has the stem well-branched, the leaves are coarsely toothed or almost lobed, the flowers in cymules (they may be solitary in seedlings), and the balloon-tipped hairs have a remarkably large apical cell (c.0.1mm in diam.); balloon-tipped hairs are usually scarce on the lower part of the stem, where the hairs are gland-tipped and up to 0.3–0.4mm long. The area of *M. pusilla* lies mainly north and east of that of *M. cheiranthus*; but the two species may meet in the vicinity of Robertson, with consequent hybridization. The following four specimens seem to be intermediate in character (leaves subentire, inflorescence tending to thyrsoid but apical cell of balloon-tipped hairs small, bracts small, posticous filaments with 1– many hairs): Robertson div., 3319DA, Breede River valley, farm Alfalfa, *Snijman* 731 (NBG). 3319DD, Vrolijkheid Nature Reserve, *van der Merwe* 2526 (PRE); a few miles S of Robertson, *Levyns* 10146 (BOL). Bushman's River, *Lewis* sub BOL 47676 (BOL).

Manulea pusilla grows in sandy places, sometimes in sand-filled depressions over rock sheets or in other rocky places, flowering in August and September.

51. Manulea aridicola Hilliard in Notes RBG Edinb. 46:338(1990).
Type: Cape, Richtersveld, 2816BD, halfway up Numees Peak, W of S entrance Helskloof, 27 ix 1981, *Hugo* 2821 (holo. STE, iso. PRE).
Annual herb, stems few to many from the base, c.50–270mm tall, erect, eventually well branched from the base and above, puberulous with a mixture of balloon- and gland-tipped hairs up to 0.2mm long, apical cell of balloon-tipped hairs up to 0.05mm in diam., remotely leafy. *Leaves* at base of plant opposite, bases connate, becoming alternate upwards and gradually passing into bracts, main leaves c.10–36 × 3–12mm, elliptic tapering into a petiolar part c.4–30mm long (nearly equalling to roughly half the length of the blade), rapidly smaller upwards and ± sessile, apex obtuse to subacute, margins obscurely to distinctly serrate, both surfaces minutely glandular-puberulous, hairs up to c.0.2mm long, occasional balloon-tipped hairs also present. *Flowers* many in narrow thyrses soon cymosely branched into panicles, cymules 2–many-flowered, mostly dichasial. *Bracts* (lowermost) leaf-like in all but the smaller plants. *Pedicels* up to c.5mm long, but mostly very short. *Calyx* obscurely bilabiate, tube 0.2–0.4mm long, lobes 2–2.4 × 0.4–0.6mm, linear-lanceolate, acute, puberulous all over with a mixture of gland- and balloon-tipped hairs up to 0.1–0.2mm long. *Corolla* tube 2.4–2.8 × 0.7–0.9mm in throat, cylindric, expanded above, limb bilabiate, 5–8mm across lateral lobes, which ascend, leaving the anticous lobe somewhat isolated, posticous lobes 2.6–5 × 0.5–0.9mm, anticous lobe 2.1–4 × 0.3–0.8mm, all lobes long-lanceolate, subulate to eye, margins revolute, dull pale yellow to orange, back of tube yellow/orange often with a dark violet patch between the posticous filaments, tube and backs of lobes puberulous outside with balloon-tipped hairs less than 0.1mm long, inside with 3 longitudinal posticous bands of clavate hairs. *Stamens* 4, anticous anthers 0.2–0.3mm long, just visible in mouth, posticous anthers 0.5–0.7mm long, included, filaments glabrous. *Stigma* 0.6–1mm long, included. *Style* 0.1–0.3mm long, shorter than stigma. *Ovary* 0.7–1 × 0.5–0.6mm. *Capsules* c.2.5–4 × 1.75–2.5mm. *Seeds* 0.5–0.8 × 0.3–0.4mm, violet-blue or pallid.

Citations:

Namibia. Sperrgebiet 1, 2816BA, Obibwasser, 17 ix 1973, *Giess* 13017 (PRE, S).
Cape. Richtersveld, 2816BD, mountains SW of Kuboos, 11 ix 1973, *Lavranos* 11026 (PRE). 2817CA, W foothills Cornellsberg, 13 ix 1983, *Williamson* 3291 (BOL); 2817CB, south of Van Zylsrus, 680m, 4 ix 1977, *Thompson & le Roux 320* (STE).

Manulea aridicola, like so many other annuals, will flower when little more than a seedling, but well-grown plants are bushy, the stem branching both from the base and above, and then forming cymose-panicles, the individual branches of which are thyrsoid, with the flowers in 2–many-flowered cymules. This habit will distinguish it from both its allies, *M. decipiens* and

M. pusilla. In *M. decipiens*, the inflorescence is thyrsoid and may branch into a panicle, but the *stems* are simple; in *M. pusilla*, the stems are rarely branched and then only at the base, and the flowers are arranged in simple racemes. *Manulea aridicola* differs further from *M. decipiens* in its mostly shorter calyx lobes (2–2.4mm versus 2.5–4.5mm), shorter corolla tube (2.4–2.8mm versus 3.5–4.5mm) and smaller limb (5–8mm across the lateral lobes versus 9–15mm). Both *M. aridicola* and *M. pusilla* have small flowers, but in *M. aridicola* the corolla tube is often shorter (2.4–2.8mm versus 2.8–4mm) and so is the stigma (0.6–1mm versus 1–1.7mm, measured in mature flowers). *Manulea aridicola* grows much taller than does *M. pusilla*.

Manulea aridicola has been recorded from southernmost Namibia at Obib, not far from the Orange River, then across the river in the Richtersveld. It has been recorded as growing in a sandy river bed and on rocky slopes; all collections, some well on in fruit, were made in September.

52. Manulea minuscula Hilliard in Notes RBG Edinb. 46:340(1990).
Type: Namibia, Fläche östl. der Buchuberge, 2 vii 1929, *Dinter* 6445 (holo. E; iso. BOL, S, SAM, STE, Z).

Syn.: *M. pusilla* auct. non Benth.; Roessler in Prodr. Fl. SWA 126: 32(1967).

Annual herb, stems one to many tufted from the base, 15–150mm tall, erect, simple or cymosely branched at base, very minutely glandular-puberulous particularly near the base, soon glabrescent, sessile glistening glands also present, leafy only near base. *Leaves* opposite, bases connate, crowded at ground level, blade c.3–22 × 1.5–12mm, ovate to elliptic tapering to a flat petiolar part c.3.5–20mm long, often about equalling blade, occasionally only ± half as long, apex obtuse, margins entire to obscurely or distinctly toothed, both surfaces with sessile glistening glands, sometimes very minutely glandular-puberulous as well especially when young. *Flowers* many in terminal racemes or thyrses, nearly the whole of the stem floriferous, flowers solitary in the axil of each bract in tiny plants, otherwise in 2–12-flowered monochasial cymules. *Bracts* (lowermost) 1–3 × 0.5–0.8mm, lanceolate. *Peduncles* or pedicels (lowermost) 1–5mm long. *Calyx* obscurely bilabiate, tube 0.2–0.4mm long, lobes c.1.8–3 × 0.5–1mm at the markedly spathulate tips, margins very minutely glandular-puberulous, occasional sessile glistening glands also present there and on the backs. *Corolla* tube 2.5–3.5 × 0.6–0.9mm in throat, cylindric, abruptly expanded near apex, limb bilabiate, 3.4–6mm across lateral lobes, posticous lobes 1.8–3.5 × 0.5–1mm, anticous lobes 1.5–2.5 × 0.5–0.8mm, all lanceolate, acute, margins ± revolute, yellow to orange, the colour running down back of throat, with a dark violet patch there, tube minutely and sparsely puberulous to glabrous outside, glistening glands on backs of lobes, inside with 3 longitudinal bands of clavate hairs on posticous side. *Stamens* 4, anticous anthers 0.2–0.3mm long, visible in mouth, posticous anthers 0.4–0.6mm long, included, posticous filaments usually glabrous, very rarely with 1–3 hairs. *Stigma* 0.4–1mm long, included. *Style* 0.3–0.5mm long, shorter than stigma. *Ovary* 0.75–1 × 0.5–0.7mm. *Capsules* 1.5–3.5 × 1–2.5mm. *Seeds* 0.6–0.8 × 0.3–0.5mm, violet-blue to white.

Citations:

Namibia. Lüderitz-Süd, 2715BC–BD, Klinghardtgebirge, 16 ix 1922, *Dinter* 3904 (BOL, SAM, Z); Buntfeldschuh, 8 ix 1922, *Dinter* 3768 (BOL, K, PRE, Z). 2715DC, 10 miles south of Chameis, 11 ix 1977, *Williamson* 2651 (BOL). 2816BA, Schakal Mountain, 2 viii 1977, *Müller* 794 (PRE).
Cape. Namaqualand, 2816DA, near the mouth of the Gariep [Orange], 1836, *Drège* (K, mounted with the lectotype of *M. gariepina*); 4 miles S of Oppenheimer Bridge, c.300ft, 19 v 1969, *Leistner* 3446 (PRE). 2816DC, summit of Witbank, x 1926, *Pillans* 5142 (BOL, K).

Dinter gave the manuscript name *M. minuscula* to this plant, and Range published it as a *nomen nudum* in Fedde Repert. 38: 264 (1935). But the species was originally collected by Drège in 1836, together with *M. gariepina*. Bentham, when the two Drège collections came into his hands, apparently failed to discriminate between them, and published for them the name *M.*

gariepina. It was left to Roessler, in 1967, to lectotypify the name *M. gariepina* by the 'b' collection; the 'a' collection was determined by Roessler as *M. pusilla*. But *M. minuscula* is easily distinguished from *M. pusilla* by its very different indumentum: minute 2-celled glandular hairs as well as more conspicuous ± sessile glistening glands, readily seen on the calyx and on the backs of the corolla lobes; in *M. pusilla* on the other hand, the most conspicuous hairs are balloon-tipped and clothe stems, pedicels, calyx and the outside of the corolla. The two species differ further in the inflorescence (thyrsoid in *M. minuscula*, racemose in *M. pusilla*), and in the shape of the calyx lobes (markedly spathulate in *M. minuscula*, tips acute or subacute in *M. pusilla*). Their areas overlap in the vicinity of the Orange River (Gariep).

Manulea minuscula ranges from the extreme southwestern part of Namibia, from roughly the Klinghardtberge southwards to the Orange river and so into the extreme north western part of Namaqualand, as far south as Witbank, between Holgat and Port Nolloth. It has been found in sandy watercourses, in 'deep red sand' and on 'coarse sandy flats', while Pillans recorded 'on recently exposed lime soil around mouse holes'! The plants flower between July and October.

53. Manulea gariepina Benth. in Hook., Comp. Bot. Mag. 1: 384 (1836) & in DC., Prodr. 10: 366 (1846); Hiern in Fl. Cap. 4(2): 242 (1904) p.p.; Merxmüller & Roessler, Prodr. Fl. SWA 126: 31 (1967) p.p. excluding *M. gariepina* subsp. *namibensis* and *M. leptosiphon*.
Lectotype: Cape, Namaqualand, near the mouth of the Orange River (Gariep), below 600ft, 1836, *Drège* 7918 (K).

Syn.: *M. campestris* Hiern in Fl. Cap. 4(2): 240 (1904). Type: Cape, Griqualand West, Dec. 1876, *Mrs Barber* 7 (holo. K).
    M. gariepina subsp. *campestris* (Hiern) Roessler in Mitt. Bot. München 6: 15 (1966) & in Prodr. Fl. SW Afr. 126: 31 (1967).

Annual herb, stems 1 to many tufted from the base, 70–360mm long, erect, simple below, often branching above into a cymose panicle, glandular-puberulous in lower part, hairs less than 0.1mm long, occasionally with a few eglandular hairs to 0.2mm as well, becoming glabrous to very sparsely glandular upwards, leafless. *Leaves* opposite, bases connate, crowded at base of plant, blade 8–33 × 4–18mm, oblanceolate to elliptic or ovate tapering into a flat petiolar part 6–35mm long (roughly equal to twice as long as blade) apex obtuse to subacute, margins entire to serrate, both surfaces glandular-puberulous, hairs up to 0.1mm long, sessile glistening glands as well. *Flowers* many in terminal racemes, simple at first, later cymosely branched into lax panicles. *Bracts* (lowermost) 1–2.2 × 1–2mm, ovate, acute. *Pedicels* (lowermost) 0.75–3(–5)mm long. *Calyx* obscurely bilabiate, tube 0.3–1mm long, lobes 1.2–2 × 0.4–0.7mm, oblong to subspathulate, margins minutely glandular-puberulous, occasional ± sessile glistening glands present, also on tube. *Corolla* tube 5–7 × 0.8–1mm in throat, cylindric, inflated near apex, limb bilabiate, 6–10.5mm across lateral lobes, posticous lobes 3–6.5 × 0.75–1.1mm, anticous lobe 2.8–5 × 0.7–1mm, all lanceolate, acute, margins strongly revolute, varying from shades of yellow (cream, fawn, mustard yellow) to brown or reddish brown, the tube often blackish-violet on the posticous side (or rarely wholly dark violet) with a yellow to orange patch on the anticous side of the inflated part, usually glabrous outside, occasionally minutely and sparsely glandular, glistening glands on backs of lobes, inside either glabrous or with a posticous longitudinal band of clavate hairs, rarely flanked by two small lateral patches. *Stamens* 4, anticous anthers 0.4–0.5mm long, tips visible in mouth, posticous anthers 0.7–1mm long, included, posticous filaments bearded. *Stigma* 1.2–2mm long, tip sometimes bifid, included. *Style* 0.6–1.2mm long, much shorter than stigma. *Ovary* 1.3–2 × 0.6–1mm. *Capsules* c.3.5–6 × 2–3.3mm. *Seeds* 0.5–1 × 0.3–0.5mm, violet-blue.

Selected citations:

Namibia. 2716DA, 27km N of Rosh Pinah, 12 x 1978, *Hardy & Venter* 5094 (PRE). 2716DD, farm Namuskluft (LUS 88), 27 ix 1977, *Merxmüller & Giess* 32346 (PRE). 2717DD, Kannebis, x 1923, *Dinter* 5034 (BOL, PRE, SAM, Z). 2718BC, farm Us (KEE 162), 17 v 1972, *Giess & Müller* 12012 (PRE). 2818BC, Warmbad, v, *Fleck* s.n. (Z).

Cape. Gordonia, 2620BC, Kalahari Gemsbok National Park, 23 miles NE of Twee Rivieren, 2800ft, 19 ix 1959, *Leistner* 1489 (PRE). Kimberley div., 2824CA, 3 miles E of Schmidtsdrift, 3400ft, 28 ix 1957, *Leistner* 929 (PRE). Prieska div., 2820AC, Aughrabies, 21 viii 1954, *Compton* 24466 (NBG). Namaqualand, 2817AC, Richtersveld, Rosyntjiesberg, 30 viii 1977, *Oliver et al* 272 (PRE, STE). 2816DA, between Arris Drift and Anisfontein, x 1926, *Pillans* 5043 (BOL). 2917DB, 2 miles E of Springbok, 8 ix 1950, *Compton* 22052 (NBG). 2917DD, Mesklip, 2000ft, 16 ix 1897, *Schlechter* 11282 (E, PRE, S).

The type of *M. gariepina* is a well-grown plant with cymose-paniculate inflorescences; that of *M. campestris* comprises several young plants with apparently simple racemes just coming into flower. However, a bud is interposed between each bract and primary flower, giving the potential to branch, and a wide range of material shows all degrees of development in the branching of the inflorescence. The other supposed difference between *M. gariepina* and *M. campestris* (given by Roessler in his key in Mitt. Bot. München 6: 13 and Prodr. Fl. SW Afr. 126: 30), namely length of corolla tube, is also untenable.

Manulea leptosiphon, reduced to subsp. *campestris* by Roessler, is a distinct species with a long style just reaching the mouth of the corolla tube; its relationship lies with *M. burchellii* in section *Dolichoglossa*. Subsp. *namibensis* has been raised to specific rank: see under that name for distinguishing characters.

Manulea gariepina is widely distributed across southern Namibia, ranging from about Tiras in the west and Koes in the east south to the Orange river (the Gariep) thence into the Richtersveld in the Cape, south to the environs of Sprinkbok and Aalwynsfontein thence east in a broad band almost as far as Kimberley. The plants have been recorded in red Kalahari sand, on the flats or between dunes, sometimes over calcrete or gravel, also in sandy river beds, flowering between April and October.

54. Manulea deserticola Hilliard in Notes RBG Edinb. 46:338(1990).
Type: Orange Free State, Fauresmith district, 2925AD, Koksfontein, 24 ix 1935, *Pont* 2816 (holo. PRE).

Annual herb, stems several to many from the crown, tufted in discrete rosettes, c.30–110mm long, erect or decumbent, simple, minutely puberulous with a mixture of many gland-tipped and few balloon-tipped hairs, leafless except at base. *Leaves* opposite, bases connate, crowded at ground level, blade c.4–25 × 2.5–12mm, elliptic to oblanceolate tapering to a flat petiolar part c.6–20mm long (about equalling to up to twice as long as blade), apex rounded to very obtuse, margins entire to obscurely serrate or crenate, both surfaces minutely glandular-puberulous, with a few minute balloon-tipped hairs as well. *Flowers* many in terminal racemes, distant. *Bracts* (lowermost) c.1–2 × 1mm, broadly lanceolate. *Pedicels* (lowermost) c.1–1.5mm long. *Calyx* obscurely bilabiate, tube 0.2–0.5mm long, lobes 2.2–2.9 × 0.4–0.6mm, oblong-lanceolate, margins minutely glandular-puberulous, balloon-tipped hairs less than 0.1mm long also present, particularly on backs of lobes and outside of tube. *Corolla* tube 4.5–5.2 × 0.8–0.9mm in throat, cylindric, expanded at apex, limb bilabiate, 6.2–8.5mm across lateral lobes, posticous lobes 4–4.8 × 0.7–1mm, anticous lobes 3.5–3.8 × 0.6–1mm, all lobes lanceolate, tip ± acute or retuse, margins flat to ± revolute, 'brown yellow', back of tube orange/yellow, corolla tube and lobes minutely pubescent outside, hairs less than 0.1mm long, balloon-tipped, inside 4 longitudinal bands of clavate hairs in throat. *Stamens* 4, anticous anthers 0.3–0.4mm long, just included, posticous anthers 0.7–1mm long, included, posticous filaments bearded. *Stigma* 1–1.7mm long, deeply included. *Style* 0.2–0.6mm long, much shorter than stigma. *Ovary* 1.4–1.6 × 0.8–1mm. *Capsules* c.3.5–5 × 2–3mm. *Seeds* c.0.3–0.5 × 0.3mm, blue-violet.

Citations:

Orange Free State. Modder River, vii 1907, *Gilfillan* s.n. (PRE).
Cape. Kimberley div., 3024AA?, Alexandersfontein, Willowbank, viii 1923, *Stowe* 2331 (BOL).

Manulea deserticola much resembles *M. fragrans* in facies but differs in the presence of balloon-tipped hairs on the stem and inflorescence axis (sometimes glabrescent), also on the calyx and outside of the corolla, mostly shorter pedicels (lowermost 1–1.5mm versus 1–6mm) and in the lanceolate (not oblong) and larger corolla lobes (posticous lobes 4–4.8 × 0.7–1mm versus 1.75–3.5 × 0.8–1.6mm, anticous lobes 3.5–3.8 × 0.6–1mm versus 2–3 × 0.8–1.8mm). *Manulea fragrans* has an almost regular limb with a yellow star-shaped patch around the mouth; in *M. deserticola*, the limb is distinctly bilabiate and in the specimens before me no trace is evident of a differently coloured patch around the mouth: this is in any case highly improbable in a limb of this shape.

The true relationship of *M. deserticola* possibly lies with *M. gariepina*. In that species, the inflorescence is often branched (probably always simple in *M. deserticola*) but they differ further in indumentum (balloon-tipped hairs wanting in *M. gariepina* and the inflorescence axis ± glabrous, outside of corolla tube also ± glabrous), calyx lobes longer in *M. deserticola* (2.2–2.9mm versus 1.2–2mm), corolla tube often shorter (4.5–5.2mm versus 5–7mm), also the style (0.2–0.6mm versus 0.6–1.2mm).

The area of *M. deserticola* apparently lies immediately east of that of *M. gariepina* and *M. fragrans* in the south western part of the Orange Free State and neighbouring part of the Cape. However, the species is known from only three collections, two of which cannot be localized precisely. Pont recorded 'plenty near windmill' and Mrs Stowe 'abundant on sand'. Flowering takes place between July and September.

55. Manulea sp.

Annual herb, stem loosely branched from the base, c.90–200mm tall, glandular-puberulous, hairs up to c.0.3mm long, on inflorescence axis scattered balloon-tipped hairs as well, apical cell up to 0.1mm in diam., leafy only near base. *Leaves* opposite, bases connate, blade c.8–22 × 4–11mm, elliptic tapering to a flat petiolar part c.3–11mm long (roughly ⅓–½ as long as blade), apex obtuse, margins obscurely to distinctly serrate, both surfaces glandular-pubescent, hairs up to 0.5mm long. *Flowers* many, in terminal racemes, distant. *Bracts* (lowermost) c.1.5–2 × 0.4–0.5mm, linear-lanceolate. *Pedicels* (lowermost) c.3–13mm long. *Calyx* obscurely bilabiate, tube c.0.2–0.3mm long, lobes c.2.8–3.5 × 0.5mm, oblong-lanceolate, puberulous all over with a mixture of gland-tipped and balloon-tipped hairs up to 0.2mm long. *Corolla* tube c.5.5–6 × 1.1–1.2mm in throat, cylindric, expanded in upper half, limb bilabiate, c.9mm across the lateral lobes, posticous lobes c.3.5–4 × 2.2–3mm, anticous lobe c.2.5–3.5 × 2.2–3mm, all lobes oblong-elliptic, obtuse, margins flat, colour unknown, tube minutely puberulous outside, hairs balloon-tipped, ± sessile glistening glands on backs of lobes, inside posticous and anticous longitudinal bands of clavate hairs between stamens thence a transverse band round throat. *Stamens* 4, anticous anthers c.0.3mm long, included, posticous anthers c.0.8–1mm long, included, posticous filaments bearded. *Stigma* c.1.2–1.4mm long, deeply included. *Style* 0.6–0.7mm long. *Ovary* c.1.2 × 0.8mm. *Capsules* c.3–4 × 2.25–2.5mm. *Seeds* c.0.7 × 0.5mm, pallid.

Griqualand West div., 2924BA, Modder River, dry veldt, xii, *Lamb* 2024 (BOL).

This plant is known only from Lamb's collection, which consists of two similar specimens. It may be an undescribed species but it is also possible that it is of hybrid origin with *M. plurirosulata* as one parent. There is a strong superficial resemblance between Lamb's plant and *M. plurirosulata*, but it differs in having longer glandular hairs on stem and calyx and more conspicuous balloon-tipped ones, longer pedicels, shorter corolla tube clad in balloon-tipped hairs, smaller anthers and much shorter stigma. *Manulea plurirosulata* belongs in section Dolichoglossa; the style in Lamb's plant is typical of that found in section Manulea. Three species in this section (*M. deserticola*, *M. fragrans*, *M. gariepina*) occur in the same general area as Lamb's plant, while *M. plurirosulata* has been recorded immediately to the east of

Griqualand West, in the southern Orange Free State and thence south into the north east Cape. Of the three species in section Manulea, *M. deserticola* has the conspicuous balloon-tipped hairs that are a feature of Lamb's plant, which differs from *M. deserticola* in its longer pedicels, somewhat longer calyx lobes clad in longer hairs, longer corolla tube and shorter, broader, obtuse corolla lobes.

Further collecting in Griqualand West and the western Orange Free State is highly desirable.

56. Manulea tenella Hilliard in Notes RBG Edinb. 46:341(1990).

Type: Namibia, Windhoek div., 2318AB, Gameros, xii 1912, *Dinter* 2677 (holo. SAM).

Annual herb, stems several to many tufted from the base, c.210–300mm long, erect (sometimes ± prostrate?), usually simple, rarely a branch from the inflorescence axis, glandular-puberulous, hairs up to 0.2mm long, becoming shorter and sparser upwards, leafy only near base. *Leaves* opposite, bases connate, crowded near base of plant, blade c.11–27 × 3–10mm, elliptic tapering to a flat petiolar part c.16–30mm long (about equalling to twice as long as blade), apex obtuse to subacute, margins entire to very obscurely toothed, both surfaces glandular-puberulous, hairs up to 0.25–0.3mm long, sessile glistening glands as well. *Flowers* few to many in lax terminal racemes, usually simple, rarely with one branch. *Bracts* (lowermost) 1.5–2.5 × 0.6–1mm, oblong-lanceolate. *Pedicels* (lowermost) 1–2mm long. *Calyx* obscurely bilabiate, tube 0.3–0.4mm long, lobes 1.4–2 × 0.4–0.5mm, oblong-lanceolate, margins glandular-puberulous, hairs up to 0.2mm long, a few sometimes on backs as well, sessile glistening glands also present. *Corolla* tube 4.5–7.8 × 0.5–0.8mm in throat, cylindric, abruptly expanded in upper half, limb bilabiate, 5–8mm across lateral lobes, posticous lobes 2–4.3 × 0.5–0.7mm, anticous lobe 2–3.5 × 0.5–0.6mm, all lanceolate, acute, margins strongly revolute, brown (*fide* Dinter) (probably shades of yellow to brown), tube light to dark violet on posticous side with a yellow to orange patch on anticous side in throat, minutely and sparsely glandular-puberulous outside, glistening glands on backs of lobes, glabrous inside. *Stamens* 4, anticous anthers 0.3mm long, just included, posticous anthers 0.6–0.7mm long, deeply included, posticous filaments bearded. *Stigma* 0.7–1mm long, included. *Style* 0.8–1mm long, about equalling stigma. *Ovary* 1–1.2 × 0.7mm. *Capsules* c.4.5 × 3mm. *Seeds* not seen.

Citation:

Namibia. Kalahari, 1890, *Fenchel* 448 (Z).

Manulea tenella is distinguished from *M. gariepina* by its longer glandular hairs, those on stem and calyx up to 0.2mm, on the leaf up to 0.25–0.3mm (less than 0.1mm in *M. gariepina*), simple racemes, and oblong-lanceolate (not ovate) bracts (lowermost 1.5–2.5 × 0.6–1mm versus 1–2.2 × 1–2mm). It is a thoroughly ill-known species, the two meagre collections having been made in 1890 and 1912, that of Dinter near Gameros, that of Fenchel not precisely localized but his no. 445, type of *M. leptosiphon*, and no. 446, *M. burchellii*, were collected at or near Rietfontein, just inside the Cape border en route Keetmanshoop. If nos 445, 446 and 448 were indeed collected in more or less the same locality, then *M. tenella* has a linear range of at least 175km.

The flowering stems are very delicate (whence the specific epithet), the oldest, fruiting, ones less than 1mm in diameter (comparable stems in *M. gariepina* are c.1.2–1.5mm in diam.). In the type specimens (what appear to be two pieces of the same plant mounted on separate sheets) the stems are so delicate that one appears to have been decumbent and produced a single branch where it turned upwards, in contrast to the erect panicles of *M. gariepina*.

57. Manulea namibensis (Roessler) Hilliard in Notes RBG Edinb. 46: 340(1990).

Type: Namibia, Aus, an der Strasse nach Lüderitzbucht, *Merxmüller & Giess* 2936 (holo. M, n.v.).

Syn.: *M. gariepina* subsp. *namibensis* Roessler in Mitt. Bot. München 6: 14 (1966) & in Prodr. Fl. S. W. Afr. 126: 32 (1967).

M. gracillima [Dinter ex] Range in Feddes, Repert. 38: 263 (1935), nomen nudum.

Annual herb, stems 1–c.8 tufted from the base, 60–350mm long erect (or occasionally ± prostrate?), simple below, sometimes branching above into a cymose panicle, glandular-puberulous near base, hairs up to 0.1–0.2mm long, rapidly glabrous upwards or with very sparse scattered hairs, leafless. *Leaves* opposite, bases connate, crowded at base of plant, blade c.8–30 × 4–18mm, ovate to elliptic tapering into a flat petiolar part 3–26mm long (about equalling to nearly three times as long as blade), apex obtuse to subacute, margins entire to serrate, both surfaces glandular-puberulous, hairs up to 0.2mm long, sessile glistening glands as well. *Flowers* few to many in a terminal thyrse, cymules few- to many-flowered, very congested, thyrse sometimes branched into a lax panicle, branches up to 60mm long. *Bracts* (lowermost) 1–2 × 0.5–1mm, ovate, acute (branches occasionally subtended by a pair of greatly reduced leaves). *Peduncles* up to c.1mm long, both cymules and flowers usually sessile. *Calyx* obscurely bilabiate, tube 0.3–0.4mm long, lobes 1.2–2.2 × 0.4–0.6mm, oblong, margins glandular-puberulous, hairs up to 0.1mm long, occasional hairs on backs, sessile glistening glands present, also on tube. *Corolla* tube (4.6–)5–7.5 × 0.9–1.2mm in throat, cylindric, swollen in upper part, limb bilabiate, 5–7mm across lateral lobes, posticous lobes 2.2–3.1 × 0.8–1.5mm, anticous lobe (1.6–)2.2–3 × 0.8–1.3mm, all lobes oblong-elliptic, margins somewhat revolute, varying from greenish-brown to orange, dark brown or reddish brown, tube often dark violet, yellowish to orange on anticous side of inflated part, puberulous on outside of upper part, hairs minutely balloon-tipped, glistening glands on backs of lobes, inside glabrous or with a few hairs in posticous position. *Stamens* 4, anticous anthers 0.5–0.7mm long, tips visible in mouth, posticous anthers 0.7–1.1mm, included, posticous filaments glabrous or with a few hairs. *Stigma* 0.6–1.5mm long, included. *Style* 0.9–1.8mm, longer than stigma. *Ovary* 1–1.25 × 0.6–1.2mm. *Capsules* c.4–5 × 3mm or occasionally only 1.5 × 1mm and containing only 4 seeds. *Seeds* c.0.8 × 0.3mm, violet-blue.

Citations:

Namibia. Lüderitz-Süd, 2615CA, bei Lüderitzbucht, 1300 m, ix 1911, *Range* 1136 (SAM, fragment); 2616CA, Farm Klein Aus, 26 vi 1949, *Kinges* 2278 (PRE); 2616CB, Schakalskuppen, 1550m, 13 vii 1913, *Range* 1773 (SAM, fragment); 2616CC, Tsirub, 1300m, 18 viii 1918, *Range* 1849 (SAM, fragment); ibidem, 21 iii 1885, *Schinz* 361 (Z, fragment). 2715BC, Klinghardt Mts, 28 vii 1977, *Müller* 822 (PRE). 2716DD, nördlich Rosh Pinah, 15 viii 1976, *Giess* 14662 (PRE). 2816BA, 1.5km nördlich Obib Wasser, 17 ix 1973, *Giess* 12998 (PRE, S). Not precisely localized: Kleinfontein, 1300m, 3 vii 1922, *Dinter* 3596 (BOL, PRE, Z); Halenburg, 14 x 1922, *Dinter* 4076 (PRE, S, SAM, Z). Klein Kaus (Karis?) - Groendoorn, 6 vi 1931, *Örtendahl* 349 (PRE, S, UPS); Garub, 900 m, x 1907, *Range* 505 (SAM).

Manulea namibensis differs from *M. gariepina* in the longer glandular pubescence on stem and leaves, flowers in cymules, and stigma shorter than the style. This last character is unusual in Section *Manulea* but is characteristic of *M. tomentosa* and its allies in Section *Thyrsiflorae*. The species of Section *Thyrsiflorae* all have a conical ovary developing into an almost beaked capsule; the capsule of *M. namibensis* is decidedly obtuse, and there is little doubt that it is correctly placed in Section *Manulea*.

Manulea namibensis has been recorded only from the district of Lüderitz Süd in south west Namibia, where it grows in river beds and on sand flats, flowering between June and October. It seems possible that *M. namibensis* can reproduce cleistogamously; the evidence for this is tiny capsules with four normal seeds produced in addition to normal capsules (*Giess* 14662, cited above).

58. Manulea fragrans Schltr. in Bot. Jahrb. 27: 179 (1899); Hiern in Fl. Cap. 4(2): 225 (1904) quoad typus et *Johanssen* 4 tantum. **Plate 3M.**
Lectotype (chosen here): Van Rhynsdorp div., [3118BC] Zoutrivier, c.450 ped., 12 vii 1896, *Schlechter* 8110 (K; isolecto. BOL, E, PRE, S).

Syn.: ? *M. pusilla* var. *insigniflora* Diels in Bot. Jahrb. 23: 479 (1897). Type: Cape, Calvinia div., Hantam Mountains, *Meyer* (B†).

Annual herb, stems several to many from the crown, tufted in discrete rosettes, 10–150mm long, erect to more or less prostrate, simple, minutely glandular-puberulous, often glabrescent, occasional sessile glistening glands as well, leafless except at base. *Leaves* opposite, bases connate, crowded at base in a ± spreading rosette, blade 2.5–20 × 2–10mm, elliptic to subrotund tapering to a broad flat petiolar part 2–35mm long (roughly equalling to twice as long as blade), apex very obtuse, margins entire to obscurely or distinctly serrate or crenate, both surfaces minutely glandular-puberulous, soon glabrescent but sessile glistening glands often present. *Flowers* many in terminal racemes, crowded initially, distant in fruit. *Bracts* (lowermost) c.1–4 × 0.4–2mm, lanceolate or linear-lanceolate. *Pedicels* (lowermost) c.1–6mm long. *Calyx* very obscurely bilabiate, tube (0.1–)0.3–0.5(–0.6)mm long, lobes (1.6–)2.5–4.5 × 0.4–0.75mm, oblong-lanceolate, margins minutely glandular-puberulous, a few sessile glistening glands also present and on calyx tube. *Corolla* tube 3.5–4.5 × 0.75–1mm in throat, cylindric, inflated in upper part, limb obscurely bilabiate, 5.5–7.5mm across lateral lobes, posticous lobes 1.75–3.5 × 0.8–1.6mm, anticous lobe 2–3 × 0.8–1.8mm, all lobes oblong, tips either ± acute to obtuse or shortly (rarely deeply) bifid, margins flat to ± revolute, white, or pale yellowish to pale or deep mauve with a yellow star-shaped patch around mouth and down back of throat, minutely puberulous to glabrous outside, glistening glands on backs of lobes, inside a posticous longitudinal band of clavate hairs together with 2–3 lesser lateral bands. *Stamens* 4, anticous anthers 0.25–0.4mm long, just visible in mouth, posticous anthers 0.6–0.8mm long, included, posticous filaments weakly bearded. *Stigma* 0.8–1.6mm long, sometimes bifid, deeply included. *Style* 0.2–0.8mm long, much shorter than stigma. *Ovary* 0.8–1.3 × 0.6–1.3mm. *Capsules* 3–4.5 × 2–3mm. *Seeds* 0.6–0.8 × 0.3–0.4mm, blue-violet or pallid.

Selected citations:

Cape. Namaqualand, 2917DB, Springbok, 14 ix 1961, *Strey* 3917 (PRE); 3118BC, Zout River, 21 ix 1929, *Lewis Grant* 4767 (PRE). Kenhardt div., 2921AC, near Kenhardt, 21 viii 1954, *Barker* 8300 (NBG). Hopetown div., 2923DC, Strydenburg, 30 viii 1935, *Taylor* 870 (BOL). Calvinia div., 3019CC?, Volwe Graafawater, c.2000ft, 23 viii 1956, *Acocks* 18957 (PRE); 3119BD, ± 28 miles N of Calvinia, 25 ix 1952, *Maguire* 1955 (NBG); 3119DC, De Bosch, 26 vii 1941, *Esterhuysen* 5346 (BOL); 3120CC, 13½ miles NW of Middelpos, c.3900ft, 16 ix 1955, *Leistner* 297 (PRE). Williston div., 3120BD, 28km from Williston to Calvinia, 17 viii 1987, *Batten* 771 (E). Victoria West div., 3122BD, Melton Wold, 4000ft, x 1935, *Thorne* sub SAM 51944 (SAM); 3122AD, 40km SE of Loxton, 1180m, 23 viii 1988, *Batten* 872 (E). Carnarvon div., 3121DC–3022CC, between Fraserburg and Carnarvon, 20 ix 1938, *Häfstrom & Acocks* 1205 (PRE). De Aar div., 3024CA, De Aar, 23 vii 1925, *Moss* 11557 (BOL, J); Quaggafontein, c.4000ft, 6 v 1946, *Acocks* 12602 (PRE).

It is only with very slight reservation that I have reduced *M. pusilla* var. *insigniflora* to *M. fragrans*. Meyer's specimen was destroyed in the Berlin fire and appears to have been a unicate. However, *M. pusilla* bears a strong resemblance to *M. fragrans* in habit, stature and leaf form, and Diels's description of his variety as having flowers almost twice as large as those of the type [of *M. pusilla*] and with the laciniae [corolla lobes] much less acuminate fits *M. fragrans* very well; also, *M. fragrans* is common around Calvinia. The varietal name has no nomenclatural significance and the identity of the plant is really immaterial.

Manulea fragrans is widely distributed on the arid plateau of the northern Cape. It appears to colonize bare areas and has been variously recorded as growing on sandy flats, sandy gravel, stony rocky clay, and sand on top of a quartzite hill, and sometimes dominant over considerable areas. The principal flowering time is May to October.

Collectors have often recorded variation in flower colour, from white to yellowish to mauve. The species has been grown here in Edinburgh (from seed collected by Mrs Batten (no. 771) near Williston) and the flowers were a pale mustard yellow turning greyish mauve, reminiscent of the colour variation we ourselves noted, in the field, in *M. crassifolia* subsp. *thodeana*. The flowers have a strong slightly unpleasant but nevertheless sweet scent, clearly noted by Schlechter too.

The type specimen of *M. fragrans* came from Zout Rivier, between Van Rhynsdorp and Nieuwerust, on the southern margin of Namaqualand; there is only one other record from

Namaqualand (at Springbok). The area of the species stretches eastwards from Zout Rivier through Calvinia to Victoria West and northwards to Kenhardt, Prieska, Douglas, De Aar and Hopetown divisions (I have seen no record between Springbok and Kenhardt). In the western part of its range, in Namaqualand and Calvinia divisions, the tips of the corolla lobes may be entire to retuse or more deeply lobed, often showing variation in a single collection (as in the type). These specimens also tend to have the longest pedicels. It may not be without significance that it is only where *M. fragrans* is sympatric with *M. praeterita* (with lobed corolla lobes and longish pedicels) that this is so: further east, outside the range of *M. praeterita*, the corolla lobes are always entire and the pedicels short to very short. On the other hand, the switch from entire to retuse or bifid lobes seems to be a genetically simple step: see for example *M. aridicola*.

Specimens with forked corolla lobes are often confused with *M. silenoides*, but *M. fragrans* is easily distinguished from both *M. silenoides* and *M. praeterita* by the minute glandular pubescence on stem and leaves (not hairs 0.1–0.3mm long, with balloon-tipped hairs also present especially on the inflorescence axes). The petiolar part of the leaf is often much longer in *M. fragrans* than in the other two species, and the pedicels are mostly shorter (see also *M. diandra* and *M. chrysantha*).

A specimen collected by Acocks (no. 12672, PRE) at Angelierspan in Prieska division appears to be a hybrid between *M. fragrans* and *M. schaeferi* (which Acocks collected at the same place under his no. 12673). The plant has much the facies of *M. fragrans*, but the stigma is too long for that species (and not long enough for that of *M. schaeferi*) while the corolla tube is intermediate in length between that of the two species (5.8mm versus 3.5–4.5mm in *M. fragrans*, 7.5–10.5mm in *M. schaeferi*). Acocks noted of no. 12672 that it was rare, and as there is only one small plant mounted on the sheet, one deduces that it was probably the only specimen. This is in contrast to no. 12673, where Acocks recorded 'locally frequent'.

59. Manulea diandra Hilliard in Notes RBG Edinb. 46:339(1990).
Type: Cape, Sutherland div., 2 miles S of Sutherland, 6 ix 1968, *Hall* 3278 (holo. NBG, iso. STE).

Annual herb, stems c.3–50 tufted from the base, 25–100mm tall, erect, simple, glandular-puberulous, hairs up to 0.1–0.2mm long, leafy only near base. *Leaves* opposite, bases connate, crowded at base of stem, blade c.5–20 × 4–10mm, elliptic, ovate or subrotund tapering to a broad flat petiolar part c.7–23mm long (about equalling to twice as long as blade), apex obtuse to rounded, margins entire to serrate, both surfaces densely glandular-pubescent, hairs up to 0.1–0.2mm long. *Flowers* many in terminal subcapitate racemes scarcely elongating (to 20mm) in fruit. *Bracts* (lowermost) 2–3.5 × 0.8–1mm, lanceolate or oblong-lanceolate. *Pedicels* (lowermost) 1–3mm long, shorter upwards. *Calyx* obscurely bilabiate, tube 0.3–0.4mm long, lobes 3–4 × 0.4–0.8mm, lanceolate-oblong, glandular-puberulous all over, hairs up to 0.1–0.2mm long, occasional ± sessile glistening glands also present. *Corolla* tube 4.5–6.2 × 0.8–1.2mm in throat, cylindric, inflated in upper half, limb bilabiate, 5–7.5mm across lateral lobes, posticous lobes 1.7–2.6 × 1.4–2.1mm, anticous lobe 2–3.6 × 1–2.2mm, all lobes oblong-elliptic, obtuse, entire, yellow, outside with glistening glands on backs of lobes, tube minutely glandular-puberulous, inside 5 longitudinal bands of clavate hairs in throat. *Stamens* often 2 only by reduction of the anticous pair, these sometimes represented by 1 or 2 staminodes, or sometimes with fertile thecae 0.1mm long, posticous anthers 0.7–1mm long, included, posticous filaments bearded. *Stigma* 1.2–2mm long, deeply included. *Style* 0.3–1.1mm long, much shorter than stigma. *Ovary* 0.8–1.2 × 0.6–0.8mm. *Capsule* c.3.2–4 × 2.5–3mm. *Seeds* c.0.5–0.8 × 0.3–0.4mm, pallid.

Citations:

Cape. Sutherland div., 3220AD, Jakhals Fontein, 7 viii 1811, *Burchell* 1324 (PRE). 3220BA, between Sutherland and Middlepost, 9 x 1928, *Hutchinson* 711 (PRE). 3220BC, Jakhals Valley, 1500m, x 1920, *Marloth* 9653 (BOL, PRE); 23½ miles SSE of Sutherland, c.5300ft, 15 ix 1955, *Leistner* 278 (PRE); 11 miles S of Sutherland, c.5200ft, 31 vii 1953,

Acocks 16836 (PRE); 3km N of Sutherland, farm Matjies Rivier 80, 1500m, 27 viii 1986, *Cloete & Haselau* 61 (STE); 3220DD, Komsberg, 1500 m, x 1920, *Marloth* 9781 (PRE); 'sent from Sutherland', iv 1921, sub *Marloth* 9765 (PRE). Fraserburg div., 3221BA, Nieuweveldberg c.60km south of Fraserburg, 1650m, 26 ii 1986, *Moffett & Steensma* 4003 (STE).

This is the only species of *Manulea* in which the stamens are frequently reduced to two by loss of the anticous pair; when the anticous stamens are present, the anthers are either very small (c.0.1mm) or they are reduced to staminodes. This character will immediately distinguish *M. diandra* from *M. fragrans* with which it has hitherto been confused, but it differs further in its persistently glandular-puberulous stems and leaves, subcapitate inflorescence, calyx lobes glandular-puberulous all over, corolla tube mostly longer, and corolla limb yellow. *Manulea diandra* is also allied to *M. chrysantha*, with which it shares the character of yellow corolla, but it may be distinguished not only be its reduced stamens, but also by the lack of balloon-tipped hairs (not to be confused with glistening glands) on stem, calyx and corolla.

Manulea diandra is known principally from the Roggeveld, where it was first collected by Burchell in 1811, but there is one record from further east, in the Nieuweveld Mountains. Its area thus lies south of that of *M. fragrans*, west of that of *M. chrysantha*. Acocks recorded the plant as growing 'in shallow frozen soil', Leistner 'in swampy leegte', Moffett & Steensma 'moist shallow soil', while Hall noted 'abundant along roadside', so it appears to be a colonizer of bare damp areas, and flowers between July and October.

60. Manulea minor Diels in Bot. Jahrb. 23: 478 (1896); Hiern in Fl. Cap. 4(2): 232 (1904) p.p. Type: Hex River valley, 1700ft, i x 1881, *Tyson* sub Herb. Norm. Austr. Afr. 668 (holo. B†; iso. BOL, K, SAM, UPS, W) (*Tyson* 633 is the same collection: BOL, SAM).

Annual herb, stems 1–c.20 tufted from the base, c.60–300mm tall, erect, simple, minutely glandular-puberulous on upper part including inflorescence axis, hairs on lower part up to 0.2–0.3mm long, mainly glandular, a few balloon-tipped, leafless. *Leaves* opposite, bases connate, crowded at base, blade c.6–28 × 2.5–16mm, oblanceolate, elliptic or ovate tapering to a flat petiolar part roughly ½–⅔ length of blade, apex obtuse, margins subentire to serrate, both surfaces glandular-puberulous, hairs up to c.0.3–0.4mm long, occasional balloon-tipped hairs also present. *Flowers* many in long terminal racemes or narrow thyrses, either solitary or in 2–4-flowered cymules, solitary flowers and cymules often occurring together in one inflorescence. *Bracts* (lowermost) 1–3 × 0.5–0.8mm, linear-lanceolate. *Pedicels* or peduncles (lowermost) c.2.5–9mm long, shorter upwards but all flowers with noticeable pedicels. *Calyx* obscurely bilabiate, tube 0.4–0.5mm long, anticous lobes 2–2.6 × 0.6mm, posticous lobes 1.75–2.5 × 0.5–0.6mm, all lobes oblong-lanceolate, tips sometimes subspathulate, margins glandular-puberulous, hairs up to 0.1–0.2mm long, sessile glistening glands as well. *Corolla* tube 3.5–4.5 × 0.75–1.1mm in throat, cylindric, inflated near apex, limb ± bilabiate, 4.5–7mm across lateral lobes, posticous lobes 1.5–2.2 × 1.75–2.2mm, anticous lobe 1.75–2.8 × 1.3–2.1mm, all lobes subrotund to elliptic-oblong, with rounded tips, white, yellow star-shaped patch around mouth and extending down tube, glabrous to minutely puberulous outside, glistening glands on backs of lobes, 4 longitudinal bands of clavate hairs in throat. *Stamens* 4, anticous anthers 0.25–0.4mm long, just included, posticous anthers 0.6–0.8mm long, included, posticous filaments bearded. *Stigma* 0.7–1.1mm long, deeply included. *Style* 0.25–0.5mm long. *Ovary* 1–1.2 × 0.6–0.7mm. *Capsules* 3–3.5 × 1.5–2mm. *Seeds* c.0.5–0.6 × 0.3–0.4mm, pallid.

Citations:

Cape. Worcester div., 3319BD, Hex River Pass, 1700ft, 1 x 1893, *Bolus* sub Guthrie 3071 (NBG); 3319BC, near Hex River East Station, 1600ft, x 1893, *Bolus* 7887 (BOL); Orchard, vii 1944, *Esterhuysen* 10303 (BOL); ibidem, 1400ft, viii 1915, *Rogers* 16519 (J). Stellenbosch div., Bellville, i 1914, *Rogers* 16519 (PRE).

Manulea minor is known with certainty only from the Hex River valley; its occurrence at Bellville needs confirmation: Rogers used the same number, 16519 (cited above) for specimens alleged to be from two different localities. The specimen in PRE bears a ticket in Roger's own hand; that in J has a printed ticket. The specimens could well represent but one collection, part of which has been mislabelled. July to October seems to be the principal flowering season and Miss Esterhuysen recorded 'in sand'.

Manulea minor has been much confused with *M. exigua*, from which it is easily distinguished by its scapose stems, much less markedly toothed leaves and glandular pubescence. Its relationship seems to lie with *M. praeterita*, from which it is distinguished by the longer hairs on the stems (up to 0.2–0.3mm versus 0.1mm) and leaves (up to c.0.3–0.4mm versus 0.25mm), and the entire corolla lobes (not distinctly lobed).

61. Manulea praeterita Hilliard in Notes RBG Edinb. 46:340(1990).
Type: Cape, Clanwilliam div., 3218BB, 1 mile N of Clanwilliam, 23 viii 1966, *Barker* 10445 (holo. NBG).

Annual herb, stems 1–c.12 tufted from the base, c.30–300mm tall, erect, simple below the inflorescence, inflorescence sometimes weakly paniculate, hairs on lower part of stem up to 0.1mm long, mostly gland-tipped, a few balloon-tipped, short-stalked balloon-tipped hairs often plentiful on inflorescence axes, sometimes few or wanting, leafless. *Leaves* opposite, bases connate, crowded at base, c.4.5–21 × 3.5–15mm, elliptic to ovate tapering to a flat petiolar part roughly ½–⅔ length of blade, apex obtuse to subacute, margins subentire to serrate, both surfaces glandular-puberulous, hairs up to 0.25mm long, occasional balloon-tipped hairs also present. *Flowers* many in long terminal racemes or narrow thyrses, either solitary or in 2–7-flowered cymules, solitary flowers and cymules often occurring together in one inflorescence. *Bracts* (lowermost) 1.5–4 × 0.4–0.6mm, linear-lanceolate. *Pedicels* or peduncles (lowermost) c.3–18mm long, shorter upwards but all flowers with noticeable pedicels. *Calyx* very obscurely bilabiate, tube 0.2–0.5mm long, lobes 1.4–2.5 × 0.4–0.5mm, lanceolate-oblong, margins glandular-puberulous, hairs up to 0.1mm long, glistening ± sessile glands as well, a few balloon-tipped hairs sometimes on backs. *Corolla* tube 3–4 × 0.7–0.8mm in throat, cylindric, inflated near apex, limb almost regular, 4.5–6mm across lateral lobes, lobes c.1.8–2.2 × 1.6–2.3mm, elliptic to deltoid in outline, apex bilobed, sometimes each lobe lobed again, or occasionally 1 or 2 lobes out of the five entire, white or mauve with a yellow star-shaped patch around mouth and extending down back of tube, occasionally sparsely minutely glandular-puberulous outside, glistening glands on backs of lobes, inside a posticous longitudinal band of clavate hairs and a transverse band in throat. *Stamens* 4, anticous anthers 0.2–0.3mm long, posticous anthers 0.5–0.7mm long, all included, posticous filaments bearded. *Stigma* 0.6–1mm long, deeply included. *Style* 0.3–0.5mm long, much shorter than stigma. *Ovary* 0.6–1 × 0.5–0.7mm. *Capsules* c.2–3 × 1.75–2mm. *Seeds* c.0.3–0.6 × 0.3mm, blue-violet or pallid.

Citations:

Cape. Calvinia div., 3119AC, Nieuwoudtville, *Leipoldt* s.n. (BOL); 3119CA, Lokenburg, c.2100ft, i viii 1953, *Acocks* 16858 (PRE). Van Rhynsdorp div., 3118DA, Van Rhynsdorp, 500ft, 30 v 1938, *Compton* 7232 (NBG); 3118DB, Widouw farm, 15km S Van Rhynsdorp, 500ft, 9 ix 1976, *Hugo* 451 (STE). Clanwilliam div., 3118DC, Nardouw Kloof, ix 1947, *Stokoe* sub SAM 63912 (SAM). 3219AC, northern Cedarberg, Nardouwsberg plateau, 27 vii 1974, *Goldblatt* 2197 (NBG). 3218BA, Zeekoe Vlei, 1400ft, ix 1925, *Levyns* 1294 (BOL). 3218BB, 5 miles W of Clanwilliam on Lambert's Bay road, vii 1948, *Lavis* 3035 (SAM); ibidem, 8 ix 1938, *Gillett* 4046 (BOL, PRE); Clanwilliam, 400ft, 6 ix 1933, *Compton* 4308 (BOL, NBG); ibidem, 20 vi 1897, *Leipoldt* 586 (NBG, SAM); ibidem, *Leipoldt* 168 (SAM); ibidem, 20 x 1897, 300ft, *Bolus* s.n. (BOL); ibidem, 350ft, 27 viii 1894, *Schlechter* 5058 (BOL); ibidem, 300ft, 4 vii 1896, *Schlechter* 8021 (E, S); ibidem, 19 x 1930, *Galpin* 11504 p.p. (PRE): see discussion below.

Manulea praeterita has hitherto hidden under the cloak of *M. silenoides*, which suggested the trivial name. It shares with *M. silenoides* the character of bifid corolla lobes, but differs in the shorter indumentum on stems and leaves (hairs up to c.0.1mm long versus 0.2–0.5mm), hairs on calyx up to 0.1mm long and almost confined to margins (not up to 0.2–0.5mm long and well distributed), corolla tube glabrous or almost so outside (not with hairs up to 0.15mm long).

It is very closely allied to *M. gariesiana*, which differs principally in having the flowers mostly solitary in the axil of each bract (not mostly in 2–7-flowered cymules), larger bracts (mostly more than 5mm long and sometimes leaflike, not just 1.5–4mm long), and mostly longer calyx lobes (2.3–3.8mm versus 1.4–2.5mm). The flowers of *M. praeterita* may be mauve or white, those of *M. gariesiana* are always white.

The two species are allopatric, the area of *M. praeterita* lying south of that of *M. gariesiana*, from Nieuwoudtville and Van Rhynsdorp south to Clanwilliam (the southernmost record for *M. gariesiana* is at Bitterfontein, roughly 80km north of Van Rhynsdorp).

Manulea praeterita grows in sandy places and flowering plants have been collected in all months between May and December.

Galpin collected *M. praeterita* 'on sandy flats' at Clanwilliam in October 1930 (*Galpin* 11504, K, PRE). The PRE sheet carries two specimens. One is typical *M. praeterita*; the other appears to be a hybrid between *M. praeterita* and *M. arabidea*, which is known only from a small area near Clanwilliam. The aberrant specimen resembles *M. arabidea* in the indumentum on calyx and corolla tube and in the pattern of clavate hairs in the throat and around the mouth; the stigmas are long (c.1.7–3mm), too long for those of *M. praeterita*, but not long enough to be exserted as in *M. arabidea*. The corolla limb is also intermediate between that of *M. praeterita* and *M. arabidea* in that the lobes may be either entire (*M. arabidea*) or retuse (*M. praeterita*), while the posticous filaments are bearded (as in *M. praeterita*) and the anthers are distinctly unequal in size (nearly equal in *M. arabidea*).

62. **Manulea gariesiana** Hilliard in Notes RBG Edinb. 46:340(1990). **Plate 3F.**

Type: Cape, Namaqualand, 3018CC, 35km before Garies on way north, 4 ix 1986, *Batten* 739A (holo. E).

Annual herb, stems 1–15 from the base, c.30–150mm long, erect or decumbent, simple or with a pair of branches near the base, sometimes branching at base of inflorescence, puberulous, hairs up to 0.1mm long, mainly gland-tipped, balloon-tipped hairs also present mainly on inflorescence axis, either leafless or with a pair of leaves at each forking. *Leaves* opposite, bases connate, mostly crowded near base, blade c.11–30 × 4.5–15mm, elliptic to oblanceolate tapering to a petiolar part c.5–17mm long, roughly ½–¾ length of blade, apex obtuse, margins subentire to serrate, both surfaces glandular-puberulous, hairs up to 0.4mm long, an occasional balloon-tipped hair also present. *Flowers* many in terminal racemes, sometimes weakly paniculate, individual flowers commonly solitary, sometimes in 2(–3)-flowered cymules. *Bracts* (lowermost) c.2.5–10 × 0.4–1mm, linear-lanceolate, or sometimes leaflike and up to c.13–25 × 2–3.5mm. *Pedicels* (lowermost) 3.5–9(–13)mm long, slightly shorter upwards. *Calyx* very obscurely bilabiate, tube 0.2–0.4mm long, lobes 2.3–3.8 × 0.3–0.6mm, lanceolate-oblong, tips sometimes subspathulate, glandular-pubescent mainly on margins, hairs up to 0.1–0.3mm long, balloon-tipped hairs wanting or very few, sessile glistening glands also present. *Corolla* tube 3–4.5 × 0.7–1.2mm in throat, cylindric, expanded in upper half, limb almost regular, 5.5–7mm across lateral lobes, lobes 2–3 × 1.6–3.5mm, deltoid in outline, apex usually bilobed, frequently each lobe lobed again, occasionally apex merely retuse, white with a yellow/orange star-shaped patch around mouth and extending into throat, minutely puberulous outside, inside a posticous longitudinal band of clavate hairs and a transverse band in throat. *Stamens* 4, anticous anthers 0.25mm long, posticous anthers 0.6mm long, all included, posticous filaments bearded. *Stigma* 0.75–1.5mm long, deeply included. *Style* 0.25–0.4mm long, much shorter than stigma. *Ovary* 0.75–1 × 0.6–0.8mm. *Capsules* c.2.5–3 × 1.5–2mm. *Seeds* c.0.4 × 0.3mm, blue-violet or pallid.

Citations:

Cape. Namaqualand, 3017BD, between Kamieskroon and Garies, 20 vii 1967, *van der Schijff* 7012 (PRE); 15km N of Garies, 2 ix 1974, *Rösch & le Roux* 638 (PRE); near Garies, 950ft, 21 ix 1929, *Grant* 4785 (PRE). 3017DB, Zwischen Zwartdoornrivier und Groenrivier, unter 1000 Fuss, *Drège* (E, K, LE, S, TCD, distributed as *M. silenoides* β *minor*); 11 miles S of Garies, vii 1950, *Lewis* 3037 (SAM); W of Garies, between roads to Groenrivier and Hondeklipbaai, 24 viii

1987, *Batten* 803 (E); Garies, 'Aanloop', viii 1987, *Batten* 797 (E). 3018CA, c.10km S of Garies, 29 ix 1981, *Hugo* 2892 (STE); 13 miles SSE of Garies, c.1000ft, 19 vii 1957, *Acocks* 19305 (PRE); near Garies, xi 1939, *Esterhuysen* 5111 (BOL); 11 miles E of Garies, 25 vii 1950, *Barker* 6208 (NBG); 17 miles S of Garies, 1300ft, 6 vii 1954, *Schelpe* 134 (BOL). Van Rhynsdorp div., 3018CC, between Garies and Nieuwerust, 2000ft, 30 ix 1939, *Levyns* 6962a (BOL); 5.6 miles from Bitterfontein to Pofadder, 26 viii 1967, *Marsh* 446 (STE). 3118AB, Bitterfontein, 1000ft, 1 ix 1897, *Schlechter* 11020 (BOL, E, PRE, S); ibidem, 23 vii 1941, *Esterhuysen* 5410 (BOL); ibidem, 23 vii 1941, *Compton* 11054 (NBG); ibidem, 23 viii 1941, *Compton* 11321 (NBG); ibidem, 20 ix 1930, *Henrici* 2131 (PRE).

Manulea gariesiana appears to have a linear range of about 80km, from somewhere north of Garies south to Bitterfontein, most collections having been made along the main south-north road to Namaqualand. Its range westwards (and eastwards?) is unknown, but Mrs Batten found it roughly 25km west of the highway, north of the Groen river. Little is known of its ecology: 'short open scrub', 'open places', 'Namaqualand Broken Veld of granite hills' have been recorded, and the plants flower between July and September.

Manulea gariesiana and its close ally *M. praeterita* are often confused with *M. silenoides*: all three species have bifid corolla lobes. *Manulea gariesiana* can be distinguished from *M. silenoides* by the shorter hairs on the stem (up to 0.1mm, not 0.15–0.3mm), mostly shorter pedicels (seldom exceeding 9mm, not often at least 10mm), calyx pubescent mainly on the margins of the lobes, the hairs gland-tipped, up to 0.1–0.3mm long (versus pubescent all over with a mixture of gland-tipped and balloon-tipped hairs up to 0.2–0.5mm), and often smaller corolla limb (5.5–7mm versus 6–12mm), which is white (not shades of blue and mauve, very rarely cream). The area of *M. silenoides* lies mainly north of that of *M. gariesiana*, but where they are sympatric around Garies they may hybridize (see further under *M. silenoides*).

See under *M. praeterita* for the distinguishing characters of that species.

63. Manulea silenoides [E. Mey. ex] Benth. in Hook., Comp. Bot. Mag. 1: 381 (1836) & in DC., Prodr. 10: 363 (1846); Hiern in Fl. Cap. 4(2): 225 (1904) p.p.
Type: Cape, Namaqualand [2917BC] Karakuis, *Drège* c.(holo. K).

Syn.: *M. incisiflora* Hiern in Fl. Cap. 4(2): 228 (1904). Type: Cape, Namaqualand, without precise locality, *Whitehead* s.n. (holo. TCD).
　　M. silenoides Benth. var. *minor* [Benth. ex] Hiern in Fl. Cap. 4(2): 225 (1904). Lectotype (chosen here): Namaqualand, 2917 DB?, Modderfontein, *Drège* b (K; isolecto. S, W).

Annual herb, stems 1–25 from the base, c.30–300mm long, erect to decumbent or nearly prostrate, simple to laxly branched mainly in upper part, puberulous, hairs up to 0.15–0.3mm long, both gland- and balloon-tipped, balloon-tipped hairs often sparse on lower part of stem, either leafless or with 2 to a few tufted at base of branchlets. *Leaves* opposite, bases connate, mostly crowded near base, blade c.6–40 × 3.5–15mm, elliptic to oblanceolate tapering to a petiolar part c.4–30mm long, roughly equalling to half as long as blade, apex obtuse, margins obscurely to distinctly serrate, both surfaces pubescent, hairs up to 0.3–0.5mm long, mostly glandular. *Flowers* many in terminal racemes, sometimes weakly paniculate, nearly always solitary in the axil of each bract. *Bracts* (lowermost) c.1.5–10 × 0.5–1mm, linear-lanceolate, or occasionally larger and leaflike. *Pedicels* (lowermost) c.5–25mm long, slightly shorter upwards. *Calyx* very obscurely bilabiate, tube 0.25–0.5mm long, lobes 2.5–5.5 × 0.4–1mm, lanceolate-oblong, tips sometimes subspathulate, pubescent all over, hairs up to 0.2–0.5mm long, both gland- and balloon-tipped, ± sessile glistening glands as well. *Corolla* tube 3.5–5.5 × 0.8–1mm in throat, cylindric, inflated near apex, limb almost regular, 6–12mm across, lobes 2–4.5 × 2.2–5.5mm, deltoid in outline, apex bilobed, frequently each lobe lobed again, usually mauve to blue, very rarely cream-coloured, with a yellow/orange star-shaped patch around mouth and extending down tube, puberulous outside, hairs up to 0.15mm long, both gland- and balloon-tipped, inside a short posticous longitudinal band of clavate hairs and a transverse band in throat. *Stamens* 4, anticous anthers 0.2–0.4mm long, posticous anthers 0.5–0.8mm long, all included, posticous filaments bearded. *Stigma* 0.6–1.5mm long, deeply included. *Style* 0.2–

0.5mm long, much shorter than stigma. *Ovary* 0.75–1.5 × 0.6–1mm. *Capsules* c.2–4 × 1.75–2.5mm. *Seeds* c.0.5–0.6 × 0.3mm, pallid.

Selected citations:

Cape. Namaqualand, 2917DA, Spektakelberg, 2000ft, 20 ix 1971, *Thompson* 1290 (PRE, STE); 2917DB, 5 miles W by S of Springbok, c.2600ft, 22 ix 1957, *Acocks* 19546 (K, PRE); 1 mile S of Springbok, 25 viii 1954, *Barker* 8386 (NBG); Modderfontein, 3200ft, ix 1883, *Bolus* 667 (BOL, SAM, UPS, Z). 2917DC, 10 miles N of Komaggas, 4 ix 1951, *Maguire* 990 (NBG). 2917DD, Mesklip, 4 ix 1945, *Barker* 3664 (NBG). 3017BB, 14 miles N by W of Kamieskroon, c.2000ft, 19 vii 1957, *Acocks* 19325 (PRE). 3018AC, Eselsfontein, viii 1949, *Hall* 288 (NBG).

The type of *M. silenoides* came from Karakuis, which, judging from Drège's map, lies about 30km north west of Springbok, while the type of *M. incisiflora* probably came from Modderfontein, Rev. Whitehead's mission station just east of Springbok, and this too seems to be the provenance of Drège's collection of *M. silenoides* var. *minor*, which is indistinguishable from Whitehead's plant. Typical *M. silenoides* has simple stems terminating in racemes; in typical *M. incisiflora* the stems may be branched into weakly paniculate inflorescences, the branching being particularly pronounced when the stem is decumbent or prostrate. Plants of *M. incisiflora* are nearly always taller than plants of *M. silenoides* and the conclusion that *M. incisiflora* represents no more than well grown plants of *M. silenoides* is inescapable, especially as the distinction, simple versus branched, is not absolute: plants of *M. silenoides* often have a bud between the bract and lowermost pedicel, and thus have the potential to branch.

Manulea silenoides ranges from Nababeep and Springbok south to Garies, and grows in sandy or stony soils often on the slopes of granite hills, flowering in August and September. Among its close allies it is distinguished by its flowers on relatively long pedicels, these borne singly in racemes. Typically, the flowers are shades of blue or mauve with a yellow or orange 'eye'; twenty five collections record this, and in several more the colours can still be seen. Only one collector records 'cream' (*Hugo* 3047 (PRE) from Nababeep) and the plants were gathered in April, which is also odd. However, Bolus (no. 667, cited above) recorded '*flores flavi*' though here it is the colour of the 'eye' that may have taken his attention. *Manulea gariesiana*, which has hitherto been confused with *M. silenoides*, has white flowers, and the two species may hybridize where their areas meet around Garies: *Acocks* 16462 (PRE) and *Acocks* 16459 (PRE) with mauve flowers and *Esterhuysen* 5417 (BOL) with flowers 'violet and white' seem to show hybrid influence; clearly, field investigation is needed. See also under *M. acutiloba*.

64. Manulea acutiloba Hilliard in Notes RBG Edinb. 46:337(1990).
Type: Cape, Namaqualand, 2917DB, Springbok, 25 viii 1941, *Esterhuysen* 5880 (holo. BOL - specimens marked A).

Annual herb, stems 1–c.7 from the base, 30–110mm tall, erect, simple, glandular-puberulous in lower part, hairs up to 0.2mm long, becoming sparse upwards, inflorescence axis sometimes nearly glabrous, but often with tiny balloon-tipped hairs, leafless. *Leaves* opposite, bases connate, few, crowded at base of plant, blade c.4–12 × 2.5–6.5mm, elliptic tapering to a broad petiolar part c.2–5mm long, roughly ½–⅓ as long as blade, margins entire to obscurely toothed, both surfaces glandular-puberulous, hairs up to 0.3mm long. *Flowers* many, crowded in terminal racemes. *Bracts* (lowermost) 1.6–2 × 0.8–1mm, lanceolate. *Pedicels* (lowermost) 1.5–6(–10)mm long, shorter upwards. *Calyx* obscurely bilabiate, tube 0.3–0.5mm long, lobes 2.2–3.2 × 0.4–0.5mm, linear-oblong, pubescent mainly on margins (a few median hairs sometimes on back), hairs up to 0.2–0.4mm long, mostly gland-tipped, occasionally balloon-tipped hairs also present, as well as ± sessile glistening glands. *Corolla* tube 4–5.5 × 0.8–1.2mm in throat, cylindric, upper half inflated, limb almost regular, 8–14mm across, lobes 3.6–5.6 × 2–4mm, ± deltoid in outline, apex deeply and sharply bilobed, some lobes again lobed or retuse, white (sometimes blue or mauve?) with a yellow star-shaped patch around mouth and extending down back of tube, puberulous outside, hairs minute, inside a posticous longitudinal band of clavate hairs and a transverse band in throat. *Stamens* 4, anticous anthers 0.25–0.5mm long, posticous anthers 0.6–1.1mm long, all included, posticous filaments glabrous, or sometimes bearded (see comments below). *Stigma* 0.6–1.3mm long, tip sometimes bifid, deeply included. *Style* 0.3–0.5mm long, much shorter than stigma. *Ovary* 0.8–1.1 × 0.5–0.7mm. *Capsules* c.3–3.5 × 2mm. *Seeds* 0.5–0.6 × 0.3mm, pallid.

Citation:

Cape. Namaqualand, between Garies and Springbok, 7 ix 1986, *Batten* 736 (E).

Manulea acutiloba differs from its close ally *M. silenoides* in its mostly narrower calyx lobes (anticous lobes 0.4–0.5mm broad versus 0.4–1mm) with hairs nearly confined to the margins (not all over the backs) and corolla lobes with rather more deeply and sharply lobed tips, a character almost impossible to quantify, but strikingly apparent to the eye. In *M. acutiloba* the posticous filaments may be glabrous to weakly or strongly bearded whereas they are always well bearded in *M. silenoides*. Bearding of the filaments is not usually a variable character, and two plants with glabrous filaments have deliberately been chosen as the holotype. Nine plants are mounted on the sheet: four are *M. silenoides* (marked B), one is possibly *M. silenoides* × *acutiloba* (marked A × B), two are *M. acutiloba* (marked A), while the remaining two (A₁) have bearded filaments though in other respects they seem to be typical *M. acutiloba*. Miss Esterhuysen noted 'flowers pink, white, mauve' but did not indicate which was which on the herbarium sheet. Mrs Batten recorded white for her collection, which also comprises specimens with glabrous to weakly or strongly bearded filaments. Whether or not *M. acutiloba* normally has white flowers and colour is introduced by hybridization remains to be determined. (The flowers of *M. silenoides* are usually mauve or blue).

The geographical area of *M. acutiloba* is not clear: the linear distance between Garies and Springbok is roughly 100km. Nothing is known of the ecology of the species but the plants flower in August and September. Obviously, they may grow mixed with *M. silenoides*, but white-flowered plants with sharply divided corolla lobes should be easy to spot among mauve/blue-flowered plants with bluntly divided lobes and form the basis for an investigation into possible hybridization.

Section Medifixae

65. Manulea androsacea [E. Mey. ex] Benth. in Hook., Comp. Bot. Mag. 1: 381 (1836) & in DC., Prodr. 10: 363 (1846); Hiern in Fl. Cap. 4(2): 232 (1904) excluding *Bolus* 666; Merxmüller & Roessler, Prodr. Fl. S.W.A. 126: 30 (1967).
Type: Cape, Namaqualand, 2917AB, sands near Noagas, *Drège* (holo. K; iso. E, LE, S, TCD, W).

Annual herb but probably sometimes a short-lived perennial, stems c.65–280mm tall, solitary to many from the crown, simple except sometimes below the inflorescence, glandular-puberulous at the base becoming glabrous upwards, sessile glistening glands on inflorescence axis, nude. *Leaves* all radical, opposite, bases connate, 15–60 × 3–15mm, oblanceolate to broadly elliptic, apex obtuse to rounded, base gradually or abruptly narrowed to a petiolar part up to roughly half the length of the blade, margins entire to serrulate, both surfaces minutely glandular-puberulous. *Flowers* usually crowded in small terminal capitate racemes, oblong and up to c.20mm long in fruit, remaining crowded but a few flowers sometimes straggling back down the stem, or rarely inflorescence elongated to c.90mm, subsidiary inflorescences on short peduncles often developing below the main inflorescence giving old fruiting plants an odd spiky look. *Bracts* (lowermost) c.2.5–6 × 0.5–1mm, subsidiary inflorescences sometimes subtended by a greatly reduced leaf. *Pedicels* (lowermost) c.0.5–2mm, otherwise very short or flowers sessile. *Calyx* very obscurely bilabiate, tube 0.2–0.5mm long, lobes 2.1–3.75 × 0.5–0.8mm, subspathulate, glandular-puberulous all over. *Corolla* tube 6.75–7 × 1–1.3mm in throat, cylindric, abruptly expanded in upper half, limb almost regular, 6.5–8mm across lateral lobes, posticous lobes 2.25–3.5 × 1.5–2.5mm, anticous lobe 2.25–3 × 1.5–2.4mm, all elliptic-oblong, very obtuse, white, yellow to orange around mouth and inside upper half of tube, outside of tube glandular-puberulous, hairs sometimes sparse, sessile glistening glands on backs of lobes, inside with 3 small longitudinal posticous bands of clavate hairs in throat. *Stamens* 4, anticous anthers 0.3–0.5mm long, just included, posticous anthers 0.75–1mm long, deeply included, posticous filaments bearded with clavate hairs. *Stigma* 0.5–1mm long, shorter than or about equalling style, deeply included. *Style* 0.75–1.3mm long. *Ovary* c.1–1.2 × 0.8mm. *Capsules* 3–4.5 × 2–3mm. *Seeds* c.0.5–0.8 × 0.4mm, pallid.

Selected citations:

Namibia. Lüderitz-Süd, 2616CB, Aus, 15 vii 1925, *Moss* 11649 (J); Klinghardtgebirge, 16 ix 1922, *Dinter* 3893 (S, Z); ibidem, 16 ix 1977, *Merxmüller & Giess* 32036 (PRE). Buchuberge, 28 vii 1929, *Dinter* 6549 (BOL, E, PRE, SAM, STE, Z).

Cape. Namaqualand, 2816DA, east of Groot Derm, x 1926, *Pillans* 5279 (BOL). 2816DC, Holgat, between Port Nolloth and Alexander Bay, 10 ix 1961, *Hardy* 608 (PRE mixed with *M. nervosa*). 2917AB, Eksteenfontein road 3–4km N of Port Nolloth-Steinkopf road, 8 ix 1980, *Goldblatt* 5719 (PRE). 2917DA, Spektakel Pass, 19 ix 1971, *Eliovson* 38 (J, PRE).

Manulea androsacea ranges from southern Namibia to the Richtersveld and Namaqualand about as far south as Kleinsee and Spektakel Mountain near Springbok. It grows in deep sand, flowering between July and September.

The shape of the corolla tube, the almost regular limb, the great disparity in size of the anthers and the deep inclusion of the posticous pair, as well as the very short style and stigma, all indicate close relationship with *M. altissima*, especially with subsp. *glabricaulis*, but *M. androsacea* is at once distinguished by the round (not laterally compressed) corolla mouth and longitudinal band of clavate hairs in the throat. In herbaria, it is often misidentified as *M. corymbosa* [= *M. benthamiana*] and *M. bellidifolia*, both of which are easily distinguished by the very long and often shortly exserted stigma.

Under *M. androsacea*, Hiern cited only one specimen other than the type, namely *Bolus* 666 (BOL, UPS). This plant came from 'Eleven Mile Station' in Namaqualand, which I have failed to localize. It has the facies of *M. androsacea* but the corolla is larger (tube 7–8.2 × 1.2mm, limb 9–10mm in diam., posticous lobes 3–4.5 × 2.5–3.2mm, anticous lobe 2.8–4.5 × 2–3.2mm) and Bolus described the flower as '*carnei*'. I am unwilling to admit this plant into a broad concept of *M. androsacea*: more needs to be known about it.

66. Manulea altissima L.f., Suppl. 286 (1782); Thunb., Prodr. 102 (1800) & Fl. Cap. ed. Schultes 472 (1823); Benth. in Hook., Comp. Bot. Mag. 1: 382 (1836) & in DC., Prodr. 10: 363 (1846); Hiern in Fl. Cap. 4(2): 228 (1904); le Roux & Schelpe, Namaqualand ed. 2: 148, 149 cum ic. (1988). **Plate 3D.**

Type: Cape of Good Hope, *Thunberg* (holo. No. 14337 in herb. Thunberg, UPS, iso. S).

Syn.: *M. scabra* [Wendl. ex] Steud., Nomencl. Bot. ed. 2, 2: 99 (1841), nomen.

Three subspecies are recognized:

subsp. **altissima**

Perennial herb (sometimes flowering in seedling stage), taproot becoming very stout and woody, up to c.20mm diam. at crown, stems (150–)200–1200mm tall, up to c.6mm in diam., few to many from the crown, simple or laxly branched low down, occasionally almost twiggy, densely glandular-puberulous with few to many long (0.5–1.5mm) glandular hairs as well, these simple or dendroid, often leafy mainly in the lower part, pedunculoid upwards, but in branching specimens leafy on upper parts. *Leaves* opposite, bases connate, uppermost often alternate and passing into distant bracts, 15–150 × 2–18mm, oblanceolate or narrowly elliptic, apex ± acute, base tapered to a petiole-like part, margins subentire to distinctly serrate, both surfaces glandular-puberulous, usually with many, sometimes few, long (c.0.3–1mm) simple or dendroid glandular hairs as well. *Flowers* many in a dense terminal raceme, capitate at first, oblong in fruit and up to c.150mm long, sometimes with 1 or 2 subsidiary inflorescences arising some distance below the main one and sometimes overtopping it. *Bracts* (lowermost) c.3–10 × 0.5–1mm, oblong. *Pedicels* (lowermost) 1.5–4mm long. *Calyx* obscurely bilabiate, tube 0.4–1mm long, anticous lobes 3.6–5.5 × 0.8–1.2mm, posticous lobes 4–6 × 0.8–1mm, all enlarging in fruit, densely pubescent all over with both glandular and eglandular hairs, some hairs up to 0.2–1mm long. *Corolla* tube 8–10 × 1.3–1.75mm in throat, cylindric, abruptly expanded in upper half, mouth strongly compressed laterally, limb almost regular, c.7–13mm in diam., posticous lobes 3–4 × 3–5.5mm, anticous lobe 3–5 × 3–5.5mm, all lobes suborbicular, white with greenish to yellow patches at the bases, or the three anticous lobes wholly yellow, the colour coalescing in the throat and extending down the inflated part of the tube, outside of tube and backs of lobes densely pubescent, hairs up to 0.2–0.75mm long, glabrous inside except for minute glandular hairs sometimes present below insertion of posticous filaments. *Stamens* 4, anticous anthers 0.4–0.6mm long, just included, posticous anthers 1.4–1.6mm long, deeply included, posticous filaments bearded with clavate hairs. *Stigma* 0.6–2mm long, about equalling or somewhat longer than style, deeply included, tip sometimes minutely bifid. *Style* 0.6–1.5mm long. *Ovary* 1.3–2.5 × 1–1.5mm, elliptic. *Capsules* 4–6.5 × 2.6–3mm. *Seeds* c.0.7 × 0.2–0.3mm, violet-blue.

Selected citations:

Cape. Clanwilliam div., 3118DC, Windhoek, 300ft, 31 vi 1896, *Schlechter* 8359 (BOL, E, PRE); Nardouw, 14 ix 1947, *Compton* 19998 (NBG). 3218BA, Graafwater, 28 viii 1987, *Batten* 824 (E). 3218BB, Clanwilliam, 350ft, 27 viii 1894, *Schlechter* 5053 (BOL, E, J, STE (very poor), UPS). Malmesbury div., 3218CD, ix 1944, *Lewis* 1550 (SAM). Piquetberg div., 3218DC, Sauer, *Compton* 15124 (NBG). 3318 AA, Meeuklip, N of Langebaan, *Wisura* 2931 (NBG).

Manulea altissima subsp. *altissima* ranges from about Doringbaai and Klaver south to the environs of Langebaan and inland to Clanwilliam and Sauer, a plant of sandy flats and slopes up to c.425m above sea level, flowering principally between July and September.

It is the only species of *Manulea* in which the corolla mouth is strongly compressed laterally and thus shaped like a keyhole. Subsp. *altissima* can grow into a robust plant over a metre tall but it may flower when little more than a seedling; the leaves are often almost all radical, and branching specimens leafy on the upper parts look very different. However, the laterally compressed mouth is diagnostic of the species as a whole, while subsp. *altissima* itself is characterized by long glandular hairs on the stem, inflorescence axis, and leaves, varying in density but always present, and the long hairs are underlain by much shorter glandular hairs. In the environs of Klawer and Vredendal, at the northern limit of its area, it is sympatric with subsp. *longifolia*, which is more or less intermediate in character between subsp. *altissima* and subsp. *glabricaulis*, whose area lies still further north, from about Nuwerus to Springbok. Field studies, particularly in the area of subsp. *longifolia*, might clarify the relationship between these three entities.

subsp. **longifolia** (Benth.) Hilliard in Notes RBG Edinb. 46:49(1989).
Type: Cape, Van Rhynsdorp div., 3118CA, on sandy hills at Ebenezer, *Drège* (lecto. K; isolecto. LE, S).

Syn.: *Manulea longifolia* Benth. in Hook., Comp. Bot. Mag. 1: 382 (1836) & in DC., Prodr. 10: 364 (1846).
    M. altissima var. *longifolia* (Benth.) Hiern in Fl. Cap. 4(2): 229 (1904).

Probably a short-lived perennial but often behaving as an annual, taproot eventually woody and up to c.8mm diam. at crown, stems 70–400mm tall, solitary to many from the crown, often simple, sometimes laxly branched low down, minutely glandular-puberulous at least at the base, longer glandular hairs (c.0.3–1mm) always present (though sparse) immediately below the inflorescence, sometimes extending further back down stem, sometimes to base, leafy mainly at the base. *Leaves* opposite, bases connate, mostly ± radical, a few reduced, alternate, leaves or bracts sometimes present on upper part of stem, main leaves c.10–90 × 2.5–20mm, narrowly to broadly elliptic, apex obtuse, base narrowed to a petiolar part, margins subentire to serrulate, both surfaces minutely glandular-puberulous. *Flowers* many in a dense terminal raceme, capitate at first, oblong in fruit and up to 20–45mm long, with a few distant capsules sometimes straggling down the peduncle. *Bracts* (lowermost) 3–13 × 0.5–1mm, oblong. *Pedicels* (lowermost) 0.5–1.5mm long. *Calyx* obscurely bilabiate, tube 0.25–0.75mm long, anticous lobes 2.75–4.75 × 0.5–1mm, posticous lobes slightly narrower, all enlarging in fruit, densely pubescent all over, hairs up to 1mm long, both glandular and eglandular. *Corolla* tube 7–9 × 1–1.6mm in throat, limb 8–13mm across the lateral lobes, posticous lobes 2–4.75 × 1.4–4.3mm, anticous lobe 2.2–5 × 1.5–4.5mm. *Stamens* as in typical plant. *Stigma, style, ovary* as in typical plant. *Capsules* c.4–7 × 2.5–3.5mm. *Seeds* 0.4–0.8 × 0.3–0.5mm, violet-blue.

Selected citations:

Cape. Van Rhynsdorp div., 3118BC, Knechtsvlakte, 3 ix 1948, *Compton* 20859 (NBG). 3118CB, 25km on road out of Vredendal to Strandfontein, 22 viii 1987, *Batten* 787 (E). 3118CC, N side of Doorn Bay, c.100ft, 7 viii 1969, *Acocks* 24078 (PRE, STE). 3118DA, Van Rhynsdorp, 3 ix 1945, *Compton* 17172 (NBG); ibidem, *Leighton* 1121 (BOL). 3118DC, Windhoek, 300ft, 8 vii 1896, *Schlechter* 8070 (BOL, E, PRE).

The type of *M. longifolia* is only a scrap, mostly in fruit, and was broken off from a branching plant; no radical leaves are present, but only upper stem leaves. The specimen therefore differs in aspect from most of the plants I have associated with it, but I judge them to be all the same thing. The epithet '*longifolia*' is unfortunate; it is appropriate to the type specimen, but many specimens have broadly elliptic leaves. It has been noted above that subsp. *longifolia* is intermediate between subsp. *altissima* and subsp. *glabricaulis*. The tendency to branch and to produce reduced leaves on the upper part of the stem, and the long hairs that are always present there, are all characteristic of subsp. *altissima*. Leaf indumentum, and to some extent leaf shape, as well as the short fruiting inflorescence, the mostly short pedicels, and calyx lobes not or scarcely overtopping the capsule, are all characteristic of subsp. *glabricaulis*.

It is difficult to judge merely from a selection of herbarium specimens, but it seems likely that subsp. *longifolia* is wholly sympatric with subsp. *altissima* in the northern part of the range of that entity, from Nieuwerust and the Knersvlakte (Knechtsvlakte) south to around Klaver; a specimen of subsp. *longifolia* in BOL and PRE (*Wilman* 820) purportedly came from almost the southern limit of subsp. *altissima*, near Sauer, but its presence there needs confirmation. Subsp. *glabricaulis* on the other hand seems to be wholly allopatric, though its area and that of subsp. *longifolia* must meet somewhere between Nieuwerust (Nuwerus) and Garies.

Subsp. *longifolia* is frequently confused with *M. corymbosa* [*M. benthamiana*] but that species is easily distinguished by its round hairy mouth and long, shortly exserted, stigma.

subsp. **glabricaulis** (Hiern) Hilliard in Notes RBG Edinb. 46:49(1989).
Lectotype: Cape, 2917DA, Little Namaqualand, Spektakel Mt, x 1878, *Morris* sub BOL 5744 (K, isolecto. BOL).

Syn.: *M. altissima* var. *glabricaulis* Hiern in Fl. Cap. 4(2): 229 (1904).

Resembles subsp. *longifolia* in facies, but the stems are mostly nude, rarely with one or two very small, distant bracts, glabrous above (though the inflorescence axis may be glandular-puberulous), either glabrous or minutely glandular-puberulous near the base, leaves radical, bracts always short (lowermost c.3–4 × 0.5–1mm), pedicels always short (up to 1mm long), flowers sometimes smaller (corolla tube 5.5–9mm long, limb 4.5–11.5mm across the lateral lobes), capsule rarely overtopped by the calyx lobes, and the seeds sometimes pallid.

Selected citations:

Cape. Namaqualand, 2917DA, Spektakel, 25 viii 1951, *Compton* 11521 (NBG); ibidem 25 viii 1941, *Esterhuysen* 5728 (BOL); 2917DB, 8½ miles W by S of Springbok, c.2300ft, 22 ix 1957, *Acocks* 19560 (PRE). 3017AD, between Wallekraal and Hondeklipbaai, 28 ix 1976, *Goldblatt* 4218 (PRE). 3017BB, Kamieskroon, ix–x 1930, *Thorne* sub SAM 48814 (SAM). 3017DC, 42km from Garies to Groenriviersmond, 50m, 18 ix 1987, *Linder* 4260 (E). 3017DD, 4.8 miles NNW of Kotzesrus, c.600ft, 19 viii 1963, *Acocks* 23308 (PRE).

Subsp. *glabricaulis* appears to be confined to Namaqualand, from about Springbok (the type locality being Spektakel Mountain) south to the environs of Garies. The corolla seems usually to be white or creamy-white with a pale yellow to bright orange-yellow patch around the mouth, the colour suffusing out onto the bases of the lobes, but there is one record of the mouth being pale greenish-yellow, the lobes themselves pale yellow (all these states are found in the typical plant and in subsp. *longifolia*). The lower lip can be entirely yellow or with the lateral lobes white-tipped, or entirely yellow with only the upper half of the posticous lobes white. However, a description of "yellow-brown flowers", possibly refers to fading specimens.

67. Manulea juncea Benth. in Hook., Comp. Bot. Mag. 382 (1836) & in DC., Prodr. 10: 364 (1846).
Type: Cape, Van Rhynsdorp div., 3118DC, Giftbergen, *Drège* 7921 (holo. K).

Herb, possibly perennial, taproot woody, up to c.7mm diam. at crown, stems up to c.600–750mm tall, up to 3–4mm diam., eventually woody, several from the crown, simple to laxly branched low down, initially glandular-puberulous, soon glabrescent on lower part but basal cells of larger hairs often persisting as ± scabrid projections, inflorescence axis sometimes persistently glandular, stems laxly leafy to bracteate throughout. *Leaves* alternate, often with dwarf axillary shoots, distant, c.15–75 × 2–10mm, rapidly smaller upwards and passing into linear bracts, oblanceolate, apex ± acute, base narrowed into a petiolar part decurrent on the stem in two very inconspicuous ridges, margins with c.2–9 pairs of somewhat falcate teeth, prominent in larger leaves, becoming obscure in smaller ones, which then pass into entire bracts, both surfaces at first glandular-puberulous, hairs up to 0.1–0.3mm long, soon glabrescent though basal cells may persist as ± scabrid projections. *Flowers* solitary in terminal moderately dense racemes, the lower flowers with a bud interposed between bract and pedicel, this growing out into a short simple bracteate branch terminating in a 2–several-flowered raceme. *Bracts* (lowermost) c.3–4 × 0.3–0.4mm. *Pedicels* c.2–3mm long, uppermost c.1mm. *Calyx* very obscurely bilabiate, tube 0.1–0.2mm long, anticous lobes 2.2–3.5 × 0.5–0.6mm, enlarging to 4.5mm in fruit, posticous slightly narrower, lanceolate, minutely glandular-puberulous all over, a few stouter hairs sometimes present on back, the basal cell of each hair persisting as a ± scabrid projection, margins with eglandular hairs up to 0.2–0.4mm long. *Corolla* tube (8.5–) 10–15.5 × 1.2–1.5mm in throat, cylindric, upper half abruptly expanded, limb almost regular, 7–10mm across lateral lobes, posticous lobes 3–3.75 × 2–2.5mm, anticous lobe 3–4.5 × 2.5–3mm, all lobes broadly elliptic to subrotund, margins often somewhat revolute, dull yellow-orange to almost red, outside of lobes and tube puberulous with somewhat retrorse hairs up to 0.25mm long, minutely glandular as well, upper surface of lobes with tiny white clavate hairs forming a star-shaped pattern, inside of tube glabrous in upper half, minutely glandular-puberulous below point of insertion of posticous stamens. *Stamens* 4, anticous anthers 0.6–1mm long, included, posticous anthers 1.3–2mm long, very deeply included, posticous filaments bearded with clavate hairs. *Stigma* 0.6–1.5mm long, often ± bifid, deeply included. *Style* 0.7–1.8mm long. *Ovary* c.1.5 × 0.8mm. *Capsules* c.5–6 × 3–3.5mm, minutely glandular-puberulous all over. *Seeds* c.0.8 × 0.5mm, pallid.

Citations:

Cape. Van Rhynsdorp div., 3118DC, Doorn Rivier, 950ft, 7 vii 1896, *Schlechter* 8060 (BOL). Clanwilliam-Piquetberg, 3218DB(?), Bosch Kloof, 3500ft, 11 ii 1936, *Compton* 6217 (NBG); 3219AA, Cedarberg, Krakadouw Poort, c.3500ft, 16 ii 1977, *Haynes* 1266 (STE); Koude Bokkeveld, 3219AC, Wagensdrift, 5400ft, 21 i 1897, *Schlechter* 10077 (BOL,

E); 3219CA, southern Cedarberg, Die Hok, c.1160m, 10 i 1984, *Taylor* 10904 (PRE, STE); ibidem, Donkerkloof Kop, i 1945, *Stokoe* sub SAM 57675 (SAM). 3219CB, Baliesberg, 13 v 1970, *Hanekom* 1384 (K). 3219CD, east of Citrusdal, Skurfdebergen, ii 1928, *Primos* sub SAM 45717 (SAM). Ceres div., 3319AD (?), Old Elands Kloof, 24 iii 1951, *Compton* 22676 (NBG).

Hiern (in Fl. Cap. 4(2): 235, 1904) reduced *M. juncea* to synonymy under *M. cephalotes*, but that species can at once be distinguished by its short dense thyrse, quite different in aspect from the slender racemes of *M. juncea*, which differs further in its much longer corolla tube and glandular-puberulous capsules. The last two characters distinguish *M. juncea* from its close ally, *M. multispicata*.

The original material of *M. juncea* came from the Gifberg south of Van Rhynsdorp and it has since been recorded south south east over the mountains (Cedarberg, Skurfdeberg near Citrusdal, Koudebokkeveld Mountains and possibly the southernmost part of the Olifants River mountains) to the mountains at Ceres. Only one collector gave any indication of habitat: 'base of krantz in rocky rubble' and all specimens were in both flower and fruit between January and April.

There is a pressing need for good collections of this species; the collection made by Schlechter and cited above is particularly poor and the determination needs confirmation.

Manulea juncea is one of a close-knit group of species (66–71) with alternate decurrent leaves, which character sets them apart from all their congeners including those that share the peculiar characters of minute clavate hairs on the upper surface of the corolla lobes (66–73) and minute glandular hairs on the inside of the corolla tube below the insertion of the posticous filaments.

68. Manulea multispicata Hilliard in Notes RBG Edinb. 46:51(1989).

Type: Cape, Montagu div., 3320CC, Baden, Pampoen Kloof, 14 vii 1946, *Walgate* sub BOL 23356 (holo. BOL, iso. PRE).

Herb, probably perennial, taproot woody, up to c.10mm diam. at crown, stems up to c.300–700mm tall, up to c.3mm diam, eventually woody, several from the crown, occasionally simple, mostly laxly branched throughout, branches ascending at angle of c.45°, initially glandular-pubescent, with coarse hairs up to 0.2–1mm long, glabrescent, but basal cells of larger hairs persisting as ± scabrid projections, much smaller glandular hairs also present, leafy throughout. *Leaves* possibly opposite with connate bases very low down on stem but soon alternate, often with dwarf axillary shoots, crowded on lower part, often distant upwards and passing into inflorescence bracts, main leaves c.15–55 × 3–10(–18)mm, oblanceolate, apex ± acute, base distinctly narrowed and decurrent in two very narrow ridges, margins pinnatifid to pinnatipartite with up to c.9 pairs of somewhat falcate teeth of uneven size, the larger ones themselves sometimes toothed, both surfaces glandular-pubescent, hairs as on stem. *Flowers* in a long lax terminal raceme, often solitary, sometimes in 2–3-flowered dichasial cymules, often ± panicled through branching of the upper part of the stem. *Bracts* (lowermost) c.4.5–15 × 0.6–5mm, sometimes ± leaflike, or the lowermost flowers in the axils of the uppermost leaves. *Pedicels* or peduncles (lowermost) c.2–10(–25)mm long. *Calyx* very obscurely bilabiate, tube 0.1–0.4mm long, anticous lobes 2.6–4.5 × 0.5–0.7mm, posticous lobes slightly narrower, all lanceolate, minutely glandular-puberulous all over with some stouter hairs on the backs, the bases of these persisting as scabrid projections, margins with eglandular hairs up to 0.4mm long. *Corolla* tube (5.5–)7.5–8.5 × 1.2–1.5mm in throat, cylindric, upper half abruptly expanded, limb almost regular, 5.5–7.5mm across lateral lobes, posticous lobes 1.5–3 × 1.2–2.75mm, anticous lobe 2.5–3 × 1.6–2.6mm, all lobes broadly elliptic to subrotund, margins often somewhat revolute, dull yellow to deep orange-yellow, orange-brown or reddish, outside of lobes and tube pubescent with somewhat retrorse hairs up to 0.1–0.5mm, minutely glandular as well, upper surface of lobes with tiny white clavate hairs forming a star-shaped pattern, inside of tube glabrous in upper half, minutely glandular-puberulous below point of insertion of posticous filaments. *Stamens* 4, anticous anthers 0.5–0.75mm long, included, posticous anthers 1–1.4mm long, very deeply included, posticous filaments bearded with clavate hairs. *Stigma* 0.5–1.75mm long, often ± bifid, deeply included. *Style* 0.3–1.1mm long. *Ovary* c.1–1.5 × 0.8–1mm. *Capsules* c.3.5–5 × 2.5–3.5mm. *Seeds* c.1 × 0.5mm, pallid.

Citations:

Cape. Ceres div., 3219CC, Cold Bokkeveld Tafelberg, 5600ft, i 1941, *Stokoe* sub SAM 57674 (SAM). 3319AD, Koudebokkeveld, Gydouwberg, 5800ft, 19 i 1897, *Schlechter* 10048 (BOL); Ertjiesland Kloof, 28 ix 1944, *Compton* 16095 (NBG). 3319BA, Baviaansberg, 5000ft, 2 i 1942, *Compton* 12849 (NBG). 3319BC, Matroosberg near Laaken Vlei, 4500ft, 27 xi 1917 and 22 i 1917, *Phillips* 2036 a, (SAM). 3319 CB, Brandwacht Mt, Fairy Glen, c.2500ft, 8 xi 1948, *Acocks* 15279 (PRE). 3319DB, southern foothills of Voetpadsberg, farm Doringkloof, 2000ft, 27 iv 1985, *Morley* 392 (STE). Laingsburg div., 3320AB, Tweedside, 3400ft, 1 vi 1925, *Marloth* 12097 (STE). Worcester div., 3320AC, near Touws River railway station in Karroo, 2800ft, xii 1885, *Bolus* 7362 (BOL) and sub Herb. norm. austr. afr. 1089 (BOL, SAM). Montagu div., 3320CC, Baden, 2000ft, 21 ix 1946, *Levyns* 7933 (BOL), ibidem, 22 ix 1946, *Walgate* sub BOL 23466 (BOL), ibidem, 22 ix 1946, *Compton* 18344 (NBG), ibidem, ix 1946, *Lewis* 1949 (SAM).

Manulea multispicata occurs in the mountains of the south western Cape, from Cold Bokkeveld Tafelberg in the north southwards in an arc over the mountains east of Ceres, Worcester and Robertson to Montagu and eastwards on the central plateau to Touws River and Laingsburg (on the main road/rail route to the interior; any lateral distribution is unknown). It has been found on 'rocky hills', 'on a quartzite range', and in the streambeds of kloofs, also 'under shady rocks', from c.600 to 1750m above sea level. Both flowering and fruiting specimens have been collected between June and January.

The area of *M. multispicata* lies mainly south and east of that of its ally, *M. juncea*, but they are sympatric in at least part of Ceres division. For example, *Compton* 22676, from Old Elands Kloof (exact locality not traced) is *M. juncea*, while *Compton* 16095, from Ertjiesland Kloof (also not traced) is *M. multispicata*.

Manulea multispicata may be distinguished from *M. juncea*, which has similar foliage and indumentum, by its shorter corolla tube (up to 8.5mm versus 10–15.5mm) and mostly smaller limb (posticous lobes 1.5–3mm long versus 3–3.75mm, anticous lobe 2.5–3mm versus 3–4.5mm) and by the glabrous (not glandular-puberulous) capsules. The more or less paniculate arrangement of the upper branches suggested the epithet *multispicata*.

69. Manulea rigida Benth. in Hook., Comp. Bot. Mag. 1: 382 (1836) & in DC., Prodr. 10: 364 (1846); Hiern in Fl. Cap. 4(2): 238 (1904) p.p. min., excluding description. **Fig. 58A.**
Lectotype (chosen here): Cape, Clanwilliam div., am Fluss Olifants rivier [3218BD] und bei Villa Brakfontein [3219AC], 1836, *Ecklon* [& *Zeyher*] (K; isolecto. LE, S, SAM, TCD, NY).

Coarse herb, branching from base, probably perennial and c.1m tall, only upper part of stems seen, these up to 700mm long, c.3.5mm diam., woody, divaricately branched, densely glandular-puberulous (also on inflorescence axes), occasional long hairs (0.5–1mm) present on lowest part, leafy. *Leaves* alternate, c.25–45 × 5–15mm, gradually diminishing in size upwards and passing into conspicuous, leaf-like bracts, lanceolate, apex ± acute, base broad, half-clasping and shortly decurrent in two very narrow ridges, margins coarsely irregularly and sharply toothed, both surfaces glandular-puberulous with occasional longer broad-based hairs also present particularly on lower surface. *Flowers* many in large panicles, either solitary or in 2–3-flowered dichasial cymules. *Bracts*: those subtending main branches leaf-like, those subtending individual flowers or cymules smaller but always conspicuous and reaching at least halfway up the calyx of the nearest flower, either toothed or entire. *Pedicels* of individual flowers c.1–4mm long. *Calyx* obscurely bilabiate, tube c.0.4–0.75mm long, anticous lobes c.4–5 × 0.6–1mm, posticous lobes a little smaller, linear-lanceolate, minutely glandular-puberulous all over, sometimes with a few hairs to c.0.3mm on margins. *Corolla* tube c.7.5–9 × 1.6–2mm in throat, cylindric, upper half abruptly expanded, limb almost regular, c.6.5–8mm across lateral lobes, posticous lobes c.2–3 × 2–3mm, anticous lobe c.2.5–3 × 2mm, all lobes broadly elliptic, very obtuse, white, tube yellowish in throat, and forming a yellow star around mouth, outside of lobes and tube puberulous, hairs to 0.1mm, ± retrorse, upper surface of lobes with tiny clavate hairs forming a star-shaped pattern, inside of tube glabrous in upper half, minutely glandular-puberulous below insertion of posticous filaments. *Stamens* 4, anticous anthers 0.5–0.8mm long, included, posticous anthers 1.2–1.5mm long, very deeply included, posticous filaments bearded with clavate hairs. *Stigma* 0.6–1mm long, sometimes ± bifid, very deeply included. *Style* 0.6–0.75mm long. *Ovary* 1–1.5 × 0.8mm. *Capsules* c.3–5 × 2–3mm. *Seeds* c.0.8 × 0.5mm, pallid.

Citations:

Cape. Clanwilliam div., 3218BD, Olifantsrivier, 300ft, 2 vii 1896, *Schlechter* 7996 (BOL, K (very poor) and det. needs confirmation). 3219AC, north of Citrusdal, Brakfontein, Hotwegskloof, 4 vi 1982, *Van Jaarsveld* 6636 (NBG). 3219CA, Cedarberg, Duiwelskop, 4000ft, i 1945, *Stokoe* sub SAM 58551 (SAM); Heksrivierkloof, Riempie se Gat, c.650m, 16 xii 1984, *Taylor* 11187 (STE).

The two syntypes of Bentham's name *M. rigida* represent two different species, and his description unequivocally embraces both plants. I have chosen the Ecklon & Zeyher specimen as the lectotype because the Drège specimen is *M. cephalotes*.

The salient features of *M. rigida* are its divaricately spreading branches, often nearly at right angles to the main axis of the panicle, these conspicuously leafy or bracteate, the leaves broad-based and coarsely and irregularly toothed. The whole plant is minutely glandular-puberulous without any development of the coarse, broad-based glandular hairs that give rise to the scabrid projections commonly seen in its allies, though some long eglandular hairs may be present on the lower part of the stem; the basal parts are unknown to me. In *M. cephalotes* on the other hand, the indumentum is coarser and longer and some hairs become scabrid with age, the leaves are less conspicuously toothed, the bracts much smaller, and the inflorescences generally denser. The flowers of *M. rigida* may always be white (those of *M. cephalotes* are yellow to orange-brown), but collections are few and collector's notes meagre or wanting, so this needs verification.

The two species are probably wholly sympatric, though *M. rigida* appears to occupy a much smaller area than *M. cephalotes*, being known with certainty only from Brakfontein, north of Citrusdal, and from two sites in the Cedarberg, a little further to the north. Plants appear to begin flowering in December, flowers and fruits were present in January, mostly fruits by March to June. Mr H. C. Taylor recorded 'sandy stream flood plain, grassy opening between stream thicket and boulder thicket'.

70. Manulea ramulosa Hilliard in Notes RBG Edinb. 46:52(1989).
Type: Cape, Van Rhynsdorp div., 3118DC, Klaver, c.150ft, 13 iii 1926, *Smith* 2610 (holo. PRE).

A rigid twiggy shrublet up to c.450mm tall, main stems c.4.5mm diam. at base, branches spreading at base of plant, upper branchlets ascending at angle of c.45°, densely scabrid-pubescent with broad-based hairs up to c.0.4mm long, minutely glandular as well, leafy. *Leaves* opposite at base, soon subopposite to alternate, c.5–25 × 1–6mm, passing into smaller bracts, oblanceolate to oblong, apex ± acute, often recurved, base somewhat narrowed and decurrent on stem in two very narrow ridges, margins with 1–7 pairs of small teeth mainly in upper half, smallest ones and bracts entire, both surfaces densely pubescent, hairs as on branches. *Flowers* few in a short (up to c.20mm) raceme terminating each twig. *Bracts* c.4 × 0.6–1mm. *Pedicels* c.1–2.5mm. *Calyx* obscurely bilabiate, tube c.0.2mm long, anticous lobes c.3 × 0.7mm, posticous lobes slightly smaller, minutely glandular-puberulous all over together with coarser broad-based ± scabrid hairs, eglandular hairs to c.0.4mm on margins. *Corolla* tube c.7 × 1.3mm in throat, cylindric, abruptly expanded in upper half, limb almost regular, c.6mm across lateral lobes, posticous lobes c.2.2 × 2.2mm, anticous lobe c.2 × 2.2mm, all lobes subrotund, 'creamy yellow to yellowy, almost brownish', outside of lobes and tube puberulous, hairs to 0.3mm, ± retrorse, upper surface of lobes with tiny clavate hairs forming a star-shaped pattern, inside of tube glabrous in upper half, minutely glandular-puberulous below insertion of posticous filaments. *Stamens* 4, anticous anthers c.0.7mm long, included, posticous anthers c.1.2mm long, deeply included, posticous filaments bearded with clavate hairs. *Stigma* c.0.6mm long, shortly bifid, deeply included. *Style* c.0.6mm long. *Ovary* c.1.2 × 0.8mm. *Capsules* c.4 × 2.5mm, minutely glandular-puberulous. *Seeds* c.0.8 × 0.5mm, pallid.

Citation:

Cape. Van Rhynsdorp div., 3118DC, Windhoek, 1000ft, 31 vii 1896, *Schlechter* 8350 (BOL).

Manulea ramulosa is a twiggy little shrublet, scabrid all over, the tips of the small, sparingly toothed leaves often recurved, and passing upwards into small entire bracts that in turn pass into the bracts of the short terminal racemes. All these characters will distinguish it from its ally *M. rigida*, a taller plant, glandular-puberulous, with jaggedly cut leaves passing imperceptibly into prominent leaf-like bracts, and divaricately branched panicles with some of the flowers in cymules.

    Manulea ramulosa is known only from the type collection and one other. Smith found the plant 'amongst rocky hills to the east of Klaver Station growing in sandy waste'. It was just coming into flower in mid March. Schlechter's collection came from Windhoek, roughly 5km east of Klaver, but at considerably higher altitude than Smith's. At the end of July, only old capsules were present.

71. Manulea cephalotes Thunb., Prodr. 101 (1800) & Fl. Cap. ed. Schultes 470 (1823): Hiern in Fl. Cap. 4(2): 235 (1904), p.p. min., type only, excluding description.
Type: Cape of Good Hope, without precise locality, *Thunberg* (no. 14351 in UPS).

Syn.: *M. densiflora* Benth. in Hook., Comp. Bot. Mag. 1: 382 (1836) & in DC., Prodr. 10: 365 (1846); Hiern in Fl. Cap. 4(2): 236 (1904), p.p. Type: Clanwilliam div., Olifantrivier [3218BD] and Brakfontein [3219AC], *Ecklon [& Zeyher]* (holo. S, fragments SAM, W).
    M. rigida Benth. in Hook., Comp. Bot. Mag. 1: 382 (1836) & in DC., Prodr. 10: 364 (1846), excluding lectotype.
    Sutera cephalotes (Thunb.) O. Kuntze, Rev. Gen. Pl. 2: 467 (1891), excl. syn. Benth.

Coarse herb, probably perennial, stems probably several from the base, up to c.1m tall, c.3–5mm diam. at base, either simple or laxly branched from base, simple to branched above, pubescent throughout (including inflorescence axes) with glandular hairs up to c.0.3mm long, many of these broad-based and eventually ± scabrid, together with longer (to c.1mm) eglandular hairs, leafy particularly in lower part becoming more distantly bracteate upwards. *Leaves* alternate, up to c.25–50(–60) × 4–6(–10)mm, those on dwarf axillary shoots smaller, oblong to oblanceolate, apex ± acute, base somewhat narrowed and decurrent on stem in two very narrow ridges, margins sometimes entire (and usually so when passing into bracts), often with 2–4 pairs of teeth in upper half, both surfaces pubescent with broad-based glandular hairs and longer eglandular ones to 0.5–0.75mm. *Flowers* many in short (up to c.80mm) narrow thyrses, either solitary and terminal or paniculately arranged, flowers sometimes solitary (and often aborted?) and distant at base of thyrse, otherwise crowded and arranged in 3–7-flowered dichasial cymules. *Bracts* small, c.2–7 × 0.5–1mm. *Pedicels* (or peduncles) c.1.5–2mm long. *Calyx* obscurely bilabiate, tube c.0.2–0.5mm long, anticous lobes 2.4–4(–4.5)×0.5–0.8mm, posticous lobes slightly smaller, lanceolate, minutely glandular-puberulous all over together with coarser broad-based glandular hairs to c.0.25mm on backs and eglandular hairs to 0.4–0.6mm long on margins. *Corolla* tube c.5.5–7.5(–9) × 1.2–1.8mm in throat, cylindric, upper half abruptly expanded, limb almost regular, c.5–8(–9)mm across lateral lobes, posticous lobes c.1.75–2.5(–3.75) × 1.75–2.2(–3)mm, anticous lobe c.1.6–2.5(–3.25) × 1.4–2.2(–3)mm, all lobes broadly elliptic to suborbicular, very obtuse, yellow to orange-brown, outside of lobes and tube pubescent, hairs up to c.0.2–0.4mm, somewhat retrorse, upper surface of lobes with tiny white clavate hairs forming a star-shaped pattern, inside of tube glabrous in upper half, minutely glandular-puberulous below insertion of posticous filaments. *Stamens* 4, anticous anthers 0.4–0.75(–1)mm long, included, posticous anthers 1–1.5mm long, deeply included, posticous filaments bearded with clavate hairs. *Stigma* 0.3–0.8(–1.2)mm long, often bifid, very deeply included. *Style* 0.3–0.8mm long. *Ovary* c.1–1.5 × 0.8mm. *Capsules* c.3–3.5 × 2mm. *Seeds* c.1 × 0.5mm, pallid.

Citations:

Cape. Nama'land Minor, *Scully* 262 (6392 in herb. BOL) (BOL, NBG, Z). Calvinia div., 3119CA, Lokenburg, c.2100ft, 23 i 1954, *Acocks* 17574 (PRE). 3119CC?, south of Brakrivier, bed of Doorn River, 11 xii 1908, *Pearson* 3884 (BOL, K). Piquetberg div., 3218DB, The Rest, iii 1940, *Esterhuysen* 5110 (BOL); Pickeniers Kloof, 3 i 1926, *Thieler* 47 (PRE); ibidem, *Zeyher* 1295 (BOL, K, SAM). Clanwilliam div., 3219AA, Wupperthal, *Wurmb* sub *Drège* 7915 (K, syntype of *M. rigida*). 3219CC, slopes of Olifants River Mountains along Thee River, 17 iv 1949, *Esterhuysen* 15309 (BOL, NBG), ibidem, 17 iv 1949, *Esterhuysen* 15315 (BOL). Ceres div., 3319AC, Elands Kloof, 17 iv 1947, *Compton* 19450 (NBG); 3319BC, Matroosberg near Laken Vlei, 3500ft, i 1917, *Phillips* 2036b (SAM); between Ceres and Leeuwfontein, 28 xi 1908, *Pearson* 3247 (BOL, SAM), included with reservation, see comments below.

Thunberg's specimen of *M. cephalotes* is not localized, but two collections (cited above) from Pikenierskloof (now Grey's Pass) match it precisely, and as Thunberg travelled over the pass, it is possible that this is where he found it. Bentham did not see Thunberg's specimen and he re-described the species as *M. densiflora*. Both types possibly came from the same quarter-degree square, and both epithets draw attention to a salient feature of the species, namely the congested thyrse. In Thunberg's specimen, this is solitary at the top of a simple stem; in Bentham's plant several are arranged in a small panicle, but they agree so well in foliage, indumentum and flower size that they surely constitute but a single species.

Bentham cited 'Olifants River and Brackfontein, *Ecklon*' under his new name *M. densiflora*. Hiern (in Fl. Cap. 4(2): 237) cited *Ecklon* 411. I have not seen a collection bearing this number, but in Stockholm (S) there is a sheet, clearly from Ecklon & Zeyher's own herbarium, which Bentham had on loan, bearing the name *M. densiflora* in Bentham's hand; I take this to be the holotype as I can find nothing in the Kew herbarium annotated by Bentham. The Stockholm sheet now carries the usual printed label with a reference to Drège in Linnaea 19 of 1847 and it is therefore post Bentham's original description.

Manulea cephalotes appears to range from the environs of Lokenburg south to Wupperthal and the Olifants River Mountains. It is also in the mountains around Ceres; two collections (cited above) indicate this, but the plant collected by Pearson (no. 3247, cited above with reservation) is larger in all its parts than typical *M. cephalotes* (the measurements in parentheses in the formal description) and the inflorescence is laxer.

Four other collections need special mention: *Goldblatt* 5527 (E, PRE, S), 3120CC, 56km from Calvinia on Blomfontein to Middelpos road; *Perry* 3406 (NBG, PRE), 3120CC, N of farm de Hoop on Middelpos to Calvinia road; *Schmidt* 565 (PRE), 3119BD, Calvinia; *Cloete & Hasselau* 244 (STE), 3220DA, 4km from Klein Roggeveld on way down pass. They resemble *M. cephalotes* in the dense thyrses paniculately arranged, but they are far less densely pubescent and the flowers are sometimes larger.

Another specimen that may belong here is *Pearson* 6305 (BOL) from 'Kopje SW of Leliefontein Mission Station' near Khamiesberg, 3018AB. Collected on 16 i 1911, it comprises two pieces of stem, one bearing some young flowers. The presence of *M. cephalotes* in Namaqualand needs confirmation on good material, precisely localized. Scully's collection (cited above), which purportedly came from Namaqualand is a puzzle: his no. 262 and the specimens numbered 6392 in the Bolus herbarium are clearly all part of the same gathering, yet no. 262 was cited by Hiern as *M. altissima* var. *glabricaulis* (which does occur in Namaqualand); no. 6392 as *M. rigida*. There may well have been some muddle over the provenance of the specimens.

Bentham's confusion of *M. cephalotes* and *M. rigida* is elaborated under *M. rigida*, where the distinctions between the two species are given.

There is scant information on either the habit or habitat of *M. cephalotes*. The plants probably begin flowering in December or early January; by April they are mostly in fruit.

72. Manulea linearifolia Hilliard in Notes RBG Edinb. 46:340(1990).

Type: Cape, Ceres div., between Bokkeveld Sneeuwkop [3219CD] and Winkelhaaks River [3219DC], 'Zuurvlakte', 20 iv 1946, *Esterhuysen* 12704 (holo. BOL).

Coarse herb, probably perennial, stems possibly several from the base or branching there, up to c.1m (?) tall, at least 4mm diam. near base, laxly branched above into a loosely paniculate inflorescence, glabrous, leafy. *Leaves* alternate or sometimes opposite, ± appressed, distant, up to c.12–22 × 1–1.5mm (much smaller on the inflorescence branches), linear, apex subacute, base scarcely narrowed, margins entire or rarely with a single tooth, thick-textured, the midvein probably impressed above, glabrous. *Flowers* many in narrow, crowded thyrses paniculately arranged, the compound inflorescence sometimes ± twiggy, cymules up to 5-flowered, lowermost flowers often solitary and distant. *Bracts* (lowermost) c.4–8 × 0.5–1mm, leaf-like. *Peduncles* (lowermost) c.2–4mm long, pedicels short. *Calyx* obscurely bilabiate, lobes subequal, 2.7–4.5 × 0.6–0.7mm, linear-lanceolate or upper half subspathulate, minutely glandular-puberulous all over, together with coarser glandular hairs to c.0.3mm on backs, the large basal cell persisting as a scabrid projection, eglandular hairs to 0.3–0.4mm on margins. *Corolla* tube 7–9 × 1.3–1.5mm in throat, cylindric, upper half abruptly expanded, limb almost regular, 6.6–9mm across lateral lobes, posticous lobes 2–3.5 × 2–3.6mm, anticous lobe 2.5–3.5 × 2.5–3.5mm, all lobes suborbicular, orange, outside of lobes and tube pubescent, hairs up to 0.1–0.2mm long, somewhat retrorse, upper surface of lobes with tiny white clavate hairs forming a stellate pattern, inside of tube glabrous in upper half, minutely glandular-puberulous below insertion of posticous filaments. *Stamens* 4, anticous anthers 0.6–0.8mm long, included, posticous anthers 1.2–1.6mm long, deeply included, posticous filaments bearded with clavate hairs. *Stigma* 0.6–1.7mm long, often bifid, deeply included. *Style* 0.5–1.7mm long. *Ovary* 1.3–1.4 × 0.8–1mm. *Capsules* c.2–4 × 1.75–3mm. *Seeds* not seen.

Citations:

Cape. Ceres div., 3219CD, Zoo Ridge, c.3500ft, 18 ix 1964, *Taylor* 5899 (STE); ibidem, 3000ft, 3 ii 1980, *Hugo* 2241 (STE); lower south slopes of Sneeuwkop, 12 xii 1910, *Pearson & Pillans* 5866 (K). Worcester div., Karadouws Mts near Orchard, NW slopes, 26 vii 1944, *Esterhuysen* 10341 (BOL).

Manulea linearifolia appears to be confined to the arc of mountains stretching from the southern end of the Cold Bokkeveld Mountains to the Karadouws Mountains east of Worcester. Its area at least abuts on that of its ally, *M. cephalotes*, which it much resembles in general facies but from which it can be distinguished by its entirely glabrous stems (including the inflorescence axes) and leaves. The leaves are very narrow and thick-textured, unlike the relatively broad herbaceous leaves of *M. cephalotes*; also, the leaves of *M. linearifolia* are nearly always entire (an occasional tooth may appear) while all but the smallest leaves of *M. cephalotes* are conspicuously toothed.

The four collections of this species are poor, comprising mostly pieces of inflorescences; good collections are needed to show the habit of the plant and the leaves on the lowermost part of the stem.

There is a fragment in the Kew herbarium, without collector or locality; it was annotated by N. E. Brown as being part of the type of *M. densiflora*, but that is not so.

73. Manulea robusta Pilger in Bot. Jahrb. 48: 437 (1912); Merxm. & Roessler, Prodr. Fl. S.W. Afr. 126: 32 (1967).

Lectotype (chosen here): Namibia, Gross-Namaqualand, Aus, 1400m, i 1910, *Dinter* 1069a (SAM).

Syn.: *Sutera remotiflora* [Dinter ex] Range in Feddes, Repert. 38: 265 (1935), nom. nud. Specimen cited: Aus, *Dinter* 6075 (E, SAM, S, STE, Z).

Shrublet, rootstock eventually stout, woody, up to 10–15mm in diam. at crown, stems up to c.300–600mm tall, 3–4mm diam. at base, woody, many from the crown, each with long rod-like branches from the base, densely glandular-puberulous, hairs up to 0.25mm long, some broad-based and eventually scabrid, these sometimes very few or wanting and minute hairs predominating, eventually glabrescent, distantly leafy in lower part. *Leaves* opposite, bases connate, becoming alternate upwards as they pass into bracts, often with dwarf axillary shoots, primary leaves c.8–40 × 2–9mm, smaller and more distant upwards, oblanceolate to elliptic, apex ± acute, base narrowed to a short petiolar part, margins with 2–5 pairs of teeth, these sometimes rather obscure, upper small leaves often entire, both surfaces minutely glandular-puberulous, longer broad-based hairs forming scabrid projections also sometimes present. *Flowers* in long lax terminal racemes, these sometimes branched, branches stiffly ascending at an angle of c.45°, flowers usually solitary, very occasionally paired. *Bracts* (lowermost) c.3.5–10 × 0.5–2mm, linear. *Pedicels* c.3–6mm long. *Calyx* obscurely bilabiate, tube 0.1–0.5mm long, lobes c.3–3.5 × 0.6–0.8mm, lanceolate, enlarging a little in fruit, glandular-pubescent all over, hairs on margins up to 0.1–0.3mm long. *Corolla* tube 11–14 × 1.2–1.9mm in throat, cylindric, abruptly expanded in upper half, limb almost regular, c.7.5–9.5mm across lateral lobes, posticous lobes 2.7–3.5 × 2.2–4mm, anticous lobe 3–4.5 × 2.4–3.5mm, all lobes ± rotund, margins often revolute, orange to red-brown, outside of tube and lobes glandular-pubescent, hairs up to 0.1–0.2mm long, upper surface of lobes with tiny clavate hairs forming a star-shaped pattern, inside of tube either glabrous or minutely glandular-puberulous below insertion of posticous filaments. *Stamens* 4, anticous anthers 0.6–1mm long, included, posticous anthers 1.5–1.7mm long, very deeply included, posticous filaments bearded. *Stigma* 0.5–1mm long, often ± bifid, very deeply included. *Style* 0.5–1mm long. *Ovary* 1.4–1.8 × 0.8–1.5mm. *Capsules* c.3.5–5 × 2–3mm. *Seeds* c.0.5 × 0.3mm (few seen), pallid.

Citations:

Namibia. Lüderitz distr., 2516CD, Farm Garub-Urus: LU6, 4 iv 1968, *Giess* 10305 (PRE). 2616CB, Aus, 1400m, 2 vi 1922, *Dinter* 3560 (BOL, K, SAM, Z); ibidem, x 1910, *Marloth* 5008 (BOL); ibidem, 13 x 1978, *Hardy & Venter* 5074 (PRE). 2616CA, W of Aus, Farm Klein Aus, 9 viii 1959, *Giess & Van Vuuren* 827 (PRE).
Cape. Richtersveld, 2817CA, gorge at the base of the Cornellsberg, 17 viii 1986, *Williamson* 3563 (BOL, NBG); Roosterbank, 11 viii 1983, *Archer* 380 (NBG).

Manulea robusta was long thought to be confined to the mountains around Aus in southernmost Namibia, but it now seems likely that it occurs in the Richtersveld: Williamson's and Archer's specimens (cited above) appear to have softer leaves than plants from Namibia, and have dried darker, but I can find nothing else to distinguish them.

Like its ally *M. dubia*, *M. robusta* favours rocky habitats, and they appear to be sympatric around Aus. *Manulea robusta* appears to be a more rigid plant than *M. dubia*, less leafy, with differently shaped leaves (much narrower in relation to their length than those of *M. dubia* and with a much less well marked petiolar part) and these rapidly smaller upwards, the lowermost floral bracts always linear (never leaf-like) and the flowers often smaller, particularly in length of calyx lobes and size of corolla limb; the flowers may also differ somewhat in colour, but this needs confirmation.

74. Manulea dubia (Skan) Roessler in Mit. Bot. Staatss. München 6: 16 (1966); Merxm. & Roessler, Prodr. Fl. S.W. Afr. 126: 31 (1967).
Type: Namibia, 2317AC, Rehoboth, *Schinz* 51 (holo. (?) K).

Syn.: *Sutera dubia* Skan in Dyer, Fl. Trop. Afr. 4(2): 304 (1906).
    Manulea dinteri Pilger in Bot. Jahrb. 48: 438 (1912). Lectotype (chosen here): Namibia, Gross-Namaqualand, Jakalskuppe, vii 1897, *Dinter* 1120 (Z).
    Manulea namaquana L. Bol. in Ann. Bol. Herb. 1: 99, t. 13 (1915). Type: Namibia, Keetmanshoop distr., 2718 BB, Great Karasberg, summit of Lord Hill, 17 i 1913; *Pearson* 7953 (holo. BOL, fragment PRE; iso. K).

Aromatic shrublet but will flower when still single-stemmed, rootstock eventually stout, woody, up to 10–15mm diam. at crown, stems up to 300mm–1.5m long, up to c.3mm diam. at base, woody below, herbaceous above, many from the crown, erect, decumbent, ascending or almost sprawling, the plant sometimes as much across as high, each stem subsimple to well branched, densely glandular-puberulous, hairs up to 0.1–0.3mm long, some broad-based and eventually ± scabrid, leafy throughout. *Leaves* opposite, bases connate, becoming alternate upwards, often with dwarf axillary shoots, primary leaves c.12–57 × 5–30mm, roughly ¼ to ½ the length petiolar, smaller upwards and passing imperceptibly into bracts, blade ovate to broadly elliptic in outline, apex ± acute, base cuneate and passing into the petiolar part, margins coarsely and often rather irregularly serrate, in larger leaves often doubly serrate, both surfaces glandular-pubescent, hairs as on stem. *Flowers* in long racemes or very narrow thyrses terminating each branchlet, often solitary, sometimes in 2–3-flowered dichasial cymules, ± distant. *Bracts* (lowermost) c.5–24 × 1–9mm, usually leaf-like, or flowers in axils of uppermost leaves. *Pedicels* or peduncles (lowermost) c.2.5–12mm long. *Calyx* obscurely bilabiate, tube 0.2–0.5mm long, lobes c.(3.2–)4–5.5 × 0.5–0.9mm, enlarging a little in fruit, lanceolate, glandular-pubescent all over, hairs on margins up to 0.4–0.7mm long. *Corolla* tube (9–)10–15 × 1.2–2mm in throat, cylindric, abruptly expanded in upper half, limb almost regular, c.9–13mm across lateral lobes, posticous lobes 3–4.75 × 2.5–4.5mm, anticous lobe (3–)4–5.5 × 2–4mm, all lobes broadly elliptic to subrotund, margins often revolute, yellow turning orange, outside of tube and lobes glandular-pubescent, hairs up to 0.2–0.3mm long, upper surface of lobes often with tiny clavate hairs forming a star-shaped pattern, these hairs sometimes only near the sinuses or wanting, inside of tube either wholly glabrous or sometimes minutely glandular-puberulous below insertion of posticous filaments, rarely with a longitudinal posticous band of clavate hairs level with the posticous anthers. *Stamens* 4, anticous anthers 0.75–1mm long, included, posticous anthers 1.25–1.8mm long, very deeply included, posticous filaments heavily bearded with clavate hairs. *Stigma* 0.75–1.5mm long, often ± bifid, deeply included. *Style* 0.5–1.5mm long. *Ovary* 1.4–2 × 1–1.3mm. *Capsules* c.4–7 × 2–4mm. *Seeds* c.0.5–0.75 × 0.3–0.5mm, pallid.

Selected citations:

Namibia. Otjiwarongo distr., 2017AC, Waterberg, 5300ft, 14 vii 1954, *Schelpe* 183 (BM, BOL). Windhoek distr., 2216DB, Farm Keres: WIN 39, 5 xii 1963, *Giess* 7673 (PRE, WIND). 2217AD, Neudamm Experimental Farm 20 miles E of Windhoek, 1800–2100m, 18 ii 1955, *de Winter* 2370 (K, NBG). 2217CA/CB, 8 miles S of Kapp's Farm, 29 iv 1949, *Wilman* 430 (BOL, K, PRE). 2217CC, Lichtenstein, 15 x 1934, *Dinter* 7895 (BM, BOL, K). 2217CD, Auas Mts, Farm Langbeen, 16 iv 1949, *Strey* 2560 (K, SAM, Z). Bethanien distr., 2516DC, farm Kunjas: BET 14, 23 vi 1974, *Giess* 13357 (PRE); 2616CB, Jakalskuppe, i xi 1922, *Dinter* 4178 (BM, S, SAM, Z). 2616BA, Tiras on road between Aus and Helmeringhausen, 8 v 1976, *Oliver & Müller* 6459 (K, PRE).

Manulea dubia ranges over the mountains and high plateaux of Namibia, from the Brandberg and Waterberg south to Aus and the Great Karasberg, c.1400–2100m above sea level. It seems to occur most frequently in shady crevices of cliffs or under boulders, sometimes as a constituent of scrub, but there is also a record 'in damp sand'. Both flowers and fruits can be found in any month.

It is closely allied to *M. robusta*: see under that species.

Manulea aurantiaca Jarosz: Type not traced. Genus? 'leaves alternate, bracts pinnatifid, corolla orange-coloured'.

M. indiana Lour. = *Adenosma indianum* (Lour.) Merrill

M. indica Willd. = *Adenosma indianum* (Lour.) Merrill

M. intertexta Herb. Banks in Benth., Comp. Bot. Mag. 1: 372 (1836), nomen = *Cromidon decumbens* (Thunb.) Hilliard

M. longiflora Jarosz: Type not traced. Genus? 'leaves scattered, subalternate, corolla yellow, tube ventricose'.

M. plantaginea Thunb. = *Cromidon plantaginis* (L.f.) Hilliard

M. plantaginis L.f. = *Cromidon plantaginis* (L.f.) Hilliard

M. sopubia Hamilton in D. Don, Prodr. Fl. Nepal 88 (1825), nomen = *Sopubia trifida* D. Don

8. MELANOSPERMUM

Melanospermum is a small genus of mostly annual herbs centred on the Transvaal and Swaziland. Six species are enumerated here, but field work would possibly reveal the existence of one or two more.

One species, *M. foliosum*, is widely distributed from northern Namibia to western Zimbabwe and western Transvaal, an annual herb of open sandy places. It is closely allied to *M. transvaalense*, which is more restricted in its distribution (mainly south eastern Transvaal) and possibly more demanding in its habitat: damp sandy or gritty sites associated with rock outcrops.

The other four species all occupy habitats associated with rock outcrops and appear to be narrowly endemic (they are undoubtedly under-collected). *Melanospermum rupestre*, the only perennial, grows under moist rock overhangs and appears to be restricted to the Witwatersrand, the Steenkampsbergen and the mountains along the escarpment. An allied, annual, species *M. rudolfii*, is known only from Botsabelo near Middelburg; field work around Middelburg should prove profitable.

Melanospermum swazicum grows in moist peaty soil under granite boulders in the environs of Mbabane in Swaziland. Plants from Ngome in northern Natal here equated with it may represent a distinct species; field work is needed in western Swaziland, neighbouring Transvaal and northern Natal, which is also the area of *M. italae*, a plant of gritty or sandy sites associated with outcropping rocks.

Three of the species were originally described in the genera *Polycarena* and *Phyllopodium*. The three genera may have similar-looking flowers but *Melanospermum* differs so markedly in seed structure and in the adnation of bract to pedicel and calyx that the relationship is probably not close. *Melanospermum* also occurs well outside the area of these two genera.

Melanospermum Hilliard in Notes RBG Edinburgh 45: 482 (1989). **Figs 15, 27A, 39B, 59.**

Small annual or perennial herbs. *Stems* leafy, pubescence usually patent, ± retrorse in one species, hairs gland-tipped or not. *Leaves* opposite, or occasionally alternate below the inflorescence. *Inflorescences* racemose, elongate or capitate, always erect. *Bract* adnate to pedicel only or sometimes to extreme base of calyx as well. *Calyx* bilabiate, tube shorter than anticous lip, anticous lip deeply to shallowly divided, membranous, persistent. *Corolla* membranous, persistent, tube glabrous outside but backs of lobes sometimes glandular-puberulous, splitting at base as capsule ripens, limb bilabiate, lobes 5, entire, orange/yellow patch at base of posticous lip or two posticous lobes wholly orange and there either glabrous or bearded with clavate hairs. *Stamens* 4, all exserted or 2 in throat, 2 exserted, posticous filaments decurrent to base of corolla tube, anthers synthecous. *Stigma* lingulate with two marginal bands of stigmatic papillae, passing gradually into the long filiform style, exserted. *Ovary* elliptic in outline, base slightly oblique with a small nectariferous gland on the shorter side, either glandular-puberulous or glabrous on upper part; ovules many in each loculus. *Fruit* a septicidal capsule with a short loculicidal split at the tip of each valve. *Seeds* 0.25–0.75mm long, seated on a round, centrally depressed pulvinus with irregular margins (fig. 15C), seeds elliptic-oblong, coarsely and deeply reticulate (3–8 round pits in each longitudinal band), black, testa tightly investing endosperm, under the SEM the reticulations clearly not cellular in origin (fig. 27A).

Type species: *Melanospermum transvaalense* (Hiern) Hilliard
Distribution of genus: South Africa, mainly Transvaal, Swaziland and northernmost Natal with one species ranging westwards through Botswana and the northern Cape to northern Namibia and north to western Zimbabwe.

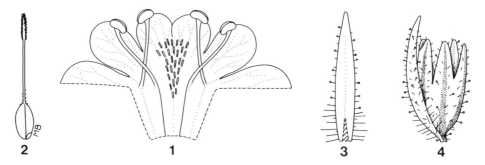

Fig. 59. *Melanospermum transvaalense* (*Hilliard & Burtt* 14164): 1, corolla opened out to show hairs on posticous side and decurrent posticous filaments; 2, stigma, style and ovary with small adnate nectariferous gland; 3, bract showing adnation scar; 4, calyx with bract adnate to pedicel (sometimes to extreme base of tube as well). All × 7.

Key to species

1a. Perennial herb well branched from the base, branches diffuse, glandular-villous, hairs on stems, leaves and bracts up to 0.75–1.5mm long .... **1. M. rupestre**
1b. Annual herbs, stems simple to branching from the base or higher, hairs mostly up to 1mm long, rarely 1.25mm 2

2a. Hairs on stems, leaves, bracts and calyx eglandular (a few minute almost sessile glands may be present below the inflorescence and on the inflorescence axis) .... **4. M. italae**
2b. Hairs either all glandular or mixed with eglandular hairs 3

3a. Hairs all glandular; cauline leaves distinctly petiolate, blade broad (c.3–13mm) in relation to its length 4
3b. Hairs mixed glandular and eglandular; cauline leaves not distinctly petiolate (though they are often narrowed to the base), blade narrow (c.0.5–2(–4)mm) in relation to its length .. 5

4a. Corolla tube c.4.5mm long, limb 5–6mm across the lateral lobes; leaf blade glandular-puberulous on both surfaces, very occasional hairs on lower margins 0.4–0.5mm long .... **3. M. rudolfii**
4b. Corolla tube 2.6–3.25mm long, limb 2–3.5mm across the lateral lobes; leaf blade sparsely glandular-hairy especially on lower surface and margins, hairs up to 0.5–1.25mm long
2. M. swazicum

5a. Eglandular hairs predominating on stems, usually more or less retrorse, up to 0.5–1mm long, tapering to the tips .... **5. M. foliosum**
5b. Glandular hairs often predominating on stems, both they and the eglandular hairs patent, eglandular hairs up to c.0.4mm long, the apical cell often inflated and balloon-like, much larger than the stalk cells .... **6. M. transvaalense**

1. Melanospermum rupestre (Hiern) Hilliard in Notes RBG Edinb. 45: 483 (1989).
Type: Transvaal, among rocks on the north side of Bezuidenhout's valley near Johannesburg,
May 1903, *Rand* 1326 (holo. BM).

Syn.: *Phyllopodium rupestre* Hiern in Fl. Cap. 4(2): 317 (1904).
    Polycarena rupestris (Hiern) Levyns in J. S. Afr. Bot. 5: 37 (1939).

Perennial herb, stems c.60–200mm long, up to 1mm in diam., well branched from the base,
stems diffuse, sometimes rooting from the lowermost parts, villous, hairs patent, glandular, up
to 0.75–1.25mm long. *Leaves* opposite, blade c.5–11 × 2.5–7mm, slightly smaller upwards and
passing imperceptibly into floral bracts, elliptic to ovate tapering into a flat petiolar part
1.5–6mm long, roughly ¼–½ the length of the blade, upper margins with 2–4 pairs of coarse
teeth, thinly villous on both surfaces, hairs up to 0.5–1.25mm long. *Flowers* few in terminal
leafy racemes, sometimes very lax. *Bracts* leaflike, the lower ones c.5–15mm long (including
the petiolar part), 1.75–5mm broad, hairs c.0.75–1.5mm long, bract adnate to pedicel only, or
sometimes to extreme base of calyx in uppermost flowers. *Pedicels* up to 0.75mm long. *Calyx*
bilabiate, tube 0.75–1(–2)mm long, anticous lobes 1.75–3(–5) × 0.6–1mm, anticous lip split
1.1–2(–3)mm, posticous lobes narrower, whole calyx glandular villous, hairs up to c.0.75mm
long. *Corolla* tube 3.5–5(–10) × 0.6–0.8(–1)mm, broadening a little in throat, limb 2.5–5(–
8)mm across the lateral lobes, posticous lobes 0.6–1.5(2.7) × 0.4–1(2)mm, anticous lobe
1–2(–3) × 0.6–1(2.8)mm, all lobes oblong-elliptic, white, orange patch at base of posticous lip,
glabrous. *Stamens* 4, anthers 0.3–0.6(–0.75)mm long, posticous pair in throat, anticous pair
shortly exserted. *Stigma* exserted. *Capsules* 3–3.5(–5) × 2–2.5mm. *Seeds* c.0.4 × 0.4mm.

Citations:

Transvaal. 2530AB, Lydenburg, *Wilms* 1060 (BM); 2530 AB, Hartebeesvlakte, 2000m, 12 xi 1979, *Kluge* 2038 (PRE);
Steenkampsbergen, 2530AC, Dullstroom, 7000ft, *Galpin* 13028 (BOL, K, PRE); 2530CA, c.8 miles from Belfast on the
Stoffberg road, *Burtt* 3112 (E); Carolina distr., 2530CC, 11 miles SE of Sewefontein, 5700ft, *Codd* 8110 (PRE). 2430DC,
Ohrigstad Dam Nature Reserve, 5200ft, 9 xi 1971, *Jacobsen* 1789 (PRE). 2430DD, Pilgrim's Rest, Clewer Falls, 5 vii
1930, *Moss* 18839 (J); 2430DC, Mount Sheba, 1 xi 1975, *Van Jaarsveld* 941 (NBG); ibidem, x 1975, *Getliffe et al.* 87
(J).

Melanospermum rupestre has been recorded from the Witwatersrand and from the Steenkampsbergen at four sites between Lydenburg and Carolina, as well as further east on the Drakensberg,
near Ohrigstad and Pilgrim's Rest, between c.1700 and 2100m above sea level. It is a delicate
straggling perennial herb, clad in remarkably long delicate glandular hairs; these hairs set it
apart from all its congeners except *M. swazicum*, in which the hairs may sometimes be nearly
as long as those of *M. rupestre*, but *M. swazicum* is an erect annual herb with the solitary stem
branching from above the base. *Melanospermum rupestre* grows on rock faces in forest and in
deep moist shade under outcropping rocks, sometimes hidden from view, which may account
for the paucity of records. It flowers mainly between October and January, but flowering and
fruiting has also been recorded in July.

   In the formal description, the measurements in parentheses refer to *Wilms* 1060, *Moss* 18839,
Van Jaarsveld 941, *Getliffe et al.* 87, and *Kluge* 2038, cited above. Wilms's specimen is dated
December 1887, and only old withered corollas remain clinging to the capsules; both the corolla
tubes and the capsules are remarkably long, but the habit and indumentum are those of *M.
rupestre*. The specimen bears only Wilms's printed label with locality given as 'bei der Stadt
Lydenburg', but it is possible that the specimen came from the nearby Steenkampsbergen rather
than from the vicinity of the town itself. The other specimens are precisely localized. This plant
should be sought again, and carefully collected to record habit, flower colour and variation, if
any, in flower size.

2. Melanospermum swazicum Hilliard in Notes RBG Edinb. 45: 483 (1989).
Type: Swaziland, Mbabane, 2631AC, Ukutula, c.4000ft, 20 ii 1956, *Compton* 25620 (holo. NBG).

Annual herb, stem c.40–270mm tall, 1–1.5mm diam. at base, simple or branching above, erect, leafy, glandular-hairy, hairs patent, up to 0.25–1mm long. *Leaves* opposite, occasionally alternate immediately below the inflorescences, blade 4–16 × 2.5–13mm, ovate-elliptic, base tapering to a flat petiolar part 1–11mm long, roughly ⅓ to ½ the length of the blade, upper margins finely to coarsely serrate with 3–6 pairs of teeth, thinly glandular-hairy especially on lower surface and margins, hairs up to 0.5–1.25mm long. *Flowers* up to c.15 in very lax racemes terminating each branchlet. *Bracts* at base of raceme c.2.5–12 × 0.5–4.5mm, smaller upwards, more or less leaflike (with 1 or 2 teeth) to spathulate, lanceolate or linear-lanceolate and then entire, adnate to pedicel and extreme base of calyx, glandular-hairy, hairs up to 0.6mm long. *Pedicels* up to 1mm long. *Calyx* bilabiate, tube 0.6–1.1mm long, anticous lobes 1.6–2.5 × 0.7–0.8mm, anticous lip split 1–2.25mm, posticous lobes narrower, whole calyx glandular-hairy, hairs up to c.0.4mm long. *Corolla* tube c.2.6–3.5 × 0.6–0.8mm, broadening a little in the throat, limb c.2–3.5mm across lateral lobes, posticous lobes 0.5–1.25 × 0.5–0.75mm, anticous lobe 0.75–1.5 × 0.5–0.9mm, all lobes oblong-elliptic, white or mauve with an orange patch at base of posticous lip and there either glabrous or thinly bearded, hairs extending down back of tube. *Stamens* 4, anthers 0.3–0.5mm long, all exserted. *Stigma* exserted. *Capsules* 2–3 × 1.5–2mm. *Seeds* c.0.3–0.6 × 0.3–0.5mm.

Citations:

Swaziland. Mbabane distr., 2631AC, hill NE of Mbabane, c.4500ft, *Compton* 26769 (NBG, PRE); ibidem, *Compton* 25351 (NBG, PRE, and see comment below); ibidem, 4200ft, *Kemp* 675 (PRE); hill 2 miles N of Mbabane, 4500ft, *Codd* 9519 (PRE); Makusini Hills, c.4000ft, *Compton* 31602 (NBG); Sipokasini, Tunnels road, c.3500ft, *Compton* 31900 (NBG, PRE); Ukutula, *Compton* 24925 (NBG).
Natal. Ngotshe distr., 2731CD, Ngome, c.3500ft, *Hilliard & Burtt* 8439 (E); ibidem, *Hilliard & Burtt* 9893 (E).

Melanospermum swazicum has been recorded principally from several sites on the hills around Mbabane, where it grows in moist shaded peaty soil in the shelter of granite rocks, between c.1050 and 1375m above sea level. It flowers mainly in January and February. Prof. Compton recorded that the plants smell foetid.

It is with some reservation that I include the specimens from Natal cited above: they have a rather longer calyx tube (0.75–1.1mm versus 0.6–0.75mm) and shorter lobes, the anticous ones 1.6–2mm with the lip split 1–1.6mm versus 2–2.5mm with the lip split 2–2.25mm. More collections from the area between Mbabane and Ngome are needed to assess the significance of this difference. The flowers of the Natal plant are mauve, those from Swaziland white. The constancy of this difference too needs to be ascertained. The Natal plant grows on bare wet earth around rock sheets.

Melanospermum swazicum is allied to *M. rupestre* from which it differs in its annual habit, the stem branching only above the base (not perennial, the stems diffuse and richly branched from the base), in its more floriferous racemes with less leaflike bracts, and in floral details, the calyx being more deeply divided, the corolla tube shorter, the limb often smaller, and the posticous lip sometimes bearded. The hairs on stems, bracts and calyx are also shorter than in *M. rupestre*. The two species are allopatric, *M. rupestre* occurring further west in the Steen-kampsbergen and along the Witwatersrand.

Specimens collected by Prof. Compton under his number 25351 (cited above) need special mention. They appear to include two different species. Those I have marked B are *M. swazicum*; those I have marked A have fewer flowers and these are in the axils of the upper leaves; in other words, the bracts are far more leaflike than in *M. swazicum* (the lowermost ones measure c.7.5 × 6.5mm). The flowers scarcely differ. I have seen no other specimens, either from Swaziland or elsewhere, to match them and further collections are clearly needed. The plants were in flower in mid January, when the first capsules had already formed.

3. Melanospermum rudolfii Hilliard in Notes RBG Edinb. 45: 483 (1989).
Type: Transvaal, Middelburg distr., 2529CB, Botsabelo, 1600m, 28 xii 1893, *Schlechter* 4078 (holo. E; iso. BOL, K, PRE, S, UPS).

Annual herb, stem erect to c.40–120mm, sparingly branched just above the base, leafy, glandular-puberulous with very occasional longer hairs up to 0.25–0.5mm long. *Leaves* oppo-site, blade c.6–20 × 3–13mm, usually elliptic, sometimes ovate, tapering gradually or more abruptly into a flat petiolar part 3–20mm long, about equalling to half the length of the blade, upper margins coarsely serrate with 2–5 pairs of teeth, both surfaces glandular-puberulous, very occasional hairs on lower margins up to 0.4–0.5mm long. *Flowers* up to c.9 in very lax terminal racemes. *Bracts* at base of raceme c.6–16 × 1.5–7mm, smaller upwards, leaflike, adnate to pedicel only or to extreme base of calyx as well, glandular-puberulous, very occasional hairs on lower margins up to 0.5mm long. *Pedicels* up to 0.8mm long. *Calyx* bilabiate, tube c.1.25–1.5mm long, anticous lobes c.2.25–2.5 × 0.75mm, anticous lip split c.1.6–1.75mm, posticous lobes slightly narrower, whole calyx glandular-puberulous, hairs up to 0.3mm long. *Corolla* tube c.4.5 × 0.75mm broadening slightly in throat, limb c.5–6mm across the lateral lobes, posticous lobes c.1–1.5 × 1–1.25mm, anticous lobe c.2–2.5 × 1.25–1.5mm, all lobes elliptic, white with an orange patch at base of posticous lip and there bearded. *Stamens* 4, anthers 0.6mm long, dark violet, all exserted. *Stigma* exserted. *Capsules* c.1.5–2.5 × 1.5mm. *Seeds* up to c.20 in each loculus, c.0.3–0.4 × 0.3–0.4mm.

Melanospermum rudolfii is known only from the type collection made by Rudolf Schlechter nearly 100 years ago at Botsabelo near Middelburg in south eastern Transvaal. The only ecological note he gave was 'in hum[idis]', but it can be deduced that the plants were growing in moist shady places under rocks, and the gorge at Botsabelo would be a likely place to begin looking for the plant. When Schlechter collected his specimens at the end of December they were mostly in flower, but a few capsules had already formed.

   Melanospermum rudolfii is clearly allied to both *M. rupestre* and *M. swazicum* but can be distinguished from both by its much shorter indumentum and by floral details (length of calyx tube and size of limb in particular).

4. Melanospermum italae Hilliard in Notes RBG Edinb. 45: 482 (1989).
Type: Natal, Ngotshe distr., Itala Nature Reserve, c.5000ft, 4 iv 1977, *Hilliard & Burtt* 10025 (holo. E; iso. NU, PRE, S).

Annual herb, stems c.40–200mm long, 1–2mm diam. at base, simple to well branched from the base and above, erect or the outer branches sometimes decumbent and rooting, leafy with axillary leaf tufts often well developed, pubescent throughout with patent or somewhat retrorse eglandular hairs up to 0.1–0.5mm long, the apical cell sometimes much bigger than the stalk cells and balloon-like, sometimes very minutely glandular-puberulous as well near the tips of the stems. *Leaves* opposite, sometimes alternate towards the tips of the stems, blade of main stem leaves (and radical leaves when present) c.5–13 × 1.5–8mm, broadly to narrowly elliptic tapering to a flat petiole-like base 1–12mm long, up to half the length of the blade or occasionally almost equalling it, margins in upper half of leaf with 2–3 pairs of teeth or entire in small leaves, glandular-punctate, nearly glabrous except for eglandular hairs on lower margins, upper surface in lower half of leaf, and sometimes on the midline below, hairs up to 0.4–0.8mm long, obtuse, apical cell sometimes inflated. *Flowers* many in crowded head-like racemes soon elongating and sometimes very lax in the lower part. *Bracts* at base of racemes c.5.5–8 × 1–4mm, lanceolate to elliptic-lanceolate, upper margins sometimes with 1 or 2 teeth, adnate to pedicel and base of calyx, hairy mainly on margins and upper surface near base, hairs mostly 0.3–0.6mm long, apical cell sometimes inflated. *Pedicels* 0–1mm long. *Calyx* bilabiate, tube 1.5–2mm long, anticous lobes 2–2.5 × 1mm, anticous lip split c.1mm, posticous lobes narrower and more deeply divided, whole calyx pubescent outside or hairs more or less confined to keels and margins, c.0.3mm long, obtuse, apical cell often more or less inflated. *Corolla* tube 5–5.2 × 1.2–1.5mm, broadening a little in the throat, limb 4.5–6mm across the lateral lobes, posticous lobes 1.2–1.6 × 0.75–1.5mm, anticous lobe 1.6–2.5 × 1–1.5mm, whole corolla sparsely puberulous outside particularly on back of posticous lip, orange patch at base of posticous lip inside and bearded there, the hairs running down back of tube, corolla lobes either white or white flushed mauve or wholly mauve. *Stamens* 4, anthers 0.6–0.75mm long, dark-coloured, all well exserted. *Stigma* well exserted. *Capsules* c.3–3.5 × 2.5mm, many-seeded. *Seeds* c.0.5–0.75mm long.

Citations:

Natal. Ngotshe distr., 2731CA, Itala Nature Reserve, near Louwsburg gate, c.4500ft, *Hilliard & Burtt* 8518 (E, NU); ibidem, *Hilliard & Burtt* 17722 (E, NU).
Swaziland. Mbabane, 2631AC, Nduma, c.5400ft, *Compton* 25376 (NBG, PRE).
Transvaal. Piet Retief distr., c.1400m, farm Mooi Hoek, *Devenish* 1765 (PRE).

Melanospermum italae is known from Mbabane in Swaziland, a site in south east Transvaal on the Natal border, and another in Natal, just south of the Transvaal border. The plants grow in sandy or gritty places around or on rock sheets, flowering between November and April. Well-grown plants form small bushy tufts smothered in flowers, and the range of colour variation from white to pale mauve can be seen in separate colonies around the extensive rock outcrops near the Louwsburg gate into the Itala Nature Reserve in northern Natal.

The primary cauline leaves in all but the most starved specimens are relatively broad and mostly distinctly petiolate, which at once sets the species apart from its ally, *M. transvaalense*. They differ too in indumentum: *M. italae* is almost entirely eglandular (there may be minute almost sessile glands on the upper parts of the stems and inflorescence axes) while *M. transvaalense* is conspicuously glandular on stems, bracts and calyces, though eglandular hairs are often present as well. Also, the corolla tube in *M. italae* is longer than it is in *M. transvaalense*.

The two species occupy similar habits but appear to be allopatric, *M. transvaalense* occurring west and north of the area of *M. italae*. On the other hand, *M. italae* is sympatric with *M. swazicum*, though Prof. Compton (to whom Botany is indebted for his extensive work on the flora of Swaziland) never collected them at the same site, but only in the same general area around Mbabane. *Melanospermum swazicum* resembles *M. italae* in its broad, petiolate cauline leaves, but differs in its glandular pubescent, more distinctly serrate leaves without axillary leaf tufts, and flowers in very lax racemes and smaller than those of *M. italae* (for example, corolla tube c.2.6–3.25mm long versus 5–5.2mm and limb c.2–3.5mm across the lateral lobes versus 4.5–6mm).

5. Melanospermum foliosum (Benth.) Hilliard in Notes RBG Edinb. 45: 482 (1989).
Type: Transvaal, Magaliesberg, *Burke* 192 (holo. K, iso. PRE, n.v.).

Syn.: *Polycarena foliosa* Benth. in DC., Prodr. 10: 351 (1846); Hiern in Fl. Cap. 4(2): 327 (1904).
    Polycarena discolor Schinz in Verh. Bot. Ver. Brandenb. 31: 191 (1890); Hiern in Fl. Cap. 4(2): 327 (1904); Hemsley & Skan in Fl. Trop. Afr. 4(2): 297 (1906); Merxmuller & Roessler in Prodr. Fl. S. W. Afr. 126: 40 (1967). Type: S.W.A., western margin of the Kalahari, Uri, *Schinz* 360 (holo. Z, iso. K).
    P. transvaalensis auct. non Hiern; Philcox in Fl. Zamb. 8(2): 26 (1990).

Annual herb, stems c.40–250mm long, up to 2.5mm diam. at base, simple to branching from the base and above, erect or decumbent, leafy, glandular-puberulous particularly on the upper parts, pubescent throughout with eglandular hairs up to 0.5–1mm long, obtuse, usually more or less retrorse. *Leaves* opposite, often becoming alternate on the inflorescence branches, blade of radical leaves (which are often wanting) c.4–13 × 2–6mm, elliptic, tapering to a flat petiolar part c.3–8mm long, from about half as long to about equalling the blade; primary stem leaves c.7–30 × 0.5–4mm, linear-oblong to oblanceolate, often with axillary tufts of smaller leaves, margins entire to toothed (2–5 pairs of teeth in the upper half) particularly in the larger leaves, upper surface with coarse tapering eglandular hairs up to 0.6–1mm long, mostly along the margins and on the lower half of the blade, lower surface glandular-puberulous, eglandular hairs confined to midrib. *Flowers* many in crowded rounded heads terminating all the branchlets, which are nude towards the tips, the heads elongating in fruit and sometimes becoming very lax. *Bracts* at base of racemes (2–)4–12 × (0.5–)1–1.75mm, smaller upwards, linear to oblong-lanceolate, adnate to pedicel only or to extreme base of calyx as well, glandular-puberulous all over, eglandular hairs as well, mainly on margins, up to 0.5–1mm long. *Pedicels* up to 0.5–1.5mm long. *Calyx* bilabiate, tube 1–1.6mm long, anticous lobe 1.5–2.5 × 0.5–0.8mm, anticous lip split 1–1.8mm, posticous lobes narrower and more deeply separated, whole calyx glandular-puberulous with longer eglandular hairs on margins. *Corolla* tube 2.5–4 × 0.75–1.2mm, widening slightly in throat, limb 4–6mm across the lateral lobes, posticous lobes 0.75–1.75 × 0.75–1.6mm, anticous lobe 2–3 × 1–2mm, all lobes elliptic, posticous lip bearded, hairs sometimes extending to the anticous lobes, orange patch at base of posticous lip or lower half of posticous lobes orange, rest of limb blue or mauve, perhaps sometimes creamy-coloured. *Stamens* 4, anthers 0.6–0.8mm long, dark-coloured, all well exserted. *Stigma* well exserted. *Capsules* 2.5–3 × 2–2.5mm. *Seeds* many in each loculus, 0.25–0.5mm long.

Selected citations:

Zimbabwe. 2028AB, 5 miles south of Bulawayo, ± 4800ft, *Miller* 7974 (K).
Transvaal. Pietersburg distr., 2328BD, 'Maxabeu' [= Magabene], 1160m, *Schlechter* 4660 (BOL, E, K, PRE, S, UPS). Waterberg, 2427BC, Kransberg, farm Geelhoutbos, *Mauve* 5053 (K). 2528CC, Halfway House between Pretoria and Johannesburg, *Mogg* 11451 (PRE).
Cape. Vryburg, 2624DC, Armoedsvlakte, *Mogg* 7933 (PRE). 2723 BB, near Litakun at the Moshowa river, *Burchell* 2252-6 (K).
Botswana. 2425DB, 3 miles north of Gaborone, 3350ft, *Mott* 312 (K, PRE).
Namibia. Otjiwarongo, 2016BC, *Bagot-Smith* 8 (NU). Okahandja, 2116DD, 1100m, *Dinter* 612 (SAM); ibidem, *Dinter* 7774 (WIND).

There are no characters to separate *Polycarena discolor* from *Melanospermum foliosum* (= *Polycarena foliosa*). Hiern (in Fl. Cap. 4(2)) sought to distinguish them in his key to the species of *Polycarena* by differences in habit and leafiness, but this does not hold when a range of material is examined. The Burchell specimen from Litakun that he cites under *P. discolor* comprises four tiny flowering seedlings, in strong contrast to the well grown specimen that typifies the name *P. foliosa*.

Melanospermum foliosum is widely distributed, though infrequently collected, from northern Namibia (recorded from Otjiwarongo, Okahandja and Gobabis, as well as Uri, the type locality) through Botswana and the northern Cape to the western Transvaal (not recorded east of Premier Mine, Pretoria district), and with a single record from western Zimbabwe, near Bulawayo. It is a plant of open sandy places; flowering has been recorded between March and October with most records (14 out of 21 collections seen) in August and September.

It is clearly allied to *M. transvaalense*, with which it is frequently confused. They can be distinguished by differences in indumentum: in *M. transvaalense*, the eglandular hairs on the stem are patent, up to c.0.4mm long, and sometimes the apical cell is inflated and balloon-like; glandular hairs often predominate. In *M. foliosum*, eglandular hairs always predominate on the stems and are up to 0.5–1mm long, taper to the tips, and are usually more or less retrorse. The flowers of the two species are similar in size but differ in colour: limb white in *M. transvaalense* (apart from the orange blotch), mauve, pink or blue in *M. foliosum*. This seems to be a constant difference, but it needs further checking.

Melanospermum foliosum flowers in spring; the plants presumably appear on sandy sites soon after a rain shower. This is in contrast to both the habitat and flowering time of *M. transvaalense*, which grows in sandy or gritty seepage areas on rock sheets or among rock outcrops, and flowers mainly between January and March. The area of *M. transvaalense* lies east of that of *M. foliosum*, possibly always above c.1500m; *M. foliosum* seems to occur only below c.1500m. Their areas meet somewhere between Premier Mine (*M. foliosum*) and the Wilge Rivier and Brug Spruit on the Pretoria–Middelburg road (type localities for *M. transvaalense* and *Phyllopodium calvum*, which is synonymous with *M. transvaalense*); field work hereabouts might prove instructive.

6. Melanospermum transvaalense (Hiern) Hilliard in Notes RBG Edinb. 45: 483 (1989). **Fig. 59, Plate 1I.**
Lectotype: Transvaal, [between Pretoria and Witbank], Wilge Rivier, 1530m, 3 i 1894, *Schlechter* 4125 (K; isolecto. BM, BOL, C, E, J, S).

Syn.: *Polycarena transvaalensis* Hiern in Fl. Cap. 4(2): 330 (1904).
> *Phyllopodium calvum* Hiern in Fl. Cap. 4(2): 315 (1904). Type: Transvaal, between Middelburg and Pretoria, highlands at Brug Spruit, c.4500ft, *Bolus* 7674 (holo. BOL).
> *Polycarena calva* (Hiern) Levyns in J. S. Afr. Bot. 5: 37 (1939).

Annual herb, stems c.55–220mm tall, up to c.1mm diam. at base, simple or branching both above and below, erect or the outer branches somewhat decumbent, leafy, glandular-puberulous with hairs up to c.0.3mm long, patent eglandular hairs also present, few to plentiful, less than 0.5mm long, the apical cell sometimes swollen and balloon-like. *Leaves* opposite, sometimes alternate towards the tips of the branches, blade of lowermost ones (which may be withered or wanting at flowering) c.5–20 × 2.5–5.5mm, elliptic, tapering to a flat petiolar part c.3.5–17mm long, from twice as long to half as long as the blade, upper margins either entire or with 1–3 pairs of teeth, indumentum variable, often both surfaces nearly glabrous except for a few coarse eglandular hairs up to 0.4mm long on the lower margins and midline, or sometimes glandular-puberulous all over or with a mixture of short glandular and eglandular hairs, leaf surface always glandular-punctate; stem leaves much narrower than the basal leaves, c.5.5–40 × 0.5–2mm, linear to narrowly oblanceolate, narrowed below but not distinctly petiolate, margins entire or with a very few tiny teeth, indumentum as in radical leaves. *Flowers* many in crowded rounded heads terminating all the branchlets, which are nude towards the tips, the heads elongating in fruit and often becoming very lax. *Bracts* at base of racemes c.5–15 × 0.5–1.5(–2)mm, smaller upwards, linear to oblong-lanceolate, usually entire, sometimes with 1 or 2 small teeth, adnate to pedicel only or to extreme base of calyx as well, glandular-puberulous, usually with few to many eglandular hairs as well, these up to c.0.4mm long at base of bract. *Pedicels* up to 0.5–1mm long. *Calyx* bilabiate, tube 0.75–1.5mm long, anticous lobes 1.5–2 × 0.5–0.75mm, anticous lip split 0.5–0.75mm, posticous lobes narrower and more deeply separated, whole calyx glandular-puberulous, often with eglandular hairs as well. *Corolla* tube 2.5–4 × 0.6–1.25mm broadening slightly in throat, limb 4–6mm across the lateral lobes, posticous lobes 1–1.6 × 0.9–1.4mm, anticous lobe 2–2.6 × 1–1.8mm, all lobes elliptic, either glabrous or glandular-puberulous outside, bearded at base of posticous lip and down back of tube, anticous lip white, posticous lip either orange or with an orange patch at the base. *Stamens* 4, anthers 0.6–0.8mm long, dark-coloured, all well exserted. *Stigma* well exserted. *Capsules* 2–3 × 2–3mm. *Seeds* many in each loculus, 0.3–0.5 × 0.3–0.5mm.

Selected citations:

Transvaal. Pilgrim's Rest distr., 2430DC, 12km from Pilgrim's Rest to Lydenburg, *Stirton* 1815 (PRE); 2530BA, Sabie-Lydenburg, Long Tom Pass, c.7000ft, *Hilliard & Burtt* 14354 (E, NU). Belfast distr., 2530AC, Dullstroom, Wanhoop Farm, 7000ft, *Drews* 251 (PRE); 2530CA, 8km from Belfast on Stoffberg road, *Hilliard & Burtt* 14432 (E, NU); 2530CD, 18km from Machadodorp on road to Badplaas, Uitkomst Farm, *Germishuizen* 3806 (PRE); 2530CB, Elandshoogte near Machadodorp, c.5500ft, *Hilliard & Burtt* 14164 (E, NU). Barberton distr., 2530DB, Kaapschehoop, *Hilliard & Burtt* 14291 (E, NU). Ermelo distr., 2630AD, Knock Du farm near Lothair, *Welman* 393 (PRE).
Orange Free State. Heilbron, 2727BD, *Gilfillan* in herb. Bolus 15507 (BOL).

Melanospermum transvaalense is relatively widespread in SE Transvaal from Pilgrim's Rest to Ermelo, with a single record from the Orange Free State at Heilbron, about 70km due south of the Vaal river. The plants favour damp or wet sandy or gritty sites associated with rock outcrops. They flower mainly between January and March, but there is one record for September, another for October, when the plants were in fruit; it seems likely therefore that they can continue to flower through winter when the soil is kept damp by seepage. We have ourselves collected the species several times and noted that the individual heads of flowers are flat-topped and form pseudanthia with the orange patches on the posticous lips turned towards the centre of the inflorescence and simulating anthers.

Melanospermum transvaalense is closely allied to *M. foliosum*: see discussion under that species.

9. POLYCARENA

Polycarena is essentially a genus of the western and south western Cape. Of the 17 species enumerated here, only four range north of Van Rhynsdorp and most are found no further east than 20° of longitude, only one ranging as far as 22° east.

The species fall into two natural groups: first, those species (ten in number) in which the corolla tube is narrow in relation to its length and is either pubescent or glabrous, and any yellow or orange patterning on the limb takes the form of either a blotch below the posticous lip or a flush all round the mouth. This corresponds to Bentham's (1846) unranked group *Longiflorae*. There are normally four stamens, the anticous pair exserted and the posticous pair more or less included with the anthers appearing in the mouth (in *P. aemulans*, the one species in which the stamens may be reduced to two, it is the exserted (anticous) stamens that disappear).

Secondly, there are the species (seven, or possibly eight in number) in which the corolla tube is broad in relation to its length, is usually glabrous (rarely a few minute hairs in one species) and there are usually four stamens, all well exserted (in this group, two species, *P. comptonii* and *P. tenella*, show reduction to one, two or three stamens, (the posticous stamens aborted) and one species, *P. rariflora*, may be cleistogamous, and then all four stamens remain within the unopened corolla). In this second group, yellow or orange colour-patterning varies mainly in relation to corolla size. Two species (*P. filiformis* and *P. tenella*) have very small, wholly white flowers; in *P. comptonii*, the flowers are also very small but the two posticous corolla lobes are orange, the anticous ones white. In the other species, all of which have white or cream flowers, the posticous lobes normally have yellow or orange tips in addition to a similarly coloured patch at the base of the lip. The exceptions are *P. aurea* in which there is a yellow-flowered form lacking any patterning, and *P. rariflora*, where the cleistogamous (and probably also the self-fertilized but not cleistogamous) flowers lack patterning.

Within Group 2, (which corresponds in part to Bentham's informal group *Parviflorae*), the species are clearcut and relatively easy to distinguish. *Polycarena rariflora* seems to stand alone. *Polycarena pubescens*, *P. aurea* and *P. formosa* are closely allied, and so are *P. comptonii*, *P. filiformis* and *P. tenella*, the latter showing extreme reduction in corolla size with reversion to a regular limb and frequent reduction in the number of stamens. *Polycarena comptonii* (known from only two collections at White Hill near Laingsburg) may have four stamens, but often there are only two, with further reduction to one (in contrast to *P. aemulans*, it is the posticous stamens that abort). *Polycarena comptonii*, *P. filiformis* and *P. tenella* differ from other species of *Polycarena* in having seeds only obscurely trigonous and weakly winged. All three are exceedingly delicate annuals.

Group 1 comprises ten species, seven of which are newly described. Of the three historic species, *P. silenoides* appears to be confined to Lion's Head in the Cape Peninsula; its circumscription presents no difficulty. The other two species are *P. capensis* and *P. gilioides*. There has clearly been a great deal of confusion about the correct use of these two names. In addition, it seems highly probable that extensive hybridization and back-crossing has taken place between several different species with consequent blurring of species limits. There is no concrete evidence to support this hypothesis, but the circumstantial evidence is detailed below.

Polycarena capensis is characterized by its corymbose branching, marked leafiness (the leaves are close-set, dwarf axillary leafy shoots are well developed in all but the most depauperate specimens, and the primary leaves scarcely diminish in length upwards though they do become narrower) and its heads of flowers (at least 10 in the main heads) with pubescent corolla tube, pale yellow limb (cream in the Cape Peninsula and nearby) with an orange to deep yellow star in the throat and there well bearded all round. It appears to be confined to the 3318 degree square with an overall range from about Hopefield to the Peninsula and east to Joostenberg between Durbanville and Paarl.

Polycarena gilioides is characterized by divaricate branching and a consequent weak development of dwarf axillary shoots, few-flowered (1–7) heads that soon become lax, relatively small flowers (tube up to 10mm, limb 5–8mm across the lateral lobes in contrast to tube mostly 9.5–14mm in *P. capensis* with limb mostly 8–14mm), the limb white or cream with an orange-yellow patch below the posticous lip only and there well bearded with hairs sometimes extending to the other three lobes. This description applies to *P. gilioides* from its type locality around Paarl and northwards to Gouda and Tulbagh. To the west and north, its limits are blurred where it overlaps in distribution with *P. lilacina* and *P. gracilis*, both of which may in turn be affected. This is discussed further below.

Polycarena lilacina is closely allied to *P. capensis* with which it has hitherto been confused and with which it shares the characters of corymbose branching, leafy habit and relatively large heads of flowers. It differs in its remarkably short indumentum (hairs mostly up to 0.1mm long with the glandular head small and inconspicuous, opposed to hairs 0.25–0.3mm long with conspicuous glandular head), mostly longer bracts and corolla tube, and mauve flowers, the limb marked with a yellow patch only below the posticuous lip and there bearded with only a few hairs extending to the lateral lobes. It is largely sympatric with *P. capensis* north of the Cape Flats and the Peninsula.

The difference in indumentum between *P. lilacina* and *P. capensis* should not be dismissed as trivial. Differences in indumentum are important both at generic level within Manuleae as a whole and at species level within many of the genera, including *Polycarena* (see for example *P. pubescens*, *P. aurea* and *P. formosa* in Group 2). What has here been described as *P. lilacina* var. *difficilis* differs from *P. lilacina* principally in its longer hairs (0.25–0.4mm) with conspicuous glandular tips (some specimens also show a tendency towards divaricate branching, and white flowers may occur). This variety occurs within the area of *P. lilacina*, but also occurs further north, reaching as far as the environs of Doring Baai, west of Klaver. In the southern part of its range it is sympatric with *P. capensis* and *P. gilioides sens. lat.*, with hairs within the 0.25–0.4mm range, and in the northern part, not only with *P. gilioides sens. lat.*, but also with *P. subtilis* and *P. gracilis*, again with hairs within the 0.25–0.4mm range. The last three all branch divaricately.

Just as the indumentum of *P. lilacina* may have been affected by the presence of one or other of these species, so in turn does *P. gilioides* appear to have been affected, possibly by *P. lilacina* (introduction of mauve colour and longer corolla tube) or *P. gracilis* (introduction of purplish flush on both corolla tube and backs of corolla lobes).

It is surely not without significance that problems of species delimitation arise only where two or more species are sympatric. *Polycarena batteniana* for example presents no problems; it is confined to the higher country from about Garies to Wupperthal, isolated from the species of the western lowlands that are discussed above.

On the other hand, the area around Pakhuis Pass (3218BB), and probably Nardouw Pass too (3118DC), presents an excellent field for biological research. Here occur no less than five allied species, *P. exigua*, *P. aemulans*, *P. gracilis*, *P. gilioides sens. lat.* and *P. nardouwensis*; (*P.*

rariflora, *P. tenella*, *P. formosa* and *P. pubescens*, all Group 2, also occur in the area but do not appear to contribute to the problems). All five species frequently have a dark flush on the corolla tube and on the backs of the lobes, and are the only species so patterned. *Polycarena exigua* is highly distinctive: a tiny annual with leaves broad for the group, corolla lobes white inside, dark red outside, the tube probably white but drying ochre-yellow with no trace of a dark flush. *Polycarena aemulans* closely resembles *P. exigua*, having the same broad leaves, but differs in details of corolla size, pubescence and colouring: corolla tube and backs of lobes weakly to strongly flushed purple. The two species possibly grow within a few metres of each other (one mixed collection seen). Here (as well as elsewhere, but not consistently) both *P. gilioides sens. lat.* and *P. gracilis* have a purplish flush on corolla tube and lobes. *Polycarena nardouwensis* stands slightly apart, both geographically and in morphological detail: it is known only from Nardouw Pass (where none of the other species has been recorded) and it has a deep yellow flush all round the mouth (instead of a patch only below the posticous lip) and the mouth is glabrous (at least the base of the posticous lip bearded in the others).

It should be pointed out that *P. capensis*, *P. subtilis* and *P. nardouwensis* all have a yellow or orange flush right round the mouth (all the other species in Group 1 have only a yellow-orange patch). In addition, the limb is yellow or cream. Unfortunately, collectors notes are scanty and I have no field knowledge of the species, but it may be that the corollas of these three species tend more to the yellow part of the spectrum and attract different pollinators from the other species in Group 1, which have mauve, white or cream (only a hint of yellow?) flowers. The only biological note was given by Leipoldt on a collection of *P. batteniana*, where he recorded 'delicious scent in evening'.

The short broad corolla tubes and different colour-patterning in the species of Group 2 suggest yet another type of pollinator. Again, there is only one collector's remark that is possibly of biological significance: Mrs Batten recorded of the white-flowered form of *P. aurea* 'flowers yellow and white (in the same head), yellow uppermost'. This may be an instance of a head of flowers simulating a single flower, a well documented biological phenomenon of which we have personal knowledge in Scrophulariaceae (*Strobilopsis wrightii* and *Melanospermum transvaalense*).

Polycarena Benth. in Hook., Comp. Bot. Mag. 1: 371 (1836), in DC., Prodr. 10: 350 (1846) & in Benth. & Hook. f., Gen. Pl. 2(2): 944 (1876); Hiern in Fl. Cap. 4(2): 321 (1904), p.p.; Hilliard & Burtt in Notes Roy. Bot. Gard. Edinb. 35(2): 161 (1977), excluding *P. capillaris* (L.f.) Benth. [= *P. parvula* Schltr.]. **Figs 16B, C, 27B, C, 40B, 60.**

Annual herbs. *Stems* leafy, pubescence always patent, hairs gland-tipped. *Leaves* opposite below, alternate above. *Inflorescence* racemose, sometimes congested, always erect. *Bract* not sharply differentiated from leaves, adnate to pedicel and calyx, only anticous lobes and sometimes uppermost part of tube free. *Calyx* 5-lobed, bilabiate, membranous, slightly accrescent, persistent. *Corolla* membranous, persistent, tube glabrous or pubescent outside, splitting at the base as the capsule ripens, limb usually bilabiate (regular in one species with minute flowers), lobes entire, throat usually with orange-yellow patch at base of posticous lip or all round mouth and there either bearded with clavate hairs or sometimes glabrous. *Stamens* usually 4, sometimes reduced to 2, or 3 or 1 in chasmogamous flowers, either all exserted or anticous pair well exserted, posticous pair just visible in mouth, posticous filaments decurrent down tube in most species, anthers synthecous. *Stigma* lingulate with two marginal bands of stigmatic papillae, exserted, passing gradually into the long filiform style. *Ovary* elliptic in outline, base slightly oblique with an adnate nectariferous gland on the shorter side, glandular-puberulous on upper part, bilocular; ovules few to many in each loculus. *Fruit* a septicidal capsule with a short loculicidal split at top of each valve, glandular-puberulous at apex. *Seeds* 4–40 in each loculus, 0.75–1(–1.5)mm long, buff-coloured, seated on a round, centrally depressed pulvinus with undulate margins (fig. 16B, C), seed usually strongly, sometimes weakly, 3-winged, testa delicate, transparent, loosely enveloping endosperm, under the SEM seen to be regularly patterned with wrinkled tubercles (fig. 27B, C).

Lectotype: *Polycarena capensis* (L.) Benth.
Distribution: South Africa, western and south-western Cape.

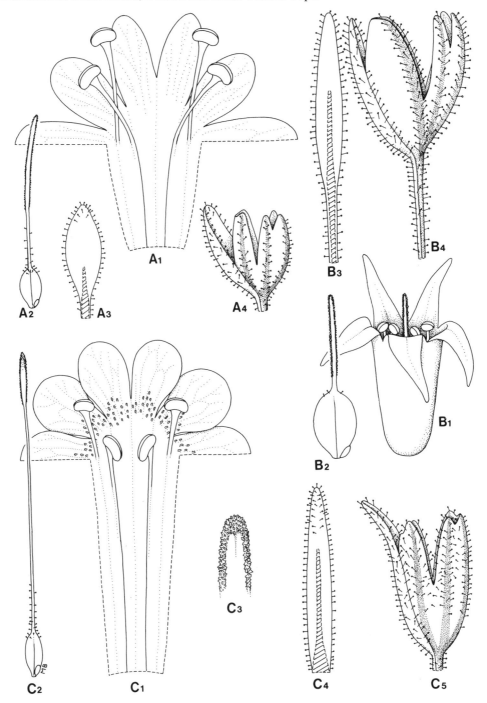

Key to species

1a. Corolla tube pubescent or puberulous 2
1b. Corolla tube either glabrous or with occasional minute hairs 10

2a. Corolla limb 3–4mm across the lateral lobes 3
2b. Corolla limb at least 5mm across the lateral lobes 4

3a. Bracts 4.5–7 × 0.4–0.75mm, calyx 2.5–4mm, corolla mouth glabrous, tube flushed purple
8. P. aemulans
3b. Bracts 6–14 × 0.75–1.25mm, calyx 4.25–5.5mm, corolla mouth bearded below the
posticous lip, tube cream-coloured . . . . **10. P. silenoides**

4a. Corolla limb yellow or cream-coloured, orange all round mouth and there bearded; plant
corymbosely branched, closely leafy right up to the inflorescence, main heads with at least
10 flowers (fewer only in depauperate specimens), compact, elongating slightly in fruit
2. P. capensis
4b. Corolla limb either mauve, white or cream-coloured; orange patch usually confined to base
of posticous lip and there bearded, but if all round mouth, then glabrous; if cream-coloured,
then plant divaricately branched, dwarf axillary shoots either absent or weakly developed,
main heads small (never more than 9 flowers) and lax 5

5a. Flowers mauve, at least 10 in each main head, crowded, becoming slightly lax in fruit, hairs
on bracts and calyx up to 0.1mm long, on stems occasional hairs reaching 0.2mm
1. P. lilacina var. lilacina
5b. Flowers either mauve, white or cream, most hairs at least 0.25mm long, often 0.4mm . . . 6

6a. Leaves oblanceolate, relatively broad in relation to their length, mostly 8–15 × 1–3mm
8. P. aemulans
6b. Leaves linear to narrowly oblanceolate, long in relation to their breadth (ignore the
lowermost 1 or 2 pairs), mostly 8–35 × 0.5–2mm 7

7a. Mouth glabrous . . . . **5. P. nardouwensis**
7b. Mouth bearded at least below the posticous lip 8

8a. Corolla lobes mostly flushed on the backs . . . . **6. P. gilioides** sens. lat.
8b. Corolla lobes concolorous 9

9a. At least 10 flowers in the main heads, bracts mostly 7.5–15mm long, calyx at least 5mm
long, corolla tube 11–16mm, capsule 5–6mm . . . . **1. P. lilacina var. difficilis**
9b. Up to 9 flowers in the main heads, bracts 5–7mm long, calyx 3.5–4.5mm, corolla tube
7.5–10mm, capsule 3–4.5mm . . . . **6. P. gilioides**

10a. Corolla tube mostly at least 8mm long; if shorter, then only c.0.5mm broad; 1 pair of
anthers well exserted, the other pair in the corolla mouth or very shortly exserted 11
10b. Corolla tube 1–6.5mm long, 0.75–2mm broad; all 4 stamens (occasionally reduced to 2
or 1) well exserted 14

Fig. 60 (opposite). A, *Polycarena aurea* (*Batten* 973): A1, corolla opened out to show decurrent posticous filaments; A2, stigma, style and ovary with small adnate nectariferous gland; A3, bract showing adnation scar; A4, calyx with bract adnate to lower part of calyx tube and pedicel. All × 7. B, *Polycarena tenella* (*Esterhuysen* 20552): B1, corolla with 5 equal lobes; B2, stigma, style and ovary with small adnate gland; B3, bract showing adnation scar; B4, calyx with bract adnate to lower half of tube and pedicel. All × 32. C, *Polycarena capensis* (*Hilliard & Burtt* 13029): C1, corolla opened out to show hairs around mouth and decurrent posticous filaments, × 7; C2, stigma, style and ovary with small adnate nectariferous gland, × 7; C3, apex of stigma showing marginal bands of stigmatic papillae, × 20; C4, bract showing adnation scar, × 7; C5, calyx with bract adnate to pedicel and halfway up tube, × 7.

11a. Corolla tube 3.5–8 × 0.5mm, limb 2.5–4mm across the lateral lobes, lobes white above, red below . . . . **9. P. exigua**

11b. Corolla tube 8–25 × 0.5–1.5mm, limb at least 6mm across, lobes white, cream or yellow, often self-coloured, flushed purple below in one species 12

12a. Corolla limb white, throat glabrous or with a very few hairs below the posticous lip
4. P. batteniana

12b. Corolla limb white, cream or yellow, at least the posticous lip bearded 13

13a. Flowers few, in lax racemes, bracts 4–7.5mm long, not or only briefly exceeding the calyx, limb white, cream or yellow with orange patch at base of posticous lip and there bearded (a few hairs may extend onto the lateral lobes) . . . . **7. P. gracilis**

13b. Flowers in compact heads, bracts 9–18mm long, conspicuously longer than the calyx, limb yellow, mouth orange and bearded all round . . . . **3. P. subtilis**

14a. Flowers crowded into rounded heads (they elongate slightly in fruit) 15

14b. Flowers in lax racemes 18

15a. Corolla limb 3–5mm across the lateral lobes . . . . **13. P. pubescens**

15b. Corolla limb 6–14mm across the lateral lobes 16

16a. Hairs on stem, bract and calyx up to 0.25mm long, bracts 5–9 × 0.5mm
See under **14. P. formosa**

16b. Hairs on stem etc. either at least 0.4mm long; or, if only 0.25mm, bracts 2.5–7.5 × 0.4–3mm, narrowly to broadly elliptic 17

17a. Pubescence short, longest hairs on stem, bract and calyx 0.25–0.4mm, bracts 2.5–7.5 × 0.4–3mm, narrowly to broadly elliptic . . . . **12. P. aurea**

17b. Pubescence longer, hairs 0.6–1.2mm, bracts 5.5–14 × 0.5–1.5 (–2)mm, linear or narrowly oblong . . . . **14. P. formosa**

18a. Corolla lobes acuminate, limb c.2mm in diam., white . . . . **17. P. tenella**

18b. Corolla lobes rounded, limb usually at least 2.4mm in diam. (smaller in flowers that are clearly cleistogamous) 19

19a. Plant with a rosette of 4 radical leaves 4.5–18 × 2.5–9mm, ovate to elliptic, corolla tube 1.5–2mm long, limb 2.4mm across the lateral lobes, posticous lobes orange, anticous white; stamens 4, 2 or 1 . . . . **15. P. comptonii**

19b. Plants without a basal rosette of leaves, corolla tube at least 2.6mm long, posticous lobes either white or *tipped* with orange; stamens 4 20

20a. Bracts 4–12 × 0.25–1mm, linear, calyx glabrous to puberulous, posticous corolla lobes often tipped with orange, anthers yellow . . . . **11. P. rariflora**

20b. Bracts 2.5–3 × 0.4–1mm, oblong, calyx with hairs up to 0.25mm long, posticous corolla lobes white, anthers almost black and very conspicuous . . . . **16. P. filiformis**

1. Polycarena lilacina Hilliard in Notes RBG Edinb. 45: 488 (1989). Type: Cape, Malmesbury div., 3318CB, Buck Bay near Darling, 16 ix 1980, *Hilliard & Burtt* 13016 (holo. E, iso. NU).

Two varieties are recognized:

1a. Hairs on stem, bract and calyx mostly up to 0.1(–0.2)mm long . . . . var. **lilacina**

1b. Hairs up to 0.25–0.4mm long . . . . var. **difficilis**

var. lilacina

Annual herb c.60–280mm tall, stem occasionally simple, mostly corymbosely branched from about or above the middle, glandular-puberulous, the hairs mostly 0.1–0.2mm long, an occasional hair reaching 0.25mm, leafy, dwarf axillary shoots often present. *Leaves* 6–35 × 0.5–2(–3)mm, diminishing in width rather than in length upwards, linear to narrowly oblanceolate, entire or with 1–3 pairs of teeth, both surfaces sparsely glandular-puberulous. *Flowers* few to many (at least 10 in the main heads) in dense terminal heads corymbosely arranged, heads elongating slightly in fruit, pedicels up to 0.5–1.5mm long. *Bracts* 9–17 × 0.5–1.5mm long, adnate almost to top of calyx, linear, lowermost often with 1 or 2 pairs of teeth, glandular-puberulous on both surfaces, hairs mostly c.0.1mm long, always less than 0.2mm, the gland very small, often narrower than the shaft of the hair. *Calyx* (5.5–)6–8mm long, anticous lobes very short, posticous one 0.75–1mm long, glandular-puberulous all over, hairs mostly c.0.1mm long, always less than 0.2mm, the glandular head inconspicuous. *Corolla* tube 13.5–16 × 1–1.75mm, glandular-puberulous, limb 8.5–12mm across, lobes broadly elliptic, posticous ones 3–5 × 1.75–3.5mm, anticous one 4–5 × 2.5–4mm, pale mauve, yellow patch at base of posticous lip and there bearded, a few hairs extending onto the other 3 lobes. *Stamens*: anticous pair well exserted, anthers 1mm long, posticous pair in throat, filaments decurrent to base of corolla tube, anthers 1.5mm long. *Stigma* exserted. *Capsule* c.4.5–6 × 3–4mm. *Seeds* many, c.1mm long, distinctly 3-winged.

Selected citations:

Cape. Piquetberg div., 3218DC, between Berg River and Sauer, 15 ix 1953, *Lewis* 3863 (SAM). Malmesbury div., 3318AA, Hopefield, Geelbek Road, 28 ix 1953, *Heginbotham* 216 (NBG); 3318AB, Hopefield, ix 1905, *Bolus* 12791 (BOL); Koperfontein, 14 ix 1953, *Barker* 8058 (NBG).

var. **difficilis** Hilliard in Notes RBG Edinb. 45: 488 (1989).
Type: Cape, Van Rhynsdorp div., 3118CD, farm Kliphoek [about 10 miles east of Doringbaai], 350ft, 12 ix 1970, *Hall* 3826 (holo. NBG; iso. K, PRE, STE).

Annual herb c.100–260mm tall, stem occasionally simple, usually corymbosely branched from above the middle, occasionally somewhat divaricate, glandular-pubescent, hairs 0.25–0.4mm long, dwarf axillary shoots sometimes present. *Leaves* 10–40 × 0.5–2(–5)mm, mostly diminishing in width rather than in length upwards, linear to narrowly oblanceolate, entire or with 1–3 pairs of teeth, both surfaces sparsely glandular-pubescent. *Flowers* few to many (at least 10 in the main heads) in dense terminal heads that elongate slightly in fruit, pedicels up to 1mm long. *Bracts* (6–)7.5–15 × 0.4–1mm, linear, lowermost often with 1 or 2 pairs of teeth, glandular-pubescent on both surfaces, hairs 0.25–0.4mm long, with conspicuous dark apical gland. *Calyx* 5–7.5mm long, anticous lobes very small, posticous one 0.75–1mm long, glandular-pubescent, hairs 0.25–0.4mm long. *Corolla* tube 11.5–17 × 0.75–1.25mm, glandular-pubescent, limb 7–13.5mm across, lobes broadly elliptic, posticous ones (2–)3–6 × 1.4–3mm, anticous one 3–5.5 × 2–4.5mm, white or pale mauve, yellow patch at base of posticous lip and there bearded, a few hairs extending onto the 2 lateral lobes. *Stamens*: anticous pair well exserted, anthers 0.75–1mm long, posticous pair in the throat, anthers 1.3–1.5mm long. *Stigma* exserted. *Capsules* 4–6 × 3–3.5mm. *Seeds* many, c.1mm long, distinctly 3-winged.

Selected citations:

Cape. Clanwilliam div., 3218AB, 1½ miles SW of Leipoldtville, c.400ft, 23 viii 1958, *Acocks* 19676 (K, PRE); 3218BA, Graafwater, *Compton* 24211 (NBG). Piquetberg div., 3218CD, Kotze, ix 1944, *Lewis* 1552 (SAM); 3218DA, 11.7km south of Redelinghuis, c.120m, 16 ix 1970, *Acocks* 24356 (K, PRE); 3218DC, near Sauer, 10 ix 1949, *Steyn* 545 (NBG, STE). Malmesbury div., 3318CB, Mamre, 28 ix 1953, *Compton* 24369 (NBG).

Typical *P. lilacina* is notable for its very short indumentum, a character that, combined with markedly different floral details, serves to distinguish it from *P. capensis* with which it shares the character of corymbose branching and leafy habit. *Polycarena lilacina* var. *difficilis* has all the characters of *P. lilacina* except the very short indumentum; instead, the hairs are up to 0.25–0.4mm long. To assign varietal rank to this plant is a matter of expediency, a means of drawing attention to a variation that may seem trivial, but it assumes significance in the context of *Polycarena* as a whole, where small differences in indumentum can often be used to distinguish species that are easily separable on other characters. It is thought that exchange of genes with other sympatric species having the same sort of indumentum may account for this variation in var. *difficilis* (see p.396), and also account for the more or less divaricate branching seen in some specimens of var. *difficilis*, and possibly the occurrence of white flowers (but also see below).

Polycarena lilacina var. *lilacina* ranges from about Piquetberg and Sauer to Buck Bay and Darling, growing in sandy places below 200m in altitude and flowering in September and October. Var. *difficilis* is found over the entire area of var. *lilacina*, but it also extends northwards through low-lying country to the vicinity of Doring Baai, where it has been collected on the farm Kliphoek. This northernmost collection was deliberately chosen as the type of var. *difficilis* because in facies it so closely resembles typical *P. lilacina*. However, the flowers are white, and the collector's notes indicate that he saw a large population of white-flowered plants. The coastal area north of Doring Baai should be explored for this white-flowered plant to check for constancy of colour in that area.

2. Polycarena capensis (L.) Benth. in Hook., Comp. Bot. Mag. 1: 371 (1836) & in DC., Prodr. 10: 351 (1846); Hiern in Fl. Cap. 4(2): 323 (1904) excluding *Schlechter* 5083, 8576; Levyns in Adamson & Salter, Fl. Cape Penins. 713 (1950). **Fig. 60C, Plate 3J.**
Type: Cap. bonae spei (LINN 790.9).

Syn.: *Lychnidea villosa foliis angustis* ... Burm., Rar. Afr. Pl. decas 5, 141, t. 50 fig. 2 (1739).
  Buchnera capensis L., Mant. 1: 88 (1767).
  Erinus umbellatus Burm. f., Prodr. Pl. Cap. 17 (1768), quotes Burm., Rar. Afr. Pl. t. 50 fig. 2.
  Manulea capensis (L.) Thunb., Prodr. 101 (1800) & Fl. Cap. ed. Schultes 467 (1823).
  M. villosa Pers., Syn. 2: 148 (1807), nom. illegit.

Annual herb 70–280mm tall, stems simple or corymbosely branched either from the base or in the upper half, glandular-pubescent, all hairs c.0.25–0.3mm long, very leafy, in part owing to the rich development of dwarf axillary shoots. *Leaves* (primary leaves) 6–28 × 0.75–1.25mm, diminishing in width rather than in length upwards except on the inflorescence branches, linear to oblanceolate, mostly with 1–2(–3) pairs of very small teeth, both surfaces glandular-pubescent. *Flowers* few to many in dense terminal heads corymbosely arranged, heads elongating somewhat in fruit, pedicels up to 1–1.5mm long. *Bracts* 5–10(–13) × 0.5–1mm, adnate almost to top of calyx, linear, very occasionally the lowermost with 1 or 2 teeth, glandular-pubescent on both surfaces, hairs 0.25–0.3mm long. *Calyx* 4–5.5(–7)mm long, anticous lobes very short, posticous one 1–1.5mm long, glandular-pubescent all over, hairs 0.25–0.3(–0.4)mm long. *Corolla* tube (7–)9.5–14 × 1–1.5mm, glandular-pubescent, limb (7–)8–14mm across, lobes broadly elliptic, posticous ones 2–4 × 2–3.5mm, anticous one 3–6 × 2.25–4.5mm, pale yellow, or creamy-coloured on the Cape Peninsula and nearby, orange-yellow at base of lobes and in throat, and there strongly bearded. *Stamens*: anticous pair well exserted, anthers 1–1.25mm long, posticous pair in throat, filaments decurrent to base of tube, anthers 1.25–1.5mm long. *Stigma* exserted. *Capsules* 4.5–7 × 3.5–5mm. *Seeds* many, c.1mm long, distinctly 3-winged.

Selected citations:

Cape. Malmesbury div., 3318AD, Darling Flora Reserve, *Lewis* 5088 (NBG); 3318BC, 18 miles NW of Malmesbury, *Lewis* 3861 (SAM); 3318CB, near Darling, *Hilliard & Burtt* 13029 (E, NU); Mamre Hills, *Wasserfall* 456 (NBG). Cape Town div., 3318CD, near Milnerton, *Salter* 8682 (BOL). 3318DC, Kanonberg, Brackenfel, *Acock* 5105 (S). Paarl div., 3318DD, Joostenberg, *Esterhuysen* 16019 (BOL, NBG).

Polycarena capensis has a narrow geographical range from about Hopefield south to the Peninsula (it appears to be confined to the 3318 degree square), and grows in sandy places, flowering between August and October, but mainly in September.

Much confusion has surrounded the correct use of the name *P. capensis*, and the species most commonly confounded with it are *P. batteniana*, *P. gilioides*, *P. gracilis*, *P. lilacina* and *P.*

subtilis, only one of which (*P. gilioides*) has hitherto been described. *Polycarena capensis* is characterized by its corymbose branching, marked leafiness and heads of yellow flowers (cream on the Cape Peninsula and nearby) marked in the corolla mouth with a deep yellow or orange star and there well bearded (see also p.396). Only *P. lilacina* resembles *P. capensis* in habit, but its mauve flowers serve as an easy distinguishing character; the significant differences between the two species are fully discussed under *P. lilacina* (above) and on p.396. *Polycarena subtilis* resembles *P. capensis* in the colour and markings of the corolla, but differs in its divaricate branching and in floral detail (see under 3, *P. subtilis*). *Polycarena batteniana* (no. 4, white flowers), *P. gilioides* (no. 6, white, cream, pink or purple flowers) and *P. gracilis* (no. 7, white or cream and possibly yellow flowers) all differ from *P. capensis* in their divaricate branching as well as in floral details other than colour. Where appropriate, the differences are discussed under those species and on p.396.

3. Polycarena subtilis Hilliard in Notes RBG Edinb. 45: 488 (1989).
Type: Cape, Piquetberg div., 3218DB, Grey's Pass, 6 ix 1949, *Steyn* 361 (holo. NBG, iso. S).

Annual herb 50–240mm tall, stems simple or divaricately branched, glandular-pubescent, hairs up to 0.25mm long. *Leaves* 9–38 × 0.4–2(–3.5)mm, linear to narrowly oblanceolate, entire or with up to 2(–4) pairs of teeth, both surfaces glandular-pubescent. *Flowers* up to 10 in small heads, lax in fruit, pedicels up to 0.5–1mm long. *Bracts* 9–18 × 0.4–0.5mm, linear, adnate nearly to top of calyx, glandular-pubescent on both surfaces, hairs up to 0.25mm long. *Calyx* (5–)6–9mm long, anticous lobes very short, posticous one c.1–1.8mm long, glandular-pubescent all over, hairs up to 0.25mm long. *Corolla* tube 11.5–26 × 0.75–1.25mm, glabrous to thinly puberulous, limb 10–13.5mm across, lobes broadly elliptic, posticous ones (2.5–)3.5–5 × 2.5–3.5mm, anticous one (2.5–)3.5–5 × (2–)3–4mm, yellow, orange all round in throat and there well bearded. *Stamens*: anticous pair well exserted, anthers c.1.25mm long, posticous pair in throat, filaments decurrent to base, anthers c.1.75mm long. *Stigma* exserted. *Capsules* c.4–5 × 3mm. *Seeds* many, c.1mm long, strongly 3-winged.

Selected citations:

Cape. Clanwilliam div., 20 miles S of Clanwilliam, Olifants River valley, 31 vii 1948, *Lewis* 3149 (SAM). Piquetberg div., 3218DA, near Het Kruis, 29 ix 1943, *Leighton* 75 (BOL, PRE); 3218DB, The Rest, ix 1930, *Gillett* 3710 (BOL, STE); top of Grey's Pass, 12 ix 1935, *Taylor* 951 (BOL); 3218DD, near Piquetberg, 2 ix 1938, *Hafstrom & Acocks* 1234 (PRE).

Polycarena subtilis has been recorded from a small area between the Olifants River valley south of Clanwilliam to Piquetberg, a linear distance of about 80km. It seems to be essentially a plant of sandy flats, and although it has been collected on top of Grey's Pass (518m), it is probably commonest at lower altitudes (up to 200m). It flowers between July and September.

Its flowers resemble those of *P. capensis*, being yellow with a bearded orange star in the mouth, but it is a much less leafy plant than *P. capensis*, divaricately rather than corymbosely branched, with longer and narrower bracts (these bracts are a conspicuous feature, always much exceeding the calyx) and (even in well-grown plants) fewer flowers in the head, which gives the plant a dainty appearance (whence the trivial name). Both calyx and corolla tube are mostly longer than in *P. capensis*, and the tube of *P. subtilis* tends to be glabrous.

The two species are allopatric, their areas meeting somewhere south of Piquetberg. On the other hand, *P. subtilis* occurs in the same geographical area as, but possibly always at lower altitudes than, *P. gracilis*, another dainty plant with even fewer flowers in the inflorescence than *P. subtilis* (up to 5(–7)) and these arranged in lax racemes rather than in heads. *Polycarena gracilis* lacks the conspicuous bracts of *P. subtilis* (4–7.5mm long, versus 9–18mm) and has a mostly shorter calyx (4–6mm, versus (5–)6–9mm). The limb of the corolla in *P. gracilis* is white to cream in colour, with the backs of the lobes as well as the tube often flushed with purple, and the orange patch in the throat is confined to the base of the posticous lip, which is more heavily bearded than the lateral lobes.

4. Polycarena batteniana Hilliard in Notes RBG Edinb. 45: 486 (1989).
Type: Cape, Niewoudtville, 3119AC, Neil McGregor's farm, 6 ix 1979, *Batten* 493 (holo. E).

Annual herb 60–180mm tall, stems simple or divaricately branched, glandular-pubescent, hairs up to 0.25–0.4mm long. *Leaves*: lower ones 12–35 × 2.25–5mm, narrowly to broadly spathulate, obscurely toothed, both surfaces glandular-pubescent; upper leaves 7.5–33 × 0.5–2(–3)mm, oblong to narrowly oblanceolate, entire or with 1–2 pairs of teeth, glandular-pubescent. *Flowers* few to many in terminal heads, elongating slightly in fruit, pedicels up to 1mm long. *Bracts* 5–13 × 0.6–1.25(–1.5)mm, adnate nearly to top of calyx, linear to oblong, both surfaces glandular-pubescent, hairs up to 0.25–0.4mm long. *Calyx* 4–6(–7)mm long, anticous lobes very short, posticous one c.1–1.5mm long, glandular-pubescent all over, hairs up to 0.25–0.4mm long. *Corolla* tube 10–17 × 1–1.5mm, glabrous or with occasional minute hairs, limb 8–13mm across, lobes broadly elliptic, posticous ones 3–4.5 × 2–3mm, anticous one 3.5–5 × 2.5–4mm, white or creamy-coloured, orange-yellow patch at base of posticous lip, usually glabrous, occasionally with a few hairs. *Stamens*: anticous pair exserted, anthers 1–1.25mm long, posticous pair in throat, filaments decurrent to base of tube, anthers 1.5–2mm long. *Stigma* exserted. *Capsules* c.4 × 3mm. *Seeds* c.16 in each loculus, c.1mm long, distinctly 3-winged.

Selected citations:

Cape. Van Rhynsdorp div., 3018CA, near Garies, 4 ix 1986, *Batten* 739 (E). Calvinia div., 3119CA, Lokenburg, c.2300ft, 17 ix 1955, *Acocks* 18467 (PRE). 3119CD, Botterkloof, 29 viii 1941, *Barker* 1955 (NBG). Clanwilliam div., 3219AA, Citadel Kop, 7 ix 1953, *Compton* 24256 (NBG). 3219AC, near Wupperthal, Storms Vlei, ix 1897, *Leipoldt* 589 (NBG, SAM).

Polycarena batteniana has been recorded from Garies, south of the Kamiesberg, south east to the environs of Wupperthal, probably always above 400m in altitude. It grows in sandy places, flowering mainly in September. Leipoldt (no. 589) recorded 'delicious scent in evening'; this, allied to the white, or at least pale, colour of the limb and the long corolla tube suggests pollination by moths (see also p. 397). Nothing appears to be known of the biology of *P. capensis*, which is allied to *P. batteniana*, but usually has the limb coloured yellow (pale on the Cape Peninsula and nearby) and marked with a deeper yellow or orange, bearded star in the mouth (the mouth of the corolla in *P. batteniana* is glabrous or nearly so, and has an orange-yellow patch only below the posticous lip). The two species are further distinguished by the corolla tube being glabrous or very nearly so in *P. batteniana*, pubescent in *P. capensis*, which is always corymbosely branched and very leafy right up to the inflorescence, partly because of the strong development of leafy dwarf axillary shoots; the branching in *P. batteniana* is more divaricate than corymbose; in other words, the axillary shoots develop into spreading branches.

Polycarena batteniana has a more northerly distribution than *P. capensis*, and grows at higher altitudes.

Mrs Batten not only collected the type material, but she has also given much time to collecting, photographing and sketching plants in the western Cape specifically to help me; naming *P. batteniana* is a small mark of my appreciation.

5. Polycarena nardouwensis Hilliard in Notes RBG Edinb. 45: 488 (1989).
Type: Cape, Clanwilliam div., 3118DC, top of Nardouws Kloof, 13 viii 1976, *Goldblatt* 3861 (holo. E, iso. PRE).

Annual herb 70–160mm tall, stems simple or corymbosely branched in the upper part, glandular-pubescent, the hairs up to 0.25–0.4mm long. *Leaves* 11–25 × 0.5–2(–3.75)mm, lowermost 1–2 pairs of leaves the broadest and often the shortest, otherwise diminishing in breadth rather than in length upwards, linear to oblanceolate, entire or with 1–2 pairs of teeth. *Flowers* up to 10 in small heads, elongating slightly in fruit, pedicels up to 1mm long. *Bracts* 8–18×0.5–0.75mm, adnate almost to top of calyx, linear, glandular-pubescent on both surfaces, hairs c.0.3–0.4mm long. *Calyx* 5.5–9mm long, anticous lobes short, posticous one c.1–1.75mm long, glandular-pubescent all over, hairs c.0.3–0.4mm long. *Corolla* tube 22–40 × 1.25–1.5mm, glandular-pubescent, limb 10–13.5mm across, lobes broadly elliptic, posticous ones 4.5–6 × 3–3.8mm, anticous one 5–6 × 3.5–4.5mm, limb cream (or pale yellow?) flushed orange all round in throat, backs of lobes and tube often flushed purplish, throat glabrous or with, literally, 1 or 2 hairs below the posticous lip. *Stamens*: anticous pair strongly exserted, anthers 1mm long, posticous pair in throat, filaments decurrent to base of tube, anthers 1.5mm long. *Stigma* exserted. *Capsules* 5–6 × 3–3.5mm. *Seeds* many, c.1mm long, distinctly 3-winged.

Citations:

Cape. Clanwilliam div., 3118DC, top of Nardouws Kloof, viii 1949, *Stokoe* sub SAM 70151 (SAM); ibidem, 6 ix 1951, *Johnson* 241 (C, NBG).

Polycarena nardouwensis is known only from the Nardouw's Kloof - Nardouw's Pass area, where it has been found by three different collectors over the period 1949 to 1976, flowering in August and September.

It is a distinctive plant, with a long corolla tube, cream-coloured limb and glabrous mouth, all of which point to an affinity with *P. batteniana*, which occurs much further north than *P. nardouwensis* as well as immediately to the east of it (the nearest recorded sites lie very roughly 30–60km away at Citadel Kop and Brandewyn River (3219AA), the top of Botterkloof Pass (3119 CD) and Lokenburg (3119CA). It differs from *P. batteniana* in its mostly longer and narrower bracts (8–18 × 0.5–0.75mm versus 5–13 × 0.6–1.5mm), longer and pubescent corolla tube (22–40mm versus 10–17mm and glabrous or very nearly so) and the orange patterning on the limb extends all round the mouth (not confined to the base of the posticous lip).

The backs of the corolla lobes and the tube are often flushed with dark purple, which is unknown in *P. batteniana*, but it does occur in *P. gracilis*, which has been recorded only south of Nardouw Pass, the nearest known locality being Pakhuis Pass (3218BB). The flowers of *P. nardouwensis* are in definite heads (not in very open racemes as in *P. gracilis*), the bracts are longer (8–18mm versus 4–7.5mm), the calyx is often longer (5.5–9mm versus 4–6mm), and the glabrous mouth is orange all round (not merely below the posticous lip and bearded there, often on the lateral lobes as well). The corolla tube in *P. gracilis* is only about 8–10mm long over most of its range; however, the two specimens collected at Pakhuis Pass have tubes about 12–25mm long, compared with tube 22–40mm long in *P. nardouwensis*. Extensive field studies are needed in this area, as has already been suggested (see p. 396).

6. Polycarena gilioides Benth. in Hook., Comp. Bot. Mag. 1: 372 (1836) & in DC., Prodr. 10: 351 (1846); Hiern in Fl. Cap. 4(2): 324 (1904), excluding *Schlechter* 10821.
Type: Cape, sands near Paarl, *Drège* (holo. K; iso. E, S, W).

Annual herb 75–150 [220]mm tall, stem either simple or divaricately branched from the base or higher, glandular-pubescent, hairs up to 0.25[0.3]mm long, dwarf axillary shoots sometimes weakly developed. *Leaves*: lower leaves 7–22 [14–38] × 0.5–2.5 [0.75–4]mm, oblanceolate, entire or with 1–2 pairs of teeth, thinly glandular-pubescent on both surfaces, particularly margins and lower part; upper leaves 10–20 [8–35]×0.5–1.5 [0.5–1(–2)]mm, linear to narrowly oblanceolate, either entire or with 1 pair of teeth. *Flowers* 2–7(–10) in small terminal clusters becoming lax in fruit, pedicels 0.5–1mm long. *Bracts* 5–7[5.5–13] × 0.4–0.75mm, adnate almost to top of calyx, linear, entire, glandular-pubescent on both surfaces, hairs up to 0.25–0.3mm long. *Calyx* 3.5–4.5[4–7.5]mm long, anticous lip shallowly lobed, posticous lobe c.1mm long, glandular-pubescent all over, hairs 0.25–0.3mm long. *Corolla* tube 7.5–10[9–13(–25)] × 0.75–1.25mm, pubescent, limb 5–8[6.5–9(–11)]mm across, lobes oblong-elliptic, posticous ones 2–3.25 [2.5–4.5] × 1–2[1.5–2.25]mm, anticous one 2–4[2.5–5] × 1–2[1.75–3]mm, all lobes white or cream [white, cream, pink, purple, both backs of lobes and tube often flushed purple], orange-yellow patch below posticous lip, bearded there, hairs often extending to the other 2 or 3 lobes. *Stamens* 4, anticous pair well exserted, anthers c.0.75mm long, posticous pair included, filaments decurrent nearly to base of tube, anthers in mouth, c.1.25mm long. *Stigma* exserted. *Capsules* 3–4.5[6] × 3[3–4]mm. *Seeds* many, c.1mm long, distinctly 3-winged.

Selected citations:
P. gilioides sens. strict.:

Cape. Paarl div., 3319AC, W foot of Vogelvlei Mts near Gouda, 9 ix 1951, *Esterhuysen* 18803 (BOL); Ceres road, 12 ix 1896, *Schlechter* 8983 (BOL, E,S); Tulbagh, Nieuwekloof, 1897, *MacOwan* s.n. (E). 3218DD(?), De Hoek Estates, ix 1936, *Lewis* s.n. (BOL). 3318DB, Paarl, *W. Brown* s.n. (E).

P. gilioides sens. lat.:

Cape. Van Rhynsdorp div., 3118DC, Gift Berg, ix 1911, *Phillips* 7364 (BOL, SAM, STE). Clanwilliam div., 3218BA, Zeekoe Vley, 19 viii 1896, *Schlechter* 8576 (E, PRE, W); 3218BB, 12 miles from Clanwilliam on road to Lambert's Bay, ix 1938, *Gillett* 4060 (BOL, K); Pakhuis Pass, 1 xii 1934, *Compton* 4773 (NBG). 3218BC, near Paleisheuwel, 5 ix 1934, *Levyns* 10163 (BOL). 3219AC, Niewoud Pass, 26 ix 1934, *Compton* 4991 (NBG). Piquetberg div., 3218AD, between Verloren Vlei and Rooikransberg, 18 x 1935, *Pillans* 7984 (BOL).

Polycarena gilioides was described from material collected by Drège near Paarl. Specimens matching the type well have come from localities between Paarl and Tulbagh, and possibly Piquetberg (De Hoek Estates, not precisely localized by the collector). It is a small annual herb, well branched from the base, with long narrow leaves and 1–7-flowered heads that soon become lax, the flowers small (corolla tube up to 10mm long, pubescent, limb 5–8mm in diam.). Collector's notes are meagre, merely a cryptic 'white' and 'cream' on two sheets, but all the material has dried with both corolla tube and lobes self-coloured (apart from the usual orange patch in the throat).

In the formal description given above, measurements and other information in square brackets refer to a plant that has been included in a broad circumscription of *P. gilioides*. This plant ranges from about Piquetberg north to the Giftberg south west of Klawer (3118DC). It has mostly larger flowers (corolla tube 9–25mm long, limb mostly 7–11mm across the lateral lobes). Again, there is no precise information about the colour of the corolla: four collectors have merely recorded 'white', 'cream', 'pink', 'purple'. But in many instances, the backs of the corolla lobes or the corolla tube, sometimes both, have dried with a purplish flush. These plants may have been influenced by *P. gracilis*, here sympatric with *P. gilioides* sens. lat.: *P. gracilis* normally has a purple flush on the outside of the corolla. In the vicinity of Sauer, Hopefield, Ysterfontein and Darling, the influence may have come from *P. lilacina*; there is a fuller discussion on p.396.

7. Polycarena gracilis Hilliard in Notes RBG Edinb. 45: 487 (1989).
Type: Cape, Clanwilliam div., 3218BB, Pakhuis Pass, 7 ix 1949, *Steyn* 400 (holo. NBG).

Annual herb 90–230mm tall, stems either simple or divaricately branched from the base, glandular-pubescent, hairs up to 0.25mm long. *Leaves*: lower leaves 9–24 × 0.5–4mm, linear to oblanceolate, with 1–3 pairs of teeth, both surfaces glandular-pubescent; upper leaves 6–30 × 0.25–1.5(–2)mm, linear to narrowly oblanceolate, either entire or with 1 pair of teeth. *Flowers* 1–5(–7) in lax terminal racemes, pedicels 0–0.5mm long. *Bracts* 4–7.5 × 0.4–0.75mm, adnate almost to top of calyx, linear, entire, glandular-pubescent on both surfaces, hairs up to 0.25mm long. *Calyx* 4–6mm long, anticous lip entire or shallowly lobed, posticous lobe 0.5–0.75mm long, glandular-pubescent all over, hairs up to 0.25–0.3mm long. *Corolla* tube 8–12(–25) × 0.75–1mm, glabrous, limb 7–7.5(–13.5)mm across, lobes oblong-elliptic, posticous ones 2.5–5.5 × 1.25–2.5mm, anticous one 2.5–3.5 × 1.5–4mm, all lobes white to cream (or yellow?) with an orange-yellow patch at base of posticous lip and there well bearded, hairs often extending onto two lateral lobes, backs of lobes and tube sometimes flushed purplish. *Stamens* 4, anticous pair well exserted, anthers c.0.75mm long, posticous pair included, filaments decurrent to base of tube, anthers in mouth, c.1.25mm long. *Stigma* exserted. *Capsules* 3.5–4 × 3mm. *Seeds* many, c.1mm long, distinctly 3-winged.

Citations:

Cape. Clanwilliam div., 3218BB, Pakhuis Pass, 27 viii 1937, *Compton* 6876 (NBG); Piquetberg div., 3218BB, Bosch Kloof, 22 ix 1940, *Compton* 9485 (NBG); 3218DA, Piquetberg Mountain Plateau, 1800ft, ix 1927, *Levyns* 2237 (BOL); ibidem, 2500–3000ft, 12 ix 1954, *Esterhuysen* 23110 (BOL, K); Hills NW of Mouton's Vlei, 6 xi 1934, *Pillans* 7452 (BOL).

Polycarena gracilis has been recorded from about Pakhuis Pass, east of Clanwilliam, south to Piquetberg Mountain, between c.550 and 900m above sea level. This is the same geographical range as *P. subtilis*, but it is not clear that the two species are actually sympatric; it seems likely that *P. subtilis* occurs only at lower altitudes than *P. gracilis* (see under no. 3, *P. subtilis*, for distinguishing characters). The collectors of *P. gracilis* have provided no information on its habitat, but it flowers between August and November.

It can be confused with *P. gilioides sens. strict.*, from which it may be distinguished by its glabrous (not pubescent) corolla tube, which is 8–12(–25)mm long (versus 7.5–10mm) with limb 7–13.5mm across (versus 5–8mm), and both tube and limb often flushed purple outside (not so in *P. gilioides*). There is some difference too in the arrangement of the flowers, the raceme being very lax in *P. gracilis*, congested in *P. gilioides* and elongating only slightly in fruit. It has already been suggested (on p.396) that *P. gracilis* and *P. gilioides* may interact where their areas overlap, but the lax inflorescence and particularly the glabrous corolla tube of *P. gracilis* are good distinguishing characters.

The relationship of *P. gracilis* possibly lies with *P. aemulans*, which has lax racemes and corolla tube and limb flushed purple outside, but *P. gracilis* may be distinguished by its relatively narrower upper leaves (6–30 × 0.25–1.5(–2)mm, linear or narrowly oblanceolate, versus 6–15 × 1–3mm, spathulate), calyx 4–6mm long (versus 2.5–4mm), corolla tube glabrous (not pubescent), limb 7–13.5mm across (versus 3–9mm), stamens always 4 (not often reduced to 2 or 3).

8. Polycarena aemulans Hilliard in Notes RBG Edinb. 45: 486 (1989).
Type: Cape, Clanwilliam div., 3219AA, near Pakhuis, 1 ix 1938, *Salter* 7515 (holo. K, iso. BOL).

Diminutive annual herb c.40–130mm tall, stems either simple or divaricately branched from the base, glandular-pubescent, hairs c.0.25–0.3mm long. *Leaves*: lower leaves (4.5–)10–19 × 2.5–4mm, spathulate to elliptic, with 2 pairs of teeth, teeth sometimes obscure, both surfaces glandular-pubescent; upper leaves similar but smaller, 6–15 × 1–3mm, mostly with 1–2(–4) pairs of teeth. *Flowers* up to 7, but mostly 3–5, in lax terminal racemes, pedicels 0.1–0.75mm long. *Bracts* 4.5–9 × 0.4–1mm, adnate almost to top of calyx, linear-oblong, entire, glandular-pubescent on both surfaces, hairs up to 0.4mm long. *Calyx* 2.5–4mm long, anticous lip shallowly lobed, posticous lobe 0.3–0.75mm long, glandular-pubescent all over, hairs up to 0.3–0.4mm long. *Corolla* tube 5.75–16 × 0.75mm, pubescent, flushed with purple, limb 3–9mm across, lobes oblong-elliptic, posticous ones 1.25–3 × 0.75–1.75mm, anticous one 1.5–3.5 × 1.5–2mm, all lobes cream-coloured above, weakly to strongly flushed purple below, yellow patch at base of posticous lobe and bearded there except in the smallest flowers. *Stamens* either 4, anticous pair well exserted, posticous pair in throat, or, in the smallest flowers, often reduced to 2 by abortion of the anticous pair, or with a third stamen, the anther much reduced, exserted; anthers of exserted stamens c.0.25mm long (if fully fertile), of included ones c.0.3mm, posticous filaments decurrent more than halfway down tube. *Stigma* exserted, or included in the smallest flowers. *Capsules* 3–3.5 × 2–3mm. *Seeds* c.15 in each loculus, c.0.75–1mm, distinctly 3-winged.

Citations:

Cape. Van Rhynsdorp div., 3119AC, top of Van Rhyn's Pass, 28 viii 1941, *Esterhuysen* 5990 (K, BOL). Clanwilliam div., 3218BB, Pakhuis Pass, eastern side, ix 1947, *Lewis* 4876 (SAM). 3219AA, near Pakhuis, ix 1933, *Leipoldt* BOL 20663 (BOL, K); Pakhuisberg, 2300ft, 23 viii 1896, *Schlechter* 8614 (E, PRE, S, W); Brandewyn River, 1000ft, 13 viii 1897, *Schlechter* 10821 (BOL, E, PRE, W); ibidem, 13 ix 1947, *Compton* 19983 (NBG); ibidem, 26 viii 1950, *Barker* 6585 (NBG). South Cedarberg, 3219AC, Krom River, 2 x 1952, *Esterhuysen* 20545 (BOL).

Polycarena aemulans grows in sandy places among rocks or in depressions in rocks, flowering between August and October. It has been recorded from Van Rhyn's Pass in the Bokkeveld mountains west of Vanrhynsdorp, the Pakhuis mountains near Clanwilliam, and the south Cedarberg, a linear range of roughly 100km. Over the entire range there occur what are clearly whole colonies of plants with very small corollas and the stamens frequently reduced to two (and then included) or there may be a third, exserted, stamen with much reduced anther lobes; occasionally the full complement of four stamens is present; the stigma is wholly or partially included. The corolla limb spreads open and it is glabrous in the throat, not bearded below the posticous lip as in larger flowers with the normal complement of stamens (the function of these hairs is not understood, but they are commonplace in Manuleae and are almost certainly of biological significance). It seems then that these small flowers, though not actually cleistogamous, may be self-fertilizing. What are probably out-crossing plants (larger corolla, four normal stamens, exserted stigma) of *P. aemulans* have been recorded around Pakhuis and Brandewynrivier. Field observations are needed, especially at Pakhuis, where other closely allied species, including *P. exigua* and *P. gilioides sens. lat.*, have been recorded; they must grow in close proximity to *P. aemulans*, and in similar habitats (see p.396 for further discussion).

The specific epithet, *aemulans*, draws attention to the close resemblance of this species to *P. exigua*. It is particularly those plants with the smallest flowers and reduced stamens that resemble *P. exigua*, but they can at once be distinguished by the corolla tube, pubescent and purplish in *P. aemulans*, glabrous or nearly so in *P. exigua* and drying ochre-yellow (white when fresh?).

The flowers of *P. aemulans* resemble those of *P. gilioides* in both pubescence and colouring. However, the limb is mostly smaller in *P. aemulans* (3–9mm across the lateral lobes versus mostly 7.5–11mm), the upper lip only is bearded in the larger flowers (not bearded all round the mouth) and the capsules are smaller (3–3.5mm long versus 5–6mm). The two species differ markedly in foliage, the leaves of *P. aemulans* being broad in relation to their length (spathulate, mostly 1–3mm broad), those of *P. gilioides* narrow (linear, mostly 0.5–1mm broad; the 2–4 lowermost leaves should be ignored as they are broad in many species of *Polycarena*). Plants of *P. gilioides* look leafier than those of *P. aemulans* because of the presence of dwarf axillary shoots (absent in *P. aemulans*).

9. Polycarena exigua Hilliard in Notes RBG Edinb. 45: 487 (1989).
Type: Cape, Clanwilliam div., S. Cedarberg, 3219AD, Wolfberg, 4000ft, 3 x 1952, *Esterhuysen* 20575 (holo. BOL, mixed with a specimen of *P. aemulans*; iso. K, PRE).

Diminutive annual herb 30–130mm tall, stems either simple or divaricately branched from the base, glandular-pubescent, hairs c.0.25mm long. *Leaves*: lower leaves 4–14 × 1.5–4mm, broadly spathulate, rarely entire, usually with 2 pairs of obscure teeth, both surfaces thinly glandular-pubescent; upper leaves similar but smaller, 3–8 × 1–2.5mm, mostly with 1 or 2 pairs of teeth. *Flowers* 1–5(–7) in very lax terminal racemes, pedicels 0.5–0.75mm long. *Bracts* 3–7 × 0.5–1.75mm, adnate almost to top of calyx, oblong to narrowly spathulate, often with 1 or 2 teeth, glandular-pubescent on both surfaces, hairs up to 0.3mm long. *Calyx* 2.5–4mm long, anticous lip scarcely notched, posticous lobe c.0.5mm long, glandular-pubescent all over, hairs up to 0.3mm long. *Corolla* tube 3.5–8 × 0.5mm, glabrous or with few minute hairs near apex, drying ochre-yellow, possibly white when fresh, limb 2.5–4mm across, lobes oblong-elliptic, posticous ones 0.75–1 × 0.5–1mm, anticous one 1–1.75 × 0.5–0.75mm, white above, red beneath, bearded at base of posticous lip and with an orange patch there. *Stamens* 4, anticous pair well-exserted, anthers 0.25mm long, posticous pair in mouth, filaments decurrent to base of tube, anthers 0.3mm long. *Stigma* exserted. *Capsules* 2.5–3 × 2–3mm. *Seeds* c.7 in each loculus, c.0.75–1mm long, distinctly 3-winged.

Citations:

Cape. Clanwilliam div., N Cedarberg, 3218BB, Pakhuis, 3000ft, 7 ix 1953, *Esterhuysen* 21766 (BOL, PRE); ibidem, 23 viii 1941, *Esterhuysen* 5903 (BOL, K); ibidem, 29 ix 1940, *Esterhuysen* 3351 (BOL). Central Cedarberg, 3219AD, near Welbedacht, Cedarhoutkop, 4000ft, 27 ix 1980, *Esterhuysen* 35511 (BOL, E, S,). 3219AC, Middelberg, 25 ix 1937, *Compton* 7066 (NBG); Tafelberg, 4000ft, 16 xii 1950, *Esterhuysen* 18145 (BOL, K). 3219AD, plateau between Wolfberg Cracks and Arch, 10 x 1976, *Esterhuysen* 34397 (BOL).

Polycarena exigua grows in damp sandy or silty places, either in depressions in rock sheets or under overhanging rocks, in both sun and shade, flowering between August and October. It appears to be endemic to the Cedarberg, where it is clearly widespread between c.900 and 1200m above sea level.

The short broad leaf blades of *P. exigua* immediately set it apart from all allied species except *P. aemulans*, which shares this character. The two species can always be separated on details of floral morphology: the mostly shorter corolla tube of *P. exigua* (3.5–8mm versus 5.75–16mm), which is glabrous or very nearly so (not pubescent) and always dries yellow (though it is probably white when fresh) while that of *P. aemulans* is purplish, and the limb is mostly smaller (2.5–4mm across the lateral lobes versus 3–9mm). It is particularly the small-flowered form of *P. aemulans* that bears a remarkable superficial resemblence to *P. exigua*; this is discussed further under *P. aemulans*.

10. Polycarena silenoides [Harv. ex] Benth. in DC., Prodr. 10: 351 (1846); Hiern in Fl. Cap. 4 (2): 324 (1904) excluding *Schlechter* 5008, 8614; Levyns in Adamson & Salter, Fl. Cape Penins. 713 (1950).
Type: Cape Town, Lion's Head, *Harvey* 31 (holo. K; iso. E, TCD).

Syn.: *Nycterinia selaginoides* (Thunb.) Benth. var. *parviflora* Benth. in Hook., Comp. Bot. Mag. 1: 370 (1836) & in DC., Prodr. 10: 350 (1846). Type: Cape, without precise locality, *Ecklon* (holo. K).
Zaluzianskya villosa F.W. Schmidt var. *parviflora* (Benth.) Hiern in Fl. Cap. 4(2): 345 (1904), excl. all cited specimens.

Annual herb, stem erect to 50–140mm, simple or branched from the base, glandular-pubescent, hairs up to c.0.5mm long. *Leaves* 7–26 × 1–4.5mm, linear to oblanceolate, apex subacute, base tapered, petiole-like, margins entire to obscurely toothed, both surfaces glandular-pubescent, the larger leaves glabrescent above. *Inflorescence* spicate, flowers few to several, often relatively distant, sessile or with pedicels to 1mm long. *Bracts* c.6–14 × 0.75–1.25mm, longest at the base, all but the uppermost overtopping the flower, adnate to entire calyx tube, linear-oblong, glandular-pubescent. *Calyx* c.4.25–4.5mm long abaxially, 5.25–5.5mm adaxially, anticous lobes c.0.5mm long, posticous lobes c.1mm, glandular-pubescent all over outside, particularly on the keels, hairs up to c.0.5mm long. *Corolla* tube c.6–8 × 1.25mm, minutely glandular-puberulous outside, limb c.3.5mm across, lobes oblong, obtuse, posticous ones 1.5 × 1mm, anticous one 1.5 × 0.75–1.25mm, posticous lip bearded at base, corolla 'deep cream colour' (Mrs Levyns). *Stamens*: anticous pair shortly exserted, anthers 0.5mm long, posticous pair included, filaments decurrent nearly to base of tube, anthers 0.75mm. *Stigma* just exserted. *Capsule* 4–4.5 × 3.5–4mm. *Seeds* many, c.1mm long, 3-winged.

Selected citations:

Cape. 3318CD, Cape Town, Lion's Head, *Bolus* 4769 (BOL); ibidem, *Wolley Dod* 3090 (BOL).

Polycarena silenoides has been recorded only from the slopes of Lion's Head in the Cape Peninsula, and flowers in August and September. Hiern (in Fl. Cap. 4(2): 325) cited as this species *Schlechter* 8614 from Pakhuis Berg and *Schlechter* 5008 from Olifants River; the former proves to be *P. aemulans*; the latter is *P. subtilis*.

The affinities of *P. silenoides* are obscure; it is here placed at the end of the linear sequence of species in Group 1; *P. rariflora*, an equally well-marked and taxonomically isolated species, is placed first in the sequence of species in Group 2. In other words, *P. silenoides* does not appear to be closely related to either *P. exigua* or *P. rariflora*, between which it falls in the numerical order adopted here.

11. Polycarena rariflora Benth. in Hook., Comp. Bot. Mag. 1: 372 (1836) & in DC., Prodr. 10: 352 (1846); Hiern in Fl. Cap. 4(2): 331 (1904), excluding var. *micrantha* Schltr.

Type: Cape, Bergrivier, Sept. Oct., *Ecklon* 336 (holo. K, iso. S).

Syn.: *P. leipoldtii* Hiern in Fl. Cap. 4(2): 326 (1904). Lectotype (chosen here): Cape, near Clanwilliam, *Leipoldt* 874 (BOL, isolecto. PRE).

Annual herb, stem erect to 40–250mm, simple or branched, often from the base, glandular-puberulous. *Leaves:* lowermost c.10–25 × 1.25–3.5mm, mostly elliptic to narrowly spathulate with a few obscure teeth, upper leaves 4–30 × 0.3–1.5mm, mostly linear, entire or with 1 or 2 small teeth each side, all narrowed to the base, glandular-puberulous, often glabrescent. *Inflorescence* spicate, borne well above the leaves on terminal filiform peduncles, flowers 1 to several, often relatively distant, pedicels c.0.5–1mm long. *Bracts* 4–12 × 0.25–1mm, decreasing in length upwards, adnate nearly to top of calyx tube, linear, glabrous inside, glandular-puberulous outside. *Calyx* 3–4mm long, lobes very small, thinly membranous, each segment with a greenish keel, nearly always flushed purple around the sinus between the two lips, either glandular-puberulous particularly on the keels and margins of lobes, or hairs very few or wanting. *Corolla* tube 3–5 × 1.5–2mm, glabrous, limb 3–5mm across, or smaller if flowers cleistogamous, lobes oblong, obtuse, posticous ones 0.75–1 × 0.75–1mm, anticous one 1.25–2.5 × 1–2mm, posticous lip bearded at base, limb cream-coloured or white with an orange patch at base of posticous lip and often each lobe tipped with orange. *Stamens* normally exserted, anthers 0.5–0.75mm long, or included in cleistogamous flowers, posticous filaments decurrent to base of tube. *Stigma* shortly exserted. *Capsule* 2.5–3.75 × 2–3mm. *Seeds* many, 0.75mm long, 3-winged.

Selected citations:

Cape. Namaqualand, 2917DA, Spektakel, *Compton* 11517 (NBG): 3017BB, Khamieskroon, *Compton* 11317 (NBG). Van Rhynsdorp div., 3118AB, Karree-bergen, *Schlechter* 8257 (E, BOL, K, S, STE, W); 3119AC, top of Van Rhyn's Pass, *Esterhuysen* 5279a (BOL). Calvinia div., 3119CA, Lokenburg, *Compton* 11539 (NBG). Clanwilliam div., 3219AA, Brandewynrivier, *Schlechter* 10822 (BOL, E, S, W); 3219CB, S Cedarberg, Krom River, *Esterhuysen* 20538 (BOL, NBG, PRE); 3218BB, Olifants River, *Schlechter* 4986 (BOL, E, PRE, S, UPS). Malmesbury div., 3318BD, Botma's Pass over Riebeeck Kasteel, *Esterhuysen* 20426 (BOL). Laingsburg div., 3320AB, Tweedside, *Compton* 22849 (NBG). Ceres div., foot of Hex River Mountains, Ezelsfontein, *Esterhuysen* 20336 (BOL). Oudtshoorn div., 3322CC, on northern foothills of Outeniqua Mountains, farm Moeasrivier, 1900ft, 24 viii 1991, *Vlok* 2479 (E).

The holotype of *P. rariflora* at Kew is labelled (in Bentham's hand) CBS., *Ecklon*; however, a sheet in S, which clearly matches the type (plants withered), bears Ecklon's own label, giving collector's number and locality.

Polycarena rariflora is widely distributed in the western and south-western Cape, from the area around O'okiep in Namaqualand south to about Malmesbury and Tulbagh and west to the flanks of the Cedarberg and the Hex River Mountains, with three isolated records from much further west, near Laingsburg, Sutherland, and on the northern flank of the Outeniqua Mountains. It is also naturalized in South Australia (100km N of Adelaide, Hoyleton, *Copley* 3220,

NBG). It grows in moist sandy places, flowering from July to October, but mainly in August and September.

Polycarena rariflora is easily recognized by its loosely spicate inflorescences with linear (almost filiform) bracts and the glabrous or almost glabrous and usually purple-tinted calyx. The corolla varies much in size, particularly in the diameter of the limb, which may be 3–5mm across, or smaller when the flowers are self-fertilized. Cleistogamy and self-fertilization appear to be commonplace. In *Compton* 17148 (NBG) nearly all the flowers on the sheet are clearly cleistogamous, with the small unopened corolla capping the ripe capsule. In a few of the flowers, the lobes of the limb have spread, but they are very small, lack any colour patterning, and the inference is that the flowers were self-fertilized. On other sheets, plants with tiny unpatterned corollas are mixed with plants with larger, patterned, corollas (e.g. *Compton* 5535, NBG) or the sheets may consist of plants with all the corollas reduced, or all the corollas well-developed.

Hiern (1904, p.321) made no attempt to discuss affinities, and the only diagnoses are those given in the key. Here, *P. leipoldtii* is separated from *P. rariflora* on 'corolla tube longer than the calyx' versus 'corolla tube about equalling the calyx'; this separates them widely in the enumeration. The type of *P. rariflora* is a single plant, in fruit, with very withered corollas, but the limb is tiny (c.2mm across). In contrast, the specimens that typify the name *P. leipoldtii* have large, patterned corollas. The types represent the two extremes of the variation range in *P. rariflora*. The species appears to have no close relatives.

12. Polycarena aurea Benth. in Hook., Comp. Bot. Mag. 1: 372 (1836) & in DC., Prodr. 10: 351 (1846); Hiern in Fl. Cap. 4(2): 326 (1904), excluding *Schlechter* 8722, *Burchell* 1260. **Fig. 60A, Plate 1G.**

Lectotype (chosen here): Cape, without precise locality, *Thom* s.n. (K).

Annual herb 40–150(–300)mm tall, stems simple or branched either from the base or in the upper half, shortly glandular-pubescent with an admixture of longer hairs c.0.4mm long. *Leaves* mainly 5–29 × 0.5–4mm, diminishing in size upwards, linear to narrowly oblanceolate, entire or with 1 or 2 pairs of obscure teeth, both surfaces glandular-pubescent. *Flowers* few to many in small terminal heads, elongating slightly in fruit but flowers always crowded, pedicels 0.5–2mm long. *Bracts* 2.5–7.5 × 0.4–3mm, narrowly to broadly elliptic, adnate roughly halfway up calyx tube (0.5–1.5mm), glandular-pubescent mainly on the backs, hairs 0.25–0.4mm long. *Calyx* 2.5–4mm long, anticous lobes very short, posticous ones c.1mm long, glandular-pubescent particularly on the greenish keels, hairs up to 0.25–0.4mm long. *Corolla* tube 3–5.5 × 1–1.75mm, glabrous, limb 6–9.5mm across, lobes elliptic, posticous ones 1.3–3 × 0.75–2mm, anticous ones 1.75–4.5 × 1.25–2.5mm, limb either wholly yellow or upper lip white tipped yellow, lower white, posticous lip either glabrous or with a few hairs on the palate. *Stamens* all well exserted, anthers 1mm long, posticous filaments decurrent to base of tube. *Stigma* exserted. *Capsule* 2.5–3.5 × 2–3mm. *Seeds* many, c.1mm long, either distinctly or obscurely 3-winged.

Selected citations:

Cape. Calvinia div., 3119AD, between Calvinia and Niewoudtville, *Batten* 746 (E); 3119BD, Akkerdam, foot of Hantam Mts, *Lewis* 5795 (NBG); 3119CA, Lokenburg, *Compton* 11540 (NBG). Sutherland div., 3120CC, Roggeveld Escarpment near Middelpost, *Levyns* 9482 (BOL); 3220AB, Voëlfontein, *Hall* 3721 (NBG); ibidem, *Hall* 4637 (NBG); 3220CB, Klipbanksrivier, *Acocks* 16967 (PRE). Worcester div., 3319AB, north of Gydoberg, *Oliver* 5107 (PRE, STE); 3319BC, Karroo Poort, *Compton* 5534 (NBG). Laingsburg div., 3320AB, Tweedside, *Compton* 22848 (NBG).

Bentham quoted unlocalized specimens collected by Thom and Masson under his new name, *Polycarena aurea*. The Masson specimen (BM) is now in poor condition. That collected by Thom is better and Bentham's description clearly refers to it: '*foliis linearibus*' and 'narrow short erect leaves' (in Masson's plant some of the leaves are relatively broad (3.75mm) and somewhat spreading). *Thom* is therefore selected as lectotype.

Polycarena aurea ranges from Niewoudtville and Calvinia in a broad arc eastwards to Sutherland and Laingsburg divisions then south west to Ceres and Worcester divisions. In the southern part of its range, the flowers are wholly yellow. Around Middelpost and south east of it (3120CC and 3220AB) the flowers may be either wholly yellow or the limb may be white with only the tips of the posticous lobes tipped with yellow. It is clear that the two forms occur together: Hall (cited above) collected them both, and so did Compton, further south around White Hill and Tweedside (3320AB, BA). In the northern part of its range (3119 degree square), *P. aurea* seems always to have white and yellow flowers, but Mrs Batten (746, cited above) made the observation 'flowers yellow and white (in the same head), yellow uppermost'.

The plants grow on sandy or clayey flats and slopes, often among shrubs, flowering mainly in September. *Polycarena aurea* is easily distinguished from *P. pubescens*, with which it is partly sympatric, by its larger limb (6–9.5mm versus 3–5mm), upper lip glabrous or very nearly so (not well bearded) and shorter indumentum (hairs on stem, bract and calyx up to 0.25–0.4mm, versus 0.5–1mm).

It is distinguished from *P. formosa* by its shorter indumentum (longer hairs 0.25–4mm versus 0.6–1.2mm), leaves only obscurely toothed, differently shaped bracts (elliptic versus linear to oblong) adnate roughly halfway up the calyx tube (not nearly to the top) and often shorter corolla tube (3–5.5mm versus 5–6.5mm). The area of *P. aurea* lies mainly west and south of that of *P. formosa*; they appear to be sympatric only in the neighbourhood of Niewoudtville.

13. Polycarena pubescens Benth. in Hook., Comp. Bot. Mag. 1: 372 (1836) & in DC., Prodr. 10: 351 (1846); Hiern in Fl. Cap. 4(2): 328 (1904) excluding *Drège* 549, b, c, *Schlechter* 8276 and *Bolus* 657.
Lectotype (chosen here): Cape [3319AC near Tulbagh], Roodezand, *Drège* 549a (K).

Syn.: *P. arenaria* Hiern in Fl. Cap. 4(2): 330 (1904). Type: Cape, Clanwilliam div., [3218BC] in sandy places near Alexander's Hoek, 100m, 2 ix 1894, *Schlechter* 5134 (holo. BOL; iso. E, J, K, S, STE, UPS).
P. glaucescens Hiern in Fl. Cap. 4(2): 325 (1904). Type: Cape, Little Namaqualand [3018CD], Modderfontein, *Whitehead* s.n. (holo. TCD, iso. S).

Annual herb 65–280mm tall, stems simple or sparingly branched either from the base or in the upper half, shortly glandular-pubescent with an admixture of longer hairs 0.25–1mm long. *Leaves* c.7–30 × 1–4(–5)mm, diminishing in size upwards, linear, oblanceolate or elliptic, with 1–3 pairs of rather obscure teeth, thinly to thickly glandular-pubescent on both surfaces. *Flowers* few to many in small terminal heads, heads elongating slightly in fruit, lowermost flowers then sometimes distant, pedicels up to 0.5mm long. *Bracts* 4.5–14 × 0.5–3mm, broadly lanceolate to ovate, adnate roughly halfway or a little more up calyx tube (1.25–3.25mm), glandular-pubescent all over backs, hairs 0.25–1mm long. *Calyx* 3–4(–4.5)mm long, anticous lobes very short, posticous one c.1.5–2mm long, glandular-pubescent particularly on the greenish keels, hairs up to 0.5–1mm long. *Corolla* tube 3.5–5 × 1–2mm, glabrous, limb 3–5mm across, rarely 6mm, lobes oblong, posticous ones 0.75–1.5 × 0.5–1.25mm, white tipped yellow, anticous ones 1.25–2.5 × 0.5–1.5mm, white, posticous lip with a yellow patch in the throat and there pubescent. *Stamens* all well exserted, anthers c.0.5mm long, posticous filaments decurrent to base of tube. *Stigma* exserted. *Capsule* 3–3.5 × 2.4–3mm. *Seeds* c.1mm long, distinctly 3-winged.

Selected citations:

Cape. Namaqualand, 2917BC, 2km from Bulletrap, *van Wyk* 6299 (PRE); 2917DD, Droedap, *Compton* 11555 (NBG); 3017BD, Kamieskroon, *Thorne* s.n. SAM 48811 (SAM); 3118AB, Bitterfontein, *Compton* 11322 (NBG). Calvinia div., 3119AC, 6km from Niewoudtville, *Batten* 780 (E); 3119CA, Lokenburg, *Acocks* 18542 (PRE). Sutherland div., 3220CA, Houthoek, *Hanekom* 1097 (PRE). Ceres div., 3319BC, Hottentots Kloof, *Walgate* 298 (NBG); 3320CC, Montagu, *Page* s.n. BOL 15621 (BOL).

Bentham, under his new name *P. pubescens*, quoted three specimens collected by Drège, at Roodesand (a), Haazenkraalsrivier (b) and Zilverfontein (c). The specimen from Roodesand (6 plants) fits the essentials of Bentham's description well: dwarf, viscous, floral leaves ovate-lanceolate, capsule pubescent. The top of the capsule is glandular-pubescent in all species of *Polycarena*, but in this particular case it is likely that Bentham was describing the capsules clearly visible on the right-hand specimen on the sheet, where the persistent, pubescent, calyx so tightly enfolds them, both before and after dehiscence, that it can easily be mistaken for the wall of the capsule itself.

The specimens from Haazenkraalsrivier and Zilverfontein differ in having long eglandular hairs mixed with shorter glandular ones on stem, leaf, bract and calyx as well as small retrorse eglandular hairs on the stem; also, the capsules are glabrous. They are *Phyllopodium collinum*.

Hiern (in his key in Fl. Cap. 4(2): 322–323) separated *P. glaucescens* from both *P. pubescens* and *P. arenaria* by describing the corolla tube as longer than the calyx, not about equalling it, but this is contradicted by the measurements given in the description of *P. glaucescens*, and it is in any case a very unreliable criterion to use in species that have a short broad corolla tube. Then *P. pubescens* is described as having 'Flowering calyx about or nearly equalling the bract', *P. arenaria* as 'Flowering calyx falling short of the bract'. This simply is not so; both criteria can be met in one inflorescence because the bracts diminish in size from base to apex.

Polycarena pubescens is however a well-marked and easily recognized species distinguished by its relatively broad, entire bracts and heads of almost sessile flowers with a small limb, a character that Bentham himself stressed in distinguishing *P. aurea* from *P. pubescens*. It is widely distributed from northern Namaqualand to Montagu and the Anysberge, where it favours moist rocky sites and flowers between July and September.

14. Polycarena formosa Hilliard in Notes RBG Edinb. 45: 487 (1989).
Type: Cape, Clanwilliam div., 3219AC, Wupperthal, *Goldblatt* 2531 (holo. NBG, iso. S).

Annual herb 70–200mm tall, stems simple or branched either from the base or in the upper half, glandular-pubescent with an admixture of longer hairs 0.6–1.1mm long. *Leaves* 7–35 × 0.5–4.5mm, diminishing in size upwards, linear to oblanceolate, sometimes entire, mostly with 1–5 pairs of teeth, both surfaces glandular-pubescent. *Flowers* few to many in terminal heads, elongating somewhat in fruit, the lowermost flowers then up to c.2.5(–5)mm apart, pedicels 0.75–1mm long. *Bracts* 5.5–14 × 0.5–1.5(–2)mm, linear or oblong, usually entire, the lower-most occasionally with 1 or 2 teeth, adnate nearly to top of calyx tube, glandular-pubescent on both surfaces, hairs up to 0.6–1.2mm long. *Calyx* 3.5–5mm long, anticous lobes very short, posticous one c.1.5mm long, glandular-pubescent mainly on the greenish keels, hairs 0.6–1.2mm long. *Corolla* tube 5–6.5 × 1.5–1.75mm, glabrous or with sparse minute glandular hairs, limb (6–)8–14mm across, lobes elliptic, posticous ones 2–3 × 1.5–2.5mm, anticous ones 3–5 × 1.5–3.25mm, limb creamy white or palest yellow, the posticous lobes tipped orange-yellow and with an orange-yellow blotch at base of lobes in throat, there either glabrous or very sparsely hairy. *Stamens* all well exserted, anthers 1mm long, posticous filaments decurrent to base of tube. *Stigma* exserted. *Capsule* 3–4.5 × 2.5–3.2mm. *Seeds* many, c.1mm long, distinctly 3-winged.

Selected citations:

Cape. Clanwilliam div., 3219AC, Wupperthal, *Leipoldt* BOL 20661 (NBG, PRE); 3219AA, Klipfontein, *Esterhuysen* 5798 (BOL); 3219AB, Bidouw, *Compton* 7750 (NBG); 3218BB, near Doornbosch, *Lewis* 3153 (NBG). Calvinia div., 3119AC, 3 miles W Niewoudtville, *Stokoe* SAM 64290 (SAM). Van Rhynsdorp div., 3118DC, Koude Berg, *Schlechter* 8722 (BOL, E, PRE, S, W).

Polycarena formosa can be recognized by its narrow, sparsely but sharply toothed leaves, linear to oblong bracts and the large creamy-white flowers vividly marked with orange (whence the specific epithet). The species has generally been confused with *P. aurea*, to which it is certainly allied; see under *P. aurea* for distinguishing characters.

The distribution of *P. formosa* lies mainly west of that of *P. aurea*, between Lambert's Bay and Wupperthal. Collectors have given some information on habitat ('sandy flats near the sea'; 'gentle sandy N slope'; 'rocky dry SW slope'; 'shale'; 'fynbos on red clay') and the plants flower in August and September.

One specimen (*Hall* 192, NBG, STE) needs special mention. It came from the bottom of Botterkloof Pass (3119CD) and was 'abundant among bushes'. It has the aspect and floral characteristics of *P. formosa*, but differs in its very short indumentum: none of the hairs on stem, bract and calyx exceed 0.25mm in length. Typical *P. formosa* has not been recorded from Botterkloof Pass, which does, however, lie within the area of that species. Field observations are needed to determine the status of the plant with short indumentum.

15. Polycarena comptonii Hilliard in Notes RBG Edinb. 45: 487 (1989).
Type: Cape, Laingsburg div., 3320BA, Whitehill, *Compton* 11230 (holo. NBG).

Delicate annual herb 20–100mm tall, stem filiform, simple in smallest plants, branched from the base in larger ones, glandular-puberulous only at the base, but some minute glandular hairs on the inflorescence axes. *Leaves*: radical leaves 4, rosetted, 4.5–18 × 2.5–9mm, ovate to elliptic, contracted to a petiolar part, glabrous, margins entire to obscurely toothed; cauline leaves few (only at the branching of the stems), mostly 6–13 × 0.6–1mm, linear, entire. *Flowers* loosely arranged in terminal racemes, peduncles filiform, pedicels c.0.5–1.5mm long. *Bracts* 1.75–7 × 0.5–1.5mm, oblong to lanceolate, glabrous except for a few minute marginal hairs at the base, adnate to lower part of calyx tube (0.6–1mm). *Calyx* 2.2–2.6mm long, lobes very small, entire calyx conspicuously keeled, glabrous except for minute glandular hairs on margins of lobes. *Corolla* tube 1.5–2 × 0.75–1.2mm, glabrous, limb c.2.4mm across, lobes oblong, obtuse, posticous ones 0.5 × 0.3–0.4mm, orange with a few minute hairs at base, anticous ones 1–1.4 × 0.6–0.75mm, white. *Stamens* either 4, or 2 by abortion of posticous pair, one of these 2 sometimes reduced to a staminode, the anticous well exserted, anthers minute, posticous filaments scarcely decurrent. *Stigma* exserted. *Capsules* c.2.5 × 2mm. *Seeds* many, c.0.75mm long, obscurely trigonous, irregularly colliculate, amber-coloured with a narrow whitish wing strongly to weakly developed.

Citation:

Cape. Laingsburg div., 3320BA, Whitehill, *Compton* 13373, mixed with *P. tenella* (NBG).

Polycarena comptonii is easily distinguished from *P. filiformis* by its basal rosette of four glabrous leaves, its glabrous (not glandular-pubescent) bracts, and much smaller flowers (corolla tube 1.5–2mm long versus 2.6–3.6mm, limb up to 2.4mm across versus 4–5.5mm) with the two posticous lobes coloured bright orange, this clearly visible in dried material, whereas in *P. filiformis* the whole limb is white, though there *may* be a pale yellow patch in the throat (no collector notes this). The very dark and comparatively large anthers are a conspicuous feature of *P. filiformis*, whereas in *P. comptonii* the anthers are minute and pale yellow.

The sheet of *P. comptonii* mixed with *P. tenella* (*Compton* 13373, cited above) is instructive: the plants are all tiny with mostly simple stems; those of *P. comptonii* have a basal rosette of broad leaves and often only one very narrow cauline leaf; those of *P. tenella* have no basal rosette but there are a few pairs of relatively broad cauline leaves. Furthermore, the flowers of *P. comptonii* are nearly sessile whereas those of *P. tenella* are on relatively long filiform pedicels. Close examination of the flowers reveals the bilabiate limb of *P. comptonii* with orange and white obtuse lobes while the limb of *P. tenella* is regular, the lobes very acute and white.

It is a pleasure to commemorate Prof. R. H. Compton in the name of this interesting little plant; not only did he collect the only material that I have seen, but his many other collections of *Polycarena* and its allies have greatly facilitated this study. He gave no ecological notes on *P. comptonii*, but the plants probably grew in the shelter of rocks; they flower in August.

16. Polycarena filiformis Diels in Engl. Bot. Jahrb. 23: 478 (1897); Hiern in Fl. Cap. 4(2): 329 (1904).
Type: Cape, Calvinia div., [3119BD] Hantam hills, *Meyer* (B†).

Syn.: *P. gracilipes* [N.E. Br. ex] Hiern in Fl. Cap. 4(2): 332 (1904). Type: Calvinia div., [3119BD] Brand Vley, *Johanssen* 17 (holo. K, iso. SAM).

Delicate annual herb 120–250mm tall, stem rarely simple, usually well-branched from the base, glandular-pubescent becoming glandular-puberulous to almost glabrous on the inflorescence axes. *Leaves* c.4.5–20 × 0.6–4mm, diminishing in size upwards, lower ones oblanceolate to linear, upper mostly linear, base tapered, petiole-like, margins few-toothed in the larger leaves, entire in the smaller, both surfaces glandular-pubescent. *Flowers* in terminal racemes, peduncles filiform, pedicels up to 0.5mm long, rather distant. *Bracts* 2.5–3 × 0.4–1mm, oblong, glandular-pubescent, adnate to lower part of calyx tube (0.5–1mm). *Calyx* 2.25–3.5mm long, anticous lobes very short, posticous lobe c.1 × 0.4mm, all lobes distinctly keeled, glandular-pubescent all over outside, often flushed with purple. *Corolla* tube 2.6–3.6 × 0.75–1mm, glabrous, limb 4–5.5mm across, white, lobes oblong or elliptic, obtuse, posticous ones 1–1.75 × 1–1.5mm, anticous ones 1.5–2.25 × 0.75–1.25mm, posticous lip either glabrous at the base inside or with a few minute hairs, and possibly with a yellow patch there. *Stamens* all well exserted, anthers c.0.8mm long, blackish, posticous filaments decurrent to base of tube. *Stigma* exserted. *Capsule* 2–3 × 1.5–2.75mm. *Seeds* 4 in each loculus, 1–1.5mm long, obscurely 3-angled, amber-coloured, irregularly colliculate, with a narrow whitish wing weakly to strongly developed.

Citations:

Cape. Calvinia div., 3119BC, Groot Toring, 8 ix 1982, *Schelpe* 8140 (BOL). 3119BD, Ekerdam, *Taylor* 2798 and 2661 (NBG); ibidem (as Akkerdam), *Barker* 9500 (NBG, STE); Moordenaarspoort, *Lewis* 4311 (SAM).

Although the type of *P. filiformis* was destroyed in the Berlin fire and I have seen no duplicate, I have no hesitation in equating *P. gracilipes* with it: Diels gave a good description with precise measurements that accord well with the material I have seen from the environs of Calvinia. All but one of the specimens, including the types of both names, came from the same quarter-degree square; one from the adjacent quarter-degree square. In Hiern's key (in Fl. Cap. 4, 2: 322–323), *P. filiformis* falls into the group with bract "1/12 to 1/3 in. long" (that is, 2–8mm), while *P. gracilipes* is segregated under "Bract 1/20 in. long" (1.2mm). But this is not so: no bract on the type of *P. gracilipes* is less than 2mm long.
  The only ecological notes given by collectors record 'in shady places; 'on slopes in shelter of small bushes' and 'shaly slopes with short open Karroid scrub'. The plants flower in August and September.
  Polycarena filiformis is at once distinguished from *P. tenella* by its nearly sessile flowers with a well-developed bilabiate limb with rounded lobes and conspicuous dark anthers (versus pedicels 1.5–5mm long, regular corolla with very pointed lobes, and minute pale-coloured anthers). They appear to be allopatric.

17. Polycarena tenella Hiern in Fl. Cap. 4(2): 333 (1904). **Fig. 60B.**
Lectotype (chosen here): Cape, 3118DC, Koudeberg, 2500ft., 28 viii 1896, *Schlechter* 8723 (BOL; isolecto. E, PRE, S, W).

Very delicate annual herb 30–190mm tall, stem erect, simple or branched, glandular-puberulous. *Leaves*: lowermost 8–15 × 2–3mm, oblanceolate or occasionally elliptic, base petiole-like, upper margins few-toothed, the teeth minute, upper surface glabrous, lower glandular-puberulous mainly at base and along midline, margins glandular-puberulous, upper leaves c.5–8 × 1–1.5mm, mostly elliptic, usually entire, otherwise as lower leaves. *Inflorescence* a raceme terminating each branch, axis filiform, 20–90mm long, flowers several to many, distant, pedicels 1.5–5mm long, filiform. *Bracts* (excluding part adnate to pedicel, which is exceptionally long) 1.5–2.75 × 0.3–0.75mm, elliptic, canaliculate, adnate to lower part of calyx tube (0.5–0.75mm), minutely glandular-puberulous outside. *Calyx* 1.25–1.6mm long, distinctly bilabiate but lobes very small, membranous but strongly keeled and glandular puberulous mainly there. *Corolla* tube 1–1.5 × 0.75–1mm, glabrous, limb probably seldom expanded and then c.2mm across; lobes all alike, 0.5–0.75 × 0.2–0.3mm at base, deltoid, very acute, white. *Stamens* either 4 or reduced to 2 or 3, then sometimes with a staminode, exserted, anthers minute, filaments not decurrent. *Stigma* exserted. *Capsule* 1.5–1.75 × 1.5–2mm. *Seeds* c.4 in each loculus, c.0.75 × 0.5mm, obscurely 3-angled, irregularly colliculate, with a narrow whitish wing at one or both ends.

Selected citations:

Cape. Richtersveld, 2817AC, Rosyntjiesberg Nek north of Lelieshoek, *Oliver et al* 293 (PRE). S Cedarberg, 3219CB, Krom River, *Esterhuysen* 20552 (BOL, NBG). Sutherland div., 3220CB, Klipbanksrivier, *Acocks* 16968 (PRE). Ceres div., 3319BC, Karoo Poort, *Esterhuysen* 5473 (BOL, STE). Laingsburg div., 3320BA, Whitehill Ridge, *Compton* 3510 (BOL).

Polycarena tenella is relatively widely distributed from just south of the Orange River, in the Richtersveld, south to the mountains near Ceres and Laingsburg, possibly always between 450 and 1100m above sea level. It may be found in the shelter of rocks and bushes, flowering between August and October. It is easily recognized by its delicate habit and relatively long racemes with minute widely-spaced flowers borne on conspicuous filiform pedicels up to 5mm long. No other species of *Polycarena* has a regular corolla limb with very acute lobes. The frequent reduction of the stamens from four to three or two suggests that selfing may take place, but in all cases both anthers and style are exserted.

EXCLUDED NAMES

Polycarena intertexta Benth. = *Cromidon decumbens* (Thunb.) Hilliard
Polycarena plantaginea (L.f.) Benth., based on *Manulea plantaginis* L.f. = *Cromidon plantaginis* (L.f.) Hilliard

10. GLEKIA

Glekia, a monotypic genus, is a plant of the summer rainfall area of southern Africa, essentially montane in its distribution, from the mountains around Graaff Reinet east to Queenstown, thence northwards to Lesotho. This area lies outside the range of any species of *Phyllopodium*, the genus in which it was originally described.

Glekia Hilliard in Notes RBG Edinb. 45: 482 (1989). **Figs 15D, 27D, 40A, 61.**

Twiggy shrublet. *Stems* leafy, pubescence confined to two narrow ridges spanning the internodes, hairs eglandular, more or less retrorse, minute patent glandular hairs sometimes present as well. *Leaves* (and bracts) usually opposite, sometimes subopposite, bases connate and decurrent in two narrow ridges. *Inflorescence* racemose, flowers solitary in upper leaf axils forming rounded terminal heads, oblong in fruit, sometimes further arranged in corymbose panicles. *Bract* adnate to pedicel and halfway up calyx tube. *Calyx* bilabiate, membranous, persistent. *Corolla* thick-textured, not persistent, tube glabrous outside, limb bilabiate, lobes 5, entire, orange patch at base of posticous lip extending down inflated part of corolla tube, heavily bearded all round mouth, hairs unicellular, clavate. *Stamens* 4, posticous pair in throat, filaments not decurrent down tube, anticous pair shortly exserted, anthers synthecous. *Stigma* lingulate with two marginal bands of stigmatic papillae passing gradually into the long filiform style, exserted. *Ovary* elliptic in outline, base slightly oblique with a nectariferous gland on the shorter side, ovules c.9–16 in each loculus. *Fruit* a septicidal capsule with a short loculicidal split at tip of each valve. *Seeds* 1–1.25mm long, seated on an irregularly rounded, centrally depressed pulvinus (fig. 15D), seeds obscurely 3-angled or flattened, irregularly wrinkled on one face, testa tightly investing endosperm, opaque, creamy or pale amber, under the SEM the cells seem to be strongly domed with a slightly flattened apex (fig. 27D).

Type species: *Glekia krebsiana* (Benth.) Hilliard
Distribution: South Africa, Lesotho and the Cape mountains from Lady Grey and Queenstown to Somerset East and Graaff Reinet.

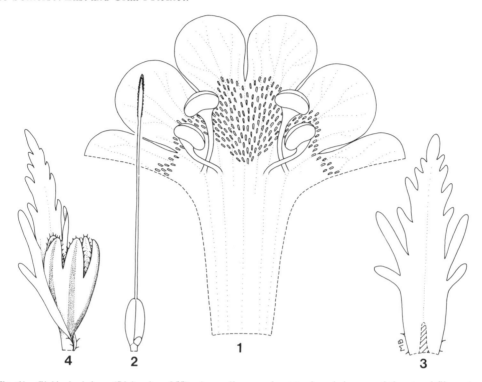

Fig. 61. *Glekia krebsiana* (*Richardson* 255): 1, corolla opened out to show hairs around throat and filaments not decurrent; 2, stigma, style and ovary with small adnate nectariferous gland; 3, bract showing adnation scar; 4, calyx with bract adnate to very short pedicel and extreme base of tube. All × 7.

Glekia krebsiana (Benth.) Hilliard in Notes RBG Edinb. 45: 482 (1989). **Figs 15D, 27D, 61, Plate 4I.**
Type: South Africa [Cape] without precise locality, *Krebs* 238 (holo. G-DC, n.v.; iso. LE, S).

Syn.: *Phyllopodium krebsianum* Benth. in DC., Prodr. 10: 353 (1846) & in Hook., Ic. Pl. 11: 63 t. 1079 (1871); Hiern in Fl. Cap. 4(2): 321 (1904).

Twiggy shrublet c.230–600mm tall, lower branches often decumbent and sometimes rooting, basal parts up to c.3.5mm in diam., stems glabrous except for 2 narrow longitudinal ciliate ridges between the nodes, the hairs patent or somewhat curved and retrorse, eglandular, obtuse, up to 0.25mm long; very leafy. *Leaves* usually opposite and decussate, rarely subopposite, 6–36 × 2.5–14mm, elliptic gradually narrowed below into a very short petiolar part, bases connate and decurrent on the stem in narrow ridges, margins finely callose-toothed, blade thick-textured, glandular-punctate, minutely glandular-puberulous as well on the upper surface, lower surface either glabrous or nearly so except for cilia on lower margins, some gland-tipped, most eglandular. *Flowers* solitary in upper leaf axils forming crowded terminal heads sometimes further arranged in corymbose panicles, becoming oblong in fruit, c.20–40mm long. *Bracts* leaflike, indumentum as on leaves, lowermost c.8–13 × 3–7mm, adnate to pedicel and up to halfway up calyx tube. *Pedicels* up to 1mm long. *Calyx* bilabiate, tube c.1.3–1.5mm long, anticous lobe 1.5–2.25 × 1.4–2mm, anticous lip split 1–2mm, 3 posticous lobes about half the width of the anticous ones, all hyaline, delicately keeled, ciliate on margins. *Corolla* tube 7–13mm long, cylindric below and there c.0.8–1.5mm in diam., abruptly campanulate above, 2–4mm across mouth, limb c.6–14mm across lateral lobes, posticous lobes 2–4 × 2–4mm, anticous lobe 3–5 × 1.75–4mm, all lobes subrotund, white inside with an orange patch at base of posticous lip running down back of inflated part of tube, heavily bearded all round mouth, each lobe with a blue-purple patch outside fading to brown. *Stamens* 4, anthers 1–2mm long, posticous pair in throat, anticous pair shortly exserted. *Stigma* exserted. *Capsules* c.6 × 3–3.5mm, c.9–16 seeds in each loculus. *Seeds* c.1–1.25 × 0.75mm, obscurely angled or flattened, irregularly wrinkled on one face, creamy-coloured or pallid amber.

Selected citations:

Lesotho. Maseru distr., 2927BD, Makhaleng valley near Molimo Nthuse Pass, c.6800ft, *Hilliard & Burtt* 12060 (E, NU); 2927BC, Botsabelo, *Dieterlen* 1034 (SAM); Roma, 5400ft, *Ruch* 1523 (PRE).
Cape. Barkly East distr., 3027DC, Barkly East to Lady Grey, above Kraai River, 6000ft, *Hilliard & Burtt* 13149 (E, NU); 3028CA, Kloppershoek Valley NE of Rhodes, c.7000ft, *Hilliard & Burtt* 16636 (E, NU). Keiskammahoek distr., 3227CA, Boma Pass, *Acocks* 9118 (PRE). Queenstown div., 3226BC, Katberg, Vulcan's Bellows, 5500ft, *Story* 2810 (PRE); 3126DD, Kama's Mountain, 5000ft, *Galpin* 7872 (BOL). Graaff Reinet div., 3124DD, S extreme of Renosterberg above Lootsberg railway halt, c.6000ft, *Hilliard & Burtt* 10652 (E, NU); 3224DC, Oudeberg, 5000ft, *Bolus* 735 (BOL, S). Somerset East div., 3225DA, Boschberg, 3500–4500ft, *MacOwan* 1363 (BOL, E, S).

Glekia krebsiana is relatively widely distributed on the mountains from western Lesotho south to the Witteberg, Cape Drakensberg, Stormberg and Amatole Mountains, thence west to Boschberg at Somerset East, the Lootsberg on the southern flank of the Sneeuwberg, and the Oudeberg north of Graaff Reinet, between c.1200 and 2150m above sea level. It grows on bare cliffs and rocky, grassy mountainsides, flowering between May and October, principally in August. Specimens from Boschberg are particularly robust, but there is nothing else to distinguish them.

Bentham described the plant in 1846 as a species of *Phyllopodium*, but it differs from that genus in details of vegetative, floral and seed morphology. The area of *Glekia* lies outside the geographical range of *Phyllopodium*: there is one record of *P. rustii* from the Karoo Nature Reserve at Graaff Reinet, but this terrain is very different from the montane habitat of *Glekia* a few kilometres north of Graaff Reinet on the Oudeberg and thence eastwards over the mountains.

11. TRIEENEA

Trieenea is a genus of at least nine species confined to the mountains of the western and southern Cape, with no less than seven species restricted to the Cedarberg or occurring there. Some of the species appear to be very locally endemic, but clearly much more field work is needed before the distribution patterns are fully established and the total number of species known. Also, it is remarkable that all the species favour similar habitats: moist shade under rocks. They are easily recognized by their bushy habit, broad deeply toothed and mostly opposite leaves, bracts mostly adnate to pedicels only, and white or mauve flowers marked with an orange patch running from the base of the posticous lip down the back of the tube.

Trieenea laxiflora, *T. lasiocephala* and *T. lanciloba* are closely allied shrublets, known only from the northern Cedarberg. They all bear their flowers in round or oblong heads, the bracts are adnate to the pedicel only, and the posticous pair of stamens are deep in the corolla tube, the anticous pair in the throat or shortly exserted. The corolla is either mauve to blue (*T. laxiflora*) or white.

Trieenea frigida also has the posticous stamens included, the anticous ones shortly exserted, but it is possibly annual, the flowers (colour unknown) are borne in lax racemes, and the bract is adnate to both pedicel and calyx tube. It is known only from the Cold Bokkeveld Mountains due south of the Cedarberg.

In *T. taylorii*, the anthers of the posticous stamens appear in the mouth of the corolla tube, while the anticous pair are fully exserted. It too may be annual; the inflorescence is capitate and the bracts are usually adnate to the pedicel only, sometimes to the extreme base of the calyx. The colour patterning of the corolla limb is remarkable in that the two posticous lobes are dark violet, contrasting sharply with the white anticous lip. The species is known only from the Cedarberg.

The remaining four species have all four stamens exserted and the bracts are usually adnate to the pedicel only or sometimes to the extreme base of the calyx tube. The most widely distributed species, *T. glutinosa*, is an annual that sometimes perennates, apparently less demanding in its habitat requirements (it will grow out on rocky peaty mountain slopes) and ranging from the Cedarberg south to the mountains around Ceres, Worcester and Robertson and east to the Great Winterhoek Mountains near Uitenhage. It has whitish to pale blue flowers in long racemes.

The other three species are all white-flowered. *Trieenea schlechteri* and *T. longipedicellata* are both perennial herbs with the flowers in elongated racemes; *T. schlechteri* is confined to the Cedarberg and Cold Bokkeveld Mountains, while *T. longipedicellata* appears to be endemic to the mountains between Paarl, Stellenbosch and Genadendal. *Trieenea elsiae* is possibly annual, bears its flowers in round heads, and is known only from the South Cedarberg and Cold Bokkeveld Mountains.

Trieenea Hilliard in Notes RBG Edinb. 45: 489 (1989). **Figs 15E, 16A, 28A, B, C, 41A, 62, Plate 1E, F.**

Annual or perennial herbs or shrublets, all eventually bushy though often flowering in the seedling stage. *Stems* leafy, pubescence always patent, hairs gland-tipped or not. *Leaves* opposite or alternate above in some species, margins deeply and coarsely toothed. *Inflorescences* racemose, capitate or elongate, often panicled, always erect. *Bracts* often adnate to pedicel only or sometimes to extreme base of calyx in uppermost flowers. *Calyx* obscurely bilabiate, lobed almost or quite to base, membranous, persistent. *Corolla* membranous, tube often glandular-pubescent outside, splitting at the base as the capsule ripens, limb bilabiate, lobes 5, entire, orange/yellow patch at base of posticous lip extending down back of tube almost to base and there either glabrous or bearded with clavate unicellular hairs. *Stamens* 4, either all exserted or the posticous pair included or appearing in the mouth of the tube, posticous filaments very briefly to strongly decurrent down corolla tube, anthers synthecous. *Stigma* lingulate with two marginal bands of stigmatic papillae, well exserted, passing gradually into the long filiform style. *Ovary* elliptic in outline, base slightly oblique with a small nectariferous gland on the shorter side, either glandular-puberulous or glabrous on upper part; ovules up to 14 in each loculus. *Fruit* a septicidal capsule with a short loculicidal split at top of each valve. *Seeds* 0.5–1mm long, seated on a round centrally depressed pulvinus (fig. 16A), seeds either elliptic or slightly angled, sinuously wrinkled in longitudinal bands, pallid, testa tightly investing endosperm, under the SEM the cells seen to be convex (fig. 28A, B, C).

Type species: *Trieenea schlechteri* (Hiern) Hilliard
South Africa, Cape mountains from the Cedarberg south and east to the Great Winterhoek Mountains.

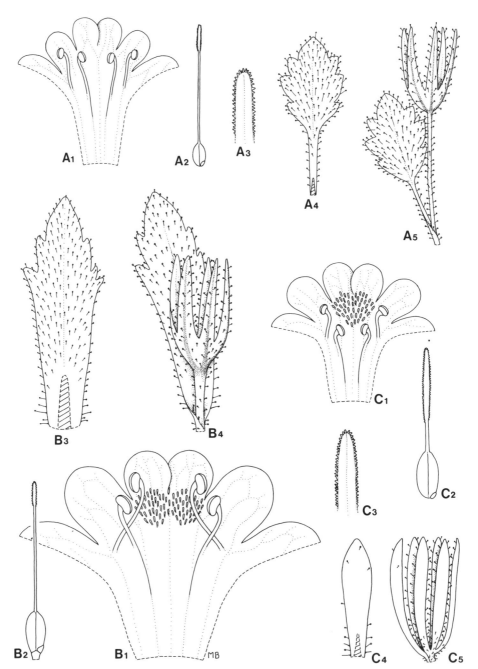

Fig. 62. A, *Trieenea longipedicellata* (*Esterhuysen* 35627): A1, corolla opened out to show decurrent posticous filaments, × 4.5; A2, stigma, style and ovary with adnate nectariferous gland, × 4.5; A3, stigma showing marginal stigmatic papillae, × 20; A4, bract showing adnation scar, × 4.5; A5, calyx with bract adnate to extreme base of pedicel, × 4.5. B, *T. glutinosa* (*Esterhuysen* 35527): B1, corolla opened out to show hairs on posticous lip, decurrent posticous filaments; B2, stigma, style and ovary with adnate nectariferous gland; B3, bract showing adnation scar; B4, calyx with bract adnate to extreme base of pedicel. All × 4.5. C, *Trieenea lanciloba* (*Taylor* 11952): C1, corolla opened out to show hairs on posticous lip and decurrent posticous filaments, × 4.5; C2, stigma, style and ovary with adnate nectariferous gland, × 4.5; C3, tip of stigma showing marginal stigmatic papillae, × 20; C4, bract showing adnation scar, × 4.5; C5, calyx with bract adnate to extreme base of calyx and pedicel, × 4.5.

Key to species

1a. Posticous lobes dark violet (visible even in dried material), contrasting with the white anticous lip .... **4. T. taylorii**
1b. Posticous lobes white, cream, orange or mauve-blue 2

2a. Plants either with long eglandular hairs at least on bracts and calyx, or glandular-puberulous to glabrous 3
2b. Plants clad in relatively long (c.0.25–1.5mm) gland-tipped hairs 5

3a. Calyx lobes lanceolate, 0.75–1mm broad, corolla tube 3.5–5mm long ..... **3. T. lanciloba**
3b. Calyx lobes linear-lanceolate, 0.4–0.75mm broad, corolla tube 5–7.5mm long 4

4a. Heads somewhat lax, elongating in fruit, mostly arranged in very open corymbose-panicles (solitary only at the tips of the smallest twiglets) .... **1. T. laxiflora**
4b. Heads compact, round, not elongating in fruit, solitary at the branch tips   **2. T. lasiocephala**

5a. Hairs on stems up to 1–1.5mm long, 2 stamens exserted, 2 included .... **5. T. frigida**
5b. Hairs on stem up to 0.75mm long, all stamens exserted although two may only reach throat 6

6a. Either flowers in axils of upper leaves or lowermost bracts leaflike and at least 1.5mm broad 7
6b. Bracts not leaflike, less than 1mm broad .... **7. T. schlechteri**

7a. Flowers in axils of upper leaves, pedicels 2.5–6mm long .... **8. T. longipedicellata**
7b. Only lowermost bracts leaflike, upper smaller and narrower, pedicels up to 4mm long   .. 8

8a. Upper leaves passing imperceptibly into bracts, these c.6–18 × 3–7mm at base of inflorescence; corolla limb often mauve, sometimes whitish .... **9. T. glutinosa**
8b. Upper leaves separated from lowermost flower by a bare peduncle; corolla limb white
6. T. elsiae

1. Trieenea laxiflora Hilliard in Notes RBG Edinb. 45: 490 (1989).
Type: Cape, Clanwilliam div., 3219AA, N. Cedarbergen, Pakhuis, 2500–3000ft, *Esterhuysen* 14972 (holo. BOL; iso. K, NBG, PRE, SAM, STE, UPS).

Shrublet up to 450mm tall (but will flower in the seedling stage), well branched from the base, branches erect or decumbent, main ones up to 4mm in diam. at base, glabrous, glandular-puberulous on inflorescence axes, leafy. *Leaves* opposite, bases connate, blade 4–30 × 5–25mm, broadly ovate, base abruptly contracted into a flat petiolar part 1–6(–10)mm long, up to c.½ the total leaf length but mostly much shorter, upper margins with 2–4(–8) pairs of coarse ± obtuse teeth, both surfaces glabrous, sometimes glaucous. *Flowers* few to several in lax heads up to 30mm long, sometimes solitary, usually corymbose-paniculately arranged on long peduncles bearing usually 1, sometimes 2, pairs of reduced leaves or bracts. *Bracts* at base of inflorescence 4–6 × 0.4–0.75mm, linear-lanceolate, glandular-puberulous, adnate to pedicels only. *Pedicels* up to 0.5–1mm long. *Calyx* very obscurely bilabiate, tube none or up to 0.5mm long, anticous lobes 3.75–6.75 × 0.4–0.75mm, posticous lobes similar, all glandular-puberulous, sometimes with a few eglandular hairs as well. *Corolla* tube 6.5–7.5mm long, funnel-shaped, c.2mm across mouth, limb 7–8mm across lateral lobes, posticous lobes 2–2.75 × 1.5–2.25mm, anticous lobe 2.75–4 × 2mm, all lobes subrotund, posticous lip well bearded at base, orange there and down back of tube, hairs sometimes extending to anticous lip as well, rest of limb pale mauve-blue or pale saxe blue. *Stamens* 4, posticous pair deep in tube, filaments decurrent nearly to base of tube, anther lobes 0.75–1.1mm long, anticous pair shortly exserted, anther lobes 0.6–0.75mm long. *Stigma* exserted. *Capsules* 2.5–4.5 × 2–2.5mm. *Seeds* up to c.9 in each loculus, c.0.75 × 0.5mm, pallid amber, wrinkled in sinuous longitudinal bands.

Citations:

Cape. Clanwilliam div., N. Cedarberg, 3219AA, near Boontjieskraal, *Esterhuysen* 34828 (BOL, K); Pakhuis, 3000ft, *Esterhuysen* 21151 (BOL, K, PRE, UPS); ibidem, *Leipoldt* 3416 (BOL).

Trieenea laxiflora is known from only two sites in the northern Cedarberg, and like all its close congeners it grows in the shelter of overhanging rocks, flowering in December and January. Its loose heads of flowers arranged in corymbose panicles at once distinguishes it from *T. lasiocephala* and *T. lanciloba*, with compact round heads at the tips of the branches.

One specimen (*Taylor* 7471, K, PRE, STE) is puzzling. It was collected at c.1375m on the higher ridges and slopes of Crevasse Peak in the northern Cedarberg, where it was 'very locally frequent among rocks' and just beginning to flower in the second week of November. It has the general facies of *T. laxiflora*: leaves with short petioles and heads borne in corymbose panicles, but the uppermost parts of the branches, the bracts and the calyx are villous as well as glandular-puberulous. Furthermore, the flowers, which are the same colour (blue or mauve-blue) as those of *T. laxiflora*, have a slightly larger limb with the posticous lip glabrous, not bearded. The plant may warrant specific rank but I am unwilling to describe it until it is better known. I have seen no other species of *Trieenea* from Crevasse Peak.

2. Trieenea lasiocephala Hilliard in Notes RBG Edinb. 45: 490 (1989).
Type: Cape, Clanwilliam div., 3219AA, Cedarberg, near Crystal Pool hut, 4500ft., *Esterhuysen* 30016 (holo. BOL, iso. S).

Shrublet up to 300mm tall, well branched from the base, branches erect or decumbent, main ones 2.5–3mm in diam. at the base, villous with patent eglandular hairs mostly 1–2.5mm long, sometimes glabrescent, sometimes minutely glandular-puberulous below the inflorescence, leafy. *Leaves* opposite, bases connate, scarcely diminishing in size upwards, except for the uppermost pair, which may be much reduced or bracteate, blade 6–24 × 6–23mm, broadly ovate, base abruptly contracted to a flat petiolar part 4–14mm long, often half or equalling the length of the blade, upper margins with 2–5 pairs of coarse, ± obtuse teeth, both surfaces thinly villous, sometimes glabrescent, hairs up to 1.25–1.5mm long especially on petiolar part, sparsely glandular-puberulous as well. *Flowers* many in rounded heads c.10–20mm in diam., not elongating in fruit, solitary at branch tips which are usually pedunculoid with a single pair of bracts or much reduced leaves. *Bracts* at base of inflorescence c.5–8 × 0.5–1.5mm, linear-lanceolate, adnate to pedicel only, villous, hairs up to 1.5mm long. *Pedicels* up to 0.75mm long. *Calyx* very obscurely bilabiate, tube none or up to 0.2mm long, anticous lobes 3.75–7 × 0.4–0.75mm, posticous lobes similar, villous with hairs up to 1.5mm long, glandular-puberulous as well at base. *Corolla* tube 5–7mm long, funnel-shaped, c.2mm across mouth, whole corolla minutely and sparsely glandular-puberulous outside in upper part, limb 5.5–8mm across lateral lobes, posticous lobes 1.5–3 × 1.75–3mm, anticous lobe 2–3 × 1.6–2.75mm, all lobes subrotund, posticous lip glabrous at base, orange there and down back of tube, rest of limb white or creamy-coloured. *Stamens* 4, posticous pair deep in tube, filaments decurrent to base, anthers 0.8–1mm long, anticous pair shortly exserted or in throat, anthers 0.5–0.75mm long. *Stigma* exserted. *Capsules* c.2.75 × 2mm (few seen). *Seeds* c.2–10 in each loculus, c.0.75 × 0.6mm, pallid amber sinuously wrinkled (only one seen).

Citations:

Cape. Clanwilliam div., 3219AA, Cedarberg Mts., between Heuning Vlei and Crystal Pool, *Esterhuysen* 7534 (BOL); Heuning Vlei, *Stokoe* sub SAM 56174 (SAM); Scorpion's Poort, *Esterhuysen* 7536 (BOL, K, NBG); Krakadouw Peak, *Stokoe* sub SAM 56173 (SAM); Klein Koupoort, *Esterhuysen* 34822 (BOL, S).

Trieenea lasiocephala is known from several different localities in the northern part of the Cedarberg, where it grows in the shelter of rocks, flowering in December and January. Fruiting material should be sought; only very old inflorescences are present on the specimens seen, but these are enough to indicate that the head does not elongate in fruit, a character that at once distinguishes *T. lasiocephala* from *T. laxiflora*, where the heads are borne in very open corymbose panicles, in contrast to the solitary heads of *T. lasiocephala*.

The specific epithet draws attention to the villous bracts and calyces, which give the compact head a woolly appearance.

3. Trieenea lanciloba Hilliard in Notes RBG Edinb. 45: 489 (1989). **Fig. 62C.**
Type: Cape, Clanwilliam div., 3219AA, Groot Krakadouw, c.1690m, 22 xi 1987, *Taylor* 11903 (holo. E, iso. STE).

Shrublet up to c.200mm tall, well branched from the base, branches erect or decumbent, main ones up to 4mm in diam. at base, glabrous, leafy. *Leaves* opposite, bases connate, blade 4–16 × 4–13mm, broadly ovate, base abruptly contracted into a flat petiolar part 2–8mm long, often about half as long as leaf blade, upper margins with 2–4 pairs of coarse ± obtuse teeth, both surfaces glabrous or occasionally with a very few hairs. *Flowers* few to several in round heads c.10mm in diam., remaining ± compact in fruit, solitary at tips of branches. *Bracts* at base of inflorescence 4–8 × 1–2.25mm, lanceolate, adnate to pedicel only, very sparsely glandular-puberulous, occasional eglandular hairs on margins c.0.1–0.5mm long. *Pedicels* up to 0.8–1.5mm long. *Calyx* very obscurely bilabiate, tube none or up to 0.4mm long, anticous lobes 3.25–6 × 0.75–1mm, lanceolate, often folded longitudinally, posticous lobes similar, sparsely and minutely hairy on margins. *Corolla* tube 3.5–5mm long, funnel-shaped, c.1.5–2.5mm across mouth, limb 4.5–6mm across lateral lobes, posticous lobes 1–1.5 × 1–1.75mm, anticous lobe 1.5–2.6 × 1–2.4mm, all lobes subrotund, white or cream-coloured, blueish outside in bud, base of posticous lip orange, the colour running back down the tube, either bearded or glabrous. *Stamens* 4, anthers of posticous pair appearing in mouth, 0.4–1.1mm long, filaments decurrent to base of tube, anticous pair shortly exserted, anthers 0.4–1.1mm long. *Stigma* exserted. *Capsules* c.3.5 × 2mm. *Seeds* c.6–7 in each loculus, c.0.75 × 0.75mm, cream-coloured, angled, obscurely wrinkled on one face.

Citations:

Cape. Clanwilliam div., Krakadouw Peak, 4500ft, *Esterhuysen* 14316 (BOL, K); ibidem, 5600ft, *Esterhuysen* 14999 (BOL, NBG); ibidem, 5700ft, *Esterhuysen* 7498 (BOL); ibidem, *Stokoe* sub SAM 60009 (SAM); ibidem, 5500ft, *Adamson* s.n. (BOL); ibidem, 26 x 1923, *Pocock* 653 (STE). Pakhuis Peak, *Esterhuysen* 7412 (BOL).

Trieenea lanciloba is known from two mountains in the northern Cedarberg: Krakadouw Peak and Pakhuis Peak, where it grows in deep rock fissures or under big rock overhangs, thus afforded moist shade. December seems to be the chief month for flowering; in late November, when Mr Taylor collected his specimens, the plants were only just coming into flower, but old capsules on these plants afforded the only seeds that I have seen. These were firmly held in bulges at the base of the capsules; the upper seeds had all been dispersed. It is desirable that further fruiting material be collected because the seeds appear to differ from those of the closely allied *T. laxiflora* and *T. lasiocephala*, both of which have elliptic seeds wrinkled in sinuous longitudinal bands while those of *T. lanciloba* are angled with only one face obscurely wrinkled. This needs confirmation on more adequate material: the apparent differences may be no more than artefacts resulting from the position of the seed in the capsule.

The trivial name *lanciloba* draws attention to the lanceolate bracts and calyx lobes, a character that at once distinguishes *T. lanciloba* from both *T. lasiocephala* and *T. laxiflora*. The inflorescences of *T. lanciloba* resemble those of *T. lasiocephala* in that they are capitate and solitary, but the bracts and calyx lack the abundant long hairs of *T. lasiocephala*. In indumentum, the bracts and calyx are more like those of *T. laxiflora*, but that species differs markedly in the more spicate heads arranged in lax corymbose-panicles.

4. Trieenea taylorii Hilliard in Notes RBG Edinb. 45: 491 (1989). **Plate 1F.**
Type: Cape, Clanwilliam div., 3219AA, N. Cedarberg, Moedersielshoek path up Groot Kraka-douw, 1450–1600m, *Taylor* 11894 (holo. E).

Perennial herb, stems well branched from the base, sprawling or diffuse, up to 80–300mm long, woody at base and there 1–2mm in diam., villous with patent eglandular hairs up to 1–2mm long, glandular-puberulous as well, and some long hairs gland-tipped, leafy. *Leaves* opposite, occasionally alternate below the inflorescences, blade 6–18 × 4–14mm, ovate to elliptic, base cuneate and tapering into a flat petiolar part 4–13mm long, often roughtly ½ to ⅓ as long as blade, upper margins with (2–)5–6 pairs of coarse teeth, both surfaces pilose, hairs up to 1–2mm long, sparsely glandular-puberulous as well. *Flowers* few to many in round heads, elongating in fruit, remaining crowded or becoming distant, the lowermost flowers usually distant, heads terminating all the branchlets, often further arranged in very open panicles, peduncles nude or nearly so. *Bracts* c.3.5–9 × 0.75–3.5mm, lowermost leaflike or broadly spathulate, narrower upwards, adnate to pedicel only or to extreme base of calyx as well, pubescent all over with eglandular hairs up to 0.6–1.25mm long, minutely glandular-puberulous as well. *Pedicels* up to 0.5–1.5mm long. *Calyx* obscurely bilabiate, tube 0.4–1mm long, anticous lobes 2.4–4 × 0.5–0.75mm, anticous lip split 2.4–4mm, whole calyx villous, hairs up to 1mm long, glandular-puberulous as well. *Corolla* tube 2.5–4mm long, cylindric in lower part and there c.0.75–1mm in diam., rapidly widening upwards, c.1.75mm across mouth, whole corolla minutely and sparsely glandular-puberulous outside, limb 3.5–5mm across the lateral lobes, posticous lobes 0.75–1.2 × 0.75–1mm, anticous lobe 1.25–2 × 1–1.75mm, all lobes subrotund, the posticous lip dark violet, with an orange-yellow patch running down back of tube, glabrous, anticous lip white or creamy-coloured. *Stamens* 4, posticous filaments decurrent to base of tube, anthers in mouth, anticous anthers exserted, all anthers 0.4–0.6mm long. *Stigma* well exserted. *Capsules* 2.5–3.5 × 1.5–2.5mm. *Seeds* c.6–10 in each loculus, 0.5–0.75 × 0.4–0.6mm, sinuously wrinkled in longitudinal bands, cream-coloured, pallid amber when fully ripe.

Citations:

Cape. Clanwilliam div., N. Cedarberg, 3219AA, Krakadouwsberg, 5000ft, *Esterhuysen* 7504 (BOL); ibidem, 3000–4000ft, *Esterhuysen* 15003 (BOL); ibidem, 5500ft., *Esterhuysen* 15000 (BOL, NBG, SAM); Koupoort, 5000ft, *Esterhuysen* 12173 (BOL); Middelberg plateau, *Compton* 12733 (NBG). Cedarbergen, 3219AC, near Crystal Pool, 4400ft., *Barnes* BOL 19479 (BOL); ibidem, *Esterhuysen* 30012 (BOL, K); ibidem, *Esterhuysen* 30015 (BOL, S); ibidem, *Weintroub* sub Moss 19776 (J). Foot of Sneeuwberg, 4500ft, *Esterhuysen* 13095 (BOL, NBG, PRE). Apollo Peak, central Cedarberg, west of Kromrivier, c.1600m, 22 x 1985, *Taylor* 11401 (STE); Laurie's Hell, 6000ft, 6 ix 1982, *Viviers* 595 (STE).

Trieenea taylorii is known only from the Cedarberg where it grows in moist shady sites under large rocks, between c.1200 and 1675m above sea level. Flowering takes place mainly between October and December.

The bicoloured limb of the corolla (upper lip dark violet, lower cream or white) and the long eglandular hairs on the vegetative parts and on the calyx make this a species very easily recognized.

It is a pleasure to associate the name of Mr H.C. Taylor with this plant: he has made a study of the vegetation of the Cedarberg and was kind enough to respond to my appeal for help and to climb Krakadouw Mountain in search of this and allied species.

5. Trieenea frigida Hilliard in Notes RBG Edinb. 45: 489 (1989).
Type: Cape, Ceres div. [3219CA?], Cold Bokkeveld Mountains at Eland's Kloof, 5000ft, *Esterhuysen* 18445 (holo. BOL).

Herb, possibly annual, well branched from the base, base woody, stems there up to 1.25mm in diam., diffuse, leafy except below the inflorescences, glandular-pilose, hairs patent, up to 1–1.5mm long. *Leaves* opposite, uppermost one on each twig sometimes solitary, blade c.7–12

× 6–9mm, elliptic or ovate, base cuneate, tapering into a flat petiolar part 3–5mm long, roughly half length of blade, upper margins with 2–3 pairs of coarse teeth, both surfaces glandular-pilose, hairs up to 2mm long. *Flowers* up to c.15 in lax racemes, each terminating a bare peduncle up to 40mm long at tips of branches. *Bracts* at base of inflorescence c.5–7 × 1–2mm, spathulate or almost leaflike, smaller and narrower upwards, adnate to pedicel and calyx tube, pubescent all over, hairs gland-tipped, up to 1–1.5mm long. *Pedicels* up to 0.75–2mm long. *Calyx* tube 0.25–0.5mm long, anticous lobe c.3.25–4 × 0.5–0.75mm, anticous lip split c.3.25–4mm, whole calyx glandular-pilose, hairs up to 0.75mm long. *Corolla* tube c.4.5 × 1mm, broadening slightly in throat, limb c.4mm across lateral lobes, posticous lobes c.1.5 × 1mm, anticous lobe c.2 × 1mm, all lobes elliptic, very sparsely and minutely glandular-puberulous outside, glabrous inside, colour unknown but possibly white with an orange patch running down back of corolla tube. *Stamens* 4, anticous pair exserted, posticous pair included, filaments decurrent nearly to base of tube, anthers c.0.5mm long. *Stigma* probably exserted. *Capsules* 3–4.5 × 2mm. *Seeds* c.6–10 in each loculus, c.0.6–0.75 × 0.5mm, sinuously wrinkled in longitudinal bands, pallid amber.

Trieenea frigida is known only from the type collection made by Miss Elsie Esterhuysen in the Cold Bokkeveld Mountains in late March when the plants were in fruit and only old corollas were present clinging to the tops of the capsules.

The very long hairs on stems, leaves and bracts (1–2mm) and the pair of included stamens at once distinguish *T. frigida* from its allies, *T. schlechteri* and *T. elsiae*. *Trieenea frigida* and *T. schlechteri* are almost certainly sympatric and as all the species in this genus favour similar habitats, careful collecting is needed: the two species are much alike in facies, but the lowermost bracts in *T. frigida* are either leaflike or spathulate; they are always ± linear in *T. schlechteri* and less than 1mm broad.

The colour of the flowers in *T. frigida* is unknown.

6. Trieenea elsiae Hilliard in Notes RBG Edinb. 45: 489 (1989).
Type: Cape, Clanwilliam div., 3219CB, S. Cedarbergen, Sandfontein Peak, 5000ft, *Esterhuysen* 13866 (holo. BOL, iso. K).

Herb, possibly annual, stems well branched from the base, c.80–200mm long, erect or sprawling, woody at base and there up to 1.5mm in diam., pubescent with patent glandular hairs up to 0.5–0.75mm long, leafy. *Leaves* opposite, becoming alternate on the inflorescence branches, blade 5–17 × 3–10mm, elliptic or ovate-elliptic, base cuneate, tapering into a flat petiolar part 2–6mm long, roughly ½–⅓ length of blade, upper margins with 3–5 pairs of coarse teeth, both surfaces glandular-pilose, hairs up to 0.5–0.75mm long. *Flowers* few to many in congested rounded heads becoming oblong in fruit, only the lowermost 1–3 flowers distant, each raceme terminating a nude peduncle up to c.40mm long, these solitary or in few-branched panicles. *Bracts* c.5–8 × 1.5–3mm, the lowermost leaflike, smaller and narrower upwards, adnate to pedicel only, hairy as leaves. *Pedicels* up to 0.5–1mm long. *Calyx* obscurely bilabiate, tube 0.5–1mm long, anticous lobes 2.5–5 × 0.5–1mm, anticous lip split 2.5–5mm, whole calyx glandular-pubescent, hairs up to 0.5–0.75mm long. *Corolla* tube 3–4.5mm long, cylindric in lower half and there c.0.75mm in diam., rapidly broadening upwards, c.1.5mm across mouth, whole corolla glandular-pubescent outside, limb 2.75–4mm across lateral lobes, posticous lobes 0.75–1 × 0.75–1mm, anticous lobe 1–1.5 × 0.75–1mm, all lobes subrotund, posticous lip bearded at base in all but the smallest (selfed?) flowers, white with an orange patch at base of posticous lip and running down back of tube. *Stamens* 4, posticous filaments decurrent well down tube, all anthers 0.3–0.5mm long, shortly exserted. *Stigma* either exserted or curled round in tiny flowers. *Capsules* 2.5–4.5 × 2–3mm. *Seeds* c.0.5 × 0.4mm, irregularly wrinkled in longitudinal bands, pallid amber.

Citations:

Cape. Clanwilliam div., S. Cedarbergen, 3219AC, Apollo Peak, 4500ft, *Esterhuysen* 18081 (BOL). Ceres div., Cold Bokkeveld Mts, Schoongezicht Peak, 5000ft, *Esterhuysen* 21294 (BOL).

Trieenea elsiae is known only from three collections made by Miss Elsie Esterhuysen in the South Cedarberg and the Cold Bokkeveld Mountains. The plants grow in the shelter of big boulders and rock overhangs, and when the collections were made, in mid December and early April, they were already mostly in fruit; good flowering material is needed from earlier in the season.

Trieenea elsiae is allied to *T. schlechteri* with which it is at least partly sympatric; they differ in facies (racemes remaining crowded in fruit in *T. elsiae*, very lax in *T. schlechteri*) as well as in floral detail: lowermost bracts in *T. elsiae* more or less leaflike (c.5–8 × 1.5–3mm versus 2–5 × 0.3–0.8mm), calyx mostly larger (anticous lobes 2.5–5mm long versus 1.75–2.5mm), and posticous lip bearded at the base in all but the smallest flowers (always glabrous in *T. schlechteri*).

Very few flowers are present on the specimens of *T. elsiae* available to me, but on the sheet of *Esterhuysen* 13866 in Bolus herbarium, some tiny flowers lack a beard at the base of the posticous lip and the stigma may be malformed, curling round the anthers; both factors indicate selfing, but further investigation is needed to confirm autogamy in *T. elsiae*.

7. Trieenea schlechteri (Hiern) Hilliard in Notes RBG Edinb. 45: 490 (1989).
Type: Cape, Ceres div., 3319AB, rocky places near Sand River in the Cold Bokkeveld, 4500ft, *Schlechter* 10113 (holo. K; iso. BM, BOL).

Syn.: *Phyllopodium schlechteri* Hiern in Fl. Cap. 4(2): 319 (1904).
  Polycarena schlechteri (Hiern) Levyns in J.S. Afr. Bot. 5: 37 (1939).

Perennial herb well branched from the base, base woody, stems there up to 2mm diam., prostrate or diffuse, the inflorescences probably held erect, very leafy but the tips scapose below the racemes, glandular-pubescent, hairs patent, up to 0.4–0.5mm long. *Leaves* opposite, occasionally the uppermost alternate, blade mostly 6–20 × 4–11mm, elliptic, base cuneate and tapering into a flat petiolar part 2–7mm long, shorter than blade, upper margins with 2–4 pairs of coarse teeth, both surfaces glandular-pubescent, hairs up to 0.25–0.4mm long. *Flowers* rarely solitary, usually 2–12 in very lax racemes terminating all the branchlets, sometimes one or a few secondary racemes developing to produce a very open panicle. *Bracts* 2–5 × 0.3–0.8mm, linear-oblong to linear-lanceolate, glandular-pubescent as in leaves, adnate to base of pedicel only. *Pedicels* up to 0.6–4mm long, gradually shorter upwards. *Calyx* obscurely bilabiate, tube 0.25–0.5mm long, anticous lobe 1.75–2.5 × 0.3–0.8mm, anticous lip split 1.75–2.5mm, calyx glandular-pubescent all over. *Corolla tube* c.2–3mm long, cylindric in lower half and there c.0.5–0.8mm broad, abruptly expanded above, c.1–1.5mm across mouth, whole corolla glandular-puberulous outside, limb c.2–4mm across lateral lobes, posticous lobes c.0.5–1 × 0.4–1mm, anticous lobe c.0.75–1.5 × 0.5–1mm, corolla smaller in autogamous flowers, all lobes subrotund, white with an orange patch at base of posticous lobe and extending down back of tube, glabrous. *Stamens* 4, shortly exserted, posticous filaments decurrent about halfway down tube, all anthers 0.3–0.5mm long. *Stigma* either exserted, or malformed and twisted round and often attached to the anthers. *Capsules* 2–2.5(–3) × 1.5–2mm. *Seeds* up to 8–12 in each loculus, c.0.5–0.75 × 0.5mm, obscurely 3-angled, sinuously wrinkled in longitudinal bands, pallid.

Citations:

Cape. Clanwilliam div., 3219AC, Cedarberg, Wolfberg, 4900ft, *Taylor* 7523 (K, STE); ibidem, 4500ft, *Esterhuysen* 29979a (BOL); ibidem, 4500ft, *Esterhuysen* 29979 (BOL, S); ibidem, *Esterhuysen* 18115 (BOL); ibidem, *Esterhuysen* 22449 (BOL); plateau between the Tafelberg and the Wolfberg, 5000ft, *Esterhuysen* 29982 (BOL, E, K, NBG, PRE, S).

Trieenea schlechteri is obviously common in the northern Cedarberg, but there are no records from the southern Cedarberg, where its ally, *T. elsiae*, occurs. However, Schlechter's original collection came from a site in the Cold Bokkeveld, roughly 85km south of the Cedarberg (I have not been able to pinpoint the precise locality). The plants grow under rock overhangs, flowering in November and December. Although many flowers open normally, selfing may be commonplace: in many of the smaller flowers the stigma is much shorter than normal and is either firmly attached to an anther or curled round the anthers. Occasionally, an unopened corolla can be seen capping the young capsule.

Trieenea schlechteri is easily recognized by its very lax racemes, sometimes forming few-branched panicles, and small narrow bracts, which are often adnate only to the extreme base of the pedicels; at the tip of the inflorescence, where the pedicels may be much abbreviated, the bract may be adnate nearly to the top of the pedicel.

8. Trieenea longipedicellata Hilliard in Notes RBG Edinb. 45: 490 (1989). **Fig. 62A.**
Type: Cape, Paarl div., 3319CC, Wemmershoek Peak, 6000ft, 31 xii 1944, *Esterhuysen* 11236 (holo. BOL; iso. K, PRE).

Perennial herb well branched from the base, branches rooting from lowermost nodes, up to 1mm diam., prostrate or diffuse, leafy throughout, glandular-pubescent, hairs patent, up to 0.4–0.5mm long. *Leaves* opposite below becoming alternate upwards and passing imperceptibly into floral bracts, blade of largest leaves c.4–10 × 4–8mm, ovate to rhomboid in outline, base cuneate and passing abruptly into a flat petiolar part c.3–7mm long, about equalling the blade or up to c.half its length, upper margins deeply and coarsely toothed, 2–3(–4) pairs each side, both surfaces glandular-pubescent, hairs up to 0.4–0.5mm long. *Flowers* solitary in upper leaf axils forming long lax leafy racemes. *Bracts* leaf-like throughout, adnate to base of pedicel only. *Pedicels*: lower ones 2.5–6mm long, slightly shorter upwards. *Calyx* obscurely bilabiate, tube 0.5–1.2mm long, anticous lobes 1.8–2.5 × 0.6–0.9mm, anticous lip split 1.5–2.6mm, calyx glandular-pubescent all over. *Corolla* tube 3–4mm long, cylindric in lower half and there 0.75–1mm broad, abruptly expanded above, c.1.5–2mm across mouth, whole corolla glandular-puberulous outside, limb 4–6mm across lateral lobes, posticous lobes 0.5–1.5 × 1–1.5mm, anticous lobe 1.25–2 × 1.25–2mm, all lobes subrotund, white with an orange patch at base of posticous lip and extending down back of tube, glabrous. *Stamens* 4, exserted, posticous filaments decurrent more than halfway down tube, all anthers 0.4–0.6mm long. *Style* well exserted. *Capsules* 2.5–3 × 2–2.5mm. *Seeds* up to 8 in each loculus, 0.75–1 × 0.6mm, obscurely 3- or 4-angled, one face obscurely wrinkled, testa cream-coloured.

Citations:

Cape. Stellenbosch div., 3318DD, Jonkershoek Forest Reserve, Dwarsberg, ?4500ft, *Esterhuysen* s.n. (BOL, S); 3419AA, Jonkershoek Forest Reserve, Victoria Peak, 4800ft, *Esterhuysen* 29296 (BOL, K). Worcester div., 3319CA, Slanghoek Needle, 5500ft, *Esterhuysen* 17807 (BOL, K, NBG, PRE, UPS); Slanghoek Mts, Witteberg, 5500ft, *Esterhuysen* 9489 (BOL, K). 3319CC, Du Toit's Peak, 6500ft, *Esterhuysen* 8582 (BOL); ibidem, *Esterhuysen* 30557 (BOL, K); ibidem, 1800m, *Marloth* 2492 (PRE). Caledon div., 3319CD, Louwshoek Peak above Kaaiman's Gat, 5500ft, *Esterhuysen* 35627 (BOL, S).

Trieenea longipedicellata is known from a few mountains in the S.W. Cape, lying between Paarl, Stellenbosch and Genadendal. It grows on the damp shady floors of rock overhangs and under big boulders, between c.1500 and 1980m above sea level; flowers can possibly be found in any month, but November to February may be the main season. It is easily recognized by its sprawling rooting branches with the flowers borne singly on long pedicels in all the upper leaf axils. These long pedicels (lower ones c.2.5–6mm) will at once distinguish *T. longipedicellata* from *T. glutinosa* (pedicels 1–3mm), but it also differs in its inflorescences (flowers never running up into definite racemes), smaller flowers (corolla tube 3–4mm long versus mostly 4–7.5mm, anticous lobe 1.25–2mm versus mostly 2.5–5mm), glabrous at the base of the posticous lip (not with clavate hairs), and smaller, rounder capsules (2.5–3 × 2–2.5mm versus

4–5 × 2.5–3mm) with fewer seeds (up to 8 in each loculus versus many). The two species appear to be allopatric.

The closest ally of *T. longipedicellata* may be *T. schlechteri*; they have similar flowers arranged in very lax inflorescences, but the bracts of *T. schlechteri* are never leaflike; they are linear and less than 1mm broad.

9. Trieenea glutinosa (Schltr.) Hilliard in Notes RBG Edinb. 45: 489 (1989). **Fig. 62B, Plate 1E.**

Type: Cape, Worcester div., on the top of Matroosberg, 3319BC/BD, 6000ft, xii 1892, *Marloth* 2216 (holo. BOL).

Syn.: *Phyllopodium glutinosum* Schltr. in J. Bot. 35: 220 (1897); Hiern in Fl. Cap. 4(2): 318 (1904), excluding *Wilms* 1060.
      Polycarena glutinosa (Schltr.) Levyns in J.S. Afr. Bot. 5: 37 (1939).

Annual herb, but sometimes perennating, stems 35–400mm long, simple to well-branched from the base, erect, decumbent or possibly sprawling, base woody, branches leafy throughout, glandular-pubescent, hairs patent, up to 0.5mm long. *Leaves* opposite becoming alternate upwards and passing imperceptibly into the floral bracts, blade of largest leaves c.6–23 × 4–14mm, gradually smaller upwards, rhomboid or ovate in outline contracted more or less abruptly to a flat petiolar part c.3–20mm long, often ⅓ to ½ the length of the blade or about equalling it, upper margins deeply and coarsely toothed, teeth in 3–7 pairs, a few teeth sometimes with a secondary tooth, both surfaces glandular-pubescent. *Flowers* many in crowded heads rapidly elongating into long somewhat lax racemes, lowermost flowers often distant in axils of upper leaves, racemes often solitary at branch tips, sometimes secondary, shorter, racemes developing below them. *Bracts* at base of racemes leaflike, often c.6–18 × 3–7mm including the petiolar part, smaller, narrower and eventually linear-oblong upwards, glandular-pubescent as in leaves, mostly adnate only to pedicel, to extreme base of calyx tube as well in uppermost flowers. *Pedicels* at base of racemes 1–3mm long. *Calyx* obscurely bilabiate, tube 0.5–1mm long, anticous lobes 2–4 × 0.6–1.1mm, anticous lip split 2–4mm, calyx glandular-pubescent all over. *Corolla* tube (3.25–)4–7.5 × 0.8–1.3mm, cylindric below, abruptly funnel-shaped above, whole corolla sparsely glandular-puberulous outside, limb (4–)5–10mm across lateral lobes; posticous lobes (1–)1.5–2.5 × 1–2.5mm, anticous lobe (1.5–)2.5–5 × (1–)1.75–3mm, all lobes elliptic-oblong, whitish to pale mauve-blue, two orange patches at base of posticous lip and there bearded with a few coarse clavate hairs. *Stamens* 4, all exserted, posticous filaments briefly decurrent on tube, all anthers 0.4–0.6mm long, very dark in colour or sometimes pale. *Style* either exserted or in small flowers appearing to curl round anthers. *Capsule* 4–5 × 2.5–3mm. *Seeds* up to c.14, 0.75–1 × 0.5mm, roughly trigonous, one face obscurely furrowed longitudinally, the other two irregularly wrinkled longitudinally, testa slate grey or greenish, probably cream or blue-grey when fully ripe.

Selected citations:

Cape. Paarl div., 3318DB, Lower Wellington Sneeuwkop, 5200ft, *Esterhuysen* 12458 (BOL, E, NBG, PRE, SAM). Tulbagh div., 3319AA, Great Winterhoek, 5500–6500ft, *Esterhuysen* 19776 (BOL, NBG, UPS). Worcester div., 3319AD, nek between Milner Peak and Milner Needle, 6400ft, *Esterhuysen* 35527 (BOL, E, S). Ladismith div., 3322AA, Swartberg near Prince Albert, 5400ft, *Bolus* 12190 (BOL). Uitenhage div., 3324BD, Great Winterhoek Mts., Cockscomb, 5500ft, *Esterhuysen* 28033 (BOL, K, PRE). Uniondale div., 3323DA, Kouga Mts., Saptokop, 5500ft, *Esterhuysen* 27982 (BOL, E, K, PRE, UPS).

Trieenea glutinosa is widely distributed on the mountains of the south western Cape from the Gifberge south of Van Rhynsdorp to the Cedarberg and the mountains about Ceres, Worcester and Robertson, then east along the Zwartberg to the Kouga Mountains and Great Winterhoek Mountains north west of Uitenhage. It grows on steep sandy-peaty rocky slopes, where it may be common after fire, or in deep rocky gullies, under rock overhangs and on rocky ledges facing south and south west at altitudes ranging from c.1600m to 2100m. Flowering begins in

September and October and by December plants are passing into fruit, but some flowers may be found as late as February. There is considerable range in the size of the corolla; the smaller flowers may be selfed as the stigma appears to curl round the anthers, which are then pale in colour instead of the usual very dark colour. This needs confirmation in the field.

Trieenea glutinosa appears to be commonest from about Ceres and Worcester south and east to the Zwartberg and Great Winterhoek Mountains. Its presence in the Cedarberg rests upon two collections made by Miss Esterhuysen. One (*Esterhuysen* 30012a, BOL) was made near Crystal Pool hut on 29 xii 1962. There are four small pieces, one of which is undoubtedly *T. glutinosa*, with the leaves running up to the base of the inflorescence and passing imperceptibly into bracts, a characteristic feature of the species. The other pieces are less easy to identify, and one is mixed with a scrap of what appears to be *T. schlechteri*. Miss Esterhuysen noted that the specimens were growing with her no. 30012, which is *T. taylorii*. On the previous day, she collected a plant that I am unable to determine satisfactorily (*Esterhuysen* 29996). It was growing on the plateau south of Tafelberg 'in shelter of overhanging rock at base of cliffs, S aspect, common'. It has the short indumentum and bearded posticous lip of *T. glutinosa*, but the bracts are sharply differentiated from the leaves. Short indumentum and narrow bracts point to *T. schlechteri*, but in that species the base of the posticous lip is glabrous. A duplicate in the Kew herbarium is a mixture of unequivocal *T. glutinosa* and the plant represented on the sheet in the Bolus herbarium. Clearly, a careful survey of the plants in this area is desirable.

12. PHYLLOPODIUM

Phyllopodium comprises 26 species, most of them confined to the western and south western Cape: 18 have their areas wholly (or very nearly so) west of 20°E, and only seven of these species are found north of the Khamiesberg (four reach southern Namibia). Of the remaining eight species, five have their areas between 20° and 25°E, and three principally east of 24°E, either on the southern Cape mountains or in the coastal districts: apart from a single record from Graaff Reinet (see under *P. rustii*) the genus appears to be absent from the central Cape.

The species fall into two subgenera distinguished by differences in the structure of their inflorescences and by differences in gross seed characters visible under a dissecting microscope or with a good hand lens. The seeds of subgen. *Phyllopodium* are colliculate and irregularly wrinkled; those of subgen. *Leiospermum* are perfectly smooth.

Within subgen. *Phyllopodium*, two main groups and several smaller subgroups can be distinguished on differences in the sculpturing of the testa as seen under the SEM (when seeds are cleared in lactic acid and examined under a compound light microscope only the outlines of the polygonal cells of the testa are visible at × 225). These differences in the fine detail of the testa are linked to other, gross, characters, and there are also differences in geographical distribution.

Subgenus Phyllopodium

Group 1

Flowers in *elongated racemes*, the lowermost often distant. *Bracts* usually *glabrous on the backs* (glandular-pubescent in two species). *Corolla* limb mostly *white* (*mauve* or marked with mauve in three species) and usually with an *orange/yellow patch* at base of posticous lip (possibly always wanting in four species). *Seeds colliculate*, irregularly wrinkled as well, *pale amber in colour, testa transparent.*

Subgroup 1A. 1. *P. cuneifolium*, 2. *P. bracteatum*, 3. *P. diffusum*, 4. *P. rustii*.

> *Annual* herbs, *hairs eglandular*, those on stems *retrorse*, up to *0.5–1mm long. Leaves* petiolate, *margins serrate. Calyx* tube shorter than anticous lip, *anticous lip deeply divided. Corolla white or mauve, orange/yellow patch at base of posticous lip. Testa*

(under SEM) cells visible, outer periclinal walls verruculose, figs 28D, 29A.

Swellendam eastwards through the coastal districts to Natal.

Subgroup 1B. 5. *P. dolomiticum*, 6. *P. elegans*, 7. *P. multifolium*.

Annual or perennial herbs, *hairs* mixed *eglandular and glandular, retrorse eglandular* hairs *0.25–0.75mm* long. *Leaves* either very *shortly petiolate or narrowed to base*, margins *entire or* with *2–4 pairs of teeth. Calyx* tube shorter than anticous lip, *anticous lip deeply divided. Corolla white or mauve, orange/yellow patch* at base of posticous lip *present or absent. Testa* (under SEM) *pusticulate*, cell outlines almost invisible, fig. 29B.

Montagu to Oudtshoorn, montane.

Subgroup 1C. 8. *P. capillare*, 9. *P. caespitosum*, 10. *P. micranthum*, 11. *P. viscidissimum*, 12. *P. tweedense*.

Annual herbs, *hairs* mixed *glandular and eglandular, retrorse eglandular* hairs up to *0.1mm* long (patent ones may be longer). *Leaves petiolate*, margins *entire or* with up to *4 pairs of teeth. Calyx* tube longer than anticous lip, *anticous lip shallowly divided. Corolla white*, orange/yellow patch possibly always absent. *Testa* (under SEM) *cell outlines visible*, outer periclinal walls convex, verruculose, fig. 29C.

SW Cape, *P. capillare* coastal (Lambert's Bay to Albertinia), the rest montane, from Clanwilliam to Ceres and inland to Laingsburg.

Phyllopodium capillare and *P. micranthum* may sometimes be self-fertilizing.

Group 2

Flowers usually in *compact round heads* (sometimes oblong in fruit) terminating nude or sparsely bracteate peduncles (two anomalous species, one with flowers in leafy racemes, one with masses of round heads terminating leafy branchlets). *Bracts* either *glabrous or pubescent* on the backs. *Corolla* limb mostly *white, creamy or yellow* (*mauve* in three species, unknown in one) usually with an *orange/yellow patch* at base of posticous lip or posticous lobes orange. *Seeds colliculate*, irregularly wrinkled as well, *testa creamy, mauve, greyish or blueish* in colour, *opaque* or semi-opaque.

Subgroup 2A. 13. *P. heterophyllum*, 14. *P. cordatum*, 15. *P. mimetes*.

Annual herbs, *hairs* predominantly *eglandular, retrorse, 0.5–1.5mm long. Leaves petiolate*, margins *entire to* obscurely or distinctly *serrate. Calyx* tube shorter than anticous lip, *anticous lip deeply divided. Corolla cream, yellow or mauve, orange/yellow patch* at base of posticous lip. *Testa* (under SEM) *cell outlines conspicuous*, outer periclinal walls slightly convex, verruculose or wrinkled, fig. 29D.

SW Cape, Van Rhynsdorp to the Peninsula and east to the Breede river, coastal.

Subgroup 2B. 16. *P. pubiflorum*, 17. *P. namaense*, 18. *P. lupuliforme*, 19. *P. collinum*, 20. *P. maxii*.

Annual herbs, *hairs* either *glandular or eglandular*, or mixed, *eglandular retrorse* hairs *0.1–0.5mm* long. *Leaves petiolate or sessile*, margins *entire to* obscurely or distinctly *serrate. Calyx* tube longer than anticous lip, *anticous lip entire to shallowly lobed. Corolla white, yellow or mauve, orange/yellow patch* at base of posticous lip or two posticous lobes orange. *Testa* (under SEM) *cell outlines conspicuous*, outer periclinal walls almost flat, minutely verruculose, fig. 30A.

Southern Namibia to Clanwilliam (Cedarberg), mostly montane.

Anomalous Species: 21. *P. anomalum*, 22. *P. alpinum*

These two disparate species are enumerated here at the end of Group 2 because of the structure of their seeds, but are isolated by the form of their inflorescences.

P. anomalum: Annual herb, *hairs* on stems *eglandular, patent, 1–1.5mm* long. *Leaves petiolate*, margins *entire to* distinctly *serrate. Flowers* in *leafy racemes. Bracts pubescent* on the backs. *Calyx* tube longer than the anticous lip, *anticous lip shallowly lobed. Corolla white*, orange/yellow patch wanting. *Stamens* 2 only. *Seeds colliculate, mauvish-grey.*

Testa (under SEM) *cell outlines conspicuous*, outer periclinal walls strongly convex, irregularly verrucate, fig. 30B.
O'okiep to Montagu.
The flowers are apparently always selfed.

P. alpinum: Shrublet (but will flower in seedling stage), *hairs* on stem *eglandular, retrorse, 0.2–0.5mm* long. *Leaves* shortly *petiolate*, margins *entire or* with up to *3 pairs of teeth. Flowers* in many *rounded heads* terminating short leafy twiglets. *Bracts glabrous or pubescent* on the backs. *Calyx* tube shorter than anticous lip, *anticous lip deeply divided. Corolla pink to mauve* or occasionally white, *orange patch* at base of posticous lip. *Seeds irregularly and sinuously wrinkled, pallid amber. Testa* (under SEM) *cell outlines conspicuous*, outer periclinal walls finely granular, fig. 30C.
Mountains from Fransch Hoek to Genadendal.

Subgenus Leiospermum

Flowers in *small round compact heads* terminating innumerable twiglets corymbosely or divaricately arranged. *Bracts puberulous to hirsute* on the backs. *Corolla* limb usually *mauve*, sometimes white, without an orange/yellow patch at base of posticous lip. *Seeds smooth, testa dark blue-grey*, opaque.

The subgenus comprises 23. *P. cephalophorum*, 24. *P. phyllopodioides*, 25. *P. pumilum*, 26. *P. hispidulum*.

Annual herbs, hairs on stems *eglandular, retrorse, 0.2–0.5mm* long. *Leaves petiolate*, margins *entire to toothed. Calyx* tube shorter than to about equalling anticous lip, *anticous lip divided halfway or less. Testa* (under SEM) *cell outlines conspicuous*, outer periclinal walls slightly convex, verruculose or wrinkled, fig. 30D.
Southern Namibia to the Peninsula.

Phyllopodium Benth. in Hook., Comp. Bot. Mag. 1: 372 (1836), in DC., Prodr. 10: 352 (1846), and in Bentham & Hooker fil., Gen. Pl. 2(2): 944 (1876); Hiern in Fl. Cap. 4(2): 311 (1904) p.p.; Hilliard & Burtt in Notes RBG Edinb. 35(2): 161 (1977), excluding *Polycarena transvaalensis* Hiern [= *Melanospermum transvaalense* (Hiern) Hilliard]. **Figs 16, 28, 29, 30, 41, 63.**

Small annual or perennial herbs. *Stems* either leafy throughout or ± nude above, pubescence characteristically eglandular, retrorse or sometimes patent, patent gland-tipped hairs sometimes present as well but rarely dominant (characteristic hair comprising 2, sometimes 3 or more, stalk cells, curved downwards, apical cell often sickle-shaped, sometimes inflated: fig. 11B, D). *Leaves* opposite below, alternate above. *Inflorescence* racemose, elongate or capitate, always erect. *Bract* adnate to pedicel and at least halfway up calyx tube, patent eglandular hairs always present at least on lower margins. *Calyx* either distinctly or obscurely bilabiate, lobed mostly halfway or more to base, membranous, persistent. *Corolla* membranous, persistent, tube cylindric to narrowly funnel-shaped, usually glabrous outside, rarely pubescent, splitting at base as capsule ripens, limb bilabiate, lobes 5, entire, with or without an orange/yellow patch at base of posticous lip and there either glabrous or bearded with unicellular clavate hairs. *Stamens* usually 4, 2 in one species, or sometimes reduced to 2 or 3 in self-fertilized flowers, well exserted in chasmogamous flowers, anthers synthecous, posticous filaments decurrent at least halfway down corolla tube. *Stigma* lingulate with two marginal bands of stigmatic papillae, exserted, passing gradually into the long filiform style. *Ovary* elliptic in outline, base slightly oblique with a nectariferous gland on the shorter side, usually glabrous, ovules few to many in each loculus. *Fruit* a septicidal capsule with a short loculicidal split at tip of each valve. *Seeds* seated on a funicle taking the form of a round, centrally depressed cushion with undulate margins (fig. 16D, E), seed elliptic or angled, mostly colliculate and more or less wrinkled as well, sometimes smooth, testa tightly investing endosperm, more or less opaque, pale amber, creamy, greyish, mauvish, blueish or dark blue-grey, under the SEM the polygonal cells often visible, variously sculptured (figs 28D, 29, 30).

Type species: *Phyllopodium cuneifolium* (L.f.) Benth.
Distribution: South Africa, mainly western and southern Cape, two species extending into Natal, four into Namibia.

Two subgenera are recognized:

1a. Inflorescence either an elongated raceme at least in fruit, or flowers crowded in rounded
    heads borne singly on nude or sparsely bracteate axes (in *P. alpinum* rounded heads
    terminating many branchlets, plant perennial); seeds colliculate, often irregularly wrinkled
    as well, amber, creamy, grey or mauve .... subgen. **Phyllopodium**
1b. Stems corymbosely or divaricately branched to bear masses of rounded heads, plants
    annual; seeds smooth, dark blue-grey .... subgen. **Leiospermum**

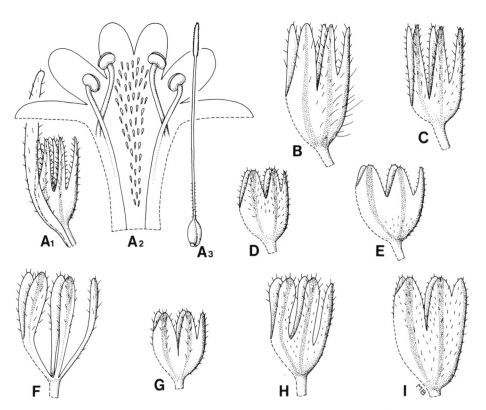

Fig. 63. *Phyllopodium*: A, *P. cuneifolium* (*Hilliard & Burtt* 10857), A1, calyx with bract adnate to it; A2, corolla opened out to show decurrent posticous filaments and clavate hairs at base of posticous lip and down back of tube; A3, lingulate stigma with marginal bands of stigmatic papillae, style, ovary with small adnate nectariferous gland. B–I: calyx with bract removed; dotted line indicates adnation; B, *P. heterophyllum* (*Davis* sub SAM 64281); C, *P. multifolium* (*Fourcade* 3255); D, *P. pumilum* (*Batten* 1028); E, *P. micranthum* (*Schlechter* 8817); F, *P. cephalophorum* (*Hilliard & Burtt* 13015); G, *P. cephalophorum* (*Harvey* 237); H, *P. alpinum* (*Rourke* 1309); I, *P. lupuliforme* (*Max Schlechter* 17). All × 10.

Key to species

1a. Flowers in elongated racemes (caution: in diminutive plants may be very short and
    congested); ripe seeds often amber-coloured, colliculate and often ± wrinkled as well .. 2
1b. Flowers in rounded heads sometimes elongating a little in fruit, or occasionally in flower,
    but not in extended racemes; ripe seeds often creamy-coloured, greyish-mauve or dark
    blue-grey, colliculate or smooth 18

2a. Hairs on stems predominantly eglandular (tiny ± sessile glands may be present, but are very
    inconspicuous) 3
2b. Both glandular and eglandular hairs present (caution: hairs may be only c.0.1mm long)   10

3a. Limb of corolla 3–7mm across the lateral lobes; stems usually at least 50mm long, simple
    or branched 4
3b. Limb of corolla 1.75–2.5mm across the lateral lobes; diminutive caespitose herb, stems
    10–50mm long . . . . **9. P. caespitosum**

4a. Many of the hairs on the stem with remarkably inflated tips, lowermost pedicels 1.5–4mm
    long . . . . **3. P. diffusum**
4b. Hairs on stems not remarkably inflated at tips, lowermost pedicels 0.25–1.5mm long   ... 5

5a. Leaf margins entire or with up to 3 pairs of teeth . . . . **22. P. alpinum**
5b. Leaf margins serrate (occasionally obscurely so) with many pairs of teeth 6

6a. Stems leafy throughout 7
6b. Stems bracteate or nude in upper part 8

7a. Flowers mauve; leaves sharply differentiated from bracts . . . . **1. P. cuneifolium**
7b. Flowers white; leaves grading imperceptibly into bracts . . . . **2. P. bracteatum**

8a. Flowers mauve . . . . **7. P. multifolium**
8b. Flowers whitish to yellowish 9

9a. Upper leaf surface thinly pilose, lower with hairs more or less confined to midrib, bracts
    glabrous or nearly so on backs, ± oblong . . . . **13. P. heterophyllum**
9b. Both leaf surfaces pilose, bracts pilose on backs, ± elliptic . . . . **14. P. cordatum**

10a. Flowers mauve or blue, limb 4.5–7mm across the lateral lobes . . . . **7. P. multifolium**
10b. Flowers white (very rarely mauve in *P. rustii*), limb up to 5mm across the lateral lobes   11

11a. Stamens 2 only, backs of bracts pubescent and glandular-puberulous as well
                                                                                 21. P. anomalum
11b. Stamens normally 4, if only 2, then bracts glabrous on the backs 12

12a. All leaves narrow in relation to length, c.18–35 × 1–3mm . . . . **6. P. elegans**
12b. Blade of at least the lowermost leaves broad in relation to length 13

13a. Stems pubescent, the eglandular hairs 0.4–1mm long 14
13b. Stems puberulous, the eglandular hairs 0.1–0.25mm long 15

14a. Lowermost leaves with petiolar part c.2mm long, blade c.7–8 × 3.5mm, stems leafy
     almost up to the racemes . . . . **5. P. dolomiticum**
14b. Lowermost leaves with petiolar part c.3–8mm long, blade c.7–15 × 3–8mm, stems with a
     distinct nude axis below the racemes . . . . **4. P. rustii**

15a. Lowermost 1–2 internodes clad only in tiny (c.0.1mm) ± retrorse eglandular hairs 16
15b. Lowermost 1–2 internodes clad primarily in patent glandular hairs c.0.25–0.5mm long   17

16a. Upper part of stem puberulous with ± retrorse eglandular hairs more or less masking any glandular hairs that may be present; corolla tube 2.5–3.5mm long, limb 1.8–5mm across the lateral lobes . . . . **8. P. capillare**
16b. Upper part of stem puberulous with patent glandular hairs c.0.1mm long, eglandular hairs few or wanting; corolla tube 1.5–2.5mm long, limb 1–2.5mm across the lateral lobes
10. P. micranthum

17a. Corolla tube c.1.8mm long, limb 2mm across the lateral lobes, anthers c.0.25mm long, yellow and inconspicuous . . . . **11. P. viscidissimum**
17b. Corolla tube c.2.75–3.5mm long, limb 5mm across the lateral lobes, anthers c.0.5–0.75mm long, dark and conspicuous . . . . **12. P. tweedense**

18a. Flowers in compact rounded heads terminating leafy stems; anticous lip of calyx deeply divided (c.2–2.75mm) . . . . **22. P. alpinum**
18b. Heads either solitary on nude or sparsely bracteate axes, or, if heads rounded and massed, anticous lip of calyx usually divided halfway or less (up to 1.75mm) 19

19a. Seeds creamy-coloured, grey or mauve, colliculate or at least wrinkled, heads solitary on nude or sparsely bracteate peduncles terminating simple or sparingly branched stems . . 20
19b. Seeds dark blue-grey, smooth; stems corymbosely or divaricately branched to bear masses of heads 28

20a. Upper part of corolla tube either finely pubescent outside with eglandular hairs, or minutely glandular-puberulous 21
20b. Corolla glabrous outside 22

21a. Corolla eglandular-pubescent outside, yellow, stamens 4, flowers in compact heads
16. P. pubiflorum
21b. Corolla glandular-puberulous outside, white, stamens 2, flowers in leafy racemes
21. P. anomalum

22a. Most conspicuous hairs on stems retrorse, often appressed 23
22b. Most conspicuous hairs on stems patent 26

23a. Calyx glabrous except for hairs on margins of lobes . . . . **17. P. namaense**
23b. Calyx pubescent to pilose at least on keels 24

24a. Corolla creamy-coloured to yellow 25
24b. Corolla mauve . . . . **15. P. mimetes**

25a. Upper leaf surface thinly pilose, lower with hairs more or less confined to midrib, bracts glabrous or nearly so on backs, ± oblong . . . . **13. P. heterophyllum**
25b. Both leaf surfaces pilose, bracts pilose on backs, ± elliptic . . . . **14. P. cordatum**

26a. Bracts glabrous on the backs . . . . **18. P. lupuliforme**
26b. Bracts pubescent or pilose on the backs 27

27a. Hairs on leaves, backs of bracts and calyx predominantly eglandular . . . . **19. P. collinum**
27b. Hairs on leaves, backs of bracts and calyx predominantly glandular . . . . **20. P. maxii**

28a. Stems corymbosely branched above, main leaves usually toothed (see also discussion under species 23 and 24) 29
28b. Stems divaricately branched, main leaves entire or obscurely toothed 30

29a. Outer bracts typically ovate-lanceolate, backs glabrous or puberulous towards the tips becoming shaggy in the lower half . . . . **23. P. cephalophorum**
29b. Outer bracts typically oblong to narrowly elliptic, puberulous on the backs, long hairs confined to margins . . . . **24. P. phyllopodioides**

30a. Leaves and bracts softly pubescent .... **25. P. pumilum**
30b. Leaves and bracts hispid .... **26. P. hispidulum**

Subgenus Phyllopodium

Group 1, subgroup 1A

1. Phyllopodium cuneifolium (L.f.) Benth. in Hook., Comp. Bot. Mag. 1: 375 (1836) & in DC., Prodr. 10: 352 (1846); Hiern in Fl. Cap. 4(2): 312 (1904). **Fig. 63A, Plate 1C.**
Type: C.B.S., *Thunberg* (holo. herb. LINN 787.10, iso. S).

Syn.: *Manulea cuneifolia* L.f., Suppl. 285 (1782); Thunb., Prodr. Pl. Cap. 101 (1800) & Fl. Cap. ed. Schultes 468 (1823).
    Polycarena cuneifolia (L.f.) Levyns in J. S. Afr. Bot. 5: 36 (1939).
    Selago herbacea Choisy in Mem. Soc. Phys. Genev. 2, 2: 110 (1823) & in DC., Prodr. 12: 17 (1848) excl. syn.;
    Rolfe in Fl. Cap. 5(1): 160 (1901), excl. syn. Type: C.B.S., herb. Burmann (G?, n.v.).

Annual herb, 45–400mm tall, stems simple to well branched from the base, lower branches often decumbent, base becoming woody, upper parts herbaceous, leafy, pubescent, the hairs 0.5–1mm long, often ± curled, retrorse. *Leaves* opposite, blade 3.5–30 × 3–20mm, broadly elliptic or ovate, contracted below into a petiolar part 1.5–20mm long, triplinerved, margins serrate, blade glandular-punctate, sparsely hairy chiefly on the margins, main veins and lower surface near the base. *Flowers* many in crowded heads rapidly elongating into long narrow dense racemes, only the lowermost flowers somewhat distant. *Bracts* 4.5–12 × 0.5–3mm, usually sharply differentiated from the leaves, spathulate, lowermost with 1–2 pairs of teeth, lower margins with patent hairs c.0.5–1mm long, glandular-punctate like the leaves, adnate to pedicel and calyx tube. *Pedicels* 0.25–1(–1.5)mm long. *Calyx* obscurely bilabiate, tube 1–1.5 (–2)mm long, anticous lobes 1.5–2 × 0.25–0.5mm, keeled and more or less conduplicate in upper part, margins with patent hairs 0.5–1mm long. *Corolla* tube 3–5 × 1–1.3mm, broadening in throat, limb (4–)5–7mm across lateral lobes, posticous lobes 1.2–2 × 0.75–1.3mm, anticous lobe 2.5–4 × 1–1.75mm, all lobes oblong, mauve, orange patch below posticous lip and bearded there, the hairs running about halfway down the tube inside, a few hairs often present at base of anticous lip. *Stamens* 4, anthers 0.75mm long. *Stigma* well exserted. *Capsules* 2.5–4 × 1.5–2mm. *Seeds* c.0.5–0.75mm long, irregularly colliculate, amber-coloured.

Selected citations:

Natal. 2931CC, Durban Flats, 50ft, 29 xi 1909, *Wood* 11543 (PRE, SAM).
Transkei. 3228CB, Kentani, coast, 50ft, 4 xii 1905, *Pegler* 1309 (PRE).
Cape. Komgha div., 3228CB, Kei Mouth, 200ft, i 1890, *Flanagan* 95 (BOL, SAM); Haga-Haga, 15 xii 1977, *Hilliard & Burtt* 11077 (E, PRE, S). Stutterheim div., 3227CB, Mt Kemp, 28 i 1979, *Hilliard & Burtt* 12423 (E). Victoria East div., 3226DD, Alice, 1700ft, 23 xi 1942, *Giffen* 1041 (PRE). East London div., 3327 BA, Kidd's Beach, 19 x 1980, *Hilliard & Burtt* 13185 (E, S). Peddie div., Hamburg, 2 ii 1957, *Taylor* 5623 (NBG). Albany div., 3326AC, road to Sidbury, x 1928, *Dyer* 1687 (PRE); 3326AD, near Seven Fountains, 17 i 1947, *Leighton* 2834 (BOL, PRE); 3326BC, Kap Valley, 4 xii 1977, *Hilliard & Burtt* 10857 (E, PRE, S). Alexandria div., 3326CB, Alexandria, 6 vi 1931, *Galpin* 10852 (BOL, PRE). Bathurst div., Bathurst, Hopewell Farm, 7 vii 1947, *Compton* 19817 (BOL, NBG). 3325DC, Port Elizabeth, vi 1871, *Bolus* 2235 (BOL); Uitenhage div., 3325CC, Vanstaadenshoogte, *Macowan* 2219 (SAM). Humansdorp div., 3324DD, near Gamtoos River, 2 xi 1950, *Maguire* 558 (NBG); 3424BB, Kabeljauws, viii 1928, *Fourcade* 3964 (BOL, NBG); 3424BA, Klipdrift, v–vi 1930, *Thode* A2500 (PRE).

Phyllopodium cuneifolium is a common weed along the southern and eastern Cape coast from Klipdrift and Humansdorp to the Kei river and inland to Grahamstown and Fort Beaufort, with a single collection from the coast near Kentani in Transkei, just north of the Kei, then a great disjunction to Durban, where Medley Wood collected it on the flats around Durban Bay in 1909; there is no modern collection from Natal and it may well have been an introduction there, just as it was surely introduced to the top of Mt Kemp (near Stutterheim, eastern Cape), where we found it flourishing on a heap of building sand. The plants favour damp sandy sites and may flower in any month.

The species is sometimes confused with the allied *P. bracteatum*, with which it is sympatric, but *P. bracteatum* differs in facies (very lax inflorescences with the lowermost flowers in the axils of the uppermost leaves with an imperceptible gradation to leaflike bracts, in contrast to the dense racemes of *P. cuneifolium* with a more or less abrupt transition from leaves to bracts; even when the lowermost flowers are widely spaced, as they sometimes are, the change from leaves to bracts is evident) as well as in floral detail: *P. cuneifolium* has bracts 0.5–3mm broad in the upper part (versus 2.5–9mm), anticous calyx lobes 1.5–2mm long (versus (2.25–)3–4mm), corolla limb (4–)5–7mm across the lateral lobes, and mauve (versus 3–4.5mm, and white), anticous lobe 2.5–4mm long (versus 1–2.1mm).

Selago herbacea (in synonymy above) proves to be *Phyllopodium cuneifolium*. Choisy, followed by Rolfe (who did not see the type) reduced *Selago lobeliacea* Hochst. (in Flora 28: 69, 1845) to synonymy under *S. herbacea*. But this cannot be; the type (*Krauss* 1094, which I have not seen) was collected on Table Mountain, soon after Krauss's arrival at the Cape, and well west of the area of *Phyllopodium cuneifolium*. Hochstetter's description of the leaves as 'obovato-lanceolatis sessilibus' is at variance with the long-petioled leaves of *P. cuneifolium*. It is possibly Choisy's own *Selago quadrangularis*!

2. Phyllopodium bracteatum Benth. in Hook., Comp. Bot. Mag. 1: 373 (1836) & in DC., Prodr. 10: 353 (1846); Hiern in Fl. Cap. 4(2): 313 (1904).
Lectotype (chosen here): [Cape, Port Elizabeth div., 3325BC] Enon, *Drège* b (K; isolecto. S, SAM, TCD, W).

Syn.: *Polycarena bracteata* (Benth.) Levyns in J. S. Afr. Bot. 5: 36 (1939).

Annual herb 50–300mm tall, stems simple to well-branched from the base, lower branches often decumbent, basal parts sometimes becoming woody, upper parts herbaceous, leafy, pubescent, the hairs c.0.5–1mm long, often ± curled, spreading to retrorse. *Leaves* opposite, blade c.7–35 × 5–23mm, broadly elliptic or ovate, contracted below into a petiolar part 3–20mm long, triplinerved, margins serrate, blade glandular-punctate, sparsely hairy chiefly on the margins, main veins and lower surface near the base. *Flowers* many in lax leafy racemes. *Bracts* 6.5–18 × 2.5–9mm, leaflike and grading imperceptibly from leaves to bracts, adnate to pedicel and calyx tube, lower margins with patent hairs 0.5–1mm long. *Pedicels* 0.25–1mm long. *Calyx* distinctly bilabiate, tube 0.75–1.5mm long, anticous lobes (2.25–)3–4 × 0.3–0.75mm, keeled, conduplicate near tips, margins with patent hairs 0.5–1mm long. *Corolla* tube 3–5 × 0.75–1.1mm, slightly broadened in throat, limb 3–4.5mm across the lateral lobes, posticous lobes 0.3–1.5 × 0.3–1mm, anticous lobe 1–2.1 × 0.75–1mm, all lobes oblong, white, orange patch at base of posticous lip and bearded there, the hairs running halfway down the tube inside, a few hairs often present at base of anticous lip. *Stamens* 4, anthers 0.3–0.6(–0.75)mm long. *Stigma* well exserted. *Capsules* 4–6 × 1.5–3mm. *Seeds* c.0.5mm long, irregularly colliculate, amber-coloured.

Selected citations:

Natal. Pietermaritzburg, 2930CB, 1968, *Garrett* 51 (E, NU). Durban distr., 2930DD, Isipingo Flats, 27 viii 1966, *Ward* 5813 (E, NU); Durban, 2931CC, 27 ix 1966, *Hilliard* 4002 (E, NU). Port Shepstone distr., 3030CD, St Michael's on Sea, 1 vii 1976, *Nicholson* 1543 (PRE); 3030DA, Southport, 25 ix 1974, *Nicholson* 1481 (PRE).
Transkei. Kabongaba, 26 x 1951, *Taylor* 3708 (NBG).
Cape. East London div., 3228CC, Kwelegha river drift, 26 xii 1900, *Galpin* 5799 (NBG, PRE). Fort Beaufort div., 3326BB, Ripplemead Estates NE of Breakfast Vlei, 12 xii 1977, *Hilliard & Burtt* 11005 (E, NU). Bathurst, 3326DB, Hopewell Farm, 7 vii 1947, *Compton* 19832 (NBG). Albany div., 3326BC, Grahamstown, 1800ft, x 1928, *Rogers* 28626 (PRE, SAM). Humansdorp div., 3324DD, Hankey, 50ft, x 1927, *Fourcade* 3317 (BOL). Uitenhage div., 3324DC, Winterhoek Mts, 12 ix 1930, *Fries et al.* 795 (PRE). Knysna div., 3422BB, Belvidere, 1926, *Duthie* 1045 (BOL); 3322CD, George, 18 iii 1893, *Schlechter* 2351 (BOL, E, PRE).

Phyllopodium bracteatum ranges from Pietermaritzburg and Durban through the coastal areas of Natal and Transkei to the Cape as far west as Mossel Bay and inland to the environs of Grahamstown and Fort Beaufort. It is a weed of damp sandy places, flowering in any month.

It is closely allied to *P. cuneifolium*; the differences between them are detailed under that species (above).

3. Phyllopodium diffusum Benth. in Hook., Comp. Bot. Mag. 1: 373 (1836) & in DC., Prodr. 10: 353 (1846); Hiern in Fl. Cap. 4(2): 314 (1904) p.p.
Type: Cape, near Uitenhage, Zwartkopsrivier, 1836, *Ecklon* s.n. (holo. K; iso. MO, S, SAM, TCD).

Syn.: *Polycarena diffusa* (Benth.) Levyns in J. S. Afr. Bot. 5: 37 (1939).

Annual herb 50–350mm tall, stems simple to well-branched from the base, lower branches often decumbent, base becoming woody, upper parts herbaceous, leafy, pubescent with retrorse hairs up to 0.75mm long, much shorter on upper parts, many tipped with an elliptic cell considerably larger than the stalk cells. *Leaves* opposite below, alternate above, blade of lower leaves c.4–18 × 2–10mm, smaller upwards, elliptic to rhomboid-ovate, contracted below into a petiolar part 2–10mm long, margins serrate, blade glandular-punctate, glabrous except for hairs on midline below and on petiolar part. *Flowers* many in crowded heads rapidly elongating into lax racemes. *Bracts* c.4.5–16 × 0.5–3mm, lowermost more or less leaflike becoming spathulate upwards, margins with patent hairs 0.3–0.5mm long, often with apical cell much enlarged, adnate to pedicel and calyx tube. *Pedicels* at base of inflorescences 1.5–4mm long, shorter upwards. *Calyx* obscurely bilabiate, tube 0.6–1mm long, anticous lobes 1.75–1.8 × 0.5–0.6mm, margins with patent hairs up to c.0.4mm long. *Corolla* tube 2.75–3.75mm long, 0.75–1mm diam. in lower part, quickly broadening upwards, limb 4.5–6mm across the lateral lobes, posticous lobes 0.75–1.5 × 0.8–1.3mm, anticous lobe 2.25–3 × 0.75–1.5mm, all lobes elliptic, white, violet outside at least in bud, orange patch at base of posticous lip with a violet patch below that in tube, bearded at base of posticous lip, the hairs running down into the tube, sometimes extending to lateral lobes as well. *Stamens* 4, anthers 0.75mm long, violet. *Stigma* well exserted. *Capsule* 2.5–4 × 1.75–2mm. *Seeds* 0.5–0.75 × 0.3–0.5mm, irregularly colliculate and wrinkled, amber-coloured.

Selected citations:

Cape. King William's Town div., 1300ft, *Tyson* 998 (E, SAM); ibidem, *Tyson* 846 (BOL, SAM, W). Uitenhage div., Uitenhage, 250ft, *Schlechter* 2511 (BOL, J, W, WU); Winterhoek Mts, Hellsgate Kloof, *Fries et al.* 889 (PRE).

Phyllopodium diffusum has been recorded from the environs of Uitenhage and Port Elizabeth, then about 150km away to the east north east at Peddie and King William's Town, a disjunction that is surely not real. It grows in sandy grassland, flowering between July and January.

The species is easily recognized by its peculiar hairs with a greatly enlarged apical cell, which occur particularly on the stems, inflorescence axes and pedicels, but often on the petiolar part of the leaves, the margins of the bracts and on the calyx tube as well. On the upper parts of the plant, these hairs may consist of no more than the relatively large apical cell and one very small stalk cell, but on the lower part of the stems and on the petiolar parts, the hairs may be much longer, the stalk cells sometimes moniliform.

Phyllopodium diffusum is most likely to be confused with *P. rustii* (below).

4. Phyllopodium rustii (Rolfe) Hilliard in Notes RBG Edinb. 45: 486 (1989).
Type: Cape, Riversdale, *Rust* 100 (holo. B†, iso. K).

Syn.: *Selago rustii* Rolfe in Fl. Cap. 5(1): 158 (1901).

Annual herb 70–200(–300)mm tall, stems simple to well branched from the base, lower branches often decumbent, base becoming woody in large plants, upper parts herbaceous, leafy, glandular-puberulous especially on upper parts, pubescent as well with retrorse eglandular hairs 0.5–1mm long. *Leaves* opposite below, alternate above, blade of lower leaves (2–3 pairs) c.7–15(–25) × 3–8(–10)mm, elliptic, rapidly smaller and narrower upwards, uppermost ± linear, petiolar part 3–8mm long, margins serrate, long patent hairs on petiolar part and over the veins on lower surface, upper surface glabrous or nearly so, occasionally whole leaf nearly glabrous. *Flowers* many in crowded heads rapidly elongating into somewhat lax racemes terminating the nude pedunculoid branch tips. *Bracts* 3.5–5 × 0.6–1mm, narrowly to more broadly spathulate, adnate to pedicel and calyx tube, margins with patent hairs 0.4–0.6mm long. *Pedicels* at base of racemes 1–1.5(–2)mm long, shorter upwards. *Calyx* obscurely bilabiate, tube 0.75–1.25mm long, anticous lobes 1.25–1.75 × 0.5–0.6mm, margins with patent hairs up to c.0.3mm long. *Corolla* tube 2–3mm long, 0.75–1mm diam. in lower part, quickly broadening upwards, limb (2.5–)3.25–5mm across the lateral lobes, posticous lobes 0.5–0.8 × 0.75–1mm, anticous lobe 1.5–3 × 0.75–1.5mm, all lobes elliptic, white or rarely mauve with orange-yellow patch at base of posticous lip, bearded there and short way down back of tube, hairs sometimes extending to lateral lobes. *Stamens* 4, anthers 0.5–0.75mm long. *Stigma* well exserted. *Capsules* 2.5–3 × 1.5–2mm. *Seeds* c.0.75 × 0.5mm, irregularly colliculate and wrinkled, amber-coloured.

Citations:

Cape. Swellendam div., 3320CD, below Langeberg Range at Clock Peaks, 800ft, *Wurts* 210 (NBG); 3420AA, Hessaqua's Kloof, *Zeyher* 3497 (S, SAM, W)*. Riversdale div., 3421AB, Riversdale, *Muir* 2565 (BOL); ibidem, *Bolus* s.n. (BOL); ibidem, *Rogers* 4347 (J); Oakdale, *Levyns* 3164 and 3545 (BOL); 3421BA, Great Vals River, *Burchell* 6549 (K). Mossel Bay div., 3422AA, Klein Berg near Mossel Bay, c.800ft, *Galpin* 4371 (PRE). Oudtshoorn div., 3322BC, De Rust, 1900ft, *Vlok* 2333 (E); ibidem, *Viviers & Vlok* 478 (E). George div., 3322CB, farm 'Die Krans', c.400m, *Dahlstrand* 2417 (J); 3322CD, hills E of Great Brak river, 150ft, *Fourcade* 4070 (BOL, K); 3322DC, western side of Kaiman's Gat, *Burchell* 5802 (K). Humansdorp div., 3324DD, west of Gamtoos river drift on road to Zuurbron, 150ft, *Fourcade* 4000 (BOL).

Rolfe himself discovered his mistake in describing this plant as a *Selago*, as shown by his annotation on the fragment of the type in the Kew herbarium, but he re-determined Rust's specimen as *Phyllopodium diffusum*. It is with this species that *P. rustii* has consistently been confused, but it is easily distinguished by its lack of the peculiar hairs that are a feature of *P. diffusum*. The stems of *P. rustii* are glandular-puberulous at least on the upper parts (*P. diffusum* is eglandular), and there are long acute retrorse hairs as well. Also, the racemes of *P. rustii* are borne on the long nude pedunculoid branch tips and the lowermost pedicels are mostly much shorter than those of *P. diffusum* (1–1.5(–2)mm versus 1.5–4mm), giving the two species a different facies.

The area of *P. rustii* lies mainly further west than that of *P. diffusum*, from about Swellendam to the Gamtoos river east of Humansdorp. The presence of *P. rustii* east of the Gamtoos needs confirmation: a specimen collected in Addo National Park (*Hall-Martin* 5955, PRE) may be *P. rustii* but it has far fewer long eglandular hairs than is usual.

Although most records are from the southern Cape coast below c.250m, there are several collections from the environs of Oudtshoorn at an altitude of c.600m, and one (*Allardice* 1548, PRE) from the Karoo Nature Reserve at Graaff Reinet, which is remarkable. On the Kamanassie Mountains near De Rust (3322DA), the flowers are mauve-blue, white in the throat and with the usual orange patch at the base of the posticous lip (*Vlok* 2507, E).

Ecological notes are scanty, but loamy sand and deep sand in disturbed Karoo vegetation near De Rust have been recorded; the plants flower between August and November.

* Zeyher used 3497 for several different species (*P. diffusum*, *P. bracteatum*, *P. cuneifolium*, *P. rustii*) collected at widely separated localities.

Group 1, subgroup 1B

5. Phyllopodium dolomiticum Hilliard in Notes RBG Edinb. 45: 484 (1989).
Type: Cape, Oudtshoorn div., Rust en Vrede, c.2500ft, 19 v 1959, *Acocks* 20463 (holo. PRE).

Perennial herb c.120–300mm tall, stems branched from the base, base woody, branches twiggy, leafy, glandular-puberulous, also with eglandular, retrorse, somewhat curved hairs c.0.4–0.75mm long, these more or less confined to two broad bands. *Leaves* opposite, lowermost c.7–8 × 3.5mm (including 2mm petiolar part), diminishing in size upwards, blade elliptic to ovate-elliptic tapering into the petiolar part, thick-textured, base expanded, half-clasping, margins with 2–4 pairs of teeth, with a few patent hairs c.0.5–1mm long, a few sometimes scattered on the leaf surface. *Flowers* few in somewhat lax racemes up to c.15mm long. *Bracts*: lowermost sometimes leaflike and then c.4.5 × 1.5mm, otherwise spathulate to linear-oblong, c.3–4 × 0.75–1.5mm, glabrous except for a few marginal hairs c.0.5mm long, adnate to pedicel and calyx tube. *Pedicels* up to 1mm long. *Calyx* obscurely bilabiate, tube c.0.75mm long, anticous lobes c.1.25 × 0.5mm, margins with short patent hairs otherwise glabrous. *Corolla* tube c.2.5–3 × 0.6–1mm, cylindric, widening in throat, limb c.3–3.25mm across lateral lobes, posticous lobes 0.75–1 × 0.75mm, anticous lobe 1.5–1.75 × 1mm, all lobes elliptic-oblong, white to light mauve-blue, yellow/orange patch at base of posticous lip, there either glabrous or bearded. *Stamens* 4, anthers c.0.5mm long, dark-coloured. *Stigma* well exserted. *Capsules* 2.5–3 × 1.5mm. *Seeds* c.0.75 × 0.5mm, colliculate, irregularly wrinkled, pale amber-coloured.

Citation:

Cape. 3322AD, lower southern foothills of Great Swartberg, near entrance to Rust en Vrede, 8 viii 1990, *Vlok* 2348 (E, K, MO, S).

Phyllopodium dolomiticum is known from only two collections: the type made near Oudtshoorn in '*Olea* scrub on dolomite hills, fairly frequent on N aspect' and the other 'in pockets of humic, loamy soil, between dolomitic boulders, on north-facing slope'. By mid-May, when the type collection was made, the plants were mostly in fruit; the second collection, made in August, has both old and young inflorescences in fruit as well as heads in full flower.

Phyllopodium dolomiticum appears to be without close allies, and it is placed for convenience next to *P. elegans*, which has been recorded in the same general area, but from which *P. dolomiticum* differs markedly in its twiggy habit, smaller and differently shaped leaves, few-flowered racemes and smaller flowers, with the base of the posticous lip either glabrous as in the type collection (which is unusual in this group), or bearded.

6. Phyllopodium elegans (Choisy) Hilliard in Notes RBG Edinb. 45: 484 (1989).
Type: Cape, Oudtshoorn div., Cango, *Mund* (holo. B†, iso. K).

Syn.: *Selago elegans* Choisy in DC., Prodr. 12: 14 (1848); Rolfe in Fl. Cap. 5(1): 160 (1901).
   Phyllopodium linearifolium H. Bolus in Trans. S. Afr. Phil. Soc. 16: 398 (1906). Type: Cape, near Prince Albert [3322AA], Zwartbergen, 850m, Dec., *Bolus* 12189 (holo. BOL; iso. BM, K).
   Polycarena linearifolia (Bol.) Levyns in J. S. Afr. Bot. 5: 37 (1939).

Annual(?) herb 150–450mm tall, stems simple to well-branched from the base, leafy becoming bracteate upwards, base woody, branches rod-like, glandular-puberulous in upper part, also with eglandular, retrorse, somewhat curved hairs c.0.25–0.5mm long, in the lower part these more or less confined to two broad bands. *Leaves* opposite and often quasi-fasciculate on lower part, becoming alternate and distant upwards, passing into bracts, up to c.18–35 × 1–3mm, linear to narrowly oblanceolate, apex acute, base narrowed then abruptly expanded and half-clasping, margins entire, minutely denticulate or with up to 4 pairs of small teeth, blade glabrous except for a few hairs on lower leaf margins. *Flowers* many in crowded heads rapidly elongating into long (up to 150mm) narrow crowded spikes, lowermost flowers sometimes distant. *Bracts* 4–5 × 0.8–1.5mm, lanceolate to elliptic-lanceolate, adnate to the short calyx tube, often glabrous on the back, sometimes minutely glandular-puberulous particularly near the base, always with eglandular marginal hairs mostly 0.75–1mm long. *Pedicels* 0–0.5mm. *Calyx* obscurely bilabiate, deeply lobed, tube 0.75–1mm long, anticous lobes 2.2–3.25 × 0.4–0.9mm, whole calyx minutely glandular-puberulous, margins with patent hairs c.0.5–0.75mm long. *Corolla* tube 3.5–5 × 0.75–1.2mm, broadening in throat, limb 4–5mm across the lateral lobes, posticous lobes 1–1.5 × 0.75–1.25mm, anticous lobe 1.75–2 × 0.75–1.6mm, all lobes elliptic-oblong, white, posticous lip heavily bearded inside, hairs often present on lower lip as well. *Stamens* 4, anthers 0.75–1mm long. *Stigma* well exserted. *Capsule* 3.5–4 × 1.75–2mm. *Seeds* c.1 × 0.5mm, obscurely trigonous, somewhat colliculate and wrinkled, pale amber.

Selected citations:

Cape. Prince Albert div., 3322AC, Swartberg Pass, *Stokoe* 8699 (BOL). Uniondale div., 3323AC, top of Prince Alfred's Pass, 3400ft, *Fourcade* 1296 (BOL, NBG, STE); 3323DA, foothills of Kouga Mts near Smutsberg, *Esterhuysen* 10759 (BOL, K, NBG). George div., 3323 DC, Keurboom's River, Long Kloof, 1800ft, *Fourcade* 4685 (BOL). Montagu div., 3320CC, Langeberg near Montagu, *Esterhuysen* 23852 (BOL, K, PRE, S). 3419BD, Caledon-Bredasdorp div. boundary, Sondag's Kloof between Stanford and Napier, c.900ft, *Taylor* 3669 (PRE, STE). Swellendam div., 3420AB, Bushman's River, *Walgate* 350a (BOL, NBG). Bredasdorp div., 3420BC, Potteberg, *Esterhuysen* 23200 (BOL, NBG).

It is fortunate that Rolfe was given part of the type of *Selago elegans*, and this is now preserved in the Kew herbarium. The ovary proves to be multiovulate and the plant is clearly conspecific with *Phyllopodium linearifolium*. The types of both names came from the Great Swartberg, *S. elegans* from near Cango, *P. linearifolium* from not far away, near Prince Albert. The species is distributed over the southern Cape mountains from about Montagu and Napier to Joubertina, embracing at least the Langeberg, Potteberg, Swartberg, Kouga mountains, Outeniqua mountains, and Lange Kloof mountains. Collector's notes are meagre, but Miss Esterhuysen recorded 'sandy slopes, common after fire'. All collectors give flower colour as 'white' but there is almost certainly a yellow patch on the posticous lip. Flowering has been recorded between September and December, also in May. Specimens in September already have long spikes of dehisced capsules as well as young flowering heads, as do those collected in May; it therefore seems likely that flowering takes place throughout the year. The plants have a simple woody taproot, and possibly persist for more than one season.

Phyllopodium elegans is closely allied to *P. multifolium* and is also frequently mistaken for *Selago*: see comments below under *P. multifolium*.

7. Phyllopodium multifolium Hiern in Fl. Cap. 4(2): 315 (1904). **Fig. 63C.**
Type: Cape, George div., Montagu Pass, 1200ft, *Young* in herb. Bolus 5527 (holo. K, iso. BOL).

Syn.: *Polycarena multifolia* (Hiern) Levyns in J. S. Afr. Bot. 5: 37 (1939).

Annual herb 80–370mm tall, stems simple to well-branched from the base, leafy becoming bracteate upwards, base woody, branches virgate, usually glandular-puberulous in upper part, or eglandular, retrorse, somewhat curved hairs c.0.25–0.5(–1)mm long predominating there as they do lower down the stem, where they are sometimes more or less confined to two broad bands. *Leaves* opposite and often quasi-fasciculate on lower part becoming alternate and distant upwards, passing into bracts, c.6–35 × 0.5–12mm, sometimes linear, often narrowly oblanceolate, the lowermost broadly oblanceolate to ovate and narrowed below into a distinct petiolar part, apex acute, base narrowed, margins usually obscurely to distinctly toothed, teeth in 3–5 pairs, blade usually glabrous except for patent hairs on margins and midrib on lower surface, sometimes glandular-puberulous as well on lower part. *Flowers* many, in crowded heads rapidly elongating into long narrow crowded spikes, lowermost flowers often distant. *Bracts* 4–5(–12) × 0.4–0.8–(1.5)mm, linear-lanceolate to narrowly spathulate, usually glandular-puberulous all over, or these hairs sometimes sparse, long (0.4–1mm) eglandular hairs as well particularly on margins, or these sometimes wanting, adnate to the short calyx tube. *Pedicels* 0–0.5mm. *Calyx* obscurely bilabiate, deeply lobed, tube 0.5–1mm long, anticous lobes 2–2.75 × 0.3–0.6mm, whole calyx glandular-puberulous, margins of lobes often with long (c.0.5mm) patent eglandular hairs as well. *Corolla* tube 3.75–5 × 0.75–1mm, broadening in throat, limb 4.5–7mm across the lateral lobes, posticous lobes 1.2–1.75 × 0.8–1.5mm, anticous lobe 2–3.2 × 1–2mm, all lobes elliptic-oblong, mauve or blue with a yellow patch at base of posticous lip, which is heavily bearded, hairs often on lower lip as well. *Stamens* 4, anthers 0.75–0.8mm long. *Stigma* well exserted. *Capsule* 2.5–3 × 1.5–2mm. *Seeds* 0.75–1 × 0.5mm, obscurely trigonous, somewhat colliculate and wrinkled, amber-coloured.

Selected citations:

Cape. George div., 3322CC, Ruiterbos State Forest, 1600ft, 16 ix 1990, *Vlok* 2405 (E); 3322CD, Montagu Pass, 2250ft, *Fourcade* 3255 (BOL, NBG, STE); 3322DC, Berg Plaats, 1000ft, *Fourcade* 3454 (BOL, PRE); 3422AB, Silver River, 6 xi 1894, *Penther* 3037 (W). Swellendam div., 3320DC, near Barrydale, *Esterhuysen* 24608 (BOL). Bredasdorp div., 3420BC, Potteberg, *Compton* 19512 (NBG, STE).

Phyllopodium multifolium, like its close ally *P. elegans*, is a plant of the southern Cape mountains. The type came from Montagu Pass through the Outeniqua mountains north of George; the species has also been collected to the west along the Langeberg, and on the Potteberg to the south. A collection made further to the north west (3319AA) in the Great Winterhoek Mountains (*Phillips* 1832, PRE) may be *P. multifolium*, but the largest leaves are c.20 × 8mm, broader in relation to length than is usual for the species. The plants grow in mountain scrub, flowering between September and November.

Phyllopodium multifolium and *P. elegans* grow together on the Potteberg, where the obvious difference in flower colour (mauve in *P. multifolium*, white in *P. elegans*) has induced collectors to take both. In this locality (and also further east, at Silver River) *P. multifolium* tends to have much broader basal leaves than *P. elegans* (up to 12mm broad, seldom exceeding 3mm in *P. elegans*). There are also differences in indumentum, *P. multifolium* being more glandular than *P. elegans* and this often shows up well on the bracts; the corolla limb of *P. multifolium* tends to be larger than that of *P. elegans* (4.5–7mm across versus 4–5mm), reflected in the size of the lobes (e.g. anticous lobe 2–3.2mm long in *P. multifolium*, 1.75–2mm in *P. elegans*).

Both species are readily mistaken for *Selago*, especially 'S. heterophylla E. Mey.' of Rolfe's account in Flora Capensis 5(1), but they are of course easily distinguished by their ovaries, multiovulate in *Phyllopodium* resulting in a many-seeded capsule, biovulate in *Selago*, which produces two 1-seeded cocci.

Group 1, subgroup 1C

8. Phyllopodium capillare (L.f.) Hilliard in Notes RBG Edinb. 46: 341 (1990).
Lectotype: C.B.S., sheet no. 14347 in herb. Thunberg (UPS).

Syn.: *Manulea capillaris* L.f., Suppl. 285 (1782); Thunb., Prodr. Pl. Cap. 101 (1800) & Fl. Cap. ed. Schultes 468 (1823).
    Polycarena capillaris (L.f.) Benth. in Hook., Comp. Bot. Mag. 1: 372 (1836) & in DC., Prodr. 10: 351 (1846); Hiern in Fl. Cap. 4(2): 329 (1904) p.p.
    P. parvula Schltr. in Bot. Jahrb. 27: 181 (1899); Hiern in Fl. Cap. 4(2): 332 (1904). Type: Cape, Clanwilliam div., in sandy places near Zuurfontein, 150ft, *Schlechter* 8534 (holo. B†; iso. BOL, E, K, W).

Annual herb c.50–260mm tall, stems simple to well branched from the base, base becoming woody, branches erect or decumbent, leafy, glandular-puberulous but these hairs often hidden by the dominant indumentum of very short (up to c.0.1mm) curved, retrorse eglandular hairs. *Leaves* opposite becoming alternate above, lowermost c.7–35 × 1.75–6(–9)mm (including petiolar part up to roughly ⅓ to ½ total length), diminishing in size upwards, blade elliptic to oblanceolate narrowed to the half-clasping base, margins often with 2–4 pairs of teeth, or sometimes entire or subentire, blade minutely puberulous on margins and midrib on lower surface, sometimes with longer hairs (to c.0.5mm) on lower margins. *Flowers* few to many in compact heads that rapidly elongate into racemes up to 40mm long, lowermost few flowers often distant, otherwise crowded, borne on nude peduncles terminating the branchlets. *Bracts* c.2.5–6 × 0.3–1.3mm, lowermost sometimes leaflike, otherwise oblong to linear, adnate to pedicel and calyx tube, sparsely puberulous especially on the lower margins and down the midline, short (c.0.2mm) patent hairs on lower margins. *Pedicels* up to 0.5–1.4mm long. *Calyx* bilabiate, tube 0.5–1.75mm long, anticous lobe 0.75–1.75 × 0.5–0.75mm, margins with short patent hairs otherwise glabrous or with minute glandular hairs on the tube, tips of lobes (and top of capsule) often flushed with purple. *Corolla* tube 2.5–3.5 × 0.75–1mm, widening in the throat, limb 1.8–5mm across the lateral lobes, posticous lobes 0.4–1 × 0.4–1mm, anticous lobe 0.75–2.5 × 0.5–1mm, all lobes elliptic-oblong, white, possibly with a yellow patch at base of posticous lip, which is usually bearded, sometimes glabrous in very small flowers. *Stamens* 4, anthers (0.1–)0.25–0.6mm long. *Stigma* well exserted but often curled back. *Capsules* 2–3.5 × 1.5–2.5mm. *Seeds* c.0.5 × 0.3mm, colliculate and irregularly wrinkled, amber-coloured.

Selected citations:

Cape. Malmesbury div., 3218CD, Kotze, *Lewis* 1555 (SAM). Piquetberg div., 3218DC, Berg River near Piquetberg, 66m, *Schlechter* 5271 (BOL, E, J, PRE, STE, UPS). Malmesbury div., 3317BB, Saldanha Bay, *Parker* 4622 (BOL, NBG). Hopefield div., 3318AB, 8.8km WNW of Hopefield, *Acocks* 24379 (PRE); ibidem, *Schlechter* 5295 (BM, BOL, E, J, K, S, STE, UPS). Malmesbury div., 3318CB, near Bokbaai, *Esterhuysen* 3838 (BOL). Cape Town div., 3318CD, near Cape Town, Blumendahl, *Bolus* 659 (BOL, E, SAM, UPS, W). Bredasdorp div., 3420DC, Potteberg, *Esterhuysen* 23225 (BOL, PRE).

Schlechter remarked of his new species, *Polycarena parvula*, that it was distinguished from *P. capillaris* by its much smaller flowers. The corolla limb in Schlechter's specimens measures c.2mm across the lateral lobes, the posticous lip may be glabrous or have but a few hairs, and the anthers are also very small, with one pair only just exserted from the mouth. Other specimens over the geographical range of the species show similar traits, which probably indicate selfing. Despite variation in the size of the flowers, *Phyllopodium capillare* is easily distinguished from its allies by the very short ± curled white hairs thickly clothing the stems and inflorescence axes and more or less masking the tiny glandular hairs that are also present. There appears to be a local strain on the Cape Peninsula (e.g. *Bolus* 659, cited above) with very leafy stems, which appear to be prostrate; the typical plant is also on the Peninsula, and although the two differ in facies, I can find nothing else to distinguish them.

Phyllopodium capillare has a relatively wide geographical range from about Lambert's Bay (west of Clanwilliam) south to the Peninsula and east to Albertinia, always in sandy or stony places, possibly never more than about 120m above sea level. It flowers between July and October, principally in September.

9. Phyllopodium caespitosum Hilliard in Notes RBG Edinb. 45: 484 (1989).
Type: Cape, Clanwilliam div., 3219AD, S Cedarberg, Wolfberg, 4000ft, 3 x 1952, *Esterhuysen* 20574 (holo. BOL; iso. K, PRE).

Annual herb, stems c.10–50mm long, caespitose, erect, decumbent or spreading, leafy, pubescent with more or less patent eglandular hairs up to 0.4mm long as well as very small (c.0.1mm) retrorse hairs. *Leaves* opposite below, alternate above, blade of lowermost few pairs c.3–12 × 1.75–7mm, elliptic, tapering to a petiolar part c.1.5–15mm long, upper leaves smaller and becoming ± sessile, margins obscurely to distinctly serrate, both surfaces thinly pubescent at first, hairs later confined to margins and midline, up to 0.5–0.75mm long. *Flowers* up to c.15 crowded in heads rapidly elongating into racemes, lowermost flowers in largest racemes distant. *Bracts* c.3–6 × 1.2–2.25mm, upper ones smaller, broadly to narrowly elliptic-oblong, adnate to pedicel and calyx tube, upper surface puberulous, lower glabrous, hairs on margins up to 0.4–1mm long. *Pedicels* of lowermost flowers 0.5–1mm long, shorter upwards. *Calyx* obscurely bilabiate, tube c.0.75–1mm long, anticous lobe c.1.25–1.6 × 1mm, puberulous mainly on keels and margins with patent eglandular hairs up to c.0.4mm long, lobes (and top of capsule) often flushed purple. *Corolla* tube c.1.25–2 × 1mm, limb c.1.75–2.5mm across the lateral lobes, posticous lobes c.0.25–0.3 × 0.25–0.3mm, anticous lobe c.0.8–1 × 0.5–0.75mm, all lobes elliptic-oblong, white or cream, posticous lip bearded at the base and possibly with a yellow patch there. *Stamens* 4, anthers 0.25–0.3mm long. *Stigma* exserted. *Capsules* 2.5–3 × 2–2.25mm. *Seeds* c.0.5–0.75 × 0.3–0.5mm, irregularly colliculate and wrinkled, amber-coloured.

Citations:

Cape. Clanwilliam div., 3219AD, Cedarberg, summit of Tafelberg, 6400ft, *Esterhuysen* 13045 (BOL, NBG, PRE); 3219AC, Cedarhoutkop, 4000ft, *Esterhuysen* 35510 (BOL). Ceres div., Baviaansberg, 6200ft, *Esterhuysen* 29837a (BOL); 3319BC, Roodeberg, 5000ft, *Esterhuysen* 29722 (BOL, PRE). Worcester div., 3319DA, Keeromsberg, 6800ft, *Esterhuysen* 9246 (BOL); ibidem, c.5500ft, *Esterhuysen* 33676 (BOL); 3319CB, near Fonteintjiesberg, ridge above Meiring's Plateau, ?5800ft, *Esterhuysen* 30421 (BOL, S).

Phyllopodium caespitosum is on the mountains from the Cedarberg near Clanwilliam south to the mountains around Ceres and Worcester. It grows in shallow rock basins or on rock ledges between c.1200 and 2075m above sea level, flowering in September and October, fruiting only by November. Yet again, Miss Esterhuysen's love of mountains has led to the discovery of an interesting plant.

Its densely tufted habit gives *P. caespitosum* a distinctive facies, different from that of its congeners, which are much more sparingly branched. It is possibly allied to *P. micranthum* from which it differs not only in habit but also in its indumentum (stems with eglandular hairs only, up to c.0.4mm long, not both glandular and eglandular hairs c.0.1mm long) and in its generally broader bracts (1.2–2.25mm versus 0.5–1(–1.5)mm).

10. Phyllopodium micranthum (Schltr.) Hilliard in Notes RBG Edinb. 45: 485 (1989). **Fig. 63E.**
Type: Clanwilliam div., 3219AC, Ezelbank, 4500ft, *Schlechter* 8817 (holo. B†; iso. BOL, E, K, S, W).

Syn.: *Polycarena rariflora* var. *micrantha* Schltr. in Bot. Jahrb. 27: 181 (1899); Hiern in Fl. Cap. 4(2): 331 (1904).

Annual herb c.40–150mm tall, stem simple to sparingly branched from the base, always sparingly branched above except in very small plants, moderately leafy, glandular-puberulous on upper parts, hairs patent, c.0.1mm long, lower 2–3 internodes with equally small, retrorse, more or less curved eglandular hairs. *Leaves* opposite below, alternate above, lowermost c.7–16 × 1.75–4mm, oblanceolate to elliptic, narrowed to a petiolar part up to ⅓ total leaf length, diminishing rapidly in size upwards, uppermost narrowly elliptic to linear, margins entire to obscurely or sharply toothed, both surfaces sparsely puberulous, lower more densely than upper, hairs prominent on margins. *Flowers* up to c.12 in slender racemes terminating short bare peduncles, lowermost flowers in largest racemes usually distant. *Bracts* 1.75–4(–7) × 0.5–1(–1.5)mm, mostly oblong, adnate to pedicel and calyx tube, minutely puberulous on inner face otherwise glabrous except for marginal hairs 0.3–0.75mm long. *Pedicels* up to 0.5–1mm long. *Calyx* obscurely bilabiate, tube 0.75–1.4mm long, puberulous and with sessile glands as well, anticous lobes 1–1.5 × 0.5–0.75mm, often flushed with purple particularly on keels and below sinuses, small patent hairs on margins. *Corolla* tube 1.5–2.5 × 0.75–0.8mm, broadening in throat, limb (1–)1.5–2.5mm across the lateral lobes, posticous lobes c.0.5 × 0.4–0.5mm, anticous lobe 0.75–1 × 0.5–0.75mm (lobes sometimes reduced to 3 or 4), all lobes elliptic-oblong, white or cream, a few hairs at base of posticous lip or hairs wanting. *Stamens* normally 4, all exserted, sometimes 2 exserted, 2 in mouth, or showing various degrees of reduction to 2 fertile stamens, exserted, 1 or 2 staminodes or stamens with reduced anthers, these either in mouth or included, anthers up to 0.3mm long. *Stigma* exserted but often curled back. *Capsules* 2–3 × 2mm, often flushed purple at apex. *Seeds* c.0.5 × 3mm, colliculate, irregularly wrinkled, amber-coloured.

Citations:

Cape. Clanwilliam div., 3219CB, south Cedarberg, Krom River, *Esterhuysen* 20469a (BOL). Ceres div., 3319AB, 7 miles W of Gydouw Pass, *Hutchinson* 1049 (K); 3319AD, E foot of Schurweberg, *Esterhuysen* 20626 (BOL). Jakhals Nest, *Compton* 16233 (NBG). Laingsburg div., 3320BC, Anysberg Nature Reserve, Farm Tap Fontein 260, *Lloyd* 1121 (STE).

When Schlechter erected the varietal name *micrantha* he remarked that it was not impossible that his plant would deserve specific rank when more material had been investigated, but he could at that time see no worthwhile difference from *Polycarena rariflora* except the much reduced corolla. *Polycarena rariflora* is however a true *Polycarena* with smooth markedly trigonous seeds, while Schlechter's plant is a *Phyllopodium* with warty seeds and I have no hesitation in raising it to specific rank. The smaller flowers often have reduced stamens and this is almost certainly associated with selfing; *Polycarena rariflora* exhibits cleistogamy associated with great reduction in flower size, but Schlechter was clearly unaware of this.

Phyllopodium micranthum may at once be distinguished from *P. capillare*, to which it seems allied, by its different indumentum: the upper part of the stem is clad in small patent glandular hairs to the total or almost total exclusion of retrorse eglandular ones, while the entire stem of *P. capillare* is clad in this type of hair, more or less masking tiny glandular hairs that may also be present. There are other differences too: in *P. micranthum*, the lower leaf surface in particular is puberulous while the hairs are confined to the midrib and margins in *P. capillare*, and the corolla tube of *P. micranthum* is mostly shorter (1.5–2.5mm versus 2.5–3.5mm). *Phyllopodium capillare* is a plant of the lowlands; *P. micranthum* is a plant of the inland plateau: it has been recorded from the Cedarberg south through the Cold Bokkeveld to the Skurweberg and east to the Anysberg. It grows in sandy rocky places or in rock crevices on cliff faces, flowering in September and October.

11. Phyllopodium viscidissimum Hilliard in Notes RBG Edinb. 45: 486 (1989).
Type: Cape, Ceres div., 3319BA, Baviaansberg, 6200ft, 4 xi 1962, *Esterhuysen* 29837 (holo. BOL).

Diminutive annual herb 30–55mm tall, stem simple to branched from the base, sparsely leafy, glandular-puberulous, hairs patent, up to 0.25mm long, eglandular hairs as well, but often sparse. *Leaves* opposite below, alternate above, blade of lowermost 1–2 pairs c.5–7 × 2–4mm, elliptic tapering to a petiolar part c.3.5–5mm long, upper leaves smaller, linear-lanceolate, margins entire to sparsely toothed, glandular-puberulous mainly on margins. *Flowers* 2–6 in crowded heads sometimes elongating into racemes, then lowermost flowers often distant. *Bracts* up to c.2.5–3 × 0.5mm, oblong, adnate to pedicel and lower part of calyx tube, glandular-pubescent on margins and inner face, patent eglandular hairs to c.0.5mm on lower margins. *Pedicel* up to c.0.4mm long. *Calyx* obscurely bilabiate, tube c.1mm long, anticous lobes c.1.2 × 0.6mm, whole calyx puberulous with glandular and eglandular hairs, lobes often flushed with purple. *Corolla* tube c.1.8 × 0.8mm, limb c.2mm across the lateral lobes, posticous lobes c.0.4 × 0.3mm, anticous lobe c.0.75 × 0.4mm, all lobes white, patch of minute hairs at base of posticous lip. *Stamens* 4, anthers c.0.25mm long. *Stigma* exserted. *Capsules* c.2.5 × 2mm. *Seeds* c.0.75 × 0.5mm, colliculate and irregularly wrinkled, amber-coloured.

Phyllopodium viscidissimum is known only from the type collection. It was growing 'on ledges at base of high cliffs, south aspect', and Miss Esterhuysen collected it together with *P. caespitosum*, under the same number, and segregated them only later. The two species differ in habit, as well as in indumentum, and *P. viscidissimum* dries darker than *P. caespitosum*. In November, when Miss Esterhuysen collected them together, *P. caespitosum* was only beginning to flower while *P. viscidissimum* was already mostly in fruit. Several of the plants on the type sheet are galled at the base, disguising the true habit of the plant.

The relationship of the species probably lies with *P. micranthum* from which it can be distinguished by its stems with longer glandular hairs extending to the base of the plant (not the lowermost 2–3 internodes with short (0.1mm) eglandular hairs only), and by the glandular-pubescent bracts, which are possibly mostly narrower than those of *P. micranthum*.

12. Phyllopodium tweedense Hilliard in Notes RBG Edinb. 45: 486 (1989).
Type: Cape, Laingsburg div., 3320AB, Tweedside, 27 ix 1951, *Compton* 22850 (holo. NBG).

Annual herb c.50–155mm tall, stem simple to sparingly branched from the base or higher, moderately leafy, glandular-puberulous (hairs c.0.1mm) on peduncles and uppermost nodes, lower down replaced by tiny eglandular retrorse hairs, lowermost 1 or 2 nodes with patent glandular hairs up to c.0.5mm long. *Leaves* opposite below, alternate above, lowermost 1 or 2 pairs c.5–16 × 2.5–4mm, ⅓ to ½ the length petiolar, blade narrowly to broadly elliptic, tapering at the base, diminishing rapidly in size upwards, uppermost linear, margins entire to few-toothed, whole leaf glandular-puberulous when young, soon glabrescent except for hairs up to c.0.5mm long on margins and midrib below. *Flowers* up to c.20 in small heads soon elongating into short (up to c.30mm) racemes, lowermost flowers somewhat distant, otherwise crowded. *Bracts* c.3.5–5.5 × 0.4–1mm, oblong, adnate to pedicel and calyx tube nearly or quite to apex, glabrous except for patent marginal hairs up to 0.5mm long. *Pedicels* up to 0.5–1mm long. *Calyx* obscurely bilabiate, tube 1.3–1.5mm long, puberulous with tiny eglandular hairs, gland-dotted as well, anticous lobes 1.25–1.3 × 0.75–1mm, small patent hairs on margins. *Corolla* tube 2.75–3.5 × 0.75–1mm, widening slightly in mouth, limb c.5mm across the lateral lobes, posticous lobes c.1–1.5 × 1mm, anticous lobe 2–2.75 × 1.25–1.5mm, all lobes elliptic, white, possibly with a yellow patch at base of posticous lip, bearded there. *Stamens* 4, anthers 0.5–0.75mm long, dark-coloured. *Stigma* well exserted. *Capsules* c.3 × 2mm. *Seeds* c.0.75 × 0.5mm, colliculate and irregularly wrinkled, amber-coloured.

Citations:

Cape. Laingsburg div., 3320AB, Cabidu, *Compton* 12103 (NBG); ibidem, Cobita [orthographic variant?], *Compton* 3762 (NBG); Tweedside, 3200ft, *Compton* 3013 (BOL). Ceres div., 3219DC, Zwartruggens, near Stompiesfontein, Groenfontein, 2 x 1991, *Bean & Trinder-Smith* 2701 (BOL, E).

Phyllopodium tweedense is known with certainty only from a small area around Tweedside, near Laingsburg; the specimen from Zwartruggens, cited above, has more glandular leaves and bracts. It flowers in September and October, but little is known of its ecology: 'open sandy slopes' were recorded at Groenfontein. It is allied to *P. micranthum*, which grows on the mountains to the west of Laingsburg, and can be distinguished by the relatively long (up to 0.5mm) patent glandular hairs at the base of the stem (tiny retrorse eglandular hairs in *P. micranthum*), and by its larger flowers with prominent dark anthers (anthers very small and yellow in *P. micranthum*).

Group 2, subgroup 2A

13. Phyllopodium heterophyllum (L.f.) Benth. in Hook., Comp. Bot. Mag. 1: 373 (1836) & in DC., Prodr. 10: 352 (1846); Hiern in Fl. Cap. 4(2): 316 (1904) p.p. **Fig. 63B.**
Type: Cape, Zwartland, *Thunberg* (sheet 14363 in herb. Thunb., UPS).

Syn.: *Manulea heterophylla* L.f., Suppl. 285 (1782); Thunb., Prodr. 101 (1800) & Fl. Cap. ed. Schultes 469 (1823).
   M. capitata L.f., Suppl. 286 (1782); Thunb., Prodr. 101 (1800) & Fl. Cap. ed. Schultes 469 (1823). Lectotype (chosen here): Cape, *Thunberg* (sheet no. 14349, UPS, bottom specimen).
   Phyllopodium capitatum (L.f.) Benth. in Hook., Comp. Bot. Mag. 1: 373 (1836) & in DC., Prodr. 10: 352 (1846); Hiern in Fl. Cap. 4(2): 316 (1904) p.p. and excl. syn. *Selago cordata* Thunb.
   Polycarena capitata (L.f.) Levyns in J. S. Afr. Bot. 5: 37 (1939) & in Adamson & Salter, Fl. Cape Penins. 714 (1950) (as *P. heterophylla*).
   P. heterophylla (L.f.) Levyns in J. S. Afr. Bot. 5: 37 (1939) & in Adamson & Salter, Fl. Cape Penins. 714 (1950) (as *P. capitata*).

Annual herb, stems 50–300mm long, simple and then erect or branched from the base, branches decumbent, ascending, or sometimes prostrate, pilose, hairs mostly 0.75–1.5mm long, retrorse, sometimes appressed on upper parts, tending to spread on lower parts, sometimes glandular-puberulous as well, these hairs very small, patent, leafy at least below, sometimes becoming scapose upwards. *Leaves* opposite below, alternate above, blade of lowermost few pairs (8–)13–34 × (2–)4–8(–12)mm, elliptic tapering ± gradually to a petiolar part c.3–15(–24)mm long, diminishing in size (particularly width) upwards, becoming narrowly elliptic to linear and sessile, margins distinctly to obscurely serrate, upper surface thinly pilose, hairs on lower surface nearly confined to midrib, hairs c.0.5–1.5mm long. *Flowers* up to c.9(–15) in small heads elongating into short racemes (the lowermost flowers then often distant), solitary at tips of nude or bracteate peduncles. *Bracts* c.3.5–9 × 0.5–1.3mm, more or less oblong or subspath-ulate, adnate to pedicel and calyx tube, glabrous or nearly so on the backs, thinly pilose on inner surface, long (1–2mm) patent hairs on margins. *Pedicels* at base of inflorescences 0.5–1.25(–2)mm long, shorter upwards. *Calyx* obscurely bilabiate, tube 0.75–1mm long, anticous lobes 2–3 × 0.5–1mm, pilose, hairs c.0.5–1mm long, mainly on margins and keels. *Corolla* tube c.2.25–4 × 1–1.5mm, limb 3.25–7mm across the lateral lobes, posticous lobes 0.4–1.5 × 0.4–1.3mm, anticous lobe 1.3–3.25 × 0.75–1.75mm, all lobes elliptic-oblong, probably always creamy-coloured to pale yellow, base of posticous lip and inside of tube marked with patches of orange or yellow, outside of lobes often with dark blotching or feathering, base of posticous lip bearded, hairs sometimes extending to 2 or 3 of the anticous lobes. *Stamens* 4, base of anticous filaments sometimes bearded, with or without an arc of hairs linking bases of filaments, anthers 0.4–1mm long, yellow or orange. *Stigma* exserted. *Capsules* 3.5–5.5 × 2–3mm. *Seeds* c.0.5 × 0.3mm, obscurely angled and colliculate, mauveish underlaid by the coppery-coloured endosperm.

Selected citations:

Cape. Van Rhynsdorp div., 3118DA, 8 miles north of Van Rhynsdorp, *Wilman* 709 (BOL, PRE). Clanwilliam div., 3218AD, Elandsbaai, *Batten* 827 (E); 3218 BC, Lange Vallei, *Compton* 5533 (NBG); 3218BD, Blaauwberg, 2500ft, *Schlechter* 8445 (E, PRE, S, W); 3218DB, Piqueniers Kloof, 700ft, *Schlechter* 10747 (BOL, E, PRE, S, W). Piquetberg div., 3218BC, 21 miles N of Piquetberg on Het Kruis road, *Barker* 6358 (NBG). Malmesbury div., 3318AD, south west of Darling, *Hugo* 2424 (PRE); 3318BC, Vredenburg, *Compton* 15887 (NBG). Ceres div., 3319AC, Elands Kloof, *Compton* 16180 (NBG).

Typification of the names *Manulea capitata* and *M. heterophylla* has proved difficult. Both names were based on Thunberg's specimens and the only ones now extant are in UPS with some duplication in S. There are two sheets in Thunberg's herbarium labelled *Manulea capitata*: sheet no. 14350 proves to be *Polycarena aurea*; sheet no. 14349 is a mixture of *Phyllopodium capillare* (top right hand specimen) and *Manulea capitata* (top left and bottom). Linnaeus the younger's description (*foliis ovatis serratis villosis, capitulis globosis, ramis diffusis*) seems to relate to all three specimens on sheet 14349: the top right hand specimen has decidedly ovate leaves and globose heads; in the other two specimens, the leaves are more elliptic than ovate, but the heads are ± globose and the bottom specimen is diffuse. I therefore propose to lectotypify the name by this last specimen.

Sheet no. 14363 is labelled *M. heterophylla*; it comprises three small plants, which are conspecific. Again, Linnaeus the younger's description (*foliis linearibus sparsis villosis integ-ris dentatisve*) does not fit the material particularly well: only the scattered cauline leaves are linear; the radical leaves are elliptic-ovate and the toothing is very obscure. Nevertheless, in the absence of any other material, this should be accepted as the type.

Manulea heterophylla and *M. capitata* are conspecific. I elect to use the name *M. heterophylla* because Bentham, who transferred the epithet *capitata* to *Phyllopodium*, introduced a further complication by quoting under that name an Ecklon specimen that proves to be *Selago cordata* (that is, *Phyllopodium cordatum*), and a Drège specimen that proves to be *P. cephalophorum*. Since then, the name *P. capitatum* has frequently been misapplied.

It seems appropriate to consider the name *Selago cordata* here. It is typified by sheet no. 13877 in Thunberg's herbarium. Rolfe (1883, pp.354, 358) referred it to *Phyllopodium heterophyllum*; Hiern (1904, p.316), referred it to *P. capitatum*. However, both were mistaken; *Selago cordata* is a species distinct from, though closely allied to, *Phyllopodium heterophyllum*, and is enumerated below.

Phyllopodium heterophyllum ranges from about Van Rhynsdorp south to the Cape Peninsula and east to about the Breede River in Bredasdorp division, and inland as far as the Hex River valley and Montagu, up to c.750m above sea level. It grows on sandy flats and slopes, flowering mainly in August and September, in fruit by October. Collectors have recorded no more of flower colour than 'whitish', 'yellow' or 'pale yellow', once 'yellow with brown stripes'. However, one specimen (*Batten* 827, E) was fresh enough to show a complicated colour patterning: the corolla basically pale yellow with an orange blotch at the base of the posticous lip, with four small orange blobs, two below the sinuses between the posticous and anticous lips, two deep inside the tube at the base of the posticous filaments; below the insertion of the anticous filaments, the whole tube is flushed light purplish-pink. On the outside of the corolla there are three lozenge-shaped purplish patches on the midline below the three anticous lobes. As the flower fades, the whole corolla turns orange, with some purpling of the anticous lobes. In some specimens (14 out of 37 examined) dark feathering, streaking and blotching is visible on the outside of the anticous lobes. It seems likely that some form of dark patterning is always present on these three lobes, although it is often destroyed in the drying process (the purple colour seems to be a pigment in the cell sap, while the orange pigment on the posticous lip is probably in the plastids and remains visible in all but the most ill-dried specimens).

The specimens seen fall into two groups: those with hairs at the base of the anticous filaments, those without. The flowers with hairy filaments tend to be larger than those without (e.g. tube mostly 3–4mm long, limb 4.5–7mm across the anticous lobes versus tube 2.25–3mm, limb 3.25–5mm). There is no geographical separation of the two forms and it is likely that the smaller flowers without hairs at the base of the anticous filaments have a different pollination biology from those with larger flowers with hairs at the base of the filaments; these hairs often extend in an arc between the filaments.

Phyllopodium heterophyllum bears a strong superficial resemblance to *P. cordatum*, from which it can most easily be distinguished by its leaves and bracts, both glabrous or nearly so on the backs (hairs are present on the margins) whereas in *P. cordatum* the leaves and bracts are pilose on both surfaces. In large plants of *P. heterophyllum* the side branches are strongly decumbent or prostrate, whereas they may always be erect in *P. cordatum* (this needs confirmation); also, the stems of *P. heterophyllum* tend to be more leafy than those of *P. cordatum*, the leaves are mostly elliptic tapering gradually at both ends (generally ovate and abruptly contracted at the base in *P. cordatum*, though this does not always hold, just as leaves in *P. heterophyllum* can occasionally be ovate). The flowers of *P. heterophyllum* always have an orange or yellow blotch at the base of the posticous lip and are further patterned with purplish or brownish lines or blotches, whereas in *P. cordatum* the corolla seems to lack any patterning. There may be a constant difference too in basic flower colour, that of *P. heterophyllum* being shades of yellow, that of *P. cordatum* pale cream or white. The ripe seeds of *P. heterophyllum* are a peculiar shade of mauve overlying amber, while those of *P. cordatum* are creamy-coloured. This may be a real difference, although it is commonplace for unripe seeds that will eventually turn blue or mauve to be cream-coloured at first.

14. Phyllopodium cordatum (Thunb.) Hilliard in Notes RBG Edinb. 46: 341 (1990).
Type: Cape, *Thunberg* (sheet no. 13877 in herb. Thunb., UPS, probable isotype S).

Syn.: *Selago cordata* Thunb., Prodr. 100 (1800) & Fl. Cap. ed. Schultes 464 (1823).

Annual herb, stems c.50–300mm long, simple to branched from the base, first stem erect, others decumbent, pilose, the hairs 0.5–1.5mm long, retrorse to appressed on upper part, sometimes tending to spread on lower part, often glandular-puberulous as well on upper part, mostly leafy only in lower part. *Leaves* opposite or uppermost alternate, blade of lowermost few pairs 6–20(–30) × 3–12mm, ovate to ovate-elliptic, abruptly or sometimes more gradually contracted into a petiolar part (2–)3–10(–13)mm long, smaller upwards and then sometimes narrowly elliptic or linear, margins obscurely to distinctly serrate, both surfaces pilose, hairs c.1–2mm long. *Flowers* up to c.20 in small heads elongating into crowded racemes (lowermost flowers sometimes distant) solitary at the tip of nude scapes. *Bracts* c.3–8 × 0.6–1.5(–3)mm, elliptic, oblong or subspathulate, adnate to pedicel and calyx tube, pilose, hairs 0.5–2mm long, longest hairs on lower margins. *Pedicels* at base of inflorescence c.0.5–1mm long. *Calyx* obscurely bilabiate, tube 0.5–1mm long, anticous lobes 1.5–2.75 × 0.4–0.75mm, pilose all over, hairs c.0.4–1mm long. *Corolla* tube 2–3 × 0.75–1mm, limb 1.5–4.5mm across the lateral lobes, posticous lobes 0.4–1 × 0.4–1mm, anticous lobe 1–2 × 0.4–1.25mm, all lobes elliptic oblong, palest cream-colour or white, possibly without any colour patterning other than the 2 upper lobes yellow, base of posticous lip bearded, the hairs sometimes extending to 2 or all 3 lobes of the anticous lip. *Stamens* 4, base of anticous filaments sometimes bearded with some hairs on the tube between the filaments, anthers 0.3–0.6mm long, yellow. *Stigma* exserted. *Capsules* c.3.5 × 2.5mm. *Seeds* c.0.5–0.75 × 0.3–0.5mm, obscurely angled, irregularly colliculate and wrinkled, pale cream-colour.

Selected citations:

Cape. Malmesbury div., 3318AB, Hopefield, 50m, *Schlechter* 5297 (BOL, E, PRE, S, UPS); 3318CB, Mamre, Groenekloof, 300ft, *Bolus* 4315 (BOL). Cape Town div., 3318CD, Kirstenbosch N of Window stream, *Esterhuysen* 23225 (BOL); 3318DC, Langverwacht above Kuils River, 725ft, *Oliver* 4673 (PRE); 3418AB, slopes of mountains above Clovelly, c.400ft, *Goldblatt* 6956 (PRE); between Fernwood and Liesbeek river, *Barker* 7552 (NBG). Caledon div., 3419AB, Zwartberg, *Hilliard & Burtt* 13069 (E, NU).

Phyllopodium cordatum ranges from the neighbourhood of Van Rhynsdorp south to the Peninsula and east to about Albertinia, and inland as far as the Cedarberg and the Hex River valley; this is also the distributional range of its ally, *P. heterophyllum*. It grows in sandy or stony places, up to c.750m above sea level, and flowers between July and October, mainly in September. It is adventive in W Australia (Cunderdin, *Whisson* 1 and 2 (K) and Northam, *Peirce* s.n., K).

Phyllopodium cordatum typically has a few pairs of ovate petiolate leaves confined to the lower part of the stem, the upper part running out into a long nude scape bearing the terminal inflorescence. Sometimes the leaves tend to be more elliptic than ovate, and sometimes they may be more numerous and run up the scape, diminishing rapidly in size. The leaves and bracts are pubescent on both surfaces and this, together with the pale flowers devoid of colour patterning as well as the creamy-coloured seeds, will distinguish the species from *P. heterophyllum* (see also under that species).

Phyllopodium cordatum is easily confused with *P. mimetes*, which also has the leaves crowded near the base of the plant and solitary heads on long peduncles; discriminating characters are given under *P. mimetes* (below).

15. Phyllopodium mimetes Hilliard in Notes RBG Edinb. 45: 485 (1989).
Type: Cape, Piquetberg div., 3218DA, near Het Kruis, 22 ix 1940, *Compton* 9518 (holo. NBG).

Annual herb c.90–200mm tall but will flower as a tiny seedling, stem simple to very sparingly branched from the base, outer branches erect or decumbent, terminating in long nude peduncles, each bearing a solitary head, pubescent with retrorse ± appressed hairs up to c.0.5mm long, glandular-puberulous as well on upper parts, leafy only at base. *Leaves* opposite becoming alternate and bracteate upwards, main leaves crowded at base, only 2–3 pairs, blade c.5–10 × 3–5mm, ovate to elliptic tapering into a petiolar part c.3.5–9mm long, this about equalling or up to twice as long as blade, rapidly shorter, narrower and distant upwards, margins entire to obscurely toothed, both surfaces thinly pubescent, hairs on petiolar part c.0.5–1mm long. *Flowers* few to many in rounded heads c.5–8mm in diam., elongating to c.10–12mm in fruit, lowermost 2 or 3 flowers very occasionally distant. *Bracts* at base of inflorescence c.2.5–4 × 0.5–1mm, smaller upwards, oblong, adnate to pedicel and calyx tube, glabrous except for patent hairs on margins, these up to 0.5–0.75mm long. *Pedicels* up to 0.5mm long. *Calyx* bilabiate, tube c.0.5–1mm long, anticous lobes 1–1.5 × 0.4–0.6mm, anticous lip split c.1–1.25mm, pubescent on margins and keels. *Corolla* tube 2.1–3 × 0.6–0.75mm broadening in throat, limb 3.25–5mm across the lateral lobes, posticous lobes c.0.75–1.75 × 0.75–1mm, anticous lobe c.1.6–3 × 0.5–0.8mm, all elliptic-oblong, mauve, bearded at base of posticous lip and on the two lateral lobes, hairs running down back of tube, probably a yellow-orange patch at base of posticous lip. *Stamens* 4, either well exserted or occasionally reduced in size and only reaching the mouth, anther lobes normally 0.5–0.6mm long, c.0.2mm when reduced and sometimes almost lacking pollen. *Stigma* well exserted. *Capsule* c.2.5–3 × 1.5mm. *Seeds* c.0.5mm long, ± elliptic in outline, almost smooth, testa greyish-mauve, endosperm amber-coloured, visible through testa.

Citations:

Cape. Piquetberg div., 3218CB, 12.8km W of Aurora, c.46m, *Acocks* 24372 (K). Malmesbury div., 3318AB, north of Hopefield, *Liebenberg* 8342 (PRE); Baarhuis/Zonquasfontein boundary, *Thompson* 3507 (STE); 3318DA, near Pella, Burger's Post Farm, *Boucher & Shepherd* 4630 (STE).

The long bare peduncles and leaves crowded towards the base of the stem give *P. mimetes* a strong superficial resemblance to *P. cordatum* (and so suggested the trivial name), but *P. mimetes* can at once be distinguished by its bracts, glabrous on the backs (not hairy) and with shorter hairs on the margins (up to 0.5–0.75mm long versus up to 2mm), and by its mauve, not creamy-white, flowers. It is also allied to *P. heterophyllum*, in which the bracts are glabrous on the backs, but they are mostly larger than those of *P. mimetes* (3.5–9mm long versus 2.5–4mm) with longer hairs on the margins (up to 1–2mm long versus 0.5–0.75mm). The flowers of *P. heterophyllum* are pale shades of yellow marked with dark lines and blotches while those of *P. mimetes* are mauve. *Phyllopodium mimetes* is sympatric with both these species, but it seems to occupy a much more restricted area, from about Het Kruis south through Aurora to Hopefield and Pella near Malmesbury. It grows in sand, flowering in September and October.

Group 2, subgroup 2B

16. Phyllopodium pubiflorum Hilliard in Notes RBG Edinb. 45: 485 (1989).
Type: Cape, Clanwilliam div., Pakhuis, 29 ix 1940, *Esterhuysen* 3363 (holo. BOL).

Annual herb, stems c.40–150mm tall, simple, either solitary or 2–4 from the base, pilose with coarse ± retrorse hairs up to 1–1.5mm long, leafy only at the base, terminating in a long nude or remotely bracteate peduncle bearing a head of flowers. *Leaves* in 1–3 pairs near base of stem, in larger plants becoming alternate and bracteate upwards, c.9–20 × 1.5–5mm, elliptic or oblanceolate tapering to a petiolar part c.⅓–½ the total leaf length, upper, alternate, leaves much smaller, sessile, margins entire to obscurely few-toothed, both surfaces pilose, hairs up to 1mm long. *Flowers* sessile or almost so, crowded into solitary terminal heads c.7–12mm in diam. *Bracts* at base of inflorescence c.5–9 × 0.8–2mm, smaller upwards, subspathulate, apex either broader or narrower than base, adnate to calyx tube, shortly hispid in upper half becoming hirsute in lower half, these hairs up to 1.25–1.5mm long, strongly patent to somewhat reflexed. *Calyx* obscurely bilabiate, tube 1–2.5mm long, anticous lobes 1.25–2 × 0.5–1mm, anticous lip split c.1–1.5mm, hispid on keels, hairs up to 1mm long, ± reflexed. *Corolla* tube 3.25–6 × 1–1.25mm, widening in the throat, limb c.5–8mm across the lateral lobes, posticous lobes 1–3.5 × 0.6–2mm, anticous lobe 3–4.5 × 1.25–2.1mm, corolla yellow, finely pubescent outside on tube and lower half of lobes, strongly bearded at base of posticous lip (and probably orange there) with some hairs on lateral lobes and running right down back of tube, hairs up to 0.5mm long. *Stamens* 4, anther lobes 0.5–0.75mm long. *Stigma* well exserted. *Capsules* and *seeds* unknown.

Citations:

Cape. Calvinia div., 3119AC, Niewoudtville, *Leipoldt* s.n. (BOL). Clanwilliam div., 3218BB, Pakhuis Pass, *Compton* 9629 (NBG); ibidem, 'Klein Klip Huis', *Hardy* 815 (PRE).

This is the only species of *Phyllopodium* with flowers hairy on the outside of the corolla tube (though the corolla tube of *P. anomalum* is minutely glandular in the upper half) and on the backs of the corolla lobes. The flowers also seem to be clear yellow in colour, which is uncommon in the genus. The hairs on the lower half of the bracts and on the calyx are remarkably stout and tend to be reflexed. These features, together with its habit (one to four simple stems from the base, a few leaves at the base, otherwise nude or nearly so below the solitary heads) make *P. pubiflorum* an altogether distinctive species. Its affinities will remain obscure until its seeds are known. Meanwhile, I have placed it near *P. mimetes*, which it at least resembles in habit.

The only ecological information that collectors have given is 'sandy slope'. All three collections from Pakhuis Pass were made at the end of September, when the plants were only just in flower; Leipoldt's collection from Niewoudtville bears a label in Bolus's hand and is undated, but it too is in first flower.

17. Phyllopodium namaense (Thellung) Hilliard in Notes RBG Edinb. 45: 485 (1989). **Plate 1B.**

Lectotype: Namibia, Anibebene, *Schenk* 285 (Z).

Syn.: *Polycarena namaensis* Thellung in Vierteljahrss. Nat. Ges. Zürich 60: 411 (1915); Merxmüller & Roessler in Prodr. Fl. S.W.A. 126, Scrophulariaceae: 40 (1967).
Phyllopodium rangei Engl. in Bot. Jahrb. 57: 611 (1922). Type: S.W.A., 2616 CA, Garub, *Range* 509 (holo. B,; iso. SAM).
Polycarena rangei (Engl.) Levyns in J. S. Afr. Bot. 5: 37 (1939).

Annual herb c.20–150mm tall, stems in all but the smallest plants several from the base, erect, either simple, nude and terminating in an inflorescence or with one to several pairs of branches, each terminating in an inflorescence, these branches again regularly branching, to produce a twiggy rounded leafy clump, stems sparsely to densely pubescent with ± obtuse retrorse hairs c.0.2–0.5mm long, those below the lower pair of leaves very short (c.0.1mm). *Leaves*: radical leaves 4, opposite and decussate, c.9–28 × 5–20mm, broadly elliptic to rhomboid-elliptic tapering to a broad flat petiolar part, entire, very thinly pubescent to glabrous on both surfaces, hairs on margins of lower half of leaf 1–1.5mm long or sometimes very few and shorter, upper leaves always narrower and progressively shorter upwards. *Flowers* many in crowded rounded heads becoming oblong in fruit. *Bracts* at base of inflorescence c.3.25–7 × 0.75–2.2mm, oblong to elliptic, adnate to pedicel and lower half of calyx tube, glabrous except for patent hairs on margins, up to 1mm long. *Pedicels* up to 0.8mm long. *Calyx* distinctly bilabiate, tube 1–1.5mm long, anticous lobe 1.5–2.5 × 0.75–1.5mm, glabrous except for small hairs on margins. *Corolla* tube 2.75–3.75 × 1.2–1.5mm, broadening in the throat, limb 4–6mm across the lateral lobes, posticous lobes 0.4–1.25 × 0.5–0.8mm, orange, pubescent, anticous lobe 2–3 × 1–1.75mm, anticous lip white or mauve, glabrous or with a few hairs near the base of each lobe. *Stamens* 4, anthers 0.5–0.75mm long. *Stigma* well exserted. *Capsules* c.3.5 × 3mm. *Seeds* c.0.75 × 0.5mm, obscurely trigonous, colliculate and wrinkled, pallid.

Citations:

Namibia. 2615CA, Lüderitzbucht, 1300m, *Range* 1134 (SAM); Schakalskuppen, 1500m, *Range* 1782 (SAM). 2616CA, farm Klein Aus, *Kinges* 2300 (PRE); Ausweiche, *Moss & Ottley* 11650 (BM). 2616CB, Aus, *Moss & Ottley* 11651 (BM); farm Augustfelde, *Erni* sub Giess 2392 (PRE, WIND). 2616CC, Tsirub, 1200m, *Range* 1844 (SAM); ibidem, *Range*

1106 (SAM). 2715BD–BC, Klinghardtgebirge, *Dinter* 3964 (BM, K, S, SAM, PRE). 2615CB, Halenberg, [Haalenberg], *Dinter* 6626 (BM, BOL, E, PRE, SAM, STE). 2716DA, farm Witpütz Nord, 16km W Polizeistation, *Giess* 13756 (WIND).

Cape. Little Namaqualand, 5 miles south of Grootderm [2816DA, Grootdoorn ?], *Pillans* 5596 (BOL).

When Thellung described *Polycarena namaensis*, he cited three specimens: *Dinter* 1117 (Gubub), *Pohle* s.n. (between the Orange River and Lüderitz) and *Schenk* 285 (Anibebene). I have not seen the Pohle specimen, but *Dinter* 1117 and *Schenk* 285 are both in the Zürich herbarium. Neither is very good, but Schenk's specimen bears Thellung's drawings and has therefore been chosen as the lectotype.

Phyllopodium namaense is known from a small area of Namibia, from the environs of Lüderitz and Aus southwards to the Orange River, where it has also been found immediately south of the river, in the Richtersveld. It has been collected between July and October, in sandy places including a dry river bed.

It is a distinctive species, with a basal rosette of four large leaves and erect stems, which range from a nude 'peduncle' terminating in a single head to a regular pattern of branching well displayed in the largest plants, where each stem may bear a pair of leaves from whose axils develop a pair of branches each terminating in an inflorescence, with further pairs of leaves and branches sometimes developing, in the manner of a cymose inflorescence.

It is often confused with *P. hispidulum* but that species can at once be distinguished by its hispid leaves, bracts and calyx, as well as by floral differences and its smooth blueish seeds. *Phyllopodium namaense* is without close affinity but its general relationship lies with *P. heterophyllum*.

18. Phyllopodium lupuliforme (Thellung) Hilliard in Notes RBG Edinb. 45: 485 (1989). **Fig. 63I.**

Type: Cape, Little Namaqualand, [2917BD], Zabies, 7 viii 1898, *Max Schlechter* s.n. (holo. Z; iso. BM, BOL, E, K, PRE, S, W).

Syn.: *Polycarena lupuliformis* Thellung in Vierteljahrsschr. Nat. Ges. Zürich 61: 433 (1916).

Annual herb c.65–150mm tall, stem simple to branched from the base and higher, branches erect or somewhat decumbent, pilose with patent pointed hairs up to 1mm long, below the lowest pair of leaves the hairs very short and retrorse, upper part of stem glandular-pubescent as well as pilose, in small plants stem tending to be leafy only at the base, in bigger, branching plants, stems leafy in lower two thirds or half, often scapose below the inflorescence. *Leaves* opposite below, sometimes alternate above, lowermost 2 or 3 pairs of leaves c.12–27 × 4–9mm including a short petiolar part, rhomboid to elliptic, becoming smaller and sessile upwards, margins entire to obscurely or distinctly serrate, blade thick-textured, pilose on both surfaces, hairs up to 1mm long, 1.5mm on petiolar part, short (c.0.1mm) stout pointed unicellular hairs on upper leaf margins. *Flowers* many in crowded rounded heads becoming oblong in fruit, solitary at the tip of each branch. *Bracts* at base of inflorescence c.3.5–5 × 3–4.5mm, smaller upwards, rhomboid-ovate, thick-textured, adnate to pedicel and lower half of calyx tube, lower surface (back) glabrous, upper surface and margins pilose with hairs to 1mm. *Pedicels* up to 0.5mm long. *Calyx* obscurely bilabiate, tube c.1.5mm long, anticous lobes c.1.75–2 × 1mm, pubescent mainly on margins and keels. *Corolla* tube c.3.5–4mm long, 1.25mm broad, widening in the upper half, limb c.5mm across the lateral lobes, posticous lobes c.1–1.5 × 0.75–1.5mm, anticous lobe c.2.5–3 × 1.25–1.5mm, all lobes elliptic-oblong, puberulous at base of posticous lip, colour unknown. *Stamens* 4, anthers c.1mm long. *Stigma* well exserted. *Capsules* c.4 × 3mm. *Seeds* c.0.6 × 0.5mm, obscurely trigonous, colliculate, dark blue-grey.

Citation:

Cape. Little Namaqualand, 2917BB, Karoechas, [near Steinkopf], 3000ft, *Schlechter* 11396 (BOL).

Phyllopodium lupuliforme is known only from the collections made by Max and Rudolf Schlechter at Karoechas and Zabies (now spelt Sabies) near Steinkopf in the northern part of Namaqualand. The only information given is '*in collibus*' (on hills) at c.900m. The type collection, made in August, is in flower, the second collection, made a year earlier in September, is in fruit. Neither collector noted the colour of the flowers. Hiern (1904, p.317) cited both collections as *P. heterophyllum*; *P. lupuliforme* and *P. heterophyllum* are somewhat similar in habit but *P. lupuliforme* is easily distinguished by its hairier leaves, and many-flowered and crowded heads with rounder bracts; there are floral differences too.

19. Phyllopodium collinum (Hiern) Hilliard in Notes RBG Edinb. 45: 484 (1989).
Type: Cape, Little Namaqualand, [3017BD] on hills at Brakdam, 2000ft, *Schlechter* 11156 (holo. BOL; iso. E, K, W).

Syn.: *Polycarena collina* Hiern in Fl. Cap. 4(2): 325 (1904).
   Polycarena pubescens Benth. in Hook., Comp. Bot. Mag. 1: 372 (1836) p.p. quoad *Drège* b, Haazenkraals rivier, and *Drège* c, Zilverfontein, excluding lectotype.

Annual herb 30–300mm tall, stem simple to branched from the base and higher, branches usually ascending, occasionally decumbent, pubescent with short (c.0.1mm) ± retrorse eglandular hairs plus patent glandular hairs to 0.4mm long and very occasional eglandular hairs c.0.4–1mm long, leafy with a strong tendency towards the development of dwarf axillary shoots. *Leaves* opposite becoming alternate upwards, lowermost primary ones c.15–30 × 1–2(–5)mm, smaller upwards, linear-oblong to narrowly elliptic, narrowed to the base, entire or with 1–4 pairs of teeth, both surfaces pilose, hairs up to 1mm long, up to 2mm on margins near base, plus very occasional glandular hairs. *Flowers* few to many in rounded heads terminating all the upper branchlets, shortly scapose below the heads, heads oblong in fruit, up to c.20mm long, all but the lowermost 1 or 2 flowers crowded. *Bracts* c.3–9 × 1.2–3mm, smaller upwards, elliptic to ovate-lanceolate, entire or the lowermost occasionally with 1 or 2 teeth, adnate to pedicel and base of calyx tube, pilose on both surfaces, a very few glandular hairs as well, hairs on margins up to 0.5–1.5mm. *Pedicels* of lowermost flowers c.0.5–1mm long, shorter upwards. *Calyx* distinctly bilabiate, tube 1.5–2mm long, anticous lobes 1.5–2 × 0.5–1mm, forming a briefly notched lip, pilose all over, hairs up to c.0.4–0.75mm, a few gland-tipped hairs as well. *Corolla* tube 3–4.5 × 1–1.5mm, broadening in the throat, limb (4–)5–7mm across the lateral lobes, posticous lobes (1–)1.5–2 × 1–1.75mm, orange-yellow, bearded at base of lip and down back of tube, hairs often extending to the 2 lateral lobes, anticous lobe 2–3.25 × (1–)1.25–1.75mm, anticous lip creamy-coloured or possibly yellow (see discussion under *P. maxii*). *Stamens* 4, anthers 0.75–1mm long. *Stigma* well exserted. *Capsules* c.3.5–4 × 2.5mm. *Seeds* c.0.75 × 0.5mm, 3- or 4-angled, colliculate and wrinkled on 1 or 2 faces, testa pallid (possibly eventually mauve-blue?), endosperm amber-coloured, visible through the testa.

Selected citations:

Cape. Namaqualand, 2917DB, 13½ miles N by W of O'okiep, 3000ft, *Acocks* 19341 (PRE); near Springbok, *Lewis* 1556 (SAM). 3017BD, Brakdam, *Lewis* 1558 (SAM); Khamieskroon, *Hall* 280 (NBG). 3118AB, Kareebergen, 1800ft, *Schlechter* 8276 (BOL, E, S, W).

Phyllopodium collinum appears to be confined to Namaqualand, from about O'okiep in the north, south to the Kareebergen near Nieuwerust, between c.550 and 900m above sea level, a plant of sandy or gravelly slopes and flats, flowering between July and September. It is very closely allied to *P. maxii*; see under that species for discussion.

20. Phyllopodium maxii (Hiern) Hilliard in Notes RBG Edinb. 45: 485 (1989).
Type: Cape, Little Bushmanland, Keuzabies, *Max Schlechter* 102 (holo. K; iso. BOL, E, S).

Syn.: *Polycarena maxii* Hiern in Fl. Cap. 4(2): 328 (1904).
    Polycarena pubescens auct., non Benth.; Merxmüller & Roessler in Prodr. Fl. S. W. A. fam. 126: 40 (1967).

Annual herb 40–300mm tall, stem simple to branched from the base and higher, branches decumbent or ascending, pubescent with patent glandular hairs up to c.0.4mm long, eglandular hairs c.0.1mm long only on lower parts of stem, often only below the first pair of leaves, leafy, with a tendency towards the development of dwarf axillary shoots. *Leaves* opposite becoming alternate upwards, lowermost primary ones 14–37 × (1–)2.5–5mm, smaller upwards, narrowly elliptic, narrowed to the base, entire or with 1–3 pairs of teeth, both surfaces pubescent, hairs glandular, up to c.0.4mm long, long (up to 1–1.25mm) eglandular hairs on margins near base. *Flowers* few to many in rounded heads terminating all the branchlets, shortly scapose below the heads, heads oblong in fruit, scarcely exceeding 10mm long, crowded. *Bracts* c.3–13 × 1.25–3.75mm, smaller upwards, lowermost often leaflike, otherwise ovate or ovate-oblong, frequently with 1–3 teeth, adnate to pedicel and base of calyx tube, glandular-pubescent on the backs, hairs c.0.4mm long, eglandular hairs on inner face and on margins, these up to 0.75–1mm long. *Pedicels* of lowermost flowers up to c.1mm long, shorter upwards. *Calyx* distinctly bilabiate, tube 1.1–1.75mm long, anticous lobes 1.5–2 × 0.5–1.2mm, forming an entire or very briefly notched lip, glandular-pubescent mainly on the keels, marginal hairs eglandular. *Corolla* tube 2.5–3 × 1–1.5mm, broadening in the throat, limb 3.5–4mm across the lateral lobes, posticous lobes 0.75–1.5 × 0.75–1mm, orange-yellow, either glabrous at base of lip or thinly bearded, anticous lobe 1–2 × 0.75–1.5mm, anticous lip probably white (see discussion below). *Stamens* 4, anthers 0.4–0.5mm long. *Stigma* well exserted. *Capsules* c.4 × 2.5mm. *Seeds* c.0.75 × 0.5mm, 3- or 4-angled, obscurely colliculate on one face, testa pallid over amber-coloured endosperm.

Selected citations:

Namibia. 2616CB, farm Kubub, *Giess* 12860 (NBG, PRE, WIND).
Cape. Springbok div., 2917DB, Verseputs, *Van der Schijff & Schweickerdt* 5763 (PRE). 2918AB, 60 miles west of Pofadder, *Schlieben* 9013 (K, PRE, S, W). 2919AB, 31 miles WSW of Pofadder, 3100ft, *Leistner* 2501 (K, PRE). 3018AC, Khamiesberg, *Pearson* 6696 (BOL, K).

Phyllopodium maxii is so closely allied to *P. collinum* that I have reservations about maintaining it as a separate species. However, they can be separated on differences in indumentum: eglandular hairs predominate in *P. collinum*, glandular in *P. maxii*, where hairs similar to the very short ± retrorse hairs that clothe the stems in *P. collinum* are confined either to the lowermost internodes or occur only below the lowermost pair of leaves. Long eglandular hairs predominate on the leaves, bracts and calyx in *P. collinum*, glandular ones in *P. maxii*. Also, the flowers of *P. maxii* tend to be smaller than those of *P. collinum* (for example, corolla tube 2.5–3mm long versus 3–4.5mm, limb 3.5–4mm across versus (4–)5–7mm, anthers 0.4–0.5mm long versus 0.75–1mm) and the posticous lip is either glabrous or very sparsely hairy, while in *P. collinum* both posticous lip and lateral lobes are well bearded.

The area of *P. maxii* lies mainly north and east of that of *P. collinum* (from the environs of Aus in southern Namibia to Pofadder and the Khamiesberg in Namaqualand), but they are sympatric around O'okiep and Kamieskroon. Where their areas meet, some specimens are difficult to place and it is clearly there that any investigation into their relationship should begin. They favour similar habitats and flower at the same time, July to September.

Collectors who have noted flower colour often merely record 'yellow'. But it is evident even from dried specimens that the posticous lip may be orange-yellow, differing in colour from the rest of the limb. In other specimens, the whole limb or just the lower lip is white, and the whole corolla becomes yellowish as the flower withers after anthesis.

Anomalous Species

21. Phyllopodium anomalum Hilliard in Notes RBG Edinb. 45: 484 (1989).
Type: Cape, Namaqualand, 1500–2000ft, August, *Drège* distributed as *Polycarena plantaginea* Benth. a (holo. E; iso. S, TCD, W).

Syn.: *Polycarena plantaginea* auct., non *Manulea plantaginis* L.f. nec *M. plantaginea* Thunb.; Bentham in Hook., Comp. Bot. Mag. 1: 372 (1836) and all subsequent authors.

Annual herb 35–250mm tall, stem simple to branched from the base and higher, branches mostly erect or ascending, occasionally decumbent, branches, leaves and bracts all minutely glandular-puberulous, thinly pilose as well with patent eglandular hairs mostly 1–1.5mm long, leafy. *Leaves* opposite below becoming alternate upwards and passing imperceptibly into floral bracts, blade of lowermost few pairs c.5–13 × 2–6mm, ovate to elliptic tapering to a petiolar part 2–7mm long, smaller upwards, margins distinctly to obscurely serrate or subentire to entire, both surfaces pubescent, hairs up to 1.25–1.5mm long, minutely glandular as well, particularly on petiolar part. *Flowers* many in more or less leafy racemes, upper flowers crowded, lower distant. *Bracts* c.4–12 × 0.6–3.5mm, lower ones leaflike becoming spathulate to linear upwards, adnate to pedicel and lower half of calyx tube, indumentum as on leaves, marginal hairs in particular up to 1.5mm long. *Pedicels* of lowermost flowers c.0.5–0.75mm long, shorter upwards. *Calyx* obscurely bilabiate, tube 1.5–2.75mm long, anticous lobes 0.5–0.75 × 0.3–0.75mm, glandular-puberulous and with long patent eglandular hairs as well. *Corolla* tube c.2.5–4 × 0.75mm, minutely glandular-puberulous outside in upper part, limb c.1.3–2mm across the anticous lobes, posticous lobes c.0.3–0.5 × 0.2–0.4mm, anticous lobe c.0.6–0.8 × 0.2–0.5mm, all lobes white. *Stamens* 2 only (the anticous pair), anthers c.0.2mm long, appearing in the mouth of the tube. *Stigma* curled round the anthers. *Capsules* 3–4 × 1.25–2mm. *Seeds* 0.4–0.75 × 0.3–0.5mm, colliculate, testa mauveish-grey, endosperm amber-coloured, showing through the thin testa.

Citations:

Cape. Little Namaqualand, 2918CC, Zilverfontein, 2000–3000ft, *Drège* distributed as *Polycarena plantaginea* Benth. b (E); Kamieskroon, 3017BD, *Thorne* SAM 48811 (mixed with *Polycarena pubescens*) (SAM); Klipfontein, 3000ft, *Bolus* 658 (SAM, W); 3018CA, 11 miles south of Garies, *Lewis* 3144 (SAM); c.10km south of Garies, *Hugo* 2896 (PRE); between Garies and Nieuwerust, 2000ft, *Levyns* 6962 (BOL). Van Rhynsdorp div., Karee Bergen, 1200ft, *Schlechter* 8195 (BOL, E, W); ibidem, *Wall* s.n. (S). Calvinia div., 3119AC, Niewoudtville Reserve, *Perry & Snijman* 2220 (NBG). Clanwilliam div., 3218BB, Alpha, *Compton* 10939 (NBG). Sutherland div., 3220CA, Houthoek, *Hanekom* 1090 (PRE, STE). Montagu, 3320CC, hill behind Baths, 2000ft, *Levyns* 3904 (BOL).

Bentham's misuse of the name *Manulea plantaginis* L.f. for this plant is inexplicable. Brief though it is (*foliis ovatis subdentatis integrisque glabris, capitulis ovatis, ramis diffusis*), Linnaeus the younger's description clearly does not fit the plant to which Bentham and all subsequent authors have applied the name. True *Manulea plantaginis* proves to have only one ovule in each loculus and is a species of *Cromidon*.

The generic placement of what should now be called *Phyllopodium anomalum* is dictated by indumentum and seed structure. The species is isolated within the genus, and is easily recognized by its long leafy racemes, tiny flowers and possession of only two stamens. There is no indication of cleistogamy, but the reduction in number of stamens appears to be constant and to be associated with selfing: the stigma is never exserted, but curls round the anthers in the mouth of the corolla tube.

Phyllopodium anomalum has a relatively wide geographical range, from about O'okiep in northern Namaqualand to Montagu in the southern Cape, between c.370 and 950m above sea level. It grows in gravelly and sandy places, flowering between July and September. Despite its wide range, it is seldom collected, though apparently plentiful in its chosen habitats.

22. Phyllopodium alpinum N. E. Br. in Kew Bull. 1901: 128 (1901); Hiern in Fl. Cap. 4(2): 319 (1904). **Fig. 63H.**

Type: Cape, Caledon div., summit of Genadendal Mt, 5000ft, *Galpin* 4407 (holo. K).

Syn.: *Polycarena alpina* (N. E. Br.) Levyns in J. S. Afr. Bot. 5: 37 (1939).

Perennial herb, branches several to many from the base, simple initially, soon branching, main branches sprawling or prostrate, bases becoming woody, up to 5mm diam., sometimes rooting from lower parts of main branches, pubescent, hairs ± retrorse, up to 0.2–0.5mm long, glandular-puberulous as well particularly on the upper parts, very leafy. *Leaves* opposite becoming alternate towards the tips, 4–14 × 1.5–6mm, decreasing little in size upwards, spathulate or elliptic tapering gradually to a flat petiolar part, thick-textured, glandular-punctate, margins entire or with up to 3 pairs of small teeth, often glabrous except for patent eglandular hairs up to 0.5–1mm long on the lower margins, or a few hairs on the midline below, occasionally very thinly pubescent on both surfaces. *Flowers* many in crowded rounded or oblong heads, in fruit up to c.30(–50) × 8mm, terminating all the branchlets, sometimes panicled, the plant very floriferous. *Bracts* at base of inflorescence c.3.75–9 × 1–3mm, leaflike, spathulate or oblong-elliptic, adnate to pedicel and base of calyx, often glabrous on the backs except for marginal hairs up to 0.5–1mm long, sometimes thinly pubescent. *Pedicels* up to 1.5mm long. *Calyx* obscurely bilabiate, tube 0.25–0.6mm long, anticous lobes 2–3 × 0.5–0.8mm, anticous lip split 2–2.75mm, whole calyx pubescent, hairs up to 0.5mm long, glandular-puberulous as well. *Corolla* tube 4–4.5mm long, cylindric below and there 0.75–1mm in diam., abruptly expanded above and c.2mm across the mouth, limb 4.5–6.5mm across the lateral lobes, posticous lobes 1–1.75 × 1–1.75mm, anticous lobe 2–3 × 1.3–2.25mm, all lobes elliptic, usually pink or mauve, occasionally white, with an orange patch at the base of the posticous lip and there heavily bearded, the hairs usually extending right round the mouth. *Stamens* 4, 2 exserted, 2 in mouth or shortly exserted, anthers 0.6–0.75mm long. *Stigma* well exserted. *Capsules* 3–4.5 × 2.25–2.5mm. *Seeds* c.5–8 in each loculus, 0.75–1 × 0.5–0.6mm, elliptic, irregularly and sinuously wrinkled, pallid amber.

Citations:

Cape. Worcester div., 3319AD, Waaihoek Peak, SE side, 5500ft, *Esterhuysen* 28898 (K, S); ibidem, 5000ft, *Esterhuysen* 8299 (BOL); Schurfteberge near Bertsberg, 2000–3000ft, *Esterhuysen* 32679a (BOL). 3319CB, Brandwacht Peak, 5800ft, *Esterhuysen* 28357 (BOL, S); Chavonnesberg-Brandwacht Peak, 5000ft, *Esterhuysen* 29921 (BOL, K). 3319DC, Robertson Microwave Station, 4800ft, *Lawder* s.n. (NBG); Jona's Kop, 5350ft, *Rourke* 1309 (NBG, PRE); ibidem, 4950ft, *Beyers* 162 (STE); summo monte prope Genadendal, *Gibbs* sub BOL 10035 (BM, K, BOL).

Phyllopodium alpinum has no close allies. It has been placed at the end of Group 2 because of its seed structure but the masses of compact rounded heads on leafy twiglets, which are seen in well grown specimens, set it apart from any species within this group (its seed structure precludes its inclusion in subgen. *Leiospermum*, where there are characteristically masses of rounded heads).

It is known with certainty from the arc of mountains stretching from Fransch Hoek east to Stormsvlei, between c.600 and 1750m above sea level. It grows in sandy depressions among rocks, flowering between October and December, mostly in fruit by February.

The type specimen came from the mountains above Genadendal; this material and subsequent collections from the same general area are mostly very compact in habit, well-branched, and have the bracts hairy on the backs, but *Beyers* 162 from Jona's Kop are clearly young plants, with tufts of unbranched stems each terminating in a head. The collections from farther west are all lax in habit and the bracts are glabrous on the backs. A collection from much further east, on the Swartberg, main peak in the ridge east of Blesberg (*Thompson* 2254, STE) may be *P. alpinum*; it comprises a few tiny depauperate twiglets terminating in few-flowered heads with bracts narrower than usual. Good specimens are needed from this area.

Subgen. Leiospermum Hilliard in Edinb. J. Bot. 50: 97 (1993).
Type species: *Phyllopodium cephalophorum* (Thunb.) Hilliard

23. Phyllopodium cephalophorum (Thunb.) Hilliard in Edinb. J. Bot. 50: 97 (1993). **Fig. 63F, G.**
Lectotype: C.B.S., *Thunberg* (right-hand specimen on sheet 13874, UPS).

Syn.: *Selago cephalophora* Thunb., Prodr. 100 (1800).
    Polycarena cephalophora (Thunb.) Levyns in J. S. Afr. Bot. 5: 37 (1939) & in Adamson & Salter, Fl. Cape Penins.
      714 (1950).

Annual herb c.55–300mm tall, stem simple to branched from the base, these branches often decumbent, all branches branching above into corymbose compound inflorescences, stems pilose with eglandular hairs up to 1–2mm long, hairs *either* retrorse in upper part becoming ± patent towards base *or* patent throughout, small glandular hairs often present as well especially on upper parts, ± sessile to shortly stipitate when eglandular hairs retrorse, longer (up to 0.2mm) when hairs patent, stems very leafy, dwarf axillary shoots often present. *Leaves* opposite becoming alternate upwards, lower primary leaves c.15–50(–70) × 2–5(–7)mm, oblanceolate tapering to a petiolar part best developed in large leaves, upper and axillary leaves quickly smaller, margins with 2–7 pairs of teeth, uppermost leaves sometimes entire, both surfaces pilose, hairs (especially on lower margins) up to 0.75–2mm long. *Flowers* many in crowded heads c.7–13mm in diameter, terminating all the branchlets. *Bracts* 3–7 × 0.75–1.6mm, smaller upwards, outermost mostly lanceolate-ovate, adnate to pedicel and ± halfway up calyx, ± glabrous to puberulous in apical part becoming pubescent to hirsute in lower half, these hairs 0.75–2mm long. *Pedicels* of lowermost flowers up to 0.5mm long, shorter upwards. *Calyx* distinctly bilabiate, *either* calyx tube 0.4–1.5mm long, the anticous lobes 1.5–3 × 0.5–0.8mm with the lip split 0.5–1.75mm, *or* 3 posticous lobes sometimes free to the base, the anticous lobes then c.3.5–5mm long, the lip split 1.75–3mm, pilose to pubescent. *Corolla* tube 2.5–4.5 × 0.75–1mm broadening slightly in the throat, limb 4–7mm across the lateral lobes, posticous lobes 1–2.5 × 0.5–1.25mm, anticous lobe (1.75–)2.5–4 × 0.5–1.2mm, all lobes oblong, usually mauve, occasionally pink or white, bearded on posticous lip and down back of throat, hairs often extending onto the lobes of the anticous lip. *Stamens* 4, anthers 0.5–0.8mm long. *Stigma* well exserted. *Capsules* c.3–3.5 × 1.75mm. *Seeds* c.0.4mm long, elliptic in outline, smooth, dark blue-grey.

Selected citations:

Cape. Van Rhynsdorp div., 3118DB, Zandkraal, *Steyn* 414 (NBG). Piquetberg div., 3216AD, Verloren Vlei, *Compton* 15060 (NBG). Clanwilliam div., 3218BB, Lange Vallei, *Acocks* 2955 (S). Malmesbury div., 3318AD, north of Modder River, 12 xi 1986, *O'Callaghan* 1326 (STE); 3318CB, Buck Bay, *Hilliard & Burtt* 13015 (E, NU); between Bokbaai and Darling, *Esterhuysen* 3849 (PRE); 3318DC, Potsdam, 16 x 1988, *Bokelmann & Paine* 49 (E). Cape Peninsula, 3418AB, Zand Vlei, *Pillans* 4013 (BOL, PRE); Chapman's Bay, *Lewis* 515 (SAM).

Phyllopodium cephalophorum is relatively widely distributed from about Van Rhynsdorp south to the Peninsula, up to c.300m above sea level. It appears to be a very common plant on sandy flats, flowering mainly in September and October.

In the northern part of its range, from Van Rhynsdorp south to Piquetberg, the eglandular hairs on the stems are always patent and any glandular hairs that may be present are up to c.0.2mm long. The flowers are usually mauve, but white or cream-coloured ones have also been recorded. From Piquetberg southwards, the hairs on the upper parts of the stems are retrorse and are often strongly appressed, and any glands that may be present are either sessile or very shortly stalked; such is the indumentum on the type specimen. Flower colour appears nearly always to be shades of mauve, though pink has been recorded at Langebaan.

There is some variation too in the degree to which the calyx lobes are united. In the area around Hopefield, Darling and Buck Bay the three posticous calyx lobes may be free to the base or very briefly united by a delicate membrane; the two anticous lobes are united for roughly half their length.

Nevertheless, *P. cephalophorum* can be recognized by its shaggy white indumentum, narrow toothed leaves and masses of small round heads corymbosely arranged, the bracts characteristically glabrous to puberulous at the tips and becoming shaggy in the lower half. It is often confused with *P. capitatum*, this going back to Bentham, who miscalled *P. cephalophorum P. capitatum*; the true *P. capitatum* is synonymous with *P. heterophyllum* (no. 13).

Phyllopodium cephalophorum is closely allied to *P. phyllopodioides* (below) and in areas where they are sympatric they may be difficult to distinguish: see under *P. phyllopodioides*.

24. Phyllopodium phyllopodioides (Schltr.) Hilliard in Edinb. J. Bot. 50: 97 (1993).
Type: Cape, Van Rhynsdorp div., 3118DB, Drooge River, 1200ft, *Schlechter* 8322 (holo. B†; iso. BOL, E, S, SAM, W).

Syn.: *Selago phyllopodioides* Schltr. in Bot. Jahrb. 27: 190 (1900); Rolfe in Fl. Cap. 5(1): 161 (1901).
    Polycarena selaginoides [Schltr. ex] Hiern in Fl. Cap. 4(2): 326 (1904). Type as above.

Annual herb 40–250mm tall, or occasionally tiny seedlings flowering, stem simple to branched from the base, these branches often decumbent (sometimes prostrate?), all branches branching above into corymbose compound inflorescences, stem pilose with eglandular hairs up to 0.5–1.5mm long, retrorse to patent, often glandular-puberulous as well on upper parts, dwarf axillary shoots either wanting or ill-developed. *Leaves* opposite becoming alternate upwards, blade of lowermost leaves c.9–27 × 4–7mm, oblanceolate tapering into a petiolar part c.6–30mm long, often roughly equalling the blade in length, upper leaves quickly shorter and narrower with the petiolar part much reduced, margins entire to sparingly toothed, both surfaces pubescent, pilose on the midrib below and on margins of petiolar part, these hairs up to 1.5–1.75mm long. *Flowers* few to several in crowded heads c.4–6(–7)mm in diam., ± oblong in fruit, terminating all the branchlets. *Bracts* 3–5 × 0.5–1.4mm, smaller upwards, oblong to elliptic, adnate to pedicel and about halfway up calyx, puberulous on the backs, hairs on margins up to 0.5–1.25mm long. *Pedicels* of lowermost flowers up to 0.5mm long, shorter upwards. *Calyx* bilabiate, tube 1–1.5mm long, anticous lobes 1.25–2 × 0.5–0.8mm, anticous lip split 0.5–0.75mm, rarely the posticous lobes free to the base, pubescent. *Corolla* tube 2.5–3.5 × 0.75–1mm, broadening slightly in throat, limb 4–5mm across the lateral lobes, posticous lobes 1–1.5 × 0.6–1.2mm, anticous lobe 2–3 × 0.75–1.3mm, all lobes oblong, usually mauve, sometimes white, bearded on posticous lip and down back of tube, the hairs often extending onto the lobes of the anticous lip. *Stamens* 4, anthers 0.5–0.75mm long. *Stigma* well exserted. *Capsules* c.2–3 × 1.5–2mm. *Seeds* c.0.4–0.5mm long, elliptic in outline, smooth, dark blue-grey.

Selected citations:

Cape. Van Rhynsdorp div., 3118BC, 3 miles N of Van Rhynsdorp, Knersvlakte, *Lewis* 1554 (SAM). 3118CB, 25km out of Vredendal on Strandfontein road, *Batten* 790 (E). 3118DA, 10 miles SSW of Van Rhynsdorp, c.400ft, *Acocks* 19631 (K, STE); Van Rhynsdorp, *Esterhuysen* 5386 (BOL, PRE). 3118DC, Klaver, *Compton* 20697 (NBG, STE). Clanwilliam div., 3218AD, Elandsbaai, *Batten* 826 (E). 3218AB, near Lambert's Bay, *Van Breda* 4292 (K).

Hiern's name, *Polycarena selaginoides*, is based on the same collection (*Schlechter* 8322) as Schlechter's name, *Selago phyllopodioides*. It seems that all the distributed sheets of the Schlechter collection bore Schlechter's manuscript name, *Polycarena selaginoides*, which Hiern validated. Meanwhile, Schlechter must have changed his mind about the generic placement of his plant, and published for it the name *Selago phyllopodioides*, diagnosing his species against *Selago cephalophora* (now *Phyllopodium cephalophorum*). Rolfe enumerated *Selago phyllopodioides* in his account of the genus in Flora Capensis 5(2), saying he had not seen the type, unaware that an isotype was already in the Kew herbarium, but under *Polycarena*.

Unfortunately, Schlechter's highly inappropriate epithet takes precedence, and the plant must be called *Phyllopodium phyllopodioides*.

Schlechter distinguished his new species from *Selago cephalophora* 'durch Habitus, Blätter und der lockeren Blutenkopchen', but as his specimens are mostly in bud, there is no question of the heads being more compact than those of *P. cephalophorum*; the species are also similar in habit, but there are subtle differences in foliage. The two species are very closely allied and it is difficult to formulate the distinctions between them. Typical *P. cephalophorum* is a more hoary-looking plant than typical *P. phyllopodioides* because of a somewhat denser indumentum almost impossible to define, axillary leaf tufts are well developed (absent or weakly developed in *P. phyllopodioides*) and the heads are mostly larger (7–13mm in diameter versus 4–6(–7)mm). But the most useful character on which to separate the two species is the shape and indumentum of the bracts: typically ovate-lanceolate in *P. cephalophorum*, the back glabrous to puberulous towards the tip, becoming shaggy in the lower half; in *P. phyllopodioides* typically oblong to narrowly elliptic, puberulous on the backs, long hairs confined to the margins.

Phyllopodium phyllopodioides ranges from the neighbourhood of Van Rhynsdorp south to Saldanha Bay and Piquetberg; over the whole of its range it is sympatric with the somewhat atypical form of *P. cephalophorum* with patent indumentum as well as with typical *P. cephalophorum* around Piquetberg and Saldanha Bay. Where the two species are sympatric, the heads of *P. cephalophorum* tend to be smaller than those in plants outside the range of *P. phyllopodioides* and the bracts may be less hairy than usual. Some specimens are difficult to place and it is not impossible that the two species influence each other where they are sympatric; they flower at the same time and, from meagre collectors's notes, appear to occupy similar habitats.

25. Phyllopodium pumilum Benth. in Hook., Comp. Bot. Mag. 1: 373 (1836) & in DC., Prodr. 10: 353 (1846); Hiern in Fl. Cap. 4(2): 317 (1904), excluding *Schlechter* 5059 and 8619. **Fig. 63D.**
Type: Cape, Clanwilliam [div.], Zwischen Zwartdoorn rivier und Groenrivier, unter 1000 Fuss, August, *Drège* (holo. K; iso. E, S, W).

Syn.: *Polycarena pumila* (Benth.) Levyns in J. S. Afr. Bot. 5: 37 (1939).

Annual herb c.40–190mm tall, stem simple to well branched from the base, branches divaricate, eventually producing a rounded twiggy bushlet, each twig terminating in a capitate inflorescence, stems pubescent with eglandular hairs c.0.3–0.75mm long, retrorse and appressed on upper parts, tending to spread lower down, often minutely glandular-puberulous as well on upper parts, leafy. *Leaves* opposite becoming alternate upwards, blade of lowermost leaves 2.5–17 × 1.75–13mm, ovate to elliptic, tapering abruptly or more gradually into a petiolar part c.1.5–20mm long (up to c.twice as long as blade), upper leaves rapidly smaller and petiolar part less conspicuous, oblanceolate, margins entire to obscurely few-toothed, both surfaces pubescent, hairs on lower margins up to 0.5–0.75mm long. *Flowers* few to several in crowded heads c.3–7mm in diam., elongating slightly in fruit. *Bracts*: lowermost c.2.5–4 × 1–1.75mm, rarely slightly longer and leaflike, mostly broadly to narrowly elliptic, shorter and narrower upwards, adnate to pedicel and about halfway up calyx tube, puberulous on backs, patent hairs up to 0.5–1.5mm on margins. *Pedicels* of lowermost flowers up to 0.5mm long, shorter upwards. *Calyx* bilabiate, tube 0.75–1.5mm long, anticous lobes 1–1.5 × 0.5–0.75mm, pubescent, hairs up to 0.25–0.4mm long. *Corolla* tube 2–2.75 × 0.75–1.2mm, widening slightly in throat, limb 4–5mm across the lateral lobes, posticous lobes c.0.6–1 × 0.5–1mm, anticous lobes c.2–2.5 × 0.75–1.3mm, all lobes elliptic-oblong, white to lilac-mauve, bearded at base of posticous lip, a few hairs sometimes on lateral lobes. *Stamens* 4, anthers 0.5–0.75mm long. *Stigma* well exserted. *Capsules* 2–3.5 × 1.5–1.75mm. *Seeds* 0.4–0.5mm long, elliptic in outline, smooth, dark blue-grey.

Selected citations:

Cape. Namaqualand, 2816BB, Andaus Poort, *Marloth* 12242 (BOL). 2816DA, near Alexander Bay, *Hall* 568 (NBG). 2916 BB, between Port Nolloth and Holgat River, *Pillans* 5633 (BOL). 2916BD, Steinkopf-Port Nolloth road, *Steiner* 1312 (NBG). 2917AB, Eksteenfontein road, *Goldblatt* 5720 (E). 3017BA, between Komaggas and Soebatsfontein, *Maguire* 408 (NBG, STE). 3017AD, 3 miles east of Hondeklipbaai on road to Garies, *Thompson* 1084 (K). 3017 DB, west of Garies, between road to Groenrivier and Hondeklipbaai, *Batten* 804 (E). 3018AC, De Kom, 3 miles from Leliefontein in the Khamiesbergen, *Leipoldt* 3730 (BOL); 3018CA, Langkloof Pass, 26 viii 1990, *Batten* 1028 (E, K, S).

Bentham described *P. pumilum* from flowering seedlings, and the name has since been misused for young plants or dwarfed specimens of very diverse species. In recent years, the species has been much confused with *P. phyllopodioides* (as *P. selaginoides*) from which it can be distinguished by its different habit: *P. pumilum* is divaricately branched from the base and soon forms a rounded twiggy bushlet with a head of flowers terminating every twig; *P. phyllopodioides* is corymbosely branched and each stem branches above into a compound corymbose inflorescence.

Phyllopodium pumilum is confined to Namaqualand, from the Richtersveld south to about Garies, Groenrivier and Tietiesbaai, from near sea level to c.1300m. It grows on sand dunes and other sandy or sometimes gravelly sites, flowering between July and October. Its area does not overlap with that of *P. phyllopodioides*, which ranges from about Van Rhynsdorp south to Saldanha Bay and Piquetberg.

Two specimens need special mention: *Compton* 11031 (NBG) and *Esterhuysen* 5412 (BOL). They were made on the same day (23 vii 1941), the first at Bitterfontein, the second between Bitterfontein and Garies (between 3018CA and 3118AB). The Compton specimens are much bigger and better grown than those collected by Miss Esterhuysen, so it is likely that they did come from slightly different localities; both collectors described the flowers as white. The plants have the divaricate branching and entire leaves typical of *P. pumilum*, but they look hoary because the hairs on the stems are more numerous and rather longer than in *P. pumilum*; the heads on the robust Compton specimens are c.8mm in diam. (up to 7mm in *P. pumilum*) and the outermost bracts are often longer (up to 6.5mm). This plant may represent no more than a local race of *P. pumilum*, at the southern limit of the range of that species. Further investigation in the area around Garies, Groenrivier (type locality of *P. pumilum*) and Bitterfontein is needed to see if formal recognition is necessary.

26. Phyllopodium hispidulum (Thellung) Hilliard in Notes RBG Edinb. 50: 97 (1993). Type: S.W.A. [Namibia], 2715BD–BC, Klinghardtberge, *Dinter* 3888 (holo. Z; iso. BM, K, SAM).

Syn.: *Polycarena hispidula* Thellung in Vierteljahrss. Nat. Ges. Zürich 74: 118 (1929); Merxmüller & Roessler in Prodr. Fl. S.W.A. fam. 126 Scrophulariaceae: 40 (1967).

Annual herb c.35–80mm tall, stem branched from the base, then divaricately branched above, each twiglet terminating in a capitate inflorescence, stems pubescent with retrorse appressed eglandular hairs, hairs up to c.0.5–0.6mm long, a few minute glandular hairs sometimes on upper parts. *Leaves* opposite below, alternate above, lower ones c.10–18 × 2.5–5mm, elliptic or ovate-elliptic tapering into a petiolar part roughly half the total leaf length, upper leaves rapidly smaller and sessile, margins entire or with 1 or 2 obscure teeth, both surfaces hispid, patent hairs on lower margins up to 1mm long. *Flowers* few in crowded heads c.4–6mm in diameter, elongating slightly in fruit. *Bracts* at base of inflorescence c.3–4.5 × 1–2.5mm, elliptic, smaller and narrower upwards, adnate to pedicel and lower part of calyx tube, backs minutely hispid, margins with patent hairs up to c.0.6–1mm long. *Pedicels* up to 0.5mm long. *Calyx* bilabiate, tube 0.75–1mm long, anticous lobes 1–1.5 × 0.6–0.75mm, pubescent, hairs up to c.0.25mm long. *Corolla* tube 1.75–2.5 × 0.75mm, broadening in the throat, limb 3.5–4mm across the lateral lobes, posticous lobes 0.5 × 0.3–0.5mm, anticous lobe 1.25–2 × 0.75–1mm, all lobes elliptic-oblong, mauve, bearded at base of posticous lip, hairs often extending to lateral lobes as well. *Stamens* 4, anthers 0.5mm long. *Stigma* well exserted. *Capsules* c.3 × 2mm. *Seeds* 0.4–0.5mm long, elliptic in outline, smooth, dark blue-grey.

Citation:

Cape. Namaqualand, 2816DA, Grootderm, *Pillans* 5287 (BOL).

Phyllopodium hispidulum is known only from the type collection made in the Klinghardtberge east of Bogenfels in southernmost Namibia not far from the coast, and from one other on hills east of Grootderm on the Orange River, not far from Alexander Bay. It flowers in September and October. The species is closely allied to *P. pumilum*, with which it is sympatric in the Richtersveld. They closely resemble each other in habit and in their entire (or nearly so) leaves; they differ principally in indumentum, that of *P. hispidulum* being coarser and harsher than that of *P. pumilum* though difficult to define precisely. There may also be some difference in flower size, evident in the corolla lobes: posticuous lobes 0.5mm long in *P. hispidulum* (versus 0.6–1mm), anticous lobe 1.25–2mm (versus 2–2.5mm).

Excluded names

Phyllopodium baurii Hiern = *Selago baurii* (Hiern) Hilliard
Phyllopodium rudolphii Hiern = *Selago humilis* Rolfe (homotypic)

13. Zaluzianskya

The characteristic features of *Zaluzianskya* are its flowers in terminal spikes, these often capitate or at least dense, sometimes lax or becoming lax in fruit; the bract adnate to the calyx; calyx strongly ribbed and plicate; long narrow corolla tube only slightly inflated at the apex; filaments decurrent to the base of the corolla tube forming a channel embracing the style; lingulate stigma with marginal papillae; the ± beaked capsule, and pallid or mauve-grey colliculate seeds, obscurely angled or distinctly winged.

Although the genus is relatively easy to recognize, many of the species are not, and it is common to find two species mixed on an herbarium sheet. The species fall into four sections (elaborated below). Those comprising sect. *Nycterinia* are particularly difficult to discriminate. They are the species with corolla lobes bifid, white above, red below, and expanding at night or in dull light (only one species has its flowers opening in sunshine), and their discrimination depends largely on differences in habit and rootstock, not always apparent in dried material; the species from Transvaal and Zimbabwe in particular are still inadequately known. Many of these species are perennial and their areas lie mainly in the eastern half of southern Africa. In the western half, from Namibia south to the Cape, all the species are annual, including the few species of sect. *Nycterinia* that occur there. Nearly all these more western species (excluding those in sect. *Nycterinia*) have flowers that open in sunshine and the upper surface of the corolla lobes commonly ranges from white to shades of pink and mauve; shades of yellow or orange are rare except in sect. *Holomeria* and sect. *Zaluzianskya* subsect. *Noctiflora* (where the flowers open at dusk or in dull light). Differences in colour patterning around the mouth of the corolla tube are important taxonomically, and possibly so too are differences in colour between the upper and the lower surfaces of the corolla lobes. The latter difference is difficult or impossible to see in dried specimens and is very inadequately documented by collectors. The patterns around the mouth can often be detected because the tissue is anatomically different from that of the rest of the lobes, and the colour is presumably held in the plastids rather than in the cell sap of the epidermal cells.

The biological significance of the colour patterns has yet to be investigated and it is potentially of great interest. Hilliard & Burtt (1983 p.4), in writing about sect. *Nycterinia* (as sect. *Zaluzianskya*), have already commented on the striking contrast between actinomorphic flowers opening at dusk and then becoming fragrant, and zygomorphic odourless flowers opening in sunshine; the limbs of actinomorphic corollas are held horizonatally, those of zygomorphic corollas vertically. In sect. *Zaluzianskya* subsect. *Noctiflora*, it has been noted by several collectors that the flowers of *Z. peduncularis* (yellow) become fragrant when they open at dusk, while in sect. *Macrocalyx*, Marloth recorded of *Z. acutiloba* (saffron yellow flowers open in sunshine) 'strong scent'. But nothing is known of the role played by the yellow/orange/scarlet star-shaped patch around the mouth; sometimes the colour changes from yellow to red as the flower ages (e.g. *Z. villosa* and *Z. affinis*) but in *Z. rubrostellata* (subsect. *Noctiflora*) for instance, it is always bright scarlet; in *Z. peduncularis* (subsect. *Noctiflora*) there is no star, but the *throat* is scarlet or not; there is possibly a change in colour as the flower ages. In *Z. chrysops*, *Z. ovata* and *Z. distans* (sect. Nycterinia), there may be a bright orange (or scarlet) zone around the mouth, circular in outline (not star-shaped); it does not appear to be a constant feature except in *Z. chrysops*. Precisely the same orange band is found in the corolla of *Jamesbrittenia dentatisepala*, which is sympatric with *Z. chrysops* and *Z. ovata* in the high Drakensberg.

The biological significance of the circlet of unicellular hairs around the mouth of the corolla tube also needs investigation. This is present in many species. In sect. *Zaluzianskya* subsect. *Zaluzianskya,* the hairs are relatively long and acute (in subsect. *Noctiflora*, only *Z. rubrostellata* sometimes has a ring of hairs). In sect. *Nycterinia,* the hairs are obtuse and the circlet may be only partially developed; in more or less zygomorphic corollas it is sometimes wanting or weakly developed on one side only. In sect. *Holomeria*, the hairs are also obtuse, as they are in sect. *Macrocalyx*, though here they may be very short, and glands are always present in a broad band around the mouth, sometimes to the exclusion of hairs. Glands sometimes occur in sect. *Nycterinia* as well.

The breeding patterns in some of the species give promise of rewarding research. Cleistogamous flowers occur occasionally in *Z. benthamiana* but cleistogamy is commonplace in three species, *Z. peduncularis* and *Z. kareebergensis* (sect. *Zaluzianskya* subsect. *Noctiflora*) and *Z. divaricata* (sect. *Holomeria*). All the flowers in a head can be cleistogamous, or sometimes a few of the flowers in such a head will expand an abnormally small limb and appear to be autogamous; or sometimes the lowermost one or a few flowers in a head are cleistogamous, the rest being apparently normal chasmogamous flowers. Autogamy appears to occur in at least three other species, all in sect. *Zaluzianskya*: *Z. affinis* produces a form with a small limb and included stigma and field investigation is needed to see if this is linked to loss of colour patterning around the mouth; *Z. parviflora* has insignifcant flowers lacking any colour pattern and the stigma is ± included; *Z. cohabitans* also has tiny flowers, the two anticous stamens have been aborted, there is no colour pattern around the mouth, and the stigma is often included.

Prolific seed-set suggests that the species of *Zaluzianskya* are self-compatible. On the other hand, the strongly reflexed corolla lobes and strongly exserted anticous anthers and stigma seen in many species, as well as the emission of fragrance, suggest outcrossing. It is possible that the anthers shed their pollen before the style elongates to push the stigma out of the corolla tube. Hybridization seems to be rare, though *Z. violacea* and *Z. bella* perhaps introgress, and I have suggested (p.517) that *Z. pusilla* and *Z. acrobareia* may hybridize.

Zaluzianskya F.W. Schmidt, Neue Selt. Pflanzen. 11 (1793); Walpers, Repert. Sp. Nov. 3: 306 (1844); Hiern in Fl. Cap. 4(2): 333 (1904); Hilliard & Burtt in Notes RBG Edinb. 41: 1 (1983); nom. cons., non *Zaluzianskia* Necker (= *Marsilea* L.). **Figs 17, 31, 32, 42A, 64.**

Syn.: *Nycterinia* D. Don in Sweet, Brit. Fl. Gard. ser. 2, 239 (1834); Benth. in Hook., Comp. Bot. Mag. 1: 369 (1836)
    & in DC., Prodr. 10: 348 (1846). Type: *N. lychnidea* D. Don [= *Zaluzianskya maritima* (L.f.) Walp.]

Annual or perennial herbs. *Stems* either leafy throughout or mostly at base, pubescent, hairs often acute, patent or retrorse, sometimes glandular hairs as well. *Leaves* opposite, often subopposite or alternate upwards, all radical or radical and cauline, simple, entire to variously

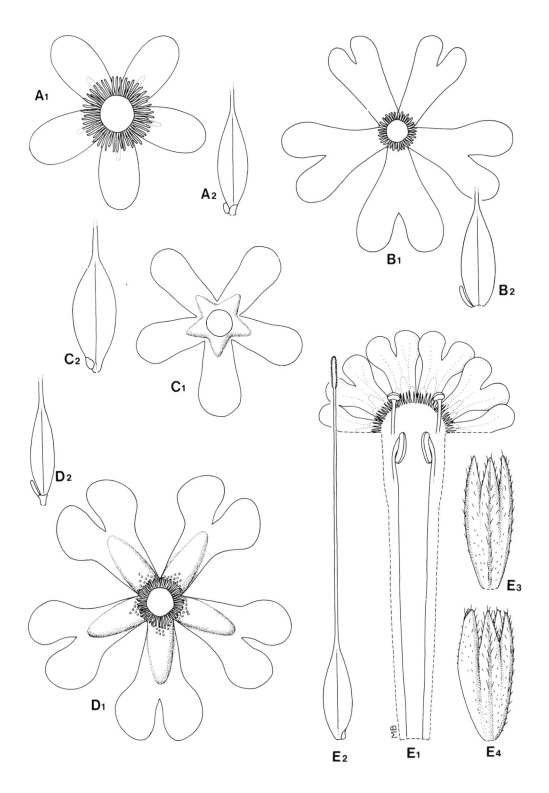

toothed. *Inflorescence* a terminal spike, often congested, sometimes lax (a minute pedicel is often present, but is so short as to be insignificant). *Bract* adnate to calyx. *Calyx* bilabiate, strongly 5-ribbed and plicate, anticous lip usually 2-lobed, occasionally ± entire, posticous lip 3-lobed, lobes ± deltoid, pubescent at least on margins. *Corolla tube* narrowly cylindric, slightly expanded in throat, pubescent outside, hairs sometimes sparse or occasionally wanting, glabrous inside but mouth often bearded with long acute or clavate hairs; limb often ± regular, sometimes bilabiate, lobes rotate or reflexed (but cleistogamy not uncommon), either suborbicular, spathulate, elliptic or oblong, then entire or retuse, or Y-shaped and then once or twice bifid, usually glandular outside, either glabrous inside or glandular particularly near mouth. *Stamens* 4 or 2 by reduction of either the anticous or the posticous pair (staminodes rarely present and never consistently so), didynamous, inserted near apex of tube, posticous pair included, filaments decurrent to base of tube forming a channel for the style, anticous anthers usually smaller, often exserted or tips appearing in mouth, all anthers synthecous. *Stigma* lingulate with 2 marginal bands of stigmatic papillae, longer than style, often exserted. *Ovary* ± elliptic in outline tapering upwards into the style, base often slightly oblique, with a nectariferous gland on the shorter side, gland either small, rounded, adnate to base of ovary, or peg-like and free from ovary, ovules many in each loculus. *Fruit* a ± distinctly beaked septicidal capsule with a short loculicidal split at tip of each valve, usually glabrous, rarely a few minute glands at apex. *Seeds* seated on a funicle in the form of a raised centrally depressed cushion, in some sections the cushions coalescing and ill-defined (fig. 17), seeds obscurely angled, the angles sometimes marked by very narrow wings, well developed only in section *Macrocalyx*, testa thin, pallid or mauve-grey, tightly investing the endosperm, under the SEM the polygonal cell outlines clearly visible, the outer periclinal wall plane or slightly convex, verruculose, in section *Macrocalyx* radially wrinkled with a central tubercle (figs 31, 32A, B).

Type species: *Zaluzianskya villosa* F.W. Schmidt
Distribution: mainly southern Arica, annual species concentrated in the western and south western Cape, thinning out northwards and eastwards, one isolated on the high mountains of East Africa; perennial species concentrated in eastern southern Africa, thinning out northwards through the Transvaal to Zimbabwe.

Bentham (1836) was the first to study the genus, under the then correct name *Nycterinia* D. Don (the name *Zaluzianskya* F.W. Schmidt, 1793, is now conserved against *Zaluzianskia* Necker); in 1846, he knew 16 species, and subdivided the genus into section *Zaluzianskya* (corolla lobes bifid) with 12 species, and section *Holomeria* (corolla lobes entire) with four species. He further subdivided section *Zaluzianskya* into two unranked groups, *Lychnideae* (leaves obovate, oblong, lanceolate or linear; corolla tube pubescent; ten species) and *Selaginoideae* (leaves spathulate, corolla tube glabrous; two species). Hiern (1904) did not recognize sections; he enumerated 32 species (four now transferred to other genera, 10 other names relegated to synonymy). Hilliard & Burtt (1983) took Bentham's sectional names back into use and chose *Z. divaricata* as the lectotype of sect. *Holomeria*. However, they were dealing with only part of the genus, mainly those species confined to the summer rainfall area of southern Africa. Now that the whole genus has been studied and 55 species recognized, it is clear that

Fig. 64 (opposite). A, *Zaluzianskya* sect. *Holomeria*, *Z. pusilla* (*Batten* 737): A1, corolla limb with orange bar on each lobe, circlet of hairs around mouth; A2, ovary with small adnate nectariferous gland. Both × 8. B, *Zaluzianskya* sect. *Nycterinia*, *Z. glareosa* (*Hilliard & Burtt* 9617): B1, corolla limb with circlet of hairs around mouth; B2, ovary with free peg-like nectariferous gland. Both × 4.5. C, *Zaluzianskya* sect. *Zaluzianskya* subsect. *Noctiflora*, *Z. rubrostellata* (*Hilliard & Burtt* 14674): C1, corolla limb with scarlet star-shaped patch around mouth; C2, ovary with small adnate nectariferous gland. Both × 4.5. D, *Zaluzianskya* sect. *Macrocalyx*, *Z. pumila* (*Batten* 1012): D1, corolla limb with orange star-shaped patch around mouth, circlet of hairs and sessile glands; D2, ovary with free peg-like nectariferous gland. Both × 4.5. E, *Zaluzianskya* sect. *Zaluzianskya* subsect. *Zaluzianskya*, *Z. violacea* (*Batten* 774): E1, corolla opened out to show circlet of hairs around mouth, decurrent posticous filaments, × 4.5; E2, ovary with small adnate gland, × 4.5; E3, calyx, posticous side, × 5; E4, calyx in lateral view, × 5.

sect. *Zaluzianskya* and sect. *Holomeria* need redefinition and that further subdivision is warranted. Consequently, four sections and two subsections are now recognized.

Key to sections and subsections

1a. Corolla limb *either* snow-white above (sometimes with an orange band around the mouth), pale to dark red below, opening at dusk or in dull light, *or* (one species only) creamy to greenish-white above, pale to dark red below, zygomorphic, opening in sunshine

    sect. **Nycterinia**

1b. Corolla limb variously coloured, but never contrasting white and red, mostly opening in sunshine, but if opening at dusk or in dull light, then creamy coloured to bright yellow above 2

2a. Calyx 7–17mm long, stamens 4, nectariferous gland conspicuous, peg-like, free from ovary

    sect. **Macrocalyx**

2b. Calyx 3–9mm long, *either* stamens 4 or 2 (anticous pair) and the nectariferous gland tiny, rounded, adnate to base of ovary, *or* stamens 2 (posticous pair), nectariferous gland tiny, peg-like, adnate to extreme base of ovary 3

3a. Corolla lobes oblong to elliptic, entire, white to bright yellow, orange or brick red above, expanding in sunshine; stamens 4 or 2 (anticous pair) .... sect. **Holomeria**

3b. Corolla lobes spathulate or oblong-spathulate, either deeply divided (Y-shaped) or retuse to entire, lobes variously coloured, including yellow; stamens 4 or 2 (posticous pair) 4

4a. Corolla lobes creamy yellow to bright yellow above, purple-brown below, tips rounded; flowers opening at night or in dull light .... sect. **Zaluzianskya** subsect. **Noctiflora**

4b. Corolla lobes white or creamy white to various shades of pink or mauve, tips often bifid or retuse; if rounded, then lobes pink or mauve or changing colour from white to pink or mauve, or sometimes white or orange, but always coppery orange below, tips deeply bifid to retuse or sometimes rounded; flowers opening in sunshine

    sect. **Zaluzianskya** subsect. **Zaluzianskya**

Zaluzianskya sect. **Nycterinia** (D. Don) Hilliard, comb. et stat. nov. **Fig. 64B.**
Type species: *N. lychnidea* D. Don [= *Z. maritima* (L.f.) Walp.].

Syn.: *Nycterinia* D. Don in Sweet, Brit. Fl. Gard. ser. 2, 239 (1834).
    Nycterinia sect. *Zaluzianskya* group *Lychnideae* Benth. in DC., Prodr. 10: 348 (1846) p.p. excl. *N. pumila*.

Annual or perennial herbs. *Corolla* limb usually actinomorphic, weakly to strongly zygomorphic in a few species, lobes spathulate in outline, deeply bifid, deep pink to red below, usually white above and opening at dusk (creamy- or greenish-white in one species and opening in sunshine), rarely with a bright orange circular band around the mouth, mouth often with a circlet of unicellular obtuse hairs, these sometimes wanting or wanting on the anticous side. *Stamens* 4. *Nectariferous gland* peg-like, conspicuous, free from ovary. *Placentae* with funicles coalescing to produce irregular longitudinal striae. *Seeds* collicate, angled. *Testa* with polygonal cell outlines clearly visible, outer periclinal walls slightly convex.

Section *Nycterinia* comprises species 1–20, eight of them annual or sometimes surviving for more than one season, the rest strongly perennial. Most of the species are confined to the eastern half of southern Africa, where rain falls mainly in the summer months, from Zimbabwe (*Z. tropicalis*) south to the Eastern Cape; here, in the area where rain falls all the year round, perennial species begin to fade out, and in the southern and western Cape (rainfall in winter) only three species occur, all annual (*Z. maritima* , *Z. capensis*, *Z. muirii*).

Zaluzianskya sect. **Macrocalyx** Hilliard, sect. nov. **Fig. 64D.**

Herbae annuae, floribus diurnis. *Calyx* insigniter conspicuus, costis exceptis papyraceus et inflatus. *Corollae* limbus actinomorphicus; lobi ambitu spatulati, profunde bifidi et interdum iterum lobati, plerumque albi vel malvini vel rosei in pagina superiore, inferne malvini vel rosei vel rubri (in una specie crocei et demum brunnescens), interdum area flava vel rubra stellata semper annulo glandularum circa oram praediti, interdum insuper annulo interiore pilorum obtusorum brevium praestante. *Stamina* 4. *Glans nectarifera* cylindrica conspicua ab ovario libera. *Placentae* funiculis pulviniformibus medio depressis discretis praeditae. *Semina* late alata, profunde rugosa; cellulae polygonales testae manifestae pariete exteriore mamillata.

Type species: *Z. inflata* Diels

Annual herbs. *Calyx* remarkably prominent, papery (except for the ribs) and inflated. *Corolla* limb actinomorphic, lobes spathulate in outline, deeply bifid and sometimes lobed again, usually white to mauve or pale to deep pink above, mauve to pink or red below, saffron yellow ageing brown in one species, opening in sunshine, with or without a yellow to orange, scarlet or crimson star-shaped patch around the mouth, always with a circlet of glands around the mouth, sometimes an inner circlet of short obtuse hairs as well. *Stamens* 4. *Nectariferous gland* peg-like, conspicuous, free from ovary. *Placentae* with discrete, cushion-like, centrally depressed funicles. *Seeds* broadly winged, deeply wrinkled and dimpled. *Testa* with polygonal cell outlines clearly visible, outer periclinal walls strongly to weakly mamillate.

Section *Macrocalyx* comprises 21. *Z. pumila*, 22. *Z. inflata*, 23. *Z. sutherlandica*, 24. *Z. mirabilis*, 25. *Z. acutiloba*, which are confined to the western and southern Cape, from Van Rhynsdorp and Calvinia divisions south through Sutherland and east along the mountains to Willowmore division. Good collections from Sutherland and Willowmore are much needed.

Zaluzianskya sect. **Zaluzianskya**

Type species: *Z. villosa* F.W. Schmidt

Annual herbs. *Corolla* limb actinomorphic, lobes spathulate or oblong-spathulate, bifid, emarginate or entire, white, creamy, mauve, pink, red-violet or yellow above, coppery orange to purple-brown below, opening in sunshine or in dull light, with or without a star-shaped yellow-orange-red patch around mouth, circlet of unicellular acute hairs present or absent. *Stamens* 4 or 2 (posticous pair). *Nectariferous gland* either small, peg-like, adnate to extreme base of ovary, or tiny, rounded, adnate to base of ovary. *Placentae* with ± discrete, often somewhat flattened, cushion-like centrally depressed funicles. *Seeds* colliculate, angled, sometimes with very narrow partially developed wings. *Testa* with polygonal cell outlines clearly visible, outer periclinal walls scarcely raised.

Section *Zaluzianskya* comprises species 26–45. It can be divided into two subsections:

subsect. Zaluzianskya. Fig. 65E.

Syn.: *Nycterinia section Zaluzianskya* group *Selaginoideae* Benth.in DC., Prodr. 10: 349 (1846) p.p.

Flowers opening in sunshine. *Corolla* lobes deeply bifid to retuse, rarely rounded, white or creamy white to various shades of pink or mauve (sometimes changing as the flowers age), rarely yellow or orange. Sometimes lower surface similar in colour to upper, sometimes shades of coppery orange (more information needed), mouth usually marked by a yellow to orange or scarlet star-shaped patch and a circlet of acute erect hairs. *Stamens* 4 or 2. *Nectariferous gland* small, peg-like in species with 2 stamens, rounded, adnate to base of ovary in species with 4 stamens.

Subsection *Zaluzianskya* embraces species 26–41 in the enumeration. They are widely distributed in the western, southern, and central Cape (though only one occurs in the Peninsula) and tail out in the eastern Cape. One (*Z. venusta*) reaches as far north as Kuruman and just gets into the western Orange Free State, but is also found as far south as Ladismith division; another (*Z. karrooica*) ranges from Carnarvon and Beaufort West to Bloemfontein and Fort Beaufort; *Z. crocea* is confined to the mountains around Barkly East, Naude's Nek and Queenstown, while *Z. vallispiscis* is in and around the Fish River valley near Grahamstown; *Z. synaptica* stretches right across the Karroo from Karroo Poort to Graaff Reinet. Some species are ill known and appear to be of very limited distribution: *Z. marlothii* near Sutherland, *Z. parviflora* around Saldanha Bay, and *Z. gracilis* in Bredasdorp and Riversdale divisions.

subsect. **Noctiflora** Hilliard, subsect. nov. **Fig. 64C.**

Flores crepusculo vel sub lumine obscuro aperientes. *Corollae* lobi integri, apicibus rotundatis, cremei vel vivide flavi in pagina superiore, inferne purpureo-brunnei, ore interdum colore altero notato, annulo pilorum plerumque absente in una specie interdum praestante. *Stamina* 4. *Glans nectarifera* minuta, rotundata, ad basin ovarii adnata.

Type species: *Z. peduncularis* (Benth.) Walp.

Flowers opening at dusk or in dull light. *Corolla* lobes entire, tips rounded, creamy white to bright yellow above, purple-brown below, mouth with or without a contrasting colour pattern, circlet of hairs sometimes present in one species, wanting in the others. *Stamens* 4. *Nectariferous gland* tiny, rounded, adnate to base of ovary.

Subsection *Noctiflora* includes four species, *Z. peduncularis*, *Z. elgonensis*, *Z. rubrostellata* and *Z. kareebergensis*. *Zaluzianskya peduncularis* has the widest range of any species, from southern Namibia south to Montagu and east through the Karroo to the Orange Free State and eastern Cape. *Zaluzianskya elgonensis* is closely allied to *Z. peduncularis*, but there is a remarkable disjunction to its only known localities, on Elgon and Kilimanjaro. *Zaluzianskya rubrostellata*, which is confined to the mountains of Lesotho and the nearby Drakensberg and Witteberg, completes the close-knit trio. *Zaluzianskya kareebergensis* is known only from Kareeberg in Van Rhynsdorp division, and stands apart by virtue of its leafy stems.

sect. **Holomeria** (Benth.) Hilliard & Burtt in Notes RBG Edinb. 41:7 (1983). **Fig. 64A.**
Lectotype: *Z. divaricata* (Thunb.) Walp.

Syn.: *Nycterinia* sect. *Holomeria* Benth. in DC., Prodr. 10: 350 (1846), p.p., excl. *Z. peduncularis*.

Annual herbs. *Corolla* limb ± actinomorphic, lobes oblong to elliptic, entire, creamy white to bright yellow or orange (rarely brick-red) above, sometimes purple-brown or maroon below (more information needed), opening in sunshine, often with a yellow to orange or scarlet median bar on each lobe or a similarly coloured star-shaped patch around the mouth (more information needed), all but one species (*Z. diandra*) with a circlet or broad band of obtuse hairs as well. *Stamens* 4 or 2 (anticous pair). *Nectariferous gland* tiny, rounded, adnate to base of ovary. *Placentae* with well flattened cushion-like centrally depressed funicles. *Seeds* colliculate, angled, testa with polygonal cell outlines clearly visible, outer periclinal walls plane.

Section *Holomeria* embraces ten species (46–55), three of which (*Z. divaricata*, *Z. pusilla*, *Z. benthamiana*) were included by Bentham in his original concept of *Nycterinia* sect. *Holomeria*; Bentham's fourth species, *Z. peduncularis*, is now included in sect. *Zaluzianskya* subsect. *Noctiflora*. The other species of sect. *Holomeria* were unknown to Bentham.

Zaluzianskya benthamiana (*Nycterinia villosa* Benth.) has a wide distribution from southern Namibia south to Saldanha Bay and Ceres then east to Oudtshoorn, while *Z. diandra* ranges from southern Namibia south to the Orange River and east to Prieska. These two species stand apart in the Section in having only two stamens; *Z. diandra* differs further in its strongly reflexed narrow corolla lobes with revolute margins and disproportionately large far-exserted anthers.

The other eight species, each having four stamens, are confined to the western Cape. *Zaluzianskya pusilla* is the most widespread, with a range from Kamieskroon to Hessequaspoort near Swellendam. Its unnamed ally (no. 53) is known only from Sneeuwkop near Kamieskroon. All but one of the other species seem to be of very limited distribution, but field work would possibly extend their ranges: *Z. collina* (Kamiesberg and Kamieskroon area); *Z. acrobareia* (Van Rhyn's Pass); *Z. glandulosa* (northern Cedarberg at Pakhuis); *Z. isanthera* (mountains north of Ceres); *Z. lanigera* (Karoo Poort). The type of the section, *Z. divaricata*, is obviously a common plant ranging from the northern Cedarberg to the Peninsula and east to Albertinia.

Key to species

1a. Corolla limb *either* snow-white above (in 3 species with an orange band around the mouth), pale to dark red below, opening at dusk or in dull light, *or* (1 species only) creamy to greenish-white above, opening in sunshine, plant perennial with a mass of vegetative buds on the crown (**section Nycterinia**) 2
1b. Corolla limb variously coloured (white, shades of pink and mauve, or yellow) but not contrasting snow-white and red, mostly opening in sunshine, if opening at dusk or in dull light, then creamy-coloured to bright yellow above 26

2a. Upper surface of corolla lobes glandular at least in lower part, mouth either not bearded or rarely a few long unicellular hairs present on the posticous side 3
2b. Upper surface of corolla lobes either glabrous or glandular, if glandular, then mouth well bearded 4

3a. Corolla limb zygomorphic, held vertically, a few hairs sometimes present on anticous side of mouth; perennial herb with a thick clump of vegetative buds on the crown
1. Z. microsiphon
3b. Corolla limb regular, held more or less horizontally, beard wanting; annual or weakly perennial, but lacking vegetative buds on the crown . . . . **12. Z. muirii**

4a. Stems either glabrous or with coarse retrorse eglandular hairs 5
4b. Stems with coarse spreading hairs with or without a glandular tip, sometimes short fine spreading glandular hairs as well 21

5a. Corolla lobes glabrous above (ignore tiny glands that may be present near the sinuses of the corolla lobes adjacent to the circlet) 6
5b. Corolla lobes glandular above at least in the lower part 16

6a. Corolla limb slightly zygomorphic (the anticous lobe slightly isolated from the other four), held vertically 7
6b. Corolla limb regular, held more or less horizontally 8

7a. Leaves coarsely toothed, pilose at least on the lower surface; corolla tube usually 17–24mm long . . . . **13. Z. oreophila**
7b. Leaves either entire or with a few small teeth, usually glabrous or with a few hairs on margins and midline; corolla tube usually 30–40mm long . . . . **7. Z. elongata**

8a. Upper surface of corolla lobes bright orange in the lower half . . . . **14. Z. chrysops**
8b. Corolla lobes wholly white above 9

9a. Plant pulvinate with many crowded green leaf rosettes forming dense cushions or small mats . . . . **4. Z. pulvinata**
9b. Plant not pulvinate though leaf rosettes sometimes present on the crown, but few, not forming cushions nor mats 10

10a. Plants strongly perennial, with either a thick clump of vegetative buds on the crown or a thickened woody taproot, which may bear a few small buds or loose leaf tufts on the crown (specimens flowering in the seedling stage will not key out). E Cape to Natal . . 11
10b. Plants either annuals or short-lived perennials with the taproot not exceeding c.4mm in diam. and lacking basal vegetative buds. Cape to E Cape 15

11a. Cauline leaves mostly up to 2mm broad 12
11b. Cauline leaves mostly at least 3mm broad 13

12a. Inflorescences usually with 10-25 flowers, corolla lobes usually 3–5mm long
6. Z. schmitziae
12b. Inflorescences usually with up to 15 flowers, corolla lobes usually 5–8mm long
5. Z. angustifolia

13a. Plant developing a thick clump of vegetative buds on the crown, stem simple
2. Z. spathacea
13b. Plants without a thick clump of vegetative buds on the crown, at most a few small ones,
stems either simple or branched (or with axillary leaf tufts) 14

14a. Stem simple; leaves glabrous or nearly so; corolla tube with acute hairs up to 0.4–0.7mm
long, glandular hairs up to 0.3mm .... **9. Z. pachyrrhiza**
14b. Stems either branched or with axillary tufts, simple mainly in young specimens and then
leaves hairy; corolla tube minutely glandular-puberulous .... **15. Z. glareosa**

15a. Leaves almost fleshy, elliptic, more or less entire, glabrous or with a few hairs on margins
and midline below .... **10. Z. maritima**
15b. Leaves herbaceous, linear to elliptic, often toothed, frequently only narrow leaves entire,
usually sparsely to densely pilose, rarely glabrous .... **11. Z. capensis**

16a. Cauline leaves mostly linear, spreading, mostly 60–90 × 0.75–3mm ... **5. Z. angustifolia**
16b. Cauline leaves mostly elliptic, at least the upper ones ascending; if spreading, then either
c.7–17mm broad or up to c.50mm long 17

17a. Stems either branched or with small axillary leaf tufts (incipient branches) 18
17b. Stems simple 19

18a. Crown with a thick clump of vegetative buds, whole plant pilose, hairs all eglandular
8. Z. pilosa
18b. Crown without a thick clump of vegetative buds, plant only moderately hairy, glandular
hairs often mixed with the eglandular ones .... **16. Z. tropicalis**

19a. Crown without a thick clump of vegetative buds, root becoming woody and carrot-like;
corolla tube with acute hairs up to 0.4–0.7mm long, glandular hairs up to 0.3mm
9. Z. pachyrrhiza
19b. Crown with a thick clump of vegetative buds; corolla tube usually glandular-pubescent,
only rarely with a few acute hairs 20

20a. Inflorescence remaining short and thick in fruit, not elongating to reveal the axis (upper
surface of corolla lobes glandular on claw) .... **3. Z. natalensis**
20b. Inflorescence elongating in fruit to reveal the axis; upper surface of corolla lobes either
glandular or eglandular but always eglandular where sympatric with Z. *natalensis* (Natal
and nearby) .... **2. Z. spathacea**

21a. Corolla lobes glabrous above (ignore tiny glandular hairs adjacent to the circlet of long
hairs) 22
21b. Corolla lobes glandular above at least in lower half 24

22a. Corolla tube 21–27mm long; leaves and bracts rhomboid-ovate, tips recurved
17. Z. turritella
22b. Corolla tube 27–58mm long; leaves oblong, elliptic or ovate, tips not recurved 23

23a. Plant herbaceous, often flowering in the seedling stage, later several-stemmed from the
crown; leaves usually more than twice as long as broad .... **15. Z. glareosa**

23b. Plant with many twiggy branches forming low interwoven clumps; leaves usually only twice as long as broad .... **19. Z. ovata**

24a. Most leaves ovate or cuneate in outline, abruptly contracted to petiolar part
18. Z. katharinae

24b. Leaves oblong-elliptic to elliptic-ovate, gradually narrowed to the petiolar part 25

25a. Leaves rapidly diminishing in size upwards; inflorescence crowded .... **16. Z. tropicalis**

25b. Leaves scarcely diminishing in size upwards; inflorescence lax, often markedly so
20. Z. distans

26a. Calyx 7–17mm long, stamens 4, nectariferous gland conspicuous, peg-like, free from ovary (**section Macrocalyx**) 27

26b. Calyx 3–9mm long, *either* stamens 4 or 2 (anticous pair) and the nectariferous gland tiny, rounded, adnate to base of ovary, *or* stamens 2 (posticous pair), nectariferous gland tiny, peg-like, adnate to extreme base of ovary 31

27a. Hairs on stem retrorse, claw of each corolla lobe with a thick-textured yellow/orange/scarlet bar, forming a raised star around the mouth visible even in dried material .... **21. Z. pumila**

27b. Hairs on stem patent, coloured star around mouth, if present, not thick-textured 28

28a. Mouth of corolla tube encircled by a band of clavate hairs giving way to tiny glands on shaft of lobes, lobes 2.5–4.5mm long 29

28b. Mouth of corolla tube encircled by a band of glands only, lobes 5–7mm long 30

29a. Lowermost bracts 20–35mm long, corolla tube 29–35mm long, anticous anthers 0.4–0.8mm, posticous anthers 1.5–1.8mm .... **22. Z. inflata**

29b. Lowermost bracts c.10mm long, corolla tube c.22mm long, anticous anthers c.1.25mm, posticous anthers c.2.5mm .... **25. Z. acutiloba**

30a. Hairs on stem gland-tipped, up to 1–1.5mm long, very occasional acute hairs to 2mm also present, corolla tube 30–38mm long, corolla lobes pink to mauve above, margins plane
24. Z. mirabilis

30b. Hairs on stem a mixture of gland-tipped and acute, or acute only, up to 1mm long, corolla tube 23–27mm long, corolla lobes saffron yellow turning brown, margins revolute, the lobes very acute to the eye .... **25. Z. acutiloba**

31a. Corolla lobes spathulate or oblong-spathulate, either deeply bifid (Y-shaped) or retuse to entire, lobes *either* white or mauve or shades of pink to red-violet above (rarely yellow) expanding in sunshine, stamens 4 or 2 (posticous pair), *or* creamy coloured to bright yellow above expanding at dusk or in dull light, stamens 4 (**section Zaluzianskya**) ... 32

31b. Corolla lobes oblong to elliptic, entire, white to bright yellow, orange or brick red above, expanding in sunshine, stamens 4 or 2 (anticous pair) (**section Holomeria**) 52

32a. Corolla lobes white or creamy-white to various shades of pink and mauve, tips often bifid or retuse; if rounded, then lobes pink or mauve or changing colour from white to pink or mauve, or sometimes white or orange, but always coppery orange below; flowers expanding in sunshine (**section Zaluzianskya subsect. Zaluzianskya**) 33

32b. Corolla lobes creamy coloured to bright yellow above, purple-brown below, tips rounded; flowers expanding at night or in dull light (**section Zaluzianskya subsect. Noctiflora**) 49

33a. Stamens 2 (posticous pair); tips of corolla lobes always bifid 34

33b. Stamens 4; tips of corolla lobes bifid, retuse or rounded, sometimes in a single flower 40

34a. Hairs on stem patent, acute; minute glandular hairs also present 35

34b. Hairs on stem ± retrorse, acute, or sometimes stems almost glabrous 38

35a. Corolla tube (9–) 15–30mm long (take care to measure fully developed flowers) 36
35b. Corolla tube 6.5–10mm long 37

36a. Leaves and bracts oblanceolate to spathulate, ± obtuse, margins toothed, densely
     pubescent all over (see further in discussion under both *Z. villosa* and *Z. affinis*)
     26. Z. villosa
36b. Leaves and bracts lanceolate, ± acute, margins entire or nearly so, leaves sparsely hairy,
     hairs mainly on margins and petiolar part, bracts glabrous or very nearly so dorsally in
     distal half . . . . **27. Z. affinis**

37a. Stems bearing flowers almost from the base, leaves oblanceolate, ± acute. Saldanha Bay
     28. Z. parviflora
37b. Stems leafy in the lower part, floriferous only in upper part, leaves oblanceolate, apex
     obtuse to subacute. Bredasdorp and Riversdale divisions . . . . **29. Z. gracilis**

38a. Hairs on corolla tube eglandular . . . . **33. Z. violacea**
38b. Hairs on corolla tube glandular 39

39a. Stems either glabrous or with hairs up to 0.2mm long, backs of leaves and bracts glabrous
     or very nearly so . . . . **30. Z. sanorum**
39b. Stems pilose, hairs up to 0.75–1mm long, backs of leaves and bracts pilose with hairs up
     to 0.5–1mm long . . . . **31. Z. pilosissima**

40a. Corolla tube 10–35mm long 41
40b. Corolla tube 4–8.5mm long 48

41a. Hairs on stem ± retrorse (tiny patent glands often present as well), hairs on corolla tube
     acute, c.0.3–0.4mm long, rarely tube glabrous or only partly hairy (see notes under *Z.
     violacea*) . . . . **33. Z. violacea**
41b. Hairs on stem patent (patent to ± retrorse in one species), hairs on corolla tube glandular,
     0.1–0.5mm long 42

42a. Bracts (and leaves) clad in acute hairs only; corolla tube 11–15mm long, lobes distinctly
     bifid . . . . **32. Z. bella**
42b. Bracts (and leaves) with minute glandular hairs as well as acute ones; corolla tube mostly
     more than 15mm long, only 13–15mm in *Z. marlothii* but then lobes rounded 43

43a. Bracts spathulate, entire, corolla tube 13–15mm long, lobes rounded . . . **34. Z. marlothii**
43b. Either bracts not spathulate, or distinctly toothed, or corolla tube longer, or lobes bifid . 44

44a. Rays of star-shaped patch around mouth reaching well up shaft of corolla lobes 45
44b. Rays of star-shaped patch around mouth very short, scarcely reaching base of shaft of
     lobes 46

45a. Corolla lobes distinctly bifid; calyx 5–6.5mm long . . . . **35. Z. venusta**
45b. Corolla lobes rounded to slightly retuse; calyx 6.5–8mm long (see comments under *Z.
     karrooica*) . . . . **36. Z. karrooica**

46a. Lowermost bracts spathulate, 2–3 times as long as broad . . . . **39. Z. crocea**
46b. Lowermost bracts oblong-lanceolate, 3–6 times as long as broad 47

47a. Eglandular (acute) hairs on stems up to 1–1.5mm long, at least the lowermost bracts
     distinctly and sharply toothed (1–4 pairs of teeth) . . . . **40. Z. vallispiscis**
47b. Eglandular hairs on stems up to 0.2–0.8mm long, bracts entire or subentire, rarely with a
     few tiny teeth . . . . **41. Z. synaptica**

48a. Hairs on stem patent, stamens 4, corolla lobes often distinctly bifid (ranging from bifid to
     entire in a single flower) . . . . **37. Z. minima**

48b. Hairs on stem retrorse, appressed, stamens 2, corolla lobes entire or occasionally retuse
38. Z. cohabitans

49a. Mouth of corolla tube surrounded by a bright scarlet star-shaped patch, corolla tube
21–31mm long .... **44. Z. rubrostellata**
49b. Mouth of corolla tube without a scarlet star-shaped patch (though it may be scarlet in the
throat), corolla tube up to 25mm long 50

50a. Upper part of stem leafless 51
50b. Stem leafy throughout, corolla tube c.7.5–9mm long in chasmogamous flowers (shorter in
cleistogamous ones) .... **45. Z. kareebergensis**

51a. Corolla tube 17–25mm long (may be shorter in cleistogamous flowers), lobes creamy
yellow to lemon yellow above, darker yellow to bright scarlet ring in throat
42. Z. peduncularis
51b. Corolla tube 10–12mm long, lobes creamy coloured with an orange spot at the base of
each lobe .... **43. Z. elgonensis**

52a. Stamens 4 53
52b. Stamens 2 60

53a. Hairs on stem ± retrorse, acute, or hairs sparse or wanting (minute glandular hairs may
also be present) 54
53b. Hairs on stem patent, mostly gland-tipped 59

54a. Bracts lanceolate, deeply and sharply toothed .... **46. Z. divaricata**
54b. Bracts linear to narrowly elliptic, oblong-lanceolate or spathulate, either entire or with 1–2
pairs of small teeth 55

55a. Bracts ± spathulate to ovate 56
55b. Bracts linear oblong or narrowly elliptic 57

56a. Bracts and calyx white-woolly, corolla lobes orange-yellow .... **48. Z. lanigera**
56b. Bracts and calyx with acute hairs up to 1–1.8mm long, corolla lobes creamy white above
49. Z. isanthera

57a. Corolla tube c.17–20mm long, lobes c.3.5–6mm, bright yellow, mouth encircled by a
broad band of hairs running out onto bases of lobes. .... **50. Z. collina**
57b. Corolla tube 6–17mm long, lobes 1–4.5mm, white to creamy, mouth with a circlet of hairs
but these not extending out onto the lobes 58

58a. Primary stem eventually branching from the base, but stems otherwise usually simple; if
branched above, then each branch subtended by a single leaf; bracts with acute hairs up to
0.7–1.5mm long, often on margins and dorsal midline only, corolla tube 6–15mm long
52. Z. pusilla
58b. Stem well branched from base and above, tuft of leaves in axil of each primary leaf
(incipient branch), bracts glabrous, corolla tube 17mm long ..... **53. Z. sp. aff. Z. pusilla**

59a. Corolla tube c.10mm long, lobes c.2.5–3mm, posticous anthers 0.6–0.8mm
47. Z. glandulosa
59b. Corolla tube c.22-27mm long, lobes c.3.5–5mm, posticous anthers 1.4–1.7mm
51. Z. acrobareia

60a. Bracts lanceolate, corolla lobes elliptic, margins plane, white, creamy or yellow on upper
surface marked with a yellow to orange star around mouth, mouth either encircled by
hairs or glabrous .... **54. Z. benthamiana**
60b. Bracts spathulate, corolla lobes oblong, margins revolute, the lobes strongly reflexed,
orange to brick red above, no colour patterning, glabrous .... **55. Z. diandra**

Section Nycterinia

1. Zaluzianskya microsiphon (O. Kuntze) K. Schum. in Just, Jahresb. 26 (1): 395 (1900); Hiern in Fl. Cap. 4(2): 344 (1904); Hilliard & Burtt in Notes RBG Edinb. 41: 9 (1983); F. N. Hepper, The Kew Magazine 5: 69 t. 101 and fig. (1988).
Type: Natal, Van Reenen's Pass, 1800m, 20 iii 1894, *Kuntze* (iso. K).

Syn.: *Nycterinia microsiphon* O. Kuntze, Rev. Gen. Pl. 3(2): 238 (1898).
  Zaluzianskya maritima var. *breviflora* Hiern in Fl. Cap. 4(2): 336 (1904). Lectotype: Natal, Van Reenen's Pass, 6000ft, 5 iii 1895, *Schlechter* 6988 (K; isolecto. BOL, G, GRA, S).
  Z. maritima var. *atro-purpurea* Hiern, loc. cit. Lectotype: Natal, East Griqualand, mountains around Kokstad, 5500ft, ii 1883, *Tyson* 1354 (K; isolecto. BOL mixed with *Z. spathacea*, NBG, PRE).
  Z. maritima var. *grandiflora* Hiern, loc. cit. Type: Natal, East Griqualand, Mount Currie, 5200ft, ii 1884, *Tyson* 1733 (BOL, NBG, PRE).

Perennial herb with a thick clump of partly subterranean vegetative buds at the crown; stems often solitary, sometimes 2 or 3, usually simple, sometimes producing 1–3 secondary spikes below the main spike, erect, up to c.400mm long, pilose with retrorse white hairs, leafy. *Leaves*: radical often more or less rosulate at flowering, oblanceolate or elliptic, mostly 35–90 × 8–20mm; cauline ascending, elliptic or oblong, mostly 20–65 × 4–9 (–20)mm, becoming slightly smaller upwards and passing imperceptibly into floral bracts; all leaves with margins entire, obscurely toothed, or callose-toothed particularly in upper part, hairy at least on margins and midline. *Flowers* many in a dense terminal spike, opening in sunshine, spike often very long, accounting for about half the total stem length, shorter in weak specimens. *Bracts* adnate to calyx for 1.5–3mm, mostly 15–30mm long, elliptic or lanceolate, margins entire or with 1 or 2 pairs of callose teeth near the tip, hairy at least on margins and midline. *Calyx* 8–14mm long, lips 3–4mm, hairy at least on margins. *Corolla* tube (16–) 20–40 (–52)mm long, glandular-puberulous, hairs up to 0.1–0.2mm long, limb held vertically, zygomorphic, mouth often glabrous, sometimes with a few large unicellular hairs, rarely thinly bearded on the anticous side; lobes deeply notched, the secondary lobes themselves sometimes notched, pink, scarlet or crimson and glandular-puberulous outside, creamy- or greenish-white inside, glandular-puberulous at least on claw of lobes. *Stamens* 4, all visible in the mouth, anticous anthers 1–1.3mm long, usually shortly exserted, the two posticous ones 2–2.5mm, often partly included, sometimes far exserted. *Stigma* 4–5mm long, exserted. *Capsule* 12–15 × 4mm. *Seeds* c.1–1.5 × 0.7–1.5mm, angled, colliculate, pallid.

Selected citations:

Transvaal. Wakkerstroom distr., 2730AD, farm Oshoek, 28 xii 1975, *Devenish* 1620 (E, NU).
Natal. Bergville distr., 2828DB, Royal Natal National Park, Basuto Gate area, c.7000ft, 1 ii 1978, *Stewart* 2036 (E, K, NU). Underberg distr., 2929CB, Sani Pass, 21 iii 1977, c.8000ft, *Hilliard & Burtt* 9740 (E, NU). Alfred distr., 3029DA, top of new Ingeli Pass, c.5300ft, *Acocks* 22008 (K, PRE). Transkei-Cape border, 3128BA, Kwenke [Ntywenka], 5200ft, i 1896, *Bolus* 8756 (K, PRE).
Lesotho. Maseru distr., 2928AC, Blue Mountain Pass, 8500ft, 10 i 1979, *Hilliard & Burtt* 12023 (E, NU).
Orange Free State. Harrismith distr., 2828DB, Bester's Vlei, 6000ft, *Flanagan* 2033 (PRE, SAM).

Zaluzianskya microsiphon appears to be rare in the Highlands of the Eastern Transvaal (records from Iron Crown Mountain and Makapansgat), becoming common on the low Drakensberg along the Transvaal-Natal border, thence along the high Drakensberg and Lesotho to the mountains around Engcobo in Transkei, between 1525 and 2745m above sea level. It grows in rocky grassland, flowering mainly between January and March.

There is remarkable variation in the size of the corolla limb, which is nevertheless the most distinctive feature of the species. The flowers open in sunshine and the limb is held vertically; it is strongly zygomorphic, glandular both above and below, and the mouth is glabrous or very nearly so. This contrasts with the actinomorphic limb of *Z. natalensis* and *Z. spathacea*, which is held horizontally and opens at dusk. These two species, like *Z. microsiphon*, produce a single stem from a mass of undeveloped buds on the crown of the rootstock, and favour similar habitats; *Z. microsiphon* and *Z.spathacea* are partly sympatric.

2. Zaluzianskya spathacea (Benth.) Walp., Repert. 3: 306 (1844); Hilliard & Burtt in Notes RBG Edinb. 41: 11 (1983), excl. *Z maritima* var. *pubens*.
Lectotype: Cape [Lady Grey distr.] at the top of the Witteberg, 7000–8000ft, 1 i 1833, *Drège* (K, isolecto. P).

Syn.: *Nycterinia spathacea* Benth. in Hook., Comp. Bot. Mag. 1: 369 (1836).
    [*Z. maritima* auct. non (L.f.) Walp.; Gumbleton in Gard. Chron. ser. 3, 42: 161 fig. 64 (1907); Skan in Bot. Mag.
      t. 8215 (1908). Cultivated plant from Barberton, SE Transvaal (K)].

Perennial herb with a thick clump of partly subterranean vegetative buds at the crown; stems solitary to several, simple, erect or ascending, up to 400mm long, usually with retrorse white hairs, sometimes glabrous, leafy. *Leaves* at base of stem elliptic or oblanceolate, often narrowed to a petiole-like base, 30–100 × 8–18mm, becoming narrowly elliptic to elliptic-oblong upwards and sharply ascending, 30–70 × 5–12 (–18)mm, a little smaller below the inflorescence, margins usually entire or obscurely toothed, rarely coarsely toothed, hairy at least on margins and midline. *Flowers* opening at dusk or in dull light, arranged in a terminal spike elongating in fruit. *Bracts* adnate to calyx for 1–5mm, 18–23mm long, lanceolate, margins entire or with one or two pairs of callose teeth near tips, hairy at least on margins and midline. *Calyx* 12–15 (–20)mm long, lips 4–6mm long, either glabrous except for hairy margins, or thinly hairy all over. *Corolla* tube (36–) 40–60mm long, glandular-puberulous, hairs up to 0.1–0.2 (–0.4)mm long, very rarely eglandular hairs as well, limb held horizontally, regular; mouth with a thick circlet of unicellular hairs, sometimes developed on upper side only; lobes 6–10mm long, deeply notched, crimson and glandular-puberulous outside, white inside, either glabrous or glandular particularly in lower half. *Stamens* 4, anticous anthers 1.3–2.4mm long, shortly exserted, posticous ones 3–5mm long, included. *Stigma* 6.5–10mm long, exserted. *Capsule* c.12–15 × 4mm. *Seeds* c.1 × 0.6mm (few seen), angled, colliculate, pallid.

Selected citations:

Transvaal. Pilgrim's Rest distr., 2530BB, Bakenkop, 8 iii 1981, *Hilliard & Burtt* 14272 (E, NU). Barberton distr., 2531CC, Summit Saddleback Mt, 5000ft, 22 ii 1890, *Galpin* 829 (GRA, K, PRE, SAM).
Natal. Underberg distr., 2929CC, vicinity of Tarn Cave above Bushman's Nek, c.7800ft, 18 i 1984, *Hilliard & Burtt* 17319 (E, NU).
Transkei. Baziya Mt., 3128CB, c.4700ft, 10 ii 1981, *Hilliard & Burtt* 13887 (E, NU). Insizwa to Umzimhlava, 3029CC, 28 i 1895, *Krook* 3044 (W).
Cape. Stutterheim div., 3227AC, Thomas River, 3000ft, i 1893, *Flanagan* 1715 (BOL, PRE, SAM). Stockenstrom div., 3226DA, Katberg Pass, 24 i 1979, *Hilliard & Burtt* 12357(E).
Orange Free State. Witzieshoek distr., 2828DB, path to The Sentinel, c.8700ft, 27 xii 1975, *Hilliard & Burtt* 8661 (E, NU).

Zaluzianskya spathacea has a wide geographical range from the eastern highlands and southeastern Highveld of the Transvaal, western Swaziland and NE Orange Free State through eastern Lesotho, Natal (between 1525 and 2650m), Transkei and the Eastern Cape roughly as far west as Somerset East. It grows in grassland, often among rocks, flowering mainly between December and February.

In Natal, the corolla lobes are always eglandular above and in the Transkei and Eastern Cape this seems to be the commoner condition, while in the Transvaal both glandular and eglandular lobes are equally common. *Zaluzianskya spathacea* is partly sympatric with *Z. microsiphon*, which differs in its zygomorphic corolla limb, expanded in sunshine, the flowers held vertically. Its relationship seems to lie with *Z. natalensis*, which differs in the inflorescence, congested even in fruit (not elongating to reveal the axis) and the corolla lobes always glandular above (eglandular in *Z. spathacea* where it is sympatric with *Z. natalensis* in Natal). See also under *Z. elongata*, p.478.

3. Zaluzianskya natalensis [Bernh. ex] Krauss in Flora 27(2): 834 (1844); Hilliard & Burtt in Notes RBG Edinb. 41: 13 (1983).
Type: 'In m. Bosjemann Rand, Natal' [World's View above Pietermaritzburg], 3000ft, Dec., *Krauss* (n.v.).

Syn. *Nycterinia natalensis* [Bernh. ex] Krauss, loc. cit. Type as above.

Perennial herb with thick clump of partly subterranean vegetative buds at the crown, one occasionally developing into a leaf rosette; stem usually solitary, simple, stout, erect to 350mm, pubescent, hairs retrorse, leafy becoming bracteate upwards. *Leaves* mostly 60–90 × 15–25mm, rapidly decreasing in size and passing imperceptibly into bracts, upper leaves sharply ascending, all elliptic, tapering at both ends, margins entire or callose-toothed, thinly hairy or hairs confined to margins and veins. *Flowers* opening at dusk or in dull light, in a short, dense spike, up to c.90 × 25mm, not elongating in fruit, the axis usually remaining invisible. *Bracts* adnate to calyx for c.4mm, resembling the leaves but smaller, imbricate, obscuring part of the bract above, mostly 20–40 × 6–15mm, margins entire or with an occasional callose tooth. *Calyx* c.15–30mm long often glabrous except for hairy margins, sometimes keels hairy as well, lips 4.5–7mm long. *Corolla* tube 32–50mm long, glandular-puberulous, hairs 0.1–0.25mm long, rarely a few eglandular hairs also present, limb held horizontally, regular; mouth with thick circlet of long unicellular hairs; lobes (5–) 6–8mm long, deeply notched, crimson and glandular-puberulous outside, white inside with minute scattered glandular hairs particularly on lower half. *Stamens* 4, anticous anthers c.2mm long, shortly exserted, posticous ones c.5–5.7mm long, included. *Capsule* c.16 × 5mm. *Seeds* c.1.5 × 1mm, angled, colliculate, pallid.

Selected citations:

Swaziland. Mbabane distr., 2631AC, Ukutula, 12 xii 1954, *Compton* 24800 (NBG, PRE).
Natal. Ngotshe distr., 2731CD, Ngome, c.4000ft, 2 ix 1977, *Hilliard & Burtt* 9940 (E, NU). New Hanover distr., 2930BD, Little Noodsberg, Laager Farm, 23 i 1983, *Hilliard & Burtt* 16312 (E). Alfred distr., Zuurberg, 4000ft, ii 1884, *Tyson* 1730 (BOL, NBG, PRE).
Transkei. Umzimkulu distr, 3029BD, Clydesdale, 3000ft, iii 1886, *Tyson* 863 (BOL, K, SAM).

Zaluzianskya natalensis ranges from western Swaziland south through Natal and just enters Transkei. It grows in grassland, between 600 and 1700m above sea level, flowering mostly between January and March. Its distinguishing features are the mass of undeveloped buds on the crown and the tightly congested inflorescence, the axis invisible even in fruit. This latter character will at once distinguish it from its ally, *Z. spathacea*; its growth habit distinguishes it from *Z. maritima*, an annual herb with which it has been much confused.

4. Zaluzianskya pulvinata Killick in Kirkia 1: 105 (1965); Hilliard & Burtt in Notes RBG Edinb. 41: 14 (1983).
Type: Natal, Bergville distr., between Sentinel Gate and base of Sentinel cliffs, c.9000ft, 3 xii 1953, *Killick & Marais* 2204 (holo. PRE).

Cushion-forming perennial herb, taproot eventually up to c.10mm diam., woody, 1 to several leaf rosettes crowded on the crown, or spreading rhizomatously into small mats. *Stems* 1 from each rosette, terminal, simple, erect or ascending, mostly 30–100mm long, exceptionally up to 150–300mm, clad in retrorse white hairs, either leafy or almost wholly bracteate. *Radical leaves* mostly 10–25 (–30) × 1–5 (–9)mm, linear, elliptic or spathulate tapering to a broad, flat petiolar part accounting for up to half the total leaf length, bases closely imbricate, apex obtuse to acute, margins entire or obscurely toothed, sometimes glabrous, usually a few coarse hairs on margins and midline, rarely sparsely hairy all over, often thick-textured; *cauline leaves* broad-based, otherwise more or less resembling the radical leaves, ascending. *Flowers* crowded in a short or long spike, usually remaining crowded in fruit, opening at dusk or in dull light and then sweetly scented. *Bracts* adnate to calyx for 2–5mm, usually shorter and much broader than the leaves, mostly (9–) 12–20 (–26) × 4–9mm, elliptic, obtuse to acuminate, broad-based and clasping the calyx, margins entire or with 1 or 2, rarely more, pairs of callose teeth near apex, hairy as the leaves. *Calyx* 8–12mm long, lips 3.2–4.5mm long, thinly hairy on upper margins and sometimes on keels, occasionally hairs wanting. *Corolla* tube 30–50mm long, glandular-puberulous, hairs up to 0.1–0.25mm long, limb held horizontally, regular; mouth with circlet of long unicellular hairs; lobes mostly 5–8mm long, deeply notched, crimson and glandular-puberulous outside, white inside, either eglandular or with a few minute glands near the sinuses. *Stamens* 4, anticous anthers 1.8–2.3mm long, shortly exserted, posticous ones 3.5–4mm long, included. *Stigma* 3.5–8mm long, exserted. *Capsule* c.8–10 × 4mm. *Seeds* c.1.4 × 1mm, angled, colliculate, pallid.

Selected citations:

Transvaal: Belfast distr., 2530CA, between Belfast and Dullstroom, 4 i 1960, *van der Schijff* 4811 (W). Barberton distr., 2530DB, Kaapsche Hoop, 9 iii 1981, *Hilliard & Burtt* 14293 (E, NU).
Natal. Bergville distr., 2829CA, Bezuidenhout Pass, c.5700ft, 10 xii 1976, *Hilliard & Burtt* 9464 (E, NU). Utrecht distr., 2730AD, Altemooi, 6500ft, xii 1920, *Thode* STE 4509 (STE). Lion's River distr., 2929BD, Fort Nottingham Common-age, 5500ft, 26 x 1976, *Hilliard & Burtt* 9055 (E, NU). Underberg distr., 2929CA, Garden Castle N.R., c.6700ft, 8 xi 1977, *Hilliard & Burtt* 10511 (E).
Transkei. Tabankulu Mt., 3029CB, c.5000ft, 18 xi 1973, *Hilliard & Burtt* 7329 (E, NU).
Cape. Barkly East div., 3027CB, Three Drifts stream below Pitlochrie, c.5800ft, 5 xii 1981, *Hilliard & Burtt* 14725 (E, NU).
Lesotho. 2929 AD, 1 mile S of Giant's Castle Pass, 10000ft, 11 xii 1973, *Wright* 1612 (E, NU). 2828 CC, Leribe, 1911, *Dieterlen* 427 (P, W).
Orange Free State. Harrismith distr., 2829 AC, Platberg, 2420m, 10 i 1974, *Jacobsz* 2507 (NBG, PRE).

Zaluzianskya pulvinata ranges from Belfast and Barberton districts in the south-eastern Transvaal to Natal (above 1550m), the mountains of the NE Orange Free State, Lesotho (up to 3000m), the mountains of Transkei, and the Cape Drakensberg and Witteberg near Barkly East and Lady Grey. It favours bare stony places, often in and around rock sheets or rock outcrops, and flowers between October and January. The growth habit of the plant (a cushion of leaf rosettes, whence the trivial name) is a valuable diagnostic feature. There is considerable variation in leaf form: plants with narrowly linear leaves look strikingly different from plants with broader, spathulate leaves. Narrow-leaved plants seem particularly common in the northern part of the range (though not to the exclusion of broad-leaved plants), for example, at Leribe (Lesotho), on Qua Qua Mountain and Platberg in the Orange Free State, on parts of the low Drakensberg in northern Natal (Bezuidenhout's Pass, Manyanyeza Mountain, Itala), and between Badplaats and Lake Chrissie in the Transvaal, where the leaves are remarkably long and narrow. Some of this variation is undoubtedly genetic: *Dieterlen* 427, collected at Leribe in 1911, is exactly like *Richardson* 215, collected there in 1982. But circumstantial evidence indicates that variation may be induced by environmental differences and here specimens (*Hilliard & Burtt* 12289) collected at Saalboom Nek (near Barkly East) seem to offer an example: very dwarfed plants (50mm high in fruit) with leaves only 1mm broad, collected on dry banks; nearby, in the shelter of rocks, plants 150mm tall, with leaves 4mm broad.

5. Zaluzianskya angustifolia Hilliard & Burtt in Notes RBG Edinb. 41: 16 (1983).
Type: E Cape, 3226DA, Stockenstrom div., Katberg Pass, on summit, 5800ft, hard bare ground, 26 i 1979, *Hilliard & Burtt* 12405 (holo. NU, iso. E).

Perennial herb forming small mats or low cushions, taproot becoming thick and woody, crown often with crowded vegetative buds, these sometimes developing into small crowded leaf rosettes. *Stems* several from the crown, prostrate or decumbent, simple to well-branched, wiry, mostly 70–300mm long, exceptionally to c.450mm, clad in ± retrorse white hairs, leafy, dwarf axillary shoots often present. *Radical leaves*, if present, oblong, very short; *cauline leaves* 10–35 × 0.75–3mm, more or less spreading, linear, narrowly oblong, or rarely some lower leaves narrowly elliptic, margins entire or with a few small callose teeth, thinly hairy, or hairs very sparse, or wanting. *Flowers* 1–15, in a short dense spike elongating somewhat in fruit, opening at dusk or in dull light, then sweetly scented. *Bracts* adnate to calyx for 1.5–4mm, unlike the leaves, c.12–30mm long, lanceolate, acuminate, margins entire or with an occasional callose tooth, indumentum as on leaves. *Calyx* 8–12mm long, lips 2–4mm long, hairy on upper margins and sometimes on keels. *Corolla* tube c.27–50mm long, glandular-puberulous, hairs up to 0.2mm long, limb held horizontally, regular, mouth with circlet of long unicellular hairs sometimes weakly developed on anticous side; lobes (4–)5–8mm long, deeply notched, crimson and glandular-puberulous outside, white inside, a few minute glandular hairs near the base, or glandular hairs scattered over lower half of each lobe, rarely glands confined to sinuses. *Stamens* 4, anticous anthers 1.6–2mm long, shortly exserted, posticous ones 2.5–5mm long, included. *Stigma* c.5–12mm long, exserted. *Capsule* c.10–15 × 3–5mm. *Seeds* c.1 × 1mm, angled, colliculate, pallid.

Selected citations:

Transvaal. Zoutpansberg, 2329BB, Louis Trichardt, 23 v 1927, *Blenkiron* sub Moss 14458 (J). 2229DD, farm Harnham, N-facing slope of Zoutpansberg, 27 iv 1982, *Balkwill* 162 (E).
Natal. Port Shepstone distr., 3030CC, Umtamvuna Nature Reserve, c.1000ft, 26 iv 1977, *Hilliard & Burtt* 10236 (E, NU). 3030CB, Izotsha, 13 iii 1973, *Strey* 11116 (E, NU). 3030CA, Oribi Gorge, 500m, i 1977, *Henderson et al.* 136 (J).
Transkei. 3129DA, Port St Johns, i 1896, *Flanagan* 2596 (PRE). 3129BC, Mateku waterfall, 11 xi 1970, *Strey* 10165 (K, PRE). 3127DB, Satana's Nek between Engcobo and Elliot, c.4500ft, 13 x 1980, *Hilliard & Burtt* 13104 (E, NU).
Cape. Stockenstrom div., 3226DB, Hogsback, Siberia, 5000ft, 11 xii 1977, *Hilliard & Burtt* 10994 (E, NU). Keiskammahoek div., 3227CB, top of Kologha Range, c.1400m, 8 v 1943, *Acocks* 9794 (PRE).

Zaluzianskya angustifolia has an extraordinary distribution range. There is an isolated record from the Lootsberg near Graaff Reinet (3124DD), then the plant has been well collected in the Eastern Cape, on the Elandsberg and the Amatole Mountains, and on the nearby mountains around Cathcart, Stutterheim and King William's Town. It is undercollected in Transkei, but clearly ranges over the mountains, coming down to near sea level on the heights above Port St Johns, thence to the heights above the Umtamvuna on both the Transkeian and the Natal side. The distributional gap thence to the Zoutpansberg, Blaauwberg and Woodbush area in the northernmost Transvaal seems real; unless my taxonomy is faulty.

The plants form small mats or low cushions on bare hard earth in the crevices of outcropping rocks or around rocks in grassland, flowering between August and May. The species is characterized by the small tight knots of vegetative buds at the crown, by its wiry spreading, branching stems, narrow spreading leaves and few-flowered inflorescences. The corolla lobes are mostly glandular above at least near the base (when we described the species, we were mistaken in thinking that the type material is eglandular).

6. Zaluzianskya schmitziae Hilliard & Burtt in Notes RBG Edinb. 41: 18 (1983).
Type: Lesotho, Makhaleng Valley near Molimo Nthuse Pass, c.2070m, 11 i 1979, *Hilliard & Burtt* 12058 (holo. NU, iso. E).

Syn.: *Nycterinia capensis* var. *foliosa* Benth. in Hook., Comp. Bot. Mag. 1: 370 (1836). Type: Cape, 3026CC, Nieuwe Hantam, 4500–5000ft, ii 1833, *Drège* (holo. K; iso. P, W).

Perennial herb, taproot eventually thick and woody, crown without a tight cluster of vegetative buds. *Stems* several from the crown, simple or sparingly branched, usually erect, rarely decumbent, 150–300mm long, clad in retrorse white hairs, dwarf leafy axillary shoots often present. *Radical leaves* rarely present and then very loosely tufted, they and sometimes the lowermost leaves c.15–46 × 2–4 (–10)mm, elliptic tapering to a broad petiolar part, usually with a few small obscure callose teeth near tip; *cauline* leaves mostly 20–40 × 1–2mm, more or less spreading, linear, base slightly broadened, clasping, apex ± obtuse, margins entire or with a few small callose teeth near tip, thinly hairy on both surfaces. *Flowers* opening at dusk or in dull light, but twice recorded as open in bright light, mostly 10–25, fewer on weak branches, arranged in a spike, usually elongating in fruit. *Bracts* unlike the leaves, adnate to calyx for 2–4mm, c.15–27mm long, lanceolate-acuminate, margins usually entire, or occasionally with a few callose teeth near apex, hairy on margins and midline, occasionally hairy all over backs. *Calyx* 8–10mm long, lips 3–4mm long, hairy at least on margins and keels. *Corolla* tube 25–39mm long, glandular-puberulous, hairs c.0.1mm, limb held horizontally, regular; mouth with circlet of long unicellular hairs; lobes 3–5mm long, deeply notched, red and glandular-puberulous outside, white inside, eglandular. *Stamens* 4, anticous anthers 1.5–2mm long, shortly exserted, posticous ones 2.5–3mm long, included. *Stigma* 5–10mm long, exserted. *Capsules* c.9 × 5mm. *Seeds* c.1 × 0.7mm, angled, colliculate, pallid.

Selected citations:

Cape. Lady Grey distr., 3027CB, Witteberg, Joubert's Pass, c.7500ft, 18 i 1979, *Hilliard & Burtt* 12160 (E, NU). Barkly East distr., 3127AB, Saalboom Nek, c.6900ft, 21 i 1979, *Hilliard & Burtt* 12308 (E, NU).
Orange Free State. Ficksburg, 2827DD, Strathcona, 4 iii 1936, *Fawkes* 136 (NBG). Bethlehem distr., 2828AB, 4 miles SE Bethlehem on Kestel road, 1675m, 16 ii 1967, *Scheepers* 1428 (K, PRE).
Lesotho. Leribe, 2828CC, *Dieterlen* 407 (P, PRE, SAM). 2927BC, Roma, 1675m, 20 ii 1960, *Ruch* 1577 (PRE). 2927BD, Mountain road, between Rual and Mpao, iii 1977, *Schmitz* 7461 (PRE, ROML).

Zaluzianskya schmitziae ranges from Bethlehem and Harrismith districts of the Orange Free State south through western Lesotho and neighbouring parts of the Free State to the Witteberg and Cape Drakensberg; it has also been recorded from Graaff Reinet and the mountains around Queenstown, and should be sought in the intervening mountains. The first collection appears to have been made by Drège, who found it in February 1833, at 'Nieuwe Hantam', an area lying near the Kraai River in Barkly East division, while nearly eight years later, in December 1840, Zeyher had it near Nieuwejaarspruit, between the Gariep (Orange) and Caledon rivers, at the foot of the Witteberg.

Its closest ally is *Z. angustifolia*, which differs in its sprawling well-branched stems, vegetative buds on the crown, mostly fewer flowers in an inflorescence, and the limb of the corolla larger (lobes mostly 5–8mm long).

7. Zaluzianskya elongata Hilliard & Burtt in Notes RBG Edinb. 41: 20 (1983).
Type: Natal, Underberg distr., Bamboo Mountain, 1675–1980m, 8 iv 1977, *Hilliard & Burtt* 10080 (holo. NU, iso. E).

Perennial herb, taproot and crown becoming thick and woody, vegetative buds, when present, small, not developing into a thick clump; stems several from the crown in old plants, ascending, up to 600mm long, often simple, sometimes branched, clad in retrorse white hairs, leafy. *Leaves* on lower part of stem 35–70 × 2–8 (–10)mm, becoming smaller and often more distant upwards, uppermost generally 15–25 × 1.5mm, sometimes broader, rarely up to 40mm long, all more or less ascending, oblong-elliptic or linear, margins entire or with a few small callose teeth, often glabrous or with hairs on margins and midline, rarely hairy all over. *Flowers* opening at dusk or in dull light and then scented, many in a long narrow spike elongating as the flowers open and capsules develop, small secondary spikes sometimes developing below the main fruiting spike. *Bracts* adnate to calyx for 2–3mm, lanceolate, (9–) 10–14 (–20)mm long, margins entire, glabrous or hairy as the leaves. *Calyx* 8–13mm long, lips 3–4mm long, either glabrous except for marginal hairs, or sometimes hairs better developed, exceptionally hairy all over. *Corolla* tube 30–40 (–44)mm long, glandular-puberulous, hairs up to 0.1mm long, limb held vertically, somewhat zygomorphic; mouth with circlet of long unicellular hairs often poorly developed or wanting on anticous side; the two upper lobes fractionally more united than the 3 lower, the anticous lobe slightly isolated from the other 4 and often narrower, lobes 3.5–5 (–6)mm long, deeply notched, crimson and glandular-puberulous outside, white and glabrous inside. *Stamens* 4, anticous anthers 1.2–1.8mm long, shortly exserted, posticous ones 3–4mm long, included. *Stigma* 6–8mm long , exserted. *Capsule* 10–15 × 4–5mm. *Seeds* c.1.3 × 0.8–1mm, angled, colliculate, pallid.

Selected citations:

Transvaal. Middelburg distr., c.2529DD, Steenkampsbergen, 5000ft, 6 iii 1922, *Rudatis* sub STE 13309 (STE). 2531CC, Barberton, 3100ft, *Rogers* 30193 (S). 2628AA, Johannesburg, Parktown, 2 ii 1922, *Moss* 8045 (J).
Swaziland. Mbabane distr., 2631AC, Ukutula, c.1220m, 20 ii 1956, *Compton* 25262 (NBG, PRE).
Natal. Bergville distr., 2828DB, Royal Natal National Park, c.5500ft, 6 ii 1982, *Hilliard & Burtt* 15458 (E, NU). Ngotshe distr., 2731 CA, Itala N.R., c.5000ft, 4 iv 1977, *Hilliard & Burtt* 10014 (E, NU). Mooi River distr., 2929BB, Bray Hill, c.5600ft, 20 i 1986, *Hilliard & Burtt* 19090 (E). Port Shepstone distr., Umpambanyoni, 1 v 1914, *Rudatis* 2092 (STE).
Transkei. 3128DB, near Umtata Falls, 18 ii 1927, *Grant* 3507 (PRE).
Cape. Maclear div., 3128AA, Ugie, Pomona, 4300ft, ii 1928, *Gill* 149 (STE).

Zaluzianskya elongata ranges from the Transvaal (eastern highlands and Highveld) to western Swaziland, Natal, Transkei, and Maclear division of the Eastern Cape, from c.460 to 1800m above sea level. *Mogg* 31633 (J) is only a fragment, but appears to be *Z. elongata*, and it is said to have come from Inhaca Island in Mozambique; the occurrence of *Z. elongata* in coastal Mozambique seems improbable, and needs confirmation.

The species may be recognized by its slightly zygomorphic flowers arranged in lax elongated spikes; also, the stems are often, but not always, only laxly leafy below the spikes, and there is no thick mass of vegetative buds on the crown. These characters will distinguish *Z. elongata* from *Z. spathacea*, with which it is sympatric in the northern half of its range.

8. Zaluzianskya pilosa Hilliard & Burtt in Notes RBG Edinb. 41: 22 (1983).
Type: Natal, New Hanover distr., Table Mountain, iii 1946, *Johnstone* 84 (holo. NU).

Syn.: *Z. maritima* (L.f.) Walp. var. *pubens* Hiern in Fl. Cap. 4(2): 336 (1904). Lectotype: Zululand, dry plains, *Gerrard* 1210 (K; isolecto. TCD, W).

Perennial herb, crown eventually thick and woody with clumped vegetative buds, stems several from the crown, erect or ascending, up to 350mm long, usually branched or at least with axillary vegetative buds, pilose with long retrorse white hairs, leafy. *Leaves* spreading, mostly 35–60 × 7–17mm, biggest about the middle of the stem, elliptic, tapering at both ends, margins obscurely to distinctly callose-toothed particularly in the upper half, both surfaces pilose. *Bracts* adnate to calyx for 2–2.5mm, leaflike, but smaller and diminishing in size upwards. *Flowers* opening at dusk or in dull light. *Spike* elongating in fruit. *Calyx* 10–15mm long, lips 4–6mm long, pilose. *Corolla* tube c.34–50mm long, glandular-puberulous, hairs 0.2–0.3mm long, rarely acute hairs as well to 0.4mm, limb held horizontally, regular; mouth with circlet of long unicellular hairs; lobes 6–8mm long, deeply notched, red and glandular-puberulous outside, white and glandular-puberulous inside particularly on lower half. *Stamens* 4, anticous anthers 1.5–1.8mm long, shortly exserted, posticous anthers 4–5mm long, included. *Capsule* 11–13 × 4–5mm. *Seeds* not seen.

Selected citations:

Natal. New Hanover distr., 2930BD, Little Noodsberg, Laager Farm, 23 i 1983, *Hilliard & Burtt* 16313 (E). Pietermaritzburg distr., 2930DA, Table Mountain, ii 1946, *Killick* 7 (E).
Transvaal. Pietersburg distr., 2430AA, The Downs, Marake, iv 1945, *Crundall* s.n. (PRE).

Zaluzianskya pilosa is ill-known; there are records from Natal, mainly over Table Mountain Sandstone, and from the highlands of the eastern Transvaal, in rocky grassland, and flowering mainly in December and January.

The branching habit of *Z. pilosa* (though this is not always displayed in individual specimens), its spreading cauline leaves, and copious indumentum, will distinguish it from *Z. spathacea*, *Z. pachyrrhiza* and *Z. natalensis*, its closest allies, as well as from *Z. elongata* , which stands further apart and has slightly zygomorphic flowers.

Through a clerical error, Hilliard & Burtt (1983) placed *Z. maritima* var. *pubens* in synonymy under *Z. spathacea* instead of under *Z. pilosa*, where it now appears.

9. Zaluzianskya pachyrrhiza Hilliard & Burtt in Notes RBG Edinb.41: 23 (1983).
Type: Natal, Hlabisa distr., St Lucia Game Park, 23 v 1977, *Hilliard & Burtt* 10347 (holo. NU, iso. E).

Syn.: [*Z. natalensis* auct. non Krauss; Harvey, Thes. Cap. 1: 37 fig. 58 (1859)].

Perennial herb, taproot becoming much thickened and often carrot-like with age, up to 15mm diam., crown without a thick clump of vegetative buds though a few buds may be present; stems one or several from the crown, simple, or occasionally branched, erect or ascending, up to 450mm long, with retrorse white hairs, leafy. *Leaves* on lower part of stem mostly 40–70 × 4–19mm, decreasing in size upwards, at least upper leaves sharply ascending, all elliptic, lanceolate or oblong-elliptic, margins entire or with a few callose teeth, glabrous or with a few hairs on margins and midline. *Flowers* several to many, the spike elongating in fruit, flowers opening at dusk or in dull light. *Bracts* adnate to calyx for 2.5–4mm, lanceolate, obtuse to shortly acuminate, mostly 13–24mm long, margins entire or with an occasional callose tooth towards the apex, glabrous. *Calyx* 9–14mm long, lips 4–6.5mm, glabrous except for hairy margins at tips. *Corolla* tube 35–52mm long, pubescent, with glandular hairs up to 0.3mm long, eglandular 0.4–0.7mm, limb held horizontally, regular; mouth with circlet of unicellular hairs; lobes 5–10mm long, deeply notched, rose pink to dark crimson and glandular-puberulous outside, white inside, generally glandular-puberulous, rarely glabrous. *Stamens* 4, anticous anthers 1–1.8mm long, shortly exserted, posticous ones 3.5–4.5mm long, included. *Stigma* 5–8mm long, exserted. *Capsule* 12–13 × 4mm. *Seeds* c.1 × 0.8mm, angled, colliculate, pallid.

Selected citations:

Natal. Hlabisa distr., 2832AD, St. Lucia Game Park, 15 viii 1975, *Pooley* 1750 (E, K, NU). Durban distr., 2931CC, Port Natal [Durban], *Gueinzius* s.n. (K, PRE, TCD, W); Clairmont, 15 viii 1893, *Schlechter* 3094 (E, GRA, J, K, P, PRE, STE). Port Shepstone, 3030CB, 27 xi 1923, *Weeks* 66 (J).

Zaluzianskya pachyrrhiza has been recorded along the Natal coast from the Lebombo Mountains and Lake St. Lucia in the north to the Umtamvuna in the south, from sea level to c.500m, in sandy grassland, flowering in any month.

Harvey's figure (loc. cit.) clearly shows the elongation of the spike that is characteristic of *Z. pachyrrhiza* and helps to distinguish it from *Z. natalensis*, in which the spike remains short and thick. *Zaluzianskya pachyrrhiza* has a thick woody rootstock often without any vegetative buds on the crown, in contrast to the crown of *Z. natalensis* with its thick tuft of vegetative buds. The corolla tube of *Z. pachyrrhiza* is often shaggy with long acute hairs mixed with shorter glandular ones; sometimes glandular hairs predominate and acute hairs may occur only near the base of the tube; the tube of *Z. natalensis* is usually glandular, the hairs up to 0.25mm long, rarely with a few eglandular hairs as well, up to 0.25mm–0.4mm long.

10. Zaluzianskya maritima (Linn. f.) Walp., Repert. 3: 307 (1844); Hiern in Fl. Cap. 4(2): 335 (1904) p.p. min.; Hilliard & Burtt in Notes RBG Edinb. 41: 24 (1983).
Type: Cape, Uitenhage distr., sea coast near Zeekoe river, *Thunberg* (sheet 14408 in herb. Thunberg, UPS).

Syn.: *Erinus maritimus* Linn. f., Suppl. 287 (1782).
 [*Erinus lychnideus* auct., non (L.) L.f.; Lindl. in Bot. Reg. 9: t. 748 (1823); Sims in Bot. Mag. 51: t. 2504 (1824);
  Lodd., Bot. Cab. 10: t. 957 (1824); Geel, Sert. Bot. Cl. 14 (1832)].
 Nycterinia maritima (Linn. f.) Benth. in Hook., Comp. Bot. Mag. 1: 369 (1836), p.p.
 N. lychnidea D. Don in Sweet, Brit. Fl. Gard. ser. 2, 3: t. 239 (1835). Lectotype: Sweet, Brit. Fl. Gard. ser. 2, 3:
  t. 239 (1835), based on plant cultivated by Patrick Neill at Canonmills, Edinburgh.
 Zaluzianskya lychnidea (D. Don) Walp., Repert. 3: 307 (1844); Hiern in Fl. Cap. 4(2): 337 (1904); Hilliard &
  Burtt in Notes RBG Edinb. 37: 316 (1979).
 Z. maritima var. *fragrantissima* Hiern in Fl. Cap. 4(2): 336 (1904). Types: Cape, Knysna div., sand hills at
  Plettenberg Bay, *Burchell* 5318 (K) and mouth of the Great Fish River, *Burchell* 3726 (K).

Herb mostly 100–300mm high, annual or sometimes persisting for more than one season, taproot becoming woody, rarely exceeding 4mm diam., stem simple or branched from the base, clad in retrorse white hairs, leafy, often with axillary leaf tufts. *Leaves* mostly 15–45 (–70) × (3–) 4–8mm, spreading, elliptic, base tapered, apex obtuse to subacute, margins entire or with a few small callose teeth, often somewhat fleshy, nearly glabrous or with coarse hairs on the margins and midline below. *Flowers* several to many in a crowded head, (sometimes elongating in fruit) opening at dusk or in dull light and then sweetly scented. *Bracts* adnate to calyx for 2.5–4mm, resembling the leaves but often shorter and broader. *Calyx* 8–14mm long, lips 4–7mm long, glabrous except for hairs on margins, and occasionally a few on keels. *Corolla* tube 25–50mm long, pubescent, hairs usually glandular, up to 0.1–0.5mm long, rarely with few to many eglandular hairs as well, 0.4–0.7mm long, limb held horizontally, regular; mouth with circlet of long unicellular hairs; lobes 5–9mm long, deeply notched, crimson and glandular-puberulous outside, white inside, eglandular or with a few glandular hairs near the base. *Stamens* 4, anticous anthers 1.4–1.8mm long, shortly exserted, 2 posticous ones included, 2.8–3.8mm long. *Capsule* 10–14 × 4–5mm. *Seeds* c.1–1.3 × 1mm, angled, colliculate, pallid.

Selected citations:

Transkei, 3228BC, Qora river mouth, 29 xii 1921, *Hilner* 494 (PRE)
Cape. Komgha div., 3228CB, Haga Haga, 15 xii 1977, *Hilliard & Burtt* 11088A (E). Alexandria div., 3326DA, Kenton-on-Sea, 6 xii 1977, *Hilliard & Burtt* 10869 (E, NU). Humansdorp div., 3424AA, Witte Els Bosch, x 1936, *Fourcade* 5348 (BOL, STE). Knysna div., 3422BB, Sedgefield, 4 xii 1949, *Middlemost* 2047 (NBG, STE).

Zaluzianskya maritima is confined to the stretch of coast from the Umtentu river in Transkei south to George in the Cape. The name has frequently been misapplied to species from inland regions as far afield as Zimbabwe, but those plants are all perennials, with swollen rootstocks, often carrying a tight mass of vegetative buds on the crown. *Zaluzianskya maritima* is characterized by its thin woody taproot, branching habit, and spreading leaves. It is essentially annual and will flower when still single-stemmed; its preferred habitat is the foreshore, along the seaward margin of dune scrub, but it will also grow on sandy grassy slopes overlooking the sea, flowering in any month. Its relationship lies with *Z. capensis*, also an annual that may grow on littoral dunes, but its thin herbaceous leaves, often with toothed margins, will distinguish it from *Z. maritima*, which usually has thick almost or quite entire leaves.

The areas of *Z. maritima* and *Z. pachyrrhiza* probably meet just south of the Umtamvuna river and field work there may prove interesting. There are few collections from the Transkeian coast, but specimens from the shore near Mkambati, south of the Umtamvuna, have the thin taproot and branching habit of *Z. maritima*, but the long eglandular hairs on the corolla tube are similar to these of *Z. pachyrrhiza*: however, these long hairs occur sporadically as far west as Mossel Bay. Representative specimens are *Abbot* 1977 (E, NH) from Mgwetyana Beach near Mkambati, and a collection made by Drège that was distributed as *Nycterinia lychnidea* D. Don b (E, K, P, W). The sheet in Paris (P) bears Drège's own label, which shows that the specimens were collected near the Umsikaba river at c.400ft above sea level on 21 ii 1832.

11. Zaluzianskya capensis (L.) Walp., Repert. 3: 307 (1844); Hiern in Fl. Cap. 4(2): 338 (1904) incl. vars; Levyns in Adamson & Salter, Fl. Cape Penins. 712 (1950); Hilliard & Burtt in Notes RBG Edinb. 41: 26 (1983). **Plate 2A.**
Lectotype: LINN 789.5.

Syn.: *Erinus capensis* L., Mant. alt. 252 (1771).
    E. aethiopicus Thunb., Prodr. Pl. Cap. 102 (1800) & Fl. Cap. 473 (1823). Type: in herb. Thunb., sheet 14397 (UPS).
    E. africanus auct. non L.; Thunb., Prodr. Pl. Cap. 102 (1800) & Fl. Cap. ed. Schultes 474 (1823) p.p.
    Nycterinia capensis (L.) Benth. in Hook., Comp. Bot. Mag. 1: 370 (1836), excl. var. *foliosa* Benth.
    N. coriacea Benth. in Hook., Comp. Bot. Mag. 1: 369 (1836) & in DC., Prodr. 10: 348 (1846). Type: Cape, Mountains near Cape Town, *Ecklon* (iso. SAM).

N. dentata Benth. in Hook., Comp. Bot. Mag. 1: 370 (1836). Lectotype: Cape, Paarl Mountain, *Drège* (K; isolecto. E, P, TCD, W).

N. longiflora Benth. in Hook., Comp. Bot. Mag. 1: 370 (1836) & in DC., Prodr. 10: 340 (1846). Type: Cape, 3018AA/AC, Roodeberg, 4000ft, 12 xi 1839, *Drège* (holo. K; iso. E, P, W).

Zaluzianskya coriacea (Benth.) Walp., Repert. 3: 306 (1844).

Z. dentata (Benth.) Walp., Repert. 3: 307 (1844); Hiern in Fl. Cap. 4(2): 339 (1904).

Z. longiflora (Benth.) Walp., Repert. 3: 307 (1844); Hiern in Fl. Cap. 4(2): 339 (1904).

Herb, mostly 100–400mm high, either annual or persisting for more than one season, taproot eventually woody, up to 4mm diam., stem initially simple, erect, soon branching from the base, side branches often decumbent, simple or branched, clad in retrorse white hairs, leafy, often with axillary leaf tufts. *Leaves* extremely variable, mostly 13–40 (–60) × 1–6 (–10)mm, spreading, linear to elliptic, apex subacute, base narrowed in broader leaves and then sometimes petiole-like, margins either entire (usually but not always in the narrower leaves) or obscurely to prominently toothed, teeth patent, occasionally glabrous, usually sparsely to densely pilose. *Flowers* opening at dusk or in dull light and then sweetly scented, few to many in a spike, often short and dense, elongating in fruit. *Bracts* adnate to calyx for 1.5–4mm, resembling the upper leaves. *Calyx* 6–11mm long, lips 4–6mm long, usually with at least a few hairs on the upper margins, keels sparsely to densely pilose. *Corolla* tube 25–35 (–40)mm long, glandular-puberulous, hairs up to 0.1–0.2mm long, limb held horizontally, regular; mouth with circlet of long unicellular hairs, sometimes poorly developed or developed on one side only; lobes (3–) 4–7 (–8)mm long, deeply notched, red and glandular-puberulous outside, white inside, either eglandular or with minute glands on the claw. *Stamens* 4, 2 anticous anthers 1.2–1.8mm long, shortly exserted, 2 posticous ones 2.5–3.5mm long, included. *Stigma* 5–7mm long, usually exserted. *Capsule* 6–12 × 4mm. *Seeds* c.1–1.3 × 0.8mm, angled, colliculate, pallid.

Selected citations:

Cape. Calvinia div., 3019CD, Kubiskouw, 1350m, 8 ix 1926, *Marloth* 12873 (PRE, STE). Clanwilliam div., 3219AA, Cedarberg, Pakhuis Pass, 3093ft, 26 v 1982, *Viviers* 347 (STE). Ceres div., Michell's Pass, 3500ft, 10 ix 1896, *Schlechter* 8933 (BOL, GRA, P, PRE). Malmesbury div., 3318CB, Mamre Hills, 22 ix 1942, *Compton* 13765 (NBG). Cape Peninsula, 3318CD, Table Mt above Kirstenbosch, 21 ix 1980, *Hilliard & Burtt* 13031 (E, NU). Somerset West div., 3418BD, Betty's Bay, 1 x 1980, *Hilliard & Burtt* 13089 (E, NU). Caledon div., 3419CB, Baviaanskloof near Die Kelders, 304m, 22 v 1971, *Thompson* 1204 (STE). Bredasdorp div., 3420AC, Kykoedie, c.500ft, 21 vii 1972, *Acocks* 22394 (STE). Uitenhage, 3325CD, 16 vii 1936, *Wall* s.n. (S). Port Elizabeth, 3325DC, 11 x 1980, *Stewart* 2175 (E). Albany div., 3326BC, 4 miles SE of Grahamstown, 18 xi 1928, *Gillett* 2476 (STE).

Zaluzianskya capensis favours sandy places, often in scrub, and flowers mainly between July and October, but as early as March and as late as December. It is widely distributed from central Namaqualand through Calvinia, Clanwilliam, Ceres and Worcester divisions to the Peninsula thence east through the coastal area to Grahamstown and Bathurst, down to sea level in the southern part of its range, but only at higher elevations in the northern part (up to 1350m). The specimens from the Bankberg and Amatole Mountains mentioned by Hilliard & Burtt (1983) have been referred to *Z. glareosa*.

The type collection of *Z. longiflora* (in synonymy above) came from the Roodeberg (Rooiberg) north of the Khamiesberg in Namaqualand, at an altitude of c.1200m, but the species seems to be rare in these northern areas, becoming common only in the south west and along the southern coast.

There is remarkable variation in the width of the leaves and they may be entire to prominently toothed, irrespective of width. The degree of hairiness is also variable, from almost shaggy to nearly glabrous, with corresponding indumentum on the bracts and calyx. In the western part of the range, indumentum is usually well developed, but nearly glabrous plants also occur, while along the southern coast, hairs on leaves, bracts and calyces are nearly confined to margins and midlines. Variation can occur in a single collection, for example, *Van Wyk* 2377 (STE) from the southern foothills of the Voetpadsberg (3319 DA) and there are several instances of similar variation within the same quarter degree square.

12. Zaluzianskya muirii Hilliard & Burtt in Notes RBG Edinb. 41: 28 (1983).
Type: Cape, 3421AC, Riversdale div., south of Riversdale, Wankoe se Rante, c.700ft, 6 ix 1975, *Oliver* 5982 (holo. PRE, iso. STE).

Herb, 150–400mm high, either annual or perhaps persisting for more than one season, taproot eventually woody, up to 3mm diam., stem initially simple, later branching from the base and higher, clad in retrorse white hairs, leafy, often with axillary leaf tufts. *Leaves* mostly 15–40 × 1–3mm, spreading, linear to narrowly elliptic, subacute, tapering to a petiole-like base in larger leaves, margins entire or with a few pairs of spreading lobes, thinly hairy. *Flowers* few to many in a spike, elongating in fruit, probably opening at dusk. *Bracts* adnate to calyx for 1.5–2.5mm, resembling the upper leaves. *Calyx* 5–8mm long, lips 3–3.5mm long, pilose on upper margins and keels. *Corolla* tube 15–21mm long, glandular-puberulous, hairs 0.1–0.25mm long, limb held horizontally, regular; mouth not bearded; lobes 6–7mm long, deeply notched, red and glandular-puberulous outside, white or pale pink inside, puberulous with conspicuous stalked glands. *Stamens* 4, 2 anticous anthers 0.7–1.1mm long, shortly exserted, 2 posticous ones 1–2.5mm long, included. *Stigma* 3–6mm long, exserted. *Capsules* c.7 × 2.5mm. *Seeds* not seen.

Selected citations:

Cape. Riversdale div., 3420BC, De Hoop, Hamerkop, c.700ft, 7 viii 1984, *Fellingham* 703 (STE); De Hoop, Potteberg N.R., 60m, 7 ix 1978, *Burgers* 1084 (STE); 3421BA, Albertinia, viii 1915, *Rogers* 16752 (J).

Zaluzianskya muirii is confined to a small area of the southern Cape coast around Riversdale, Albertinia and Still Bay, where it grows in scrub up to c.210m above sea level, and flowers mainly between June and September. Its distinguishing features are its glabrous corolla mouth and lobes conspicuously glandular above (the glands are stalked); these characters, together with the shorter corolla tube, set it apart from *Z. capensis*, which it most resembles.

13. Zaluzianskya oreophila Hilliard & Burtt in Notes RBG Edinb. 41: 29(1983).
Type: Lesotho, Mokhotlong distr., 2929CB, Sani Top, east of pass, c.2900m, 14 i 1976, *Hilliard & Burtt* 8806 (holo. NU, iso. E).

Annual herb, or possibly surviving for more than one season, taproot remaining thin, no clump of vegetative buds at crown, stems solitary and erect in young plants, several from the crown and ascending in older plants, usually simple, mostly 100–300mm long, clad in retrorse white hairs, leafy. *Radical leaves* usually present only on young plants, oblanceolate, c.18–30 × 7–10mm; *cauline leaves* more or less ascending, elliptic or oblong, mostly 15–40 × 2–5mm, slightly smaller upwards, margins coarsely toothed, upper small leaves with 1 or 2 pairs of teeth near apex, upper surface sparsely pilose or almost glabrous, lower surface thinly pilose. *Flowers* many, opening at dusk or in dull light, spike elongating as the flowers open. *Bracts* adnate to calyx for 2.3–4mm, lanceolate, 11–15 (–20)mm long, margins usually with 1 or 2 pairs of teeth near the tip, rarely entire, indumentum as on leaves. *Calyx* 6–9mm long, lips 3–4mm long, margins hairy. *Corolla* tube 16–24mm long, glandular-puberulous, hairs c.0.1mm long, limb held vertically, somewhat zygomorphic; mouth with circlet of long unicellular hairs, often poorly developed on anticous side; anticous lobe slightly isolated from other 4, lobes deeply notched, 4–5mm long, crimson and glandular-puberulous outside, white and glabrous inside. *Stamens* 4, 2 anticous anthers 1.8–2mm long, shortly exserted, 2 posticous ones 2.8–3mm long, included. *Stigma* 3–5mm long, exserted. *Capsule* not seen.

Selected citations:

Lesotho. 2829CC, Cleft Peak, c.10000ft, 21 i 1956, *Edwards* 1153 (NU). 2929AD, Thabana Ntlenyana, 10500ft, 21 i 1955, *Coetzee* 591 (PRE); Sani Top, 9500ft, 7 i 1977, *Hilliard & Burtt* 9643 (E, NU).

Zaluzianskya oreophila is endemic to the Eastern Mountain Region where it grows on the summit plateau, between c.2285 and 3200m above sea level. At Sani Top, it is common in rough wet grassland, often growing in the shelter of small bushes and grass tussocks, and flowers in December and January.

It resembles *Z. elongata* in its slightly zygomorphic flowers, but that species is a coarser and more glabrous plant with larger flowers (calyx 8–13mm, corolla tube 30–44mm long).

14. Zaluzianskya chrysops Hilliard & Burtt in Notes RBG Edinb. 41: 32 (1983).
Type: Natal, Underberg distr., Garden Castle Nature Reserve, Pillar Cave valley, 1908m and above, 4 xi 1977, *Hilliard & Burtt* 10415 (holo. NU, iso. E).

Herbaceous, flowering initially in the seedling stage, then stems solitary, simple, erect, later branching from the base and surviving for more than one season, these stems simple or sparingly branched low down, erect or decumbent, eventually somewhat woody at the base but without a thick clump of vegetative buds on the crown, 35–150mm long , hairy with coarse retrorse white hairs. *Leaves* crowded towards base of stem, distant upwards, lower leaves mostly 10–40 × 5–15mm, elliptic to subrotund, apex obtuse, base narrowed to a petiole-like part, margins subentire to coarsely toothed, both surfaces thinly pilose; cauline leaves similar but smaller, sessile. *Flowers* opening at dusk or in dull light, spike few-flowered, crowded, scarcely elongating in fruit. *Bracts* adnate to calyx for 2–5mm, similar to the cauline leaves. *Calyx* 8–10mm long, lips 4–4.5mm long, margins and keels hairy. *Corolla* tube c.30–43mm long, glandular-puberulous, hairs up to 0.1–0.2mm long, limb held horizontally, regular; mouth with circlet of long unicellular hairs, lobes deeply notched, 5–7mm long, dark red and glandular-puberulous outside, glabrous inside, white in upper half, vivid orange in lower. *Stamens* 4, 2 anticous anthers 1.5–2.6mm long, slightly exserted, 2 posticous ones 2.5–3.2mm long, included. *Stigma* 4–7mm long, eventually exserted. *Capsule* c.10 × 4mm. *Seeds* not seen.

Selected citations:

Natal. Mpendhle distr., 2929BC, Highmoor F.R., headwaters of Elandshoek River, c.8100ft, 5 i 1983, *Hilliard & Burtt* 16198 (E, NU). Underberg distr., 2929CA, Garden Castle F.R., Mlambonja valley, 7000–7500ft, 8 i 1982, *Hilliard & Burtt* 15035 (E, NU).
Lesotho. Sani Top, 2929CB, NE of chalet, 10000ft, 17 i 1976, *Hilliard & Burtt* 8844 (E). 2929CC, Sehlabathebe, gorge of Tsoelikane, xi 1979, *Schmitz* 8794 (NU). 2928AC, 7km east of Blue Mt Pass, 2550m, 13 xi 1982, *Richardson* 196 (E).

Zalzianskya chrysops is possibly widespread in the eastern Mountain Region above c.2000m, but collections are few and widely scattered. The plants favour damp bare areas, particularly the boulder beds of streams, earth banks above streams, and along the foot of cliffs, flowering between October and December, mostly going into fruit by January.

The loose basal rosette of leaves will distinguish it from its ally *Z. glareosa*, and from *Z. ovata*; also *Z. glareosa* never has an orange 'eye', though *Z. ovata* sometimes does. However, *Z. ovata* differs further in the spreading, often gland-tipped, hairs on its stems.

15. Zaluzianskya glareosa Hilliard & Burtt in Notes RBG Edinb. 41: 30 (1983).
Type: Natal, Underberg distr., Cobham Forest Reserve, Upper Polela Cave area, 2300–2375m, 15 ii 1979, *Hilliard & Burtt* 12510 (holo. NU, iso. E).

Herbaceous, taproot becoming woody, up to 10mm diam., flowering initially in the seedling stage, then stems solitary, simple, erect, later branching from the base and surviving for more than one season, these stems either simple or branched, decumbent or ascending, becoming woody at the base with small vegetative buds but never with a thick clump of vegetative buds, up to c.450mm long, hairy with ± retrorse or spreading white hairs, leafy, often with axillary leaf tufts. *Leaves* spreading, mostly 15–50 (–60) × 2–8 (–15)mm, linear, oblong, or narrowly elliptic, rarely broadly elliptic, base sometimes petiole-like, margins entire or with a few small callose teeth, or occasionally more coarsely toothed, thinly hairy. *Flowers* opening at dusk or in dull light and then emitting a spicy fragrance, spike initially condensed, often few-flowered, elongating in fruit. *Bracts* adnate to calyx for 2.5–5mm, 15–20 (–30)mm long, lanceolate, entire or with 1 or 2 pairs of teeth near the apex, hairy. *Calyx* 7–13mm long, lips 3.5–6mm long, hairy. *Corolla* tube 27–50mm long, glandular-puberulous, hairs up to 0.1–0.2mm long, limb held horizontally, regular; mouth with circlet of long unicellular hairs; lobes deeply notched, 5–10mm long, crimson and glandular-puberulous outside, white inside, glabrous or sometimes with a few minute glandular hairs near the base. *Stamens* 4, 2 anticous anthers 1.5–2.2mm long, shortly exserted, 2 posticous ones 2.6–3.5mm long, included. *Stigma* 5–10mm long. *Capsule* 8–11 × 4–5mm. *Seeds* c.1–1.2 × 0.8–1mm, angled, colliculate, pallid.

Selected citations:

Orange Free State. Witzieshoek, 2828DB, road to The Sentinel, c.2225m, 26 xii 1975, *Hilliard & Burtt* 8636 (E, NU).
Natal. Bergville distr., 2828DB, Royal Natal National Park, Tugela Gorge, c.1800m, 2 ii 1982, *Hilliard & Burtt* 15392 (E). Mpendhle distr., 2929BC, Loteni Nature Reserve, 24 xii 1983, *Hilliard & Burtt* 11809 (E). Lion's River distr., 2929BB, Fort Nottingham Commonage, c.5600ft, 6 ii 1978, *Wright* 2431 (E, NU). Alfred distr., Weza, 3029DA, Zuurberg, 5500ft, 3 iii 1974, *Hilliard* 5473 (E, NU).
Transkei. 3128CB, Bazeia Mt, Mpolompo valley, c.4500ft,11 ii 1981, *Hilliard & Burtt* 13940 (E, NU).
Cape. Elliot div., 3127BB, Bastervoetpad, 7500ft, 20 i 1982, *Matthews* 826 (NBG). Graaff Reinet div., 3224BC, near Graaff Reinet, iv 1867, *Bolus* 428 (K, S).

Zaluzianskya glareosa is montane in its distribution, from the low Drakensberg on the Transvaal - Natal border along the face of the high Drakensberg in Natal and on outlying spurs and mountains (mainly between 1200 and 2750m above sea level) to the mountains of Transkei, the Cape Drakensberg and other eastern Cape mountains as far west as Graaff Reinet and possibly the Nieuweveld Mountains: Drège collected what appears to be *Z. glareosa* at Zakrivierspoort on 12 xi 1826, but the specimen is very poor (*Drège* 7898, P); Guthrie also had it at 6800ft in the Nieuweveld Mountains in 1894 (*Guthrie* 3493, NBG). Collections from the Cape are scanty and widely scattered geographically, and there are few modern ones, probably because these mountains are scarcely botanized. The plants favour bare gritty areas, often around and in the crevices of rock sheets, or in the boulder beds of streams, flowering mainly between December and April.

16. Zaluzianskya tropicalis Hilliard in Edinb. J. Bot. 47:349 (1990).
Type: Zimbabwe, Inyanga distr., summit of Mt Inyangani, 2500m, 7 iii 1981, *Philcox et al.* 8909 (holo. K, iso. E).

Perennial herb, crown woody, c.10mm diam., without a thick clump of vegetative buds, stems several, up to 400mm long, erect or decumbent, either simple and then with small axillary leaf tufts, or branched, clad in spreading or ± retrorse coarse white hairs often mixed with glandular hairs up to 0.8mm long, leafy. *Leaves* ascending or spreading, mostly 20–50 × 2–18mm, lower ones narrowly elliptic, becoming smaller and linear or oblong upwards, obtuse, margins entire or with a few pairs of callose teeth in upper half, both surfaces nearly glabrous to thinly hairy, most hairs acute, some shorter (c.0.5mm) and gland-tipped. *Flowers* opening at dusk or in dull light; spikes few- to many-flowered, elongating in fruit. *Bracts* adnate to calyx for 3.5–6mm, c.15–20mm long, oblong, or lanceolate, resembling the leaves. *Calyx* 10–13mm long, lips 3–7mm long, pilose with long acute hairs on upper margins and keels, sometimes with glandular hairs as well, up to 0.5mm long. *Corolla* tube 35–45mm long, glandular-puberulous, hairs up to 0.3mm long, rarely with acute hairs to 0.4mm as well; limb held horizontally, regular, lobes deeply notched, 6–9mm long, pink or red and glandular-puberulous outside, white inside, glandular on the lower part, circlet of long unicellular hairs around mouth. *Stamens* 4, 2 anticous anthers c.1.5–1.6mm long, shortly exserted, 2 posticous ones c.4mm long, included. *Stigma* 5–12mm long. *Capsules* c.11 × 4mm. *Seeds* not seen.

Selected citations:

Zimbabwe. Matoba distr., farm Quaringa, 4700ft, v 1954, *Miller* 2412 (K). Mazoe distr., upper slopes Iron Mask Hill, 5100ft, v 1907, *Eyles* 548 (K). Inyanga distr., near Inyanga Down, c.2000m, 29 i 1931, *Norlindh & Weimarck* 4691 (K, S). Mutare distr., Odzani river valley, 1914, *Teague* 51 (K).

Mozambique. Manica & Sofala, Tsetsera, 6500ft, 2 iii 1954, *Wild* 4476 (K).

Zaluzianskya tropicalis is widespread in the highlands of Zimbabwe, from Matoba district in the west to the eastern highlands and neighbouring heights in Mozambique, between c.1200 and 2600m above sea level, in grassland or *Brachystegia* woodland, flowering mainly between January and May, but beginning as early as September.

Hilliard & Burtt (1983) mentioned the plant under *Z. glareosa* and *Z. distans. Zaluzianskya glareosa*, which is common in the Eastern Mountain Region of southern Arica, resembles *Z. tropicalis* in habit and general facies but differs in its corolla lobes, glabrous above (not with stalked glands), and in its lack of all but the smallest glandular hairs on stems, leaves, bracts and calyx, whereas in *Z. tropicalis* glandular hairs up to 0.5–0.8mm long are often (but not always) present. Like *Z. glareosa, Z. tropicalis* shows considerable variation in the width of the leaves so that the commonly misused names *Z. capensis* and *Z. maritima* have often been applied to this most northern of the species in Section *Nycterinia. Zaluzianskya tropicalis* differs from *Z. distans,* which also has spreading glandular and eglandular hairs on its stems and glandular hairs around the corolla mouth, by its less leaflike bracts and more crowded inflorescence.

17. Zaluzianskya turritella Hilliard & Burtt in Notes RBG Edinb. 37: 318 (1979) and in Notes RBG Edinb. 41: 33 (1983).
Type: Lesotho, Mokhotlong distr., escarpment south of Sani Pass, c.2990m, 18 i 1976, *Hilliard & Burtt* 8876 (holo. NU, iso. E).

Annual herb, taproot slender, crown unthickened, stems 25–130mm long, simple or with 2–6 branches from the base, these branches simple or shortly branched near the base, erect or decumbent, clad in spreading gland-tipped hairs, closely leafy. *Leaves* mostly 8–20 × 5–15mm, thick-textured, rhomboid-ovate, contracted to a broad petiole-like part, apex obtuse to subacute, recurved-spreading, margins entire or obscurely crenate to bluntly toothed in upper half, both surfaces nearly glabrous to sparsely hairy, some hairs gland-tipped, dark green above, beetroot red below, aromatic. *Flowers* opening at dusk or in dull light, spikes short, crowded. *Bracts* adnate to calyx for 2.5–3mm, leaf-like, mostly larger than the leaves and increasing in size upwards. *Calyx* 7–10mm long, lips 4–6mm long, glandular-pubescent on upper margins and keels. *Corolla* tube 21–27mm long, glandular-puberulous, hairs c.0.1mm long, limb held horizontally, regular; mouth with circlet of long unicellular hairs; lobes deeply notched, 5–6mm long, dark red and glandular-puberulous outside, white and glabrous inside. *Stamens* 4, 2 anticous anthers 1.8–2mm long, shortly exserted, 2 posticous ones 2.5–2.7mm long, included. *Stigma* c.4mm long, exserted. *Capsule* (immature) 9 × 5.5mm.

Selected citations:

Lesotho–Natal border, 2929CB, Sani Top, saddle on escarpment N of pass, c.9900ft, 8 i 1977, *Hilliard & Burtt* 9661 (E, NU). Upper Injasuti, 3300m, 28 i 1966, *Trauseld* 546 (NU).

Zaluzianskya turritella is known only from a small area of the high Drakensberg, along the edge of the escarpment between Upper Injasuti and Giant's Castle south to Thaba Ntšo (Devil's Knuckles) above Bushman's Nek in southern Natal. It grows in gravel and silt patches overlying bare basalt rock sheets, between c.2940 and 3300m above sea level. Despite its diminutive size, it has relatively large and somewhat fleshy rhomboid leaves and bracts increasing in size upwards; this feature, as well as the spreading glandular hairs on the stems, make it a most distinctive species.

18. Zaluzianskya katharinae Hiern in Fl. Cap. 4(2): 341 (1904); Lucas & Pike, Wild Flowers of the Witwatersrand 78, pl. XIV d (1971); Hilliard & Burtt in Notes RBG Edinb. 41: 34 (1983).
Lectotype: Transvaal, Johannesburg, Jeppestown Ridge, 6000ft, ii 1898, *Gilfillan* in herb. *Galpin* 1478 (K, isolecto. PRE).

Shrubby, loosely branched, stems straggling, sometimes prostrate or decumbent, up to 600mm long, clad in coarse ± spreading hairs, some gland-tipped, leafy. *Leaves* mostly 10–30 × 4–25mm, elliptic, ovate or cuneate in outline, margins coarsely toothed, apex acute, base cuneate, narrowed into the petiole, petioles mostly 4–10mm long, both blade and petiole coarsely pubescent with glandular and eglandular hairs. *Bracts* adnate to calyx for 1–2mm, leaf-like. *Flowers* few to many in lax spikes, nearly always opposite, opening at dusk or in dull light and then scented. *Calyx* 12–17mm long, lips 4–7mm long, coarsely pubescent. *Corolla* tube 35–52mm long, puberulous, hairs either glandular, c.0.2–0.4mm long, or mixed with eglandular hairs up to 0.5mm long, limb held horizontally, regular; mouth with circlet of long unicellular hairs; lobes deeply notched, 7.5–12mm long, pink to crimson and glandular-puberulous outside, white and glandular-puberulous inside. *Stamens* 4, 2 anticous anthers 1–2.2mm long, shortly exserted, 2 posticous ones 4–4.5mm long, included. *Capsule* 12–15 × 5–6mm. *Seeds* not seen.

Selected citations:

Transvaal. Zoutpansberg, 2229DD, 5 miles W of Wylie's Poort, 4800ft, 22 viii 1930, *Hutchinson & Gillett* 4381 (BM, K). Pilgrim's Rest distr., 2430DB, Mariepskop, 6 vii 1961, *van der Schijff* 5620 (K, PRE, W). Belfast distr., 2530CB, Elandshoogte, 16 ii 1978, *Richardson* 55 (E). Johannesburg distr., 2628AA, Observatory, 3 x 1971, *Moss* 4854 (J). Potchefstroom distr., 2627BC, Elandsfontein, 6 iv 1954, *Louw* 1423 (PRE).

Zaluzianskya katharinae is endemic to the Transvaal, where it is confined to the mountains from the Zoutpansberg south to Carolina and west along the Witwatersrand and Magaliesberg, reaching its southern limit on the hills near Heidelberg and Potchefstroom. It grows in partially shaded damp places among rock outcrops and rock tumbles, often quartzite or conglomerate, and flowers can be found in any month. It is easily recognized by its coarsely toothed patent leaves, the uppermost ones nearly always remaining opposite, distant, and producing a flower from each axil.

19. Zaluzianskya ovata (Benth.) Walp., Repert. 3: 307 (1844); Hiern in Fl. Cap. 4(2): 340 (1904); Trauseld, Wild Flow. Natal Drakensberg 172 (colour plate), 173 (1969); Hilliard & Burtt in Notes RBG Edinb. 41: 34 (1983). **Plate 4D.**
Type: Cape [Lady Grey distr.], Wittebergen, 7500ft, 15 i 1833, *Drège* s.n. (holo. K; iso. P, also W, but very poor).

Syn.: *Nycterinia ovata* Benth. in Hook., Comp. Bot. Mag. 1: 370 (1836).
 Zaluzianskya montana Hiern in Fl. Cap. 4(2): 342 (1904). Type: [Lesotho], summit Mont aux Sources, 9500ft, Jan., *Flanagan* 2032 (holo. K; iso. PRE, SAM).

Strongly aromatic twiggy shrublet forming loose clumps, stems sometimes prostrate or decumbent, up to c.450mm long, brittle, much branched, clad in long coarse spreading white hairs, gland-tipped or not, or these hairs occasionally wanting, as well as short spreading gland-tipped hairs, leafy. *Leaves* mostly 15–60 × 4–35mm, narrowly to broadly elliptic or ovate, apex obtuse to subacute, base cuneate, tapering into a short petiole-like part, margins subentire to coarsely and often somewhat irregularly toothed or lobulate, both surfaces shaggy with gland-tipped hairs. *Flowers* opening at dusk or in dull light; spikes 1 to several flowered, the flowers then usually crowded. *Bracts* adnate to calyx for 1.5–4mm, resembling the leaves but smaller. *Calyx* 8–13mm long, lips 4–6mm long, glandular-pubescent. *Corolla* tube 30–58mm long, glandular-puberulous, hairs up to 0.2–0.3mm long, limb held horizontally, regular; mouth with circlet of long unicellular hairs, sometimes weakly developed on anticous side, or wanting; lobes deeply notched, c.6–11mm long, pink to crimson and glandular-puberulous outside, glabrous inside, usually white, sometimes brilliant orange near base. *Stamens* 4, 2 anticous anthers 1.5–2.5mm long, shortly exserted, 2 posticous ones 3–3.5mm long, included. *Stigma* 4.5–6mm long, exserted. *Capsule* 7–11 × 4–6mm. *Seeds* c.1.8 × 1.2mm, angled, colliculate, pallid.

Selected citations:

Orange Free State. Harrismith distr., 2829AC, Platberg, c.700ft, 13 xii 1976, *Hilliard & Burtt* 9510 (E, NU).
Lesotho. 2927BD, Blue Mt Pass, c.8500ft, 10 i 1979, *Hilliard & Burtt* 12026 (E, NU). 2929CA, Black Mts, 10400–10600ft, 13 i 1976, *Hilliard & Burtt* 8766 (E, NU). 2928CC, below Maletsunyane Falls, 19 x 1946, *Esterhuysen* 13180 (K, PRE).
Natal. Bergville distr., 2829CC, Cathedral Peak F.R., foot of Tlanyaku Pass, 24 x 1973, *Hilliard & Burtt* 6913 (E, NU). Underberg distr., 2929CD, Upper Umzimouti Valley, 6500ft, 27 xi 1976, *Hilliard & Burtt* 9401 (E, NU). Alfred distr., 3029DA, Mt Ngeli, c.5600ft, 4 i 1969, *Hilliard & Burtt* 5837 (E, K, NH, NU).
Transkei. 3028BB, Ramatseliso's Beacon, 2350m, 23 x 1976, *Boardman* 172 (PRE).
Cape. Lady Grey div., 3027CB, Witteberg, Joubert's Pass, 7500–7700ft, 18 i 1979, *Hilliard & Burtt* 12188 (E, NU). Graaff Reinet div., 3224AA, Koudeveldberge SE of Doornbosch, 6000ft, 6 xi 1974, *Oliver* 5204 (K, PRE, STE). Ladismith div., 3321BD, Swartberg, Toverkop, 6500ft, 16 xii 1956, *Esterhuysen* 26799 (K). Calvinia div., 3119BD, Akkerendam, c.4800ft, 14 xi 1955, *Acocks* 18627 (K, PRE). Worcester div., 3319CA, Du Toit's Peak, 6000ft, 15 xi 1981, *Esterhuysen* 35707 (E).

Zaluzianskya ovata is widely distributed on the mountains from Platberg near Harrismith in the Orange Free State through Lesotho, Natal and Transkei (between 1750 and 3230m above sea level) to the Cape mountains, from the Witteberg and Cape Drakensberg in the north east and the mountains around Queenstown in the south east to the Hantam Mountains near Calvinia and the Cape fold mountains as far west as Worcester and Tulbagh divisions, between c.1460 and 2130m above sea level. The plants favour partly shaded cliff faces, but at very high altitudes they will thrive in bare silty or gravelly patches.

Zaluzianskya ovata shows considerable variation in leaf size, attendant upon growing conditions; much more interesting is the occasional presence of a bright orange or red band around the mouth of the corolla. This colour zonation may also be found in the allied *Z. distans*, and it is a constant feature of *Z. chrysops*. These three species are sympatric over part of their ranges, and the biological significance of the phenomenon needs investigation.

20. Zaluzianskya distans Hiern in Fl. Cap. 4(2): 341 (1904); Hilliard & Burtt in Notes RBG Edinb. 41: 37 (1983).
Lectotype: Natal, Van Reenen, 5–6000ft, 3 iii 1898, flowers chocolate outside, white within, *Wood* 7906 (K).

Syn.: *Z. latifolia* [Schinz ex] O. Hoffm. & Muschler in Ann. Nat. Hofmus. Wien 24: 323 (1910), nomen, written on *Schlechter* 6944 (ex K, P, W, Z) syntype of *Z. distans*.

Herbaceous, flowering in seedling stage, then stem solitary, simple, erect, later loosely and laxly branched from the base and surviving for more than one season, main branches then becoming woody at the base but without a thick clump of vegetative buds, stems up to c.400mm long, ascending, loosely branched, somewhat hoary with coarse spreading white hairs, gland-tipped or not, as well as much shorter glandular hairs, leafy. *Leaves* spreading, mostly 20–60 × 9–30mm, elliptic-ovate, the base sometimes petiole-like, margins obscurely and irregularly toothed or the toothing more pronounced and regular, both surfaces glandular-hairy. *Flowers* opening at dusk or in dull light. *Spike* few- to many-flowered, lax, often remarkably so. *Bracts* adnate to calyx for 3–5.5mm, 14–30 × 5–15mm, becoming smaller upwards, the lower ones in particular leaf-like and distant, elliptic or ovate, toothed and hairy as in the leaves. *Calyx* 8–15mm long, lips 3–4mm long, hairy. *Corolla* tube 24–50mm long, glandular-pubescent, hairs up to 0.2–0.4mm long, rarely acute hairs as well to 0.4mm, limb held horizontally, regular; mouth with circlet of long unicellular hairs; lobes deeply notched, 4–10(–13)mm long, crimson and glandular-puberulous outside, either white inside or scarlet to bright orange to pale yellow in lower half, and there usually glandular-puberulous. *Stamens* 4, 2 anticous anthers 2mm long, rarely reduced to 0.5–1mm, shortly exserted, 2 posticous ones 3–3.5mm long, included. *Stigma* 2.5–5mm long, eventually exserted. *Capsule* c.10–12 × 4–5mm. *Seeds* not seen.

Selected citations:

Transvaal. Lydenburg distr., 2530BA, Mt Anderson, 7300ft, 24 xii 1932, *Smuts & Gillett* 2358 (PRE). Wakkerstroom distr., Paardekop range, c.7500ft, 2 ii 1930, *Galpin* 9787 (K, PRE).
Natal. Utrecht distr., 2730AD, Naauwhoek, 6800ft, 7 i 1962, *Devenish* 800 (K, PRE). Polela distr., 2929DD, farm Glengariff, upper Nkife Gorge, 5800ft, 27 xii 1981, *Rennie* 1280 (E, NU).
Orange Free State. Harrismith distr., 2829AC, Platberg, One Man Pass, 13 xii 1976, *Hilliard & Burtt* 9515 (E).

There are scattered records of *Z. distans* in the highlands of the eastern Transvaal, from Lydenburg southwards to the Low Drakensberg on the Natal-Transvaal-Orange Free State border, where it seems to be relatively common, then an isolated record from Mawahqua Mountain in southern Natal. The plants grow in damp partially shaded places, often in the shelter of rocks or montane scrub, and in scrub-filled watercourses, between c.1765 and 2200m above sea level, flowering between December and March.

Zaluzianskya distans is closely allied to *Z. ovata*, from which it differs in habit (tufts of simple or subsimple stems, not a well branched shrublet) and in its generally much laxer inflorescence. The two species are sympatric on Platberg, where these differences, as well as the differences in the type of habitat that is favoured, are strikingly demonstrated.

Section Macrocalyx

21. Zaluzianskya pumila (Benth.) Walp., Repert. 3: 307 (1844); Hiern in Fl. Cap. 4(2): 340 (1904).
Type: Cape, 'In the Nieuweveld or Kowp' (Bentham), zwischen Dweka [Dwyka river] und Zwartbulletje, 2500–3000 Fuss, Sept., (Drège, Zwei pflanzengeographische Documente), *Drège* (iso. P).

Syn.: *Nycterinia pumila* Benth. in Hook., Comp. Bot. Mag. 1: 370 (1836) & in DC., Prodr. 10: 349 (1846).
    Erinus africanus auct., non L.; Thunb., Prodr. 102 (1800) & Fl. Cap. ed. Schultes 474 (1823) p.p.
    Nycterinia africana auct., non (L.) Benth.; Benth. in Comp. Bot. Mag. 1: 371 (1836), quoad descr. et spec. excl. syn.
    Zaluzianskya falciloba Diels in Bot. Jahrb. 23: 481 (1896); Hiern in Fl. Cap. 4(2): 343 (1904). Type: Cape, Calvinia div., Hantam Mts, 1869, *Meyer* s.n. (B†).
    Z. africana Hiern in Fl. Cap. 4(2): 342 (1904), p.p. Lectotype (chosen here): Cape, without precise locality, *Thunberg* (sheet no. 14402, left hand specimen, herb. Thunberg, UPS).
    Z. pseudafricana Paclt in Taxon 21(4): 539 (1972), nom. illegit.

Annual herb 20–230mm tall, primary stem erect, simple or sparingly branched from the base and above, branches ascending or decumbent, pubescent with long (up to c.1mm) delicate acute hairs, retrorse, ± appressed. *Leaves* few, blade of radical leaves 5–30 × 3–15mm, elliptic to oblong tapering into a broad flat petiolar part c.3–15mm long (roughly half as long as to almost equalling blade), cauline leaves c.8–40 × 3–11mm, elliptic tapering into a short petiolar part, all leaves with margins almost entire to distinctly toothed, thinly pubescent, hairs denser towards the base, long (to 2mm), delicate. *Flowers* up to c.20, at first crowded in a head, elongating into a crowded spike, the lowermost few flowers sometimes distant, open in sunshine. *Bracts* adnate to calyx for c.6–10mm, lowermost 16–40 × 3–10mm, oblong, 3–5 pairs of deep narrow patent teeth in upper half, softly pubescent, hairs up to 1.3–2mm long. *Calyx* 7–13mm long, lips 2–3mm long, sparsely pubescent, mainly on the ribs, hairs delicate, acute, up to c.1mm long, minutely glandular as well. *Corolla* tube (17–)21–32mm long (throat 1.5–2mm in diam.), sparsely and very minutely glandular, limb c.10–22mm in diam., regular, lobes 4.5–10 × 2–7mm, deeply bifid (occasionally each lobe lobed again), Y-shaped, distally white or shades of mauve above, mauve to pink or red below, the tissue very thin, proximally thick-textured, the patch of thick tissue elliptic in outline, occupying nearly the whole of the shaft of the lobe, forming a yellow to orange or bright scarlet star around the mouth, base of each patch densely clothed in tiny sessile glands, these giving way to short (c.0.4mm) blunt hairs forming a wide band around the mouth. *Stamens* 4, anticous pair exserted, anthers 1.1–1.5mm long, posticous pair included, anthers 2–2.5mm long. *Stigma* 2.5–4mm long, shortly exserted. *Style* 14–26mm long. *Ovary* c.3–5 × 0.7–2mm, nectariferous gland peglike, remarkably long (c.1mm), lightly attached to extreme base of ovary. *Capsules* 10–12 × 5–6mm. *Seeds* c.1.5–1.7 × 1.2–1.5mm, palest mauvish-grey, irregularly 3–4winged and deeply and very coarsely 'dimpled'.

Selected citations:

Cape. Van Rhynsdorp div., 3118BC, Knechtsvlakte, 23 viii 1941, *Esterhuysen* 5939 (BOL, K); 3118BD, between Van Rhynsdorp and Van Rhyn's Pass, 7 viii 1974, *Goldblatt* 2281 (E, NBG, PRE); 3118BD, Vuurfontein, 900ft, 24 viii 1897, *Schlechter* 10978 (E, K, PRE, S, SAM, W). 3118DC, Klaver, 300ft, 31 vii 1920, *Andreae* 441 (PRE, STE). Calvinia div., 3119BD, 15 miles from Calvinia, Loeriesfontein, 26 ix 1952, *Middlemost* 1782 (NBG). 3119BD, Calvinia Commonage, c.3700ft, *Leistner* 388 (PRE). Sutherland div., 3220CB, Klipbanksrivier, c.3000ft, 26 viii 1953, *Acocks* 16944 (PRE); 3220CC, near Yuk River, 19 vii 1811, *Burchell* 1259 (K, now in poor condition). 3319BB, 26km from Karoopoort to Sutherland, 18 viii 1975, *Wisura* 3465 (NBG). Laingsburg div., Whitehill, 12 viii 1929, *Compton* 3498 (BOL, NBG).

Bentham's holotype of *Z. pumila* is not in the Kew herbarium but there is an isotype in Paris (P), with a determinavit slip in Bentham's hand. This sheet bears another label (presumably Drège's own) giving the precise locality: Wilgeboschfontein, 2500ft, 24 ix 1826 (in the citation above, I have given the localities quoted by Bentham and Drège). This farm still appears on the 1:500000 map at 3321AA, about 5km north west of Koup and 30km west of Dwyka.

Diels's holotype of *Z. falciloba* was destroyed in the Berlin fire, but his description of the deeply toothed upper leaves and bracts and 'the peculiar appearance of the corolla and its glabrous tube' (my translation) are decisive. The peculiar appearance of the corolla derives from the structure of the lobes; after anthesis, these are reflexed straight back, the thickened basal portion then lying parallel to the corolla tube while the thin distal portion curls up and back towards the thickened part – whence the trivial name *falciloba*.

The typification of Hiern's name, *Z. africana*, is problematical. Hiern had on loan from Uppsala six sheets from Thunberg's herbarium, all bearing the name *Erinus africanus* L. but none equivalent to Linnaeus's plant (which is *Z. villosa*). They are a very mixed bag: several specimens of *Z. capensis* (determined by Hiern as either *Z. dentata* or *Z. longiflora*), several more of *Bellardia trixago* (L.) All. (*Bartsia trixago* L.), a Crucifer, and, on sheet no. 14402, a whole plant on the left, a fragment on the right. Hiern designated this sheet *Z. africana* (*Nycterinia africana* D. Don); the left hand specimen is *Z. pumila*; the fragment on the right is a different species, but I have failed to determine it. Hiern wrote a composite determinavit label, which he headed '*Erinus africanus* Thunb. Hb. = *Zaluzianskya dentata* Walp.' and then enumerated the six sheets, labelling them alpha to zeta, with the appropriate determinations.

It is not clear why Hiern's description of *Z. africana* should consist largely of a translation of Thunberg's (1823) description of *Erinus africanus*, when he knew that much of Thunberg's material under that name is *Zaluzianskya capensis*. But it *is* clear that Hiern was renaming a plant that Thunberg had misdetermined as *Erinus africanus* and that *Z. africana* should be lectotypified by the left hand specimen on sheet 14402, not only because it was so determined by Hiern, but also because it is the only specimen that could have yielded the information on duration, root and habit.

The other specimens cited by Hiern (1904) are an astonishing mixture of species (not all *Zaluzianskya*) and all except *Burchell* 1259 (cited above) should be ignored.

Zaluzianskya pumila is widely distributed from the environs of Van Rhynsdorp and Calvinia south east across the Tanqua Karoo and Roggeveld to Laingsburg, then east to Dwyka river. It grows on gravelly, shaly or sandy flats and slopes, c.100–1125m above sea level, flowering between July and September, principally in August.

In herbaria, *Z. pumila* has been much confused with *Z. inflata* but it can be recognized at once by the remarkably thick-textured (and therefore raised) 'star' around the mouth; allied species may have a constrasting 'star' but the tissue is always thin. All the species in this section have a rather papery and much inflated calyx, delicate pubescence, and very odd, ± winged seeds.

22. Zaluzianskya inflata Diels in Engl. Bot. Jahrb. 23: 481 (1896); Hiern in Fl. Cap. 4(2): 342 (1904).

Type: Clanwilliam distr., Hantam Mts, 1869, *Meyer* s.n. (B†).

Annual herb, main stem 30–250mm long, weak, more or less erect, often simple, sometimes sparingly branched from the base and a little higher, branches decumbent or ascending, villous with long (up to 1–2mm) delicate acute patent hairs. *Leaves* few, blade of lowermost leaves c.6–22 × 3.5–10mm, ovate to elliptic tapering into a broad flat petiolar part c.3–12mm long (roughly half as long as blade), cauline leaves 12–35 × 4–14mm, elliptic tapering into a short petiolar part, all leaves with margins subentire to distinctly toothed, thinly villous, hairs acute, up to 2–3mm long especially on lower margins. *Flowers* up to c.10 in a lax or crowded spike, open in sunshine. *Bracts* adnate to calyx for 5–8mm, lowermost 20–35 × 7–16mm, oblong-lanceolate, 3–4 pairs of small teeth on upper part, thinly villous with acute hairs up to 2mm long, tiny (0.1mm) glandular hairs also present. *Calyx* 12–17mm long, lips 3–4mm long, thinly villous mainly on the ribs, hairs delicate, acute, up to 1.5–2mm long, minutely glandular as well. *Corolla* tube 29–35mm long (throat 1.5–2mm in diam.), glabrous or with a few scattered minute glands, limb c.7–12mm in diam., regular, lobes c.2.5–4.5 × 1.8–4mm, deeply bifid, Y-shaped, 'red' or 'magenta' (a crimson patch can be detected at the base of each lobe in some dried specimens), mouth encircled by a broad band of obtuse hairs up to c.0.4mm long, these giving way to tiny (less than 0.1mm) glands on shafts of lobes. *Stamens* 4, anticous pair exserted, anthers 0.4–0.8mm long, posticous pair included, 1.5–1.8mm long. *Stigma* 2–4mm long, just included or slightly exserted. *Style* 21–27mm long. *Ovary* c.5 × 1mm, nectariferous gland peglike, c.0.4mm long. *Capsules* c.14 × 7mm. *Seeds* c.2 × 1.7mm, pallid, irregularly 3–4 winged, colliculate.

Citations:

Cape. Calvinia div., 3119BD, Akkerdam, lower slopes of Hantam Mts, 22 vii 1961, *Lewis* 5806 (NBG); ibidem (as Ekerdam), 27 ix 1947, *Taylor* 2819 (NBG). 3119AD, Matjesfontein, 2200ft, 20 viii 1897, *Schlechter* 10921 (BOL, K). 3119BC, 20km from Calvinia on road to Loeriesfontein, hills north of the Hantam Mts, 27 viii 1983, *Coetzer* 851 (PRE). 3119AC, near Elandsfontein Farm, 28 ix 1929, *Lewis Grant* 4895 (PRE).

The type of *Z. inflata* was destroyed in the Berlin fire, and no isotype has been traced. Nevertheless, it seems certain that the name is being correctly applied to the plant described here. Diels gave the size of the calyx, corolla and limb, and his measurements accord with those in the description above. Meyer's specimen came from the Hantams, and it is only thereabouts that the plant has been re-collected. Collectors have given no ecological information but the plants flower between July and September.

Zaluzianskya inflata is an odd-looking plant with the relatively huge bracts and calyces seeming to weigh down the stems, which, when they attain any height, have dried contorted and bent. The small corolla limb is all out of proportion to these large organs and the long corolla tube. The species has been much confused with *Z. pumila*, which has a mostly smaller calyx, and larger corolla limb with a distinctive raised orange 'star' around the mouth, and larger anthers.

23. Zaluzianskya sutherlandica Hilliard in Edinb. J. Bot. 47: 349 (1990).

Type: Cape, Sutherland div., Kruis Rivier, 5400ft, ix 1926, *Levyns* 1661 (holo. BOL).

Annual herb, stem c.30mm high, branched from the base, branches decumbent, villous with long acute patent hairs. *Leaves*: blade c.7–10 × 5–10mm, ovate tapering into a broad flat petiolar part 7–8mm long (roughly equalling the blade), margins serrate, both surfaces pilose with acute hairs up to 3mm long. *Flowers* c.10 in a short crowded spike, open in sunshine. *Bracts* adnate to calyx for c.4.5mm, c.10 × 5mm, oblong with 2–3 pairs of long narrow teeth towards apex, shaggy with acute hairs up to c.2.5mm long. *Calyx* c.7–8mm long, lips 1.5mm, villous mainly at tips of lobes, hairs acute, up to 1.5mm long. *Corolla* tube c.22 × 1.5mm in throat, very minutely and sparsely glandular, limb c.7.5mm across, regular, lobes c.2.5 × 2.5mm, bifid, Y-shaped, each 'arm' lobed again, 'pinkish', mouth fringed with obtuse hairs up to c.0.5mm long, these giving way to tiny glands on shafts of lobes. *Stamens* 4, anticous pair exserted, anthers 1.25mm long, posticous pair included, anthers 2.5mm long. *Stigma* 1mm long, just reaching mouth. *Style* 18mm long. *Ovary* swelling into fruit, nectariferous gland c.0.5mm long. *Capsules* not seen.

Zaluzianskya sutherlandica is known only from a single tiny plant collected at Kruis Rivier. There are at least two places of this name in Sutherland division, one about 10km north of Sutherland (3220BC), the other to the north east (3220AD) on the Vis River. The plant was found in 'open places' and was well in flower (but not in fruit) in September.

The indumentum around the mouth, a band of clavate hairs giving way to tiny glands on the shaft of the lobes, is similar to that in *Z. pumila* and *Z. inflata*. The plant has the facies of *Z. pumila*, but can at once be distinguished by the lack of a thick-textured orange 'star' around the mouth. It differs from *Z. inflata* not only in aspect but also in its much smaller calyx and corolla, as well as, possibly more significantly, in its longer anthers (anticous ones 1.25mm long (versus 0.4–0.8mm), posticous ones 2.5mm (versus 1.5–1.8mm).

24. **Zaluzianskya mirabilis** Hilliard in Edinb. J. Bot. 47: 348 (1990).
Type: Cape, Sutherland div., 3220CA, Houthoek, 14 viii 1968, *Hanekom* 1073 (holo. STE; iso. K, PRE).

Annual herb, primary stem 10–80mm long, erect, simple or branched once from the base and sometimes another pair of branches higher up, branches decumbent, pubescent with patent glandular hairs up to c.1mm long. *Leaves* few, c.12–30 × 3–12mm, elliptic tapering to a flat petiolar base sometimes almost as long as the blade in the radical leaves, much shorter upwards, ± acute, margins subentire to distinctly and sharply toothed, both surfaces glandular-pubescent, hairs up to 1–1.5mm long, very occasional acute hairs to 2mm long also present especially on lower margins. *Flowers* up to c.10, crowded at first, elongating into a dense spike, open in sunshine. *Bracts* adnate to calyx for 3.5–5mm, lowermost c.22–32 × 6–9mm, oblong-lanceolate, 3–6 pairs of sharp narrow patent teeth in upper half, softly villous with glandular hairs up to 0.7–1.5mm long, acute hairs as well to 1.5–2mm. *Calyx* 11–15mm long, lips 3–3.5mm, pubescent mainly on the ribs with delicate glandular hairs up to c.0.7mm long, some acute hairs as well mainly at tips of lobes. *Corolla* tube c.30–38mm long (throat 2–2.3mm in diam.), sparsely and very minutely glandular, limb c.14–18mm in diam., regular, lobes c.5.5–7 × 5–7mm, deeply bifid, Y-shaped, the lobes lobed again, reddish below, pinkish-mauve above, yellow in throat sometimes with an orange (or scarlet?) star-shaped patch around mouth, densely glandular there, the glands (up to 0.1mm long) extending up the shaft of the lobe. *Stamens* 4, anticous pair exserted, anthers 1–1.4mm long, posticous pair included, anthers 2–3mm long. *Stigma* 2–7mm long, just reaching mouth or shortly exserted. *Style* 18–28mm long. *Ovary* c.3 × 0.8mm, nectariferous gland peglike, c.1.1mm long. *Capsule* c.11 × 5mm. *Seeds* all immature but winged and colliculate.

Citation:

Cape. Sutherland div., 3220DA, Verlaten Kloof, 4300ft, 8 ix 1926, *Levyns* 1646 (BOL, K).

Zaluzianskya mirabilis is known with certainty only from Sutherland division, one collection from gravelly ground in and near the dry course of the Houthoek river, the other from Verlaten Kloof, flowering in August and September. The plants have the aspect of *Z. pumila* but are easily recognized by the mainly glandular pubescence on stems, leaves and bracts, and by the lack of a raised orange star-shaped patch around the mouth. In *Z. pumila* there is a band of clavate hairs around the mouth giving way to sessile glands on the shafts of the lobes; in *Z. mirabilis*, only tiny stalked glands are present.

A specimen from much further east (Willowmore div., 3323BC, Baviaanskloof, Nuwekloof Pass, *Immelman* 341, PRE) may be *Z. mirabilis* but it is only a tiny scrap, with bracts c.15 × 4mm, calyx 9mm, corolla tube 20mm, limb c.8mm, lobes c.2.5 × 2.5mm and lobed only once, anticous anthers 0.5mm long, posticous anthers 1.2mm. These measurements are all smaller than those of *Z. mirabilis* and the mouth appears to be glandular. Good collections from this area are much needed.

25. Zaluzianskya acutiloba Hilliard in Edinb. J. Bot. 47: 345 (1990). **Plate 2J.**
Type: Cape, Sutherland div., Waterkloof, 1430m, ix 1921, *Marloth* 10409 (holo. STE; iso. PRE).

Annual herb, primary stem 30–150mm long, erect, simple or branched from the base and above, branches decumbent or ascending, densely pubescent with delicate patent hairs up to c.1mm long, either minutely gland-tipped or acute, moderately leafy. *Leaves* c.17–30 × 6–12mm, elliptic or ovate-elliptic, subacute, tapering into a broad flat petiolar part, this accounting for up to half the total length, margins serrate, both surfaces villous with hairs up to 2.5–3mm long, either gland-tipped or acute. *Flowers* up to c.20 crowded in a short spike, open in sunshine. *Bracts* adnate to calyx for 4–6mm, lowermost c.11–22 × 3.5–7mm, oblong, acute, 1–3 pairs of narrow sharp teeth near apex, softly villous with acute hairs up to 2–4mm long, usually mixed with glandular hairs up to 0.8–1.5mm long. *Calyx* 8.5–12mm long, lips 2.5–3mm long, villous mainly on the ribs, hairs gland-tipped or acute, up to 0.8–1.25mm long. *Corolla* tube 23–27mm long (throat 1.4–2mm in diam.), glabrous, limb 11–15mm in diam., regular, lobes 5–7 × 2.5–5mm, deeply bifid, Y-shaped, margins revolute, the segments very acute to the eye, saffron-yellow ageing brown, band of minute sessile globular glands around mouth. *Stamens* 4, anticous pair exserted, anthers 1.25–1.6mm long, posticous pair included, anthers 2.2–3.2mm long. *Stigma* 1.5–7mm long, shortly exserted. *Style* 14–21mm long. *Ovary* 3.5–4 × 0.8–1mm, peglike nectariferous gland 0.5mm long. *Capsules* c.8 × 4mm. *Seeds* c.2 × 1.5mm, more or less 3-winged, deeply 'dimpled', pallid.

Citations:

Cape. Calvinia div., 3119DB, farm Vlakfontein, Rebunie, 28 viii 1974, *Hanekom* 2341 (K, PRE); 3119 BC, Hantamsberg, upper south eastern slopes, 4 ix 1991, *Goldblatt* 9179 (E); Akkerendam, Hantamsberg, c.4800ft, 22 ix 1955, *Leistner* 396 (K). Sutherland and Fraserburg, 19 ix 1938, *Wall* s.n. (S).

Zaluzianskya acutiloba is known from Calvinia and Sutherland divisions, where it grows in gravelly or stony ground or on shale screes and flowers in August and September. It has the soft indumentum, inflated calyx and peculiar seeds of *Z. pumila*, *Z. inflata* and *Z. mirabilis*, but it is at once distinguished by its corolla limb, deeply divided into two very acute lobes (this partly optical due to the revolute margins) and yellowish to brownish in colour, apparently lacking a contrasting 'eye' and with a band of globular glands around the mouth. Marloth noted 'young flowers saffron, older dark brown, strong scent'. The indumentum on both Marloth's and Wall's specimens is very glandular, but Hanekom's and Leistner's are eglandular and Hanekom noted that the white hairs gave the green leaves a greyish appearance.

sect. Zaluzianskya subsect. Zaluzianskya

26. Zaluzianskya villosa F. W. Schmidt, Neue Selt. Pfl. 11 (1793); Usteri, Ann. Botanik. 6: 116 (1793); Hiern in Fl. Cap. 4(2): 345 (1904) excl. vars.; Levyns in Adamson & Salter, Fl. Cape Penins. 712 (1950) p.p. **Plate 3B.**
Neotype: (chosen here): Cape of Good Hope, False Bay, vii 1882, *Bolus* in herb. Norm. Austro-Africanum 652 (E; isoneo. W).

Syn.: *Lychnidea villosa foliis ex alis floriferis* ... Burm., Rar. Afr. Pl. decas 5: 139 t. 50, fig. 1. (1739).
   Erinus africanus L., Sp. Pl. ed. 1: 630 (1753). Type: in herb. van Royen, sheet no. 908234607 (L).

E. selaginoides Thunb., Prodr. Pl. Cap. 102 (1800) & Fl. Cap. ed. Schultes 475 (1823). Type: CBS, crescit in arena mobile inter Leuwestart et littus, floret Augusto, *Thunberg* (holo. sheet no. 14410, UPS, iso. S).

E. villosa Thunb., Prodr. Pl. Cap. 102 (1800) & Fl. Cap. ed. Schultes 474 (1823). Type: CBS, *Thunberg* (holo. sheet no. 14414, UPS, iso. S).

Nycterinia selaginoides (Thunb.) Benth. in Hook., Comp. Bot. Mag. 370 (1836) excl. var. *glabra* and var. *parviflora*.

N. africana (L.) Benth. in Hook., Comp. Bot. Mag. 371 (1836) quoad syn. tantum.

Zaluzianskya selaginoides (Thunb.) Walp., Repert. 3: 308 (1844) excl. var. *glabra* and var. *parviflora*.

Annual herb, often flowering in the seedling stage, 30–300mm tall, stem initially simple, erect, soon branching from the base and above, branches erect, ascending or decumbent, villous, hairs up to 1–1.5mm long, acute, patent, underlain by minute glandular hairs as well, leafy throughout. *Leaves:* lowermost 14–50 × 2–9mm, slightly smaller upwards and passing imperceptibly into bracts, oblanceolate to linear, apex obtuse, or sometimes subacute, base narrowed into a petiolar part up to ⅓–¼ the total leaf length but shorter upwards, margins with 2–4 pairs of teeth in upper half, these rarely obscure, both surfaces densely pubescent, longest hairs on the petiolar part, up to 1mm long. *Flowers* in crowded, eventually much elongated, spikes, open in sunshine. *Bracts* adnate to calyx tube for c.3.5–5mm, lowermost c.11–25 × 2–9mm, smaller upwards, either narrowly to broadly spathulate or oblanceolate, less commonly oblong above broadening in lower half, densely pubescent on both surfaces with longer hairs (1–2.25mm) on lower margins and sometimes on lower back and midline as well. *Calyx* 6.5–9mm long, anticous lip 1.5–3mm long, posticous lip 1.6–4mm long, pubescent all over, lobes tipped with longer hairs to c.1mm, these sometimes also on the keels. *Corolla* tube 14–23mm long (throat 1.25–1.5mm in diam.), minutely glandular-puberulous, occasional acute hairs as well, limb c.10–14mm in diam., regular, held horizontally, lobes eventually reflexed, mouth encircled by stiff erect acute hairs, lobes 4–6mm long, bifid, broadly Y-shaped, white to mauve above, marked with a yellow to red star around mouth, glabrous. *Stamens* 2 (posticous pair), included, tips of the anthers sometimes just visible in mouth, anthers 1.25–2mm long. *Stigma* 1.5–3mm long, exserted. *Style* 11–15mm. *Ovary* 2.3–4 × 0.8–1.5mm. *Capsule* 7–9 × 2.5mm. *Seeds* c.0.5–0.7 × 0.4–0.6mm, pale mauve-grey, irregularly wrinkled, obscurely 3-angled, the angles marked by very narrow pallid wings, sometimes wanting.

Selected citations:

Cape. Cape Town div., 3318CD, Paarden Island, 2 viii 1940, *Bond* 426 (NBG); 3318DC, Blackheath, 18 x 1973, *Oliver* 4764 (K, PRE, STE). 3418AB, Rondevlei, 13 x 1940, *Compton* 9839 (NBG); 3418AB, False Bay, vii 1882, *Bolus* 652 (BOL, E, PRE, SAM, W); 3418BA, Cape Flats near Zwartklip, 8 ix 1980, *Hilliard & Burtt* 13073 (E, K, S). Bredasdorp div., 3419CB, beyond Strandkloof, 4 ix 1943, *Barker* 2467 (BOL, NBG).

As no type of *Z. villosa* is known to be extant, a neotype has been chosen. Typical *Z. villosa*, that is, plants according with the types of both names, *villosa* and *selaginoides*, ranges from about Langebaan to the Peninsula (including Robben Island) and east to Pearly Beach in Bredasdorp division. It is distinguished by its bifid corolla lobes, two stamens, and leaves and bracts more or less obtuse, margins of at least the leaves usually distinctly toothed, and both surfaces clad in short dense pubescence, longer hairs also being present particularly on the lower margins and lower backs. The whole plant generally, but not always, dries dusky. From Langebaan north to Velddrift, Aurora and Sauer, the hairs on leaves and bracts may be sparser and longer and the tips of both rather more acute (e.g. *Thompson* 3523, STE; *Compton* 9409, NBG; *Leipoldt* 3738, BOL; *Theron* 2041, K, NBG, PRE; *Goldblatt* 6013, E; *Thompson* 794, K, PRE, STE). Some of these plants dry dusky, others are greenish: they begin to approach *Z. affinis* in the shape of leaf and bract, but the leaf margins are toothed as in *Z. villosa* and the indumentum is closer to that of *Z. villosa* than that of *Z. affinis*. They are therefore included in a broad circumscription of *Z. villosa*. See further under *Z. affinis*, *Z. sanorum* and *Z. pilosissima*, all of which have been confused with *Z. villosa* because they share the characters of bilobed corolla lobes and stamens reduced to two.

Zaluzianskya villosa is a common plant, usually growing in large stands, on sand dunes and in other sandy places, along the coast and up to c.200m above sea level, flowering between July and November. The corolla limb may be white to various shades of mauve and usually has a yellow 'eye' that darkens to orange or red as the flower ages.

27. Zaluzianskya affinis Hilliard in Edinb. J. Bot. 47: 345 (1990). **Plate 5A.**
Type: Cape, Van Rhynsdorp div., 3118DA, farm Liebendal, 7 miles N of Vredendal, 5 viii 1970, *Hall* 3697 (holo. NBG, iso PRE).

Syn.: *Nycterinia selaginoides* (Thunb.) Benth. var *glabra* Benth. in Hook., Comp. Bot. Mag. 1: 370 (1836). Lectotype: Cape, bei Holrivier, auf Karrooartigen Höhe, unter 1000 Fuss, Aug., *Drège* (K; isolecto. E, P, S, TCD, W).
    Zaluzianskya villosa F. W. Schmidt var. *glabra* (Benth.) Hiern in Fl. Cap. 4(2): 345 (1904).

Annual herb often flowering in seedling stage, 20–300mm tall, stem initially simple, erect, soon branching from base and above, branches erect, ascending or decumbent, usually villous, hairs up to 0.4–2mm long, acute, patent, minute glandular hairs as well, laxly leafy. *Leaves*: lowermost 10–45 × 1–8mm, slightly smaller upwards and passing imperceptibly into bracts, lanceolate, ± acute, base narrowed to a petiolar part up to ½ total leaf length (or sometimes ± equalling blade in lowermost pair of leaves), margins ± entire to obscurely few-toothed, sparsely hairy, hairs mostly on lower margins and petiolar part, up to 0.75–2mm long, much shorter upwards. *Flowers* in crowded, eventually much elongated, spikes, open in sunshine. *Bracts* adnate to calyx tube for (2–)3.5–5mm, lowermost c.7.5–30 × 1.1–5(–8)mm, smaller upwards, lanceolate, the upper ones in particular often much narrowed and strongly folded in the upper part, ± acute, usually sparsely pilose on lower half especially on margins, hairs up to 1.5–3mm long, much shorter and sparser upwards, upper half often glabrous except on midline below, minute glandular hairs also present, rarely short fine pubescence present on lower back. *Calyx* (4–)5.5–9.5mm long, anticous lip 0.75–3mm long, posticous lip 1.25–3.5mm long, minutely pubescent, long hairs at tips and often on posticous keels as well, up to 1–2mm long. *Corolla* tube (9–)15–30mm long (throat (0.8–1)–1.1–1.5mm in diam.), minutely glandular-puberulous, hairs sometimes sparse, limb (4–6)–8.5–14mm in diameter, regular, held horizontally, lobes eventually reflexed, mouth encircled by stiff acute erect hairs, lobes (1.5–2)–3.75–6mm long, bifid, broadly Y-shaped, above white to mauve or yellow, marked with a yellow to orange or red star around mouth, glabrous. *Stamens* 2 (posticous pair), included, anthers (0.8–1)–1.5–2mm long. *Stigma* (1–)1.5–7mm long, exserted (except in autogamous flowers). *Style* (4–8)–8.8–20.5mm. *Ovary* 1.75–5 × 0.75–1.2mm. *Capsules* 6–9 × 2–2.5mm. *Seeds* c.0.5–0.75 × 0.3–0.5mm, pale mauve-grey, irregularly wrinkled, obscurely angled, angles marked by very narrow pallid wings, sometimes wanting.

Selected citations:

Cape. Richtersveld, 2816BD, Numees, 252m, 21 ix 1981, *McDonald* 715 (PRE, STE). 2916BD, Port Nolloth, viii 1883, *Bolus* 653 (SAM, W). Calvinia div., Wolwe Graafwater, c.2000ft, 23 viii 1956, *Acocks* 18948 (K, PRE). Van Rhynsdorp div., 3118DC, 6 miles north of Klawer, 22 viii 1950, *Barker* 6433 (NBG); Heerenlogement, 3118DC, 21 vii 1941, *Esterhuysen* 5578 (BOL, PRE); 3118DC, Windhoek, 300ft, 9 vii 1896, *Schlechter* 8080 (E, K, P, S, W). Vredendal div., 3118CB, Vredendal, 22 viii 1987, *Batten* 792 (E). Clanwilliam div., 3218BB, near Clanwilliam, 20 vii 1971, *Compton* 11007 (NBG).

Bentham (loc. cit. 1836) described this plant as a variety of *Nycterinia selaginoides* (that is, *Z. villosa*): he saw three specimens collected by Drège. When he published his account of Scrophulariaceae in De Candolle's Prodromus 10 (p.349, 1846), he no longer kept up his var. *glabra* (though Hiern, 1904, chose to do so). Now that plentiful material is available, it is clear that 'var. *glabra*' is a well-marked entity widely distributed in the western Cape from just south of the Orange River to the environs of Lambert's Bay, Clanwilliam and Citrusdal. I chose to recognise it as a full species distinct from *Z. villosa*, though closely related to it (whence the epithet *affinis*), and despite the fact that in the area about Langebaan, Velddrift, Aurora and Sauer, where their areas meet, the distinctions between them are slightly blurred. This blurring could stem from introgression or from incomplete speciation, a problem whose resolution lies in the field and laboratory.

Bentham cited under his var. *glabra* 'On the Olifants and Zwartdoorn rivers, *Drège*'. In Drège's Zwei pflanzengeographische Documente, three collections of *Nycterinia* var. β are listed: 'a' 'Bei Olifantrivier, auf Karrooartigen Höhen, unter 1000 Fuss, August' (which proves to be *Z. violacea* Schltr.), 'b' 'Bei Holrivier, auf Karrooartigen Höhen, unter 1000 Fuss, August', and 'c', 'Bei Mierenkasteel, Karrooartigen Höhen, 1000–2000 Fuss, August'. I have lectotypified var. *glabra* (not *glabrior* as on the tickets of the distributed specimens) by the specimen from Holrivier in the Kew herbarium (the village of that name lies about 5km NE of the Olifants River); the duplicates in E and S each have a specimen of *Z. benthamiana* mixed with the specimens of var. *glabra*. For this and other reasons I have not used Drège's specimens to typify the name *Z. affinis* but have chosen instead good recent specimens from the farm Liebendal, about 5km S of Holrivier. They match precisely the Drège specimens from that locality.

The two species, *Z. affinis* and *Z. villosa*, differ primarily in the shape and indumentum of the leaves and bracts: lanceolate and ± acute with margins entire or nearly so in *Z. affinis*, oblanceolate to spathulate and ± obtuse with margins toothed in *Z. villosa*, the leaves of *Z. affinis* sparsely hairy, hairs mainly on lower margins and the petiolar part, those of *Z. villosa* densely pubescent all over as are the bracts, while in *Z. affinis* the bracts are usually glabrous or very nearly so in the distal half: when hairs are present, particularly on the upper surface, they are scattered, minute and inconspicuous. The bracts of *Z. affinis* are usually much narrower in the distal half than those of *Z. villosa*, a reflection of the basic difference in shape between the leaves and bracts of the two species.

Specimens of *Z. affinis* usually dry greenish, those of *Z. villosa* dusky, but the distinction is not absolute.

What may be an autogamous form of *Z. affinis* has been recorded in the southern part of the range of the species, from roughly Heerenlogement south to Porterville. The following specimens have been seen: Van Rhynsdorp div., von Olifantsrivier bis Langevalei, meistens Sandflachen in welchen Knakasberg und Heerenlogementberg ihre felsigen Gipfel erheben, 500–2000 Fuss, *Zeyher* 1283b (P, W). Clanwilliam div., 3219AC, Brackfontein, 500ft, 8 viii 1897, *Schlechter* 10780 (BOL, E, K, P, PRE, W); ibidem, 3 ix 1936, *Wall* s.n. (S). Piquetberg div., 3218DD, Piquetberg, 350ft, x 1895, *Bolus* s.n. (PRE); De Hoek, 19 vii 1941, *Compton* 10914 (NBG); ibidem, 19 vii 1941, *Esterhuysen* 5516 (K, BOL); 3218DC, Sauer, 30 ix 1943, *Compton* 15080 (NBG). Malmesbury div., 3318BB, Porterville, 800ft, 20 viii 1894, *Schlechter* 4903 (BOL); 3318CB, Groenekloof, *Zeyher* 1283 (BOL, K, P, PRE). The last locality needs confirmation: *Zeyher* 1283a came from Groenekloof, *Zeyher* 1283b from much further north; no discriminating letters appear on the sheets in BOL, K and PRE.

The specimens quoted above all have very small flowers: the figures in parentheses in the formal description refer to them. Autogamy is suggested by the small size of the flowers and by the stigmas often being included. The plants are readily confused with *Z. parviflora*, but differ in the stems being leafy in the lower part (not floriferous) and the leaves and bracts being much less hairy than in *Z. parviflora*. They appear to be allopatric but the area between Langebaan and Porterville should be carefully explored.

Zaluzianskya affinis is frequently recorded on sandy flats and slopes clad in karroid vegetation, usually in large colonies in open places among the bushes, from sea level to about 900m, flowering between June and November, at its peak in August and September. The upper surface of the corolla limb is most often white, but some collectors have recorded 'mauve to white', others only mauve, while there are a few collections from scattered localities that record pale to dark yellow flowers. The yellow to orange or red star-shaped pattern around the mouth seems to be a constant feature, with the orange colour running down the back of the throat (this colour patterning is possibly absent in the autogamous form; field observations are needed). Plants from the Bidouw valley are unusual in that the hairs on the stem are sometimes very short and the lower backs of the bracts are very shortly pubescent, all long hairs sometimes confined to the margins.

28. Zaluzianskya parviflora Hilliard in Edinb. J. Bot. 47: 348 (1990).
Type: Cape, Saldanha Bay, Hoetje's Bay, ix 1905, *H. Bolus* 12794 (holo. BOL; iso. K, PRE).

Annual herb 100–310mm tall, main stem erect, soon sparingly branched from the base and above, branches erect, ascending or decumbent, villous with patent acute hairs up to 1mm long, minutely glandular as well, laxly leafy throughout, all but the lowermost 1 or 2 pairs bearing an axillary flower. *Leaves*: lowermost 11–30 × 3–7mm, scarcely shorter but often narrower upwards, oblanceolate, apex ± acute, base narrowed but scarcely petiolar except in the lowermost pair, margins with 2–6 pairs of small teeth in upper half, uppermost leaves (bracts) sometimes ± entire, both surfaces thinly villous, hairs up to c.1.5–3mm long particularly on lower margins and backs. *Flowers* open in sunshine, solitary in the leaf axils, forming long, lax, many-flowered spikes. *Bracts* adnate to calyx tube for c.2–3mm, lowermost c.10–30 × 3.5–7mm, lanceolate, leaf-like. *Calyx* 5–6mm long, both anticous and posticous lips 1–3mm long, shortly pubescent all over, hairs at tips and on keels up to c.0.8–1mm long. *Corolla* tube 8.2–10mm long (throat 0.7–1mm in diam.), minutely glandular-puberulous, limb 2.5–3.6mm in diam., regular, lobes eventually reflexed, mouth encircled by stiff acute erect hairs, lobes 0.8–1.4mm long, bifid, broadly Y-shaped, creamy-white fading to ochre-yellow, no coloured 'eye' reported and none visible in dried material. *Stamens* 2 (posticous pair), anthers 0.6–0.7mm long, included. *Stigma* 1.5–2.5mm long, included. *Style* 3.7–5.3mm long. *Ovary* 2–2.5 × 0.8–1mm. *Capsules* 5–7 × 2–2.5mm. *Seeds* c.0.5–0.7 × 0.4–0.5mm, pale mauve-grey, irregularly wrinkled, obscurely angled, angles marked by very narrow pallid wings, sometimes wanting.

Citations:

Cape. Malmesbury div., 3317BB, Saldanha Bay, ix 1949, *MacNae* 1112 (SAM); 3318AA, peninsula W of Langebaan, x 1933, *Pillans* 6965 (BOL); Postberg, 8 ix 1957, *Lewis* 5242 (NBG); ibidem, 9 ix 1966, *Pamphlett* 119 (NBG, STE). Steenberg Cove, ix 1944, *Lewis* 1940 (SAM); ibidem, 2 ix 1944, *Compton* 15931 (NBG).

The sheet of *Nycterinia selaginoides* var. *parviflora* Benth. in Bentham's own herbarium (now in K) bears two separate collections, one, made by Ecklon and written up by Bentham as *N. selaginoides* β *parviflora*, is clearly the holotype of the variety, because the second collection, made by Zeyher and determined by Bentham, came to Bentham from Wallich in 1847, long after he had published the name. Ecklon's specimen proves to be *Polycarena silenoides* (which the hawk-eyed N. E. Brown noted on the sheet); Zeyher's is *Zaluzianskya affinis*. The only other collection determined by Bentham as *Nycterinia selaginoides* var. *parviflora* and seen by me is in the Natural History Museum in Stockholm (S): the sheet bears three specimens in a row, with three collector's labels and Bentham's determinavit slip gummed along the lower margin of the sheet, so it is impossible to judge exactly which specimen Bentham saw. Two of the labels were written by Zeyher (224 and 150) and the specimens came from Saldanha Bay; they are the species here recognized as *Z. parviflora*; the third specimen, collected by Sieber, is *Z. affinis*. Hiern's (1904) transfer of the varietal name *parviflora* to *Zaluzianskya villosa* compounded the muddle, and the name inevitably became attached to the plant now to be known as *Z. parviflora*. It differs from *Z. villosa* not only in its much smaller flowers, but also in the shape of the leaves and bracts (oblanceolate, acute, not oblanceolate, obtuse), and in the stems, floriferous almost to the base in *Z. parviflora*, leafy in the lower part in *Z. villosa*. It is this last character above all that confirmed my decision on its rank.

Zaluzianskya parviflora is easily confused with the small-flowered form of *Z. affinis*. This plant has the stems leafy in the lower part and the leaves and bracts are much less hairy than those of *Z. parviflora*; most of the specimens cited by Hiern (1904) as *Z. villosa* var. *parviflora* are *Z. affinis*. *Zaluzianskya parviflora*, like the small-flowered form of *Z. affinis*, may be autogamous: the insignificant flowers and included style (both exceptional in the genus) suggest this, also the prolific seed-set. The species appears to be confined to the environs of Saldanha Bay, flowering in August and September.

See also *Z. gracilis* (below).

29. Zaluzianskya gracilis Hilliard in Edinb. J. Bot. 47: 346 (1990).
Type: Cape, Riversdale div., 3421AC, Wankoe south of Riversdale, 400ft, 6 ix 1975, *Oliver* 5981 (holo. STE, iso. PRE).

Annual herb c.50–250mm tall, main stem erect, soon branching from the base and above, branches erect, ascending or decumbent, villous with patent acute hairs up to c.1mm long, minutely glandular as well, laxly leafy throughout. *Leaves* 8–24 × 1.75–4.5mm, scarcely shorter upwards and passing imperceptibly into bracts, oblanceolate, apex obtuse to subacute, base narrowed to a short petiolar part, margins with 1–4 pairs of small teeth in upper half, both surfaces densely pubescent, hairs on lower margins up to 1.5mm long, to c.1mm on blade, minutely glandular as well. *Flowers* open in sunshine, in long terminal spikes, initially crowded, becoming lax in fruit. *Bracts* adnate to calyx for 2.5–4mm, lowermost c.10–16 × 2–3.5mm, leaflike. *Calyx* 5.2–7mm long, both lips 1–2mm long, pubescent all over, lobes tipped with longer hairs to c.1mm. *Corolla* tube 6.5–10mm long (throat 0.8–1mm in diam.), minutely glandular-puberulous, limb c.2–4mm in diam., regular, lobes eventually reflexed, mouth encircled by stiff acute erect hairs, lobes 0.8–2.8mm long, bifid, broadly Y-shaped, white to pale mauve or pink, yellow to orange around mouth, glabrous. *Stamens* 2 (posticous pair), included, anthers 0.6–1mm long. *Stigma* 1.4–2.5mm, included or shortly exserted. *Style* 2–3.6mm long. *Ovary* c.2–2.5mm long. *Capsules* c.5.5 × 2mm. *Seeds* c.0.5–0.6 × 0.4mm, pale mauve-grey, irregularly wrinkled, obscurely 3-angled, angles marked by very narrow pallid wings, sometimes wanting.

Citations:

Cape. Bredasdorp div., Brandfontein, 13 x 1951, *Esterhuysen* 18994 (BOL); 3420BC, Potteberg, 19 ix 1954, *Esterhuysen* 23343 (BOL, K); 3420AD, De Hoop–Potberg Nature Reserve, 30m, Dronkvlei, 11 ix 1979, *Burgers* 2181 (STE); 3420CB, De Hoop, Buffelsfontein, 60–80m, 9 viii 1984, *Fellingham* 761 (STE); 3420CA, Arniston, c.100ft, 24 viii 1962, *Acocks* 22607 (PRE). Riversdale div., 3421AD, Stil Baai, 18 viii 1929, *Nel* sub STE 9620 (STE); ibidem, 60m, 16 viii 1986, *Bohnen* 8720 (STE); ibidem, 26 viii 1978, *Bohnen* 401.3 (K); Takkiesfontein, c.500ft, 25 ix 1978, *Hugo* 1230 (PRE, STE); on the Stil Baai–Blombos road, 7 ix 1957, *Wurts* 1546 (NBG).

Zaluzianskya gracilis appears to be confined to calcareous sands in Bredasdorp and Riversdale districts, on littoral dunes or in open patches in scrub on ridges, up to c.150m above sea level, flowering between August and October. In its very small flowers it resembles *Z. parviflora*, but is easily distinguished by its leafy stems (those of *Z. parviflora* bear flowers almost to the base) and more obtuse leaves. The two species are possibly vicariads, developed since the drowning of the coast in the Middle Pleistocene, isolating *Z. parviflora* around Saldanha Bay from *Z. gracilis* along the southern coast; *Manulea thyrsiflora* shows a similar distribution, but the two now isolated populations have not diverged so sharply.

Zaluzianskya gracilis resembles *Z. villosa* in habit and foliage, but is easily distinguished by its much smaller flowers. The two species are allopatric, *Z. villosa* not having been recorded further east than Pearly Beach between Danger Point and Quoin Point.

30. Zaluzianskya sanorum Hilliard in Edinb. J. Bot. 47: 349 (1990).
Type: Cape, Little Namaqualand, 2917DB, near Springbok, 25 viii 1941, *Esterhuysen* 5888 (holo. BOL, iso. K).

Annual herb often flowering in seedling stage, stems 10–140mm long, erect, ascending, decumbent or prostrate, initially simple, soon branching from the base and more sparingly above, pubescent to almost glabrous particularly near base, hairs up to 0.2mm long, acute, ±

retrorse, stems leafless when simple, later a leaf developing at each forking. *Leaves* mostly radical, few, 12–40 × 3–9mm, spathulate, ½–⅔ total length petiolar, apex ± obtuse, upper surface sparsely and minutely pubescent, lower glabrous, margins entire to obscurely toothed with sparse hairs up to 1.1mm long. *Flowers* in crowded, eventually much elongated, spikes, the primary stem sometimes floriferous nearly to the base, flowers open in sunshine. *Bracts* adnate to calyx tube for 2.5–5mm, lowermost c.8–25 × 2.8–9mm, smaller upwards, leaf-like, mostly ± ovate above contracted into an oblong basal half, subacute, upper surface pubescent with acute hairs up to 0.25mm long, lower surface either glabrous or with scattered minute hairs particularly in lower half, lower margins with hairs up to 0.4–1mm long. *Calyx* 5.5–8mm long, both lips 1–2mm long, minutely pubescent, tips with hairs up to 0.4mm long, these sometimes on posticous keels as well. *Corolla* tube 14–24mm long, (throat 1.3–1.7mm in diam.), pubescent with both glandular and eglandular hairs, either predominating, glandular hairs less than 0.1mm long, acute hairs up to 0.2–0.25mm long, limb 9–16mm in diam., regular, held horizontally, lobes eventually reflexed, mouth encircled by stiff acute erect hairs, lobes 3.5–7mm long, bifid, broadly Y-shaped, rich mauve to red-violet above, yellow to bright orange star-shaped patch around mouth, glabrous. *Stamens* 2 (posticous pair), included, 1.5–2mm long. *Stigma* 1.5–4mm long, exserted. *Style* 11–20mm long. *Ovary* 3–4 × 0.8–1mm. *Capsules* 6–7 × 2.5–3mm. *Seeds* c.0.5–0.6 × 0.4–0.5mm, pale mauve-grey, irregularly wrinkled, obscurely angled, angles marked by very narrow pallid wings, sometimes wanting.

Selected citations:

Cape. Namaqualand, 2818CD, 27 miles S of Goodhouse, 27 vii 1950, *Barker* 6300 (NBG); 2917BB, ± 10.5km from Steinkopf on road to Vioolsdrift, 27 viii 1983, *Van Wyk* 6570 (PRE); 2917BD, Goechas, 3000ft, 21 ix 1897, *Schlechter* 11372 (BOL); 2917DB, 25 miles N of O'Okiep, 25 viii 1959, *Barker* 9050 (NBG); 2917DD, Droedap, 26 viii 1941, *Compton* 11537 (NBG); 2918AA, 48 miles NE of Springbok, 29 v 1961, *Schlieben* 9082 (K, S, W); 2918AB, Little Bushmanland, Keuzabies [=Koisabis], 18 vi 1898, *Max Schlechter* s.n. (BOL, E, K, P, PRE, S, W). Pofadder div., 2919BC, farm Houmoed on road between Pofadder and Kenhardt, 23 ix 1961, *Hardy* 763 (K, PRE).

All species of *Zaluzianskya* with bifid corolla lobes and only two stamens have consistently been lumped under *Z. villosa*, so despite the fact that *Z. sanorum* was collected as early as 1897 by Rudolf Schlechter, and a year later by his brother Max, the species has hitherto gone unrecognized. Both these early collections were cited by Hiern (1904: 346) as *Z. villosa* var. *glabra*.

    Zaluzianskya sanorum is readily distinguished from *Z. affinis* (*Z. villosa* var. *glabra*) by its glabrous to sparsely pubescent stems, the hairs up to 0.2mm long and ± retrorse (not villous with patent hairs up to 0.4–2mm long), hairs on margins of bracts up to 0.4–1mm long (not 1.5–3mm) and corolla tube distinctly pubescent with a mixture of glandular and eglandular hairs, one or the other often predominating, glandular hairs less than 0.1mm long, acute hairs up to 0.2–0.25mm (not minutely puberulous to nearly glabrous). The corolla limb of *Z. sanorum* seems always to be coloured rich shades of red-violet, contrasting with the bright yellow or orange 'eye'; that of *Z. affinis* is frequently white, sometimes mauve, also with a contrasting 'eye'. The area of *Z. affinis* lies west and south of that of *Z. sanorum*, mostly at lower altitudes. Their areas appear to meet west of Steinkopf. *Zaluzianskya sanorum* is essentially a plant of the interior plateau, more than 900m above sea level (sea level to 900m in *Z. affinis*). Ecological information is scanty, but the plants have been recorded as growing in red Kalahari sand and in red sandy loam on gravel slopes, and they flower between May and October. The epithet commemorates the San (Bushmen) in whose territory, Little Bushmanland, the plant has mostly been collected, though records stretch as far east as Pofadder and south to Kamieskroon; the most northerly record was made south of Goodhouse (Gudaos) on the Orange river (the Gariep).

    See also *Z. pilosissima* (below).

31. Zaluzianskya pilosissima Hilliard in Edinb. J. Bot. 47: 348 (1990).
Type: Cape, Fraserburg div., 3221AB, 38 miles SW of Fraserburg, c.4300ft, 22 viii 1953, *Acocks* 16904 (holo. PRE).

Annual herb often flowering in seedling stage, stems 10–140mm long, erect to prostrate, initially simple, soon branching from the base and more sparingly above, pilose, hairs up to 0.75–1mm long, ± retrorse, stems leafless when simple, later a leaf developing at each forking. *Leaves* mostly radical, few, 7–30 × 2.5–7mm, spathulate, ½–⅔ total length petiolar, apex ± acute, both surfaces pilose with acute hairs up to 1mm long, minutely and sparsely glandular as well, margins entire or with a few small teeth in upper half, hairs up to 1–2mm long particularly on petiolar part. *Flowers* in crowded spikes, open in sunshine. *Bracts* adnate to calyx for 2–4mm, lowermost 7–16 × 2–5mm, smaller upwards, leaf-like, ovate to elliptic above contracted into an oblong basal half, ± acute, both surfaces pilose with hairs up to 0.5mm long, hairs on margins up to 0.75–2mm long. *Calyx* 5–6.5mm long, both lips 1.2–2mm long, pilose, hairs on keels up to 1–1.5mm long, minutely glandular as well. *Corolla* tube 12.5–19mm long, (throat 1.3–1.7mm in diam.), pubescent with acute hairs 0.25–0.4mm long, minutely glandular as well, exclusively so on backs of lobes, limb 7.8–12mm in diam., regular, held horizontally, lobes eventually reflexed, mouth encircled by stiff acute erect hairs, lobes 2.8–5mm long, bifid, broadly Y-shaped, pink to mauve, yellow to bright orange star-shaped patch around mouth, glabrous. *Stamens* 2 (posticous pair), anthers 1.8–2mm long, included, tips sometimes just visible in mouth. *Stigma* 1–4mm long, exserted. *Style* 9–15mm. *Ovary* 1.75–3.5 × 0.6–1mm. *Capsules* 5–7 × 2.5–3mm. *Seeds* c.0.5–0.6 × 0.4–0.5mm, pallid faintly tinged mauve-grey, irregularly wrinkled, obscurely angled, angles marked by very narrow pallid wings sometimes wanting.

Citations:

Cape. Calvinia div., 3020CC, Breekbeenkolk to Kotzeskolk, 2 ix 1986, *Burger & Louw* 173 (STE). Carnarvon division, 3022CC, Carnarvon, ix 1935, *Vermeulen* s.n. (BOL). 3121BA, 13 miles E of Scorpion's Drift, c.4200ft, 27 viii 1952, *Acocks* 16416 (PRE). Between Fraserburg (3121DC) and Carnarvon (3022CC), 4000ft, 20 ix 1938, *Wall* s.n. (S); ibidem, *Häfstrom & Acocks* 1249 (PRE, S); 3220BC, near Sutherland, 5000ft, *Wall* s.n. (S); between Sutherland and Middlepost, 9 x 1928, *Hutchinson* 704 (K). 3220BD?, between Jackals Fountain and Kuilenberg, 8 viii 1811, *Burchell* 1342 (K). 3120CD, Roggeveld Escarpment, Kookfontein 868, 1163m, 30 viii 1986, *Cloete & Haselau* 115 (STE). 3120DD, farm Uitkoms on Middelpos–Calvinia road, 1100m, 3 ix 1986, *Burger & Louw* 222 (STE).

Zaluzianskya pilosissima has hitherto been confused with *Z. villosa* from which it is distinguished by its stems, leafless or almost so, clad in ± retrorse hairs up to 0.75–1mm long (not leafy throughout with spreading hairs up to 1–1.5mm long), mostly shorter bracts, the longest 7–16mm (versus 11–25mm), mostly shorter calyx (5–6.5mm versus 6.5–9mm), and distinctly pubescent corolla tube with acute hairs up to 0.4mm long (versus minutely glandular-puberulous). Its closest ally is possibly *Z. sanorum*, a much more glabrous plant (stems glabrous or with hairs up to 0.2mm long, lower surface of leaves and bracts glabrous or very nearly so, hairs on margins up to 0.4–1mm long, hairs on corolla tube seldom more than 0.2mm long).

Zaluzianskya pilosissima occurs south and east of the areas of both *V. villosa* and *V. sanorum*, in the arid Karoo of the central Cape, from Kotzeskolk, Middelpos and Sutherland east to Carnarvon and Fraserburg, between 1100 and 1500m above sea level. It grows in sandy places including dry stream beds, and flowers in August and September.

32. Zaluzianskya bella Hilliard in Edinb. J. Bot. 47: 346 (1990).
Type: Cape Province, Sutherland div., 3220CA, Houthoek, 14 viii 1968, *Hanekom* 1074 (holo. K; iso. PRE, STE).

Annual herb, primary stem 15–150mm tall, simple or with 2–3 branches from the base, these ascending or decumbent, simple or with a pair of branches, pubescent with patent acute hairs up to 0.3–0.8mm long, more or less minutely glandular as well. *Leaves* few, usually only the radical pair and one other, blade 2–12 × 2–6mm, tapering into a flat petiolar part 2–11mm long (roughly as long as the blade or occasionally shorter in the upper leaves), ovate to elliptic, subacute, margins entire to very obscurely toothed, both surfaces pubescent, hairs acute, longest on margins, particularly of petiolar part, up to 1–2mm long. *Flowers* up to c.20, at first crowded in a head, elongating into a spike in fruit, open in sunshine. *Bracts* adnate to calyx for 2–4mm, lowermost c.7–15 × 3–6mm, smaller upwards, elliptic to ovate in upper part, contracting into a broad shaft, pubescent on both surfaces, hairs all acute, those on backs of bracts up to 0.8–1.2mm long, up to 1.5–2mm on margins. *Calyx* 4.2–6mm long, lips 1–1.5mm long, pubescent with acute hairs up to 1mm long. *Corolla* tube 11–15mm long (throat 0.8–1.3mm in diam.), minutely glandular-puberulous (hairs up to 0.1mm long), limb 8–10mm in diam., regular, mouth encircled by long acute erect hairs, lobes eventually reflexed, 3.5–4 × 2.5–4mm, Y-shaped, bifid, pale to deep mauve above, deep red below, deep yellow star-shaped patch around mouth, the rays extending well up shaft of lobe, sometimes with a small crimson blotch at the base of each lobe. *Stamens* 4, posticous pair included, anthers 1–1.4mm long, anticous anthers well exserted, 0.4–0.7mm long. *Stigma* 0.6–2mm long, exserted. *Style* c.8.4–9mm long. *Ovary* c.1.5 × 0.8mm. *Capsules* 5–6 × 2.4–3mm. *Seeds* c.0.4 × 0.3mm, pale amber colour, strongly colliculate, obscurely angled, angles ± marked by very narrow pallid wings.

Citations:

Cape. Ceres div., 3220AC, Ceres Karoo, farm Tandschoonmaak, 4 vii 1975, *Van Breda* 4336 (K, PRE). Sutherland div., Tanqua Karoo (3219DD?), 1500ft, 10 vii 1938, *Compton* 7250 (NBG). 3220BD, Roggeveld, Wilgenboschfontein, ix 1909, *Worsdell* s.n. (K). 3220CA, Koedoesberg, 2500ft, 5 ix 1926, *Levyns* 1564 (BOL). 3220DA, Klein Roggeveld, 1100m, x 1920, *Marloth* 9601 (PRE). 3220DB, 20 miles SSW of foot of Komsberg Pass, c.3700ft, 15 ix 1955, *Leistner* 267 (PRE). 3220DC, at turnoff to De Plaat, 1200m, 14 ix 1986, *Fellingham* 1191 (STE).

This pretty little plant has been found in the Tanqua and Ceres Karroos north and north west of the Koedoesberg, thence east to the Komsberg and Klein Roggeveld, at c.500 to 1200m above sea level. It grows in sandy or gravelly places among shrublets, and flowers between July and September, being well on into fruit in October.

Its area lies south east of that of *Z. violacea* and west of that of *Z. venusta*, though it appears to be sympatric with *Z. venusta* in the environs of the Koedoesberg. It is at once distinguished from *Z. violacea* by its stems with patent (not retrorse, appressed) hairs and corolla tube minutely glandular (not with delicate acute hairs up to 0.3–0.4mm long). *Zaluzianskya bella* resembles *Z. venusta* in the indumentum on its stems and corolla tube, but its bracts lack minute glandular hairs, and it has rather smaller flowers (corolla tube 11–15mm long versus 15–26mm, posticous anthers 1–1.4mm versus 1.5–2mm, anticous anthers 0.4–0.7mm versus 0.6–1.2mm).

33. Zaluzianskya violacea Schltr. in Bot. Jahrb. 27: 183 (1899); Hiern in Fl. Cap. 4(2): 344 (1904), excluding *Schlechter* 10780.
Type: Van Rhynsdorp div., sandy hills near the Zout River, 450ft, 13 vii 1896, *Schlechter* 8113 (holo. B†; iso. E, K, PRE, S, W).

Annual herb 20–150mm tall, primary stem erect, soon branching from the base, branches ascending or decumbent, simple to sparingly branched, pubescent with retrorse ± appressed acute hairs up to 0.3–0.4mm long, distantly leafy, sometimes only 1 pair above the radical pair and the cotyledons. *Leaf* blade c.3.5–20 × 2–10mm tapering into a flat petiolar part often as long as the blade, but shorter on stem leaves, ovate in lowermost pair, elliptic-ovate to elliptic upwards, apex subacute, margins entire or obscurely dentate, both surfaces shortly pubescent, minutely glandular as well, longer hairs to c.1mm on margins, particularly of petiolar part. *Flowers* up to 20, at first crowded in a head, elongating into a spike in fruit, open in sunshine. *Bracts* adnate to calyx for 3–4.5mm, lowermost c.7–18.5 × 2–5mm, elliptic to ovate in upper part, contracting into a broad membranous shaft, usually shortly pubescent on backs, hairs sometimes densest on lower half, very rarely backs almost glabrous, densely pubescent on inner face, margins with long hairs to 1.5mm, occasional long hairs (to 1mm) sometimes also on midline. *Calyx* 4.5–8mm long, lips 1–3mm long, finely pubescent, minutely glandular as well, long (up to c.0.4–0.6mm) acute hairs mainly on keels and at tips of teeth. *Corolla* tube 10–23 (–35)mm long (throat 1.2–1.8mm in diam.), usually densely pubescent with delicate acute hairs up to c.0.3–0.4mm long, (sometimes minute glandular hairs as well), very rarely hairs confined to swollen apical part or whole tube glabrous; limb 7–12mm in diam., regular, lobes eventually reflexed, mouth usually encircled by stiff acute erect hairs, hairs rarely few or wanting, lobes 3–7.5 × 2–7mm, Y-shaped, bifid, usually shades of mauve or rosy-mauve (very rarely pale yellow), deep yellow star-shaped patch around mouth, rays extending well up shaft of lobe, often with a small crimson blotch at base of each lobe. *Stamens* usually 4, rarely the anticous pair aborted, posticous pair included or very rarely exserted, anthers 1.3–2mm long, anticous pair exserted, anthers 0.3–0.8mm long. *Stigma* 3–6mm long, exserted. *Style* 11–18mm long. *Ovary* c.2–3.5 × 0.8mm. *Capsules* c.5–7 × 3mm. *Seeds* 0.75–0.8 × 0.4–0.5mm, pale mauvish-grey, colliculate, obscurely 3-angled, angles ± marked by very narrow pallid wings, or wings wanting.

Selected citations:

Cape. Calvinia div., 3119AD, Vlakfontein, 21 viii 1954, *Barker* 8290 (NBG); 3119BD, Akkerdam, foot of Hantam Mts, 22 vii 1961, *Barker* 9305 (NBG); ibidem, *Acocks* 17745 (K, PRE); ibidem, 17 viii 1987, *Batten* 774 (E); 3119DA, Kareeboomfontein, 9 ix 1974, *Hanekom* 2425 (K, PRE); 3119DB, Bloukrans Mts, farm Kneu, 3800ft, 21 viii 1975, *Thompson* 2503 (STE).

Schlechter's specimens of *Z. violacea* came from Zoutrivier, on the road between Vanrhynsdorp and Neuwerus, and Drège found it somewhere along the Olifants River west of Vanrhynsdorp; all subsequent collections have been made around Loeriesfontein and Calvinia and south east as far as Middelpos, on the road to Sutherland (though possibly the species extends south to Ceres division; see comments below). The species is clearly common around Akkerdam at the foot of the Hantam Mountains, and it was possibly hereabouts that Thunberg found it. There is no specimen of *Z. violacea* in Thunberg's own herbarium, but he gave one to Montin (as a species of *Eranthemum*, in Acanthaceae), which is now in Stockholm (S).

Zaluzianskya gilgiana Diels (in Bot. Jahrb. 23: 480, 1896) is probably conspecific with *Z. violacea*, which it antedates by three years. The type came from the Hantams, collected there by Meyer, a medical practitioner in Calvinia in the 1860s, who sent his specimens to Berlin, where they were destroyed in the great fire of the 1940s. Diels described the calyx of his new species as 5–10mm long, corolla tube 17–25mm (4.5–8mm and 10–23mm respectively in *Z. violacea*, which is an acceptable discrepancy), and the corolla tube as '*tenuissime pubescente*', which I interpret as very finely pubescent, a precise description of the tube in *Z. violacea* (Hiern, in his translation of Diels's description, writes 'very thinly pubescent'). A discordant note is struck by Diels's description of the bracts as '*albovillosis*'; those of *Z. violacea* have really long hairs only on the margins, with shorter hairs on the blade. On the other hand, I have seen nothing from the Hantams with truly whitevillous bracts; also, it seems inconceivable that Meyer lived there and failed to collect a common and very attractive plant.

The practical solution seems to be to maintain *Z. violacea*, well typified by specimens in several different herbaria, rather than make a doubtful reduction to a name that will remain without a type.

Zaluzianskya violacea is normally easily recognized by the combination of retrorse hairs on the stems, finely pubescent corolla tube, mauve limb with bifid lobes, a deep yellow 'eye', and four stamens. The two anticous (exserted) stamens are occasionally completely aborted, apparently throughout a whole population; the two posticous stamens are then shortly exserted (*Lewis* 3866, SAM, from Moordenaarspoort near Agter Hantamberg, and *Middlemost* 1768, NBG, from ± 30 miles north of Calvinia, which cannot be far from Moordenaarspoort). This specimen is mixed with *Z. peduncularis*).

Several collections from the Loeriesfontein-Calvinia area show various degrees of reduction of the hairs on the corolla tube: these may be well developed, or confined to the apical part of the tube, or completely wanting, all in one collection (*Hugo* 510, PRE, STE, Rheeboksfontein, south west of Loeriesfontein; *Lewis* 3867, SAM, near the Hantams River; *Burger & Louw* 286, STE, Calvinia-Loeriesfontein). These specimens all have mauve flowers; but *Hugo* 511, PRE, STE, Vosfontein, south of Loeriesfontein, and *Burger & Louw* 281, STE, Kareeboom, Calvinia-Loeriesfontein, have yellow flowers (pale limb, darker 'eye'), and the corolla tube glabrous to hairy on the upper part only. Hugo noted that the mauve-flowered plant was 'fairly frequent', the yellow-flowered plant 'occasional'.

Several more collections need special mention. These are: Ceres div., 3219DB, Die Bos, *Stayner* s.n. (NBG, STE); Gansfontein, 3219DA, *Compton* 5528 (NBG); Ceres-Calvinia Karoo near Katbakkies Rock [3219DC?], *Levyns* 11366 (BOL). Calvinia div., Vogelstruis Vlakte (not traced), *Compton* 11139 (NBG); ibidem, *Esterhuysen* 5326 (BOL, K). These all have stems with retrorse hairs, an important characteristic of *Z. violacea*, as well as its flower colour; however, the corolla tube is glabrous, and the backs of the bracts are very nearly so.

The specimens I have been able to localize precisely came from well south of the main area of *Z. violacea*, and abut on the western limits of *Z. bella* (easily distinguished by the long spreading hairs on the stems). The possibility of introgression at the geographical interface between the species should be investigated.

34. Zaluzianskya marlothii Hilliard in Edinb. J. Bot. 47: 347 (1990).
Type: Cape, Sutherland div., 3220AD, Uitkyk, Sneeuwkrans, 1700m, x 1920, *Marloth* 9903 (holo. STE).

Annual herb, primary stem c.20–50mm long, simple, soon branching from the base, branches simple, decumbent, pubescent with patent acute hairs up to c.0.5mm long, minutely glandular as well. *Leaves* few, blade c.3–8 × 2–6mm, ovate to elliptic tapering into a broad flat petiolar part 3–8mm long (roughly equalling blade in length), subacute, margins entire, both surfaces pubescent with acute hairs up to 1–1.5mm long, minutely glandular as well. *Flowers* crowded in a dense terminal spike, open in sunshine. *Bracts* adnate to calyx for c.2.5mm, lowermost c.7–11 × 2.5–5mm, ± spathulate, obtuse to subacute, pubescent all over, acute hairs up to 1.5–2.6mm long, minutely glandular as well, both short glandular and short acute hairs forming a fine dense pubescence underlying the long coarse hairs. *Calyx* 5–6mm long, lips 1mm, pubescent with acute hairs up to 1mm long, minutely glandular as well. *Corolla* tube c.13–15mm long (throat c.1.3–1.5mm in diam.), glandular-puberulous, hairs up to c.0.15mm long, limb c.8–11mm in diam., ± regular, mouth encircled by acute hairs and there with a crimson star (visible in dried state), the rays extending well up the shaft of the lobes, lobes c.3.8–4.5 × 2.8–4.5mm, broadly spathulate, entire, mauve. *Stamens* 4, anticous pair exserted, anthers 0.8mm long, posticous pair included, anthers 1.5–1.6mm long. *Stigma* 2–3mm long, exserted. *Style* 10–11mm. *Ovary* c.3 × 0.8mm. *Capsules* (few seen) c.5 × 2.2mm. *Seeds* c.1 × 0.6mm, obscurely angled, colliculate, mauvish-grey.

Citation:

Cape. Sutherland div., Roggeveld, 1500m, x 1920, *Marloth* 9772 (BOL).

Zaluzianskya marlothii has the aspect of *Z. bella* but can be distinguished not only by its entire corolla lobes, but also by the dense fine pubescence underlying the long coarse hairs particularly on the backs of the bracts and the petiolar part of the leaves, this composed of both glandular and eglandular hairs (both leaves and bracts in *Z. bella* lack any glands); the anthers of *Z. marlothii* may be longer than those of *Z. bella* but this needs confirmation on more material (anticous anthers 0.8mm versus 0.4–0.7mm, posticous anthers 1.5–1.6mm versus 1–1.4mm). Its true relationship possibly lies with *Z. minima*, from which it is easily distinguished by its longer corolla tube and larger limb with broadly spathulate lobes, and well-bearded mouth.

Zaluzianskya marlothii is known only from Marloth's two collections, made at 1500 and 1700m; *Z. bella* also grows near Sutherland, but it has been recorded only west and south of the site where *Z. marlothii* was found and at lower altitudes; *Z. minima* is sympatric with *Z. marlothii*.

35. Zaluzianskya venusta Hilliard in Edinb. J. Bot. 47: 350 (1990).
Type: Cape Province, Prince Albert div., 3221DC, 3½ miles N of Prince Albert Road station, c.2200ft, 18 ix 1953, *Acocks* 17103 (holo. PRE).

Annual herb, primary stem 10–150mm tall, simple or with 2–6 branches from the base, these ascending or decumbent, simple or with a pair of branches, pubescent with patent acute hairs up to 1–1.4mm long, minutely glandular-puberulous as well. *Leaves* few, only 1 or 2 pairs above radical pair, blade 4–20 × 1.5–11mm, tapering into a flat petiolar part 2–16mm long (rarely about equalling the blade, usually roughly ½–⅓ as long), usually elliptic, rarely lowermost pair ovate, subacute, margins entire to very obscurely toothed, both surfaces pubescent with acute hairs, those on margins up to 1–2mm long, glandular-puberulous as well. *Flowers* up to c.30, often fewer, initially crowded in a head, elongating into a spike in fruit, open in sunshine. *Bracts* adnate to calyx for 2–4mm, lowermost c.7–15 × 2.5–5mm, smaller upwards, lanceolate-elliptic, ± entire, pubescent on both surfaces, glandular-puberulous as well, acute eglandular hairs on margins up to 1.5–2.3mm long, those on backs of bracts up to 0.8–1.2mm long. *Calyx* 5–6.5mm long, lips 1.5–2mm long, pubescent with acute hairs up to 1–1.5mm long, glandular-puberulous as well. *Corolla* tube 15–26mm long (throat 1.2–2mm in diam.), glandular-puberulous (hairs up to 0.1–0.2mm long), limb (5.5–)8.5–15mm in diam, regular, mouth encircled by long acute hairs, lobes eventually reflexed, (2–)3.2–6 × (1.6–)2.2–6mm, bifid, broadly Y-shaped, mauve to bright purplish-pink above, orange below (*Acocks* 547; *Leistner* 931 and 2785), orange/yellow star-shaped patch around mouth, rays extending well up shaft of lobes. *Stamens* 4, posticous pair included, anthers 1.5–2mm long, anticous pair exserted, anthers 0.6–1.2mm long. *Stigma* 1–3mm long, exserted. *Style* 12.5–21.5mm long. *Ovary* 2–3 × 0.8mm. *Capsules* 4–6 × 2–3mm. *Seeds*: ripe seeds not seen.

Selected citations:

Orange Free State. Boshof distr., 2825CA, Bett-el-Pillar, 11 ix 1925, *Burtt Davy* 10764 (BOL, SAM).
Cape. Upington div., near Upington (?), 2821AC, 10 vi 1965, *Mostert* 1607 (PRE). Griqualand West, 2822BB, 21 miles SW of Olifantshoek, c.3900ft, 28 viii 1961, *Leistner* 2785 (K, PRE); without precise locality, xii 1876, *Barber* 13(K). Kimberley div., 2824CA, 3 miles E of Schmidtsdrift, 3400ft, 28 ix 1957, *Leistner* 931 (PRE). Prieska div., 2922AD, Asbestos Hills, vii 1894, *Marloth* 2031 (PRE). Sutherland div., 3220CA, near Koedoesberg, Gelukshoop Farm, 500–550m, 15 viii 1985, *Hilton-Taylor & Midgley* 30 (STE). Beaufort West townlands, 3222BC, west of Walker Dam, 1100ft, 14 ix 1979, *Bohnen* 6493 (STE). Ladismith div., 3321CA, Little Karoo, Mannshoop, 395m, 16 vii 1982, *Laidler* 210 (PRE, STE); Gamka Poort Nature Reserve, 3321BC, 1300ft, 3 viii 1982, *Cattell* 105 (STE).

Zaluzianskya venusta is widely distributed in the arid northern and central Cape, from Upington in the west to Kuruman and Kimberley in the east (and just entering the Orange Free State), south to Strydenburg and Kraankuil in Hopetown division, then a distributional gap (which is filled by its close ally, *Z. karrooica*) to Beaufort West division, extending west and south to the Koedoesberg (SW of Sutherland), Ladismith and Laingsburg.

In the northern part of its range it grows in sandy places, on sand dunes or in bare ground between grass tufts; in the Karoo, it favours sandy or gravelly places, from ridges down to river beds, between c.350 and 1500m above sea level. It flowers chiefly between July and September, though there is a record as early as May, and another as late as December.

It has been much confused with *Z. violacea*, whose area lies to the west and north west of that of *Z. venusta* (mainly around Calvinia) and which is easily distinguished by its stems with retrorse appressed hairs, and corolla tube finely pubescent with long (up to 0.3–0.4mm) acute hairs. It is closely allied to *Z. bella*: see under that species. Its relationship to *Z. karrooica* needs further investigation; details are given under that species.

36. Zaluzianskya karrooica Hilliard in Edinb. J. Bot. 47: 347 (1990).
Type: Cape, Graaff Reinet div., 3124CD, Sneeuwbergen, Zuureplaats, 5000 Fuss, 12 ix 1829, *Drège* (holo. S; iso. E, K, P, TCD, distributed as *Nycterinia africana* Don b.).

Annual herb, stems c.10–130mm long, primary one simple, erect, soon branching from the base and sometimes higher, branches decumbent, simple to sparingly branched, very minutely glandular, coarse patent acute hairs up to 0.7–2mm long as well, distantly leafy. *Leaf* blade c.7–25 × 2.5–12mm, elliptic to ovate-elliptic, subacute, base tapering into a broad flat petiolar part c.3–17mm long (roughly equalling to about half as long as blade), margins entire to obscurely or distinctly toothed in upper part, young leaves thinly villous, hairs almost confined to petiolar part in older leaves, acute, up to 1–2mm long, minutely glandular as well. *Flowers* open in sunshine, in dense terminal spikes, elongating in fruit, the lowermost flowers then often distant, spike seldom accounting for more than half total stem length. *Bracts* adnate to calyx for 2–4.5mm, lowermost 7–23 × 1.7–6mm, mostly 3–6 times as long as broad, leaf-like, acute hairs up to 2–3mm long, minutely glandular as well. *Calyx* 6.5–8mm long, lips 1.5–3mm, minutely glandular, coarse patent acute hairs up to 1–2.4mm long particularly on ribs and tips. *Corolla* tube 22–27mm long (throat 1.25–1.6mm in diam.), glandular-puberulous, hairs up to 0.1–0.2mm long, limb 6.5–11mm in diam., regular, mouth encircled by stiff acute hairs, lobes 2.4–5 × 1.4–3mm, spathulate, apex rounded to slightly retuse, white to bright mauve above, yellow to red 'star' around mouth, the rays extending well onto the shaft of the lobe, orange to dark red-brown below. *Stamens* 4, anticous pair exserted, anthers 0.8–1.2mm long, posticous pair included, anthers 1.8–2.5mm long. *Stigma* 1–3.5mm long, shortly exserted. *Style* 18–20mm long. *Ovary* 3–4 × 1–1.2mm. *Capsules* c.5–7 × 3mm. *Seeds* c.0.4–0.6 × 0.3–0.4mm, obscurely angled, colliculate, pallid.

Selected citations:

Orange Free State. Bloemfontein distr., 2926AA, Tempe Farm, 26 viii 1922, *Potts* 3514 (PRE).
Cape. De Aar div., 3024CA, De Aar, 23 vii 1925, *Moss* 11552 (BOL, J). Hanover div., 3124DB, S of Middelburg, Dwarsvlei, c.4200ft, 1 x 1964, *Acocks* 23524 (K, PRE). Richmond div., 3124CA, Rhenosterfontein, c.6000ft, 26 iv 1950, *Acocks* 15818 (PRE). Middelburg div., 3125AC, Bangor Farm, 4000ft, ix 1917, *Bolus* s.n. (BOL). Beaufort West div., 3222AD, Karoo National Park, 1945m, 31 x 1984, *Bengis* 406 (PRE).

Zaluzianskya karrooica will be found in herbaria under a variety of names including *Z. africana* and *Z. crocea*. The confusion with *Z. africana* dates back to Bentham: *Drège* from the Sneeuwbergen (chosen here as the type of *Z. karrooica*) and *Ecklon* from the Winterbergen, were cited by him as *Z. africana* (Hiern too cited the *Drège* specimen, as no. 5846). Also, many of the specimens cited by Hilliard & Burtt (1983) as *Z. crocea* are *Z. karrooica*. The species is at once distinguished from *Z. crocea* by the yellow to red star-shaped patch around the mouth: in *Z. karrooica*, the rays of the 'star' reach well up the shaft of the corolla lobes; in *Z. crocea* they are so short that they scarcely reach the base of the shaft; also, the tips of the corolla lobes in *Z. karrooica* are often narrower in relation to their length than are those of *Z. crocea*. The two species differ further in leaves and bracts, those of *Z. crocea* being more markedly spathulate than those of *Z. karrooica* (e.g. lowermost bracts of *Z. karrooica* mostly 3–6 times as long as broad, those of *Z. crocea* 2–3 times) and they are usually more distinctly toothed in *Z. crocea* than in *Z. karrooica*.

The area of *Z. karrooica* appears to stretch from the south central part of the Orange Free State south and west to Carnarvon and Beaufort West, south and east to Middelburg, Cradock and the environs of the Winterberg; it thus lies west of the area of *Z. crocea*. However, the area of *Z. karrooica* occupies the apparent distributional gap between the northern and southern parts of the range of *Z. venusta*, from which it seems to differ little except in its obtuse to emarginate corolla lobes. Most specimens can be assigned to one or the other name on this basis, but *Acocks* 547 (PRE) and *Collins* 43 (J, PRE) from Hay Division in the northern Cape have lobes that vary from entire to retuse or shallowly bifid, sometimes in a single corolla. Field investigations are needed, and the area around Kimberley, Boshof, Bloemfontein and De Aar is central to the problem.

Zaluzianskya karrooica grows in sandy patches in grassland or scrub, c.1200 to 1950m above sea level, flowering mainly between July and October, but as early as April. This is also when *Z. venusta* flowers; *Z. crocea* is later.

37. Zaluzianskya minima (Hiern) Hilliard in Edinb. J. Bot. 47: 348 (1990).
Type: Cape, Calvinia div., Hantam, *Meyer* (n.v.).

Syn.: *Phyllopodium minimum* Hiern in Fl. Cap. 4(2): 318 (1904).
    Polycarena minima (Hiern) Levyns in J. S. Afr. Bot. 5: 37 (1939).

Annual herb, primary stem 10–100mm tall, simple or with a pair of branches from the base, these ascending or decumbent, they and the primary stem in all but the very smallest plants with a pair of branches, each subtended by a leaf, about midway or lower, pubescent with patent acute hairs up to 0.5–1mm long, minutely glandular-puberulous as well. *Leaves* few, usually two radical and only one pair on each branch, blade 3–11 × 2–6mm tapering into a flat petiolar part 1.5–10mm long (roughly equalling blade), ovate to elliptic, apex subacute, margins entire to obscurely toothed, pubescent with acute hairs, longest (up to 1–1.5mm) on margins especially of petiolar part. *Flowers* up to c.15, often fewer, initially crowded in a head, elongating into a spike in fruit, open in sunshine. *Bracts* adnate to calyx for 2–3mm, lowermost 4–15 × 2–7mm, smaller upwards, ovate in upper half then contracted into a broad shaft, pubescent on both surfaces, hairs often sparser on posticous face, longest hairs on lower margins, up to 1–1.7mm long. *Calyx* 3–5mm long, lips 0.6–1.2mm, pubescent with acute hairs up to 0.5–1mm long. *Corolla* tube 4–8.5mm long (throat 0.7–1mm in diam.), very minutely glandular-puberulous, limb 2.5–4mm in diam., nearly regular, mouth either glabrous or partially encircled by a few acute hairs, lobes 1–1.8 × 0.6–1mm, narrowly oblong with expanded tip, shortly bifid to entire (in one flower), bright purplish-pink above, with a yellow/orange star-shaped patch around mouth, sometimes with a crimson fleck as well at base of each lobe. *Stamens* 4, posticous pair included, anthers 0.6mm long (very rarely reduced to a single tiny anther), anticous pair exserted, anthers 0.2mm long. *Stigma* 0.8–1.5mm long, exserted. *Style* 2.5–5mm long. *Ovary* c.1.2 × 0.6mm. *Capsules* 4–5 × 2.5–2.7mm. *Seeds* c.0.4 × 0.3mm, subglobose when fully ripe, pale amber-colour, strongly colliculate.

Selected citations:

Cape. Calvinia div., 3119BD, below Hantam Peak, Akkerdam Nature Reserve, 3300ft, 19 viii 1975, *Thompson* 2395 (STE); ibidem, 17 viii 1987, *Batten* 772A (E); south of Keiskie, 3119DB, 4 ix 1986, *Oliver* 8905 (STE). Fraserburg div., 3121DC, 38 miles SW of Fraserburg, c.4300ft, 22 viii 1953, *Acocks* 16907 (PRE). Sutherland div., 3220BC, 15 miles N by W of Sutherland, c.5000ft, 28 viii 1953, *Acocks* 16996 (K, PRE). 3220CC, near Yuk river, 19 vii 1811, *Burchell* 1256 (K). Laingsburg div., 3320BA, Whitehill, 18 viii 1941, *Compton* 11247 (NBG). Ladismith div., 3320CA, Nougashoogte, road between Touwsrivier and Montagu, 950m, 5 viii 1989, *Viviers & Vlok* 405 (E).

Although I failed to trace the type, I am satisfied that Hiern's *Phyllopodium minimum* is the species of *Zaluzianskya* enumerated here. His specimen was clearly tiny (stem only 12–18mm tall), and the measurements given by him in Flora Capensis can be matched by very small specimens now before me; the biggest specimen known to me (*Burchell* 1256, cited above) was seen by Hiern but he failed to place it. However, the range of material now available demonstrates the great variation in size (particularly stature) that the annual species of *Zaluzianskya* can display.

Zaluzianskya minima is quite widely distributed from Calvinia south to the Roggeveld, Witteberg at Laingsburg, and Nougashoogte due south of Touw's River, and east to the environs of Fraserburg, between 800 and 1500m above sea level. It favours stony places over shale or clay, on both flats and slopes, flowering between July and September.

Its relationship lies with *Z. bella* and *Z. venusta*, with both of which it is partly sympatric, but it is easily recognized by its much smaller flowers, particularly corolla and anthers. The almost hispid patent hairs on the stems, which nearly always carry at least one pair of leaves above the radical ones, will at once distinguish it from *Z. cohabitans*, with which it is very easily confused. The two species differ further in a number of details, including the shape of the bracts and the number of stamens: four in *Z. minima*, two in *Z. cohabitans*; see further under that species.

38. Zaluzianskya cohabitans Hilliard in Edinb. J. Bot. 47: 346 (1990).
Type: Cape, Calvinia div., 3119BD, Hantam Mts, Akkerendam, 17 viii 1987, *Batten* 772 (holo. E, iso. K).

Annual herb, primary stem 5–100mm tall, simple or branching once or twice at the base, branches sharply ascending, pubescent with retrorse acute hairs up to 0.25–0.5mm long, minutely glandular as well, scapose. *Leaves* in 2 to several pairs, crowded at base of plant, blade c.2–15 × 1.3–15mm, ovate to elliptic tapering to a flat petiolar part c.1–10mm long (roughly a half to two thirds as long as blade), apex subacute, margins entire to rather obscurely denticulate, glabrous dorsally or a few hairs on the midline, thinly pubescent ventrally, hairs acute, up to 0.7–1mm long. *Flowers* open in sunshine, up to c.15 crowded in a terminal head, either remaining congested in fruit or the lowermost few capsules distant. *Bracts* adnate to calyx for 3–7.5mm, lowermost 5.5–17 × 1–4mm, linear-lanceolate to oblong, usually entire, occasionally with a pair of minute teeth, pubescent on both surfaces, hairs acute, often sparse on dorsal face, lowermost short and ± retrorse, up to 0.7–1.5mm on margins. *Calyx* 3.6–8mm long, lips 0.5–1.5mm long, pubescent particularly on the ribs, hairs acute, up to 0.7–1.5mm long. *Corolla* tube 6–8.5mm long (throat c.0.8–1mm in diam.), minutely glandular-puberulous particularly on upper part, limb c.3–5mm in diam., nearly regular, mouth encircled by relatively long clavate hairs, lobes 1–1.8 × 0.7–1.2mm, elliptic, usually entire, rarely one or two lobes retuse, rose-pink, yellow in throat, but apparently no pattern on lobes. *Stamens* 2 (the posticous pair), anthers 0.5–0.8mm long, included or tips visible in mouth. *Stigma* c.1–1.5mm long, either exserted or ± included and attached to anthers. *Style* c.2.5–4.5mm long. *Ovary* c.2–2.5 ×0.8mm. *Capsules* 5.5–8 × 2–3mm. *Seeds* c.0.5–0.8 × 0.5mm, obscurely angled, colliculate, pallid.

Citations:

Cape. Calvinia div., 3119BD, Moordenaarspoort, 27 ix 1952, *Lewis* 3865 (SAM); 5 miles SSE of Calvinia, c.3400ft, 20 viii 1956, *Acocks* 18942 (PRE); 3119DB, Dorp se Kop S of Keiskie, 4 ix 1986, *Oliver* 8905A (STE); 3119DD, Karigatbosch Fontein, c.4000ft, 20 viii 1975, *Thompson* 2463 (STE). Ripjoeni Mts, 1200m, viii 1921, *Marloth* 10291 (PRE). Sutherland div., 3220BC, 15 miles NW by N of Sutherland, c.5000ft, 28 viii 1953, *Acocks* 17000 (PRE); Roggeveld Escarpment, Swaarweerberg, farm Geelhoek 103, 1844m, 4 ix 1986, *Cloete & Haselau* 212 (STE). 3220AA, Roggeveld, Uitkyk, 23 ix 1981, *Goldblatt* 6363 (E, PRE).

Zaluzianskya cohabitans is so named because it often grows with *Z. minima*: four of the six collections seen from the area around Calvinia are a mixture of the two species. However, the keen-eyed Acocks collected both species near Calvinia and north of Sutherland, and did not fail to distinguish them. The corolla limb in *Z. cohabitans* is possibly self-coloured (mauve to rose-pink); the flowers of *Z. minima* are similarly coloured but also have a yellow to orange star-shaped 'eye' and sometimes there is as well a small crimson patch at the base of each lobe. The corolla lobes are mostly entire in *Z. cohabitans*, mostly bifid in *Z. minima*. But they differ further in a number of characters: stem of *Z. cohabitans* with retrorse hairs and leaves crowded at base; stem of *Z. minima* with patent hairs and, in all but the smallest plants, a pair of leaves well up the stem, each subtending a branch; bracts of *Z. cohabitans* linear-lanceolate to oblong (not ovate-spathulate); mouth of corolla tube well bearded (not glabrous nor with just a few hairs), and stamens 2 (not 4).

Zaluzianskya cohabitans has been recorded from the foot of the Hantam Mountains near Calvinia south to the Roggeveld, growing in sandy, clayey or shaly soils in karroid scrub, c.1000 to 1850m above sea level, and flowering in August and September. The flowers may often be autogamous: the included stigma attached to the anthers suggests this, also the reduction to two, included, stamens and the lack of patterning on the corolla, rare in this group. But some outcrossing is also probable, as the stigma may be well exserted.

39. Zaluzianskya crocea Schltr. in J. Bot. 35: 221 (1897); Hiern in Fl. Cap. 4(2): 346 (1904); Hilliard & Burtt in Notes RBG Edinburgh 41: 40 (1983), p.p. excl. many of the cited specimens. **Plates 3H, 5B.**
Type: Cape, Queenstown div., near Bailey, summit of Andriesberg, 6700ft, iv 1895, *Galpin* 1927 (holo. K; iso. BOL, GRA, PRE).

Annual herb, stems c.30–120mm long, primary one simple, erect, soon branching from the base or a little higher, branches often simple, occasionally sparingly branched, decumbent, very minutely glandular, coarse patent acute hairs up to 1–1.5mm long as well, moderately leafy. *Leaves* 7–25 × 3–10mm, spathulate, subacute, upper margins usually with 2–3 pairs of teeth, lowermost pair of leaves sometimes subentire, both surfaces thinly villous or hairs nearly confined to margins, hairs up to 3mm long particularly on lower margins, very minutely glandular as well. *Flowers* in dense terminal spikes often accounting for ± ⅔ total stem length, open in sunshine. *Bracts* adnate to calyx for 2.5–4mm, lowermost 9–17 × 3.2–7mm, 2–3 times as long as broad, leaf-like but often hairier, hairs up to 1.5–4mm long. *Calyx* 6–7mm long, lips 2mm long, minutely glandular, coarse patent acute hairs up to 1–2mm long particularly on ribs and at tips. *Corolla* tube 18–32mm long (throat c.1.5mm in diam.), glandular-puberulous, hairs up to c.0.15mm long, limb 8–11mm in diam., regular, mouth somewhat thickened, encircled by stiff ± acute hairs, lobes 3–4 × 3–4mm, spathulate, apex ± rounded to slightly retuse, white above fading to pinkish-mauve, yellow very short-rayed 'star' around mouth fading to orange/red, coppery-orange below. *Stamens* 4, anticous pair well exserted, anthers 1–1.25mm long, posticous pair included, anthers 1.5–2.5mm long. *Stigma* 1–4mm long, just reaching mouth or shortly exserted. *Style* c.14–27mm long. *Ovary* c.2–3 × 0.8–1.2mm. *Capsules* 4.5–7 × 2–4mm. *Seeds* c.0.8–1 × 0.5–0.6mm, slightly angled, colliculate, greyish-mauve.

Selected citations:

Lesotho. Buffalo river waterfall, 3028CA, 8500ft, 13 iii 1904, *Galpin* 6791 (K, BOL).
Cape. Lady Grey div., 3027DA, Witteberg, Joubert's Pass, i 1925, *Thode* A506 (K, PRE); top of Witteberg, 1861, *Cooper* 614 (TCD). Barkly East div., 3027DB, Rhodes to Naude's Nek, 7800ft, 21 ii 1971, *Hilliard & Burtt* 6687 (E, K, MO, NBG, NU, PRE); Ben McDhui, c.8400ft, 3 ii 1983, *Hilliard & Burtt* 16389 (E, K, NU). 3028CA, Naude's Nek Pass, c.8600ft, 13 i 1979, *Puff* 790113-5/2 (J). Molteno div., 3126AD, Broughton, 6300ft, xii 1892, *Flanagan* 1619 (BOL, K, SAM).

Zaluzianskya crocea appears to be nearly confined to the mountains of the north eastern Cape, namely the Stormberg, Witteberg and Cape Drakensberg, thence just over the Lesotho border near Ben McDhui and Naude's Nek Pass, and Andriesberg, due south of the Stormberg, between c.1920 and 2600m above sea level. It grows in damp gravelly patches around bare rock sheets, flowering between December and April. Hilliard & Burtt (1983) included in their circumscription of *Z. crocea* two other species, *Z. karrooica* and *Z. vallispiscis*, hence their comments on odd distribution and flowering times. See further under these two names.

40. Zaluzianskya vallispiscis Hilliard in Edinb. J. Bot. 47: 350 (1990).
Type: Cape, Albany div, 3326BB, near Breakfast Vlei, 8 xi 1938, *Häfstrom & Acocks* 1251 (holo. S, iso. PRE).

Syn.: *Z. divaricata* auct. non (Thunb.) Walp.; Batten & Bokelmann, Wild Fl. E Cape 130 pl. 103, 7 (1966).

Annual herb, stems c.30–150mm long, primary one simple, erect, soon branching from the base and higher, branches decumbent or ascending, simple to branched, very minutely glandular, coarse patent acute hairs up to 1–1.5mm long as well, moderately leafy. *Leaf* blade c.6–20 × 2.5–10mm, elliptic, ± acute, tapering into a broad flat petiolar part c.3–15mm long (about equalling to roughly one third as long as blade), margins subentire to distinctly toothed, both surfaces thinly villous, hairs acute, up to 1–2mm long, minutely glandular as well. *Flowers* open in sunshine, up to c.30 in dense terminal spikes, elongating as they pass into fruit, capsules then laxly arranged. *Bracts* adnate to calyx for 3–5mm, lowermost c.12–19 × 2.5–6mm, 3–5 times as long as broad, oblong-lanceolate, 1–4 pairs of teeth in upper part, thinly villous with acute hairs up to 1.5–2.5mm long, glandular hairs as well up to 0.3–0.5mm long. *Calyx* 5.5–8mm long, lips 1.8–2.5mm long, glandular hairs up to 0.3–0.5mm long, coarse acute hairs as well mainly on ribs and tips, up to 1–1.5mm long. *Corolla* tube 18–28mm long (throat 1.3–1.8mm in diam.), glandular, hairs on lower part especially up to 0.4–0.5mm long, limb c.9–10mm in diam., regular, mouth encircled by acute hairs, lobes 3–4 × 3–3.5mm, spathulate, rounded to retuse or shallowly bifid, white to pale mauve or pinkish-mauve above with a yellow to orange/red very short-rayed 'star' around mouth, pale to deep coppery-orange below. *Stamens* 4, anticous pair exserted, anthers 0.7–1mm long, posticous pair included, anthers 1.5–2.2mm long. *Stigma* 1–2.5mm long, shortly exserted. *Style* c.17–20.5mm long. *Ovary* 2–3 × 0.8mm. *Capsules* 5–6 × 2.5mm. *Seeds* not seen.

Selected citations:

Cape. Albany div., 3326BB, Fish River Valley near Committees, 1000ft, v 1928, *Dyer* 1541 (K, PRE); ibidem, ix 1963, *Bokelmann* s.n. (NBG); ibidem, Heatherton Towers, 1500ft, 18 vi 1961, *Jacot Guillarmod* 3994 (PRE).

Zaluzianskya vallispiscis is allied to *Z. crocea* from which it is distinguished by its flower spikes, laxer in fruit than those of *Z. crocea*, the bracts oblong-lanceolate and roughly 3–5 times as long as broad (not spathulate and 2–3 times as long as broad; examine the lowermost bracts); it is also more glandular than *Z. crocea* with hairs up to 0.4–0.5mm long on the corolla tube (up to 0.15mm in *Z. crocea*).

The earliest collection of the species appears to have been that of Ecklon & Zeyher, from Hermannskraal on the Great Fish River, which was cited by Hiern (1904) under *Z. africana* as *Ecklon* 221. The plant may be confined to the valley of the Great Fish River, where it has variously been recorded as growing 'in damp ditches', 'roadside in disturbed ground' and 'in *Euphorbia* karroid scrub', flowering between May and November.

41. Zaluzianskya synaptica Hilliard in Edinb. J. Bot. 47: 349 (1990).
Type: Cape, Willowmore div., 3323AC, slopes east of Toorwater, 2500ft, 5 x 1971, *Oliver* 3655 (holo. STE).

Annual herb, primary stem c.10–240mm long, simple to laxly branched, eventually branching from the base, the branches simple to laxly branched, ascending or decumbent, pubescent with ± retrorse to patent acute hairs up to 0.2–0.8mm long, minutely glandular-puberulous as well, laxly leafy. *Leaf* blade c.4.5–20 × 2.5–13mm, narrowly to broadly elliptic, occasionally lowermost ovate-elliptic, tapering into a broad flat petiolar part c.2.5–15mm long (nearly equalling to about half as long as blade), subacute, margins entire, subentire, or with a few pairs of very small teeth, both surfaces thinly pubescent with acute hairs up to 1–1.5mm long, minutely glandular as well. *Flowers* open in sunshine, up to c.20 in dense spikes elongating in fruit, the lowermost few capsules often distant. *Bracts* adnate to calyx for 3–5mm, lowermost 7.5–21 × 1.8–3(–7)mm, oblong-lanceolate, entire to shortly and rather obscurely toothed, pubescent on both surfaces, glandular-puberulous as well, acute hairs on margins up to 0.6–2.5mm long, on backs c.0.25–0.8mm long. *Calyx* 6–9mm long, lips 1.3–2mm long, pubescent with acute hairs up to 0.5–2mm long, finely puberulous as well, some hairs glandular. *Corolla* tube 17–26mm long (throat 1.3–1.7mm in diam.), glandular-puberulous (hairs up to 0.1–0.4mm long), limb 7.5–10mm in diam., regular, mouth encircled by long acute hairs, lobes 3–6 × 2–4mm, spathulate, bifid to retuse or subentire, white above (or sometimes orange), yellowish to coppery-orange below, yellow/orange patch around mouth with very short rays scarcely reaching base of lobes. *Stamens* 4, anticous pair exserted, anthers 0.6–1mm long, posticous pair included 1.4–2mm long. *Stigma* 1.4–5mm long, shortly exserted. *Style* 11–20mm long. *Ovary* 3–4 × 0.8–1mm long. *Capsule* c.5–8 × 2.5–3mm. *Seeds* c.0.5–0.6 × 0.4mm, obscurely angled, colliculate, pallid.

Selected citations:

Cape. Worcester div., 3319DC, Ribbokkop, 300–600m, 29 ix 1982, *Bayer* 3359 (NBG); 3320AC, c.20km ESE Touws River, farm Avondrust, c.750m, 11 ix 1985, *Glen* 1522 (PRE). Robertson div., 3319DA, near Robertson, 4 viii 1949, *Steyn* 209 (NBG). Ladismith div., 3321AD, road to Waterkloof in Klein Swartberg Mts, c.600m, 23 x 1981, *Mauve, Reid & Wikner* 96 (STE). Graaff Reinet div., 3224AC, near Aberdeen, 3100ft, 1 x 1973, *Bayliss* 6084 (K); 3224BC, near Graaff Reinet, ix 1868, *Bolus* 1869 (K, S).

Zaluzianskya synaptica has been recorded from Hex River Kloof and the mountains south of Touws River and Worcester east to Ladismith, Willowmore, Graaff Reinet and Steytlerville divisions, in sandy or shaly places in scrub, between 600 and 1200m above sea level, flowering between August and October.

It may be distinguished from *Z. venusta*, which it most resembles in facies, by the shorter indumentum on the stems (acute hairs up to 0.8mm long versus 1–1.4mm, the glandular hairs also being shorter than those of *Z. venusta*), corolla lobes white (not mauve to purplish pink) above, with the rays of the yellow to orange star-shaped patch around the mouth very short and scarcely reaching the base of the lobes, whereas in *Z. venusta* the rays reach well up the shaft of the lobes. The two species appear to be sympatric only in the Ladismith area, whence *Z. venusta* spreads north north east, *Z. synaptica* west and east.

In *Z. venusta*, the corolla lobes are mauve to bright purplish-pink above and possibly always shades of orange below; in *Z. synaptica* they appear to be mostly white above, occasionally orange, and shades of orange below. Collector's notes are scanty and field observations are needed; the bright orange specimens have been recorded from 3324AB, Klipplaats, and 3224AC, near Aberdeen.

If the colour patterning of the corolla limb is a guide to relationships, then the true relationship of *Z. synaptica* lies with *Z. crocea* and *Z. vallispiscis*, all three having a very short-rayed 'star' around the corolla mouth: *Z. crocea* differs in its spathulate well-toothed leaves and bracts and longer indumentum, *Z. vallispiscis* in the longer hairs on the stem and more sharply toothed leaves and bracts.

Bolus 1869 (cited above) was referred by Hiern to *Z. africana*, but material can be found in herbaria under many different names. The epithet *synaptica* draws attention to the variation in the lobing of the corolla, which helps to blur the distinction between species with entire lobes and those with decidedly bifid lobes.

Section Zaluzianskya subsect. Noctiflora

42. Zaluzianskya peduncularis (Benth.) Walp., Repert. 3: 308 (1844); Hiern in Fl. Cap. 4(2): 349 (1904); Hilliard & Burtt in Notes RBG Edinb. 41: 39 (1983).
Type: Cape, Albany near Theopolis, on the Kowie [Kovi] River, *Ecklon* (holo. K; iso. E, P, S, SAM, TCD, W).

Syn.: *Nycterinia peduncularis* Benth. in Hook., Comp. Bot. Mag. 1: 371 (1836) & in DC., Prodr. 10: 350 (1846).
  N. peduncularis var. *hirsuta* Benth., loc. cit. Type as for *Z. peduncularis*.
  N. peduncularis var. *glabriuscula* Benth., loc. cit. Type: Namaqualand, Haazenkraals river, 2000ft, 24 viii 1830, *Drège* (holo. K, iso. P).
  Zaluzianskya peduncularis var. *hirsuta* (Benth.) Walp., loc. cit.
  Z. peduncularis var. *glabriuscula* (Benth.) Walp., loc. cit.
  Z. gilioides Schltr. in Bot. Jahrb. 27: 182 (1899); Hiern in Fl. Cap. 4(2): 350 (1904); Merxmüller & Roessler in Prodr. Fl. S.W.A. 126. Scrophulariaceae: 59 (1967). Type: Clanwilliam distr., 3219AB, Bidouw Berg, 3800ft, 26 viii 1896, *Schlechter* 8681 (holo. B†; iso. BOL, E, K, P, PRE, S, W).
  Polycarena dinteri Thell. in Vierteljahrss. Nat. Ges. Zürich 60: 413 (1918). Type: S.W.A., Gubub, *Dinter* 1141 (holo. Z).

Annual herb, primary stem 13–250mm tall, solitary and simple, or several stems eventually developing from base, these erect or decumbent, often simple, but often with leaf tufts indicating ability to branch, occasionally branches developed, pubescent with acute retrorse appressed hairs up to 0.5–0.7mm long, minutely glandular as well, often nude between radical leaves and inflorescence, sometimes with a few pairs of leaves. *Leaves*: radical leaves several, loosely rosetted, blade c.3–30 × 2–15mm, ovate to elliptic tapering into a broad flat petiolar part c.2–15mm long (about equalling blade to ⅔ as long), cauline leaves c.4–27 × 1.2–10mm, elliptic tapering to a short petiolar part, all leaves subentire to obscurely or distinctly toothed, upper surface sometimes glabrous, usually thinly pubescent with acute hairs up to c.1mm long, lower surface glabrous except on midline, acute hairs up to 1–1.5mm long on margins. *Flowers* up to c.10 in a terminal head not or scarcely elongating in fruit, opening at dusk but sometimes cleistogamous. *Bracts* adnate to calyx for 3.5–7mm, lowermost c.8–20 × 1–4.5mm, linear-lanceolate to lanceolate, entire or with 1 or 2 pairs of teeth, variable in pubescence but always with long (up to 1.2–1.8mm) acute hairs on the margins, the inner face pubescent, the outer glabrous to shortly pubescent, the hairs acute, up to 0.7mm long (but often much shorter), retrorse. *Calyx* 4.5–8mm long, lips 0.6–1mm long, usually puberulous to pubescent all over. *Corolla* tube 17–25mm long (throat 1–1.8mm in diam.), glandular-puberulous, limb (of chasmogamous flowers) 5–10mm in diam., almost regular, mouth glabrous, yellow to bright scarlet in the throat, the colour in an almost regular ring around the mouth, lobes c.1.5–4.3 × 1–3.2mm, broadened at the tips, creamy- to lemon-yellow above, maroon to purplish-brown below. *Stamens* 4, anticous pair exserted, anthers 0.4–1.4mm long, posticous anthers included, anthers 0.6–1.8mm long. *Stigma* 1–4mm long, scarcely exserted. *Style* 11–20mm long. *Ovary* 3.5–4.5 × 1mm. *Capsules* c.6–9 × 3–4mm. *Seeds* c.0.4–0.6 × 0.3–0.4mm, obscurely angled, colliculate, pallid.

Selected citations:

Namibia. 2616CB, north of Aus, 8 ix 1973, *Giess* 12828 (PRE, S).
Cape. Namaqualand, 2917BA, near Klipfontein, 3000ft, viii 1883, *Bolus* sub herb. norm. austr. afr. 654 (BOL, K, SAM, W); 3018AC, Kamiesberg near Karas, 25 viii 1987, *Batten* 794 (E). Van Rhynsdorp div., 3119AC, Van Rhyn's Pass, 25 vii 1941, *Esterhuysen* 5275 (BOL). Calvinia div., 3119BD, Akkerdam, foot of Hantam Mts, 22 vii 1961, *Lewis* 5798 (NBG). Clanwilliam div., 3219AD, S Cedarberg, Wolfberg, 3 x 1952, *Esterhuysen* 20567 (BOL, K, PRE). Sutherland div., 3220CA, Houthoek, viii 1968, *Hanekom* 1210 (K, PRE, STE). Laingsburg div., 3320BA, NE of Matjiesfontein, Ngaapkop, 1000m, 15 ix 1971, *Thompson* 1241 (PRE, STE); 3320AB, Constable, 21 ix 1931, *Compton* 3758 (BOL, NBG). Worcester div., 3319BC, Matroosberg, 5000ft, 2 xii 1947 [fruiting, cleistogamous], *Esterhuysen* 14201 (BOL). Albany div., c.3326B–D, Fish River Heights, 1880, [cleistogamous], *Hutton* s.n. (K). Stutterheim div., 3227CB, Dohne Hill, 5000ft, 1897, [cleistogamous], *Sim* 20360 (E, PRE). De Aar, 3024CA, 23 vii 1925, *Moss* 11668 (J). Richmond div., 3124CA, Rhenosterfontein, c.5600ft, 27 iv 1950, [cleistogamous and chasmogamous], *Acocks* 15837 (K, PRE).
Orange Free State. Ficksburg distr., 2827DB, Nebo, 6200ft, 29 x 1934, *Galpin* 13971 (BOL, PRE).
Lesotho. 2828CC, Leribe, *Dieterlen* 191 (BOL, PRE, SAM). 2927BA, Teyateyaneng, 18 viii 1954, *Martin* 1020 (NBG).

On the type sheet of *Z. peduncularis* at Kew, Bentham has transposed the collector's names and localities: the Drège specimen (type of var. *glabriuscula*, collected in Namaqualand) is top right, the Ecklon specimen (type of var. *hirsuta*, collected near Grahamstown) is bottom left. The varietal names must refer to the bracts, shaggy on the backs in Ecklon's collection, almost glabrous on the backs in Drège's. However, the degree of hairiness shows a complete gradation from glabrous through puberulous to pubescent even within a single collection, and no varieties are recognized here.

Zaluzianskya peduncularis has a remarkably wide geographical range from southern Namibia south through Namaqualand to Van Rhynsdorp, Sutherland and Laingsburg divisions thence east and north as far as Stutterheim in the eastern Cape and De Aar and Colesberg to the north east, thence through the Orange Free State as far north as Kroonstad and east to the borders of Lesotho.

Zaluzianskya peduncularis also has a very wide altitudinal range, from c.100 to 1900m above sea level, favouring sandy or stony places, often in karroid vegetation. Flowering takes place mainly between July and September, but there are records as early as April and as late as November. It is clearly a very good weed: cleistogamy appears to be commonplace, sometimes the whole head being cleistogamous, occasionally only the first one or two flowers (the type material comprises wholly cleistogamous and what appear to be normal outbreeding plants, though the latter are not represented in every isotype). Cleistogamous flowers are easy to spot because a tiny bud with abnormally small corolla lobes caps the fruit; it is not possible to judge if flowers with a small expanded limb are autogamous or outbreeding, but my guess is that autogamy is also common. The calyx, only very shallowly bilabiate and almost always puberulous or pubescent, is a good distinguishing character; allied to the glabrous mouth, it will at once distinguish *Z. peduncularis* from *Z. collina*, with which it has been much confused and with which it is partly sympatric in Namaqualand. However, its relationship is with *Z. elgonensis* and *Z. rubrostellata*: *Z. elgonensis* has much smaller flowers (corolla tube only 10–12mm long) and different colour-patterning; the patterning in *Z. rubrostellata* (a distinct star-shaped scarlet 'eye') also marks it off from *Z. peduncularis*, but it is further distinguished by differences in foliage and corolla (see further under *Z. rubrostellata* (no. 44).

43. Zaluzianskya elgonensis Hedberg in Bot. Not. 123: 512 (1970); Hilliard & Burtt in Notes RBG Edinb. 41: 6 (1983) in obs.
Type: Uganda, Bugishu distr., Mt Elgon, 3800m, 5 xii 1967, *Hedberg* 4478 (holo. UPS; iso. K, PRE, also EA, MHU, n.v.).

Annual herb, primary stem 15–70mm tall (–150mm in cult. fide Hedberg), solitary or branching once or twice from the extreme base, simple, pubescent with acute ± retrorse hairs up to 0.7–1mm long, minutely glandular as well, leafless in the upper part. *Leaves* crowded at and near the base of the stem, c.3–15 × 0.8–5mm, ovate or elliptic to oblong-elliptic, entire, lowermost narrowed to a broad flat petiolar part c.1–3mm long, pubescent with acute hairs up to 1mm long particularly on lower margins. *Flowers* up to c.6 in a terminal head, possibly opening at dusk. *Bracts* adnate to calyx for c.4mm, lowermost c.5–11 × 1.6–2.5mm, oblong-lanceolate, entire, pubescent, hairs on margins up to c.1mm, on backs c.0.4mm, ± retrorse. *Calyx* 5–5.6mm long, lips c.1mm, pubescent all over, hairs c.0.3mm long. *Corolla* tube 10–12mm long (throat c.0.7–1.2mm in diam.), minutely glandular- puberulous, limb c.4–5mm in diam., almost regular, mouth glabrous, lobes c.1.5–2 × 0.9mm, broadened at tips, cream-coloured above with an orange spot at base of each lobe, purplish-brown outside (Hedberg). *Stamens* 4, anticous pair exserted, anthers c.0.4mm long, posticous pair included, c.0.7mm long. *Stigma* c.2mm long, scarcely exserted. *Style* c.6mm long. *Ovary* not seen. *Capsules* c.5–6 × 2.7mm. *Seeds* c.0.4 × 0.3mm, obscurely angled, colliculate, pallid.

Citation:

Tanzania. Kilimanjaro, Shira plateau, 3800m, 4 vii 1970, *Friis* 241 (K).

Zaluzianskya elgonensis seems to be known from only two collections, the type material from Elgon, the other from Kilimanjaro, both at 3800m, in thin soil on rocky ground. Hedberg (loc. cit.) suggested that it is at least facultatively autogamous because 'its flowers seem to remain mostly closed', but the photographs (loc. cit., fig. 2) suggest that the closure may simply be in response to light intensity; in other words, the plant may be nightflowering like its allies, *Z. peduncularis* and *Z. rubrostellata*. The disjunction of over 3000km between *Z. elgonensis* and its two close allies in southern Africa is most remarkable.

44. Zaluzianskya rubrostellata Hilliard & Burtt in Notes RBG Edinb. 37: 317 (1979) & op. cit. 41: 39 (1983); Trauseld, Wild Flow. Natal Drakensberg 172 (colour photograph), 173, as *Z. pulvinata* (1963); Manning in Fl. Pl. Afr. 49 pl. 1950 (1987). **Plate 2F.**
Type: Lesotho, 2929CB, above Sani Pass, east of chalet, 2920m, 17 i 1976, *Hilliard & Burtt* 8849 (holo. E, iso. NU).

Annual herb, primary stem at first solitary but soon branching from the extreme base, branches c.20–140mm long, simple or branching very low down, erect or decumbent, pubescent with acute ± retrorse hairs up to c.0.5–0.7mm long, laxly leafy. *Leaves* c.7–20 × 1.5–6mm, spathulate, elliptic or oblong, the lowermost in particular tapering into a broad flat petiolar part, subentire to distinctly toothed, acute hairs up to c.1mm long on margins, otherwise glabrous or very nearly so. *Flowers* up to c.10 in a terminal capitate cluster, not elongating in fruit, opening at dusk or in dull light. *Bracts* adnate to calyx for 4–5mm, lowermost c.9–14 × 2–4mm, oblong-lanceolate, toothed, tips recurved, acute hairs up to 1.5–2mm long on margins, backs glabrous to puberulous. *Calyx* 5.5–8mm long, lips 1mm long, puberulous all over. *Corolla* tube 21–31mm long (throat c.1.6–2mm in diam.), glandular-puberulous, limb 10–15mm in diam., almost regular, mouth glabrous to partially or fully ringed with small clavate hairs, bright scarlet in throat extending out onto bases of lobes to form a slightly raised star around the mouth, lobes c.4.3–6 × 3.2–5mm, broadened at the tips, bright yellow above, chocolate-brown below. *Stamens* 4, anticous pair exserted, anthers 1.4–1.7mm long, posticous pair included, anthers 2.3–2.8mm long. *Stigma* c.3–6mm long, scarcely to well exserted. *Style* c.15–30mm long. *Ovary* c.3.5–5 × 1mm. *Capsules* c.7–8 × 3.5–4mm. *Seeds* not seen.

Selected citations:

Cape. Lady Grey div., 3027DB, Witteberg, Joubert's Pass, c.7700ft, 18 i 1979, *Hilliard & Burtt* 12183 (E). Barkly East distr., 3028CA, Doodman's Krans, c.8500ft, 7 iii 1904, *Galpin* 6794 (PRE).
Lesotho. 2929AD, Th. N [Thabana Ntlenyana], 10800ft, 20 i 1955, *Jacot Guillarmod* 2356 (K, PRE). 2927BD, Molimo Nthuse, 2300m, 6 xi 1975, *Schmitz* 6293A (PRE).

Zaluzianskya rubrostellata is known with certainty only from high altitudes in Lesotho and along the Drakensberg on the Natal–Lesotho border as well as the Cape Drakensberg from Naude's Nek to Ben McDhui, and the Witteberg near Lady Grey. It grows in basalt grit, at altitudes ranging from 2350 to 3300m, flowering mainly between December and March. There is one collection purportedly from Hogsback in the eastern Cape (*Young* sub Moss 18174, J), but its presence there needs confirmation.

Zaluzianskya rubrostellata is allied to *Z. peduncularis* and their areas abut on the western margins of Lesotho, but *Z. peduncularis* is always at lower altitudes than *Z. rubrostellata*, on the sandstone, not the basalt. The leaves of *Z. rubrostellata* are thicker in texture than those of *Z. peduncularis* and usually extend further up the stems, the colour patterning around the mouth is markedly different (always starshaped in *Z. rubrostellata*, merely a neat ring in the throat of *Z. peduncularis*), the limb mostly larger, with broader lobes (10–15mm in diam. versus 5–10mm with lobes 3.2–5mm broad at the tips versus 1–3.2mm), and the anthers are larger (anticous anthers 1.4–1.7mm versus 0.4–1.4mm, posticous ones 2.3–2.8mm versus 0.6–1.8mm). Unlike *Z. peduncularis*, *Z. rubrostellata* gives no indication of either cleistogamy or autogamy.

45. Zaluzianskya kareebergensis Hilliard in Edinb. J. Bot. 47: 347 (1990).
Type: Cape, Van Rhynsdorp div., 3118BA, Kareebergen, 1500ft, 20 vii 1896, *Schlechter* 8233 (holo. E; iso. K, P, PRE, W). See notes below.

Annual herb, primary stem 60–150mm tall, erect, solitary and simple, or branching from the base and above, branches sharply ascending, pubescent with acute retrorse hairs up to 0.5–1mm long, leafy and often with axillary leaf tufts (incipient branches). *Leaves*: radical in one or two pairs, blade c.10–15 × 4–7mm, ovate to elliptic-ovate, subentire, tapering into a broad flat petiolar part c.5–8mm long (roughly half as long as blade), primary cauline leaves c.13–25 × 1.5–6mm, elliptic to oblong-elliptic, acute, base narrowed, margins denticulate, both surfaces thinly villous, hairs acute, up to 1–1.5mm long. *Flowers* c.10 in a compact terminal head not elongating in fruit, often cleistogamous. *Bracts* adnate to calyx for 5.5–7mm, lowermost c.10–14 × 1.75–2.5mm, linear-oblong, ± acute, entire, acute hairs up to 1.5–2mm long on margins, shortly pubescent on backs, hairs ± retrorse. *Calyx* c.5.5–8mm long, lips c.1mm long, puberulous all over. *Corolla* tube (of chasmogamous flowers) c.7.5–9mm long (throat c.1mm in diam.), very minutely glandular-puberulous especially at apex, limb c.3.5mm in diam, almost regular, mouth glabrous, lobes c.1–1.2 × 0.5–0.7mm, slightly broadened at tips, colour unknown but probably white or yellow. *Stamens* 4, anticous pair scarcely exserted (included in cleistogamous flowers), anthers 0.3–0.5mm long, posticous pair included, anthers 0.6–0.8mm long. *Stigma* c.3mm long, shortly exserted in chasmogamous flowers. *Style* c.6mm long. *Ovary* c.2.5 × 0.8mm. *Capsules* 6–7 × 2.5–3mm. *Seeds* c.0.7–0.8 × 0.4–0.5mm, obscurely angled, colliculate, pallid.

Citation:

Cape. Van Rhynsdorp div., 3118AB, Karee Bergen, 28 viii 1937, *Wall* s.n. (S).

Zaluzianskya kareebergensis is known only from 'Karee Bergen'. In the published account of Schlechter's itinerary (Jessop, 1964), the farm Karreeberg is said to be about 13 miles south east of Bitterfontein; however, the 1:500000 map (1980 edition) shows only a mine, Kareeberg, about 18 miles (30km) south east of Bitterfontein, and this is probably Schlechter's site. Unfortunately, Wall's collection too gives only 'Karee Bergen', but I judge it came from the same place as Schlechter's.

The sheet of *Schlechter* 8233 in E (the holotype) is mixed with *Z. benthamiana* (the lectotype of *Z. bolusii*, that is, *Z. benthamiana*, is a Schlechter specimen from Karee Bergen), those in P, PRE and W are mixed with *Z. pusilla* (*Schlechter* 8193 from Karee Bergen is *Z. pusilla*), while the sheets of *Schlechter* 8233 in BOL and K are wholly *Z. pusilla*. *Zaluzianskya benthamiana* is easily distinguished by its very glandular stems, leaves and bracts, mixed with long patent hairs on the stems; *Z. pusilla* has retrorse hairs on the stems, but lacks them on the backs of the bracts (present in *Z. kareebergensis*) and the calyx is much more deeply divided than it is in *Z. kareebergensis*. Although small-flowered forms of *Z. benthamiana* and *Z. pusilla* bear a superficial resemblance to *Z. kareebergensis*, its relationship lies with *Z. peduncularis*, with which it shares the characters of retrorse hairs on the stems and on the backs of the bracts and a shallowly divided calyx; it differs in its leafy stems (upper part of the stem nude in *Z. peduncularis*), chasmogamous flowers with a much shorter corolla tube (c.9mm versus 17–25mm) and mostly smaller anthers. The flowers of *Z. kareebergensis* are very small, and mostly cleistogamous; however, some clearly have a spreading limb and long-exserted stigma.

No ecological information was given by the collectors.

Section Holomeria

46. Zaluzianskya divaricata (Thunb.) Walp., Repert. 3: 308 (1844); Hiern in Fl. Cap. 4(2): 347 (1904); Levyns in Adamson & Salter, Fl. Cape Penins. 712 (1950).
Type: Cape of Good Hope [sandy places near Cape Town fide Walpers], *Thunberg* (holo. sheet no. 14362, UPS; prob. iso. S).

Syn.: *Manulea divaricata* Thunb., Prodr. 101 (1800) & Fl. Cap. ed. Schultes 468 (1823).
    Nycterinia divaricata (Thunb.) Benth. in Hook., Comp. Bot. Mag. 1: 371 (1836) & in DC., Prodr. 10: 350 (1846).
    N. rigida Benth. in DC., Prodr. 10: 350 (1846), lapsu pro *N. divaricata* sec. I.K.
    N. divaricata var. ? *parviflora* Benth. in DC., Prodr. 10: 350 (1846). Type: C.B.S., *Harvey* (K).

Annual herb, primary stem 30–250mm tall, simple at first, soon with few to several branches from the base and above, branches decumbent, pubescent with ± retrorse acute hairs up to c.1mm long, moderately leafy. *Leaf* blade c.5–25 × 2–14mm, elliptic to (rarely) ovate, acute or subacute, tapering into a broad flat petiolar part c.4–17mm long (roughly equalling to half as long as blade), margins subentire to sharply toothed, thinly villous particularly on margins and midline on dorsal face, hairs acute, up to 1.5mm long. *Flowers* open in sunshine, many in each spike, at first crowded, becoming lax in fruit, primary spike sometimes floriferous almost to base. *Bracts* adnate to calyx for c.5–7mm, lowermost c.11–28 × 5–8mm, lanceolate, acute, deeply toothed, villous on lower part, often almost glabrous above, hairs acute, up to 1.3–1.6mm long. *Calyx* 6–7mm long, lips 3–4mm long, thinly villous with acute hairs up to 1.5–1.8mm long. *Corolla* tube 18–20mm long (throat c.0.8–1.3mm in diam.), glabrous or with a few minute glandular hairs, limb of chasmogamous flowers 6–7mm in diam. (cleistogamous flowers frequently present), almost regular, mouth encircled by long clavate hairs, lobes 2.5–2.8 × 1.5–1.8mm, oblong, apex rounded, upper surface yellow with an orange to scarlet median bar extending ⅔ of the way up the lobe, backs of lobes dark red to chocolate-brown. *Stamens* 4, anticous pair exserted, anthers 0.25–0.4mm long, posticous pair included, anthers 0.8–1.25mm long. *Stigma* 3–4mm long, exserted. *Style* 13–14mm long. *Ovary* c.3.5–4 × 1mm. *Capsules* 6–10 × 3–3.5mm. *Seeds* c.0.5–0.8 × 0.4–0.5mm, obscurely angled, colliculate, pallid.

Selected citations:

Cape. Clanwilliam div., 3219AA, Pakhuisberg, 2500ft, 24 viii 1896, *Schlechter* 8638 (E, K, P, S, W); 3218DB, Piekenierskloof, 1 ix 1986, *Burger* 124 (STE). Cape Town div., 3318CD, Table Mt, 700ft, 16 ix 1892, *Guthrie* 1226 (BOL). Paarl div., 3318DB, Donker Kloof, 11 ix 1949, *Esterhuysen* 15712 (NBG). Stellenbosch div., 3418BB, Somerset West, 11 ix 1949, *Parker* 4446 (K, NBG, PRE, SAM). Worcester div., 3319CA, Du Toit's Kloof, 23 viii 1953, *Esterhuysen* 21710 (BOL, PRE). Bredasdorp div., 3420CA, farm Nachtwacht, c.150ft, ix 1926, *Smith* 3015 (PRE).

When Bentham (1846) described var. *parviflora* (in synonymy above) he queried its status and suggested that the corollas might be abnormal. He was looking at three young plants collected by Harvey, just beginning to flower, and all the flowers cleistogamous with the unopened corollas capping the half-formed capsules (not in young bud as an annotation by N. E. Brown claims). It is clear from the large number of specimens now before me that it is commonplace for a spike to produce first cleistogamous flowers then chasmogamous ones; more rarely are all the flowers cleistogamous. Sometimes the corolla limb is abnormally small, but expands; these flowers may be autogamous.

Zaluzianskya divaricata ranges from Pakhuis near Clanwilliam south to the Peninsula and inland to Tulbagh and Worcester, then through the coastal areas as far east as Albertinia in Riversdale division. It is obviously a common plant at least on the Peninsula and in its hinterland, growing on gravelly or stony slopes, sometimes on sand (but there are no records from the Cape Flats), up to c.750m above sea level, flowering between July and October. The oblong, apically rounded corolla lobes of *Z. divaricata* allied to its sharply toothed bracts make it an easily recognizable species.

It is established as a weed in Australia (South Australia, *Symon* 2830 (K), Victoria, *Melville* 1179B (K)).

47. Zaluzianskya glandulosa Hilliard in Edinb. J. Bot. 47: 346 (1990).
Type: Cape, Clanwilliam div., N Cedarberg, Pakhuis, 3000ft, 7 ix 1953, *Esterhuysen* 21737 (holo. BOL, iso. PRE).

Annual herb, stem 4–40mm tall, often simple, sometimes with a pair of branches near base, occasionally a second pair higher up the stem, branches decumbent, pubescent with patent glandular hairs up to 0.4–0.5mm long, occasional acute hairs as well up to 0.4–0.8mm long, often leafy mainly at the base and then with a pair of leaves immediately below the inflorescence. *Leaf* blade c.2–10 × 1.5–6mm, ovate-elliptic to elliptic tapering into a broad flat petiolar part 0.5–7mm long (often ± half as long as blade, sometimes nearly equalling it), subacute, margins entire to rather obscurely toothed, hairy mainly on the petiolar part and midrib on dorsal face, acute hairs up to 1.8mm long, glandular to 0.8mm. *Flowers* 1–4 in a crowded terminal spike, open in sunshine. *Bracts* adnate to calyx for 3–3.5mm, lowermost c.11–12 × 1.8–2.5mm, ± oblong, ± obtuse, margins entire to obscurely few-toothed, acute hairs up to 1–1.5mm long almost confined to lower margins, glandular hairs up to 0.4–0.8mm long on dorsal midline. *Calyx* c.4.5–6mm long, lips 2.2–3mm long, pubescent with acute hairs up to 0.8–1.5mm long. *Corolla* tube c.10mm long (1–1.2mm diam. in throat), very minutely and sparsely glandular, limb c.6mm in diam., regular, mouth encircled by long acute hairs, lobes 2.5–3 × 1–1.8mm, oblong, apex rounded, yellow above, brown below. *Stamens* 4, anticous pair exserted, anthers 0.4–0.6mm long, posticous pair included, anthers 0.6–0.8mm long. *Stigma* c.1.5mm long, exserted. *Style* 4.5–7mm long. *Ovary* c.2 × 0.6mm. *Capsules* (immature) 6 × 2.5mm. *Seeds* immature.

Zaluzianskya glandulosa is known only from the type collection; the plants were growing in sand at c.900m, and in the first week in September they were in flower and very young fruit. Its relationship seems to lie with *Z. divaricata*, from which it is at once distinguished by the conspicuous patent glandular hairs on stems, leaves and bracts, by the almost entire bracts, and flowers with a much shorter corolla tube and limb lacking any colour patterning.

48. Zaluzianskya lanigera Hilliard in Edinb. J. Bot. 47: 347 (1990).
Type: Cape, Ceres div., Karoo Poort, Doorn River, ix 1933, *Acock* 1648 (holo. S).

Annual herb, stem c.18–35mm tall, either simple or producing a pair of short (c.7mm long) simple branches at the apex, pubescent with acute, retrorse appressed hairs up to c.0.8mm long, leafless except at base and apex. *Leaves*: basal pair cotyledonary, blade c.2–3 × 2–3mm, ovate, subacute, entire, tapering into a broad flat petiolar part c.1.5–3mm long, apical pair (or perhaps sometimes 2 pairs) subtending inflorescence, blade c.4–5 × 3mm, elliptic or elliptic-ovate, subacute, entire, tapering to a broad flat petiolar part c.2–4mm long, all leaves shortly pubescent with acute hairs, longer hairs (to c.1mm) on margins of petiolar part. *Flowers* up to c.10 crowded in tight head, open in sunshine. *Bracts* adnate to calyx for c.1.4mm, lowermost c.7–8.5 × 1.8–2.4mm, spathulate, expanded apical half very shortly pubescent with acute hairs, green to eye, lower half densely white woolly. *Calyx* c.4.5–4.8mm long, lips 1.8–2mm long, densely white woolly in upper half. *Corolla* tube c.11–12mm long (throat 1.2–1.4mm in diam.), glabrous or with a few minute scattered glands, limb c.6–7.5mm in diam., regular, mouth encircled by long acute hairs, lobes c.2.5–3 × 2.5mm, elliptic, entire, apex rounded, orange-yellow with a deeper orange median longitudinal bar on each lobe. *Stamens* 4, anticous pair exserted, anthers 0.7–0.8mm long, tips of posticous pair just visible in mouth, anthers 0.8–1mm long. *Stigma* c.2.5mm long, exserted. *Style* c.7.5mm. *Ovary* c.2 × 0.6mm. *Capsules* c.4.3 × 2.4mm. *Seeds* c.0.6 × 0.4mm, obscurely angled, colliculate, pallid.

Zaluzianskya lanigera is a most distinctive species, known only from the type collection made (in Acocks' student days, when he spelt his name Acock) on sandy flats along the east bank of the Doorn River at Karoo Poort. The white-woolly calyx and bracts white-woolly on the lower half, green and very shortly pubescent on the upper, are unmatched in the genus. The orange-yellow, entire corolla lobes each marked with an orange bar point to its being in the general affinity of *Z. collina*, but the relationship is not close.

49. Zaluzianskya isanthera Hilliard in Edinb. J. Bot. 47: 347 (1990).
Type: Cape, Ceres div., 3319AB, Schurweberg (next to Bok Tafelberg), 3500ft, 12 x 1952, *Esterhuysen* 20636 (holo. BOL, iso. E).

Annual herb, primary stem 10–40mm tall, often simple, sometimes sparingly branched from the base or higher, pubescent with ± retrorse acute hairs up to 1–1.4mm long. *Leaves* few (1–3 pairs), c.4–12 × 1.7–4mm, spathulate, obtuse to subacute, margins entire, both surfaces pubescent, upper more densely than lower, hairs acute, up to c.1mm long particularly on lower margins. *Flowers* c.2–15 in a dense terminal spike, open in sunshine. *Bracts* adnate to calyx for 2–3mm, lowermost c.5–15 × 2.8–5.8mm, ovate to spathulate, upper surface densely pubescent, hairs acute, c.0.6–1mm long, up to 1–1.8mm on margins, lower surface glabrous to pubescent, often showing a gradation in pubescence from outer to inner bracts. *Calyx* c.4–5.5mm long, lips 1.7–2mm long, pubescent, hairs acute, up to 1mm long. *Corolla* tube c.8.5–11mm long (throat 0.8–1mm in diam.), minutely glandular-puberulous, limb c.3.5–5mm in diam., nearly regular, mouth encircled by a band of acute hairs, sometimes orange-coloured, lobes 1.5–2 × 0.8–1.2mm, elliptic, entire, creamy-white above, mauve below, yellow/orange 'eye'. *Stamens* 4, anticous pair exserted, anthers 0.5–0.7mm long, posticous pair at least partially exserted, anthers 0.5–0.7mm long. *Stigma* 1mm long, partly exserted. *Style* 5–8mm long. *Ovary* c.1.3 × 0.5mm. *Capsules* c.4 × 2mm. *Seeds* c.0.5 × 0.3mm, obscurely angled, colliculate, bright mauve when young, whitish at maturity.

Citations:

Cape. Ceres div., 3319AB, E foot of Schurweberg, 12 x 1952, *Esterhuysen* 20627 (BOL). 3219DC, top of Katbakkies Pass, 1300m, 22 ix 1985, *Viviers* 1523 (BOL).

Zaluzianskya isanthera is known from only two sites in the mountains north of Ceres, where it was found growing in sand over rock sheets, between c.900 and 1300m above sea level, flowering and fruiting in September and October. It seems to be allied to *Z. lanigera*, which it resembles in facies and which also has nearly equal anthers, but from which it can at once be distinguished by its bracts and calyx, which are clothed in acute hairs up to 1–1.8mm long (not white-woolly as in *Z. lanigera*), and corolla lobes creamy-white above (not bright orange-yellow).

50. Zaluzianskya collina Hiern in Fl. Cap. 4(2): 346 (1904), p.p. **Plate 3E.**
Lectotype (chosen here): Cape, Namaqualand [3017BD, between Garies and Kamieskroon], Waterklip, 2300ft, 10 ix 1897, *Schlechter* 11177 (BOL; isolecto. E, K, P, PRE, S, W).

Annual herb, primary stem 30–250mm tall, simple to well branched from the base and above, lowest branches decumbent or ascending, either glabrous, glabrous in the upper half pubescent in the lower with ± retrorse acute hairs up to 0.25–0.5mm long, or sparsely pubescent throughout, smaller plants sometimes leafless between radical leaves and inflorescence, larger ones remotely leafy. *Leaves*: radical leaves in 1 or 2 pairs, blade c.3–20 × 2–8mm, ovate to elliptic tapering into a flat petiolar part 1–20mm long (mostly ± equalling blade), cauline leaves c.5.5–40 × 0.8–3mm, linear, ± obtuse, margins of all leaves entire to very obscurely denticulate, nearly glabrous, a few hairs up to c.1mm long on lower margins and dorsal midline. *Flowers* 1–c.15 in a terminal capitate spike elongating in fruit but still crowded, open in sunshine. *Bracts* adnate to calyx for 4–5mm, lowermost c.10–25 × 0.8–1.1mm, linear, entire, almost glabrous, a few acute hairs up to 0.6–1.4mm long on margins only. *Calyx* 5–7mm long, lips 2.5–3mm long, glabrous except for fine acute hairs up to 0.3–0.6mm long fringing the lobes. *Corolla* tube 17–20mm long (throat 1.5–2mm in diam.), very minutely puberulous with scattered glands or these confined to swollen part of tube only, limb c.8–16mm in diam., almost regular, mouth thickly bearded with long clavate hairs extending out onto lower part of each lobe, lobes 3.5–6 × 2.5–4.2mm, elliptic, bright yellow with a dark orange median bar extending about ⅓ of the way up the lobe. *Stamens* 4, anticous pair exserted, anthers, 0.7–1.8mm long, posticous pair included, anthers 1–2mm long. *Stigma* 2.5–5mm long, exserted. *Style* 10.5–14mm long. *Ovary* c.3–4 × 0.8–1mm. *Capsules* c.7–12 × 3–4mm. *Seeds* not seen.

Selected citations:

Cape. Namaqualand, 3017BB, south of Kamieskroon, 27 viii 1941, *Esterhuysen* 5976 (BOL); Kamieskroon, Grootvlei, 16 viii 1967, *Van Breda* 4099 (PRE). 3017BD, along Karkams–Spoegrivier road, 716m, 20 ix 1981, *van Berkel* 422 (NBG). 3018CA, ±20km N of Garies, Garagams, 4 ix 1989, *Batten* 982 (E); Garies, 27 viii 1941, *Compton* 11448 (NBG). 3018AC, near Leliefontein, 5100ft, 27 ix 1932, *Levyns* 4024 (BOL).

Hiern based the name *Z. collina* on two different species; his description embraces both, but the name has become attached to the plant conspecific with *Schlechter* 11177, which has therefore been chosen as the lectotype. The other syntype, *Schlechter* 8193, is *Z. pusilla*.

Zaluzianskya collina has been recorded from only a small area of Namaqualand, roughly encompassed by lines drawn between Kamieskroon, the Kamiesberg, Garies and the Spoeg River. It favours damp sandy or gravelly places and is in full flower in August and September.

The remarkably broad band of hairs around the mouth allied to the bright yellow elliptic corolla lobes marked with an orange median bar and the linear, almost glabrous, cauline leaves and bracts make recognition easy, and distinguish it at once from *Z. peduncularis* (= *Z. gilioides*) (with which it has been much confused), which has the calyx only very shortly divided into two lips, the mouth of the corolla tube glabrous, and mostly broader cauline leaves. Also, the flowers of *Z. collina* open during the day, those of *Z. peduncularis* at dusk.

The closest ally of *Z. collina* is possibly *Z. pusilla*, which is a hairier plant with the corolla lobes white or creamy-coloured above, the mouth bearded but the hairs not extending up the lobes as they do in *Z. collina*. They are sympatric over the whole range of *Z. collina*.

51. Zaluzianskya acrobareia Hilliard in Edinb. J. Bot. 47: 345 (1990).
Type: Cape, Calvinia div., 3119AC?, Nieuwoudtville, farm Uitkoms, 820m, 9 viii 1983, *Van Wyk* 1350 (holo. STE).

Annual herb, stem 5–17mm tall, solitary (but clearly will branch from base later), simple, pubescent with patent mainly glandular hairs up to c.0.7mm long, leafy. *Leaves*: radical leaves in 1 or 2 pairs, blade c.2.5–7 × 2–3.7mm, ovate or elliptic-ovate tapering into a broad flat petiolar part 2–5mm long (somewhat shorter than blade), cauline leaves c.3–6 × 1.2–2.5mm, lanceolate-elliptic tapering to a short petiolar part, margins of all leaves obscurely to distinctly toothed, both surfaces with acute hairs up to 1mm long plus glandular hairs to c.0.5mm. *Flowers* 1–5 in a congested terminal spike, open in sunshine. *Bracts* adnate to calyx for c.4mm, lowermost c.8–14 × 2–2.2mm, oblong to oblong-lanceolate, margins with 1 or 2 pairs of tiny teeth, both surfaces densely pubescent with acute hairs up to 1.5mm long, some glandular hairs to 0.8mm. *Calyx* c.5–6mm long, lips c.3mm long, densely pubescent all over with acute hairs up to 1.2mm long. *Corolla* tube 22–27mm long (throat c.2mm in diam.), very minutely and sparsely glandular-pubescent, limb c.10–12mm in diam., almost regular, mouth bearded with long clavate hairs extending to base of lobes, lobes c.3.5–5 × 3–4.3mm, elliptic, yellow above with a median orange bar, purple below. *Stamens* 4, anticous pair exserted, anthers 0.6–0.9mm long, posticous pair included, anthers 1.4–1.7mm long. *Stigma* c.2mm long, shortly exserted? *Style* c.20–22mm long. *Ovary* c.2.5–3 × 0.8mm. *Capsules* not seen.

Zaluzianskya acrobareia is known only from the type material, which consists of a number of tiny plants, obviously in the first stages of growth, with the very short stems terminating in disproportionately large flowers (the epithet means top heavy). They were found growing in moist patches on a rocky plateau; the farm Uitkoms is not marked on the 1:500000 map but it is possibly not far from the top of Van Rhyn's Pass, whence came two collections that may represent hybrids between *Z. acrobareia* and *Z. pusilla* (*Lewis* 5897, NBG, STE, and *Esterhuysen* 5284, BOL).

The relationship of *Z. acrobareia* seems to lie with *Z. pusilla* from which it can be distinguished by the patent glandular pubescence on the stems (not acute ± retrorse hairs), longer corolla tube (22–27mm versus 6–15mm) and limb yellow above (not white nor creamy-white). The plants from the top of Van Rhyn's Pass (collected in 1941 and 1961) were noted by both collectors as having yellow flowers. *Esterhuysen* 5284 has flowers with short corolla tubes (two measured, 8.5 and 16mm) and the hairs on the stem appear to be eglandular; it thus seems very close to *Z. pusilla*. *Lewis* 5897 on the other hand has flowers with longer corolla tubes (19.5 and 22mm) and the hairs on the stems are mostly gland-tipped; it is thus very close to *Z. acrobareia*, but the material does not look homogeneous, and someone has written on the sheet in NBG '? mixed collection'. *Zaluzianskya acrobareia* should be sought again, and the colonies on Van Rhyn's Pass scrutinized.

52. Zaluzianskya pusilla (Benth.) Walp., Repert. 3: 308 (1844); Hiern in Fl. Cap. 4(2): 349 (1904), p.p.

Type: Cape [Van Rhynsdorp div., c.3118CB] between Hol river and Mierenkasteel, 5 viii 1830, *Drège* (holo. K; iso. E, P, S, TCD, W).

Syn.: *Nycterinia pusilla* Benth. in Hook., Comp. Bot. Mag. 1: 371 (1836) & in DC., Prodr. 10: 350 (1846).
    Zaluzianskya collina Hiern in Fl. Cap. 4(2): 346 (1904) p.p. excl. lectotype.

Annual herb, primary stems c.20–155mm tall, solitary or up to 7 stems from base, usually simple, sometimes with 1–3 branches, often pubescent with ± retrorse acute hairs up to 0.5–1mm long, these sometimes almost wanting, minute glands also often present particularly on inflorescence axis and at apex of stem, often leafless between radical leaves and inflorescence, occasionally up to c.5 leaves present. *Leaves*: radical leaves in 1 or 2 pairs, blade c.4.5–22 × 4–13mm, ovate to elliptic tapering into a broad flat petiolar part 4.5–18mm long (very roughly equalling blade), cauline leaves often wanting, when present c.3.5–28 × 1–5mm, linear to elliptic slightly tapering to the base, margins of all leaves usually entire or very obscurely denticulate, rarely distinctly toothed, thinly pubescent mainly on margins and midline, hairs acute, up to c.1mm long. *Flowers* 1–c.30, initially crowded, forming a lax spike in fruit, open in sunshine. *Bracts* adnate to calyx for 3.5–5.5mm, lowermost 11–25 × 1–4mm, linear to narrowly elliptic, entire to very obscurely toothed, pubescent all over with acute hairs up to 0.7–1.5mm long, hairs sometimes sparse and confined to margins and dorsal midline. *Calyx* 4–7mm long, lips 3.5–5.5mm, hairy all over though hairs sometimes sparse, acute, up to 0.5–1.3mm long. *Corolla* tube 6–15mm long (throat 0.8–1.8mm in diam.), glabrous or very minutely and sparsely glandular, limb 3–11mm in diam., almost regular, mouth bearded with long clavate hairs extending to base of lobes, lobes 1–4.5 × 0.6–3.1mm, oblong, white to creamy yellow above, yellow in throat with a median orange bar extending c.⅓ of way up lobe, maroon below. *Stamens* 4, anticous pair exserted, anthers 0.25–0.9mm long, posticous pair included, 0.5–1.2mm long. *Stigma* 1.5–3mm long, shortly exserted. *Style* 2–11mm long. *Ovary* 2.5–3.5 × 0.8–1mm. *Capsules* c.6–7 × 2.4–3mm. *Seeds* c.0.4–0.5 × 0.3–0.5mm, obscurely angled, colliculate, pallid.

Selected citations:

Cape. Namaqualand, 3017BB, S of Kamieskroon, 27 viii 1941, *Esterhuysen* 5975 (BOL). 3018AC, Studers Pass, 25 viii 1987, *Batten* 810 (E). 3018CA, Garies, 27 viii 1941, *Compton* 11447 (NBG). Van Rhynsdorp div., 3118BA, Kareebergen, 1200ft, 18 vii 1896, *Schlechter* 8193 (E, K, P, PRE, W). 3118DC, Heerenlogement, 21 vii 1941, *Esterhuysen* 5577 (BOL). Calvinia div., Lokenburg, c.2400ft, 25 ix 1955, *Acocks* 18541 (PRE). Clanwilliam div., 3219CB, S Cedarberg, Krom River, 2 x 1952, *Esterhuysen* 20549 (BOL, PRE); 3218BB, Olifants River Barrage, 22 vii 1941, *Compton* 11041 (NBG). Ceres div., 3319BC, top of Theronsberg Pass, 18 ix 1938, *Häfstrom & Acocks* 2232 (PRE, S). Laingsburg div., 3320AB, Tweedside, 27 ix 1951, *Compton* 22851 (NBG). Swellendam div., Hessekwas Poort, 19 ix 1962, *Taylor* 3957 (STE).

Zaluzianskya pusilla is widely distributed in the western Cape, from the environs of Kamieskroon in Namaqualand south east through Van Rhynsdorp, Clanwilliam and Ceres to Laingsburg and Hassequa's Poort near Stormsvlei. It favours rocky or sandy sites on slopes and flats, in arid fynbos or other scrubby vegetation, from c.150 to 950m above sea level, flowering between July and October.

It is a difficult plant to characterize, partly because it shows such a remarkable range in the size of the corolla limb and there is some geographical patterning to this. From Hassequa's Poort north to the Ceres Karoo, the limb is only c.3–4.5mm across; northwards, it is mostly larger, though plants from, for example, Lokenburg (Calvinia div.) and the Knechtsvlakte (Van Rhynsdorp div.) may have the limb only c.4mm in diam. (the type material, from somewhere near Holrivier, has the limb c.6mm in diam.), and it may be only c.4.5mm in the Kamiesberg area, though there the limb is more usually 8–10mm across. There is no indication of cleistogamy, but autogamy cannot be ruled out.

The corolla of even the smallest flowers is always bearded around the mouth; this character, as well as the white or creamy-white limb expanding in sunshine, will at once distinguish *Z. pusilla* from *Z. peduncularis* (glabrous mouth, bright yellow limb expanding at dusk or in dull light) with which it has been much confused. Its relationship, however, lies with *Z. collina*, but it differs from that species in the colour of the limb (bright yellow in *Z. collina*), less heavily bearded mouth (the hairs do not extend onto the lobes as they do in *Z. collina*), the shorter corolla tube, and hairier bracts and calyx (the hairs, though they may be sparse, are rarely confined to the margins of the bracts and the calyx lobes as they are in *Z. collina*).

Zaluzianskya pusilla appears to hybridize with *Z. acrobareia* on Van Rhyn's Pass: see under that species.

53. Zaluzianskya sp.

Annual herb, about 50mm tall, well branched from the base and above, glabrous except for ± sessile glands on upper parts, laxly leafy, with leaf tufts (incipient branches) in axils of primary leaves. *Leaves* (primary) c.6–11 × 1–2mm, elliptic tapering to a petiolar part accounting for about half the length of the lower leaves, shorter upwards, entire, glabrous. *Flowers* solitary at the tip of each branchlet then overtopped by a slender stem again terminating in a flower. *Bracts* adnate to calyx for c.3mm, lowermost c.6–11 × 1mm, oblong, obtuse, entire, glabrous. *Calyx* c.4mm long, lips c.2.3mm long, glabrous except for very small hairs on margins of lobes. *Corolla* tube c.17mm long (throat c.1mm in diam.), glabrous, limb c.6mm in diam., almost regular, mouth bearded, lobes c.2.5 × 1.2mm, elliptic, colour unknown. *Stamens* 4, anticous pair exserted, anthers 0.7mm long, posticous pair included, anthers 1mm long. *Stigma* c.1.5mm long. *Style* c.9mm. *Ovary* not seen. *Capsules* c.5–6 × 2.6–3mm. *Seeds* not seen.

Citation:

Cape. Namaqualand [3017BB NE of Kamieskroon], damp places in rock crevices, upper western slopes Sneeuwkop, 11 xii 1910, *Pearson & Pillans* 5812 (K).

This is a very odd little plant, its growth habit (solitary flowers overtopped by a younger flower terminating a branch arising immediately below the first flower) not matched elsewhere in the genus. The bracts and calyx are very like those of *Z. pusilla*. The material is inadequate to typify a name, and the plant should be sought again.

54. Zaluzianskya benthamiana Walp., Repert. 3: 309 (1844); Hiern in Fl. Cap. 4(2): 348 (1904); Merxmüller & Roessler, Prodr. Fl. S.W. Afr. 126. Scrophulariaceae: 58 (1967).
Type: Cape, Namaqualand, Haazenkraalsrivier, 2500ft, 24 viii 1830, *Drège* (holo. K; iso. E, P, S, TCD, W).

Syn.: *Nycterinia villosa* Benth. in Hook., Comp. Bot. Mag. 1: 371 (1836) & in DC., Prodr. 10: 350 (1846), excl. syn. Type as for *Z. benthamiana*.
 Zaluzianskya aschersoniana Schinz in Verhandl. Bot. Ver. Brandenb. 31: 190 (1890); Hiern in Fl. Cap. 4(2): 348 (1904); Merxm. & Roessler, Prodr. Fl. S.W. Afr. 126. Scrophulariaceae: 58 (1967). Lectotype (chosen here): Namibia, 2616CC, Tsirub, v 1885, *Schenk* 134 (Z, isolecto. PRE).
 Z. bolusii Hiern in Fl. Cap. 4(2): 352 (1904). Lectotype (chosen here): Cape, Van Rhynsdorp div., 3118BA, Karee Bergen, 1500ft, 23 vii 1896, *Schlechter* 8272 (K; isolecto. BOL, E, P, PRE, S, W).
 Z. ramosa Hiern in Fl. Cap. 4(2): 353 (1904); Levyns in Adamson & Salter, Fl. Cape Penins. 712 (1950). Type: Cape, near Houtsbay (see comment below), c.100ft, 12 vi 1892, *Schlechter* 968 (holo. not found; iso. P, PRE, W, WU).
 Manulea buchubergensis [Dinter ex] Range in Feddes, Repert. 38: 263 (1935), nom. nud.

Annual herb, primary stem 20–330mm tall, erect, simple to well branched from the base and above, glandular-puberulous, also ± villous with patent acute hairs up to 1–2mm long, these hairs occasionally sparse, leafy, often with axillary leaf tufts (incipient branches). *Leaves* (primary ones) c.5–35 × 1–6mm, broadly to narrowly elliptic, subacute, base narrowed to a petiolar part, particularly in lowermost leaves, margins entire to obscurely dentate, glandular puberulous all over, thinly villous as well (or hairs occasionally sparse), hairs up to 1–2mm long. *Flowers* open in sunshine, many in a terminal spike, crowded at first, becoming lax and spike greatly elongated in fruit. *Bracts* adnate to calyx for 3–4.5mm, lowermost 10–27 × 2–5(–10)mm, lanceolate, entire or occasionally obscurely toothed, glandular-puberulous, villous as well with acute hairs up to 1.5–2.5mm long. *Calyx* (5–)6–8mm long, lips 1.5–3mm long, glandular-puberulous, villous as well with acute hairs up to 1–1.5mm long. *Corolla* tube (9–)14–20mm long (throat 0.8–1.4mm in diam., scarcely wider than tube), glandular-puberulous, hairs sometimes sparse, or rarely up to 0.2mm long, limb 4–8.25mm in diam., almost regular, mouth sparsely bearded or glabrous, lobes 1.5–3.5 × 1.2–2mm, broadly elliptic, variously recorded as white, cream, creamy-yellow, yellow and pale lemon yellow, deeper yellow/orange in throat radiating out as a star-shaped 'eye', backs of lobes once recorded as maroon. *Stamens* 2 (anticous pair), anthers 0.8–2.3mm long, included to shortly exserted. *Stigma* c.2–6mm long, included to shortly exserted. *Style* c.6.5–15mm long. *Ovary* c.2.5–3.5 × 1mm. *Capsules* c.5–7 × 2.5–3mm. *Seeds* c.0.6–1 × 0.4–0.6mm, obscurely angled, colliculate, pallid.

Selected citations:

Namibia. 2616CB, Aus, 1500m, 4 vi 1922, *Dinter* 3562 (BOL, K, S, SAM, Z). 2715DD, Buchubirge, 28 vii 1929, *Dinter* 6551 (BOL, PRE, S, SAM, Z).
Cape. Richtersveld, 2817AA, Kodaspiek, 900m, 2 ix 1977, *Oliver, Tölken & Venter* 467 (PRE). 2816BD, Hottentots-paradysberg, head of Helskloof, 700m, 28 viii 1977, *Thompson & le Roux* 122 (PRE, STE). Namaqualand, 2917DA, Spektakelberg, 2000ft, 20 ix 1971, *Thompson* 1292 (STE); 3018CA, Garies, 24 vii 1941, *Esterhuysen* 5434 (BOL); Kenhardt div., 2919AB, 10 miles SE of Pofadder, 3400ft, 22 v 1961, *Leistner* 2477 (K, PRE); 2918BB, Wortel, 29 vi 1936, *Wall* s.n. (S). Van Rhynsdorp div., 3118DA, 54km N of Van Rhynsdorp, 22 viii 1987, *Batten* 800 (E); 3118AB, Bitterfontein, 23 viii 1941, *Compton* 11324 (NBG). Calvinia div., Wolwe Graafwater, c.2000ft, 23 viii 1956, *Acocks* 18959 (PRE, Z). Ceres div., 3219DC, Skittery Kloof, 29 viii 1971, *Wisura* 2211 (NBG). Sutherland div., 3220CA, Houthoek, 14 viii 1968, *Hanekom* 1072 (K, PRE, STE). Oudtshoorn div., 3322CB, farm 'Die Krans', 21 v 1973, *Dahlstrand* 2418 (J).

When publishing *Nycterinia villosa*, Bentham (1836) described a Drège specimen (here cited as type), but he also included *Erinus villosus* Thunb. as a synonym, with a question mark. His doubts were perfectly justified: Thunberg's plant is not the same as Drège's. Therefore, despite Bentham's use of the epithet *villosus*, one is not justified in taking *E. villosus* as the type of *Nycterinia villosa*, thereby forcing Bentham into a formal misidentification that he did not make. *Erinus villosus* Thunb. is a heterotypic synonym of *Zaluzianskya villosa* Schmidt (see there); *Nycterinia villosa* was renamed in *Zaluzianskya* as *Z. benthamiana* Walp. and that is its correct name. Hiern (1904) also used the name in this sense, excluding the Thunberg synonym.

Bentham implied that *Nycterinia villosa* (that is, *Zaluzianskya benthamiana*) has four (not two) stamens, and this was accepted without question by Hiern, who proceeded to describe *Z. bolusii* Hiern as having two stamens, and thereby separating *Z. bolusii* from *Z. benthamiana* in his key. Schinz separated his *Z. aschersoniana* from *Nycterinia villosa* on the breadth of the bracts and their degree of adnation to the calyx, but this does not hold over the wide range of material now available.

Zaluzianskya ramosa Hiern is also inseparable from *Z. benthamiana*. The type purportedly came from Hout Bay (Houts Bay of Schlechter), but this seems improbable as the species has never been recollected on the Peninsula, nor anywhere nearby. It is interesting that one of the two beautiful sheets of *Schlechter* 968 (type of *Z. ramosa*) in Vienna (W), as well as the sheet in Paris (P), is mixed with *Z. affinis*. The two species are sympatric from the Richtersveld south to about Van Rhynsdorp division, which reinforces my suspicion that Schlechter's specimen came from somewhere in that area. The full geographical range of *Z. benthamiana* is from southern Namibia (Lüderitz and Aus southwards) to Namaqualand, thence south to Vanrhynsdorp, Calvinia and Williston divisions, then bearing south and east to the Tanqua Karoo, the Anysberg and Prince Albert; there are few collections from the southern and eastern part of the range. The plants grow on sandy or gravelly flats and slopes, flowering between June and August, after early rains.

Zaluzianskya benthamiana is very glandular on stems, leaves, bracts and calyx, and this feature, in combination with the long patent acute hairs, relatively long, narrow, entire or subentire leaves, and rounded corolla lobes make it easily recognizable. Its closest ally appears to be *Z. pusilla*, with which it is partly sympatric; *Z. pusilla* differs not only in having four stamens, but it is also a much less glandular plant, with ± retrorse hairs on the stems, and a shorter corolla tube.

55. Zaluzianskya diandra Diels in Bot. Jahrb. 23: 482 (1896); Hiern in Fl. Cap. 4(2): 352 (1904); Merxmüller & Roessler in Prodr. Fl. S.W.A. 126 Scrophulariaceae: 58 (1967). **Plate 2K.**

Type: Calvinia div., Hantam Mts, 1869, *Meyer* (B†); see note below.

Annual herb, primary stem 50–300mm tall, soon branching from the base and above, branches prostrate, decumbent or ascending, pubescent with coarse patent acute hairs up to 1–1.5mm long, minutely glandular as well, leafy, with leaf tufts and eventually more than one branch developing from the axil of each primary leaf. *Leaves*: radical c.17–45 × 2–10mm, elliptic tapering to a petiolar part accounting for up to half total leaf length, entire or obscurely toothed, very thinly pubescent, hairs acute, up to 1mm long, primary cauline leaves c.6–26×1.5–4.5mm, narrowly obovate to narrowly elliptic tapering to the base, entire or obscurely toothed, glabrous on the upper half dorsally, thinly villous on lower part, villous ventrally, hairs up to 1–1.8mm long particularly on lower margins. *Flowers* open in sunshine, up to c.30 in congested terminal spikes, only the lowermost 2–3 capsules lax in fruit. *Bracts* adnate to calyx for 1.5–2.5mm, lowermost c.4–15 × 1.8–5mm, spathulate, subacute, entire, glabrous or nearly so dorsally on the upper half, villous on lower half, villous ventrally, hairs acute, up to 1–2mm long, minutely glandular as well. *Calyx* 4.5–6mm long, lips 1.2–2mm long, villous mainly on the ribs, hairs acute, up to 0.5–1mm long, minutely glandular as well. *Corolla* tube 10–13mm long (throat c.1–1.3mm in diam.), glabrous or with sparse minute glandular hairs, limb c.5–6mm in diam., slighly zygomorphic, lobes c.2–3 × 0.8–1mm, strongly reflexed, oblong, margins revolute, orange to brick-red, yellow in throat. *Stamens* 2 (anticous pair), anthers (1–)2mm long, well exserted. *Stigma* c.2.5–4.5mm long, well exserted. *Style* c.7–11.5mm long. *Ovary* c.2.5 × 0.8mm. *Capsules* 5 × 2.5mm. *Seeds* 0.6–0.75 × 0.4–0.5mm, angled, colliculate, pallid.

Selected citations:

Namibia. Bethanien distr., 2616BA, farm Tiras, 16 viii 1963, *Merxmüller & Giess* 2842 (PRE). Keetmanshoop distr., 2717DB, farm Holoogberg, 30 v 1972, *Giess & Müller* 12305 (PRE). Seeheim distr., 2718CA, Klein Karas, 3 viii 1923, *Dinter* 4790 (BOL, K, PRE, SAM, STE, Z).
Cape. Namaqualand, 2817DC, just south of Vyfmylspoort, 2 vii 1986, *Fellingham* 1101 (PRE, STE). 2818CC, 9 miles S of Goodhouse, 27 vii 1950, *Barker* 6271 (NBG). Kenhardt distr., 2820DC, Kakamas, 2300ft, viii 1939, *Acocks* s.n. (PRE); 2919AC, 22 miles W of Pofadder, 'Namies', 21 ix 1961, *Hardy* 748 (K, PRE); 2921AC, 19 miles ENE of Kenhardt, c.3000ft, 4 v 1961, *Leistner* 2324 (K, PRE).

The type of *Z. diandra* was destroyed in the Berlin fire; it was said to have come from the Hantam Mountains near Calvinia, where Meyer practised medicine. However, this provenance is unlikely as the area of *Z. diandra* lies well to the north of Calvinia, with a range from the Tiras Mountains in southern Namibia south to the Orange river at Vioolsdrif and east to Prieska. Diels's description leaves no doubt that the name is being used correctly. The plants grow in sandy or gravelly places, flowering between May and September.

Zaluzianskya diandra is a most distinctive species, both in habit (well-grown plants are bushy) and in the flowers: the strongly reflexed narrow orange lobes with revolute margins are more like those of many species of *Manulea* than any *Zaluzianskya*, but the two well exserted, disproportionately large, anthers are unique. It has no obvious relationships. In the linear sequence I have placed it next to *Z. benthamiana*, which is itself an isolated species, and the pair make a rather uncomfortable appendage to Sect. *Holomeria*.

INSUFFICIENTLY KNOWN SPECIES

Erinus gracilis Lehm. in Del. Sem. Hort. Hamb. Bot. 4, 8 (1833) (not seen) and in Linnaea 10(2): 76 (1836).

Lehmann's Scrophulariaceae are in Stockholm (S), but the type of the name *Erinus gracilis* could not be found when Prof. Nordenstam kindly made a thorough search on my behalf. The plant was grown from seed sent to Hamburg by Ecklon and it is possible that no herbarium specimen was made. Lehmann's description reads '*Planta hirsuta, caule herbaceo ramoso, foliis petiolatis spathulato-lanceolatis dentatis, bracteis lanceolatis ciliatis, tubo corollae pubescente, limbi laciniis emarginatis*' and the plant could be a species of *Zaluzianskya* as Hiern (1904, p.337) suggested.

Erinus pulchellus Jarosz, Pl. Nov. Cap. 20 (1821).

This name was placed by Hiern (1904, p.353) at the end of *Zaluzianskya* as an imperfectly known species. The whereabouts of Jarosz's herbarium is unknown and I have seen no specimen that I could relate to the name. The description (in Fl. Cap. 4(2): 353) reads:

> Not perennial; stem shrubby, erect, terete, pubescent; leaves opposite, linear, obtuse, sessile, somewhat sheathing, subpilose, but little dentate; spike terminal, imbricate; bracts ovate-lanceolate, sessile, ciliate, longer than calyx; calyx campanulate, 5-partite, persistent, ½ inch long, segments linear, acute; corolla funnel-shaped, glabrous, orange-coloured, 1½ inches long; limb equally 5-lobed; lobes oblong, obcordate, ⅙ inch long, anthers 2, half-exserted.

The only *Zaluzianskya* with flowers 1½ inches long and 'orange-coloured' is *Z. pumila*, with conspicuous thickened orange to scarlet bars forming the lower part of each corolla lobe, but *Z. pumila* is a sparingly branched annual herb with elliptic leaves and 4 stamens. I doubt that Jarosz's plant was a *Zaluzianskya* (both the shape of the corolla tube and the linear calyx segments are discordant), but I am unable to place it elsewhere.

14. REYEMIA

The generic name *Reyemia* commemorates Dr H. Meyer, who practised medicine in Calvinia in the 1860s, and whose collections from that area are the basis of many names.

When Diels described *Zaluzianskya nemesioides*, he could suggest no affinity for the plant, but noted that the morphology of the corolla placed it amongst the most advanced of the Manuleae. The corolla is indeed peculiar in that the corolla limb (in both *R. nemesioides* and *R. chasmanthiflora*) is resupinate: the corolla tube curves in towards the axis with the bract at the back of the curve, whereas usually a corolla tube curves away from the axis so that the bract is under the curve. The one-lobed anticous lip thus stands erect and the 4-lobed posticous lip inclines downwards. The flowers bear no resemblance to those of any species of *Zaluzianskya* and specimens of *R. nemesioides* may be found in herbaria under *Nemesia*, *Sutera* and *Phyllopodium*, or nameless at the end of the family. However, the affinity of *Reyemia* appears to be with *Zaluzianskya*: they have similar calyces, which are distinctive in the Manuleae by virtue of being plicate, and their seeds are similar. *Reyemia* is easily distinguished from *Zaluzianskya* by its loose panicles of flowers, resupinate corolla-limb, clavate hairs inside corolla throat on posticous side, and two stamens plus two staminodes (no staminodes in *Zaluzianskya*).

Although Burchell collected the first specimens of *R. nemesioides* in 1811 (it was determined by Hiern as *Phyllopodium* ?), the first collection of *R. chasmanthiflora* seems to have been made only in 1986. The Middelpos-Fraserburg-Sutherland area still awaits thorough botanical exploration, and it will be interesting to see what further discoveries are made.

Reyemia Hilliard in Edinb. J. Bot. 49: 297 (1992). **Figs 18A, 32D, 33A, 65, Plate 2C.**

Small annual herbs. *Stems* laxly leafy, glandular above, retrorse eglandular hairs below. *Leaves* opposite below, alternate above, mainly radical. *Inflorescence* a lax raceme, often panicled. *Bract* adnate to pedicel and most of calyx tube. *Calyx* persistent, bilabiate, ± plicate at anthesis, lips short, anticous lip scarcely lobed, posticous lip shortly 3-toothed, membranous, midrib of each lobe thickened. *Corolla* resupinate by curvature of upper part of tube, tube narrowly cylindric below, widening towards the throat, limb bilabiate, anticous lip entire, posticous lip 4-lobed, bearded with clavate hairs at base and down back of throat. *Stamens* 2 (the posticous pair), filaments decurrent to base of tube forming a channel enveloping the style, anthers synthecous, well exserted, connective glandular; 2 small staminodes. *Stigma* lingulate with two marginal bands of stigmatic papillae, exserted, tapering into the style. *Ovary* oblong-elliptic in outline, small nectariferous gland completely adnate to base of ovary on one side, glandular or not on the sutures, ovules many (more than 20 in each loculus). *Fruit* a septicidal capsule with a short loculicidal split at the tip of each valve, valves apiculate. *Seeds* seated on a funicle taking the form of a round centrally depressed cushion (fig. 18A), seed slightly longer than broad, somewhat angled, irregularly colliculate, greyish when young, later pallid (figs 32D, 33A).

Type species: *Reyemia chasmanthiflora* Hilliard
Distribution: South Africa, Cape, in Calvinia, Sutherland and Fraserburg divisions.

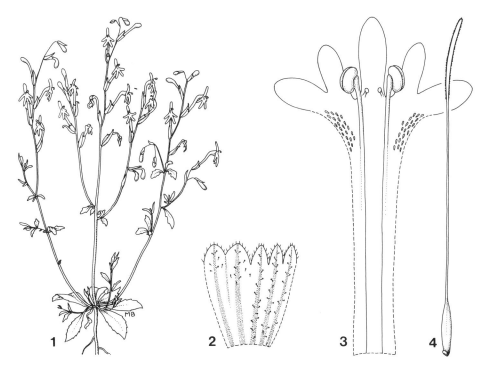

Fig. 65. *Reyemia chasmanthiflora* (*Batten* 1016): 1, habit of plant, flowers with upper part of corolla tube curving towards axis and so bringing the anticous corolla lobe into the posticous position, × 0.3; 2, calyx opened out, × 4.5; 3, corolla opened out to show anticous staminodes, decurrent posticous filaments and exserted anthers, clavate hairs in throat, × 4.5; 4, lingulate stigma with marginal bands of stigmatic papillae, style, ovary with small adnate nectariferous gland, × 4.5.

Key to species

1a. Corolla tube c.23–30mm long, limb cream-coloured .... **1. R. chasmanthiflora**
1b. Corolla tube c.7–8.5mm long, limb yellow and white (anticous lobe white, posticous lobes
    yellow) .... **2. R. nemesioides**

1. Reyemia chasmanthiflora Hilliard in Edinb. J. Bot. 49: 297 (1992). **Fig. 65, Plate 2C.**
Type: Cape, Calvinia div., 3120CB, farm Annexe Kransfontein 721, 1123m, 2 ix 1986, *Cloete & Haselau* 172 (holo. STE).

Annual herb, primary stem c.80–200mm tall, soon branching from the base and sparingly so above, branches sharply ascending, glandular-puberulous, hairs up to 0.15mm long, these hairs disappearing towards the base of each stem and there replaced by retrorse eglandular hairs up to 0.25mm long, distantly leafy, often with axillary leaf tufts (incipient branches). *Leaves*: radical with blade c.6–20 × 3–10mm, elliptic tapering to a broad flat petiolar part c.4–15mm long (roughly equalling to half as long as blade), margins entire to sparsely toothed, they and the upper surface with scattered eglandular hairs up to 1–2mm long, scattered minute sessile glands as well, lower surface glabrous; cauline leaves c.4–14 × 1–2.8mm, elliptic, briefly narrowed to the base, otherwise as the radical leaves. *Flowers* loosely panicled, inflorescence branches up to c.50mm long, 1–4-flowered. *Pedicels* up to 1–3mm long, filiform. *Bracts* adnate nearly to apex of calyx, c.10–15 × 1.2–2.5mm, oblong, entire or rarely with a tooth, coarse eglandular hairs up to 1mm long on margins and tip of inner face or these hairs wanting, sparsely glandular-puberulous outside. *Calyx* c.5–10mm long, lips c.1.5mm long, posticous lobes c.1 × 1mm, margins shortly pubescent with eglandular hairs, glandular-puberulous elsewhere but especially on the ribs. *Corolla tube* c.23–30mm long (throat c.3–3.5mm in diam.), curved just below the apex, minutely and sparsely glandular on upper part of tube and outside of lobes, limb zygomorphic, anticous lip c.7.5–8 × 1.5–2mm, narrowly elliptic, posticous lip c.10 × 10mm, 4-lobed, lobes roughly 7 × 2.2–2.8mm, narrowly spathulate, the two posticous lobes united for about ⅓ their length, cream, bearded at base with clavate hairs c.0.1mm long running down back of tube to point of insertion of stamens. *Stamens* 2, filaments c.3–4mm long, glabrous or with a few minute glandular hairs, anthers c.2.8–3.5mm long, well exserted; staminodes 2, c.0.3–1mm long, exserted. *Stigma* c.6–12mm long, well exserted. *Style* c.16–18mm long. *Ovary* c.5 × 2mm. *Capsules* c.10 × 3.5mm. *Seeds* c.0.8 × 0.5mm.

Citation:

Cape. Calvinia div., 3120CA, farm Leeudrift, 1121m, 20 vii 1990, *Batten* 1016 (E, K, NBG, S).

This striking plant is known from only two collections, both made between Calvinia and Middelpos: Cloete & Haselau first collected it in September 1986 and Mrs Batten, at my request, kindly sought it again in 1990. She found it in karroid vegetation on very dry stony flats, localized but then growing in profusion. In late August, the plants were in flower but not in fruit, so Mrs Batten potted up a few, hand-pollinated them, and made the interesting observation that in the afternoon (2pm) the two lateral lobes of the functionally lower lip (corolla resupinate) are folded back ('often touching, in hands-behind-its-back position'), but by 8 pm are spread out laterally, and the flowers then emit a heavy scent of carnations; the scent and narrow corolla lobes, made narrower by revolute margins, suggest pollination by moths (see plate 2C).

    Hand-pollination resulted in seed-set (there are no capsules on the type collection). Unfortunately, seed of *Reyemia* has proved difficult to germinate (Dr Steiner sent seed of *R. nemesioides*) and information on chromosomes is therefore lacking. It is noteworthy that seed of the apparently closely allied genus *Zaluzianskya* germinates readily.

    The corolla tube of *R. chasmanthiflora* is much longer and more slender than that of *R. nemesioides* and the plant is often more glandular; in other respects, the two species are remarkably similar, even in details of the indumentum on the stem. The trivial name *chasman-*

thiflora was suggested by the superficial resemblance that the corolla bears to a flower of the iridaceous genus *Chasmanthe*.

2. Reyemia nemesioides (Diels) Hilliard in Edinb. J. Bot. 49: 297 (1992).
Type: Cape, Calvinia div., Hantam Mts, *Meyer* s.n. (holo. B†).

Syn.: *Zaluzianskya nemesioides* Diels in Bot. Jahrb. 23: 482 (1897); Hiern in Fl. Cap. 4(2): 353 (1904).

Annual herb, primary stem 20–300mm long, erect, soon branching from the base and above, branches erect or decumbent, glandular-puberulous on upper parts (inflorescence axis), hairs up to 0.1mm long, pubescent below, hairs up to c.0.25mm long, eglandular, retrorse, distantly leafy, often with axillary leaf tufts (incipient branches). *Leaves*: radical with blade c.5–18 × 2–8mm, elliptic tapering into a broad flat petiolar part c.3–15mm long (nearly equalling to up to half as long as blade), margins entire to sparsely toothed, they and the upper surface with scattered eglandular hairs up to 0.4–0.6mm long, lower surface glabrous; cauline leaves c.4–15 × 0.8–3.2mm, elliptic briefly narrowed at the base, otherwise as the radical leaves. *Flowers* loosely panicled, inflorescence branches up to 50mm long. *Pedicels* up to 1.5–3mm long, filiform. *Bracts* adnate nearly to apex of calyx, c.5–12 × 0.8–2.2mm, oblong-lanceolate, entire, coarse eglandular hairs up to 0.4–0.6mm long on margins and tip of inner face, a few tiny almost sessile glands as well, backs glabrous. *Calyx* c.5–6.25mm long, lips c.1mm long, margins very shortly pubescent, otherwise glabrous except for minute ± sessile glands. *Corolla* tube c.7–8.5mm long (throat c.2–3mm in diam.), minutely glandular on upper part and on outside of lobes, limb zygomorphic, anticous lip c.3.5–6 × 3.6–5mm, broadly elliptic, white, posticous lip c.6–10 × 8–10mm, 4-lobed, lobes roughly 3.5 × 3mm, oblong-elliptic, the two posticous ones united for about half their length, all yellow, palate orange, bearded at the base with clavate hairs c.1mm long, running down back of tube to point of insertion of stamens. *Stamens* 2, filaments c.2–5mm long, bearded at base with clavate hairs c.0.3mm long, anthers 2–2.75mm long, well exserted; staminodes 2, c.0.5–0.7mm long, exserted. *Stigma* c.4–5.5mm long, exserted. *Style* 3.5–6mm long. *Ovary* 3–4 × 1.5–2mm. *Capsules* 7–8 × 3.5–4mm. *Seeds* c.0.8–1 × 0.6–0.8mm.

Citations:

Cape. Calvinia div., 3119BD, foot of Hantam Mts. between Hantams Peak and Akkerdam, 29 vii 1948, *Lewis* 3191 (SAM). c.3120C, between Middlepos-Calvinia at Elandsfontein, 4000ft, 15 ix 1926, *Levyns* 1697 (BOL). Sutherland div., at the Great Reed river, by Piet Mulders, 10 viii 1811, *Burchell* 1374 (K); between Sutherland and Middlepos, 9 x 1928, *Hutchinson* 713 (BM, BOL, K, PRE). 3220AD, Uitkyk, Sneeuwkrans, 1600m, x 1920, *Marloth* 9868 (PRE). 3220BC, 6 miles N of Sutherland, c.4800ft, 23 viii 1953, *Acocks* 16923 (K, PRE); farm Rooikloof, 1450m, 27 ix 1984, *Steiner* 800 (NBG). 3220DB, Theronsrust, c.5300ft, 18 x 1954, *Acocks* 17802 (PRE). Fraserburg div., 3221AB, 38 miles SW of Fraserburg, c.4300ft, 22 viii 1953, *Acocks* 16906 (PRE).

Although the type of the name *Zaluzianskya nemesioides* was lost in the Berlin fire, Diels's detailed description leaves no doubt that the name is being used correctly. The species was unknown to Hiern (the description in Flora Capensis is a translation from Diels), who suggested it should be compared with *Lyperia diandra* E. Mey. In this revision, the identity of *L. diandra* is discussed under *L. violacea*.

Reyemia nemesioides seems to be confined to a small area of the west central Cape, from Calvinia to Sutherland and Fraserburg, above c.1350m in altitude. Acocks described it as growing in Marginal Arid Karroo and Western Mountain Karroo of rugged dolerite hills. It flowers between July and October. Steiner (in litt.) noted that the flowers open around noon and close again in late afternoon; this is in contrast to the flowers of *R. chasmanthiflora*, which open at night. The flowers of the latter species are creamy-coloured; in *R. nemesioides* only the physically upper (anticous) lobe is white, while the four lower lobes are shades of yellow. Steiner (in litt.) noticed that the upper lip varied from white to pale yellow in a single population. This is possibly a phenomenon related to pollination. The two lateral lobes of the lower lip are a paler yellow than the two median lobes while the palate is orange.

15. GLUMICALYX

Glumicalyx is a small well-marked genus of 6 closely allied but readily distinguishable species confined to the Eastern Mountain Region of southern Africa (the Drakensberg Centre of Weimarck, 1941, and Nordenstam, 1969). The core of this region is Lesotho which is clearly the centre of diversity for the genus: one species is endemic, 3 more are widely distributed there, while a fifth, *G. apiculatus*, will surely be found at least in southernmost Lesotho when that area is thoroughly explored botanically. Only *G. goseloides* may truly be confined to the eastern side of the Drakensberg outwith the borders of Lesotho. The species all occupy similar habitats and indeed more than one species may be found in the same locality and flowering at the same time. The nodding congested heads of similar-looking flowers suggest a common suite of pollinators, but hybrids have been found only once, between *G. nutans* and *G. goseloides*.

The species are arranged in a sequence displaying progressive decrease in the length of the corolla tube, but, perhaps more importantly, the first four species in the enumeration have roughly 35–80 ovules in each loculus, while there are 20–25 (–35) in *G. montanus* and 7–12 in *G. apiculatus*; these last two species also show reduction in the calyx (see fig. 66).

The phylogenetic significance of reduction in the number of ovules is discussed on p.67.

Glumicalyx Hiern in Hook. Ic. Pl. 28, tab. 2769 (Nov. 1903) & in Fl. Cap. 4, 2: 369 (July 1904); Marloth, Fl. S. Afr. 3, 1: 129 (1932); Junell in Svensk Bot. Tidskr. 55: 172 (1961), Hilliard & Burtt in Notes RBG Edinb. 35: 155 (1977). **Figs 18C, 33B, C, D, 66, Plate 2E.**

Perennial herbs or dwarf shrublets. *Stems* leafy throughout, pubescent, hairs retrorse, minutely glandular as well. *Leaves* opposite below, alternate above, simple, usually toothed, glandular-punctate, sometimes glandular and/or eglandular hairs as well. *Inflorescence* a terminal cone-shaped, very congested, head, nodding in flower, erect and elongated in fruit. *Bracts* broad, sharply differentiated from leaves, adnate to pedicel, free from calyx. *Calyx* unequally 5-lobed, rarely 2 anticous lobes obsolete, split nearly or quite to base at least on anticous side, thin-textured. *Corolla* glabrous, thick-textured, leathery when dry, tube funnel-shaped, limb slightly bilabiate, lobes spreading, oblong to suborbicular, entire, creamy-yellow to orange or orange-red on upper surface with pale thin-textured margins. *Stamens* 4, didynamous, inserted near top of corolla tube, either 2 or 4 exserted, posticous filaments decurrent almost or quite to base of tube, anthers synthecous. *Stigma* lingulate, exserted. *Ovary* ± oblong in outline, nectariferous gland at base on posticous side, ovules 7 to many in each loculus. *Fruit* a septicidal capsule with a short loculicidal split at tip of each valve, glabrous. *Seeds* seated on a funicle taking the form of a raised, centrally depressed cushion (fig. 18C), reddish-brown, angled by pressure, sinuously colliculate, under the SEM cells of testa seen to be polygonal, outer periclinal walls convex or not, verruculose (fig. 33B, C, D).

Type species: *G. montanus* Hiern
Distribution: Eastern Mountain Region of southern Africa.

Key to species

Note: hybrids not catered for

1a. Stamens all fully exserted at anthesis 2
1b. Two stamens included or just visible in mouth at anthesis 5

2a. Leaves shortly pubescent .... **3. G. flanaganii**
2b. Leaves glabrous, or with some hairs on lower margins and over main vein 3

3a. Calyx of 5 equal spathulate lobes, free to the base or rarely briefly connate **5. G. montanus**
3b. Calyx of 3–5 lobes clearly fused below 4

4a. Corolla tube up to 4mm long . . . . **6. G. apiculatus**
4b. Corolla tube at least 7mm long . . . . **4. G. lesuticus**

5a. Corolla tube up to 16mm long, lobes oblong . . . . **2. G. nutans**
5b. Corolla tube at least 20mm long, lobes suborbicular . . . . **1. G. goseloides**

1. Glumicalyx goseloides (Diels) Hilliard & Burtt in Notes RBG Edinb. 35: 172 (1977) &
Botany of the Southern Natal Drakensberg 200, plate 26E (1987). **Fig. 66F, Plate 2E.**
Type: Natal, Estcourt distr., Injasuti valley, 1800–2100m, *Thode* 70 (B†).

Syn.: *Zaluzianskya goseloides* Diels in Bot. Jahrb. 23: 480 (1879); Hiern in Fl. Cap. 4, 2: 351 (1904).

Perennial herb, stems one or several from the crown, erect or decumbent then erect to c.450mm,
usually simple, occasionally forking once or twice above or below, sometimes with a few short,
slender (sterile?) branches near the base, pubescent with short gland-tipped and long acute hairs,
closely leafy. *Leaves* opposite below becoming alternate upwards, mostly 20–65 × 4–15mm,
decreasing in size upwards; radical leaves (often wanting at flowering) elliptic, obtuse,
narrowed to a petiole-like half-clasping base; stem leaves oblong to elliptic-oblong, scarcely
narrowed below, sessile, clasping, obtuse to subacute at apex, entire or obscurely toothed to
crenate or serrate on margins, thick-textured, upper surface pubescent with acute and gland-
tipped hairs, lower similar or hairs confined to main veins and margins. *Inflorescence* turbinate

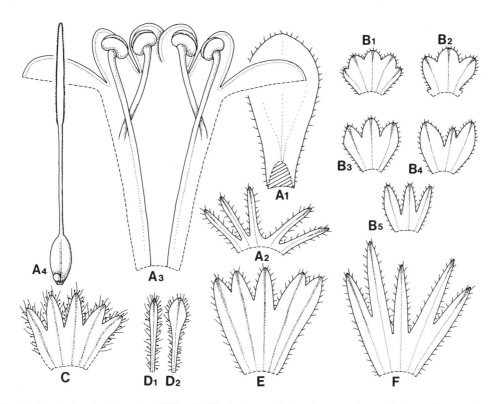

Fig. 66. *Glumicalyx*: A, *G. lesuticus* (*Hilliard* 5451), A1, bract, × 6; A2, calyx opened out, × 4; A3, corolla opened out
to show decurrent posticous filaments and lack of clavate hairs, ×6; A4, lingulate stigma with marginal bands of stigmatic
papillae, style, ovary with small adnate nectariferous gland, × 6. B1–5, *G. apiculatus* (1–3, *Drège* (type); 4, *Galpin* 6812;
5, *Hilliard & Burtt* 16397) showing variation in form of calyx. C, *G. flanaganii* (*Hilliard* 5418), calyx. D, *G. montanus*
(1, *Hilliard & Burtt* 6707; 2, *Marais* 1322) showing variation in calyx segments. E, *G. nutans* (*Hilliard* 5450), calyx.
F, *G. goseloides* (*Thode* STE 6342), calyx. B–F × 4.

initially, c.30–100mm long, rapidly becoming oblong, nearly always solitary, rarely with subsidiary heads, elongating in fruit. *Bracts* 12–19 × 7–17mm, ovate, acute and apiculate or obtuse, entire, pubescent outside with acute and gland-tipped hairs, minutely glandular-puberulous inside. *Calyx* 5–7(–8)mm long, thin, membranous, obscurely bilabiate, lobes free nearly to the base or fused up to ⅓–½ their length, free to the base anticously (fig. 66F), delicate hairs on bracts and margins, many gland-tipped. *Corolla* tube 20–29mm long, c.1.25mm broad, slightly dilated under the limb, creamy; limb nearly regular, 7–9mm diam., lobes suborbicular, orange to orange-red inside, creamy outside. *Stamens* with anthers markedly unequal, posticous pair with long vertical anthers, included, anticous pair with short horizontal anthers, exserted. *Capsule* c.7–8 × 3–4mm, seeds many, c.1 × 0.7mm.

Selected citations:

Orange Free State. Witzieshoek, Elands River valley, *Flanagan* 2034 (BOL, NH); road to Sentinel, *Hilliard & Burtt* 8637 (E, NU).

Natal. Bergville distr., Royal Natal National Park, *Trauseld* 254 (PRE); Mnweni area, *Esterhuysen* 21647 (BOL); Ndedema river, *Hilliard & Burtt* 6915 (E, NU); Estcourt distr., Giant's Castle, *Symons* 88 (PRE). Impendhle distr., Loteni river, *Wright* 1344 (NU). Underberg distr., Drakensberg Garden, *Hilliard & Burtt* 7741 (E, NU); Sani Pass, *Hilliard & Burtt* 8757 (E, NU). Alfred distr., Mt Ngeli, *Hilliard & Burtt* 5813 (E, NU); ibidem, *Tyson* 1323 (BOL).

Transkei. Mt Ayliff distr., Mt Insizwa, *Hilliard & Burtt* 7300 (E, NU); ibidem, *Strey* 10819 (E, NH, NU, PRE); ibidem, 27 i 1895, *Krook* 3044 (W).

Glumicalyx goseloides ranges down the face of the Drakensberg from The Sentinel and nearby Elands River valley above Witzieshoek south to the border of Natal and Transkei; it has also been recorded on the outlying massifs of Ngeli on the Natal-Transkei border and Mount Insizwa in Transkei. The plants grow in the boulder beds of mountain streams and in other bare gravelly areas between c.1600 and 2800m above sea level, flowering between October and July but mainly in December and January.

It hybridizes with *G. nutans*: see under that species.

2. Glumicalyx nutans (Rolfe) Hilliard & Burtt in Notes RBG Edinb. 37: 315 (1979). **Fig. 66E**. Lectotype (chosen here): Cape, Aliwal North distr., Wittebergen, 7000–7500ft, *Drège* (K).

Syn.: *Selago nutans* Rolfe in Journ. Linn. Soc. 20: 354, 358 (1883) & in Fl. Cap. 5, 1: 159 (1912).
   Zaluzianskya alpestris Diels in Bot. Jahrb. 23: 480 (1897); Hiern in Fl. Cap. 4(2): 350 (1904). Type: Natal, Mont aux Sources, 2700–3000m, Feb. 1893, *Thode* 71 (B†; prob. iso. *Thode* s.n., STE 8300).
   [*Selago cephalophora* auct., non Thunb.; E. Meyer, Comment. Pl. Afr. Austr. 256 (1837)].
   Glumicalyx alpestris (Diels) Hilliard & Burtt in Notes RBG Edinb. 35: 169 (1977).

Perennial herb, woody at the base and with a woody taproot; stems solitary in young plants, many from the crown in older, decumbent then erect, 100–450mm long, simple, pubescent with long acute hairs and sessile glands, leafy. *Radical leaves* 12–30 × 4-11mm, oblanceolate, obtuse, base petiole-like, crenate-serrate in upper part, thick textured, hairy mainly on margins, particularly in lower part, sometimes over the main vein as well, rarely all over; *cauline leaves* sometimes opposite below, bases connate, alternate upwards, 11–40 × 1.5–4mm, oblong or elliptic-oblong tapering below, sessile, otherwise as radical leaves. *Inflorescences* turbinate initially, c.20–30mm long, solitary or, in luxuriant specimens, with smaller subsidiary heads in the upper leaf axils, elongating in fruit. *Bracts* 6–12 × 3–6mm, broadly elliptic to ovate, obtuse to subacute, membranous, margins entire, they and the backs ciliate with long acute hairs and short-stalked glands, sparsely pubescent inside. *Calyx* 4–7mm long, membranous, obscurely bilabiate, lobes ⅓ or less the length of the tube, free to the base anticously, or sometimes shortly connate at base (fig. 66E), long cilia on margins and backs, short-stalked glands as well, minutely glandular-puberulous within. *Corolla* tube 12–16mm long, 1mm broad, widening slightly under the limb, creamy; limb bilabiate, 7–8mm across, lobes narrow, oblong, eventually strongly reflexed, orange to deep brick-red inside, buff outside or sometimes dull violet. *Stamens* with anthers subequal, posticous pair included, anticous pair exserted. *Capsule* c.5–6.5 × 2.5mm; seeds many, c.1 × 0.7mm.

Selected citations:

Orange Free State. Harrismith distr., Harrismith, *Sankey* 212 (BOL, K); Platberg, *Hilliard & Burtt* 8703 (E, NU); Witzieshoek, road to Sentinel, *Hilliard & Burtt* 8614 (E, NU).

Natal. Bergville distr., Mponjwane Mt, *Thode* s.n. (STE 8299); Cathedral Peak area, upper Tsanatalana valley, *Schelpe* 7230 (BOL); Estcourt distr., Giant's Castle Game Reserve, Langalibalela Pass, *Trauseld* 496 (K, NU, PRE); Kamberg, Gladstone's Nose, *Wright* s.n. (NU). Underberg distr., Garden Castle Forest Reserve, Mlambonja Valley, 4 i 1982, *Hilliard & Burtt* 14880(E).

Cape. Barkly East distr., Doodman's Krans Mt, *Galpin* 6799 (BOL, K, NH, PRE); Maclear distr., top of Naude's Nek Pass, *Acocks* 12335 (PRE). Lady Grey distr., Witteberg, Joubert's Pass, 18 i 1979, *Hilliard & Burtt* 12180 (E). Elliot distr., Fetcani Pass, 22 i 1979, *Hilliard & Burtt* 12315 (E).

Lesotho. Butha Buthe distr., Oxbow area, *Roberts* 3532, 3621 (PRE). Caledon River Pass, *Thode* s.n. (STE 6341); ibidem, *Thode* 43 (BOL, K). Pone Valley, Mothae Mts, *Coetzee* 823 (PRE). Berea, Mamalapi, *Marais* 1293 (K, PRE). Maseru distr., Blue Mt. Pass, c.8500ft, 10 i 1979, *Hilliard & Burtt* 12025 (E). Makhaleng Valley, *Ruch* 1602 (PRE). Sani Top, *Hilliard & Burtt* 8814, 8815, 8797 (E, NU). Sehlabathebe, *Guillarmod, Getliffe & Mzamane* 242 (PRE).

Glumicalyx nutans is widely distributed in the Eastern Mountain Region, from Platberg above Harrismith in the north eastern Orange Free State, through Lesotho and the Drakensberg in Natal to the Witteberg and Cape Drakensberg, between c.1800 and 3350m above sea level.

The plants grow on grassy mountain slopes, often at the foot of cliffs and among rocks, in boulder beds along streams, and in other gritty places, including roadsides, flowering between December and March. Hilliard & Burtt (1977) reported in some detail hybridization between *G. nutans* and *G. goseloides* along the then newly-cut road from Witzieshoek Rest Camp to the foot of The Sentinel. The hybrids appeared to be backcrossing to both parents (*Hilliard & Burtt* 8617, 8669).

3. Glumicalyx flanaganii (Hiern) Hilliard & Burtt in Notes RBG Edinb. 35: 168 (1977). **Fig. 66C**.

Type: Orange River Colony (Orange Free State), summit of Mont aux Sources, 9500ft, *Flanagan* 2036 (holo. K; iso. BOL, NH, PRE).

Syn.: *Zaluzianskya flanaganii* Hiern in Fl. Cap. 4, 2: 351 (1904).

Tufted perennial herb or subshrub, stems erect or decumbent then erect, 150–600mm high, mostly simple but with dwarf axillary shoots that sometimes elongate, these usually sterile, pubescent with long acute hairs and short-stalked glands, closely leafy. *Leaves* opposite, subopposite or alternate above, mostly 15–37 × 5–16mm, decreasing in size upwards, roughly half the length petiolar, blade narrowly to broadly ovate or oval, obtuse to subacute, coarsely crenate-serrate to almost lobulate on margins, at base narrowed into the flat petiolar part, expanded and half-clasping, thick-textured, both surfaces pubescent with long acute hairs and short-stalked glands or, more usually, upper surface thinly pubescent, hairs confined to margins and main veins below. *Inflorescences* turbinate initially, c.25–30mm long, mostly solitary, sometimes subtended by smaller spikes on short leafy axillary shoots, spikes elongating in fruit. *Bracts* 8–15 × 2.5–5mm, oblong, abruptly acute, entire, backs and margins pubescent with long acute hairs and short-stalked glands, less hairy within. *Calyx* 4.5–6.5mm, membranous, obscurely bilabiate, lobes c.⅓–½ the length of the tube, free nearly to the base anticously (fig. 66C) with long delicate acute hairs and short-stalked glands on backs and margins, glandular-puberulous within. *Corolla* tube 13–17.5mm long, c.2mm broad, dilated under the limb, creamy; limb slightly irregular, 6–8mm diam., lobes oblong to suborbicular, bright orange inside, buff to brownish outside. *Stamens* all exserted, anthers equal. *Capsule* c.6 × 3mm, seeds many.

Selected citations:

Natal. Bergville distr., Mont aux Sources, gorge in Beacon Buttress, *Galpin* 10363 (BOL, PRE). Estcourt distr., Giant's Castle, *Symons* 265A (PRE). Mpendhle distr., Highmoor Forest Reserve, headwaters of Elandshoek river, c.8100ft, 5 i 1983, *Hilliard & Burtt* 16193 (E). Underberg distr., upper tributaries S of Mkomazi river (feeders of Ka-Ntubu), c.8000ft,

2 xii 1982, *Hilliard & Burtt* 15773 (E). Bushman's Nek, Thamathu Pass, *Hilliard & Burtt* 8909 (E, NU); Sani Pass, *Hilliard* 5329, 5418 (E, NU).
Cape. Barkly East distr., Ben Mcdhui, *Galpin* 6792 (BOL, K, NH, PRE).
Lesotho. Oxbow, Fanana valley, *Williamson* 369 (K). Mamalapi, *Jacot Guillarmod* 21395, 617 (PRE). Berea distr., top of Nsututse Pass, *Marais* 1323 (PRE). Sani Top, *Hilliard & Burtt* 8792 (E, NU). Black Mts, *Hilliard & Burtt* 8776 (E, NU). Maseru distr., Blue Mountain Pass, c.8600ft, 10 i 1979, *Hilliard & Burtt* 12027 (E).

Glumicalyx flanaganii appears to be confined to the high mountains of Lesotho and the High Drakensberg in Natal, from c.1900–3350m above sea level, just entering the Cape on the southern border of Lesotho. The plants grow in damp rocky places, often along the foot of cliffs, in rock tumbles, and in stream gullies, flowering mainly between November and March.

4. Glumicalyx lesuticus Hilliard & Burtt in Notes RBG Edinb. 35: 166 (1977). **Fig. 66A.**
Type: Lesotho, Sani Top, 2850m, 2 i 1974, *Hilliard* 5451 (holo. NU, iso. E).

Perennial herb developing a woody taproot and woody crown; stems 100–350mm long, woody below (up to 5mm diam.), solitary to many from the crown, erect or decumbent then erect, usually simple, rarely with short axillary shoots near the base, pubescent with long acute hairs and short-stalked glands, hairs sometimes nearly confined to 2 longitudinal bands, leafy. *Leaves* opposite becoming alternate upwards, radical leaves, when present, up to 25 × 4mm, blade elliptic, tapering to a broad flat petiolar part accounting for most of the length, apex subacute, blade crenate-serrate to almost lobulate on margins, entire on petiolar part, leathery, glabrous except for hairs on margins of petiolar part and sometimes over main veins; cauline leaves mostly 9–15 × 1.5–4mm diminishing slightly upwards, oblong or elliptic-oblong, subacute, leathery, coarsely toothed to lobulate in upper half, entire in lower, base with a distinct pulvinus, long acute and short gland-tipped hairs nearly confined to lower margins or both surfaces pubescent. *Inflorescences* globose initially, c.20 × 20mm, usually solitary, or rarely subtended by smaller heads on axillary shoots, elongating in fruit. *Bracts* 6–9.5 × 2.5–6mm, oblong to ovate, apex abruptly apiculate, pubescent with long acute and short gland-tipped hairs outside, glandular inside and with a few long hairs. *Calyx* 3–3.5mm long, lobes linear-oblong (fig. 66A2), connate only at the base, delicately membranous, ciliate with acute and gland-tipped hairs. *Corolla* tube 7–11.5mm long, c.1.5mm broad, slightly dilated under the limb, creamy, occasionally suffused dull violet; limb bilabiate, c.6–8mm across, lobes elliptic to suborbicular, orange above, creamy or yellowish below, occasionally suffused dull violet. *Stamens* all exserted, anthers equal. *Capsule* c.5 × 2mm, seeds many, c.0.8 × 0.4mm.

Citations:

Lesotho. Sani Top, *Hilliard* 5313, 5336, 5397 (E, NU), *Hilliard & Burtt* 8791, 8813, 8813A (E, NU); *Ruch* 2407 (PRE); Black Mts, *Hilliard & Burtt* 8764 (E, NU). Khalong-la-Mashulu, *Jacot Guillarmod* 5900 (PRE), mixed with *G. montanus*. Mamalapi, *Jacot Guillarmod* 616 (PRE). Meniaming Pass, *Coetzee* 493 (PRE). Little Bokong, *Jacot Guillarmod* 112 (PRE), Cathedral Peak area, Cleft Peak, *Edwards* 1159 (NU, mixed with *G. nutans*). Maseru distr., Blue Mountain Pass, c.9000ft, 13 i 1979, *Hilliard & Burtt* 12111 (E); ibidem, c.8500ft, 10 i 1979, *Hilliard & Burtt* 12025A (E).

Glumicalyx lesuticus appears to be endemic to Lesotho, where it is widely distributed between c.2550 and 3180m above sea level. Like all its congeners, it favours gritty, rocky places, flowering between December and February.

5. Glumicalyx montanus Hiern in Hook. Ic. Pl. 28, tab. 2769 (Nov. 1903) and in Fl. Cap. 4, 2: 369 (1904); Hilliard & Burtt in Notes RBG Edinb. 35: 164 (1977) and Botany of the southern Natal Drakensberg 200, plate 26 D (1987). **Fig. 66D.**
Type: Orange River Colony (Orange Free State), on the slopes of Mont aux Sources, 7000–8000ft, Jan. 1894, *Flanagan* 2018 (BOL).

Tufted perennial herb becoming woody at the base; stems mostly 150–300mm long, many from the crown, decumbent then erect, simple, or very rarely with short sterile (?) axillary shoots, pubescent with both acute and gland-tipped hairs, sometimes nearly confined to two longitudi-nal bands, leafy. *Leaves* opposite, becoming alternate upwards, 5–20 × 2–8mm, scarcely diminishing upwards, oblong-elliptic to obovate, tips obtuse to rounded, base narrowed with a distinct pulvinus, bases connate when leaves opposite, margins crenate-serrate, leathery, glabrous except for minute gland-tipped hairs on lower margins. *Inflorescences* globose initially, up to c.15 × 15mm, usually solitary, rarely subtended by smaller spikes on very short axillary shoots, elongating in fruit. *Bracts* 5(–6) × 4.5–10mm, suborbicular, tips rounded or abruptly apiculate, backs minutely glandular-puberulous, margins entire or laciniate, with long delicate hairs. *Calyx* 3.5–4.5mm long, segments narrowly to broadly spathulate (fig. 66D), free or connate at the very base, delicately membranous, ciliate with acute and gland-tipped hairs. *Corolla* tube 4–6mm long, c.1.5mm broad, creamy; limb bilabiate, c.4–5mm diam., lobes oblong to elliptic, creamy to pale yellow above, creamy below. *Stamens* all exserted, anthers equal. *Capsule* c.4.5 × 3mm, seeds many (c.20–25 ovules in each loculus, perhaps only half of which develop), c.1 × 0.7mm.

Selected citations:

Orange Free State. Bethlehem distr., Golden Gate National Park, Generaalskop, *Roberts* 3198 (PRE, mixed with *G. nutans*). Harrismith distr., road to Sentinel, *Hilliard & Burtt* 8638 (E, NU).
Natal. Bergville distr., Royal Natal National Park, W slope of The Sentinel, *Trauseld* 148 (PRE); MnWeni area, *Esterhuysen* 21664 (BOL); Cathedral Peak area, upper Tsanatalana valley, *Schelpe* 7231 (K, BOL); Ndedema valley near Schoongezicht Cave, 16 iv 1978, *Hilliard* 8136 (E). Estcourt distr., Giant's Castle Game Reserve, upper Injasuti, *Trauseld* 526 (NU, PRE). Underberg distr., Cobham Forest Reserve, Upper Polela Cave area, 7700ft, 13 ii 1979, *Hilliard & Burtt* 12469 (E); Lakes Cave area, 7900ft, 15 xii 1982, *Manning, Hilliard & Burtt* 16079 (E).
Cape. Maclear distr., top of Naude's Nek Pass, *Hilliard & Burtt* 6707 (E, K, MO, NU, PRE). Barkly East distr., base of Doodman's Krans, *Galpin* 6812 (BOL, K, PRE mixed with *G. apiculatus*; NH).
Lesotho. Butha Buthe distr., Oxbow Camp, Tsehlanyane Valley, *Jacot Guillarmod* 3682 (PRE); Berea distr., top of Nsututse Pass, *Marais* 1322 (K, PRE). Sehlabathebe, *Guillarmod, Getliffe & Mzamane* 110 (PRE). Sani Top, *Hilliard* 5392 (E, NU). Ridge 1½ miles NW of entrance to Bushman's River Pass, *Wright* 338 (NU). Maseru distr., Blue Mountain Pass, c.9000ft, 13 i 1979, *Hilliard & Burtt* 12096(E).

Glumicalyx montanus grows on the high mountains of Lesotho, extending southwards onto the Cape Drakensberg around Naude's Nek and eastwards down the face of the high Drakensberg into Natal, mostly between c.2350 and 3050m above sea level, rarely as low as c.1800m in the major stream valleys of the Natal Drakensberg.

The plants favour gritty or gravelly places on rock sheets or in the boulder beds of streams, flowering between December and March. Plants in the rock garden of the Royal Botanic Garden, Edinburgh, are still flowering well after 25 years in cultivation. *Glumicalyx goseloides* on the other hand thrives only in the milder climate of the west coast of Scotland, in the RBG's satellite garden at Logan.

6. Glumicalyx apiculatus (E. Mey.) Hilliard & Burtt in Notes RBG Edinb. 35: 164 (1977). **Fig. 66B**.
Type: Cape, summit of the Wittebergen, 7000-7500ft, *Drège* (E, K, PRE fragment).

Syn.: *Selago apiculata* E. Mey., Comm. 256 (1837); Walp., Rep. 4: 151 (1845); Choisy in DC., Prodr. 12: 16 (1848).
    Walafrida apiculata (E. Mey.) Rolfe in Fl. Cap. 5, 1; 121 (1901).

Tufted perennial herb, woody at the base, eventually developing a thick woody stock; stems 60–250mm long, many from the crown, erect, simple, pubescent with both acute and gland-tipped hairs, hairs sometimes nearly confined to 2 longitudinal bands, leafy. *Leaves* opposite becoming alternate upwards, c.5–15 × 3–6mm, diminishing slightly upwards, broadly or narrowly elliptic, subacute, base slightly narrowed with a distinct pulvinus, bases connate when leaves opposite, margins crenate-serrate in upper half, leathery, glabrous. *Inflorescences* globose initially, less than 10 × 10mm, solitary, elongating to 15–45mm in fruit. *Bracts* 3–5 × 2–4mm, broadly elliptic to ovate, abruptly acute, margins entire, ciliate with long acute hairs and short-stalked glands, inside glandular-puberulous and with a few long hairs, leathery. *Calyx*

2.5–3.5mm long, membranous, obscurely bilabiate, lobing variable, always with 3 posticous segments fused for c.⅔ their length, 2 anticous segments nearly equalling the posticous ones or minute, or wanting, free nearly or quite to the base anticously, margins ciliate (fig. 66B). *Corolla* tube 3–4mm long, c.1.5mm broad, creamy; limb bilabiate, c.4mm diam., lobes elliptic, orange-yellow to pale cream inside, pale outside. *Stamens* all exserted, anthers equal. *Capsule* 3 × 2.5mm, seeds few (c.7–12 ovules in each loculus), c.1 × 0.7mm.

Citations:

Cape. Lady Grey distr., 3027CB, Witteberg, Joubert's Pass, c.7700ft, 18 i 1979, *Hilliard & Burtt* 12199 (E). Barkly East distr., base of Doodman's Krans, c.2500m, *Galpin* 6812 (BOL, K, PRE, mixed with *G. montanus*); 3027DB, Ben McDhui, c.8400ft, 4 ii 1983, *Hilliard & Burtt* 16397 (E); Kraalberg, near Barkly Pass, c.2400m, *Rattray* in herb. *Galpin* 7313 (PRE). Elliot-Maclear distr. boundary, 3127BB, Bastervoetpad, c.7250ft, 15 ii 1983, *Hilliard & Burtt* 16666 (E).

Glumicalyx apiculatus is known from only a small part of the Witteberg and Cape Drakensberg, within an area encompassed by lines drawn from Lady Grey (at the foot of the Witteberg) east to Doodman's Krans (where the borders of Lesotho, Cape and Transkei meet) and south to the Bastervoetpad across the Cape Drakensberg north-east of Barkly Pass (Kraalberg has not been traced); these mountains are ill-explored botanically.

The plants grow in gravel on wet rock sheets, between c.2200 and 2560m above sea level, flowering between January and March. *Glumicalyx apiculatus* and *G. montanus* bear so strong a resemblance to one another that Galpin made a mixed collection; they differ principally in the lobing of the calyx and the number of ovules in the ovary.

16. STROBILOPSIS

Strobilopsis Hilliard & Burtt in Notes RBG Edinb. 35: 172 (1977) & in Notes RBG Edinb. 45: 206 (1989). **Figs 18B, 34A, 67.**

Annual or perennial herb. *Stems* leafy throughout, pubescent, hairs retrorse, minutely glandular as well. *Leaves* opposite below, alternate above, simple, entire or toothed, glandular-punctate, a few long hairs as well on margins. *Inflorescence* terminal, capitate, solitary or with few to several smaller heads on short leafy axillary shoots below the main one, globose or oblong-cylindric, erect, scarcely elongating in fruit. *Bracts* broad, sharply differentiated from leaves, adnate to minute pedicel, free from calyx. *Calyx* unequally 5-lobed, thin-textured. *Corolla* tube funnel-shaped, limb bilabiate, lobes spreading, oblong-elliptic, entire, orange patch at base of posticous lip and there with clavate hairs, tube either glabrous or very minutely glandular outside. *Stamens* either 4, didynamous, or 3 or 2 by reduction of the anticous pair, inserted near top of corolla tube, exserted, posticous filaments decurrent to base of tube, anthers synthecous. *Stigma* lingulate with marginal papillae, exserted. *Ovary* ± oblong-elliptic in outline, nectariferous gland at base on posticous side, ovules up to 6 in each loculus on upper part of placenta, few developing. *Fruit* a septicidal capsule with a short loculicidal split at tip of each valve, minutely glandular on sutures. *Seeds* each seated on a massive, pulvinate funicle (fig. 18B), 1–4 in each loculus, elliptic in outline, flat on inner face, convex on outer, sinuously wrinkled, pallid, eventually blackish-brown, under the SEM the cells of the testa seen to be polygonal, outer periclinal walls convex, finely granular (fig. 34A).

Type species: *Strobilopsis wrightii* Hilliard & Burtt
Distribution: Natal Drakensberg and Sehlabathebe in Lesotho. Monotypic.

Strobilopsis wrightii Hilliard & Burtt in Notes RBG Edinb. 35: 173 (1977) & Botany of the southern Natal Drakensberg 200, plate 26 F (1987). **Fig. 67, Plate 1K.**
Type: Natal, Mpendhle distr., Whiterocks, summit of Little Berg, c.2100m, growing in damp soil, 9 i 1967, *Wright* 388 (holo. E, iso. NU).

Annual or perennial herb, taproot becoming woody. *Stems* up to 450mm tall, either simple or sparingly branched below, sparingly branched above, pubescent with retrorse whitish hairs c.0.5–1mm long, minutely glandular as well. *Leaves* up to c.15–45 × 2–5mm, narrowly elliptic to linear-oblong, subacute, scarcely to distinctly narrowed to the base and there with a distinct pulvinus, decurrent in two narrow ridges, obscurely to distinctly serrate in upper part, thick-textured, glandular-punctate, a few coarse white hairs up to c.1mm long on lower margins and midrib below, a few tiny glandular hairs as well. *Flowers* ± sessile, crowded in a dense rounded or somewhat oblong terminal head, often with smaller heads on short simple or branched stems below the main one, heads c.15–20 × 10–13mm, elongating slightly in fruit. *Bracts* sharply differentiated from leaves, adnate to minute pedicel, free from calyx, c.5–7 × 4–6mm, broadly ovate to suborbicular, abruptly acute or subacute, margins ciliate with hairs up to 1mm long, sometimes minute glandular hairs as well, a few minute glands on lower backs, otherwise glabrous. *Calyx* tube c.1mm long, lobes c.1.4 × 0.7mm, the posticous one often longer and narrower, glandular-puberulous all over, margins ciliate with delicate hairs up to 0.7mm long. *Corolla* tube c.3.5–4 × 2mm in mouth, sparsely and very minutely glandular-puberulous outside, limb bilabiate, c.6–8mm across lateral lobes, posticous lobes c.2.5 × 2–3mm, anticous lobe c.3–4 × 2–2.8mm, all lobes oblong-elliptic, posticous lip creamy-colour to pale mauve with an orange blotch at base, this thickly clad in clavate hairs, a few hairs extending round mouth, anticous lip shades of mauve. *Stamens* either 4 or reduced to 2 or 3 by abortion of anticous ones, didynamous, arising about two thirds of the way up the tube, well exserted,

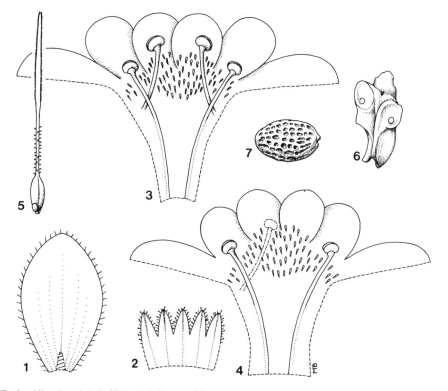

Fig. 67. *Strobilopsis wrightii (Hilliard & Burtt* 12553): 1, bract; 2, calyx opened out; 3, corolla opened out to show decurrent posticous filaments, exserted anthers, clavate hairs around mouth; 4, corolla opened out to show either one or both anticous stamens aborted; 5, lingulate stigma with marginal bands of papillae, style, ovary with small adnate nectariferous gland; 6, placenta showing two funicles with small circular attachment scars; number of ovules in each loculus reduced to two; 7, seed. 1–5 × 6, 6,7 × 15.

filaments c.2.8–3mm, posticous pair decurrent to base of tube, anthers c.0.8mm. *Stigma* nearly as long as style. *Style* minutely glandular-puberulous. *Ovary* c.1.4 × 0.8mm. *Capsule* c.2.5–3 × 2mm. *Seeds* c.1.5 × 1mm, very few.

Citations:

Lesotho. Sehlabathebe, 2300–2500m, i 1973, *Guillarmod, Getliffe & Mzamane* 31 (PRE); ibidem, xii 1976, *Schmitz* 6987 (ROMA); ibidem, Matša a Mafikeng, c.2450m, 2 xi 1976, *Hoener* 1688 (E).
Natal. Bergville distr., 2828DD, Mponjwane Mt, 7–8000ft, i 1891, *Thode* sub STE 7749 (STE). Mpendhle distr., Kamberg area, Storm Heights, c.7000ft, 16 xii 1978, *Hilliard & Burtt* 11776 (E); ibidem, 28 i 1978, *Hilliard & Burtt* 11260 (E); Mulangane ridge above Carter's Nek, 7000–7300ft, 2 xii 1983, *Hilliard & Burtt* 17031 (E, NU); ibidem, fruiting material of 17031, 3 ii 1984, *Hilliard & Burtt* 17540 (E). Underberg distr., Cobham Forest Reserve, Upper Polela Cave area, c.6800ft, 14 ii 1979, *Hilliard & Burtt* 12553 (E, NU); end of ridge above Pillar Cave, c.6700ft, 8 xi 1977, *Hilliard & Burtt* 10510 (E, NU).

Strobilopsis wrightii is known mainly from Cave Sandstone outcrops in the southern Natal Drakensberg above c.2000m, and from Sehlabathebe in south-eastern Lesotho where the Cave Sandstone rises to an altitude of over 2400m. There is one collection from the northern part of the Drakensberg, near Cathedral Peak, made by Justus Thode in 1891. The plants grow on grassy or scrubby sites and flower between November and February.

Strobilopsis resembles *Glumicalyx* in the indumentum of stems and leaves, in its broad bracts, and flowers in tightly congested heads, though these are always erect in *Strobilopsis*, nodding in *Glumicalyx* at anthesis, straightening up only in fruit. The genera differ in details of calyx, in the colouring, texture and indumentum of the corolla, and in their seeds. *Strobilopsis* may have two, one or no anticous stamens; they are either present or absent; there are no staminodes. The most northerly collection, that of Thode, has only two stamens; travelling thence southwards, collections from Storm Heights may have four or two stamens, Mulangane four, three or two, Garden Castle four, Upper Pholela four, Sehlabathebe four. There seems therefore to be some geographical pattern to reduction in the number of stamens.

17. TETRASELAGO

Tetraselago has an extraordinary ovary: it is bilocular with two ovules in each loculus developed from the centre of the axile placenta; one ovule is erect, one pendulous, each held by a massive funicle that partly embraces it (fig. 68). The leaves are more or less fascicled, the inflorescence is a corymb, and the corolla is mostly shades of blue or mauve, sometimes white. The genus is thus very easily recognized. The species were originally described under *Selago*, some species of which are superficially similar to *Tetraselago* (mixed collections are not unknown), but the distinctive ovary, which ripens into a tiny capsule, contrasts with the 1-seeded indehiscent cocci of *Selago*.

Tetraselago is confined to the highlands of the northern and eastern Transvaal, western Swaziland, and the Midlands of Natal (there between c.600 and 1400m above sea level, c.1200–2000m further north). Although the genus is readily circumscribed, the species are not; only an intensive field and laboratory study will sort out the complexities in the Transvaal. *Tetraselago natalensis* has relatively few-flowered, lax, partial inflorescences, the stems mostly lack long hairs on the lower part, the leaves are long and narrow. Its area lies south of that of the rest of the genus in the Transvaal and Swaziland. Specimens from as far north as the Zoutpansberg have, in herbaria, been assigned to *T. natalensis*, presumably because they have narrow leaves, but they differ in the greater number of flowers in the partial inflorescences and in their stems, mostly with long coarse hairs throughout. They belong in the complex centred on *T. nelsonii*, *T. wilmsii* and *T. longituba*. *Tetraselago nelsonii* is distinguished primarily by its very congested ultimate partial inflorescences (glomerules); it seems to be confined to a small area stretching from the Makapaansberg east to the Transvaal Drakensberg around Woodbush and the Wolkberg. A very broad species concept has of necessity been adopted here for *T. wilmsii* and *T. longituba*, distinguished from *T. nelsonii* by their lax ultimate partial inflorescences. The epithet *longituba* was given originally to a plant with corolla tube 6–8mm long and relatively short broad leaves, while the epithet *wilmsii* was given to a plant with corolla tube c.5mm long, and leaves similar in shape to those of *longituba*. Over a wide range of material, the distinction in length of corolla tube disappears, and leaves vary tremendously in both length and width, at one extreme resembling those of *T. natalensis*. It is just possible to maintain the epithets *longituba* and *wilmsii* if the presence or absence of long coarse hairs on the leaves is taken into account (present in *longituba*, absent in *wilmsii*), but this is a dubious distinction because hairiness varies from total coverage of both surfaces of the leaf to a very few hairs on the margins and on the lower surface of the leaf along the midrib. However, using this criterion, specimens assigned to *T. longituba* sens. lat. range from the Barberton Mountains (type locality) west to the environs of Belfast and Ermelo and north to Lydenburg, Sabie and Mount Sheba above Pilgrim's Rest. Specimens assigned to *T. wilmsii* sens. lat. occur along the Zoutpansberg, then from Blyde River south to Ohrigstad, thence westwards to the environs of Lydenburg (type locality) and Belfast, with two isolated records from the environs of Heidel-berg (Suikerbosrand). Further details are given under *T. longituba* and *T. wilmsii*. A field investigation could usefully begin in the 2530 degree square where specimens assigned to both *T. longituba* and *T. wilmsii* grow and where the greatest variation seems to occur, but much of the grassland has been lost to afforestation.

Tetraselago Junell in Svensk Bot. Tidskr. 55, 1: 190 (1961); Hilliard & Burtt in Notes RBG Edinb. 35: 175 (1977). **Figs 18D, 34B, 68, Plate 1J.**

Tufted woody perennial herbs. *Stems* pubescent, leafy. *Leaves* alternate fascicled, simple, reduced in size below inflorescence, glandular-punctate. *Inflorescence* corymbose, terminal, occasionally lateral corymbs lower down. *Bract* adnate to pedicel and calyx tube. *Calyx* slightly bilabiate, shortly toothed. *Corolla* tube funnel-shaped, limb slightly bilabiate, lobes 5, spread-ing, glabrous outside, often clavate hairs around mouth. *Stamens* 4, didynamous, exserted, anterior pair inserted in throat, posticous pair a little lower down with filaments decurrent to base of tube, anthers synthecous. *Stigma* lingulate, exserted. *Ovary* oblong, ovules 2 in each loculus from centre of axis, funicles pulvinate, hollowed, one turned up, one down (fig. 18D). *Fruit* a septicidal capsule with a short loculicidal split at tip of each valve. *Seeds* amber-coloured turning dull black, ornamented with c.8 irregularly sinuous longitudinal bands, under the SEM testa seen to be composed of polygonal cells with flat or somewhat convex verruculose outer periclinal walls (fig. 34B).

Type: *T. natalensis* (Rolfe) Junell
Distribution: Transvaal, Swaziland, Natal.

Key to species

1a. Leaves glabrous except occasionally a few minute hairs on undersurface of lower part of
  midrib 2
1b. Leaves either hairy all over or with at least a few coarse hairs on margins and undersurface
  of midrib (caution: hairs may be very few) . . . . **4. T. longituba** sens. lat.

2a. Ultimate partial inflorescence a small ± rounded glomerule, pedicels short (up to
  0.5–0.8mm long), at least the lowermost bracts relatively broad (0.7–1.6mm) **2. T. nelsonii**
2b. Ultimate partial inflorescence an extended raceme, pedicels often longer (0.5–1.8mm),
  bracts often narrower (c.0.4–0.8mm) but sometimes up to 1–1.2mm 3

3a. Partial inflorescences with up to 6–14 flowers (examine the racemes terminating the minor
  corymbs making up the whole corymb), bracts narrow (0.4–0.5mm), coarse hairs on

inflorescence axes and upper parts of stem rarely extending to base of stem and then confined to a band from below each leaf base .... **1. T. natalensis**
3b. Partial inflorescences terminating a lateral corymb with up to 10–45 flowers, bracts often (but not always) 0.6–1.2mm broad, coarse hairs usually present over whole stem

3. T. wilmsii sens. lat.

1. Tetraselago natalensis (Rolfe) Junell in Sv. Bot. Tidskr. 55, 1: 190 (1961); Hilliard & Burtt in Notes RBG Edinb. 35: 176 (1977). **Fig. 68, Plate 1J.**
Lectotype: Natal [Camperdown distr.], hillside near Botha's Railway Station, 2–3000ft, 29 iv 1892, *Medley Wood* 4863 (K; isolecto. E, PRE).

Syn.: *Selago natalensis* Rolfe in Fl. Cap. 5, 1: 151 (1901).

Stems c.350mm–1m tall, ± erect, mostly simple or branching from base, occasionally corymbosely branched above, minutely glandular-puberulous, coarse hairs up to 0.3–0.5mm long as well particularly on upper parts including inflorescence, only occasionally running back down stem in bands below each leaf-base. *Primary leaves* up to 15–38 × 1–3(–4)mm, linear-lanceolate, acute, gradually narrowed to a petiolar part, margins plane or slightly revolute, c.1–16 tiny teeth on upper part (axillary leaves smaller and sometimes entire), glandular-punctate otherwise glabrous except occasionally a few minute hairs on lower surface at base of midrib. *Flowers* up to c.6–14 in each partial inflorescence, loosely arranged. *Bracts* up to 2–3.2 × 0.4–0.5mm, glabrous except for a few hairs c.0.15–0.25mm long on margins. *Pedicels* up to c.0.5–1.2mm long. *Calyx* tube 1.4–2mm long, posticous lobes 0.3–0.5 × 0.3–0.4mm, hairs up to 0.15–0.2mm long on margins. *Corolla* tube 3–4.5 × 1.2–1.7mm in throat, limb 3–4mm across lateral lobes,

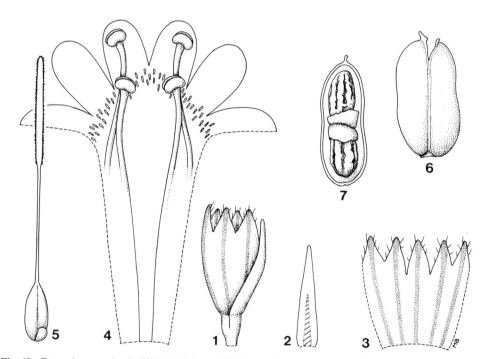

Fig. 68. *Tetraselago natalensis* (*Hilliard & Burtt* 10016): 1, calyx with bract attached; 2, bract showing degree of adnation to calyx; 3, calyx opened out; 4, corolla opened out to show decurrent posticous filaments, well exserted anthers, clavate hairs around mouth; 5, lingulate stigma with marginal bands of papillae, style, ovary with small adnate nectariferous gland; 6, capsule beginning to dehisce; 7, half capsule (septum removed) showing the two opposed seeds still attached to their massive funicles. All × 16.

posticous lobes 1–1.6 × 0.6–1mm, anticous lobe 1.3–1.8 × 0.7–1mm, shades of blue and mauve, occasionally a white sport, clavate hairs around mouth or sometimes confined to posticous lip, or wanting. *Stamens*: posticous filaments 1.5–2.5mm long, anthers 0.5mm, anticous filaments 1.7–2.5mm, anthers 0.5mm. *Stigma* 1.5–2.2mm long. *Style* 2–3mm. *Ovary* 0.8 × 0.3–0.4mm. *Capsules* c.2–2.5 × 1–1.2mm. *Seeds* c.0.8 × 0.3–0.4mm.

Selected citations:

Transvaal. Piet Retief distr., 2730BB, Iswepe, 6 iii 1949, *Sidey* 1604 (PRE).
Swaziland. Mankaiana distr., 2631CA, 10 miles W of Mankaiana, c.4500ft, 19 iii 1959, *Compton* 28676 (PRE). Hlatikulu distr., 2631CD, near Hlatikulu, c.3500ft, 2 iii 1962, *Compton* 31333 (PRE).
Natal. Ngotshe distr., 2731CA, Itala Nature Reserve, c.4500ft, 4 iv 1977, *Hilliard & Burtt* 10016 (E, PRE). Paulpietersburg distr., 2731AC, 26 miles E of Paulpietersburg, farm Mielieplaat, 3500ft, 22 vi 1948, *Codd* 4315 (PRE). Nkandla distr., 2831CA, Nkandla Forest, 3400ft, 26 iii 1956, *Edwards* 1325 (PRE). Pietermaritzburg distr., 2930DA, Table Mountain, 9 iii 1926, *McClean* 169 (PRE). Camperdown distr., 2930DC, Botha's Hill, iii 1985, *Edwards* 28 (E). Pinetown distr., 2930DD, Kloof, 16 i 1948, *Dohse* 34 (E). Richmond distr., 2930CB, 5 miles Thornville–Eston, Tala, 3000ft, 31 i 1966, *Moll & Morris* 613 (PRE).

Tetraselago natalensis ranges from southern Swaziland and the south-eastern corner of the Transvaal through Natal as far south as Eston, Botha's Hill and the Groeneberg at Inanda, between c.600 and 1400m above sea level. It favours open grassland, often on slopes and among rocks, flowering between January and June.

Characteristic features of the species are stems tending to be merely glandular-puberulous on their lower parts, principal primary leaves seldom more than 1–3mm broad and usually at least 20mm long, relatively few-toothed, glabrous, bracts narrow (up to 0.4–0.5mm broad), the partial inflorescences lax and relatively few-flowered (up to 14 flowers in the largest racemes, which are usually those terminating each minor corymb of the whole inflorescence).

2. Tetraselago nelsonii (Rolfe) Hilliard & Burtt in Notes RBG Edinb. 35: 176 (1977).
Lectotype: E Transvaal, Houtbosch [Woodbush], *Nelson* 439 (K, isolecto. PRE).

Syn.: *Selago nelsonii* Rolfe in Fl. Cap. 5, 1: 152 (1901).
    S. transvaalensis Rolfe in Fl. Cap. 5, 1: 153 (1901). Type: Transvaal, Houtbosch, *Rehmann* 6211 (holo. K, iso. BM).

Stems c.250–800mm long, erect or decumbent, often simple, sometimes corymbosely branched above, minutely glandular-puberulous, usually coarse spreading hairs as well, up to 0.5–1mm long, to base of stem (these completely wanting in specimens from Makapan's Gat). *Primary* leaves up to 9–25(–30) × (1–)1.5–3(–5)mm, linear-lanceolate to narrowly oblanceolate, ± acute, gradually narrowed to a petiolar part, entire or a few tiny teeth, glandular-punctate, otherwise glabrous except sometimes a few minute hairs on lower surface at base of midrib. *Flowers* up to c.14–25 in each partial inflorescence, crowded. *Bracts* up to 2.5–3.5 × 0.7–1.6mm, glabrous except for a few hairs c.0.2mm long on margins. *Pedicels* up to 0.5–0.8mm long. *Calyx* tube 1.5–1.8mm long, posticous lobes 0.5–0.8 × 0.3–0.5mm, hairs up to 0.15–0.2mm long on margins. *Corolla* tube 2.6–3.8 × 1.2–1.5mm in throat, limb 3–4mm across lateral lobes, posticous lobes 1–1.6 × 0.7–1mm, anticous lobe 1.2–1.7 × 0.8–1.1mm, shades of blue and mauve, occasionally a white sport, clavate hairs all round mouth. *Stamens:* posticous filaments 1.2–1.8mm long, anthers 0.4–0.6mm, anticous filaments 1.2–2.2mm long, anthers 0.4–0.6mm. *Stigma* 1.4–3mm long. *Style* 1–2mm. *Ovary* 0.7–0.8 × 0.3–0.4mm. *Capsules* c.1.5 × 1mm. *Seeds* c.0.8 × 0.4mm.

Selected citations:

Transvaal. Potgietersrus distr., 2429AA, Makapan's Gat, M'Swazis ridge, 30 iii 1970, *Maguire* 8053 (E, J). Pietersburg distr., 2329DD, Houtboschberg, 2180m, 9 ii 1894, *Schlechter* 4449 (E, J, PRE); 2330CC, New Agatha, vi 1916, *Rogers* 18895 (J, K, PRE); 2430AA, Wolkberg, N slopes Serala Peak, 1800m, 11 iii 1982, *Venter* 9375 (PRE).

The leaves of *Rehmann* 6211 (type of *Selago transvaalensis*) are glabrous, not hispidulous as claimed by Rolfe, and the specimen bears only buds, thus Rolfe's measurement of the corolla tube (1 lin.) is misleading.

Tetraselago nelsonii appears to be confined to a small area of the north-eastern Transvaal, with records from the Makapaansberg east of Pietersburg, Woodbush, the Wolkberg, Seralaberg and Marakeberg, the last four localities all part of the Transvaal Drakensberg. The plants grow in grassland, often on rocky sites and on forest margins, flowering mainly between January and June.

The distinguishing marks of the species are the narrow leaves, flowers in small glomerules c.3–6mm long (these aggregated into the terminal corymb), pedicels scarcely reaching 0.5mm (occasionally 0.8mm) in length, and the relatively broad bracts. The stems usually have long hairs throughout their length, but these hairs are entirely wanting in specimens from the environs of Makapan's Gat. The species is often confused in herbaria with *T. natalensis*, which also has narrow leaves, but the short pedicels, crowded glomerules and broader bracts will at once distinguish it.

An investigation is needed into the relationship of *T. nelsonii* with a plant occurring along the whole length of the Zoutpansberg. This plant differs in its laxer partial inflorescences, often longer pedicels, and often narrower bracts, all characters that point to *T. wilmsii*: see further under that species.

3. Tetraselago wilmsii (Rolfe) Hilliard & Burtt in Notes RBG Edinb. 35: 177 (1977).
Type: Transvaal, Lydenburg distr., Paardeplaats, *Wilms* 1163 (holo. K; iso. E, fragment PRE).

Syn.: *Selago wilmsii* Rolfe in Fl. Cap. 5, 1: 151 (1901).
     S. aggregata Rolfe in Fl. Cap. 5, 1: 152 (1901). Lectotype (chosen here): Transvaal, near Lydenburg, *Wilms* 1165 (K, isolecto. E).

Note: Data in square and in round brackets refer to specimens included in a broad species concept; further information given in discussion.

Stems c.300–650mm long, erect or decumbent, simple to branched particularly above, minutely glandular-puberulous, coarse hairs up to 0.4–1mm long as well, often over whole length of stem. *Primary leaves* up to 11–21 × 3–5.5mm, ratio of length to breadth roughly 3–6:1 [9–30 × 1.1–5mm, ratio roughly 4–10:1] (8–25 × 1–2.5mm, ratio roughly 4.5–25:1), elliptic to narrowly obovate [narrowly obovate to linear-lanceolate] (narrowly obovate to linear-lanceolate), acute, gradually narrowed to a petiolar part, margins entire to c.10-toothed, glandular-punctate otherwise glabrous except sometimes a few very minute hairs on back of midrib near base of leaf. *Flowers* up to 8–25 in each partial inflorescence [15–28] (20–45), loosely arranged. *Pedicels* up to 0.5–1.5mm long [1–1.8mm] (0.8–1.5mm). *Calyx* tube 1.7–2.3mm long [1.4–2mm] (1–1.7mm), posticous lobes 0.5–1 × 0.4–0.7mm [0.4–0.8 × 0.4–0.6mm] (0.3–0.8 × 0.3–0.4mm), nearly glabrous or hairs up to 0.2–0.4mm long on margins. *Corolla* tube 4.2–6.5 × 1.2–1.8mm in throat [2.5–5 × 1.2–1.6mm] (2–4 × 1–1.3mm), limb 3.6–5.5mm across lateral lobes [3–4.1mm] (2.5–4.3mm), posticous lobes 1.2–2 × 0.8–1.1mm [1–1.8 × 0.8–1mm] (0.7–1.5 × 0.5–1mm), anticous lobe 1.4–2.5 × 0.8–1.2mm [1.2–2.2 × 0.8–1.1mm] (1–2 × 0.5–1mm), shades of blue and mauve, rarely a white sport, clavate hairs around mouth, sometimes confined to posticous lip. *Stamens*: posticous filaments 1.5–2.2mm [1.1–2.2mm] (1.2–1.8mm), anthers 0.4–0.7mm [0.4–0.6mm] (0.3–0.4mm), anticous filaments 1.3–2.3mm [1.2–2.3mm] (1.4–2.2mm), anthers 0.4–0.8mm [0.4–0.6mm] (0.3–0.4mm). *Stigma* 1.5–5mm [1.5–3mm] (1.4–4mm). *Style* 2–3mm [0.8–2.2mm] (0.7–1.7mm). *Ovary* 0.7–1 × 0.3–0.4mm. *Capsules* c.1.7–2 × 1–1.5mm. *Seeds* c.0.6–1 × 0.4mm.

Selected citations of *T. wilmsii* sens. strict.:

Transvaal. Pilgrim's Rest distr., 2430DB, summit Blyde River gorge, 12 xii 1937, *Galpin* 14634 (PRE). 2430DC, Ohrigstad Nature Reserve, 20 i 1976, *Theron* 3458 (PRE). Belfast distr., 2530AC, Dullstroom, farm Klipbankspruit,

2060m, 15 ii 1988, *Burgoyne* 55 (PRE); farm Kleinzuikerboschkop, 1890m, 18 iii 1987, *de Villiers* 95 (PRE). Lydenburg distr., 2530BA, Long Tom Pass, 1700–2000m, *Werdermann & Oberdieck* 2093 (K, PRE); near Lydenburg, 8km on Sabie road, 5 iii 1981, *Hilliard & Burtt* 14190 (E).

T. wilmsii sens. lat., data in square brackets in formal description:
Transvaal. Pilgrim's Rest distr., 2430DB, Mariepskop, 5 xii 1957, *Merxmüller* 598 (PRE); 2430DD, Bourke's Luck mine, 24 ii 1937, *Galpin* 14295 (PRE); Belfast distr., 2530DC, Machadodorp, 3 iii 1917, *Pole -Evans* H16955 (PRE). Heidelberg distr., 2628AD, Heidelberg Kloof, 25 ii 1937, *Repton* 962 (PRE).

T. wilmsii sens. lat., data in round brackets in formal description:
Transvaal. Zoutpansberg, 2329AB, farm Lejuma, 1636m, 23 ii 1982, *Venter* 7553 (PRE); 2329BB, Muchindudi Falls, 1 vi 1947, *Bruce & Kies* 3 (J). 2230CD, Tate Vondo Forest Reserve, 1219m, 29 iv 1977, *Hemm* 115 (J, PRE). 2230CD, Lake Funduzi, 14 vii 1935, *Galpin* 14916 (PRE).

The original material of *T. wilmsii* (and its synonym *Selago aggregata*) came from the Lydenburg area; the corolla tube is c.5mm long, the largest leaves measure c.15–20 × 5–6mm and are glabrous. The distinction from *T. longituba* sens. *strict.* (below) is thus very slight, being no more than somewhat shorter corolla tube and glabrous leaves. Material seen from the following quarter degree squares has leaves similar to those of typical *T. wilmsii* but the corolla tube may be c.4–6.5mm long (as long as in some specimens of *T. longituba*): 2430DC, 2530AA, AB, AC, BA, BB, 2529BD, DB. In the formal description, data in square brackets refer to plants that sometimes have the leaves longer and narrower than in typical *T. wilmsii*, and corolla tube sometimes shorter. They have been seen from 2430DB, DD, 2530BB, BC, 2628AD, CA. Why the leaves should be narrower and the corolla tube shorter (and compare the distribution of narrow-leaved plants assigned to *T. longituba*) is not clear: the area of the narrow-leaved species *T. natalensis* and *T. nelsonii* lie to the south and north and nowhere are they sympatric with either *T. wilmsii* or *T. longituba* even in the broadest circumscription of those names.

Plants from the Zoutpansberg (data in rounded brackets in the formal description) will be found in herbaria under *T. natalensis* and *T. nelsonii*. They differ from *T. natalensis* in having the stems clad throughout in long coarse hairs and more flowers in the partial inflorescences, from *T. nelsonii* in having looser partial inflorescences, longer pedicels, and often narrower bracts. However, they cannot be distinguished from some specimens assigned to *T. wilmsii* sens. lat., and the variation is continuous.

The favoured habitat and flowering season of these plants are similar to those of other species of *Tetraselago*. The altitudinal range is c.1200 to 2000m.

4. Tetraselago longituba (Rolfe) Hilliard & Burtt in Notes RBG Edinb. 35: 176 (1977).
Type: E Transvaal, Barberton, 2800–3000ft, July-August, *Galpin* 398 (holo. K, iso. PRE).

Syn.: *Selago longituba* Rolfe in Fl. Cap. 5, 1: 151 (1901).
   S. montis-shebae Brenan in Kew Bull. 40: 81 (1985). Type: Transvaal, 2430DC, Mount Sheba, 14 ii 1982, *Brenan* 14966 (holo. K; iso. E, PRE).

Note: Data in square and in round brackets refer to specimens included in a broad species concept; further information given in discussion.

Stems 200mm–1m long, erect or decumbent [200–600mm, possibly straggling] (250–600mm, erect or decumbent), simple to well-branched particularly above, minutely glandular-puberulous, coarse hairs up to 0.6–1mm long as well over whole length of stem. *Primary* leaves up to 10–28 × 3–11mm, ratio of length to breadth roughly 2–4:1 [8–16 × 3–9mm, ratio 1–3:1] (10–24 × 2–7.8mm, ratio 2–5:1), elliptic to broadly elliptic or narrowly obovate [broadly ovate to suborbicular] (mostly narrowly obovate), acute [subacute to obtuse], gradually [abruptly] narrowed to a petiolar part, margins sharply serrate, sometime entire in secondary leaves, glandular-punctate, few to many coarse hairs on margins and midrib on lower surface of leaf, rarely hairy all over both surfaces [hairy on both surfaces] (hairs only on margins and midrib, sometimes very sparse). *Flowers* up to 5–20 in each partial inflorescence [12–23] (11–25), loosely arranged. *Bracts* up to 3–4.3 × 1–1.7mm [3–4 × 1–1.2mm] (3–4 × 1–1.3mm), almost glabrous or with hairs up to 0.2–0.4mm long nearly always confined to margins [0.3–0.8mm, backs and margins] (0.2–0.4mm, margins). *Pedicels* up to 0.5–1.5mm long [0.5–1.2mm] (0.8–2mm). *Calyx* tube 1.6–2.3mm long, posticous lobes 0.5–1 × 0.4–0.7mm, hairs up to 0.2–0.4mm long on margins. *Corolla* tube 6–9 × 1.2–1.5mm in throat [3.7–4.2 × 1.4–1.6mm] (4–6.8 × 1.2–1.8mm), limb 4–6mm across lateral lobes [3.6–5mm] (3.5–5mm), posticous lobes 1.5–2.2 × 0.9–1.5mm [1.2–2.2 × 0.8–1.3mm] (1.2–2 × 0.8–1.4mm), anticous lobe 1.7–2.7 × 0.8–1.4mm [1.4–2.5 × 0.9–1.25mm] (1.5–2 × 0.8–1.5mm), shades of blue and mauve [white] (shades of blue and mauve, sometimes almost white), clavate hairs around mouth, sometimes confined to posticous lip, sometimes wanting. *Stamens*: posticous filaments 1–1.8mm long, anthers 0.5–0.8mm long, anticous filaments 1–2mm, anthers 0.5–0.8mm. *Stigma* 2.2–4.7mm [2–2.2mm] (2.3–4mm). *Style* 3.6–6.3mm [1.7–2.7mm] (1.5–3.5mm). *Ovary* 0.8–1 × 0.3–0.4mm. *Capsules* 2.5–2.7 × 1.5mm. *Seeds* c.1 × 0.5mm.

Selected citations of *T. longituba* sens. strict.:

Transvaal. Barberton distr., 2531CC, mountain behind Barberton village, ± 4500ft, 18 ii 1931, *Liebenberg* 2425 (PRE); about 21km south east of Barberton, Kangwane, 1700m, 13 i 1985, *Balkwill & Cadman* 2643 (E,J). 2531CB, between Louw's Creek and Maid of the Mist Mountain, 5 i 1929, *Hutchinson* 2419 (K, PRE).
Swaziland. Pigg's Peak distr., 2531CC, Emlembe Mountain, c.4500ft, 9 ii 1961, *Compton* 30526 (PRE); Havelock, 3800ft, iii 1960, *Miller* 7253 (PRE).

T. longituba sens. lat., data in square brackets in formal description:
Transvaal. Belfast distr., 2530AC, Kleinzuikerboschkop, 1890m, 7 i 1987, *de Villiers* 13 (PRE). 2530AC, Dullstroom, 2000–2100m, 30 i 1959, *Werdermann & Oberdieck* 2018 (PRE). 2530CA, Belfast, xii 1909, *Leendertz* 2909 (PRE). The only collections seen.

T. longituba sens. lat., data in round brackets in formal description:
Transvaal. Pilgrim's Rest distr., 2430DC, Mt. Sheba Nature Reserve, 1800m, i 1976, *Forrester & Gooyer* 29 (J, PRE), ibidem, *Brenan* 14966 (type of *Selago montis-shebae*, E, K, PRE). Lydenburg distr., 2530BA, Mt. Anderson, 16 i 1952, *Prosser* 1802 (J, PRE). Belfast distr., 2530CB, Elandshoogte near Machadodorp, c.5500ft, 3 iii 1981, *Hilliard & Burtt* 14179 (E). Carolina distr., 2630BB, 33km from Jessievale Forest Station, Sonstraal, 11 i 1984, *Welman* 425 (K, PRE). 2630BA, 9 miles from Warburton p.o. on Mbabane road, *Hilliard* 4782 (E, PRE). Ermelo distr., 2629DB, P.O. The Gem, 3 iv 1933, *Walker* 136 (PRE).
Swaziland. Mbabane distr, Ukutula, 13 ii 1955, *Compton* 24922 (PRE). 2631AA, 25km NW of Mbabane, Ngwenya Hills, Stag Peak, 5 iv 1966, *Maguire* 7454 (J); Malolotja Nature Reserve, 1524m, *Braun* 669 (PRE).

The original material of *T. longituba* came from the Barberton Mountains. In this area (quarter degree squares 2530 DA, 2529 CB, CC) the length of the corolla tube varies from 6–9mm, the leaves are broad in relation to their length (for example, 10 × 4mm, 25 × 10mm, 27 × 8mm), and they have anything from a very few coarse hairs on the lower margins and back of the midrib (as in the type collection) to many hairs, and in one collection there are hairs all over both surfaces of the leaf. On the high ground to the south of the Barberton Mountains (from about Forbes Reef south to Mbabane in Swaziland, 2631AA, AC), westwards to the environs of Ermelo and Carolina (2629DB, 2630AA, AB) and north to Lydenburg, Sabie and Pilgrim's Rest (2529DB, 2530AB, BA, BB, BD, CB, 2430DC), the length of the corolla tube is 4–6.8mm, the leaves are often rather narrower than in the typical plant (for example, 9 × 2mm, 29 × 5mm, but also 8 × 4mm, 21 × 7.8mm) and there are few to many coarse hairs on the margins and back of the midrib. In the formal description, data in rounded brackets refer to these plants. Measurements in square brackets refer to a plant that grows around Belfast and Dullstroom (2530AC, CA). It is very hairy (leaves hairy all over, bracts hairy on the backs as well as on the margins), the leaves are short and broad, the corolla tube short (up to 4.2mm), and the limb white, though in dried material the tube is clearly flushed with mauve. White flowers have not been recorded in *T. longituba* (though white sports occur in *T. natalensis* and *T. wilmsii*). It might well be worth taxonomic recognition when more is known about it.

All the plants grow in grassland, often favouring rocky sites, and flower between January and June.

PHYTOGEOGRAPHY

All but two of the 344 species of Manuleae are confined to Africa south of the Sahara. *Camptoloma canariense* is endemic to Grand Canary, and *Jamesbrittenia dissecta* is found in the Nile valley, Sudan and the Indian subcontinent. *Jamesbrittenia* (83 species) is the most widely distributed genus, its main area being Angola (four species), Zambia (two species) and Malawi (two species) southwards, with a rich representation in Namibia (35 species) and nearly as many (30) in the Transvaal, Natal, Lesotho, Transkei and Eastern Cape; the genus is poorly represented in the Cape Floral Region except along the southern Cape coast and adjacent mountains, but there are a good many species in Namaqualand. In contrast, *Sutera* (49 species), with an overall range from Namibia, Botswana and Zimbabwe southwards, is poorly represented in Namibia (two species) and richly developed in the Cape Floral Region (28 species), while there are 15 species in the Transvaal-Eastern Cape area. *Manulea* (74 species) has its main area in the western Cape (31 species in the Cape Floral Region, 15 in Namaqualand); there are 12 species in Namibia, nine in the Transvaal-Eastern Cape including the widespread *M. parviflora* that is the only species in Zimbabwe and Angola. *Zaluzianskya* (55 species) is nearly confined to the area south of the Orange and Limpopo rivers; there is one species on Elgon and Kilimanjaro, one in Zimbabwe, three in Namibia, but 14 in the Cape Floral Region and 21 in the Tranvaal-E Cape area. The genus is also well represented in Namaqualand and the karroid areas of Cape Province. One genus (*Trieenea*, nine species) is confined to the Cape Floral Region, *Polycarena* (17 species) is very nearly confined to the western Cape, *Reyemia* (two species) is known only from the Calvinia-Roggeveld area, *Lyperia* (six species) is concentrated in the western and south western Cape, with *L. tristis* ranging northwards into Namibia and eastwards into the Karroo and southern Cape fold mountains. *Phyllopodium* (26 species) is also concentrated in the western, south western and southern Cape, with three species in the Eastern Cape, two of them ranging into Natal; three species occur in Namibia. The monotypic genus *Manuleopsis* is confined to Namibia. *Tetraselago* (four species) and *Melanospermum* (six species) are centred on the Transvaal. There is one species of *Tetraselago* in Natal; one species of *Melanospermum* ranges to Zimbabwe, Botswana and Namibia while two occur in Natal and Swaziland. *Glekia* (monotypic) is confined to the elevated parts of the eastern Cape and Lesotho, while *Glumicalyx* (six species) and *Strobilopsis* (monotypic) are endemic to the Eastern Mountain Region. The three species of *Camptoloma* introduce a disjunct Afro-arid element to the phytogeography of Manuleae. A detailed analysis of species-distribution follows, summarized in Table 5.

The phytogeographical features of the Cape Floral Region are comparatively well known, but outside that Region, the patterns of distribution of species are mostly obscure due in no small part to lack of monographic work. Also, many areas are either ill-known or unknown botanically. Nordenstam (1969) gave an account of the phytogeography of the genus *Euryops* (Compositae), which has its main area in southern Africa, then a disjunct distribution northwards over the mountains to Ethiopia and Arabia. Within the Cape Floral Region, he recognized 10 phytogeographical groups, these being a modified version of Weimarck's (1941) classic analysis of the Cape flora. Outwith the Cape Floral Region, he proposed a further 17 groups, based on his knowledge, not only of *Euryops* but also of other genera. Although he did not anticipate a general validity for his scheme, I find it fits very well the distribution patterns of those members of Manuleae sympatric with *Euryops*. I have therefore adopted Nordenstam's scheme with modification (both geographical and nomenclatural) where necessary and a good many additions. There are, for example, far more members of Manuleae than there are species of *Euryops* in Namibia, Zimbabwe and extra-montane south eastern Africa. Some of the patterns of distribution recognizable in Namibia are commented on by Nordenstam (1974, pp.56 et seq.) in his account of the flora of the Brandberg.

Table 5. Phytogeographical groups and the number of species of Manuleae occurring within them.

Group	A. Cape Floral Region										B. Africa S of Cunene and Zambezi Rivers, excl. Cape Floral Region																											
	1	2	3	4	5	6	7	8	9	10	11	12	13	14	15	16	17	18	19	20	21	22	23	24	25	26	27	28	29	30	31	32	33	34	C	D	E	F
1. *Antherothamnus*																													1									
2. *Manuleopsis*																										1												
3. *Camptoloma*																										1											1	1
4. *Jamesbrittenia*				5				3	1	2	1		1	3	1	1	1	7	2		7		6	5	3	7	2	4	3	7	4	1	5	2	1		1	
5. *Lyperia*	1			1	1								1						1																			
6. *Sutera*	3	1	4	2	1	2	3	4	6	1			2	4				2	2				1	4	5				1	4	5	1	1					
7. *Manulea*	2	1	1	1	1	7	18	2	1		2		1	2	5		2	4	2	3	6		1	4	3				2	2	1	1						
8. *Melanospermum*																													1	5								
9. *Polycarena*						5	5			1	1	1								4																		
10. *Glekia*																1																						
11. *Trieenea*	1			1			7																															
12. *Phyllopodium*			3	3		4	4			2											1	4	2								3							
13. *Zaluzianskya*		2						3	2		11		1		4	1	4	4	3	3	2	1		1					1	3	8		1			1		
14. *Reyemia*											2																											
15. *Glumicalyx*																		6																				
16. *Strobilopsis*																		1																				
17. *Tetraselago*																														4								

A. Cape Floral Region: 1, Cape Ubiquists; 2, Karroo Mt-Western; 3, Southern; 4, Western; 5, Western and Southern; 6, South western; 7, North western; 8, Bredasdorp-Riversdale; 9, Southern Cape Mts; 10, South eastern.

B. Africa S of Cunene and Zambezi Rivers, excl. Cape Floral Region: 11, Western Upper Karroo; 12, Roggeveld-Cape Karroo; 13, Upper Karroo; 14, E Upper Karroo-grassland; 15, Sneeuwbergen; 16, Sneeuwbergen-Drakensberg; 17, Sneeuwbergen-Cape; 18, Drakensberg; 19, Karroo Ubiquists; 20, Namaqualand-Cape; 21, Namaqua; 22, Vanrhynsdorp Karroo; 23, Gariep; 24, N Cape-Namibia; 25, Namibian central and southern Highlands; 26, Namibian western escarpment; 27, Kaokoveld; 28, Waterberg-Otavi Mts; 29, Interior savanna; 30, Highveld; 31, E Cape-Transkei-Natal; 32, Eastern Ubiquists; 33, SE Tropical Africa; 34, Tropical Africa Ubiquists.

Nordenstam also recognized, within southern Africa, nine phytogeographical centres of outstanding species-concentration and endemism (Nordenstam, 1969, pp.62–67). Some of these are important in relation to the species of Manuleae, and there are other centres in geographical areas outwith the area covered by Nordenstam (Africa south of the Orange and Limpopo rivers). An account of these centres follows the analysis of species-distribution.

It is not practicable to publish the large number of maps showing the distributions of individual species of Manuleae but these are housed at the Royal Botanic Garden Edinburgh. Figures 69–77 show the overall distribution of each genus, fig. 78 the phytogeographical groups, fig. 79 the centres of endemism (see pp.558–564).

PHYTOGEOGRAPHICAL GROUPS

GROUP A.
Species with their main distribution in the Cape Floral Region (as defined by Bond & Goldblatt, 1984) but including outliers in the Eastern Cape.

1. The Cape Ubiquists (five genera, nine species).
The species are ubiquitous or nearly so within the Cape Floral Region.

    5. Lyperia
        3. *antirrhinoides* (outlier in Spektakelberg)
    6. Sutera
        3. *foetida* (outlier in Khamiesberg)
        22. *decipiens* (extends slightly north)
        40. *caerulea* (extends slightly north)
    7. Manulea
        33. *laxa* (montane)
        48. *cheiranthus* (absent from the area of Group 9)
    11. Trieenea
        9. *glutinosa*
    13. Zaluzianskya
        11. *capensis* (extends east to Grahamstown, isolated records north to the Khamiesberg)
        46. *divaricata* (absent from the area of Group 9)

2. The Karroo mountain – Western Group (one genus, one species).
The species occur in the area of the Southern Cape Mountains Group (9) as well as in one or both of the western Groups (6, 7).

    6. Sutera
        14. *comptonii* (Karroopoort to Witteberg)

3. The Southern Group (five genera, 14 species).
The species are found mostly east of the Breede river as far as Alexandria division, but excluding the limestone endemics in Bredasdorp and Riversdale divisions. They are either coastal or penetrate inland to the mountains.

    4. Jamesbrittenia
        25. *pinnatifida* (Langekloof to Grahamstown; ill-known)
        26. *argentea* (Cogman's Kloof to Port Elizabeth)
        29. *tenuifolia* (Mossel Bay to Plettenberg Bay, inland to the Long Kloof, Baviaans Kloof, upper Gamtoos river valley)
        32. *microphylla* (Knysna to Port Alfred, coastal)
        33. *aspalathoides* (Elim, in Bredasdorp div., to Addo)
    5. Lyperia
        4. *violacea* (Peninsula, then Swellendam to Uniondale div.)
    6. Sutera
        17. *integrifolia* (Albertinia to Humansdorp)
        23. *langebergensis* (Langeberg)

43. *cordata* (George to East London, along the coast but penetrating inland to the Outeniquas, Zuurberg and Grahamstown)

49. *campanulata* (Still Bay and Albertinia to Port Alfred)

7. Manulea

44. *obovata* (roughly Plettenberg Bay to Port Alfred; sand dunes and coastal scrub)

12. Phyllopodium

4. *rustii* (Swellendam to Gamtoos river, with scattered records from Oudtshoorn, Kamanassie Mts and possibly Graaff-Reinet)

6. *elegans* (Montagu to Joubertina, montane)

7. *multifolium* (Langeberg to Gt Winterhoek Mts)

4. The Western Group (four genera, seven species).

The species occur in the areas of both the North western (7) and South western (6) Groups and may extend slightly east of the Breede river.

6. Sutera

10. *glabrata* (montane; Worcester-Barrydale-De Doorns)

18. *uncinata* (montane; Giftberg (Klaver) south to Somerset West and the Peninsula, east to Montagu)

7. Manulea

32. *leiostachys* (montane; Cedarberg to Piquetberg Mt, Ceres and Worcester)

11. Trieenea

8. *longipedicellata* (montane, environs Paarl, Stellenbosch, Genadendal)

12. Phyllopodium

8. *capillare* (coastal; Lamberts Bay south to the Peninsula, east to Albertinia)

13. *heterophyllum* (lowlands; Van Rhynsdorp div. south to the Peninsula, east to Breede river)

14. *cordatum* (as *heterophyllum*)

5. The Western and Southern Group (three genera, three species).

The species occur in mainly coastal areas from the Great Berg river to the Gouritz river.

5. Lyperia

1. *lychnidea* (Saldanha Bay south to the Peninsula, east to environs of Still Bay; coastal)

6. Sutera

29. *aethiopica* (mainly Caledon, Swellendam, Bredasdorp and Riversdale divs.; mountains and flats)

7. Manulea

28. *thyrsiflora* (Veldrift to Bloubergstrand, disjunction to Bredasdorp and Riversdale divs.; coastal)

6. The South western Group (five genera, 20 species).

This is basically the South-western subregion of Weimarck and Nordenstam (the area south and west of the Great Berg and Breede rivers) but the area is here extended to include the coastal area from the Great Berg north to the Olifants river and the lower Olifants river valley, and excludes the limestone endemics of Bredasdorp division. The species may be coastal or montane, or both.

6. Sutera

20. *multiramosa* (Olifants river to Lambert's Bay; coastal)

24. *hispida* (Paarl, Jonkershoek, Peninsula to Bredasdorp; coast and lower mountain slopes)

7. Manulea

15. *corymbosa* (Great Berg river to Camps Bay and Cape Flats)

17. *psilostoma* (Graafwater; lowlands)

18. *augei* (Saldanha Bay to Hopefield; lowlands)

26. *tomentosa* (Great Berg river to Pearly Beach (Bredasdorp div.); coastal)

27. *rubra* (Great Berg river to Somerset West and inland to Paarl, Fransch Hoek and Swellendam; lowlands)

34. *pillansii* (Lambert's Bay to Verloren Vlei, coastal)

46. *exigua* (mainly Caledon div., lowlands)

9. Polycarena

1. *lilacina* (roughly Olifants river south to Darling; lowlands)

2. *capensis* (Hopefield south to Peninsula; lowlands)

3. *subtilis* (Olifants river valley and Piquetberg area; lowlands)

6. *gilioides* sens. strict. (Tulbagh to Paarl, insufficiently known; lowlands)

10. *silenoides* (Lion's Head only)

12. *Phyllopodium*

15. *mimetes* (Het Kruis to Malmesbury, lowlands)

22. *alpinum* (Fransch Hoek Mts and Rivierzondereinde Mts, doubtful disjunction to the Zwartberg)

23. *cephalophorum* (Van Rhynsdorp to Saldanha Bay and Piquetberg; lowlands)

24. *phyllopodioides* (as *P. cephalophorum*)

13. *Zaluzianskya*

26. *villosa* (just north of the Great Berg river to the Peninsula and east to Pearly Beach, Bredasdorp div.; coastal)

28. *parviflora* (environs of Saldanha Bay)

7. The North western Group (seven genera, 41 species)

This is the north west mountainous area of the Cape Floral Region (the coastal lowlands are referred to Group 6), from the Bokkeveld Mts and Giftberg south through the Cedarberg, Olifants River Mts, Piquetberg Mt to the mountains around Touws River and Montagu in the east, Ceres, Robertson and Worcester in the west.

5. *Lyperia*

6. *formosa* (Voetpadsberg; insufficiently known)

6. *Sutera*

11. *paniculata* (Sneeuwkop and Khamiesberg then a disjunction to the Bokkeveld Mts south to the Cedarberg)

19. *longipedicellata* (Van Rhyn's Pass and nearby; insufficiently known)

21. *subsessilis* (Cedarberg south to ?; insufficiently known)

7. *Manulea*

16. *adenocalyx* (roughly Wupperthal to Karroopoort, west to Piquetberg)

19. *paucibarbata* (Van Rhynsdorp to the Cedarberg)

20. *annua* (N Cedarberg to Piquetberg and Porterville)

21. *arabidea* (Pakhuis-Clanwilliam area; insufficiently known)

30. *turritis* (Piquetberg, Great Winterhoek, Tulbagh, Ceres, Worcester; submontane)

31. *ovatifolia* (Piquetberg)

35. *montana* (Langeberg (Clanwilliam div.), Middelberg, Cedarberg)

36. *glandulosa* (Giftberg, Nardouwsberg, N Cedarberg)

39. *virgata* (Nardouwsberg, Cedarberg, Muishoekberg)

40. *stellata* (mountains west of Clanwilliam)

47. *adenodes* (N Cedarberg)

60. *minor* (Hex river valley)

61. *praeterita* (Nieuwoudtville and Van Rhynsdorp south to Clanwilliam)

67. *juncea* (Giftberg, Cedarberg, Skurfdeberg, Koudebokkeveld Mts, Ceres)

68. *multispicata* (Cold Bokkeveld Tafelberg to mountains east of Ceres, Worcester and Robertson to Touws River and Laingsburg)

69. *rigida* (Cedarberg)

71. *cephalotes* (Lokenburg (3119CA) to Ceres, possibly a disjunction northwards to the Khamiesberg; ill-known)

72. *linearifolia* (Cold Bokkeveld Mts to Nardouwsberg; ill-known)

9. *Polycarena*

5. *nardouwensis* (Nardouw's Pass area; ill-known)

7. *gracilis* (Pakhuis Pass to Piquetberg Mt)

8. *aemulans* (Bokkeveld Mts, Pakhuis Mts, S Cedarberg)

9. *exigua* (Cedarberg)

14. *formosa* (roughly Nieuwoudtville south to Wupperthal, but extending westwards towards the sea)

11. *Trieenea*

1. *laxiflora* (N Cedarberg; ill-known)

2. *lasiocephala* (N Cedarberg; ill-known)

3. *lanciloba* (N Cedarberg; ill-known)

4. *taylorii* (Cedarberg)

5. *frigida* (Cold Bokkeveld Mts; ill-known)

6. *elsiae* (S Cedarberg, Cold Bokkeveld Mts; ill-known)

7. *schlechteri* (N Cedarberg, Cold Bokkeveld; ill-known)

12. *Phyllopodium*

9. *caespitosum* (Cedarberg south to mountains near Ceres and Worcester)

10. *micranthum* (Cedarberg south to Skurweberg and Anysberg, east to the Karroo mountains)

11. *viscidissimum* (Baviaansberg, 3319BA; ill-known)

16. *pubiflorum* (Nieuwoudtville and Pakhuis Pass)

13. *Zaluzianskya*

47. *glandulosa* (N Cedarberg; ill-known)

48. *lanigera* (Karroo Poort; ill-known)

49. *isanthera* (environs of Skurweberg, 3219DC, 3319AB; ill-known)

8. The Bredasdorp-Riversdale Limestone Group (four genera, 11 species).

The species are confined to calcareous substrates mainly in these two districts; Nordenstam (1969, p.26) advocated the recognition of this Group and Oliver et al. (1983) did so.

4. *Jamesbrittenia*

34. *calciphila* (Hagelkraal to Albertinia)

35. *stellata* (Bredasdorp and Riversdale divisions with a remarkable disjunction to a limestone cliff at Buffels Baai on the Peninsula)

36. *albomarginata* (Stanford to Still Bay)

6. *Sutera*

27. *subspicata* (Stanford to Cape Infanta)

28. *titanophila* (cliffs at De Hoop vlei; ill-known)

30. *calciphila* (Bredasdorp to Still Bay)

31. *placida* (Still Bay; ill-known)

7. *Manulea*

22. *calciphila* (Bredasdorp to Still Bay, but one record from Namaqualand)

29. *caledonica* (Kleinrivier lagoon to Still Bay)

13. *Zaluzianskya*

12. *muirii* (limestone area south of Riversdale and Albertinia)

29. *gracilis* (Struys Bay to Still Bay)

9. The Southern Cape Mountains Group (five genera, eleven species).

The species are confined (or nearly so) to the mountainous tract on the southern flank of the Karroos: Witteberg (near Laingsburg), Anysberg, Klein and Groot Zwartberg and neighbouring mountains, thence east to mountains around Willowmore and Steytlerville (Grootrivierberge, Klein Winterhoekberge, Baviaans Kloof Berge) south to the northern flanks of the Outeniqua and Langekloof mountains.

4. *Jamesbrittenia*

18. *tortuosa* (roughly Gamka river east to the Klein Winterhoekberge, submontane)

6. *Sutera*

25. *marifolia* (E end Outeniqua Mts to Vanstaadensberg near Uitenhage)

26. *cinerea* (southern flank Baviaanskloof Mts; ill-known)

32. *denudata* (mountains around Willowmore and Uniondale east to Great Winterhoek Mts and Suuranysberge)

33. *subnuda* (Klein and Groot Zwartberg thence south to northern flank of Langeberg and Outeniqua Mts, submontane)

34. *affinis* (E end Outeniqua Mts to W end Langekloof; ill-known)

35. *tenuicaulis* (Zwartberg Pass; ill-known)

7. *Manulea*

25. *derustiana* (southern flank of Groot Zwartberg, near Le Roux; ill-known)

9. *Polycarena*

15. *comptonii* (Witteberg, Whitehill; ill-known)

12. *Phyllopodium*

5. *dolomiticum* (southern flank Groot Zwartberg, near Rust en Vrede; ill-known)

12. *tweedense* (Tweedside near Laingsburg; insufficiently known)

10. The South eastern Group (two genera, three species).
The species are confined to a small area east of the Sundays river, from the Zuurberg to the environs of Grahamstown.

4. *Jamesbrittenia*

21. *albanensis* (Grahamstown NE to Alice, SE to Kowie, W to Ado, but one record from 'the Long Kloof')

31. *zuurbergensis* (Zuurberg; ill-known)

6. *Sutera*

6. *racemosa* (Zuurberg; ill-known)

GROUP B.

Species with their main distribution in southern Africa (south of the Cunene and Zambezi rivers) outside the Cape Floral Region. Several species penetrate the Cape Floral Region but seldom as a constituent of 'fynbos'; several overleap the Limpopo interval to reach Zimbabwe, and a few reach Angola.

11. The Western Upper Karroo Group (five genera, 17 species).
This group comprises species endemic to an area including the Hantam Mountains, Roggeveld and Nieuweveld.

4. *Jamesbrittenia*

17. *incisa* (Calvinia and Middle Roggeveld)

6. *Manulea*

43. *incana* (Middle Roggeveld; ill-known)

59. *diandra* (Roggeveld, Nieuweveld)

9. *Polycarena*

16. *filiformis* (Calvinia; ill-known)

13. *Zaluzianskya*

21. *pumila* (Van Rhynsdorp, Calvinia, Tanqua Karroo, Roggeveld, Moordenaars Karroo)

22. *inflata* (Hantam Mts and nearby)

23. *sutherlandica* (Middle Roggeveld; ill-known)

24. *mirabilis* (Middle Roggeveld; ill-known)

25. *acutiloba* (Calvinia, Middle Roggeveld; ill-known)

31. *pilosissima* (Roggeveld to Carnarvon and Fraserburg)

32. *bella* (Tanqua and Ceres Karroos, Komsberg, Klein Roggeveld)

33. *violacea* (Barren Karroo, Bokkeveld Karroo, Roggeveld, possibly Tanqua Karroo)

34. *marlothii* (Roggeveld; ill-known)

38. *cohabitans* (Hantams, Roggeveld)

51. *acrobareia* (Bokkeveld Mts?; ill-known)

14. *Reyemia*

1. *chasmanthiflora* (between Calvinia and Middelpos; ill-known)

2. *nemesioides* (Calvinia, Roggeveld)

12. The Roggeveld-Cape Karroo Group (five genera, six species).
The species occur in the combined areas of the Western Upper Karroo Group (no. 11) and the Southern Cape Mountain Group (no. 9).

5. *Lyperia*

5. *tenuiflora* (hills N of Ceres to northern flank of Groot Zwartberg and Karroo north of these mountains)

6. *Sutera*

12. *violacea* (Nieuwoudtville south to Karroo Poort, thence east to the Komsberg and Groot Zwartberg)

13. *revoluta* (Calvinia and Sutherland divisions south to Touwsriver, east across the mountains to Uniondale, down almost to sea level west of Mossel Bay)

7. *Manulea*

24. *latiloba* (southern Tanqua Karroo to Ceres Karroo and area north of the Langeberg)

9. *Polycarena*

12. *aurea* (Nieuwoudtville and Calvinia south west to Ceres and Worcester, east to the Roggeveld and Laingsburg Witteberg)

13. *Zaluzianskya*
 37. *minima* (Calvinia and Roggeveld south to Laingsburg Witteberg and heights south of Touws River)

13. The Upper Karroo Group (three genera, seven species).

The species in this group are found in the combined areas of the Western Upper Karroo Group (no. 11) and the Sneeuwbergen Group (no. 15). Some may extend into the Orange Free State. Disjunctions may occur in the distributions of the montane species.

 4. *Jamesbrittenia*
  19. *tysonii* (northern Middle Roggeveld and Nieuweveld fanning out to reach the Barkly West-Hay area in the north east, Cradock and the border of the Orange Free State in the east)

 6. *Sutera*
  15. *archeri* (Nieuweveld, Moordenaars Karroo, Laingsburg Witteberg to Tierberg near Prince Albert and northern flanks of Groot Zwartberg west of Vondeling; ill-known)
  16. *macrosiphon* (Nieuweveld to southern Orange Free State, Stormberg and Andriesberg)
  41. *rotundifolium* (mountains around Victoria West east to the Sneeuwberg, a few records north to the Asbestos Hills; ill-known)
  42. *pauciflora* (Roggeveld and Nieuweveld to Three Sisters Mountain, then a disjunction (Great Karroo) to the Groot Zwartberg)

 7. *Manulea*
  23. *karrooica* (Karroo, from Fraserburg and Graaff Reinet south to Prince Albert; ill-known)
  45. *chrysantha* (Karroo, from about Beaufort West, Graaff Reinet and Pearston in the north to roughly Oudtshoorn and Uitenhage; ill-known)

14. The Eastern Upper Karroo - grassland Group (two genera, eight species).

The species occur in the karroid and grassland areas of the north eastern Cape (Middelburg-Colesberg-Lady Grey-Barkly East area), sometimes northwards to the border of Botswana, the Orange Free State, western Lesotho and southern Transvaal.

 4. *Jamesbrittenia*
  28. *albiflora* (mainly southern half of Orange Free State, extending slightly westwards into the Cape)
  63. *stricta* (SE Transvaal, eastern half of O.F.S., western Lesotho, around Aliwal North and Indwe in the Cape)
  77. *aurantiaca* (widely distributed in Transvaal, excluding the Lowveld, and extending into neighbouring Botswana, O.F.S., western Lesotho, northern and eastern Cape from the Malopo river south to Prieska and south east to the Stormberg-Cape Drakensberg area; moist places)

 7. *Manulea*
  7. *flanaganii* (Bethulie only; ill-known)
  9. *paniculata* (southern Transvaal Highveld, eastern O.F.S., western Lesotho, Cape Drakensberg, Stormberg, disjunction (real?) to Mt Currie and the Zuurberg near Kokstad)
  10. *buchneroides* (southernmost Transvaal, low Drakensberg in Natal, eastern O.F.S., western Lesotho, Witteberg at Lady Grey)
  12. *plurirosulata* (southern half of O.F.S. to Colesberg-Middelburg-Steynsburg area of Cape)
  54. *deserticola* (SW Orange Free State and nearby parts of Cape; ill-known)

15. The Sneeuwbergen Group (2 genera, 2 species).

The species are found in the Koudeveldberge and Sneeuwberg east to the Stormberg, Winterberge and Amatole Mountains, all montane.

 4. *Jamesbrittenia*
  62. *crassicaulis* (Koudeveldberg and Sneeuwberg east to the Bamboesberg, Andriesberg, Bankberg and Wildeschutsberg)

 6. *Sutera*
  14. *glandulifera* (Stormberg, mountains around Queenstown, Amatole Mountains, Kologha Range, Pirie Mountain)

16. The Sneeuwberg-Drakensberg Group (four genera, eight species).

The species are montane and occur in the combined areas of the Sneeuwbergen and Drakensberg Groups (nos 15, 18), mostly reaching their western limit in the Koudeveldberge and

Sneeuwberg, rarely with disjunctions to the southern Cape fold mountains and other Cape mountains.

4. *Jamesbrittenia*

20. *filicaulis* (Koudeveldberge north east to the mountains along the Orange Free State-Lesotho border, south east to the Elandsberg and Amatole Mts, thence over the mountains to western Transkei as far north as the Natal border at Coleford)

7. *Manulea*

4. *crassifolia* subsp. *crassifolia* (main area in Lesotho and neighbouring Orange Free State south to Witteberg and Cape Drakensberg and west to Bankberg, Sneeuwberge and Koudeveldberge, with a few records from Groot Zwartberg and Kouga Mts)

5. *rhodantha* subsp. *rhodantha* (western Lesotho to Cape Drakensberg and Stormberg)

10. *Glekia* (monotypic)

krebsiana (western Lesotho south to Witteberg, Cape Drakensberg, Stormberg and Amatole Mts, west to Boschberg, Lootsberg, Oudeberg)

13. *Zaluzianskya*

6. *schmitziae* (mountains about Graaff Reinet and Queenstown north to Cape Drakensberg, Witteberg, Lesotho and neighbouring mountains in the O.F.S.)

13. *crocea* (Stormberg, Andriesberg, Witteberg, Cape Drakensberg)

15. *glareosa* (Drakensberg and its outliers from Transvaal-Natal border to Cape Drakensberg, thence to mountains around Graaff Reinet, possibly a disjunction to the Nieuweveld Mts; under-collected in Cape)

19. *ovata* (well-collected in Drakensberg, Lesotho and neighbouring O.F.S., then scattered records from Cape mountains at Queenstown, Graaff Reinet, Calvinia, Ladismith, Worcester, Tulbagh)

17. The Sneeuwbergen - Cape Group (two genera, two species).

The species occur in the combined areas of the Sneeuwbergen Group (no. 15), the Southern Cape Mountain Group (no. 9) and the South Eastern Group (no. 10).

4. *Jamesbrittenia*

30. *foliolosa* (Karroo around Aberdeen, Graaff Reinet and Cradock thence south east to Grahamstown, south west to the Cape fold mountains as far west as the east end of the Langeberge, almost to sea level near Uitenhage and Alexandria)

13. *Zaluzianskya*

41. *synaptica* (Karroo around Aberdeen and Graaff Reinet south and west to the southern Cape Mts, from Hex River Kloof east to the northern flank of the Baviaanskloofberge)

18. The Drakensberg Group (Eastern Mountain Region) (six genera, 24 species).

This group embraces montane species occurring in the Drakensberg in Natal, Lesotho and Cape, with neighbouring areas in Transkei and the Orange Free State, also the Witteberg at Lady Grey.

4. *Jamesbrittenia*

11. *dentatisepala* (northern part of Lesotho and Natal Drakensberg, above c.1980m)

64. *pristisepala* (widespread above c.1500m)

65. *lesutica* (environs of Mokhotlong, c.2400m; ill-known)

67. *beverlyana* (Sehlabathebe National Park, c.2300m; ill-known)

73. *breviflora* (widespread above c.1400m)

74. *jurassica* (Lesotho, above c.2500m; ill-known)

75. *aspleniifolia* (southern Lesotho and neighbouring Cape mountains, above c.1700m; ill-known)

6. *Sutera*

1. *cooperi* (environs of Bethlehem south along O.F.S.-Lesotho border to foot of Witteberg near Lady Grey; sandstone cliffs)

5. *polelensis* subsp. *polelensis* (Cave Sandstone in Lesotho and adjacent Natal and O.F.S., Witteberg, Cape Drakensberg, above c.1500m)

7. *Manulea*

4. *crassifolia* subsp. *thodeana* (Natal Drakensberg from Mont aux Sources to Bushman's Nek, above c.2100m)

8. *florifera* (foothills of Natal Drakensberg from Highmoor and Mooi River south to the Transkei border, above c.1500m; ill-known)

41. *platystigma* (Sani Top to Mokhotlong, above c.2700m; ill-known)
42. *dregei* (Witteberg and Cape Drakensberg, above c.2200m; ill-known)

13. *Zaluzianskya*
13. *oreophila* (escarpment between Natal and Lesotho, above c.2300m; ill-known)
14. *chrysops* (scattered localities in Lesotho and Natal above c.2000m; ill-known)
17. *turritella* (escarpment between Natal and Lesotho, above c.3000m; ill-known)
44. *rubrostellata* (scattered localities in Lesotho, Cape and Natal Drakensberg, Witteberg, above c.2350m)

15. *Glumicalyx* (whole genus endemic)
1. *goseloides* (Drakensberg and outliers in Natal and Transkei)
2. *nutans* (Platberg in O.F.S., Lesotho, Natal and Cape Drakensberg, Witteberg)
3. *flanaganii* (Lesotho, Natal Drakensberg, just reaching Cape Drakensberg)
4. *lesuticus* (Lesotho)
5. *montanus* (Lesotho, Natal Drakensberg, just reaching Cape Drakensberg)
6. *apiculatus* (Witteberg and Cape Drakensberg)

16. *Strobilopsis* (monotypic)
wrightii (southern Natal Drakensberg, Sehlabathebe in Lesotho, on Cave Sandstone; ill-known)

19. The Karroo Ubiquists (five genera, ten species).

The species are widely distributed in the karroid areas of the Cape and may spread into the Orange Free State, low-lying parts of Lesotho, and Namibia.

4. *Jamesbrittenia*
14. *atropurpurea* subsp. *atropurpurea* (widespread in Cape, extending into southern and eastern O.F.S., western Lesotho, just entering Transkei)
46. *thunbergii* (most records from western Cape, Calvinia div. south to Ceres div., east to Groot Zwartberg, also in Kenhardt div.)

5. *Lyperia*
2. *tristis* (widespread in the west, from west central Namibia to the Peninsula, east to Kenhardt div. in the north, Graaff Reinet and Avontuur in the south)

6. *Sutera*
36. *halimifolia* (Klein Karas in Namibia, then Griqualand West south to Fraserburg, Prince Albert and Humansdorp divisions, thence east and north east to Cathcart, fringes of the Cape Drakensberg, southern O.F.S.)
39. *patriotica* (Rehoboth and Windhoek districts in Namibia then a disjunction (real?) to Hay, Barkly West and Prieska divisions (Cape) thence western Transvaal, O.F.S., Lesotho, south to east central Cape from Murraysburg to Cathcart)

7. *Manulea*
13. *schaeferi* (Aus and Keetmanshoop in Namibia south to Warmbad, thence east to Kakamas, Upington, Kenhardt and Prieska in the Cape)
58. *fragrans* (Springbok, Van Rhynsdorp and Middelpos east to Prieska, De Aar, Victoria West)

13. *Zaluzianskya*
35. *venusta* (Upington to Kuruman, Kimberley and just entering O.F.S., south to Hopetown division, then a disjunction to Beaufort West div., thence west and south to the Koedoesberg, Laingsburg and Ladismith)
36. *karrooica* (south central O.F.S., south and west to Carnarvon and Beaufort West, south and east to Middelburg, Cradock and northern flank of Great Winterberg)
42. *peduncularis* (southern Namibia, thence through Namaqualand south and east to Van Rhynsdorp, Sutherland and Laingsburg divisions, thence east and north to Stutterheim, Colesberg and De Aar, O.F.S., western Lesotho)

20. The Namaqualand-Cape Group (four genera, 11 species).

The species have a western distribution from Namaqualand south into the Cape Floral Region; they may cross the Orange into Lüderitz Süd and Warmbad districts of Namibia, or extend slightly eastwards into karroid areas.

7. *Manulea*
49. *decipiens* (northern Namaqualand to roughly Verloren Vlei and Clanwilliam)

50. *pusilla* (Namaqualand south to the Olifants river thence south through the Bokkeveld, Tanqua and Ceres Karroos to the Witteberg and Warmwaterberg)

66. *altissima* (Springbok south to Langebaan)

9. *Polycarena*

4. *batteniana* (Garies to Wupperthal)

11. *rariflora* (O'okiep to Malmesbury and Tulbagh, with isolated records from Sutherland, Laingsburg and northern flank of Outeniqua Mts)

13. *pubescens* (northern Namaqualand to Montagu and the Anysberg, a few records from western Roggeveld)

17. *tenella* (Richtersveld south to Ceres, Laingsburg, also fringes of Roggeveld, possibly always montane; ill-known)

12. *Phyllopodium*

21. *anomalum* (O'okiep to Montagu, one record from Sutherland div.)

13. *Zaluzianskya*

27. *affinis* (Richtersveld south to Great Berg river, inland to Bokkeveld Karroo and environs of Wupperthal)

52. *pusilla* (Kamieskroon south east through Van Rhynsdorp, Clanwilliam and Ceres to Laingsburg and Hassequa's Poort)

54. *benthamiana* (Lüderitz-Süd district of Namibia south to Saldanha Bay in the west; east to Pofadder, Calvinia and the Anysberg, thence east to Prince Albert; under-collected in the southern and eastern part of its range)

21. The Namaqua Group (four genera, 19 species).

The species are distributed in the area from Lüderitz-Süd and Warmbad districts of Namibia south through Namaqualand to the Olifants river. Endemics in the Van Rhynsdorp Karroo and in the Richtersveld are treated separately as Groups 22 and 23.

4. *Jamesbrittenia*

16. *namaquensis* (Richtersveld south to Leliefontein)

37. *merxmuelleri* (Lüderitz to Kleinsee in Namaqualand, coastal)

38. *fruticosa* (Lüderitz to Doorn river south of Van Rhynsdorp)

44. *amplexicaulis* (Richtersveld south to Eenkokerboom and Loeriesfontein)

45. *racemosa* (Henkries south to the Olifants river)

47. *aridicola* (Lüderitz-Süd and Warmbad districts, thence Springbok to Upington and Kenhardt)

72. *pedunculosa* (O'okiep south to the Khamiesberg)

7. *Manulea*

14. *nervosa* (Springbok north east to Pella, south east to the Khamiesberg area)

37. *cinerea* (Witbank and Hondeklip Bay, coastal dunes; ill-known)

62. *gariesiana* (environs of Garies south to Bitterfontein; ill-known)

63. *silenoides* (Nababeep and Springbok south to Garies)

64. *acutiloba* (Springbok to Garies; ill-known)

65. *androsacea* (Lüderitz-Süd district south to environs of Springbok)

12. *Phyllopodium*

18. *lupuliforme* (near Steinkopf; ill-known)

19. *collinum* (O'okiep to Karreebergen near Nieuwerust)

20. *maxii* (Aus south to Pofadder and the Khamiesberg; under-collected)

25. *pumilum* (Richtersveld south to Garies and the Groen river)

13. *Zaluzianskya*

30. *sanorum* (Goodhouse on the Orange south to Kamieskroon, east to Pofadder)

50. *collina* (Kamieskroon to Garies; ill-known)

22. The Vanrhynsdorp Karroo Group (two genera, two species).

The species are confined to a small area of quartzitic and other rock outcrops in Van Rhynsdorp division, especially the Knersvlakte.

7. *Manulea*

70. *ramulosa* (environs of Klaver; insufficiently known)

13. *Zaluzianskya*

45. *karreebergensis* (farm Karreebergen, 3118BA; insufficiently known)

23. The Gariep Group (three genera, 12 species).

The species are endemic to the country adjoining the lower reaches of the Orange (Gariep) river, including the Richtersveld, and parts of Lüderitz-Süd, southernmost Bethanien and Warmbad districts in Namibia.

4. *Jamesbrittenia*

40. *sessilifolia* (Namibia, Kuibis south to the Orange; ill-known)
41. *major* (roughly Karios south to the Richtersveld, but possibly a disjunction to the north, 2014CB, CC, 2013DB, which would then indicate placement in Group 26; under-collected)
42. *megaphylla* (environs of Vioolsdrift, on both sides of the Orange; ill-known)
43. *bicolor* (Namibia, Klinghardtberge, Tsausberge, Pockenbank south to Witpütz)
49. *glutinosa* (Namusberge east to Klein Karasberge in Namibia, south to the Richtersveld and along the Orange to Goodhouse)
68. *ramosissima* (Spitskop, Namuskluft and Rosh Pinah (Namibia) and the mountains of the Richtersveld (Cape) east to environs of Aughrabies Falls)

7. *Manulea*

51. *aridicola* (Richtersveld and across the Orange around Obib; ill-known)
52. *minuscula* (Klinghardtberge, Chameis and Schakalsberg south towards Port Nolloth; ill-known)
57. *namibensis* (environs Aus, Klinghardtberge, Huib Plateau)
73. *robusta* (mountains around Aus, then Cornellsberg in Richtersveld; ill-known)

12. *Phyllopodium*

17. *namaense* (Lüderitz and Aus south to the Orange, just entering the Richtersveld, but under-collected there)
26. *hispidulum* (known only from the Klinghardtberge and Grootderm on the Orange river)

24. The arid northern Cape-Namibia Group (three genera, nine species).

The species occur in the arid northern Cape, from roughly the Orange (rarely as far south as the Khamiesberg) north to Vryburg and Gordonia, and in the southern half of Namibia.

4. *Jamesbrittenia*

27. *integerrima* (Cape, from the Richtersveld east to Hopetown, north to Griqualand West, Vryburg, Gordonia, scattered records in Namibia from Windhoek south to Aus)
48. *megadenia* (mainly Keetmanshoop and Warmbad districts of Namibia, one record from western Gordonia in the Cape)
61. *canescens* var. *canescens* (south central Namibia and scattered records in the Cape along the Orange from the Richtersveld to Prieska)
70. *tenella* (western Namibia, mainly Brandberg to Waterberg south to the Naukluft Mts, one record from the Kaokoveld (Marienfluss), scattered records from around Aus and Great Karasberg, thence a disjunction (real?) to the Asbestos Mts in Griqualand West)
81. *adpressa* (scattered records in southern half of Namibia, one record from Namaqualand (Platbakkies), another from environs of Upington; river beds, pans, marshes)

7. *Manulea*

1. *leptosiphon* (Namibia, between Keetmanshoop and Rietvlei, on edge of Kalahari, c.2619CB; insufficiently known)
2. *burchellii* (Cape mainly north of Orange river, extending to eastern fringe of Namibia)
53. *gariepina* (Namibia south of the 26th parallel, the Cape as far south as the Khamiesberg and east to Kimberley)

13. *Zaluzianskya*

55. *diandra* (Namibia, from the Tiras Mts south to the Orange river at Vioolsdrift, thence east to the Great Karasberge, and Prieska in the Cape)

25. The Namibia central and southern Highland Group (two genera, six species).

The species are mostly confined to the central and southern highlands of Namibia, including the Brandberg, Erongo Mts, Khomashochland, Auas Mts, Gamsberg, Naukluft Mts, Tiras Mts, Karasberge and surrounding high ground, usually above c.900m; sometimes a species may extend into Botswana on the high ground east of Gobabis. The limestone area centred on the Waterberg is treated separately (Group 28).

4. *Jamesbrittenia*

50. *primuliflora* (Auas Mts south almost to the Orange)
55. *pallida* (Erongo Mts, Giftkuppe and Omatako Mt south to Khomashochland)
57. *lyperioides* (Karibib and Okahandja south to the Great Tiras Mts, but also Etosha Pan)

7. *Manulea*

3. *conferta* (Swakopmund to Okahandja and Windhoek, thence east to Gobabis and just across the Botswana border; ill-known)
56. *tenella* (known with certainty only from Gameros)
74. *dubia* (Brandberg and Waterberg south to Aus and Great Karasberge)

26. The Namibia western escarpment Group (three genera, nine species).

The species occur mainly in the transitional zone between the eastern fringe of the Namib and the escarpment of the interior plateau; they may extend into southern Angola and northern Namaqualand.

2. *Manuleopsis* (monotypic)

dinteri (Kaokoveld south to the Naukluft Mts, but extending to nearly 18°E)

3. *Camptoloma*

2. *rotundifolium* (Benguela and Mossamedes districts of Angola (two collections seen) then western Namibia as far south as the Naukluft Mts)

4. *Jamesbrittenia*

39. *maxii* (Mossamedes in Angola, south through western Namibia to the northern Cape roughly between Anenous and Pofadder)
51. *fimbriata* (Vreemdelingspoort only, 2415DD; insufficiently known)
56. *fleckii* (Gamsberg area; insufficiently known)
58. *pilgeriana* (Naukluft Mts only)
59. *barbata* (Namibia only, roughly from the Omaruru river south to the Kuiseb, thence a few sites in Maltahöhe and Bethanien districts on gravel flats; specimens from these two districts are leafier and more glandular, and need further investigation)
60. *chenopodioides* (Namibia only, Kaokoveld (1913B) south to the Brandberg)
69. *hereroensis* (Namibia only, Brandberg south to the Kuiseb, east to Karabib; a plant from the Naukluft Mts included under *J. hereroensis* needs further investigation)

27. The Kaokoveld Group (one genus, two species).

The species are confined to southern Angola and northern Namibia. This is an interesting phytogeographical group that contrasts with what I have called the Eastern Ubiquists (Group 32), where the link to Angola is through the eastern part of southern Africa.

4. *Jamesbrittenia*

54. *heucherifolia* (Angola, in Mossamedes district, sea level to c.825m, then in the mountains along the lower Cunene in Namibia, Otjihipa Mts and Baynes Mts, south to Sanitatas, c.700–1000m)
61. *canescens* var. *laevior* (Kaokoveld and Karstveld of north west Namibia, one record from Angola, near source of Cunene in Huila district; river beds, pans, sponges, often in calcareous soils)

28. The Waterberg-Otavi Mountains Group (one genus, four species).

The species are probably endemic to these mountains.

4. *Jamesbrittenia*

2. *giessii* (south west end of Waterberg, on quartzite, and adjacent Etjo plateau; ill-known)
52. *acutiloba* (Waterberg, limestone cliffs and rocks)
53. *dolomitica* (Otavi to Grootfontein, limestone cliffs and rocks)
71. *fragilis* (Otavi Mountains, limestone cliffs and rocks)

29. The interior savanna Group (four genera, six species).

The species are often widely distributed in the savannas and woodlands of Namibia, Botswana, northern Cape (Gordonia to Griqualand West), western Transvaal, western Zimbabwe.

1. *Antherothamnus* (monotypic)

pearsonii (Huib plateau and Great Karasberge in SE Namibia, a few sites around Kakamas on the Orange, then eastern border of Botswana, western Transvaal, Matopas in Zimbabwe)

4. *Jamesbrittenia*
14. *atropurpurea* subsp. *pubescens* (mainly Botswana, 2 records from Gobabis district in Namibia, Cape north of the Orange river, possibly always on calcareous soils)
61. *canescens* var. *seineri* (central Namibia between the 21st and 24th parallels, a few records from central Botswana and Kalahari Gemsbok Park in the northern Cape)
79. *concinna* (northern Namibia, from lower Cunene south east to Okahandja and east to central Botswana, there only scattered records; marshes, pans and river banks, often calcareous)

6. *Sutera*
2. *griquensis* (Asbestos Hills and Langeberg in Griqualand West, hills N of Prieska, SW Transvaal from Zeerust to Rustenburg; under-collected)

8. *Melanospermum*
5. *foliosum* (western Transvaal, then scattered records from north west Cape, Botswana, Namibia, western Zimbabwe)

30. The Highveld Group (six genera, 23 species).

The species in this Group are found principally on the Transvaal Highveld, but may extend into western Swaziland and Natal, or more rarely bridge the Limpopo interval to reach Zimbabwe, or reach the eastern fringe of Botswana. This Group corresponds to Group 6 of Hilliard (1978, p.417) who found 43 species in 18 genera of Compositae occurring in Natal belong there; it includes 5 species of *Euryops* (Nordenstam, 1969, p.48).

4. *Jamesbrittenia*
5. *candida* (small area of NE Transvaal, Strydpoortberge to Transvaal Drakensberg around The Downs; margins of forest patches)
6. *grandiflora* (eastern escarpment, Mt Sheba to Barberton and neighbouring parts of Swaziland; margins of forest patches, scrubby grassland)
7. *macrantha* (small area of NE Transvaal, Sekukunis and Lulu Mts; scrubby grassland)
9. *burkeana* (Transvaal from Zoutpansberg south almost to border with O.F.S., western Swaziland, northern Natal, eastern Botswana; forest margins, scrubby grassland)
10. *accrescens* (Zoutpansberg and eastern highlands of Transvaal; scrubby grassland, forest margins)
66. *silenoides* (Transvaal from environs of Potgietersrus south to Belfast, Hlobane Mt in northern Natal; rocky grassland; ill-known)
78. *montana* (one site in SW Zimbabwe, western Transvaal, N Natal; damp or marshy places)

6. *Sutera*
5. *polelensis* subsp. *fraterna* (SE Transvaal and W Swaziland; rock outcrops)
8. *debilis* (southern Zimbabwe, northern Transvaal; rock outcrops and cliffs, in damp shade)
37. *neglecta* (SE Transvaal, western Swaziland, low Drakensberg on Transvaal-Natal-O.F.S. border, O.F.S.-Lesotho border to Fouriesburg and Leribe; rocky grassland)
38. *levis* (Transvaal Highveld from Magaliesberg to the Vaal, east to Middelburg, west into the fringes of Botswana; rocky grassland)

7. *Manulea*
5. *rhodantha* subsp. *aurantiaca* (E and SE Transvaal, from Haenertsburg to Standerton and Wakkerstroom, extreme northern Natal)
6. *parviflora* var. *limonioides* (Witwatersrand south to the Vaal and east to Piet Retief)

8. *Melanospermum*
1. *rupestre* (Transvaal: Witwatersrand, Steenkampsbergen, eastern escarpment, rock faces in forest, deep under rocks in moist shade; insufficiently known)
2. *swazicum* (Swaziland, on hills around Mbabane, Natal at Ngome; moist peaty soil under and around rocks; insufficiently known)
3. *rudolfii* (Transvaal, at Botsabelo near Middelburg, moist shade under rocks; insufficiently known)
4. *italae* (Swaziland around Mbabane, Transvaal near Piet Retief, Natal at Itala near Louwsburg, damp sand; insufficiently known)
6. *transvaalense* (SE Transvaal, from Pilgrim's Rest to Ermelo, one site in northern O.F.S. at Hebron; damp sand)

13. *Zaluzianskya*
18. *katharinae* (Transvaal only, Zoutpansberg, eastern escarpment, Magaliesberg, Witwatersrand, hills near Potchefstroom and Heidelberg; rock outcrops)

17. *Tetraselago*
1. *natalensis* (extreme SE Transvaal, southern Swaziland, Natal Midlands)
2. *nelsonii* (NE Transvaal, Makapaansberg east to edge of escarpment)
3. *wilmsii* (Zoutpansberg, then Blyde river south to Heidelberg)
4. *longituba* (Barberton Mts (both Swaziland and Transvaal) south to Mbabane and Carolina, north to Belfast, Pilgrim's Rest and Lydenburg districts)

31. The Eastern Cape-Transkei-Natal Group (five genera, 16 species).
The species are more or less coastal (up to c.900m above sea level) in their distribution, from the southern Cape coast to Natal, though few cover the whole area. The group contains some narrowly endemic species; see further under Centres of endemism 2b and 2c (below).

4. *Jamesbrittenia*
22. *phlogiflora* (E Cape, from about the Great Fish river to the Kei and inland to King William's Town and Stutterheim; grass and scrub on margins of forest patches)
23. *maritima* (E Cape, Alexandria to East London, grass and scrub on margins of forest patches)
24. *kraussiana* (E Cape, environs of East London to the Umfolozi river in Natal, inland to c.800m; grass and scrub on forest margins)
76. *multisecta* (E Cape around Stutterheim and Komgha, Transkei, grassland; ill-known)

6. *Sutera*
9. *platysepala* (Natal, Entumeni to Uvongo; sandstone cliffs or overhanging rocks, margins of or near forest patches)
45. *impedita* (E Cape around Stutterheim and Komgha, Transkei near Kentani, grassland and scrub on forest margins; ill-known)
46. *calycina* (most records from southern Transkei and around Komgha, but also records from Cala and Barkly Pass; ill-known)
47. *polyantha* (littoral dunes from George eastwards to drier parts of E Cape: lower Gamtoos valley, Zwartkops valley, inland to Fort Beaufort, Peddie and Cookhouse on the Great Fish river)
48. *roseoflava* (E Cape, from Port Alfred to East London, inland to King William's Town and Stutterheim)

7. *Manulea*
11. *bellidifolia* (E Cape coast between Alexandria and Port Alfred, inland to the valleys of the Kabusie, Toise, Zwart Kei and Komgha rivers)

12. *Phyllopodium*
1. *cuneifolium* (roughly Humansdorp to the Kei river, inland to Grahamstown and Fort Beaufort, isolated records from Transkei, then Durban Bay – possibly introduced there; weed)
2. *bracteatum* (Mossel Bay through coastal areas to Natal, inland to Grahamstown and Fort Beaufort; weed)
3. *diffusum* (Uitenhage to Port Elizabeth, then Peddie and King William's Town; ill-known)

13. *Zaluzianskya*
9. *pachyrrhiza* (Natal coast, from Lebombo Mts and Lake St Lucia south to the Umtamvuna)
10. *maritima* (coastal dunes, George to Umtentu river in Transkei)
40. *vallispiscis* (valley of Great Fish river)

32. The Eastern Ubiquists (four genera, 11 species).
The species have a wide or sometimes more restricted range, from the eastern Cape through Natal to the elevated parts of the Transvaal and adjoining parts of Swaziland. Because many of them range along the low Drakensberg, they occur in the Orange Free State near its border with Natal and along the northern part of the O.F.S.-Lesotho border. Some may overleap the Limpopo interval and range northwards to at least Zimbabwe with a disjunction to Huila district of Angola. Hilliard (1978), in analysing species of Compositae found in Natal, subdivided this unit into Groups 4 (15 genera, 59 species), 4a (eight genera, 19 species), 5 (four genera, three species), 8 (20 genera, 55 species). It has been suggested that elements of the Cape flora have migrated along this eastern route to Angola via stations in Zimbabwe (see, inter alia, Nordenstam, 1974, p.62). Although one species of *Jamesbrittenia* and one of *Manulea* show this distribution pattern, neither genus is a typical element of the Cape Floral Region.

4. *Jamesbrittenia*
　15. *huillana* (E Cape, low-lying parts of Natal, western Swaziland, eastern Transvaal, Zimbabwe, a few records from Zambia, Huila in Angola)
6. *Sutera*
　4. *floribunda* (widely distributed in Natal and Transvaal and neighbouring areas of Swaziland and O.F.S., south to the Kei river, north to the high ground in Zimbabwe, mainly along the eastern escarpment)
7. *Manulea*
　6. *parviflora* var. *parviflora* (Transkei, Natal below c.1800m, coastal Mozambique, Swaziland, Transvaal, Zimbabwe, Angola)
13. *Zaluzianskya*
　1. *microsiphon* (a few records from NE Transvaal, then low Drakensberg on Transvaal-Natal-O.F.S. border, mountains of Lesotho, Natal, Transkei, above c.1500m)
　2. *spathacea* (high ground in Transvaal, W Swaziland, low and high Drakensberg and outliers in Natal, Cape Drakensberg, Witteberg, mountains of Transkei, eastern Cape roughly as far west as Somerset East)
　3. *natalensis* (western Swaziland south through Natal to Transkei border, 600–1700m above sea level)
　4. *pulvinata* (SE Transvaal, low and high Drakensberg and outliers in Natal, Cape Drakensberg and Witteberg, Lesotho)
　5. *angustifolia* (apparently disjunct: Zoutpansberg, Blaauwberg, Woodbush in Transvaal, southernmost coastal Natal, scattered sites in Transkei, Cape on Amatole mountains and nearby mountains, Lootsberg near Graaff Reinet)
　7. *elongata* (Transvaal Highveld and eastern highlands, western Swaziland, widespread in Natal, south to border of Transkei)
　8. *pilosa* (a few records in Natal and highlands of E Transvaal; insufficiently known)
　20. *distans* (scattered records from highlands of E Transvaal, low Drakensberg, then isolated record from Mawahqua Mt in southern Natal; insufficiently known)

33. The South east Tropical Africa Group (three genera, seven species).

The species have their main distribution in Zimbabwe, but may extend to the mountains of Mozambique and Malawi.
4. *Jamesbrittenia*
　1. *fodina* (endemic to the Great Dyke in Zimbabwe)
　4. *carvalhoi* (eastern highlands of Zimbabwe from Inyanga to the Chimanimani mountains, also Gorongoza in Mozambique)
　8. *albobadia* (highlands of Zimbabwe, overleaping the Zambezi interval to the heights around Dedza and Lilongwe in Malawi; almost certainly in Mozambique on Tambanchipere Mt, Domwe Mt and the Kirk Mts)
　13. *zambesica* (Victoria Falls only, on the Zimbabwe side; insufficiently known)
　82. *myriantha* (NW and W Zimbabwe, scattered sites, always damp or marshy; insufficiently known)
6. *Sutera*
　7. *septentrionalis* (NE Zimbabwe, at c.1500m; granite boulders and cliffs; ill-known)
13. *Zaluzianskya*
　16. *tropicalis* (widespread in highlands of Zimbabwe and neighbouring heights in Mozambique)

34. The tropical African Ubiquists (one genus, two species).

The species are widely distributed in the savannas and grasslands of tropical Africa, where they are confined to moist places.
4. *Jamesbrittenia*
　12. *elegantissima* (Angola, Botswana, Zambia, Zimbabwe, Namibia, confined to the Cunene-Cubango-Cuanavale-Cuando-Okavango-Zambezi drainage system)
　80. *micrantha* (Zambia, Tanzania, Malawi, Mozambique, Zimbabwe, Botswana, N and E Transvaal lowveld, Swaziland lowveld, NE Natal)

GROUP C.

Species endemic to Angola.
4. *Jamesbrittenia*
　3. *angolensis* (Huila distr., Sà da Bandeira, Serra da Chela; ill-known)

GROUP D.
Species endemic to North East Tropical Africa.
13. *Zaluzianskya*
> 43. *elgonensis* (Uganda, Mt Elgon; Tanzania, Kilimanjaro, at both sites in thin soil on rocky ground at 3800m; ill-known). There is a disjunction of over 3000km between *Z. elgonensis* and its two closest allies, *Z. rubrostellata* and *Z. peduncularis*.

GROUP E.
NE Africa-Arabia-India species.
The species now have a highly disjunct distribution and are the only ones to occur outside Africa and its islands.
3. *Camptoloma*
> 3. *lyperiiflorum* (Somalia, Socotra, Kuria Muria Islands, Hadhramaut; crevices of cliffs and similar rocky sites)
4. *Jamesbrittenia*
> 83. *dissecta* (Nile valley, Sudan, Gangetic plain south to Madras; weed of moist places, river banks, gullies, reservoirs)

GROUP F.
Species endemic to Canary Islands.
3. *Camptoloma*
> 1. *canariense* (endemic to Gran Canaria)

The disjunct Afro-arid element.
The three species of *Camptoloma* are vicariads that form part of the disjunct Afro-arid element in African phytogeography. This interesting group is becoming increasingly known as revisionary work reveals the links, for example, *Calliandra gilbertii* (NE Kenya, Somalia) and *C. redacta* (Namibia) which was originally described in *Acacia* (see Thulin, Guinet & Hunde, 1981), and *Camptoloma*, hitherto lost in *Sutera*, with one species in Namibia-Angola, another in the Canary Islands, the third in a small area around the Horn of Africa. Goldblatt (1978, pp.395–398, table 2) lists genera known at that date to have a north-south distribution and gives useful references to more detailed treatments.

PHYTOGEOGRAPHICAL CENTRES

The phytogeographical groups outlined above are based on the broad distributions of all the species of Manuleae (and many other taxa as well), but it is clear that in some areas there is a rich concentration of species, many of which are endemic. Nordenstam (1969) recognized nine geographical areas in southern Africa (south of the Orange and Limpopo rivers) as centres of outstanding species concentration and of endemism. His scheme was based on his experience of *Euryops* as well as other genera, and many of these centres are important in Manuleae. I have therefore followed his sequence of Centres, adding four in the Cape Floral Region as well as several more in areas either without significance in *Euryops* or outside the area dealt with by Nordenstam. These centres of endemism are on an entirely different scale from those used, for example, by White (1983) in considering Africa as a whole.

1a. The Caledon Centre
The Cape Floral Region is not important as a centre of endemism in *Euryops*, but it is in Manuleae. Within this Region, Nordenstam (1969, p.62) was able to recognize only one, the Caledon Centre, which corresponds roughly with the South-western and Peninsula Centres of Oliver et al. (1983, p.437), who were dealing entirely with 'Cape' genera. My Group 6, the South western Group, corresponds roughly with the West Coastal, Peninsula and South-western Centres of Oliver et al., and it is the coastal parts of this area that are particularly important as

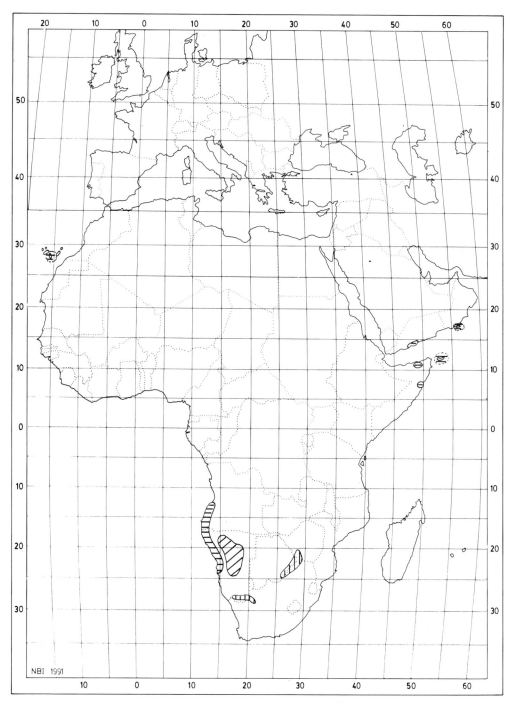

Fig. 69. Map indicating overall distribution of *Camptoloma* ▤ , *Manuleopsis* ▨ , *Antherothamnus* ⦚ .

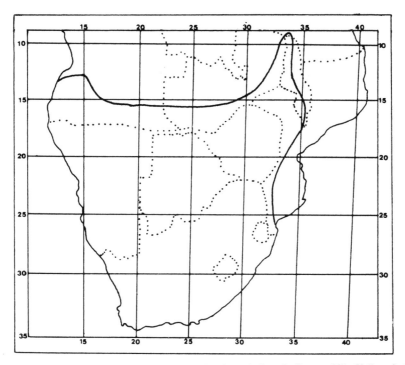

Fig. 70. Map indicating overall distribution of *Jamesbrittenia* excluding *J. dissecta* (Nile Valley, Sudan, Indian subcontinent).

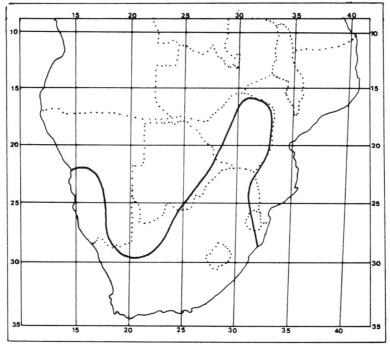

Fig. 71. Map indicating overall distribution of *Sutera*.

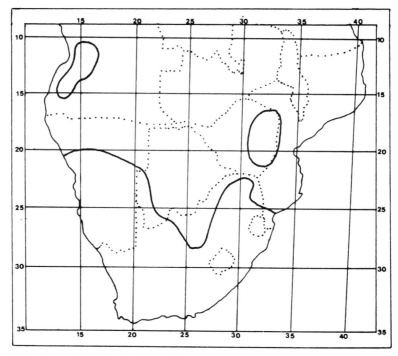

Fig. 72. Map indicating overall distribution of *Manulea*.

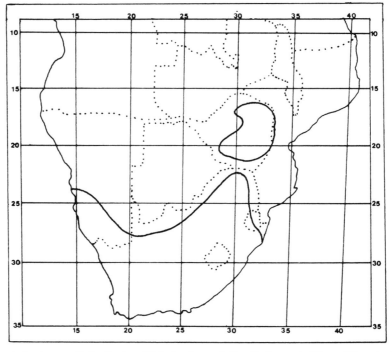

Fig. 73. Map indicating overall distribution of *Zaluzianskya* excluding Z. *elgonensis* (Elgon, Kilimanjaro).

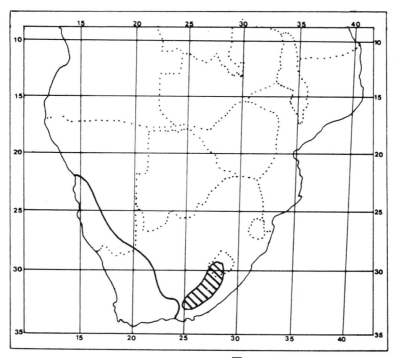

Fig. 74. Map indicating overall distribution of *Lyperia* and *Glekia* ▧ .

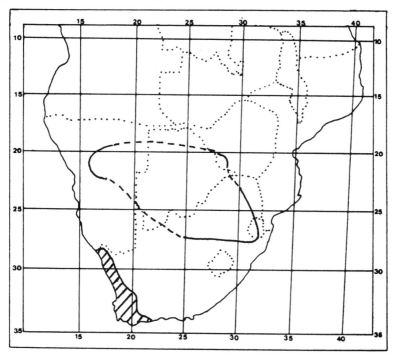

Fig. 75. Map indicating overall distribution of *Melanospermum* (dashed lines indicate no records) and *Polycarena* ▨ .

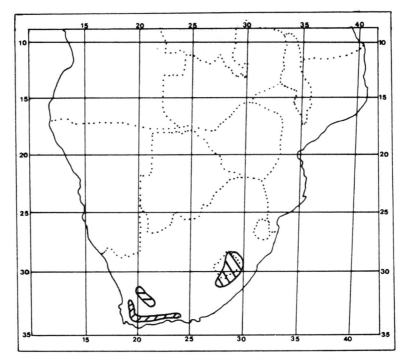

Fig. 76. Map indicating overall distribution of *Trieenea* , *Reyemia* , *Glumicalyx* .

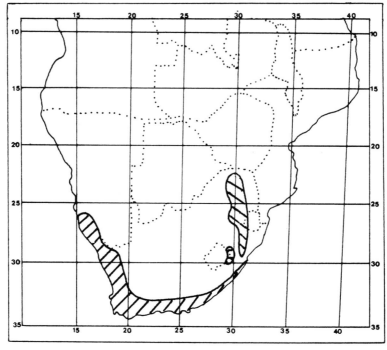

Fig. 77. Map indicating overall distribution of *Phyllopodium* , *Strobilopsis* , *Tetraselago* .

Fig. 78. Map indicating approximate boundaries of phytogeographical groups: 3, Southern; 6, South western; 7, North western; 8, Bredasdorp-Riversdale; 9, Southern Cape Mountains; 10, South eastern; 11, Western Upper Karroo; 14, Eastern Upper Karroo – grassland; 15, Sneeuwbergen; 18, Drakensberg; 21, Namaqua; 22, Vanrhynsdorp Karroo; 23, Gariep; 25, Namibian central and southern highland; 30, Highveld; 31, Eastern Cape-Transkei-Natal. Groups 26, 27, 28, 29, 33 are not shown, being partly (26, 29) or wholly off the map; the northernmost part of Group 25 is also off the map.

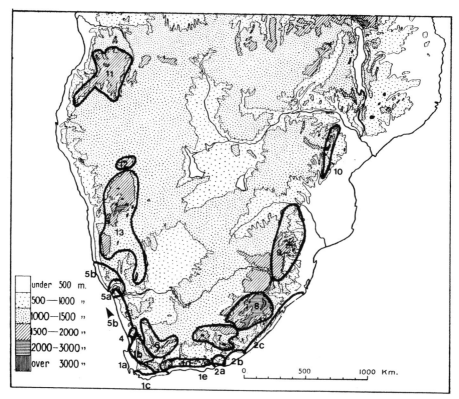

Fig. 79. Map indicating centres of endemism: 1a, Caledon Centre; 1b, North western Centre; 1c, Bredasdorp Centre; 1d, Southern Cape Mountain Centre; 1e, Southern Cape Coastal Centre; 2a, Albany Centre; 2b, Alexandria-East London Centre; 2c, Pondoland Centre; 3, Little Karroo Centre (not mapped); 4, Vanrhynsdorp Centre; 5a, Gariep Centre; 5b, Namaqua Centre; 6, Western Upper Karroo Centre; 7, Sneeuwbergen Centre; 8, Drakensberg Centre; 9, Barberton Centre; 10, Inyangani Centre; 11, Angolan Centre; 12, Waterberg-Otavi Mountains Centre; 13, Namibia central and southern highlands Centre.

a centre of endemism in Manuleae: *Manulea* (seven species), *Polycarena* (five), *Phyllopodium* (three), *Sutera* (two), *Zaluzianskya* (two).

Oliver et al. (1983) recognized a Northern Centre, equivalent to Weimarck's and Nordenstam's Northwestern Subregion and my North western Group (no. 7), a Bredasdorp Centre, equivalent to my Group 8, and a Southern Centre, equivalent to my Group 9. These three geographical areas, as well as the area of my Group 3, are all centres of endemism in Manuleae. To keep in line with Nordenstam's enumeration, they are numbered 1b, etc. This device also serves to unite under one number centres of endemism within the Cape Floral Region.

1b. The North western Centre

This Centre corresponds to Group 7 (above), where its geographical area is outlined. It is particularly important in Manuleae, and the species are mainly on the mountains: *Manulea* (18 species), *Trieenea* (seven out of a total of nine species, many known only from the Cedarberg), *Polycarena* (five), *Phyllopodium* (four). There are also three species of *Zaluzianskya*, but collections are too few to be certain of their distribution, and three species of *Sutera* that are also insufficiently known: *S. subsessilis* and *S. longipedicellata* with very few records, *S. paniculata* on the Khamiesberg and nearby then a disjunction to the Bokkeveld Mts and Cedarberg. *Lyperia formosa* (Voetpadsberg) is known only from the type collection.

1c. The Bredasdorp Centre

This centre, based on the limestone areas mainly in Bredasdorp and Riversdale divisions, is becoming well known for its endemism and revisionary work will make its importance ever clearer: all but one of the 11 species of Manuleae endemic there are newly described! However, that *Manulea calciphila* appears to occur in Namaqualand as well as on the southern limestone emphasises the need for caution in declaring a species endemic to a particular area.

Jamesbrittenia (three species, *J. stellata* disjunct to a limestone cliff on the Peninsula), *Sutera* (four), *Manulea* (two, with the caveat above), *Zaluzianskya* (two). See Group 8 for details.

1d. The Southern Cape Mountain Centre

This centre covers the geographical area of Group 9 (above). These mountains are not well explored, and most collections have come from the major passes. Some ranges appear never to have been traversed by a botanist. Six species of *Sutera* are endemic there, but three are known from very few collections. Then there is one species in each of the genera *Jamesbrittenia, Manulea, Polycarena* and *Phyllopodium*, all but *Jamesbrittenia tortuosa* known from only one or two sites.

1e. The Southern Cape coastal Centre

This centre covers the geographical area of Group 3 (above), roughly from the Breede River to Alexandria, excluding the limestone. The species are not necessarily confined to the coast but may ascend the mountains or penetrate inland up the river valleys. *Phyllopodium elegans* and *P. multifolium* are montane, and *Sutera langebergensis* is possibly confined to the Langeberg. The centre is important for endemism in *Jamesbrittenia* (five species), slightly less so in *Sutera* (four species), *Phyllopodium* (three), *Manulea* (one), *Lyperia* (one).

2a. The Albany Centre

This Centre corresponds roughly to the area of my Group 10 (Zuurberg to the environs of Grahamstown) and to the Zuurberg Subcentre of Weimarck (1941, p.75). It is unimportant for Manuleae. *Jamesbrittenia zuurbergensis* and *Manulea racemosa* may be confined to the Zuurberg, but both are under-collected; similarly, the distribution of *Jamesbrittenia albanensis* is not clear.

Two more areas in the Cape-Transkei-Natal coastal region are centres of endemism; they are detailed below under 2b and 2c.

2b. The Alexandria-East London Centre

A small area of the E Cape, from roughly the Sundays river to the Kei and neighbouring Kentani district in Transkei, is a minor centre of endemism for Manuleae. The eastern Cape is not well known botanically and the importance of this Centre in a wider context is therefore obscure (though two Composites, *Vellereophyton vellereum* (R. A. Dyer) Hilliard and *Helichrysum praecinctum* Klatt come to mind, endemics in the littoral zone, as well as *Diascia personata* Hilliard & Burtt). Its species are included in my much wider Group 31. The endemics are: *Jamesbrittenia phlogiflora, J. maritima, J. multisecta, Sutera impedita, S. roseoflava, Manulea bellidifolia, Phyllopodium diffusum* (insufficiently known), *Zaluzianskya vallispiscis*.

2c. The Pondoland Centre

Another centre of endemism is the sandstone region of Natal and Pondoland (Transkei) that runs from about Hlabisa in Natal (Zululand) to Port St Johns in Transkei. The southernmost part in particular of this sandstone area (between the Umzimkulu and Umzimvubu rivers) is noted for its endemism and its links to the Cape Floral Region (see, for example, Hilliard, 1978, p.422, and Van Wyk, 1989). Van Wyk (loc. cit., p.245) called this the Pondoland Centre and I have adopted the term though as used here it covers the whole of the sandstone area. This centre is not important for Manuleae; *Sutera platysepala* is confined to sandstone cliffs and outcrops between Entumeni in Zululand and Uvongo in southern Natal.

3. The Little Karroo Centre

Nordenstam (1969, p.63) wrote 'This limited Karroo area forms an enclave within the Cape Region surrounded by fynbos vegetation'. It appears to be unimportant in Manuleae and in any case lack of ecological information on most herbarium specimens makes it difficult to judge if a species is confined to it.

4. The Vanrhynsdorp Centre

This centre corresponds to Nordenstam's phytogeographical group 21, The Vanrhynsdorp Karroo Endemics (my Group 22). It is not important in Manuleae, and only two species may be endemic there, *Manulea ramulosa* and *Zaluzianskya karreebergensis*, each known from only two collections.

5a. The Gariep Centre

This Centre corresponds to my Group 23, embracing the country adjoining the lower reaches of the Gariep (Orange). Nordenstam emphasized the importance of the Centre for the succulent Karroo flora. It is significant for *Jamesbrittenia* (six species); four species of *Manulea* and two of *Phyllopodium* may also be endemic there, but all are under-collected.

The Centre includes the Richtersveld, the northernmost part of Namaqualand. The rest of Namaqualand, and the area south of it to the Olifants river is unimportant for *Euryops* except in the Van Rhynsdorp enclave (no. 4, above), but it is a significant centre of endemism for Manuleae and is intercalated here as 5b.

5b. The Namaqua Centre

This centre covers the geographical area of Group 21 (above), from southernmost Namibia south through Namaqualand to the Olifants river. The endemics comprise seven species of *Jamesbrittenia*, six of *Manulea*, four of *Phyllopodium*, two of *Zaluzianskya*.

6. The Western Upper Karroo Centre

This Centre encompasses the geographical area of Group 11 (above) and includes the Hantam Mountains, Roggeveld and Nieuweveld. Nordenstam (1969, p.66) found it to be one of the most marked centres in *Euryops*, and gave examples of other endemics, including *Cliffortia hantamensis* and *C. arborea, Secale africanum*, and the monotypic genus *Daubenya*. It is also important for the Manuleae. The ditypic genus *Reyemia* is endemic there, also 11 species of *Zaluzianskya*, all annuals, half of them under-collected. Other endemics include two species of *Manulea*, one each of *Jamesbrittenia* and *Polycarena*.

7. The Sneeuwbergen Centre

This corresponds to my Group 15. It is very important in *Euryops*, but is insignificant in Manuleae: only one species of *Jamesbrittenia* and one of *Sutera* are endemic there.

8. The Drakensbergen Centre

Nordenstam (1969, p.67) stressed the general importance of this Centre, which corresponds to Group 18 (above). Hilliard (1978, p.411), in analysing the distribution of Compositae in Natal, found 111 species (in 23 genera) to be endemic there, while Hilliard & Burtt (1987) gave a much fuller discussion on endemism in the Eastern Mountain Region (that is, the Drakensberg Centre), and listed 394 species, subspecies and varieties over a wide range of families. The Centre is important for Manuleae: two genera, *Glumicalyx* (six species) and *Strobilopsis* (one) are endemic there as well as species in several more genera: *Jamesbrittenia* (seven), *Zaluzianskya* (five), *Manulea* (four), *Sutera* (two).

9. The Barberton Centre

Nordenstam (1969, p.67) defined this Centre as including the regions around Barberton and Lydenburg. My Group 30, The Highveld, encompasses the enlarged Centre, which covers the Zoutpansberg, the eastern highlands of the Transvaal and neighbouring parts of Swaziland, and the mountains of northernmost Natal. The genus *Tetraselago* (four species) is virtually endemic to it (*T. natalensis* extends south to the Natal Midlands) and so is *Melanospermum* (five out of six species). Other endemics are found in *Jamesbrittenia* (five species), *Sutera* (one subspecies) and *Manulea* (one subspecies).

10. The Inyangani Centre

This corresponds to the Inyangani Subcentre of Weimarck (1941, p.78) and falls within my Group 33. It encompasses the mountains along the eastern border of Zimbabwe and neighbouring parts of Mozambique. The endemics confined to the quartzites of the Chimanimani Mountains are well known (see Wild, 1964), but no member of Scrophulariaceae figures among them, and the Inyangani Centre as a whole is unimportant for Manuleae. Only *Jamesbrittenia carvalhoi* appears to be strictly endemic there; *J. albobadia* bridges the Zambezi interval to reach the mountains near Dedza in Malawi.

11. The Angola Centre

Weimarck (1941, p.80) recognized an Angolan Subcentre composed of 'several plateaux and mountains in Angola ..... The isolated montane areas in SW Africa may be included here'. The flora of Angola is too ill-known for me to be confident about recognizing a centre of endemism there, and although the flora of Namibia is comparatively well-known, the full distribution of many species is unclear because of the inaccessibility of many areas and consequent under-collecting. I prefer, for the reason given below, to restrict my Angolan Centre to the highlands in Huambo and Huila provinces in SW Angola, which lie above 2000m, and are clad in a mosaic of woodland and grassland (Barbosa, 1970) and probably constitute a centre of endemism. The only member of Manuleae endemic there is *Jamesbrittenia angolensis*.

Among the Manuleae, all but one of the species common to southern Angola and Namibia are confined to Mossamedes district of Angola and do not occur in the highlands (see Groups 26 and 27). *Welwitschia* is another striking example of such a 'lowland' connexion across the Cunene.

12. The Waterberg-Otavi Mountains Centre

This north eastern montane area of Namibia is mostly limestone. It has yielded four endemic species of *Jamesbrittenia* (see Group 28 for details).

13. The Namibia central and southern highlands Centre

This centre includes the mountains and plateaux of Namibia from the Brandberg, Erongoberge and Omatako peaks south to the Huis Hoch Plateau and Karasberge. The endemics belong in two phytogeographical groups, nos 25 and 26. The following species appear to be sufficiently well known to be included with some confidence in the Centre: *Jamesbrittenia primuliflora, J. pallida, J. fleckii* (?), *J. pilgeriana, J. hereroensis, Manulea dubia* (also on the Waterberg).

That the Centre is of real significance is borne out by the number of species (11) listed by Nordenstam (1974, p.56) as endemic to the Brandberg, with the caveat that some would in due course probably be found elsewhere. The list does not include any members of Scrophulariaceae, but, in the discussion following, Nordenstam mentions several more possible endemics, including '*Sutera* cf. *atropurpurea*' (which proves to be *Jamesbrittenia huillana*, widespread and extremely variable; the type of the name came from Huila in Angola), and '*Sutera canescens*', a misdetermination of the species now recognized as *Jamesbrittenia chenopodioides*. The area of this species extends from the Kaokoveld south to the Brandberg, and it is listed in my Group 26, the species occurring in the escarpment zone stretching from southern Angola to Namaqualand.

REFERENCES

ARGUE, C. L. (1993). Pollen morphology in the Selagineae, Manuleae (Scrophulariaceae), and selected Globulariaceae, and its taxonomic significance. *American Journal of Botany* 80(6): 723–733.

BAILLON, H. E. (1888). *Histoire des Plantes IX, Scrophulariacées*. Paris. Hachette.

BARBOSA, L. A. Grandvaux. (1970). *Carta fitogeográfica de Angola*. Instituto de Investigaçao científica de Angola. Luanda. 323 pp. Map.

BARRINGER, K. A. (1984). Seed morphology and classification of the Scrophulariaceae. *American Journal of Botany* 71 (Abstracts): 156.

BARRINGER, K. A. (1993). Five new tribes in the Scrophulariaceae. *Novon* 3: 15–17.

BARTHLOTT, W. (1984). Microstructural features of seed surfaces. In Heywood, V. H. & Moore, D. M. eds. *Current concepts in plant taxonomy*. Academic Press. London & Orlando.

BENTHAM, G. (1835) (June). In *Edwards, Botanical Register* 21, sub t. 1770.

BENTHAM, G. (1836). Synopsis of the Buchnerae, a tribe of Scrophulariaceae. In Hooker, W. J., *Companion to the Botanical Magazine* 1: 356–384.

BENTHAM, G. (1846). Scrophulariaceae. In De Candolle, *Prodromus* 10. Paris.

BENTHAM, G. (1876). Scrophulariaceae. In Bentham, G. & Hooker, J. D. *Genera Plantarum* 2. London.

BERGIUS, P. J. (1767). *Descriptiones plantarum ex Capite Bonae Spei*. Stockholm.

BOND, P. & GOLDBLATT, P. (1984). Plants of the Cape Flora. A descriptive catalogue. *Journal of South African Botany* Supplementary volume 13.

BROWN, R. (1810). *Prodromus Florae Novae Hollandiae et Insulae van-Diemen*. London.

BRUCE, E. A. (1941). Scrophulariaceae. *Kew Bulletin of Miscellaneous Information* 1940: 63–64.

BURTT, B. L. (1971). From the south. In Davis, P. H. et al. (eds.), *Plant Life of South West Asia* 135–149.

CARLQUIST, S. (1992). Wood anatomy of sympetalous dicotyledon families: a summary with comments on systematic relationships and evolution of the woody habit. *Annals of the Missouri Botanical Garden* 79: 303–332.

CRONQUIST, A. (1981). *An integrated system of classification of flowering plants*. Columbia University Press.

DAFNI, A. et al. (1990). Red bowl-shaped flowers: convergence for beetle pollination in the Mediterranean region. *Israel Journal of Botany* 39: 81–92.

DAHLGREN, R. M. T. (1980). A revised system of classification of the angiosperms. *Botanical Journal of the Linnean Society* 8: 91–124.

DALLA TORRE, K. W. VON & HARMS, H. (1904). *Genera Siphonogamarum ad systema Englerianum conscripta*. Leipzig. W. Engelmann.

DE WINTER, B. (1971). Floristic relationships between the northern and southern arid areas in Africa. *Mitteilungen der botanischen Staatssammlung München* 10: 424–437.

DIELS, L. (1897). Beiträge zur Kenntnis der Scrophulariaceen Afrikas. *Botanische Jahrbuch* 23: 471–517.

DON, D. (1835). Nycterinia lychnidea. In Sweet, *The British Flower Garden*, second series, 3, t. 239.

DYER, R. A. (1975). *The genera of southern African flowering plants*. 1. Dicotyledons. Pretoria

ENDLICHER, S. L. (1839). *Genera Plantarum secundum Ordines Naturales disposita*. Vienna. 1836–1840.

FAEGRI, K. & PIJL, L. VAN DER. (1971). *The principles of pollination ecology*, ed. 2. Pergamon Press.

GAERTNER, J. (1788). *De fructibus et seminibus plantarum*. Stuttgart.

GIBBS RUSSELL, G. E. et al. (1987). *List of species of southern African plants*, ed. 2, part 2, Dicotyledons. Pretoria.

GOLDBLATT, P. (1978). An analysis of the flora of southern Africa: its characteristics, relationships and origins. *Annals of the Missouri Botanical Garden* 65: 369–436.

GRABIAS, B., SWIATEK, L. & SWIETOSLAWSKI, J. (1991). The morphology of hairs in Verbascum species. *Acta Societatis Botanicorum Poloniae* 60: 191–208.

HARTL, D. (1959). Das alveolierte Endosperm bei Scrophulariaceen, seine Entstehung, Anatomie und taxonomische Bedeutung. *Beiträge zur Biologie der Pflanzen* 35(1): 95–110.

HARTL, D. (1975). Scrophulariaceae. In Hegi, G. *Illustrierte Flora von Mitteleuropa*, ed. 2, Bd. VI/1. Berlin and Hamburg. Paul Parey.

HEMSLEY, W. B. & SKAN, S. A. (1906). Scrophulariaceae. In Thiselton-Dyer, *Flora of Tropical Africa* 4(2): 261–462.

HIERN, W. P. (1904). Scrophulariaceae. In Thiselton-Dyer, *Flora Capensis* 4(2): 121–420.

HILLIARD, O. M. (1978). The geographical distribution of Compositae native to Natal. *Notes from the Royal Botanic Garden Edinburgh* 36: 407–425.

HILLIARD, O. M. (1990). A brief survey of Scrophulariaceae-Selagineae. *Edinburgh Journal of Botany* 47: 315–343.

HILLIARD, O. M. & BURTT, B. L. (1977). Notes on some plants of southern Africa chiefly from Natal: VI. *Notes from the Royal Botanic Garden Edinburgh* 35: 155–177.

HILLIARD, O. M. & BURTT, B. L. (1982). Notes on some plants of southern Africa chiefly from Natal: IX. *Notes from the Royal Botanic Garden Edinburgh* 40: 247–298.

HILLIARD, O. M. & BURTT, B. L. (1983). Zaluzianskya (Scrophulariaceae) in south eastern Africa. *Notes from the Royal Botanic Garden Edinburgh* 41: 1–43.

HILLIARD, O. M. & BURTT, B. L. (1987). The botany of the southern Natal Drakensberg. *Annals of Kirstenbosch Botanic Garden* vol. 15.

HUTCHINSON, J. (1926). *The families of the flowering plants* I. London.

JESSOP, J. P. (1964). Itinerary of Rudolf Schlechter's collecting trips in southern Africa. *Journal of South African Botany* 30: 129–142.

JONG, K. (1993). Variation in chromosome number in the Manuleae (Scrophulariaceae) and its cytotaxonomic implications. *Edinburgh Journal of Botany* 50(3): 365–379.

JUNELL, S. (1961). Ovarian morphology and taxonomical position of Selagineae. *Svensk Botanisk Tidskrift* 55(1): 168–192.

KOOIMAN, P. (1970). The occurrence of iridoid glycosides in the Scrophulariaceae. *Acta Botanica Neerlandica* 19: 329–340.

KUNTZE, O. (1891). *Revisio Generum Plantarum* 2. Leipzig.

KUNTZE, O. (1898). *Revisio Generum Plantarum* 3(2). Leipzig.

LEVYNS, M. R. (1939). Some changes in nomenclature. *Journal of South African Botany* 5: 35–38.

LINNAEUS, C. (1759). *Systema Naturae* 2, ed. 10. Stockholm.

LINNAEUS, C. (1767). *Mantissa plantarum*. Stockholm.

MABBERLEY, D. J. (1987). *The Plant-Book*. Cambridge University Press.

MARLOTH, R. (1932). *The Flora of South Africa*, 3. Cambridge.

MELCHIOR, H. ed. (1964). A. Engler, *Syllabus der Pflanzenfamilien* II. Berlin.

MERXMÜLLER, H. & ROESSLER, H. (1967). Scrophulariaceae, fam. 126 in Merxmüller, ed., *Prodromus einer Flora von Südwestafrika*. Lehre.

METCALFE, C. R. & CHALK, L. (1950). *Anatomy of the Dicotyledons* II. Oxford. Clarendon Press.

MEYER, E. (1849). Annotationes ad hortum seminiferum Regimontanum 1848. *Annales Sciences Naturelles* sér. 3, 11: 254.

NORDENSTAM, B. (1969). Phytogeography of the genus Euryops (Compositae). *Opera Botanica* 23. 77 pp.

NORDENSTAM, B. (1974). The flora of the Brandberg. *Dinteria* 11: 1–67.

OLIVER, E. G. H., LINDER, H. P. & ROURKE, J. P. (1983). Geographical distribution of present-day Cape taxa and their phytogeographical significance. *Bothalia* 14: 427–440.

PFEIFFER, L. G. C. (1874). *Nomenclator botanicus*. Casselis. Fischer.

PHILLIPS, E. P. (1951). The genera of South African flowering plants. *Botanical Survey Memoir* 25. Pretoria.

PIJL, L. VAN DER. (1961). Ecological aspects of flower evolution II. Zoophilous flower classes. *Evolution* 15: 44–59.

PIJL, L. VAN DER. (1982). *Principles of dispersal of higher plants*. Springer Verlag.

PROCTER, M. & YEO, P. (1973). *The pollination of flowers*. Collins. London.

RAMAN, S. (1989–1990). The trichomes on the corolla of the Scrophulariaceae I–IX. *Beiträge Biol. Pflanzen* 64, 65.

ROLFE, R. A. (1883). On the Selagineae described by Linnaeus, Bergius, Linnaeus fil. and Thunberg. *Journal of the Linnean Society* 20: 338–358.

ROLFE, R. A. (1901). Selagineae. In Thiselton-Dyer, W. T., *Flora Capensis* 5(1): 95–180. London.

ROTH, A. G. (1807). *Botanische Bemerkungen und Berichtigungen*. Leipzig.

ROTH, A. G. (1821). *Novae plantarum species praesertim Indiae orientalis*. Halberstad.

SCHINZ, H. & THELLUNG, A. (1929). Scrophulariaceae. *Vierteljahrsschrift der Naturforschenden Gesellschaft in Zürich* 74: 113–123.

SCOTT ELLIOT, G. F. (1891). Notes on the fertilisation of South African and Madagascan flowering plants. *Annals of Botany* 5: 368.

SUTTON, D. A. (1988). *A revision of the tribe Antirrhineae*. Oxford University Press.

TAKHTAJAN, A. L. (1980). Outline of the classification of flowering plants (Magnoliophyta). *Botanical Review* 46(3): 225–359.

THEOBALD, W. L., KRAHULIK, J. L. & ROLLINS, R. C. (1979). Trichome description and classification. In Metcalfe, C. R. & Chalk, L., *Anatomy of the Dicotyledons* ed. 2, 1. Clarendon Press. Oxford.

THORNE, R. F. (1983). Proposed new realignments in the angiosperms. In Ehrendorfer, F. & Dahlgren, R., eds. New evidence of relationships and modern systems of classification of the angiosperms. *Nordic Journal of Botany* 3(1): 85–117.

THULIN, M., GUINET, P. & HUNDE, A. (1981). Calliandra (Leguminosae) in continental Africa. *Nordic Journal of Botany* 1: 27–34.

URBAN, I. (1912). *Symbolae Antillanae* 7. Berlin. Borntraeger.

VAN WYK, A. E. (1989). The sandstone regions of Natal and Pondoland: remarkable centres of endemism. *Palaeoecology of Africa and the surrounding islands* 21: 243–257.

VOGEL, S. (1954). Blütenbiologische Typen als Elemente der Sippengliederung dargestellt anhand der Flora Südafrika. *Botanische Studien* Heft 1. Gustav Fischer Verlag. Jena.

VOGEL, S. (1990). *The role of scent glands in pollination. On the structure and function of osmophores*. English translation. Smithsonian Institution Libraries and the National Science Foundation, Washington, DC.

WEBER, A. (1989). Family position and conjectural affinities of Charadrophila capensis Marloth. *Botanische Jahrbuch* 111: 87–119.

WEIMARCK, H. (1941). Phytogeographical groups, centres and intervals within the Cape Flora. *Lunds Universitets Årsskrift N.F. Afd.* 2, 37(5): 143 pp.

WEISS, M. R. (1991). Floral colour changes as cues for pollinators. *Nature* 354: 227–229 cum ic.

WETTSTEIN, R. VON. (1891). Scrophulariaceae. In Engler, A. & Prantl, K. A. E. *Die Natürlichen Pflanzenfamilien* 4(3b). Leipzig. Engelmann.

WHITE, F. (1983). *The vegetation of Africa: a descriptive memoir to accompany the UNESCO/AET-FAT/UNSO vegetation map of Africa*. Paris: UNESCO.

WILD, H. (1964). The endemic species of the Chimanimani Mountains and their significance. *Kirkia* 4: 125–157.

WILLDENOW, C. L. (1800). *Species plantarum* 3: Berlin.

WILLDENOW, C. L. (1809). *Enumeratio plantarum Horti Regii Berolinensis* ..... Berlin.

INDEX